THE HIGH NITROGEN COMPOUNDS

THE HIGH NITROGEN COMPOUNDS

FREDERIC R. BENSON

Retired
ICI Americas Inc.
Wilmington, Delaware

A Wiley-Interscience Publication

JOHN WILEY AND SONS,

New York · London · Sydney · Toronto

Library of Congress Cataloging in Publication Data:

Benson, Frederic R.
The high nitrogen compounds.

"A Wiley-Interscience publication."
Includes index.
1. Cyclic compounds. 2. Organonitrogen compounds.
I. Title.

QD401.B457 1983 547'.04 83-10476
ISBN 0-471-02652-2

Printed in the United States of America

10 9 8 7 6 5 4 3 2 1

PREFACE

What is the nature of chemicals with backbones of nitrogen? What structural possibilities have been realized? What physical, chemical, biological, and use properties result from linking nitrogen atoms in chains bearing organic and inorganic appendages?

This book, with a focus on nitrogen chains, has been written to try to provide at least partial answers to these questions. The conceptual parents of compounds with sequential nitrogen atoms were called hydronitrogens by the late Professor L. F. Audrieth in analogy to the hydrocarbons. It may be said at once that the derivatives of hydronitrogens exhibit distinctive chemical and physical properties and have diverse uses.

An explanation of the title of this book is in order. During World War II, I had the good fortune to work at Picatinny Arsenal under the direction of Dr. Audrieth on a project called Compounds of High Nitrogen Content in Propellant Powders. Most of the molecules under consideration had one or more nitrogen-nitrogen links. In daily usage, the materials of the project were called high nitrogen compounds. Extensions of this work by various groups after the war included investigation of derivatives of aminoguanidine and other hydrazines, triazoles, and tetrazoles.

Although substances with two linked nitrogen atoms are of scientific interest, and find a variety of uses, the distinctive ranges of characteristics of compounds with sequences of nitrogen appear to a greater extent with chain lengths of three or more nitrogen atoms. Further, the publications on hydrazine, azo, and pyrazole derivatives are so numerous that prudence dictated exclusion of these and other compounds with two-nitrogen units. There remains a large quantity of literature on acyclic and cyclic compounds with three or more contiguous nitrogen atoms; I have designated these chemicals and this book *The High Nitrogen Compounds.*

Uses are emphasized to a greater extent than is customary in most treatises on chemical classes; those chemicals that are useful and saleable help pay the dues for the pursuit of chemistry and other sciences. Information about each class of high nitrogen compounds is illustrative and representative but not exhaustive. It is hoped that the organized information on hydronitrogen derivatives with three or more linked nitrogen atoms will inform, intrigue, stimulate, and reward the reader.

Many friends and colleagues assisted in one or more aspects of the preparation of *The High Nitrogen Compounds.* In particular, I am indebted to Dr. J. W. Le Maistre for his penetrating and constructive criticisms of the manuscript. I also gratefully acknowledge the helpful comments of J. R. Almond, R. O. Beauchamp, Jr., Dr. C. H. Nelson, Dr. H. Skolnik, and Dr. E.-J. Tillmanns-Skolnik on portions of the manuscript. The on-line computer searches conducted by Dr. C. H. Nelson and J. R. Almond were of tremendous value, both in identifying relevant references and, within the time limits of the data bases used, in determining the existence or absence of information pertaining to potential high nitrogen classes and compounds. For obtaining countless articles, patents, and other references from near and far, I thank Velga Rukuts and her staff in ICI Americas' Atlas Library, especially Karen Zeitler and Karol Hannover. I also thank Mary Ann Gregson for typing the entire manuscript, deciphering my handwriting, and making many useful suggestions of format. Thanks also are due F. A. Stepanek for ably typing the lengthy reference lists. Finally, I cannot thank my wife enough for her continual encouragement, patience, and understanding while *The High Nitrogen Compounds* was being planned, organized, and written.

FREDERIC R. BENSON

Wilmington, Delaware
June 1983

CONTENTS

1

INTRODUCTION

The high nitrogen compounds are defined operationally as those with chains of three or more contiguous nitrogen atoms covalently bound. Presently 10 linked nitrogen atoms are the upper limit. Representatives of all these nitrogen chain lengths, except nine, have been synthesized. The nitrogen chains may be entirely acyclic; distributed within and without a ring (acyclic-cyclic); entirely within one ring system (cyclic); or distributed between two singly joined, nonfused rings (cyclic-cyclic). In a cyclic chain, one or more nitrogen atoms may be shared by two or three rings, and hence are embedded. Nitrogen is present in the chains at oxidation levels from −2 to +4. Thus the high nitrogen compounds start with the acyclic triazanes **(I)** discovered by Wohl (23) in 1900 and extend to the acyclic-cyclic 1,1′-azobis(tetrazoles) **(II)** synthesized by Stollé (21) in 1933. This longest sequence of nitrogen atoms is to be compared with the ex-

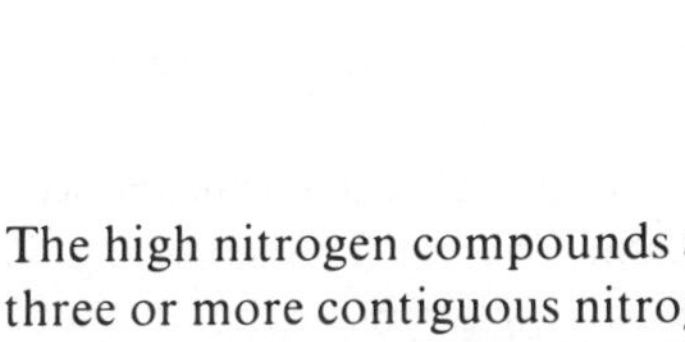

1-Formyl-2-phenyl-3-(phenyl-methylene)triazane

(I)

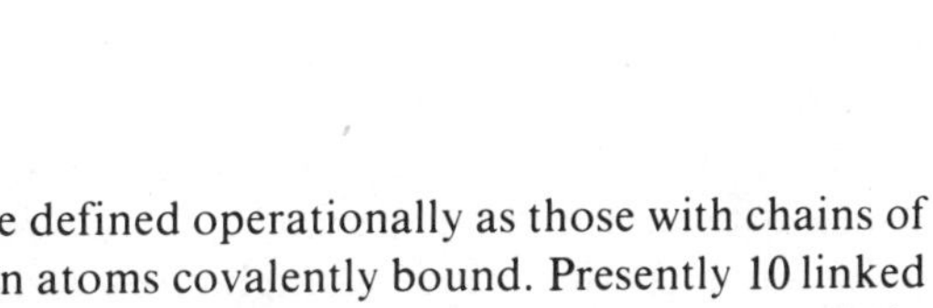

1,1′-Azobis(5-phenyltetrazole)

(II)

tent of catenation by other first-row elements in the periodic table. A borane exists with 20 linked boron atoms (7), carbon chains in polyethylene reach lengths of over 100,000 atoms (5), three oxygens are linked in ozone, and four are joined in the unstable O_4F_2 (8).

Many acyclic high nitrogen compounds fall within Audrieth's classification of compounds of hydrogen and nitrogen (1, 2), which provides a formal framework for their consideration. He grouped hydronitrogens into one saturated and three unsaturated series, as indicated. The parent compounds, hydrogen azide and 2-tetrazene, can be isolated, but other hydronitrogens with three or more linked nitrogen atoms are known only as organic substitution products.

Saturated			Unsaturated		
N_nH_{n+2}	ammonia	NH_3	N_nH_n	diimide	HN=NH
	hydrazine	H_2NNH_2		triazene	$HN{=}NNH_2$
	triazane	H_2NNHNH_2		1-tetrazene	$HN{=}NNHNH_2$
				2-tetrazene	$H_2NN{=}NNH_2$
				ammonium azide	$NH_4^+\bar{N}{=}\overset{+}{N}{=}\bar{N}$
			N_nH_{n-2}	hydrogen azide	$H\bar{N}{-}\overset{+}{N}{\equiv}N$
				diiminohydrazine	HN=NN=NH
			N_nH_{n-4}	octaztriene	HN=NNHN=NNHN=NH

Indeed, most high nitrogen compounds owe their existence to the groups attached to the nitrogen chains.

There are approximately 50 classes within the high nitrogen compounds; their existence and thermal stability are discussed in Chapter 2. A sampling of older and recently discovered high nitrogen species suggests both the diversity of compounds with multiple linked nitrogen atoms and the potential for extension of this field. Knowledge of compounds with three or more nitrogens in sequence began with the syntheses by Peter Griess of aromatic triazenes including diazoaminobenzene from 1859 to 1862 (13), followed by phenyl azide and other aryl azides in 1864 (14) during his investigations of aromatic diazo compounds. With the rush to expansion of organic chemistry in the latter part of the nineteenth century also came an extensive foundation for the chemistry of polynitrogen chains. By 1900 members of the triazanes **(I)**, nitrosohydrazines **(III)**, tetrazenes **(IV)**, pentazdienes **(V)**, and octaztrienes **(VI)** had been synthe-

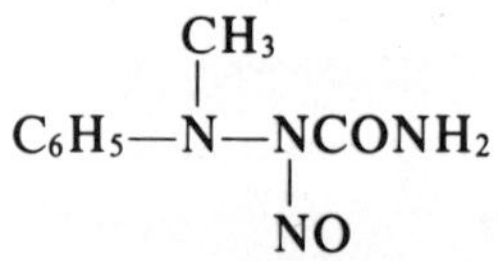

(III)

1878
Fischer (10)

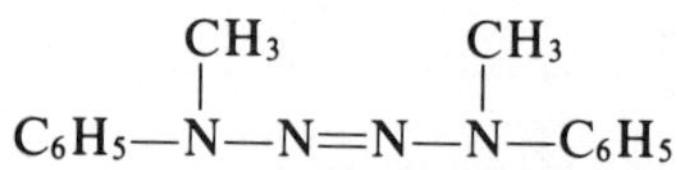

(IV)

1878
Fischer (10)

sized, shown here with the first year of their preparation and their discoverers. Some heterocycles with three or more linked nitrogens made by this time include benzotriazole **(VII)**, benzotriazinone **(VIII)**, *v*-triazole **(IX)**, tetrazole **(X)**, and tetrazolo[1,5-*a*]quinoline **(XI)** with an embedded nitrogen. Some new high nitrogen classes, first synthesized in the last half of the twentieth century, are the dipolar amine nitroimides **(XII)**, azimines **(XIII)**, and tetraazapentalenes

(XIV), the pyrazolidinedione dimers **(XV)** with a cyclic-cyclic tetrazane chain, the first authentic triaziridine **(XVI)**, and the pentazoles **(XVII)**.

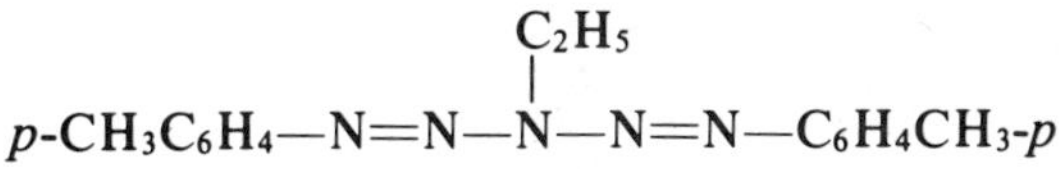

(V)

1888
Goldschmidt and Holm (12)

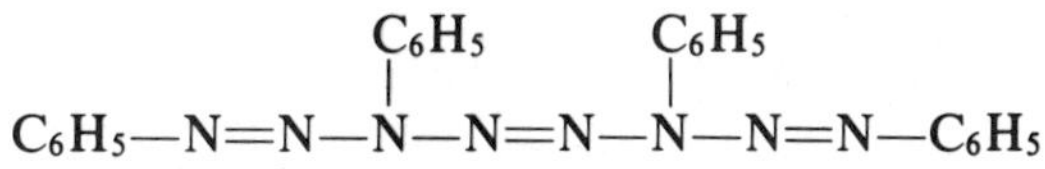

(VI)

1900
Wohl and Schiff (24)

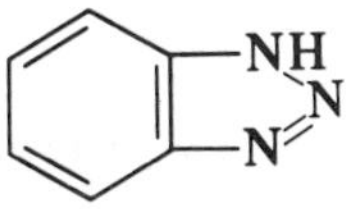

(VII)

1876
Ladenburg (16)

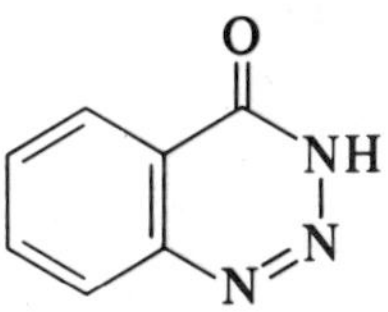

(VIII)

1887
Weddige and Finger (22)

(IX)

1891
Baltzer and v. Pechmann (3)

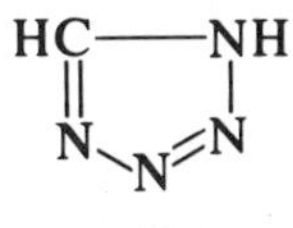

(X)

1892
Bladin (4)

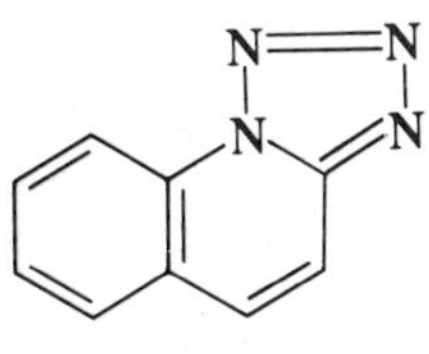

(XI)

1900
Marckwald and Meyer (18)

In comparison with corresponding derivatives of carbon chains, compounds with sequences of nitrogen atoms tend to be more fragile. Relative stabilities of these two series are partially related to the average bond energies of their single, double, and triple bonds. Thus compounds with nitrogen-nitrogen single bonds would be expected to be the least stable. The nitrogen-nitrogen double bond, however, is stronger than a carbon-carbon single bond, and the presence of N=N in a chain should enhance stability. Formation of elemental nitrogen during decomposition of many substances with contiguous nitrogen atoms is an indication of their inherent thermochemical instability. Nitrogen chains, however,

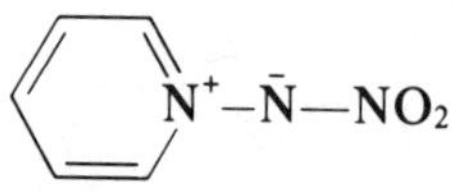

(XII)

1969
Epsztajn and Katritzky (9)

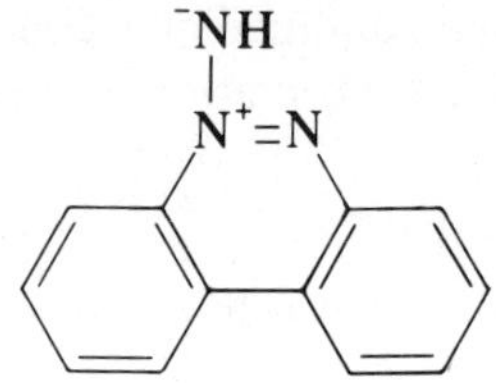

(XIII)

1971
Gait, Rees, and Storr (11)

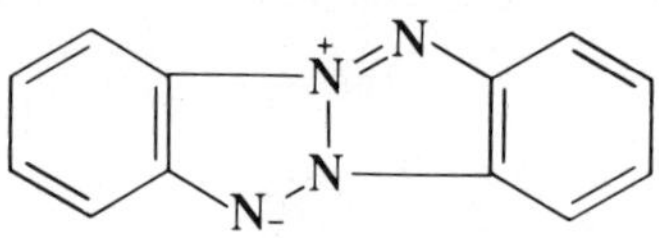

(XIV)

1962
Carboni and Castle (6)

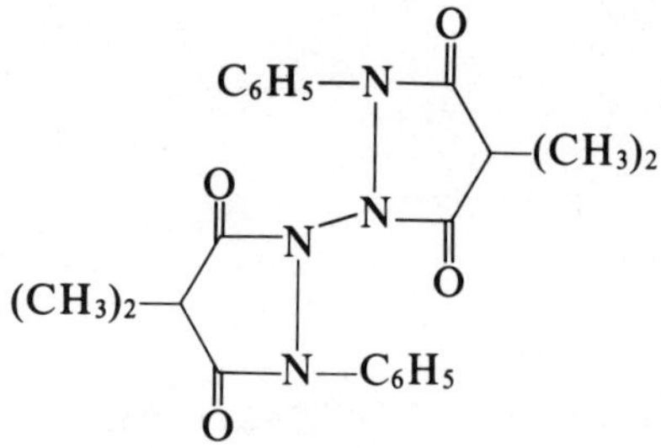

(XV)

1978
Pirkle and Gravel (20)

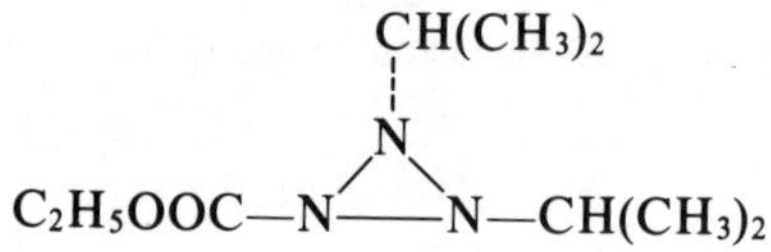

(XVI)

1980
Leuenberger, Hoesch, and Dreiding (17)

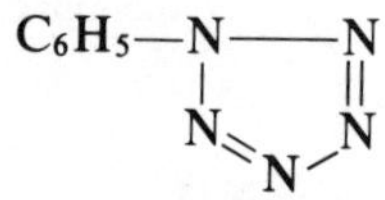

(XVII)

1956
Huisgen and Ugi (15)

often exhibit greater stability than average bond energies would suggest; this is related to the unshared electron pairs on each nitrogen atom except for ammonium type nitrogen. In the absence of adverse steric effects, unshared electron pairs on adjacent nitrogen atoms undergo resonance interaction with each other, with unsaturation in the chain and with unsaturated and aromatic substituents, increasing the effective strength of the nitrogen-nitrogen bonds.

Bond Energies
(kcal/mole) (19)

C—C	83	N—N	39
C═C	140	N═N	100
C≡C	200	N≡N	226

We shall examine the influence of nitrogen chain length, of unsaturation, and of variation in degree and kind of substitution on stability of high nitrogen compounds toward heat and chemical agents; note the ranges of their physical and physical chemical properties; and consider how they are used, their biological effects, and the numerous methods for forming nitrogen chains. It is the goal of this exploration of the high nitrogen compounds to provide a basis for further understanding and extension of their scope, not only to new compounds and classes but to materials with useful properties and applications.

REFERENCES

1. L. F. Audrieth, *J. Chem. Educ.*, **1930,** *7,* 2055–2062.
2. L. F. Audrieth and B. A. Ogg, *The Chemistry of Hydrazine,* Wiley, New York, 1951, pp. 1–12.
3. O. Baltzer and H. v. Pechmann, *Ann.*, **1891,** *262,* 314–324.
4. J. A. Bladin, *Ber.*, **1892,** *25,* 1411–1413.
5. E. D. Caldwell in *Encyclopedia of Polymer Science and Technology,* Vol. 6, Wiley-Interscience, New York, 1967, p. 336.
6. R. A. Carboni and J. E. Castle, *J. Am. Chem. Soc.*, **1962,** *84,* 2453–2454.
7. F. A. Cotton and G. Wilkinson, *Advanced Inorganic Chemistry—A Comprehensive Text,* 4th ed., John Wiley, New York, 1980, p. 305.
8. *Ibid.*, pp. 487, 494.
9. J. Epsztajn and A. R. Katritzky, *Tetrahedron Lett.*, **1969,** 4739–4742.
10. E. Fischer, *Ann.*, **1878,** *190,* 67–183.
11. S. F. Gait, C. W. Rees, and R. C. Storr, *J. Chem. Soc., Chem. Commun.*, **1971,** 1545–1546.
12. H. Goldschmidt and J. Holm, *Ber.*, **1888,** *21,* 1016–1026.
13. P. Griess, *Proc. Roy. Soc.*, **1859,** *9,* 594–597; *Ann.*, **1861,** *117,* 1–67; **1862,** *121,* 257–280.
14. P. Griess, *Phil. Trans. Roy. Soc. (London)*, **1864,** *154,* 667–731.
15. R. Huisgen and I. Ugi, *Angew. Chem.*, **1956,** *68,* 705–706.
16. A. Ladenburg, *Ber.*, **1876,** *9,* 219–223.
17. C. Leuenberger, L. Hoesch, and A. S. Dreiding, *J. Chem. Soc., Chem. Commun.*, **1980,** 1197–1198; L. Hoesch *et al.*, *Helv. Chim. Acta,* **1982,** *65,* 682–696.
18. W. Marckwald and E. Meyer, *Ber.*, **1900,** *33,* 1885–1895.
19. I. T. Millar and H. D. Springall, *Sidgwick's Organic Chemistry of Nitrogen,* Oxford University Press, 1966, p. 34.
20. W. H. Pirkle and P. L. Gravel, *J. Org. Chem.*, **1978,** *43,* 808–815.
21. R. Stollé, *J. Prakt. Chem.*, **1933,** *138,* 1–17.
22. A. Weddige and H. Finger, *J. Prakt. Chem.*, **1887,** *35,* 262–264.
23. A. Wohl, *Ber.*, **1900,** *33,* 2759–2760.
24. A. Wohl and H. Schiff, *Ber.*, **1900,** *33,* 2741–2758.

2

THE HIGH NITROGEN POPULATION—EXISTENCE, STRUCTURES, AND THERMAL STABILITY

This chapter illustrates the known structural types of high nitrogen compounds and their stability to heat.* Organization of the high nitrogen compounds is primarily by nitrogen chain length and secondarily as acyclic, mixed acyclic-cyclic, and cyclic. Within subgroups, the sequence is from most reduced to most oxidized. Examples are selected to indicate, for each structural class, the range of both substituents and variation in thermal stability. The compounds discussed here usually have been analyzed and characterized and it is reasonable to believe, on the basis of physical or chemical data or both, that they have the constitutions reported. In a few chemical categories compounds of uncertain structure are included to indicate limitations in stability or the present state of knowledge. Physical-chemical information, elucidating the structure of particular compounds or classes, is discussed under the technique used. Chemical evidence is presented in the chapters on synthesis or reactions as appropriate.

The criteria for existence used are that a chemical must be capable of isolation and handling under some attainable conditions and that physical or chemical measurements of it can be made. Thus a substance isolated at liquid nitro-

* Over fifty years ago, in his book *The Pyrolysis of Carbon Compounds,* C. D. Hurd (531) organized the then-existing knowledge of transformations of carbon compounds through the agency of heat alone. Although his focus was not on stability, he described the thermal fate of compounds in many classes with nitrogen chains, including triazenes, azides, *v*-triazoles, tetrazenes, and tetrazoles.

gen temperature that can be characterized by a physical or chemical measurement is considered to exist and is included, even if it decomposes before reaching room temperature. Intermediates that cannot be isolated, however, are not considered part of the high nitrogen population, although they may be the subject of speculation and frequently help to rationalize chemical and physical behavior.

Thermal stability is best evaluated by quantitative measurements of decomposition or lack thereof at a particular temperature and under other specified conditions, such as the presence of a solvent. Determination of kinetics, especially with knowledge of the identity of decomposition products, produces the greatest understanding, but measurement of weight loss or gas evolved in a standard period at a specified temperature is of practical value. Programmed heat input, as in differential thermal or thermogravimetric analyses, provides excellent descriptions of thermal behavior including decomposition, but unfortunately has been applied to few high nitrogen compounds. The information that a substance is unchanged after exposure to a given temperature enables categorization of its thermal stability.

For many high nitrogen compounds, the only evidence of instability is a notation of melting with decomposition, or, less often, of decomposition on attempted distillation. A report of explosion or violent decomposition, with or without melting, makes obvious the end point of stability. Occasionally the fact of stability on or after melting is reported.

The least satisfactory measure of thermal stability is a melting point without mention of decomposition. The fact of melting at a given temperature only means that intermolecular forces in the solid have been overcome. In the absence of an indication of accompanying decomposition, and with the presumption of accurate reporting in the literature used, it is considered that under the usual rapid heating conditions of a melting point determination, no intramolecular bond breaking has occurred to the melting point. There are numerous instances, however, of solid high nitrogen compounds which do not disintegrate at their melting points, but which decompose either on standing at room temperature or in solution. Although useful for characterization of a chemical, a melting point value has limited significance as an indication of thermal stability.

2.1. ACYCLIC THREE-NITROGEN CHAINS

2.1.1. Triazanes and Triazanium Salts

Triazane, H_2NNHNH_2, with a completely reduced acyclic three-nitrogen chain, is the hypothetical parent of this class. Most known triazanes carry both acyl and alkyl or aryl substituents together with one or more hydrogen atoms, and are moderately stable as indicated by their melting points, which range from 69 to 182° with decomposition (Table 2.1). Stability seems to be conferred by the

Table 2.1. Triazanes

Triazane	Melting Point (°C)	Reference
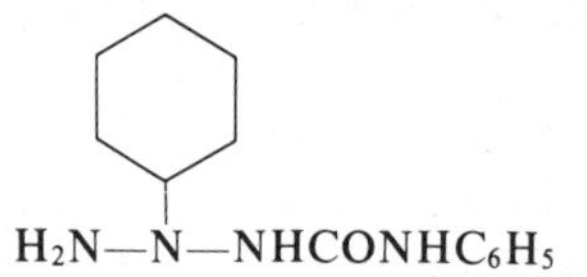	89–91 dec.	969
$C_6H_5CH{=}N{-}N(C_6H_5){-}NHCOCH_3$	162–164 dec.	1194
$(CH_3)_2N{-}N(COOC_2H_5){-}NHCOOC_2H_5$	92.5–93 dec.	583
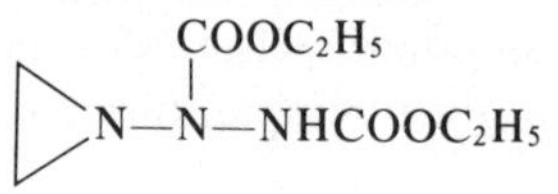	69–70 dec.	675

simultaneous presence on the triazane chain of an electron-releasing and an electron-withdrawing group. In contrast, 1,1,3-trifluoro-2,3-bis(trifluoromethyl)-triazane, with only electron-withdrawing substituents, decomposes on warming to room temperature (950).

$$F_2N{-}N(CF_3){-}N(CF_3)F \xrightarrow{-46 \text{ to } 25^\circ} CF_3N{=}NCF_3 + \tfrac{1}{2}\, N_2F_4$$

bp (calc.) 19°

The presence of methyl triazane, $CH_3NHNHNH_2$, carrying an electron-releasing substituent, in a fraction boiling at 45–85° also containing diethyl ether, is indicated by proton magnetic resonance and chemical data; it has not been obtained in a pure state (888).

1,2-Bis(ethoxycarbonyl)-3-phenyltriazane at room temperature in benzene solution, in a reversal of synthesis, liberates aniline in a quantity corresponding to about 6% decomposition (223a).

$$C_6H_5NH{-}N(COOC_2H_5){-}NHCOOC_2H_5 \xrightarrow[\text{benzene}]{26^\circ} C_6H_5NH_2 \quad (\text{3 weeks})$$

mp 138°(256)

Triazanium salts that have been prepared carry hydrogen, alkyl, or cycloalkylene substituents on the middle nitrogen, and, based on their melting points, are moderately stable. The triazanium salts with alkyl or cycloalkylene

Table 2.2. Triazanium Salts

Salt		Melting Point (°C)	Ref.
$H_2N-\overset{+}{N}H_2-NH_2$	$\bar{O}_3SCH_3$	77–79 dec.	400
$H_2N-\overset{CH_3}{\underset{CH_3}{N^+}}-NH_2$	Cl^-	124	721, 1121
	ClO_4^-	194–195	1119
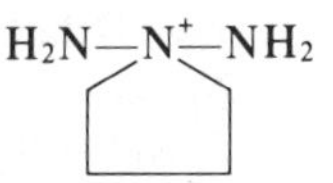	Cl^-	121–123 dec.	1120

substituents are more stable than the unsubstituted triazanium salt (Table 2.2). Thermolysis of 2,2-dimethyltriazanium chloride in a helium atmosphere at 150° led to no product, but decomposition occurs at 200°, forming ammonium chloride, nitrogen, 1,1-dimethylhydrazine, and formaldehyde dimethylhydrazone

$$H_2N-\overset{CH_3}{\underset{CH_3}{N^+}}-NH_2\ Cl^- \xrightarrow[He]{200\text{–}250°} NH_4Cl + N_2 + (CH_3)_2NNH_2 + (CH_3)_2NN{=}CH_2$$

(721). Solid dipolar triazanium compounds can be isolated, but are very unstable as indicated by the 15 min half-life of 2,2-dimethyltriazanium betaine carbanilide, which decomposes to tetramethyltetrazene and phenylurea (969).

$$H_2N-\overset{CH_3}{\underset{CH_3}{N^+}}-N^--CONHC_6H_5 \xrightarrow[THF]{\text{room temp.}} (CH_3)_2N-N{=}N-N(CH_3)_2 + C_6H_5NHCONH_2$$

2.1.2. Triazenes

The triazenes, also called diazoamino compounds, are classified as monosubstituted ($RN{=}NNH_2$), disubstituted ($RN{=}NNHR'$ and $RN{=}NN{=}Y$), and trisubstituted ($RN{=}NNR'R''$). Dipolar triazenes and triazene salts also are known.* With the exception of the lower dialkyl 1,3-disubstituted triazenes and the trialkyltriazenes, which are liquids, most known triazenes are solids.

*Triazene structures and thermal stability are reviewed by Süling (1053) and Smith (1004). Diaryl triazenes are covered by Campbell and Day (192) and a more recent review by Vaughan and Stevens covers monoalkyl monoaryl triazenes (1126).

The parent triazene, $H_2NN{=}NH$, is considered to be present after pulse radiolysis of aqueous hydrazine solution on the basis of spectral and kinetic evidence (446). Its half-life in aqueous solution at pH 9 (most stable) and 24° is calculated from its decay rate constant (1060) to be about 11.5 min. Triazene also is formed in a high-speed gas stream of hydrazine by electrodeless electrical discharge (331, 332).

Triazenes with only one substituent are unstable (Table 2.3). Both phenyl- and *p*-bromophenyltriazene are obtained in interconvertible forms; one modification of each of these compounds exhibits a remarkable spontaneous motion with conversion to its alternate form. Presumably the motion results from small quantities of nitrogen generated by decomposition of the triazenes. Other monosubstituted phenyl (266) and anthraquinone triazenes (1136) have been reported, usually without sufficient stability for characterization or analysis.

Substituents on 1,3-triazenes bearing a hydrogen atom on the diazoamino chain include alkyl, aryl, heterocyclic, and acyl groups. Unsymmetrical 1,3-disubstituted triazenes can exist, in principle, in two forms depending on the position of the hydrogen (192):

$$RN{=}NNHR' \qquad RNHN{=}NR'$$

No separate tautomers have been isolated, but evidence of hydrogen location in some triazenes has come from X-ray crystallography and equilibrium compositions have been determined from nuclear magnetic resonance, infrared, and mass spectral studies.

Thermal decomposition of disubstituted triazenes liberates nitrogen and forms free radicals which undergo further reaction with each other and with solvent, if any. In addition to the alkyl or aryl radicals from the 1- and 3-sub-

Table 2.3. Monosubstituted Triazenes

Triazene	Melting Point (°C)	Thermal Stability	Ref.
$C_6H_5N{=}NNH_2$	40 dec.	Dec. to N_2 and aniline at room temp.; expl. in presence of aniline	264
	50 dec.	Spontaneous motion with slight d and conversion to lower melting form	
p-$BrC_6H_4N{=}NNH_2$	36.5 dec.	Spontaneous motion with slight d and conversion to higher melting form	266
	39 dec.	Dec. in several days below room temp. to N_2 and *p*-bromoaniline	

stituents, substituted amino radicals also are formed (941). The alkyl group of alkyl aryl triazenes is converted to the corresponding hydrocarbon, as illustrated by the decomposition of 1-methyl-3-phenyltriazene (271, 272):

$$CH_3NHN{=}NC_6H_5 \xrightarrow{120^\circ} N_2 + CH_4 + CH_3NHC_6H_5 + (C_6H_5\cdot)$$

Representative 1,3-disubstituted triazenes with a hydrogen on the nitrogen chain are listed in Table 2.4 with examples of low, medium, and high thermal stability. The least heat-resistant 1,3-disubstituted triazenes are comparable to the triazanes in stability. Combination of an aryl and an electron-withdrawing group (e.g., acyl, trifluoromethyl) on the triazene induces relative instability. Alkyl-substituted diazoamino compounds are less stable than aryl derivatives. 1-Aryl-3-methyltriazenes, having aryl substituted with *meta* or *para* electron-withdrawing groups, are moderately stable solids which usually decompose on standing. Heating a benzene or alcohol solution of 1-*p*-nitrophenyl-3-methyltriazene leads to 1,3-bis(*p*-nitrophenyl)-3-methyltriazene (12). The analogous

$$p\text{-}O_2NC_6H_4N{=}NNHCH_3 \xrightarrow[\text{benzene}]{\text{heat in}} p\text{-}O_2NC_6H_4N{=}N{-}\underset{}{\overset{\displaystyle CH_3}{\overset{|}{N}}}{-}C_6H_4NO_2\text{-}p$$

1-*o*-nitrophenyl-3-methyltriazene is less stable and decomposes within hours of preparation, even in a vacuum desiccator. A progression of increasing thermal stability is seen in the half-lives of the 1,3-bis(*p*-tolyl), phenyl, and *p*-nitrophenyl triazenes, indicating, in symmetrical diaryl triazenes, a small destabilizing effect of the methyl substituent and a considerable stabilizing influence of the electron-withdrawing nitro group. As indicated by the melting points of the 3-pyridyl and 2-benzothiazolyl derivatives and the *s*-triazole polymer, relatively stable triazenes result from symmetrical substitution with electron-withdrawing heterocycles.

The 1,3-triazenes with monovalent substituents apparently exist in the trans form, as suggested by dipole moment data (639); this configuration has been confirmed by X-ray analysis in the case of *p,p'*-dibromodiazoaminobenzene (600).

Presently known disubstituted triazenes without hydrogen are substituted with monovalent aryl or more strongly electron-accepting radicals at N-1 and divalent cyclopropene, heterocyclic, and phosphorus groups at N-3 (Table 2.5). Triazenes with divalent carbon substituents may be looked on as 1,2,3-triazabutadienes. The phosphorus substituents utilize the *d*-orbitals of this element in bonding to nitrogen.

Some of these triazenes exist in both cis and trans forms. In the 1-aryl-3-[3'-methylbenzothiazolinylidene-(2)]triazene series, cis isomers with ortho and para electron-withdrawing substituents on the phenyl group are converted to the trans forms on heating in solution (307, 309). Analogous disubstituted tria-

Table 2.4. 1,3-Disubstituted Triazenes with Hydrogen on Nitrogen

Triazene	Melting Point (°C)	Thermal Stability	Ref.
$C_6H_5N{=}NNHCOC_6H_5$	84 dec.		112
$C_6H_5N{=}N{-}NH{-}CF_3$	91	Dec. reflux in benzene	702
$CH_3N{=}NNHCH_3$	−12	Bp 92°/1 atm, incip. dec.	263
		$t_{1/2}$ (200°, 10–80 mm) 1.9 hr[a]	902
$C_6H_5N{=}NNHCH_3$	37–37.5	Stable in dry benzene at 40°:	33
		incip. dec., 90°	261, 262
		$t_{1/2}$ (110°, cumene) 48 min	933
p-$CH_3C_6H_4N{=}NNHC_6H_4CH_3$-$p$	118	$t_{1/2}$ (150°, paraffin) 27 min	760, 823
$C_6H_5N{=}NNHC_6H_5$	100	slow N_2 evoln. in vac., 105–110°	249, 474
		$t_{1/2}$ (150°, paraffin) 55 min	823
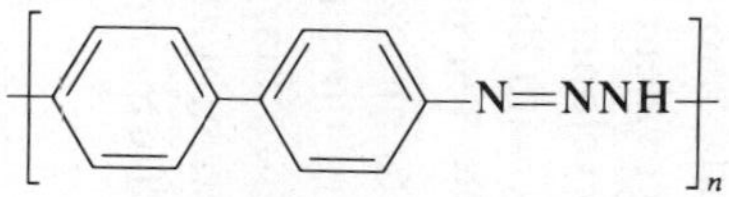	75–85	Dec. 90–100°; 5.2% weight loss at 150°	99
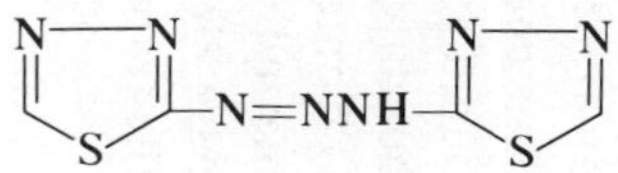	89–90 dec.		996

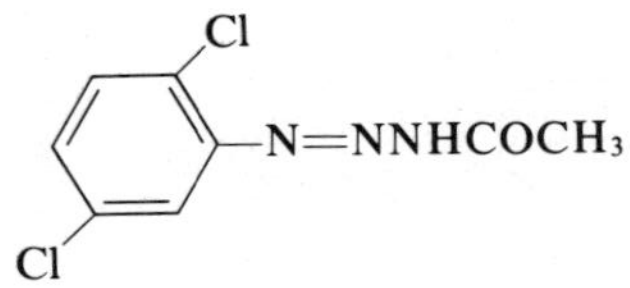	165–166 dec.		507
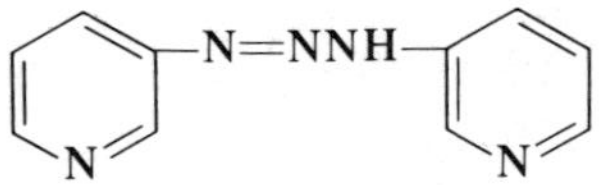	173–174 dec.		749
p-$O_2NC_6H_4N{=}NNHC_6H_4NO_2$-*p*	225–230 dec.	$t_{1/2}$ (160°, paraffin) 10.5 hr	511, 823
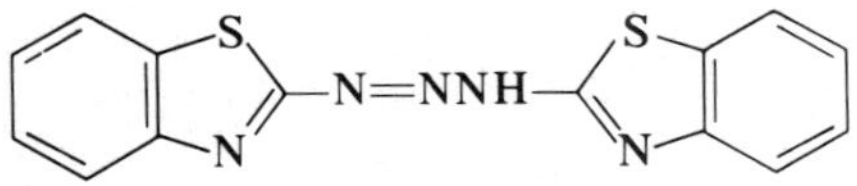	247–250		587
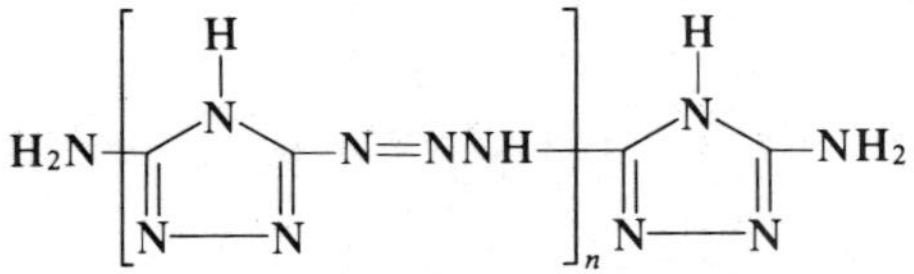	>300°		443

[a]Kinetic data reported in the literature as first-order rate constants have been converted to half-life values to facilitate comparisons of thermal decomposition. Temperatures and either pressure conditions for gas phase kinetic studies or solvent used, if in solution, are reported for each half-life value.

Table 2.5. 1,3-Disubstituted Triazenes with Divalent Substituents

Triazene	Melting Point (°C)	Thermal Stability	Ref.
$(C_2H_5)_2P(C_6H_5)=N-N=N-C_6H_5$	51–52 dec.	Stable short time at room temp.	1028
$(C_6H_5)_3P=N-N=N-SO_2C_6H_5$	87–88 dec.	Dec. on solution in benzene or $CHCl_3$	647
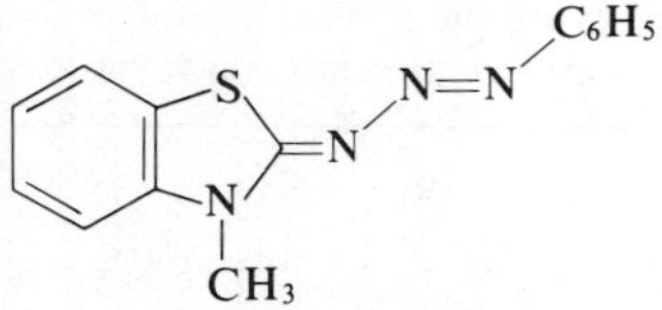	127	Dec. 170–180	307
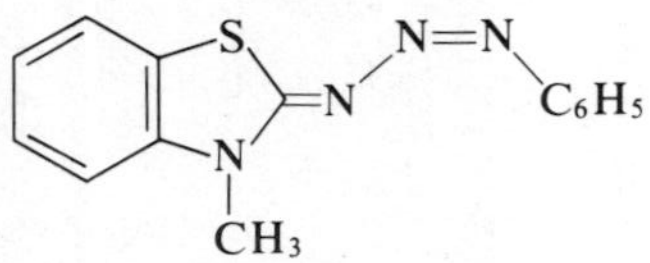	148–152	Dec. 170–180	307
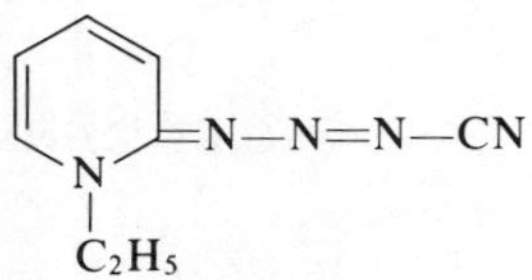	154–157 dec.		457
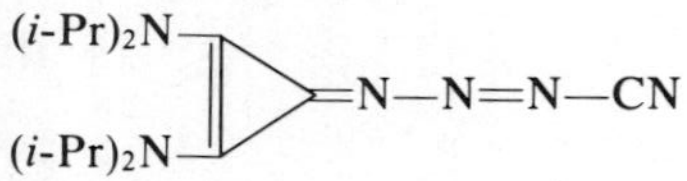	162–165	Dec. reflux in toluene	390
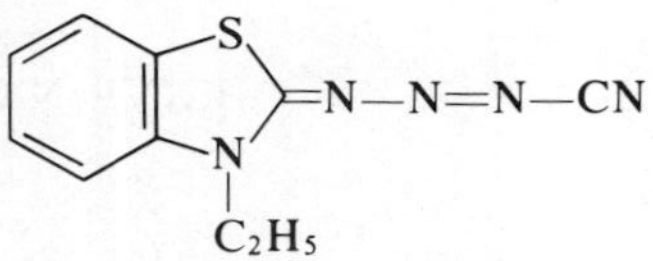	233–234 dec.		58

zenes with phenyl or with phenyl substituted with electron-releasing groups do not rearrange, but decompose at higher temperatures.

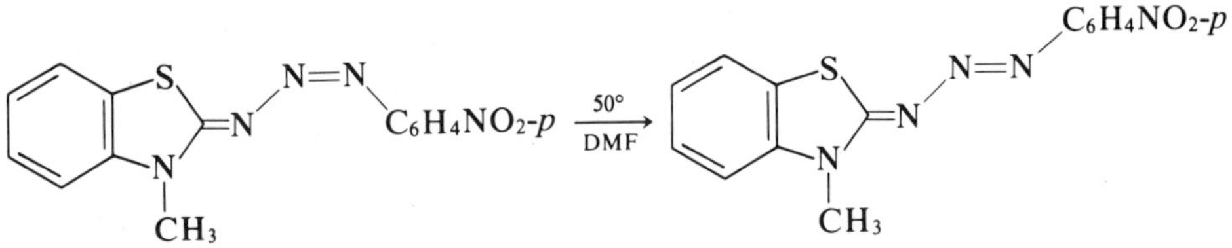

As indicated in Table 2.5, the phosphorus derivatives are less stable than other members of this subclass of triazenes, most of which tend to be slightly less stable than the 1,3-disubstituted triazenes with hydrogen.

Known trisubstituted triazenes are substituted with alkyl, aryl, heteroaryl, acyl, and noncarbon radicals. Boiling point data for the liquid trialkyltriazenes

$C_6H_5CH_2N{=}NN(CH_3)_2$

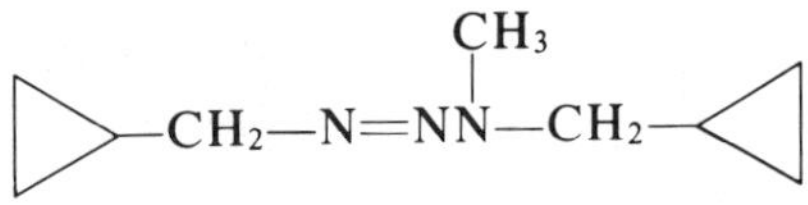

$$C_6H_5CH_2N{=}N\overset{CH_3}{\overset{|}{N}}C_4H_9\text{-}n$$

without substituents in the alkyl groups have not been determined; 1-benzyl-3,3-dimethyltriazene can be stored in an oxygen atmosphere at room temperature for several weeks without appreciable decomposition (992).

Although a solid, 1,1,3-triphenyltriazene is less stable than 1-alkyl-1,3-diphenyltriazenes or the liquid 1,1-dialkyl-3-phenyl derivatives, as indicated in Table 2.6. Thermal decomposition of the trisubstituted triazenes involves split-

$$p\text{-}ClC_6H_4N{=}N{-}\overset{CH_3}{\overset{|}{N}}{-}C_6H_4Cl\text{-}p \xrightarrow[\text{reflux in } CH_3OH]{65^\circ}$$

$$N_2 + \underset{34.2\%}{p\text{-}ClC_6H_4NHCH_3} + \underset{13.2\%}{p\text{-}ClC_6H_4\overset{CH_3}{\overset{|}{N}}C_6H_4Cl\text{-}p} + \underset{5.0\%}{p\text{-}ClC_6H_4C_6H_4Cl\text{-}p}$$

ting out of nitrogen and combination of the resulting radicals (57). It is possible to distill 1,1-di-*n*-hexyl-3-phenyltriazene at 156–158° at 0.6 mm without decomposition (935). The inherent instability of the triazene chain, however, is shown by the thermal decomposition of other aryl or heterocyclic dialkyltriazenes (Table 2.6).

Replacement of one phenyl by an acyl in triphenyltriazene renders the molecule slightly less stable. The stability properties of 1-acetyl-1,3-diphenyltriazene (Table 2.7) illustrate the poor utility of a melting point value alone as an indica-

Table 2.6. Trisubstituted Triazenes with Alkyl, Aryl, and Heterocyclic Substituents

Triazene	Melting Point (°C)	Thermal Stability	Ref.
$CF_3N{=}N{-}N(CH_3)_2$	Bp 58	Distills without dec.	702
$C_6H_5{-}N{=}NN(C_6H_5)_2$	47	Rapid dec., 100°	1130
$p\text{-}ClC_6H_4N{=}NN(CH_3){-}C_6H_4Cl\text{-}p$	72–74	Dec. 65° in refluxing CH_3OH	570, 1128
$C_6H_5N{=}NN(C_6H_5){-}C(C_6H_5)_3$	161	$t_{1/2}$ (91.2°, cumene) 1.0 hr	498, 1081
$C_6H_5N{=}NN(C_6H_5){-}CH_2C_6H_5$	74	$t_{1/2}$ (136.8, cumene) 1.1 hr	498, 1029
$C_6H_5N{=}NN(CH_3)_2$	Bp 113–114/12		474
$C_6H_5N{=}NN(C_2H_5)_2$	Bp 95–96/6	$t_{1/2}$ (230°, diphenylmethane) 24 min	498, 687
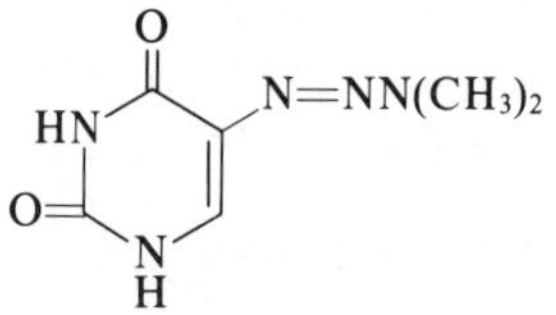	166–167 dec.	Less pure sample, expl. 162–164°	1090, 1099
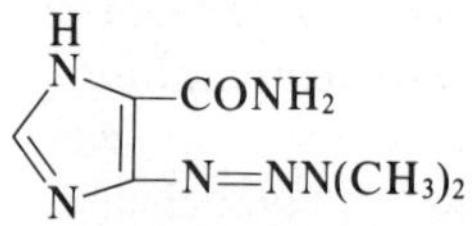	210–211 dec.		1204

Table 2.7. Trisubstituted Triazenes with Acyl Substituents

Triazene	Melting Point (°C)	Thermal Stability	Ref.
$C_6H_5{-}N{=}N{-}N(C_6H_5)COC_6H_5$	131 dec.	$t_{1/2}$ (70°, toluene) 2.2 hr	86, 498
$C_6H_5N{=}N{-}N(C_6H_5)COCH_3$	138–140	Dec., room temp., in cyclohexane in 4–5 days	398
		$t_{1/2}$ (95°, benzene) 15.1 hr	234
$NCN{=}N{-}N(CH_3)_2$	33		1039

Table 2.8. Trisubstituted Triazenes with Noncarbon Substituents

Triazene	Melting Point (°C)	Thermal Stability	Ref.
$C_6H_5{-}N{=}N{-}N[Si(CH_3)_3]_2$	Crystalline at −15	Liquid at 0°, slowly forming N_2; $t_{1/2}$ (80°, mesitylene) 16 min	1171 1174
$CH_3O{-}N{=}N{-}N[Si(CH_3)_3]_2$	−10 Bp 25/1	Dec., 65° in benzene, 1 hr	1168
$CH_3N{=}N{-}N(CH_3){-}Si(CH_3)_3$	Bp 57/50		156
$C_6H_5N{=}N{-}N(CH_3){-}Si(CH_3)_3$	Bp 73–74/high vac.	Dec., 180° in cycloheptane; thermolysis of pure material can be explosive	1172 1174
$C_6H_5N{=}N{-}N(C_6H_5){-}Sn(C_2H_5)_3$	Bp 151–153/0.001	$t_{1/2}$ (222° diphenylmethane) 13 min	498

tion of thermal stability, because of the often considerable crystal lattice energy. 1-Cyano-3,3-dimethyltriazene may be more stable than the monoacetyl or benzoyl derivatives (Table 2.7), but the absence of specific studies of thermal decomposition makes generalizations moot.

Compounds with three noncarbon radicals attached to the triazene chain are unknown. Triazenes with two trialkyl or alkoxysilyl groups decompose readily but those with one trialkyl or alkoxysilyl or one trialkylstannyl are relatively stable (Table 2.8). The 1-aryl-3,3-di(trialkylsilyl)triazenes can be isolated in interconvertible cis and trans forms (1173).

Alkoxy triazene derivatives with oxygen linked to nitrogen such as 1-methoxy-1-methyl-3-*p*-nitrophenyltriazene have been prepared (128). Other than its

$$p\text{-}O_2NC_6H_4N{=}N{-}N(CH_3)OCH_3$$

mp 66°

melting point, no data bearing on its thermal stability have been reported.

Dipolar triazenes include the triazene oxides, the aryl diazoisocyanides, and a related diazonitrilium derivative (Table 2.9). Triazene oxides with one, two, and three organic substituents are known. Those with one and two substituents formerly were considered to be hydroxytriazenes but infrared and proton magnetic resonance data show them to be dipolar triazene oxides (199, 745, 746).

Table 2.9. Dipolar Triazenes

Triazene	Melting Point (°C)	Thermal Stability	Ref.
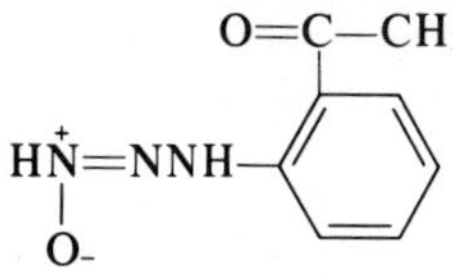	116–117 dec.	Dec. on heating in benzene	733
$CH_3\overset{+}{N}(O^-)=NNHC_6H_5$	72–73		294
$C_6H_5\overset{+}{N}(O^-)=NNHC_6H_5$	127 dec.	Can be kept indefinitely at room temperature	746, 893
$C_6H_5\overset{+}{N}(O^-)=NN(CH_3)C_6H_5$	72	Volatile with steam with slight dec.	66
$C_6H_5\overset{+}{N}(O^-)=NN(C_6H_5)_2$	128.5–129		63, 66
$p\text{-}ClC_6H_4N=N\overset{+}{N}\equiv C^-$	21	Can be distilled at 60°/0.8	539
$C_6H_5N=N\overset{+}{N}\equiv C\overset{-}{B}(C_6H_5)_3$	79–81 dec.	Unchanged at −15° under N_2	150

No specific investigations of thermal stability of these compounds have been made, but melting point data and incidental observations show no special heat resistance for the dipolar triazenes. Triazene oxides with one organic substituent are less stable than those with two or three such groups.

Both anionic and cationic salts of triazenes are known (Table 2.10). With heavy metal ions, triazene anionic salts form four-membered chelate rings (283). Little investigation has been made of the thermal stability of the alkali metal salts of the triazenes. The palladium salt of diazoaminobenzene does not decompose until above 300°, but the salts of copper I and II and silver explode on heating (283, 1145), and even the lithium salt of 1-bicycloheptyl-3-tosyltriazene is thermally unstable. Salts in which the triazene is part of the cation are generally stable, at least to 150°. Although with both anionic and cationic triazene salts, formulas with conventional imido or quaternary nitrogen atoms can be written, partial dispersal of charge over the triazene chain seems probable.

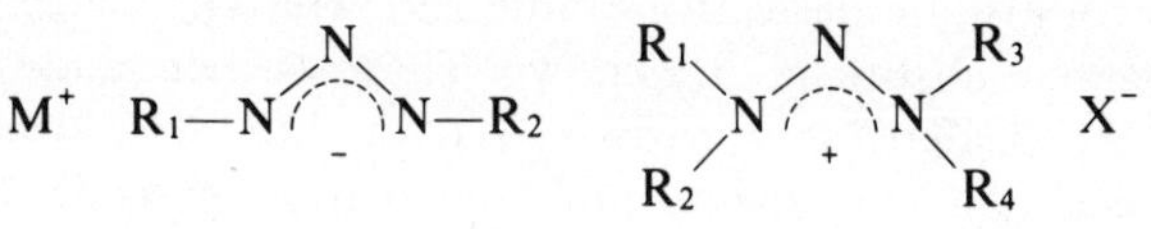

Table 2.10. Triazene Salts

Triazene	Melting Point (°C)	Thermal Stability	Ref.
$C_6H_5N{=}NN(Ag){-}C_6H_5$	Expl. 199–200		283, 1145
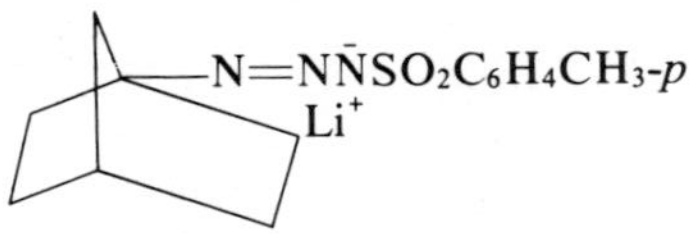	Solid	Dec. >75°/0.1; occasional expl. during pyrolysis	905
$p\text{-}CH_3C_6H_4N(CH_3){-}N{=}\overset{+}{N}[CH(CH_3)_2]_2\ \ B\overset{-}{F}_4$	96–98		425
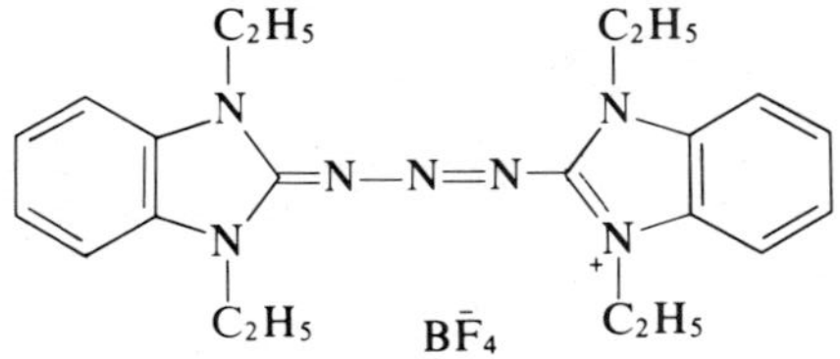	161–165 dec.	Heat 5 hr at 150° in DMF, <3% dec.	61
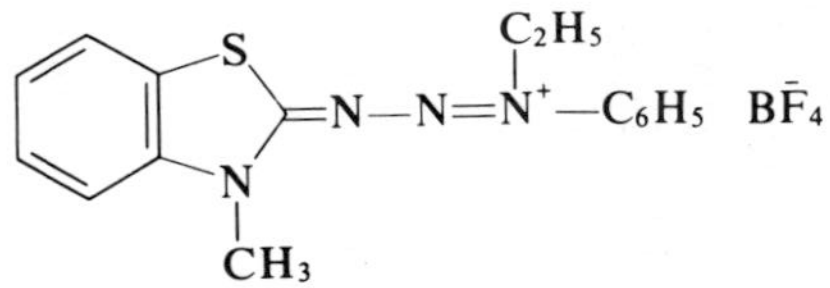	195–200		308
$(C_6H_5)_3\overset{+}{P}{-}N{=}N{-}N(CH_3)_2\ \ \overset{-}{O}_3SCF_3$	179–180		1039

2.1.3. Mononitrosohydrazines

Mononitrosohydrazines with one, two, and three alkyl or aryl substituents can be prepared as can monoacyl derivatives with one or two alkyl or aryl radicals (Table 2.11). *N*-Acetyl-*N*-arylsulfonyl-*N'*-tetraacetylglucosyl-*N'*-nitrosohydrazines contain two acyl groups.

Most nitrosohydrazines are unstable, decomposing either on storage at room temperature or at their melting points, usually below 160°. Monoalkyl derivatives appear to be more thermally stable than those with one aryl group. Heating *N*-nitrosophenylhydrazine in bromobenzene forms substantially stoichiometric quantities of aniline and nitrous oxide (219). Nitric oxide is evolved,

$$C_6H_5N(NO){-}NH_2 \xrightarrow[\text{bromobenzene}]{130^\circ} C_6H_5NH_2 + N_2O$$

nonquantitatively, on decomposition of *N*-nitrosotriphenylhydrazine at its melting point (387). The phenylmethylene derivative of *p*-bromophenylnitroso hydrazine decomposes in toluene solution overnight or on keeping the solid in a vacuum desiccator for 3 weeks (315). Oxides of nitrogen are formed as well as a denitrosated dimer, called a tetrazane, but more likely the product of carbon-nitrogen rather than nitrogen-nitrogen coupling (740a, 1196a).

$$C_6H_5CH{=}N{-}N(NO)C_6H_4Br\text{-}p \xrightarrow[\text{24 hr}]{\text{toluene, room temp.}} [NO] + \underset{12\%}{C_{26}H_{20}Br_2N_4}$$

2.1.4. Nitric Oxide-Amine Reaction Products

Salts with a chain of three nitrogen atoms result from the combination of nitric oxide and primary or secondary aliphatic amines. Their structure (683, 974) is illustrated by the product from diethylamine, which can be stored indefinitely

$$(C_2H_5)_2N{-}\overset{+}{N}(O^-){=}N{-}O^- \quad H_2\overset{+}{N}(C_2H_5)_2$$

under ether at room temperature and melts with decomposition in a sealed tube at 107° (277). Nitric oxide-primary amine reaction products are usually less stable than those from secondary amines and longer dialkyl derivatives are more thermally stable than shorter chain compounds (Table 2.12).

2.1.5. Acyclic Amine Nitroimides and Azimines

Trimethylamine nitroimide, the first acyclic example of the dipolar amine nitroimides, sublimes at 190°/0.3 mm, survives at 200° for several hours, and melts

$$(CH_3)_3\overset{+}{N}—\overset{-}{N}—NO_2$$

with decomposition at 212–213° (299, 301).

In 1963 before any members of this class of ylides had been recognized, Huisgen (516) suggested the name azimine for the three-nitrogen 1,3-dipole with

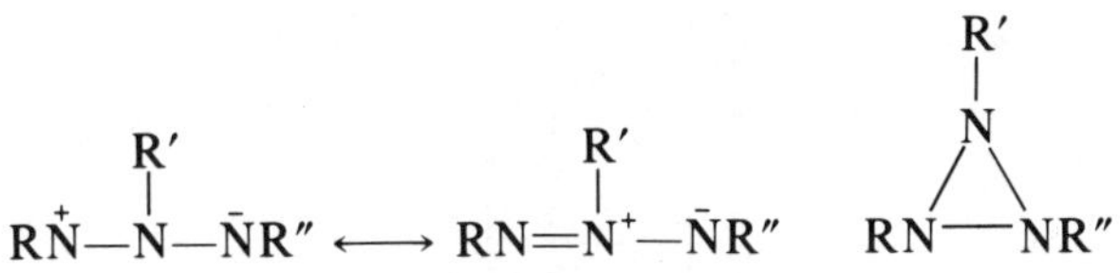

substituents at each nitrogen. Open chain azimines are isomers of the triaziridines (1023). The acyclic azimine structure was confirmed by X-ray analysis of the 1-(4,5-di(methoxycarbonyl)-2-thiazolyl)-2,3-diphenylazimine (132, 133). On the basis of the melting points of these azimines thermal stability of this class appears to be greater than the mononitrosohydrazines and similar to that of the

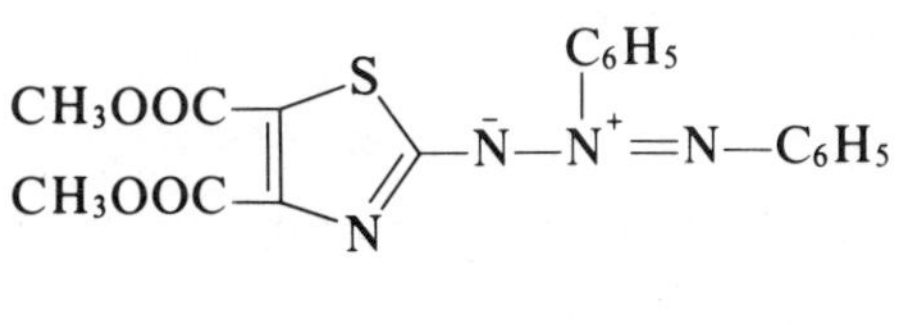

mp 178–180° (132)

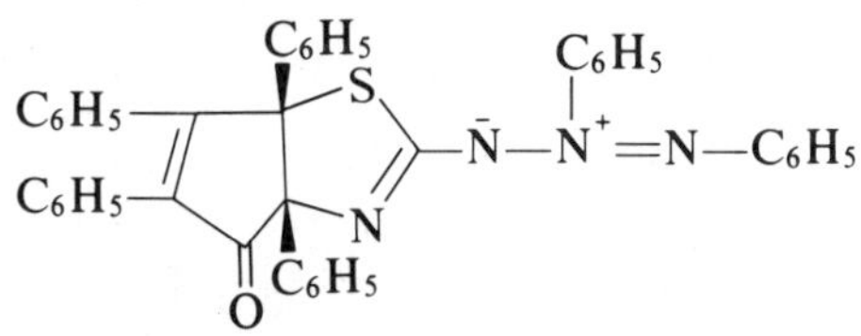

mp 166–169° dec. (133, 134)

more stable disubstituted triazenes. Azimines with alkoxycarbonyl and alkyl or aryl substituents also have been prepared (653, 654).

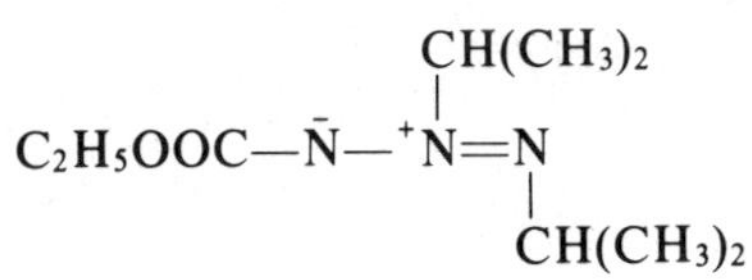

oil; bp 40°/0.01

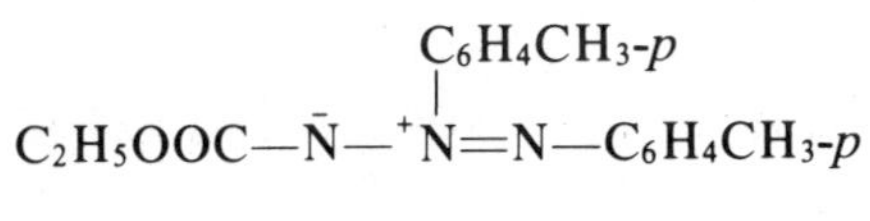

oil

2.1.6. Diazonium Amine Salts

Diazonium amine hexachloroantimonate salts, in which a diazonium group is attached to an amino group, have very little heat resistance (964). The

Table 2.11. Nitrosohydrazines

Nitrosohydrazine	Melting Point (°C)	Thermal Stability	Ref.
$H_2N-N(C_6H_5)-NO$	51	Slight dec., 8–12° for 14 days; dec. 130° in bromobenzene	64, 219 247
$H_2N-N(CH_3)-NO$	44–45		858
$C_6H_5CH{=}N-N(CH_3)-NO$	73.0–73.5; 77–78		1084, 1096
$H_2N-N(CH_2C_6H_5)-NO$	70	Does not dec. at mp	236
$(CH_3)_2CHNH-N(CH(CH_3)_2)-NO$	Bp 65–66/6–8	Unstable, room temp.; rapid dec., 160–162	680
$C_6H_5CONH-N(C_6H_5)-NO$	110 dec.	Dec. room temp.	1135

(901). Decomposition in a flowing system at 850° with cooling of the exit gases to liquid nitrogen temperature produces nonstoichiometric quantities of the same products, and in addition hydrogen and a blue solid with the composition of $(NH)_n$ which changes to ammonium azide at −125° (924).

$$HN_3 \xrightarrow{850^\circ} 1.03\ N_2 + 0.10\ H_2 + 0.19\ NH_4N_3 + (NH)_n$$

Nonexplosive decomposition to metal and nitrogen occurs near 300° for alkali metal azides and between 110° and 160° for the alkaline earth azides (Table 2.13). Once decomposition has begun, evolution of nitrogen can proceed from sodium azide and the alkaline earth azides at temperatures lower than their initial decomposition points by 10–50°. This difference between temperatures required for initiation and continuation of decomposition probably is related to an induction period before significant decomposition begins. Between 98.9 and 131.7°, barium azide requires induction periods from 340 to 20 min before decomposition accelerates strongly; toward the end, the reaction slows to a unimolecular decomposition (434). During the acceleration, centers of decomposition consisting of metallic barium are formed at an increasing rate. With barium and other inorganic azides the released

$$Ba(N_3)_2 \longrightarrow 3N_2 + Ba^0$$

$$2Ba(N_3)_2 + Ba^0 \longrightarrow Ba_3N_2 + 5N_2$$

metal induces further decomposition of the azide, forming nitrides (46, 1091).

Lead azide evolves negligible amounts of gas on heating at 150° for 40 hr in the vacuum stability test. At a heating rate of 5°/min in differential thermal analysis exothermal decomposition begins at 285–297°, with peaks from 296–308°, sometimes accompanied by detonation (49).

Table 2.13. Decomposition Points of Inorganic Azides

Azide	Initial Decomposition (°C) (1091)
NaN_3	330
KN_3	320
RbN_3	260
CsN_3	290
$Ca(N_3)_2$	110
$Sr(N_3)_2$	140
$Ba(N_3)_2$	160

Alkyl, aralkyl, and alicyclic azides (Table 2.14) generally withstand temperatures to about 150°. Based on their half-lives at 200°, methyl azide is more stable than ethyl or isopropyl azides. The decomposition products of methyl azide in kinetic studies from 155 to 200° are nitrogen, small amounts of hydrogen, and methane, and hexamethylenetetramine or a polymer of similar composition (819). An earlier kinetic investigation of methyl azide decomposition reported ammonium azide among the products (643). Aliphatic azides with a β-hydrogen decompose by splitting out hydrogen azide with formation of an olefin, as well as by loss of nitrogen. When nitrogen is released, the products formed can be rationalized by assuming formation of a nitrene, followed by further reaction of this intermediate. In the case of ethyl azide from thermolysis at 187.5–220° and an initial pressure of 25 mm, piperazine, ethylene imine, ethylene imine polymers, hydrogen azide, ethylene, butane, ethyl amine, acetonitrile, and ammonia have been identified (366).

$$CH_2{=}CH_2 + HN_3 \longleftarrow CH_3CH_2N_3 \longrightarrow N_2 + [CH_3CH_2N{:}]$$

Table 2.14. Alkyl, Aralkyl, and Alicyclic Azides

Azide	Boiling Point (°C)	Thermal Stability	Ref.
CH_3N_3	20–21	$t_{1/2}$ (200°, 133 mm) 8.4 hr $t_{1/2}$ (155°, 22–240 mm) 24 days	267, 643 819
CF_3N_3	−28.5/743	Expl. 330°	702
$CH_3CH_2N_3$	48.5–49.5	$t_{1/2}$ (240°, 133.7 mm) 5 min; $t_{1/2}$ (200°, 25 mm and >) 2.0 hr	191, 366
$(CH_3)_2CHN_3$	66.5	$t_{1/2}$ (200°, 25 mm and >) 1.2 hr	366
$CH_3COCH_2N_3$	Bp 38/1	Dec. 180–200°, trichlorobenzene	144
$(CH_3)_3CN_3$	68–71; 72		79, 706
n-$C_8H_{17}N_3$	62/3.3	Nearly unchanged at 150° in 48 hr	196, 659
$C_6H_5CH_2N_3$	62–63/4		327
$(C_6H_5)_3CN_3$	Mp 64; 64–65	$t_{1/2}$ (169.4°, nitrobenzene) 1.5 hr	105, 948
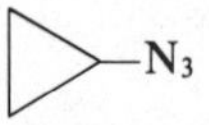	79.5	Expl. on attempted elemental analysis	589
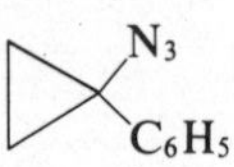	45–48/0.001	$t_{1/2}$ (102.5°, o-dichlorobenzene) 37 min	1064
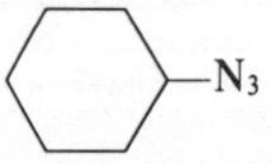	72/30	$t_{1/2}$ (194.4, ethyl benzoate) 5.1 hr	659, 1141
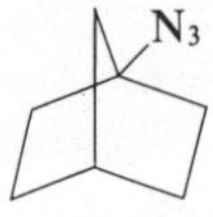	25/0.1	Stable 65° for 8 hr in methanol; dec. 140°	905

Representative temperature and pressure limits, above which ethyl azide explodes (191) are as follows:

Temp. (°C)	Critical Pressure (mm)
258.5	105.2
270.0	36.4
280.0	15.9

Cyclohexyl azide is similar to open-chain azides in stability, but azidocyclopropanes are appreciably less stable (1064). Tertiary azides appear to be slightly less resistant to heat than secondary or primary azides.

During the thermolysis of primary, secondary, and tertiary azides, hydrogen, alkyl, or aryl, but not acyl, migrates from carbon to nitrogen before or simul-

$$C_6H_5CO\overset{\overset{\displaystyle CH_3}{|}}{\underset{\underset{\displaystyle CH_3}{|}}{C}}{-}N_3 \xrightarrow{170\text{–}200^\circ} N_2 + C_6H_5CO\overset{\overset{\displaystyle CH_3}{|}}{C}{=}NCH_3$$

taneously with elimination of nitrogen (144, 145). In the rearrangement of aryldiphenylmethylazides to benzophenone anils at 170–190° (948), slightly

$$p\text{-}(CH_3)_2NC_6H_4\overset{\overset{\displaystyle C_6H_5}{|}}{\underset{\underset{\displaystyle C_6H_5}{|}}{C}}{-}N_3 \xrightarrow{185^\circ} N_2 + (C_6H_5)_2C{=}NC_6H_4N(CH_3)_2\text{-}p$$

greater migration aptitude is shown by aryl groups with electron-releasing than electron-withdrawing substituents.

Aromatic azides (Table 2.15) are less stable toward heat than alkyl azides. Phenyl azide, as well as *meta-*, *para-*, and some *ortho-*substituted phenyl azides decompose in the 100–150° range. The effect of an electron-releasing substituent meta or para to the azide is to decrease stability slightly, whereas electron-attracting substituents in these positions increase stability. Particularly when unsaturated, substituents ortho to the azide lower stability of the molecule. This neighboring group participation is confirmed by the enthalpy of activation for decomposition, ranging from 22.4 to 27.2 kcal/mole for phenyl azides with unsaturated ortho substituents in comparison with values for phenyl azide and para-substituted phenyl azides of 31.4 to 40.6 kcal/mole (285, 1011, 1141).

When no other substance is present, decomposition of aryl azides at temperatures near 350° leads to azo compounds as in the vapor phase pyrolysis of phenyl azide (1016). The initial product of thermal decomposition of phenyl azide is phenyl nitrene and its subsequent transformations result in a variety of products. Above 550° the amount of azobenzene formed decreases and cyanocyclopentadiene is first observed, becoming significant by 650° (228, 448). In

$$C_6H_5N_3 \xrightarrow{350\text{–}360°} [C_6H_5N:] \longrightarrow \underset{75\%}{C_6H_5N{=}NC_6H_5}$$

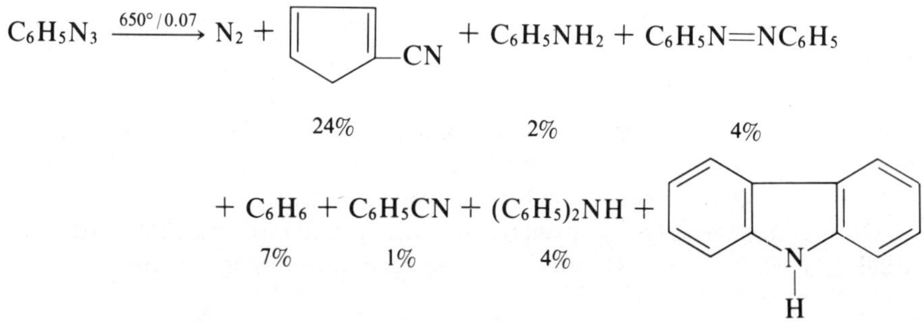

the presence of reactive solvents other products result during decomposition. Thus thermolysis of phenyl azide in *p*-xylene involves abstraction of

$$C_6H_5N_3 \xrightarrow[\text{xylene}]{138°} N_2 + \underset{25\%}{C_6H_5NH_2} + C_6H_5N{=}NC_6H_5 + p\text{-}CH_3C_6H_4CH_2CH_2C_6H_4CH_3\text{-}p$$

hydrogen from the solvent and formation of a coupling product as well as azobenzene (110). Decomposition of *o*-nitrophenyl azide, conveniently carried out at about 100°, is a preparative method for benzofurazan oxide (1007). Under

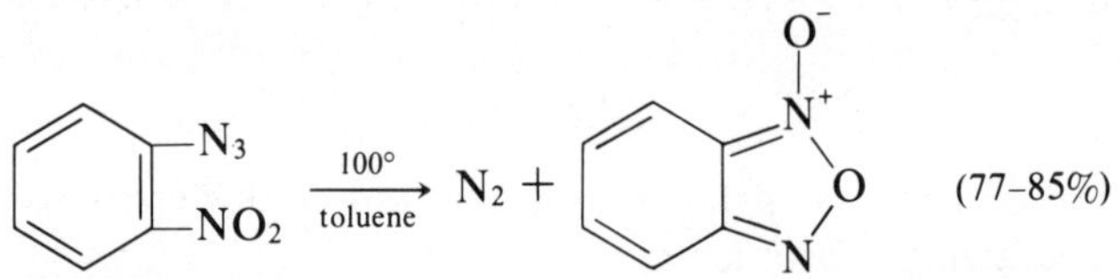

more vigorous conditions the decomposing azide group reacts with saturated ortho substituents, as in the pyrolysis of *o*-isopropylphenyl azide to dihydroskatole and the internal oxidation-reduction of *o*-azidobenzyl alcohol (1016).

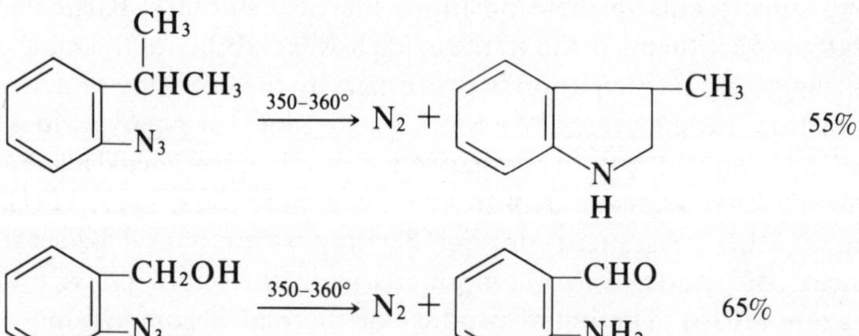

Table 2.15. Aryl Azides

Azide	Melting Point (°C)	Thermal Stability	Ref.
$C_6H_5N_3$	Bp 66–68/21	Dec. > 100° $t_{1/2}$ (140.8°, decalin) 4.6 hr	196, 674 284
$C_6F_5N_3$	Bp 35/5	Dec. begins 80°, proceeds smoothly at 120°	118
p-$CH_3OC_6H_4N_3$	34–36	$t_{1/2}$ (141.3°, decalin) 1.4 hr $t_{1/2}$ (132°, ethyl benzoate) 2.1 hr	1011 1141
p-$(CH_3)_2NC_6H_4N_3$	43–45	Dec. 110°	305, 1111 196
p-$ClC_6H_4N_3$	19.5–20.5 Bp 50–51/0.7	$t_{1/2}$ (137.9°, decalin) 3.7 hr	284, 645
o-$ClC_6H_4N_3$	Bp 66–67/3	$t_{1/2}$ (137.9°, decalin) 1.3 hr	284
p-$NCC_6H_4N_3$	65–65.5	$t_{1/2}$ (132°, decalin) 5.4 hr	284, 645
p-$O_2NC_6H_4N_3$	71	$t_{1/2}$ (141.3°, decalin) 3.0 hr $t_{1/2}$ (131.5°, decalin) 6.4 hr	1011 128, 284
m-$O_2NC_6H_4N_3$	55	$t_{1/2}$ (141.3°, decalin) 6.8 hr	809, 1011
o-$O_2NC_6H_4N_3$	51–52; 53–55	Dec. 75–80° $t_{1/2}$ (79.6°, decalin) 2.0 hr	809, 1007 284
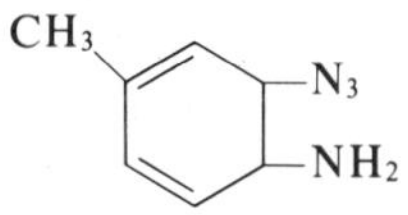	38	Dec. 80–90	418
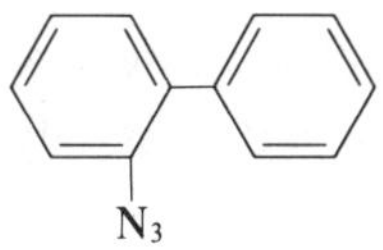	49–50	$t_{1/2}$ (148°, decalin) 1.5 hr	1008, 1011

A carbon-hydrogen linkage in a benzene ring ortho to the azide also can be attacked with formation of a carbazole (1011).

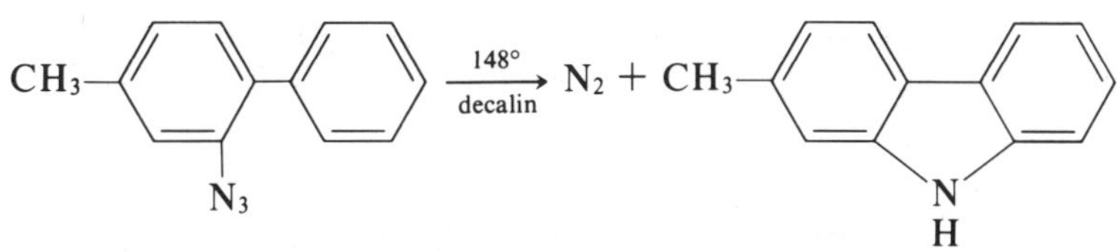

Thermal decomposition of o-azidoazobenzenes leads to the formation of 2-phenylbenzotriazoles at temperatures as low as 40° (285, 417). Electron-withdrawing groups para to azo in the nonazide ring increase, whereas electron-releasing groups decrease the decomposition rate (417). An intramolecular

1,3-cycloaddition of azide to azo forming a pentazole is postulated as the rate-determining step before loss of nitrogen.

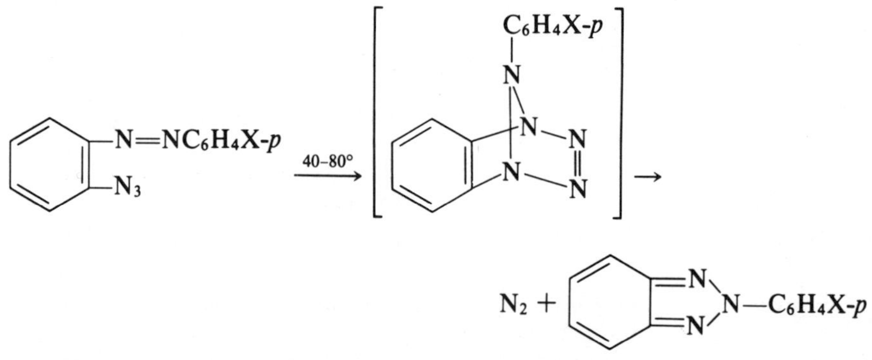

X	$t_{1/2}$, min (50°, anisole)
$N(CH_3)_2$	103
H	47
CN	28

Thermal intramolecular combinations of azides with acetylene or cyano groups suitably placed in the same molecule to form fused *v*-triazolo or tetrazolo ring systems are described in Chapter 7, Synthesis.

Most alkenyl azides are much less thermally stable than saturated alkyl or aryl derivatives, as shown by the examples in Table 2.16. In many instances only

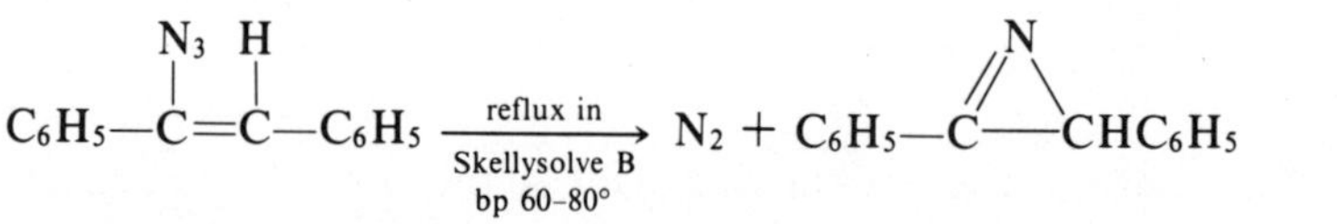

(341)

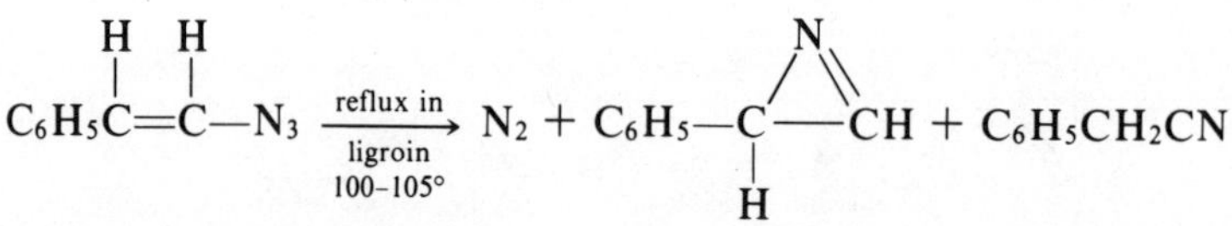

(545, 546)

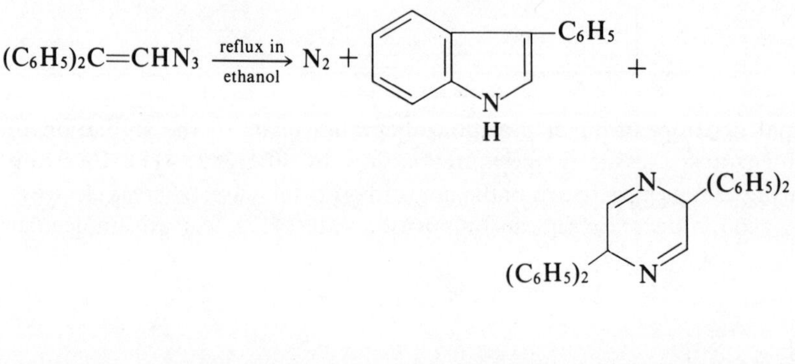

(1018)

Table 2.16. Alkenyl Azides

Azide	Boiling Point (°C)	Thermal Stability	Ref.
$CH_2{=}CH{-}N_3$	30	Expl. on warming from liquid nitrogen temp.	334, 1183
$C_6H_5C(CH_3){=}CH{-}N_3$	Oil	Dec. reflux in ethanol	1018
$C_6H_5CH{=}CH{-}N_3$	Oil	Dec. about 70°	141
$CH_2{=}C(C_6H_5){-}N_3$	Mol. dist./0.6	Dec. room temp. in 1 month	142, 1017
$C_2H_5OOC{-}C(NC){=}CH{-}N_3$	66 dec.	Relatively stable at 20°; dec., reflux in acetone-H_2O	349
$C_6H_5{-}CH{=}C(N_3){-}COOC_2H_5$	42–43		458
$p\text{-}O_2NC_6H_4CH{=}C(N_3){-}C({=}O){-}C_6H_4CH_3\text{-}p$	126–126.5 dec.		592

spectral characterization has been possible because of instability at or near room temperature, liberating nitrogen, sometimes explosively. Thus 1-azido-cyclopentene, -cyclohexene, and -cycloheptene are greenish oils whose easy decomposition precludes elemental analysis or determination of boiling point (975). The α-azidochalcones are more stable than α-azidocinnamic esters. Although α-azidochalcones melting above 80° do so with decomposition, they appear to be the most stable of the alkenyl azides (592).

Decomposition of alkenyl azides leads initially to azirines and, usually, small quantities of iminoketenes. Azirines from internal azides (no hydrogen on the olefinic carbon bearing the azide) are more stable than those from terminal azides (340, 624), which often react further to form indoles or polymers. The iminoketenes usually form nitriles or dihydropyrazines. Heating β-azidovinyl p-tolyl sulfone produces only nitrogen and p-toluenesulfonylacetonitrile (731). Chalcones carrying azide and carbonyl substituents on

$$p\text{-}CH_3C_6H_4SO_2CH{=}CHN_3 \xrightarrow[CH_3OH]{\text{reflux in}} N_2 + p\text{-}CH_3C_6H_4SO_2CH_2CN$$

the same olefinic carbon also form nitriles on heating with accompanying migration of the acyl group (591). The reaction is an acyclic example of the zwittazido cleavage (753), described further in connection with the thermolysis of cy-

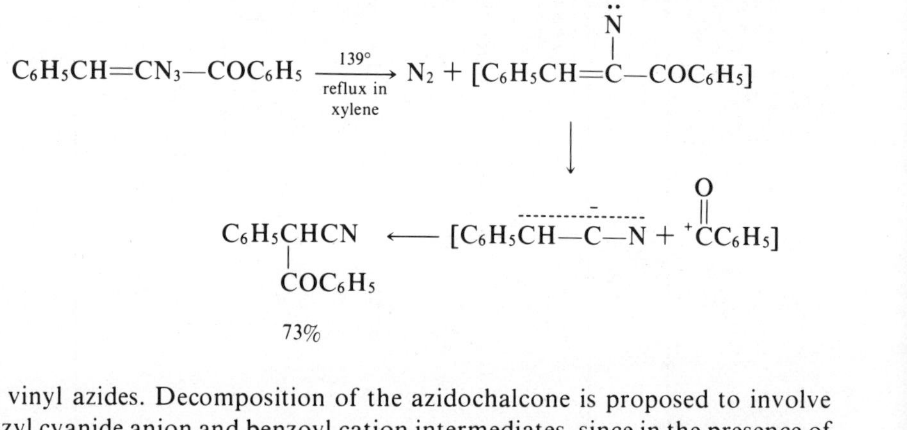

clic vinyl azides. Decomposition of the azidochalcone is proposed to involve benzyl cyanide anion and benzoyl cation intermediates, since in the presence of alcohols or benzyl amine, benzoate esters or N-benzylbenzamide are obtained, together with a lowered yield of α-cyano-α-phenylacetophenone (101).

Azides also react with olefinic linkages in the same molecule one or more atoms removed. Thus allyl azide dimerizes to the tricyclic bistriazoline (337,

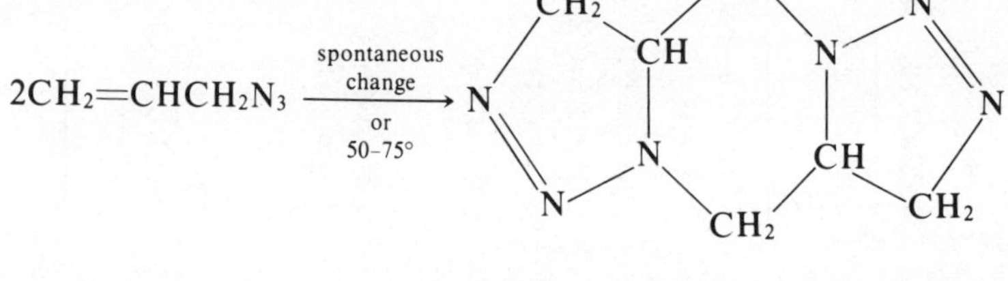

681, 864). Thermolysis of *o*-azido-*β*-alkyl, aryl, or acyl styrenes leads to 2-substituted indoles (1054).

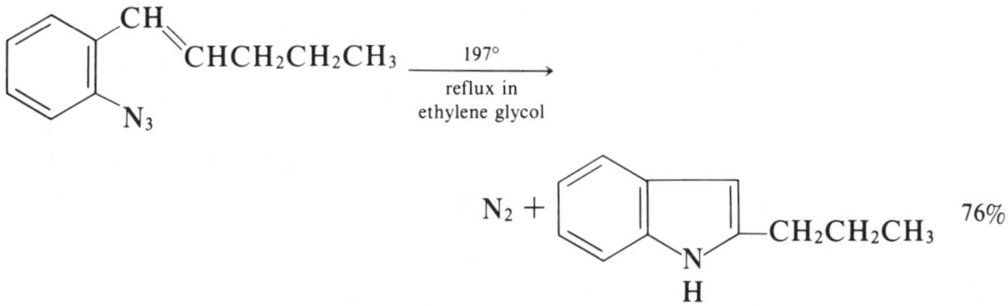

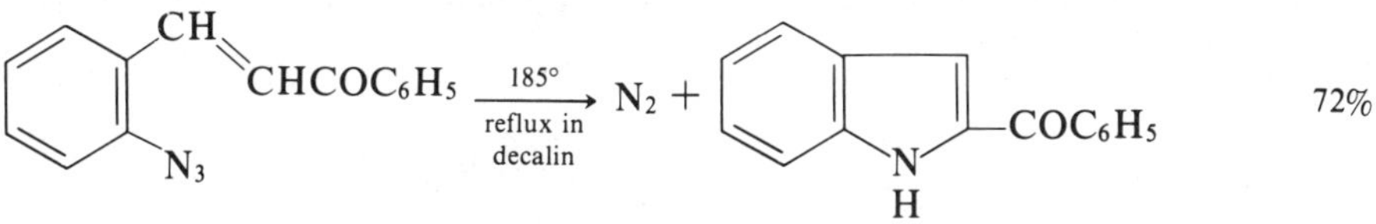

Refluxing 1-azido-4-phenylbutadiene in *n*-hexane leads to 2-phenylpyrrole, but 1-azido-2,4,4-triphenylbutadiene is converted to 2,4,4-triphenyl-2-butenonitrile (547). Although heating 4-azido-1-pentene, with two carbon atoms be-

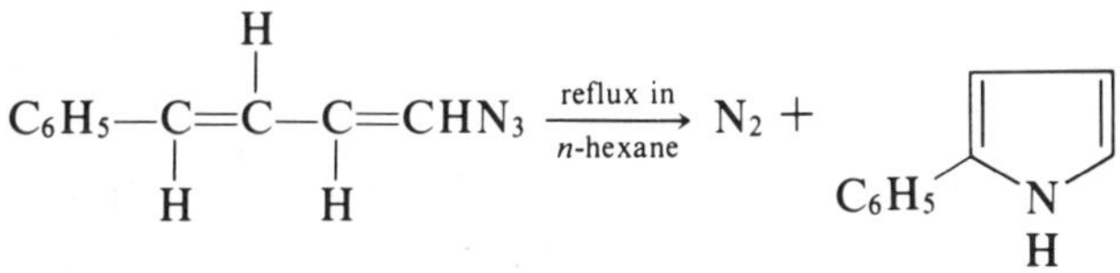

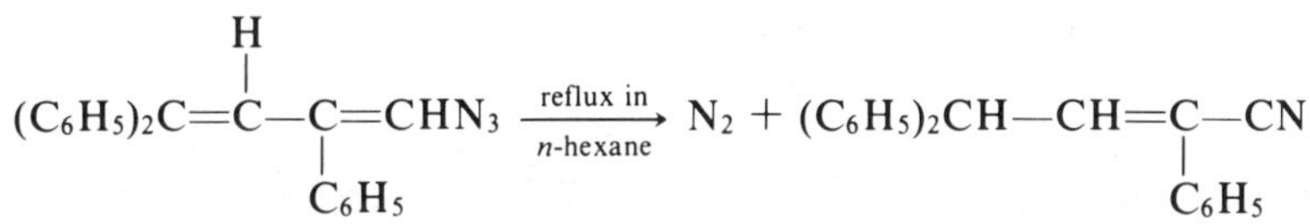

tween the azide and the double bond segment, produces no volatile products, thermal decomposition of olefinic azides with three and four intervening car-

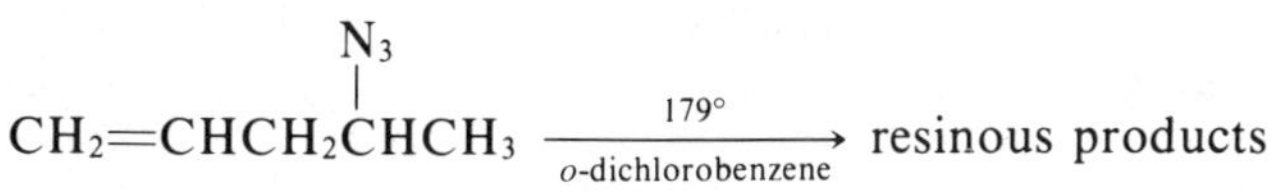

bons leads to nitrogen, cyclic imines, and usually, a fused bicyclic azirine. Either on standing or on heating at low temperatures, triazolines evidently are first formed from these olefinic azides. Decomposition of the triazolines under the same conditions used for the azides forms identical products (681).

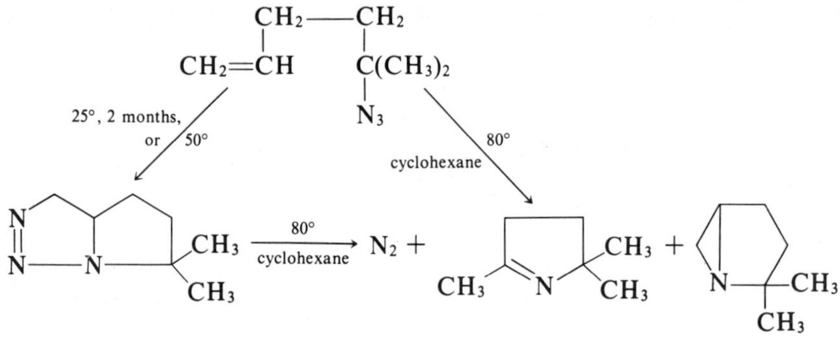

Thermolysis of *o*-azidochalcones leads to 2,1-benzisoxazoles and 3-substituted 4-quinolones (998).

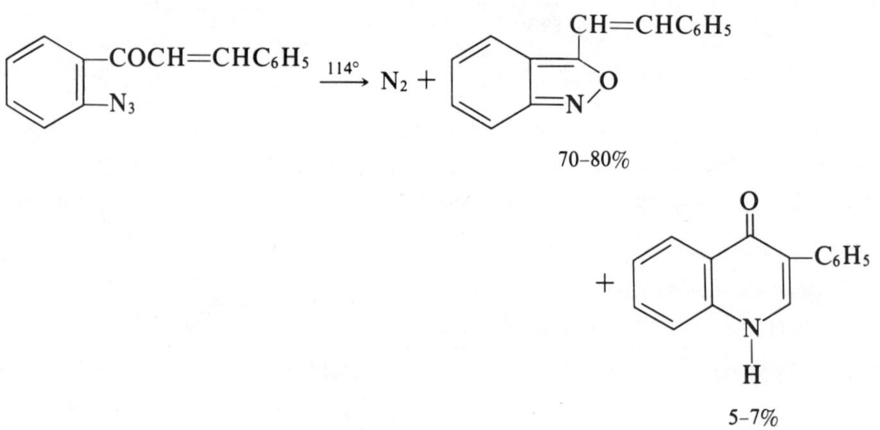

Quinone azides (Table 2.17), which generally decompose in the 80–110° range, are more thermally stable than alkenyl azides, but less so than aryl azides. The thermal conversion of 1-azido-2-vinyl aromatic compounds to indoles also is applicable to the synthesis of indolequinones from 2-azido-3-vinylquinones (367).

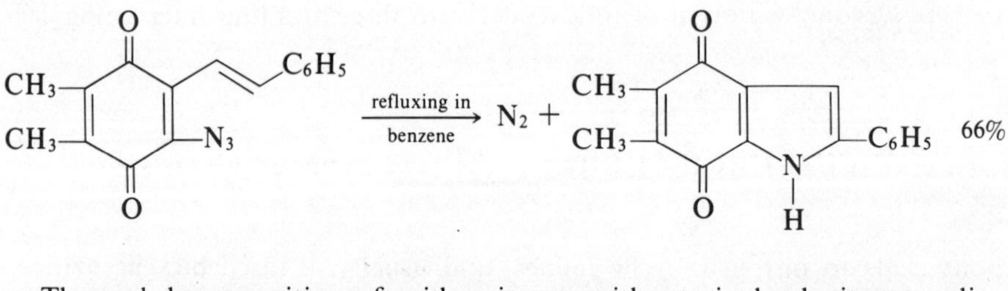

Thermal decomposition of azidoquinones without vinyl substituents adjacent to the azide involves ring contraction to 2-cyano-4-cyclopentene-1,3-diones (1159). The process of converting such cyclic vinyl azides to rearranged cyano

Table 2.17. Quinone Azides

Azide	Melting Point (°C)	Thermal Stability	Ref.
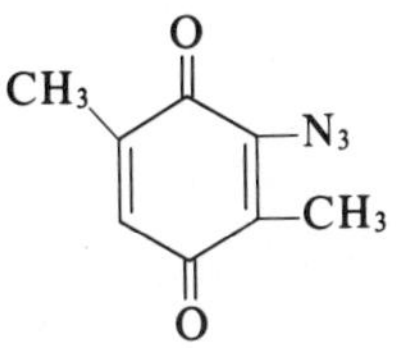	85–87 dec.; 87 dec.	$t_{1/2}$ (88.3°, chlorobenzene) 2.5 hr	248, 755, 1159
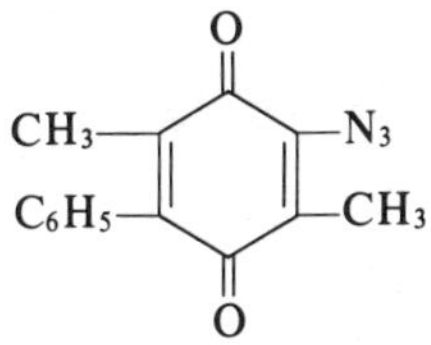	89–92 dec.	$t_{1/2}$ (105.95°, chlorobenzene) 27 min	1159
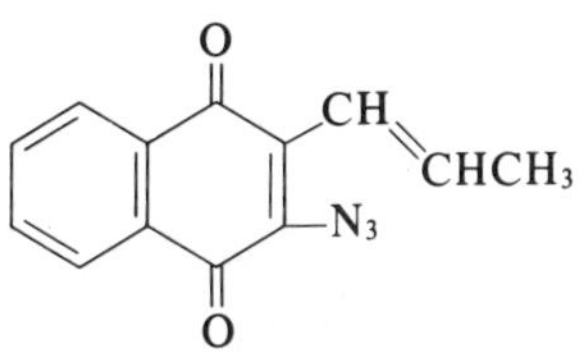	112 dec.	$t_{1/2}$ (64.3°, chlorobenzene) 53 min	367

products has been called the zwittazido cleavage by H. W. Moore (753) because of postulated zwitterionic intermediates, and embraces the decomposition of a

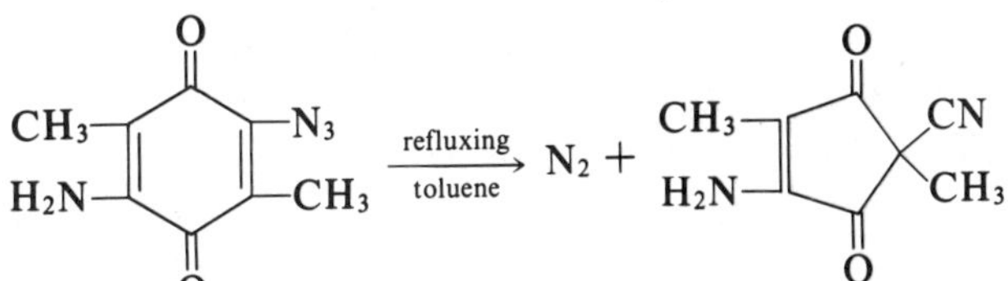

89%

variety of unsaturated carbocyclic and heterocyclic azides, usually also carrying α-carbonyl groups. The proposed mechanism is illustrated by the thermolysis of 4-azido-5-*t*-butyl-2,2-dimethyl-cyclopentene-1,3-dione (759).

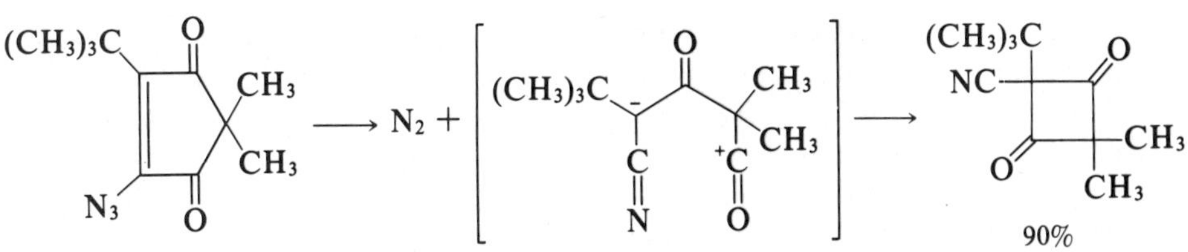

90%

Table 2.18. Heterocyclic Azides

Azide	Melting Point (°C)	Thermal Stability	Ref.
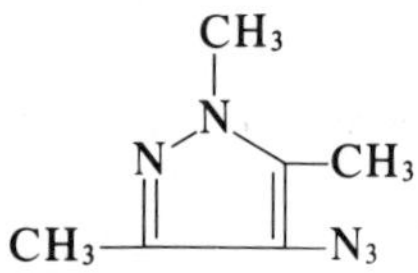	36–38	Dec. 80°, cyclohexane	1010
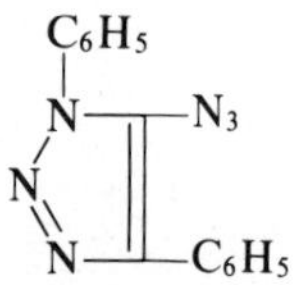	70 dec.	Dec. 50°, petroleum ether	1014
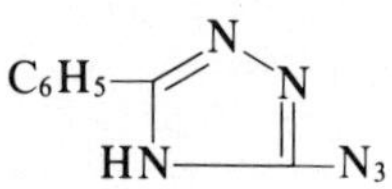	191–192 dec.	DTA exotherm 190–210°	253
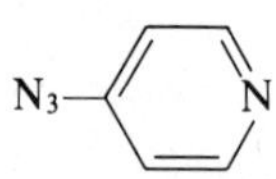	Bp 73/11		297
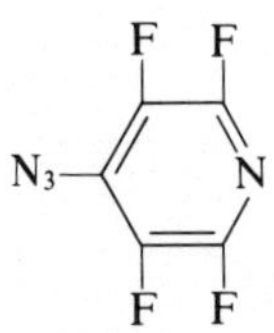	Bp 53/7.5	Dec. 130°	69
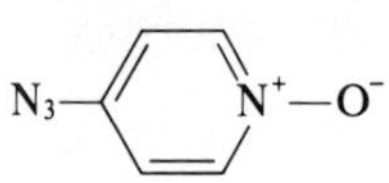	142–143 dec.		549, 1093
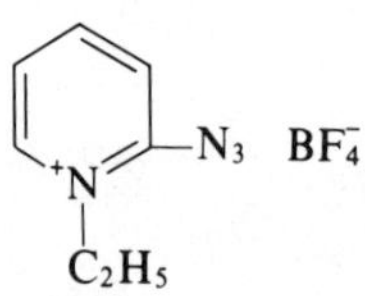	87–88 dec.	Expl. > 100°	59
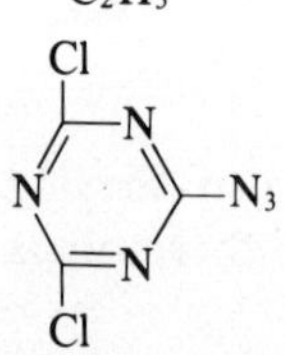	60–61	$t_{1/2}$ (184.3°, decalin) 38 min, dec. 205°	580

Heterocyclic azides (Table 2.18) exhibit a range of thermal stabilities dependent on the chemical environment of the azide group. The behavior of azide derivatives of aziridines and of a few five- and six-membered aromatic nitrogen heterocycles* illustrates their diverse responses to heat.

* All compounds reported in this section are azides, although some with adjacent ring carbon-nitrogen unsaturation in principle could exist as tetrazoles. See Section 2.16.2.

A series of 3-aryl-3-azido-2-ethoxycarbonyl aziridines has been prepared which decompose in the 80–100° range (1063). Thermolysis leads to nitrogen,

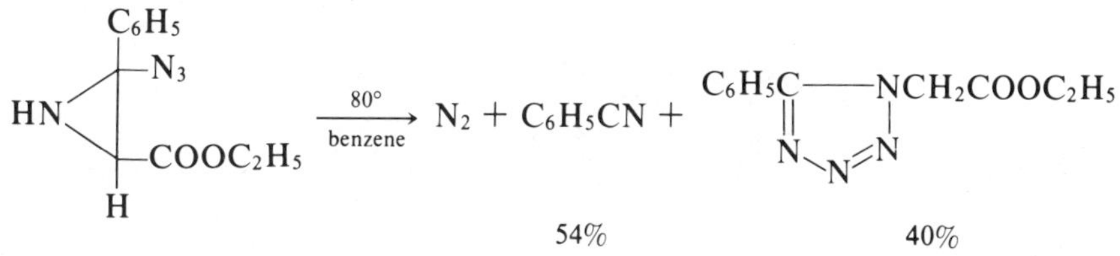

aryl nitriles, tetrazoles, and resinous material. The half-life of the 3-phenyl derivative at 80° in benzene is 1.7 hr.

In the pyrazole series, an aryl substituent at the 1-nitrogen, adjacent to an azide at the 5-position, promotes azide decomposition slightly above room temperature, forming a deep red β-azoacrylonitrile which may equilibrate with small quantities of a cyclic nitrene form (1006). In contrast, decomposition of 5-azido-3-methyl-4-phenylpyrazole, with a hydrogen on ring nitrogen, requires 110° for a reasonable rate. Nitrogen evolved from the azido group is accompanied by loss of pyrazole ring nitrogen (1010). Thermolysis of 4-azidopyrazoles occurs at 80° forming nitriles and in some cases azopyrazoles (1010).

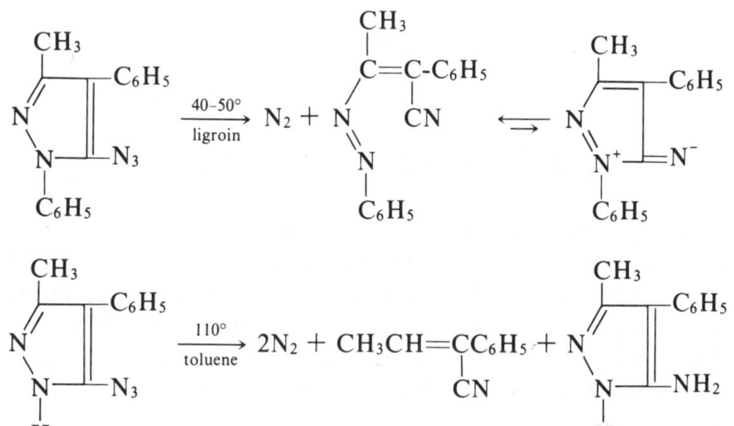

Other azido pyrazoles and triazoles illustrate the higher thermal stability associated with hydrogen on ring nitrogen adjacent to the azide, rather than with methyl or phenyl substitutions. To the extent that related compounds have been studied, pyridine azides appear to be similar in thermal stability to benzene azides. *s*-Triazine azides are relatively thermally stable, with no significant substituent effect.

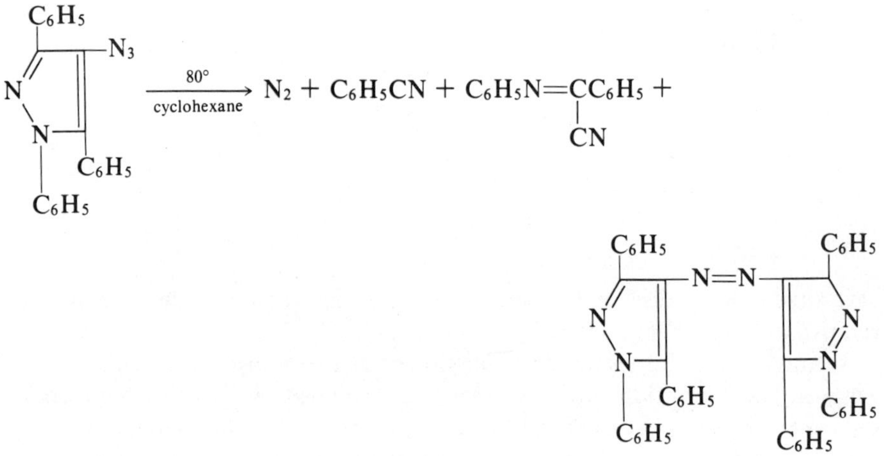

Azidooxiranes have been prepared and have half-lives of about 2 days in refluxing benzene (623). No report has been found of a 2-azidofuran, but 3-azidofurfural and related derivatives decompose in the range 110–115° (408).

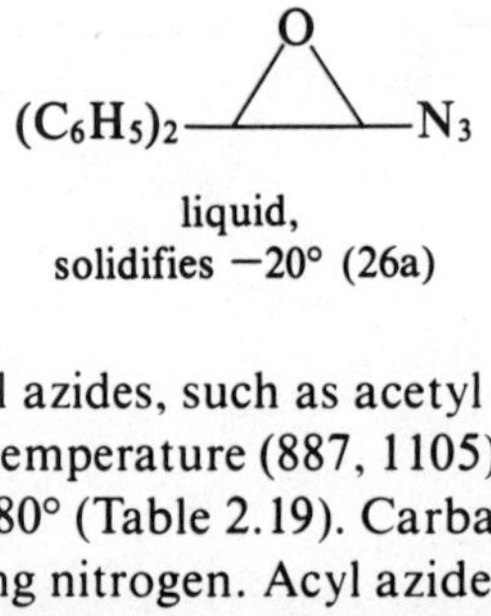

liquid,
solidifies −20° (26a)

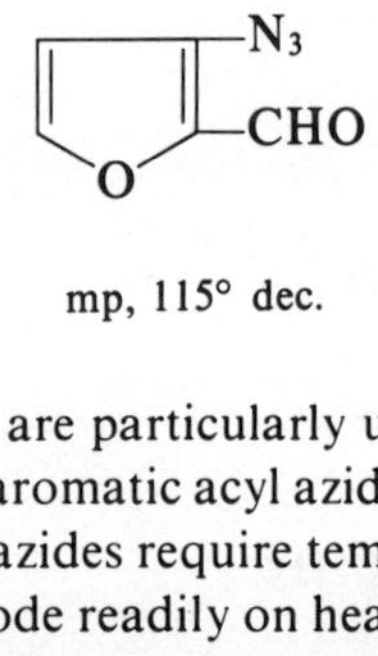

mp, 115° dec.

Aliphatic acyl azides, such as acetyl azide, are particularly unstable, decomposing at room temperature (887, 1105), and aromatic acyl azides decompose in the range of 50–80° (Table 2.19). Carbamoyl azides require temperatures above 100° before losing nitrogen. Acyl azides explode readily on heating. Pure cyanogen azide is dangerously explosive, but solutions can be handled safely; a 30% solution in acetonitrile does not detonate on instantaneous exposure to a hot bar at 250° (710).

Thermal decomposition of acyl azides involves the Curtius rearrangement, in which nitrogen release is accompanied by formation of an isocyanate, usually converted further to urethanes or amines (917, 973, 1005).

$$C_6H_5CON_3 \xrightarrow[\text{benzene}]{70^\circ} N_2 + C_6H_5NCO \qquad 87\%$$

$$n\text{-}C_{11}H_{23}CON_3 \xrightarrow[\text{toluene}]{100^\circ} N_2 + n\text{-}C_{11}H_{23}NCO \qquad 92\%$$

Table 2.19. Acyl Azides

Azide	Melting Point (°C)	Thermal Stability	Ref.
$(C_6H_5CH_2)_2CHCON_3$	51–53	Dec. room temp., 1 day	887
$C_6H_5CON_3$	27–27.5	Expl. 80° $t_{1/2}$ (25°, benzene), 66 days $t_{1/2}$ (55°, neat) 18.5 hr	1038 801
p-$O_2NC_6H_4CON_3$	71–72 dec.	 $t_{1/2}$ (65.2°, toluene) 6.1 hr	772 1201
H_2NCON_3	97	 Unchanged on reflux in benzene, 12 hr; dec. 100–115, benzene, closed tube	616, 930 242
$C_2H_5NHCON_3$	Bp 90/17 mm		689
$H_2NCONHCON_3$	193–194 dec.	Dec., 110°	47
$F_2NC(=NF)-N_3$	Bp 35°		165
$O_2NNHC(=NH)-N_3$	79 dec.	Expl. rapid heating > mp	657
$NCN=C(NHCH_3)-N_3$	122–123 dec.		814
$C_2H_5OCON_3$	Bp 40/30.5	No dec., heating at 100°, neat; expl. 160°, slow heating	690
NCN_3	Est. bp 90°	Stable indef. −20 to 0° in CH_3CN $t_{1/2}$ (25°, CH_3CN) 15 days $t_{1/2}$ (55°, cyclohexane) 1.6 hr	710

Heating carbamoyl azide by itself yields nitrogen, hydrogen azide, cyanuric acid (probably from isocyanic acid first formed), and smaller quantities of ura-

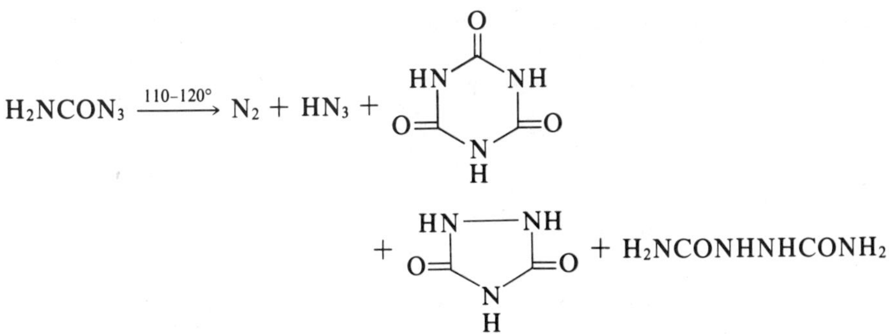

zole and hydrazodicarbamide (242). It is not necessary to invoke a rearrangement to account for these products. Other carbamoyl azides rearrange on heating, presumably first forming *N*-isocyanates, which undergo further reaction, as illustrated by formation of 1-phenylindazolinone from *N,N*-diphenylcarbamoyl azide (1040).

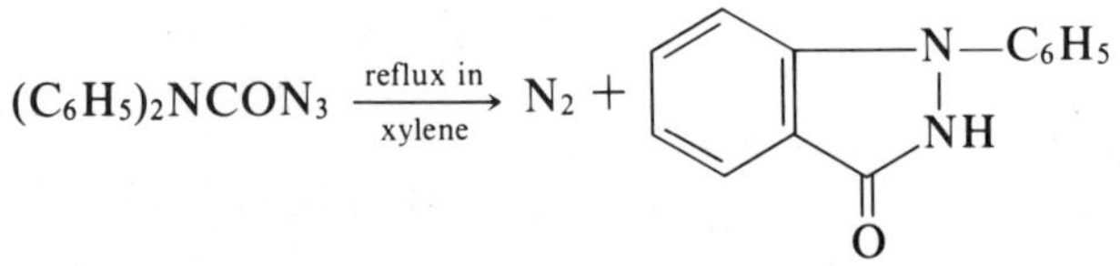

Cyanogen azide forms cyanonitrene, NCN, on heating at the relatively low temperature of 40–50°, which reacts readily with saturated hydrocarbons, olefins, and aromatic compounds. With alkanes, the nitrene inserts in carbon-hydrogen bonds, preferentially at secondary and tertiary sites, rather than at primary C—H, as indicated by the proportions of different alkyl cyanamides formed during thermolyses of cyanogen azide in *n*-hexane and 2,3-dimethylbu-

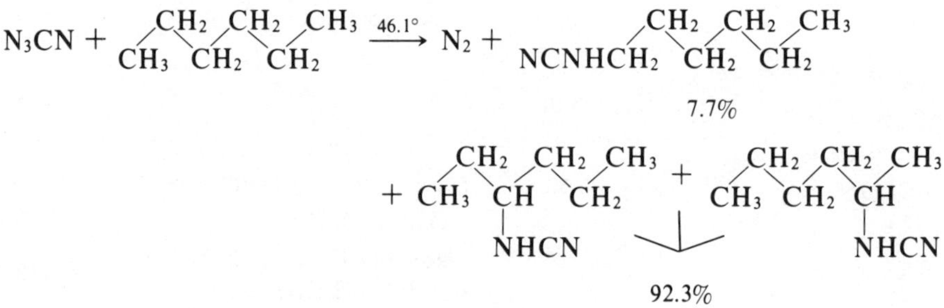

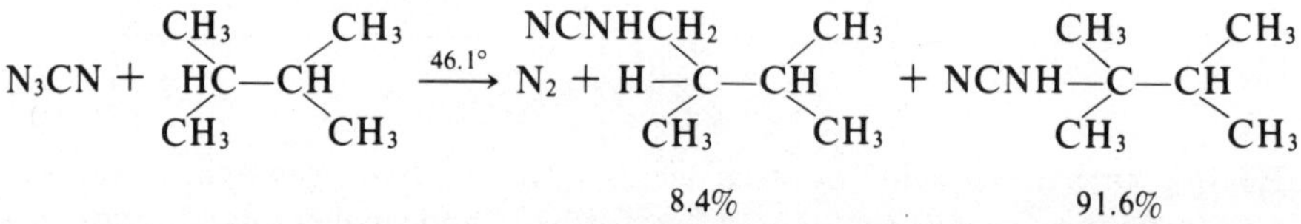

tane (32). Insertion in a tertiary C—H is highly stereospecific, evidenced by the substantially total retention of configuration in the cyanamides produced by decomposition of cyanogen azide in both *cis*- and *trans*-1,2-dimethylcyclohex-

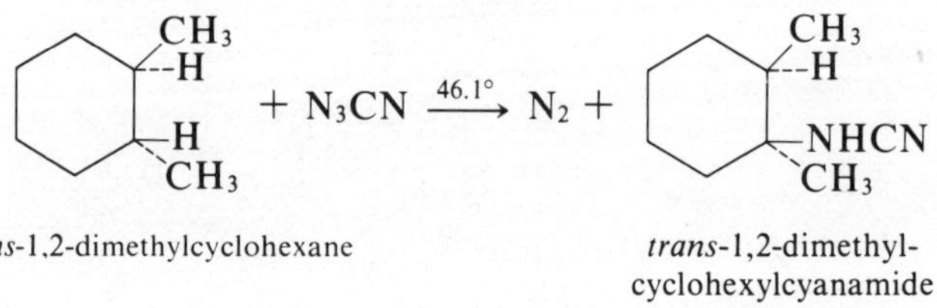

trans-1,2-dimethylcyclohexane

trans-1,2-dimethylcyclohexylcyanamide >98% *trans*

ane. Cyanonitrene reacts with olefins to produce alkylidene cyanamides and *N*-cyanoaziridines, probably through an intermediate triazoline (711), and with

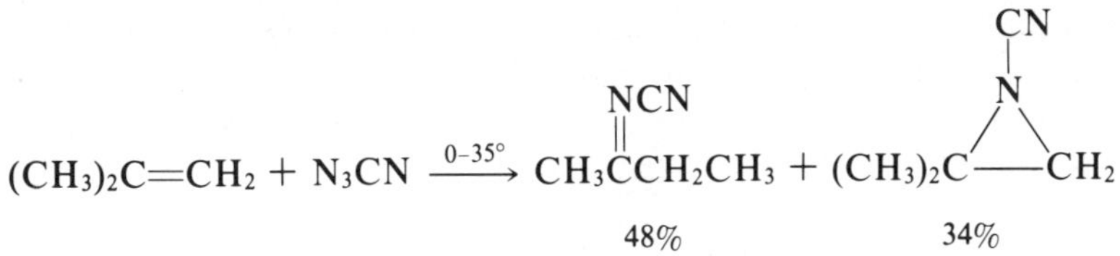

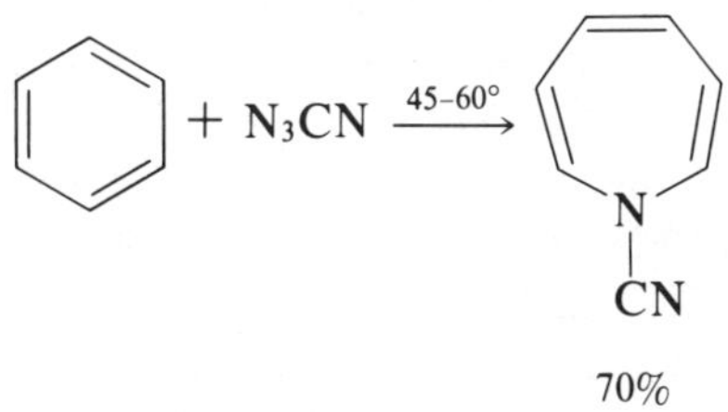

benzene derivatives to expand the ring to *N*-cyanoazepines (713). Cyanogen azide decomposes in the vapor phase to azodinitrile (712).

$$2N_3CN \xrightarrow{200^\circ} 2N_2 + NCN{=}NCN$$

54%

Alkoxy imidoyl azides, although stable at −40°, decompose slowly at room temperature (1049). Alkyl azidoformates are somewhat more stable, as indi-

$$C_2H_5OC(=NSO_2CH_3)—N_3 \qquad C_2H_5OC(=NCN)—N_3$$

mp 58–59° liquid

cated by both the reported stability at 100° of the ethyl derivative (690) and the half-life of octadecyl azidoformate in diphenyl ether at 100° of 7.2 hr (153). In cyclohexane, the initially produced nitrene from octadecyl azidoformate, probably as a singlet, inserts into C—H, forming octadecyl *N*-cyclohexylcarbamate. Nitrene, probably in the triplet state, removes two hydrogens from cyclohexane

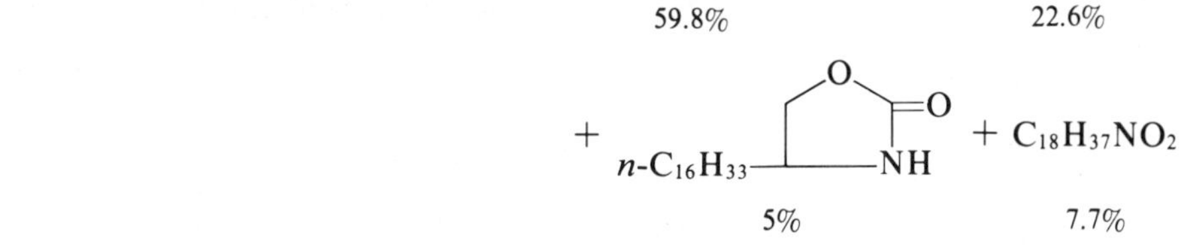

forming octadecyl carbamate, and internal condensation leads to hexadecyl oxazolidinone along with a related isomer.

Introduction of two or more azide groups in a molecule decreases thermal stability and increases explosive character. Aliphatic diazides (Table 2.20) substituted on the same or adjacent carbon atoms tend to be less stable than diazides spaced further apart. Forster, Fierz, and Joshua (338) found that ethyl diazidoacetate and 1,2-diazidoethane could be

$CH_3CH(N_3)_2$	$C_2H_5OOCCH(N_3)_2$	$N_3CH_2CH_2N_3$
bp 38°/14 mm	bp 70–72°/2 mm	bp 53°/9 mm

prepared as distillable oils which exploded when Dumas analysis was attempted; ethylidene diazide, although successfully distilled, exploded on admitting air to the apparatus. Curtius obtained carbonyl diazide as a colorless oil which exploded spontaneously and on storage (238, 241). Other *gem*-diazides, with a greater proportion of inert atoms are more stable, but decompose over

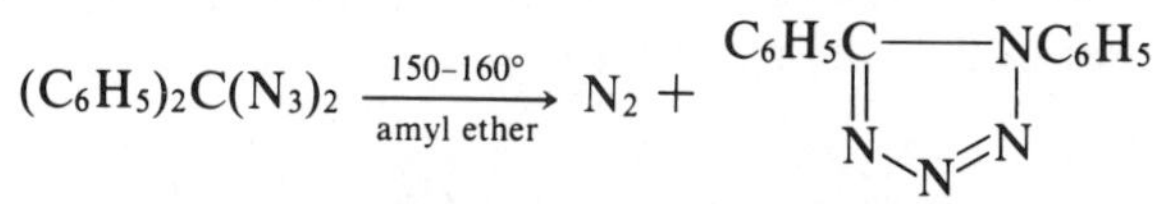

100°. Thermal decomposition of diazidodiphenylmethane liberates nitrogen and a phenyl group migrates, forming 1,5-diphenyltetrazole (973).

The α,ω-diazides with five or more atoms between the azide groups tend to be relatively stable and are comparable to mono aliphatic azides with a similar number of carbon or oxygen atoms per azide group.

Aromatic and heteroaromatic di-, tri-, and tetraazides (Table 2.21) are not significantly less thermally stable than monoazides, although many have a greater tendency to explode on heating. Aromatic *ortho*-diazides decompose to

Table 2.20. Aliphatic Diazides

Diazide	Melting Point (°C)	Thermal Stability	Ref.
$(C_6H_5)_2C(N_3)_2$	42	Dec. 120°, amyl ether; Exp. > 150°	394 973
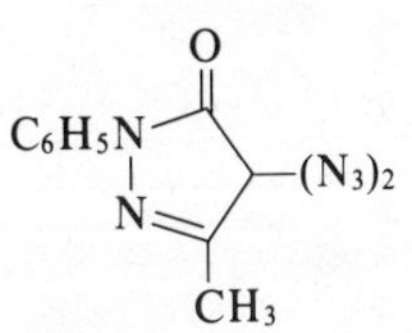	26–27	Dec. 130°, chlorobenzene	635
$N_3(CH_2)_5N_3$	Bp 100/9		1022
$N_3(CH_2)_9N_3$	Bp 142/6		1022
$N_3CH_2CH_2OCH_2CH_2N_3$	Bp 73/3		282

Table 2.21. Aryl and Heteroaryl Di- and Polyazides

Polyazide	Melting Point (°C)	Thermal Stability	Ref.
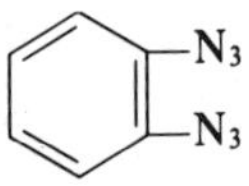	23–25	Rapid dec. 160–170° Exp. 205	506, 764 418
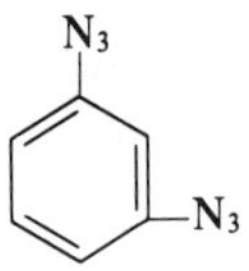	5		336
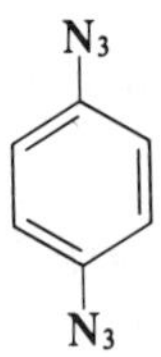	83	Expl. > mp	406
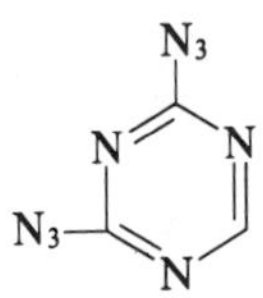	63		595
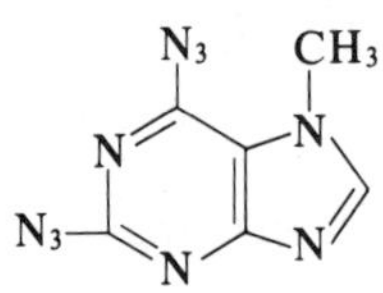	175–180 dec.		548
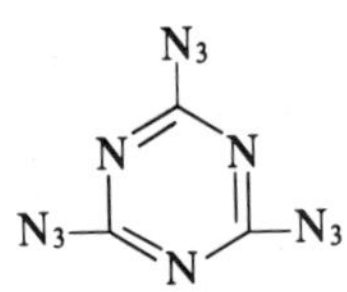	94	No dec. 112° in 17 cm argon, several hours $t_{1/2}$ (171.4°, 23 cm argon) 11.5 min Expl. temp, heated 5°/min, 170°	1071, 1199 314
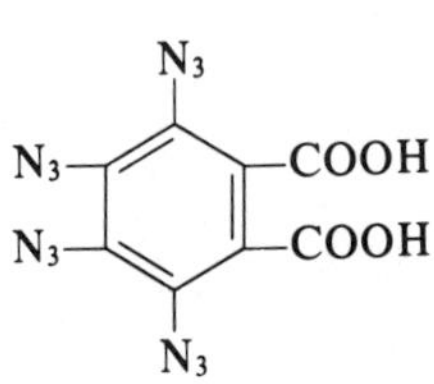	113	Dec. heating in DMF	705

dinitriles (418). Heating *p*-diazidobenzene under the same conditions forms a red oil believed to be polyazobenzene.

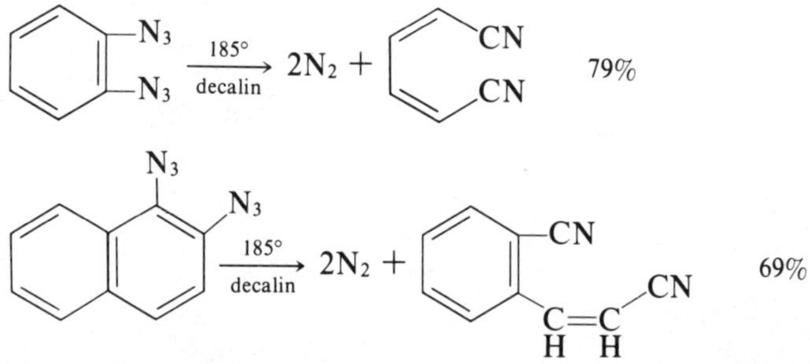

Cyanuric azide melts without decomposition and even at 150° gas formation is slow (868, 1071). In addition to nitrogen, its thermal decomposition produces cyanogen, which rapidly polymerizes.

Both 2,5-diazidohydroquinone (757) and the corresponding quinone (1206) have similar thermal stabilities. Tetraazidoquinone is extremely explosive (1189).

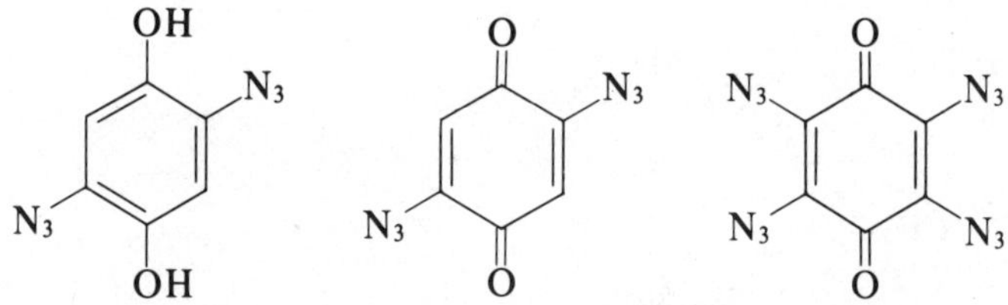

expl. 100° mp 93–95° expl. 115° expl. 120°

A variety of diazidoquinones decompose to cyano compounds (754) as illustrated by the conversion of 2,5-diazido-3,6-di-*t*-butylquinone to *t*-butylcyano-

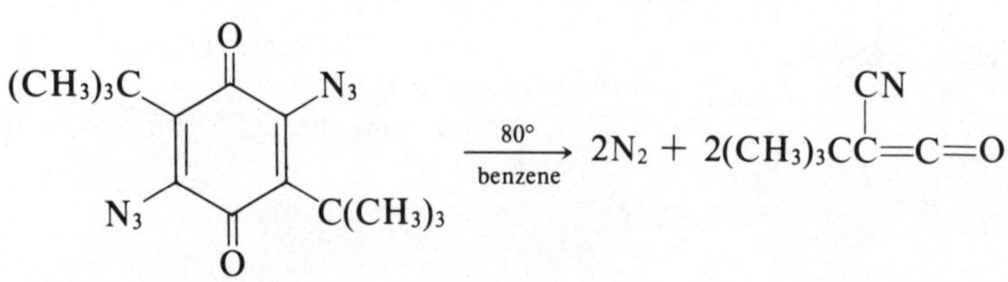

nearly quant.

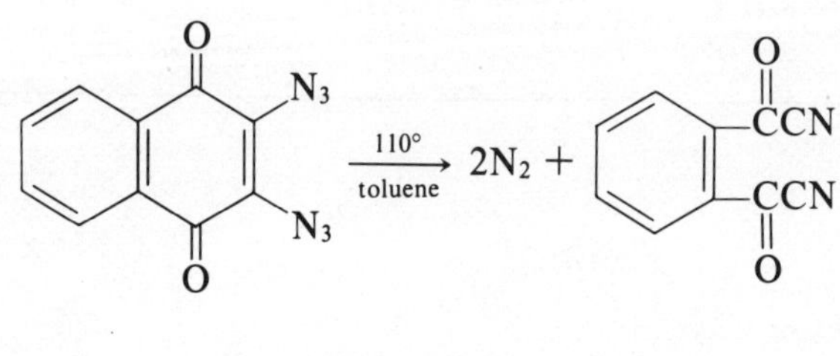

Table 2.22. Azidomethylium Salts

Salt		Melting Point (°C)	Thermal Stability	Ref.
$(H_2N)_2\overset{+}{C}N_3$	$\bar{Cl}$	145 dec.		965
	$Sb\bar{Cl}_6$	Dec. 150	No melting to 360°	
$H_2N\overset{+}{C}(N_3)_2$	$Sb\bar{Cl}_6$	Dec. ~220	Mp 332°	965
$\overset{+}{C}(N_3)_3$	$Sb\bar{Cl}_6$	142 dec.; 145 dec.		771, 965

ketene (758) and of 2,3-diazido-1,4-naphthoquinone to phthaloyl cyanide (1123).

The azide group can serve as a ligand in guanidinium-type cations, forming salts in which azide has replaced one, two, and three amino groups. Thermal stabilities of these azidomethylium salts (Table 2.22) are greater than most aryl azides.

Azides of elements other than carbon from periodic groups III, IV, V, and VI show a wide range of structural types and thermal stabilities. Mono-, di-, and triazidoboron derivatives are known (Table 2.23), which form Lewis acid complexes with amines, analogous to other borane-amine compounds. Uncomplexed boron azides are somewhat less resistant to heat than correspondingly substituted carbon azides, but their aminate complexes are appreciably more stable. Complete decomposition of the dialkylaminoboron azides requires temperatures of 270° (838).

Table 2.23. Boron Azides

Azide	Boiling Point (°C)	Thermal Stability	Ref.
$(CH_3)_2BN_3$	54/720	Expl. 140	838
$(CH_3)_2BN_3 \cdot$ N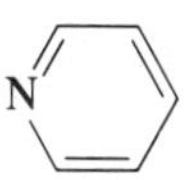	57.5/0.1	Dec. 210–220	838
$(C_6H_5)_2BN_3$	$110–115/10^{-3}$	Evolution of N_2, 110° in vac.	838
		$t_{1/2}$ (170°, hydrocarbon)	839
		25.4 min, slight expl. 180°, closed tube	840
$[(CH_3)_2N]_2BN_3$	40/12		838
$B(N_3)_3$	Cryst.	Very expl.	1163
$B(N_3)_3 \cdot$ N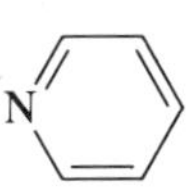	130/vac.	Dist. unchanged	838

During heating of diorganylboron azides, the release of nitrogen occurs with rearrangement of the organic radical from boron to nitrogen, analogous to the migration in trisubstituted methyl azides (646, 839). The rearrangement

$$(C_6H_5)_2BN_3 \xrightarrow{200^\circ} N_2 + (C_6H_5BNC_6H_5)_2 + (C_6H_5BNC_6H_5)_3$$

products of diarylboron azide thermolysis consist of both the four-membered tetraaryldiazadiboretane and the hexaarylborazines. At 120° only the diazadiboretane is formed (838, 839). β-Triazidoborazine detonates when heated over

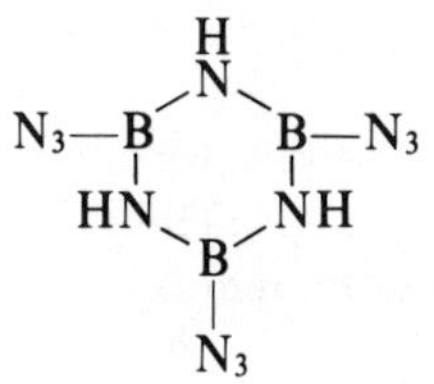

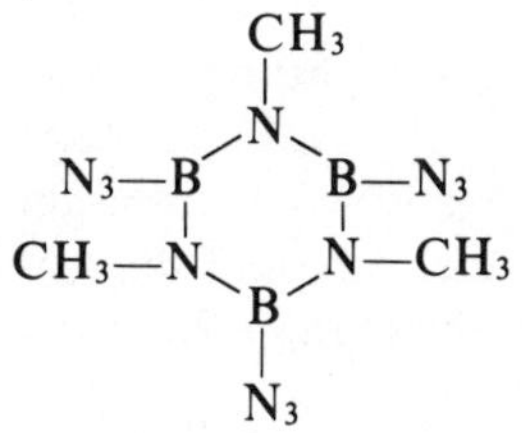

mp ~100°

$HB(N_3)_4$

liquid
−70 to −60°
dec. on raising temp

$NaB(N_3)_4$

solid

a free flame, but does not melt to 300° (777). The free acid corresponding to the salt sodium borazide can be isolated as a liquid at low temperatures, but dissociates to boron triazide and hydrogen azide on raising the temperature above −60° (1162).

Aluminum triazide and $CH_3Al(N_3)_2$ are white solids (1160, 1161); dimethylaluminum azide exists as a six-membered aluminum-nitrogen ring, $[(CH_3)_2AlN_3]_3$, melting at 65° (770, 1170).

The higher group IV azides (Table 2.24) are much more stable to heat than other covalent azides. Stability decreases as atomic weight of the central atom

Table 2.24. Higher Group IV Azides

Azide	Melting Point (°C)	Thermal Stability	Ref.
$(CH_3)_3SiN_3$	Bp 96.0–96.2	Dec. > 250°	916
$(C_6H_5)_3SiN_3$	83.5–84.5	Dec. 380°	916
$(C_6H_5)_2Si(N_3)_2$	Bp 108–110/0.03	Dec. 320°	916
$C_6H_5Si(N_3)_3$	Bp 62.5–63.5/0.01	Stable 325; dec. 400°	916, 1055
$Si(N_3)_4$	Cryst.	Highly expl.	1164
$(C_6H_5)_3GeN_3$	107–107.5	Dec. 375°	916
$(C_6H_5)_3SnN_3$	111.5–112.5	Dec. 300°	916
$(C_6H_5)_3PbN_3$	187.5–188.5	Dec. 200°	916

Table 2.25. Phosphorus Azides

Azide	Boiling Point (°C)	Thermal Stability	Ref.
$(C_6H_5)_2PN_3$	Dec. 13.6–13.8		836
$(C_6H_5)_2PON_3$	137–140/0.05	Dec. > 140°	56, 916
$C_6H_5PO(N_3)_2$	72–74/0.1	Expl. during attempted fract. dist.	55, 1118
$P(N_3)_3$	Liquid	Can expl. on warming from 0° to room temp.	167
$PO(N_3)_3$	Liquid	Can expl. on warming from 0° to room temp.	167
$P(N_3)_5$	Liquid	Dec. room temp. to 80°, in acetonitrile	167

increases. Triphenylsilicon and triphenylgermanium azides also decompose to nitrogen and products with phenyl attached to nitrogen, but the corresponding tin and lead azides dismutate to nitrogen and tetraphenyl metal derivatives (916).

Azides of phosphorus in all oxidation states (Table 2.25) are much less stable than the organosilicon azides. Diphenylphosphinous azide decomposes to nitrogen, diphenylphosphonitrile, and related linear polymers (836). The familiar rearrangement of phenyl from phosphorus to nitrogen occurs with diphenylphosphinic azide (916). Phosphorus pentaazide decomposes in solution to phosphorus triazide and nitrogen (167).

$$P(N_3)_5 \xrightarrow[\text{acetonitrile}]{80^\circ} 3N_2 + P(N_3)_3$$

Azide groups can function as ligands in phosphonium compounds, forming progressively more explosive compounds as the azide content increases (Table 2.26). Phosphonium salts with one and two azide groups are at least as stable as aryl azides.

Table 2.26. Phosphonium Azides

Azide	Melting Point (°C)	Thermal Stability	Ref.
$(CH_3)_3\overset{+}{P}N_3\ \ Sb\overset{-}{Cl}_6$	251 dec.		966
$(C_6H_5)_3\overset{+}{P}N_3\ \ Sb\overset{-}{Cl}_6$	160–162		166
$(C_6H_5)_2\overset{+}{P}(N_3)_2\ \ Sb\overset{-}{Cl}_6$	144–146 dec.		166
$C_6H_5\overset{+}{P}(N_3)_3\ \ Sb\overset{-}{Cl}_6$	68–70 dec.	Dec., several days at room temp.	166
$\overset{+}{P}(N_3)_4\ \ Sb\overset{-}{Cl}_6$	151 dec.	Expl. on heating	967

Benzenesulfinyl azide, $C_6H_5SON_3$, solidifies at $-40°$, and can be stored safely at Dry Ice temperature, but explodes at 11° (707). Benzenesulfonyl azide, $C_6H_5SO_2N_3$, melting at 13–14°, decomposes at about 135° (647, 916). Although considered to be a "rigid" azide not capable of undergoing a Curtius rearrangement, vapor phase pyrolysis produces azobenzene, indicating migration of phenyl from sulfur to nitrogen under these conditions (916). Diphenylsulfamoyl azide decomposes to diphenylamine (4).

$$C_6H_5SO_2N_3 \xrightarrow{625°} N_2 + SO_2 + C_6H_5N{=}NC_6H_5 \qquad 17.5\%$$

$$(C_6H_5)_2NSO_2N_3 \xrightarrow[\text{benzene}]{130°} N_2 + (C_6H_5)_2NH \qquad 54\%$$

Arylsulfonyl azides with an ortho substituent bearing an additional aryl group undergo various thermal intramolecular cyclizations as well as hydrogen-abstraction reactions (2). Examples of these reactions, indicative of the intermediacy of sulfonylnitrenes, are the thermolyses of 2-phenoxybenzenesulfonyl azide and diphenyl sulfide 2-sulfonyl azide in the absence of solvent.

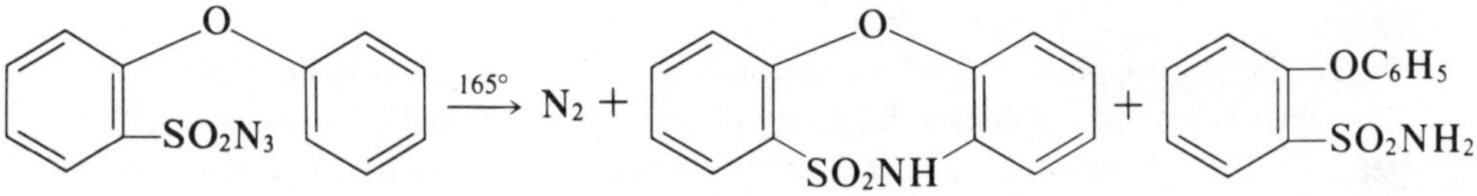

38%

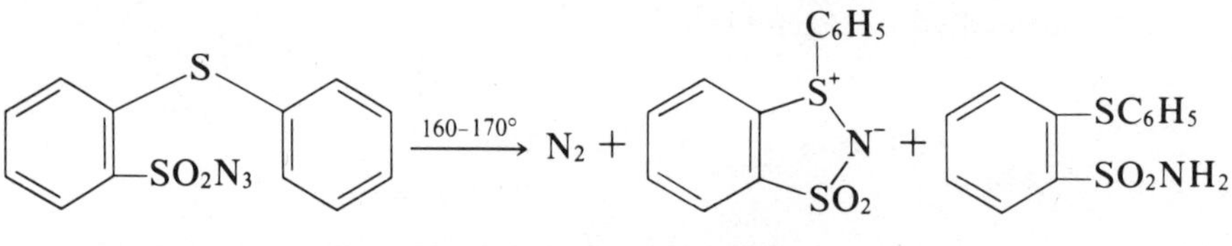

27%

2.2. ACYCLIC-CYCLIC AND CYCLIC-CYCLIC THREE-NITROGEN CHAINS

2.2.1. Two Exocyclic Nitrogen Atoms

Two-nitrogen exocyclic groups attached to a cyclic nitrogen atom to form a three-nitrogen chain include azo, nitrosamine, and nitroimide.

Some triazenes with an external aryl azo radical linked to a heterocyclic nitrogen are listed in Table 2.27. Triazenes in which this cyclic nitrogen is part of a piperidine or other six-membered ring are of the same order of thermal stability as open-chain analogs such as 1-phenyl-3,3-diethyltriazene. When the end nitro-

Table 2.27. Acyclic-Cyclic Triazenes

Triazene	Melting Point (°C)	Thermal Stability	Ref.
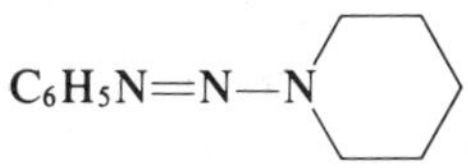	43; 44–45 Bp 97–101/0.05	Dec. 225–230°	474, 1142 944
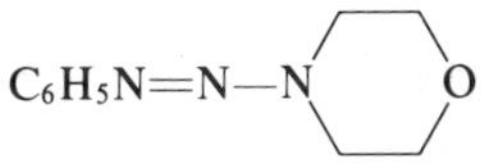	28–29; 29–30 Bp 126–128/0.08		463, 944
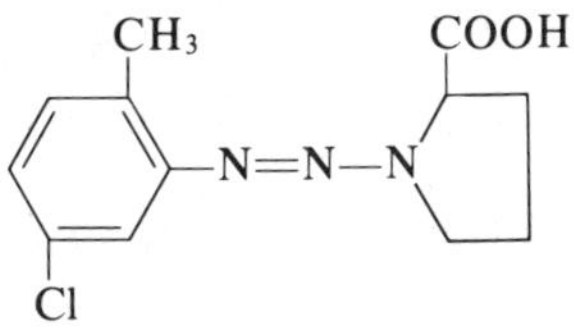	116–120 dec.		1202
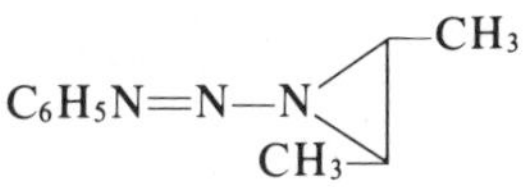	Cryst., 0–5°	Stable at −78° $t_{1/2}$ (20°, methylene chloride) 90 min	213
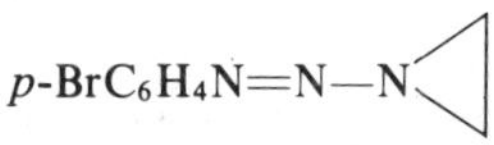	56–57		453

gen is part of an aziridine ring, the inherent instability of the three-membered ring evidently contributes to the poor heat resistance of the aryl azoaziridine, which is similar to that of the monoaryl triazenes. Heating *N*-phenylazoaziridine

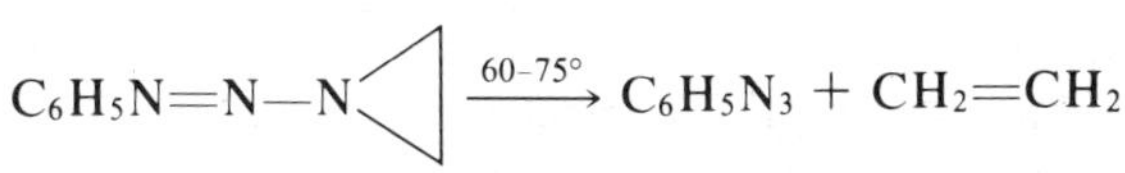

leads to ethylene and phenyl azide (935). A series of *N*-substituted phenylazopyrrolidine triazenes, with both electron-releasing and electron-withdrawing substituents, melting from 51° to 161.5° appear to be indefinitely shelf-stable (339). A triazenium derivative of benzothiazole also has been prepared (790).

(*Z*)-1-Phenyl-2-phthalimido-diazene 1-oxide **(I)** is thermodynamically more stable than its (*E*) isomer (derived from **I** by irradiation), from which it is ob-

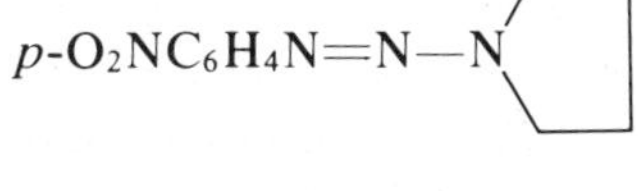

mp 161.5–163°

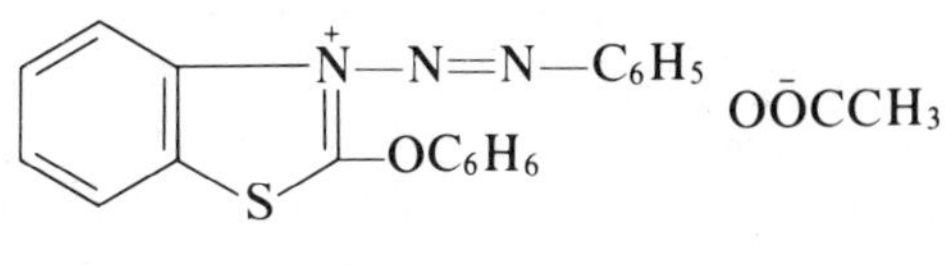

mp 89°

tained by heating (489). Thermolysis of **I** leads to products evidently arising from an intermediate phthalimide radical. Although this decomposition for-

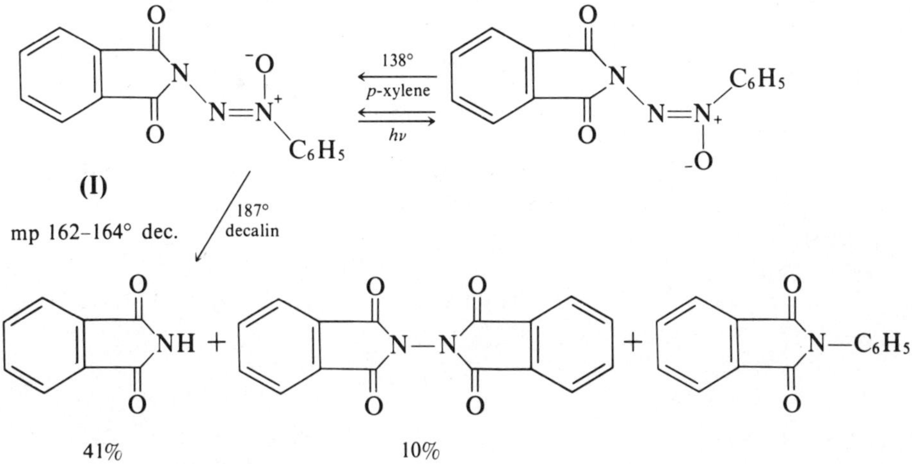

(I)

mp 162–164° dec.

mally requires loss of nitrous oxide, it was not detected. The 1-cyclohexyl derivative corresponding to **I** melts at 109–110° without decomposition.

The melting point of 4-benzylnitrosoamino-1,2,4-triazole (87) suggests that

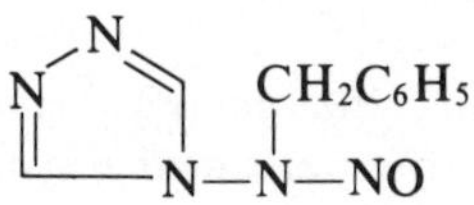

mp 115°

it is somewhat more stable than most acyclic nitrosohydrazines, possibly because of the electron-withdrawing *s*-triazole moiety.

Dipolar *N*-nitroimides of saturated nitrogen heterocycles (Table 2.28), based on melting point data, are as stable as trimethylamine *N*-nitroimide. The *N*-nitroimides of five-membered aromatic nitrogen heterocycles are somewhat less heat resistant but those derived from six-membered aromatic rings seem to be as stable as the saturated *N*-nitroimides.

2.2.2. One Exocyclic Nitrogen Atom

Single nitrogen groups, attached to two linked cyclic nitrogen atoms to form three-nitrogen chains, are amine, imide, nitroso, and nitro.

Compounds with an *N*-amino group attached to one of two cyclic contiguous, reduced nitrogen atoms may be regarded as mixed acyclic-cyclic triazanes. Stability to heat of this reduced three-nitrogen chain is greater in the acyclic-cyclic compounds, with a melting point range of 104–202°, than in the acylic triazanes. In each of the *N*-aminopyrazoles, pyrazolines, or substituted deriva-

Table 2.28. Acyclic-Cyclic Amine Nitroimides

Amine Nitroimide	Melting Point (°C)	Ref.
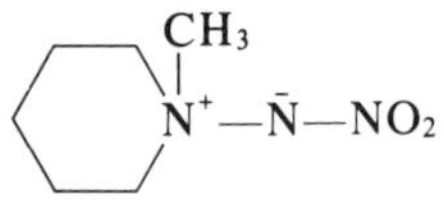	181–182	299, 301
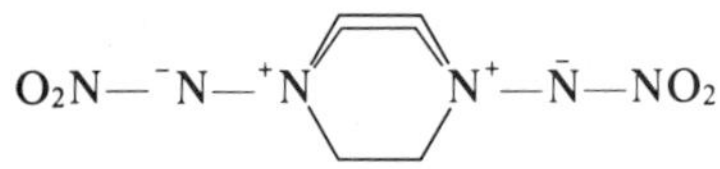	Chars >220°	299
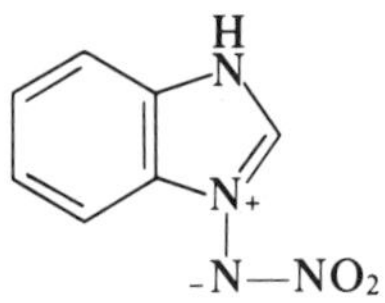	169 expl.	577
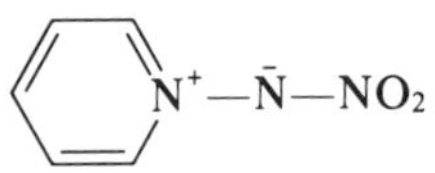	149.5–150	300
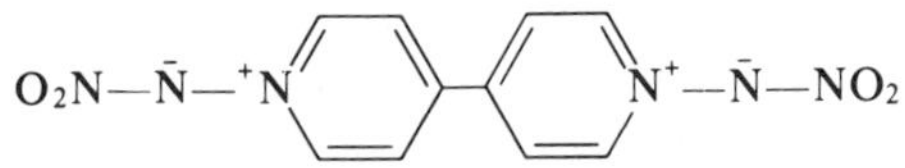	Chars >220°	299

tives (Table 2.29), there is at least one electron-withdrawing substituent (carbonyl, cyano, or ring unsaturation) directly attached to the aminohydrazine chain.

Salts of the acyclic-cyclic azimines (Table 2.30), with the middle nitrogen of the three nitrogen chain bearing a positive charge, have a higher thermal stability range (162–213°) than the open-chain triazanium salts.

The dialkylamino sydnonimine hydrochloride salts are closely related to the other azimine salts, but treatment with alkali, which would be expected to liberate the meso-ionic sydnonimine, instead leads to decomposition with ring opening (717).

Loss of the elements of a hydrogen acid from the azimine salts in a formal sense, and actually with some salts, leads to azimines. These *N*-imides* (Table 2.31) of cyclic azo compounds are analogous to the corresponding *N*-oxides. The *N*-imide of pyridazine is too unstable to be isolated, but *N*-acyl imides of pyridazine are more thermally resistant (594), as are also the dibenzopyridazines.

From the parent benzo[*c*]cinnoline *N*-imide, thermal stability increases as the electron-withdrawing character of the *N*-exocyclic substituent increases. Thermolyses of these derivatives form carbazole or *N*-substituted carbazole and benzocinnoline, except for the *N*-methyl imide, from which only benzocinnoline is obtained. Pyrolysis of the *N*-imide hydrochloride forms only carbazole (362).

* Heterocyclic *N*-imides are reviewed by Timpe (1092).

Table 2.29. *N*-Aminopyrazoles and Related Compounds

N-Aminopyrazole or Related Compound	Melting Point (°C)	Ref.
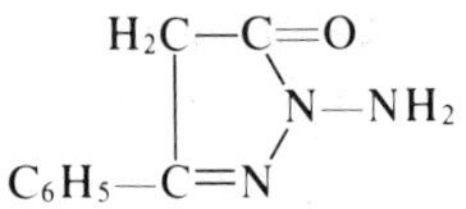	159–160 dec.	7
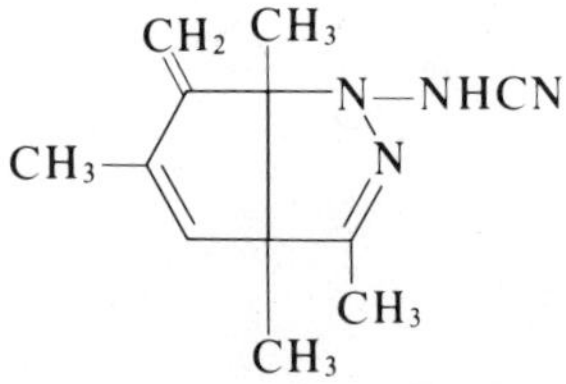	149–150	31
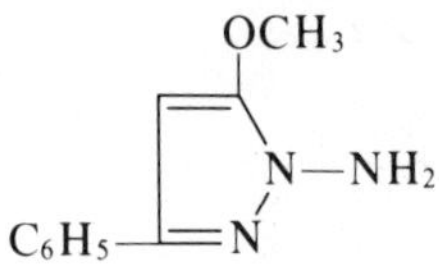	158–160 dec.	7
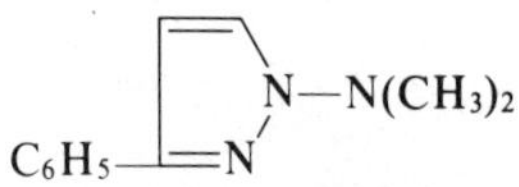	35–36 Bp 123/3.7	590
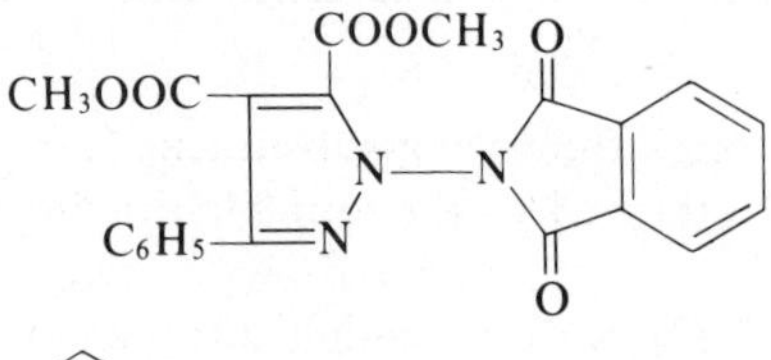	202–203.5	375
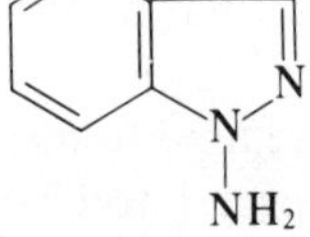	104–108	8

Heating benzo[*c*]cinnoline *N*-(*N*-phenylbenzimidoyl)imide at its melting point eliminates benzonitrile and forms benzo[*c*]cinnoline *N*-phenylimide (76). The *N*-alkylbenzimidoylimides of benzo[*c*]cinnoline are appreciably less thermally stable than the *N*-aryl derivatives.

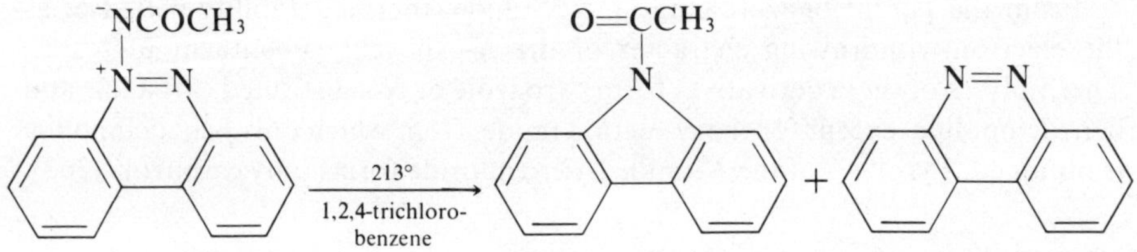

52% 17%

Table 2.30. Acyclic-Cyclic Azimine Salts

Azimine Salt	Melting Point (°C)	Ref.
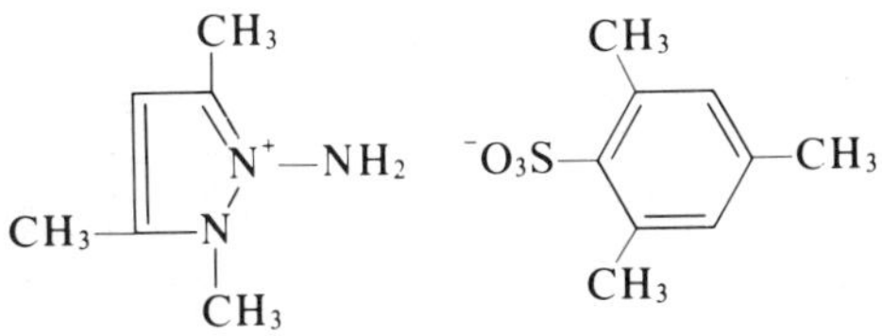	175	598
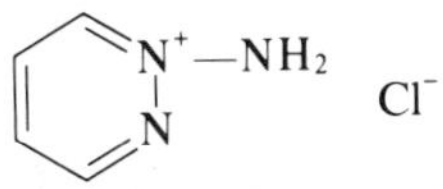	153–154	594
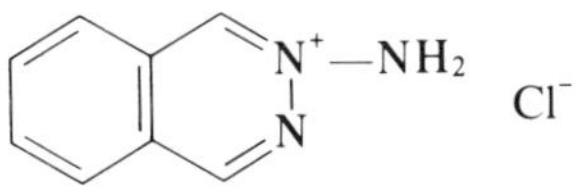	208 dec.	649
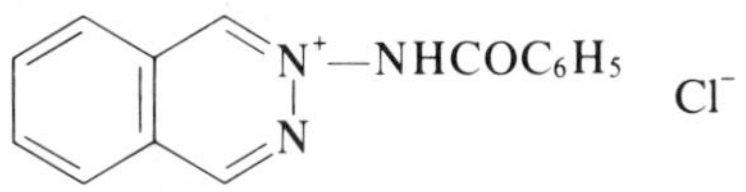	213–214 dec.	649
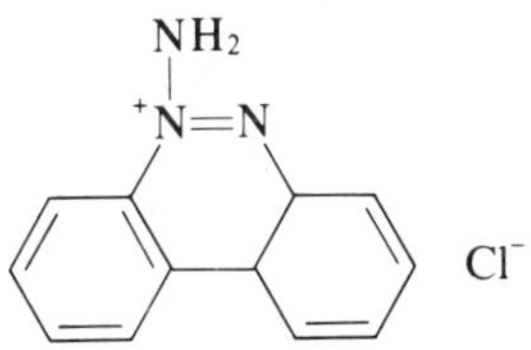	162–164 dec.	362
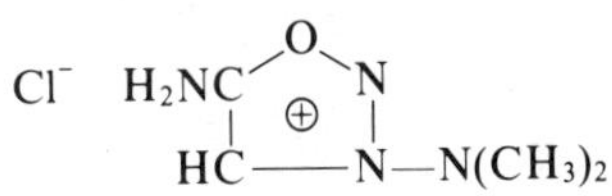	180–181 dec.	717

N-Nitrosopyrazolines and pyrazolidinones (Table 2.32) are similar in thermal stability to the open-chain nitrosohydrazines. The *N*-nitrosopyrazolidinones with phenyl on the remaining ring nitrogen are less stable than the unsubstituted nitrosopyrazolidinone. When 1-nitroso-3-aryl-Δ^2-pyrazolines are heated

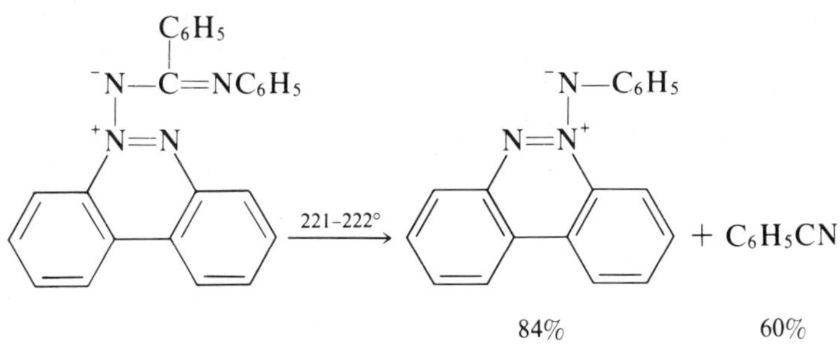

Table 2.31. Acyclic-Cyclic Azimines

Azimine	Melting Point (°C)	Thermal Stability	Ref.
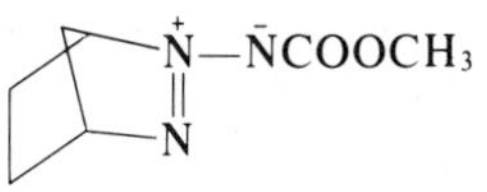	80–82		653
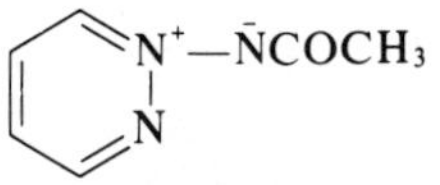	125–127 dec.		594
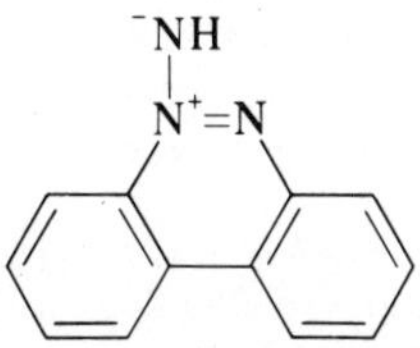	Dec. 127–128; 130	Dec., 160°, reflux in diglyme	363, 793, 362
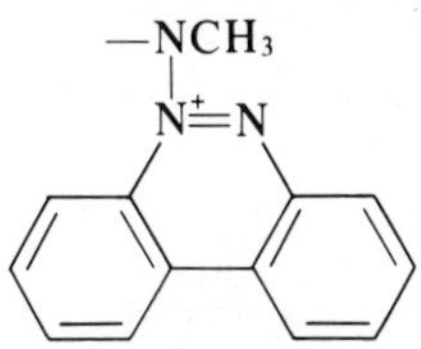	Dec. 88–89; 92		362, 793
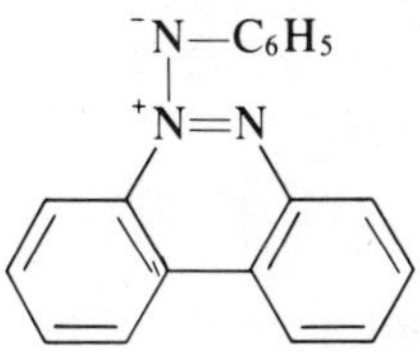	129–131; 130–132	Stable 179°, reflux in *o*-dichlorobenzene 1–2 hr, but dec. after refluxing 2–4 days	362, 1023
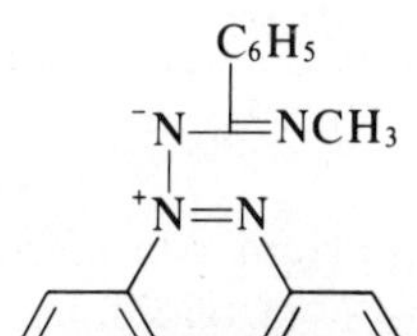	113–114 dec.		76
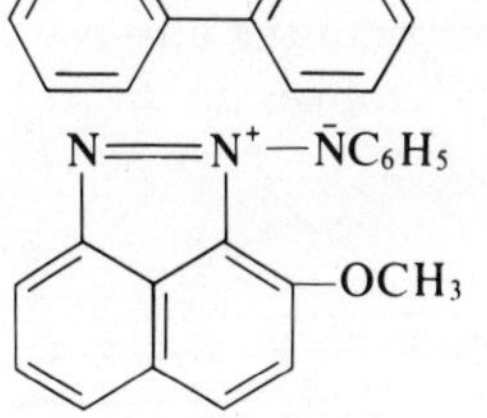	111–112	Largely unchanged, reflux in toluene; dec. 180° reflux in *o*-dichlorobenzene	1024

Table 2.32. *N*-Nitrosopyrazolines and Pyrazolidinones

Pyrazoline or Pyrazolidinone	Melting Point (°C)	Ref.
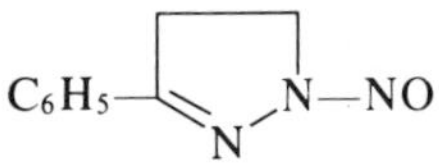	151–152 dec.	780
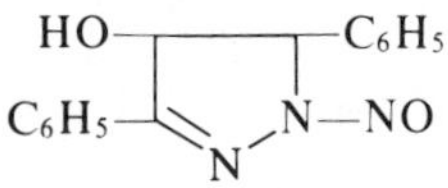	155 dec.	1177
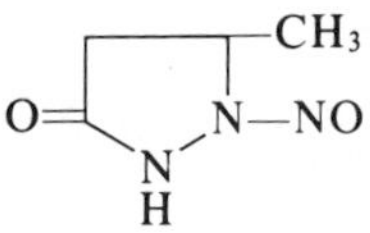	131 dec.	766, 767
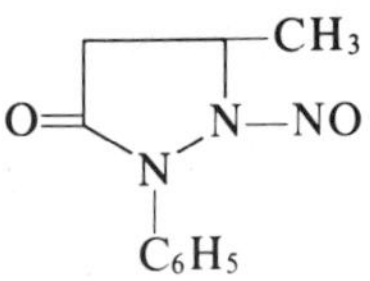	54–55, very unstable	593
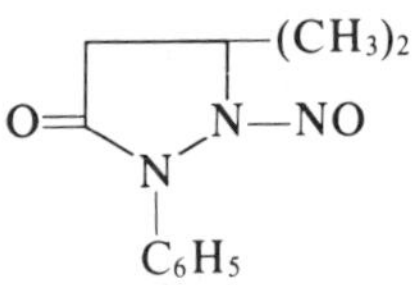	Dec. slowly on standing	889

dry at 160–170°, or refluxed in *o*-dichlorobenzene, nitrogen is evolved and 3-aryl-isoxazolines are formed (780).

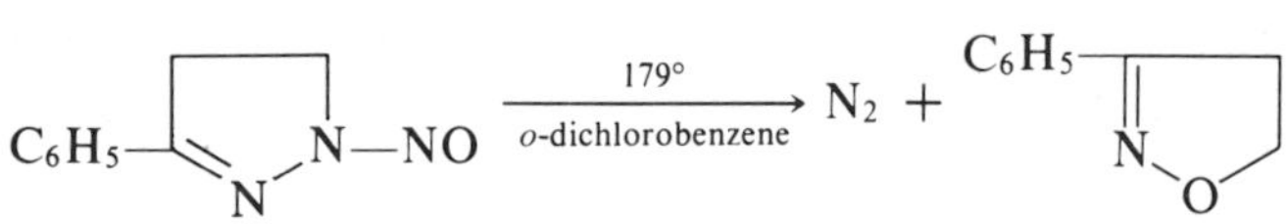

The *N*-nitropyrazoles and indazoles (Table 2.33) are stable to 130–140°, slightly below the decomposition temperature of the *N*-nitrosopyrazolines.

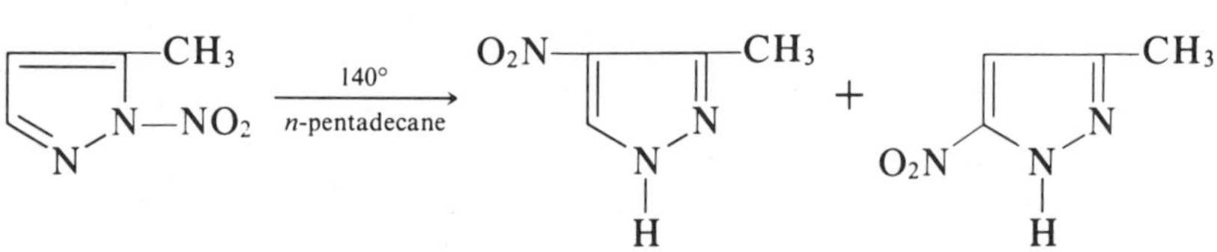

93% 7%

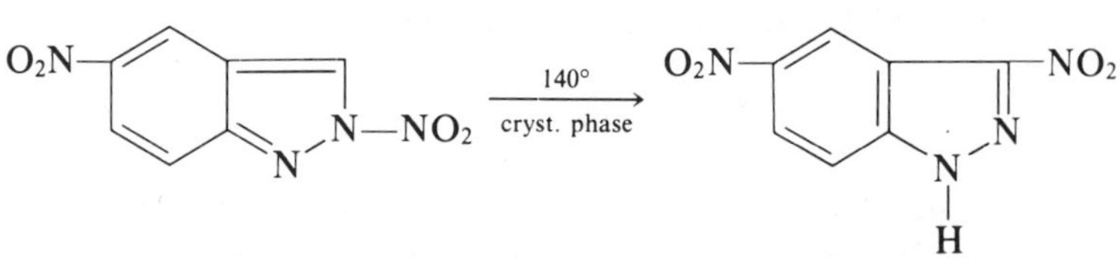

Table 2.33. *N*-Nitropyrazoles and Related Compounds

N-Nitropyrazole	Melting Point (°C)	Thermal Stability	Ref.
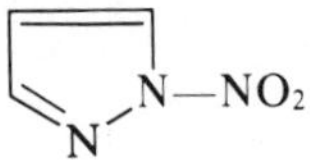	91–92; 93	Can be stored for prolonged periods; rearranges 140–150°	533, 554, 824
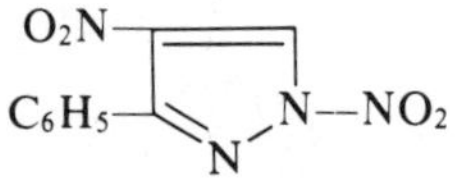	163 dec.	Rearranges 140° in benzonitrile	553
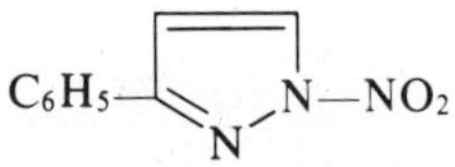	119; 122	Rearranges 130° in anisole	245, 553
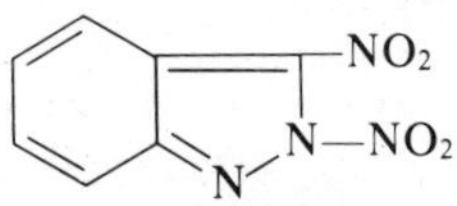	154 dec.		222

Above this temperature, the *N*-nitro group rearranges intramolecularly to *C*-nitropyrazoles (553, 554) or indazoles (222).

2.2.3. Cyclic-Cyclic Three-Nitrogen Chains

A few *N,N'*-linked bi(heteroaryls) consisting of pyrazoles or *s*-triazoles joined to other heterocycles are known which contain three-nitrogen chains. Each of these compounds is stable at its melting or reduced pressure boiling point (252).

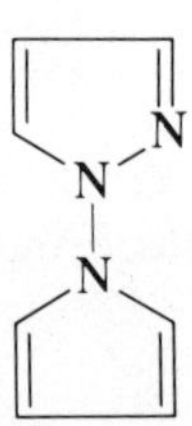

bp 104–105°/16

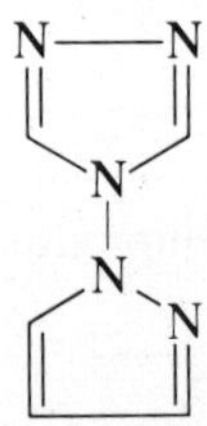

mp 136°
sublimes 90°/0.1

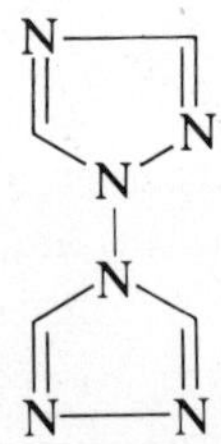

mp 179–181°
sublimes 140°/0.02

2.3. CYCLIC THREE-NITROGEN CHAINS

2.3.1. Triaziridines and Triazetidines

The first authentic compound with an all nitrogen three-membered ring was not prepared until 1980 (654). 1-Ethoxycarbonyl-*trans*-2,3-diisopropyltriaziridine decomposes into a mixture of isomeric azimines; at room temperature its half-life is 3.5 days and thus is less stable than related acyclic triazanes.

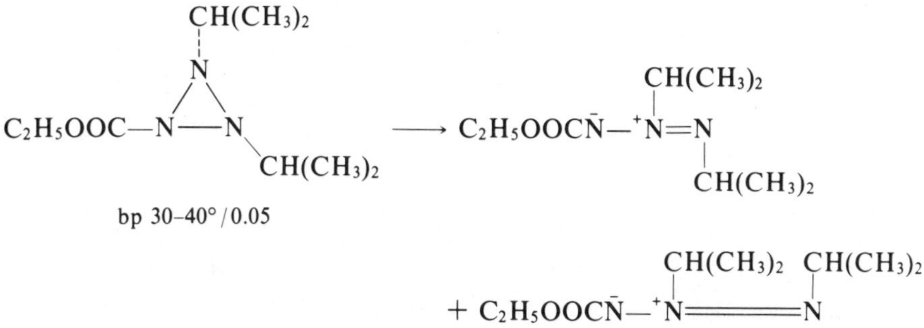

bp 30–40°/0.05

A triazetidine structure is proposed for one of the products resulting from reaction of methoxycarbonylnitrene with triethylamine, based on elemental

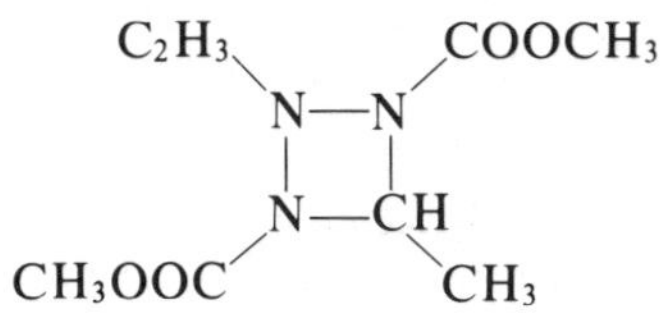

colorless liquid

analysis, infrared, NMR and mass spectral evidence (413a); alternative structures are possible.

2.3.2. Triazolines and Related Compounds

Triazolines (dihydrotriazoles) are five-membered heterocycles with a three-nitrogen chain, two carbon atoms, and one double bond in the ring. The Δ^4-triazolines **(I)** have substituents on each nitrogen, the Δ^3-class **(II)** have substituents in the 1- and 2-positions, and the Δ^2-class **(III)** have one substituent at N-1. In structures **II** and **III**, substituents W and X can be a single double-bonded group, as also Y and Z in structure **III**. Most known triazolines belong to the Δ^2 class.

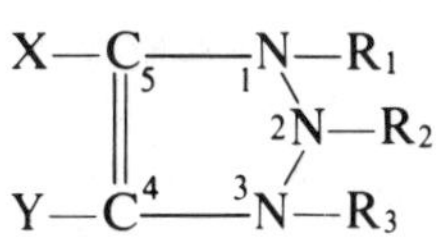

(I)

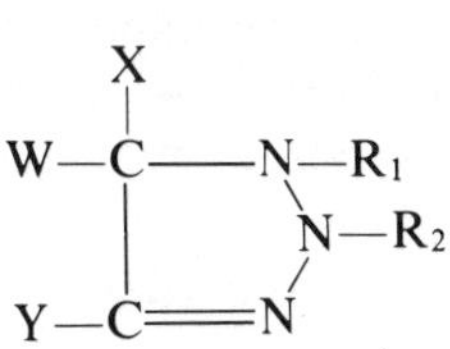

(II)

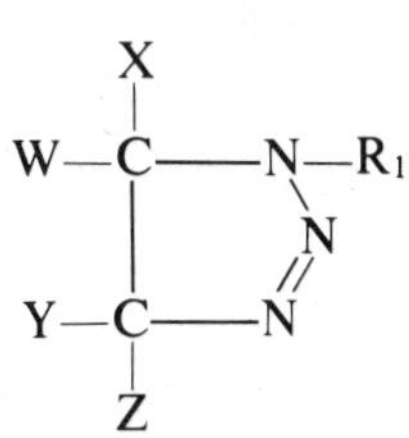

(III)

In the Δ^3- and Δ^4-triazolines, the nitrogen chain is at the triazane oxidation level. Based on melting point data, these triazolines appear to be as stable as acyclic triazanes, if not more so.

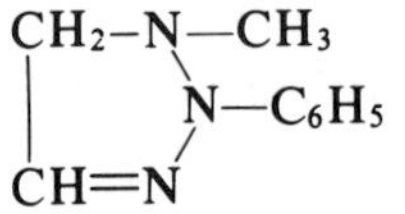

liq. (543)

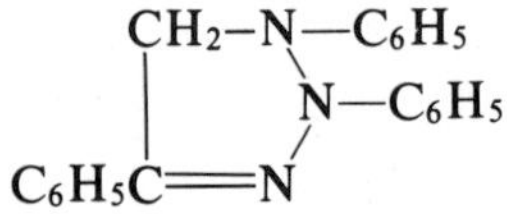

mp 136.5° (179)

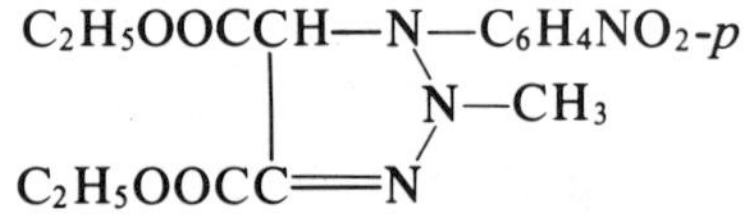

mp 162° (445)

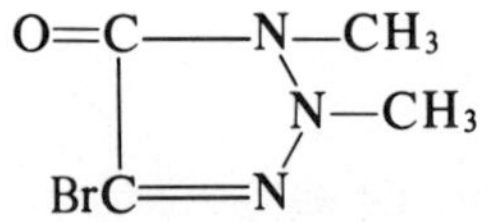

mp 96–97° (95)

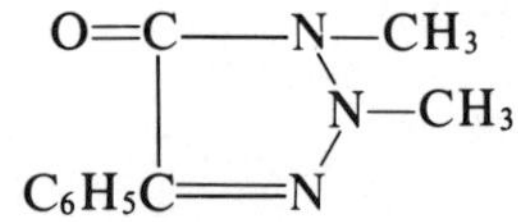

mp 120° (94)

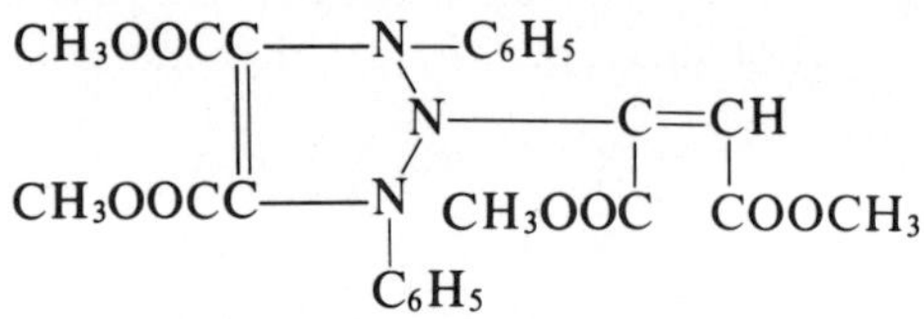

mp 188.5–189° (5)

The Δ^2-triazolines, referred to hereafter without the prefix, are cyclized triazenes, with the ends of the three-nitrogen chain connected by a two-carbon group.* Their thermal stability depends, in part, on the nature of the substituent at the 1-nitrogen, but is also affected by the substituents at the 4- and 5-carbon atoms.† A fine balance is required for existence of the triazolines, which can be upset by either excessive electron-withdrawing or electron-releasing substituents. 1-Alkyl-Δ^2-triazolines are liquids that can be distilled under reduced pressure (365).

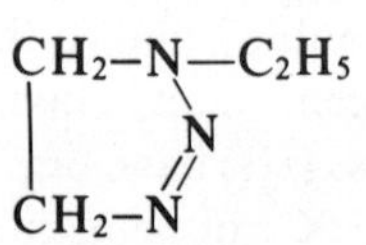

bp 32°/3

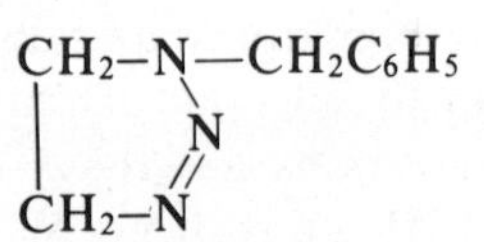

bp 120°/0.7

The routes by which triazolines decompose thermally include elimination of a molecule HX, usually small, from the 4- and 5-positions, forming *v*-triazoles; dissociation into azides and olefins, a reversal of synthesis; breaking a ring ni-

* Unfortunately melting point or other thermal stability data for the closest open-chain analogs to the 1-aryl-Δ^2-triazolines, the 1,3-dialkyl-1-aryltriazenes, do not appear to have been determined and thus no direct comparison of stability is possible.

† Two excellent reviews of the triazolines, with emphasis on decomposition reactions, are those of Scheiner (957) and Bourgois, Bourgois, and Texier (131).

trogen-nitrogen bond and elimination of nitrogen, forming aziridines, imines, or both; and forming α-diazo amines.

Triazolines unsubstituted at positions 4 and 5 are thermolyzed to aziridines, as illustrated by the decomposition of 1-*m*-chlorophenyltriazoline (453).

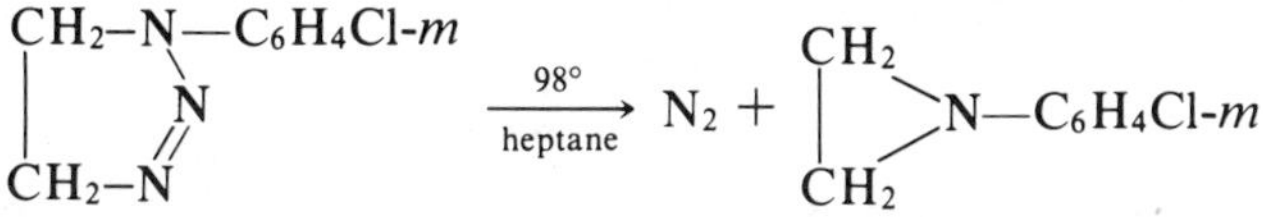

Pyrolysis of the bicyclic triazolines from cycloalkenes and aryl azides leads to both aziridines and cyclic imines (956). The 1-aryl triazolines with 5-alkyl,

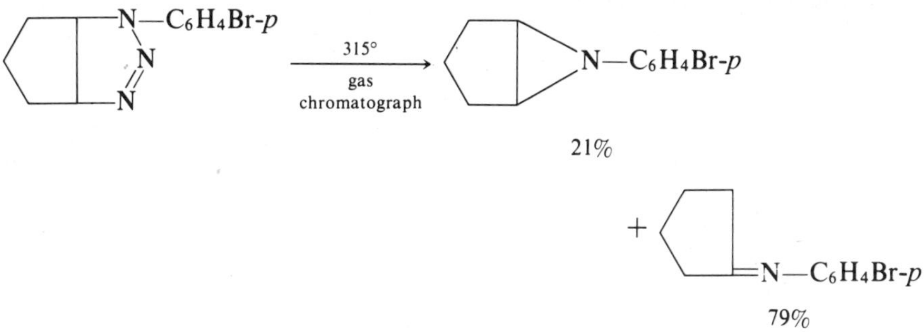

4,5-dialkyl, and cycloalkylene substituents (Table 2.34) decompose in the approximate range 110–170°.

The complexity of thermal triazoline decomposition is illustrated by the products from heating the adduct of phenylazide and norbornene(3-phenyl-

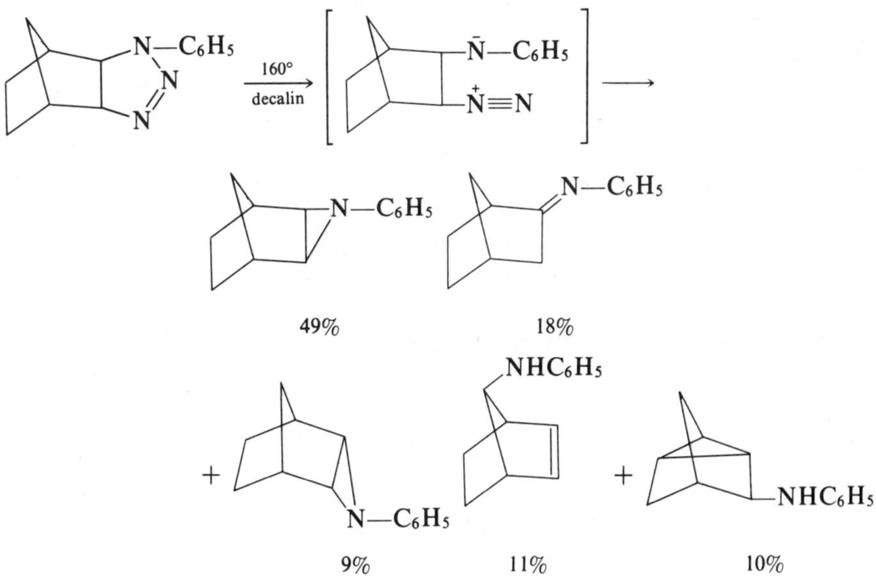

Table 2.34. 1-Aryl 5-Alkyl, 4,5-Dialkyl, or Cycloalkylene Triazolines

Triazoline	Melting Point (°C)	Ref.
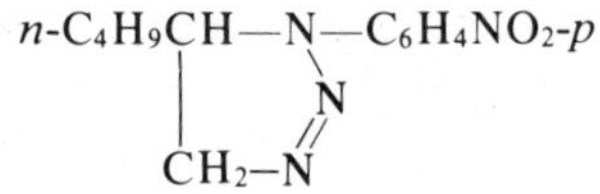	105.5–106.5	955
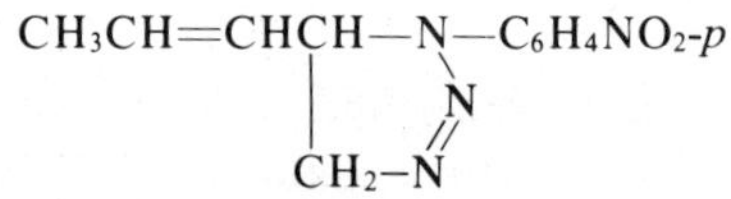	139–140 dec.	955
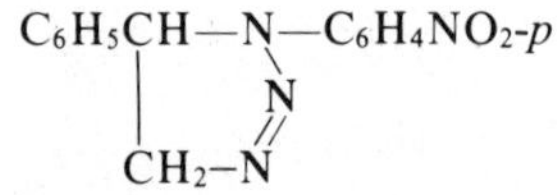	152–153 dec.	572
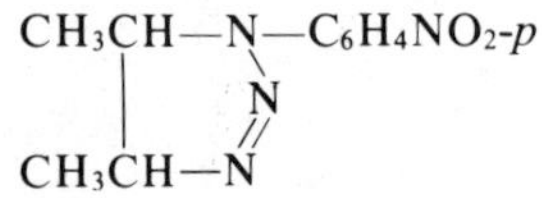	109 dec.	955
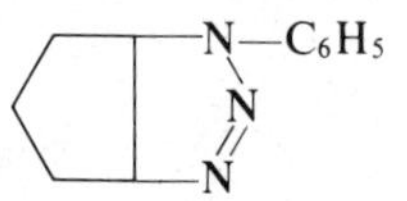	53; dec. 150	26
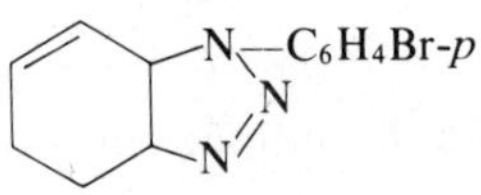	113–114 dec.	955
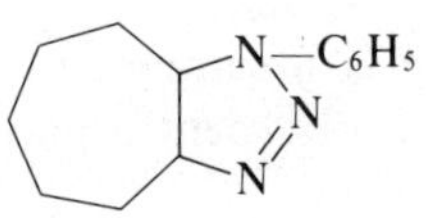	76–77; dec. 170	26

3,4,5-triazatricyclo[5.2.1.$0^{2.6}$]dec-4-ene) in which the triazoline ring has an exo orientation. At least five different products are formed (925). Because the pyrolysis is accelerated in polar solvents a diazonium betaine intermediate is proposed. Formation of the *endo*-aziridine requires a molecular rearrangement. Similar products, including an *endo*-aziridine, result from the pyrolysis of the corresponding *p*-nitrophenyl norbornene triazoline (1203).

The electronic character of the substituent on the 1-nitrogen influences the thermal stability of the norbornene triazolines (Table 2.35). Decomposition of the benzyl derivative requires the most strenuous conditions. Half-lives of the *p*-tolyl, phenyl, and *p*-nitrophenyl triazolines become progressively shorter as electron-withdrawing capability increases; the ethoxycarbonyl derivative has the poorest stability of this series. The benzoyl, tosyl (520), diethylphosphoryl, and diphenylphosphinyl (107, 108) triazolines were not isolated from interaction of the azides with norbornene, but their presence as intermediates is indicated by NMR and kinetic data, and by their decomposition products. Thermal

Table 2.35. Norbornene Triazolines

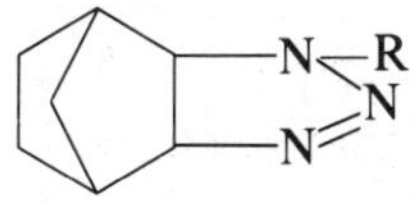

R	Melting Point (°C)	Thermal Stability	Ref.
$C_6H_5CH_2$-	74–75	Dec. 170–180°	520
p-$CH_3C_6H_4$-	79–80.5	$t_{1/2}$ (141.6, nitrobenzene) 7.9 hr	725, 958
C_6H_5-	99–100; 101–102	$t_{1/2}$ (141.6, nitrobenzene) 4.6 hr	725, 958
p-$O_2NC_6H_4$-	164–165	$t_{1/2}$ (141.6, nitrobenzene) 39 min	958, 1203 725
C_2H_5OOC-	Liq.	Dec. >70°	953

breakdown of norbornene triazolines substituted with strongly electron-withdrawing groups proceeds faster than their formation.

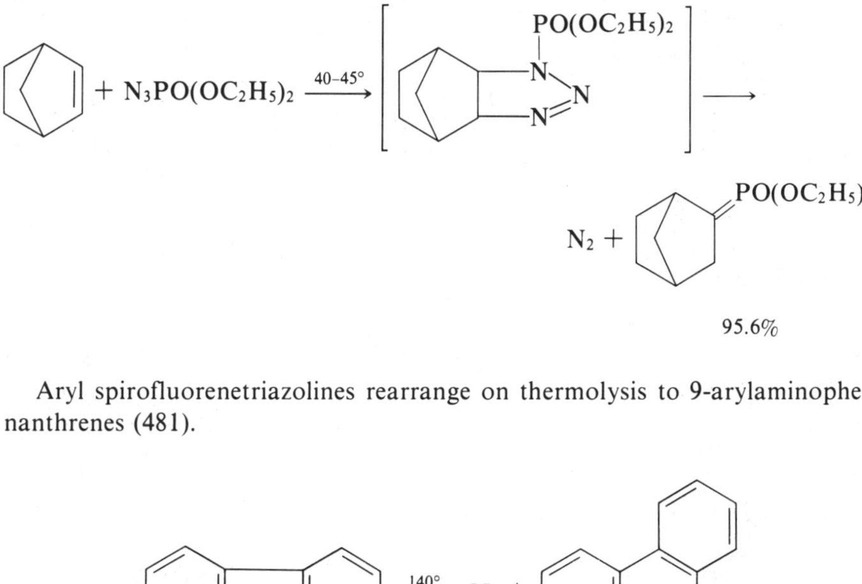

Aryl spirofluorenetriazolines rearrange on thermolysis to 9-arylaminophenanthrenes (481).

Triazolines with electron-releasing substituents on carbon are listed in Table 2.36. Those with two methoxy groups at position 5, adjacent to nitrogen substituted with an electron-withdrawing group, lose nitrogen at room temperature. The simultaneous presence of an electron-releasing group at the 4-position further decreases stability. Existence of 4,4-dimethoxy-5,5-dimethyl-1-phenyl triazoline, substituted on carbon solely by electron-releasing groups, indicates that the range of possible triazolines is considerable. The triazoline from the combination of phenylketene dimethylacetal with ethoxycarbonyl azide is not stable enough to be isolated (951). At 35°, a phenylketene dimethylacetal solution of the triazoline initially forms the diazoamine **(IV)**, with which it is in equilibrium. Further heating at this temperature leads principally to the triazole **(V)** and iminoether **(VI)**.

Triazolines from the combination of aryl azides and 1,2-diaminoethylenes are too unstable for isolation and only the aminoacetamidine decomposition products are found (210). Triazolines with a single ether or amino substituent at position 5 are more stable. Thermolysis of 4,4-dimethyl-1-(1-phenylvinyl)-5-(1-

Table 2.36. Triazolines with Electron-Releasing Substituents at Positions 4 and 5

Triazoline	Melting Point (°C)	Thermal Stability	Ref.
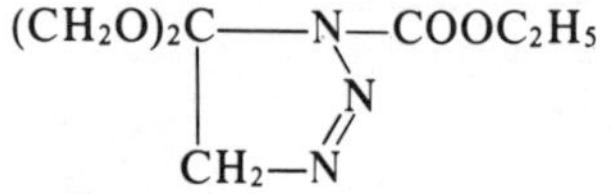	Liq.	$t_{1/2}$ (36°, CCl_4) 144 hr	403
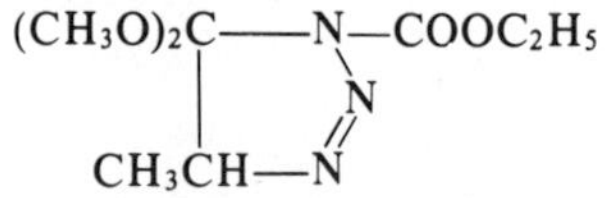	Liq.	$t_{1/2}$ (36°, CCl_4) 5.5 hr	403
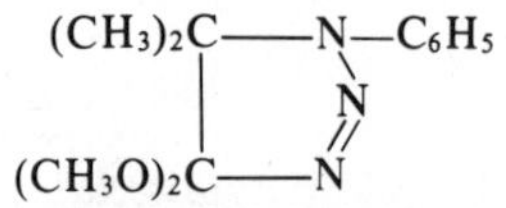	61–62	Dec. on heating	952
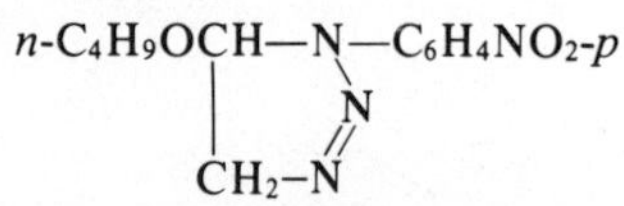	122–123	Dec. 125°	522
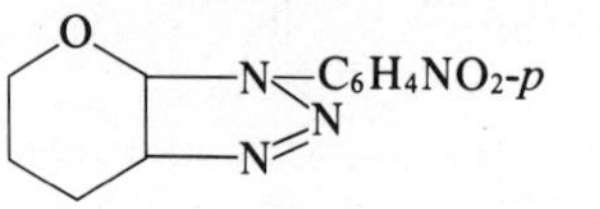	95–96 dec.	Dec. 110°, toluene	522
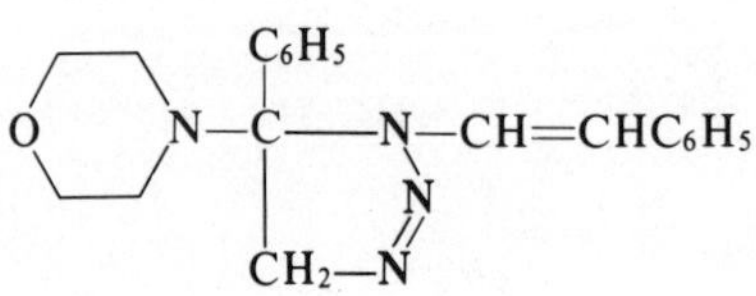	125.5 dec.		810

pyrrolidinyl)triazoline, prepared *in situ* by reaction of α-azidostyrene with 1-(2-methyl-1-propenyl)pyrrolidine, leads exclusively to the azabutadiene shown (811, 812).

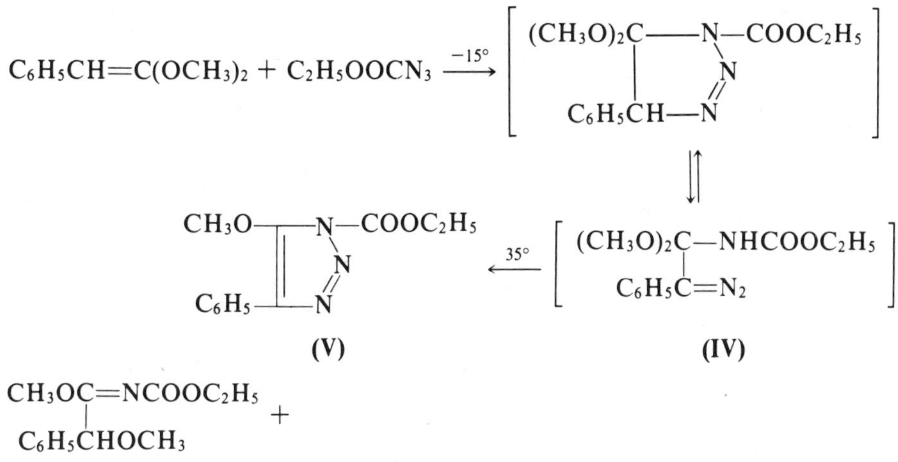

(V) (IV) (VI)

Two different paths of thermal decomposition are followed by 5-alkoxy triazolines. Thus on heating 5-ethoxy-5-methyl, 5-methoxy-5-phenyl, and 5-butoxy-1-*p*-nitrophenyl triazolines, a mole of alcohol is eliminated, forming the corresponding *v*-triazole (522). Thermolysis of 4-methyl-1-*p*-nitrophenyl-5-propoxytriazoline with only one substituent at position 5, either cis or trans,

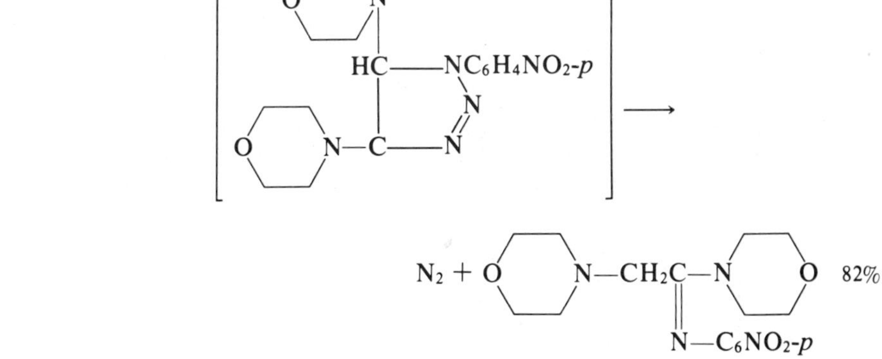

destroys the ring, leaving an imino ester (524a). Quantitative study of the decomposition of 5-octoxy-1-ethoxyethyl triazoline shows that both decomposi-

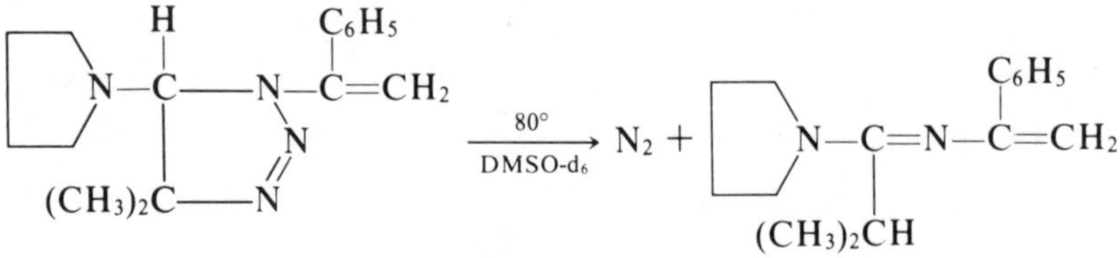

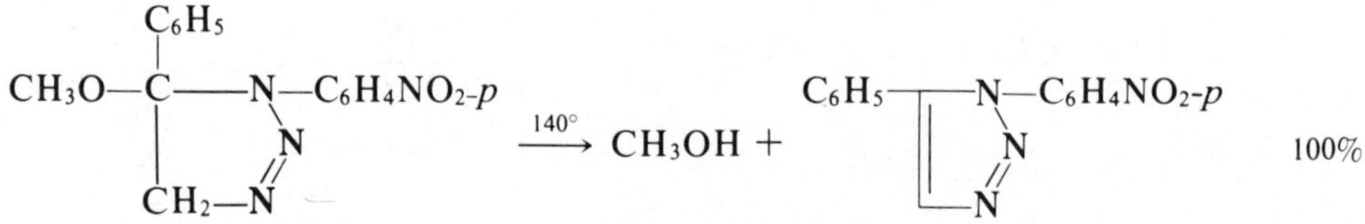

tion routes occur (636). Thermolysis of 1-aryltriazolino furans or pyrans with a cyclic ether oxygen at the 5-position leads to the acyclic-cyclic imino esters, tetrahydrofuran- or pyran-2-anils (522, 954).

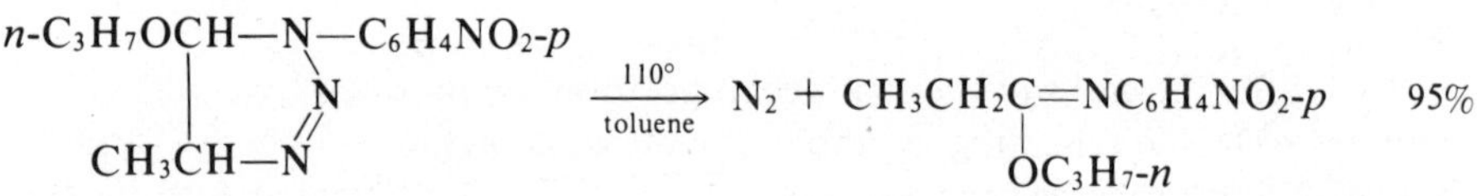

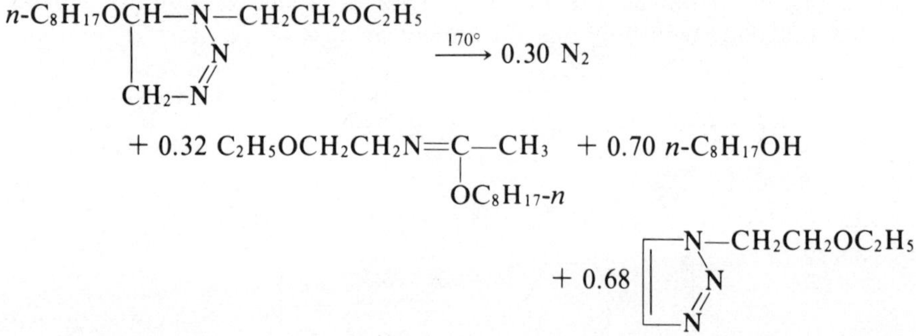

Aziridines are the only thermal decomposition products of triazolines with two strongly electron-withdrawing groups at the 5-position, such as 1-[2,2-dimethylvinyl]-5,5-bis(trifluoromethyl)- (170) or 1-methoxy-5,5-bis(methylsulfonyl)triazoline (51). The triazolines expected from the combination of aryl

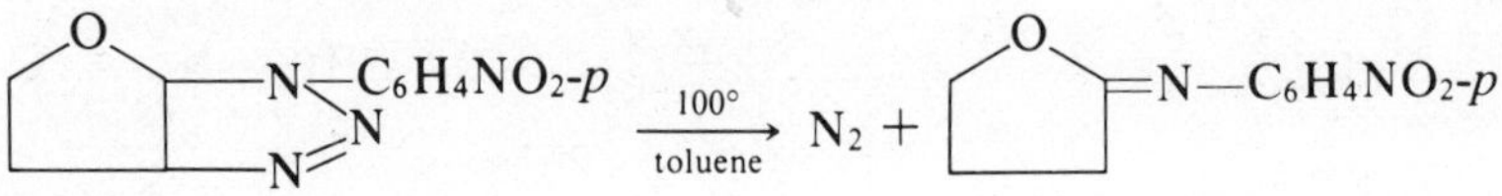

azides with α-cyano acrylates and related unsaturated compounds with two *gem*-substituted electron-withdrawing groups are too unstable for isolation; instead, pyrrolidines result from combination with 2 moles of α-cyanoacrylate,

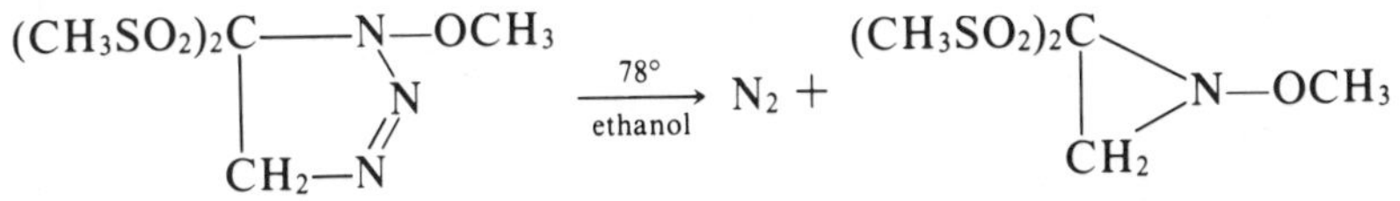

presumably through an aziridine in equilibrium with an isomeric ylide form (1079).

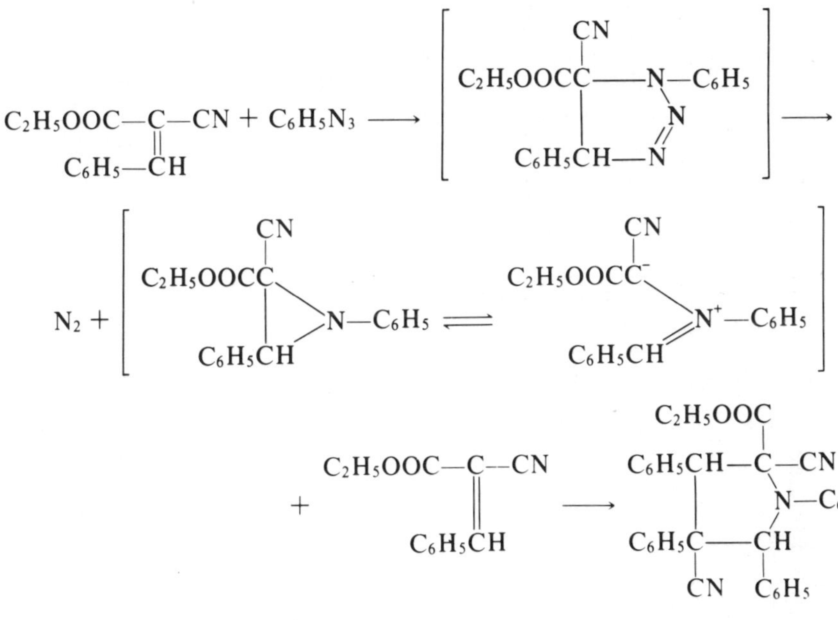

Triazolines with only electron-releasing groups at position 4 and no substituents at position 5 do not appear to have been prepared. Several triazolines with electron-attracting groups at the 4-position are known (Table 2.37); most decompose below 100°. The triazoline adduct of *n*-butyl azide and ethyl acrylate can be isolated, but spontaneously isomerizes to the unisolated α-diazoamine **(VII)**, with which it appears to be in equilibrium (157). Addition of a second mole of ethyl acrylate forms a pyrazoline.

Comparison of the 1-phenyltriazoline isomers with methyl and acetyl groups at position 4 or 5 shows that the 4-regioisomer is much less stable. From the 4-methyl-4-acetyl isomer, only the aziridine is formed, but the 5-isomer yields both the aziridine and a β-ketoenamine rearrangement product (157).

Table 2.37. Triazolines with 4-Substituted Electron-Withdrawing Groups

Triazoline	Melting Point (°C)	Thermal Stability	Ref.
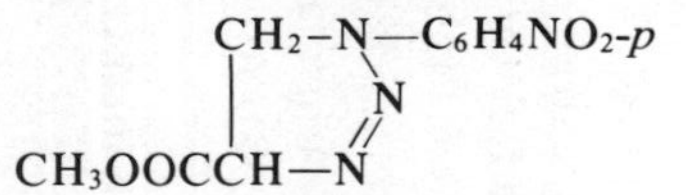	Dec. > 95	$t_{1/2}$ (85°, *o*-dichlorobenzene) 20 min	525, 1062
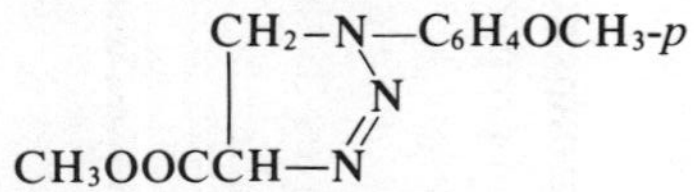	Dec. 71–72.5	$t_{1/2}$ (85°, *o*-dichlorobenzene) 5 min	525, 1062
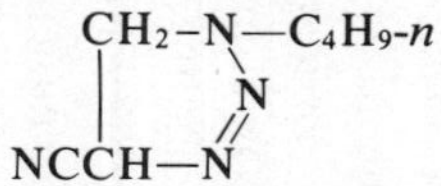	Liq.	Dec. 90°, toluene	157
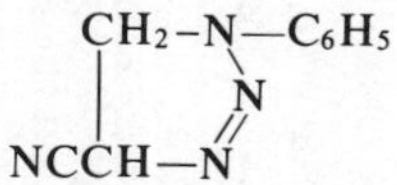	98°	$t_{1/2}$ (60°, neat) 90 hr	70, 1062
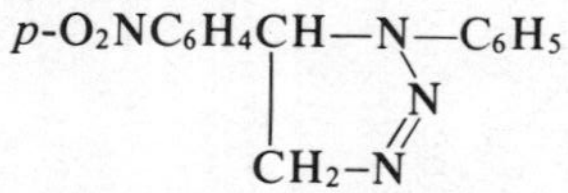	130–130.5 dec.		572
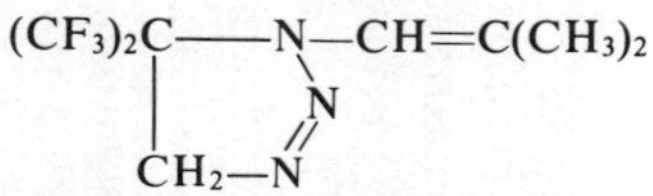	Bp 26–29/0.3	Dec. 130–140°	170

With a 5-aryl substituent, decomposition products are more varied. Thermolysis of the cis form of 4-cyano-1-*p*-methoxyphenyl-5-phenyltriazoline leads to cis and trans aziridines and an enamine, also with migration (929).

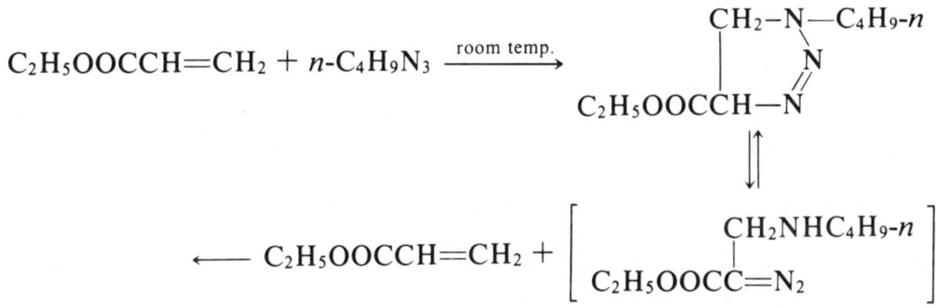

VII

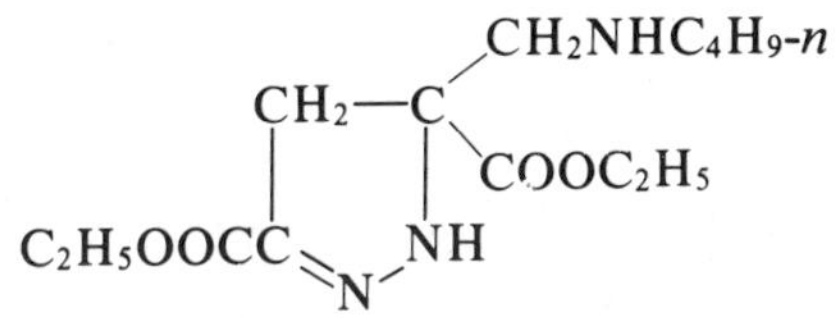

Decomposition of 1,5-diphenyl-4,4-di(methoxycarbonyl)triazoline in refluxing toluene proceeds with rearrangement through an azomethine ylide, as indicated by its interception by an isocyanide, forming an azetidine (171).

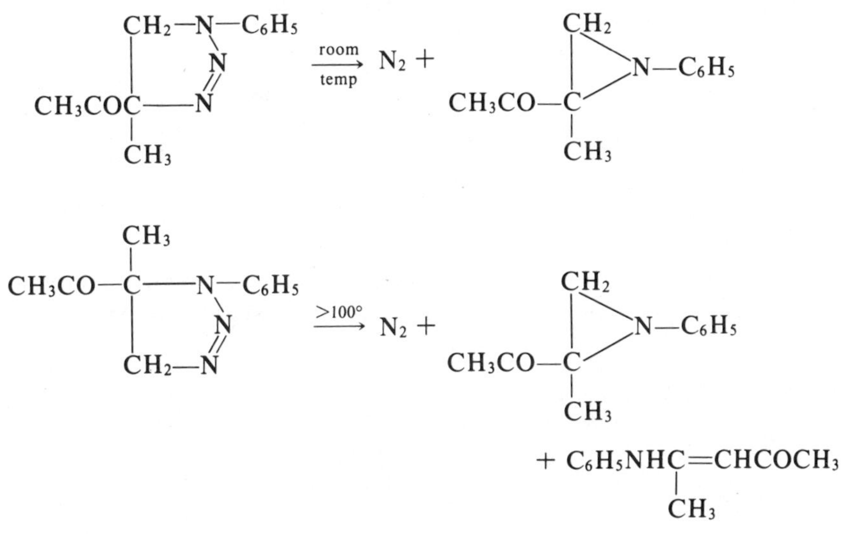

The fused ring 3-substituted 3*a*,8*a*-dihydroindene[1,2-*d*]triazol-8-ones, with an electron-withdrawing carbonyl at the 4-position of the triazoline and aryl at the 5-position (Table 2.38), are somewhat more stable (426). The relative effects of electron-withdrawing and electron-releasing groups at triazoline nitrogen are the approximate reverse of those of the norbornene triazolines.

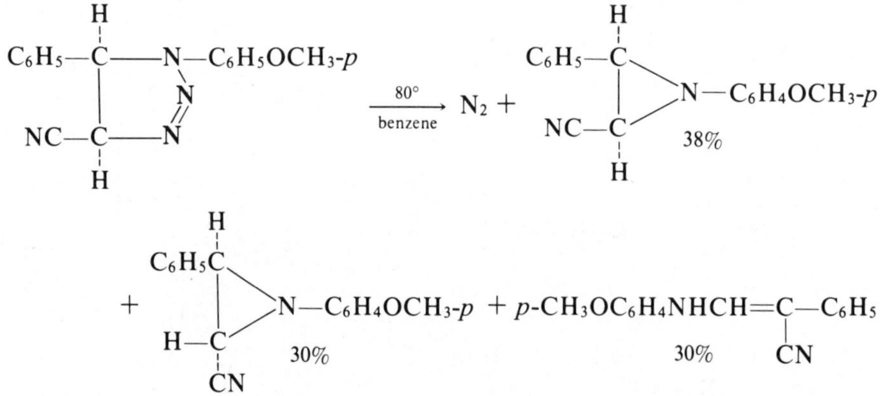

Heating the aryl and benzyl compounds gives both aziridine and imine, but only imine results from the methyl and butyl derivatives.

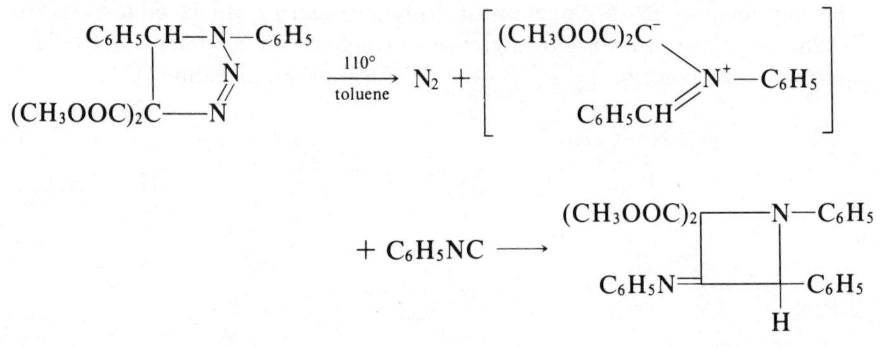

A triazoline cannot be isolated from the reactions of fumaronitrile with *p*-methoxyphenyl azide, but a triazole results from elimination of hydrogen cya-

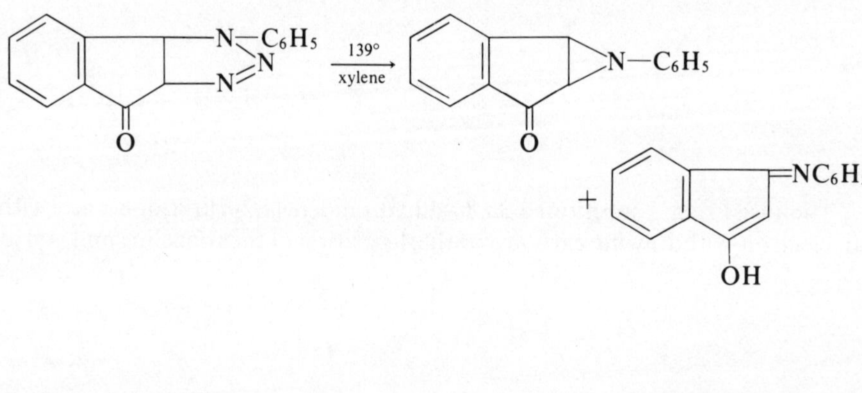

Table 2.38. Decomposition Points of 3-Substituted 3a,8a-Dihydroindene[1,2-*d*]triazol-8-ones

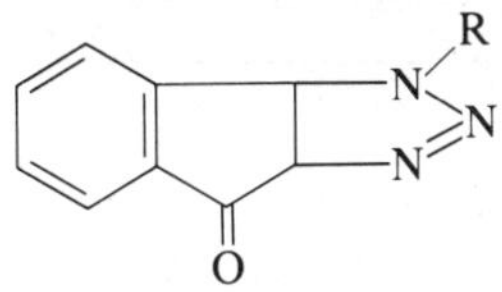

R	Decomposition Point (°C)
p-$O_2NC_6H_4$—	200
C_6H_5—	132
p-$CH_3OC_6H_4$—	140
$C_6H_5CH_2$—	146
CH_3—	115
n-C_4H_9—	70

nide from the intermediate (525). Triazolines exist, however, with other electron-withdrawing groups, such as methoxycarbonyl and the cyclic residue of

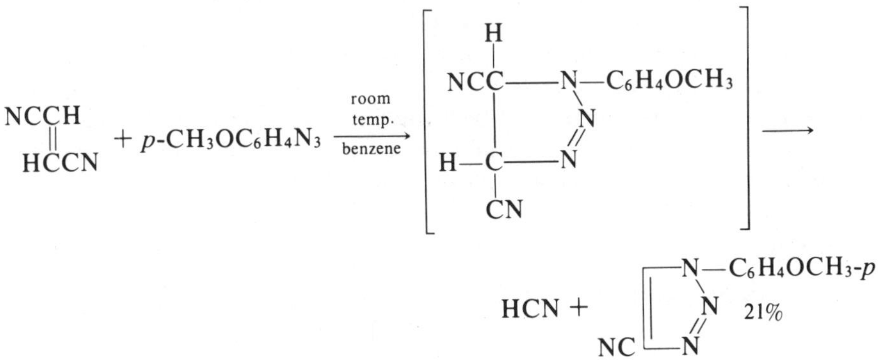

N-phenylmaleimide at the 4 and 5 positions (Table 2.39). Among the most stable triazolines are those with trifluoromethyl and fluorine substituted on the ring carbons.

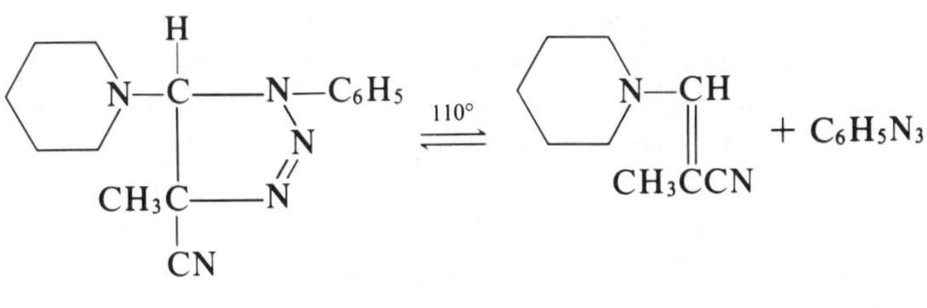

Table 2.39. Triazolines with 4,5-Disubstituted Electron-Withdrawing Groups

Triazoline	Melting Point (°C)	Thermal Stability	Ref.
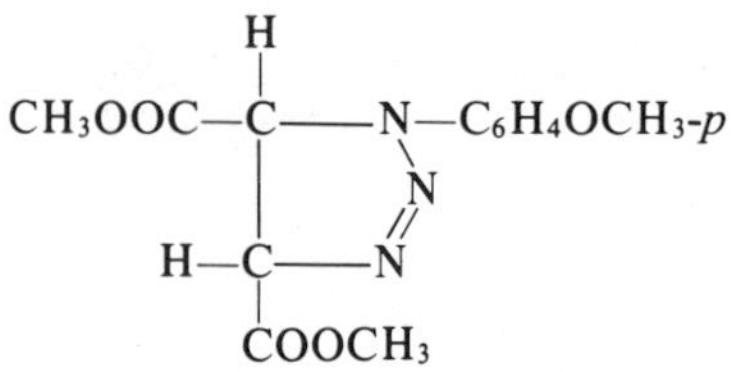	Dec. 94–95	Dec. 100°, toluene	525,1062
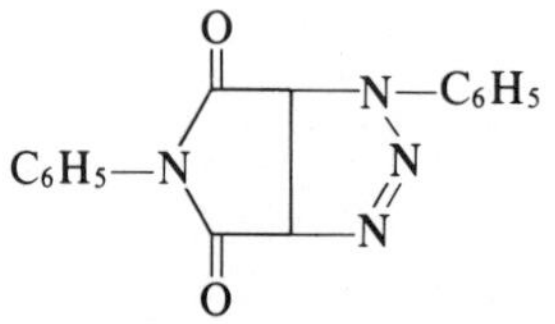	156 dec.	Dec. 130°, neat, vacuum	776
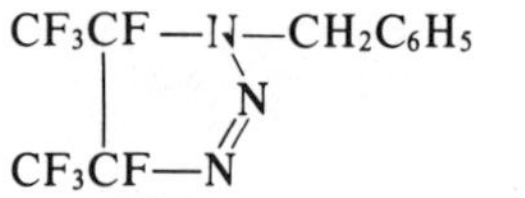	Bp 60–99/2	Dec. > 200°	197

Still different thermal behavior is found in the 5-amino-4-cyano-1-phenyl triazolines, which dissociate through a 1,3-dipolar cycloreversion (114) into enamine and phenyl azide (1078). Indeed, in the system of phenyl azide and β-piperidinomethacrylonitrile at 110° in toluene, less than 5% triazoline is present. Heating 5-cyano-1-phenyl-4,4-bis(trifluoromethyl)triazoline decomposes a portion to the related azirdine and dissociates the remainder to β,β-bis-(trifluoromethyl)acrylonitrile and phenyl azide (949).

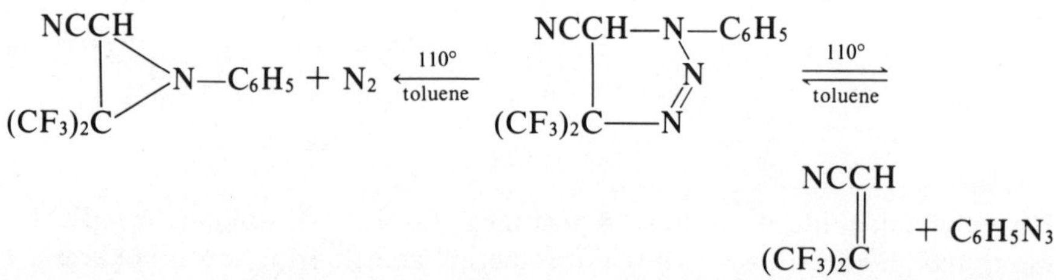

Triazolines with methylene and spiro(cyclopropane) substituents at position 4 are compared with 4,4-diphenyl derivatives in Table 2.40. 1-Aryl 5-substituted amino triazolines with methylene in the 4-position decompose from 57 to 153° (244). 5′-Amino-1′-aryl-spiro(cyclopropane-1,4′-*v*-triazoline) decompose from 130 to over 221°; derivatives with bisaryl substituents on cyclopropane do not decompose at their melting points (877). The structure of the unstable 4,4-diphenyltriazolin-5-one, with hydrogen on N-1, is indicated by a sharp NH infrared band (496, 540). Stability in the 4,4-diphenyl triazolines is increased by ring methylation and by replacing carbonyl oxygen with imine.

Table 2.40. Triazolines with 4,4-Methylene, Spiro or Diaryl Substituents

Triazoline	Melting Point (°C)	Thermal Stability	Ref.
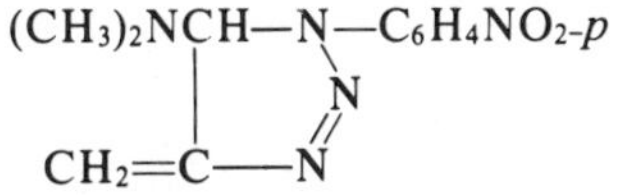	111–114 dec.		244
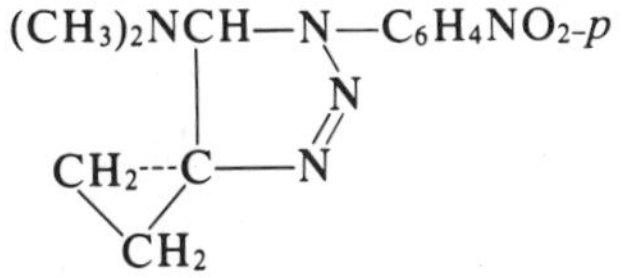	130–131 dec.		877
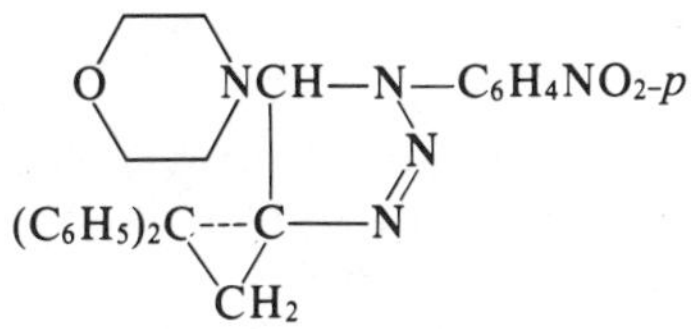	221–224		877
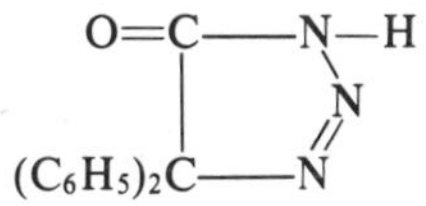	Solid	Stable at room temp.; dec starts 50°, expl. 60° in quantities > 20 mg	540, 495, 496
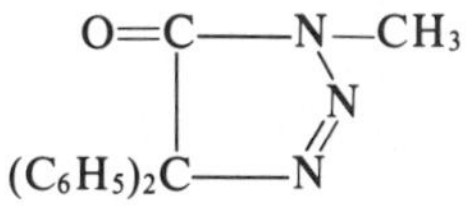	139–142 dec.		496
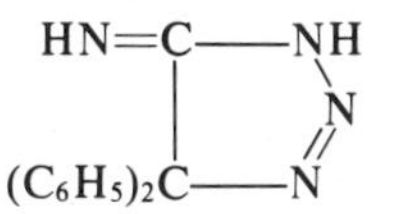	Dec. 187		494

The triphenyl 1,2,3-triazafulvene shown, with a cyclic triazene chain substituted by one divalent and one monovalent radical, is much less stable than its open-chain analogs. It is blood red in tetrahydrofuran solution, dimerizes on

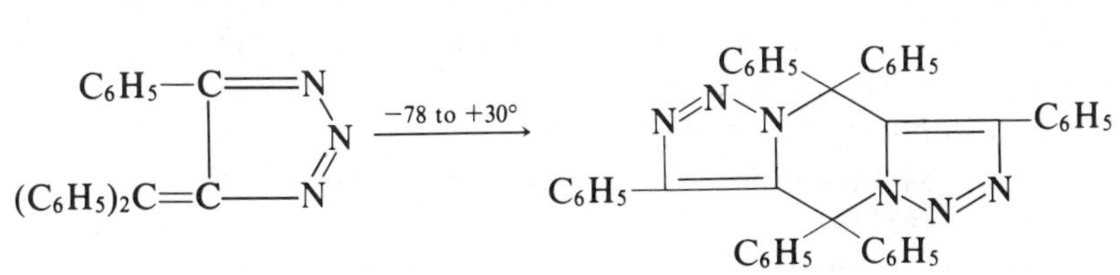

warming, and has a half-life at −78° of 4.2 hr (173, 174).

Triazolines with a 5-hydroxyl group usually exist as such in the solid state and in solution, but those bearing an aryl group on N-1 exist in equilibrium

Table 2.41. 5-Hydroxytriazoline-Ketotriazene Equilibria

Triazoline Substituents			Triazene (%)
N-1	C-4	C-5	($CDCl_3$ Solution)
C_6H_5	H, CH_3	$(CH_3)_2CH$	0
C_6H_5	H, CH_3	$(CH_3)_3C$	24
C_6H_5	CH_3, CH_3	$(CH_3)_3C$	91
p-$O_2NC_6H_4$	H, CH_3	$(CH_3)_3C$	42
p-$O_2NC_6H_4$	CH_3, CH_3	$(CH_3)_2CH$	70
p-$H_2NC_6H_4$	H, CH_3	$(CH_3)_3C$	11

with open-chain keto triazenes owing to the lability of the bond between N-1 and C-5 (Table 2.41). In these triazolines with different groups at C-5 two diastereomeric forms can exist, and are interconvertible through the keto triazene (828). Although no triazene exists in equilibrium with 5-hydroxy-5-isopropyl-4-methyl-1-phenyl-triazoline, replacement of isopropyl by the bulkier *t*-butyl

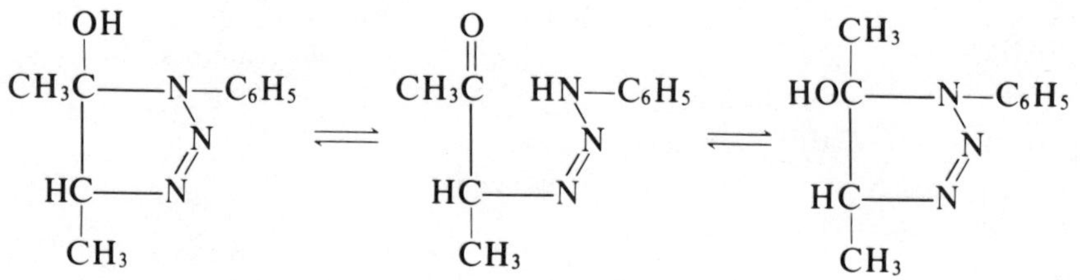

mp 84–85 dec.

group causes some ring opening. With two methyl groups at position 4, large proportions of the triazene form are present, and indeed, the infrared spectrum in potassium bromide of 5-*t*-butyl-4,4-dimethyl-5-hydroxy-1-phenyltriazoline **(VIII)** indicates the existence of some triazene, the only hydroxy triazoline thus far reported to show such ring opening in the solid state (828). When two diastereomeric forms can be present, smaller proportions of triazenes are in equilibrium. *p*-Nitroaryl on N-1 causes greater ring opening than aryl groups with electron-releasing substituents. Similar 1-alkyl-5-hydroxy triazolines do not form triazenes in amounts sufficiently large to be detected, but isomerization to diastereomers is presumed to proceed through the triazene (827).

The 1-aryl-5-hydroxy triazolines all lose nitrogen in the range 57–153°. Lowest decomposition temperatures are those of compounds with *t*-butyl or isopropyl at position 5 and two *gem*-methyl groups at the 4-position (828). The highest decomposition temperature is found with 5-phenyl and single alkyl substituents at position 4. Some 1-alkyl-5-hydroxy triazolines do not decompose at the melting point, but their decomposition temperatures usually are similar to those of their 1-aryl analogs (827).

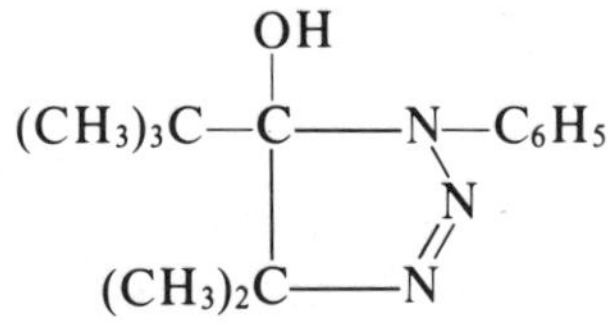

(VIII)

mp 57–59° dec.

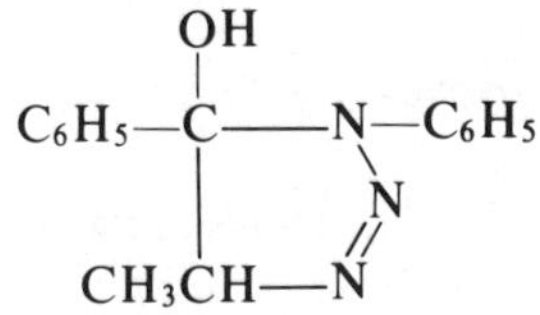

mp 149–150° dec.

Cyclic triazenium salts are disubstituted at the 1- and 3-nitrogen atoms of the triazene chain in triazolines (Table 2.42). As with acyclic triazenium cations, the

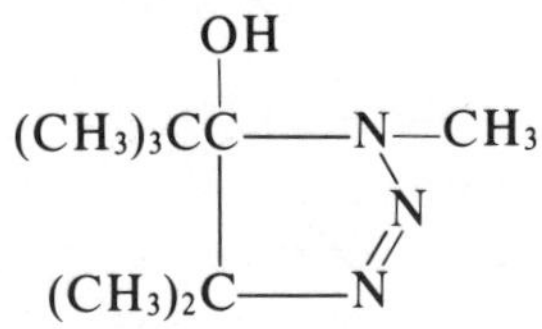

mp 118–120°

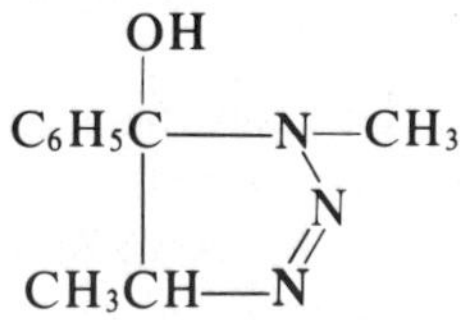

mp 138–139 dec.

Table 2.42. Cyclic Triazenium Salts

Salt		Melting Point (°C)	Ref.
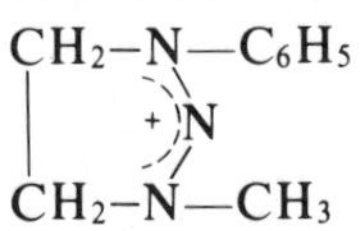	ClO_4^- BF_4^-	127–128 107–108	425
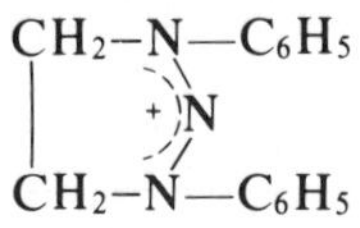	ClO_4^-	246–250 dec.	425
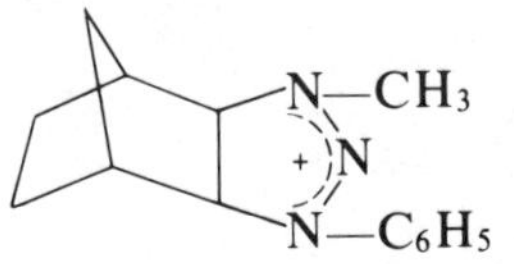	BF_4^-	109–111	425
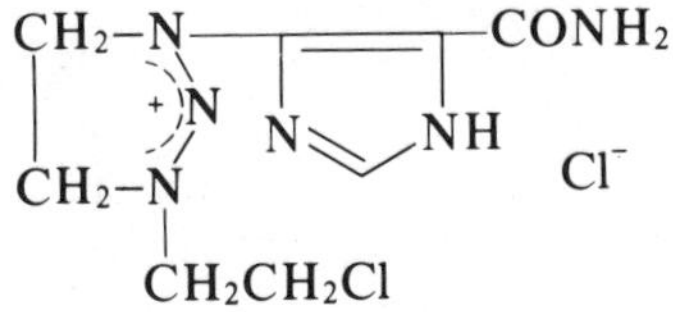	Cl^-	197–199 dec.	1, 987

positive charge is probably dispersed over the three-nitrogen chain. Decomposition points of cyclic triazenium salts are near or over 200° and much higher than those of related nonionic triazolines.

2.3.3. Monocyclic *v*-Triazoles

The monocyclic *v*-triazoles are aromatic five-membered heterocycles consisting of sequences of three-nitrogen and two-carbon atoms.* They may be subdivided into those with hydrogen on ring nitrogen and those substituted at the 1- and 2-nitrogen atoms. The hydrogen atom in the first structure is arbitrarily in-

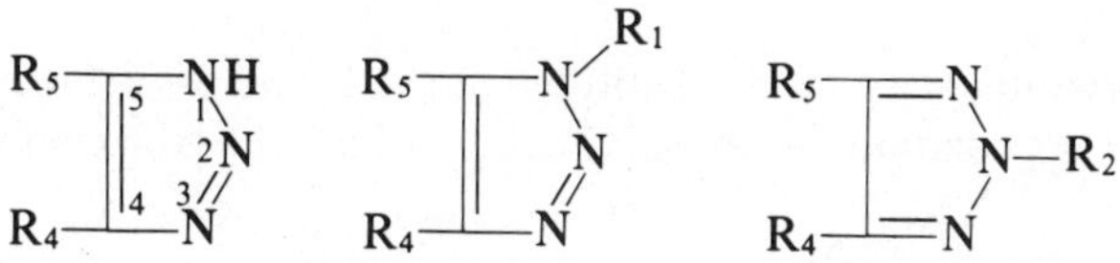

dicated at N-1, but may be located at any of the cyclic nitrogen atoms, depending on the nature of the substituents on the *v*-triazole carbon atoms, as discussed in Section 3.7, Acid-Base Properties. Virtually every type of organic radical has been substituted on both the nitrogen and carbon atoms of the *v*-triazoles.

v-Triazoles with 4- and 4,5-alkyl or aryl groups (Table 2.43) are stable at least to their melting or boiling points. Overheated vapors of the parent compound are reported to explode (62).

Table 2.43. 4- and 4,5-Alkyl and Aryl *N*-Unsubstituted Monocyclic *v*-Triazoles

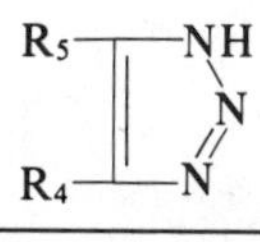

R_4	R_5	Melting Point (°C)	Ref.
H	H	22.5 Bp 208–210	62, 1182
CH_3	H	35–36; 36–36.5 Bp 108–109/25	433, 535
CF_3	H	78.5	226
CH_3	CH_3	70; 72–73	851, 769
C_6H_5	H	145–146; 148.4	433, 732
C_6H_5	C_6H_5	135–137; 136–138	374, 907

* The *v*-triazoles have been reviewed by Boyer (135), by Gilchrist and Gymer (372), and by Finley (319).

Phenylacetonitrile results from thermolysis of 4-phenyl-*v*-triazole and tetraphenylpyrazine is formed from 4,5-diphenyl-*v*-triazole (978).

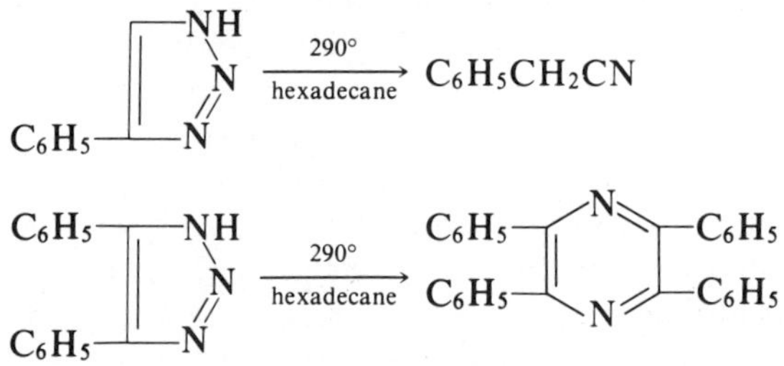

The ranges of melting points with decomposition of 4-mono- and 4,5-disubstituted *v*-triazoles is 110–235°, but most *v*-triazoles with acyl and noncarbon substituents (Table 2.44) are stable on melting or boiling under reduced pres-

Table 2.44. 4- and 4,5-Acyl and Noncarbon *N*-Unsubstituted Monocyclic *v*-Triazoles

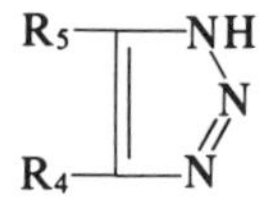

R_4	R_5	Melting Point (°C)	Ref.
C_6H_5CO	H	121–122	1127
HOOC	CH_3	Dec. 235	896
NC	H	110–112	43
H_2N	H	74–75	797, 1065
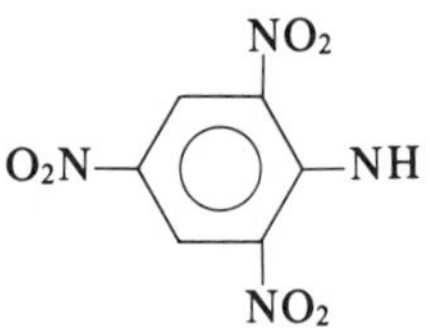	H	228 dec.; DTA exotherm 220°	797
$(CH_3)_2N$	CH_3	Bp 63–65/0.1	460
C_6H_5NH	C_6H_5	167–169	661
H_2NCO	H_2N	226–227 dec.	1146
HO	H	130–132; 135 dec., sublimes 110/0.1	240, 856
C_6H_5O	H	91–92	715
HS	H	60	384
p-$CH_3C_6H_4SO_2$	H	156–158	729, 730
Cl	H	71	226
Br	H	108	226
I	H	110–111 dec.	535

sure. The lowest melting point with decomposition is that of 4-iodo-*v*-triazole; the 4-chloro and 4-bromo derivatives appear to be stable at their melting points. Most 4-hydroxy-*v*-triazoles without nitrogen substituents decompose when melting, a change that probably is related to their conversion to α-diazoamides, discussed further in the section on open chain-cyclic isomers and equilibria. Heating *v*-triazole carboxylic acids at or above their melting points leads to decarboxylation, but the *v*-triazole ring remains intact.

The 1-substituted *v*-triazoles (Table 2.45) are somewhat more stable than their unsubstituted relatives. Except in the presence of a 5-amino or hydroxy group, which facilitates rearrangement, the ring appears to be stable to over

Table 2.45. 1-Substituted Monocyclic *v*-Triazoles

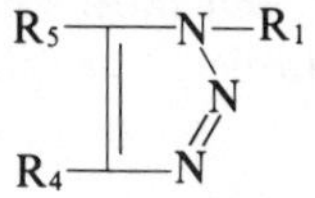

R_1	R_4	R_5	Melting Point (°C)	Ref.
CH_3	H	H	Bp 223–225	857
CH_3	CH_3	CH_3	50–51	829
CH_3	COOH	H	224 dec.	534
$(CH_3)_2CH$	H	C_6H_5NH	138–139	1061
$HOCH_2CH_2$	$(CH_3)_2COH$	$(CH_3)_2COH$	169.5–170	282
$C_6H_5CH_2$	$CONH_2$	NH_2	234	505
C_6H_5	H	H	56	111
C_6H_5	CH_3	CH_3	85–86	829
C_6H_5	C_6H_5	C_6H_5	228–229	829
C_6F_5	C_6H_5	C_6H_5	220–221; sublimes 170–180/<1	67, 68
C_6H_5	H	NH_2	110	257
C_6H_5	CHO	CHO	107	458a
C_6H_5	COOH	COOH	147–148 dec.; 149–150 dec.	738, 1210
C_6H_5	CH_3	OH	Dec. 133–134	260
p-$CH_3OC_6H_4$	H	C_6H_5	85	522
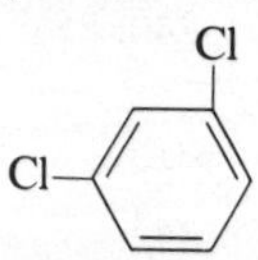	C_2H_5OOC	CH_3	270 dec.	203
CH_3CO	H	H	60–62	122
$(CH_3)_3Si$	CH_3OOC	CH_3OOC	39–41 Bp 141/0.1	120

200°. 5-Amino-1-substituted *v*-triazoles are converted thermally as well as in basic (17) and acidic (160) media to 5-substituted amino *v*-triazoles in the Dimroth rearrangement which involves breaking the nitrogen chain. An equilibrium composition results, starting from either 5-amino-4-ethoxycarbonyl-1-phenyl-*v*-triazole or 5-anilino-4-ethoxycarbonyl-*v*-triazole (258). Relatively electron-

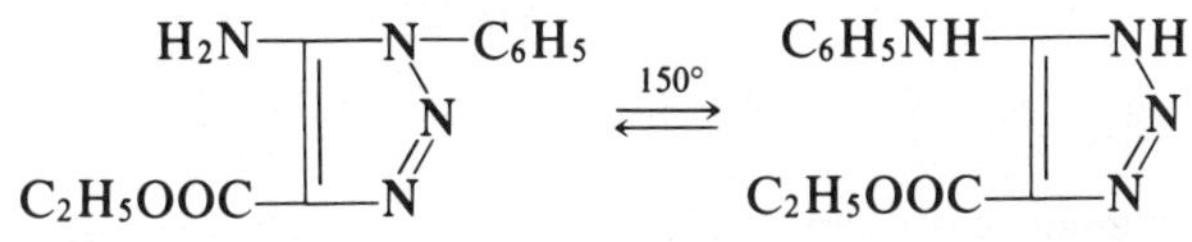

76.8% at equilibrium

withdrawing substituents on N-1 shift the equilibrium toward the isomeric 5-substituted amino-*v*-triazole, illustrated by isomerization of 5-amino-4-phenyl-1-substituted-*v*-triazoles (660, 669). The rearrangement must go through an open-chain diazoamidine intermediate, shown in one resonance form.

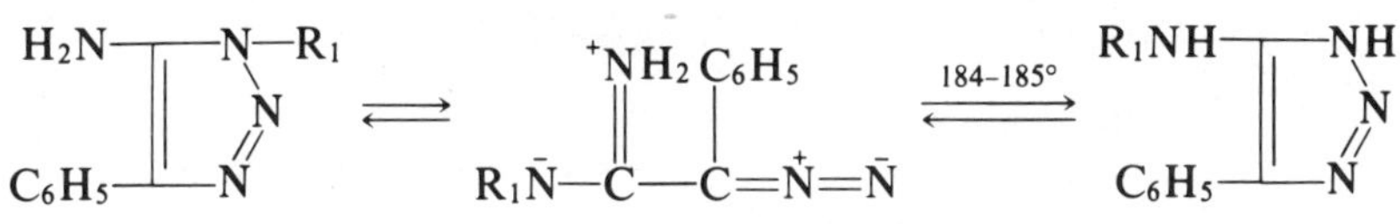

R_1	Equilibrium Composition, % 5-Substituted Amino Isomer
$C_6H_4CH_2$	7.7
m-$CH_3C_6H_4$	69.1
C_6H_5	77.2
m-$O_2NC_6H_4$	93.2

As with the *v*-triazoles unsubstituted at N-1, the 1-substituted *v*-triazoles with hydroxyl groups at position 5 can dissociate to diazoamides. The 4-hydroxy 1-substituted *v*-triazoles exhibit no special instability and exist as dipolar compounds both as solids and in solution (91). Some 5-hydroxy 1-substituted

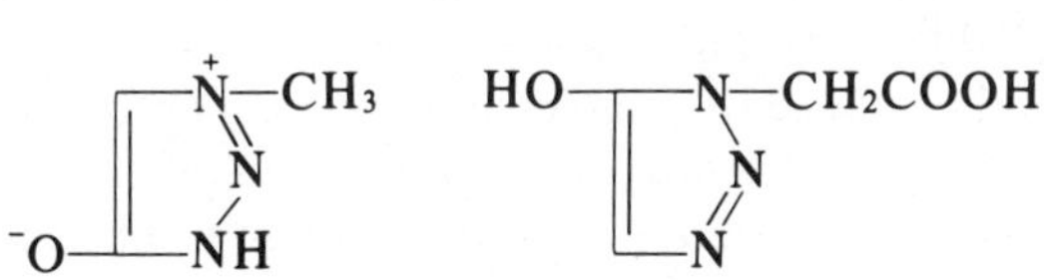

mp 140° expl. 169–170°

v-triazoles, such as the 1-acetic acid derivative, explode on heating (239). The 1-hydroxy *v*-triazoles are nearly as stable as those with 1-alkyl or aryl groups (1195).

v-Triazoles with strong electron-withdrawing groups on N-1 have poorer thermal stability than related derivatives with less electron-attracting radicals. *N*-Chloro-4,5-diphenyl-*v*-triazole decomposes, partially forming nitrogen and,

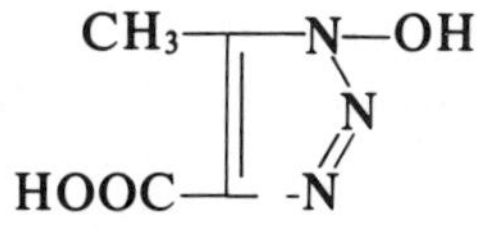

mp 200–205 dec.

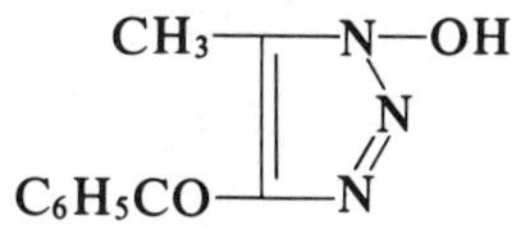

Dec. 190

with rearrangement, a diphenyl chloro aziridine which condenses with another molecule of *N*-chloro-4,5-diphenyl-*v*-triazole, liberating chlorine. An alterna-

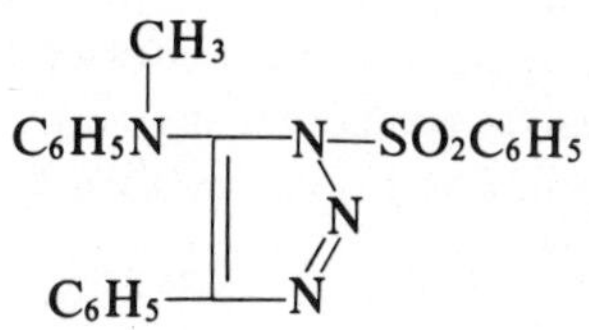

mp 66–69° dec. (478)

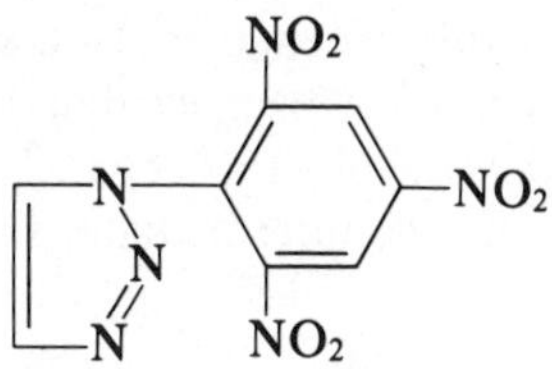

mp 224 dec.
DTA exotherm 150° (798)

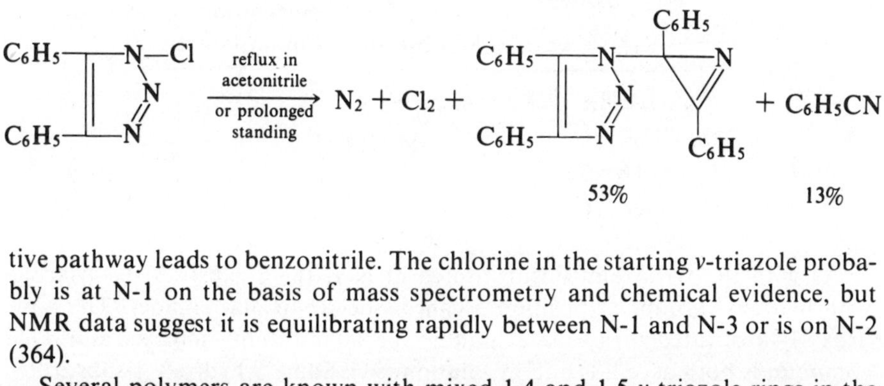

53% 13%

tive pathway leads to benzonitrile. The chlorine in the starting *v*-triazole probably is at N-1 on the basis of mass spectrometry and chemical evidence, but NMR data suggest it is equilibrating rapidly between N-1 and N-3 or is on N-2 (364).

Several polymers are known with mixed 1,4-and 1,5-*v*-triazole rings in the repeating unit (Table 2.46). Thermal stabilities of most of these *v*-triazole polymers are better than those of related monomers. The different stabilities reported for poly(phenylene *v*-triazole) may arise from different preparative methods or from polymers of different molecular weights. Thermal degradation of poly(phenylene-*v*-triazole) produces nitrogen, terephthalonitrile, benzene, and

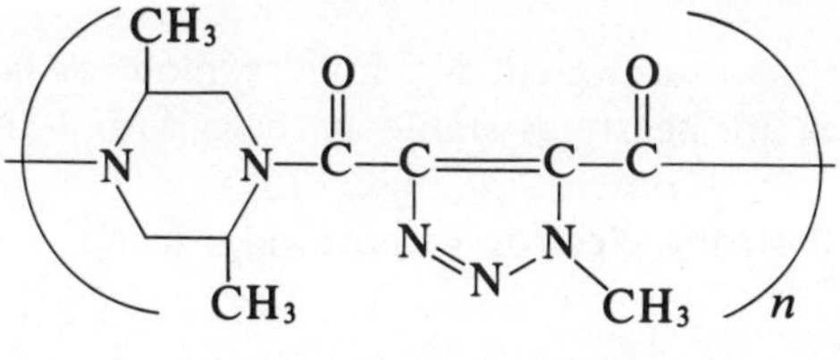

Table 2.46. 1-Substituted Monocyclic *v*-Triazole Polymers

Polymer	Thermal Stability	Ref.
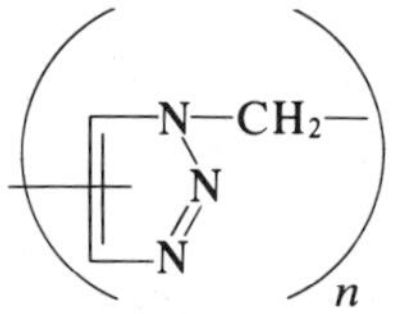	Stable to 200°; darkens > 200° with slow evolution of N_2, but does not melt	54, 510
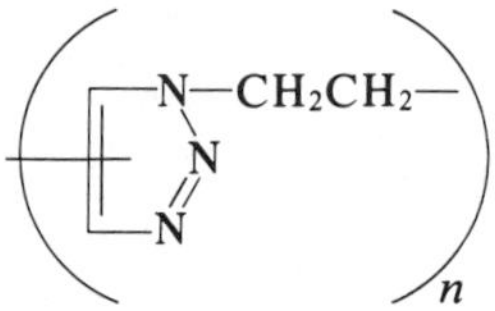	No exotherm below 320°	566
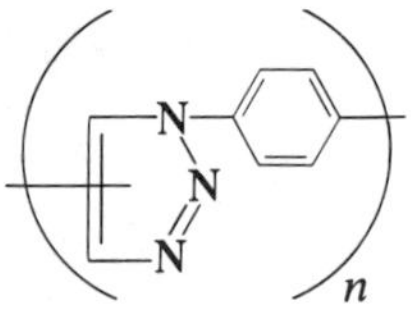	Dec. starts 250°; 90% weight retained at 300° during TGA; darkens in air at 500°	928 377 54
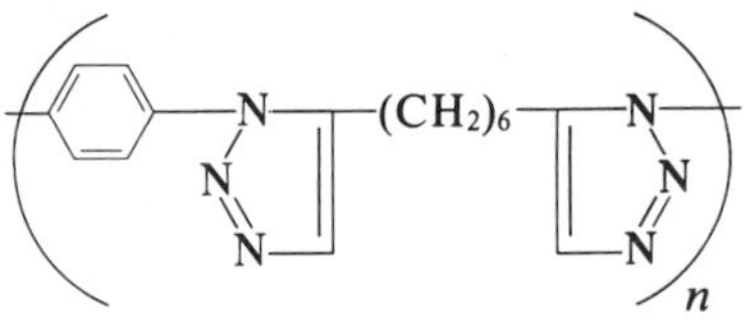	Insoluble fraction stable to 350°; 80% weight retained at 400°	1198

acetylene (928). The polyamide, consisting of a *trans*-3,5-dimethylpiperazine and 1-methyl-4,5-dicarbonyl-*v*-triazole units, requires a temperature of 362° for loss of 5% of its weight (695).

The 2-substituted *v*-triazoles (Table 2.47) usually are even more stable than

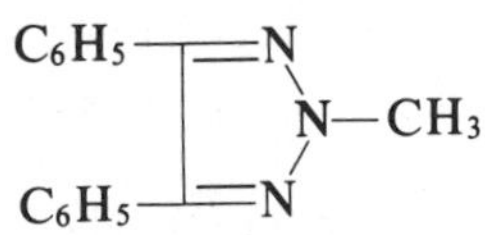

mp 61.5–63

the 1-substituted compounds. On holding 4,5-diphenyl-2-methyl *v*-triazole at 220–225° for 6 hr no change occurs in its melting point or ultraviolet or infrared spectra (344). Only 1 mole of carbon dioxide is evolved on heating 2-phenyl-*v*-

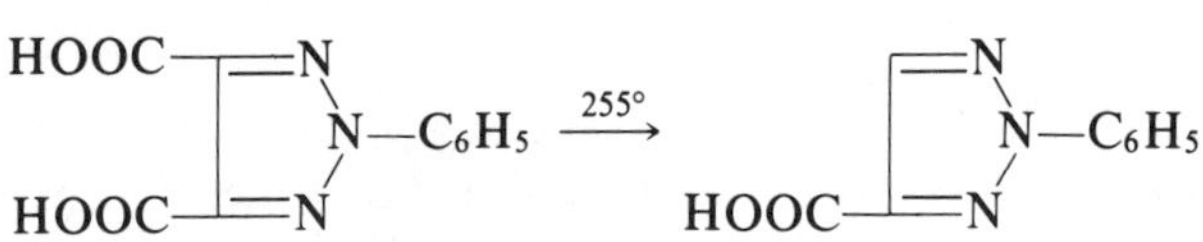

mp 255° dec.   mp 191°

Table 2.47. 2-Substituted Monocyclic *v*-Triazoles

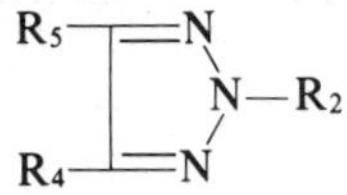

R_2	R_4	R_5	Melting Point (°C)	Ref.
CH_3	H	H	21.5–22	535
			Bp 102–103	857
CF_3CH_2	CF_3	CF_3	Bp 133	316
$CH_2{=}CH$	CH_3OOC	CH_3OOC	59–60	735
C_6H_5	H	H	Bp 228	693
C_6H_5	C_6H_5	C_6H_5	122	116
C_6H_5	CH_3	CH_3	34–35	223, 850
			Bp 253–255	
C_6H_5	CH_2OH	H	67	568
C_6H_5	CHO	H	68–69; 70	424, 568
C_6H_5	CN	H	89	92
C_6H_5	CH_3OOC	CH_3OOC	91	350
C_6H_5	NH_2	H	70	1086
C_6H_5	OH	H	123–124	89, 1086
p-$O_2NC_6H_4$	CH_3	CH_3	233–234	223
CH_3CO	H	H	33–36	122
			Bp 110°/50	
$(CH_3)_3Si$	CH_3	CH_3	Bp 65–66/13	121
$(CH_3)_3Sn$	C_6H_5	H	191–193	607
HO	COOH	CH_3	184–185	302

triazole-4,5-dicarboxylic acid, forming 2-phenyl-*v*-triazole-4-carboxylic acid which is stable under the conditions of monodecarboxylation (350). On heating either 1- or 2-acetyl-*v*-triazole at 100–130° an equilibrium mixture is established in which the 2-isomer preponderates (122). The thermal stability of 2-picryl-*v*-

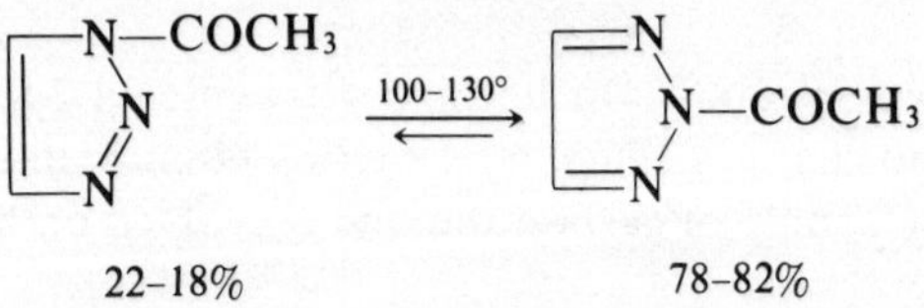

triazole is remarkable; heated above its melting point, it does not show a DTA exotherm until 335° (798). The polyamide made of 2-methyl-4,5-dicarbonyl-*v*-triazole and 2,5-dimethyl piperazine units loses 5% of its weight at 239° but is less thermally stable than the corresponding 1-methyl-*v*-triazole polymer (695).

Phosphonium *v*-triazole ylides, bearing a negative charge on the ring, and

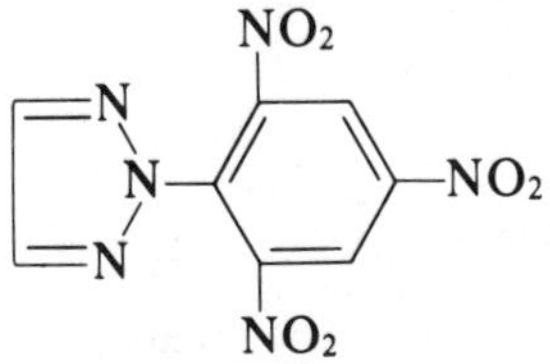

mp 211–212°

the related 2-substituted salts are as stable as most other *v*-triazoles with groups in the 2-position (1070).

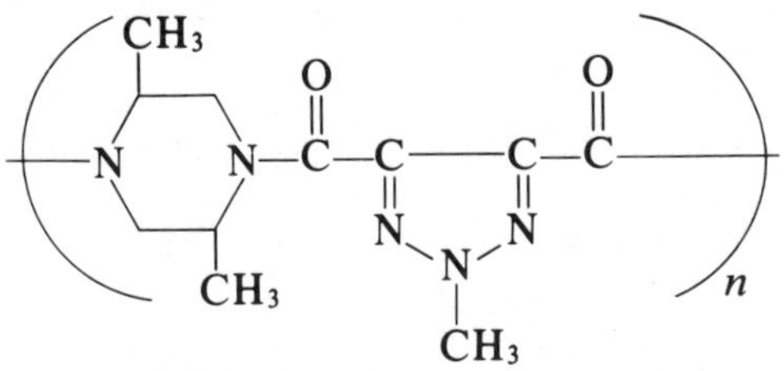

The monocyclic *v*-triazole oxides are more stable than the triazene oxides but

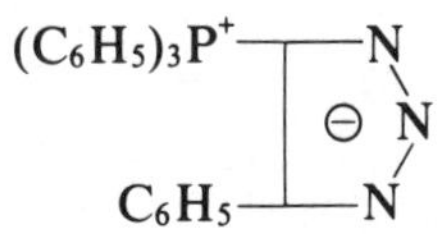

mp 240–241°

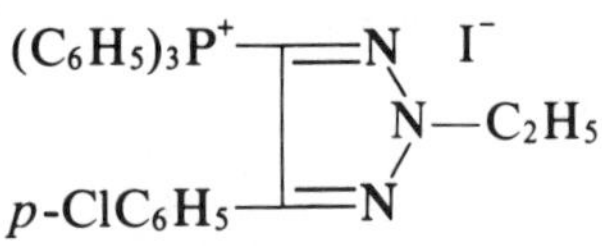

mp 219–222°

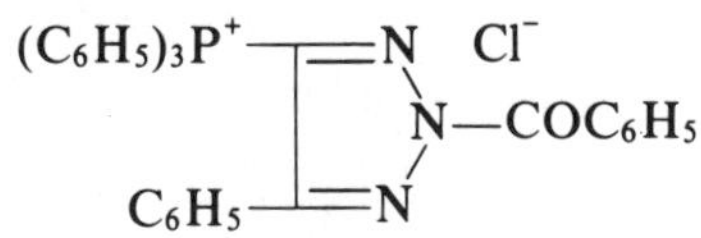

mp 299–305° dec.

insufficient data are known to categorize their thermal stability more definitely. The 2-aryl-4-hydroxy-*v*-triazole 1-oxides melt with decomposition (672).

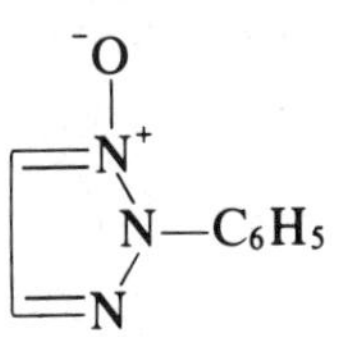

mp 60–61° (62)

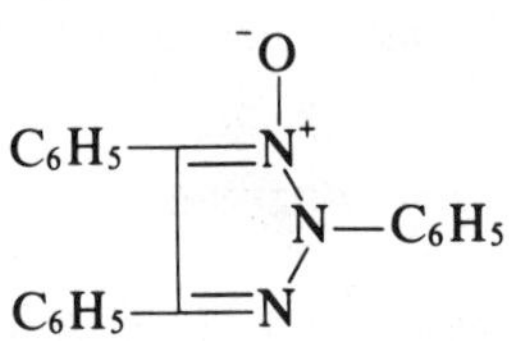

mp 168° (1068)
169° (881)

Both 1,2 and 1,3-monocyclic *v*-triazolium salts exist, as well as a series of

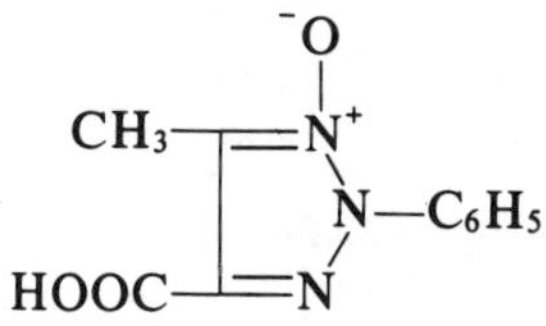

mp 93° (882)

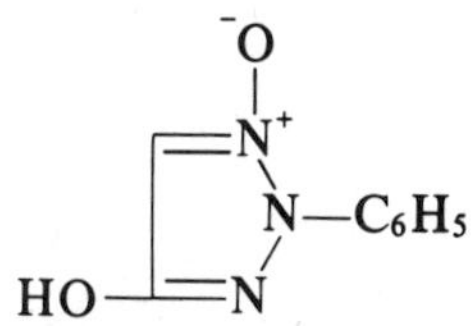

mp 155–156° dec. (672)

dipolar 1,3-disubstituted compounds with substituents bearing a negative

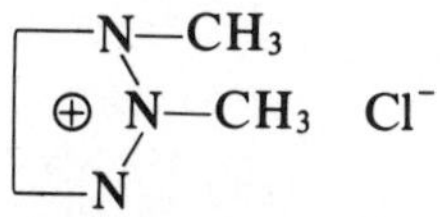

mp 211° dec. (95)

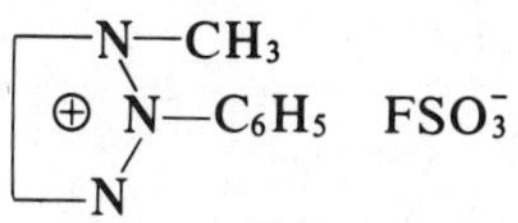

mp 170–173° (95)

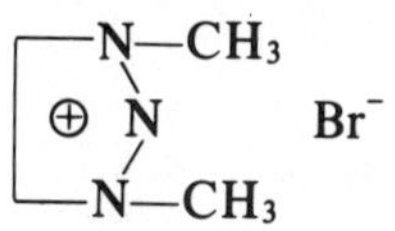

mp 177–179° (93)

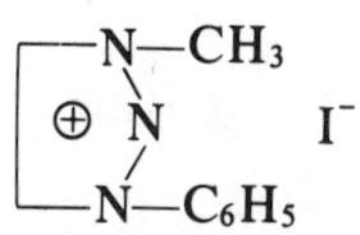

mp 134.5–135.5° (934)

charge at C-4. The melting points of these *v*-triazolium salts indicate stability

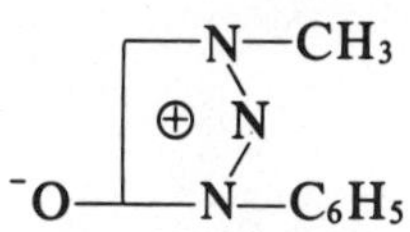

mp 69–73° (88)

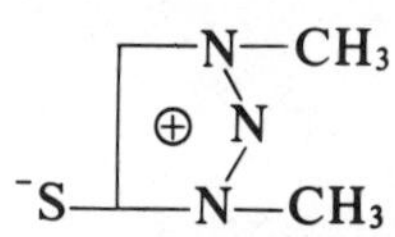

mp 171–172° (90)

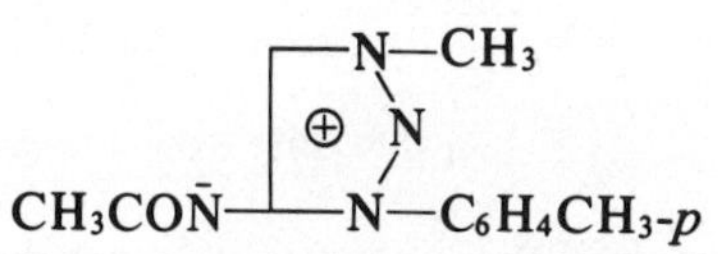

mp 214–215° (884)

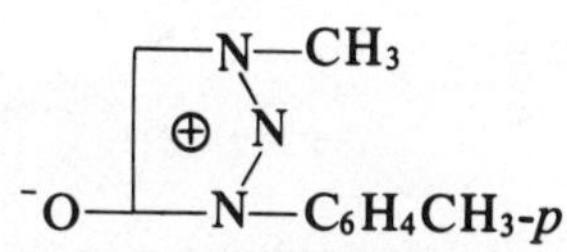

mp 141–143° (884)

commensurate with that of uncharged *v*-triazoles.

2.3.4. Fused Ring *v*-Triazoles

The *v*-triazole ring has been fused to many cyclic compounds to form polycyclic heterocycles. Those fused ring *v*-triazoles joined at the 4,5-triazole positions are discussed here, whereas those annelated at carbon-nitrogen or nitrogen-nitrogen bonds are covered in Section 2.3.9, Embedded Three-Nitrogen Chains. The fused ring *v*-triazoles may be subdivided into those without substituents on nitrogen (usually bearing a hydrogen atom) and those substituted at the 1- or 2-ring nitrogens, analogous to the monocyclic *v*-triazoles.

Cyclic 4,5-polymethylene *v*-triazoles are similar structurally to 4,5-dialkyl monocyclic *v*-triazoles and show no thermal instability to their melting or reduced pressure boiling points. Besides the 4,5-tetramethylene compounds of each subclass, a series of 1-phenyl-*v*-triazoles with larger 4,5-polymethylene chains are known. Many *N*-unsubstituted and 1-substituted derivatives of re-

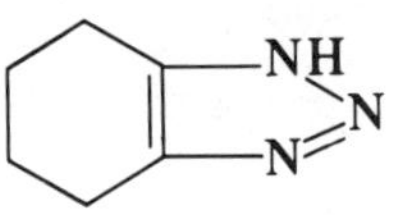

mp 74° (325)
84° (820)
bp 150°/1-2

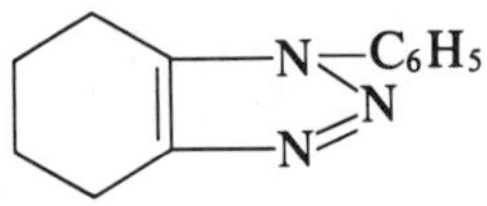

mp 117–118° (26)

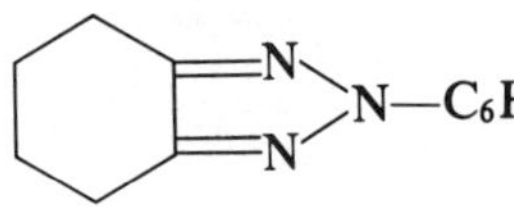

mp 90° (1052)
95° (352)

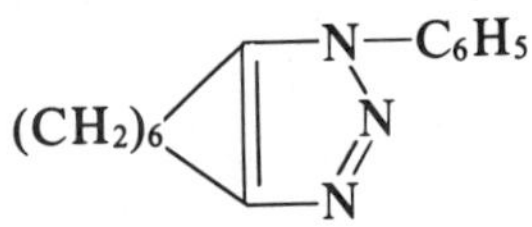

mp 84–85° (1191)

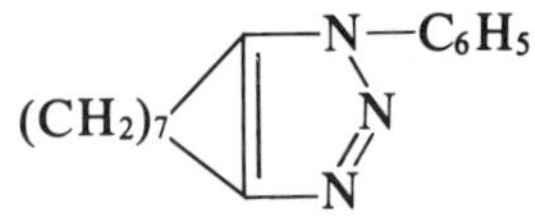

mp 79–80° (1191)

duced benzotriazoles with electron-withdrawing substituents in the six-membered ring decompose in the 100–175° range. The 2-substituted reduced benzo-

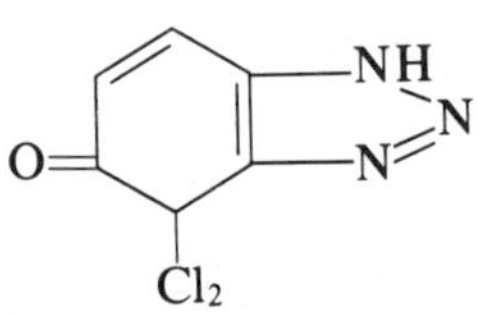

mp 132 dec. (353)

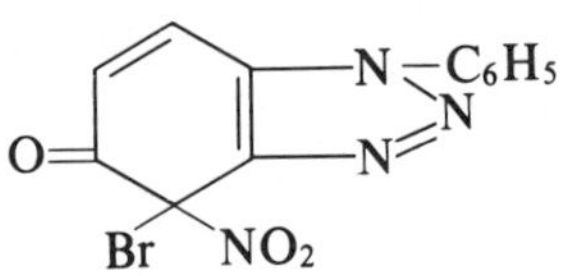

dec. < 100° (351)

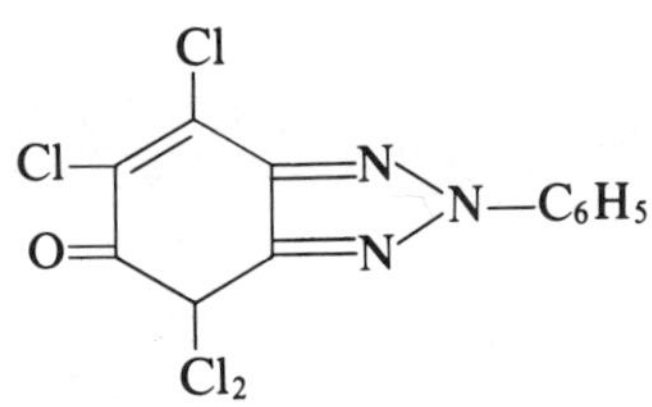

mp 168° (1211)

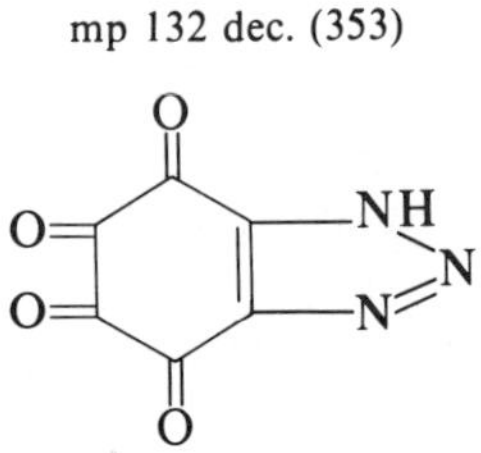

mp dec. > 160° (1209)

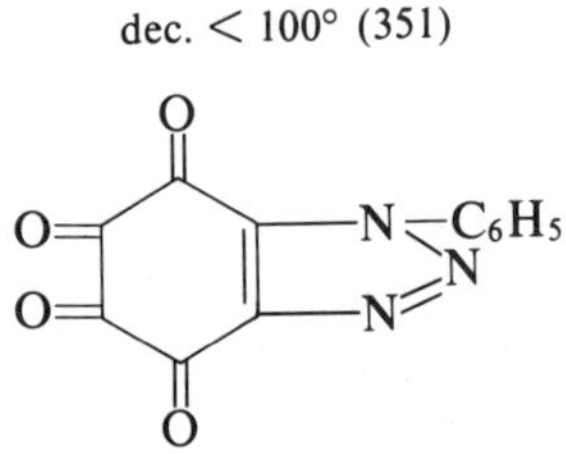

mp 175° dec. (1210)

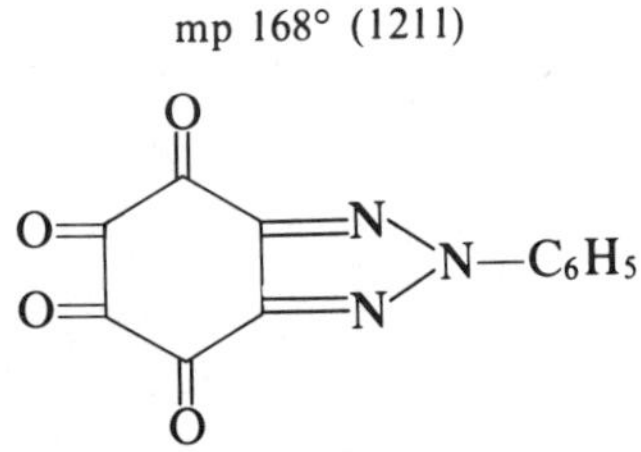

mp 165–170° dec. (1211)

triazoles do not usually decompose at their melting points. Nitric oxide is evolved from the decomposition of 4-bromo-4-nitro-5-oxo-1-phenyl-4,5-

dihydrobenzotriazole (351), suggesting that the initial decomposition does not involve fission of the triazole ring. Indeed, in most fused ring *v*-triazole decompositions, whether the triazole ring is first involved is presently unknown. The tetraoxotetrahydro benzotriazole derivatives are relatively thermally unstable regardless of the triazole substitution. With the exception of the *ortho*-quinone of 1-methyl-7,8-dihydrobenzotriazole, the dihydrobenzotriazole quinones decompose in the 160–260° range, sometimes explosively. Even higher decomposi-

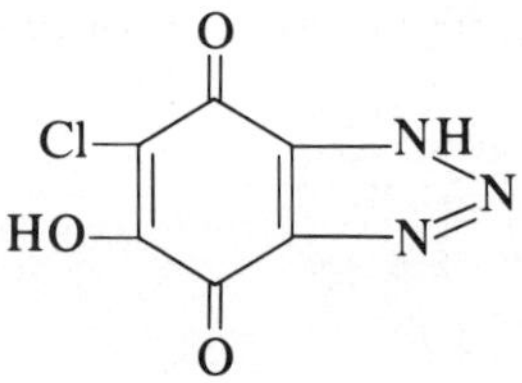

expl. 230–231° (1209)

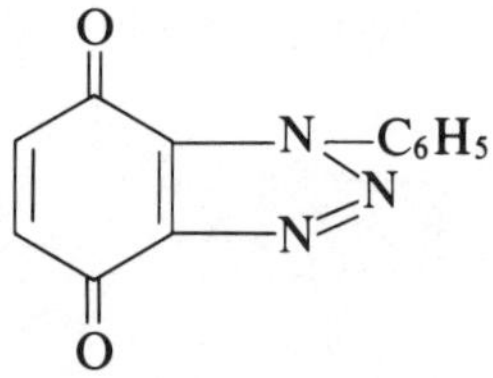

mp 180–184° dec. (1196)

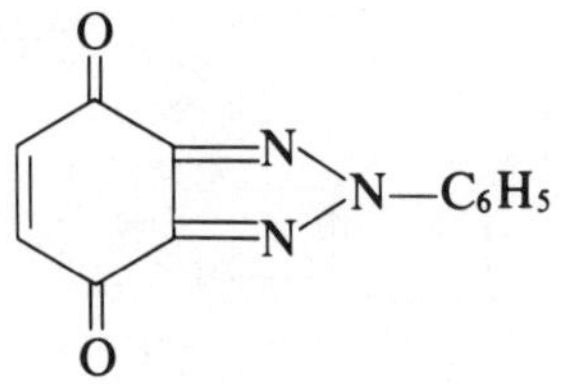

mp 224–225° dec. (38)

tion temperatures are found for bistriazolo quinones, and 2,6-dipicrylbenzo[1,2-

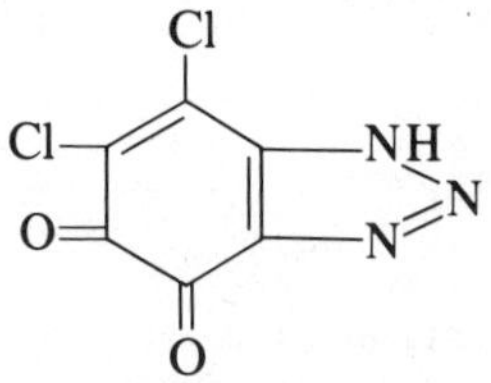

expl. > 260° (1209)

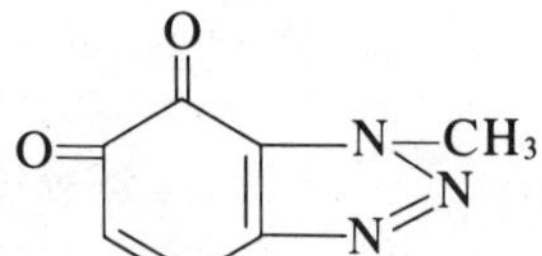

darkens > 100° (353)
gradual dec. at higher temp.

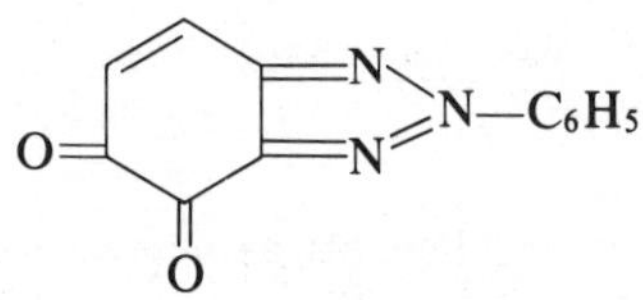

darkens 160° (346)
does not melt 340°

di-4,5-*d'*]bistriazole-4,8-dione has one of the highest explosion points known.

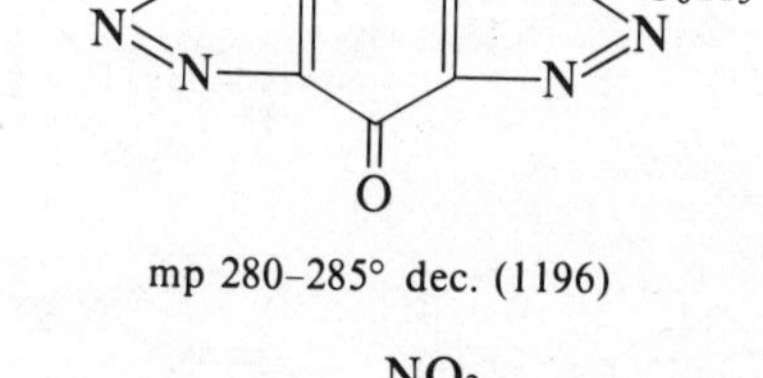

mp 280–285° dec. (1196)

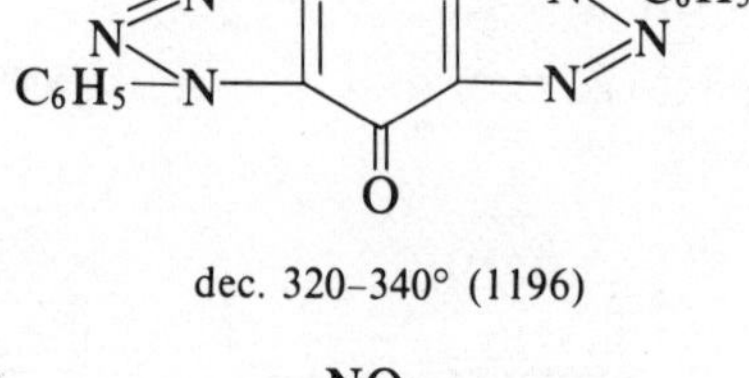

dec. 320–340° (1196)

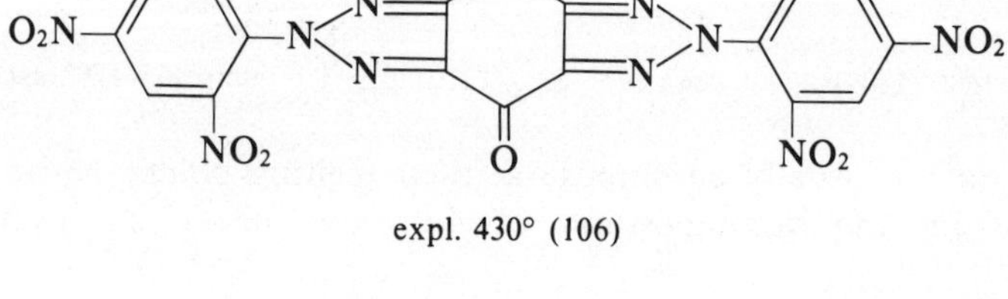

expl. 430° (106)

Fusion of the *v*-triazole ring to other aromatic rings creates extended conjugated systems. For the most part, the fully aromatic fused ring *v*-triazoles are more thermally stable than corresponding monocyclic *v*-triazoles. Fused, aromatic, carbocyclic *v*-triazoles and their derivatives, unsubstituted on nitrogen, melt with decomposition from 210 to over 360°.

Nevertheless, there is an inherent instability in the aromatic polycyclic *v*-triazoles. Although benzotriazole has been distilled without decomposition in small quantities at temperatures of 201–204° at 15 mm (246), its thermal instability is indicated by the severe explosion of a 2000 lb batch during vacuum distillation at 2 mm, with a temperature rise from 160° to over 220° (39). No information has been reported on the resistance to heat of either the linear or the angular naphthotriazole. The high temperatures required to decompose the benzobistriazoles are remarkable.

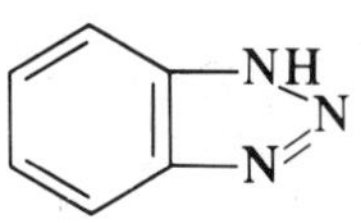

mp 98–99° (317)
bp 156–159°/2 (246)

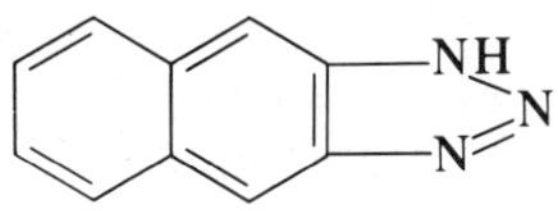

mp 187° (348)

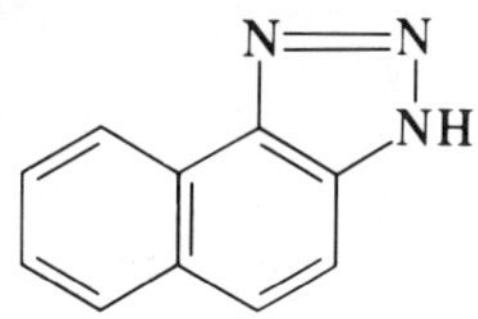

mp 188.5–189.5° (651)

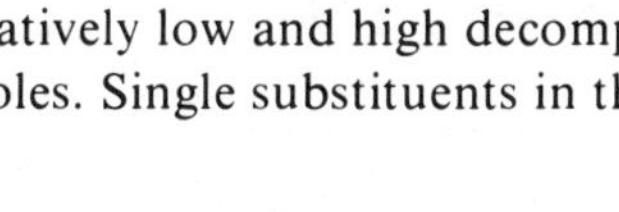

mp 315° dec. (779)
expl. 370° (20°/min) (221)

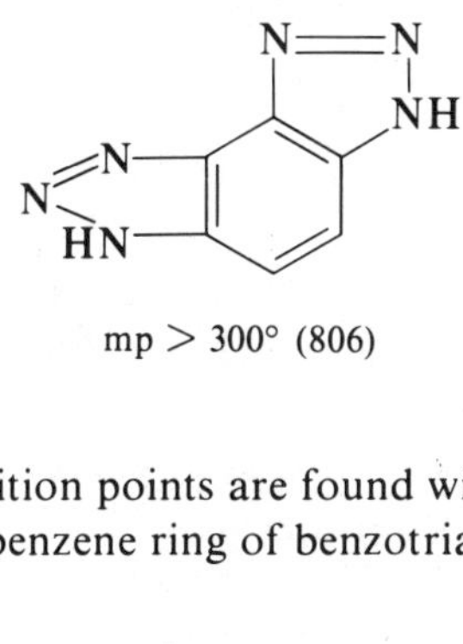

mp > 300° (806)

Both relatively low and high decomposition points are found with phenolic benzotriazoles. Single substituents in the benzene ring of benzotriazole, either electron-withdrawing or electron-releasing, confer neither thermal stability nor instability, as illustrated by the melting point data for a series of 5-substituted benzotriazoles (Table 2.48).

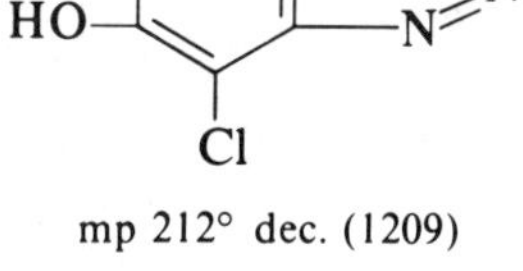

mp 212° dec. (1209)

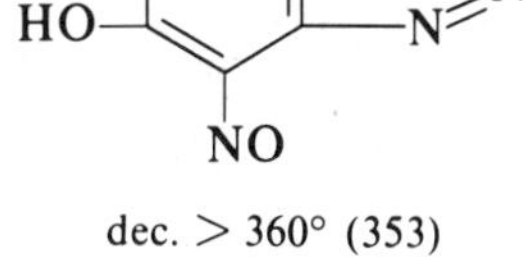

dec. > 360° (353)

Table 2.48. 5-Substituted *N*-Unsubstituted Benzotriazoles

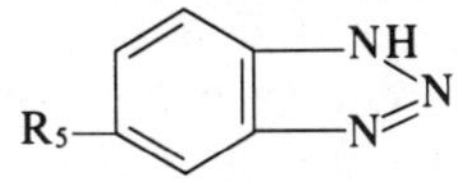

R_5	Melting Point (°C)	Ref.
CH_3	83–84	1212
OH	234–235 dec.	317
HOOC	>270	1208
CF_3	132–133; sublimes 120°/0.001	1155
NO_2	211	1209

The relatively more symmetrical 2-alkyl benzotriazoles are more volatile than the corresponding 1-alkyl compounds (Table 2.49) (611). In contrast, the melting points of the 2-aryl benzotriazoles are higher than those of the analogous 1-aryl compounds (Table 2.50), suggesting for these 2-aryl derivatives

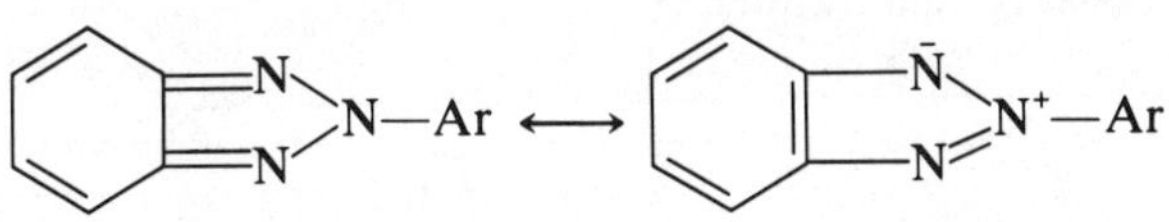

(I) (II)

some dipolarity not implied in the *o*-quinoid formula **(I)**. A contribution from structure **(II)**, requiring higher crystal lattice energies, would explain the higher

Table 2.49. Comparative Boiling Points of 1- and 2-Alkylbenzotriazoles

	Boiling Point (°C)	
R	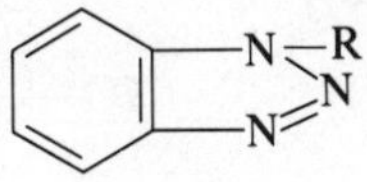	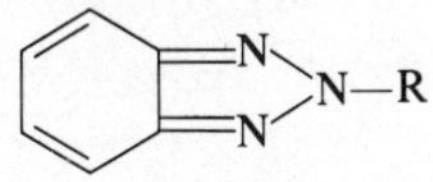
CH_3	152/14 (612) Mp 65	103.5–104/15 (612)
CH_2=CH	117–118/3	84–84.5/3
CH_2=$CHCH_2$	161–162/15	127–128/15
n-C_4H_9	170–172/17	137–139/16

Table 2.50. Comparative Melting Points of 1- and 2-Arylbenzotriazoles

R	Melting Point (°C) 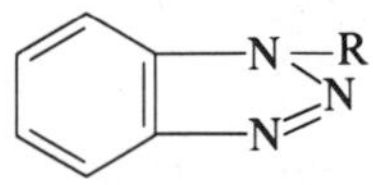	Melting Point (°C) 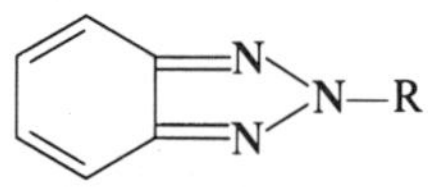
C_6H_5	89–90 (521)	109.5 (65)
p-$CH_3C_6H_4$	84–85 (129)	122 (417)
p-HOC_6H_4	170 (1114)	220.5–221.5 (415)
p-$O_2NC_6H_4$	239 (405)	286.5–287.5 (417)

melting points of the 2-aryl derivatives. Structure **II** has a more extended conjugated system than **I**, involving the 2-aryl group. This dipolar structure is the same invoked for linear fusion of *v*-triazole with other rings where uncharged forms are not possible, such as 1,2:4,5-benzobistriazole (see below).

The 1-alkyl and aryl benzotriazoles are stable to over 200° unless they contain a thermally labile group. Many 1-substituted fused ring triazoles remain intact to 350°. Decomposition of 4,7-dihydroxy-1-phenylbenzotriazole occurs at the low end of the range. Several 1-substituted benzotriazoles decompose ex-

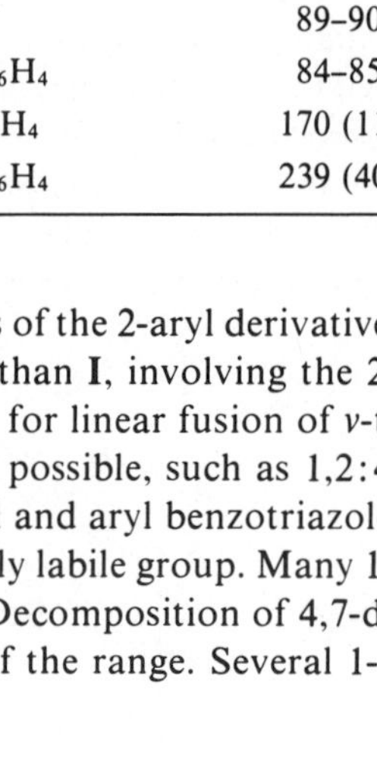

mp 203° dec. (1196)

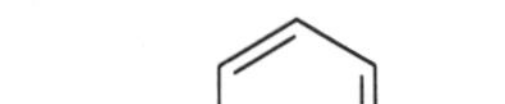

mp 180–182° (513)
expl. 250°

plosively. Both 1-alkyl and aryl benzotriazoles are among those not melting below 300°. The high decomposition temperature of 1,4-dimethylol-5-hydroxybenzotriazole is surprising, since many *N*-methylol benzotriazoles show instability below 200°, presumably liberating formaldehyde rather than breaking the triazole ring.

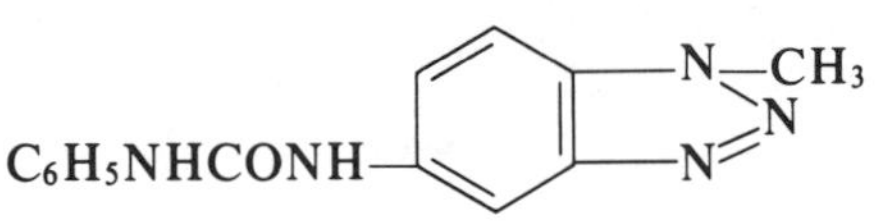

mp > 305° (869)

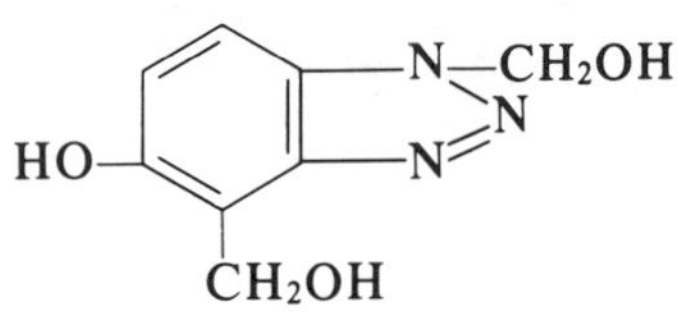

expl. 327° (353)

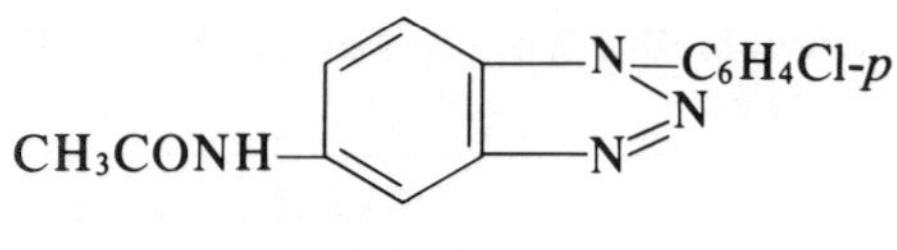

mp > 300° (125)

The thermal decomposition of 1-aryl fused ring aromatic *v*-triazoles occurs at 300–360°, expelling nitrogen from the triazole moiety, followed by ortho closure to form carbazole or a heterocyclic analog. The reaction, discovered by Graebe and Ullmann (395), has been applied to benzotriazoles and naphthotriazoles carrying alkyl, halo, and amino groups as well as *v*-triazolo pyridines. With unsaturated groups on the fused aromatic ring, yields of carbazole are lower (892). An analogous reaction of 1-benzylbenzotriazole, which does not begin to decompose until 350°, takes place in the presence of copper powder, resulting in phenanthridine (368).

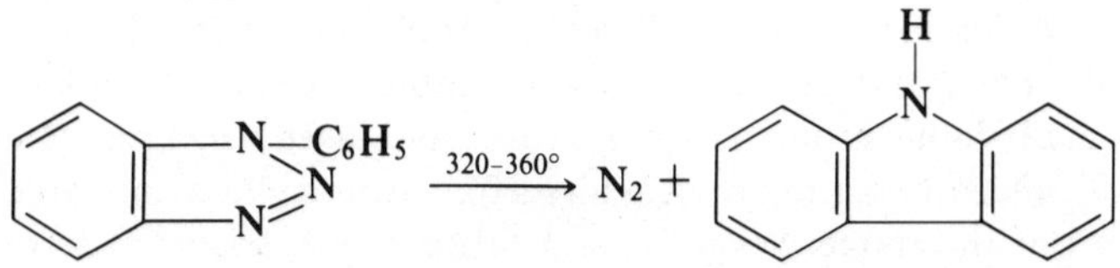

nearly quant.
(395)

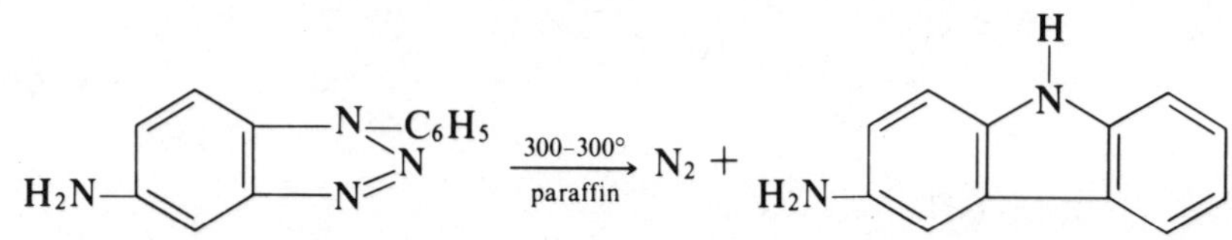

65%
(1113)

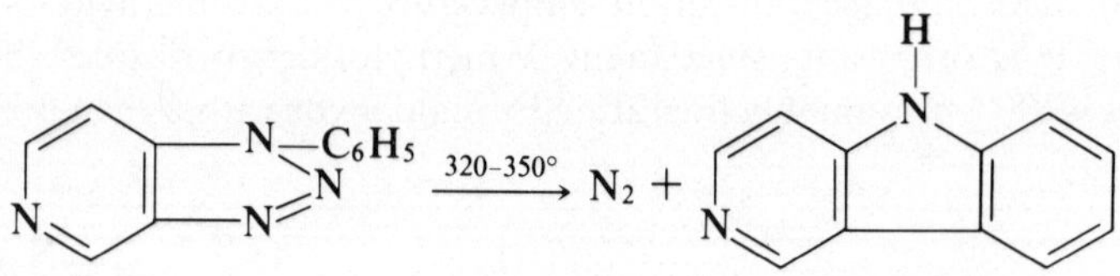

nearly quant.
(151)

The 1- and 2-*N,N*-disubstituted aminomethyl benzotriazoles readily isomerize in solution to mixtures rich in the 1-isomer (1002). Nonpolar solvents in-

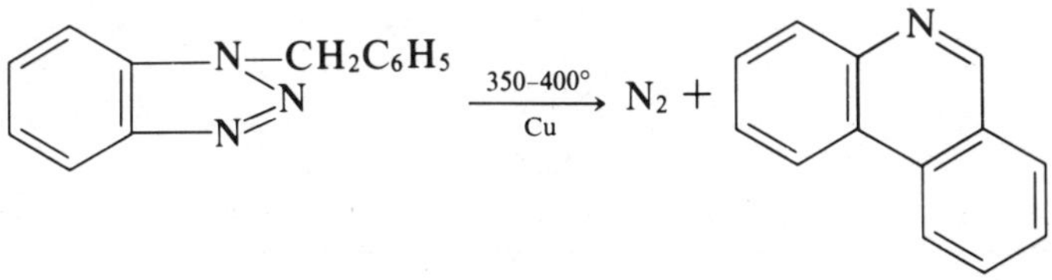

crease the amount of 2-isomer, but the equilibrium composition is insensitive to temperature change.

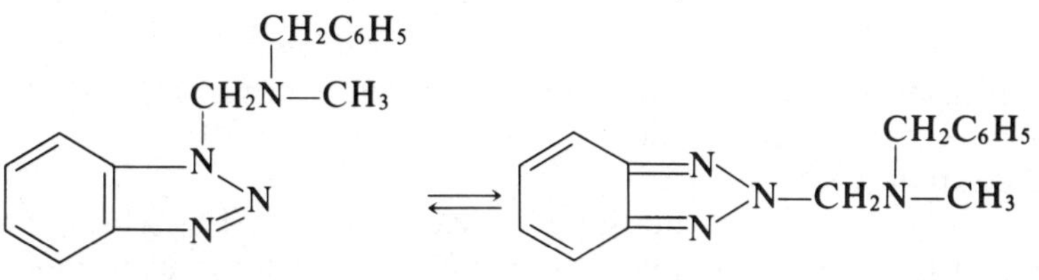

mp 61–61.5°

Solvent	% 2-isomer
$(CD_3)_2CO$	21
C_6D_6	28
CCl_4	45

Thermal decomposition of 1-benzoyl benzotriazole leads to 2-phenylbenzoxazole; the half-life of the benzoyl derivative at 250° in diphenyl ether is approximately 2.5 hr (278).

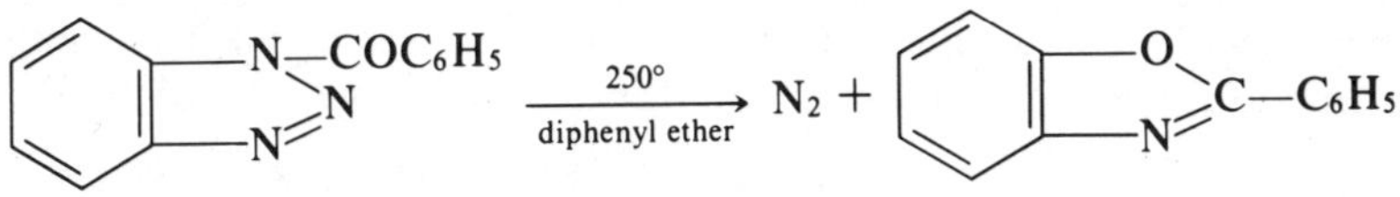

Halogen atoms substituted on the 1-nitrogen of benzotriazole are strongly electropositive, but the 1-halobenzotriazoles are stable to over 100° (Table

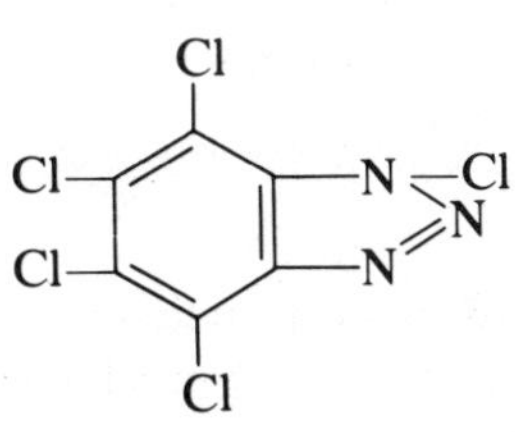

mp 169–171° dec.

Table 2.51. 1-Halobenzotriazoles

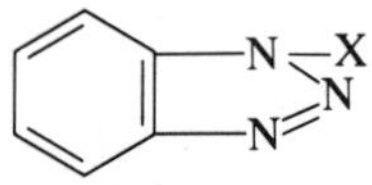

X	Melting Point (°C)	Ref.
Cl	104–106	910
Br	113–114 dec.	947
I	214–216 dec.	947

2.51). Introduction of chlorine atoms on the benzotriazole ring confers relative stability on the resulting *N*-chloro derivative, which probably is substituted at the 1-position (947).

The fully aromatic 2-substituted fused ring triazoles, which melt with decomposition, do so from 130° to over 400°. Many, but not all, phenolic benzo-

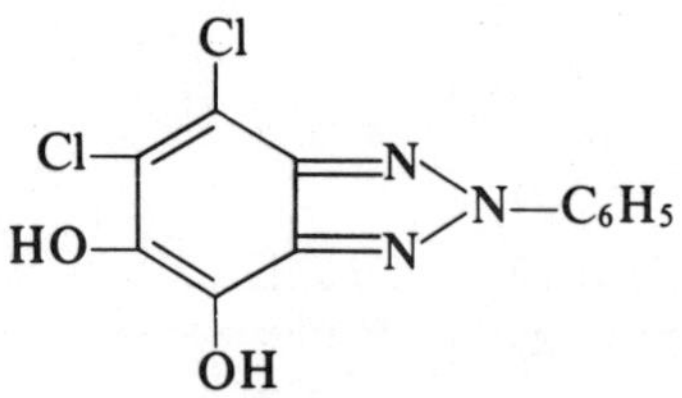

sinters 130°
mp 154–155° dec. (1211)

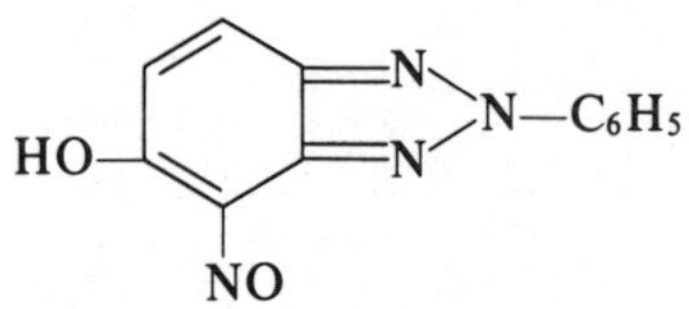

dec. 185° (346)

triazoles decompose below 200°, particularly if also substituted with electron-withdrawing groups. Low thermal stability is found in 5-phenyl triazolo[4,5-*d*]-*v*-triazole without substituents in the second triazole ring, but the 1-methyl and the dipolar 2-methyl compounds are stable at their melting points. Extended

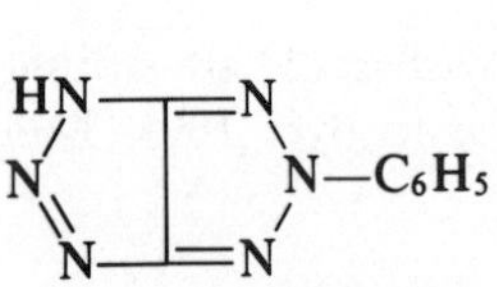

browns 130–140°
expl. 147° (1086)

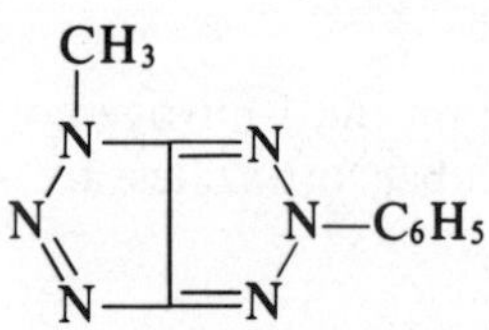

mp 139.2–139.8° (1200)

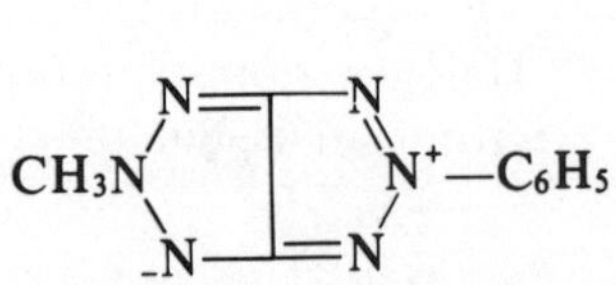

mp 149.0–149.8° (1200)

linear and cross-conjugated 2-substituted bistriazolobenzene derivatives and 2,5,8-triphenylbenzotristriazole have appreciable thermal stability with no evidence of decomposition at their melting points.

The *v*-triazole ring has been combined with many nitrogen heterocycles, including pyridine and pyrimidine. Much study has been made of the *v*-triazolo[4,5-*d*]pyrimidine ring system because of its analogy to purine, consisting

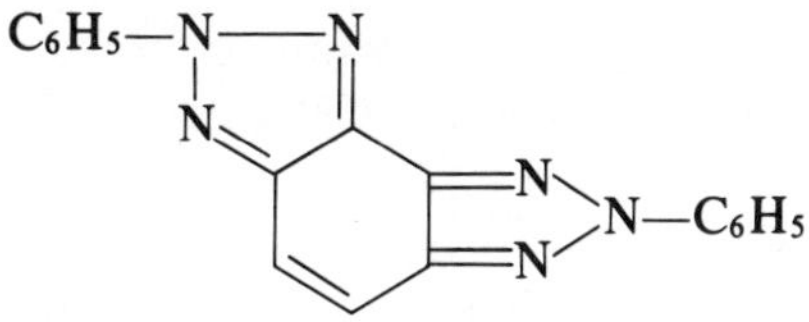

mp 221° (968)

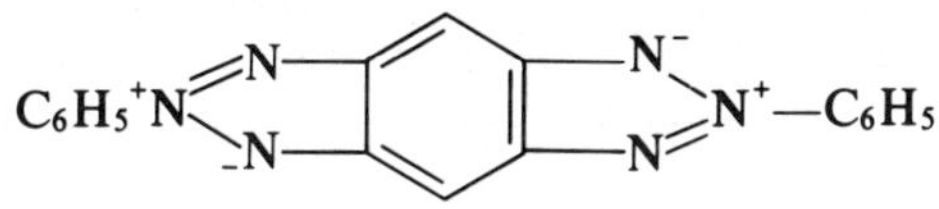

mp 334° (968)

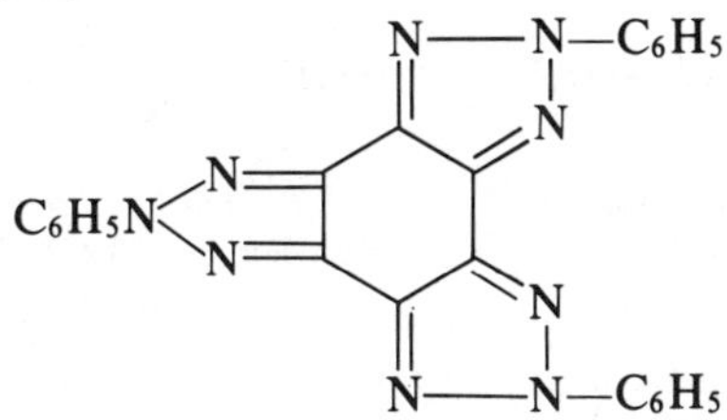

mp 414° (778)

of replacement of the 8-carbon and its hydrogen by a nitrogen atom. Analogs of

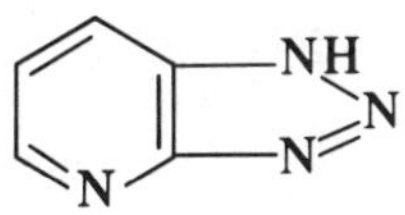

mp 205° dec. (936)

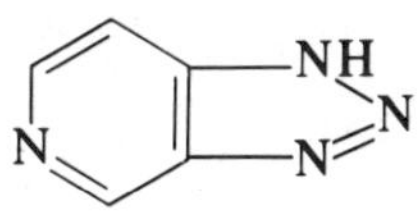

mp 240° (152)

both purine bases and nucleosides have been prepared. The decompositions of

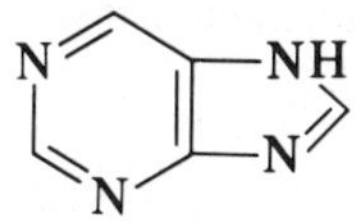

purine
mp 216–217° (325)

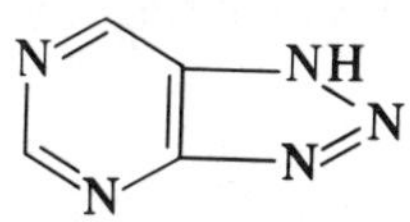

v-triazolo[4,5-*d*]pyrimidine
mp 174–175° dec. (14)
sublimes 120–130°/0.01

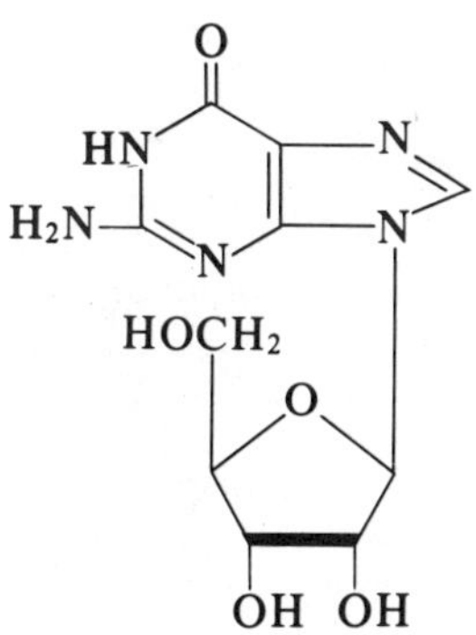

guanosine
chars without melting 235°
(251)

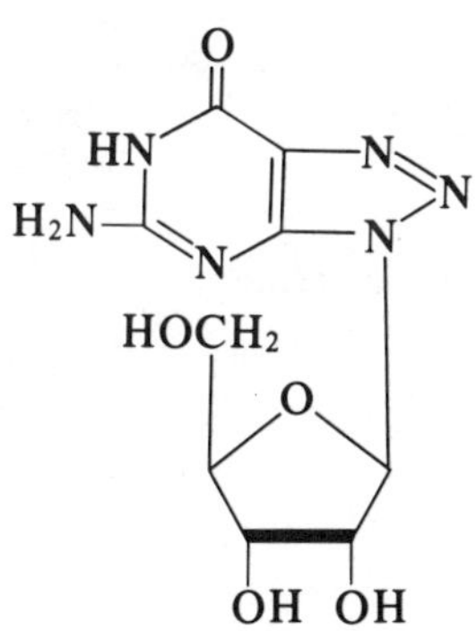

8-azaguanosine
mp 220° browns (537)
250° dec.

guanosine and 8-azaguanosine probably involve the ribose moiety rather than the nitrogen heterocycles.

Decomposition points of *v*-triazolo[4,5-*d*]pyrimidines, unsubstituted on the triazole, range from 140° to over 310°. The partially reduced 5-methyl derivative is slightly less stable than its completely aromatic analog (14). Substitution on the pyrimidine ring with nonhydrogen bonding groups such as chloro results

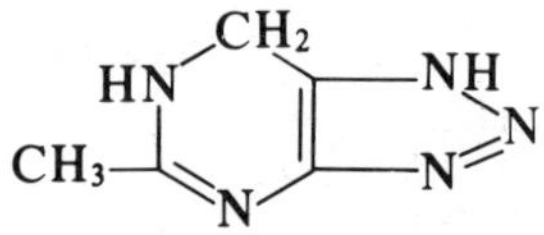

dec. slowly 145°
rapidly 220°

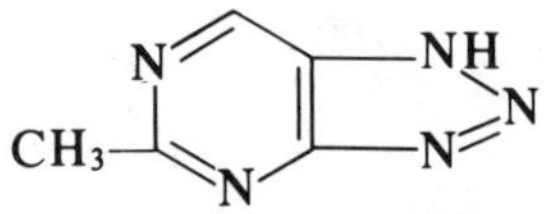

mp 160° dec.
sublimes 125°/0.01

in relatively low decomposition temperatures. *v*-Triazolo[4,5-*d*]pyrimidines with amino, hydroxyl, or tautomeric carbonyl groups, which can participate in

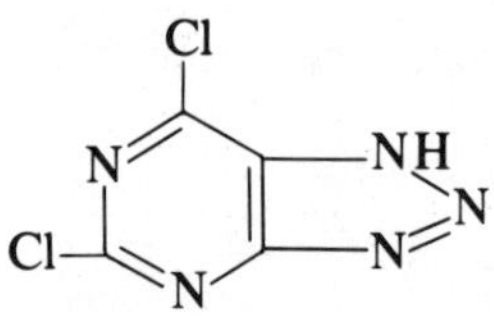

mp 140–142 dec. (123)

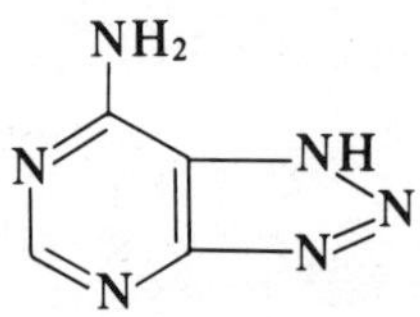

dec. > 310° (927)

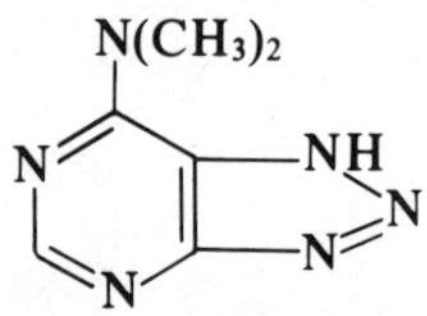

mp 288–290° (1149)

intermolecular hydrogen bonding, melt or decompose near 300°. The 6-methyl *v*-triazolo[4,5-*d*]pyrimidine has neither hydrogen nor substituent in the triazole ring.

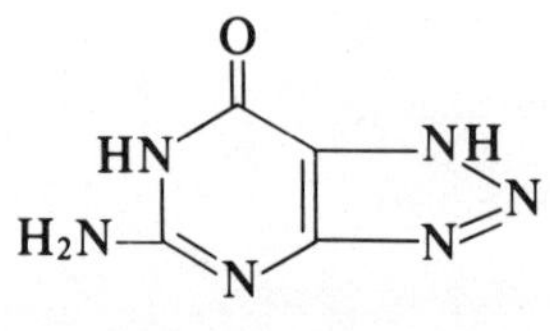

dec. > 300° (927)

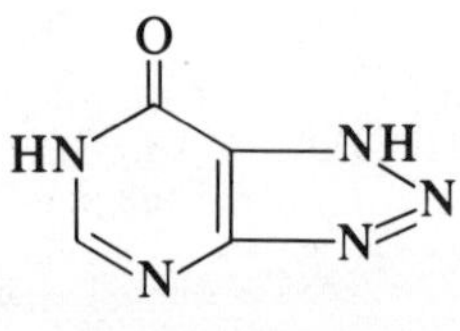

mp 306° dec. (16)
exp 308° (927)

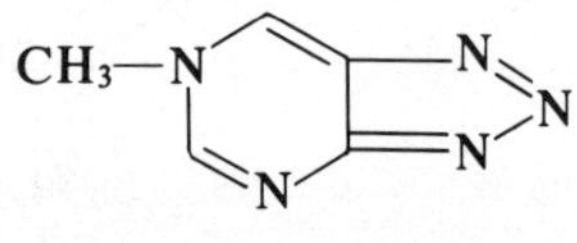

mp ~ 235° dec. (19)

Because of the asymmetry of *v*-triazolo[4,5-*d*]pyrimidine, the 1- and 3-posi-

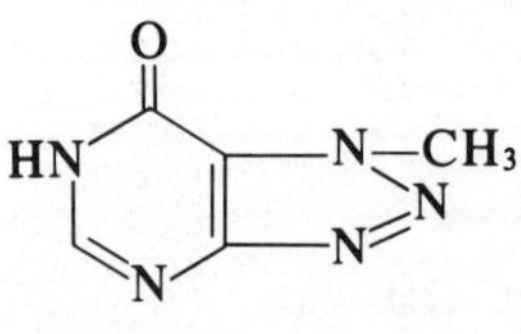

mp 264° (18)

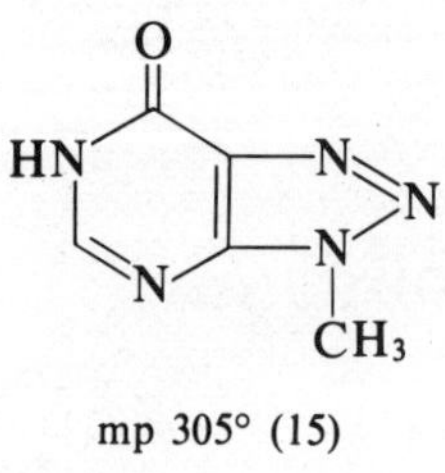

mp 305° (15)

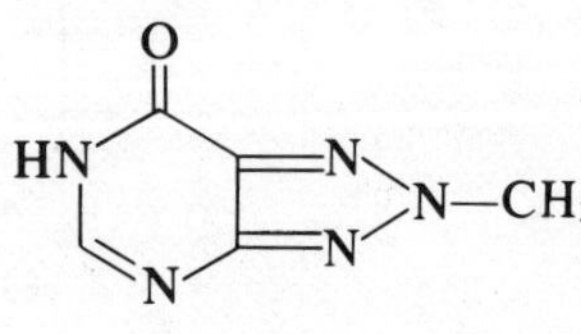

mp 261° (15)

tions on triazole are not equivalent and both 1- and 3-substituted derivatives exist as well as the 2-substituted compounds. Some samples of other 2-substituted *v*-triazolo heterocycles include derivatives of oxadiazole, pyrazine, and pyrimidine.

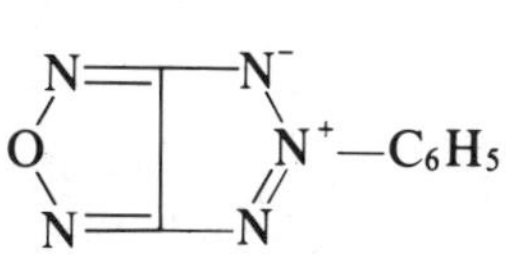

mp 165.5–166.5° (719)

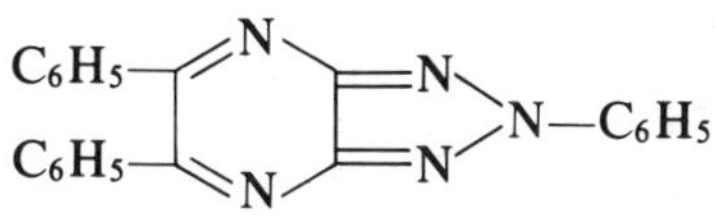

mp 217° (1086)

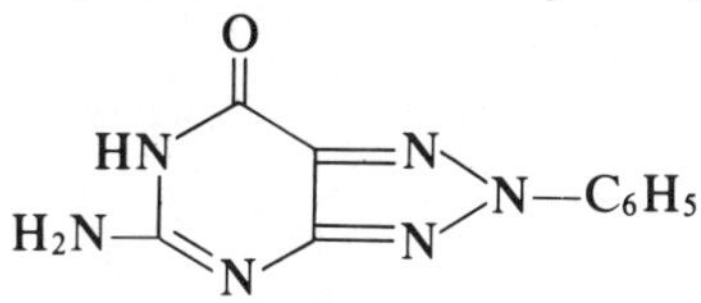

dec. > 300° (104)

A 1-hydroxyl group on a fused ring *v*-triazole confers instability and often explosive properties at or above their melting points. Ethers and esters of these hydroxybenzotriazoles, however, are more stable. Alternate tautomeric *N*-oxide or hydroxy forms are possible for the *v*-triazoles nominally substituted with a 1-hydroxy group. Based on ultraviolet spectra, and comparison with 1-methoxy and 3-methyl 1-oxide derivatives, the equilibrium favors the *N*-oxide

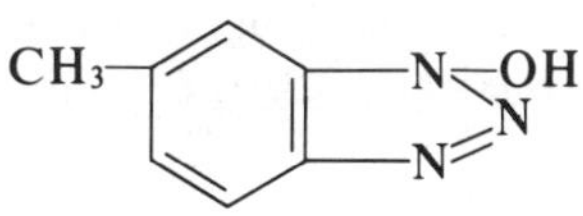

mp 176° dec. (1213)

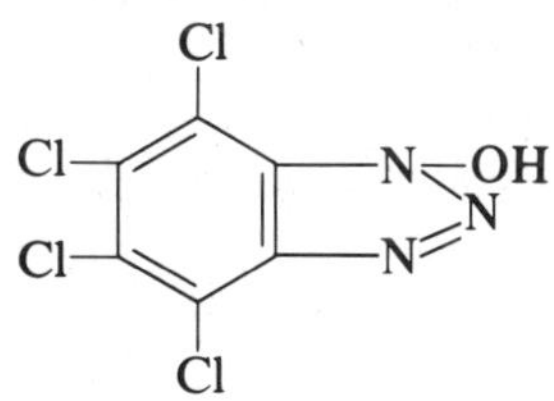

mp 213°, violent dec. (650)

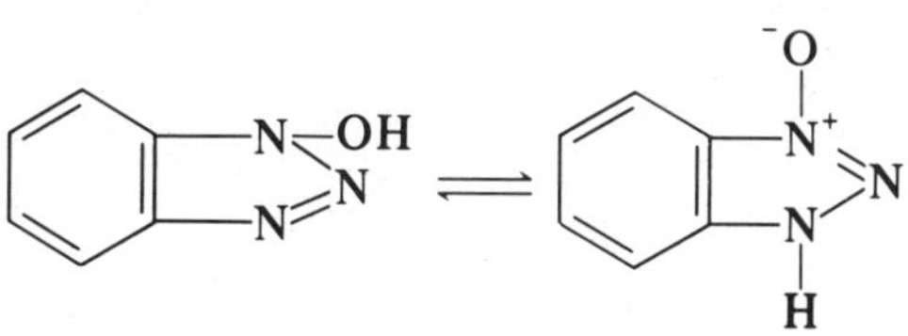

84% in H_2O
22% in C_2H_5OH

form in water and the hydroxyl form in ethanol (146). In the solid, it would be expected that intermolecular hydrogen bonding would tend to make the N—H and O—H forms indistinguishable. No X-ray crystal structural determinations have been reported for these benzotriazole derivatives. In pairs of 1-methoxy and 3-methyl 1-oxide benzotriazoles, the dipolar oxide form is the higher melt-

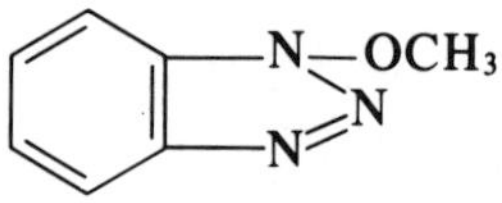

mp 87–88°

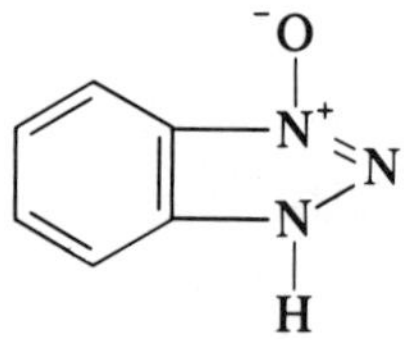

mp 158°, violent dec. (650)

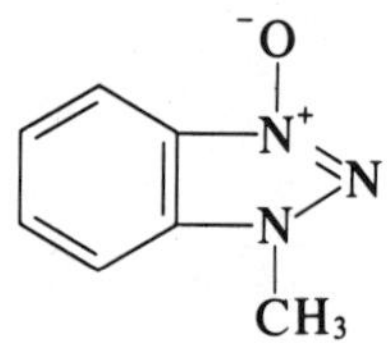

mp 144°

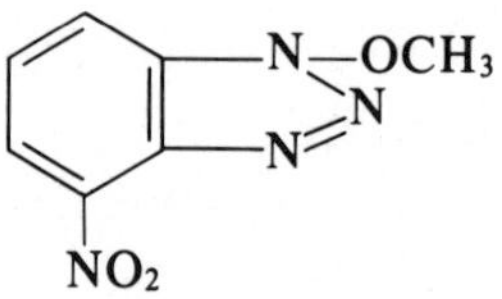

mp 148–149°

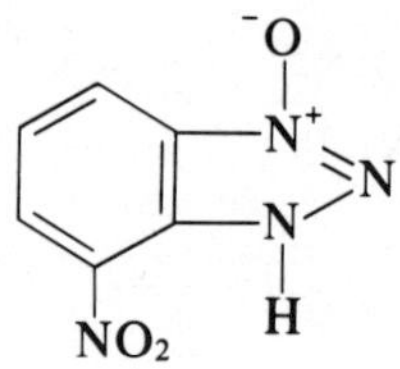

expl. 229° (130)

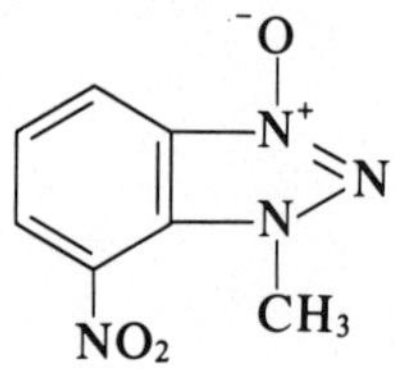

mp 242–243°

ing (146). Decomposition points of the unmethylated parents are closer to the oxide form, suggesting that the parents exist in the solid state as oxides and not hydroxy compounds.

Alkyl, aryl, and heteroaryl substituted fused ring *v*-triazole oxides are known; the oxide group has been attached at both the 1- and 2-positions.

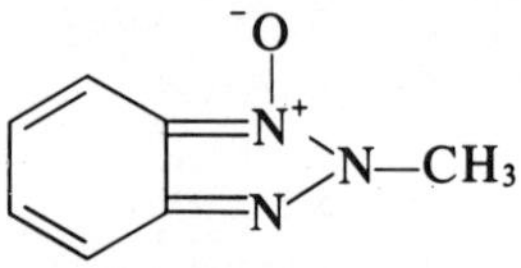

mp 96° (146)

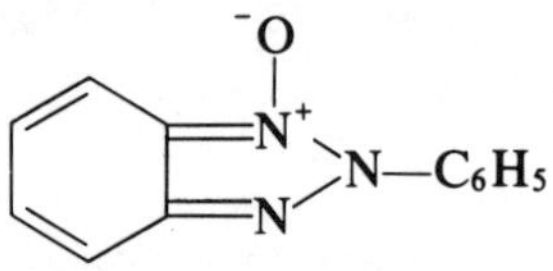

mp 88° (202)

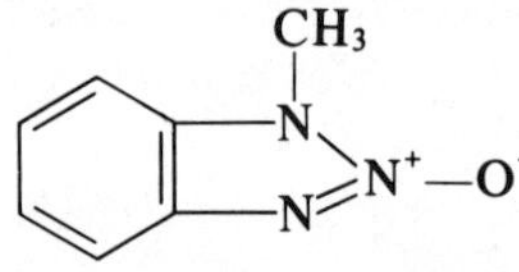

mp 125–126 (979)

Thermal decomposition of both 1-(2-pyridyl)benzotriazole 2- and 3-oxides leads to the unoxygenated 1-(2-pyridyl)benzotriazole in 50–70% yield (513).

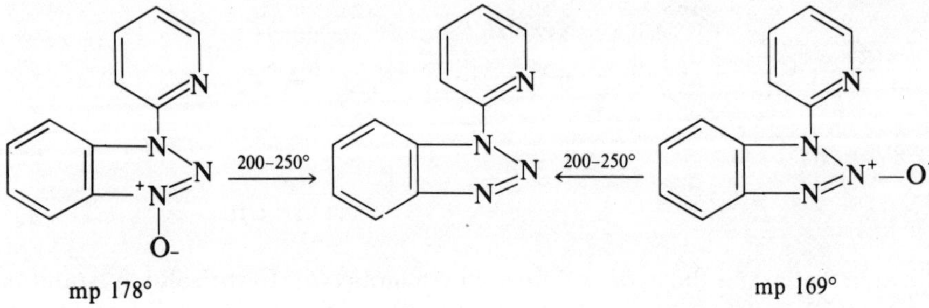

mp 178° mp 169°

Few fused ring *v*-triazole oxides decompose below 200° and many have higher melting points.

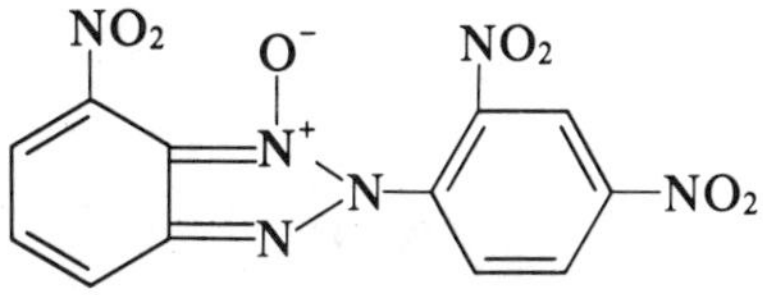

mp 224° (1186)

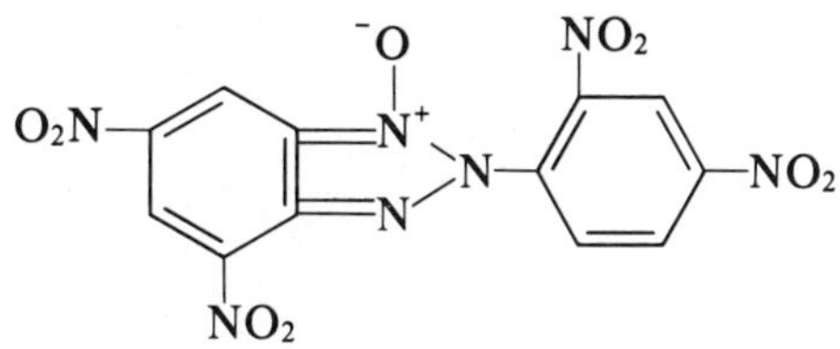

dec. 192° (1185)

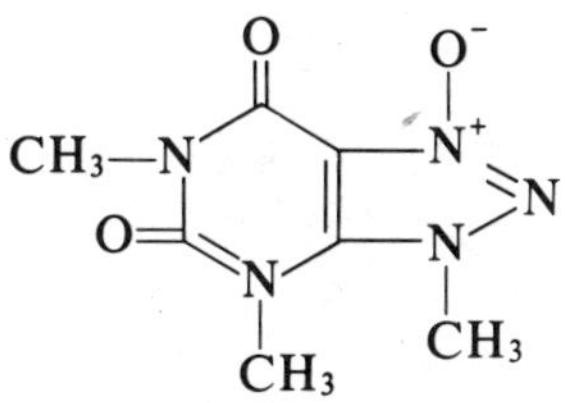

mp 293° (703)

A variety of ether and ester derivatives of 1-hydroxybenzotriazole are

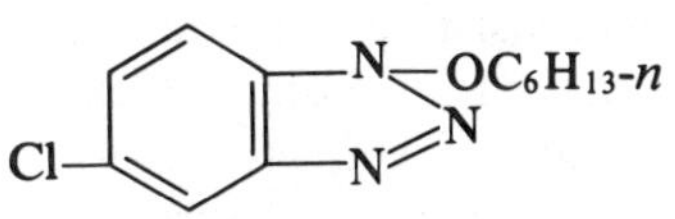

bp 156–158°/0.1 (724)

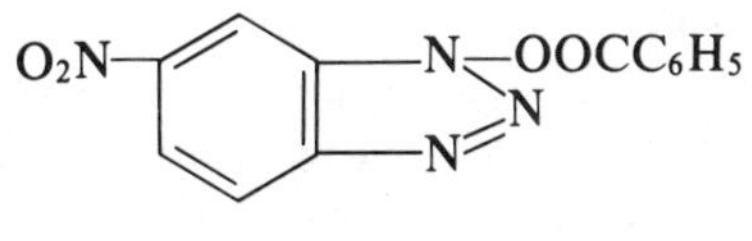

mp 165–166° (601)

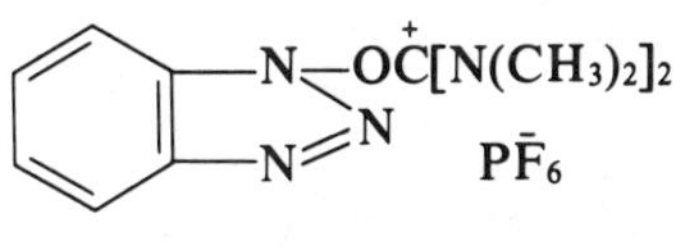

mp 254° dec. (274)

known. Pyrolysis of 1-acetoxy-4,5,6,7-tetrachlorobenzotriazole leads to 2-methoxy-3,4,5,6-tetrachlorophenyl isocyanate (650).

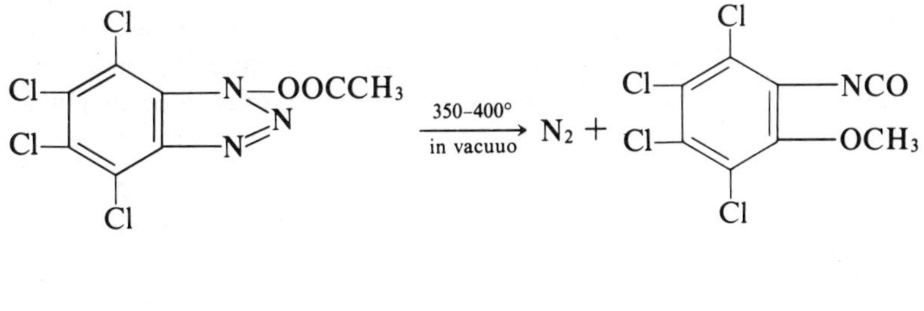

As with the monocyclic triazoles, fused ring triazoles may carry both 1,2-and 1,3-substituents, forming triazolium salts. Salts melting above 160° decompose at their melting points.

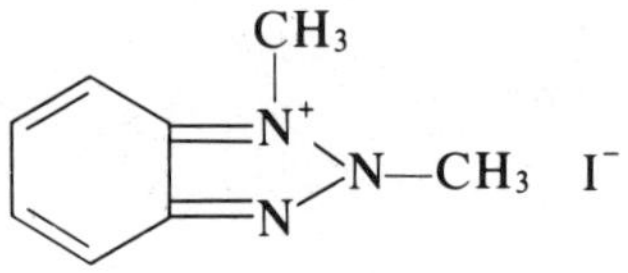

dec. 160° (612)

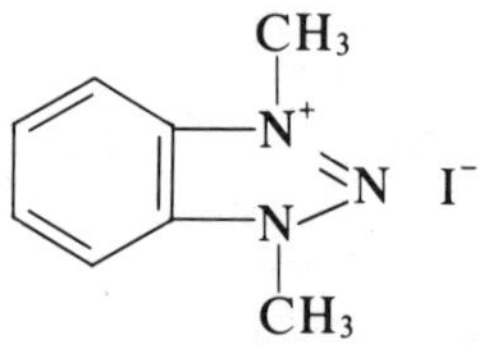

dec. 185° (612)

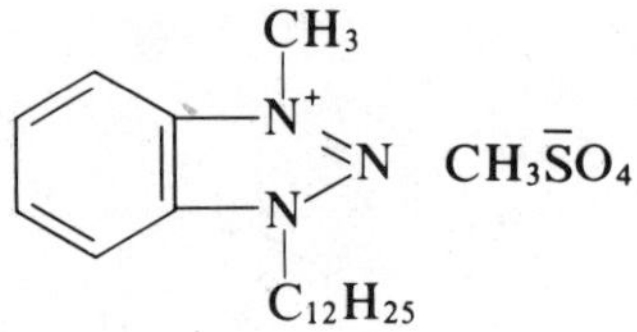

mp ~25° (621)

2.3.5. *v*-Triazines

The *v*-triazines are six-membered ring systems consisting of a sequence of three contiguous nitrogen atoms and three contiguous carbon atoms, which can exist in reduced, fully aromatic, *N*-alkylated, arylated, or oxide forms.*

Reduced *v*-triazines, where the three-nitrogen chain is at the triazane oxidation level, show no decomposition at their melting points, which extend to 187°, but studies of their thermal stability have not been made.

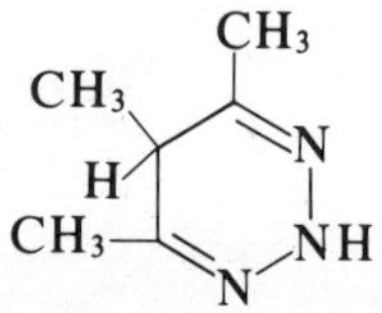

mp 54° (821)

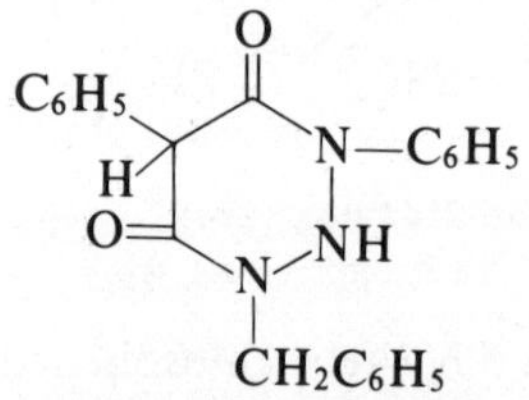

mp 136–137° (575)

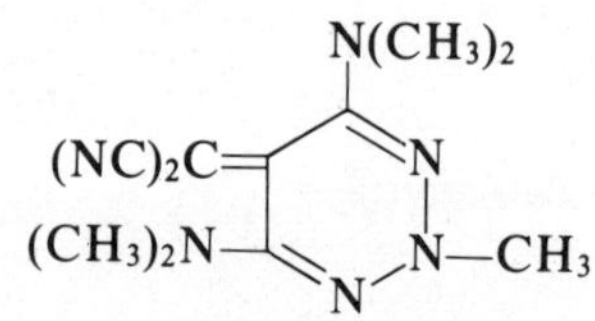

mp 184–187° (392)

Dipolar *v*-triazines at the triazene oxidation state decompose at their melting points, but cyclic six-membered triazenium salts are stable on melting.

Monocyclic, aromatic *v*-triazines have been known only since 1960 when the 4,5,6-triphenyl derivative was prepared (200), and the parent of the series was

*The *v*-triazines are reviewed by Neunhoeffer (799), by Kobylecki and McKillop (595a), and by Stevens (1034).

not synthesized until 1981 (822). Carbon substituents on *v*-triazine include alkyl, aryl, amino, alkoxy, and halo. At 300° in a flow system, trimethyl *v*-tria-

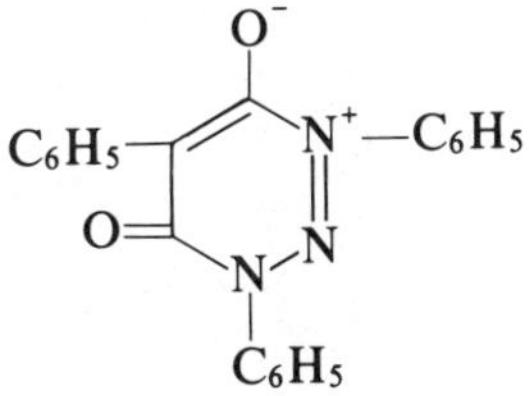

mp 171–172° dec. (575)
dec., reflux in xylene

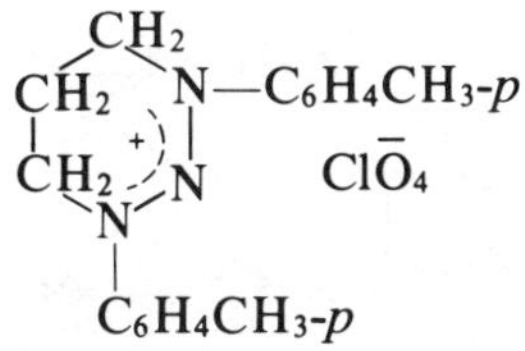

mp 182–184° (425)

zine is stable, but decomposes at 500° into 2-butyne and acetonitrile (215).

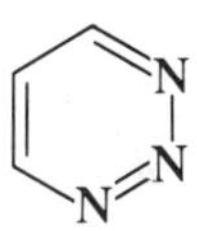

mp 69.5–71.0° (822)

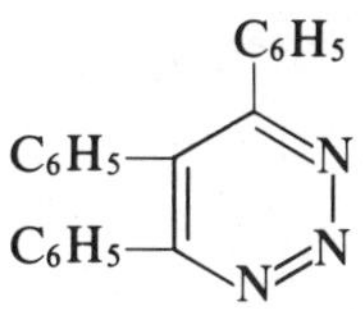

mp 280° (796)

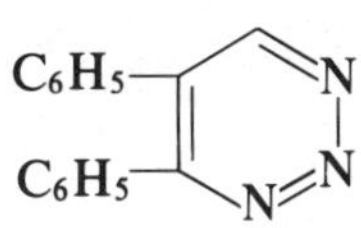

mp 176–177° (796)

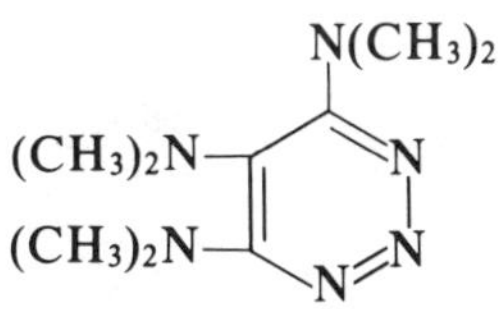

mp 93° (799)

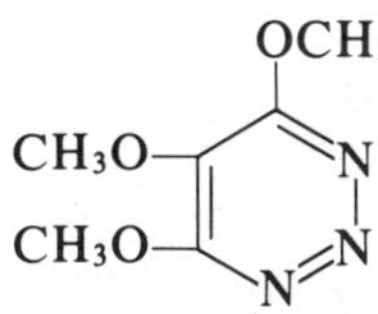

mp 115–117° (391)

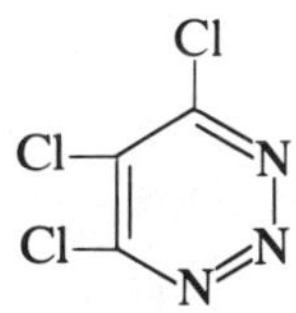

mp 110–112° (391)

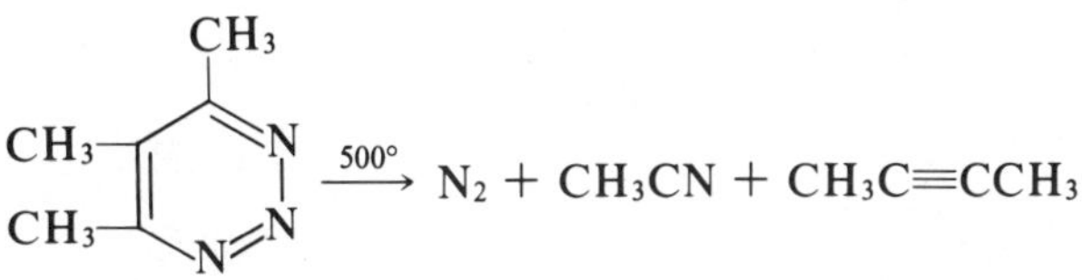

mp 146–147°

Thermolysis of triphenyl *v*-triazine above 250° leads to evolution of nitrogen and formation of indene imines (800). In other studies, at a temperature below 400°, 70% of triphenyl *v*-triazine was recovered. At 420°, 16% starting material was obtained along with nitrogen, diphenylacetylene, and benzonitrile; at 460° diphenylacetylene was the only product isolated (10). Flash vacuum pyrolysis of tris(dimethylamino)-*v*-triazine leads to tris(dimethylamine)azete, stable only below −25° (980).

The *v*-triazinium salts are alkylated and probably protonated at N-2 of the *v*-triazine ring. Most protonated *v*-triazines decompose at low temperatures, as

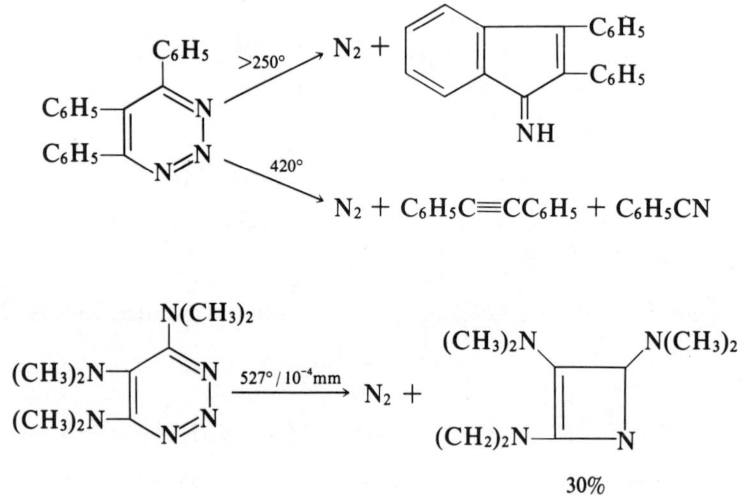

illustrated by the fluoborate salt of tris(diisopropylamino)-*v*-triazine, formed at −78°, which loses nitrogen and rearranges to a dihydroimidazolium fluoborate on warming above −20° (392). It is possible, however, to isolate the fluoborate

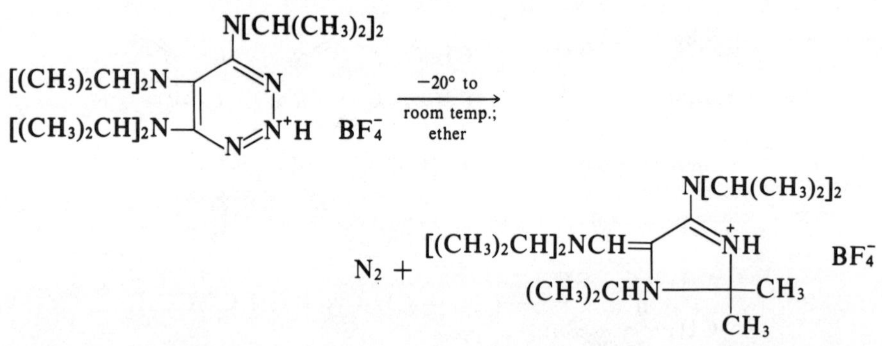

of 5-chloro-4,6-bis(dimethylamino)-*v*-triazine. The alkylated *v*-triazinium compounds are appreciably more stable than the protonated salts. Dicationic *v*-triazino-cyclopropenylium salts also can be prepared.

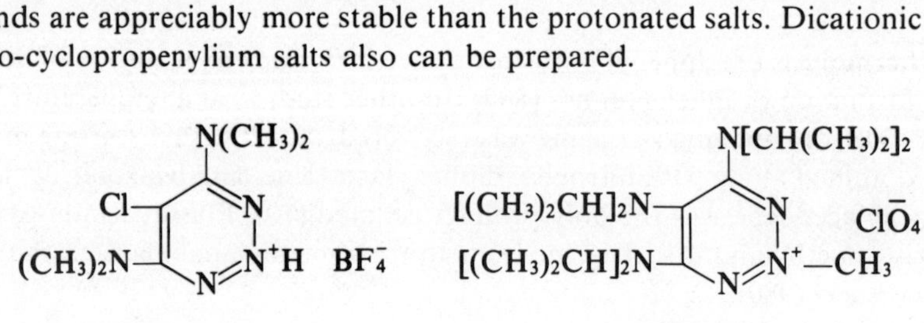

mp 159–160° mp 236–239°

The most reduced fused ring *v*-triazines are products from the combination of cyanogen azide and hexamethylbicyclo[2,2,0]hexadiene, whose only ring un-

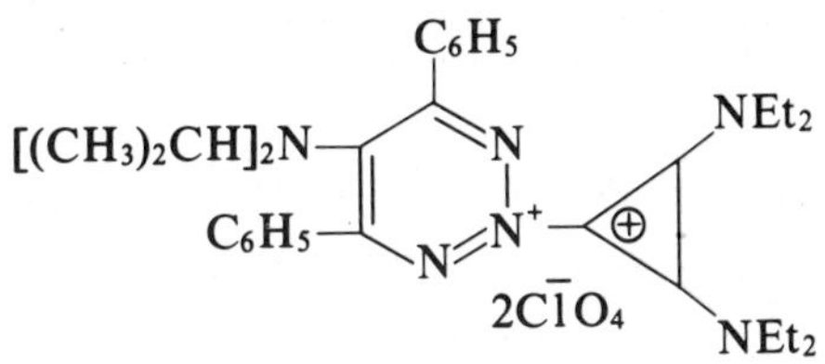

mp 207–210°

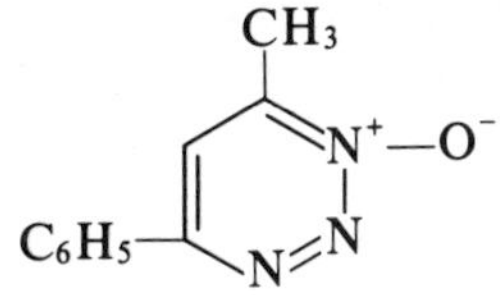

mp 197°

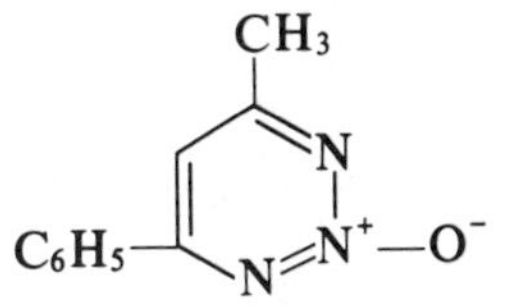

mp 178°

saturation is the azo group of the three-nitrogen chain, illustrated by the structure shown. Several 3-alkyl or aryl 3,4-dihydrobenzotriazines and a 1-phenyl-1,4-dihydrobenzotriazine have been synthesized. The mono alkyl or aryl 3,4-dihydrobenzotriazines decompose in the 120–160° range, but the 1,4,4- or 3,4,4-trisubstituted derivatives appear to be stable to their melting points. These bicyclic trisubstituted *v*-triazines are about as thermally stable as the corresponding open-chain 1-aryl-3,3-dialkyl- or 1-aryl-3-alkyl-3-aryltriazenes.

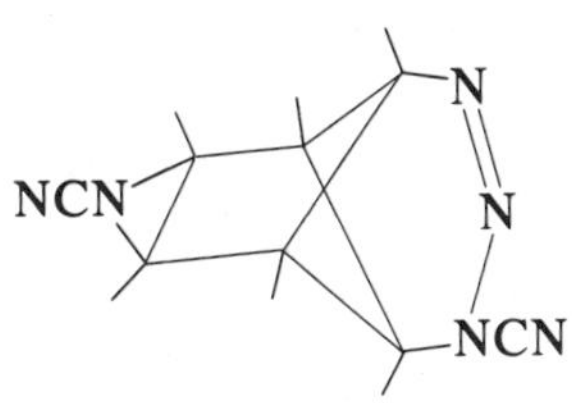

mp 160–161° (31)

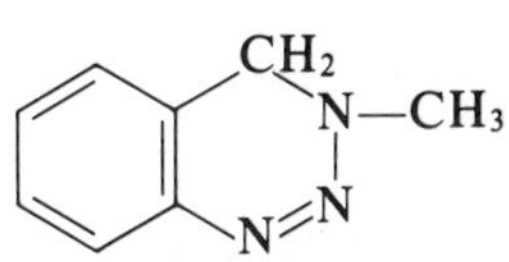

mp 72–73° (178)
dec. 120°

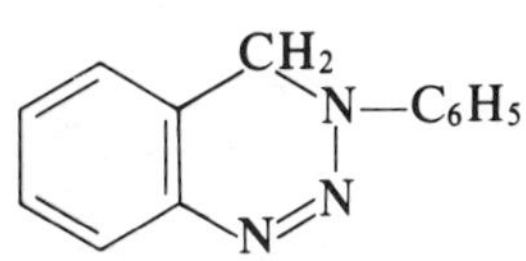

mp 128° dec. (177,447)

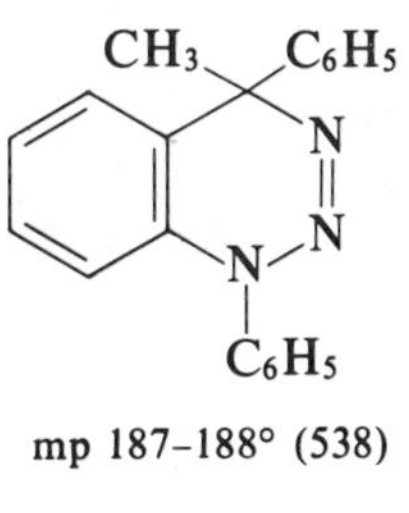

mp 187–188° (538)

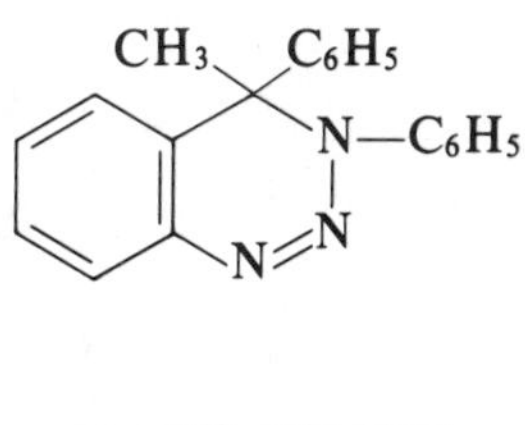

mp 162–163° (538)

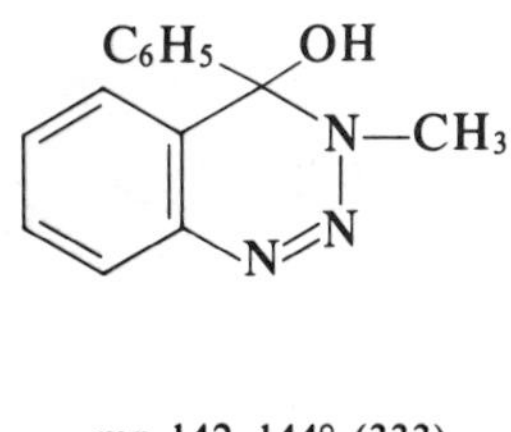

mp 142–144° (333)

Many unsubstituted fused ring *v*-triazinones are known (Table 2.52) which decompose in the 160–270° range, often explosively. They are appreciably more

Table 2.52. *N*-Unsubstituted Fused Ring *v*-Triazinones

Fused Ring *v*-Triazinone	Melting Point (°C)	Thermal Stability	Ref.
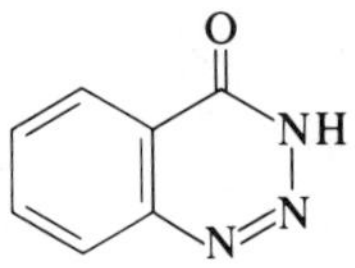	220 dec.; 242 dec.	Dec. 162–164°, diglyme	447, 775, 993
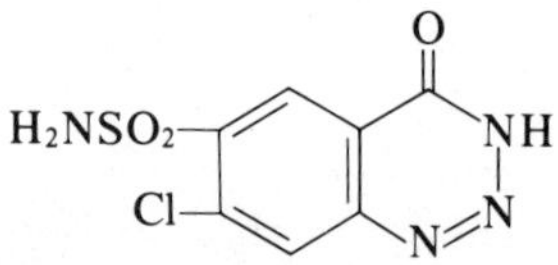	270–272 dec.		358
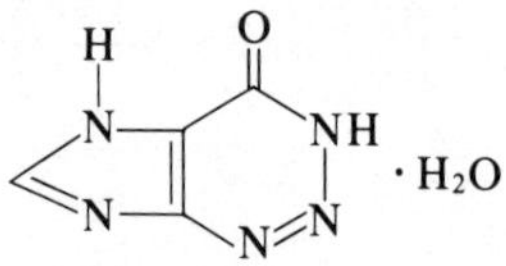	Expl. 210°	Heated gradually, darkens 150°, does not melt 260°	986
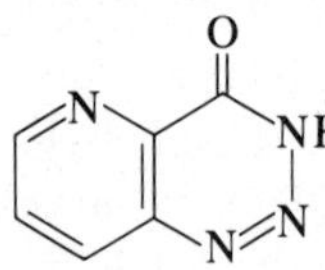	230 dec.		936
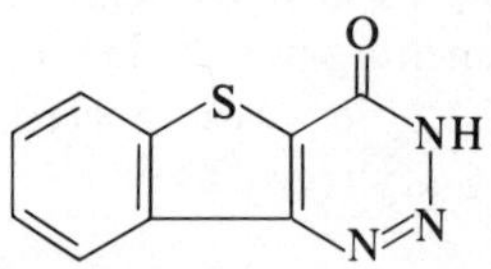	180–182 expl.		970

stable, however, than their acyclic analogues, the 1-acyl-3-aryl triazenes, which usually are destroyed at temperatures of 165° or below. Based on infrared evidence, not only are the fused ring triazinones in the keto form, but the 4-sulfur and 4-substituted imino benzotriazines also exist with a divalent group at position 4 and a hydrogen on the nitrogen chain. Distinction between hydrogen at

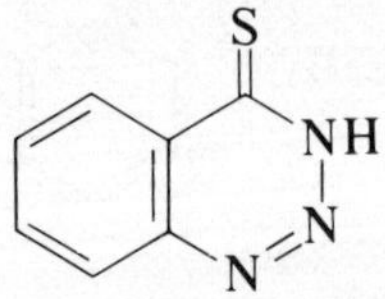

mp 205–206° (1027)

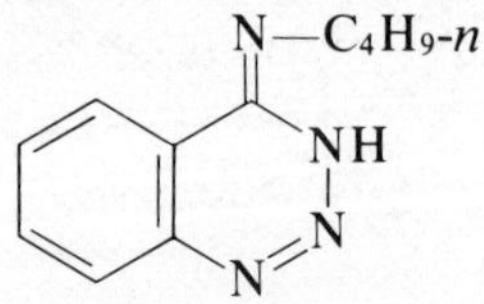

mp 153–155° (371)

N-3 or the tautomeric form at N-1 usually has not been made, but X-ray crystal analysis shows the ring hydrogen of benzotriazinone in the solid state to be at N-3 (483).

Thermolysis of benzotriazinone at 162–164° leads to expulsion of nitrogen from one molecule, condensation with a second benzotriazonone, and formation of quinazolino[3,2-*c*]benzotriazin-8-one (775). At 245°, 2-(*o*-aminophenyl)-

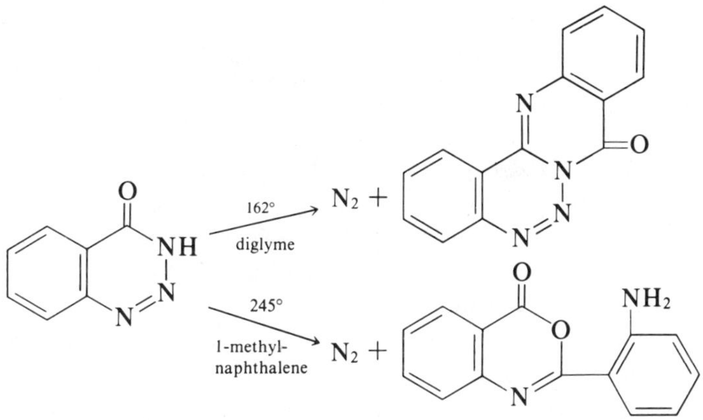

4*H*-3,1-benzoxazin-4-one is formed (225, 775, 999). Likely intermediates for these and other benzotriazinone decompositions are the benzazetidinones or their ketene imine valence isomers.

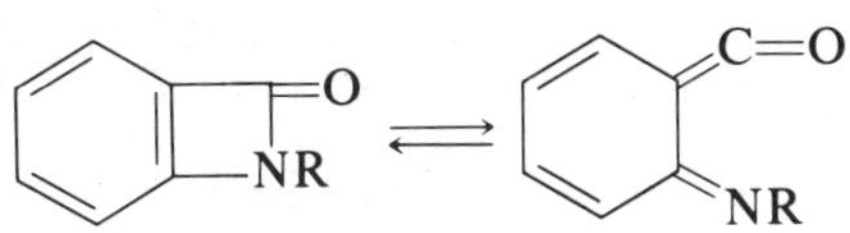

Benzotriazinones, reduced in the benzene ring and substituted at the 3-position with a lower alkyl group are stable at their melting points in the 63–75° range. The 3-butyl derivative is a liquid distilling without decomposition at 155° and 0.5 mm (188).

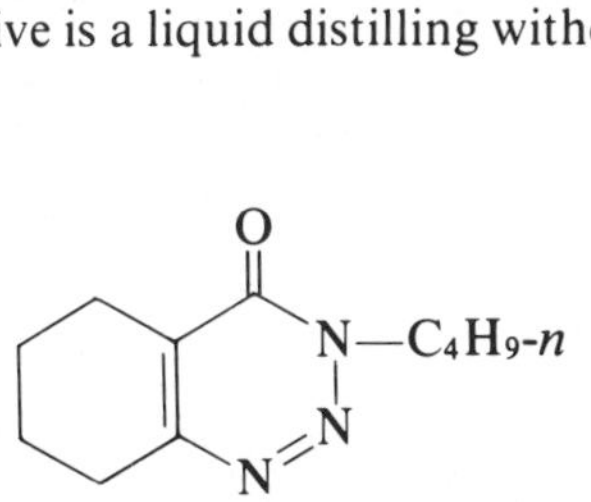

Alkyl, aryl, acyl, oxygen, and sulfur radicals have been attached to N-3 of benzotriazinone (Table 2.53). The 3-substituted benzotriazinones, except for some acyl derivatives, are more stable thermally than those with hydrogen on ring nitrogen. Heating 3-phenylbenzotriazinone alone at 282° gives rise to acridone and phenanthridone, but in paraffin only benzanilide is found (475).

Table 2.53. 3-Substituted Benzotriazinones

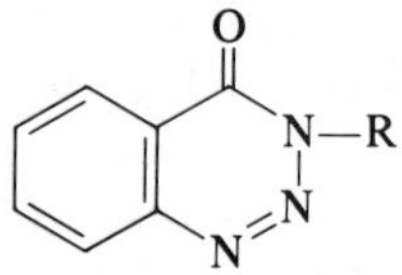

R	Melting Point (°C)	Thermal Stability	Ref.
CH_3	121–122	Stable at 206° in ethyl cyanoacetate	638, 990
$NCCH_2$	109–110		53
C_6H_5	150–151	Slow dec. at 250°, 91% unchanged after 48 hr;	369, 475
		explosive dec., 320°	476
p-$(CH_3)_2NC_6H_4$	246–247 dec.		585
p-$O_2NC_6H_4$	271–272		369
CH_3CO	165		370
HO	182–182.5 dec.	Dec., reflux in benzene 4–5 days, or toluene 20 hr	13
CH_3O	134–135	Unchanged after reflux 22 hr in p-xylene	13
CH_3COO	174–175		13
CCl_3S	125		459

Thermolysis of 3-imidoyl benzotriazinones leads principally to 1,2-diaryl-1,4-dihydroquinazolin-4-ones and phenanthridones (849). Although pyrolysis of 3-

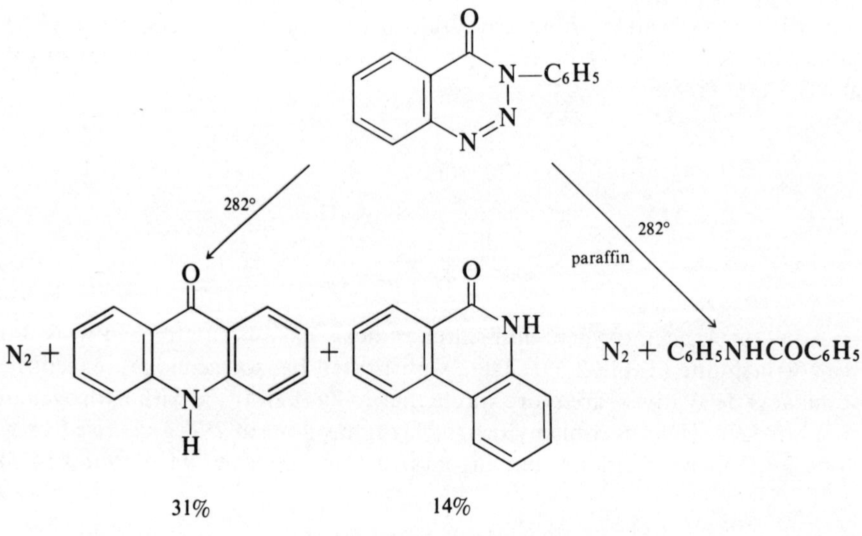

(1-adamantyl)benzotriazinone at 500° and 0.005 mm results in only minor decomposition, at 600° and 0.05 mm 1-(1-adamantyl)benzazet-2-one is obtained

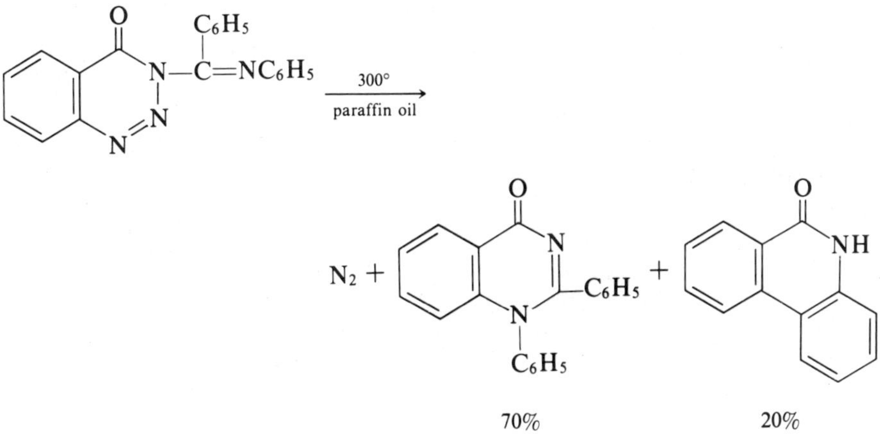

along with small quantities of adamantyl isocyanate and biphenylene (82). 3-Phenylcarbamoylbenzotriazinone, which may be considered to be a cyclic 1-aryl-3,3-diacyl triazene, is less stable, losing nitrogen at its melting point with rearrangement to 2-anilinobenzoxazinone (225).

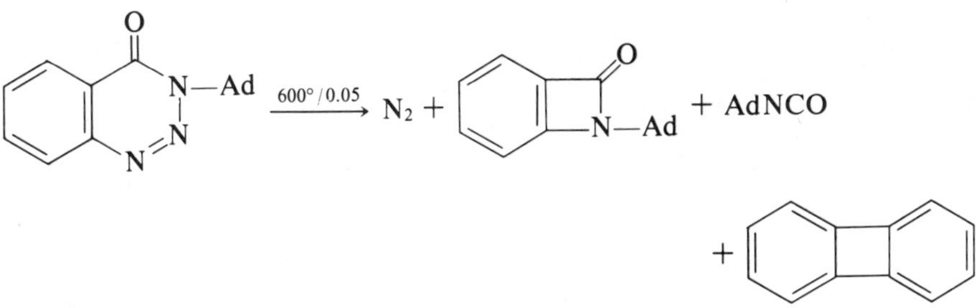

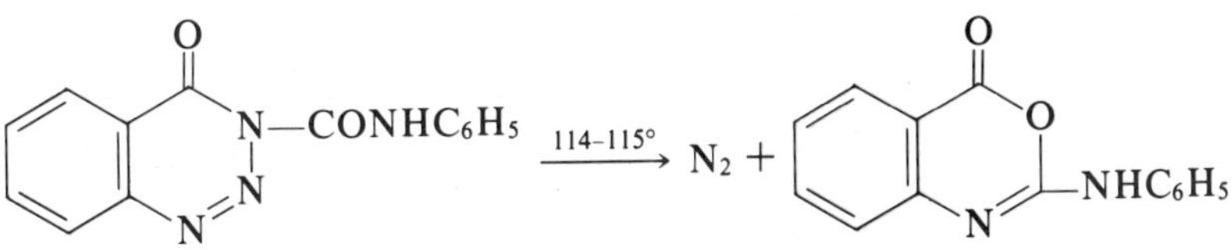

Imine and methylene analogs of the 3-substituted benzotriazinones exist. On heating in alcohol, 3-aryl-3,4-dihydro-4-imino-1,2,3-benzotriazines undergo a thermal Dimroth rearrangement to 4-arylaminobenzotriazines, a reaction facilitated by electron-withdrawing groups on the aryl ring (1031). The 3-phenyl- and *p*-anisyl-4-methylene-3,4-dihydro-1,2,3-benzotriazines were obtained only as oils, but the 3-*p*-acetylphenyl-4-methylene derivative is a crystalline solid. After refluxing in ethanol, 3-methyl-4-methylene-3,4-dihydro-1,2,3-benzotriazine is recovered unchanged, but decomposes on reflux in xylene or diglyme (333).

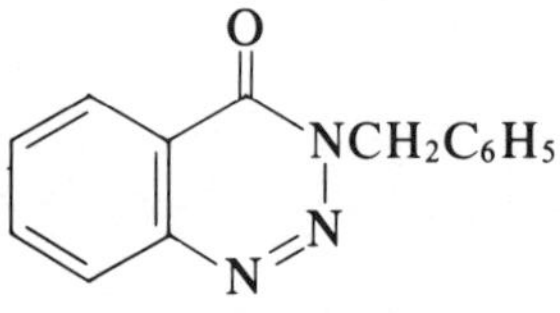

mp 114–116° (701)

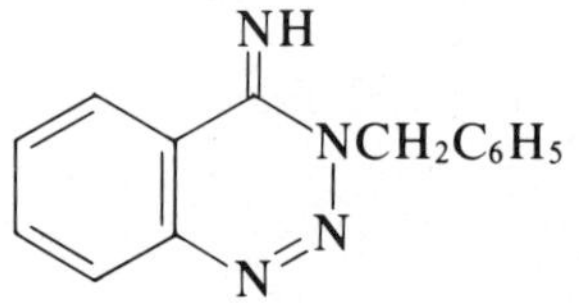

mp 119–120° (1031)

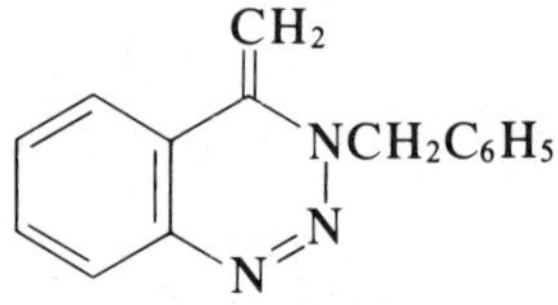

mp 80° dec. (333)

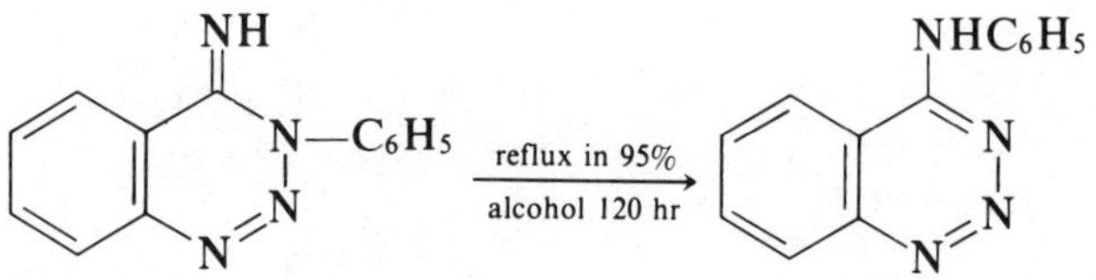

quant.

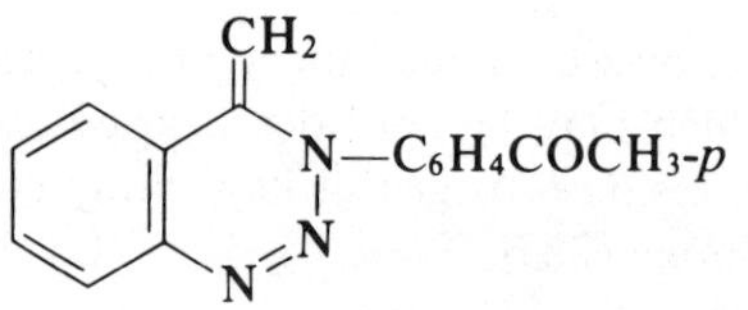

mp 121–123°

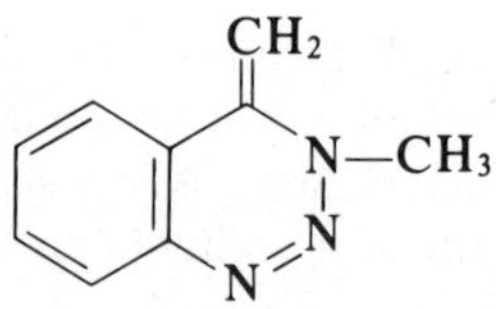

mp 64–66°

Table 2.54. 4-Substituted Benzotriazines

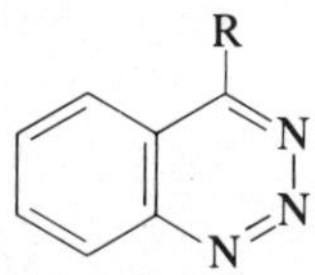

R	Melting Point (°C)	Thermal Stability	Ref.
H	119–120	Dec. 500°/0.1	8, 9
CH_3	120–121	Dec. 420–450°/0.1	8, 9
C_6H_5	156–157	Stable to 400°; dec. 420–450°/0.1	8, 10
NH_2	266 dec.		847
$C_6H_5CH_2NH$	207–209 dec.		1031
C_6H_5NH	200–201° dec.		1031
CH_3O	105.5–106.5		8
CH_3S	100–101		1027

4-Substituted *v*-benzotriazines are listed in Table 2.54. The aromatic *v*-benzotriazine and its 4-methyl and 4-phenyl derivatives are very stable thermally. Benzotriazines with an amino or substituted amino at the 4-position decompose from 200–266°. Information on the heat resistance of 4-alkoxy and alkylthio ethers of benzotriazine has not been reported.

Vapor phase pyrolysis of *v*-benzotriazine and its 4-*t*-butyl (911) and phenyl (9, 10) derivatives leads to biphenylene, the dimer of the corresponding benzazete, an acridine, and benzonitrile. Under similar conditions 4-methylbenzotriazine forms biphenylene and 9-methyl acridine, but gives little evidence for a 2-methylbenzazete dimer (911). Although scrambling of a $C_6H_5C{=}N$ unit occurs

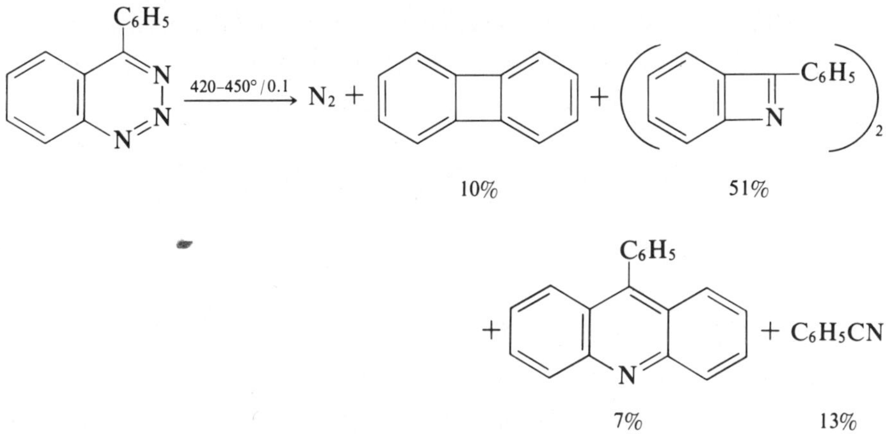

during flash vacuum pyrolysis of other heterocycles, no evidence of a C—N transposition was found in the benzazetes found in pyrolyses of *v*-benzotriazines (250). Pyrolysis of 4,5,8-triphenylpyridazino[4,5-*d*]triazine extrudes nitrogen and forms 1,4-diphenyl-5*H*-indeno[1,2-*d*]pyridazine-5-imine and diphenylbutadiyne (375). There is no indication of collapse of the triazine to a pyridazino azete.

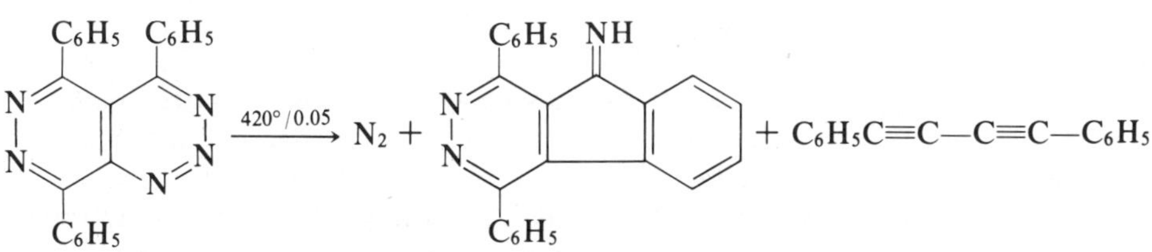

Both 2-aza and 2,8-diaza analogs of adenosine have been prepared (752).

Naphtho[1,8-*de*]triazine and its 1-methyl derivative have an unusual cross-conjugated system and are both colored red. The parent compound decomposes at or near its melting point, but no thermolysis data for the 1-methyl compound have been determined. Pyrolysis of naphtho[1,8-*de*]triazine leads to

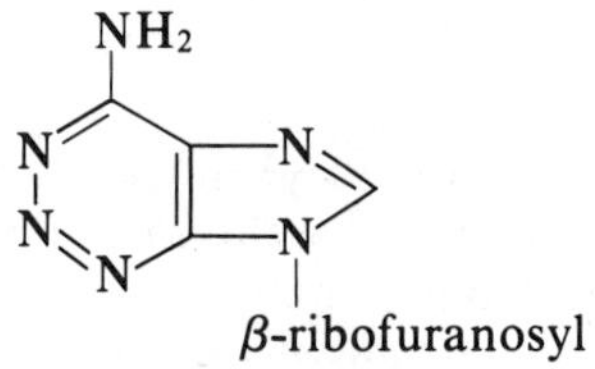

2-azaadenosine
mp 240–241°

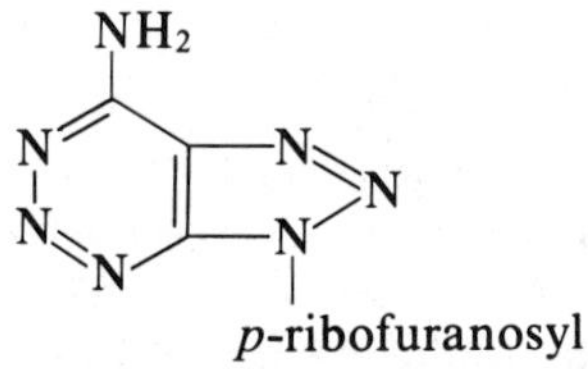

2,8-diazaadenosine
mp 193–195°

α-naphthylamine (229). The blue 2-methyl-2*H*-naphtho[1,8-*de*]triazine, a cyclic azimine, is shown in one resonance form. Both the 2-methyl derivative and the

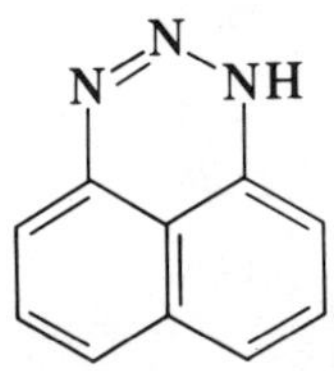

mp ~230° dec. (802)
dec. 252° (993)

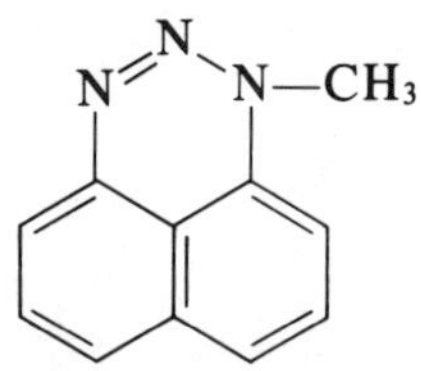

mp 91–92° (862)

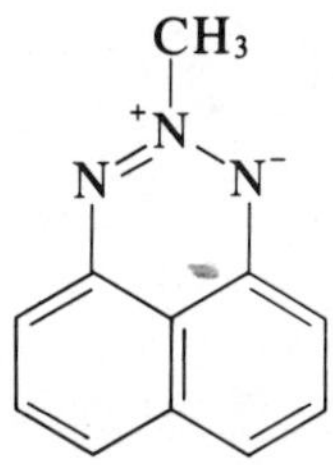

mp 132° (862)

corresponding 2′-methoxy-1′-naphthyl compound are largely unchanged after refluxing in *o*-dichlorobenzene (180°) (1024).

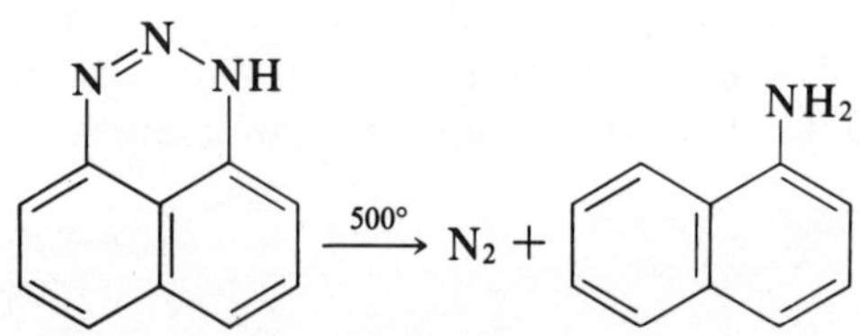

The 2-substituted benzo-1,2,3-triazonium betaines are closely related dipolar compounds, in one resonance form of which the negative charge is distributed to the oxygen or nitrogen attached at the 4-position. When 2-(*p*-tolyl)benzo-1,2,3-triazonium betaine is heated above 60°, the *p*-tolyl group migrates to the oxygen at position 4 (726). The rearrangement is reversed with 4-(tetraacetyl-β-D-glucopyranosyloxy)-1,2,3-benzotriazine, which is converted on heating at 110° to both the 2-substituted benzotriazanium betaine and the 3-substituted benzotriazine-4-one (1140).

A series of 1,2,3-benzotriazine 3-oxides are known (Table 2.55). Judging by the decomposition at their melting points, the 4-methyl and 4-phenyl com-

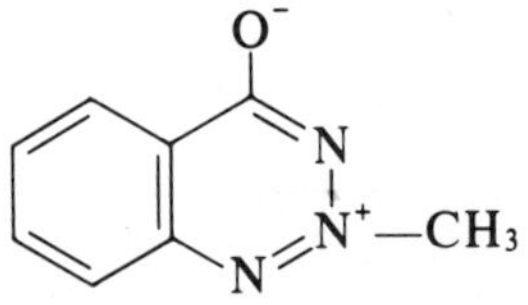

mp 122–123° (fast) (726)
143–145° (slow)

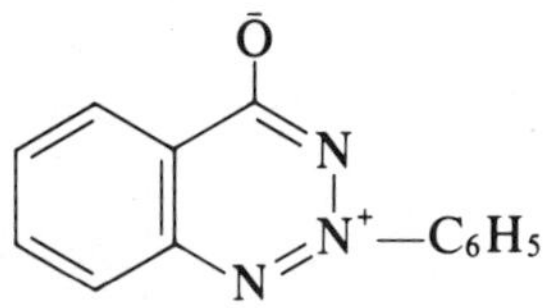

mp 116–118° (726)

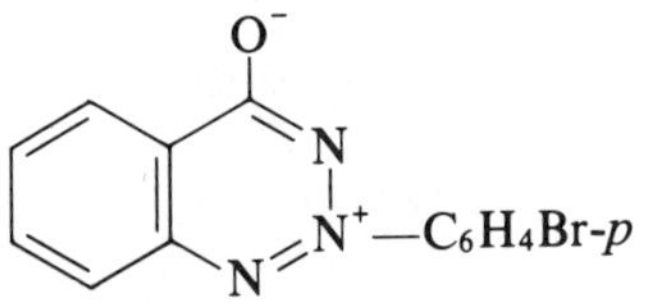

mp 199–200° (584)

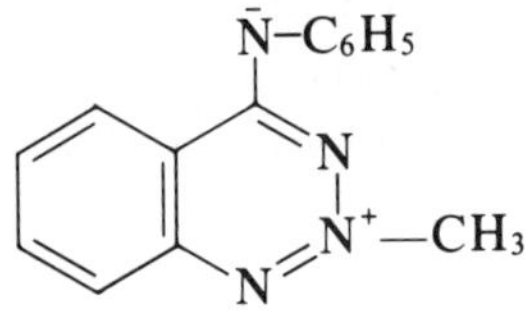

mp 130–131° (1032)

pounds are less heat resistant than the corresponding nonoxides, which are stable to over 400°. The corresponding 4-amino-7-chloro and 7-methoxy-1,2,3-

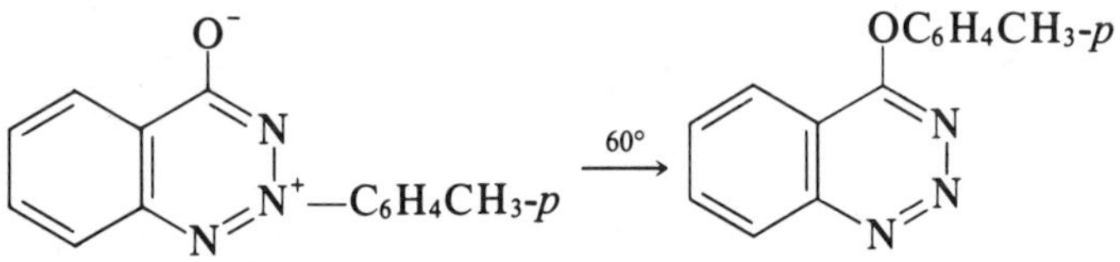

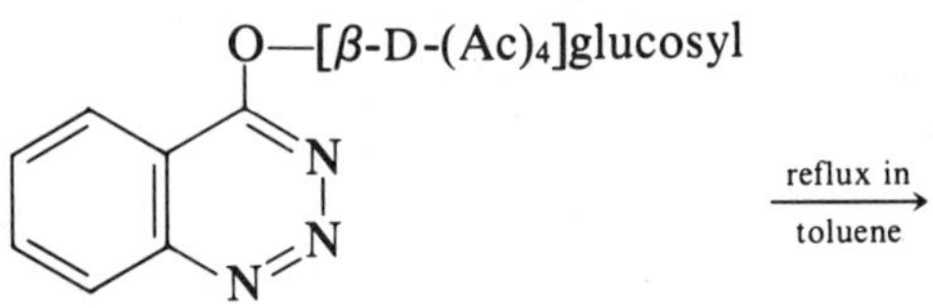

mp 142–145°

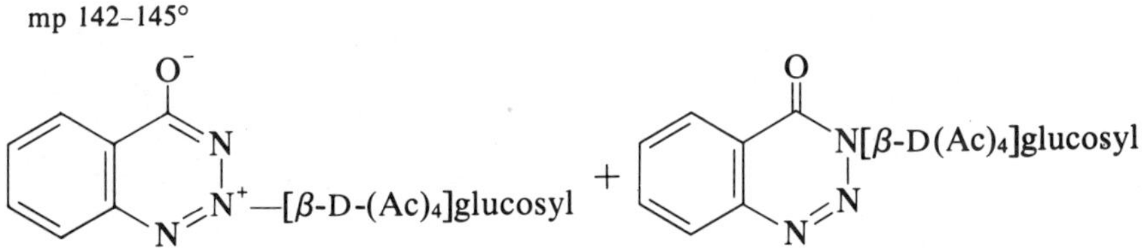

60% 20%

benzotriazine 3-oxides decompose at 205–206° and 215–216°, respectively (411, 847). Condensed triazine oxides with heterocyclic rings are illustrated by the imidazo, pyrazolo, and pyrimido compounds. Each of these heterocycles has the

Table 2.55. *v*-Benzotriazine 3-Oxides

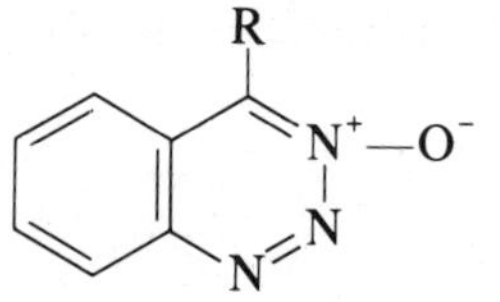

R	Melting Point (°C)	Ref.
H	158–159	733
CH_3	185–188 dec.	733
C_6H_5	154 dec.	733
NH_2	180	44, 847

capacity of multiple intermolecular hydrogen bonding and decomposes at temperatures comparable to those of the fused ring *v*-triazole oxides.

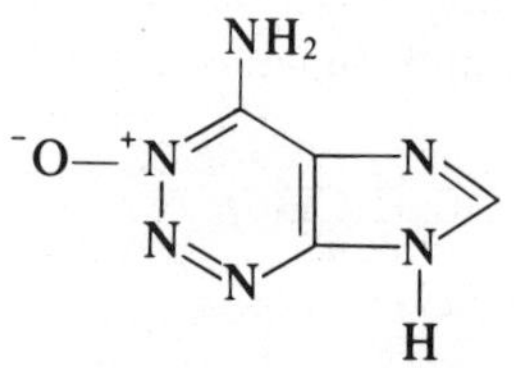

Darkens 115–120° (1033) Complete dec. > 350°

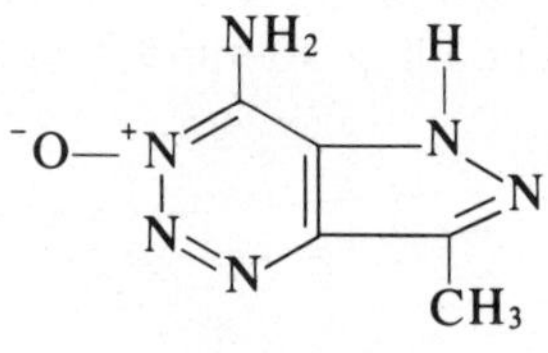

mp 252° expl. (682)

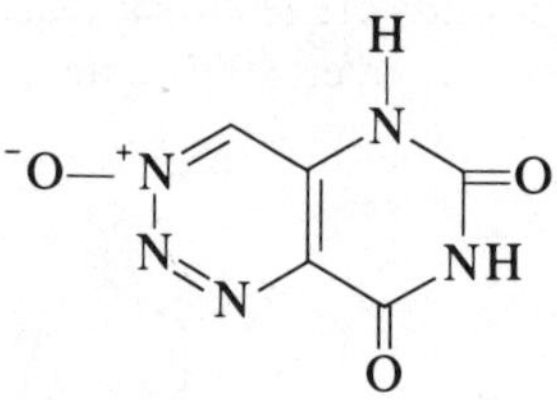

mp 232° dec. (844)

Flash vacuum pyrolysis of 4-phenyl-1,2,3-benzotriazine 3-oxide yields 3-

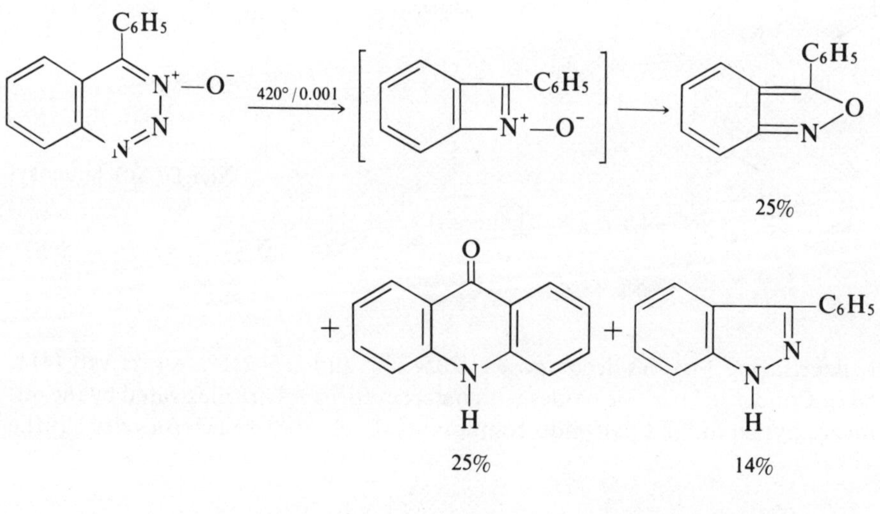

phenyl-2,1-benzisoxazole and acridone, probably through initial formation of 2-phenylbenzazete 1-oxide, together with 3-phenylindazole (10).

Several betaine 1-oxides of the 2-substituted benzotriazinones have been prepared, all of which decompose below 150°. Insufficient data are available to

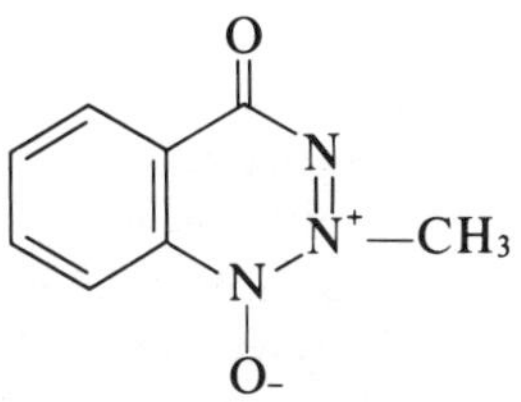

mp 145–157 dec. (726)

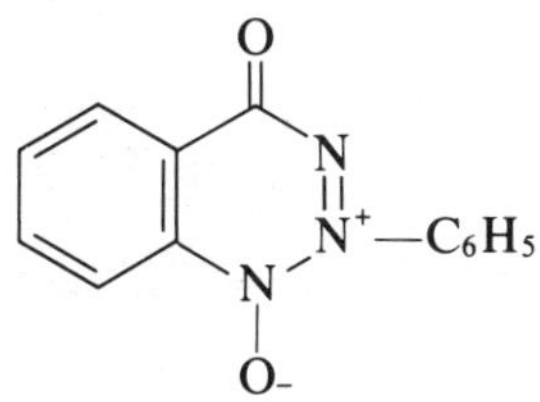

mp 147–149 dec. (726)

compare their thermal stabilities with their 2-substituted benzotriazinone betaine parents.

2.3.6. Oxatriazoles

The oxatriazoles are five-membered rings with three sequential nitrogen atoms, one oxygen, and one carbon atom. Oxatriazoles prepared thus far are mesoionic compounds, bearing a delocalized positive charge on the ring and, at the ring carbon, an exocyclic oxygen or sulfur atom or a nitrogen group carrying a negative charge (Table 2.56).* Cationic salts of oxatriazoles also are known.

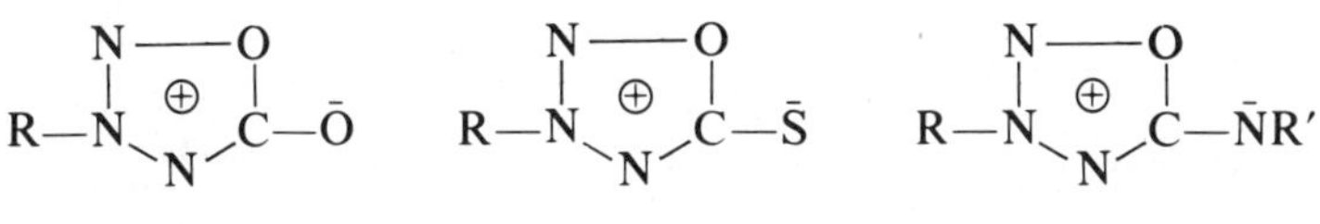

Alkyl, cycloalkyl, aryl, and saturated nitrogen heterocyclic groups have been attached to N-3 of the oxatriazole ring. On the basis of boiling and melting points and incidental observations, the oxatriazolium olates and thiolates appear to be heat resistant at least to 100°, but presently there are no data to determine an upper temperature stability limit, nor have systematic studies of their thermal stability been made. Although 3-phenyloxatriazolium-5-aminide decomposes on melting, its 5-anilide and 5-benzoylaminide appear to be stable at their melting points.

Salts of the oxatriazoles include a fluoborate of the 3-phenyl-5-ethylthiooxatriazolium cation as well as the hydrochlorides of oxatriazolium 5-aminides. The decomposition points of over 160° for the hydrochlorides indicate moderate stability, as does that of the mesoionic iodide of 3-(*N,N*-dimethylpiperidinium-4-yl)oxatriazolium-5-olate.

* The definitive review of the mesoionic compounds is that of Ollis and Ramsden (826).

Table 2.56. Oxatriazoles

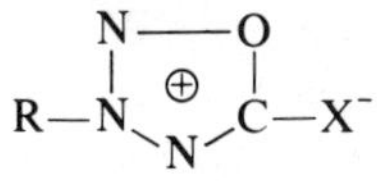

R	X^-	Melting Point (°C)	Thermal Stability	Ref.
CH_3	O	Bp 101–101.5/3 11	Unchanged on reflux in abs. ethanol 3 hr	438
$C_6H_5CH_2$	O	46.5–47.5	Unchanged on reflux in water 1 hr	321
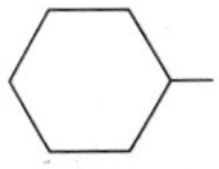	O	73.2		139
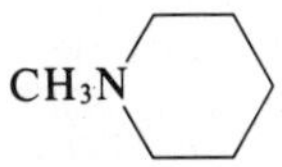	O	122–123.5		1087
C_6H_5	O	85		209, 310
C_6H_5	S	100		421
C_6H_5	NH	60–62 dec.; 80–83 dec.	Can be kept only a few hours	183, 209, 718
C_6H_5	NC_6H_5	110–111		422
C_6H_5	$NCOC_6H_5$	165–167; 166		183, 209, 718

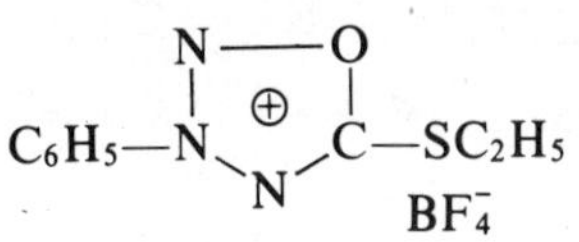

mp 90° (421)

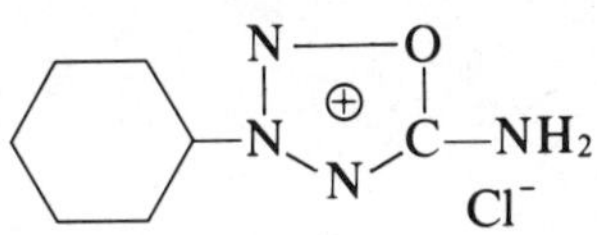

mp 162–163° dec. (208)

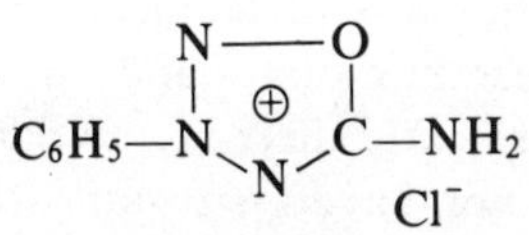

mp 191–192° dec. (718)

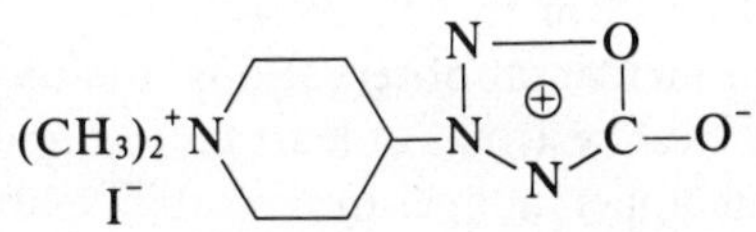

mp 218–220° dec. (1087)

2.3.7. Thiatriazolines and Thiatriazoles

The 1,2,3,4-thiatriazoles are five-membered heterocycles consisting of three sequential nitrogen atoms, a sulfur, and a carbon atom.* Both the thiatriazoles and the thiatriazolines have an unsaturated nitrogen-nitrogen linkage.

*Thiatriazoles were reviewed in 1964 by Jensen and Pedersen (562) and updated by Holm in 1976 (499).

Thiatriazolines are substituted with a divalent group at the ring carbon, numbered 5, and a monovalent radical at either the 2- or 4-position. Salts of the 4,5-thiatriazolines also are known.

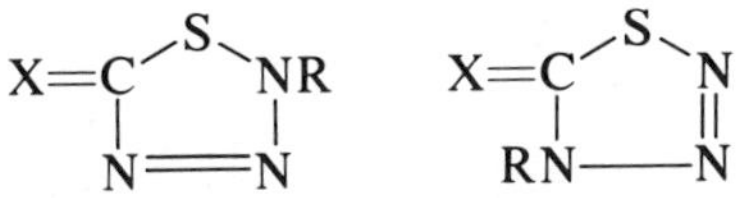

Presently known, 2,5-thiatriazolines have a hexafluoroisopropylidene group at position 5 and an aryl group at position 2. Thermolysis of the 2,5-thiatriazolines at 180–210° forms 2,1-benzisothiazoles (898).

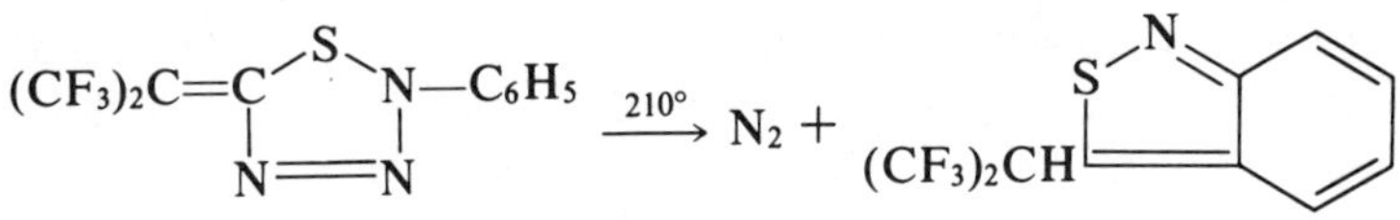

mp 125–125.5°

The 4,5-thiatriazolines prepared thus far are substituted with an imine at the

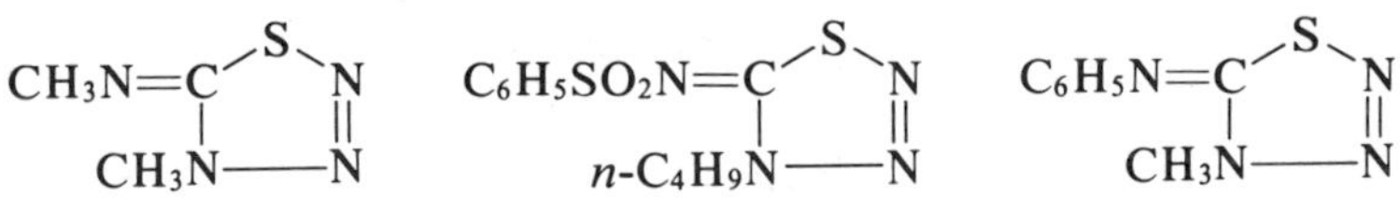

oil (1098) slow dec. 40–60° | mp 60–61° dec. (626) | mp 68–70° (631) dec. 110° (630)

ring carbon and a 4-alkyl group. They are less stable thermally than the 2,5-thiatriazolines. Heating 4-alkyl-5-alkylimino-1,2,3,4-thiatriazolines releases nitrogen

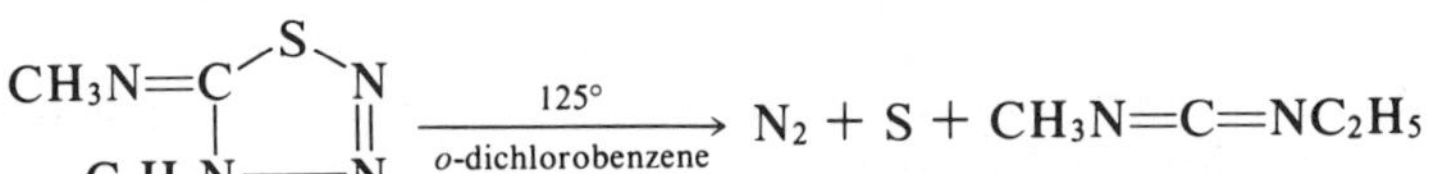

16%

rapidly at 125° and in addition forms carbodiimides, and sulfur (1098). Presence of sulfonyl carbodiimides from decomposition of 4-alkyl-5-sulfonylimino-

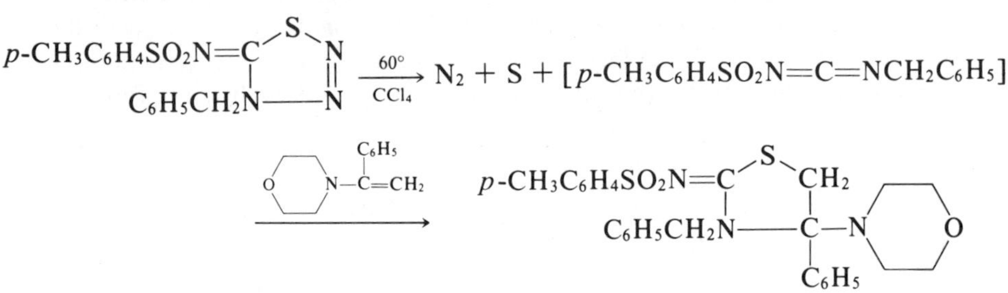

67%

1,2,3,4-thiatriazolines at 45–80° is inferred from trapping experiments with enamines. The half-life of 4-benzyl-5-*p*-tosyl-imino-1,2,3,4-thiatriazoline at 60° in carbon tetrachloride is 27.5 min (626). Thermolysis of 4-alkyl-5-arylimino-

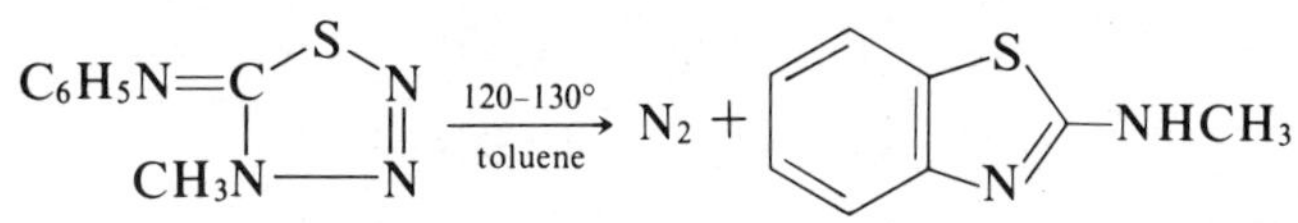

55%

thiatriazolines produces 2-alkylaminobenzothiazoles (782). Fluoborate salts of 4-alkyl-5-imino-1,2,3,4-thiatriazolines are more stable than the free bases. Although the protonation position in the salts is not specified (1098), the most probable location is the exocyclic nitrogen at position 5.

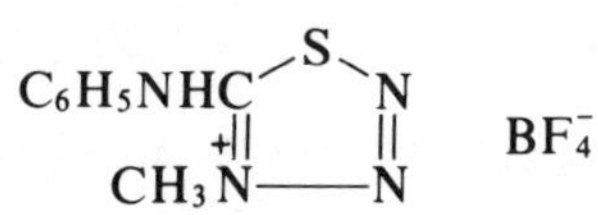

mp 162° dec.

Aromatic thiatriazoles have been substituted at the ring carbon with alkyl, aryl, amino, oxygen, and sulfur groups and chlorine. Formerly, thiatriazoles

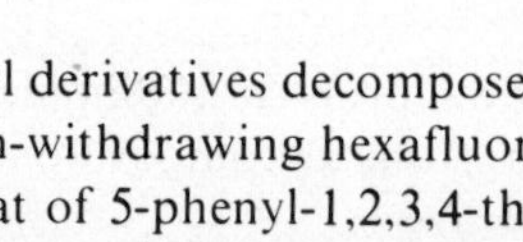

were considered to be thioacyl azides, but the absence of an azide band in the infrared, chemical evidence (666, 670), and X-ray crystallography (833) establish the cyclic structure.

The 5-unsubstituted alkyl derivatives decompose at or below room temperature (1180), but the electron-withdrawing hexafluoroisopropyl group increases heat resistance to about that of 5-phenyl-1,2,3,4-thiatriazole. Insufficient data are available on thiatriazoles bearing substituted phenyl groups to determine any effect on thermal stability of electron-releasing or electron-withdrawing substituents. Representative 5-alkyl and aryl thiatriazoles are listed in Table 2.57.

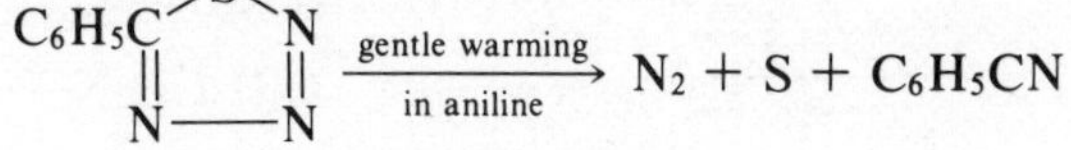

95% 63%

Table 2.57. 5-Alkyl and Arylthiatriazoles

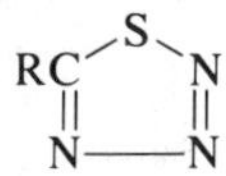

R	Melting Point (°C)	Thermal Stability	Ref.
$C_6H_5CH_2$	2	Dec., room temp.	588
$(CF_3)_2CH$	60.5–61	Dec. 100°	897
C_6H_5	94.5 dec.	Stable 4 months at room temp.; dec. 5 days in CCl_4, room temp.; $t_{1/2}$ (61.1°, bromobenzene) 7.0 hr	588, 670 502
p-$CH_3OC_6H_4$	103–104	Dec. 206° in tetralin	1013
p-ClC_6H_4	97–98		1013

Thermal decomposition of 5-phenylthiatriazole is exothermic and forms nitrogen, sulfur, and benzonitrile (588).

On the basis of ^{13}C and ^{1}H NMR, 5-amino and substituted aminothiatriazoles (Table 2.58) exist exclusively in the amino, rather than the tautomeric imino form, at least in solution in deuterated dimethyl sulfoxide, acetone, or chloroform (627). The 5-aminothiatriazoles decompose on standing at room temperature, but in differential thermal analysis do not exhibit an exotherm below their melting points.

Unsubstituted 5-aminothiatriazole is very unstable; it cannot be purified by recrystallization, because on even slight heating of its solutions, sulfur is deposited (656). Arylaminothiatriazoles appear to be more thermally resistant than alkylamino derivatives. Decomposition of the aminothiatriazoles follows the same reaction path as the 5-arylthiatriazoles, liberating nitrogen and sulfur and forming a cyanamide (329).

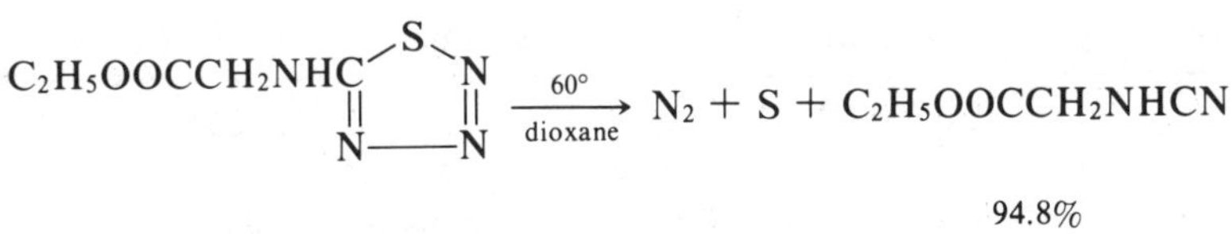

Ethers of the 5-thiatriazoles (Table 2.59) are among the least heat-resistant subclasses of this ring system. Almost all members are unstable at ambient temperature and many decompose below 0°.

Table 2.58. 5-Aminothiatriazoles

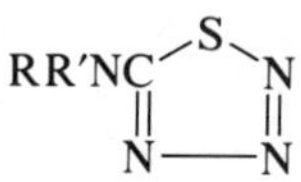

RR′N	Melting Point (°C)	Thermal Stability	Ref.
H_2N	128–130; expl. 130	DTA exotherm at mp, 80% weight loss	330, 616, 656
CH_3NH	96		616
$C_6H_5CH_2NH$	80.5–81	Stable few hrs room temp. DTA exotherm at 87°	330, 656 1019
$(CH_3)_2N$	49–51		656
$C_2H_5OOCCH_2NH$	69.0–69.5	$t_{1/2}$ (60°, dioxane) 2.2 hr	329
$(CH_3)_3C—C(C_6H_5)=N$	63–64	Dec. reflux in benzene	532
C_6H_5NH	142–143 dec.; 148–150 dec.	DTA exotherm 149°	330, 667 1134
p-$(CH_3)_2NC_6H_4NH$	140		668
p-$O_2NC_6H_4NH$	152–153		668
$C_6H_5SO_2NH$	132–133		716
$C_6H_5N(C(=O)CH_3)$	114		678

Table 2.59. 5-Thiatriazole Ethers

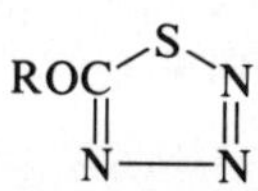

RO	Melting Point (°C)	Thermal Stability	Ref.
CH_3O	Expl. dec.	Stable few days at −30°; $t_{1/2}$ (25°, dibutyl phthalate) 1.5 hr	557 558
$(CH_3)_3CO$	Bp 40–43/0.1, sl. dec.	Can be stored months at −30°; rapid dec. room temp.	559
C_6H_5O	33–34	dec. in solution, 0° and higher; $t_{1/2}$ (25°, dibutyl phthalate) 1.5 hr	714 557
p-$C_6H_5C_6H_4O$	102–104, violent dec.	Unstable in benzene, room temp.	449

Based on ^{13}C NMR (627) and infrared (658) evidence, the structure of the nominal 5-mercaptothiatriazole is actually 5-thionothiatriazoline. It can be kept

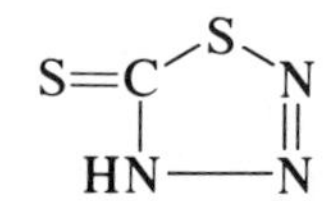

white crystals (1001)
expl. 50–60° (556)

1 or 2 days below 10° without significant decomposition. The thioethers and other thiatriazoles with sulfur at the 5-position (Table 2.60), however, are derivatives of the thiol (207). Their stability is poor at room temperature, but slightly better than that of the 5-oxygen derivatives. The considerable instability of thiatriazoles substituted in the 5-position with electronegative elements is further illustrated by the disulfide and the chloride.

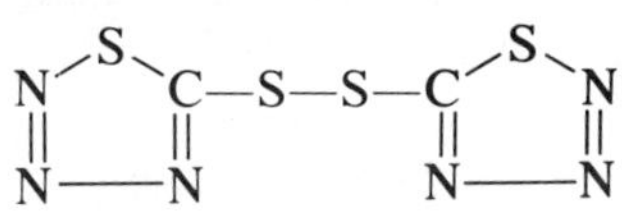

white solid
expl. > 40° (666, 1001)
gradual dec. room temp. (163)

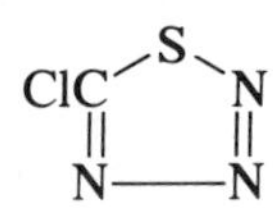

oily solid at 0°
very expl. (663)

Charge on the thiatriazole ring enhances its stability. Mesoionic thiatriazoles presently known are substituted with phenyl at position 3 and with carbon, nitrogen, oxygen, and sulfur groups bearing a negative charge at thiatriazole carbon. All are more stable than 5-monosubstituted thiatriazoles.

Known thiatriazole salts are substituted at the 3-position with alkyl and aryl groups and exhibit stability comparable to the mesoionic thiatriazoles. Holm

Table 2.60. 5-Thiatriazole Sulfur Derivatives

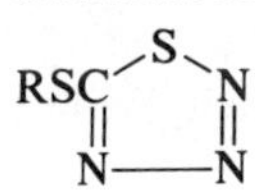

RS	Melting Point (°C)	Thermal Stability	Ref.
CH_3S	34		655
$C_6H_5CH_2S$	66; 66–67 dec.	Stable few days at 0°, few hrs at room temp.	1019, 666
C_6H_5S	64–65	Dec., heating in xylene	714
C_6H_5COS	93–94 dec.	Dec., 46° in $CHCl_3$	207
NCS	84–85 dec.	Dec., reflux in CS_2	666

(499) notes that alkylation at the 3-position blocks the ability of the system to eject nitrogen under mild conditions, and 3-arylation and oxide formation

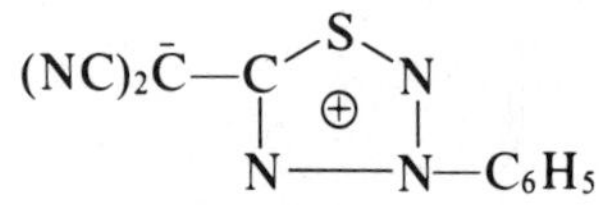

mp 200° dec. (420)

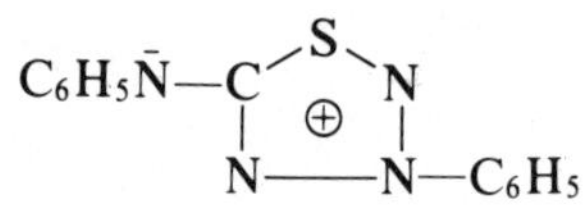

mp 125° (423)

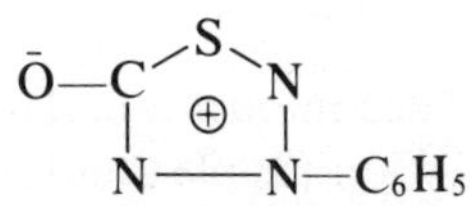

mp 99° (421)

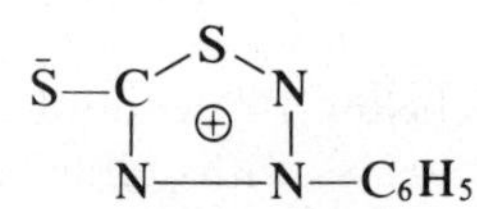

mp 128° (421)

would be expected to have a similar effect. The 3-oxide of 5-phenylthiatriazole is unstable above 185°, but is unaffected on reflux in bromobenzene (156°). Its

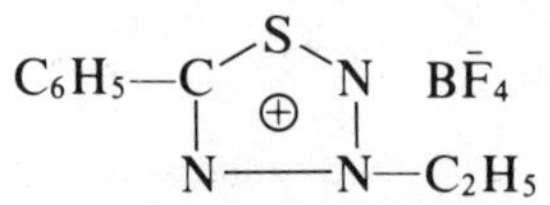

mp 87.5–88.5° (500)
stable to 180–200°

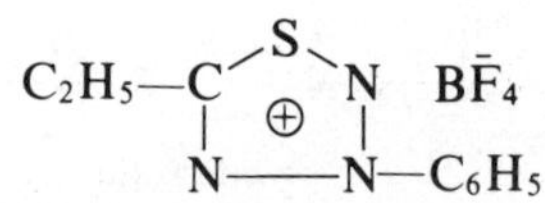

mp 106° (421)

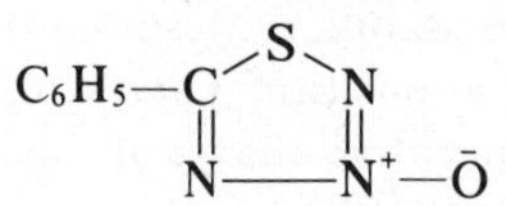

mp 133–134° (501)

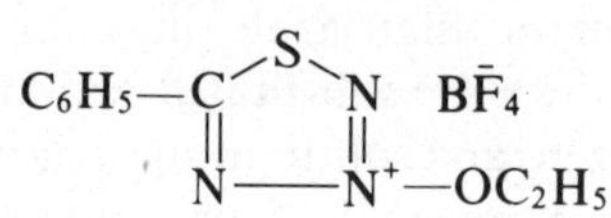

mp 102–103° (501)

O-ethyl derivative, however, decomposes on heating slightly above its melting point.

The 6*a*-λ^4-thia-1,2,3,4,6-pentaazapentalenes, fused ring thiatriazoles with tetravalent sulfur, are stable at their melting points.

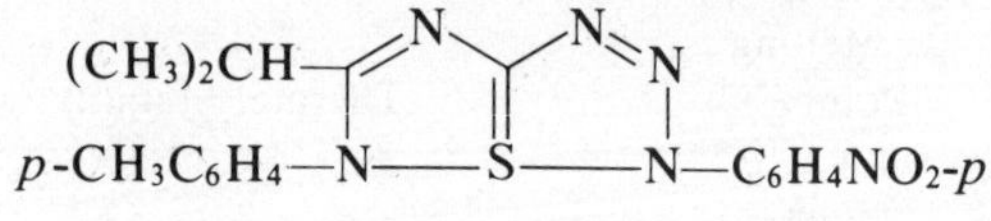

mp 78–81° (632)

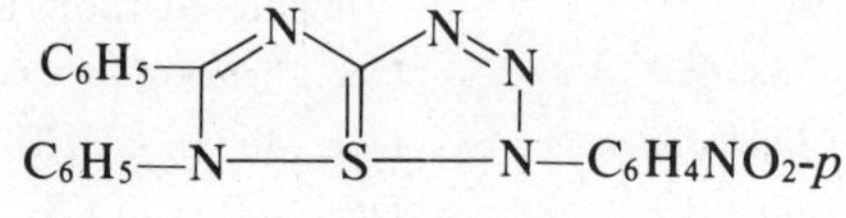

mp 223–225° (632)

2.3.8. Other Three-Nitrogen Heterocycles

Less extensively studied heterocycles with three-nitrogen chains include those with larger rings and rings with sulfur and phosphorus hetero atoms.

The presently known seven-membered heterocycles with three adjacent nitrogen atoms are the dibenzo[*d,f*][1,2,3]triazepines and a bicyclotriazanonatriene. Dibenzotriazepine isomerizes to benzo[*c*]cinnoline *N*-imide on refluxing

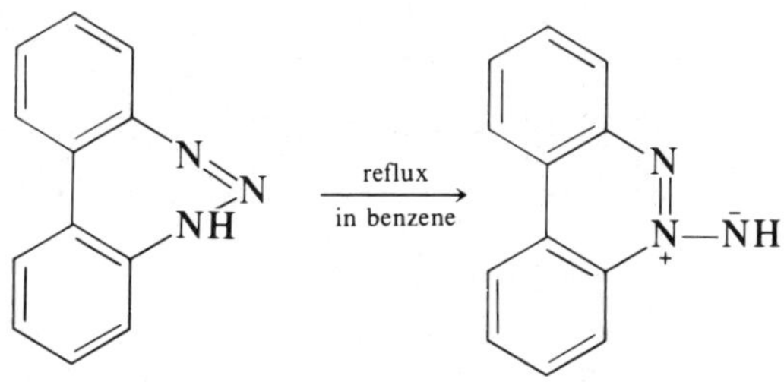

mp 99–100° (360, 361)

in benzene without disrupting the three-nitrogen chain. Heating either dibenzotriazepine or its *N*-methyl derivative in diglyme leads to the corresponding car-

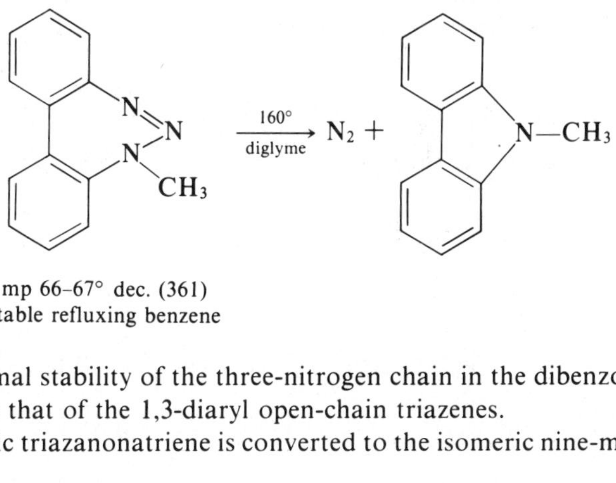

mp 66–67° dec. (361)
stable refluxing benzene

bazole. Thermal stability of the three-nitrogen chain in the dibenzotriazepines approximates that of the 1,3-diaryl open-chain triazenes.

The bicyclic triazanonatriene is converted to the isomeric nine-membered 1-

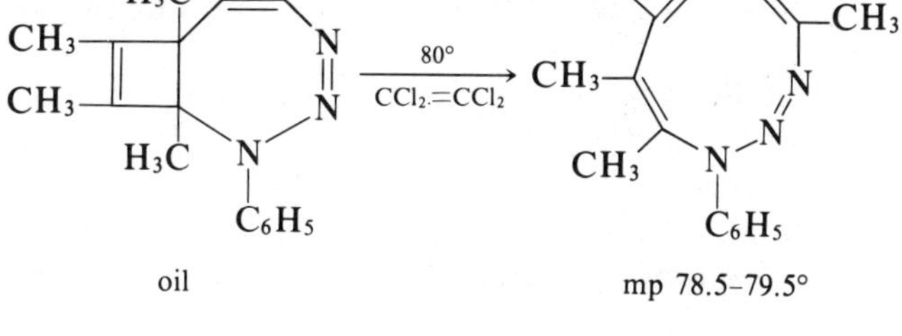

oil

mp 78.5–79.5°

phenyl-1,2,3-triazonine, which is unchanged after an hour at 195° in decalin (845, 846).

In refluxing diglyme, the 14-membered bistriazene forms carbazole in 77% yield (361).

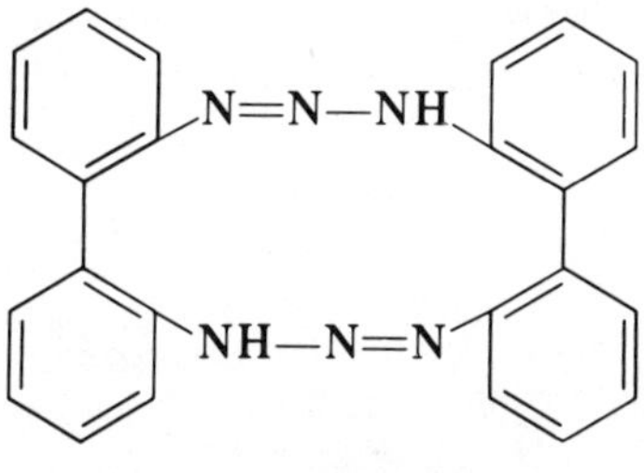

mp 210° dec.

4-Amino-6-cyanoguanidino-1,2,3,5-tetrazine is less stable than aromatic *v*-triazines, but its potassium salt does not decompose until 225° (615). Melting points have been reported for the triaryl-4*H*-1,3,4,5-oxatriazines (359), but no study of their thermal stability has appeared. The unusual 2,2-dihydro-1,3,4,5,2-thiatriazaphosphorines decompose at their melting points, but have a range of stabilities comparable to that of open-chain disubstituted triazenes without hydrogen on nitrogen (867). Dihydrothiatriazaphosphorines are stable

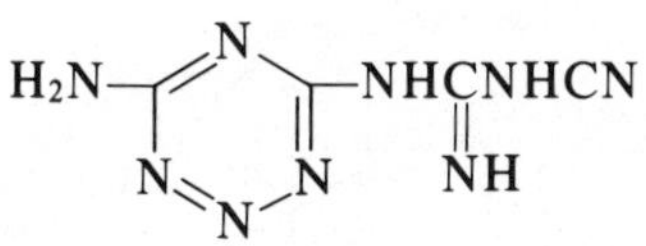

expl. 130°

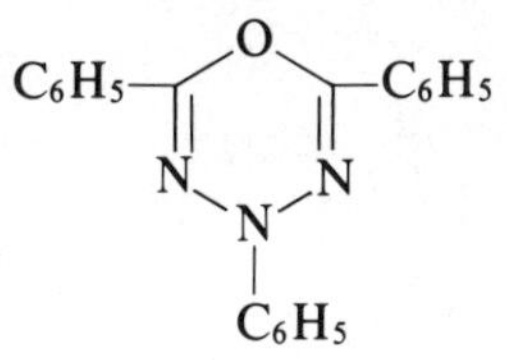

mp 172°

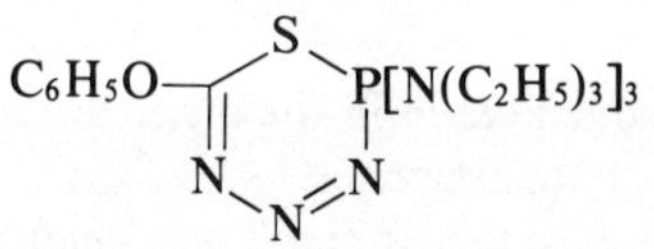

mp 71–72° dec.

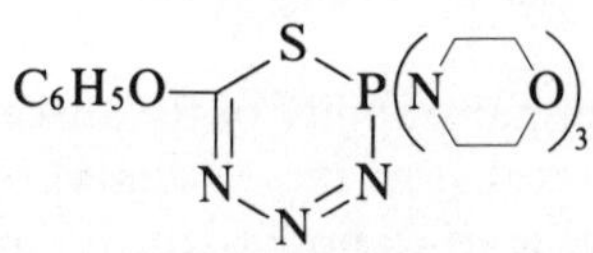

mp 164–165° dec.

at room temperature and have been stored for more than 8 years without decomposition.

The 1,2,3,4-benzothiatriazine 1,1-dioxides may be considered to be cyclic 1-sulfonyl-3-aryl triazenes and are structurally similar to the benzotriazinones. Most decompose at their melting points. Benzothiatriazine dioxides with hydrogen on the nitrogen chain have appreciably higher decomposition points than acyclic 1-aryl-3-sulfonyl triazenes, which disintegrate below 100°. Stability of 2-phenylbenzothiatriazine 1,1-dioxide is comparable to acyclic acylated diaryl triazenes but much inferior to the analogous 3-phenylbenzotriazinone.

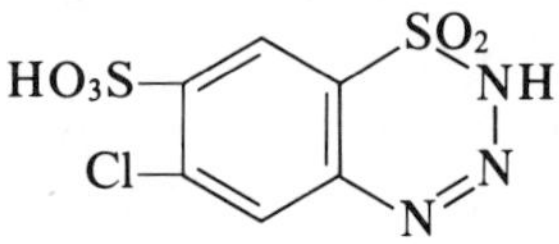

mp 260° dec. (1117)

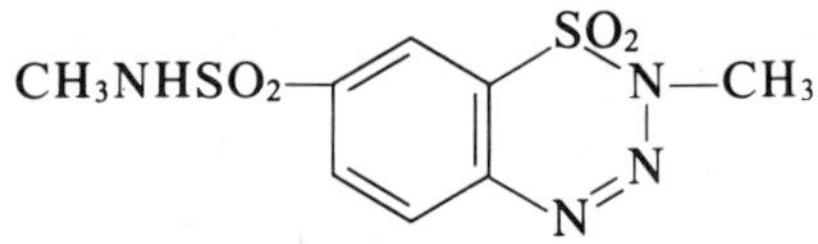

mp 176° dec. (206)

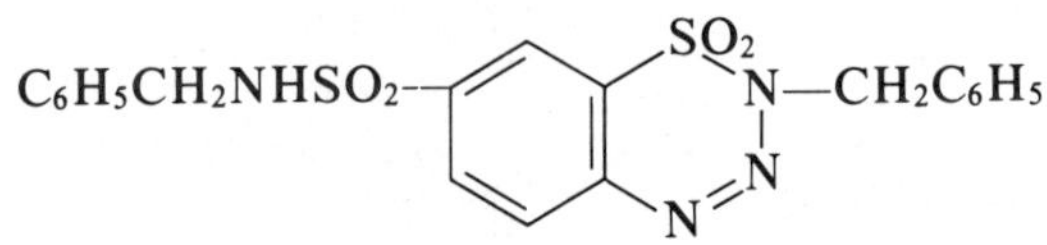

mp 158° dec. (206)

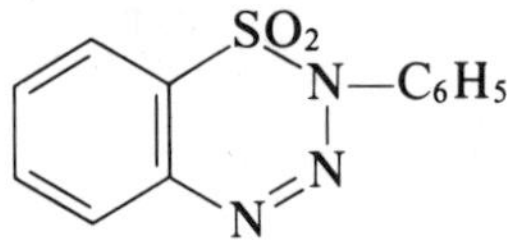

mp 111° dec. (1115)

Similarly the 2-alkyl benzothiatriazine dioxides are less thermally resistant than 3-alkyl benzotriazinones.

2.3.9. Embedded Three-Nitrogen Chains

Chains of nitrogen atoms in ring systems are more restricted in kinetic motion than those in acyclic compounds. There are further constraints on the motion of contiguous nitrogens when any of these nitrogen atoms are part of the fusion of two or more rings. Both to illustrate such structural types and to consider the effect on thermal stability, if any, of embedded or transannular nitrogen chains, these compounds are discussed separately. Chemicals with embedded nitrogen sequences consist of those with a single end or internal nitrogen atom in the chain in multiple rings, and those with two or more nitrogen atoms in the chain, common to two or more rings.

Examples of each type of embedded chain are known in ring systems with three-nitrogen sequences at the triazane oxidation level. Thus there is an end nitrogen of the three nitrogen chains in two rings in both 1,3-diphenyl-3,4-dihydro[1,2,3]triazolo[1,5-*a*]benzimidazole and 3-phenyl-2,3-dihydro[1,2,3]triazolo[3,4-*b*]benzisoxazole. Several pyrazolo and *s*-triazolo[2,3-*b*]indazoles and

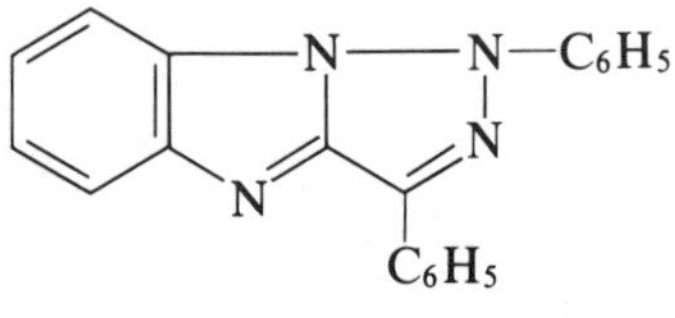

mp 148° (737)

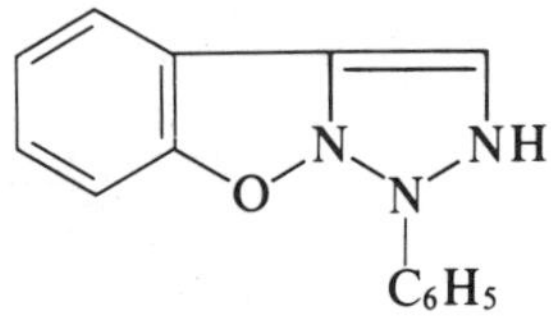

mp 178° (1192)

s-tetrazino[2,3-*b*:5,6-*b*′]diindazole contain embedded sequences of reduced three-nitrogen atoms in which the middle nitrogen is shared by two rings.

Two adjacent nitrogens of a transannular three-nitrogen chain are the points

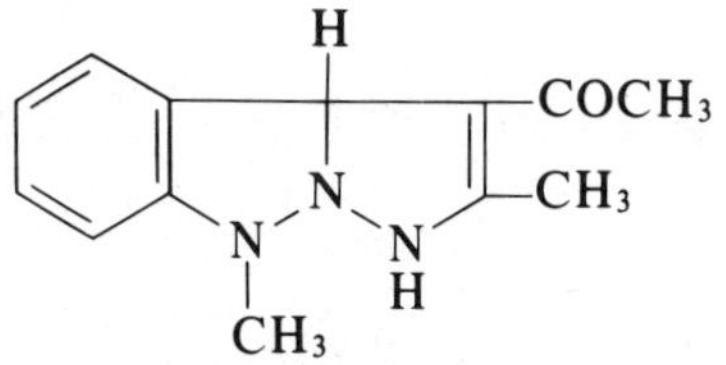

mp 72–75° (598)

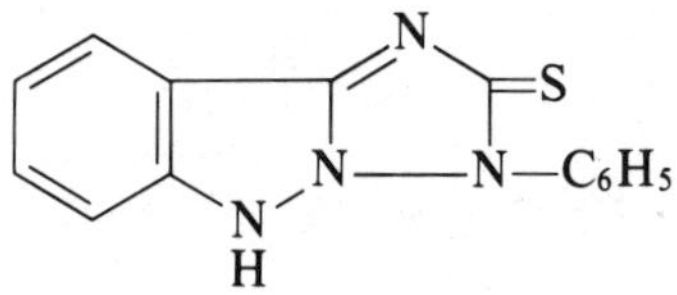

mp 268–273° (605)

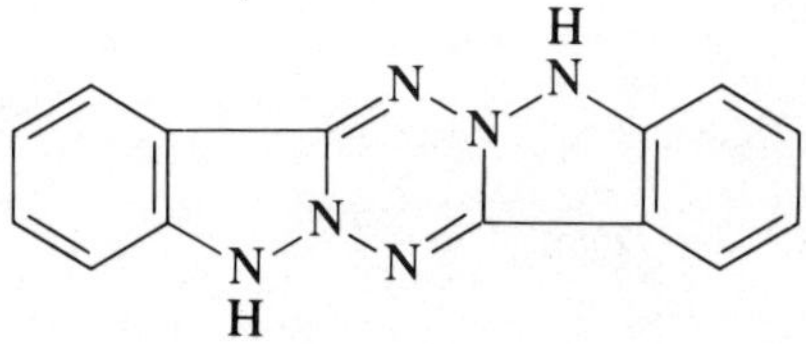

mp > 320° (1027)

of fusion of some polycyclic triazolines and all three sequential nitrogen atoms

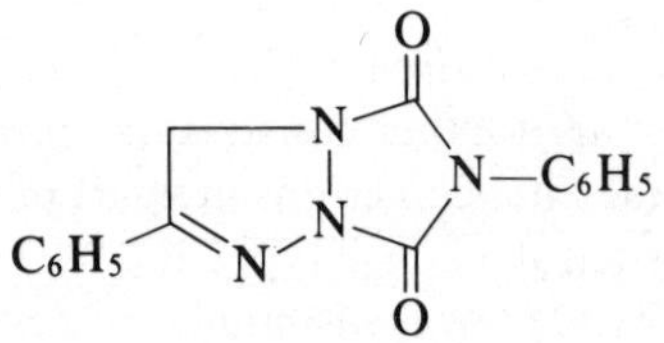

mp 173–174° (439)

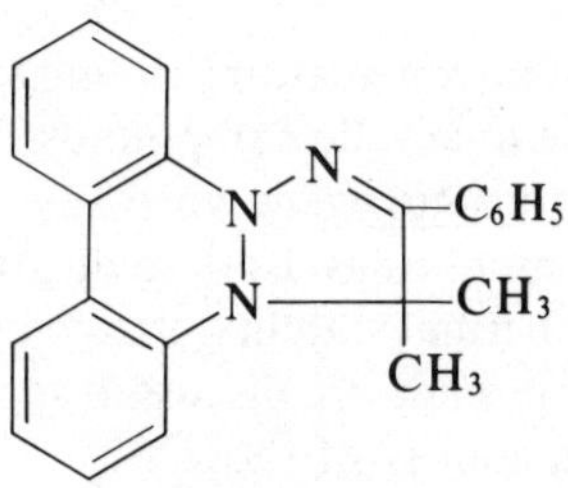

mp 163–165° (77)

are shared by multiple rings in 10-methyl-1-phenyl-2*a*,6*b*,10*c*-triazabenzo-[2,3]pentaleno[1,6-*ab*]indene.

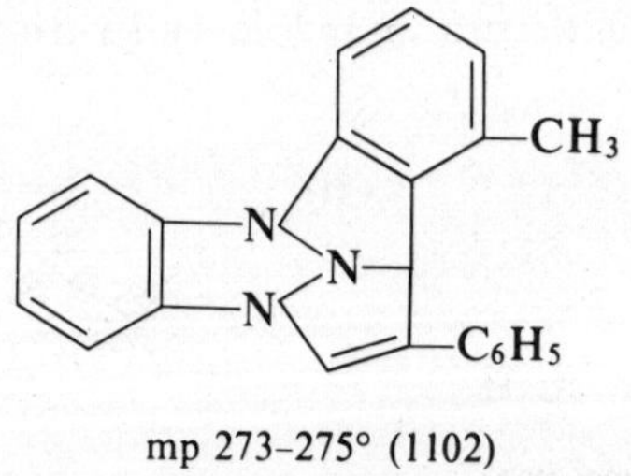

mp 273–275° (1102)

Very few thermal stability data are available for this series of reduced, embedded three-nitrogen chains. Heating 1-methyl-1,2-diphenylbenzo[*c*]-*v*-triazolo[1,2-*a*]cinnoline either in toluene (75) or at its melting point (77) eliminates benzonitrile and forms benzo[*c*]cinnoline. For most of these cyclic triazanes with one or more shared nitrogen atoms, however, there is no report

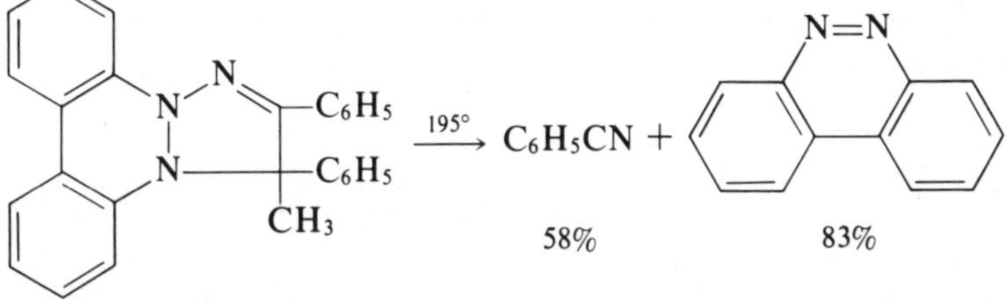

of decomposition at their melting points. Based on the generally (but not universally) higher melting points of these cyclic compounds with reduced three-nitrogen chains, and particularly when the middle nitrogen or two or three nitrogens are shared between multiple rings, it appears that these compounds with embedded triazane chains are more thermally stable than acyclic triazanes or Δ^2, Δ^3, or Δ^4 triazolines.

Embedded three-nitrogen chains in five-membered rings at the triazene or Δ^2-triazoline oxidation level are approximately as thermally stable as monocyclic Δ^2-triazolines. In refluxing benzene 5,13b-dihydro-5-methylene-1*H*-

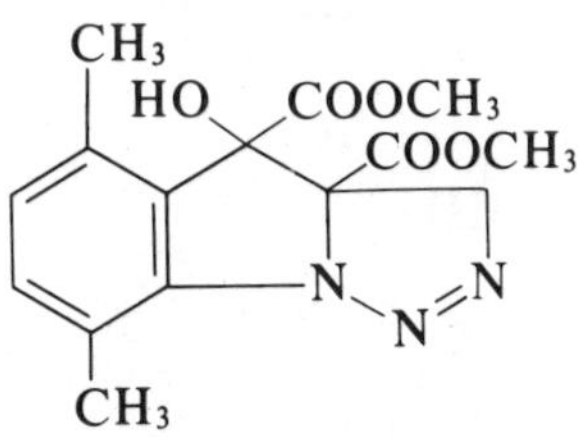

dec. ~60° (931)

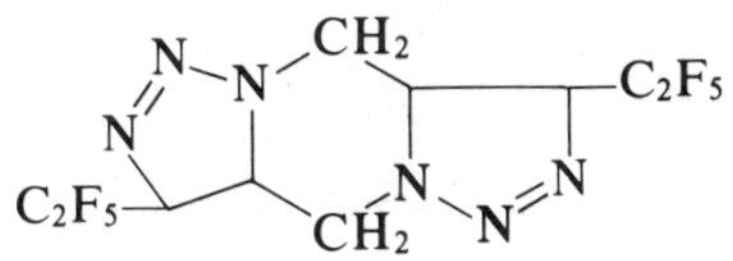

mp 186° dec. (704)

dibenzo[*c,e*]-*v*-triazolo[1,5-*a*]azepine forms the corresponding aziridine (837). The relatively lower decomposition points of these cyclic triazenes in compari-

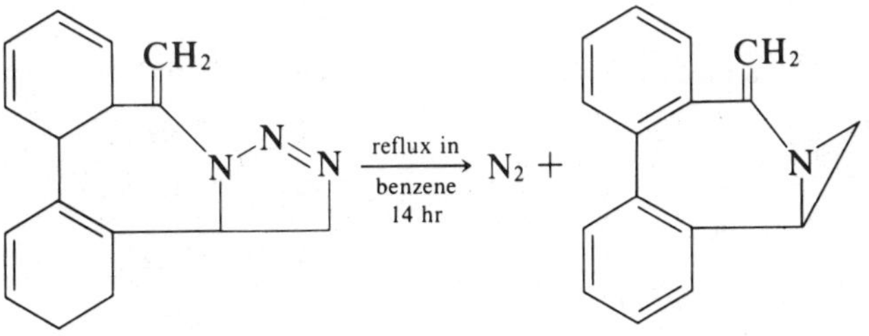

(IV)

mp 141–142°

son to those at the triazane level is related to the ease of losing nitrogen from the cyclic azo group.

Fused five-membered ring systems with two double bonds and three-nitrogen sequences with one or more nitrogen atoms shared between two rings are essentially *v*-triazoles annelated at carbon-nitrogen bonds. Fused *v*-triazoles with an end nitrogen of the embedded chain common to two rings also contain

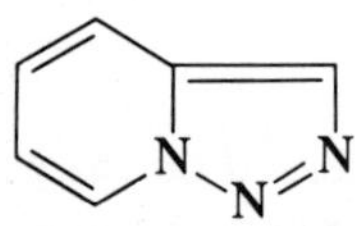

(V)

mp 39–40° (744)
bp 78°/0.09 (136)

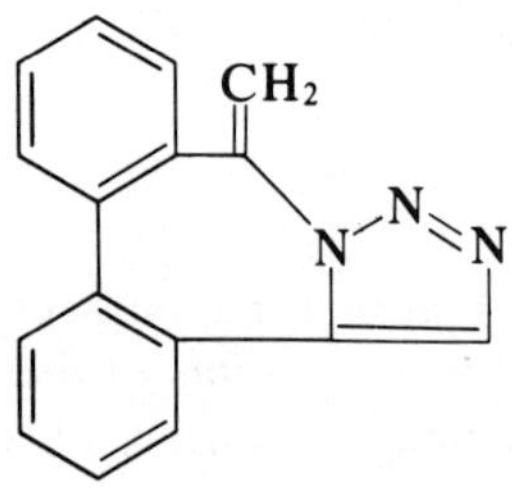

(VI)

mp 165–166° (837)

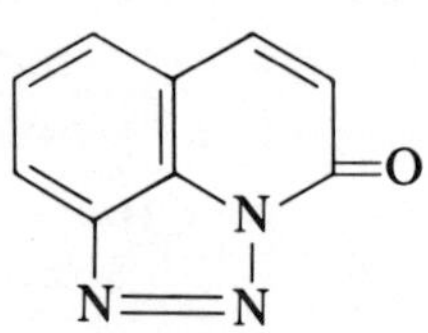

mp 208° dec. (419)

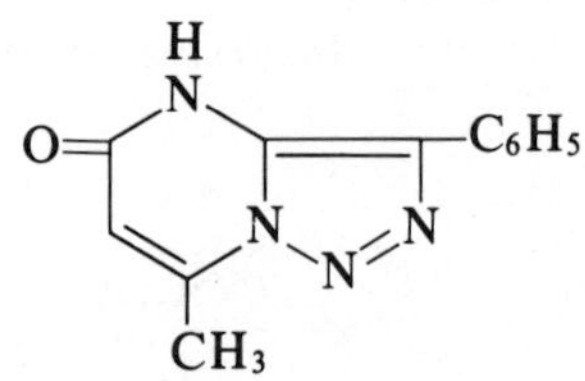

mp 240° (1059)

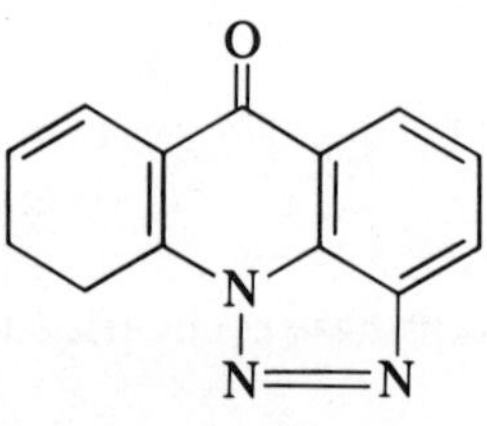

mp 266° dec. (214)

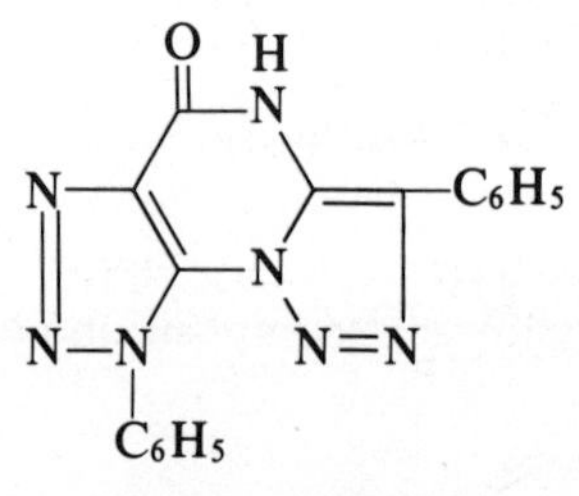

mp 327° (1057)

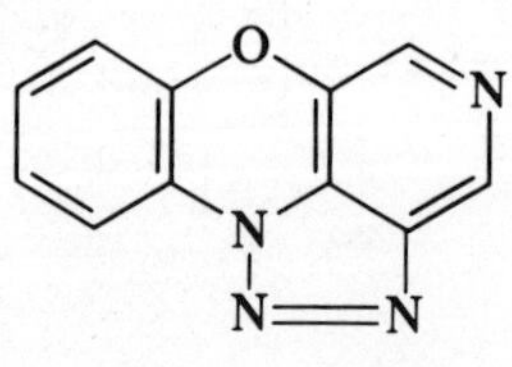

mp 171° dec. (863)

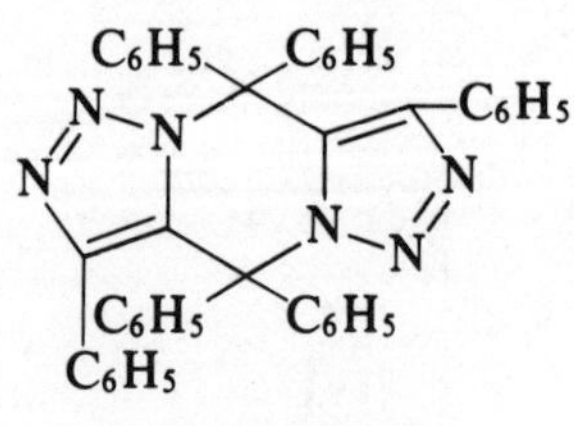

mp 277–280° (174)

an azo group conjugated with at least one double bond. They are generally more thermally resistant than triazolines with embedded nitrogen chains. Thus the reduced methylene dibenzotriazoloazepine **(IV)** decomposes on refluxing in benzene, but the aromatic analog **(VI)** survives these conditions during its preparation and gives no evidence of decomposition at its melting point (837). A lower limit of thermal stability of these aromatic systems with a shared end nitrogen of the embedded three-nitrogen chain is indicated by the half-life of 55 hours of *v*-triazolo[1,5-*a*]pyridine **(V)** at 180.5° in diphenyl ether (1151). The 1,2,3-triazolo[5,1-*d*][1,2,3,5]tetrazin-4-ones are less stable than either other fused *v*-triazoles with an end nitrogen shared by two rings, or 3-arylbenzotriazinones.

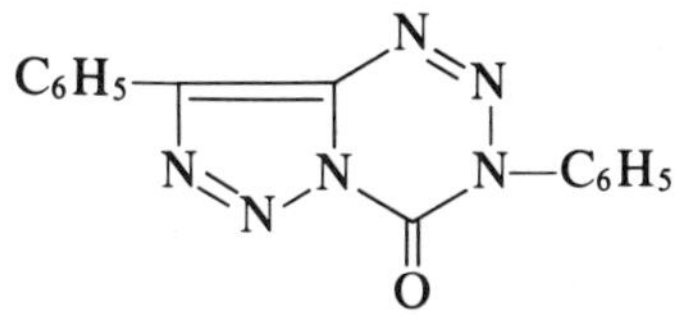

mp 140–142° dec. (288)

Compounds with the middle nitrogen of a transannular three-nitrogen chain shared between two rings and a cyclic conjugated system of four or more double bonds, also involving at least one of the end nitrogens of the chain, seem to be stable to over 250°. Both an end and the middle nitrogen of the embedded

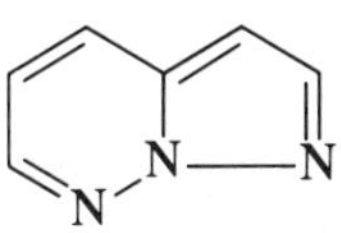

bp 270° (908)

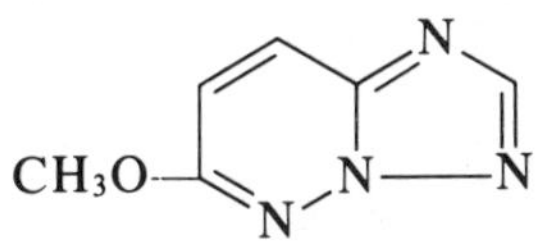

mp 178–180° (880)
sublimes 150–155°/5

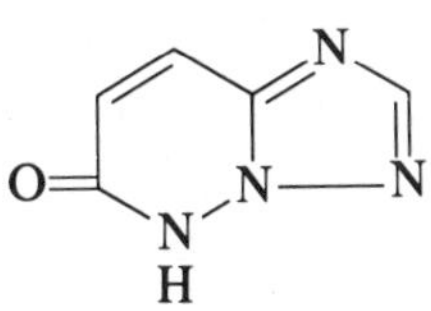

mp 156–157° (880)
boils 250–256°/5

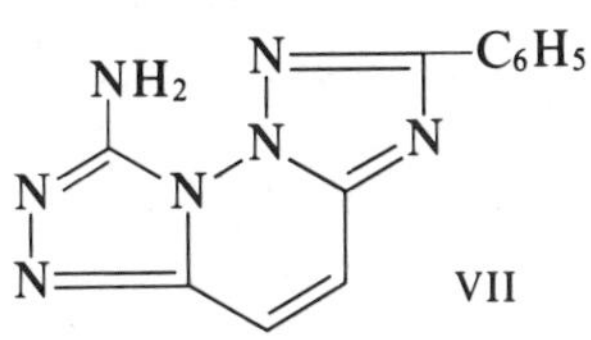

VII

mp 320–321° (1215)

three-nitrogen chain are shared between rings in the high melting, tricyclic *s*-triazolo[2,3-*b*]-*s*-triazolo[3,4-*a*]pyridazine **(VII)**.

Dihydrobenzotriazines with additional rings fused at the 3,4-positions contain an embedded three-nitrogen chain in a six-membered ring with an end nitrogen common to two rings. The presently known uncharged examples have a nitrogen at the 4-position of the benzotriazine moiety and are cyclic analogs of the 4-iminobenzotriazines. Decomposition temperatures range from 140° to over 200°, lower than the decomposition points of the 3-arylbenzotriazinones.

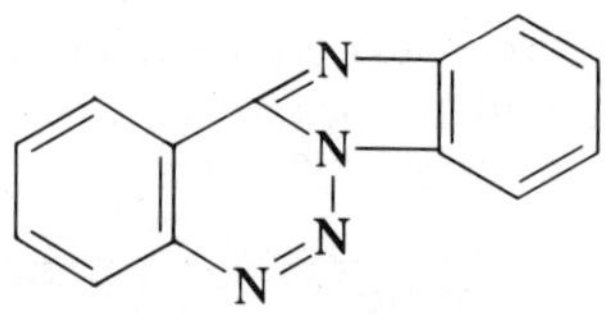

mp 207–208°, dec. > mp (805)
210.5–211° (1202a)

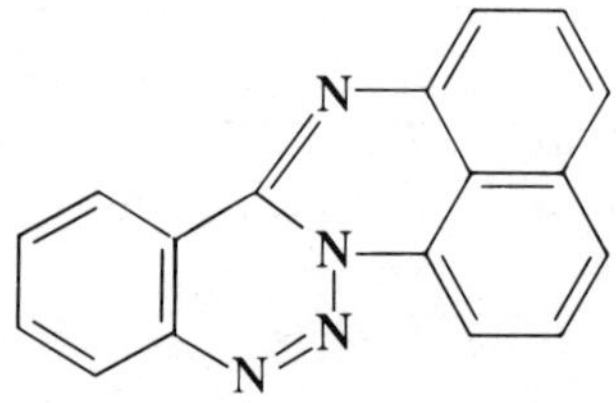

dec. 140°, weak expl. (942)

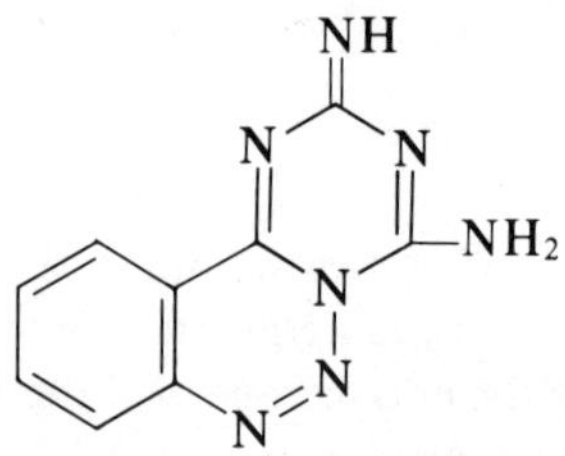

mp ~ 200° dec. (699)
dec., reflux in dioxane

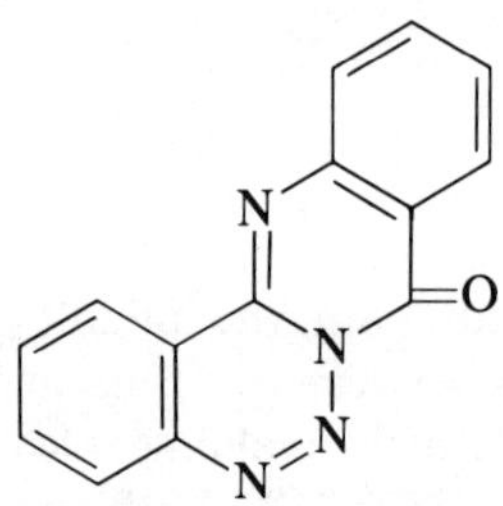

mp 216–218° (287)

In all these benzotriazine derivatives, the azo group of the nitrogen sequence is part of an extended conjugated system, which, however, does not confer any special thermal stability. Because of the constraint of the rings fused at the 3,4-positions in benzotriazine, thermal Dimroth rearrangement is improbable.

Presently known compounds with dipolar embedded three-nitrogen chains share two of the nitrogen atoms between two rings.* Although 7,8-dimethyldibenzo-1,3*a*,6*a*-triazapentalene decomposes below 200°, the 7-*p*-nitrobenzoyl derivative is resistant to over 300°; alternate resonance forms with charge on the

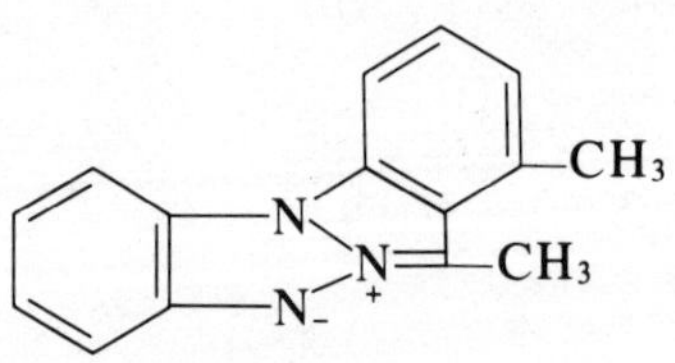

mp 190–191° dec. (1104)

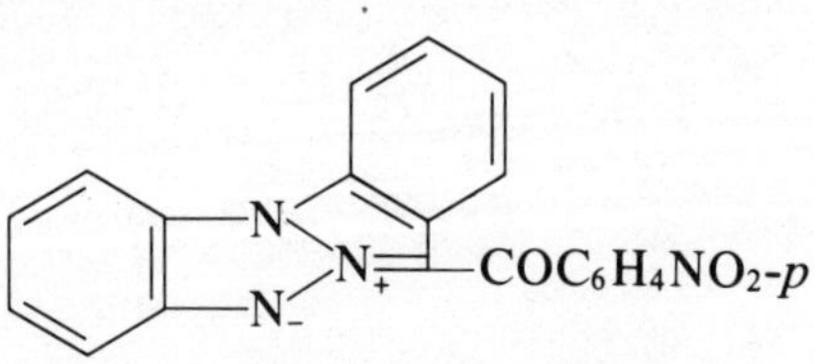

mp 304–305° dec. (1101)

* Triazapentalenes are among the mesomeric betaine derivatives of heteropentalenes reviewed by Ramsden (900).

carbonyl group no doubt contribute to the enhanced stability of the latter com-

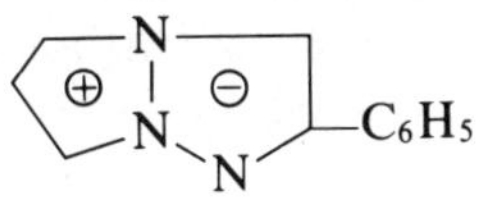

mp 86–87° (599)

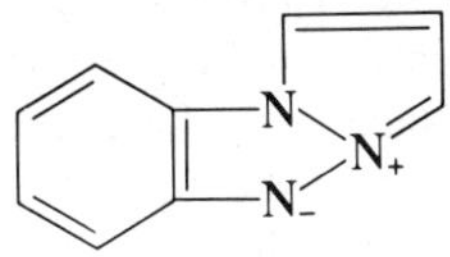

mp 102–103° (694)
103.5–104 (429)

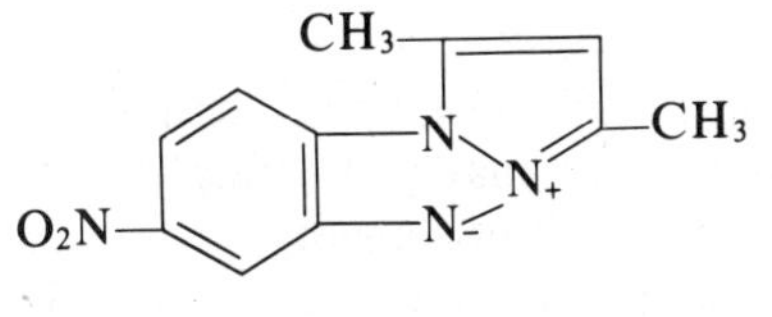

mp 227–228 (673)

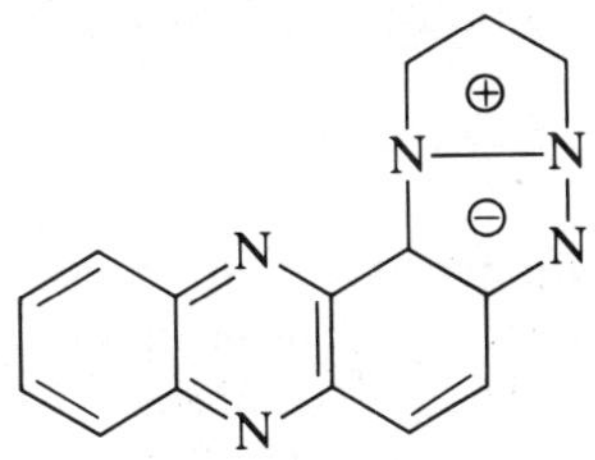

mp 255° (21)

pound. Stability data are lacking for 2-phenyl-1,3*a*,6*a*-triazapentalene, pyrazo-

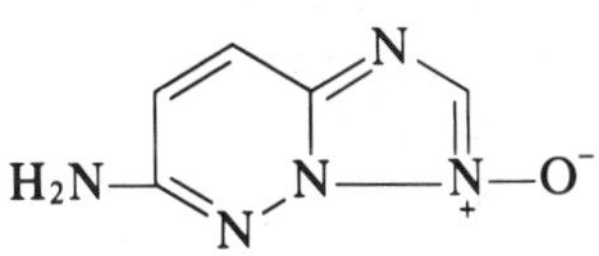

mp 263° (50)

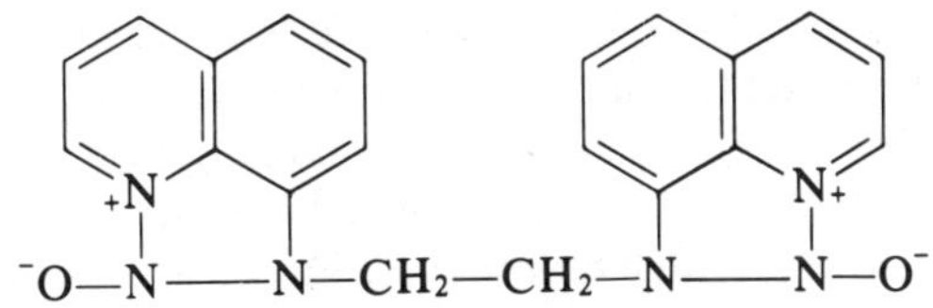

mp 182–183° (560)

lo[1,2-*a*]benzotriazole, 3′,5′-dimethylpyrazolo[1,2-*a*]-5-nitrobenzotriazole, and pyrazolo[1′,2′: 1,2]1,2,3-triazolo[5,4-*a*]phenazinyliumide.

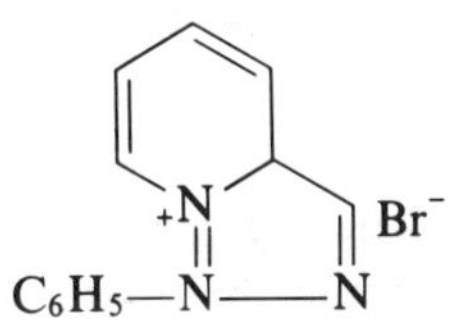

mp 163–164° (620)

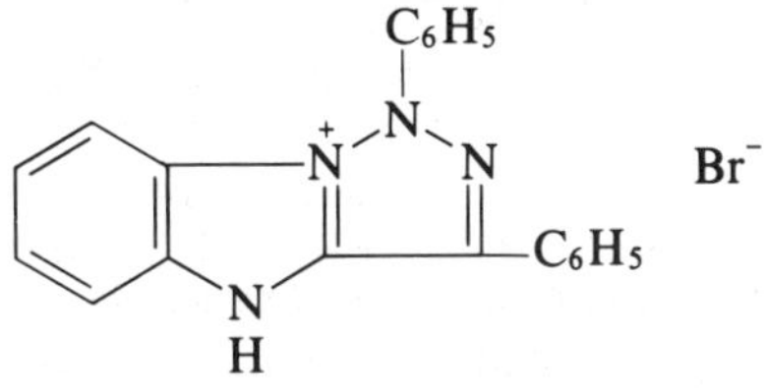

mp 195° (737)

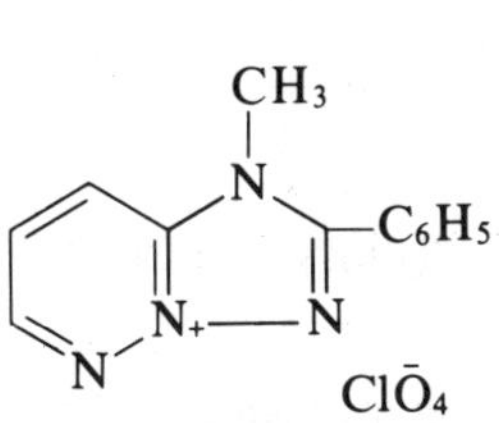

mp 195–199° (1215)

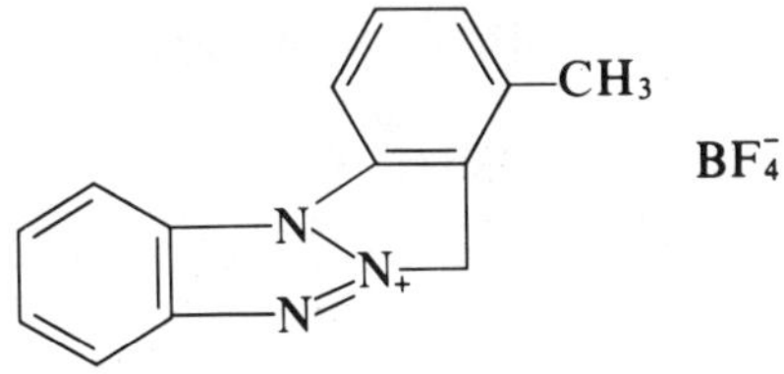

mp 252.5° dec. (1101)

Melting point data do not enable a conclusion about the stability of the few *N*-oxides of heterocycles with transannular three-nitrogen chains.

Salts of embedded three-nitrogen chain heterocycles with both an end and a middle nitrogen at the bridgehead are known. Their stability appears to be similar to that of benzotriazolium salts and to related uncharged three-nitrogen embedded heterocycles.

2.4. ACYCLIC FOUR-NITROGEN CHAINS

2.4.1. Tetrazanes

The acylic tetrazanes are constructed of four contiguous, completely reduced nitrogen atoms. Those known thus far are made of straight, rather than branched nitrogen chains, and carry methyl, trifluoromethyl, hydrogen, aryl, or acyl groups, often simultaneously with more than one substituent or substituent type from this list.

The parent compound, $H_2NNHNHNH_2$, is postulated to be formed by the bimolecular combination of hydrazyl radicals, $H_2NNH\cdot$, arising from pulse radiolysis of aqueous hydrazine (446). Since its decomposition is pH dependent in the range 8–11.5, the presence is suggested of singly and doubly protonated forms, $N_4H_7^+$ and $N_4H_8^{2+}$. The lifetime of the presumed tetrazane is very short; it decays unimolecularly to ammonia and triazene, N_3H_3.

Stability of the substituted tetrazanes (Table 2.61) ranges from hexaphenyltetrazane, which decomposes in 1 day at room temperature, to the hexa(trifluoromethyl) derivative, which is stable to 350°. Above this temperature, hexatrifluoromethyltetrazane disproportionates into tetratrifluoromethylhydrazine and perfluoroazomethane (243, 378).

$$(CF_3)_2N{-}\underset{\text{bp }100.4^\circ}{N(CF_3){-}N(CF_3)}{-}N(CF_3)_2 \xrightarrow{350\text{–}400^\circ} (CF_3)_2NN(CF_3)_2 + CF_3N{=}NCF_3$$

The 2,3-bond of many tetrazanes is easily broken with formation of two substituted hydrazyl radicals. Tetrazanes, hydrazyls, and their equilibrium conditions were extensively studied by Goldschmidt (387). Hexaphenyltetrazane is stable for several days at −80°, but at 0° is almost completely dissociated to triphenylhydrazyl, forming an intensely blue solution (389). Substitution of picryl

$$(C_6H_5)_2N{-}N(C_6H_5){-}N(C_6H_5){-}N(C_6H_5)_2 \underset{CHCl_3}{\rightleftarrows} 2(C_6H_5)_2N{-}\dot{N}{-}C_6H_5$$

groups at the middle nitrogen atoms leads to a material which exists both in the solid state and in solution only as a radical (389). Although instability prevents

Table 2.61. Acyclic Tetrazanes

Tetrazane	Melting Point (°C)	Thermal Stability	Ref.
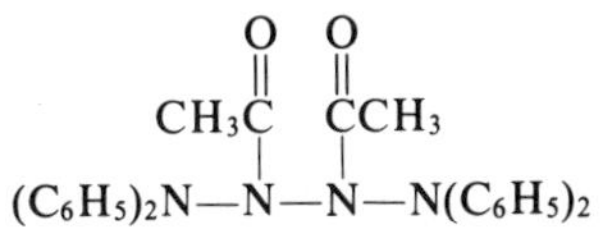	141 dec.	Sl. dissociation in $CHCl_3$ solution	386
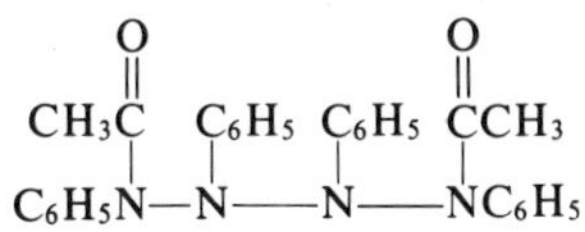	126 dec.	Sl. dissociation in $CHCl_3$ solution	388
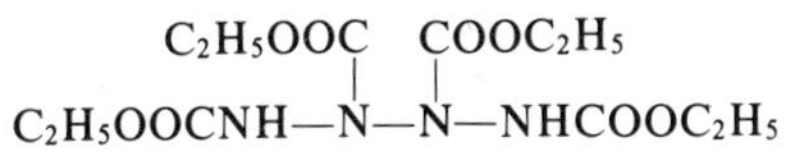	80	Dissociation > mp; disproportinates > 150°	959
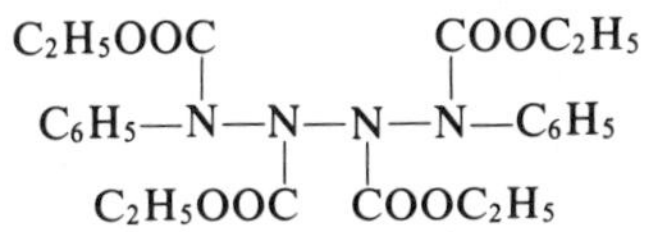	137.5–138	No dissociation at 200°	52

measurement, in chloroform 1,4-diacetyl-1,2,3,4-tetrakis(*m*-dimethylaminophenyl)tetrazane appears to be extensively dissociated (386).

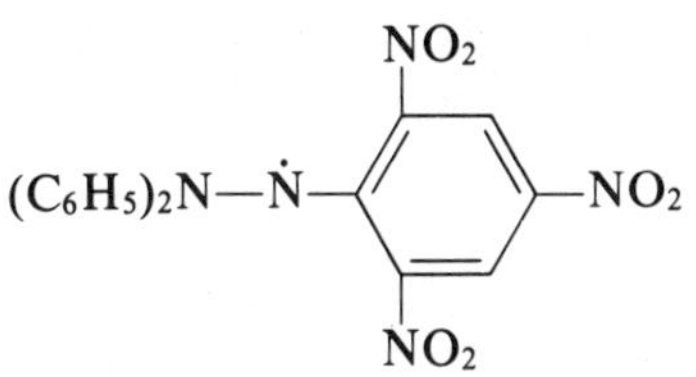

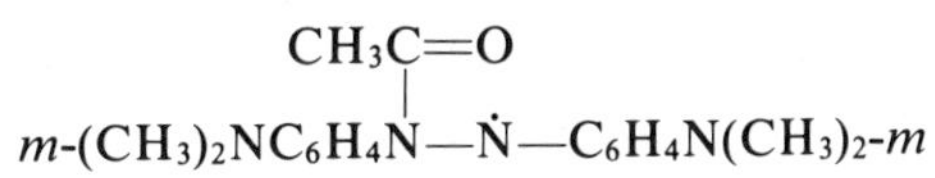

Diacyl tetraphenyl tetrazanes are more stable, but toluene solutions of 1,4-diacetyl-1,2,3,4-tetraphenyltetrazane turn bright green on heating, indicating the presence of hydrazyl radicals, and on cooling return to a weak yellow color; the color changes that occur on heating and cooling can be repeated indefinitely (388). Dissociation of the 2,3-diacyl-1,1,4,4-tetraaryltetrazanes requires less strenuous conditions. At −18° in chloroform, 2,3-dibenzoyl-1,1,4,4-tetraphenyltetrazane, at a dilution of 300, exists to the extent of 56% as a hydrazyl radical (386). The effect of increasing electron-withdrawing influence is to stabilize the tetrazane chain. Thus there is no dissociation of hexa(trifluoromethyl)tetrazane or 1,4-diphenyl-1,2,3,4-tetra(ethoxycarbonyl)tetrazane, and when the phenyl groups in the latter compund are replaced by hydrogen, forming 1,2,3,4-tetra-(ethoxycarbonyl)tetrazane, radicals only appear on heating above the melting point (959). In confirmation, quantitative study of the dissociation of the 1,1,4,4-tetraaryl-2,3-dibenzoyltetrazanes shows that as the σ constants of the aryl substituents became more positive, dissociation is less extensive; a linear Hammett ρ-σ relation is found (1188).

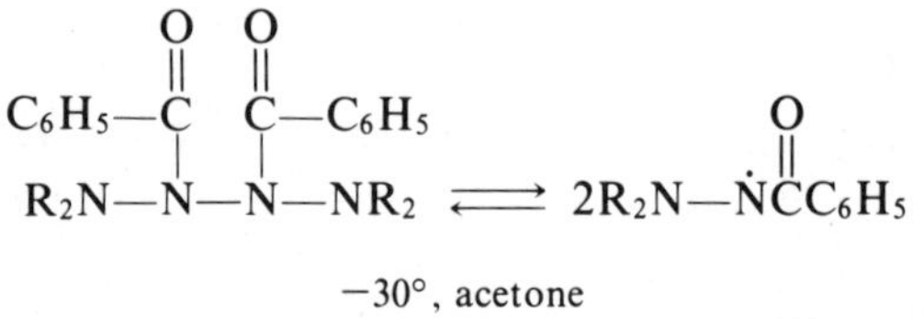

R	σ, *p*-substituent	$K \times 10^4$
p-$CH_3C_6H_4$	−0.14	53.6
C_6H_5	0	4.15
p-BrC_6H_4	+0.26	0.17

2.4.2. Tetrazenes

The two subclasses of the acyclic four-nitrogen chain with one double bond are the 1-tetrazenes, RN=NNR′NR″R‴, unsaturated at the end of the chain, and the 2-tetrazenes, RR′NN=NNR″R‴, unsaturated between the middle nitrogens. In addition, 2-tetrazenium salts and dipolar compounds also have been prepared.

The parent, 2-tetrazene, is a colorless solid, sublimable at −15° in a high vacuum and stable to 0° (1167). Since it is formed by hydrolysis of *trans*-tetrakis-(trimethylsilyl)-2-tetrazene at −78°, the resulting tetrazene probably also is in

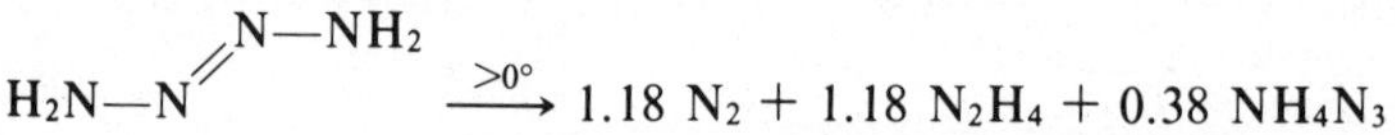

the trans configuration. Thermolysis of tetrazene proceeds along two paths, about 75% to nitrogen and hydrazine and 25% to ammonium azide.

Substituents on the 1-tetrazene chain include alkyl, aryl, heteroaryl, and acyl. Disubstitution is at the 1,3- or 1,4-positions and trisubstitution is at the 1,3,4-positions. All 1-tetrazenes (Table 2.62) decompose at their melting points, which range from 46 to 173°. The decomposition temperatures of 1,3-disubstituted 1-tetrazenes bearing an acidic group (carboxyl, tetrazole) are somewhat higher (108–165°) than those with neutral substituents (46–120°), possibly owing to dipolar structures from internal protonation of the tetrazene chain or intermolecular hydrogen bonding.

Structures of the 1,3-disubstituted 1-tetrazenes are indicated by formation of aldehyde derivatives on condensation with the terminal amino groups, a reac-

Table 2.62. 1-Tetrazenes

1-Tetrazene	Melting Point (°C)	Thermal Stability	Ref.
$C_6H_5N{=}N{-}N(CH_2C_6H_5){-}NH_2$	46–48 dec.		1193
$C_6H_5N{=}N{-}N(C_6H_5){-}NH_2$	70–73 dec.		1193
m-$HOOCC_6H_4N{=}N{-}N(C_6H_5){-}NH_2$	108–112 dec.	Dec. reflux in benzene	1193
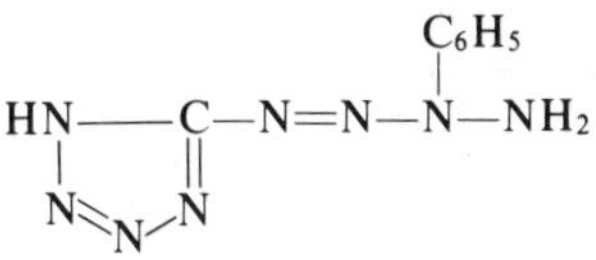	142 dec.		508
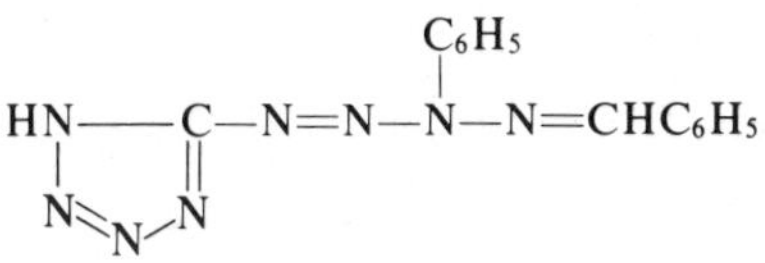	116–117 dec.		508
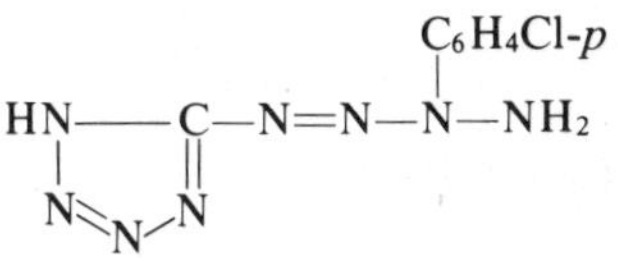	165 dec.		508
p-$O_2NC_6H_4N{=}NNHNHC_6H_4NO_2$-$p$	55 dec.		303
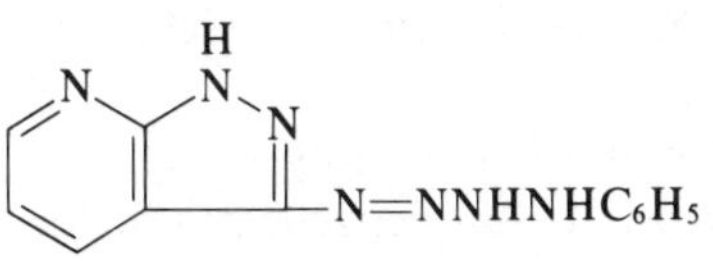	173–175 dec.		596
p-$CH_3C_6H_4N{=}NN(COCH_3){-}NHCOCH_3$	60 dec.		265
p-$O_2NC_6H_4N{=}NN(COCH_3){-}NHCOCH_3$	107 dec.	No visible change in one month when kept dry	265
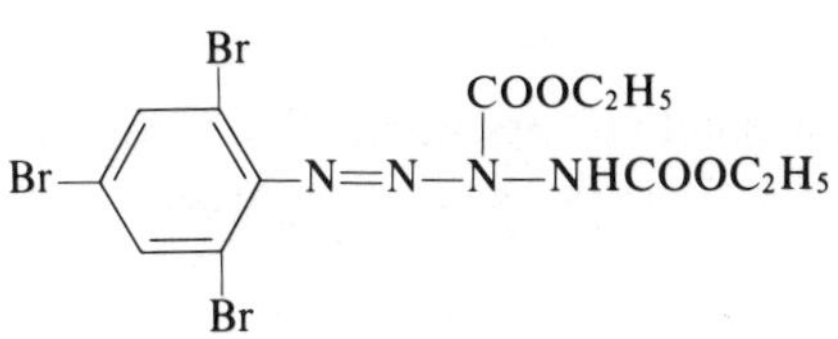	111–115 expl.		265

tion not possible with 1,4-disubstituted tetrazenes. Both 1,3- and 1,4-derivatives, however, yield azides and amines on thermal decomposition. Thus formation of phenyl azide from 1-*m*-carboxyphenyl-3-phenyl-1-tetrazene requires a

$$m\text{-}HOOCC_6H_4{-}N{=}N{-}\underset{\displaystyle |}{\overset{C_6H_5}{N}}{-}NH_2 \xrightarrow[1/2\ \text{hr}]{\text{reflux in benzene}} m\text{-}HOOCC_6H_4NH_2 + C_6H_5N_3$$

rearrangement (1193). Decomposition of 4-phenyl-1-pyrazolo[3,4-*b*]pyridin-3-yltetrazene leads to fission of all N—N and C—N bonds in the tetrazene chain and its substituents, with formation of reduction products of each substituent

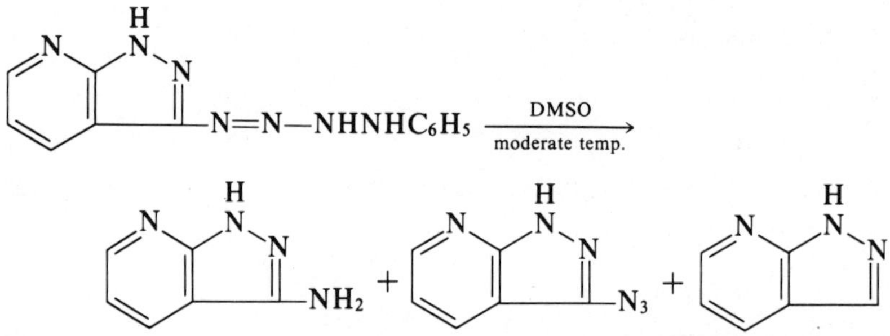

$$+ C_6H_5NH_2 + C_6H_5N_3 + C_6H_6$$

as well as amine and azide derivatives of each (596). Similar decomposition of 1,4-bis(*p*-nitrophenyl)-1-tetrazene leads to nearly quantitative yields of *p*-nitroaniline and *p*-nitrophenyl azide (303).

p-Chlorophenyl-3,4-diacetyl-1-tetrazene decomposes rapidly in refluxing benzene or chlorobenzene; isomer ratios formed in chlorobenzene of dichlorodiphenyl ($o{:}m{:}p = 61.2{:}24.6{:}14.2$) indicate a radical pathway (696). Derivatives of 1-tetrazene with a single acyl substituent are less stable than those with two acyl groups, and decompose easily to acyl azides and amides (265).

$$p\text{-}ClC_6H_4N{=}N\overset{COCH_3}{N}NHCOCH_3 \xrightarrow[131^\circ]{\text{chlorobenzene}}$$

$$\underset{60\%}{N_2 + ClC_6H_4(C_6H_4Cl\text{-}p)\text{-}o,m,p} + CH_3CONHNHCOCH_3 + \underset{10\text{–}25\%}{p\text{-}ClC_6H_4\overset{COCH_3}{N}NH_2}$$

$$p\text{-}O_2NC_6H_4N{=}NNHNHCOCH_3 \xrightarrow[\text{organic solvents}]{\text{warm in}} N_2 + p\text{-}O_2NC_6H_4N_3 + H_2NCOCH_3$$

Aldehyde hydrazones of 1,3-diaryltetrazenes rearrange to formazans in a few hours at ordinary temperature, and immediately at their melting points (182, 442). Although the rearrangement proceeds more difficultly with electron-

$$C_6H_5{-}N{=}N{-}N(C_6H_5){-}N{=}CHC_6H_5 \xrightarrow{90^\circ} C_6H_5NH{-}N{=}C(N{=}NC_6H_5){-}C_6H_5$$

releasing substituents on the 3-aryl group than with electron-withdrawing groups, the red color of formazans appears on melting all 3-aryl tetrazene aldehyde hydrazones. In contrast 1-aryl-3-alkyl or aralkyl tetrazene hydrazones do not rearrange to formazans (182, 1193).

$$C_6H_5{-}N{=}N{-}N(CH_2C_6H_5){-}N{=}CHC_6H_5$$

mp 108–109° dec.

$$p\text{-}O_2NC_6H_4N{=}N{-}N(CH_3){-}N{=}CHC_6H_5$$

mp 148° dec.

Substituents on the 2-tetrazenes include alkyl, aryl, acyl, heterocyclic, and noncarbon groups (Table 2.63). Thermolysis of 2-tetrazenes liberates nitrogen and initially forms nitrogen-centered disubstituted amino radicals. The presence of amino radicals has been established by their self-combination to form hydrazines; by disproportionation, hydrogen abstraction, and other radical reactions, including their ability to initiate polymerization of styrene and other olefines; and by measurement of electron spin resonance. The dimethylamino radical has been isolated from pyrolysis of tetramethyltetrazene as a violet solid at −196°, decomposing at −160° (923). Products from decomposition of tetramethyltetrazene in chlorobenzene, in addition to tetramethylhydrazine, include the dimethylamino disproportionation products, dimethylamine and the dimer and trimer of methylmethyleneamine, as well as a nonvolatile residue (567).

$$(CH_3)_2NN{=}NN(CH_3)_2 \xrightarrow[\text{chlorobenzene}]{151^\circ} N_2 + 2[(CH_3)_2N\cdot] \longrightarrow$$

$$\underset{21.2\%}{(CH_3)_2NN(CH_3)_2} + \underset{26.0\%}{(CH_2{=}NCH_3)_2} + \underset{3.8\%}{(CH_2{=}NCH_3)_3} + \underset{66.2\%}{(CH_3)_2NH}$$

The tetraalkyl and aralkyl 2-tetrazenes are stable to over 100°, but begin to decompose in the range 110–140°. Tetraaryltetrazenes, however, lose nitrogen at temperatures as low as 40°. More positions are available for coupling in the aromatic tetrazenes, as illustrated by the thermolysis of 1,4-bis(2-naphthyl)-1,4-diphenyl-2-tetrazene (154). No hydrazine is formed. Stable free radicals could not be detected by ESR, but the products are consistent with radical disproportionation.

Table 2.63. 2-Tetrazenes

2-Tetrazene	Melting Point (°C)	Thermal Stability	Ref.
$(CH_3)_2N—N{=}N—N(CH_3)_2$	Bp 32/15		168, 393
		$t_{1/2}$ (110°, hexane) 160 hr	817
		$t_{1/2}$ (125.5°, 10^{-3} mm) 212 hr	1144
		$t_{1/2}$ (150°, benzene) 2.4 hr	1165
$(C_2H_5)_2N—N{=}N—N(C_2H_5)_2$	Bp 79/12	Dec. reflux, ~140°	168, 1178
$(C_6H_5CH_2)_2N—N{=}N—N(CH_2C_6H_5)_2$	96.5–97; 98–99	Dec. 128°	904, 1178
$C_6H_5—N(CH_3)—N{=}N—N(CH_3)—C_6H_5$	141–142 dec.	$t_{1/2}$ (95°, hexane) 15.4 hr $t_{1/2}$ (110°, hexane) 2.5 hr	785, 832
$C_6H_5—N(C(CH_3)_3)—N{=}N—N(C(CH_3)_3)—C_6H_5$	130–131 dec.	$t_{1/2}$ (79.9, cumene) 4.5 hr	783, 786
$(C_6H_5)_2N—N{=}N—N(C_6H_5)_2$	123–124 dec.	$t_{1/2}$ (65°, DMF) 17.4 min	124, 1050
$2—C_{10}H_7—N(C_6H_5)—N{=}N—N(C_6H_5)—C_{10}H_7—2$	101 dec.	$t_{1/2}$ (40°, benzene) 4.1 hr	154, 155
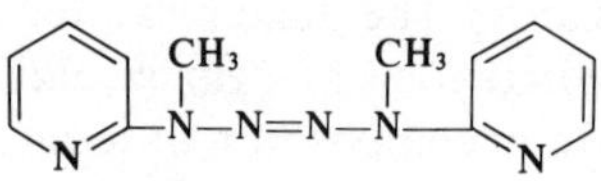	137–138 dec.		37
$CH_3CON(C_6H_5)—N{=}N—N(C_6H_5)—COCH_3$	196	Dec. 152°, cumene	567
$CH_3OOC—N(CH_3)—N{=}N—N(CH_3)—COOCH_3$	190–190.5 dec.		487

Tetrazenes with 1,4-dialkyl-1,4-diaryl substituents are intermediate in thermal stability between tetraalkyl and tetraaryl tetrazenes. Electron-withdrawing substituents on the aryl group in 1,4-diaryl-1,4-dimethyl-2-tetrazenes increase and electron-releasing substituents decrease stability, as indicated by the half-lives of the 1,4-bis-*p*-ethoxycarbonylphenyl, *p*-anisyl, and phenyl derivatives (Table 2.64) (785). There is no significant variation in activation energy for the decomposition in relation to the electronic character of the substituents. It is clear, however, that the effect on rate is related to the entropy term, thereby suggesting a conformational effect. Nelsen and Heath proposed (785) that the substituent effect reflects the ease of attaining a conformation in which the N—N

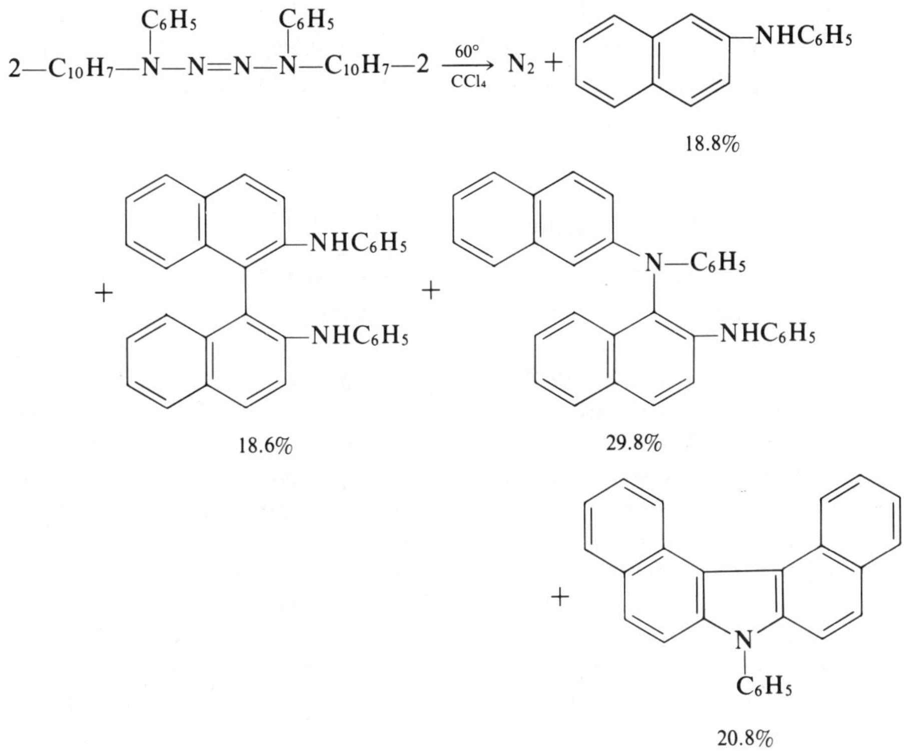

bond being broken is perpendicular to the aromatic rings and in which the lone electron pairs on N_1 and N_4 are unconjugated with the aromatic rings. Electron-withdrawing substituents on the aryl group would enhance the conjugation with the lone pair, resisting attainment of the perpendicular N—N arrangement for decomposition and increasing stability of the tetrazene. Validity of this concept was demonstrated by the relative instability of tetrazenes with substituents sterically forcing a conformation in which conjugation of the lone pair with the

Table 2.64. Half-Lives and Activation Energies of 1,4-Diaryl-1,4-dimethyl-2-tetrazene Decomposition

$$p\text{-}RC_6H_4\text{—}\overset{\displaystyle CH_3}{\overset{|}{N}}\text{—}N\text{=}N\text{—}\overset{\displaystyle CH_3}{\overset{|}{N}}\text{—}C_6H_4R\text{-}p$$

(110°, cumene)

R	$t_{1/2}$	Activation Enthalpy, $\Delta H\ddagger$ (kcal/mole)	Activation Entropy, $\Delta S\ddagger$ (eu)
CH_3O	8 min	33	15
H	3.2 hr	34	10
C_2H_5OOC	27.5 hr	31	−3

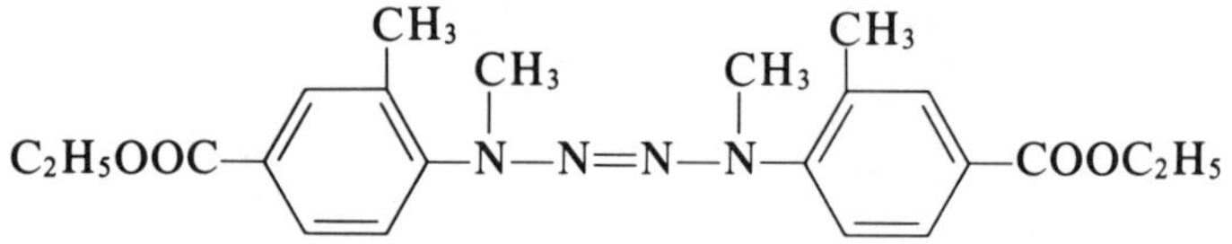

$t_{1/2}$ (110.2°, cumene) 6 min (783)

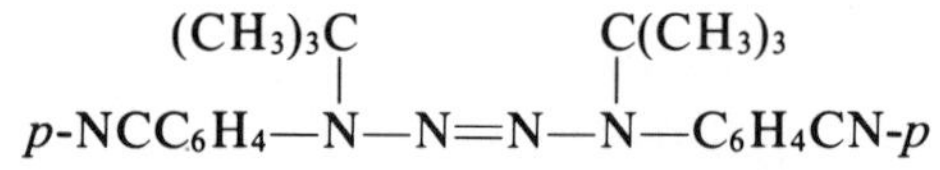

$t_{1/2}$ (107.6°, cumene) 14 min (786)

aromatic ring was minimized. Thus introduction of an ortho methyl group in the ethoxycarbonyl derivative completely eradicates the stabilizing effect of the ethoxycarbonyl. Similarly, in 1,4-di-*t*-butyl-1,4-bis-*p*-cyanophenyltetrazene, the *t*-butyl groups on positions 1 and 4 twist the *p*-cyanophenyl groups out of conformation with the lone pairs on these nitrogens and oppose stabilization of the tetrazene by cyano. Indeed, it may be inferred from X-ray crystallographic analysis of 1,4-di-*t*-butyl-1,4-bis(2,4,6-trideuteriophenyl)-2-tetrazene that, at least in the solid state, the lone pair is in conjugation neither with the aryl rings nor the azo linkage (787).

A small quantity of *N,N'*-dimethylhydrazobenzene is formed on decomposition of 1,4-dimethyl-1,4-diphenyl-2-tetrazene in cumene, but most of the prod-

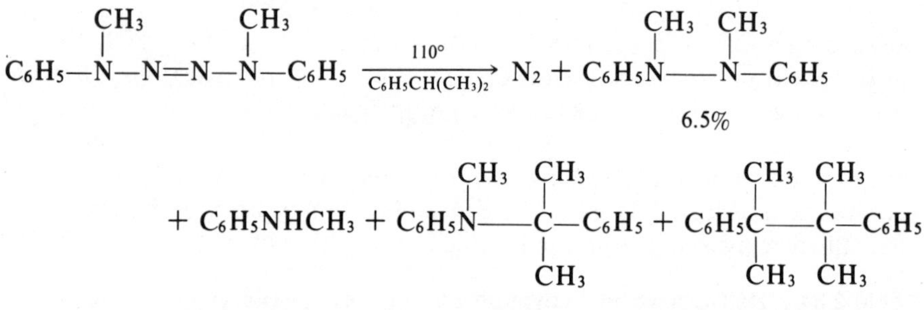

ucts are the result of hydrogen abstraction from the solvent and combination of the resulting radicals (785).

Thermal stabilities of 1,4-diacyltetrazenes are greater than corresponding dialkyl or diaryl derivatives; decomposition does not begin before 150°.

Stabilities of 2-tetrazenes carrying trimethylsilyl, germanyl, or stannyl substituents decrease with increasing atomic weight of the group IV element (Table 2.65).

The partially substituted trimethylsilyl-2-tetrazenes are liquids. Tris(trimethylsilyl)tetrazene decomposes slowly at 100° and more rapidly, but nonexplosively, at 150° to bis(trimethylsilyl)amine and trimethylsilyl azide (1175).

Table 2.65. Higher Group IV Tetrasubstituted 2-Tetrazenes

$X_2N{-}N{=}N{-}NX_2$

X	Melting Point (°C)	Half-life (150°, benzene) (hr)	Ref.
$(CH_3)_3Si$	46	53.2	1165, 1168
$(CH_3)_3Ge$	71	15.4	1165
$(CH_3)_3Sn$	88	0.9	1165

1,1-Bis(trimethylsilyl)tetrazene rearranges to the symmetrical isomer at 100° (1166).

$$[(CH_3)_3Si]_2N{-}N{=}N{-}NHSi(CH_3)_3 \xrightarrow[\text{benzene}]{150^\circ} [(CH_3)_3Si]_2NH + (CH_3)_3SiN_3$$

bp 65°/0.5 (1166)

$$[(CH_3)_3Si]_2N{-}N{=}N{-}NH_2 \xrightarrow[\text{closed tube}]{100^\circ,\ \text{pentane;}} (CH_3)_3SiNHN{=}NNHSi(CH_3)_3$$

bp 15–20°/high vacuum   bp 50°/2

A 2-tetrazene with a terminal triphenylphosphine moiety has been prepared (126).

$$(CH_3)_2N{-}N{=}N{-}N{=}P(C_6H_5)_3$$

mp 161–163°

Tetrazenes which are part of a conjugated system with benzothiazolinylidene substituents at the 1,4-positions have appreciable thermal stability. Heating the

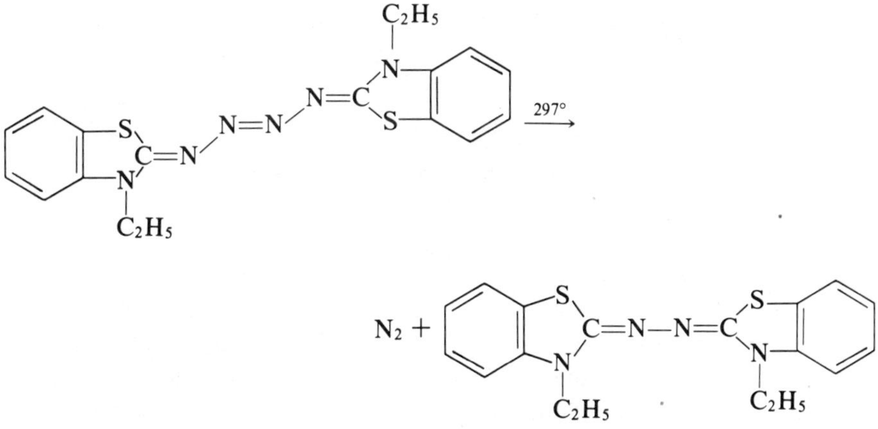

3-ethyl derivative at its melting point liberates nitrogen and forms the corresponding azine (57).

Quaternization of tetraalkyltetrazenes renders the resulting salts relatively thermally unstable. Tetrazenium ions which incorporate unsaturation on N-1,

expl. ~ 70° (698)

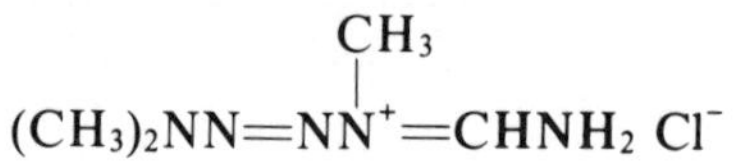

mp 189–190 dec. (1122)

conjugated with the 2-tetrazene unsaturation, are more stable. Dipolar dinitrovinyl trialkyl tetrazenes, such as the trimethyl derivative shown, are stable for several months at ambient temperature (1089). The tetramethyltetrazene cation radical is stable in chloroform, acetone, and other solvents for a few minutes at room temperature (1097).

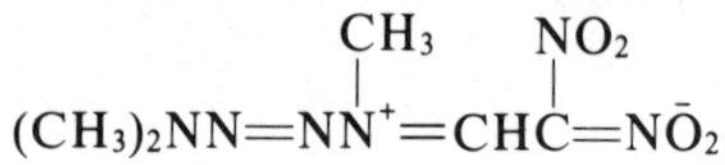

mp 131–132°

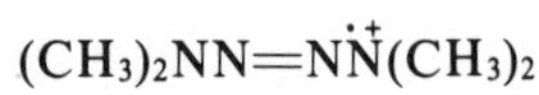

2.4.3. Nitroso, Nitro, and Azido Four-Nitrogen Chains

Compounds with four-nitrogen chains which include nitroso, nitro, or azide functions are much less stable than most of the tetrazanes or tetrazenes.

Mononitroso triazenes are labile, explosive solids, appreciably less stable than nitrosohydrazines. As indicated by their decomposition points, from −20° to 0°, the di-*p*-anisyl is the least and the di-*p*-chlorophenyl the most stable nitro-

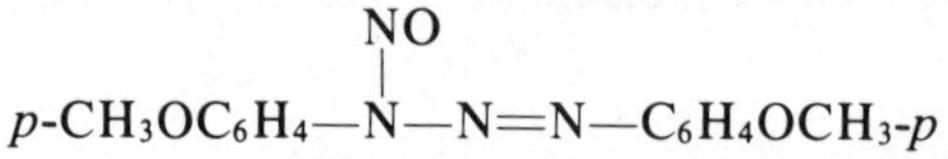

dec. −20°

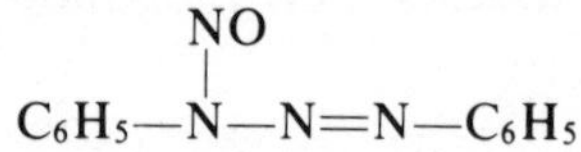

dec. −10°

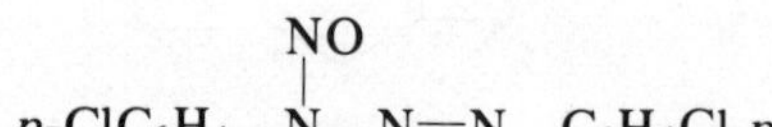

dec. 0°

sotriazene (768). At ambient temperature, most nitrosotriazenes decompose in 20–30 min. The di-*p*-chlorophenyl derivative deteriorates over 120 hr, liberating nitrogen and forming both *p,p'*-dichloroazo- and azoxybenzene as well as resinous material. Intermediacy of radicals in nitrosotriazene disintegration is indicated by the decomposition of the diphenyl derivative in nitrobenzene,

$$p\text{-}ClC_6H_4{-}N(NO){-}N{=}N{-}C_6H_4Cl\text{-}p \longrightarrow N_2 + p\text{-}ClC_6H_4{-}N{=}N{-}C_6H_4Cl\text{-}p$$

$$+\ p\text{-}ClC_6H_4N^{+}(O^{-}){-}N{=}N{-}C_6H_4Cl\text{-}p$$

forming *o, m,* and *p*-nitrobiphenyls, and also by the induction by di-*p*-chlorophenylnitrosotriazene of polymerization of acrylonitrile and other olefins.

A few dinitrosohydrazines have been described and appear to be more thermally stable than mononitrosotriazenes; their structures have not been investigated with modern techniques. Decomposition of 1,2-dibenzyl-1,2-dinitroso-

$$C_6H_5CH_2{-}N(NO){-}N(NO){-}CH_2C_6H_5 \xrightarrow{30\text{–}40^\circ} C_6H_5{-}CH_2{-}N(NO){-}N{=}CHC_6H_5$$

mp 44° (1084)

hydrazine evolves red vapors (probably from initially evolved nitric oxide reacting with atmospheric oxygen forming nitrogen dioxide) and yields 1-benzyl-1-nitroso-2-(phenylmethylene)hydrazine (237).

N-Nitrotriazenes are unstable light colored powders without melting points

$$m\text{-}O_2NC_6H_4N{=}N{-}N(NO_2){-}CH_3$$

or spectral characterization (71, 72). The methyl aryl nitrotriazenes flash before melting; 1-methyl-1-nitro-3-(*m*-nitrophenyl)triazene can be stored at room temperature about a week, but other methyl aryl nitrotriazenes liberate gas in less time, and diaryl nitrotriazenes have still poorer stability (73).

Nitrosyl azide, ONN_3, has been postulated as an intermediate in the reaction of nitrous acid or nitrosyl ion with hydrazoic acid or azide ion (1030), and its preparation as an unstable yellowish-red product, melting from −66 to −59°, and decomposing at −50° was described (686). More recent investigations (541) suggest that the presumed nitrosyl azide was instead nitrosyl chloride, whose vapor pressure and infrared spectra are very close to those reported for ONN_3. Studies of the reaction between nitrous acid and hydrogen azide, separately labeled with ^{15}N, show that the middle nitrogen in the resulting nitrous oxide

$$H\overset{*}{N}O_2 + HN_3 \longrightarrow H_2O + [ONN_3] \xrightarrow{H_2O} O\overset{*}{N}N + N_2$$

comes from nitrous acid and the end nitrogen from hydrogen azide (216). The data are consistent with, but do not prove, nitrosyl azide to be a momentary intermediate.

Although nitryl azide, O_2NN_3, has not been isolated, it apparently has been prepared in acetonitrile solution, based on infrared and mass spectral data, and its quantitative decomposition above $-10°$ to 2 moles of nitrous oxide (275).

$$O_2NN_3 \xrightarrow[\text{acetonitrile}]{-10°} 2N_2O$$

Several compounds have been prepared with acyclic four-nitrogen chains consisting of an amine nitrogen combined with azide. They have been variously termed amine azides, or tetraazo, tetrazadiene or diazo-azo compounds. *N*-Azidodimethylamine is a highly explosive liquid (126), but azidohexamethyl-

$$(CH_3)_2NN_3 \qquad [(CH_3)_3Si]_2NN_3$$

bp 30°/10 bp 40–42°/11

disilazane decomposes slowly above 150°, liberating nitrogen (1169). Azidoamine derivatives of 3-ethyl-benzothiazoline can be formed in tetrahydrofuran or dimethylformamide solution at $-50°$, but decompose above $-10°$ to nitrogen and 1,4-bis(3-ethylbenzothiazolinylidene)tetrazene (57). The azidoamino

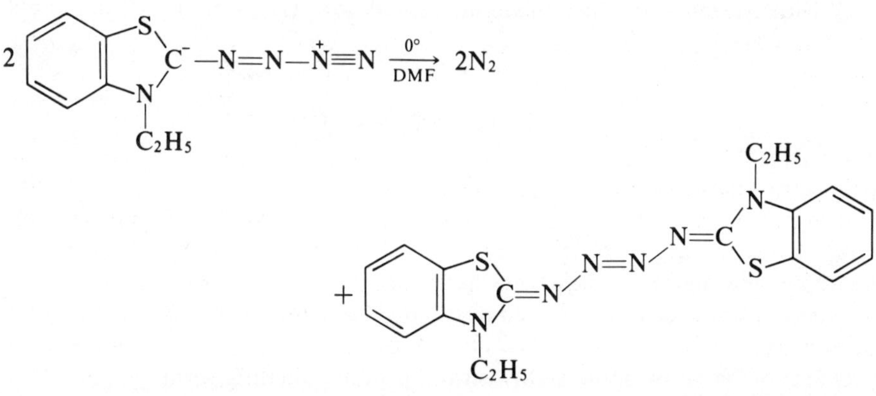

derivative of 3-ethyl-6-nitrobenzothiazoline is obtained in solid form as an explosive yellow compound (60).

2.5. ACYCLIC-CYCLIC AND CYCLIC-CYCLIC FOUR-NITROGEN CHAINS

2.5.1. Three Exocyclic Nitrogen Atoms

Presently known four-nitrogen chain compounds with one cyclic and three exocyclic nitrogen atoms are all dipolar.

A series of phthalimidoazimine derivatives with dialkyl or diaryl substituents on the two terminal nitrogen atoms were prepared by Hoesch, Dreiding, and co-workers (486). Both cis and trans isomers have been isolated with most substituents. An equilibrium mixture of the diphenyl isomers is attained from

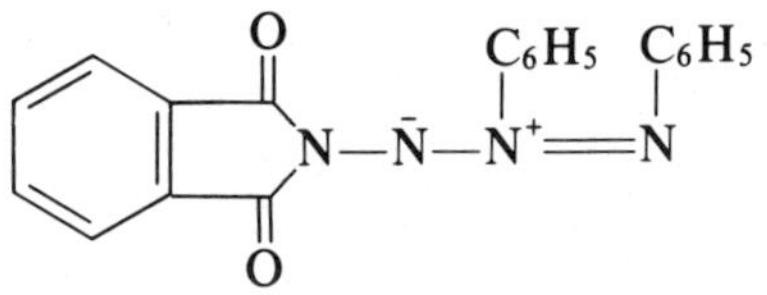

mp 133° dec.

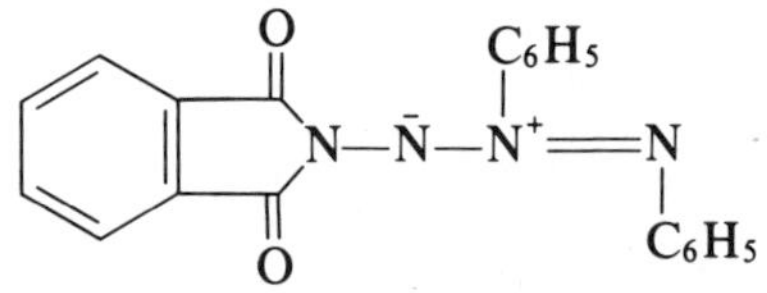

mp 131° dec.

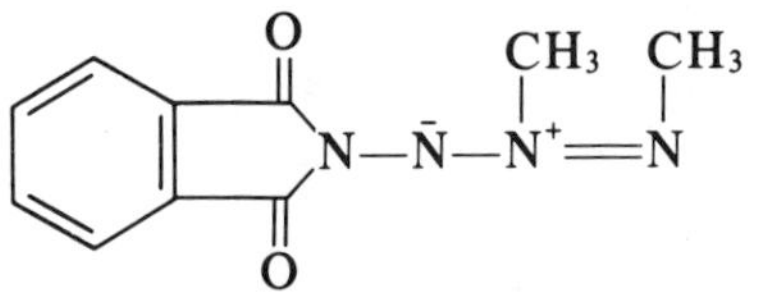

mp 138–142° dec.

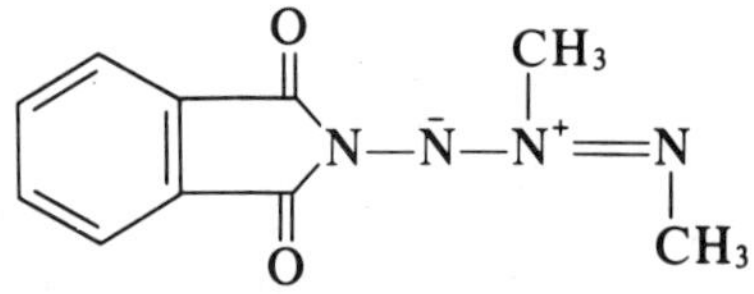

mp 149° dec.

either direction on refluxing 20–30 minutes in chloroform with a cis : trans ratio of 70 : 30. As shown by the presence of only interconverted stereoisomers and the absence of cross-products from mixtures of positional phenyl-*p*-tolylphthalimidoazimine isomers, the nitrogen chain remains intact during this unimolecular isomerization (485).

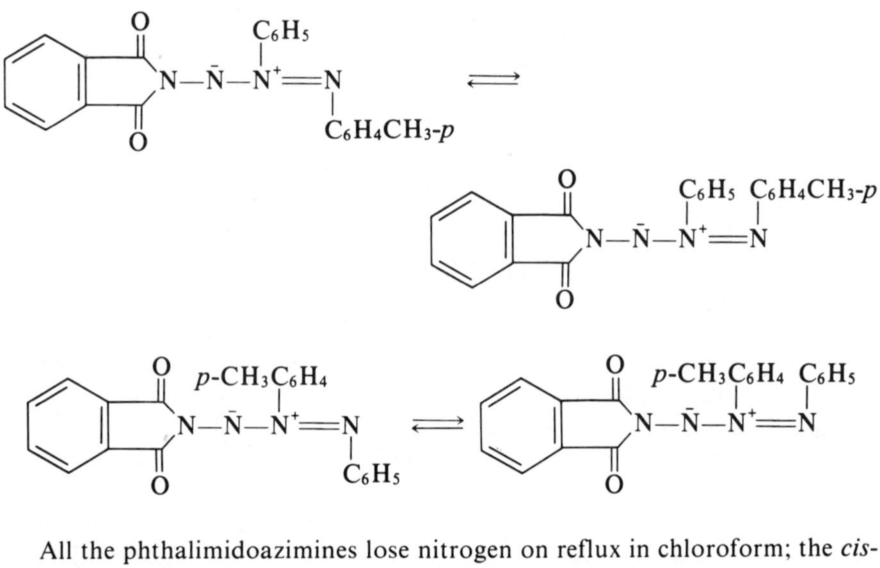

All the phthalimidoazimines lose nitrogen on reflux in chloroform; the *cis*-dimethyl isomer is completely decomposed in 5 days, but after 26 days the *trans*-dimethylazimine is 76% destroyed. Thermolysis of the *cis*-diphenyl derivative leads to loss of nitrogen and formation of *N,N'*-diphenyl phthaloyl hydrazine,

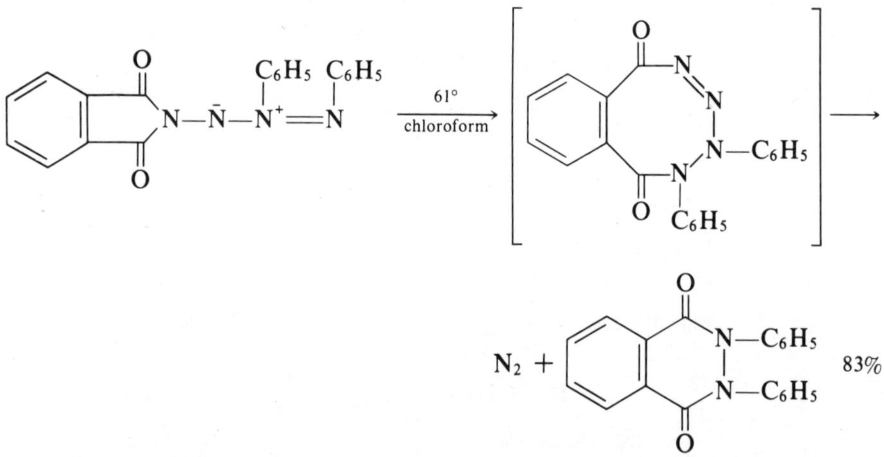

possibly through the intermediate cyclic tetrazene. Through use of ^{15}N labeling of the two terminal azimine nitrogens, it was established that all the extruded nitrogen came from the phthalimide moiety and its adjacent nitrogen (652).

Compounds with a dipolar four-nitrogen sequence consisting of a pyridinium nitrogen joined to a chain of three exocyclic nitrogen atoms terminated by

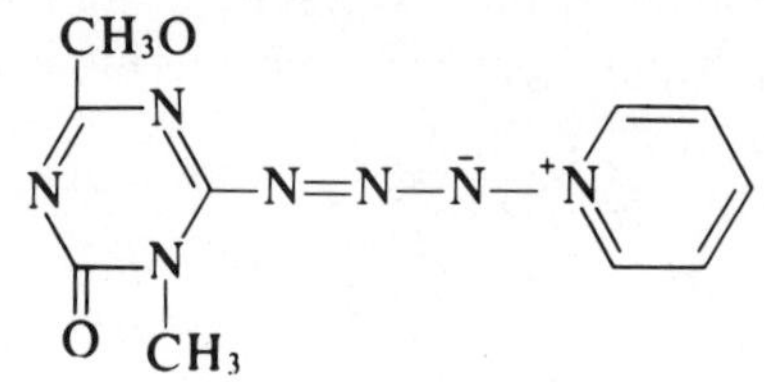

mp 133–134°

a substituted triazine show no decomposition by differential thermal analysis below 190° (354). Pyridine is lost at 220°.

2.5.2. Two Exocyclic Nitrogen Atoms

Acyclic-cyclic compounds with four-nitrogen chains at the tetrazane oxidation level and cyclic acyl groups at the 1- and 4-positions are unstable solids which explode on heating (36). These tetrazanes with hydrogen on the 2,3-nitrogen atoms give no evidence of dissociation to radicals.

The 2-tetrazenes with reduced cyclic amine moieties at each end of the chain (Table 2.66) decompose near 100°, as found for open chain tetraalkyl-2-tetrazenes. Thermolysis of azobispiperidine at 120–170° leads to piperidine and a tetrahydropyridine (1178) and heating (*E*)-1,2-bis(*cis*-9-azabicyclo[6.1.0]non-9-yldiazene) at 100° produces nitrogen, the bicyclic aziridine, *N*-*N*′-biaziridine, and cyclooctene (488).

The 2-tetrazenes with heteroaromatic amine residues as terminal groups de-

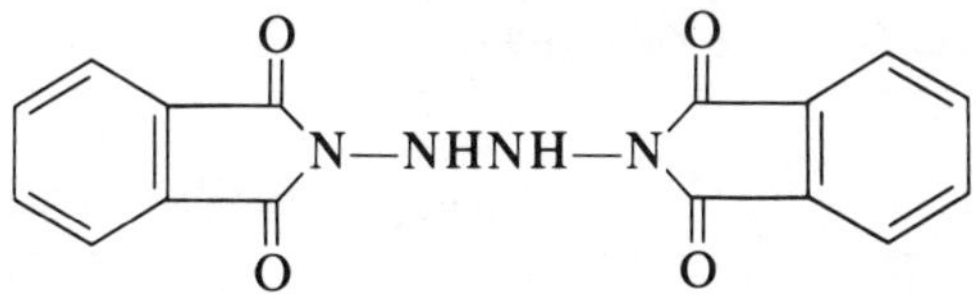

dec. 210–212°

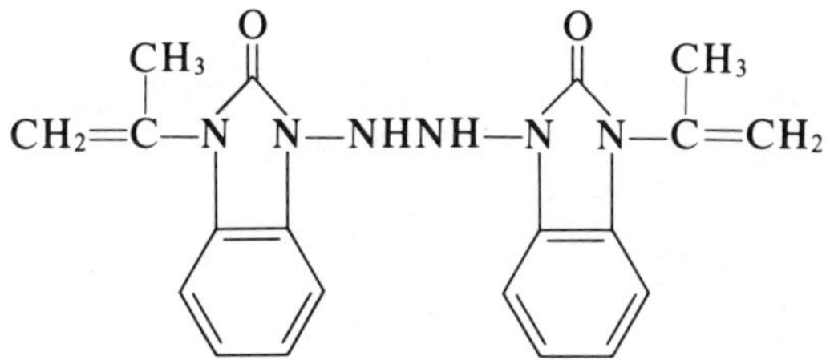

dec. 86°

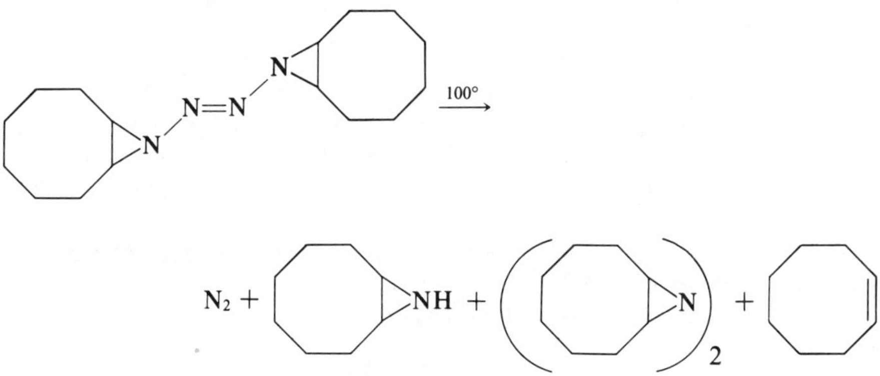

15% 40% 5–10%

compose at appreciably higher temperatures than acyclic tetraaryl-2-tetrazenes. Carbazole results from refluxing **VII** in the dimethyl ether of tetraethylene glycol at 275° (648).

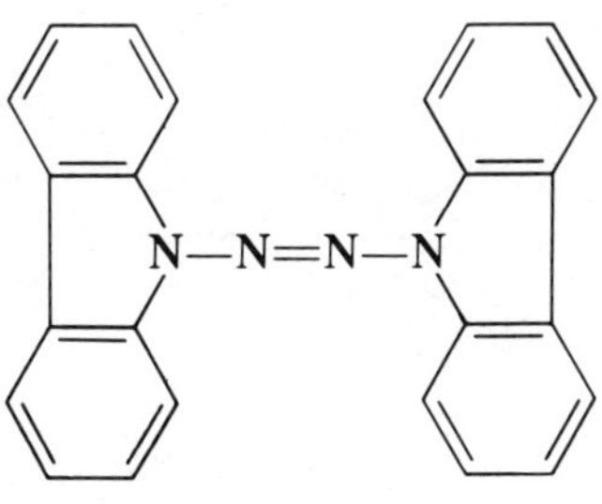

(VII)

mp 216° dec. (1179)

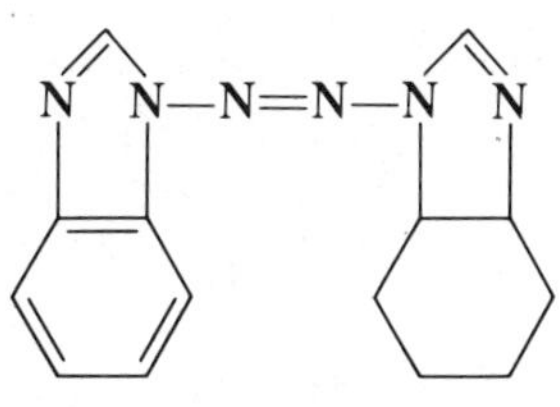

mp 270° dec. (577)

Table 2.66. Acyclic-Cyclic 2-Tetrazenes

2-Tetrazene	Melting Point (°C)	Thermal Stability	Ref.
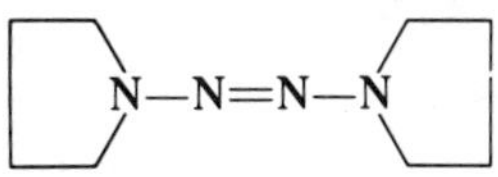	37.5–38.5		648
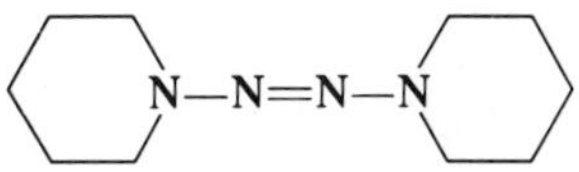	44.5; 44.6–44.7	Dec. 120°	622, 698, 1178
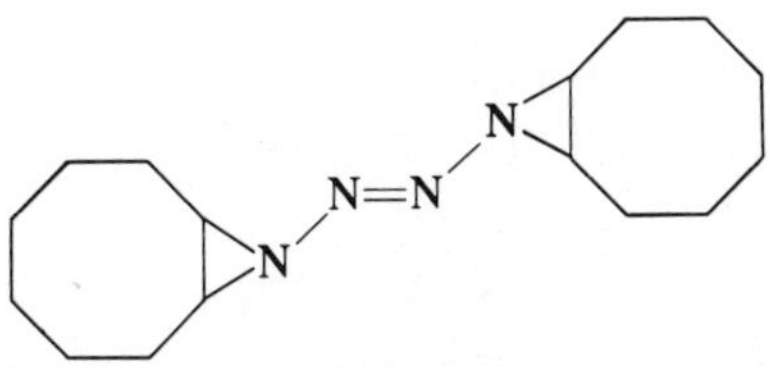	97–99 dec.	Dec. 100°	488

Bis-cyclic acyl-2-tetrazenes are as stable or more so than acyclic 1,4-diacyl analogues. *trans*-1,4-Bicyclic acyl-2-tetrazenes are more stable than their cis isomers. Loss of nitrogen occurs on heating *cis*-1,4-diphthaloyl-2-tetrazene in the melt or on refluxing in benzene with formation of *N*-(*o*-isocyanatobenzoyl)phthalimide (569).

Quaternary salts of 1,4-bis-heteroaryl acyclic-cyclic 2-tetrazenes also have considerable thermal stability.

Aminoazimine derivatives of 1,2,4-triazolidinedione have two exocyclic ni-

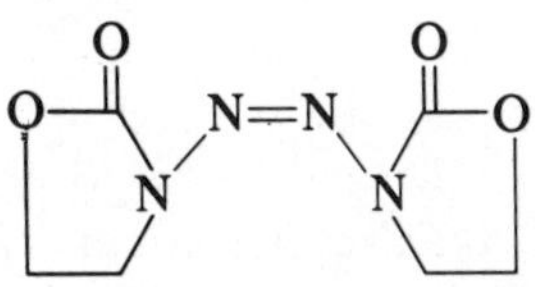

mp 170–171° expl. (335)

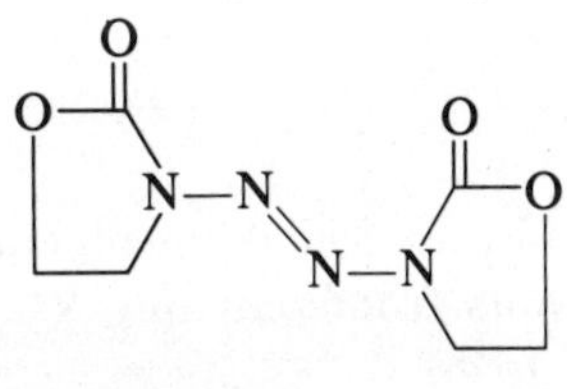

mp 298–299° (335)

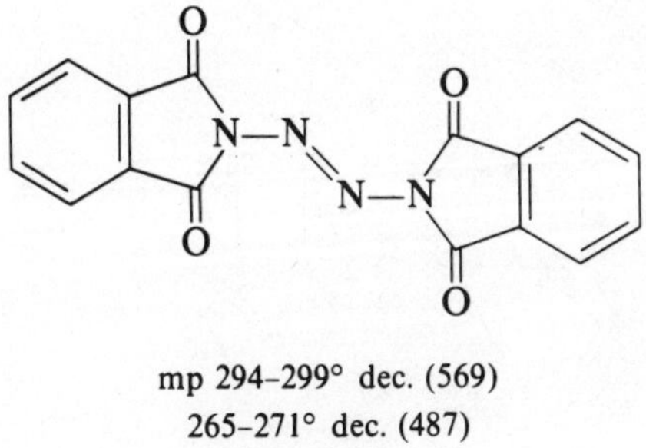

mp 294–299° dec. (569)
265–271° dec. (487)
sublimes 240°/0.01
undecomposed

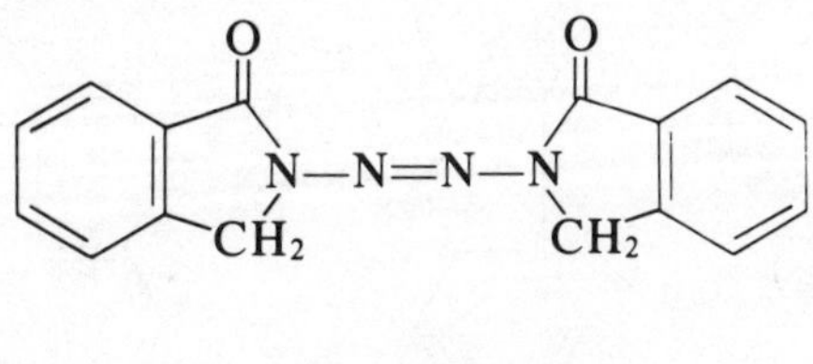

mp > 320° (35)

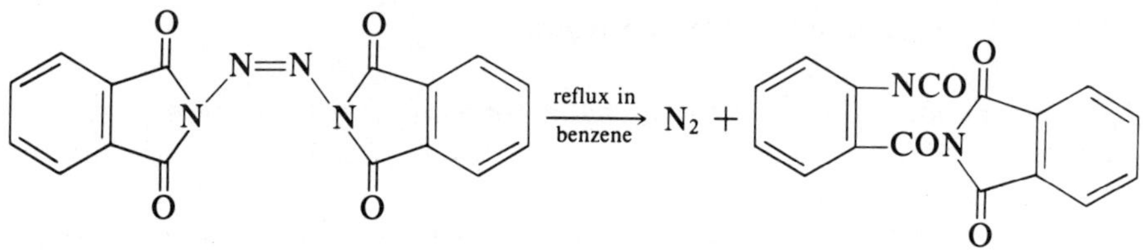

mp 160–161° dec.

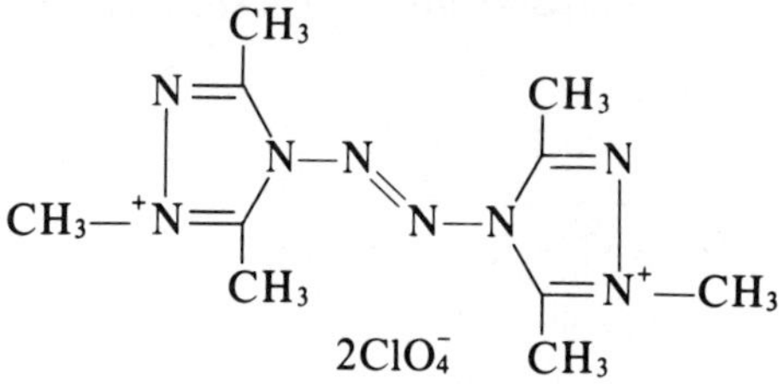

mp 274–275° (382)

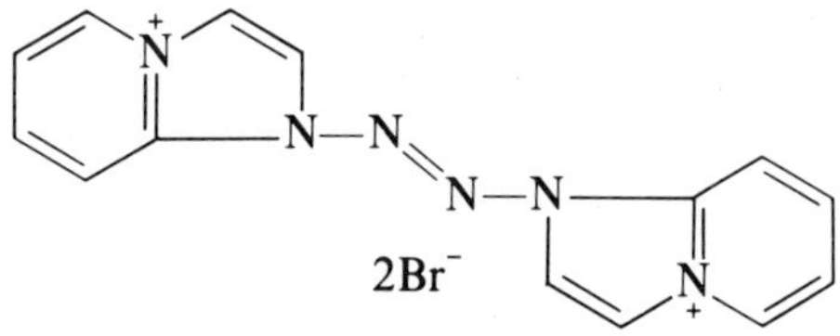

mp 309° dec. (383)

trogen atoms (597). They undergo thermal decomposition by fission of the four-nitrogen chain between the exocyclic and cyclic nitrogen atoms and between the terminal disubstituted amino and its adjacent nitrogen. Several mechanisms are involved including initial rearrangement to an allophanyl azide and dissociation to 4-phenyltriazolinedione and products of dimethylaminonitrene (306, 597).

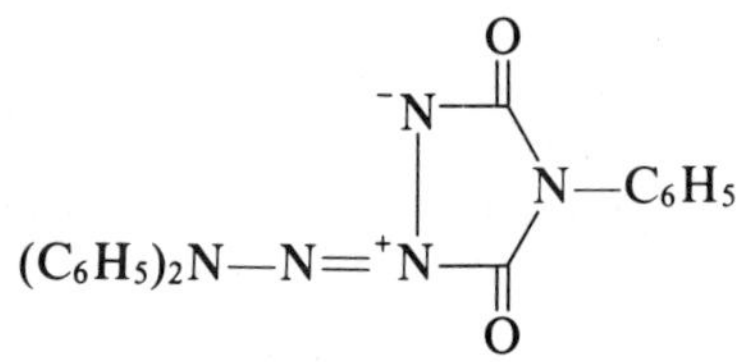

mp 140–141° dec.

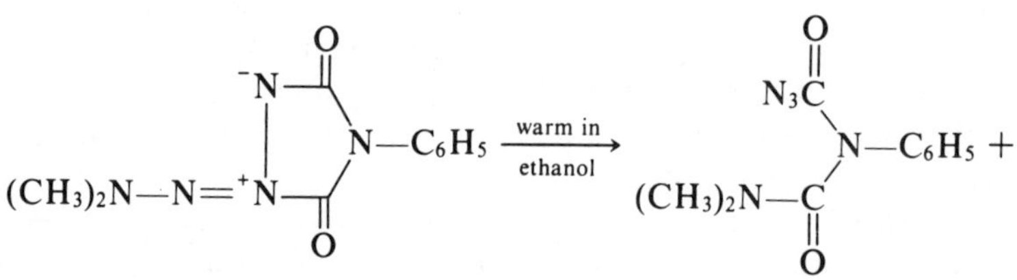

mp 133–134° dec.

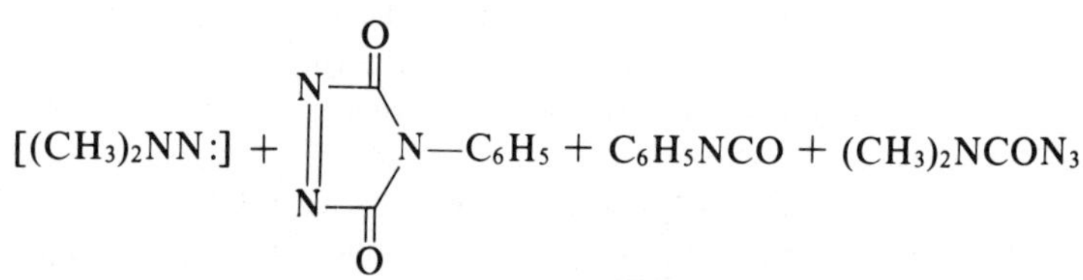

2.5.3. One Exocyclic Nitrogen Atom

The most common four-nitrogen chains compounds with one exocyclic nitrogen atom are *N*-amino three-nitrogen heterocycles, particularly *v*-triazoles and benzotriazines.

Both linear 1-amino-*v*-triazole and branched 2-amino-*v*-triazole four-nitrogen sequences are known. No obvious relationship exists between the melting points of the isomeric *N*-amino-*v*-triazoles. Amino derivatives of a wide variety

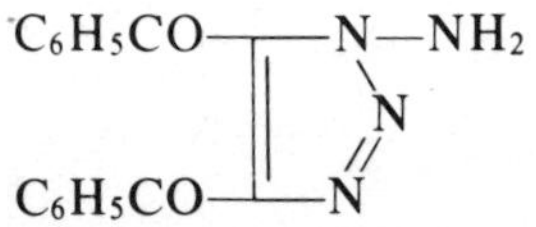

mp 158–160° (375)

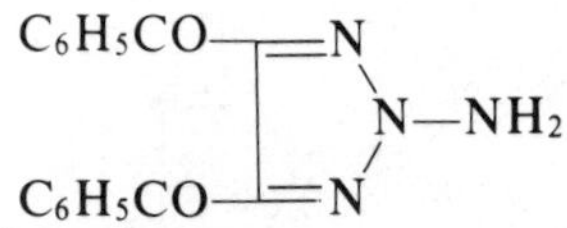

mp 100–102° (375)

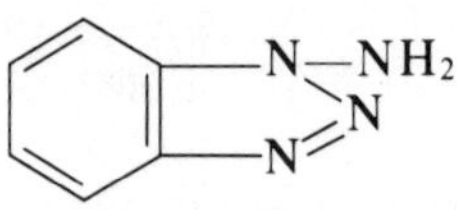

mp 84° (190)

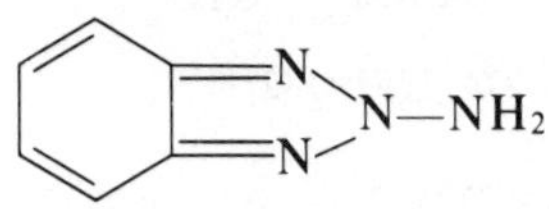

mp 121–122° (190)

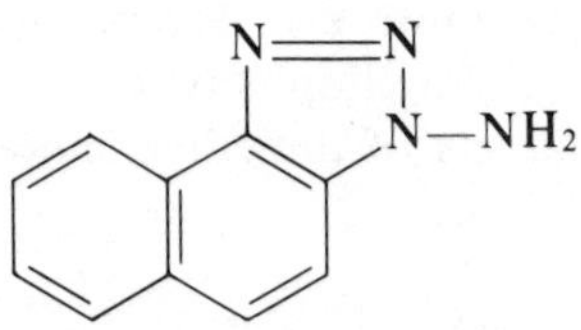

mp 152–153° (81)

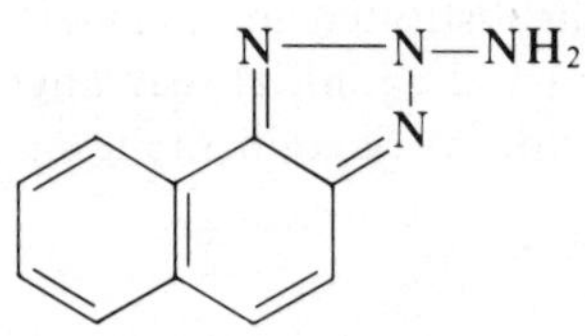

mp 166–167° (81)

of monocyclic and fused ring *v*-triazoles have been prepared. As with 2-aryl fused ring *v*-triazole quinones, the melting points of the 2-amino fused ring *v*-triazole quinones are over 300°.

Substituents on the amino group of the *N*-amino-*v*-triazoles include trityl, pentafluorophenyl, acyl, and phenylmethylene. Many, but not all, *N*-amino-*v*-triazoles melting over 175° do so with decomposition. 1-Amino-*v*-triazole decomposes explosively on heating to an unspecified temperature (851). High thermal stability is shown by the lithium salt of 1-*p*-tosylamino-4,5-dimethyl-*v*-triazole which is recovered unchanged after refluxing in ethylene glycol; at 300°, the salt slowly loses nitrogen (1184).

Compounds formerly regarded as 2,3-dihydro-1,2,3,4-tetrazines have been shown, instead, to be 1-amino-*v*-triazole derivatives. The structure of one of the most extensively investigated members of this class originally considered to be 2,3-dibenzoyl-5,6-dimethyl-2,3-dihydro-1,2,3,4-tetrazine (851) is 1-(α-benzoyloxybenzylideneamino)-4,5-dimethyl-*v*-triazole, whose structure has been established by X-ray crystallographic (84), chemical, infrared, NMR (29, 232), and mass spectral (28) evidence. On heating at 150° and above, the isoimide isomerizes to the *N*,*N*-dibenzoylamino derivative (232, 981).

Dipolar *N*-imides of 2-arylbenzotriazoles with electron-withdrawing groups

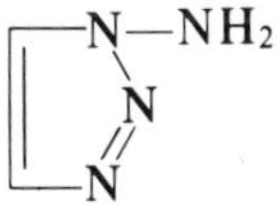

mp 51° (851, 1041)

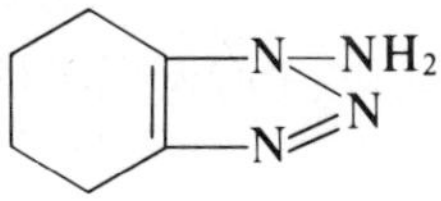

mp 94–95° (1191)
bp 128–130°/0.01

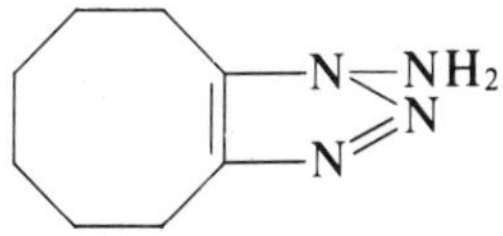

mp 61–62° (1190)
bp 181–182°/0.05

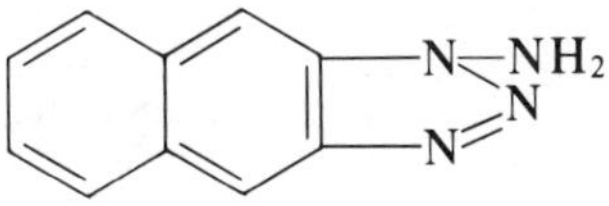

mp 206–208° (81)

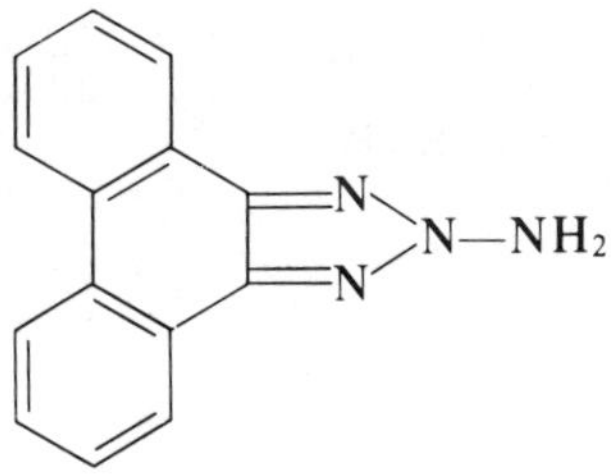

mp 177–179° dec. (80)

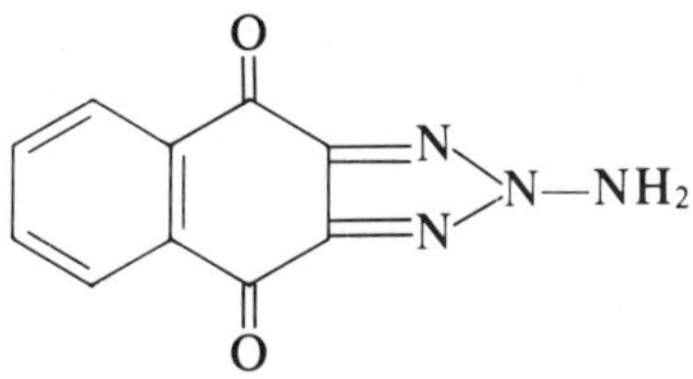

mp 313–314° dec. (764)

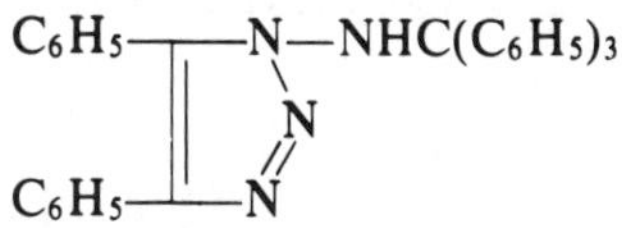

mp 270–271° dec. (233)
sublimes > 250°

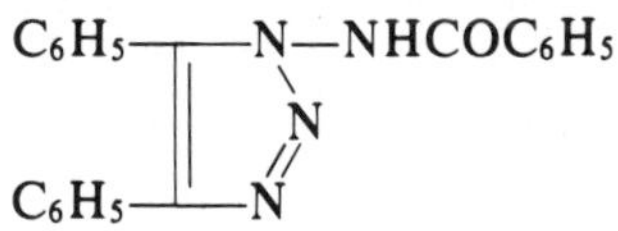

mp 255.5–257° (233)

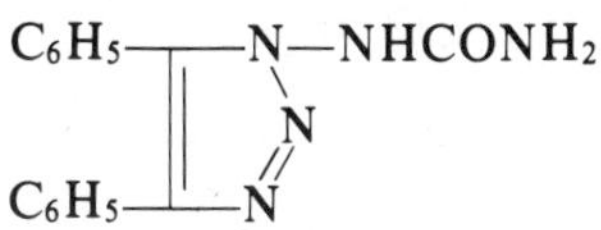

mp 209–211° (27)

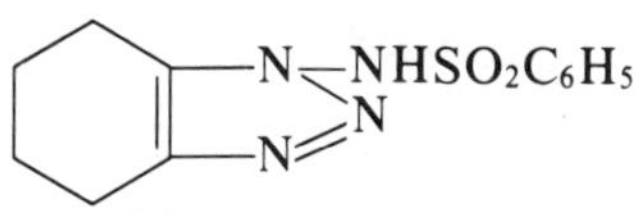

mp 176–177° dec. (1191)

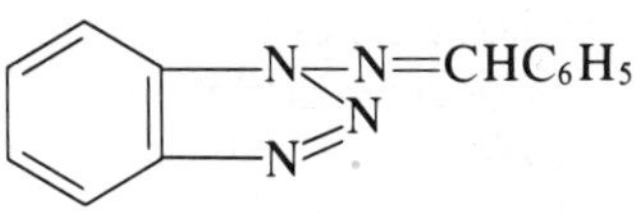

mp 112–113° (903)

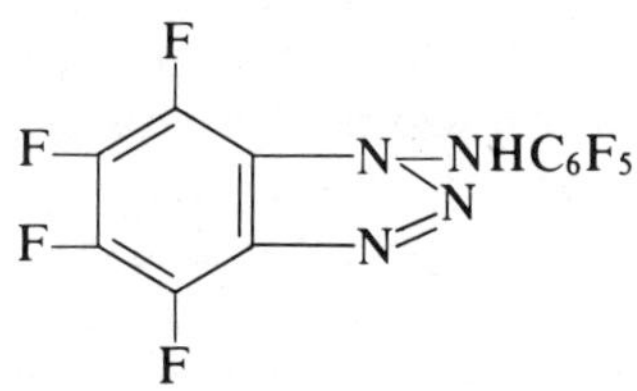

mp 158–159° (117)

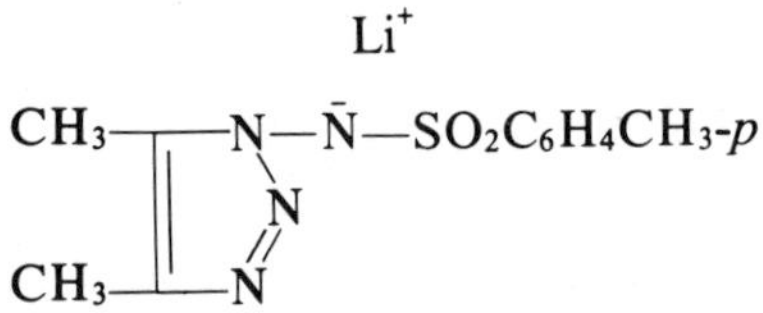

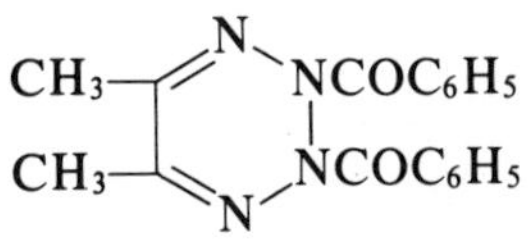

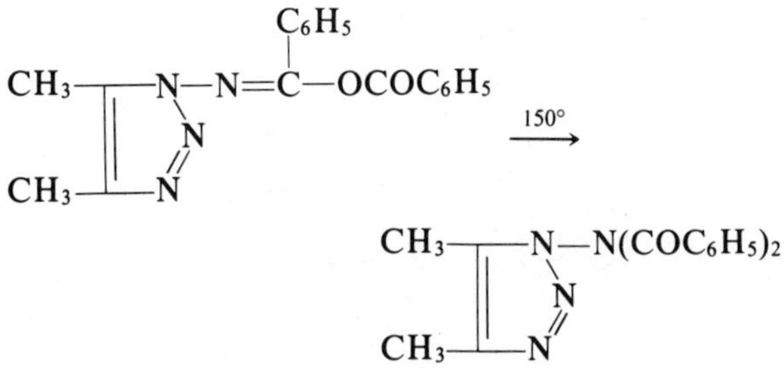

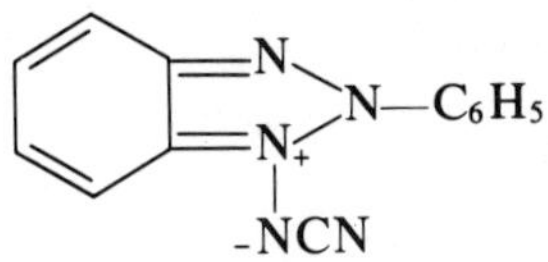

mp 165° dec.

on the imide nitrogen can be prepared, but are less thermally stable than *N*-aminobenzotriazoles (85). On heating 2-phenylbenzotriazole-1-benzenesulfonylimide in ethanol, nitrogen is evolved and 2-phenylsulfonylazobenzene results.

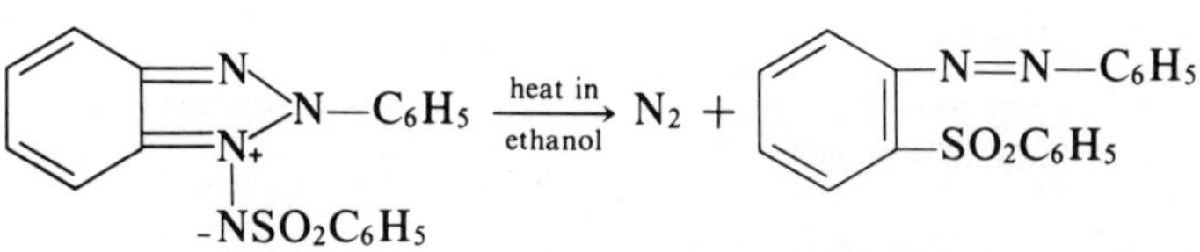

mp 84–85° dec.

Six-membered rings with one exocyclic nitrogen attached to three cyclic contiguous nitrogens include 3-aminobenzotriazinones and the 1- and 2-aminonaphtho[1,8-*de*]triazines. The aminobenzotriazinones, which may be viewed as cyclic 1-aryl-3-acyl-1-tetrazenes, are somewhat less thermally resistant than the *N*-amino-*v*-triazoles. Pyrolysis of 3-aminobenzotriazinone expels nitrogen and forms indazolin-3-one. Heating 3-arylideneaminobenzotriazinones at 300° causes rearrangement to 2-arylquinazolin-4-ones (848).

Both 1-nitroso and 1-nitrobenzotriazole have been prepared. The nitroso derivative is unstable, but no stability data have been reported for the nitro compound.

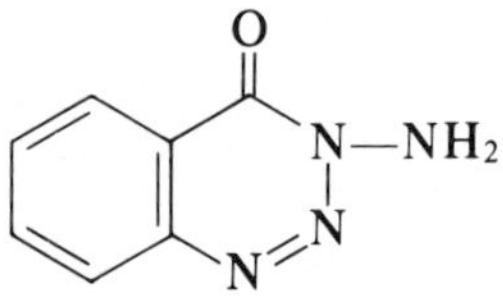

mp 154–155° dec. (6)

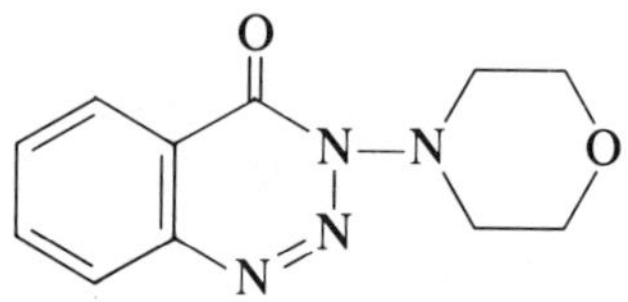

mp 199–200° dec. (471)

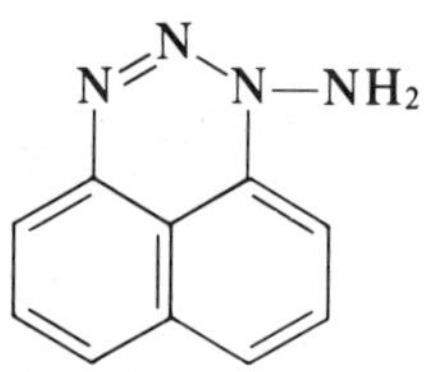

mp 154.5–155.5° (909)
158° (490)

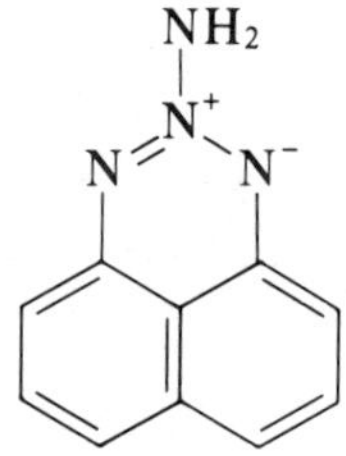

mp 179–181° (909)

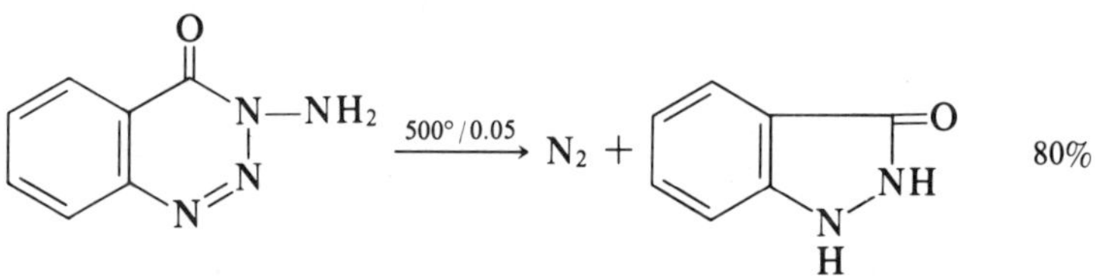

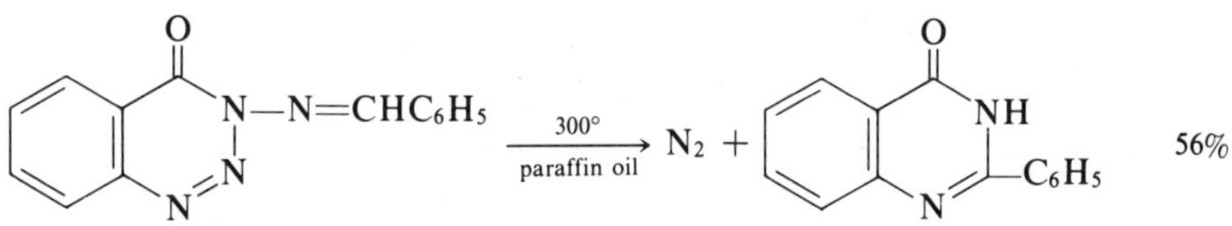

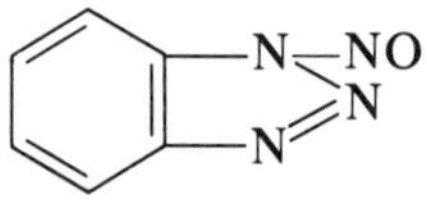

mp 40–42° (189)

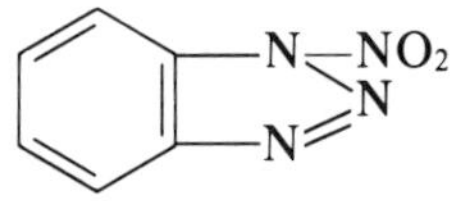

mp 74° (222)

2.5.4. Cyclic-Cyclic Four-Nitrogen Chains

Cyclic-cyclic compounds with four contiguous nitrogen atoms are constructed of either a single nitrogen atom in one ring joined to a three-nitrogen chain in a second ring or of two two-nitrogen sequences in two rings joined through nitrogen.

Presently known 1,3 cyclic-cyclic four-nitrogen chain compounds are *v*-triazole derivatives. Pyrolysis of 4,5-diphenyl-1-phthalimido-*v*-triazole at 425°

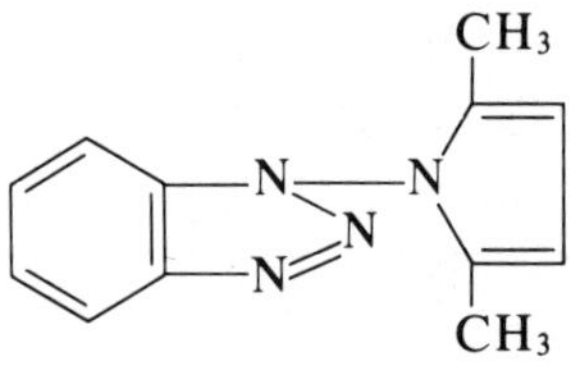

mp 90–91° (1020)

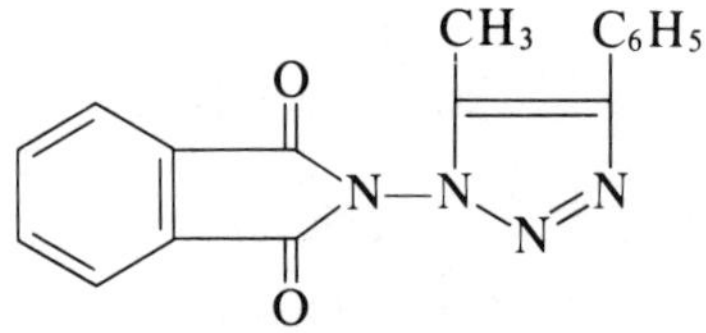

mp 215–216° (373)

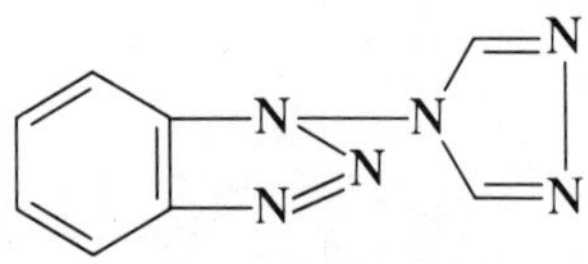

mp 175–176° (252)
sublimes 135°/0.07

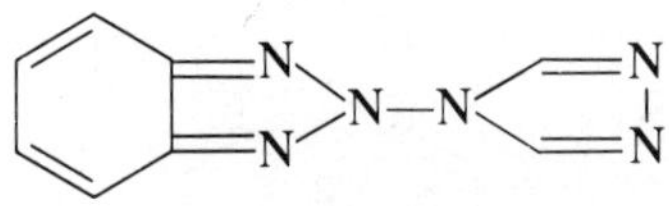

mp 259° (252)
sublimes 200°/0.02

leads exclusively to an azirine, but at 500° only to an indole (34).

Biindazole, with the two-nitrogen chains in each ring joined, shows no evidence of dissociation.

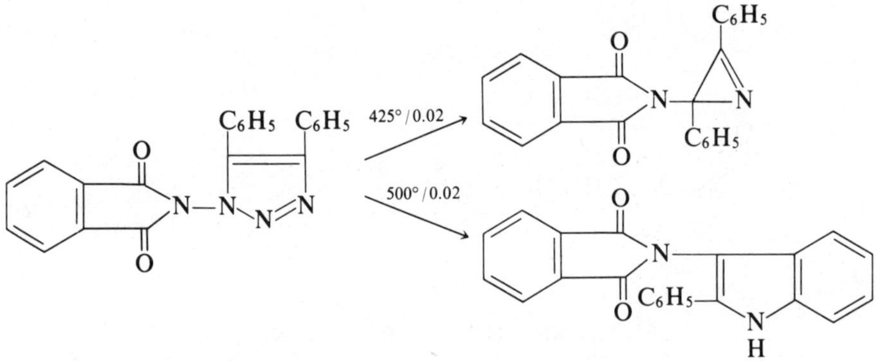

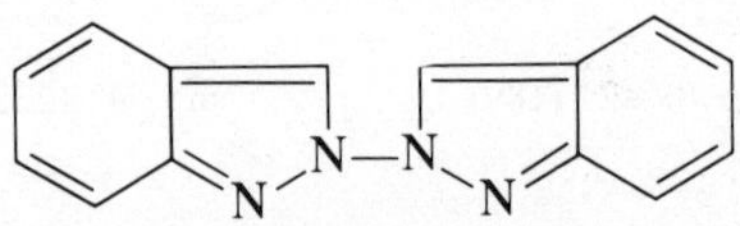

mp 267–268° dec. (943)
268° (608)

Tetrazanes derived from the combination of *N*-substituted pyrazolidinediones or 1,2,4-triazolidinediones (urazoles) are stable indefinitely as solids, but form highly colored solutions containing cyclic diacylhydrazyl radicals (871). Association of urazole radicals to tetrazanes (Table 2.67) increases as the polarity of the solvent decreases. The urazole diacylhydrazyl radicals are noteworthy in not having an aryl group directly attached to hydrazine nitrogen. Persistence

Table 2.67. 1,2,4-Triazolidinedione Radical Dimerization

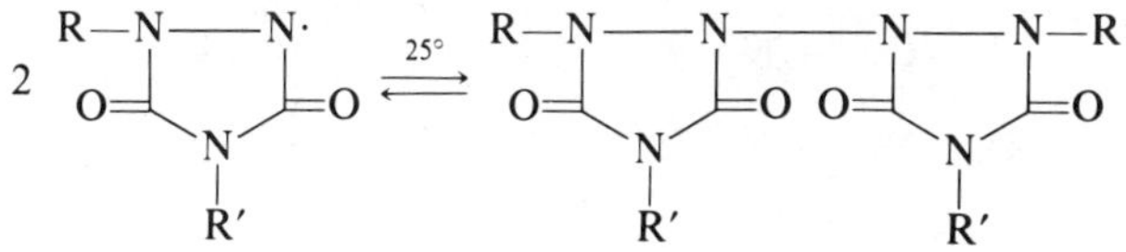

R	R′	$K_{assoc.}$		
		CCl_4	C_6H_6	CH_3CN
$C_6H_5C(CH_3)_2$	C_6H_5	6.0	1.8	0.16
$(CH_3)_3C$	C_6H_5	1.5	0.66	0.33
$(CH_3)_3C$	CH_3	5.6	3.2	1.6

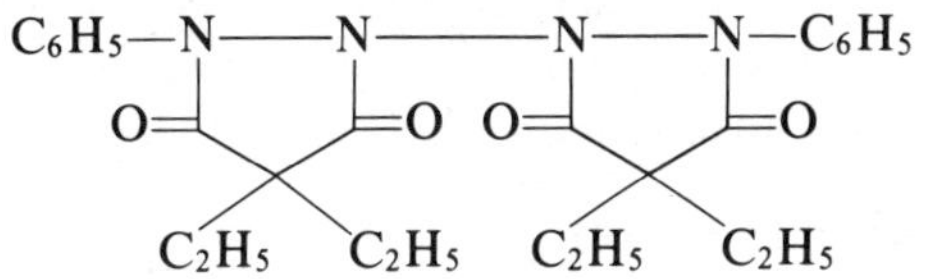

mp 106–108°

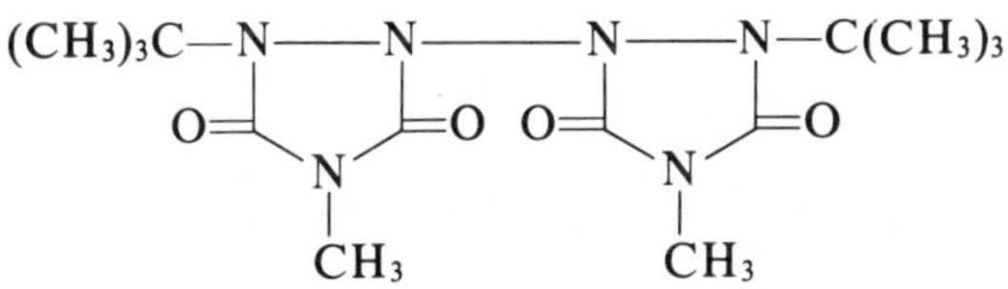

mp 94–96°

of these hydrazyl radicals results from the size of the tertiary alkyl substituents, their inability to undergo disproportionation, and delocalization of the unpaired electron over the entire heterocycle.

2.6. CYCLIC FOUR-NITROGEN CHAINS

2.6.1. Tetrazolines

The tetrazolines (dihydrotetrazoles) are monounsaturated heterocycles with one carbon and four nitrogen atoms in the ring and either two monovalent groups or one divalent substituent at the ring carbon. Presently described tetrazolines are substituted at the 1- and 4-nitrogens with a double bond between the 2- and 3-positions. Reference has been made to unfused 1,2-tetrazolines (379, 676) but without adequate characterization of the products to date. Mesoionic 1,3-disubstituted tetrazolium-5-olates, sometimes called 1,3-tetrazolines, are discussed under tetrazolium compounds.

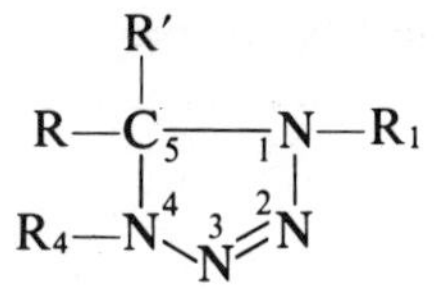 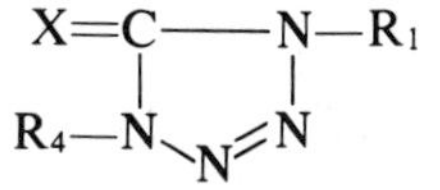

Tetrazolines with an alkyl or aryl group and hydrogen at the 5-carbon as well as methylene tetrazolines are known. The 5-hydrogen 5-substituted 1,4-tetrazo-

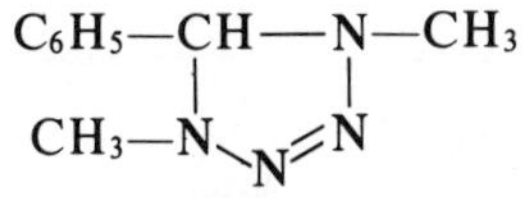

mp 53.0–53.5° (543)

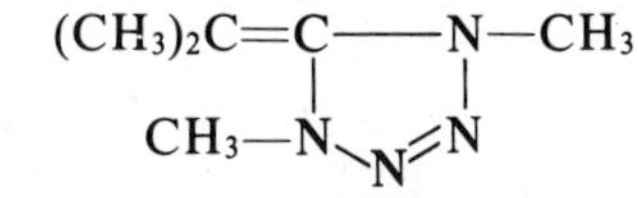

bp 30–35° / 10^{-3} (115, 894)

lines are cyclic analogs of the open-chain tetraalkyl-2-tetrazenes and decompose in approximately the same temperature range, 120–150°. Thermolysis of these tetrazolines yields azides and Schiff's bases (543). The simple methylene

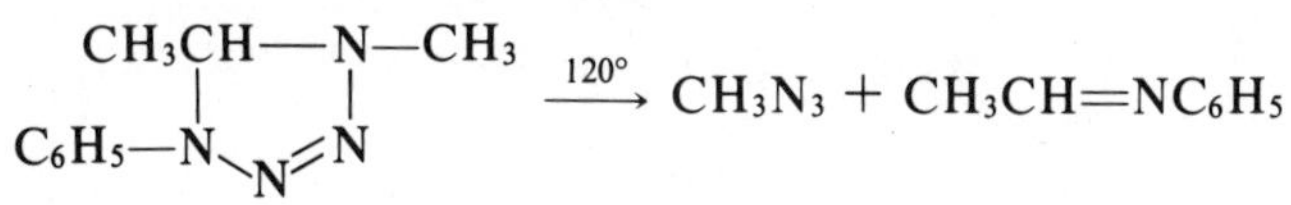

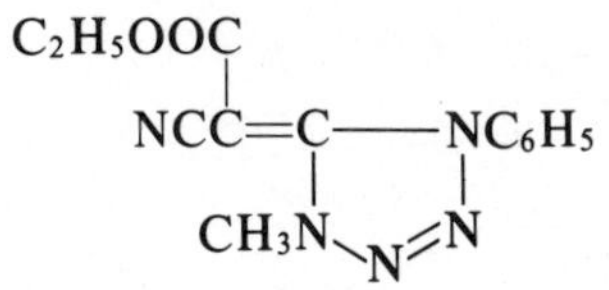

mp 194–196° (1000)

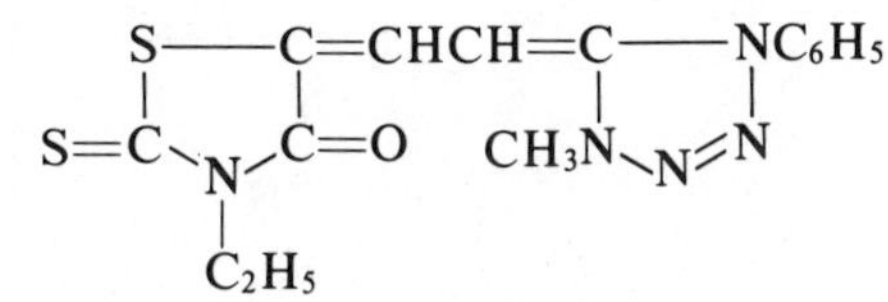

mp 226° dec. (1137)

tetrazolines are intensely yellow, extremely air- and moisture-sensitive liquids (894). Introduction of electron-withdrawing substituents on the methylene carbon enhances stability. Cyanine dyes with carbon-carbon unsaturation at the 5-tetrazoline position melt and decompose above 200°.

Iminotetrazolines are cyclic guanidines in which two nitrogens are connected by an azo group. They are substituted at the 1- and 4-positions with alkyl and aryl or two alkyl groups, and the imino moiety carries hydrogen, alkyl, or acyl or is part of an azine (Table 2.68). Except for 1,4-diphenacyl-5-iminotetrazoline, no instability for the free iminotetrazolines is indicated at their melting points. Because the iminotetrazolines are strongly basic and hygroscopic they are frequently isolated as salts, usually hydrochlorides, which melt with decomposition above 190°.

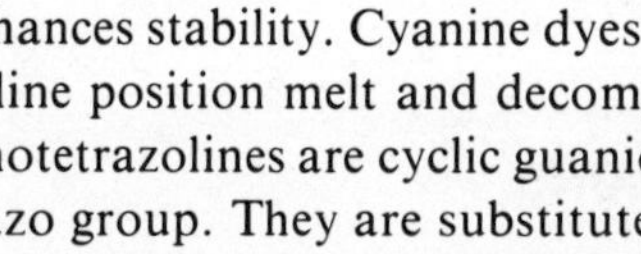

mp 242–244° dec. (466)

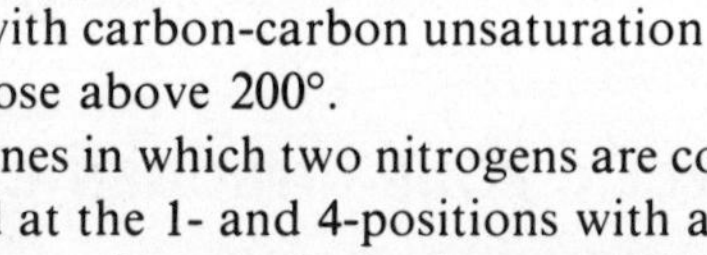

mp 216° dec. (861)

Table 2.68. 5-Imino-1,4-Disubstituted Tetrazolines

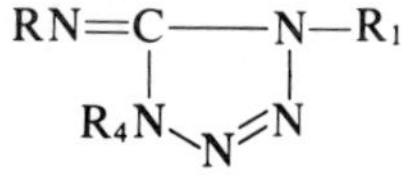

R	R_1	R_4	Melting Point (°C)	Ref.
H	CH_3	CH_3	106.4–107.4	773
			108.5–109.5	466
H	CH_3	C_6H_5	56–57	932
H	$C_6H_5COCH_2$	$C_6H_5COCH_2$	154 dec.	212
CH_3	CH_3	C_6H_5	33.5–35.3	932
			bp 120–125/0.4	
CH_3CO	CH_3	CH_3	35–36	983, 984
NC	CH_3	CH_3	112–113	814
p-$O_2NC_6H_4CH{=}N$	CH_3	CH_3	221–222	529

Alkyl, aryl, and acyl groups have been substituted at the 1- and 4-positions of tetrazolinone (Table 2.69). Thermal stability of the 1,4-tetrazolinones is greater than that of the tetrazolines with alkyl or aryl at the 5-carbon, and of their acyclic analogs, 1,4-diacyl-1,4-dialkyl or aryl-2-tetrazenes. Thus 1-cyclohexyl-4-phenoxycarbonyltetrazolinone is recovered almost quantitatively after three days at 150° (1124) and 1,4-dimethyltetrazolinone does not decompose at 300° (1139). The melting point of 1,4-dimethyltetrazolinone and that of its *N*-

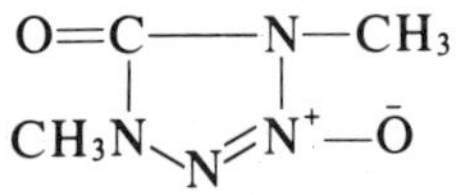

mp 113°

oxide are close (977); thermal stability of the *N*-oxide has not been investigated. 1-Alkyl-4-sulfonyltetrazolinones are less stable and, at about 100°, form an equilibrium system with their cycloreversion products, alkyl azides and sulfonylisocyanates. The half-life of 1-*n*-butyl-4-phenylsulfonyltetrazolinone at 100.8°

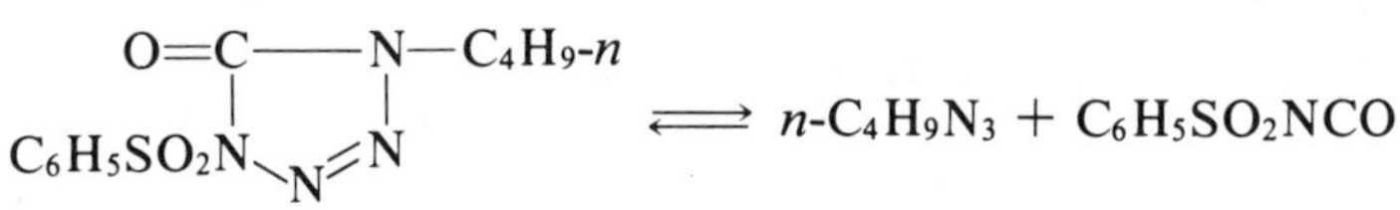

in chlorobenzene is about 72 hr (1124). Tetrazolinones with 1-alkyl and 4-chlorosulfonyl substituents are hygroscopic liquids which decompose unless kept under nitrogen (1056, 1124).

Table 2.69. 1,4-Disubstituted Tetrazolinones

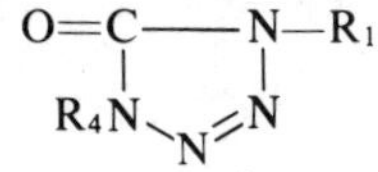

R_1	R_4	Melting Point (°C)	Ref.
CH_3	CH_3	116–117	441, 1139
CH_3	C_6H_5	71–72	932
$CH_2{=}CHCH_2$	C_6H_5	bp 120–/0.55	298
$CH_2{=}CH$	C_6H_5	35.5	254
CH_3CO	C_6H_5	103	679, 1046
C_6H_5OCO	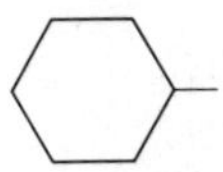	111–113	1124
$C_6H_5SO_2$	n-C_4H_9	42.5–43.5	1124

Tetrazoline-5-thiones have been prepared with alkyl, aryl, acyl, and organometallic substituents in the 1- and 4-positions (Table 2.70). The 5-thiono-1,4-tetrazolines for which some determination of heat resistance has been made are similar in thermal stability to the corresponding tetrazolinones. Thus 1-methyl-4-phenyl-5-thionotetrazoline requires a temperature greater than 300° for decomposition (1048) and the 1-benzyl-4-cyanoethyl derivative is unchanged after 2 days at 120–150° (629).

Table 2.70. 1,4-Disubstituted Tetrazoline-5-Thiones

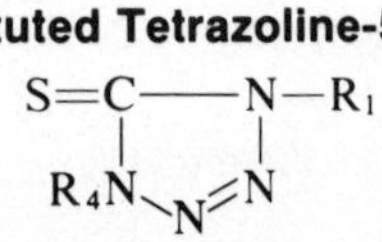

R_1	R_4	Melting Point (°C)	Ref.
CH_3	CH_3	107	1095
$C_6H_5CH_2$	$NCCH_2CH_2$	59	629
C_6H_5	$FSO_2CH_2CH_2$	89–90	613, 614
CH_3	C_6H_5	49	1048
$C_6H_5CH_2$	$C_6H_5CH_2CO$	63–66	628
C_6H_5	$(C_2H_5)_2NCH_2$	110–111	883
C_6H_5	$(C_6H_5)_3Sn$	129–137	281

2.6.2. Tetrazoles

Tetrazoles are five-membered, doubly unsaturated heterocycles consisting of one carbon and four nitrogen atoms. The 5-monosubstituted tetrazoles carry a

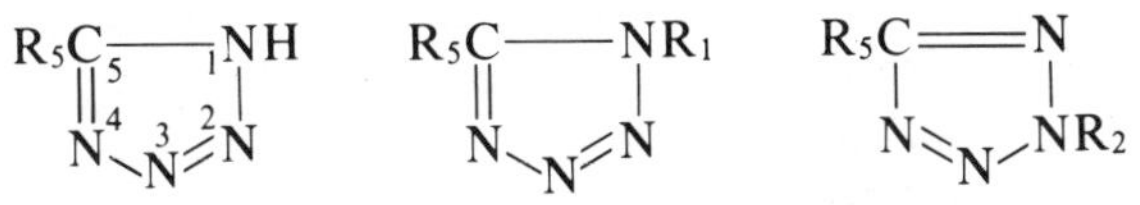

hydrogen on ring nitrogen, arbitrarily indicated at position 1. Ring nitrogen substitution is at the 1- or 2-positions. Alkyl, aryl, heterocyclic, acyl, and non-carbon substituents have been introduced at each position.* Tetrazoles fused to other rings are discussed in Section 2.6.6, Embedded Four-Nitrogen Chains.

The majority of monocyclic tetrazoles are solids, but liquids are found in each major subdivision. Liquid tetrazoles include those with a 5-trifluoromethyl or similar perfluoroalkyl group, with or without lower alkyl substitution at N-1 or N-2. Monosubstituted tetrazoles such as the 1-ethyl, allyl, and trimethylsilyl derivatives are liquids, as are those with lower alkyl or phenyl at the 2-position.

Most tetrazoles with hydrogen on ring nitrogen and alkyl, aryl, or heterocyclic groups at the 5-carbon are stable to 180° (Table 2.71). Kinetics of thermal decomposition of 5-phenyltetrazoles are not significantly influenced by the presence of either electron-releasing or electron-withdrawing substituents on phenyl.

Polymers with pendant 5-tetrazolyl groups include poly(5-vinyltetrazole)

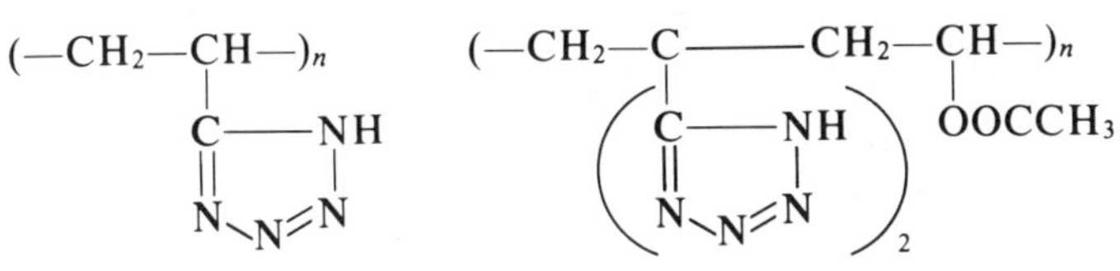

and the copolymer of vinylidene-5-tetrazole with vinyl acetate (462). Both decompose vigorously when heated rapidly to 300°.

Tetrazoles with 5-amino and substituted amino groups decompose from about 200° to 250°, but the 5-methoxy and 5-methylthiotetrazoles are less stable (Table 2.72).

Tetrazoles with strongly electron-withdrawing groups either directly attached to tetrazole carbon or substituted on aliphatic carbon or amino nitrogen

* Butler's excellent review of the tetrazoles (184), emphasizing modern physical techniques, updates Benson's earlier summary (102).

Table 2.71. *N*-Unsubstituted 5-Alkyl, Aryl, and Heteroaryl Tetrazoles

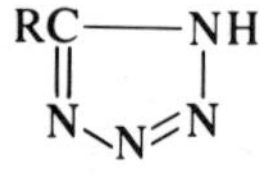

R	Melting Point (°C)	Thermal Stability	Ref.
H	157.5–158;	Dec. 180–225°	740, 834
	sublimes ~50		723
CH_3	145.2–145.7; 148–148.5; sublimes 90–100 in high vacuum		709, 740, 816
$C_6H_5CH_2$	125.5–126		700
$CH_2{=}CH$	126–127		43
CF_3	Bp 81–82/5	Dec. 180°	161, 813
C_6H_5	222 dec.; 214.5–215.5 dec.		41
		$t_{1/2}$ (180.6°, benzonitrile) 8.0 hr	708
p-$CH_3OC_6H_4$	237 dec.; 232–233 dec.		41
		$t_{1/2}$ (180.6°, benzonitrile) 8.3 hr	708
p-$O_2NC_6H_4$	225–226 dec.;		41
	213.5–214.7 dec.	$t_{1/2}$ (180.6°, benzonitrile) 7.9 hr	708
p-BrC_6H_4	172 dec.		41
p-NCC_6H_4	>300 dec.; >320 dec.		41, 324
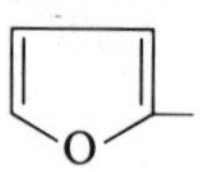	203–204		886
	250 dec.		43
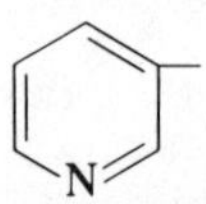	234–235 dec.		727

attached at the 5-position are less stable than other 5-monosubstituted derivatives.

Infrared spectral data show that the potential 5-hydroxy, thiol, or cyanaminotetrazoles exist, instead, as tetrazolines.

Thermal decomposition of 5-monosubstituted tetrazoles causes expulsion of hydrogen azide, nitrogen, or both. Thus hydrogen azide is evolved from the

Table 2.72. *N*-Unsubstituted Tetrazoles with 5-Nitrogen, Oxygen, or Sulfur Substitutents

R	Melting Point (°C)	Thermal Stability	Ref.
H_2N	203 dec.	Dec. 200–210° in 1-methylnaphthalene	919, 1082
C_6H_5NH	206; 209–210		1046, 1187
$(CH_3)_2N$	237–239 dec.; 240 dec.		1187 843
H_2NNH	199 dec.		1085
$C_6H_5CH{=}NNH$	235	Dec. 251	830, 1085
p-$(CH_3)_2NC_6H_4N{=}N$	150 dec.	Expl. ~155°	1082
N_3	72–73; 79.6–80.2	Expl. 217°	664, 710
CH_3O	159 dec.		1044
C_6H_5O	144; 137–138		407, 715
CH_3S	151 dec.; 144–146 dec.		345, 662

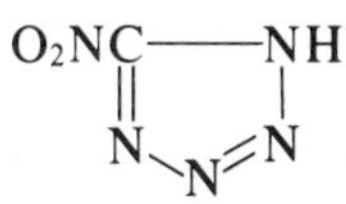

Visc. liquid (555)
expl. without external stimulus in 2–3 weeks

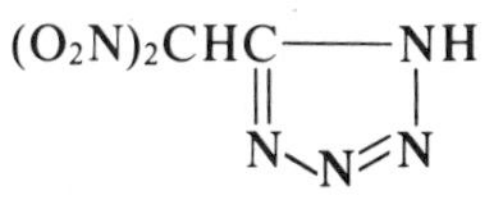

White crystals (289)
fumes 100°, darkens >150°, turns black 180°

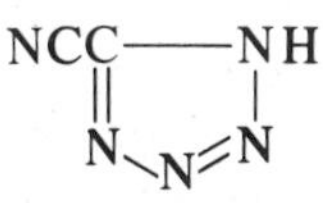

mp 103–104° dec. (722)

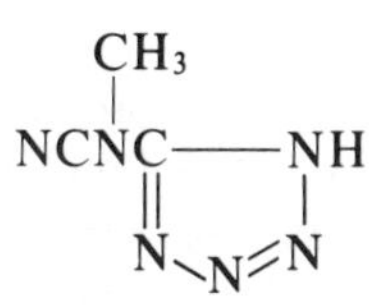

mp 129° dec. (814)

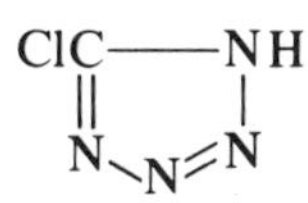

mp 73° (1042)
dec. ~ 120°

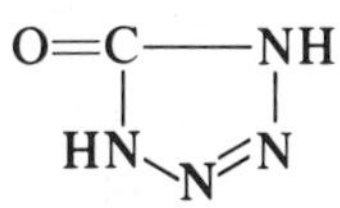

mp 254° dec. (1042)
260° dec. (441)

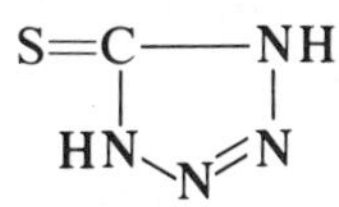

mp 205° dec. (345)

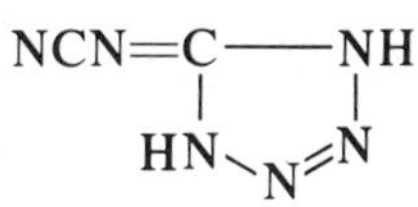

expl. 120° (814)

thermolysis of tetrazole (834) and 5-tetrazolyl thioether (662), leaving hydrogen cyanide and a thiocyanate, respectively. This decomposition path is followed in

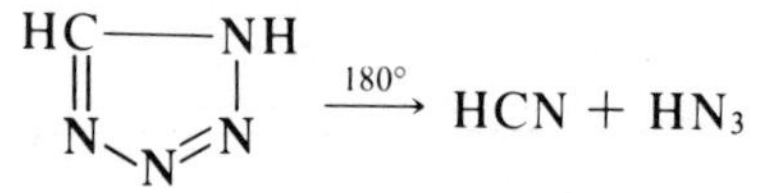

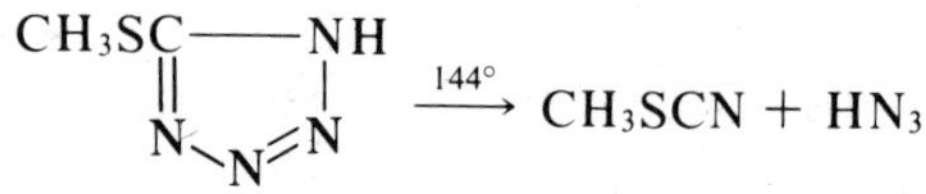

part on thermal destruction of 5-phenyltetrazole, forming benzonitrile, but in addition, nitrogen is eliminated and an intermediate phenyl nitrile imine is formed (523). Products isolated are the result of combination of benzonitrile

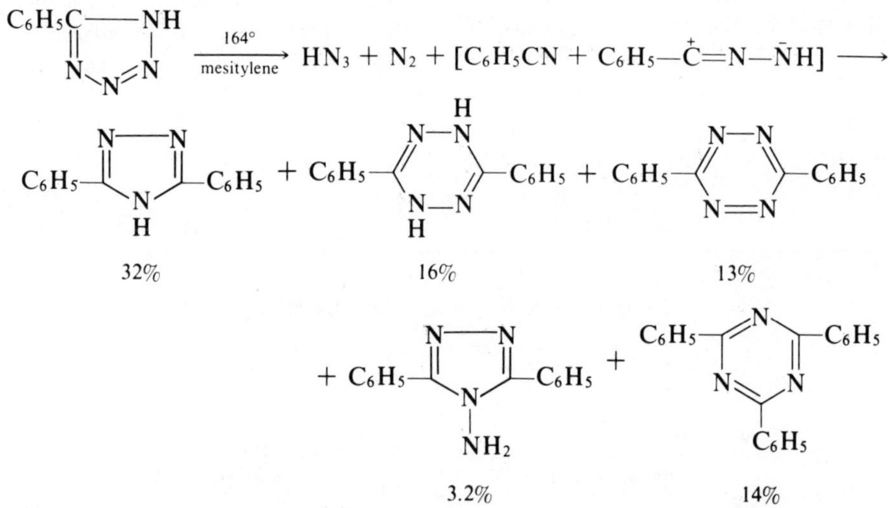

and the nitrile imine both individually and together. Pyrolysis of 5-phenyltetrazole in a vacuum, in addition to the competing fission to hydrogen azide and benzonitrile, causes the stepwise loss of nitrogen, also through the phenylnitrile imine, to phenyldiazomethane (380, 1154). At higher temperature, the remaining nitrogen is eliminated, forming phenyl carbene which undergoes ring con-

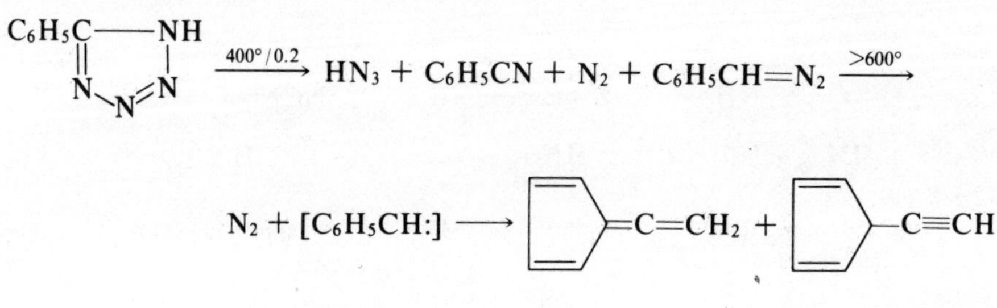

traction to fulveneallene and ethynylcyclopentadiene. On pyrolysis of 5-(2-pyridyl) and (2-pyrazinyl)tetrazole, the presumed diazomethyl intermediates cyclize to fused triazolo derivatives (1151).

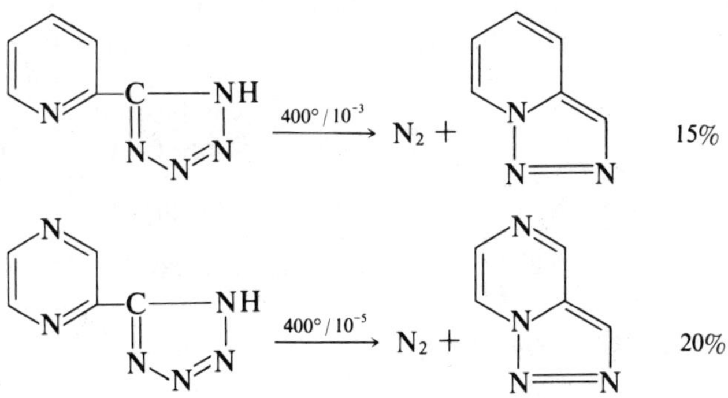

Pyrolysis of 5-^{13}C-benzyltetrazole forms benzyl cyanide and styrene, labeled respectively in the cyano group and at the β-carbon, showing no phenyl migration (227).

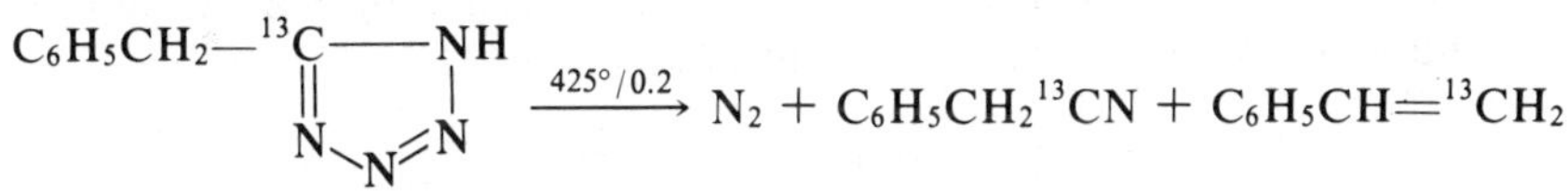

Thermolysis of a 5-substituted methyltetrazole which can eliminate a small molecule, HX, such as hydrogen chloride, water, or ammonia, with migration of another substituent on methyl, leads to an acetylene derivative (100).

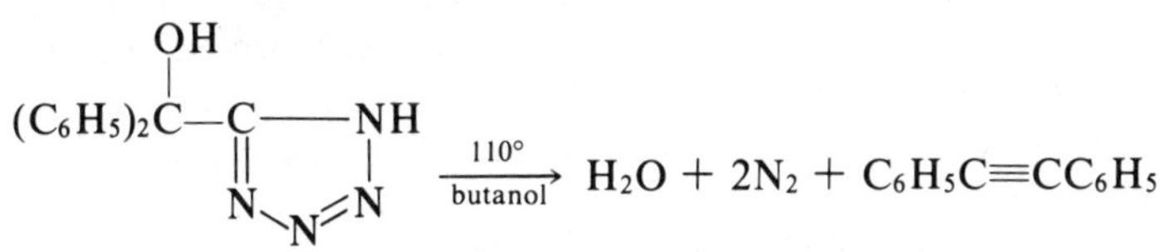

At 200–210° in 1-methylnaphthalene, 5-aminotetrazole eliminates nitrogen and hydrogen azide with formation of hydrazonium azide and an insoluble poly-

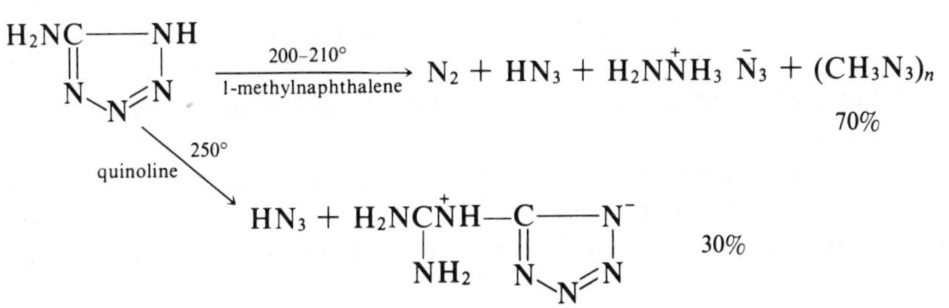

mer. The hydrazonium moiety results from migration of the amino group (919). In quinoline, hydrogen azide is split out and 5-guanidinotetrazole is formed.

Metal salts of 5-monosubstituted tetrazoles usually are more thermally stable than their parent tetrazole acids. Thus both 5-phenyl and 5-perfluoropropyltetrazole decompose at about 180°, but their alkali metal and silver salts are not destroyed until over 240°. The effect of a tetrazolate anion is illustrated by the

	Decomposition Temp. (°C)	Ref.
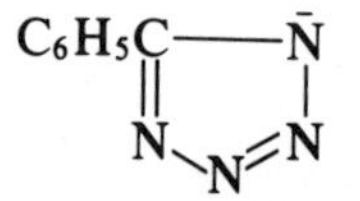 K^+	>240°	685
Ag^+	ignites 286–351°	83
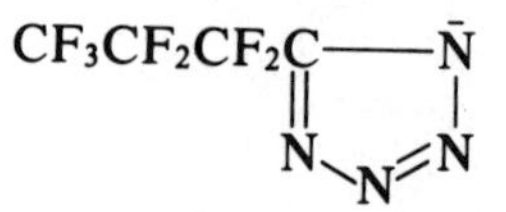 Na^+	258°	161
Ag^+	290–350°	161

difference in stability of the mono- and disodium salts of 5-dinitromethyltetrazole. Since the acidity of the dinitromethyl hydrogen is greater than that of the tetrazole hydrogen, the monosodium salt still contains hydrogen on the tetrazole ring, and is much less stable than the disodium salt (289).

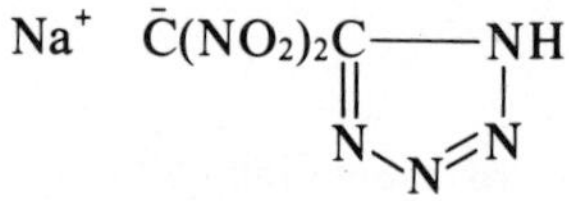

expl. 160°

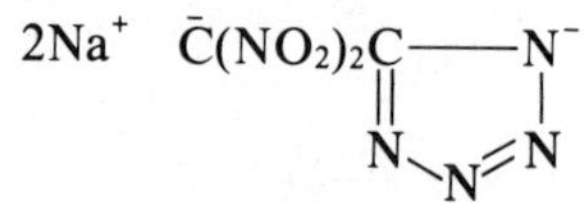

dec. > 310°

Herbst has compared the tetrazole ring with hydrogen on nitrogen to the carboxyl group (468). Acidic character of 5-monosubstituted tetrazoles is discussed in Section 3.7, Acid–Base Properties, but mention must be made of the high decomposition points of tetrazole analogs of amino acids. It must be as-

$H_3\overset{+}{N}CH_2COO^-$

glycine
mp 232–236° dec.

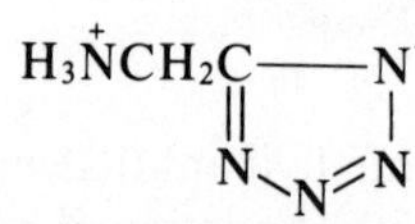

mp 268.5° dec. (728)
280–281° dec. (412)

sumed that the tetrazoles with amino on aliphatic carbon at position 5 also exist as zwitterions.

Other monosubstituted tetrazoles with a sufficiently basic group at the 5-position also exist as zwitterions, either by proton transfer or after alkylation. The high melting point of 5-guanidino tetrazole no doubt signifies a dipolar

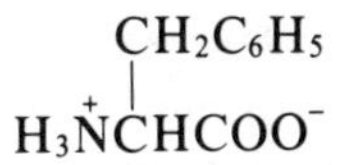

D,L-phenylalanine
mp 271–273° dec.

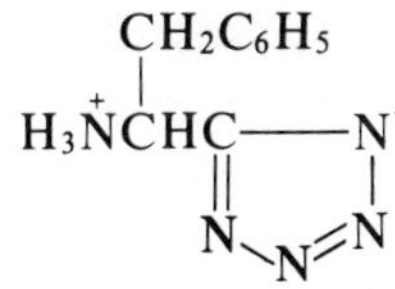

D,L-5-(α-amino-β-phenylethyl)tetrazole
mp 270.5–271.5° dec. (728)

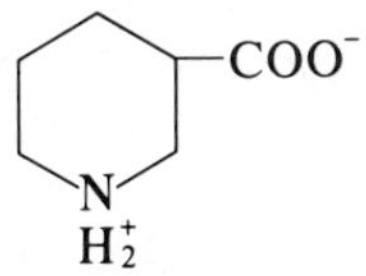

mp 249–250° dec. (634)
261° dec. (347)

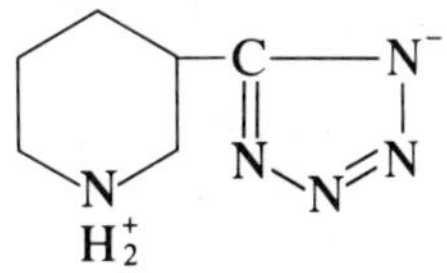

mp 296–297° dec. (727)

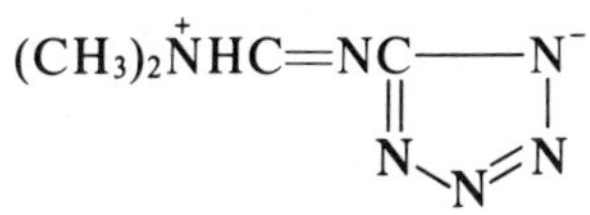

mp 200° (148)

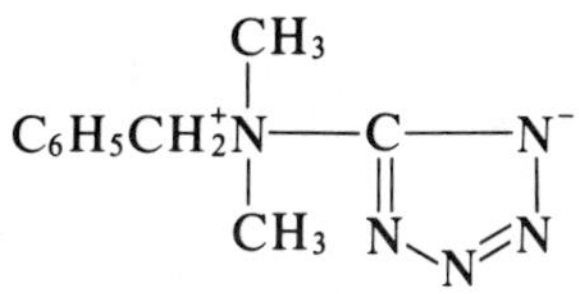

mp 204–205° (514)

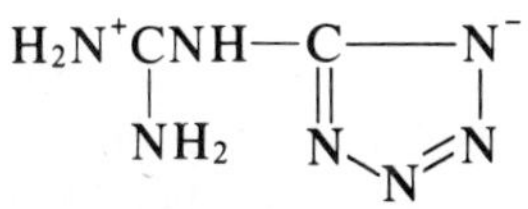

mp > 300° (1042)

form. The explosive tetracene (also called tetrazene) likewise is dipolar. Its structure, finally elucidated by X-ray analysis in 1971 (280) after investigations over 60 years (491, 493, 842), consists of a 1-tetrazene chain, substituted by 5-tetrazolate and amidinium at positions 1 and 3, respectively, together with a molecule of water, hydrogen bonded to these substituents. As indicated by differential scanning calorimetry, tetracene deteriorates at 75° after 8–14 days

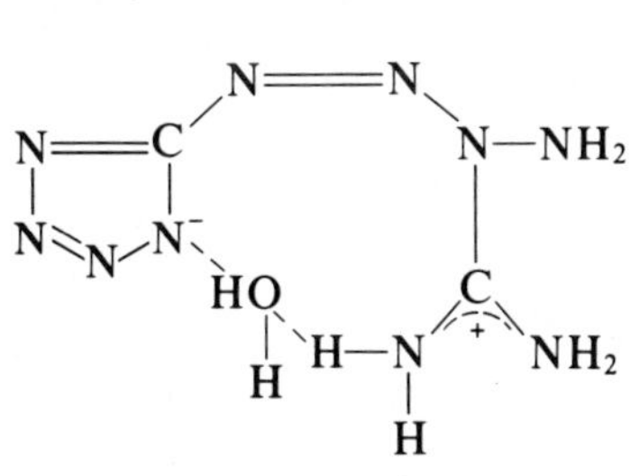

expl. 135° (842)

(633) and when heated at 0.625°/min, shows exothermic peaks at 128° and 130° (815). Thermal decomposition at 90° converts tetracene to nitrogen and 5-aminotetrazole. Based on studies of decomposition with ^{15}N-labeled material in both the tetrazole ring and the tetrazene chain, the 5-aminotetrazole derives from both parts of the tetracene molecule (119).

The least stable tetrazole is 5-tetrazolyldiazonium chloride, its acidic aqueous solutions exploding at 0° if more concentrated than 2% (1082, 1085). Even at pH 5, solutions of diazotized 5-aminotetrazole decompose violently (295). Indeed, ethereal solutions of the diazonium chloride have exploded at −78°. Although the solid is extemely unstable, it can be isolated. Its thermolysis above 80° is a source of atomic carbon (988a, 989). In contrast, the disodium salt of

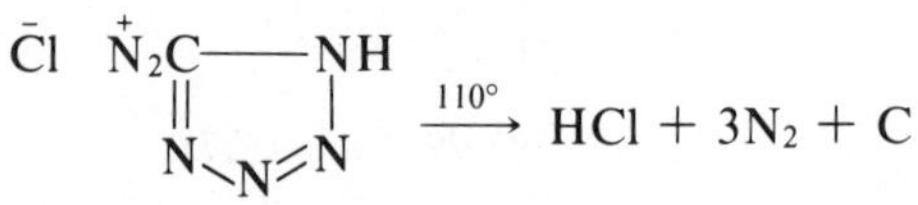

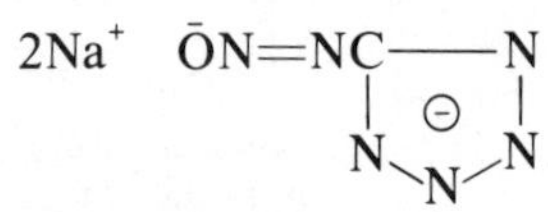

tetrazole-5-diazotate is much more stable, surviving heating at 80° in aqueous solution (1085).

Decomposition points of the 1-substituted tetrazoles (Table 2.73) range from approximately 90° to over 300°. The least stable of this subclass, 1-benzyl-5-nitrosaminotetrazole, explodes at 87° (185) but other 1-substituted 5-nitrosaminotetrazoles are more stable. The 1-aryl tetrazoles with a 5-halogen or methylthio group are less thermally resistant than 5-alkyl or aryl derivatives. As measured by the kinetics of decomposition at 175°, the most stable 1-phenyl tetrazoles are the 5-methyl, phenyl, and amino derivatives. Except for 1-*p*-nitrophenyl-5-bromotetrazole, most 1-*p*-nitrophenyltetrazoles have decomposition points over 200°. 1,5-Bis(*p*-carboxyphenyl)tetrazole melts above 300°, probably because of intermolecular hydrogen bonding. Tetrazoles with 1-hydroxyl and acyloxy groups are discussed in Section 2.16.2 on imidoyl azide-tetrazole ring-chain isomers and equilibria.

Polymers have been made with 1,5-tetrazole units in the polymer chain. The polyaminotetrazole copolymers decompose rapidly between 200° and 300° during thermogravimetric analysis, consistent with the behavior of monomeric

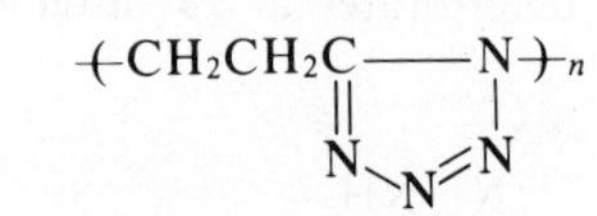

softens 100° (291)

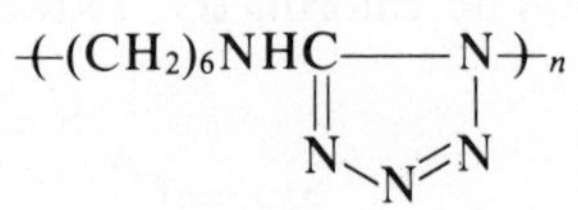

dec. 125° (286)

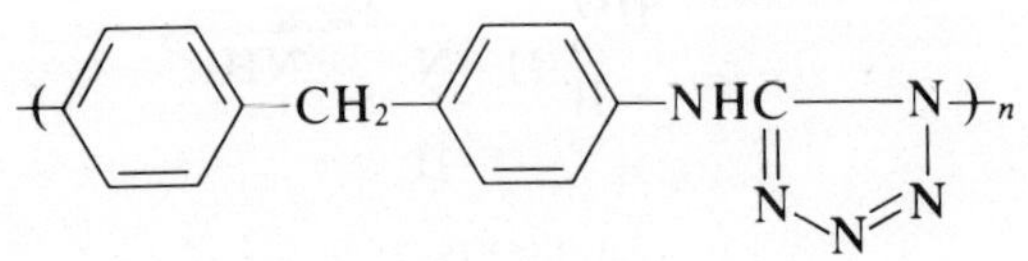

dec. > 300° (286)

Table 2.73. 1-Substituted Tetrazoles

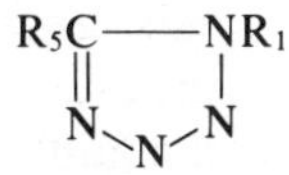

R_1	R_5	Melting Point (°C)	Thermal Stability	Ref.
CH_3	H	36		1046
CH_3	C_6H_5	104–105 Bp 144–146/0.3		436 147
CH_3	CHO	76–79		747
$CH_2{=}CH$	H	Bp 94/1		320
$O_2NCH_2CH_2$	CF_3	Bp 115/10^{-2}		734
C_6H_5	CH_3	97.5–99	$t_{1/2}$ (175°, butylcarbitol) 28.9 hr	436, 579
C_6H_5	C_6H_5	144–145	$t_{1/2}$ (175°, butylcarbitol) 14.4 hr	436, 579
C_6H_5	NH_2	160	$t_{1/2}$ (175°, butylcarbitol) 11.6 hr	579, 773
C_6H_5	CH_3S	78.5–80	$t_{1/2}$ (175°, butylcarbitol) 34 min	78, 579
C_6H_5	Cl	123.4–124.6	$t_{1/2}$ (175°, butylcarbitol) 23 min	579
m-$HOOCC_6H_4$	C_6H_5	224–226 dec.		96
p-$HOOCC_6H_4$	*p*-$HOOCC_6H_4$	310		147
p-$O_2NC_6H_4$	*p*-$O_2NC_6H_4$	262		147
p-$O_2NC_6H_4$	Br	120 dec.		1133

tetrazoles. Several vinyl polymers with 1,5-tetrazole moieties in the side chain are known.

Tetrazoles with 5-amino substituents bearing hydrogen in principle could be 5-iminotetrazolines, but both for derivatives unsubstituted in the ring and for 2-

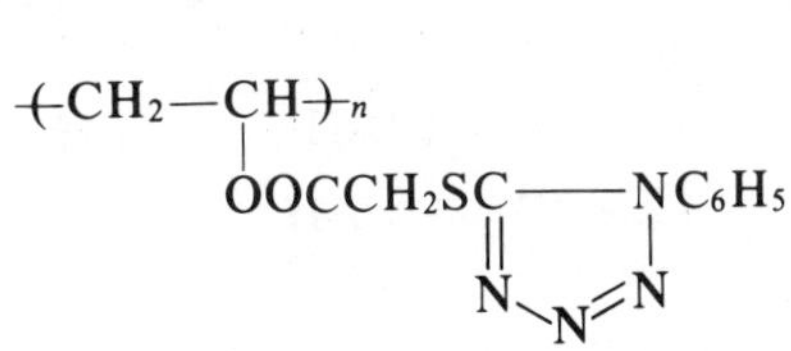

(1106)

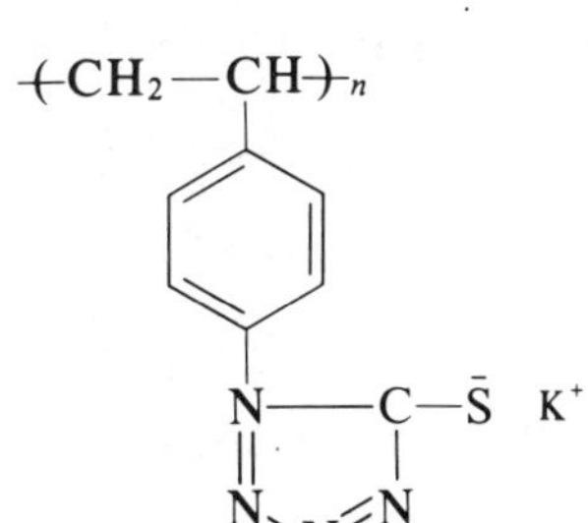

(401, 402)

substituted 5-aminotetrazoles, spectral data indicate the amino form. Most 1-substituted 5-aminotetrazoles also exist as such, but when a strongly electron-attracting substituent is present on the exocyclic nitrogen, imino forms are possible.* Spectral data indicate that an equilibrium exists between the amide and

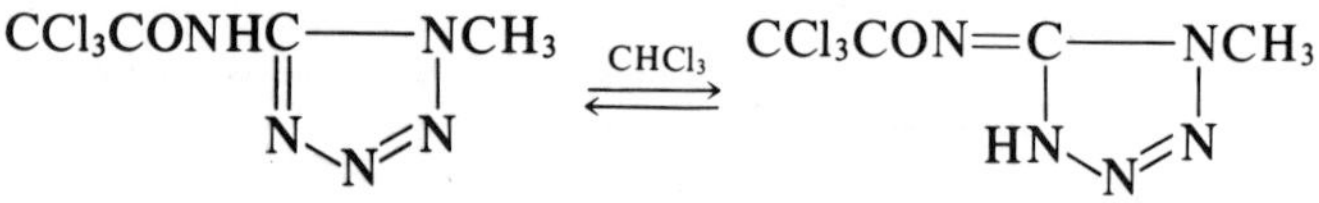

mp 138–139° (984)

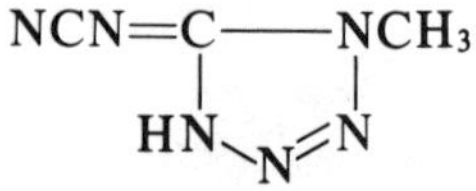

mp 162° dec. (814)

imide forms of 1-methyl-5-trichloroacetylaminotetrazole (985). In the solid state, however, 1-methyl-5-trifluoroacetyl, *p*-toluenesulfonyl (985), and cyan-aminotetrazole exist instead as iminotetrazolines. Similarly, infrared spectra of the solids indicate that the potential 5-hydroxy and 5-thiol 1-substituted tetrazoles are tetrazolines. NMR data show these forms also to predominate in deuterated chloroform and dimethyl sulfoxide solution (64).

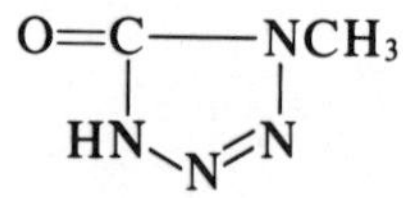

mp 122–124° (441)

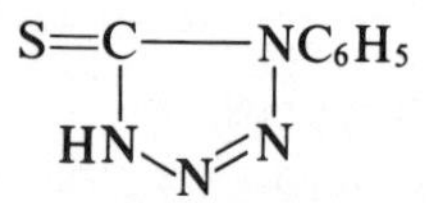

mp 147–148° dec. (667)

The tetrazole ring survives some thermal treatments in which substituents migrate internally. Thus 1-phenyltetrazoles with 5-allyloxy type radicals undergo Claisen rearrangements in high yield to 4-substituted 1-phenyltetrazolinones in half-lives of $\frac{1}{2}$ to 9 hr, illustrated by conversion of 5-crotyloxy-1-phenyltetrazole (298).

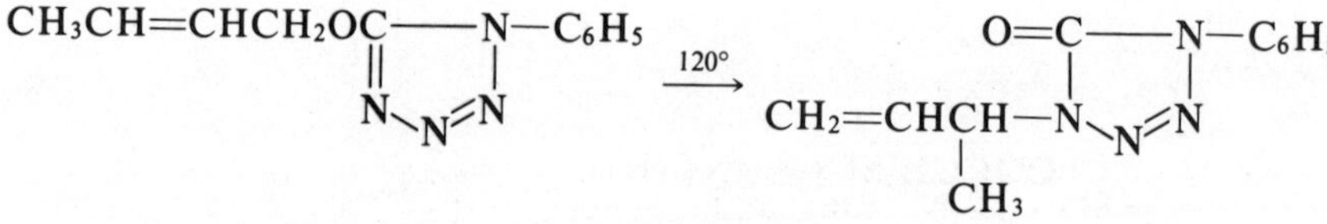

Heating 5-alkylamino, 1-aryl-5-amino (323), 1-aryl-5-hydrazino (808), and 1-aryl-5-alkylamino (323) tetrazoles effects isomerization to 1-alkyl-5-amino

*Katritzky has vigorously advocated the accurate portrayal of proper tautomeric forms, both of heterocycles in general (292, 576) and of aminotetrazoles in particular (328).

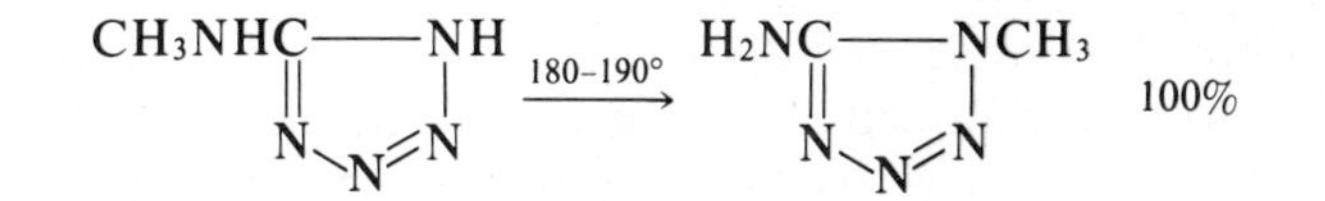

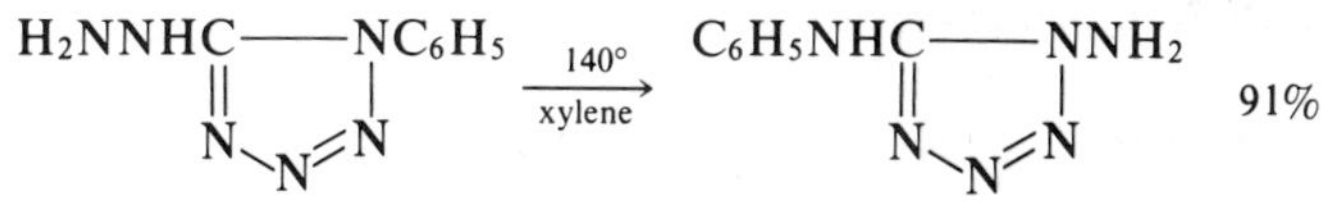

and 1-substituted-5-substituted aminotetrazoles in the Dimroth reaction. The kinetics are first order (467), attaining an equilibrium in many instances starting from either isomer. A substituted guanyl azide which can undergo ring closure in either direction appears to be the intermediate. For ring closure to occur, the azide group must be bent at the middle nitrogen. Electron-withdrawing substituents on the amino group at tetrazole carbon shift the equilibrium toward the

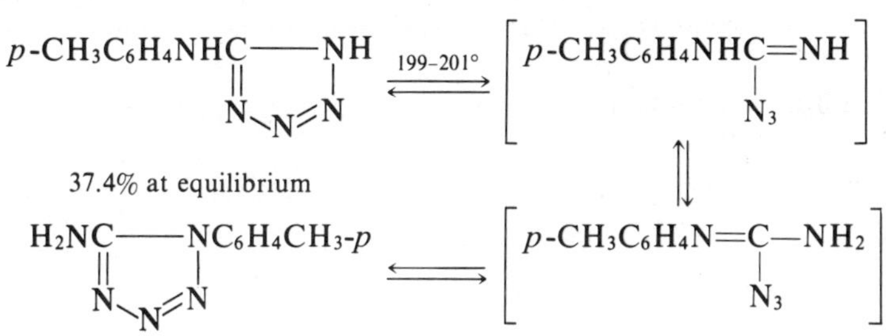

substituted amine form. An approximately linear inverse relationship is found between the pK_a of the 5-substituted aminotetrazole and the proportion of this form at equilibrium (465).

In similar manner, 1-aryl-5-thionotetrazolines rearrange to thiatriazoles (579).

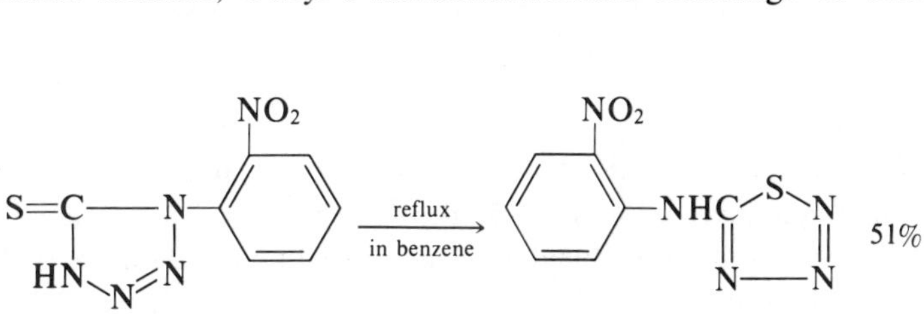

Establishment of the structures of ring-acylated tetrazoles has required NMR and infrared spectral information. The acetyl derivative of tetrazole, bis-

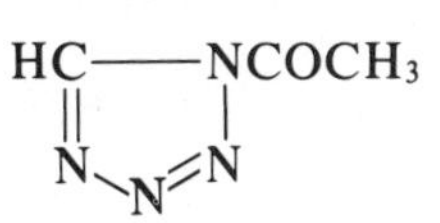

mp 58–59°
(603, 1025, 1138)

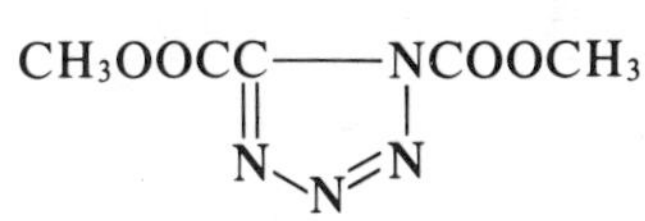

mp 77–80° (761)

(methoxycarbonyl)tetrazole, and the acyl derivatives of 5-alkylamino- or arylamino tetrazole are substituted in the 1-tetrazole position (469, 470).

A Dimroth rearrangement also occurs with 1-acyl 5-substituted aminotetrazoles, although the 1-acyl derivative usually is not isolated (469). Similarly, 1-

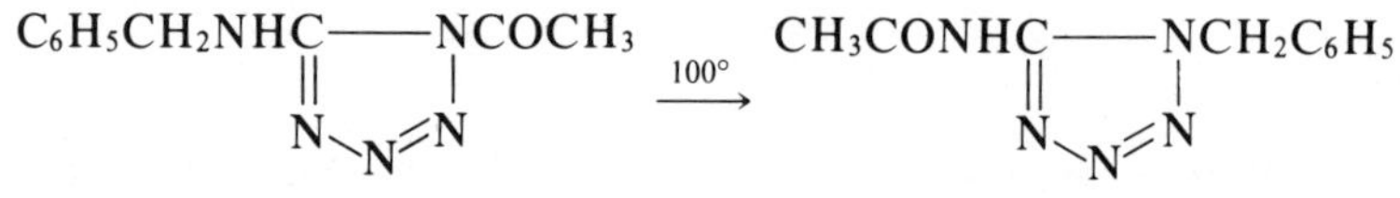

mp 106–108° (469)

alkylcarbamoyl-5-aminotetrazoles isomerize to 5-ureidotetrazoles on heating beyond their melting points; differential thermal analysis indicates an initial endotherm just above the melting point, followed by an exotherm (255). The unsubstituted 5-aminotetrazole-1-carboxamide rearranges nearly quantitatively to 5-ureidotetrazole at 180–190° in the solid phase after a broad exotherm beginning at 150°.

The 1-sulfonyltetrazoles are unstable, decomposing on storage in 4 to 20 days.

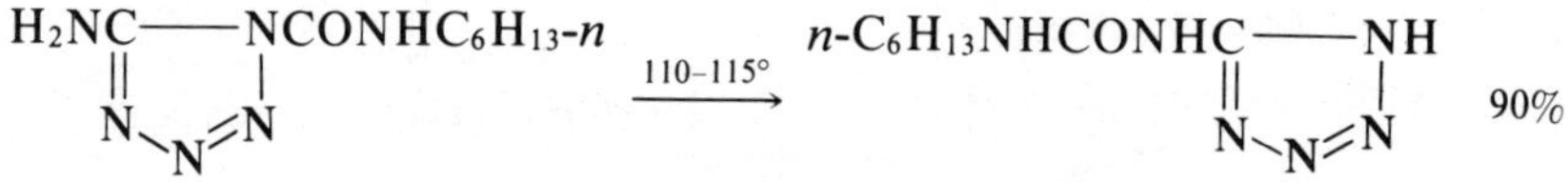

mp 108–110°

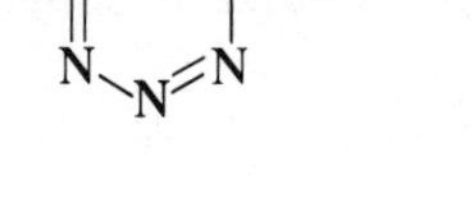

mp 88–90° (1029)

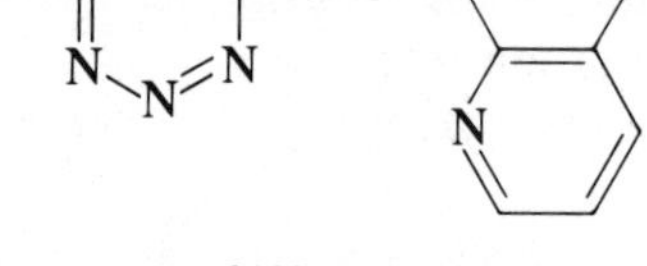

mp 210° dec. (1066)

Thermal decomposition of 1,5-di(methoxycarbonyl)tetrazole, accompanied by loss of nitrogen, gives 2-methyl-3-methoxycarbonyl-1,2,4-oxadiazolin-5-one without migration of the substituent at tetrazole carbon but with internal methylation of ring nitrogen (762).

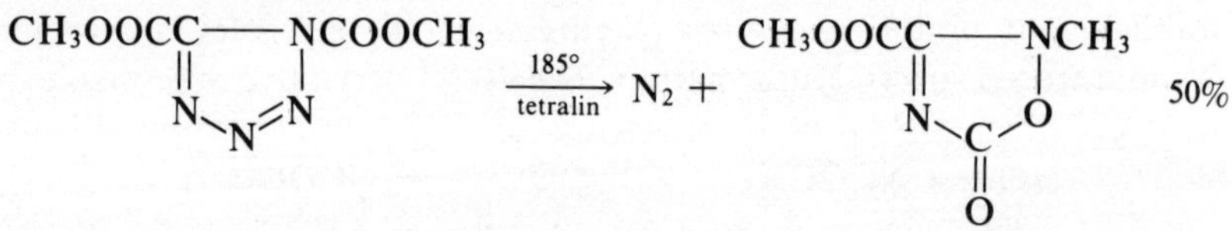

Destruction of the tetrazole ring occurs on thermolysis of 1,5-diaryl tetrazoles with loss of nitrogen and formation of carbodiimide or derivatives, 2-substituted arimidazoles, or both, as in the conversion of 1,5-diphenyltetrazole to diphenylcarbodiimide and 2-phenylbenzimidazole (1015). Electron-withdraw-

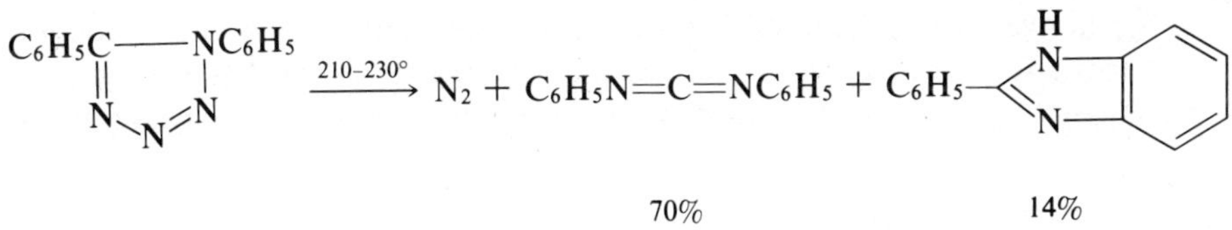

ing substituents on the 5-aryl increase the proportion of benzimidazole (1125). The reaction can be presumed to go through ring opening of the tetrazole to an imidoyl azide, followed by loss of nitrogen and migration of the carbon substi-

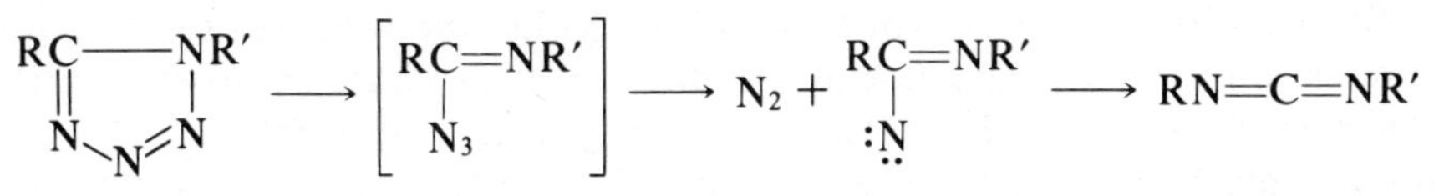

tuent to the nitrene. Addition of copper powder lowers the decomposition temperature of 1,5-diphenyltetrazole by about 60° (1125).

The same mechanism seems to be present at elevated temperatures; pyrolysis of 1-methyl-5-phenyltetrazole leads to methyl phenyl carbodiimide (326).

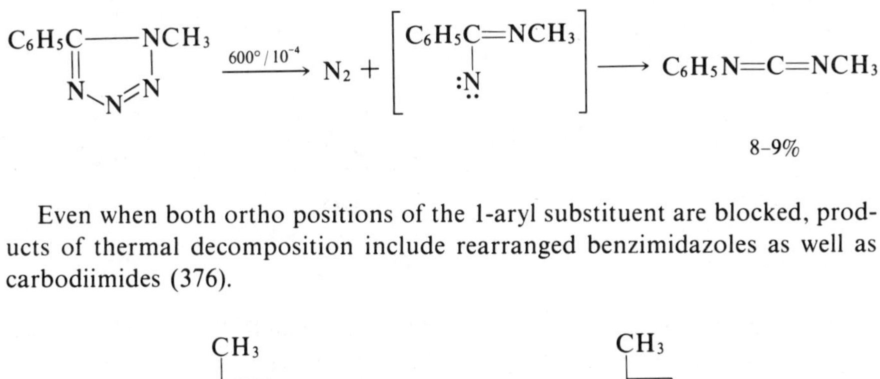

Even when both ortho positions of the 1-aryl substituent are blocked, products of thermal decomposition include rearranged benzimidazoles as well as carbodiimides (376).

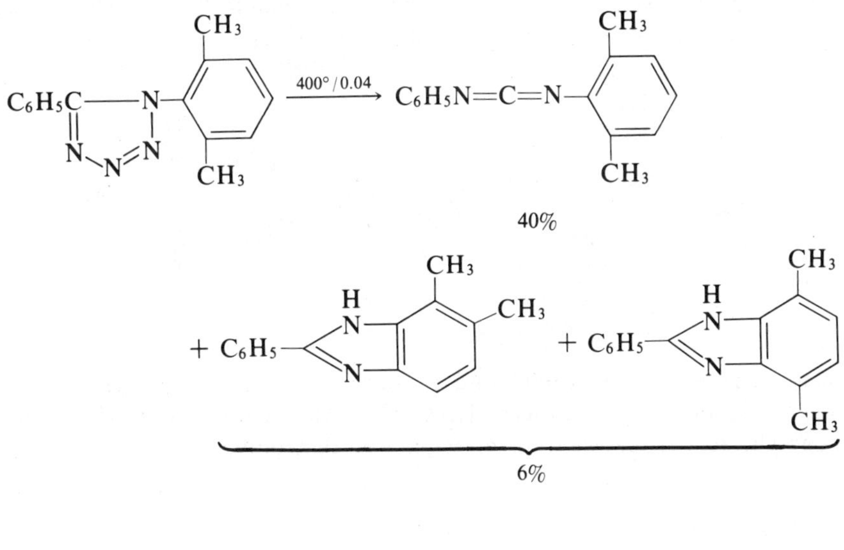

A reactive ortho substituent reduces the decomposition temperature and introduces a further reaction as illustrated by the conversion of 1-*o*-nitrophenyl-5-phenyltetrazole to 2-phenylbenzotriazole (509). Thermal decomposition of

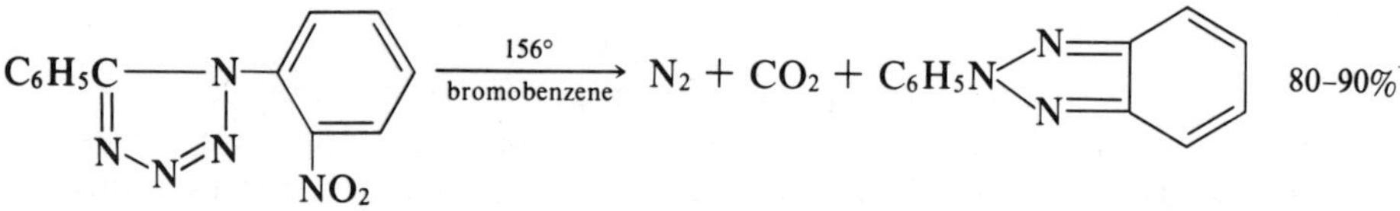

tetrazoles with an *o*-carboxyphenyl group in either the 1- or 5-position causes elimination of hydrogen azide and formation of products incorporating the elements of carboxyl. The major substance formed on thermolysis of 1-*o*-carboxyphenyl-5-phenyltetrazole is 2-phenyl-4*H*-3,1-benzoxazin-4-one (96); the 5-

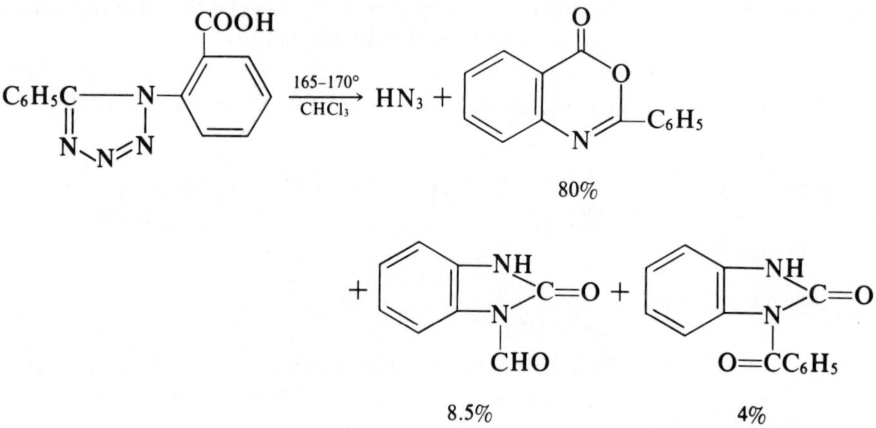

o-carboxyphenyl isomer yields only *N*-phenylphthalanil (97). In related reactions, tetrazoles with α-acrylic acid moieties at the 1-nitrogen form

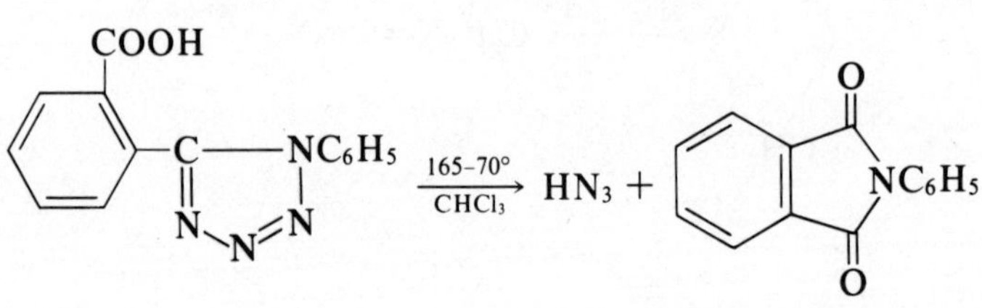

97.6%

4-methylene-5-oxazolones (691) and 1-(α-acrylamide)tetrazoles form imidazolones by copper-catalyzed thermolyses (692). Analogous decomposition of 1-thioacrylic acid tetrazoles leads principally to thiazolones.

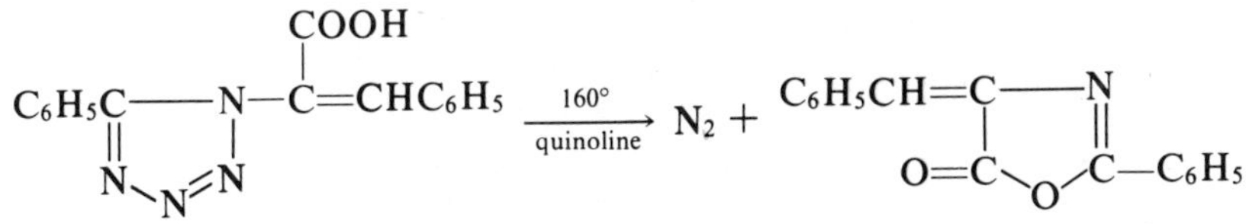

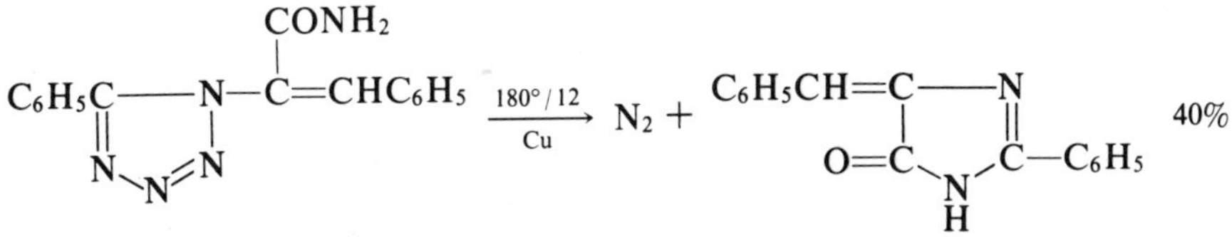

Thermolysis of 1-*p*-nitrophenyl-5-acetylaminotetrazole results in destruction of the tetrazole ring and formation of 2-methyl-5-*p*-nitrophenyl-1,3,4-oxadiazole (470). A probable mechanism is through rearrangement to an acetylamino

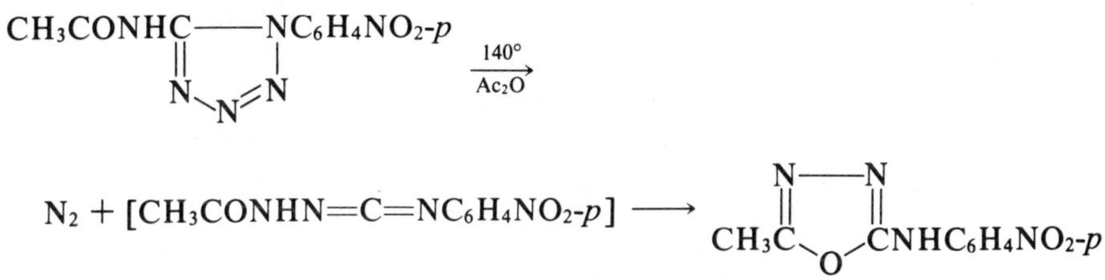

carbodiimide. The corresponding 1-phenyl derivative is stable under the same conditions.

Trimethylsilytetrazoles decompose thermally in similar but distinctive ways. Trimethylsilylcarbodiimide and a cyanamide polymer are formed in the thermolyses of both 1-trimethylsilytetrazole and 1-trimethylsilyl-5-trimethylsilyl-

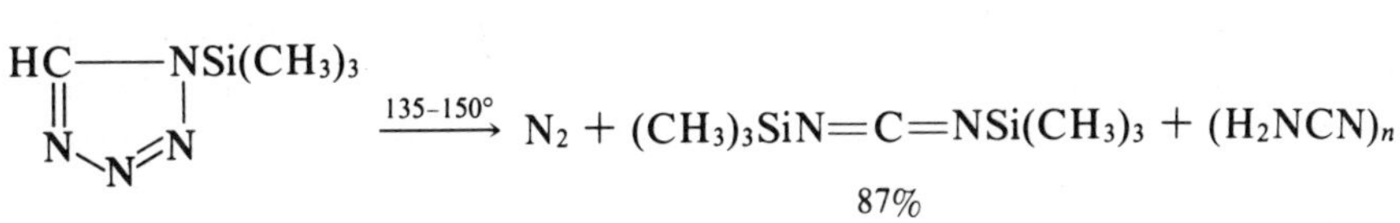

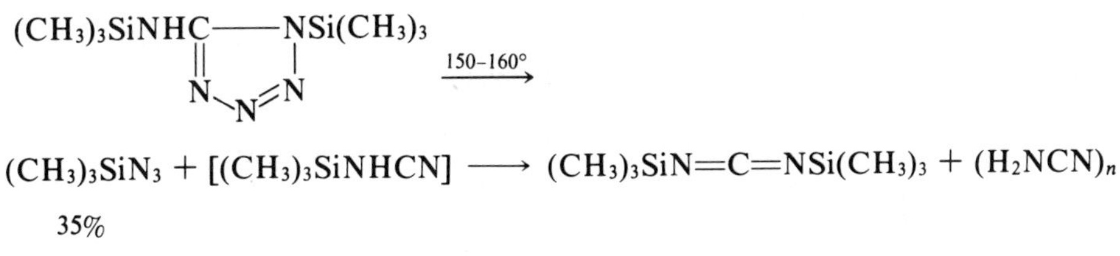

aminotetrazole (120). The initial step in the trimethylsilylaminotetrazole decomposition evidently involves fission to trimethylsilylazide and trimethylsilylcyanamide.

Heating 1-alkoxy-5-phenyltetrazoles causes rearrangement to 2-alkyl-5-phenyltetrazole 4-oxides (875).

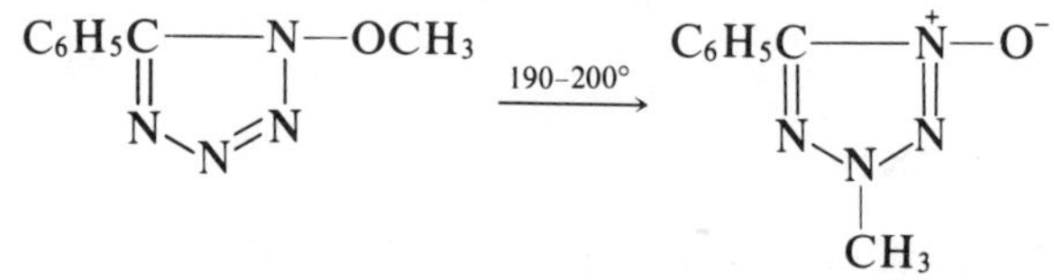

mp 33–35° mp 134.5–135.5°

Since nitrogen-substituted tetrazoles have no hydrogen on the ring available for hydrogen bonding or ion formation, 1- and 2-derivatives melt lower and are much more volatile than their 5-isomers. The 2-substituted tetrazoles boil below their isomers, (Table 2.74) in keeping with their greater molecular symmetry. Although the 2-isomers often are lower melting, exceptions are known, as illustrated. No structural regularity relating the melting points of 1- and 2-tetrazole isomers is apparent (Table 2.75).

Thermal resistance of 2-substituted tetrazoles (Table 2.76) is less than that of 1-substituted isomers. At 175°, the half-life of 2,5-diphenyltetrazole is 21 min (518) compared to 14.4 hr for 1,5-diphenyltetrazole. An electron-withdrawing group at position 2 of the tetrazole ring contributes to low thermal stability indicated by decomposition points in the 95–127° range for the 2-acetyl and 2,4-dinitrophenyl-5-phenyl derivatives. As with the 1-substituted tetrazoles, 2-carboxyphenyl tetrazoles are relatively high melting, often over 350°. Although efforts to prepare 5-carboxytetrazole and 1-substituted 5-carboxytetrazoles result in immediate decarboxylation, both 5-carboxy-2-methyl and 2-phenyltetrazoles are stable compounds.

Electron-releasing groups on 2-substituted phenyl retard the thermal decomposition rates of 2-aryl-5-phenyltetrazoles, but increase those of 2-phenyl 5-substituted phenyltetrazoles (503). Straight line relations are found between the log *k* of these reactions and the Hammett sigma constants of the substituents. The relative stabilizing influence of electron-releasing groups at the 2-position

Table 2.74. Comparative Boiling Points of 1- and 2-Alkyl Tetrazoles

		Boiling Point (°C)	
R	R_5	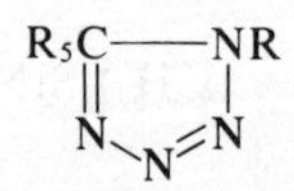	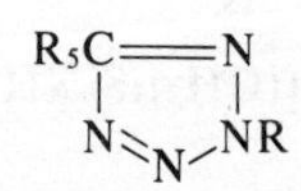
C_2H_5	H	162–164/30 (825)	72/37 (320)
CH_3	CH_3	Mp 73–74 (436)	57.0–57.2/13 (709)
CH_3	CF_3	101–102/46 (813)	67–68/46 (813)

Table 2.75. Comparative Melting Points of 1- and 2-Substituted Tetrazoles

		Melting Point (°C)			
R	R_5	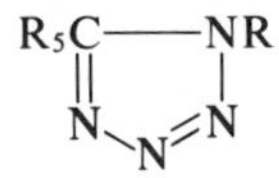		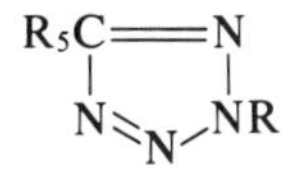	
CH_3	NH_2	232	(773)	104.5–105.5	(464)
$C_6H_5CH_2$	C_6H_5	92–93	(515)	70–71	(515)
C_6H_5	CHO	97	(747)	73–74	(748)
p-$O_2NC_6H_4$	H	202–203 dec.	(677)	145–145.5	(677)
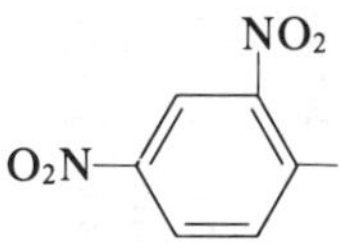	C_6H_5	183	(113)	127–128 dec.	(61)
CH_3	NO_2	57.8–58	(464)	85–86	(464)
CH_3	*m*-$O_2NC_6H_4$	145–146	(765)	174–175	(765)
$C_6H_5CH_2$	$(CH_3)_2N$	77–78	(514)	94–95	(514)
p-$O_2NC_6H_4$	C_6H_5	155–157	(571)	187–188 dec.	(551)

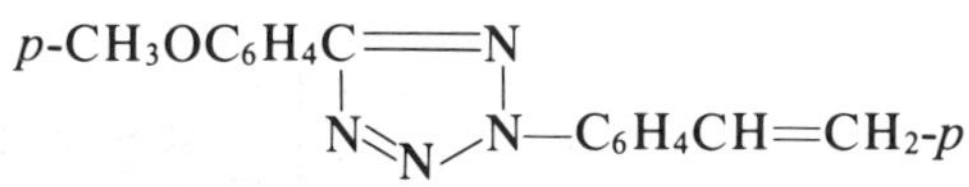

mp 153–154°
dec. 175°

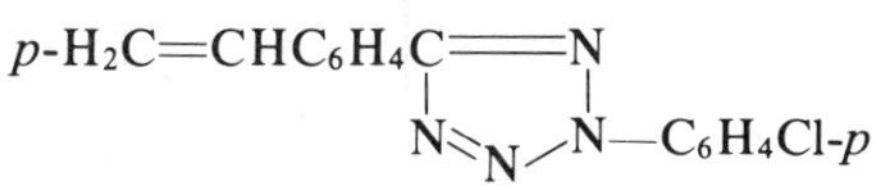

mp 103–104°
dec. 160°

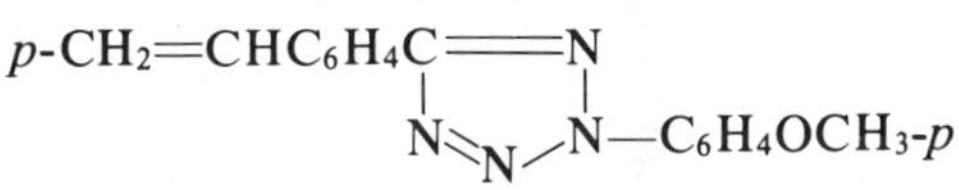

mp 81–82°
dec. 220°

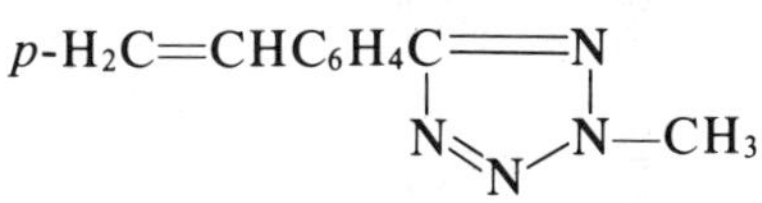

mp 61–62°
dec. 240°

Table 2.76. 2-Substituted Tetrazoles

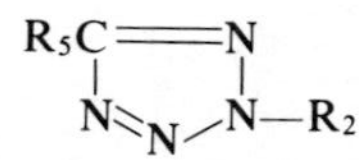

R_2	R_5	Melting Point (°C)	Thermal Stability	Ref.
CH_3	H	Bp 145–147		825
CH_3	C_6H_5	50.5–51 Bp 160–170/0.01	Dec. 190–200/0.01	461, 518
CH_2=CH	H	Bp 66–68/60		320
C_6H_5	COOH	137–138		1148
C_6H_5	C_6H_5	99.5–100; 101–103	$t_{1/2}$ (145.8°, 1-chloronaphthalene) 4.0 hr	503, 524
			$t_{1/2}$ (165.8°, 1-chloronaphthalene) 42 min	503
C_6H_5	*p*-$CH_3C_6H_4$	92–93	$t_{1/2}$ (165.8°, 1-chloronaphthalene) 39 min	503

C_6H_5	*p*-NCC_6H_4	152–154	$t_{1/2}$ (165.8°, 1-chloronaphthalene) 1 hr	503
p-$CH_3C_6H_4$	C_6H_5	103–104	$t_{1/2}$ (165.8°, 1-chloronaphthalene) 1.1 hr	503
p-ClC_6H_4	C_6H_5	120.5–121.5	$t_{1/2}$ (165.8°, 1-chloronaphthalene) 21 min	503
p-$HOOCC_6H_4$	C_6H_5	350°		982
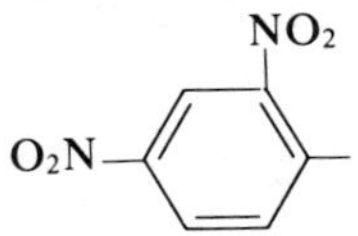	C_6H_5	127–128 dec.		601
NC	C_6H_5O	72–73		715
$(CH_3)_3Si$	C_6H_5	25–27 Bp 104–106/0.05		120
$(CH_3)_3Sn$	CH_3	215–216 with sublimation		995

is illustrated by the decomposition points of 2,5-tetrazoles with styryl substituents (1035). Copolymers of each of these styryl tetrazoles decompose at substantially the same temperature as the corresponding monomers.

Low molecular weight copolymers with repeating units consisting of *p*-xylylene and two 2,5-tetrazole groups, variously connected, have been prepared (286). The copolymer with a methylene between the two tetrazole carbons is

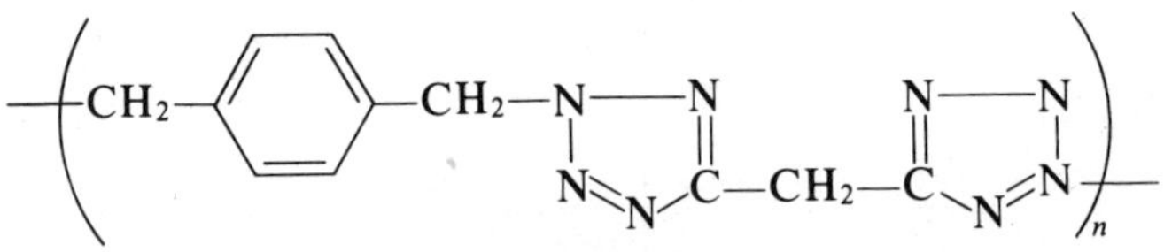

MW 2050
dec. 120°

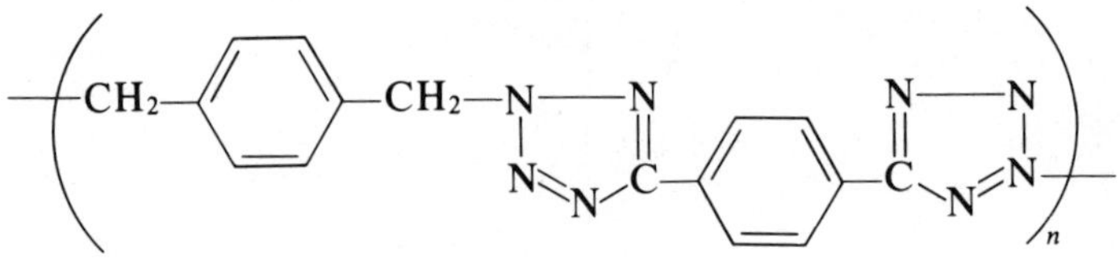

MW 1350
dec. 240°

relatively unstable in comparison with the polymer with a *p*-phenylene group. Other 2,5-tetrazole polymers with an acetamido connecting unit, probably of higher molecular weight, decompose from above 190° to over 250° (290). Poly-(2-methyl-5-vinyltetrazole), with the tetrazole group in the side chain, evolves gas linearly but not autocatalytically at 145° in the vacuum stability test; it decomposes in 2–5 sec at 300° (322).

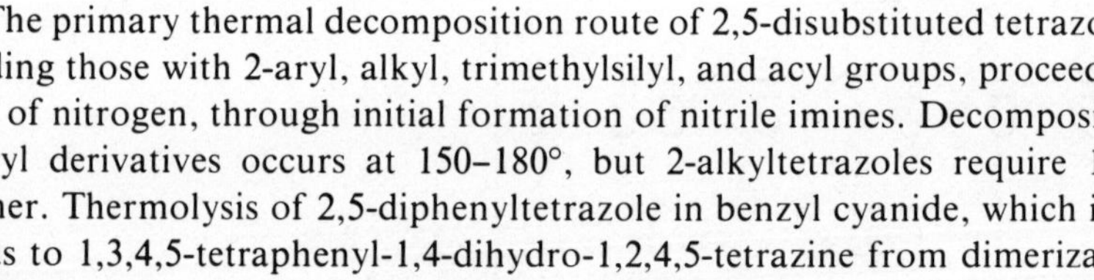

The primary thermal decomposition route of 2,5-disubstituted tetrazoles, including those with 2-aryl, alkyl, trimethylsilyl, and acyl groups, proceeds, with loss of nitrogen, through initial formation of nitrile imines. Decomposition of 2-aryl derivatives occurs at 150–180°, but 2-alkyltetrazoles require 190° or higher. Thermolysis of 2,5-diphenyltetrazole in benzyl cyanide, which is inert, leads to 1,3,4,5-tetraphenyl-1,4-dihydro-1,2,4,5-tetrazine from dimerization of the nitrile imine (524). In the presence of a reactive solvent the nitrile imine is intercepted; thus with *p*-tolunitrile, 1,3-diphenyl-5-*p*-tolyl-1,2,4-triazole is formed (518) and with aniline, α-anilinophenylmethylene phenylhydrazone results (524). From thermal decomposition of 2-methyl-5-phenyltetrazole without solvent, methyl azide and benzonitrile result by one reaction path, and nitrogen and *C*-phenyl-*N*-methylnitrile imine are formed by a second route (518). The

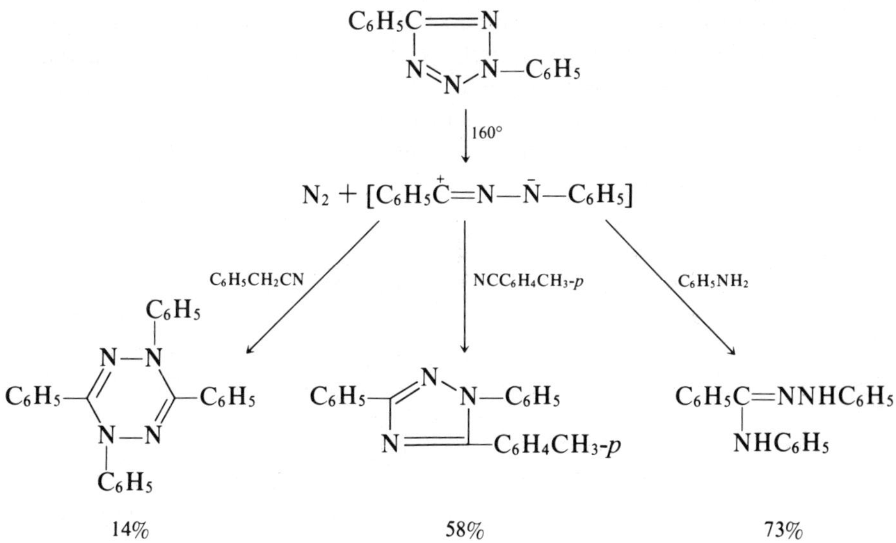

benzonitrile trimerizes to cyaphenin and combines with the nitrile imine to give 1-methyl-3,5-diphenyl-1,2,4-triazole. Flash vacuum pyrolysis of 2-methyl-5-phenyltetrazole also goes through the nitrile imine, forming phenylmeth-

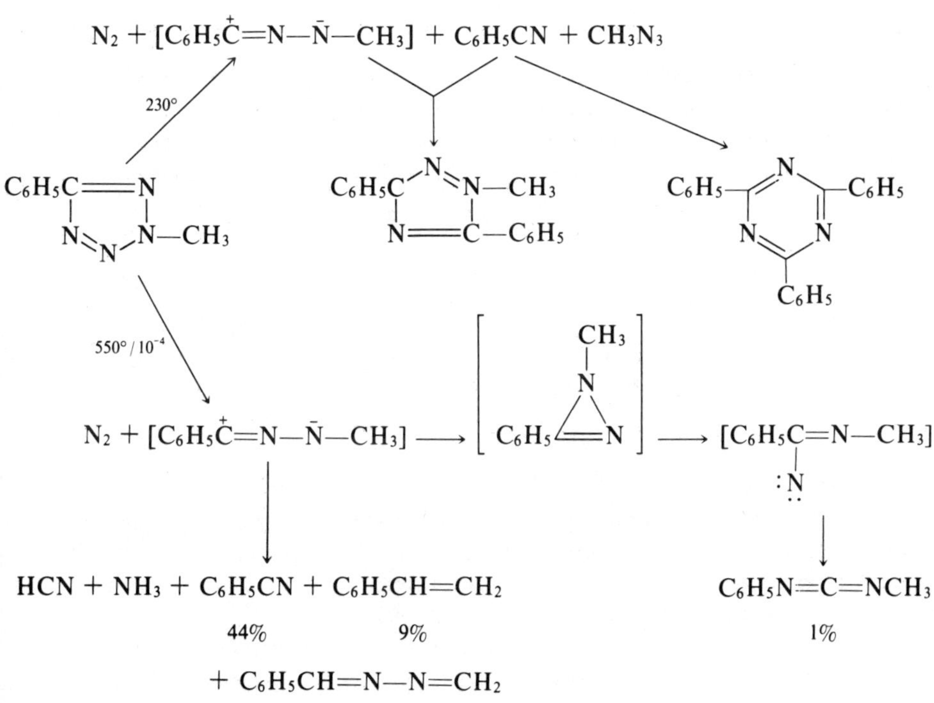

ylene-methylene azine by a 1,4-hydrogen shift, as well as other decomposition products, and by rearrangement, methyl phenyl carbodiimide (326). The thermolysis of 5-phenyl-2-trimethylsilyltetrazole also goes through a nitrile imine as shown by its interception by ethyl cinnamate to form a pyrazoline (120). In the absence of a reactive solvent, the nitrile imine dimerizes with migration of a trimethylsilyl group.

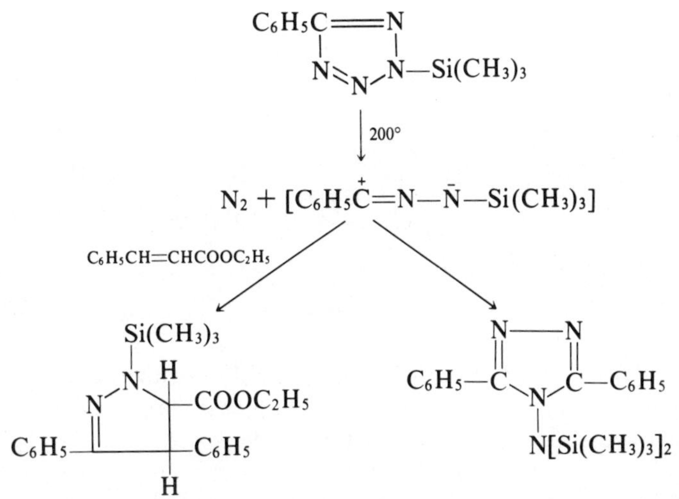

Many of the possible reactions of nitrile imines have been carried out on compounds with both a 2,5-disubstituted tetrazole and a reactive structural element in the same molecule. A polyphenylenepyrazole results from thermal decomposition of 5-*p*-ethynylphenyl-2-phenyltetrazole by combination of the

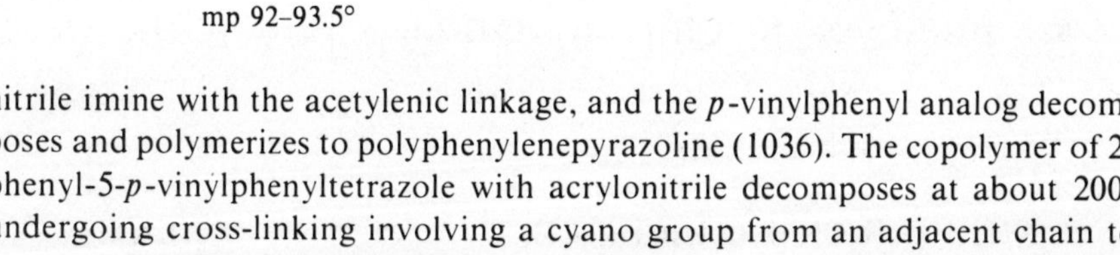

mp 92–93.5°

nitrile imine with the acetylenic linkage, and the *p*-vinylphenyl analog decomposes and polymerizes to polyphenylenepyrazoline (1036). The copolymer of 2-phenyl-5-*p*-vinylphenyltetrazole with acrylonitrile decomposes at about 200° undergoing cross-linking involving a cyano group from an adjacent chain to form a 1,2,4-triazole ring (1037).

Intramolecular cyclization by condensation of a nitrile imine with an amino group is illustrated by formation of a fervenulin from 1,3-dimethyl-5-amino-6-(5-phenyltetrazolyl)uracil (482). Internal combination with a nitro group and

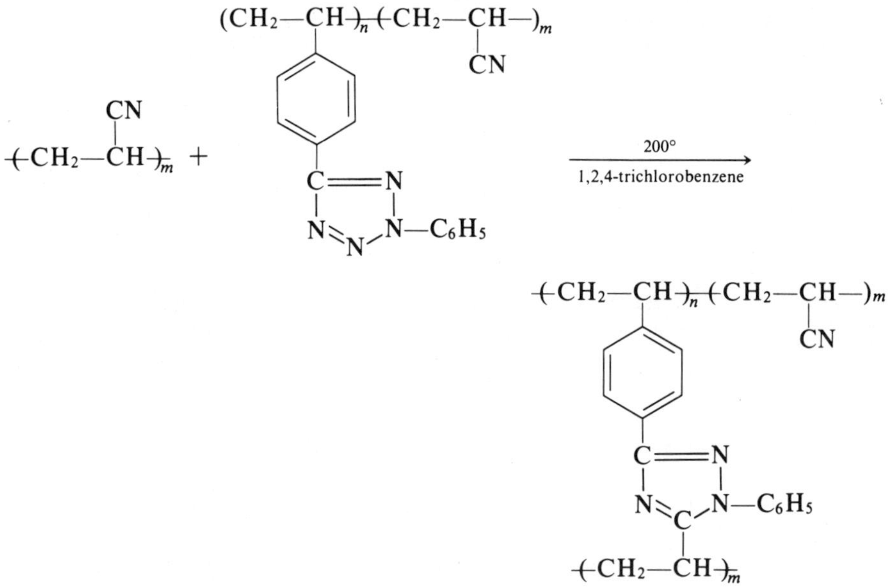

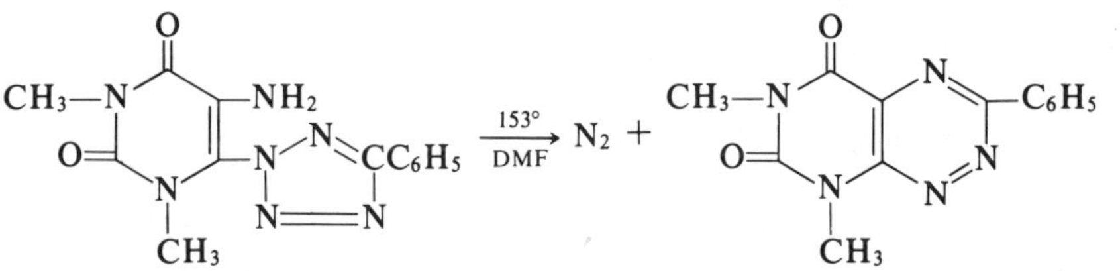

90%

migration of a benzoyl radical are involved in the conversion of 2-(2,4-dinitrophenyl)-5-phenyltetrazole to 1-benzoyloxy-6-nitrobenzotriazole (601). Condensation of an *N*-phenylnitrile imine to an indazole occurs on heating either in tet-

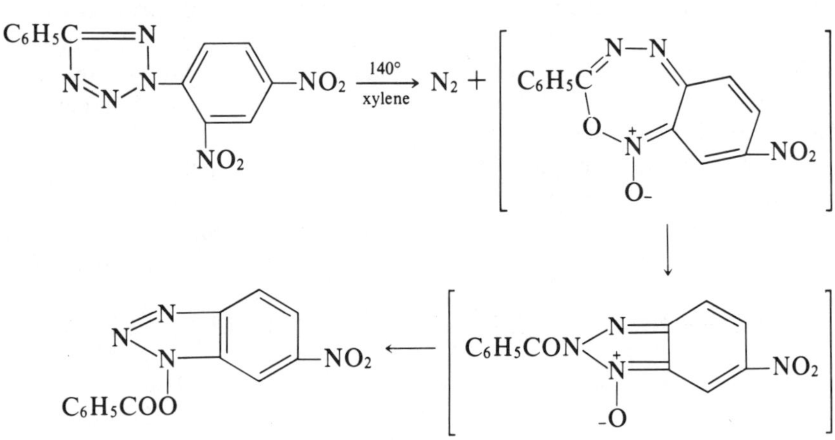

97%

ralin or under pyrolysis conditions, as illustrated by the decomposition of 2-phenyl-5-(4-pyridyl)tetrazole (1156).

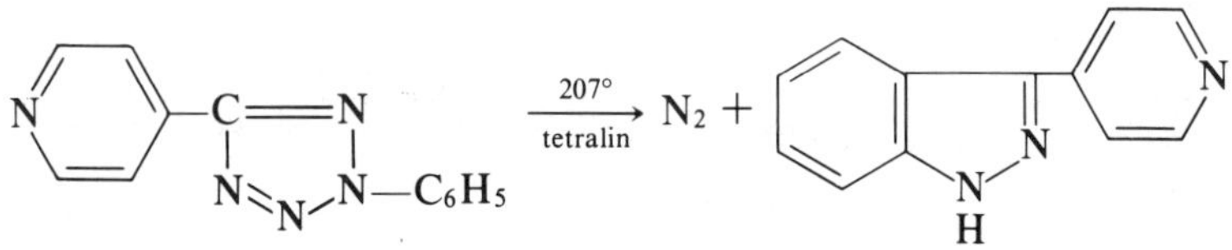

64%

The 2-acyl 5-substituted tetrazoles likewise liberate nitrogen on heating, forming an *N*-acyl nitrile imine which cyclizes to a 1,3,4-oxadiazole (519). The

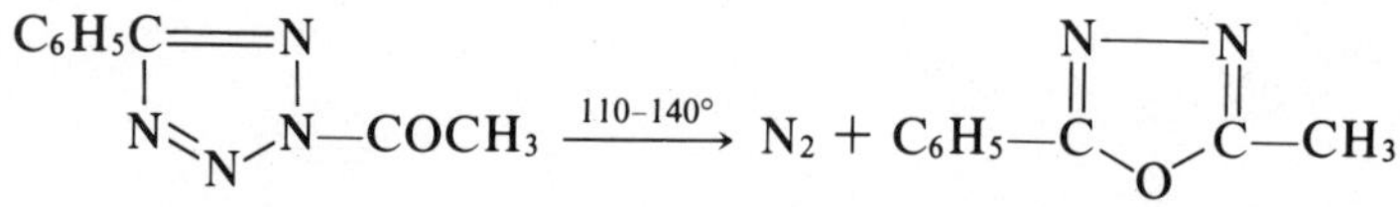

75–79%

mp 95–97°

reaction also is applicable to 2-acyl-5-aryloxytetrazoles (715). For preparative purposes the 2-acyltetrazoles are rarely isolated.

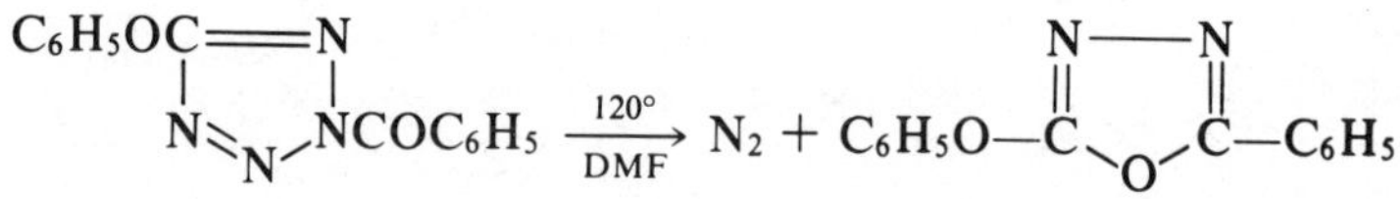

85%

2.6.3. Tetrazolium Compounds

Monocyclic tetrazolium cations are disubstituted on ring nitrogen in the 1,3-, 1,4-, or 2,3-positions, are doubly unsaturated in the five-membered tetrazole ring, and bear a single positive charge. Resonance hybrid structures can be written for each of the tetrazolium classes with positive charge located at each ring position; for convenience, and to approximate the true distributions, the charge is represented as delocalized. Both salts with separate anions and dipolar compounds have been prepared.

Alkyl and aryl groups are substituted on nitrogen in the 1,3-tetrazolium compounds. In the mesoionic compounds, negative charge is carried at the 5-position by substituent carbon, nitrogen, oxygen, or sulfur atoms. Few of the 1,3-tetrazolium compounds decompose on melting. The 1,3-bis(*m*-carboxy-

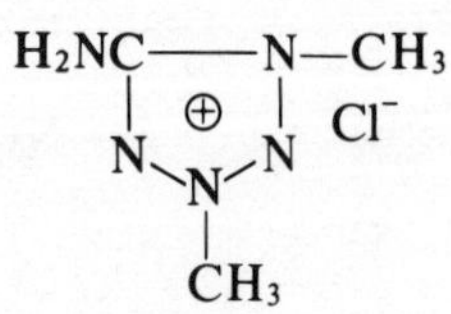

mp 208–209° (466)

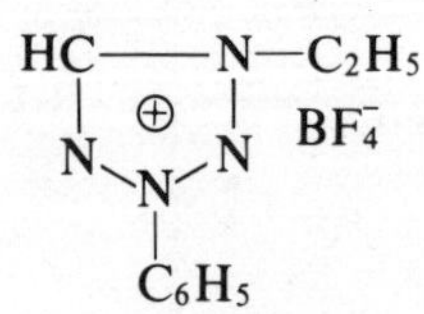

mp 96–98° (602)

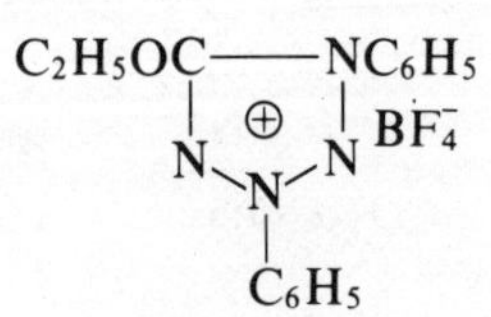

mp 174° (422)

phenyl)-5-olate has a remarkably high melting point, characteristic of many high nitrogen carboxylic acids. 3,5-Diaryl-1-(5-tetrazolate)tetrazolium betaines melt with decomposition from 259 to 312° (508).

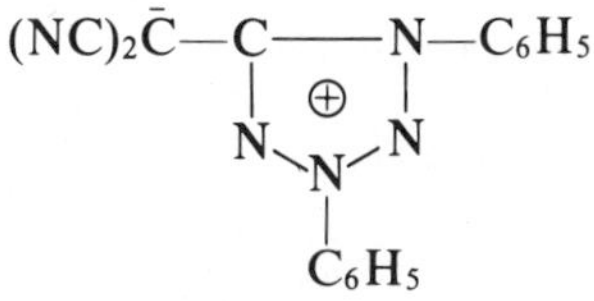

mp 234° (420)

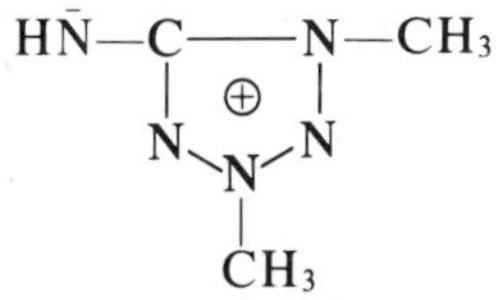

mp 41–43° (466)
bp 88/1 mm

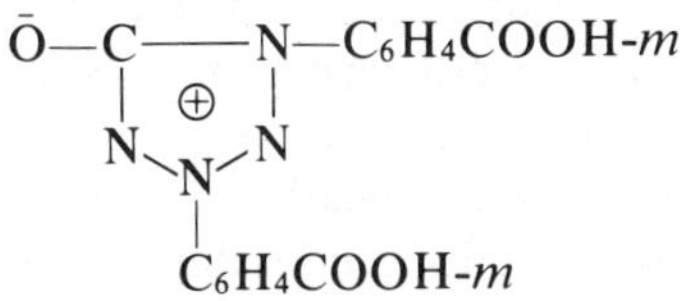

mp 400° (788)

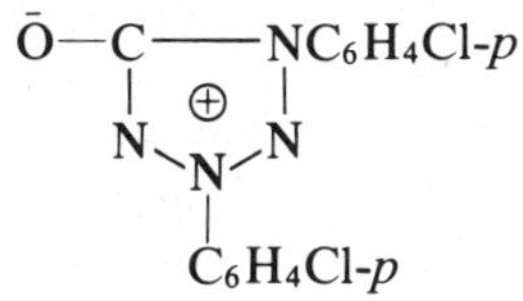

mp 231° (789)

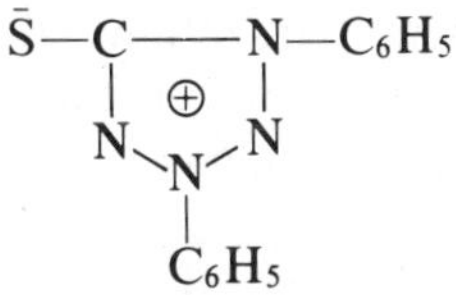

mp 153° (890)
155° (423)

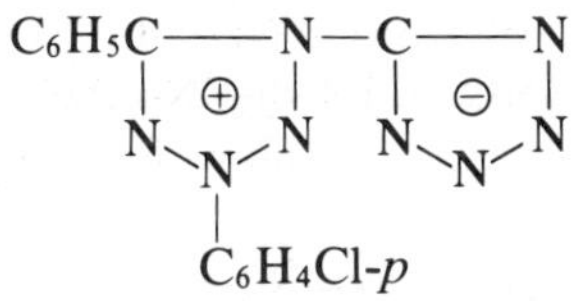

mp 312–315° dec. (508)

The 1,4-tetrazolium salts are substituted on nitrogen with alkyl and aryl. Thermal decomposition of 1,3- and 1,4-tetrazolium iodides with at least one

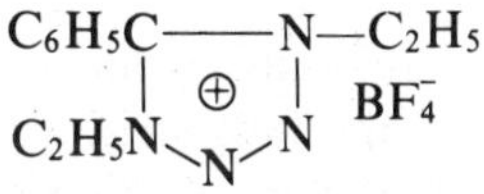

mp 131–132° (642)

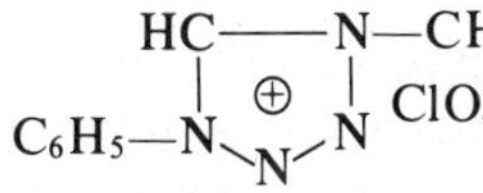

mp 164–165° dec. (602)

alkyl group on nitrogen involves elimination of alkyl iodides, the 1,4-disubstituted tetrazolium iodide being more heat resistant than the 1,3-isomer. On thermolysis, 1,3,5-trimethyltetrazolium iodide preferentially loses the 1-methyl

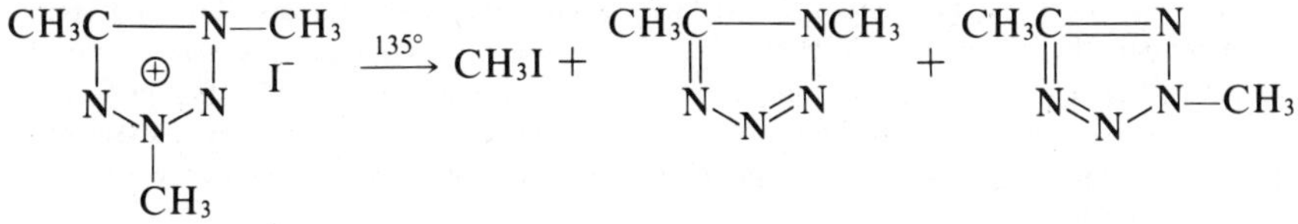

group, forming a 15:85 mixture of 1,5- and 2,5-dimethyltetrazole (542). The isomeric 1,4,5-trimethyltetrazolium iodide, which requires a higher temperature for decomposition, also forms the same dimethyltetrazoles. Since heating pure

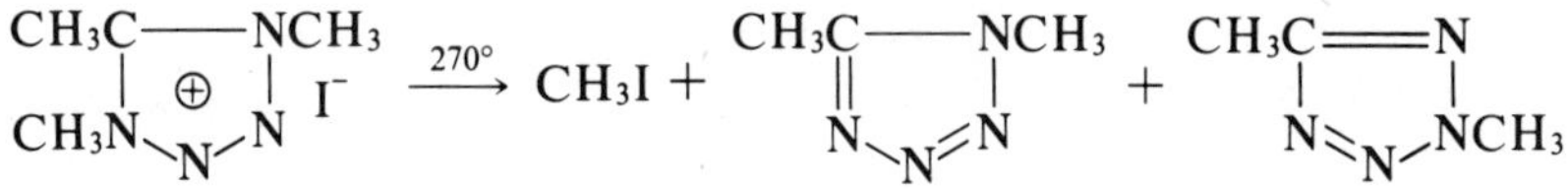

74% 21%

1,5-dimethyltetrazole at 270° produces no 2,5-dimethyltetrazole, its formation from the 1,4-tetrazolium salt must follow from alkylation of liberated 1,5-dimethyltetrazole by methyl iodide, followed by dealkylation (542, 544).

Tetrazolium compounds substituted at the 2,3-nitrogen atoms are the largest tetrazolium class.* Only aryl and heteroaryl substituents are found at the 2- and 3-locations, but alkyl, aryl, heteroaryl, acyl, and noncarbon substituents are present at the 5-position. In the 2,3-tetrazolium salts, a consequence of planarity (see Section 3.1, Molecular Dimensions) and positive charge delocalization is that the 2- and 3-positions are equivalent; with different groups at each of these positions, only one isomer should exist, as found in practice (807, 835).† Infrared spectra of 1-mono ^{15}N-labeled triphenyltetrazolium salts, in comparison with unlabeled and 1,4-di-^{15}N-labeled material, indicate that only one tetrazolium cation is present, from which it may be concluded that resonance forms,

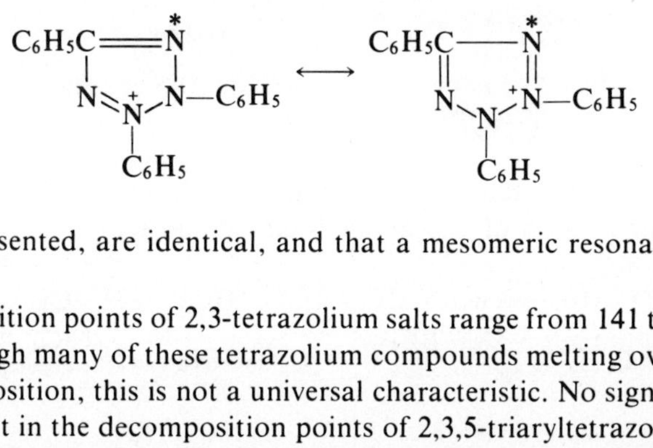

such as represented, are identical, and that a mesomeric resonance system is present (835).

Decomposition points of 2,3-tetrazolium salts range from 141 to 285° (Table 2.77). Although many of these tetrazolium compounds melting over 200° do so with decomposition, this is not a universal characteristic. No significant difference is evident in the decomposition points of 2,3,5-triaryltetrazolium salts between those with electron-releasing or electron-withdrawing substituents.

High and low melting modifications of a series of differently substituted 2,3-

*The 2,3-tetrazolium salts were comprehensively reviewed by Nineham (807) in 1955, and new advances were surveyed by Hooper (504) in 1969.

†Isomers with different substituents at 2 and 3 have been reported, diverging slightly in decomposition potential, thermogravimetric behavior, infrared spectra, and Debye-Scherrer X-ray diagrams (960–962). The isomerism is proposed to arise from orientation of the 2-, 3-, and 5-substituents above and below the tetrazolium plane. In the absence of X-ray structural analyses of the supposed isomers, indicating differing spatial arrangements of substituents, their existence remains unsubstantiated.

Table 2.77. 2,3-Disubstituted Tetrazolium Salts

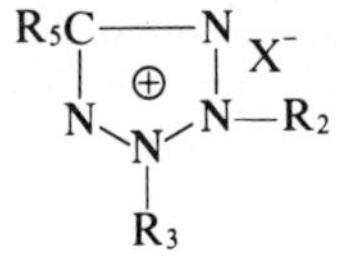

R_2	R_3	R_5	X	Melting Point (°C)	Ref.
C_6H_5	C_6H_5	H	Cl^-	268 dec.	855
C_6H_5	C_6H_5	CH_3	Cl^-	271 dec.	618, 564
C_6H_5	C_6H_5	$(CH_3)_3C$	NO_3^-	Dec. 250	795
C_6H_5	C_6H_5	$C_{11}H_{23}$	Cl^-	Dec. 141	618
C_6H_5	C_6H_5	C_6H_5	Cl^-	259–260 dec.	30
C_6H_5	C_6H_5	p-$H_2NC_6H_4$	Cl^-	285 dec.	45
C_6H_5	C_6H_5	p-$O_2NC_6H_4$	Cl^-	250 dec.	45
C_6H_5	C_6H_5	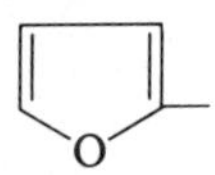	Cl^-	239–240	224
C_6H_5	C_6H_5	C_6H_5CO	Cl^-	220–225	1147
C_6H_5	C_6H_5	HO	Cl^-	194	357
C_6H_5	C_6H_5	$HOOCCH_2CH_2S$	Br^-	236–239 dec.	536
p-$H_2NC_6H_4$	C_6H_5	C_6H_5	Cl^-	235–237 dec.	45
p-$O_2NC_6H_4$	C_6H_5	C_6H_5	Cl^-	230–232 dec.	45
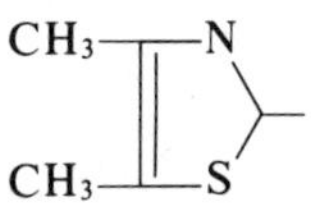	C_6H_5	C_6H_5	Br^-	223–225 dec.	30
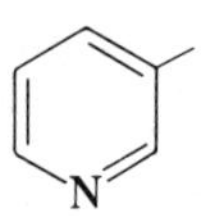	C_6H_5	C_6H_5	Cl^-	245 dec.	45

tetrazolium perchlorates can be interconverted by crystallization from methanol (high melting form) or by precipitation from methanol with ether (960).

During differential thermal analyses of a series of 2,3-diphenyltetrazolium salts with 5-p-chlorophenyl, phenyl, p-tolyl, and p-anisyl substituents, small

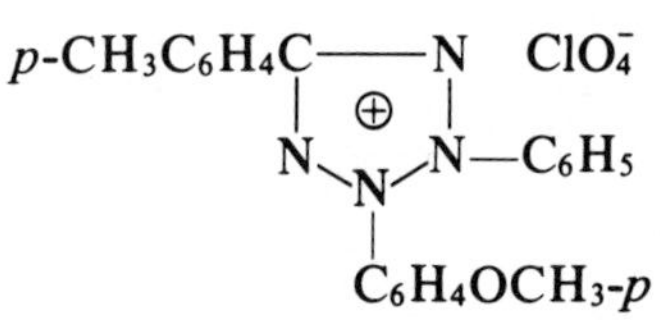

mp low, 113–115°
high, 175–177°

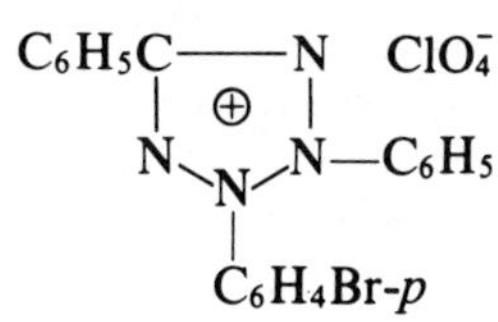

mp low, 115–117°
high, 220–221°

exotherms are observed in the solid state in the 160–200° range (962). Color changes also occur on heating 2,3-tetrazolium salts; thus 2,3-diphenyl-5-*p*-chlorophenyltetrazolium bromide changes from citron yellow to dark yellow at about 150°, to yellow at 190°, and melts to a brown liquid and evolves gas at 261–262°.

Dry distillation of a mixture of 2,3-diphenyl-5-ethoxycarbonyltetrazolium chloride and sand at an unspecified temperature forms azobenzene (855).

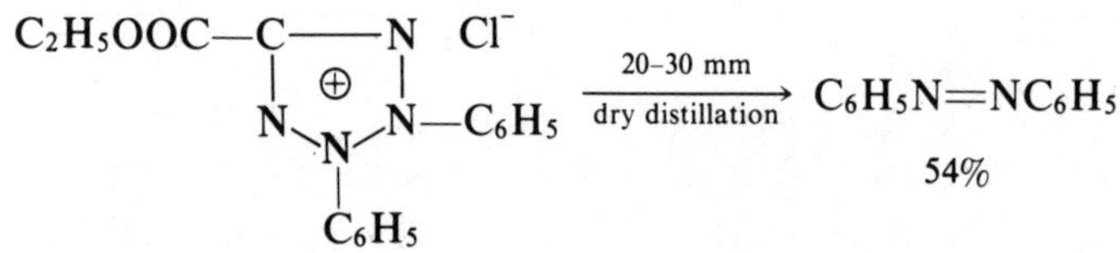

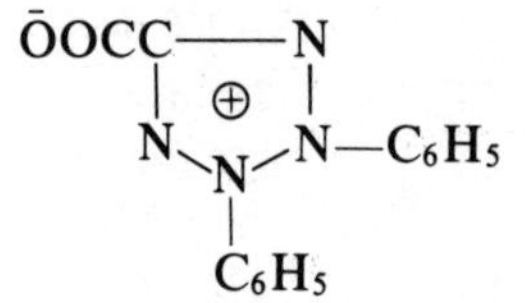

mp 161° (855)

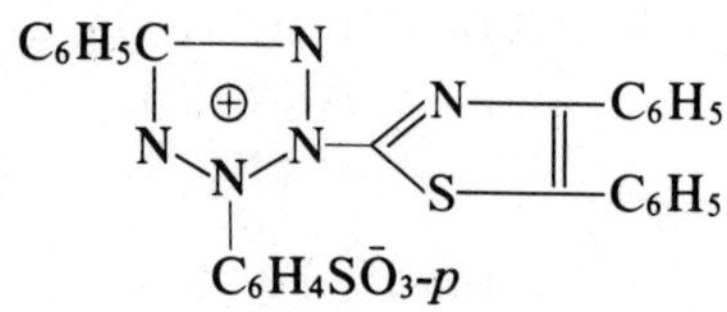

mp 242° dec. (565)

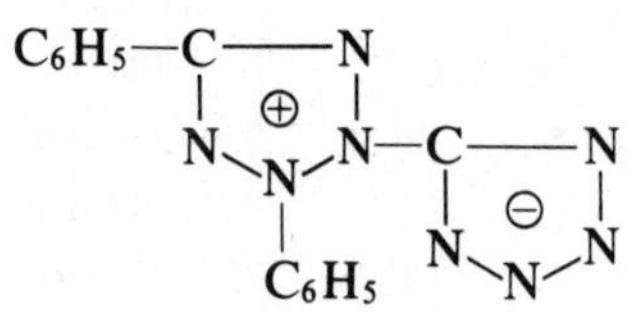

mp 209–210° expl. (619)

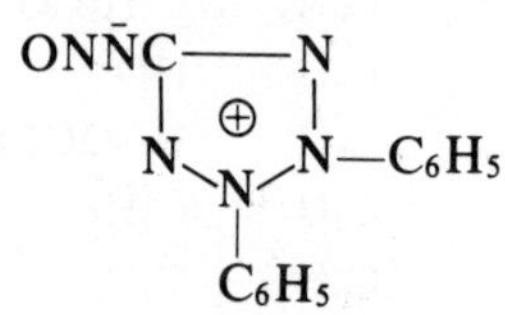

mp 177° (826)

Dipolar 2,3-tetrazolium compounds are slightly less stable than tetrazolium salts with separate anions. On heating, both 2,3-diphenyltetrazolium olates and thiolates rearrange to the corresponding mesoionic 1,3-diphenyltetrazolium compounds (891).

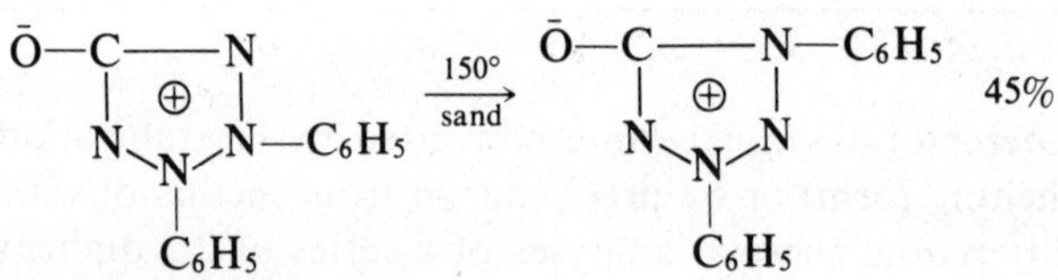

mp 168–169° expl. (1069)
175° (420)

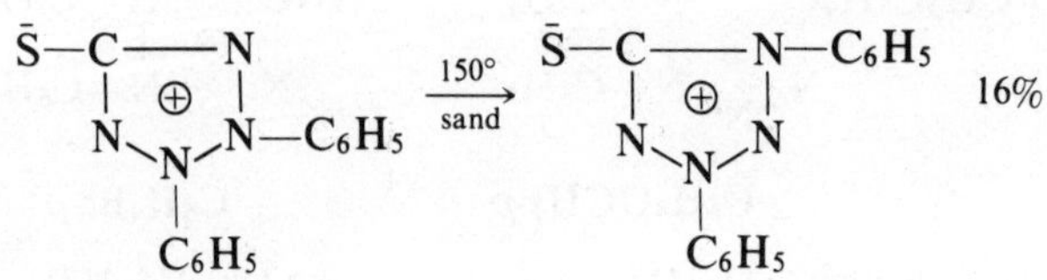

dehydrodithiazone
mp 180° (420)

2.6.4. *v*-Tetrazines

The 1,2,3,4- or *v*-tetrazines are six-membered rings with four contiguous nitrogen and two carbon atoms. No completely aromatic members of this class have been prepared; the known derivatives are tetrahydro-*v*-tetrazines with nitrogen chains at the oxidation levels of both tetrazane and tetrazene. Most *v*-tetrazines reported before 1963 were, in fact, bis(arylazo) ethylenes or derivatives of 1-amino-*v*-triazole.*

The *v*-tetrazines at the tetrazane oxidation level contain one ring carbon-nitrogen double bond and carry substituents on three of the ring nitrogen atoms, of which two are adjacent acyl groups at positions 1 and 2 (1021). The structure of this class was confirmed by X-ray analysis of **I**. Some, but not all, 1,2,3,6-tetrahydro-*v*-tetrazines disintegrate on melting. Decomposition points

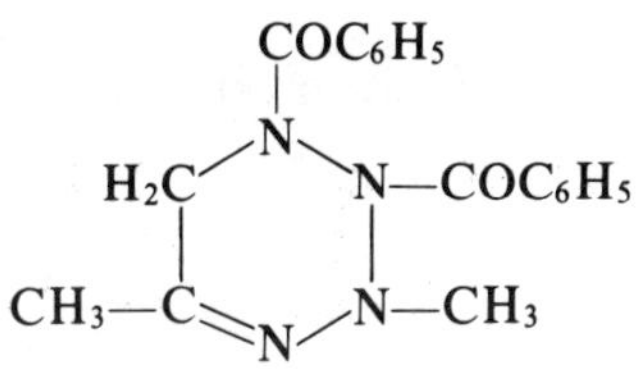

mp 133–134°

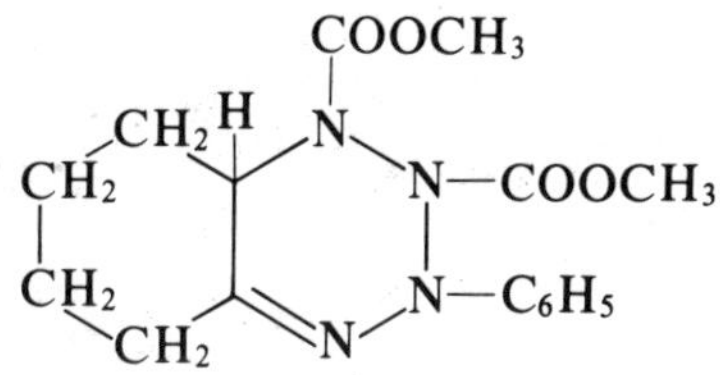

(I)

mp 132°

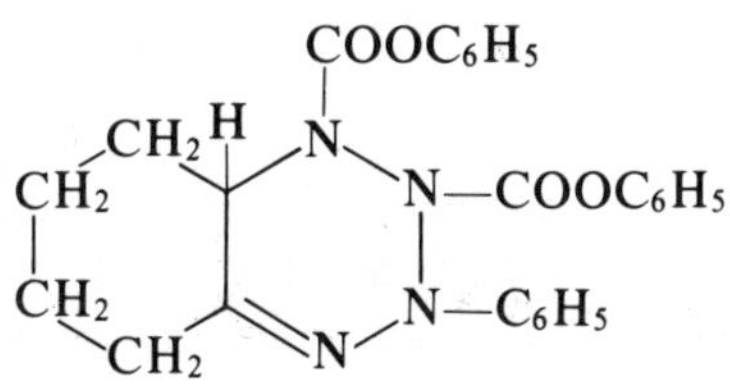

mp 161–162° dec.

range from 131 to 176°. Direct comparison with open-chain analogs cannot be made because 4-alkylidene-1,2-diacyltetrazanes are unknown.

Other tetrahydro-*v*-tetrazines have a 2,3-azo linkage in analogy to the 2-tetrazenes. Substitutents at the 1- and 4-positions are alkyl or aryl. The conformation at the azo linkage, of necessity, is cis.

Decomposition data for these 1,4,5,6-tetrahydro-*v*-tetrazines suggest them to be stable to over 100°. Cyclic 1,4-dimethyl-1,4,5,6-tetrahydro-*v*-tetrazine is less stable than the acyclic 1,1,4,4-tetramethyl-2-tetrazene, which has a trans orientation. The half-life of the closely related seven-membered 1,4-dimethyl-1,2,3,4-tetraaza-2-cycloheptene, with less conformational rigidity than the six-mem-

* Problems with the earlier *v*-tetrazine formulations are discussed by Wystrach (1197) and Wiley (1181).

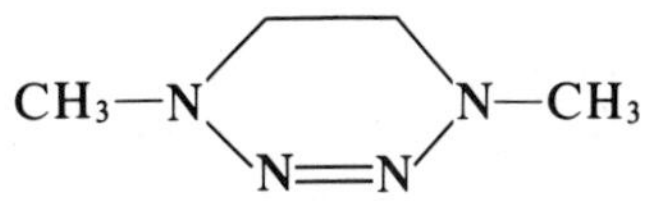

bp 90°/10 (977)
$t_{1/2}$ (129.7°, tetralin) 2.0 hr (784)

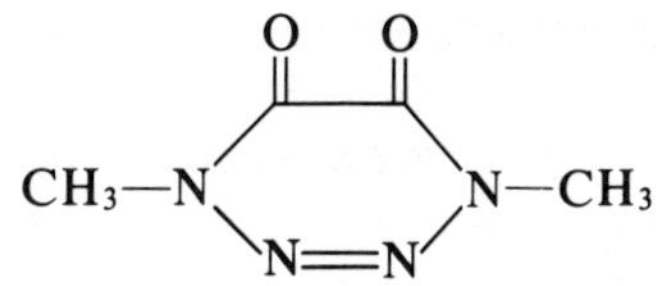

mp 111° (977)

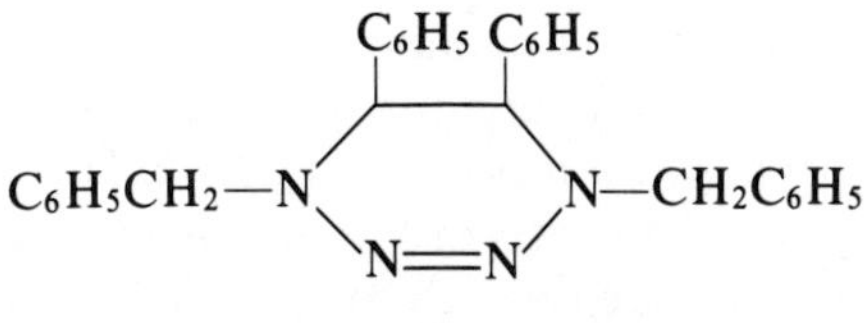

mp 161–163° dec. (720)

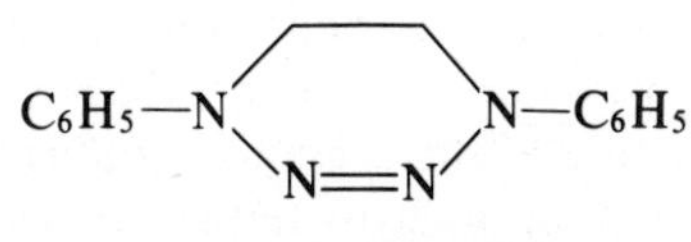

mp 176–177° dec. (609)

bered 1,4,5,6-tetrahydro-*v*-tetrazines, and with a negative entropy of decomposition, is 1.1 hr at 65° (739).

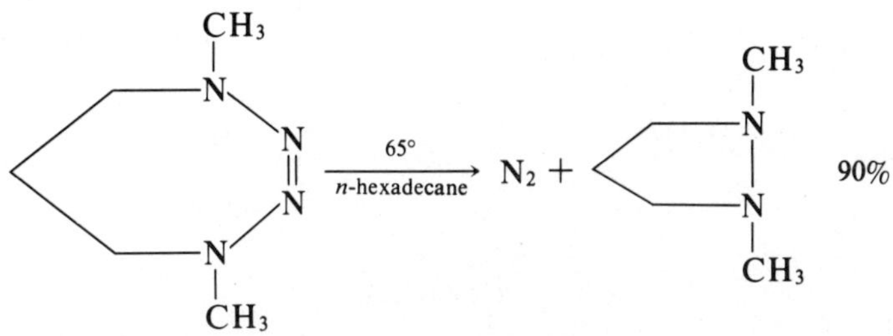

The *N*-oxides of the 1,4,5,6-tetrahydro-*v*-tetrazines have no acyclic counterparts. They do not decompose at their melting points, which range from 61 to 149°, but no study of their thermal stability has been made.

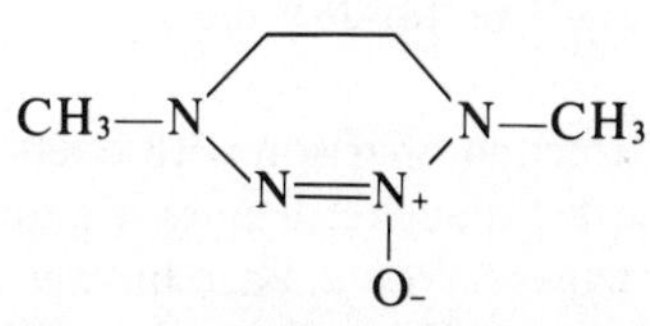

mp 61° (977)

2.6.5. Tetraaza Boron, Silicon, and Phosphorus Heterocycles

The Δ^2-tetraazaborolines are similar in structure to the 1,4-tetrazolines in consisting of a cyclized 2-tetrazene chain joined at the 1,4-positions to a nonnitrogen atom, and substituted on each of the same adjacent nitrogens. Alkyl and

Table 2.78. Tetraazaborolines

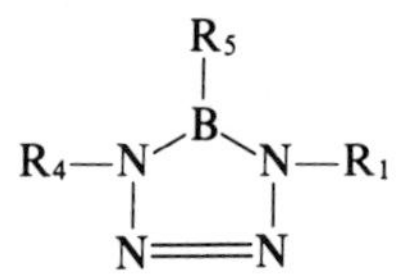

R_1	R_4	R_5	Melting Point (°C)	Ref.
CH_3	CH_3	H	11	763
CH_3	CH_3	CH_3	12.2	640
CH_3	CH_3	Cl	12.3	473
CH_3	CH_3	CN	Just below room temp.	473
CH_3	C_6H_5	H	32 Bp $80/10^{-2}$	763
C_6H_5	C_6H_5	H	116–117; 117	404, 763
C_6H_5	C_6H_5	CH_3	116	640
C_6H_5	C_6H_5	C_6H_5	186	640
p-$CH_3OC_6H_4$	C_6H_5	C_6F_5	128–130 dec.	841

aryl groups have been attached to nitrogen; alkyl, aryl, halogen, and cyano have been substituted on boron (Table 2.78).*

The 1,4-dimethyltetrazaborolines melt below room temperature and the 1,4-diaryl derivatives are solids melting over 100°. No systematic studies of the thermal stability of Δ^2-tetraazaborolines have been made, but 1-methyl-4-phenyl-Δ^2-tetraazaboroline can be distilled *in vacuo* at 80° (763). The 1,4-diphenyl derivative is stable to its melting point (117°) (404) and 1-*p*-anisyl-4-phenyl-5-perfluorophenyl-Δ^2-tetraazoboroline undergoes partial loss of nitrogen on vacuum distillation at 130° (841). Heating a mixture of dimethyltetraazaboroline and diethyltetraazaboroline at 130° results in scrambling of the nitrogen substituents with formation of a mixture of dimethyl, ethylmethyl, and diethyltetraazaboroline (641). Based on these observations, the tetraazaborolines with hydrogen, alkyl, or aryl on boron appear to be about as stable as the 5-hydrogen-5-alkyl or aryl tetrazolines or the tetraalkyl-2-tetrazenes, both of which decompose in the 120–150° range.

The compound formulated as 1,1,3,5-tetraphenyl-1-bora-2,3,4,5-tetraaza-4-cyclopentene is of a different character, involving a coordinate bond between

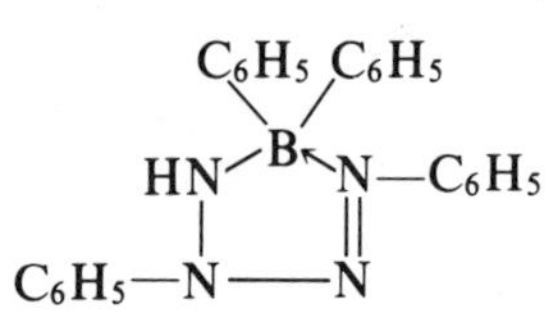

*Boron-nitrogen ring systems have been reviewed by Finch, Leach, and Morris (318).

nitrogen and boron (1116). No spectral data or X-ray crystal analysis have been obtained for this substance.

On heating, 5,5-dimethyl-1,4-bis(trimethylsilyl)silatetrazoline rearranges to azidobis(trimethylsilyl)aminodimethylsilane (1176). Its half-life in benzene is about 1 hr at 130° and 11 min at 150°.

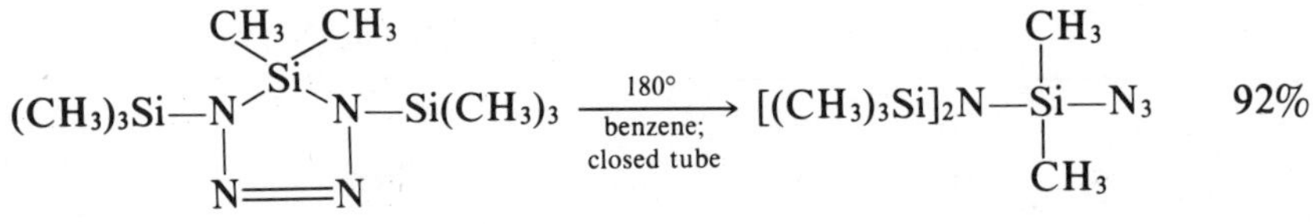

mp 88–89°

Tetraazaphospholines such as the 1,4-di-*t*-butyl-5-bis(isopropyl)amino derivative are monomeric solids (804, 879). The structure of the 5-bis(trimethylsilyl)amino analog has been confirmed by X-ray crystal analysis. Thermal sta-

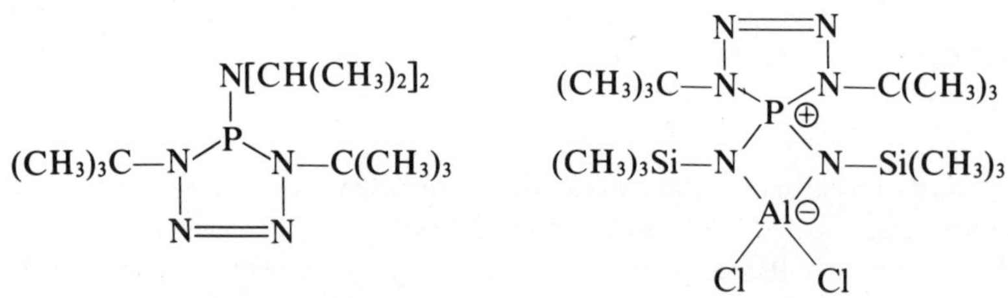

mp 78° mp 95° dec.

bility has not been determined. One of the most unusual high nitrogen compounds is the bicyclic tetraazaphosphorus diazaaluminum zwitterion. Heating above 120° liberates *t*-butyl azide and forms polymeric products (802, 803).

2.6.6. Embedded Four-Nitrogen Chains

A variety of heterocyclic systems are known with embedded four-nitrogen chains at the tetrazane oxidation level, with one, two, and four nitrogen atoms shared between multiple rings. None of these compounds with completely reduced four-nitrogen sequences, with and without cyclic unsaturation at the terminal nitrogens, and carrying alkyl, aryl, and acyl substituents on nitrogen, decompose at their melting points. Thus these embedded four-nitrogen chains appear to be more thermally stable than comparably substituted acyclic tetrazanes and acyclic-cyclic four-nitrogen chains with two exocyclic nitrogen atoms. There is no indication of dissociation of these transannular four-nitrogen chains into radicals.

At the tetrazene stage, compounds are known with embedded chains with both single end and middle nitrogen atoms shared between two rings as well as heterocycles with two nitrogens of the contiguous four-nitrogen chain common

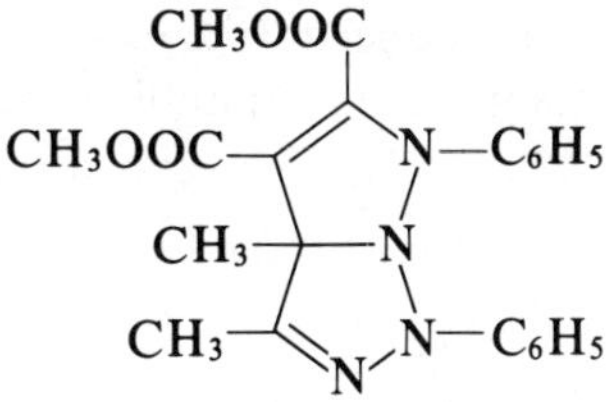

mp 136° (1052)

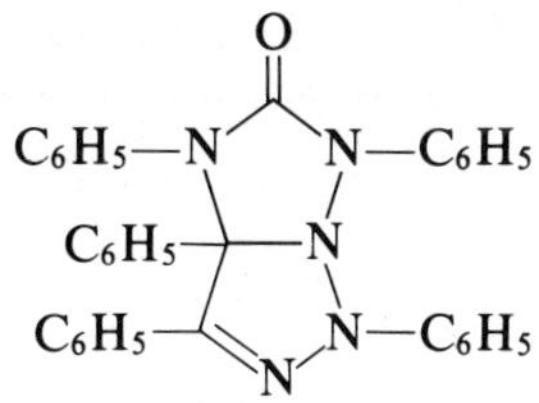

mp 238° (1051)

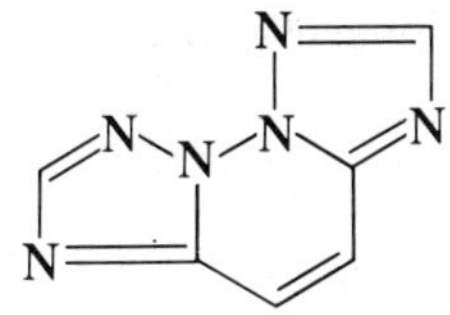

mp 208–212° (880)

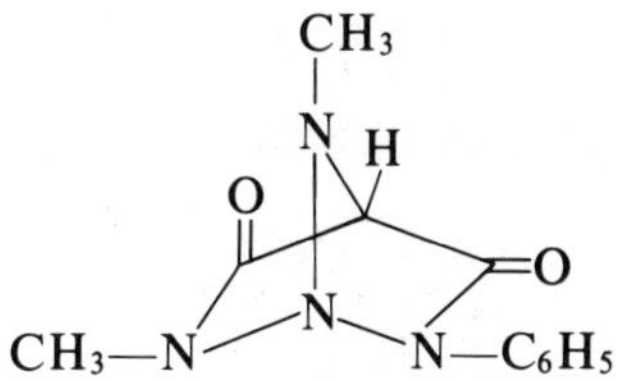

mp 167–168° (885)

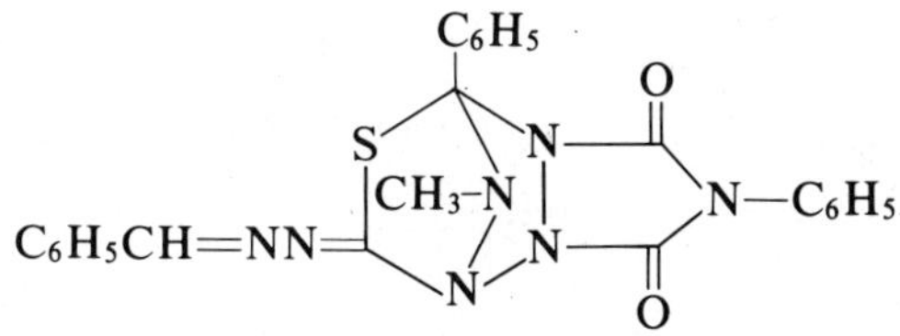

mp 202–203° (198)

to two rings. Compounds **I** and **II**, respectively a fused ring tetrahydro-*v*-tetrazine and its corresponding oxide, are similar in melting point to related monocyclic alkyl tetrahydro-*v*-tetrazines. *N*-Amino-*v*-triazoles fused to another ring,

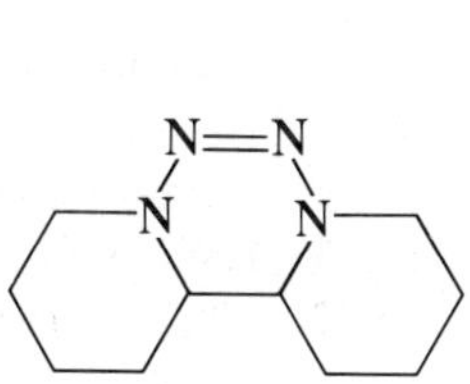

(I)

mp 66.5–67° (977)

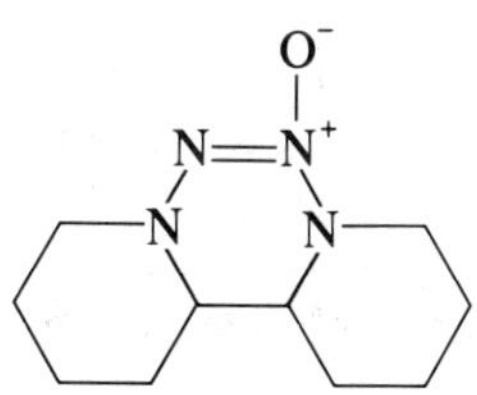

(II)

mp 79° (977)

such as **III**, resemble most acyclic-cyclic *N*-amino-*v*-triazoles in not decomposing on melting. The *v*-triazolo dihydrocinnolone **(IV)**, formally an *N*-aminotri-

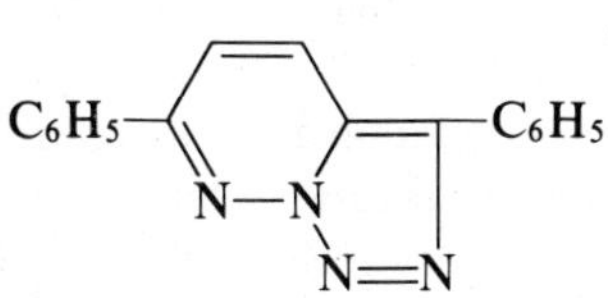

(III)

mp 254° (304)

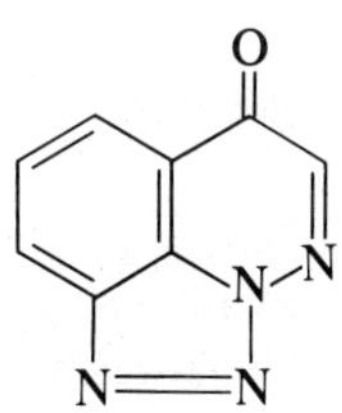

(IV)

mp 159–160° dec. (971)

azoline, is less stable. Embedded derivatives of *N*-aminobenzotriazines, as judged by their melting behavior, appear to be more stable than *N*-aminobenzo-

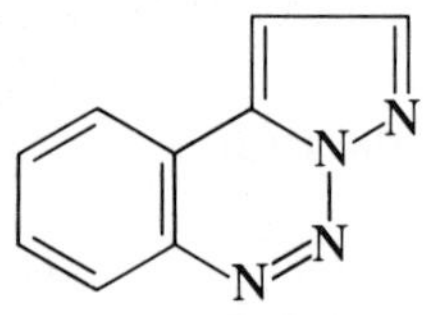

mp 133° (20)

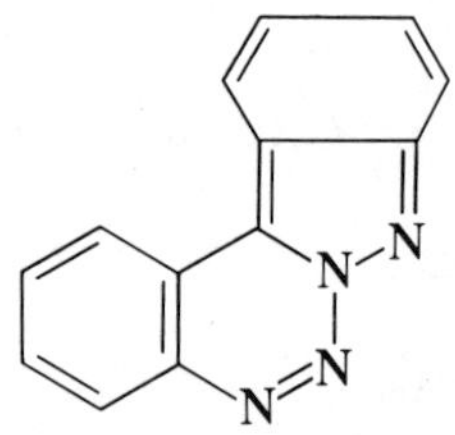

mp 246–247° dec. (1143)

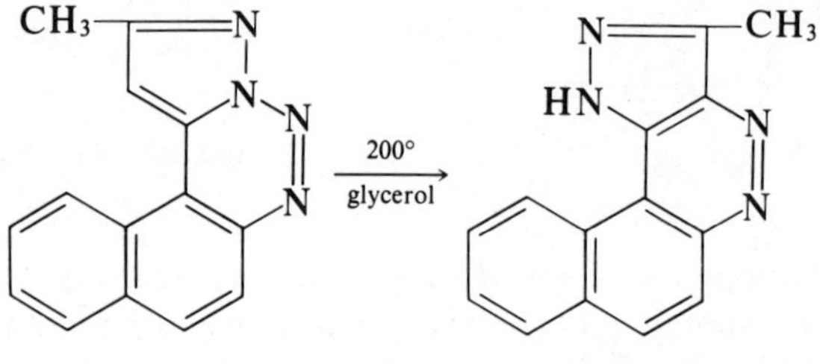

mp 180–181° mp 253–256°

triazinones, but the four-nitrogen chain in 2-methylnaphtho[1,2-*e*]pyrazolo[1,5-*c*]-1,2,3-triazine is broken during thermal isomerism to the more stable naphthopyrazolopyridazine (906). A transannular benzo-dihydro-*v*-tetrazine is stabilized by fusion with a cyanopyrazoline in **V**. The compound with two 2,5-tetrazole moieties is an unusual example of embedded 1-tetrazenes. Embedded

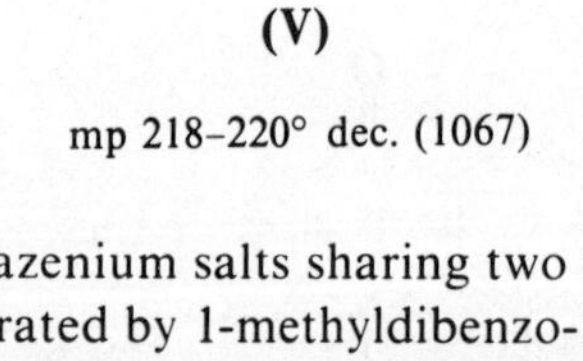

(V)

mp 218–220° dec. (1067)

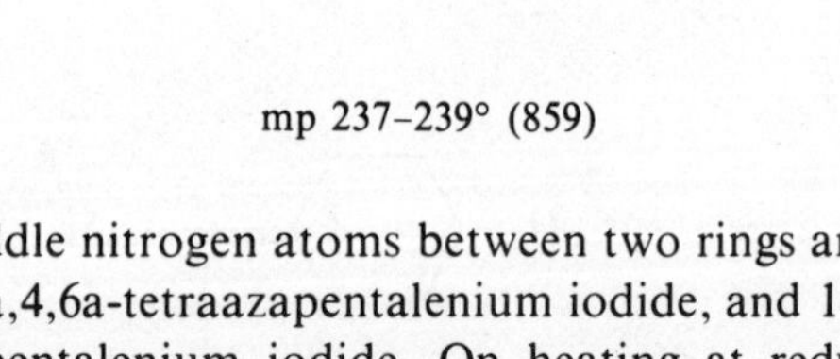

mp 237–239° (859)

tetrazenium salts sharing two middle nitrogen atoms between two rings are illustrated by 1-methyldibenzo-1,3a,4,6a-tetraazapentalenium iodide, and 1-isopropyldibenzo-1,3a,6,6a-tetraazapentalenium iodide. On heating at reduced pressure the *N*-methyl derivative loses methyl iodide and forms dibenzo-1,3a,4,6a-tetraazapentalene, discussed below in this section.

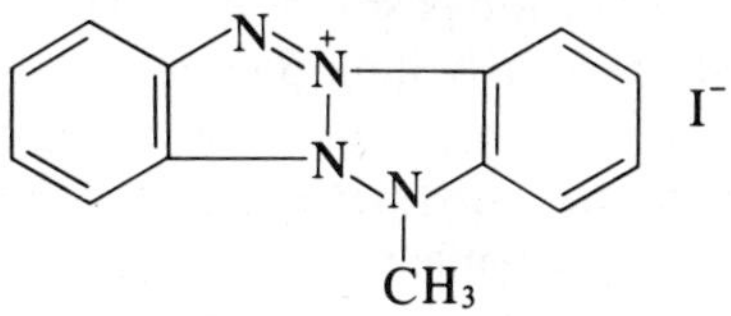

mp 191° dec. (194)

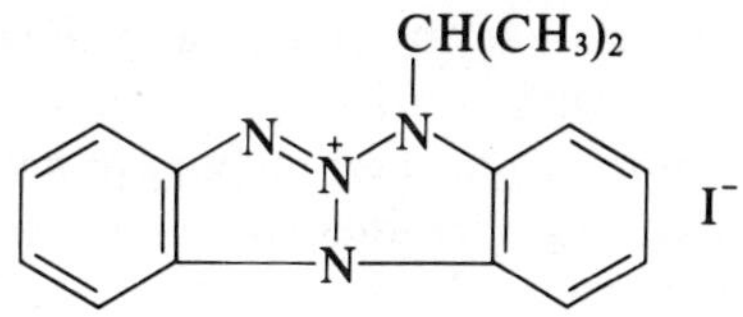

mp 174° dec. (578)

Both 1,2 and 1,4-embedded tetrazolines have been prepared. The 1,4-tetrazoline **(VI)** melts with decomposition, evolving methyl isocyanide, at a temperature below that of most monocyclic 1,4-iminotetrazolines (976).

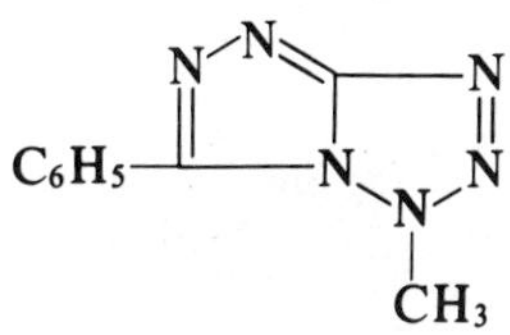

mp 173–174° (187)

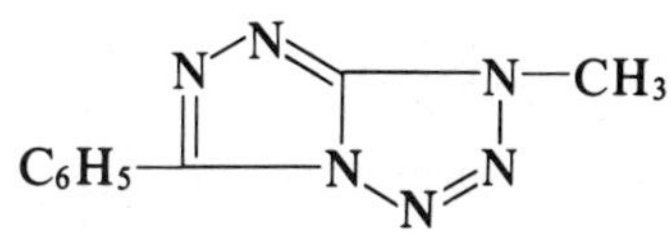

(VI)

mp 157° dec. (976)

The most numerous class of embedded four-nitrogen chains consists of tetrazoles fused to another ring at the 1,5-positions with one end nitrogen common to two rings. Compounds of this structure are potentially capable of opening to imidoyl azides as discussed in Section 2.16 on open chain-ring isomers and equilibria. All substances discussed in this part exist only as tetrazoles in

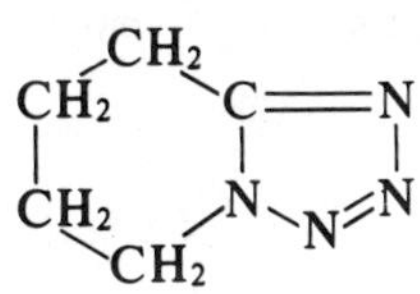

mp 117–118° (137)

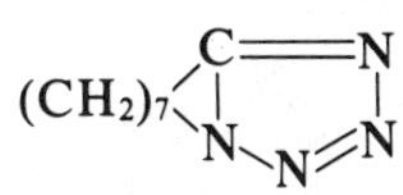

mp 42–43° (269)
bp 145–146°/0.1 (940)

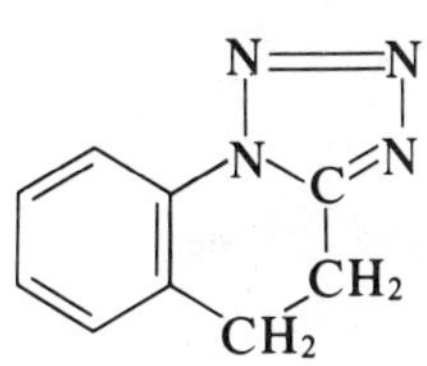

mp 101.5–102° (527)

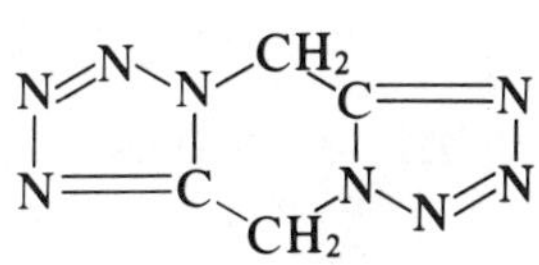

sublimes 235° (138)
discolors 355°

the solid state or in solution at ordinary temperature. None of these embedded tetrazoles with saturated groups at one or both fusion positions give evidence of decomposition at their melting or reduced pressure boiling points.

Fused ring tetrazoles conjugated with exocyclic unsaturation including other aromatic rings, decompose from 125° for tetrazolopyrazine to 300° for 9,10-tetrazolophenanthroline. Insufficient data are available to generalize on structural characteristics leading to stability or instability.

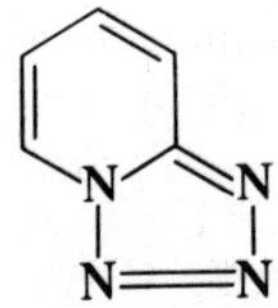

mp 157–159° (147)

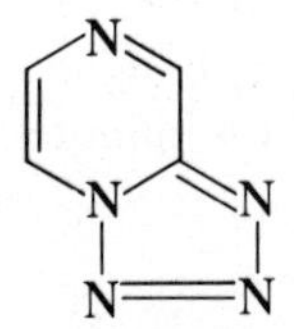

mp 90.8–91.5° (939)
dec. 125° in $CDCl_3$ (1150)

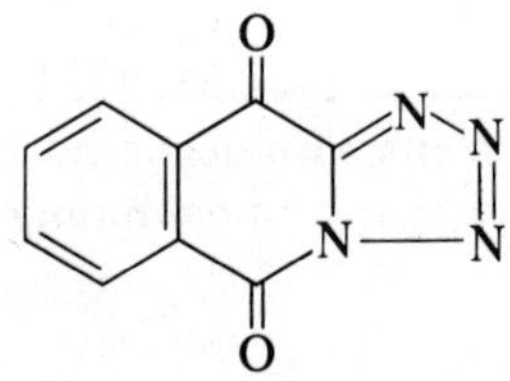

mp 216°, violent dec.
(756)

Cyano derivatives are formed as a result of both relatively low temperature decomposition and high temperature pyrolyses of fused ring tetrazoles. Thus

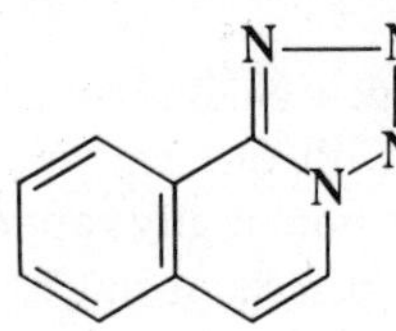

mp 140–142° (920)

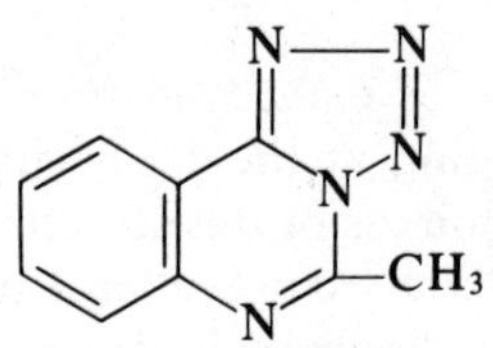

mp 180° (204)

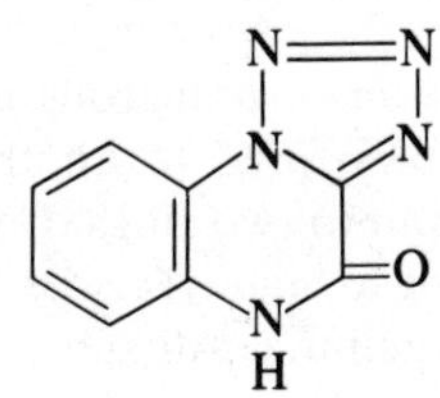

mp 297–298° dec. (205)

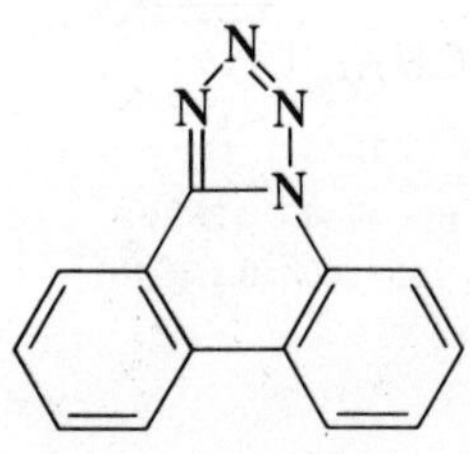

mp 221.5–223.5° (1009)
dec. 300°

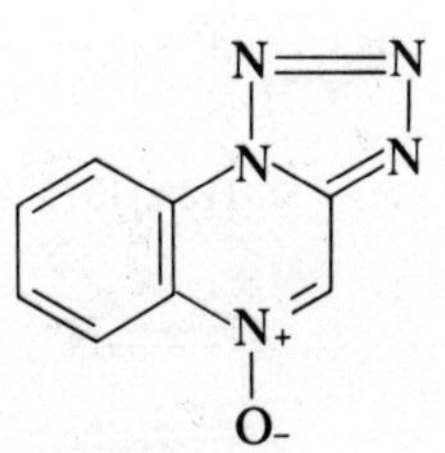

mp 189–190° dec. (231)

thermolysis of 4-phenyl-tetrazolo[1,5-*a*]quinoxaline leads to 1-cyano-2-phenylbenzimidazole (1157), and pyrolysis of polymethylene tetrazoles (1153) and

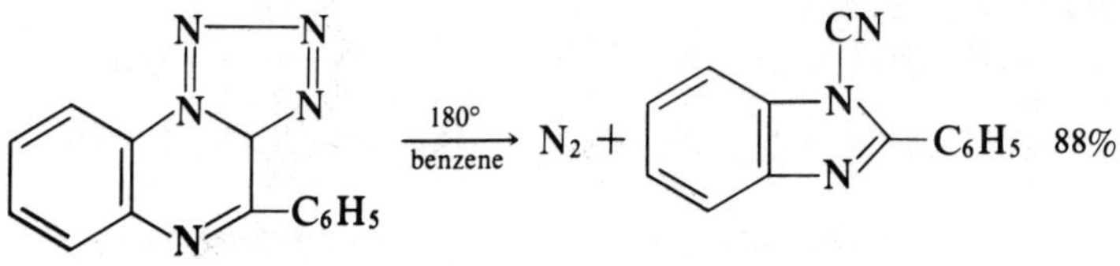

tetrazolopyridine (230) forms cyano derivatives of acyclic and cyclic amines. High-temperature decomposition of 9-phenyltetrazolo[1,5-*a*]quinoline involves

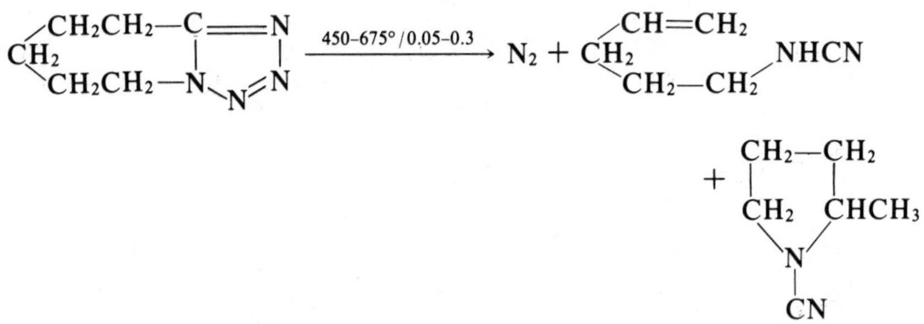

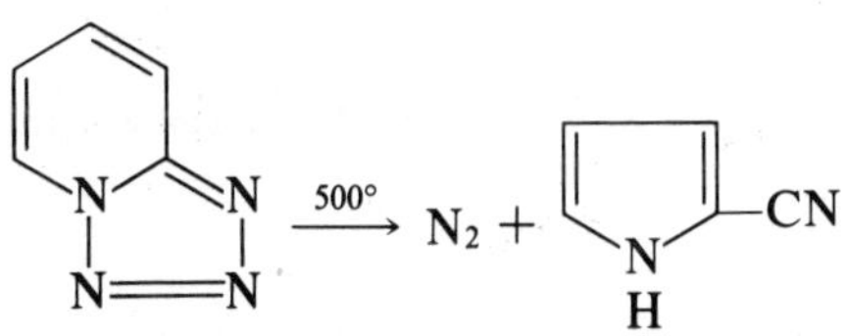

the phenyl group and a skeletal rearrangement, leading to a pyridophenanthridine in high yield (162, 1157).

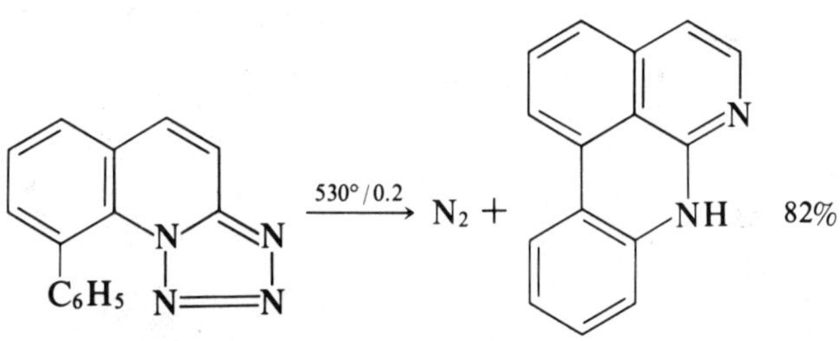

Fused tetrazolium salts with substitution in the 1,2-, 1,4-, and 2,3-positions of the tetrazole ring have been prepared. Based on the limited data available, the 1,2-fused ring tetrazolium salts appear to be somewhat more thermally stable than the 1,4-derivatives.

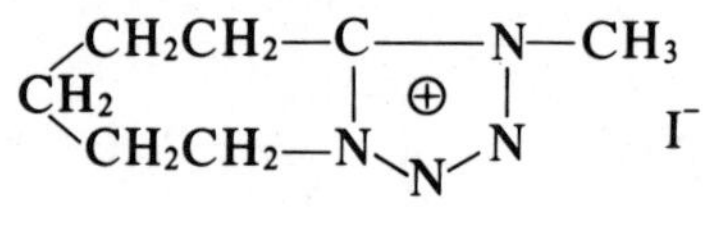

mp 175–180° (437, 279)

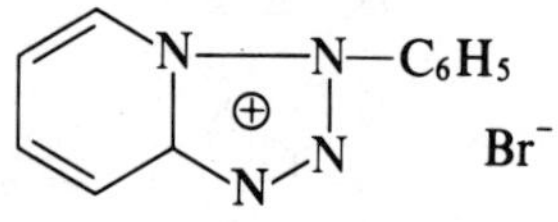

mp 278–280° (736)

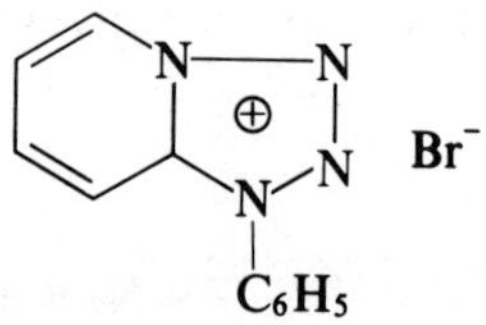

mp 234° (37)

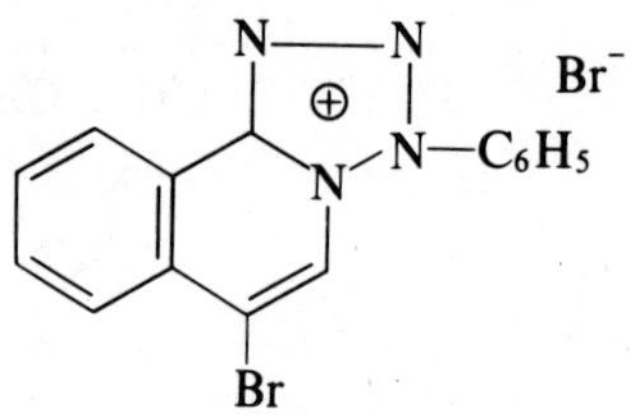

mp 292° (736)

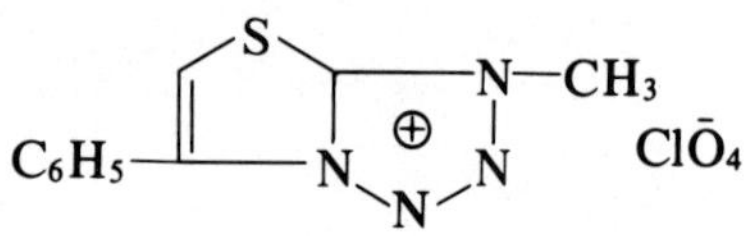

mp 185°, violent dec. (313)

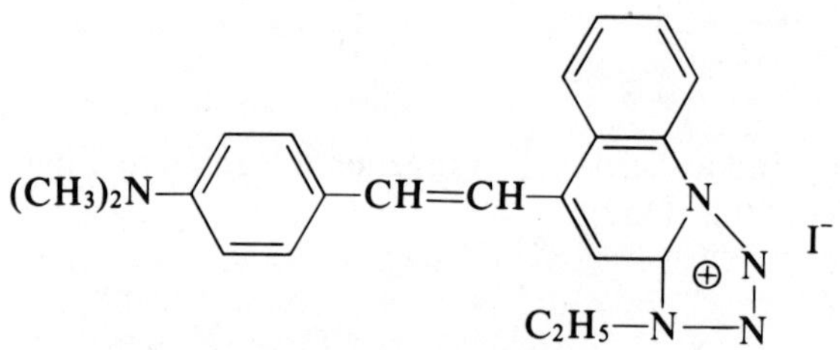

mp 259–261° dec. (158)

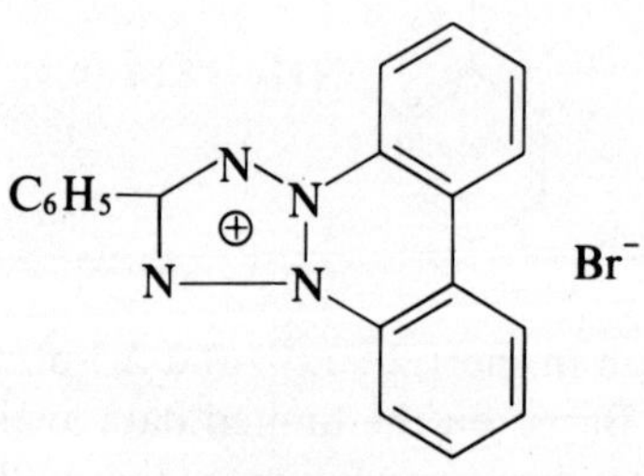

(VII)

mp 378–380° (563)

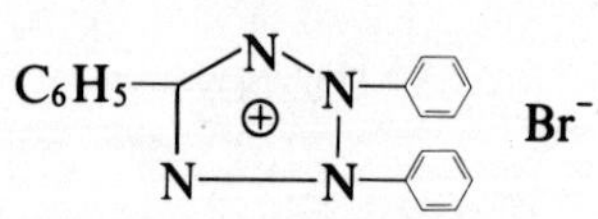

dec. 250° (962)

Annelation of the tetrazolium ring in the 2,3-positions, as in the biphenylene compound **(VII)**, appreciably increases thermal stability over that of the corresponding unfused tetrazolium salt.

v-Triazolo[1,2-*b*] and [1,2-*a*]-*v*-triazoles contain dipolar four-nitrogen chains with two nitrogen atoms shared between two fused five-membered rings. These aromatic heterocycles, illustrated by their monobenzo derivatives, **VIII** and **IX**, are more often called, respectively, 1,3*a*,4,6*a*-tetraazapentalenes (with a straight nitrogen chain) and 1,3*a*,6,6*a*-tetraazapentalenes (with a branched nitrogen chain). These compounds belong to the class of mesomeric betaines.*

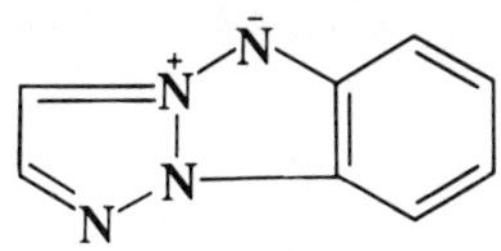

(VIII)

mp 108.8–109.7° (194)

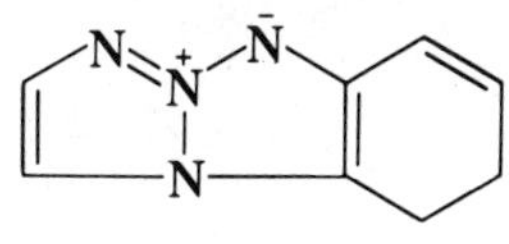

(IX)

mp 121.6–122.2° (578)

Decomposition points for bicyclic 1,3*a*,4,6*a*-tetraazapentalenes range from 208 to over 300°, but not all of these compounds decompose on melting. Dibenzo derivatives are even more stable than the bicyclic tetraazapentalenes.

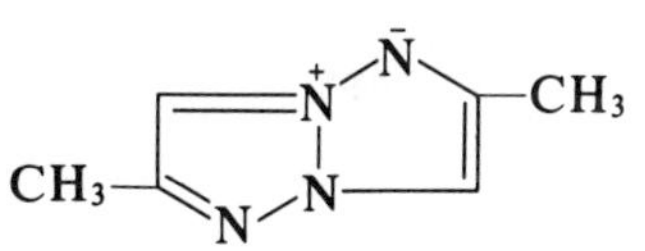

mp 93–94° (805)
bp 230°

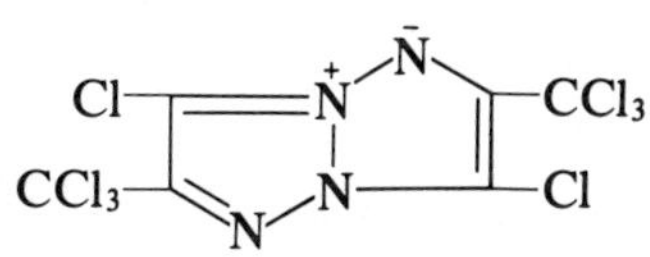

mp 208–210° dec. (865)

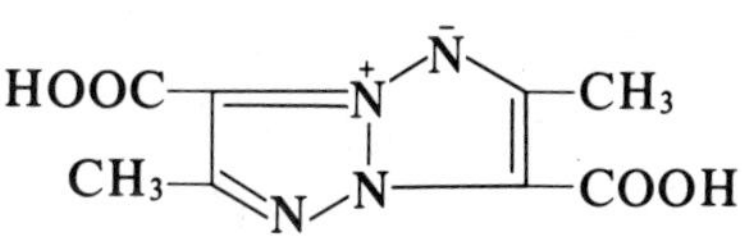

mp 310° dec. (865)

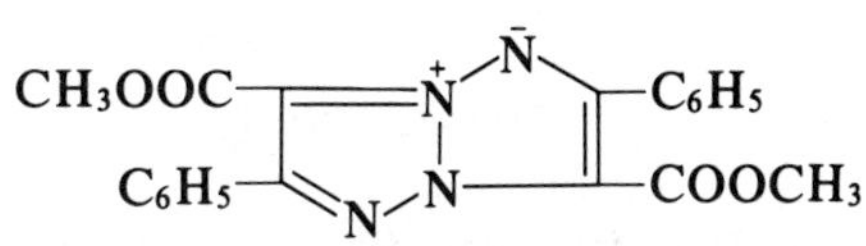

mp 247–248° (865, 866)
stable 300°

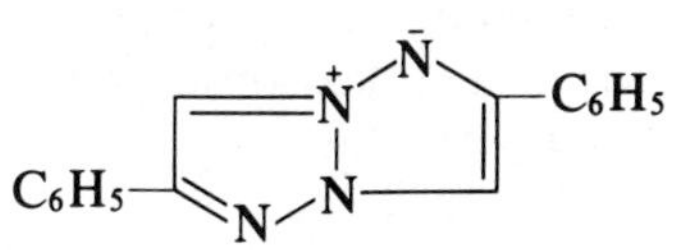

mp 257° dec. (865, 866)
sublimes unchanged
210–220°/0.05

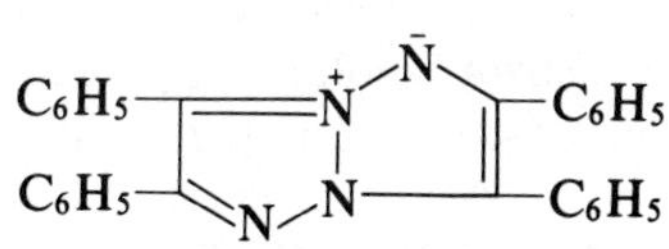

mp 285–286° (963)

* Mesomeric betaine derivatives of heteropentalenes are reviewed by Ramsden (900).

Most dibenzo-1,3*a*,6,6*a*-tetraazapentalenes melt higher than the corresponding 1,3*a*,4,6*a*-isomers, although the tetranitro derivatives are an exception. Both dibenzotetraazapentalenes are little affected by heating in solvents near 300°. The tetranitrodibenzotetraazapentalenes are remarkable in their thermal stability; indeed, tetranitrodibenzo-1,3*a*,4,6*a*-tetraazapentalene does not degrade at temperatures as high as 375° (40).

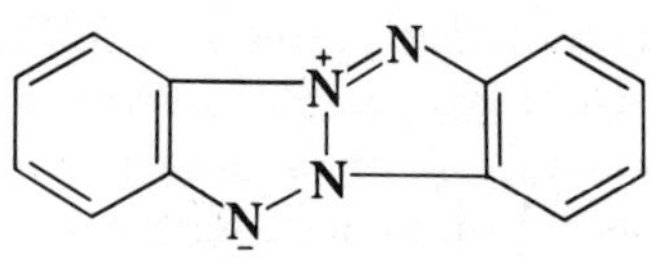

mp 237–238° (193)

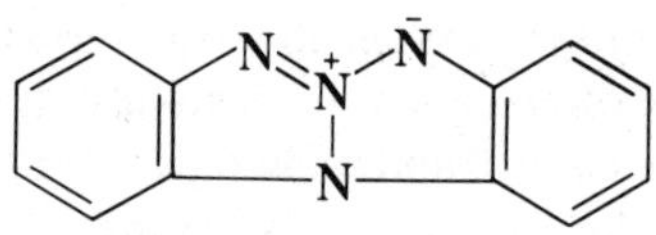

mp 254.8–255.2° (193)

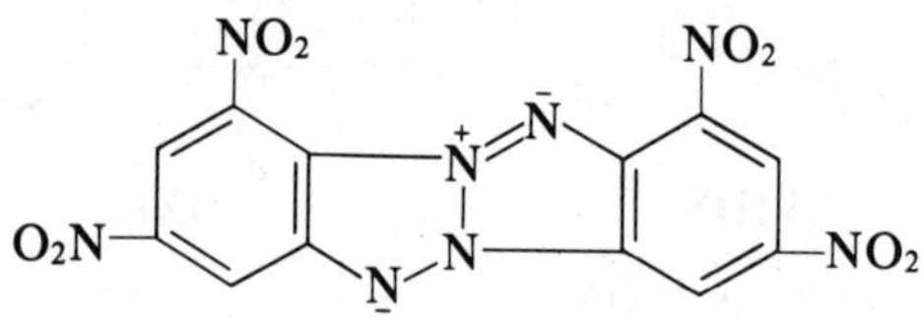

mp 410° dec. (194)

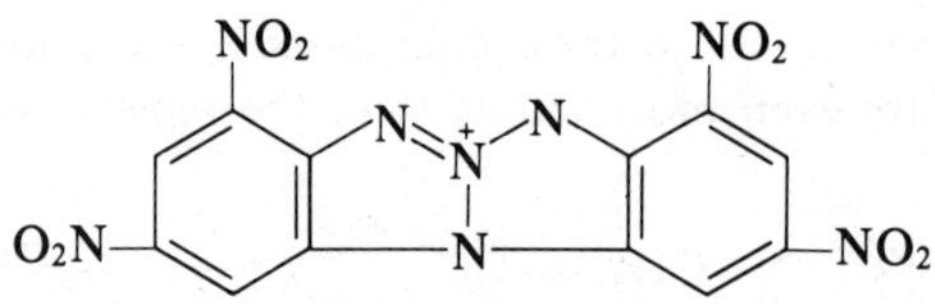

mp 400° dec. (578)

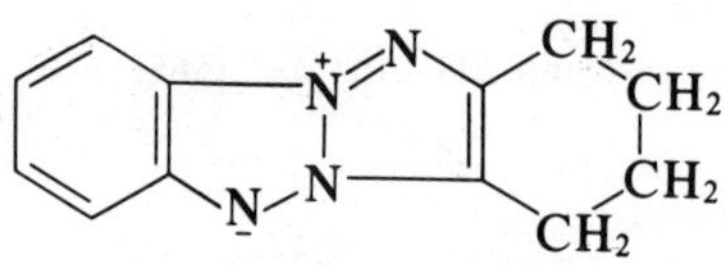

mp 131.5–133° (194)

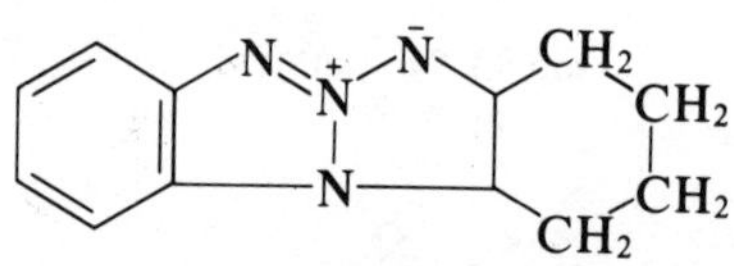

mp 176.5–177.5° (578)

Related dipolar compounds with embedded four-nitrogen chains contain a five- and a six-membered ring, annelated through two nitrogen atoms. These include benzotriazolo[2,1-*a*]naphtho[1,8-*de*]triazine **(X)**, benzotriazolo[1,2-*a*]-naphtho[1,8-*de*]triazine **(XI)**, and a benzotriazinone derivative.

2.6.7. Tetrazolinyl Radicals

Presently known tetrazolinyl radicals are 2,3-diaryl or heteroaryl substituted five-membered unsaturated rings with four contiguous nitrogen atoms, one carbon atom and a delocalized electron, arbitrarily indicated at N-1.* Their oxi-

* Tetrazolinyl and related radicals are reviewed by Neugebauer (791).

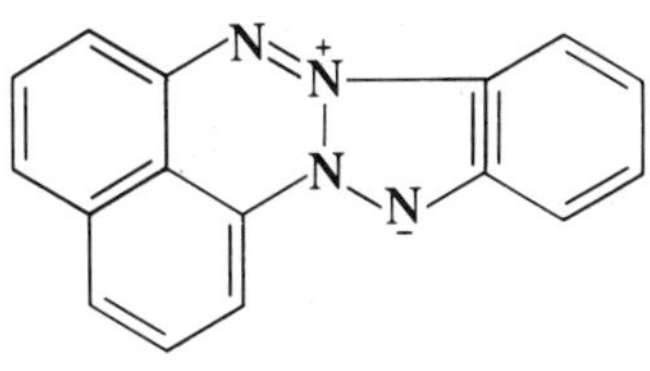

(X)

mp 214–215° (994)

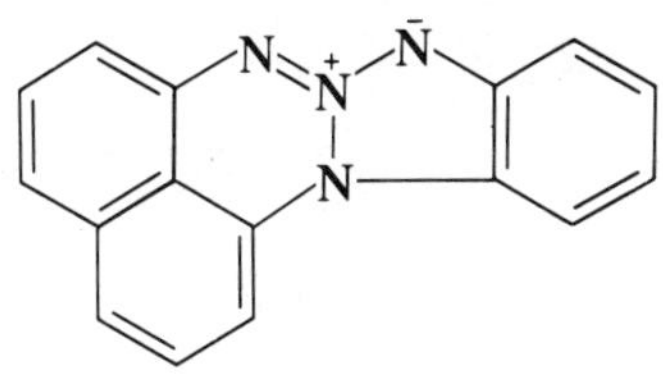

(XI)

mp 230.5–232° (994)

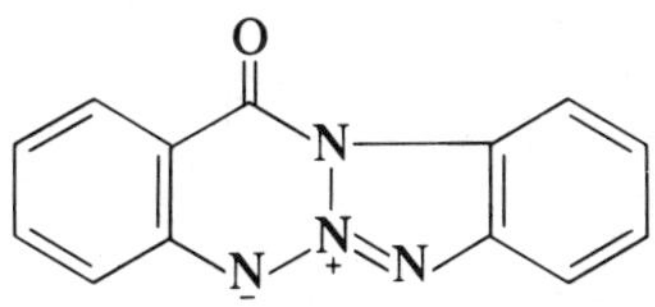

mp 261–262° (774)

dation state is intermediate between formazans and tetrazolium cations from which they differ by one electron. Electron spin resonance data confirms the cyclic rather than the isomeric acyclic structure (794).

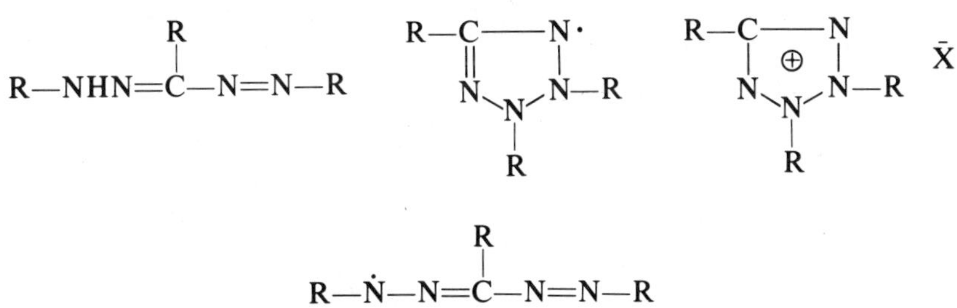

In addition to the simple tetrazolinyl radicals with one tetrazole ring, a second class consists of polycyclic tetrazolinyls with embedded four-nitrogen chains and two shared middle nitrogen atoms, the phototetrazolinyls. Although the magnetic and oxidation-reduction characteristics of many monocyclic tetrazolinyl radicals have been measured in solution (564, 792), only 5-*t*-butyl-2,3-bis(*p*-nitrophenyl)tetrazolinyl **(I)** has been isolated thus far; many tetracyclic phototetrazolinyls have been obtained in the solid state.

Decomposition temperatures of tetrazolinyl radicals range from 126 to over 300°. Higher decomposition points are associated with electron-withdrawing groups on the 2,3-biaryl substituents, reducing the electron spin densities at N-2 and N-3. Decomposition or melting points of tetrazolinyl radicals are usually much lower than those of the corresponding tetrazolium salts; an exception is the higher decomposition point of 2,3-(5′-chlorobiphenylene)-5-phenyltetrazolinyl **(III)** than its corresponding tetrazolium nitrate, which decomposes at 291–292° (564).

Crystalline tetrazolinyl radicals can be kept for months without decompos-

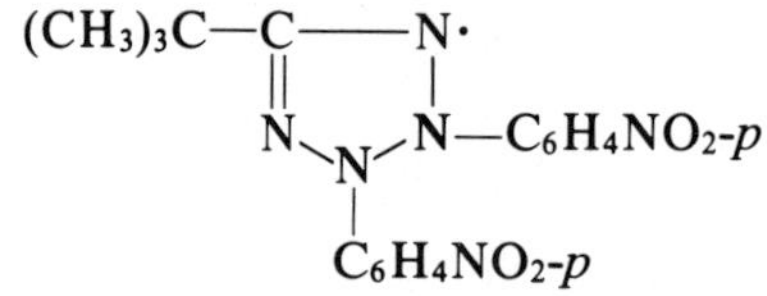

(I)

mp 126–127° dec. (792)

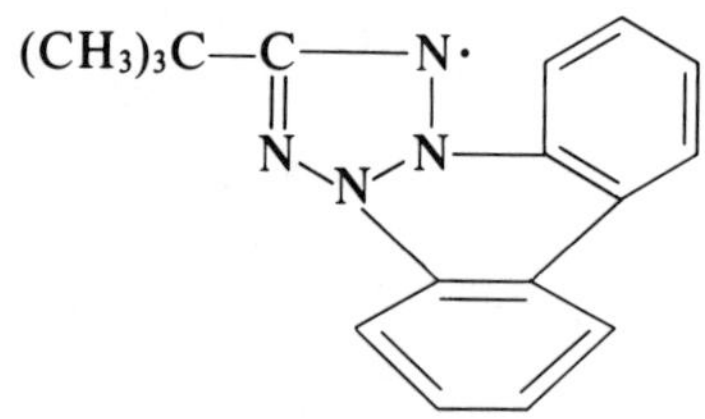

dec. 128° (795)
sublimes 120–125°/10^{-3}

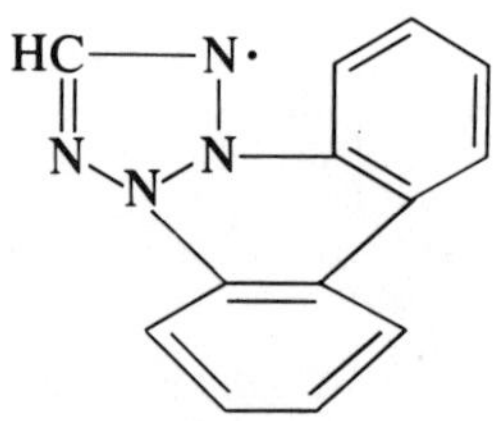

(II)

mp 119–122° (564)

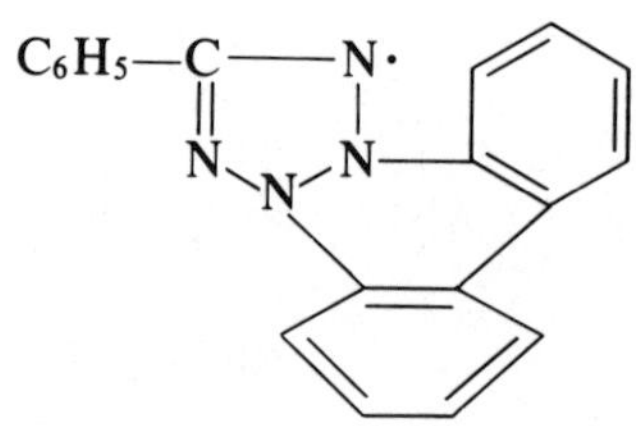

mp 160–161° (564)

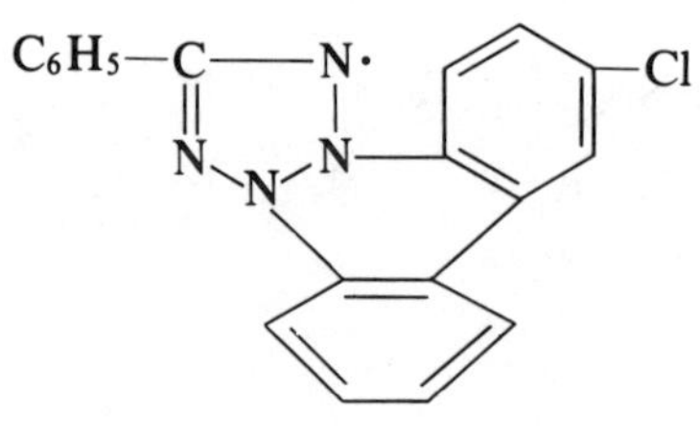

(III)

mp 310° dec. (504)

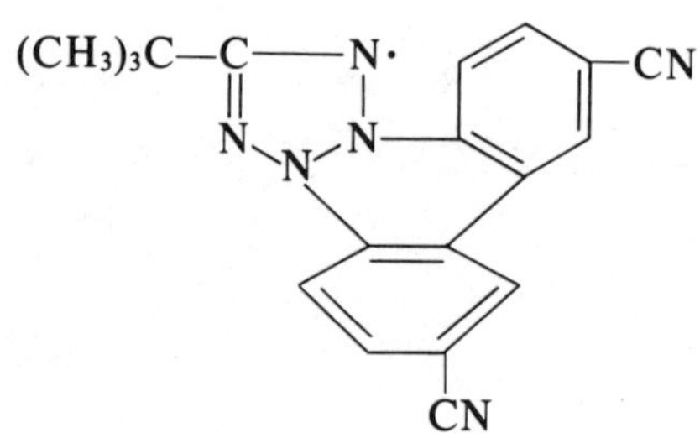

mp 289–292° dec. (792)

ing if light, carbon dioxide, and water vapor are excluded (791). Tetrazolinyl radicals are stable to oxygen; indeed, 2,3-diphenylene-5-phenyltetrazolinyl **(II)** is unchanged after passing air through a benzene solution for 4 hr (617). Both monocyclic and bicyclic tetrazolinyl radicals are monomeric in solution down to at least 90 and 70° K, respectively, and have magnetic moments nearly equal to the theoretical value for a monoradical (444, 792).

2.7. ACYCLIC FIVE-NITROGEN CHAINS

2.7.1. Pentazdienes and Related Compounds

Pentazdienes contain a chain of five contiguous nitrogen atoms with two nonconjugated double bonds, aryl groups at the terminal nitrogens, and hydrogen, alkyl, or aryl at N-3.

$$\mathrm{Ar{-}N{=}N{-}N(R){-}N{=}N{-}Ar}$$

The pentazdienes are solids which melt or decompose from 80° to over 300° (Table 2.79). Relatively higher melting points are associated with electron-withdrawing substituents (NO_2, CN, *p*-$C_6H_5N{=}N$, SO_3Na) on the aryl groups. In solution, however, in the aprotic solvent cumene, at room temperature, half-lives of 1,3,5-triphenyl and tri(*p*-tolyl)pentazdiene are less than 20 min. The small amount of data available suggest that at least 1,5-diarylpentazdienes are somewhat more stable in methanol than in cumene.

Decomposition of tri-*p*-tolylpentazdiene in cumene yields nitrogen and di-*p*-tolyltriazene as well as products of radical combination (742). Intermediate

$$p\text{-}CH_3C_6H_4\text{-}N{=}N{-}N(C_6H_4CH_3\text{-}p){-}N{=}N{-}C_6H_4CH_3\text{-}p \xrightarrow[\text{cumene}]{25.8^\circ}$$
$$N_2 + p\text{-}CH_3C_6H_4N{=}NNHC_6H_4CH_3\text{-}p + CH_3C_6H_5 + \left(C_6H_5C(CH_3)_2{-}\right)_2$$

formation of di-*p*-tolyltriazene-*p*-tolylazo radical pairs during decomposition of tri-*p*-tolylpentazdiene at 40° in chloroform has been detected by chemically induced dynamic polarization (CINDP) (497).

Thermolysis in methanol of 1,5-di-*p*-tolylpentazdiene, with hydrogen on N-3, forms both the triazene and *p*-tolylazide (741).

$$p\text{-}CH_3C_6H_4N{=}NNHN{=}NC_6H_4CH_3\text{-}p \xrightarrow{31^\circ}$$
$$N_2 + CH_3C_6H_5 + p\text{-}CH_3C_6H_4N{=}NNHC_6H_4CH_3\text{-}p + p\text{-}CH_3C_6H_4N_3$$

The only five-nitrogen compound with a nitroso group reported to date is nitrosotrimethyl-2-tetrazene, the melting point of which is above the decomposition points of diaryl nitrosotriazenes.

$$(CH_3)_2NN{=}NN(CH_3){-}NO$$

mp 40° (1088)

In the pentaaza-pentamethinecyanines, the five-nitrogen chain is at a higher oxidation level than in the pentazdienes, involving three conjugated double bonds (530). The nitrogen chain is attached to heterocyclic groups at both ends. These quaternary salts lose nitrogen at 110–150°, changing color from a greenish metallic luster to red-brown and forming triaza- and monoazacyanines (895). In solution, in the aprotic solvents acetonitrile, nitromethane, or di-

Table 2.79. Pentazdienes

$$R_1{-}N{=}N{-}N(R_3){-}N{=}N{-}R_5$$

R_1	R_3	R_5	Melting Point (°C)	Thermal Stability	Ref.
p-$CH_3C_6H_4$	H	p-$CH_3C_6H_4$	Expl. 82–83	$t_{1/2}$ (31°, methanol) 13 min	853
				$t_{1/2}$ (25°, cumene) 3 min	741
p-$O_2NC_6H_4$	H	p-$O_2NC_6H_4$	228–230		854
C_6H_5	CH_3	C_6H_5	109–110; 112–113		12, 385
p-$CH_3C_6H_4$	CH_3	p-$CH_3C_6H_4$	147 dec.	Dec. 50°, methanol	750, 935
p-$CH_3C_6H_4$	$CH_2{=}CHCH_2$	p-$CH_3C_6H_4$	85–87	$t_{1/2}$ (50°, methanol) 1 hr	387, 750
p-FC_6H_4	CH_3	p-FC_6H_4	135–136		921
o-NCC_6H_4	CH_2CN	o-NCC_6H_4	148.5–150		53
C_6H_5	CH_2COOH	C_6H_5	Expl. 124		181
p-$O_2NC_6H_4$	CH_2COOH	p-$O_2NC_6H_4$	Expl. 166–167		181
p-$NaO_3SC_6H_4$	$CH_2C_6H_5$	p-$NaO_3SC_6H_4$	Decolorizes 300		751
C_6H_5	C_6H_5	C_6H_5	Expl. 80–81	$t_{1/2}$ (25° cumene) 18.5 min	497, 852
p-$CH_3C_6H_4$	p-$CH_3C_6H_4$	p-$CH_3C_6H_4$	84.5–85 dec.	Dec. −3°, ether 36 hr	742
				$t_{1/2}$ (25.8°, cumene) 9 min	
p-$C_6H_5N{=}NC_6H_4$	p-$C_6H_5N{=}NC_6H_4$	p-$C_6H_5N{=}NC_6H_4$	184		832

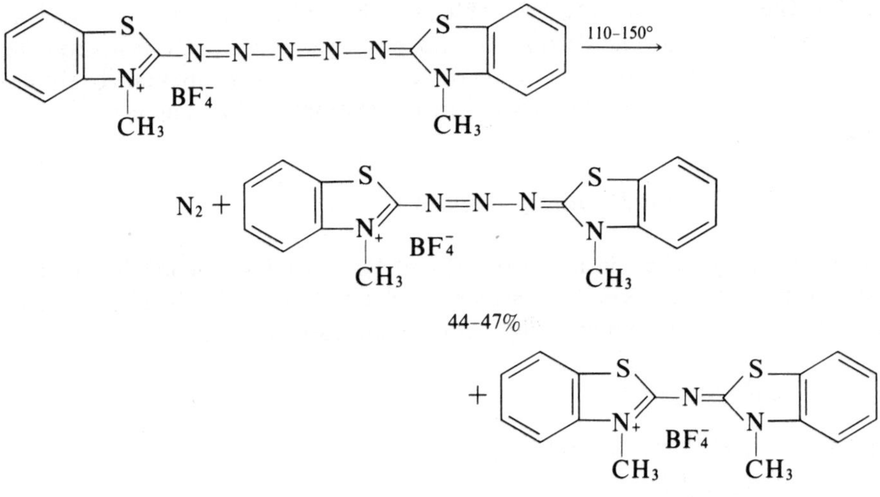

44–47%

37–40%

methylformamide at room temperature, half-lives of the pentaazacyanines are from 1 to 12 hr; the half-life of the 2-quinoline derivative is about 2.5 hr.

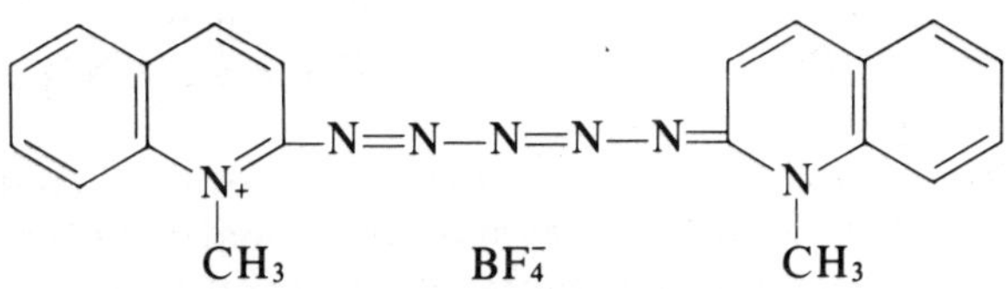

2.8. ACYCLIC-CYCLIC FIVE-NITROGEN CHAINS

2.8.1. Two Exocyclic Nitrogen Atoms

Straight and branched acyclic-cyclic five-nitrogen chains with two exocyclic nitrogen atoms occur, respectively, in the 1- and 2-nitroimides of benzotriazoles

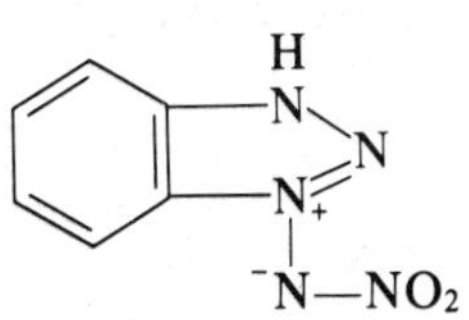

mp 74.5° expl.

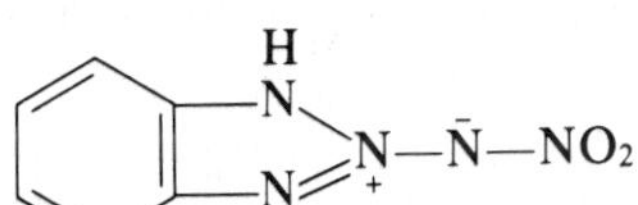

mp 81° expl.

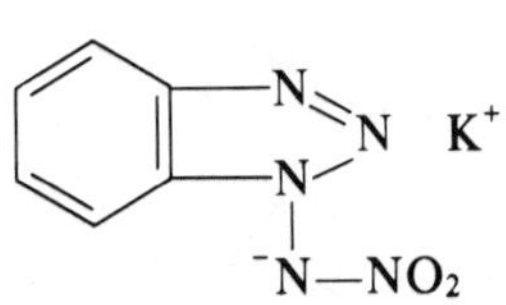

mp 181–182° dec.

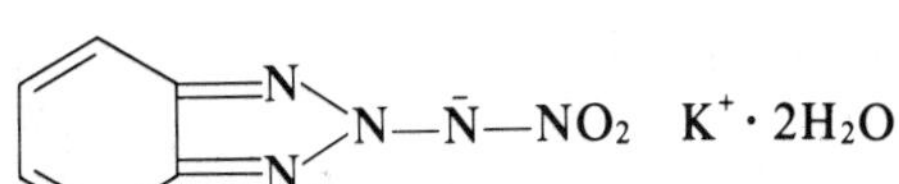

mp 258° dec.

(577). These compounds are less stable than three-nitrogen nitroimides such as benzimidazole-1-nitroimide (mp 169°, expl.). Not only do the benzotriazole nitroimides explode at lower temperatures, but they may spontaneously decompose violently, even in solution. Their potassium salts are appreciably more stable.

2.8.2. One Exocyclic Nitrogen Atom

The five-nitrogen chain in 1-[bis(trimethylsilyl)amino]-5,5-dimethyl-4-trimethylsilyl-5-silatetrazoline is at the pentazene oxidation level (1176). It survives distillation in a high vacuum at 64–66°, but decomposes at 160°.

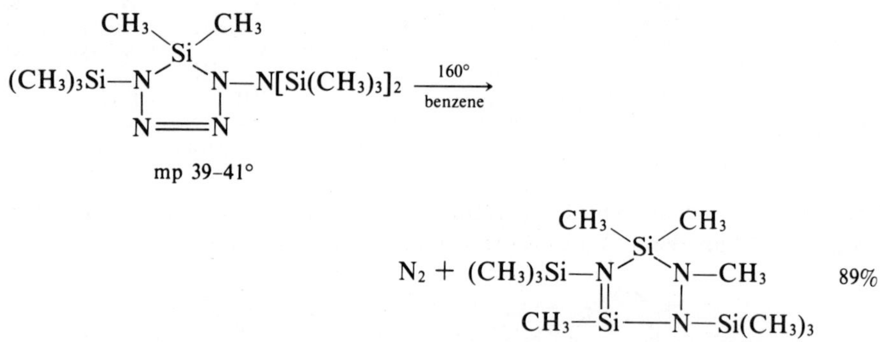

Several 1- and 2-aminotetrazoles and their derivatives with, respectively, linear and branched five-nitrogen chains have been prepared (Table 2.80). As a rule, the more symmetrical 2-aminotetrazoles boil and melt at lower temperatures than do their 1-isomers. Decomposition temperatures on melting range from 125° for 2,5-diaminotetrazole to 229° for 5-amino-1-(2-nitro-5-furfurylideneamino)tetrazole (899). Explosive properties are particularly pronounced in the liquid parent compounds; less than 4 mg of 1-aminotetrazole must be used in its carbon and hydrogen analysis because of the brisance of its explosion (413). Although no detonation results on heating 1,5-diaminotetrazole, the 2,5-isomer explodes at 210° (899). Consistent with the behavior of other high nitrogen classes, increase of carbon content decreases explosive tendency.

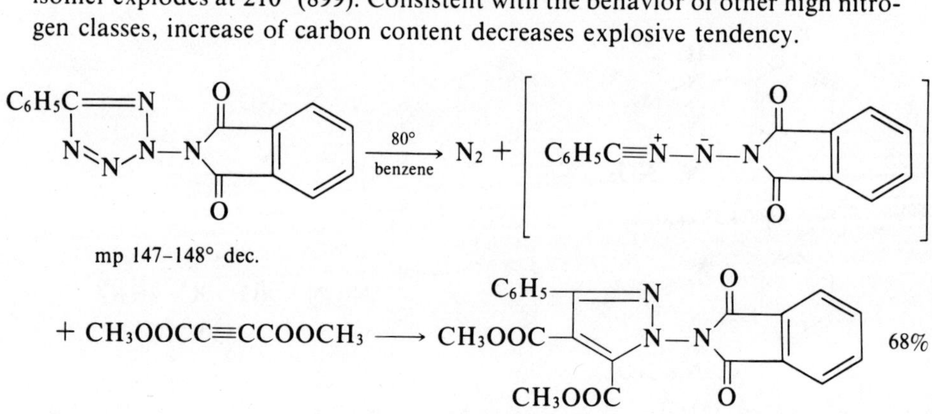

Table 2.80. 1- and 2-Aminotetrazoles

R	R′	R_5	Melting Point (°C)		
			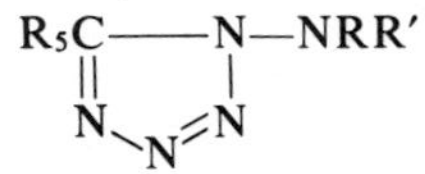		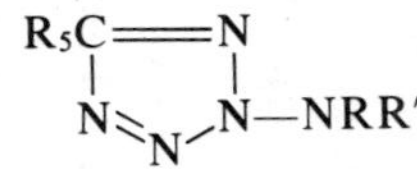
H	H	H	Bp 118–120/0.01	(413)	Bp 89–91/0.3 (899)
			137–142/0.7	(899)	
C_2H_5	C_2H_5	H	Bp 56/0.02	(149)	—
H	H	C_6H_5	155–157	(375)	109–111 (375, 899)
			156–158	(899)	
p-$O_2NC_6H_4$	H	C_6H_5	144.5–145 dec.	(872)	—
C_6H_5CO	H	C_6H_5	Dec. 136–138	(98)	—
H	H	NH_2	187–188 dec.	(899, 1100)	125–127 dec. (899)
$C_6H_5CH=$		NH_2	225–226 dec.	(899)	159–160 dec. (899)
p-$O_2NC_6H_4CH=$		p-ClC_6H_4S	111–113	(273)	—

Thermolysis of 5-phenyl-2-phthalimidotetrazole occurs in refluxing benzene, a much lower temperature than required for decomposition of other 2,5-disubstituted tetrazoles, initially forming the benzonitrile *N*-(phthalimido)imide dipole (375). As in the thermal decomposition of 2-alkyl and aryl tetrazoles, the dipole can be intercepted by olefins and acetylenes, illustrated with acetylene dicarboxylic ester. An acyclic-cyclic five-nitrogen chain occurs in the 1-(diethylamino)-4-methyl tetrazolium cation (1207).

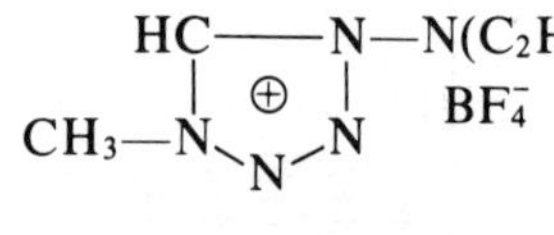

mp 103°

2.9. CYCLIC FIVE-NITROGEN CHAINS

2.9.1. Pentazoles

Pentazole, the ultimate five-membered nitrogen ring, consisting solely of nitrogen atoms, has been sought since at least the early 1900s (427). Solution of this classical problem in nitrogen chemistry came in 1956, when Huisgen and Ugi deduced the presence in solution of phenylpentazole by the kinetics of decomposition of the reaction product of azide and ^{15}N-labeled benzene diazonium ions and from the isotopic composition of their decomposition products (217, 526). Two years later Ugi reported the isolation and characterization of unstable but crystalline arylpentazoles, providing assurance of their existence (1110).* Pentazole itself has not been prepared.

Crystalline arylpentazoles can be kept below −20°, but may explode on touching above −10° (1107, 1110). The least stable arylpentazoles carry electron-withdrawing substituents on aryl and the most stable, electron-releasing groups (Table 2.81). *p*-Dimethylaminophenylpentazole, the most durable pentazole, can be kept at room temperature for several hours (1107). As with other high nitrogen classes such as azides, ortho substituents decrease stability. Phenylpentazole decomposes faster in solvents with low rather than high dielectric constant (1110).

The kinetic evidence for the existence of phenylpentazole depends on the rapid release of 65–76% of the nitrogen from the benzenediazonium-azide product in methanol, 2-methoxyethanol, or water up to about 0°, and the remainder more slowly above this temperature, suggesting intermediates of two different stabilities (526, 528). Isotopic evidence involves the distribution of decomposition products from end ^{15}N-labeled benzenediazonium ion. Approximately 85% of the phenyl azide bears the label on the middle nitrogen and about 15% in the

* The pentazoles are reviewed by Ugi (1107).

Table 2.81. Pentazoles

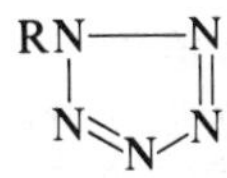

R	Decomposition Point (°C)	Thermal Stability	Ref.
C_6H_5	−5 to −3	$t_{1/2}$ (0°, *n*-hexane) 2.6 min	1110
		$t_{1/2}$ (0°, methanol) 11.8 min	
		$t_{1/2}$ (23°, H_2O) 1.2 min	925
p-$O_2NC_6H_4$	—	$t_{1/2}$ (0° methanol) 2.0 min	1109
		$t_{1/2}$ (23°, H_2O) 14 sec	925
p-ClC_6H_4	8–10	$t_{1/2}$ (0°, methanol) 9.5 min	1109, 1110
		$t_{1/2}$ (23°, H_2O) 35 sec	925
o-ClC_6H_4	—	$t_{1/2}$ (0°, methanol) 1.3 min	1109
p-$CH_3C_6H_4$	2–5	$t_{1/2}$ (0°, methanol) 20.6 min	1109, 1110
		$t_{1/2}$ (23°, H_2O) 4.4 min	925
o-$CH_3C_6H_4$	—	$t_{1/2}$ (0°, methanol) 3.7 min	1109
p-$C_2H_5OC_6H_4$	26–29	$t_{1/2}$ (0°, methanol) 38 min	1109, 1110
p-$(CH_3)_2NC_6H_4$	50–54	$t_{1/2}$ (0°, methanol) 1.1 hr	1109, 1110

liberated nitrogen (217, 220, 528). This distribution is consistent with formation of both a linear phenyldiazoazide and a phenylpentazole, each of which decomposes to phenyl azide and nitrogen, with different label distribution. In a

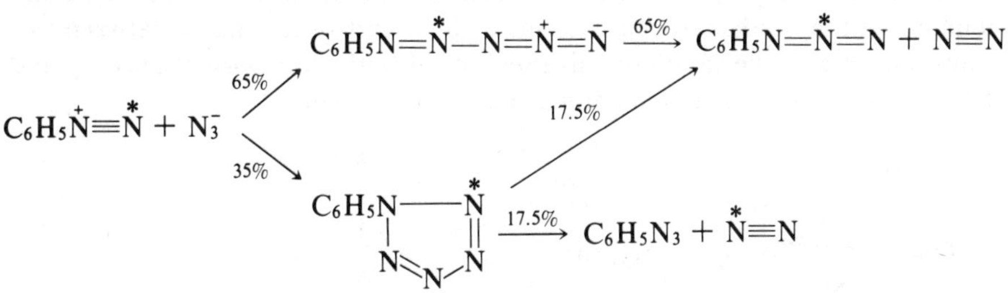

two-phase system in which phenyldiazoazide is more soluble in the nonpolar phase, decomposition of phenyldiazoazide and phenylpentazole are mutually independent (1108). Isolated, crystalline *p*-ethoxyphenylpentazole labeled at N-2 decomposes to form equal parts of labeled *p*-ethoxyphenyl azide and nitrogen (1112).

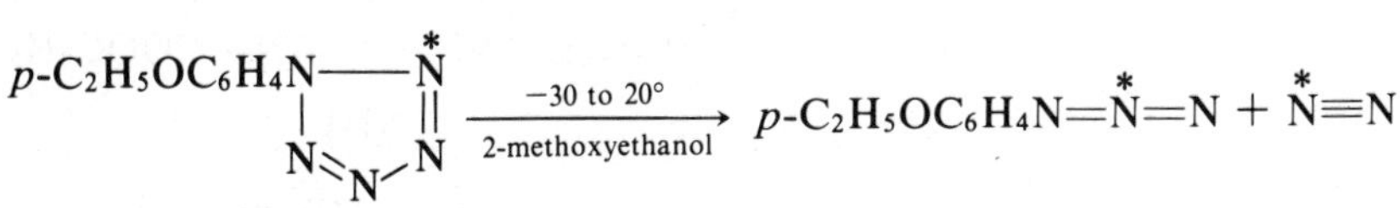

In aqueous solutions of aryldiazonium and azide ions at room temperature, the appearance of aryl azide occurs in two steps, consisting, first, of an apparent bimolecular reaction, followed by a slower reaction with first-order kinetics (925). The slower first-order reaction is believed to involve conversion of the arylpentazole to aryl azide. Rate of formation of *p*-anisylpentazole and disappearance of *p*-anisyldiazonium ion are the same, and decomposition of *p*-anisylpentazole and appearance of *p*-anisyl azide occur at the same rate. In this

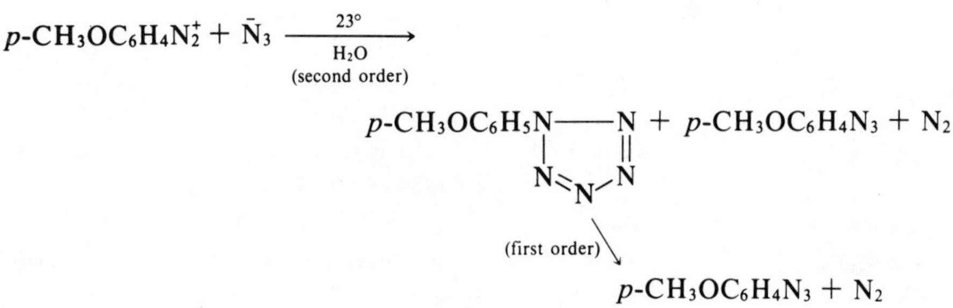

system, no aryldiazoazide was detected and, if present at all under these conditions, is a reactive intermediate.

2.9.2. Embedded Five-Nitrogen Chains

On the basis of melting behavior, compounds with embedded five-nitrogen chains seem to be more stable than other five-nitrogen classes. The branched five-nitrogen chains in **I** and **II** are at the most reduced level, perhaps owing their apparent stability not only to the restricted motions of the nitrogen chain inherent in the condensed ring systems with two nitrogens shared between two rings, but also to the simultaneous presence of both electron-withdrawing and electron-releasing substituents along the nitrogen chain.

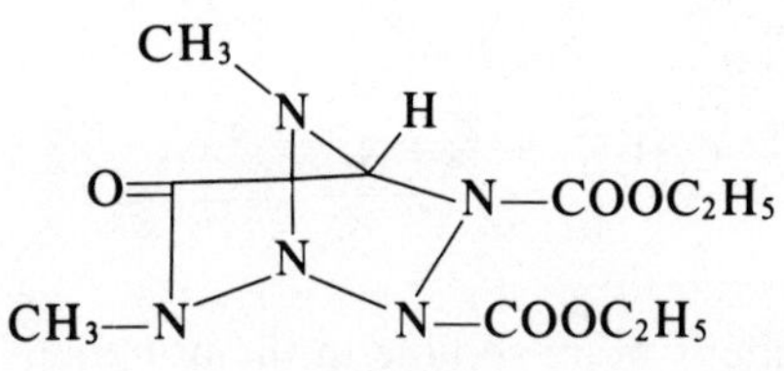

(I)

mp 166–168° (885)

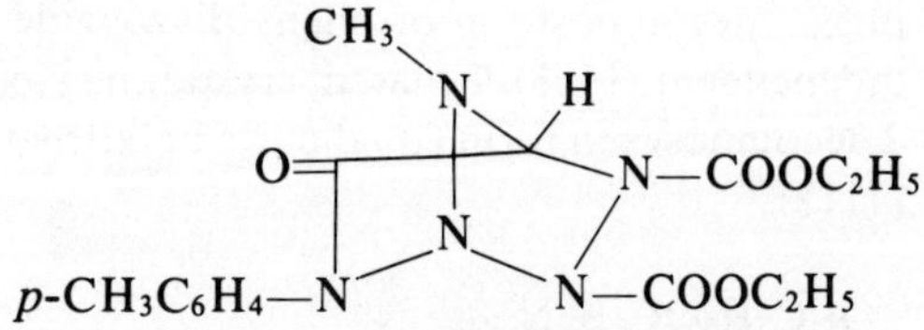

(II)

mp 172–173° (884)

Embedded compounds at the pentazene oxidation level include both those with unsaturation between N-1 and N-2 and those with a double bond between N-2 and N-3. The fourth nitrogen in the chain is shared between two rings. These dihydrotetrazolopyrazoles and *s*-triazoles may be considered as embedded 1-amino 2-substituted and 1-amino 4-substituted tetrazolines.

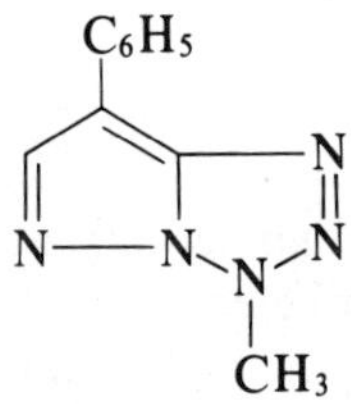

mp 138–139° (1)
139–140° (23)

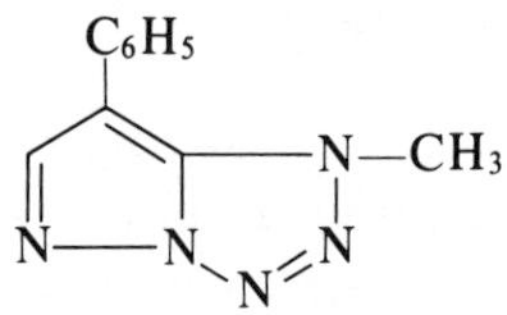

mp 101–103° (22)

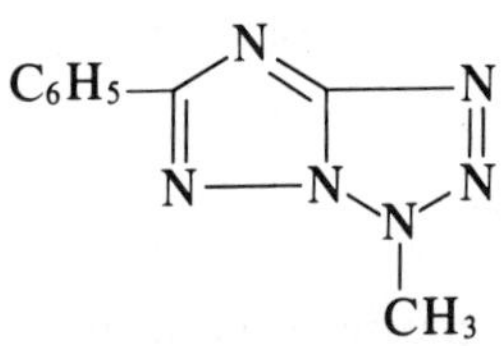

mp 182–184° (186)

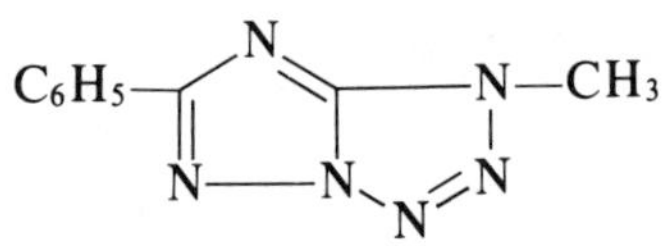

mp 150–152° (186)

The largest group of heterocycles with transannular five-nitrogen chains are derivatives of 1-aminotetrazole, sharing the nitrogen next to the end of the chain between the tetrazole and another ring. Fusion of tetrazole is to both five- and six-membered rings. These compounds melt from slightly below 100° to over 300° with few instances of decomposition. In comparison with acyclic-

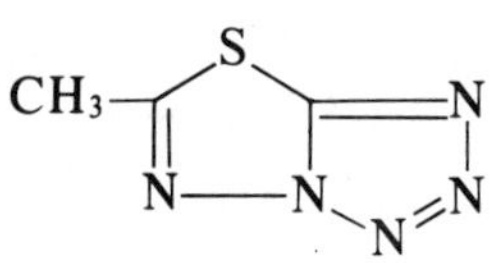

mp 92–93° (574)

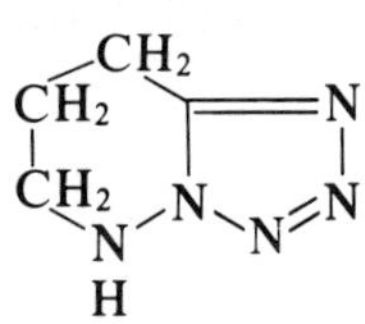

mp 158–159° (573)

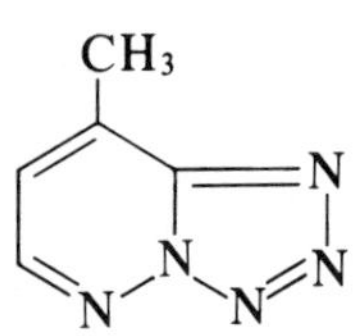

mp 126–127° (870)

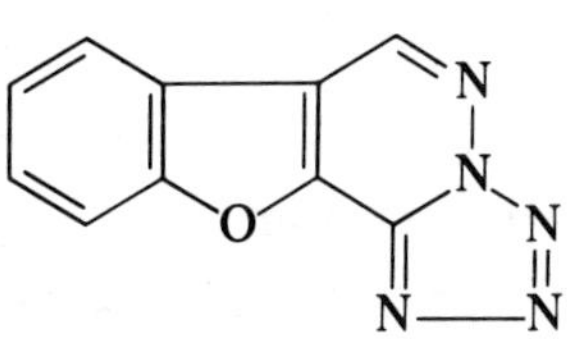

mp 298° (926)

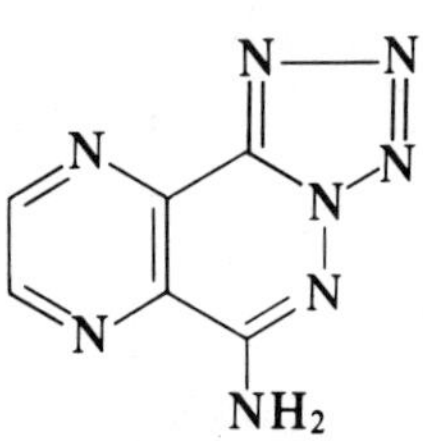

mp 315° dec. (268)

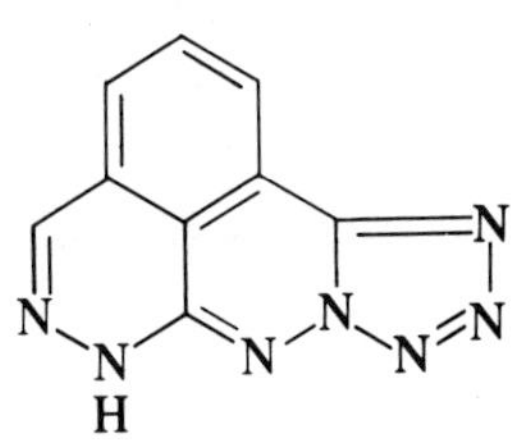

mp >350° (342)

cyclic *N*-aminotetrazoles and derivatives, these embedded five-nitrogen compounds are much more heat resistant, again indicating the strong stabilizing influence of annelation of the nitrogen chain.

2.10. ACYCLIC SIX-NITROGEN CHAINS

2.10.1. Hexazdienes

The known acyclic compounds with six contiguous nitrogen atoms contain double bonds at either end of the chain, aryl or heteroaryl groups at positions 1 and 6, and hydrogen, methyl, aryl, heteroaryl, or acyl groups at N-3 and N-4.

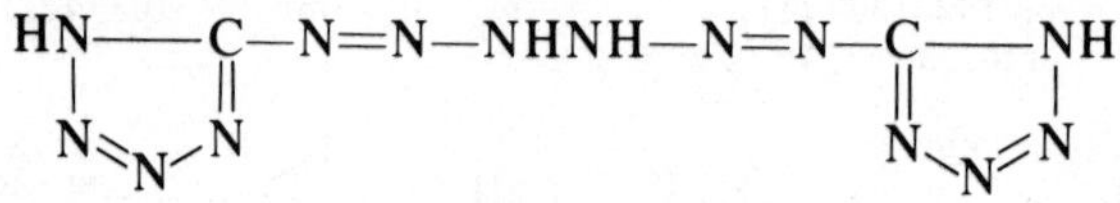

expl. 90° (492)
stable at 25° 1 month

Hexazdienes decompose on heating, often explosively, between 80 and 178°. The only hexazdiene presently described with hydrogen at N-3 and N-4 is the 1,6-bistetrazolyl derivative, originally prepared in 1911. Its structure has not

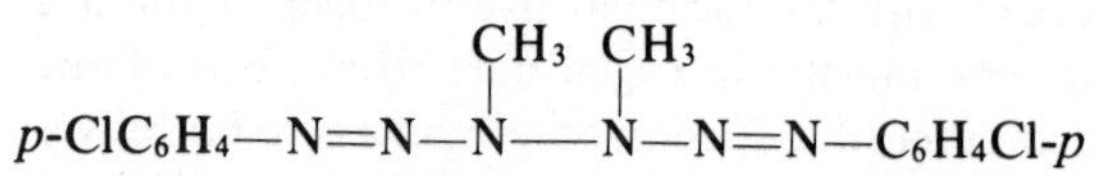

mp 86–87° dec. (397)

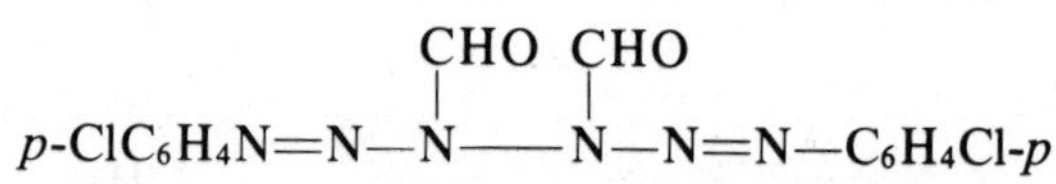

mp 124–125° dec. (507)

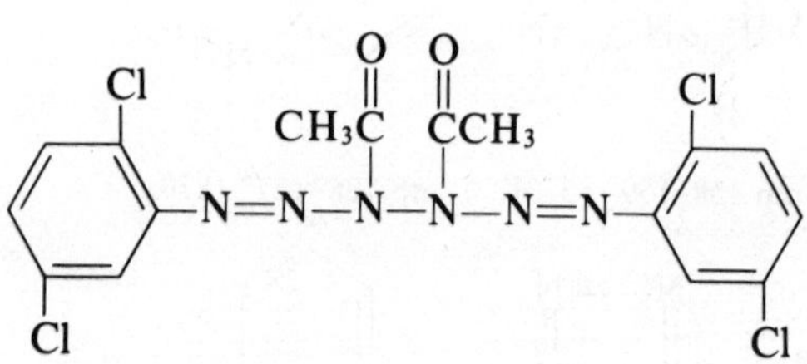

mp 178–179° dec. (507)

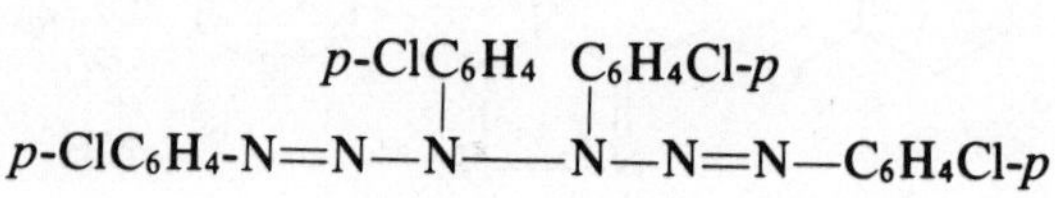

expl. 110° (1081)

been reinvestigated by modern spectroscopic techniques, nor has it been subjected to X-ray crystal analysis. The base strength of N-3 and N-4 may be sufficient for protonation by tetrazole hydrogen so that this hexazdiene might exist as a dipolar molecule, or involve intra- or intermolecular hydrogen bonding.

Known hexazdienes all are substituted with at least two electron-withdrawing groups. There is no evidence of dissociation of the tetraarylhexazdienes to triazenyl radicals (1081). On standing or on heating benzene solutions of tetraarylhexazdienes evolve nitrogen and form azo compounds. 1,6-Di-*p*-chlorophenyl-3,4-diacetyl-1,5-hexazdiene is stable indefinitely at room temperature. In refluxing benzene its half-life is about 6.4 hr; decomposition is by a radical pathway (696).

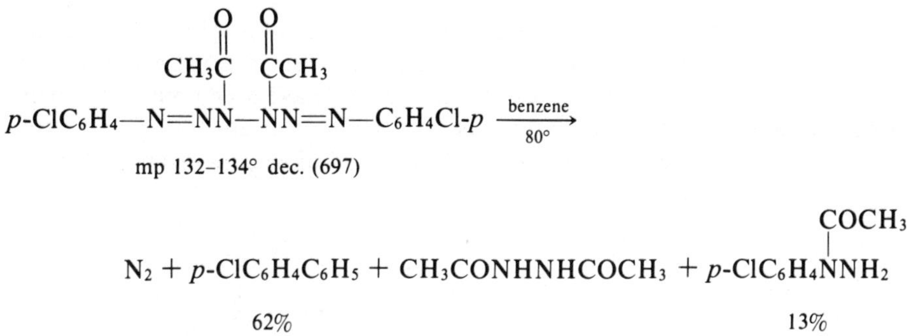

2.11. CYCLIC SIX-NITROGEN CHAINS

2.11.1. Cyclic-cyclic Six-Nitrogen Chains

The nitrogen chain in 6,6′-bi(1,2-oxoindazolo[1,2-a]benzotriazolyl) is at the hexazane oxidation level and also includes two embedded nitrogen atoms in each fused triazoline ring. This structural constraint, together with the attached carbonyl groups and benzene rings, must contribute to the stability suggested by its relatively high melting point.

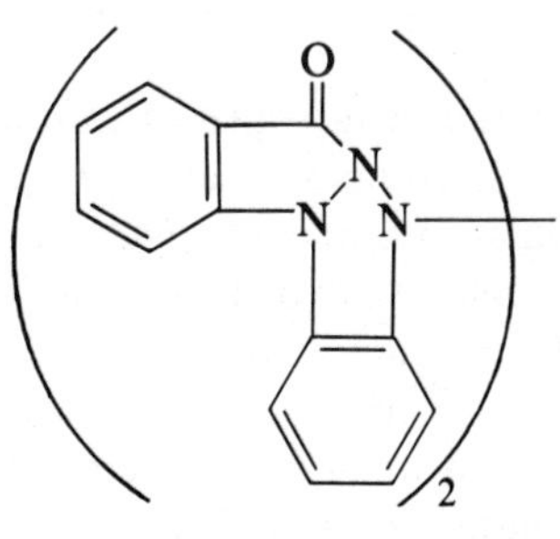

mp 256–257.5° (1103)

The simplest cyclic-cyclic hexazdiene chain is that in the *gem*-trifluoromethyl derivative of 1,1′-bitriazoline, linked to strongly electron-withdrawing groups at N-3 and N-4 but to an aliphatic carbon at each terminal nitrogen. In other

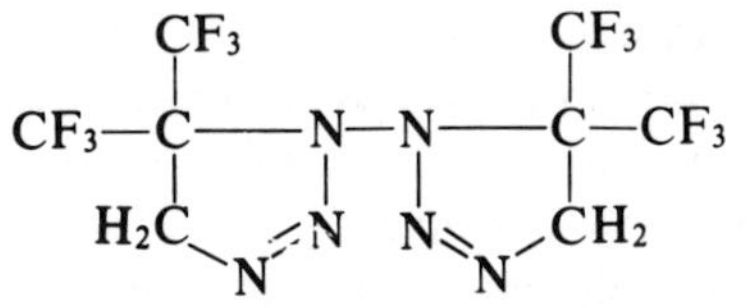

mp 96° expl. (172)

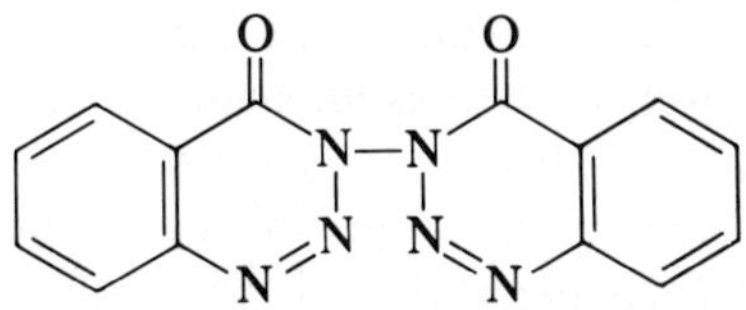

mp 240° dec. (484)
249° dec. (456)

cyclic-cyclic compounds with a hexazdiene chain such as 3,3′-bi-1,2,3-benzotriazine-4-one and the bibenzotriazoles, the azo linkages are conjugated with an aromatic ring. Both 1,1′- and 1,2′-bibenzotriazoles have been prepared, having a straight and a branched six-nitrogen chain (428). On heating, they lose nitrogen and form, respectively, straight and branched dibenzotetrazapentalenes.

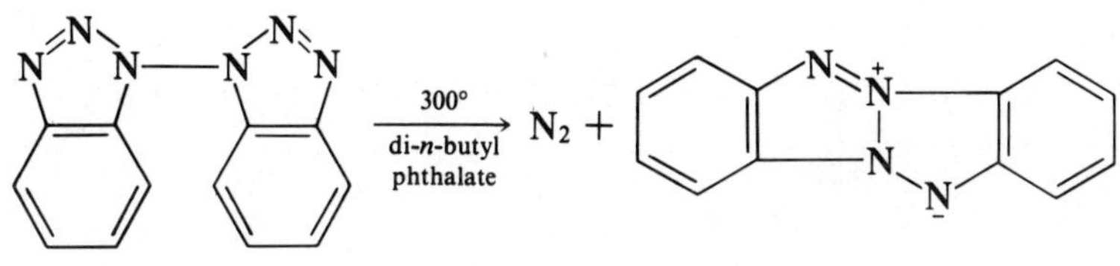

mp 235° dec. 54%

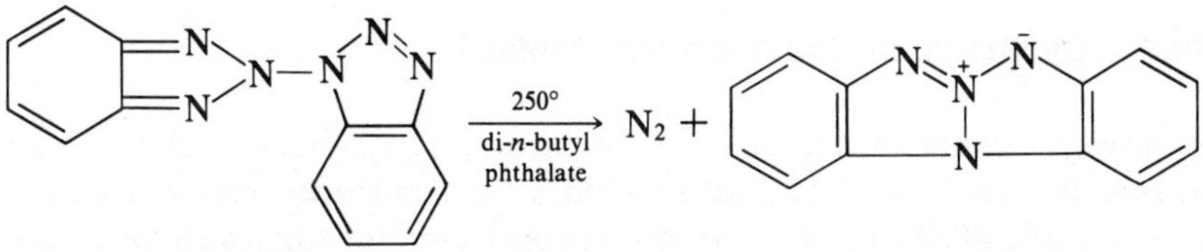

mp 123.5–124.5° 68%

2.11.2. Hexazine

Attempts to prepare a polymer of nitrogen or to indicate its presence by electrolysis of ionic azides have been made since 1895 (164, 860). A molecule consisting solely of six nitrogen atoms, in principle, could result from the combination of two azide radicals. If two azide radicals unite to form one nitrogen-nitrogen bond the product would be a linear N_6 entity, and if two nitrogen-nitrogen bonds are formed, a cyclic hexazine would be obtained. Although no choice between a linear or an annular intermediate is possible from the study of Clusius and Schumacher in 1958 (218), temporary formation of N_6 is suggested by the electrolysis of potassium azide labeled with ^{15}N at the middle nitrogen. The isotopic composition of the elemental nitrogen product, with no

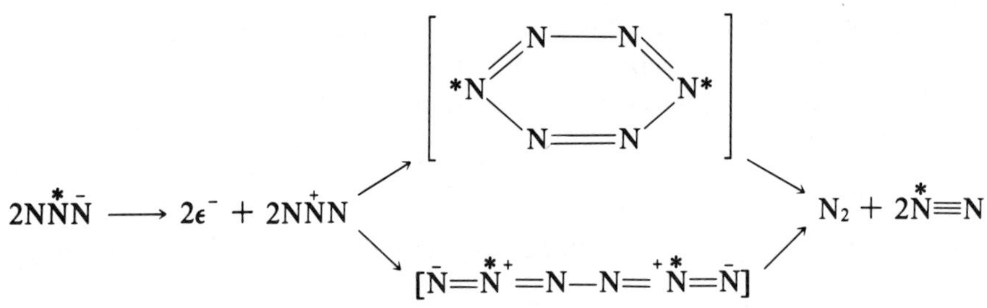

molecular nitrogen doubly labeled, and twice as much singly labeled as unlabeled, could result only through an N_6 intermediate or by the improbable dissociation of azide radical to singly labeled elemental nitrogen and unlabeled atomic nitrogen, which would have to recombine only to unlabeled N_2.

Although theoretical calculations indicate that N_6 is less stable in the gas phase than $3N_2$ (512), results suggestive of the formation of N_6 at low temperature come from the photochemical treatment of a metal coordinated diazide (1132). At 77°K, a yellow material (λ_{max} 380 nm) is formed on irradiation of an ethanol or tetrahydrofuran solution of *cis*-$Pt(N_3)_2[P(C_6H_5)_3]_2$ with no evidence by ESR of N_3 radicals, which can be detected on similar treatment of the trans isomer. As the matrix is warmed, the color disappears with simultaneous evolution of nitrogen. During flash photolysis of the cis isomer at room temperature no radicals were detected, but an intermediate absorbing at 380 nm was observed.

2.11.3. Embedded Six-Nitrogen Chains

Completely reduced six-nitrogen chains are present in the polycyclic compounds **I** and **II**. Four of the contiguous nitrogens in each of these heterocycles are shared between two rings and two are common to three rings. The end ni-

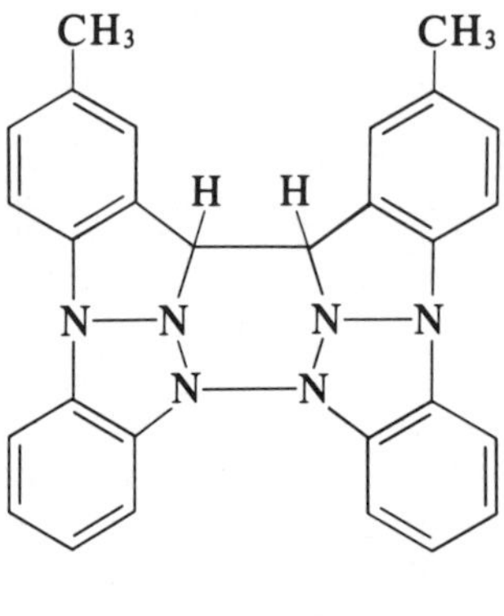

(I)

mp 248–250° dec (1104)

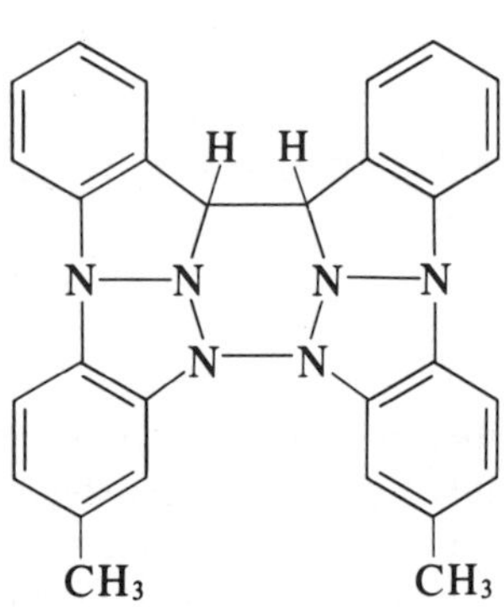

(II)

mp 199–200° dec. (1104)

trogen atoms of the six-nitrogen chain are disubstituted and the 3- and 4-nitrogens are monosubstituted by aryl, and aliphatic carbon is substituted at N-2 and N-5. Stability, as measured by decomposition on melting, is similar to compounds with cyclic-cyclic six-nitrogen chains. Fused tetrazolotriazines with one embedded nitrogen are known that contain intact six-nitrogen chains, indicated by NMR and infrared analyses. The imidazolo-tetrazolo-*v*-triazine and related compounds are discussed in Section 2.16.2 on acyclic-cyclic isomers and equilibria.

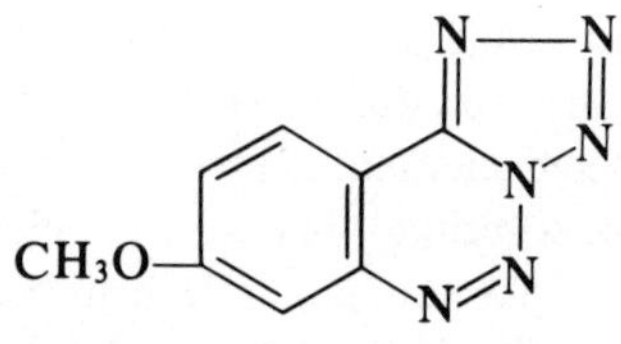

expl. solid (156a)

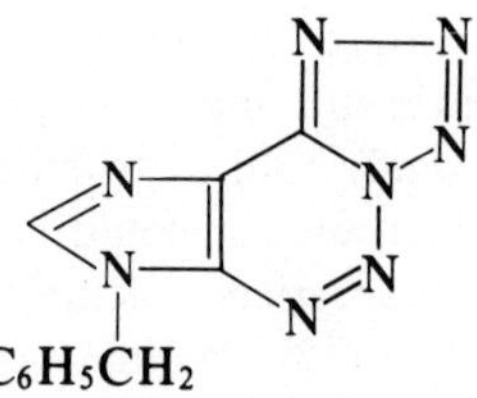

expl. 147° (1074)

2.12. ACYCLIC-CYCLIC SEVEN-NITROGEN CHAINS

2.12.1. Four Exocyclic Nitrogen Atoms

The only substance known to have seven contiguous nitrogen atoms is the lithium salt of 1-(benzotriazol-1-yl)-4-*p*-tolylsulfonyltetrazene (581). This yellow solid is too unstable for elemental analysis, decomposing unpredictably and spontaneously. At or below room temperature in acetonitrile or tetrahydro-

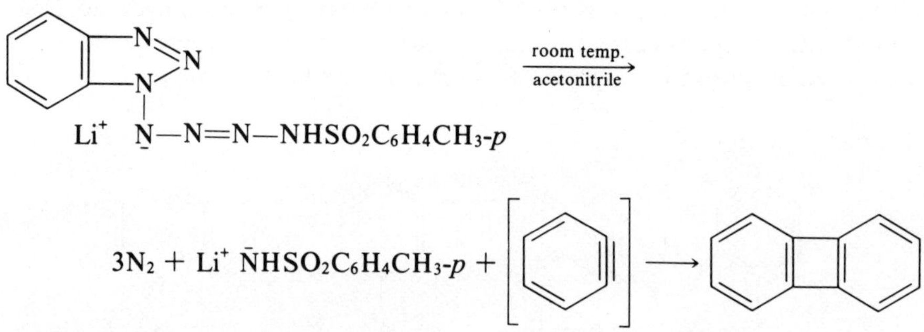

44%

furan, it liberates nitrogen and forms lithium *p*-toluenesulfonamide and benzyne, isolable either as biphenylene or as a product with a trapping material.

A compound reported earlier (671, 1083) to contain two seven nitrogen chains seems more likely to be the salt di(azidoguanidinium) azo-bis-tetrazolate, with electrovalent and not covalent bonds between the guanyl azide and azotetrazole moieties.

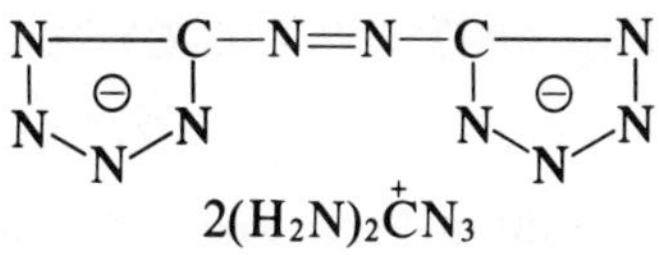

2.13. ACYCLIC EIGHT-NITROGEN CHAINS

2.13.1. Octaztrienes

In 1900, Wohl and Schiff reported a series of compounds with eight sequential nitrogen atoms, and three double bonds, substituted with aryl radicals at the 1-, 3-, 6-, and 8-positions (1193). Although the octaztriene carrying four phenyl groups was too unstable to analyze, satisfactory Dumas nitrogen analyses were obtained with the 1,8-di-*p*-bromophenyl and di-*p*-tolyl-3,6-diphenyl derivatives. Minato, Oku, and Chan in 1966 confirmed the preparation and properties of 1,8-di-*p*-tolyl-3,6-diphenyloctaztriene and found its half-life in cumene to be 46 hr at 10° and 2 hr at 30° (743). The products of thermal decomposition indicate radical intermediates, possibly initially involving the 1-phenyl-3-*p*-tolyltriazene radical, $p\text{-}CH_3C_6H_4N{=}N\dot{N}C_6H_5$.

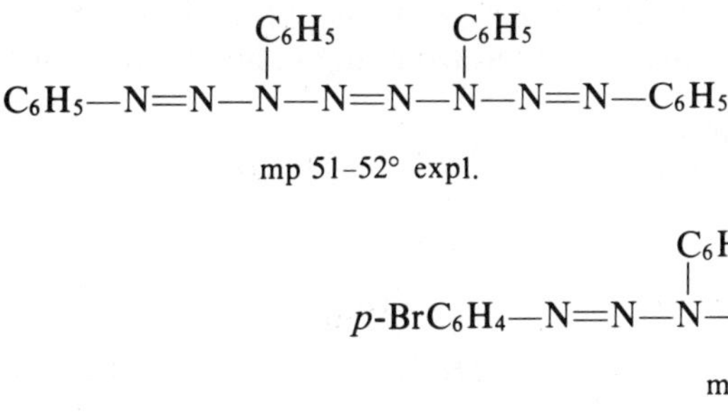

mp 51–52° expl.

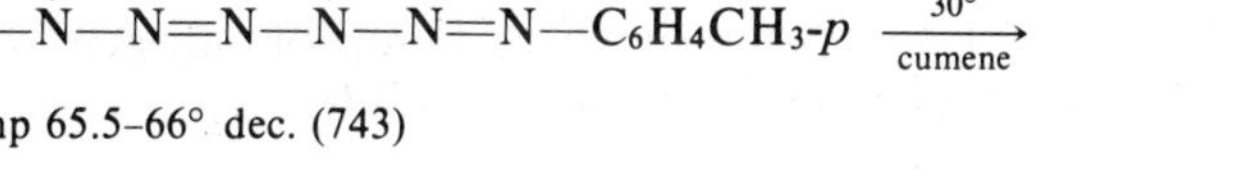

mp 60° expl.

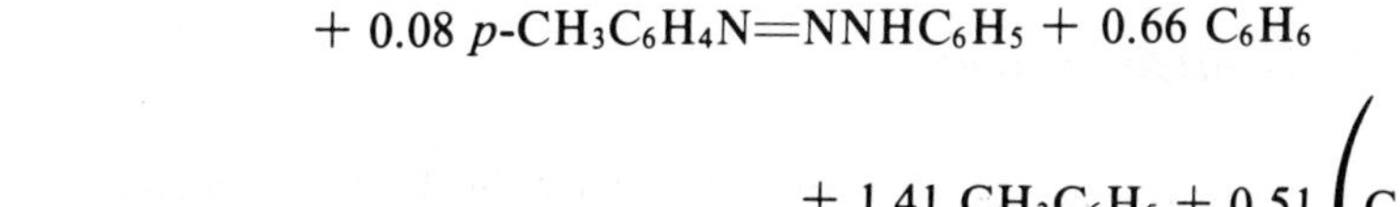

mp 65.5–66° dec. (743)

$$2.94\ N_2 + 0.44\ C_6H_5N{=}NC_6H_5 + 0.40\ p\text{-}CH_3C_6H_4N{=}NC_6H_5$$

$$+ 0.08\ p\text{-}CH_3C_6H_4N{=}NNHC_6H_5 + 0.66\ C_6H_6$$

$$+ 1.41\ CH_3C_6H_5 + 0.51$$ 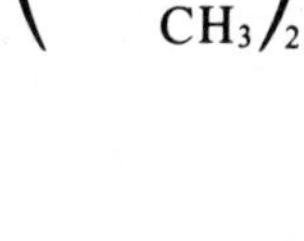

2.14. ACYCLIC-CYCLIC AND CYCLIC-CYCLIC EIGHT-NITROGEN CHAINS

2.14.1. Azobisbenzotriazolium Salts and 1,1′-Bitetrazoles

The 3,3′-dimethyl-1,1′-azobisbenzotriazolium salts are much more stable than the octaztrienes. Efforts to prepare uncharged azobisbenzotriazole were unsuc-

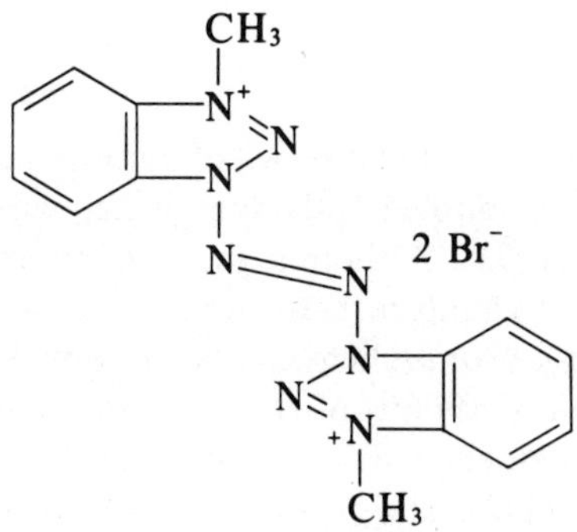

mp 156° violent dec. (381, 382)

cessful, suggesting that quaternization contributes to the stability of the eight-nitrogen chain.

At room temperature 1,1′-bi-5-phenyltetrazole shows no infrared azide band but does so at 50°, indicating opening of one tetrazole ring to an imidoyl azide

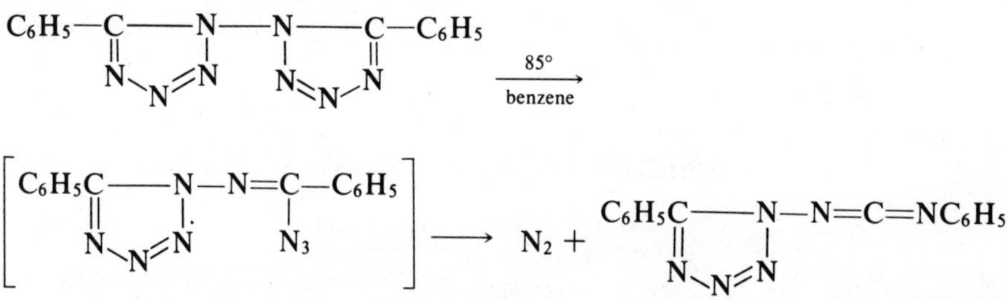

(98). Thermolysis in solution yields 1 mole of nitrogen from the rearrangement of the intermediate tetrazole imidoyl azide.

2.15. ACYCLIC-CYCLIC TEN-NITROGEN CHAINS

2.15.1. Azo-1,1′-Bistetrazoles

The longest nitrogen sequences presently known are the 10-nitrogen chains in the azo-1,1′-bistetrazoles (1043). Besides the 5,5′-bisphenyl compound, the *o*-chlorophenyl and *p*-tolyl derivatives have been prepared. All are highly explo-

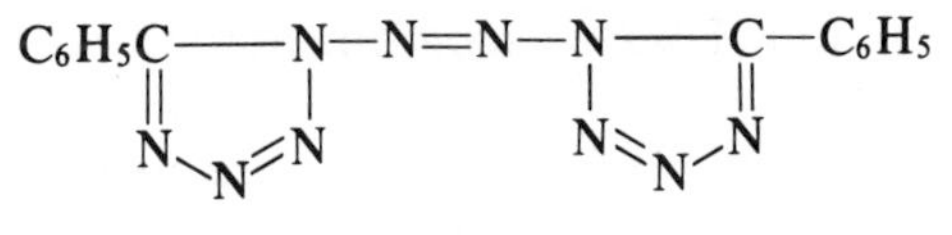

expl. 141°

sive and decompose on heating in solution, liberating nitrogen. The constitution of these azobistetrazoles is inferred from their method of preparation; elemental analysis was not attempted, nor have the materials been reinvestigated using modern techniques of structure determination.

2.16. ACYCLIC-CYCLIC ISOMERS AND EQUILIBRIA

Finally, two classes of the high nitrogen population can exist in alternate isomeric open chain or ring structures. The systems are the α-diazoimine-*v*-triazoles and the imidoyl azide-tetrazoles. Their acyclic forms are 1,3-dipoles bound to a carbon-nitrogen double bond and the cyclic isomers are aromatic five-membered heterocycles. Stability of the cyclic nitrogen form depends on

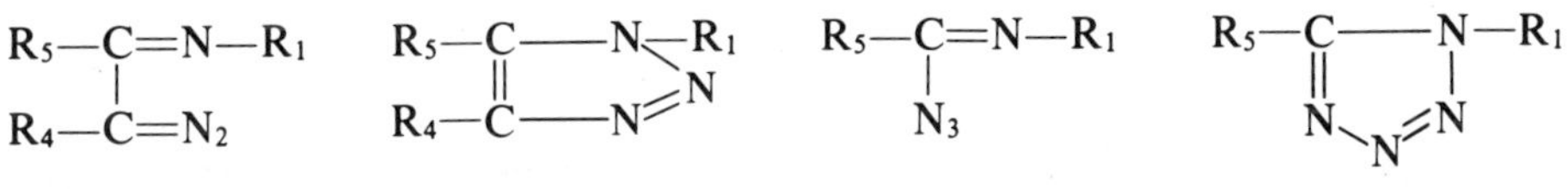

the bond between N-1 and N-2 remaining intact. Theoretical calculations indicate the ring forms to be more stable thermochemically (175, 176).

There are several possibilities for isomer relationships in these classes. Thus only one isomer may exist, or both isomers may be preparable, but are not interconvertible; isomerization may proceed in one direction only, or a mobile equilibrium may be established between the isomers. These ring closures are 1,5-electrocyclizations.* Although the thermal aspects of isomerization are of principal concern here, acidity and basicity of media and solvent polarity also influence existence or relative quantities of acyclic and cyclic forms.

Qualitatively the yellow color of the α-diazoimines frequently allows a distinction from the colorless *v*-triazoles; no comparable differentiation is available between the imidoyl azides (also called azidoazomethines) and tetrazoles, both of which are usually colorless. Infrared spectra are more definitive in identifying diazo (2050–2150 cm^{-1}) and azide (2080–2220 cm^{-1}) and ^{1}H, ^{13}C, and ^{15}N (528a) NMR data are of particular value for quantitative measurement of components of these isomer pairs. Dipole moments also have been used to calculate isomer proportions (311).

* 1,5-Electrocyclizations forming both *v*-triazoles and tetrazoles are reviewed by Taylor and Turchi (1072) and by Huisgen (517).

2.16.1. α-Diazoimine-ν-Triazole Isomerization*

The combined electron-releasing or electron-attracting character of substituents at the 1-, 4-, and 5-positions of the ν-triazole ring determines the strength of the bond between N-1 and N-2. With hydrogen, alkyl or aryl sulfonyl at N-1, and hydrogen, alkyl, or aryl at C-4 or C-5 no ring opening occurs. A stronger electron-withdrawing group, such as cyano, however, at position 1 of ν-triazole

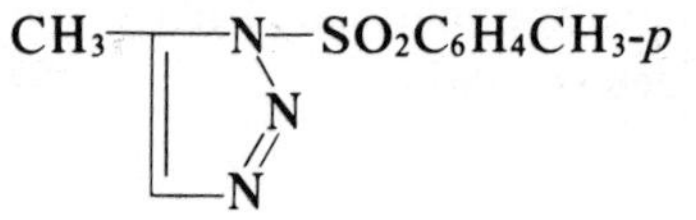

mp 119.5–121° (435)

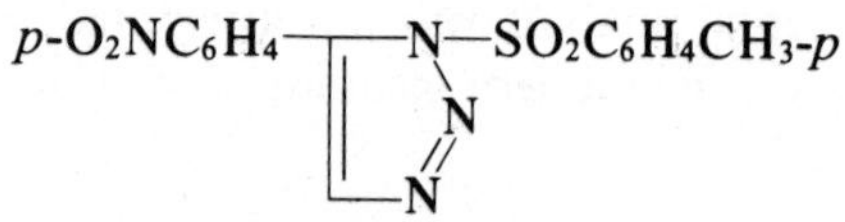

mp 172–173° (435)

tends to open the ring. 1-Cyano-ν-triazole is a colorless solid below its melting point and exists as such, but in the melt or in $CDCl_3$ solution is in equilibrium

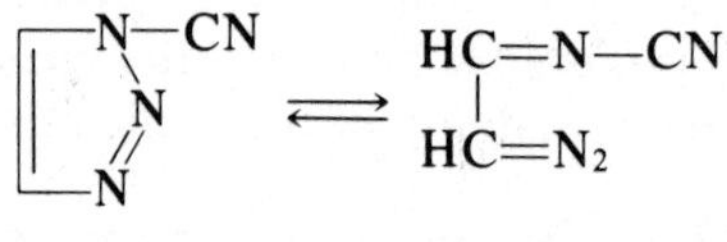

mp 33°

with its yellow isomer, α-diazo-*N*-cyanoethylideneimine (472). The same behavior is observed with 1-cyano derivatives of 4- and 5-alkyl-ν-triazoles, and the reversible thermochromism of 1-cyanobenzotriazole suggests similar isomeriza-

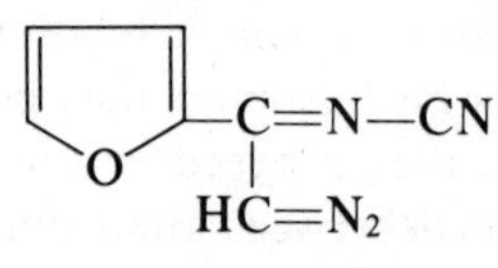

mp 90°

tion. α-Diazo-1-(2-furyl)-*N*-cyanoethylideneimine exists only in the acyclic form (42).

The 4-oxo-1-aryl-4,5,6,7-tetrahydrobenzotriazoles, even with a 1-*p*-nitro-

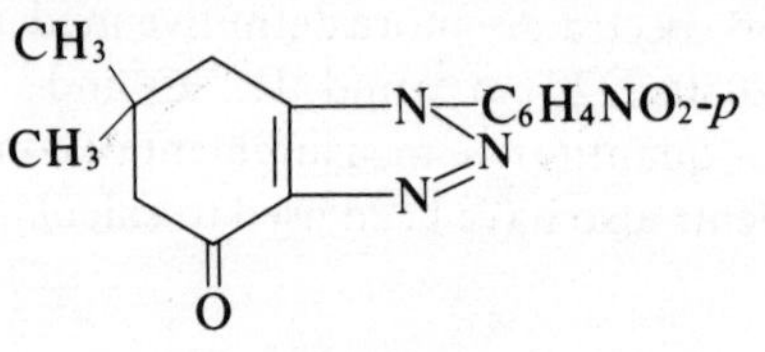

mp 219°

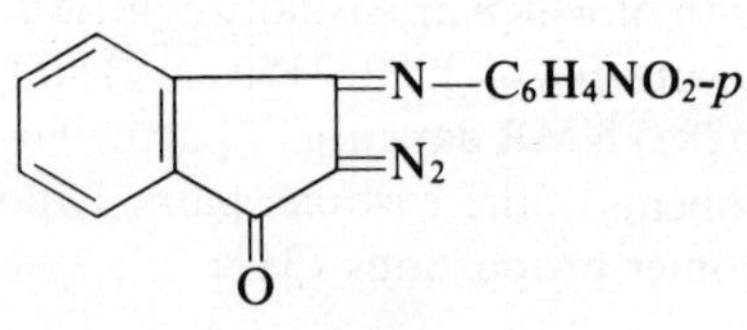

yellow
mp 203–205° dec.

*Interconversion of α-diazoimines and ν-triazoles is reviewed by Regitz (913).

phenyl and an acyl group at triazole carbon, are entirely cyclic (915). In contrast, the analogous, more aromatic, ketoindane is a diazoimine. Solid 8-oxo-indeno[1,2-*d*]triazoles with 3-aryl substituents carrying electron-releasing groups are not completely free from diazo compounds, and thermolysis effects ring opening in high yield (915). Strong acids convert both the oxoindeno triazoles and diazoimines to immonium salts. When the triazole is the stable form, it is regenerated from the immonium salt by treatment with alkali.

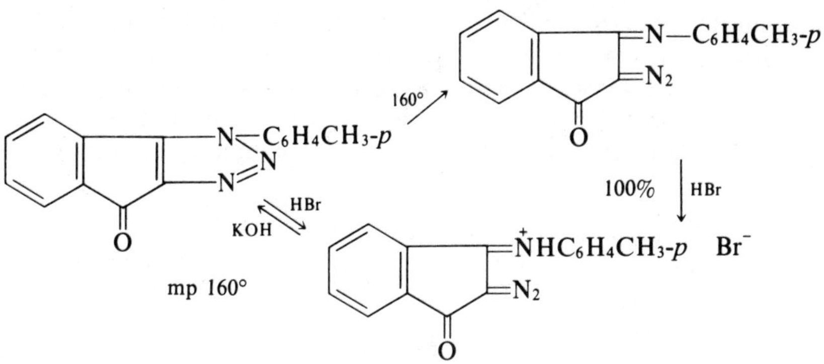

The 4- or 5-amino-*v*-triazoles with hydrogen or a 1-alkyl exhibit no ring opening, nor do 5-diarylamino or alkyl aryl amino-*v*-triazoles even with an electron-withdrawing substituent at position 1. Similarly, the balance between a 1-*p*-nitrophenyl and the electron-releasing 4-ethylthio and 5-diethylamino groups does not cause rupture of the *v*-triazole ring.

A 1-sulfonyl group on a 4-alkyl or aryl-5-dialkylamino-*v*-triazole, however, weakens the N-1–N-2 bond, and most of the resulting substances contain both

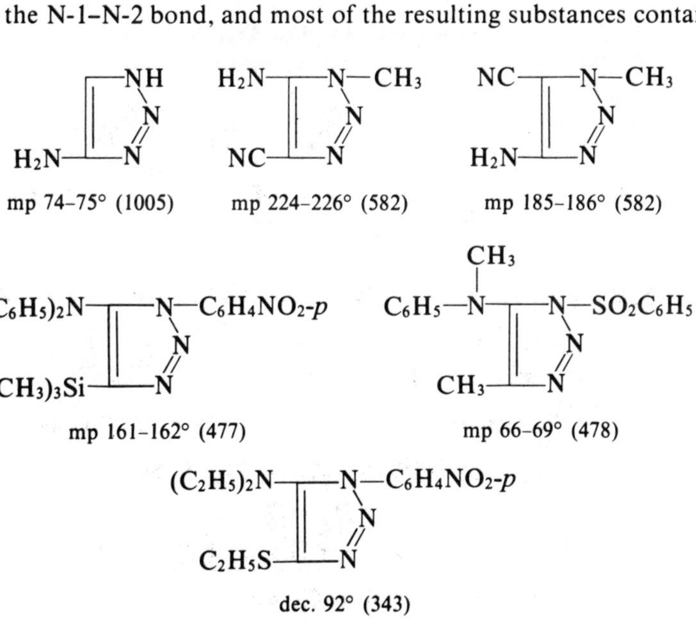

v-triazole and α-diazoamidine in equilibrium (Table 2.82), illustrated by the compositions of a series of potential 5-diethylamino-4-methyl and phenyl-*v*-triazoles (430, 431, 478). The only compound of this series to be in the triazole form exclusively is 1-*p*-dimethylaminobenzenesulfonyl-4-methyl-5-diethylamino-*v*-triazole (478). Introduction of phenyl instead of methyl at the 4-triazole position shifts the equilibrium in solution toward the diazoamidine (431, 478). The substituent attached to sulfonyl also exerts a considerable influence on the equilibrium position; as expected, electron-withdrawing substituents increase the proportion of diazoamidine. In comparison with diethylamino the less electron-releasing methylphenylamino at position 5, in combination with 1-*p*-nitrophenylsulfonyl, causes a high proportion of triazole; dissociation is greater with an *o*-nitrophenylsulfonyl (478). Most of these compounds with approximately 50% or greater triazole content in $CDCl_3$ solution are triazoles as solids.

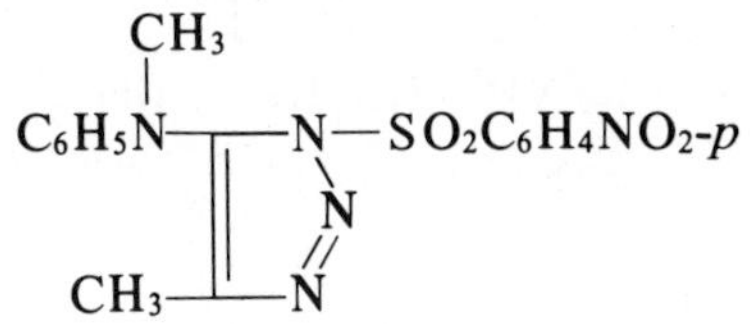

80% triazole in $CDCl_3$

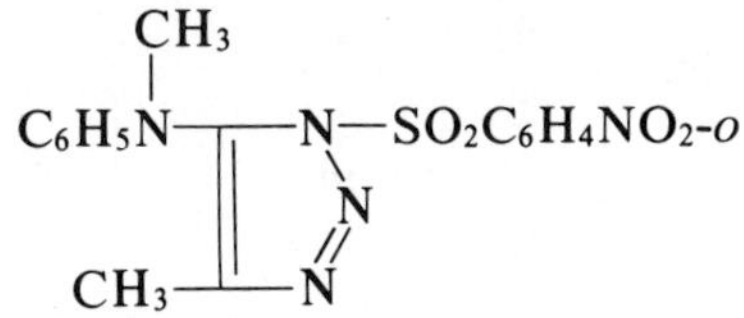

30% triazole in $CDCl_3$

Polar solvents increase the diazoamidine content; thus the amount of 5-diethylamino-4-methyl-1-(1-naphthalenesulfonyl)-*v*-triazole **(I)** in $CDCl_3$ is 64% in CCl_4 and 36% in CD_3OD (478). Lower temperatures slightly increase triazole content; in $CDCl_3$ the amount of **I** is 48% at +40° and 58% at −20° (478).

Table 2.82. 1-Sulfonyl-4-Alkyl or Aryl-5-Diethylamino-*v*-Triazole-α-Diazoamidine Equilibria

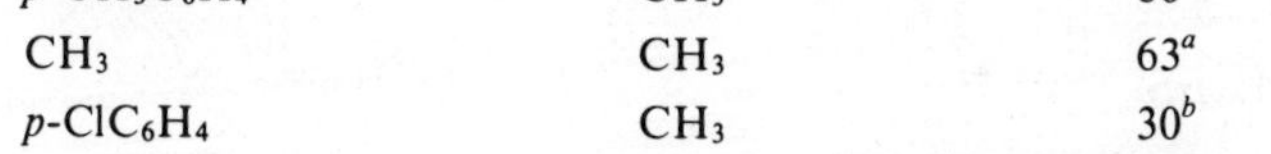

R	R_4	% Triazole
p-$(CH_3)_2NC_6H_4$	CH_3	100[a]
p-$CH_3C_6H_4$	CH_3	80[a]
CH_3	CH_3	63[a]
p-ClC_6H_4	CH_3	30[b]
p-$O_2NC_6H_4$	CH_3	0
p-$(CH_3)_2NC_6H_4$	C_6H_5	50[a]
p-$CH_3C_6H_4$	C_6H_5	10
p-$O_2NC_6H_4$	C_6H_5	0

[a] Exists as *v*-triazole in solid.
[b] Weak diazo infrared band in solid.

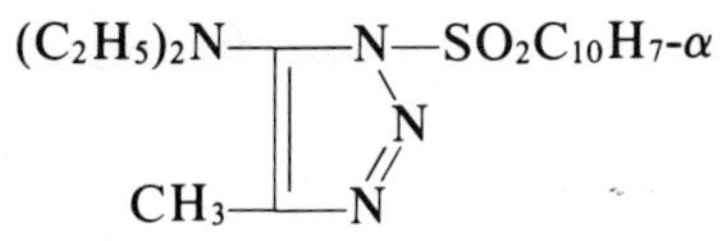

(I)

Two electron-withdrawing groups at positions 1 and 4 increase the extent of ring opening, as shown by the combined effect of sulfonyl substituents at position 1 and phosphoryl groups at position 4 (Table 2.83) (480). Introduction of a 4-benzoyl group, even with a strongly electron-releasing sulfonyl substituent,

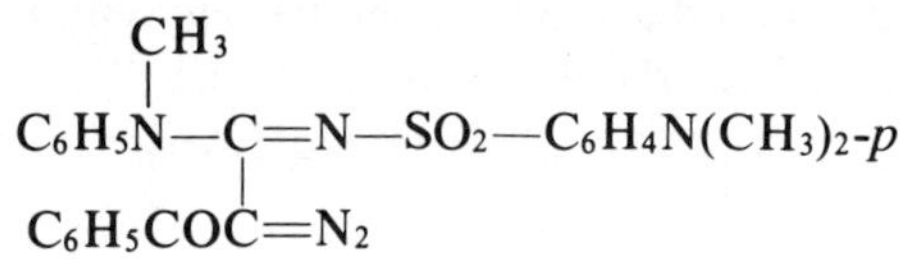

Table 2.83. 1-Sulfonyl-4-Phosphoryl-5-Methylphenylamino-*v*-Triazole-α-Diazoamidine Equilibria

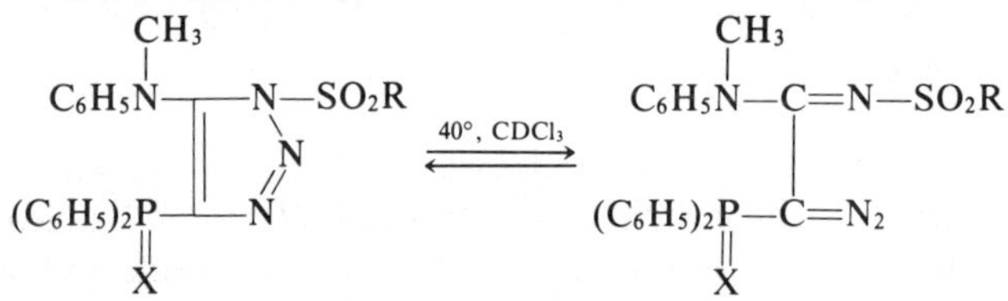

R	X	% Triazole
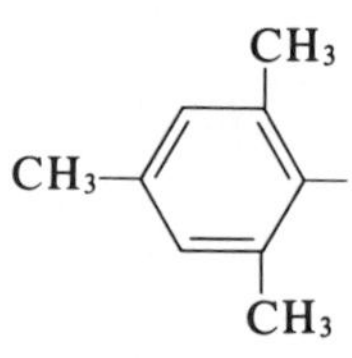	O	30
p-$CH_3C_6H_4$	O	8
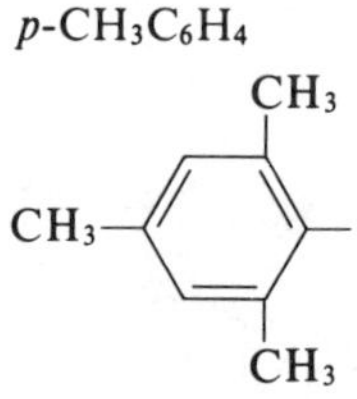	S	60
p-$CH_3C_6H_4$	S	38
p-$O_2NC_6H_4$	S	0
p-$(CH_3)_2NC_6H_4$	p-$CH_3C_6H_4N=$	85
p-$CH_3C_6H_4$	p-$CH_3C_6H_4N=$	32
p-$O_2NC_6H_4$	p-$CH_3C_6H_4N=$	0

causes the system to exist only as a diazoamidine both as a solid and in solution (480).

Potential fused 1-phenylsulfonyldihydro-*v*-triazolopyridines are solely diazo compounds (831). Acyl *v*-triazolo[3,4-*a*]pyridines with an extended conjugated

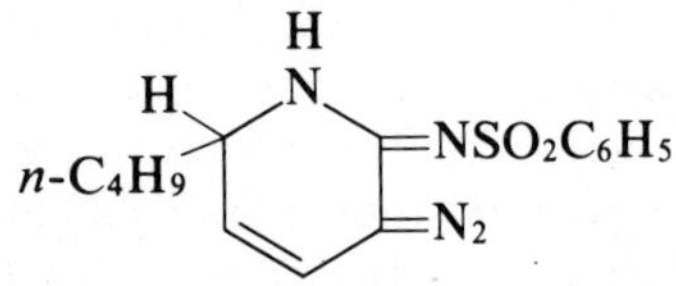

system, however, open to a diazo only when treated with perchloric acid (912); the acid is readily lost from the 2-(diazoacetylmethyl)pyridinium perchlorate, even in ethanol, regenerating acetyl *v*-triazolo[3,4-*a*]pyridine.

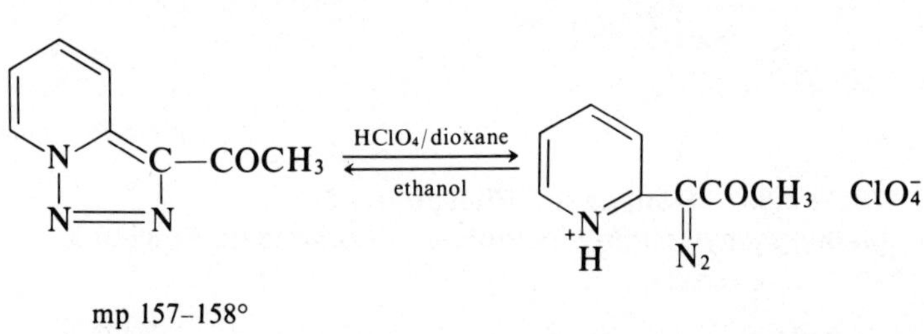

At room temperature, both as solids and in solution, *v*-triazolo[1,5-*a*]pyrimidines retain their bicyclic structure (1058). From 60 to 100°, tautomerism occurs to an equilibrium system containing diazo carboxamide and both triazolopyrimidine carboxamides (1077).

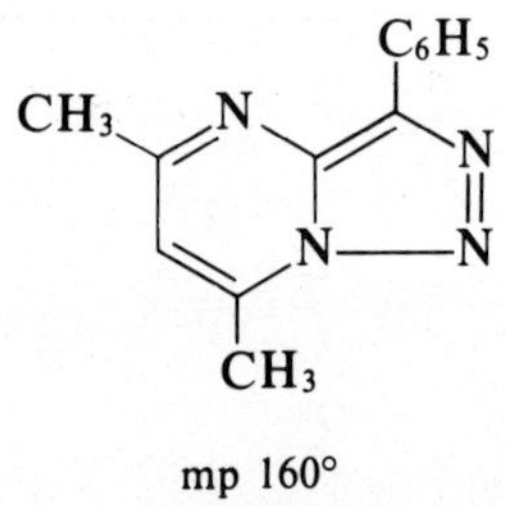

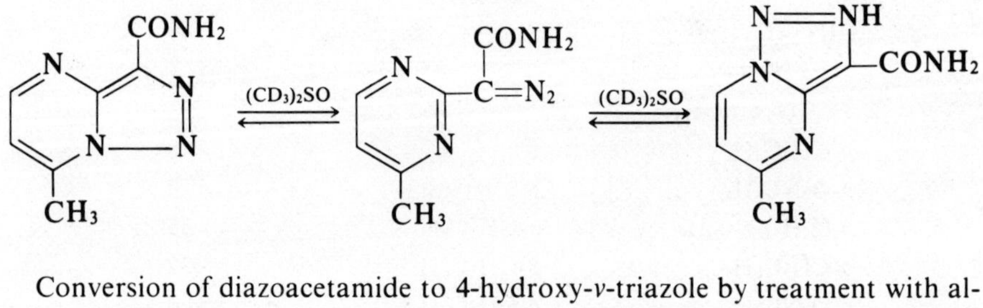

Conversion of diazoacetamide to 4-hydroxy-*v*-triazole by treatment with alkali established the existence of this type of α-diazoimine-*v*-triazole isomerism

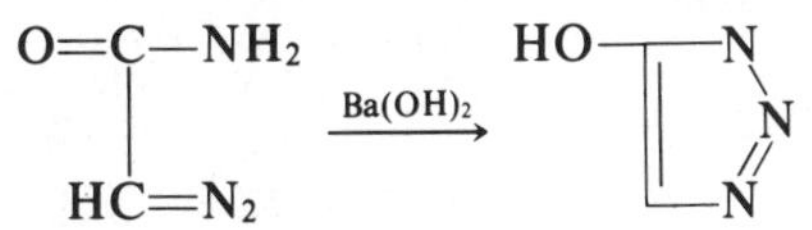

golden yellow
114° dec. (235)

mp 135°

in 1906 (240). 4-Hydroxy-*v*-triazole has no tendency to regenerate diazoacetamide. The substituted diazoacetamide **(II)** spontaneously undergoes similar unidirectional isomerization on attempted crystallization (684).

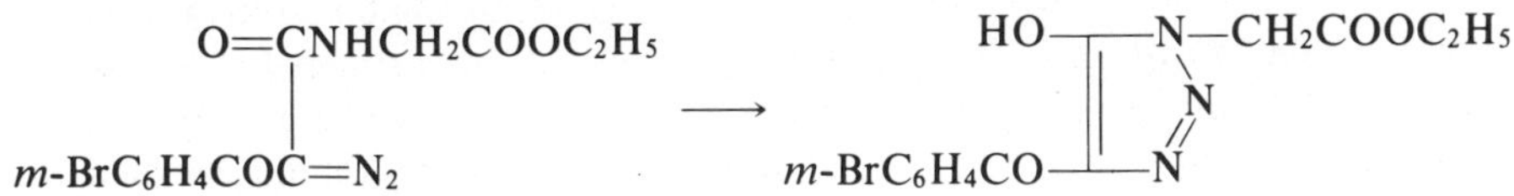

(II)

yellow oil

mp 99°

Dimroth demonstrated, however, that many α-diazoamides and hydroxy-*v*-triazoles can be interconverted and determined the equilibrium composition for the dissociation of several 1-substituted-4-alkoxycarbonyl-5-hydroxy-*v*-triazoles (Table 2.84) (259). In this series, the diazo component is present to the largest extent, particularly as the electron-withdrawing character of the N-1 aryl substituent increases. A reinterpretation of Dimroth's data shows the reaction in water to be bimolecular, involving dissociation of the 4-hydroxy group to hydrogen ion and enol anion and protonation of the 1-nitrogen (159). In the rearrangement of 1-aryl-4-alkoxycarbonyl-5-hydroxy-*v*-triazoles in acetonitrile and dimethylformamide, the same zwitterionic intermediate is proposed, presumably solvated, thereby stabilizing this dipolar form (644). In these solvents, for-

Table 2.84. 1-Alkyl or Aryl-4-Alkoxycarbonyl-5-Hydroxy-*v*-Triazole-α-Diazoamide Equilibria

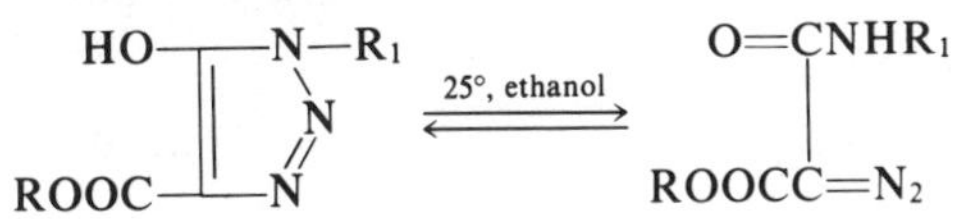

R_1	R	$K_{(\text{diazo/triazole})}$
H	CH_3	36
$C_6H_5CH_2$	CH_3	118
p-$CH_3C_6H_4$	C_2H_5	120
C_6H_5	CH_3	300
p-BrC_6H_4	C_2H_5	555

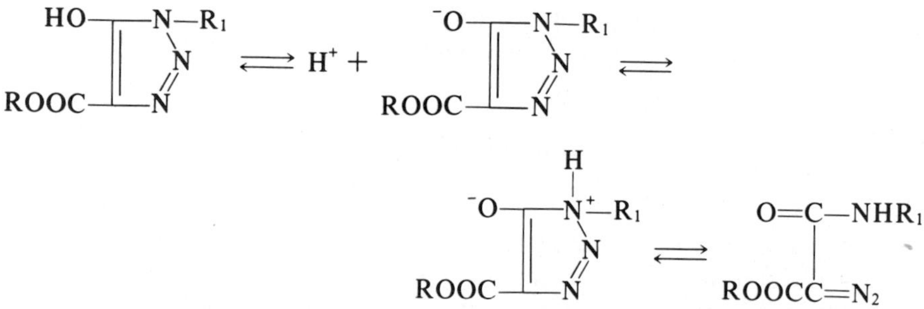

mation of diazoamide from hydroxy-*v*-triazole is unimolecular. Activation energies are shown in Table 2.85. Enthalpy of activation is somewhat less with electron-withdrawing substituents on aryl. Further decrease of enthalpy as well as a large reduction of entropy of activation results on addition of salts such as lithium chloride, believed to function in part by competing with triazole zwitterion for solvation, thus assisting in collapse of the triazole zwitterion transition state.

Under the influence of alkali, α-diazoacetoacetanilide cyclizes to the salt of 4-acetyl-5-hydroxy-1-phenyl-*v*-triazole; acidification and recrystallization from ethanol reforms the diazoamide (914). In the alkoxy derivatives, the triazole

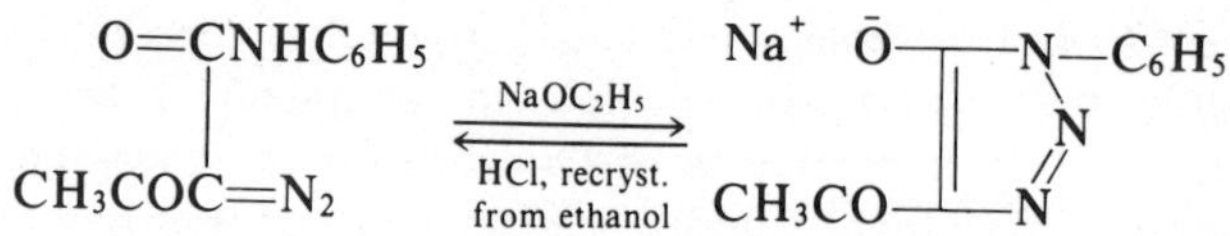

ring remains intact in 1-(2,4-dinitrophenyl)-4-ethyl-5-methoxy-*v*-triazole, but its picryl analog exists only in the open-chain form (479). No acyclic-cyclic equilibrium is observed in $CDCl_3$. Other compounds with an electron-with-

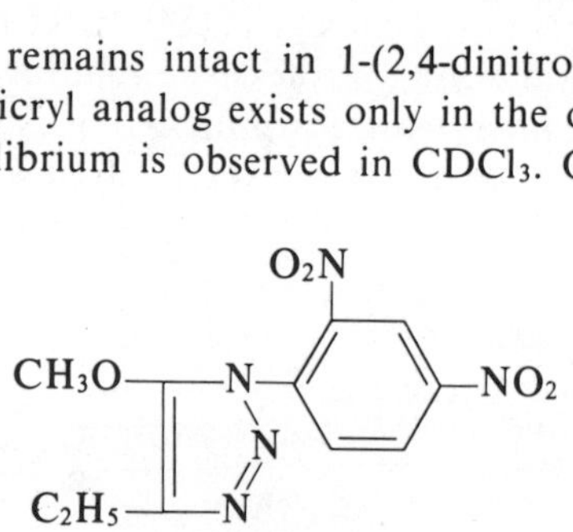

mp 78–80°

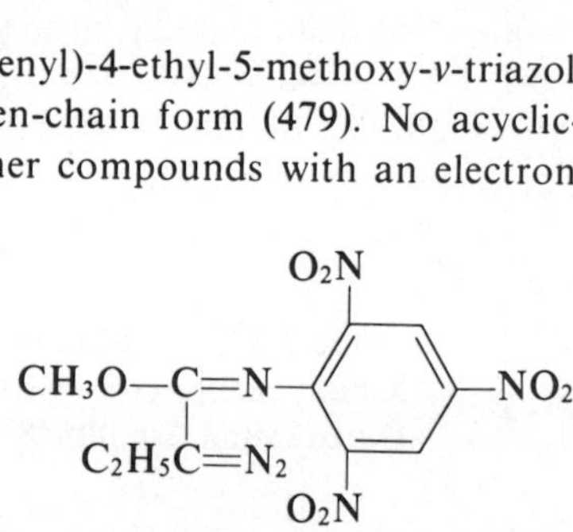

dec. 96–98°

drawing group on imine nitrogen and alkoxy on imine carbon also are diazoimines. In the potential 1-sulfonyl-5-ethoxy-4-methyl-*v*-triazoles an equilibrium

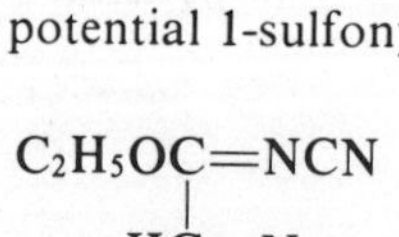

mp 49.5–52 (472)

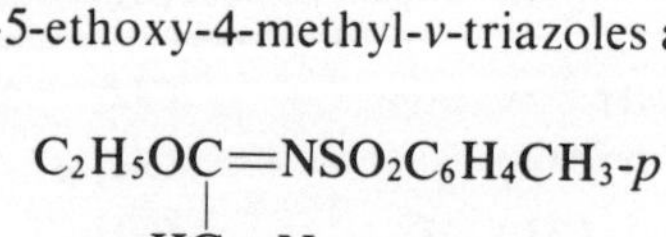

mp 73–74° (409)

Table 2.85. Activation Energies of 1-Aryl-4-Ethoxycarbonyl-5-Hydroxy-*v*-Triazole-α-Diazoamide Isomerization

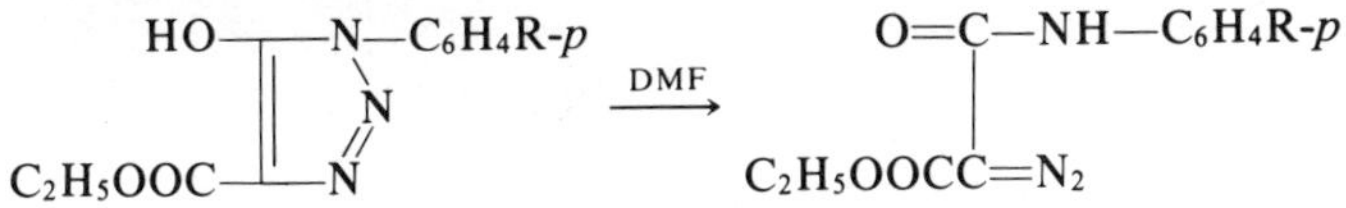

R	ΔH‡	ΔS‡
CH_3O	26.33	9.32
Br	25.62	10.4
CH_3O (+0.115 *M* LiCl)	23.58	−1.94

results (Table 2.86) (479). As with the related 5-dialkylamino-*v*-triazoles, the substituent on sulfonyl exerts a strong influence on breaking the bond between N-1 and N-2. A significant proportion of triazole is found only with the strongest electron-releasing group.

2.16.2. Imidoyl Azide-Tetrazole Isomerization*

Monocyclic tetrazoles with hydrogen or unsubstituted alkyl or aryl at positions 1 or 5 or both show no tendency to break the 1,2-nitrogen bond, and are cyclic (296). Similarly, a strongly electron-withdrawing group directly attached to

Table 2.86. 1-Sulfonyl-4-Methyl-5-Ethoxy-*v*-Triazole-α-Diazoiminoester Equilibria

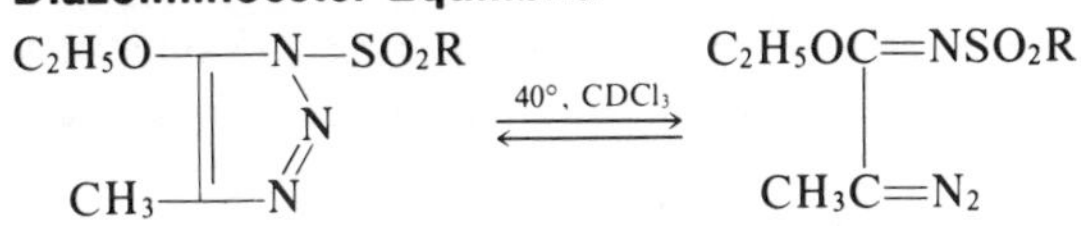

R	% Triazole
p-$(CH_3)_2NC_6H_4$	75
p-$CH_3C_6H_4$	22
C_6H_5	5
p-$O_2NC_6H_4$	0
CH_3	0

* Tautomerism between imidoyl azides and tetrazoles is reviewed by Butler (184), Pochinok *et al.* (878), and Tišler (1094).

tetrazole carbon usually does not cause ring opening; 5-cyanotetrazole and 1-alkyl derivatives of 5-trifluoromethyltetrazole show no evidence of isomeriza-

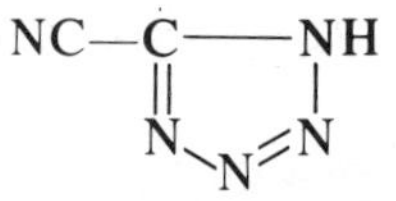

mp 103–104° dec. (722)

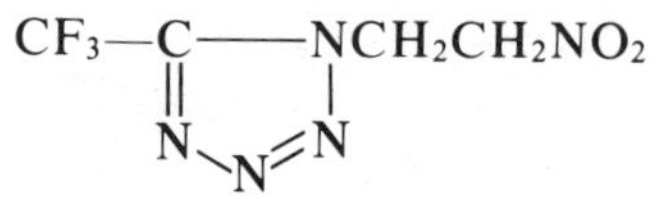

bp 115/10^{-2} (734)

tion to an open chain. Instances are known, however, of imidoyl azides with electron-withdrawing groups on imidoyl carbon which are convertible to tetrazoles. Thus the freshly distilled adduct of trimethylsilyl azide with trichloroacetonitrile shows an azide infrared band which disappears on crystallization

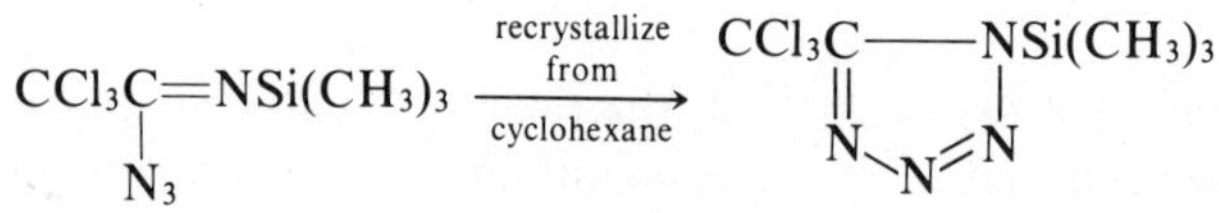

(637). Also probable formation of an imidoyl azide carrying a trifluoromethyl group on carbon is inferred by an unsuccessful attempt to prepare 1-(2-naphthyl)-5-trifluoromethyltetrazole; on work-up, only *N*-(2-naphthyl)trifluoroacetamide, the hydrolysis product of the corresponding imidoyl azide, is isolated (1205). The related 1-(2-naphthyl)-5-methyltetrazole exists only in the cyclic form.

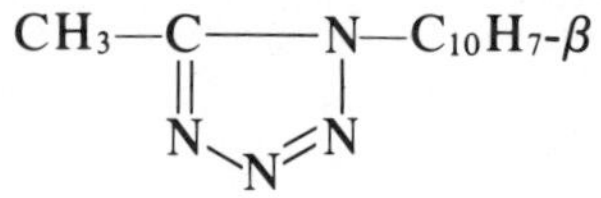

mp 164–165° (1205)

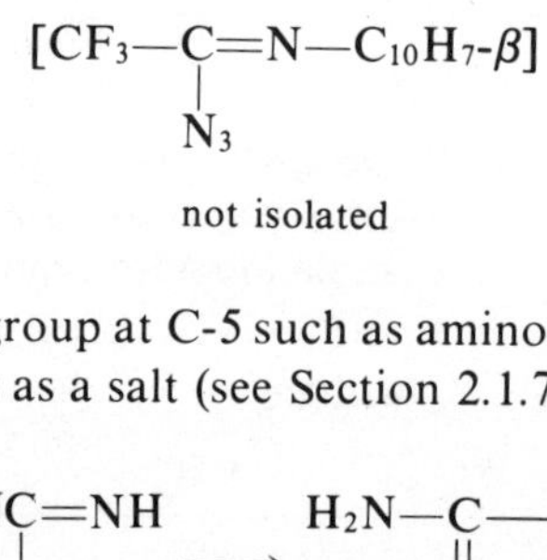

not isolated

Tetrazoles with an electron-releasing group at C-5 such as amino do not undergo ring fission. Guanyl azide is stable as a salt (see Section 2.1.7), but even

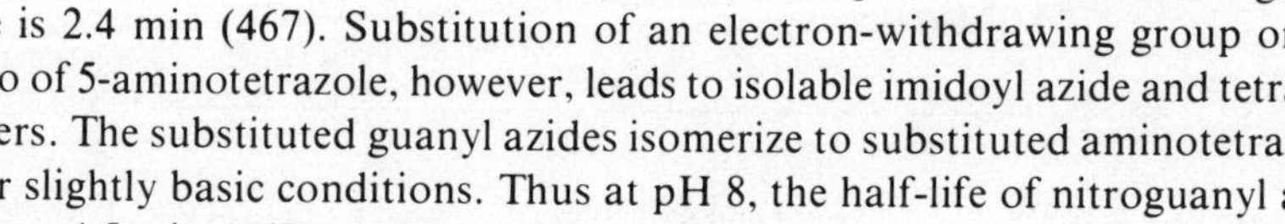

with water, as well as stronger base, it is rapidly converted to 5-aminotetrazole (1082). Indeed, at 0° in aqueous solution of about pH 9, the half-life of guanyl azide is 2.4 min (467). Substitution of an electron-withdrawing group on the amino of 5-aminotetrazole, however, leads to isolable imidoyl azide and tetrazole isomers. The substituted guanyl azides isomerize to substituted aminotetrazoles under slightly basic conditions. Thus at pH 8, the half-life of nitroguanyl azide is about 4.5 min (467), and weak bases convert *N*-cyano-*N*-methylguanyl azide and arenesulfonyl guanyl azides to the corresponding tetrazoles (561, 814).

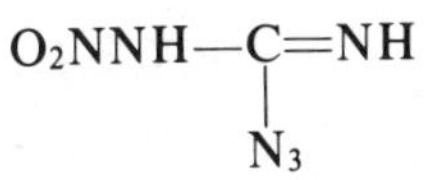

mp 79° dec. (657)

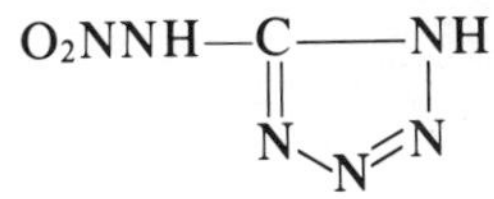

expl. 140° (657)

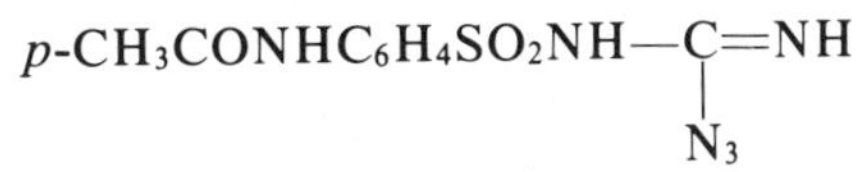

mp 166° (561, 781)

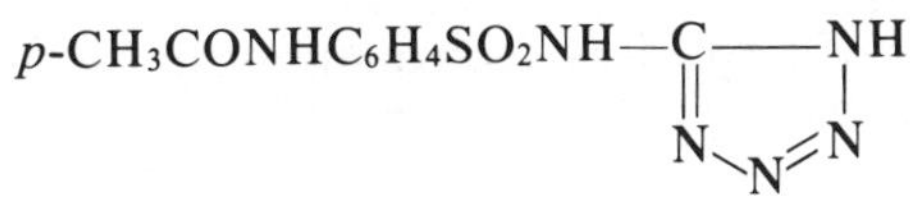

mp 205–206° (561)
217–219° dec. (781)

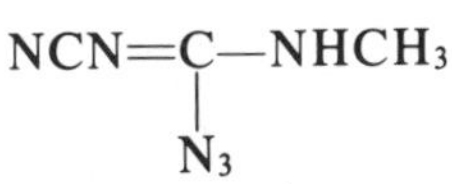

mp 122–123° dec. (814)

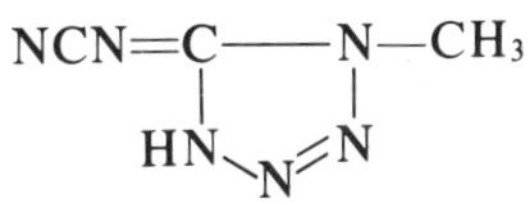

mp 162° dec. (814)

Some azidoximes such as the mesityl and pyridine derivatives do not cyclize, and none appear to be converted directly to 1-hydroxytetrazoles on heating or

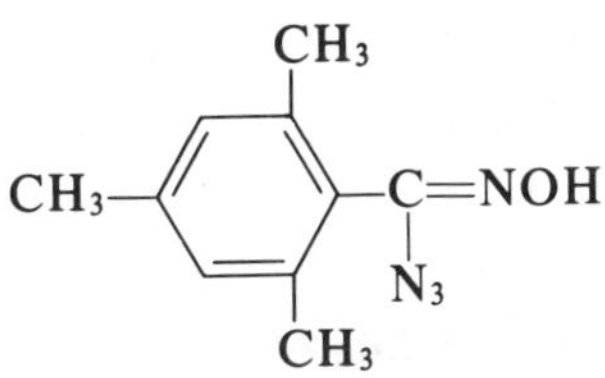

mp 89–89° dec. (410)

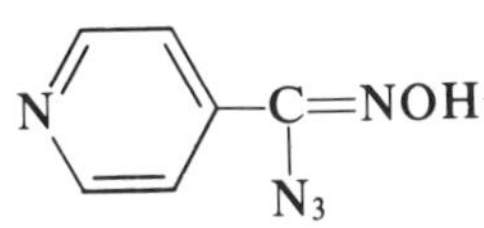

mp 173–174° dec. (201)

treatment with base (296). Reaction with acid chlorides, however, forms 1-esters of some 5-alkyl, aryl, and acylhydroxy tetrazoles, which hydrolyze to the isomeric hydroxytetrazole (873, 874). Indeed, because of its hydrolytic susceptibility, isolation of 1-benzoyloxy-5-phenyltetrazole is impractical and the hydroxytetrazole is obtained by alcoholysis in hot ethanol. In the same manner as azidoximes, aryl hydrazidic azides, which also have a hydrogen on the atom adjacent to imide nitrogen, cyclize to anilinotetrazoles on reaction with acid chlorides (872). Heating *N*-(phenylsulfonyl)benzohydrazonyl azide forms a small quantity of the corresponding tetrazole along with decomposition products

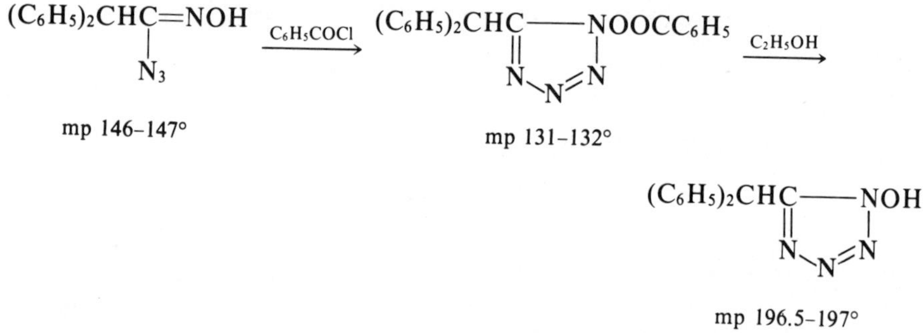

(552). Melting points of the hydroxytetrazoles usually are higher than those of the corresponding azidoximes (Table 2.87) (873).

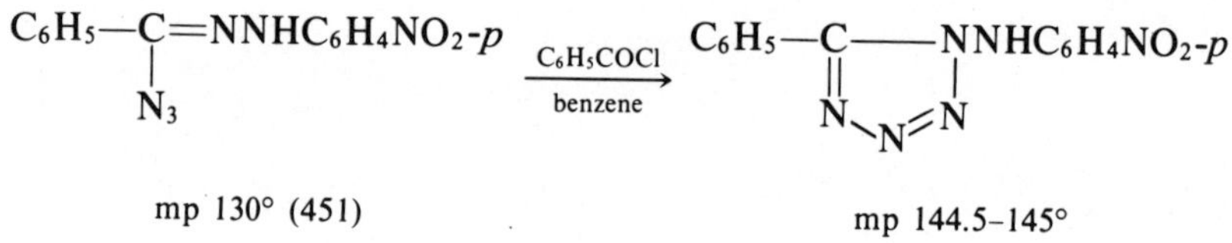

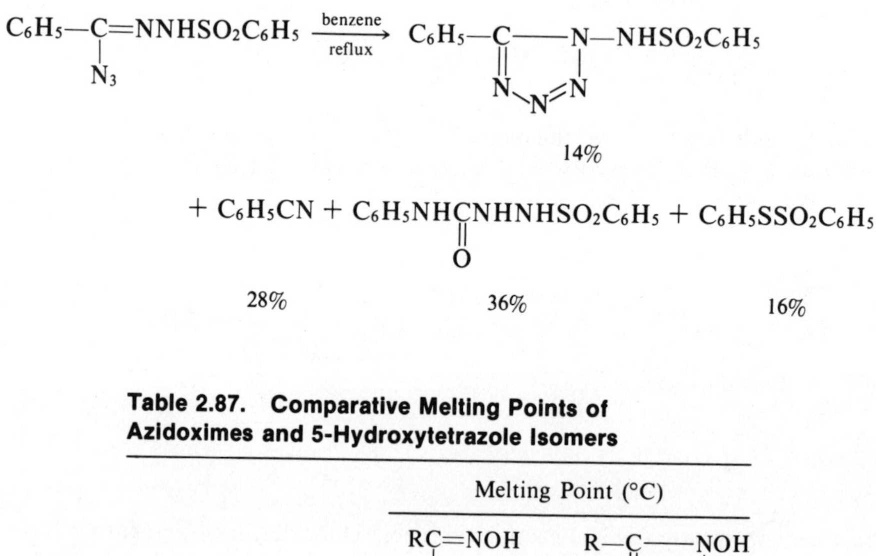

Table 2.87. Comparative Melting Points of Azidoximes and 5-Hydroxytetrazole Isomers

R	Melting Point (°C)	
	$RC(N_3){=}NOH$	$R{-}C{=}N{-}N{=}N{-}NOH$ (ring)
C_6H_5	123–124	151–152
p-$CH_3C_6H_4$	108.5–110	195–196
p-$O_2NC_6H_4$	142–144	178.5–180
C_6H_5CO (876)	78–82 dec.	106–108

O-Substituted azidoximes change to their oxytetrazole isomers on treatment with acid (873).

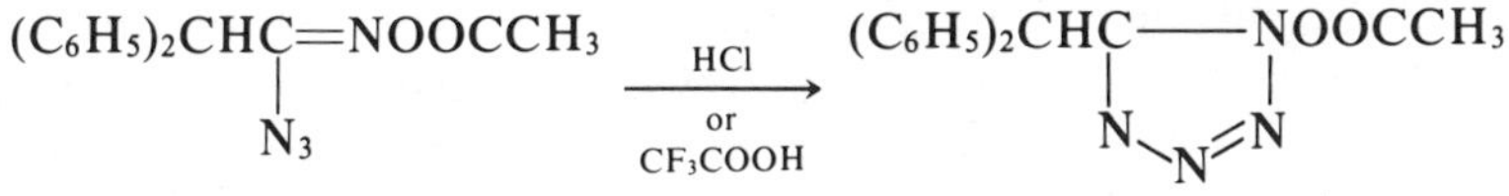

mp 65–66°   mp 132.5–133.5°

Cyclization of imidoyl azides to tetrazoles requires the lone pair of electrons on the imide nitrogen to be accessible to the terminal azide nitrogen. The lone pair in cyclic imidoyl azides is constrained to be cis to the azide, facilitating ring closure, but in open-chain analogs, the electron pair on imide nitrogen can be either cis or trans to the azide. Accordingly, to form a tetrazole, an imine with a trans pair must first isomerize to the cis form, a step that is expected to be slow (452).

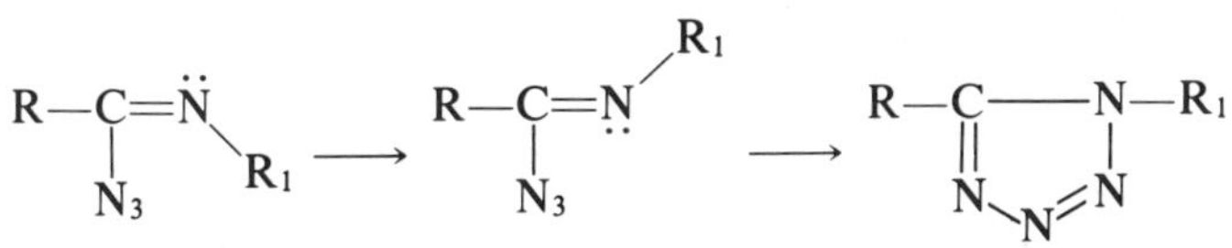

The 1,4-diaryl-1-azido-2,3-diazabutadienes **(I)** isomerize on heating to 1-phenylmethyleneamino-5-aryltetrazoles (Table 2.88) (1045, 1047). Comparative stability of these imidoyl azides results both from the *Z* configuration about the double bond from azide carbon to azine nitrogen and the slow $Z \rightarrow E$ isomerization at ordinary temperatures (452). The cyclization proceeds by first-order ki-

Table 2.88. Half-Lives of Aryl-Substituted 1-Azido-2,3-Diazabutadiene-1-Aminotetrazole Isomerizations

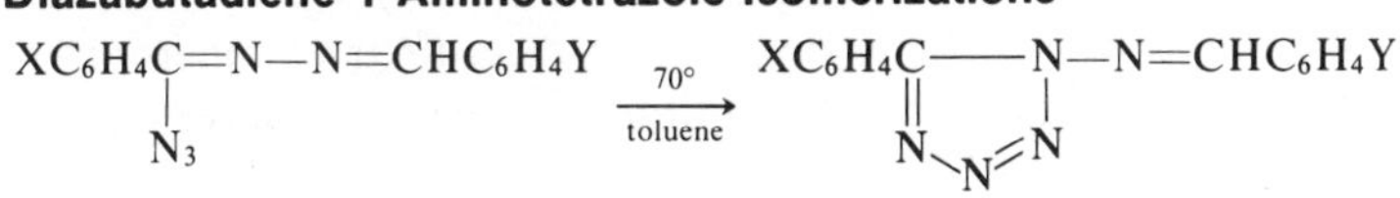

(I)

X	Y	$t_{1/2}$ (hr)
H	*p*-O_2N	0.3
H	H	3.3
H	*p*-CH_3O	10.2
m-O_2N	H	4.8
p-CH_3	H	2.6

netics and is insensitive to small quantities of acid or base or the dielectric constant of the medium. Electron-attracting groups on the phenylmethyleneamino aryl residue induce the fastest rate, and electron-releasing groups, the slowest (452). Substituents on the aryl attached to azide carbon have a relatively small influence on the rate of ring closure. These conversion rates to tetrazole are characteristic of other nitrogen inversions, consistent with the change of the lone pair on nitrogen from trans to cis being the rate-controlling step.

Further indication of the importance of the cis orientation of azide and the lone pair on imide nitrogen comes from the behavior of imidoyl azide substituents, which fix the imide nitrogen by intramolecular hydrogen bonding so that the lone pair is trans to azide. Tetrazole formation from **II** and **III** are complex

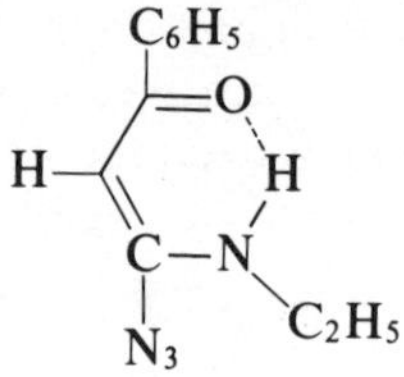

(II)

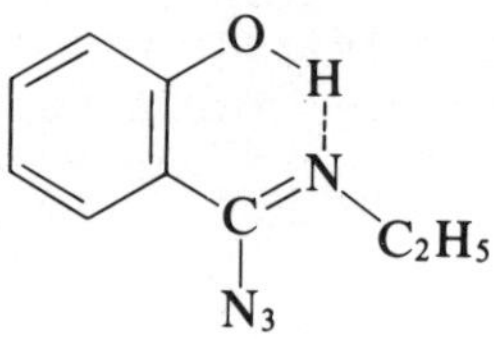

(III)

functions of pH, differing in each case, but occur only when the hydrogen bonding is interrupted, allowing inversion to the cis form (11, 450).

Azido diphenylphosphine tosyl imide exists solely in the open-chain form with no evidence of a tetraazaphosphole (127).

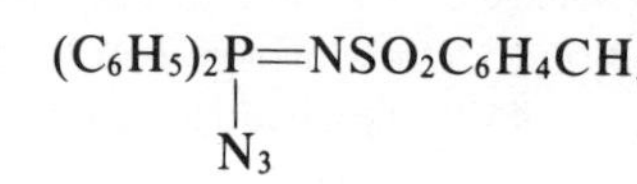

mp 84–85°

Bicyclic compounds consisting of a tetrazole fused to a saturated ring such as trimethylenetetrazole or tetrazole imidazoline (143) are as stable as related monocyclic 1,5-disubstituted tetrazoles and at ordinary temperatures, both in chloroform solution and as solids, exhibit no isomerization to imidoyl azides.

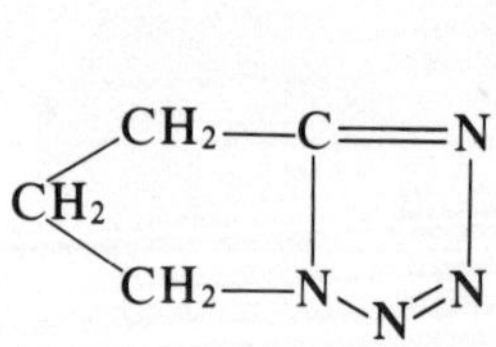

mp 109–110° (143)
110° (269)

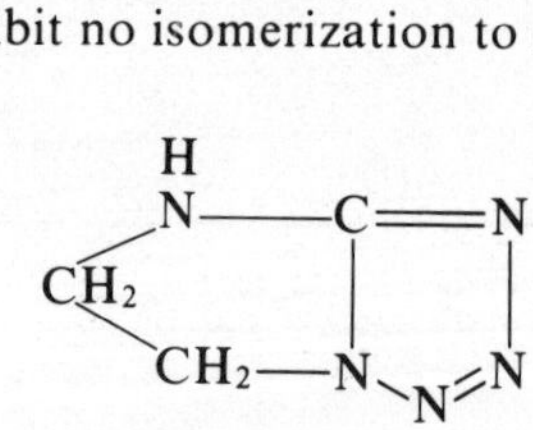

mp 163–164° dec. (323)

Attachment of azide adjacent to a ring nitrogen in five-membered aromatic heterocycles leads predominantly to cyclic imidoyl azides; some bicyclic tetrazoles exist and in some cases the two forms are in equilibria. Determining factors include the electron-withdrawing power of other ring atoms and substituents, the presence of an ionizable hydrogen, and the polar character of the solvent.

Thus the azido pyrazoles, *s*-triazole, *v*-triazole, and tetrazole shown exist only in the monocyclic azide form as solids or in solution in neutral solvents

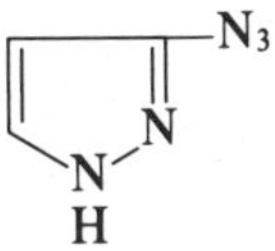

mp 60° (25)

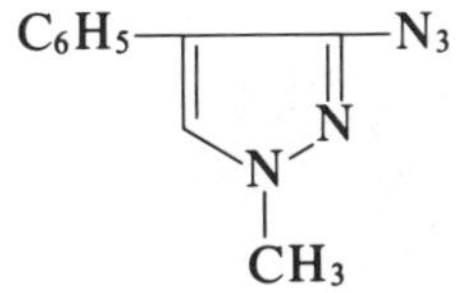

mp 70–73° (22)

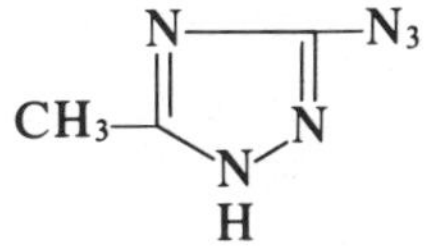

mp 144–145° (454)

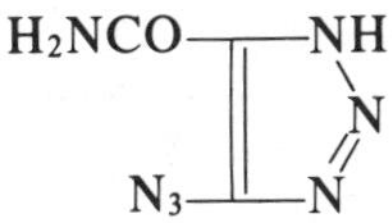

expl. 175° (988)

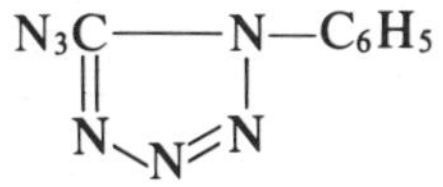

mp 96–96.7° (579)

(211). Azido pyrazoles with hydrogen on ring nitrogen are converted slowly in strong base to bicyclic pyrazolotetrazole anions. Besides possible equilibria between neutral azide and tetrazole, anionic forms of each are involved, illustrated by the interconversion of 3-azido-4-ethoxycarbonyl-5-methylpyrazole in

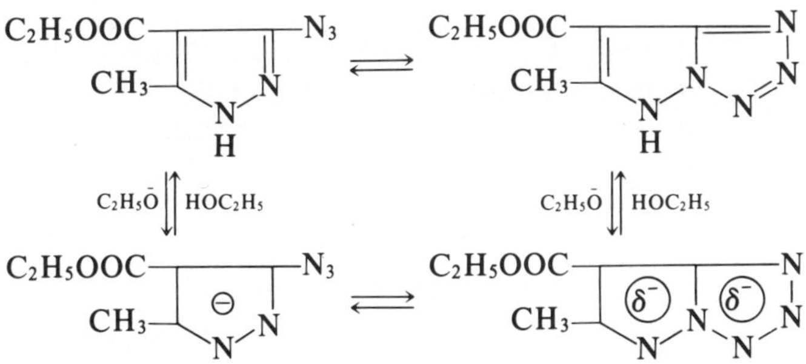

a solution of sodium ethylate in ethanol (25). The half-life of disappearance of azide is 72 hr in 0.1 *M* sodium ethylate. Cyclization of azido pyrazole anion to tetrazole anion is the slow step.

Most azide derivatives of imidazole and *s*-triazole do not cyclize, but by substituting acyl on ring nitrogen, azide-tetrazole tautomerism can be detected (399). Equilibrium positions of *N*-acylimidazo[1,2-*d*]tetrazole and the corresponding 2-azidoimidazole are shown in Table 2.89. Groups on imidazole nitrogen which are highly electron-withdrawing by the inductive effect favor azide, but bulkier substituents and lower temperatures shift the equilibrium toward the tetrazole (938). Pivaloyl imidazo[1,2-*d*]tetrazole exists as such in the solid state, at least at room temperature.

Table 2.89. *N*-Acyl Imidazo[1,2-*d*]tetrazole-2-Azidoimidazole Equilibria

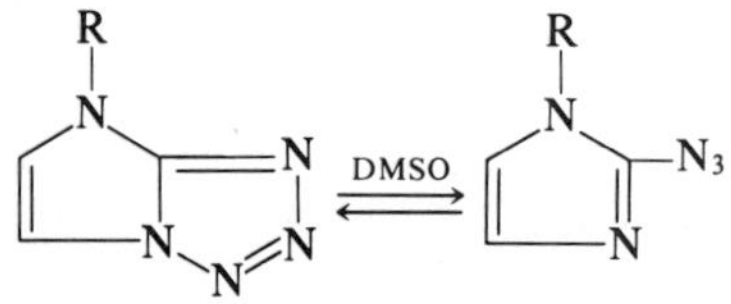

R	Temperature (°C)	$K_{A/T}$	Ref.
C_2H_5OOC	35	6.25	937
Cl_2CHCO	20	4.2	937
CH_3CO	35	2.1	399
	15	1.7	399
$(CH_3)_3CCO$	35	0.6	938

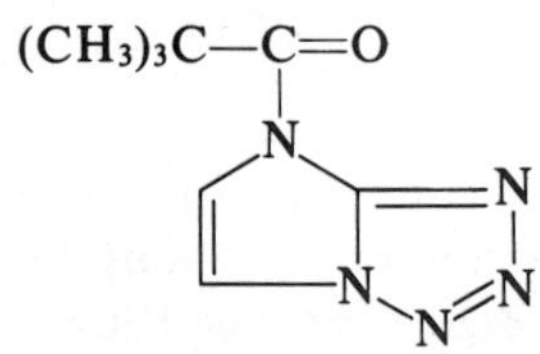

dec. 122–123° (938, 1131)

In analogy to the monocyclic imidazole azides, 2-azidobenzimidazole unsubstituted on ring nitrogen does not cyclize, but a small amount of 1-acetyl-2-azidobenzimidazole tautomerizes to the isomeric tetrazole.

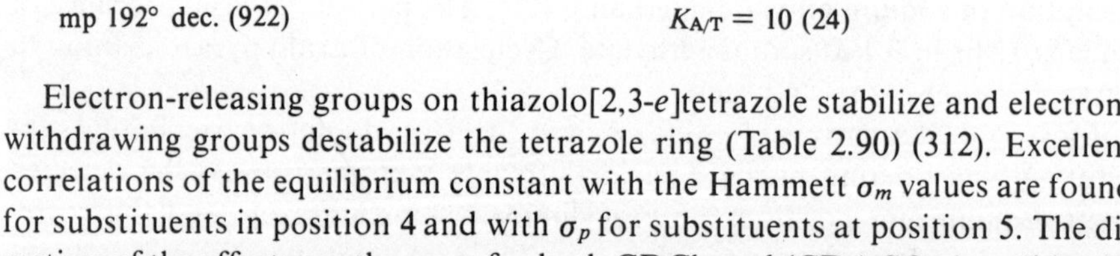

mp 192° dec. (922) $K_{A/T} = 10$ (24)

Electron-releasing groups on thiazolo[2,3-*e*]tetrazole stabilize and electron-withdrawing groups destabilize the tetrazole ring (Table 2.90) (312). Excellent correlations of the equilibrium constant with the Hammett σ_m values are found for substituents in position 4 and with σ_p for substituents at position 5. The direction of the effects are the same for both $CDCl_3$ and $(CD_3)_2SO$. As solids, the parent thiazolotetrazole-azide system and its methyl derivatives exist as bicyclic tetrazoles (48). Except for trifluoracetic acid, which strongly favors the azide form, greater solvent polarity and, to a lesser degree, basicity increase the pro-

Table 2.90. Thiazolo[2,3-*a*]tetrazole-2-Azidothiazole Equilibria

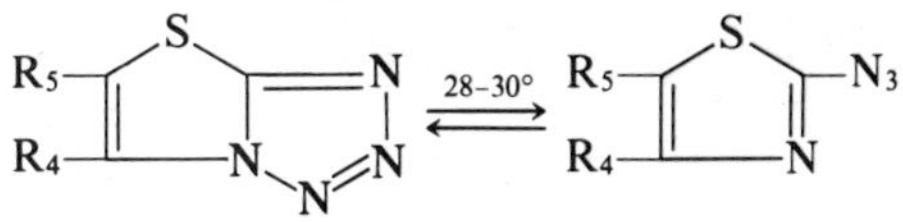

R_4	R_5	$K_{A/T}$	
		$CDCl_3$	$(CD_3)_2SO$
H	H	5.70	0.21
CH_3	H	0.85	0.03
CH_3	CH_3	0.40	$\leqslant 0.03$
H	Br	32.0	1.60
p-$O_2NC_6H_4$	H	$\geqslant 20$	$\geqslant 20$

portion of tetrazole form, illustrated by equilibrium constants for the 2-azidothiazole-thiazolo[2,3-*e*]tetrazole composition (293).

Solvent	$K_{A/T}$
CF_3COOH	>25
CCl_4	>25
C_6D_6	15.5
CD_3OD	1.2
CD_3CN	1.0

Boyer and Miller separated azide and tetrazole forms of the azidobenzothiazole-tetrazolobenzothiazole system in ethanol by paper chromatography (143). As with the thiazole derivative, the tetrazole form is predominant in deuterated

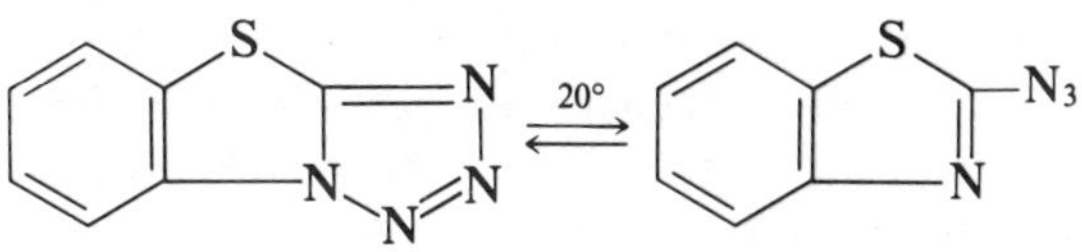

	$K_{A/T}$
$CDCl_3$	1.0
$(CD_3)_2SO$	0.08

dimethyl sulfoxide (312). Introduction of a 6-nitro group in the benzene ring of 2-azidobenzothiazole strongly shifts the equilibrium to the azide in both $CDCl_3$ and $(CD_3)_2SO$, and 3-azidobenzisothiazole 1,1-dioxide exists only as the azide

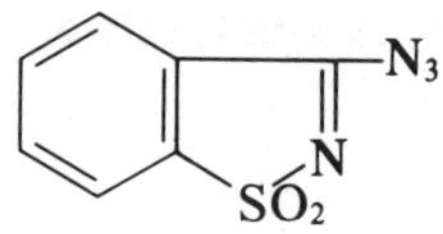

mp 153° dec. (972)

(143). Crystalline 2-methyltetrazolo[4,5-*b*]-1,3,4-thiadiazole is bicyclic, both tautomers are present in chloroform, and only the azide is present in carbon tetrachloride (997).

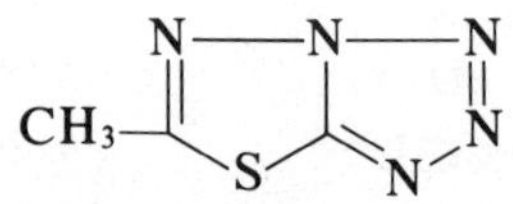

mp dec. 92–93

Tetrazolo[1,5-*a*]pyridine, the simplest six-membered ring with a fused tetrazole, is stable in the solid state and in solution, but can be converted to 2-azidopyridine by flash vacuum pyrolysis (1158). The azide is permanently stable at

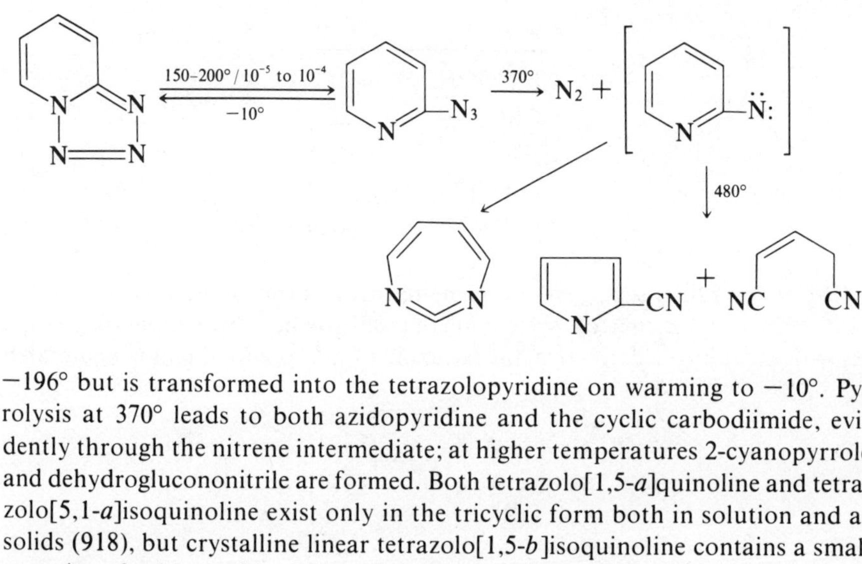

−196° but is transformed into the tetrazolopyridine on warming to −10°. Pyrolysis at 370° leads to both azidopyridine and the cyclic carbodiimide, evidently through the nitrene intermediate; at higher temperatures 2-cyanopyrrole and dehydrogluconproperties are formed. Both tetrazolo[1,5-*a*]quinoline and tetrazolo[5,1-*a*]isoquinoline exist only in the tricyclic form both in solution and as solids (918), but crystalline linear tetrazolo[1,5-*b*]isoquinoline contains a small quantity of azide, and in $CDCl_3$ the azide is the principal form (414).

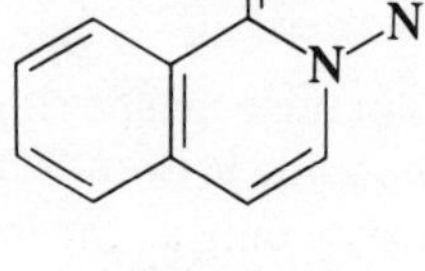

mp 140–142° (918)

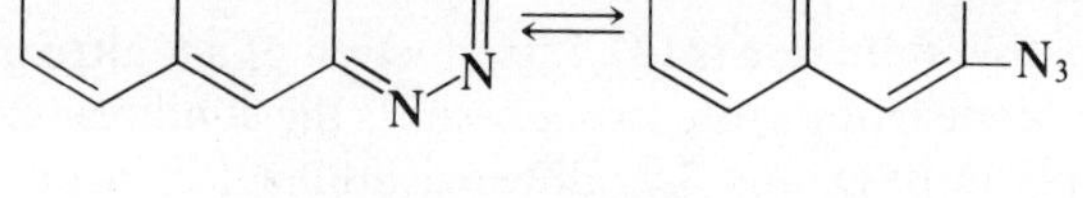

mp 122° $K_{A/T}(CDCl_3) = 1.5$

Azide is present in large quantity in $CDCl_3$ solution with both 5-chloro and 6-nitrotetrazolo[1,5-*a*]pyridine (946). The 6-bromo derivative, however, exists

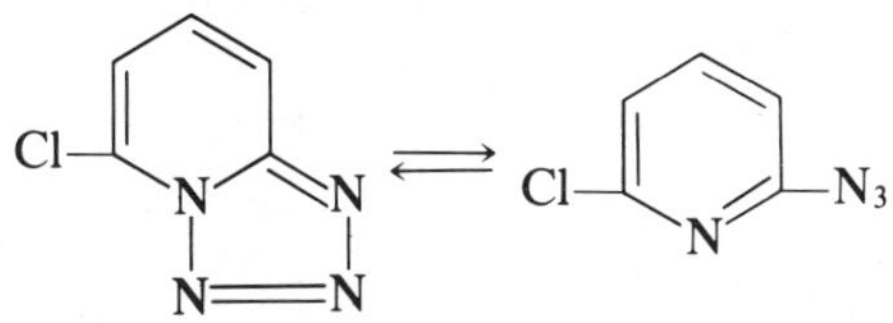

	$K_{A/T}$
23°	18.0
40°	azide

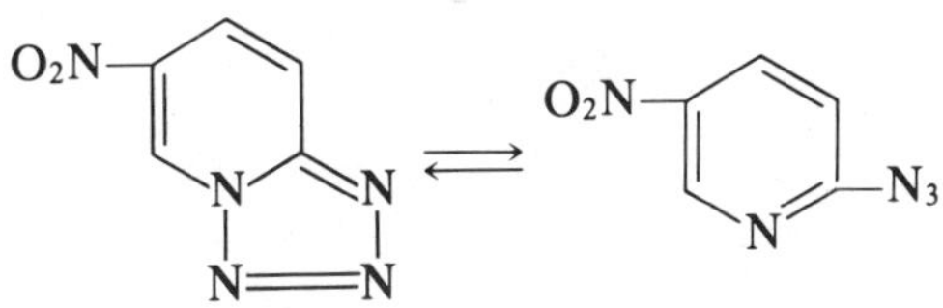

	$K_{A/T}$
23°	0.83
40°	1.04

only in the tetrazole form at 40° in $CDCl_3$ and up to 120° in $(CD_3)_2SO$. As expected, tetrazolopyridine substituted with 6- or 8-hydroxy or 8-amino are cyclized both as solids and in dimethylformamide (140).

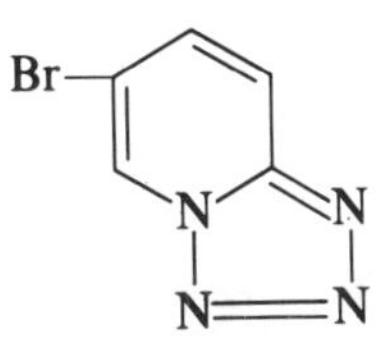

mp 155–157°

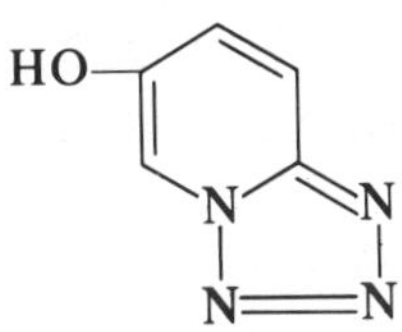

mp 215–220° dec.

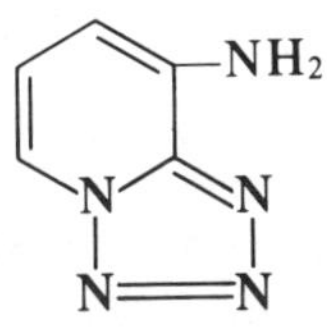

mp 180–183° dec.

On heating a Nujol suspension of tetrazolo[1,5-*a*]pyrazine above its melting point a weak azide band appears, but none is found with tetrazolo[1,5-*a*]pyrimidine. Both show azide in deuterated chloroform and trifluoracetic acid. Presumably in trifluoracetic acid, azido- or tetrazolopyrimidines and similar heterocycles are present as protonated forms. The influence of an electron-withdrawing group in increasing proportion of azide is shown by comparison of 5,7-dimethyl and 5-chloro-7-methyltetrazolo[1,5-*a*]pyrimidines (1152). Both tetrazole and azide forms of the 5-chloro-7-methyl derivative are mutually interconverted at their melting points.

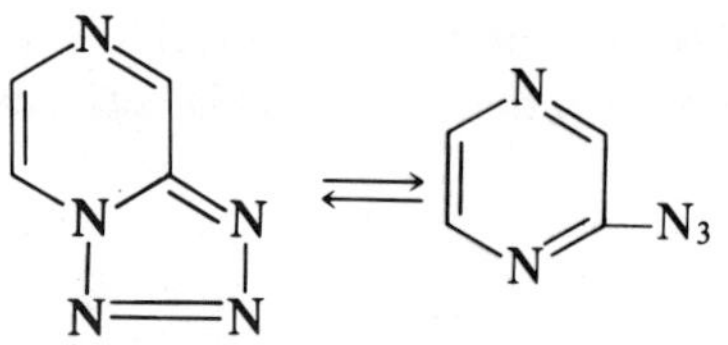

mp 90–91° (1152)

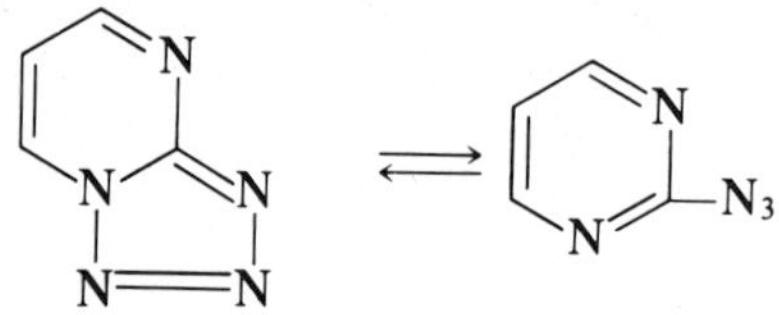

mp 119.5–120° (1152)

	$K_{A/T}$
$CDCl_3$	0.11
$(CD_3)_2SO$	tetrazole

	$K_{A/T}$
$CDCl_3$	azide (1076)
$(CD_3)_2SO$	0.08 (1073)

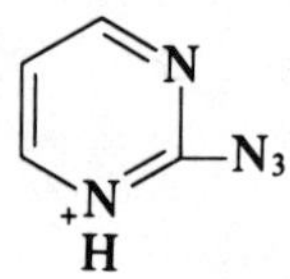

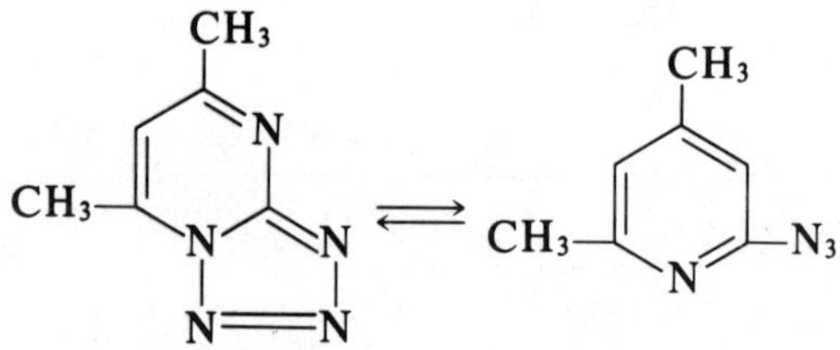

mp 151–152° (169)

	$K_{A/T}$
$CDCl_3$	0.33
$(CD_3)_2SO$	tetrazole

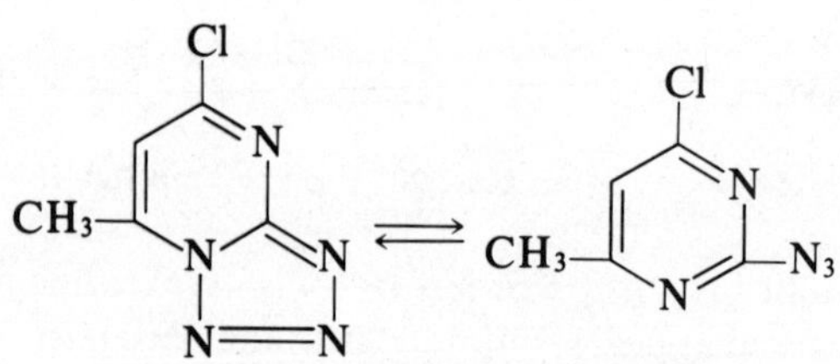

mp 108–110° mp 58–60°

	$K_{A/T}$
$CDCl_3$	11
$(CD_3)_2SO$	0.15

The major tautomer of solid tetrazolo[1,5-*c*]pyrimidine is the cyclized compound along with some azide (1075). This system readily adds water to give covalent hydrates. Although 5,7-dimethyltetrazolo[1,5-*c*]pyrimidine remains in-

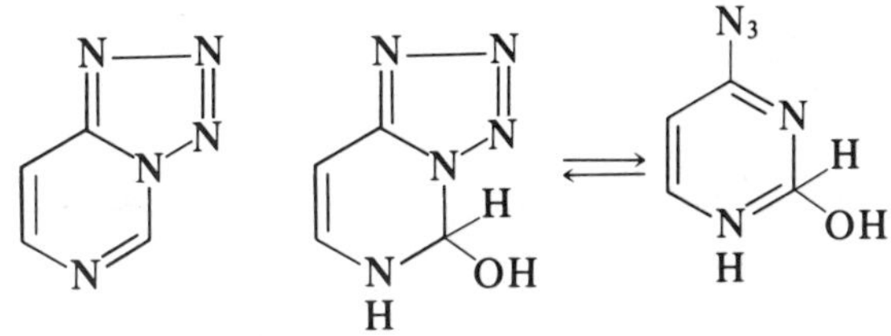

mp 77–79°

tact in both $CDCl_3$ and $(CD_3)_2SO$, the dimethoxy derivative is an azide both in these solvents and as a solid (1152). The aromatic azide group is electron donat-

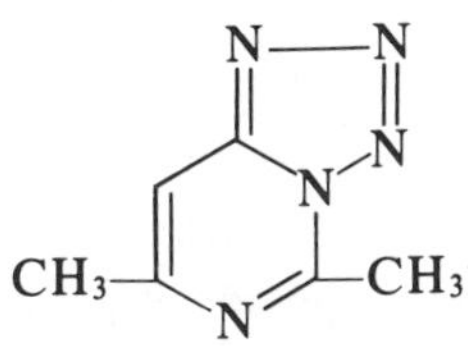

mp 84–85°

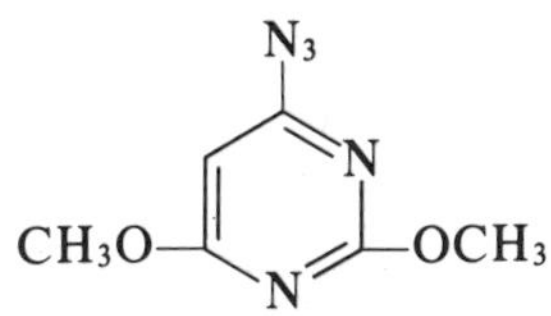

mp 37–38°

ing (1012) and the effect of introducing azide groups at the 2- and 4-positions of pyrimidine is to form a system with two different azido tetrazolopyrimidines in equilibrium with the diazide (1152). The predominant form of the solid methyl

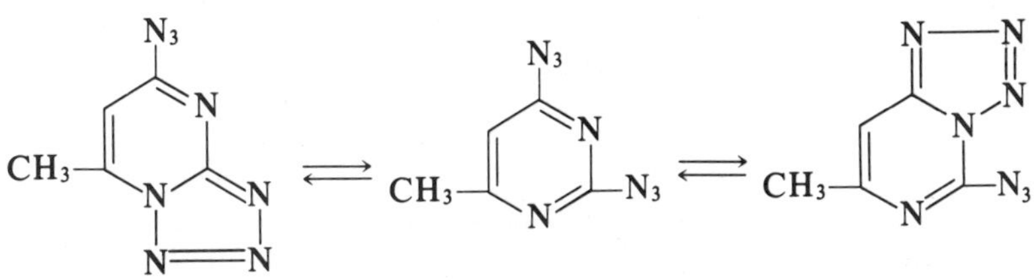

(IV) **(V)**

mp 126–127° (103, 1152) mp 40–42°

derivative is 5-azido-7-methyltetrazolo[1,5-*a*]pyrimidine **(IV)** with about 10% of another azidotetrazolopyrimidine, presumably **(V)**. On melting **(IV)** the diazide is formed; it resolidifies at 45–50° to the tetrazole mixture. Pure diazide results from adding the melt of **IV** and **V** to liquid nitrogen or from evaporation of a trifluoracetic acid solution. Both tetrazole forms and the diazide are present in $CDCl_3$.

Tetrazolo[1,5-*b*]pyridazine exhibits no azide bands either in solution or on heating above 140°. Similarly, neither electron-withdrawing nor electron-releasing groups on tetrazolo[1,5-*b*]pyridazines cause tautomerism to the azide as

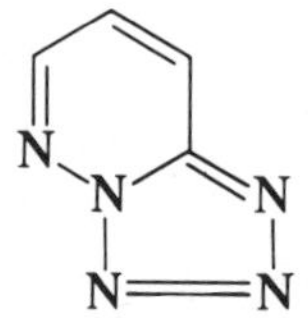

mp 110–111° (1152)

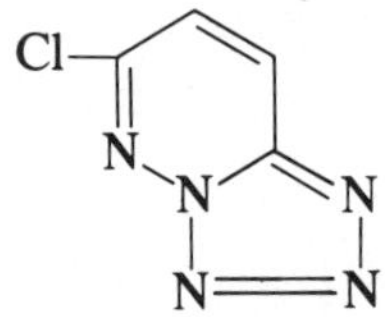

mp 108–109° (945)

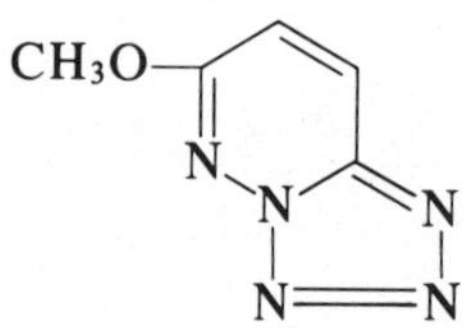

mp 155–157° (945)

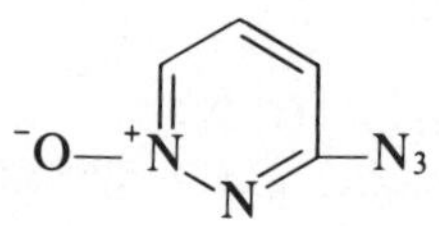

mp 151–152° (550, 1026)

solids or in solution, even in trifluoracetic acid. The *N*-oxide group on adjacent pyridazine nitrogen, however, effects complete ring opening. Fusion of an imidazole to azidopyridazine also destabilizes the tetrazole. An eight-nitrogen chain in the pyridazine series does not exist, and the potential ditetrazolo or diazido

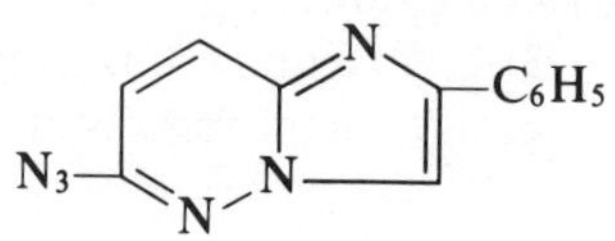

mp 200–204° (606)

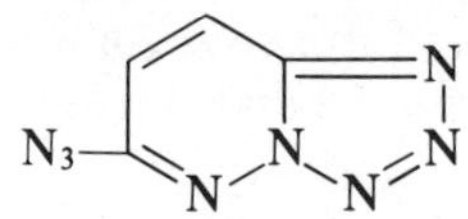

mp 130–131° (606)

product, as with the pyrimidines, is an azidotetrazolo[1,5-*b*]pyridazine. Indeed, there seem to be no instances of an annelated tetrazolo heteroaromatic with two tetrazole fusions in the same molecule. Thus potential ditetrazolonaphthyridines are, instead, azido tetrazolo[1,5-*a*][1,8]naphthyridines.

Consistent with protonation by trifluoracetic acid favoring the azide form in

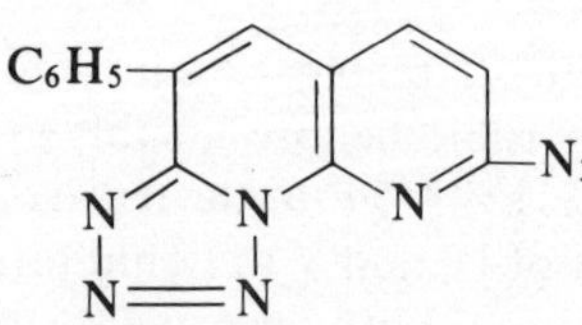

mp 204–206° dec. (195)

azide-tetrazole systems is the existence of nitrogen heterocycles containing neighboring quaternary nitrogen as azides rather than tetrazoles.

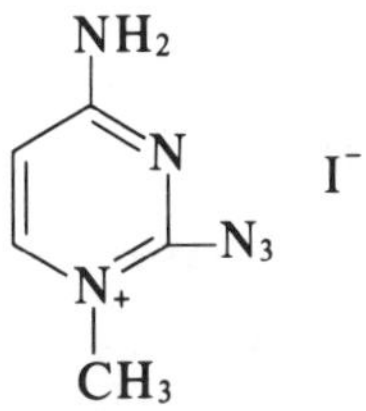

(610)

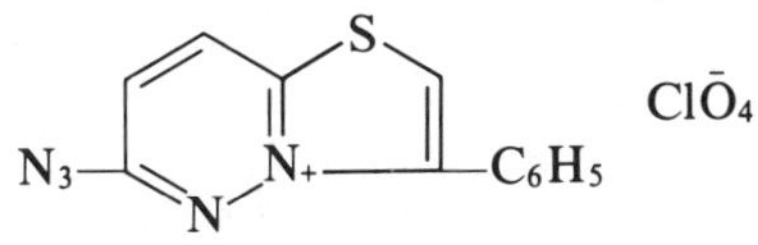

(1026)

Two tetrazolopurines and the azidopurine coexist in $(CD_3)_2SO$ solution; at 34° $K_{VII/VI} = 0.44$ and $K_{VII/VIII} = 3.5$ (1073). The linear tricyclic form **(VI)** is the

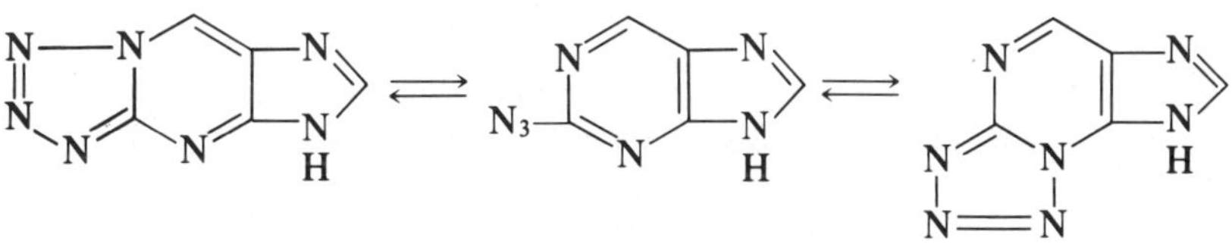

(VI) **(VII)** **(VIII)**

most favored and the angular form **(VIII)** the least favored isomer over the temperature range of 34–99°. With increasing temperature, the proportion of azide increases and at 78°, $K_{VII/VI} = 1.09$ and $K_{VII/VIII} = 6.0$. Calculated thermodynamic values indicate, in both this azido purine and 2-azidopyrimidine, that opening of the tetrazole ring to the less constrained azide increases entropy.

A related tautomerism is that of the very explosive dipolar 4-diazo-5-tetrazolyl imidazole, which may be in equilibruim with azido imidazolo[4,5-*d*]-*v*-triazine through the imidazolo-tetrazolo-*v*-triazine (1074). In $(CD_3)_2SO$ and in 1:1 $(CD_3)_2SO$ and CF_3COOH, only the diazo form is detected, as is also the case with the 1-benzyl derivative. The 3-benzyl compound, however, in the solid

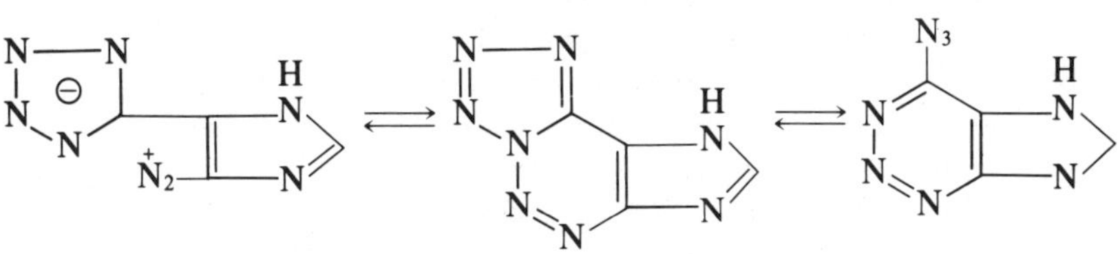

state and in $(CD_3)_2SO$ contains the embedded six-nitrogen chain, but in trifluoracetic acid exists in the diazo form.

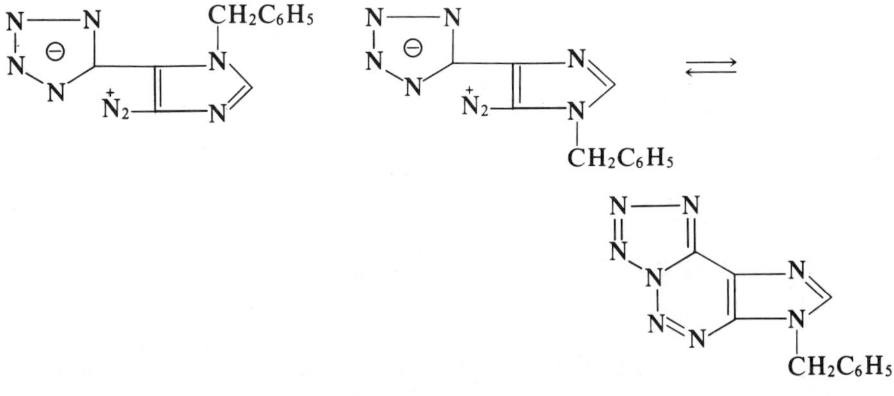

REFERENCES

1. D. J. Abraham, J. S. Rutherford, and R. D. Rosenstein, *J. Med. Chem*, **1969,** *12,* 189–190.
2. R. A. Abramovitch et al., *J. Org. Chem.*, **1978,** *43,* 1218–1226.
3. R. A. Abramovitch and E. P. Kyba, "Decomposition of Organic Azides" in S. Patai, Ed., *The Chemistry of the Azide Group,* Wiley, New York, 1971, pp. 221–329.
4. R. A. Abramovitch and K. Miyashita, *J. Chem. Soc., Perkin I,* **1975,** 2413–2416.
5. R. M. Acheson and M. W. Foxton, *J. Chem. Soc. (C)*, **1968,** 389–391.
6. J. Adamson et al., *J. Chem. Soc. (C)*, **1971,** 981–988.
7. G. Adembri, F. Ponticelli, and P. Tedeschi, *J. Heterocycl. Chem.*, **1972,** *9,* 1219–1225.
8. B. M. Adger et al., *J. Chem. Soc., Perkin I,* **1975,** 31–40.
9. *Ibid.*, 41–45.
10. B. M. Adger, C. W. Rees, and R. C. Storr, *J. Chem. Soc., Perkin I,* **1975,** 45–52.
11. E. P. Ahern, K. J. Dignam, and A. F. Hegarty, *J. Org. Chem.*, **1980,** *45,* 4302–4308.
12. T. P. Ahern, H. Fong, and K. Vaughan, *Can. J. Chem.*, **1977,** *55,* 1701–1709.
13. T. P. Ahern, T. Navratil, and K. Vaughan, *Can. J. Chem.*, **1977,** *55,* 630–639.
14. A. Albert, *J. Chem. Soc. (B)*, **1966,** 427–433.
15. A. Albert, *J. Chem. Soc. (C)*, **1968,** 2076–2083.
16. A. Albert, *J. Chem. Soc. (C)*, **1969,** 152–160.
17. A. Albert, *J. Chem. Soc. (C)*, **1970,** 230–235.
18. A. Albert, *J. Chem. Soc., Perkin I,* **1972,** 461–467.
19. A. Albert, *J. Chem. Soc., Perkin I,* **1978,** 513–519.
20. C. Alberti, *Gazz Chim. Ital.*, **1957,** *87,* 772–780.
21. A. Albini, G. F. Bettinetti, and G. Minoli, *J. Chem. Soc. Perkin I,* **1981,** 4–7.
22. E. Alcalde et al., *J. Heterocycl. Chem.*, **1978,** *15,* 395–339.
23. E. Alcalde and R. M. Claramunt, *J. Heterocycl. Chem.*, **1976,** *13,* 379–381.
24. E. Alcalde and R. M. Claramunt, *Tetrahedron Lett.*, **1975,** 1523–1524.
25. E. Alcalde, J. de Mendoza, and J. Elguero, *J. Heterocycl. Chem.*, 1974, *11,* 921–928.

26. K. Alder and G. Stein, *Ann.*, **1933,** *501,* 1–48.

26a. D. C. Alexander, private communication, Nov. 6, 1980.

27. N. E. Alexandrou and S. Adamopoulos, *Synthesis,* **1976,** 482–483.

28. N. E. Alexandrou and E. D. Micromastoras, *Tetrahedron Lett.*, **1968,** 231–234.

29. N. E. Alexandrou, N. A. Rodios, and C. P. Hadjiantoniou-Louizou, *Org. Magn. Resonance,* **1973,** *5,* 579–583.

30. F. P. Altman, *Prog. Histochem. Cytochem.*, **1976,** *9,* 1–56.

31. A. G. Anastassiou and S. W. Eachus, *Chem. Commun.*, **1970,** 429–430.

32. A. G. Anastassiou and H. E. Simmons, *J. Am. Chem. Soc.*, **1967,** *89,* 3177–3184.

33. V. Y. Andakushkin, B. A. Dolgoplosk, and I. I. Radchenko, *J. Gen. Chem. USSR, Engl. Ed.*, **1956,** *26,* 3307–3314.

34. D. J. Anderson et al., *J. Chem. Soc., Chem. Commun.*, **1971,** 1518.

35. D. J. Anderson et al., *J. Chem. Soc., Perkin I,* **1972,** 1317–1321.

36. D. J. Anderson, T. L. Gilchrist, and C. W. Rees, *Chem. Commun.*, **1971,** 800.

37. S. Anderson, E. E. Glover, and K. D. Vaughan, *J. Chem. Soc., Perkin I,* **1975,** 1232–1234.

38. A. S. Angeloni et al., *Tetrahedron,* **1972,** *28,* 303–314.

39. Anonymous, *Chem. Eng. News,* May 14, **1956,** 2450.

40. Anonymous, *Chem. Eng. News,* Nov. 5, **1962,** 49–50.

41. A. Antonowa and S. Hauptmann, *Z. Chem.*, **1976,** *16,* 17.

42. B. Arnold and M. Regitz, *Tetrahedron Lett.*, **1980,** 909–912.

43. C. Arnold, Jr. and D. N. Thatcher, *J. Org. Chem.*, **1969,** *34,* 1141–1142.

44. M. A. Aron and J. A. Elvidge, *Chem. Ind.*, **1958,** 1234–1235.

45. J. N. Ashley et al., *J. Chem. Soc.*, **1953,** 3881–3888.

46. L. F. Audrieth, *Chem. Rev.*, **1934,** *15,* 169–224.

47. L. F. Audrieth and P. G. Gordon, *J. Org. Chem.*, **1955,** *20,* 244–249.

48. L. F. Avramenko et al., *Khim. Geter. Soedin., Engl. Ed.*, **1968,** 423–427.

49. L. Avrami and J. Haberman, "Sensivity to Heat and Nuclear Radiation," in H. D. Fair and R. F. Walker, Eds., *Energetic Materials,* Vol. 2, Plenum Press, New York and London, 1977, pp. 199–211.

50. K. Babič, et al., *J. Heterocycl. Chem.*, **1976,** *13,* 487–490.

51. H. J. Backer, *Rec. Trav. Chim.*, **1950,** *69,* 1223–1232.

52. B. D. Baigrie, J. I. G. Cadogan, and J. T. Sharp, *J. Chem. Soc., Perkin I,* **1975,** 1065–1068.

53. K. M. Baines et al., *J. Org. Chem.*, **1981,** *46,* 856–859.

54. M. G. Baldwin et al., *J. Polym. Sci., Part B,* **1967,** *5,* 803–806.

55. R. A. Baldwin, *J. Org. Chem.*, **1965,** *30,* 3866–3871.

56. R. A. Baldwin and R. M. Washburn, *J. Org. Chem.*, **1965,** *30,* 3860–3866.

57. H. Balli, *Angew. Chem. Int. Ed. Engl.*, **1966,** *5,* 132.

58. H. Balli, *Helv. Chim Acta,* **1974,** *57,* 1912–1919.

59. H. Balli and F. Kersting, *Ann.*, **1961,** *647,* 1–10.

60. H. Balli and F. Kersting, *Ann.*, **1963,** *663,* 103–107.

61. H. Balli and R. Maul, *Helv. Chim. Acta,* **1976,** *59,* 148–155.

62. O. Baltzer and H. v. Pechmann, *Ann.*, **1891,** *262,* 314–324.

63. E. Bamberger, *Helv. Chim. Acta,* **1931,** *14,* 242–249.

64. E. Bamberger and H. Hauser, *Ann.*, **1910,** *375,* 316–333.

65. E. Bamberger and R. Hübner, *Ber.*, **1903,** *36,* 3822–3827.

66. E. Bamberger and A. Stiegelmann, *Ber.*, **1899,** *32,* 3554–3560.

67. R. E. Banks and A. Prakash, *J. Chem. Soc., Perkin I,* **1974,** 1365–1371.
68. R. E. Banks and A. Prakash, *Tetrahedron Lett.,* **1973,** 99–102.
69. R. E. Banks and G. R. Sparkes, *J. Chem. Soc., Perkin I,* **1972,** 2964–2970.
70. Y. Bardo, T. Nakaya and M. Imoto, *Macromol. Chem.,* **1973,** *172,* 127–134.
71. N. M. Baranchik, I. V. Grachev, and D. Z. Zavel'skii, *Dokl. Akad. Nauk. SSSR,* **1954,** *94,* 701–703.
72. N. M. Baranchik, I. V. Grachev, and D. Z. Zavel'skii, *Zh. Obshch. Khim., Engl. Ed.,* **1953,** *23,* 1651–1656.
73. *Ibid.,* **1957,** *27,* 133–141.
74. M. G. Barlow and K. W. Cheung, *J. Fluorine Chem.,* **1977,** *10,* 191–195.
75. J. J. Barr and R. C. Storr, *J. Chem. Soc., Chem. Commun.,* **1975,** 788–789.
76. *Ibid., Perkin I,* **1979,** 185–191.
77. *Ibid.,* 192–197.
78. J. R. Bartel-Keith, M. T. Burgess, and J. M. Stevenson, *J. Org. Chem.,* **1977,** *42,* 3725–3731.
79. D. H. R. Barton and L. R. Morgan, *J. Chem. Soc.,* **1962,** 622–631.
80. J. W. Barton and A. R. Grinham, *J. Chem. Soc., Perkin I,* **1972,** 634–637.
81. J. W. Barton and S. A. Jones, *J. Chem. Soc.,* (*C*), **1967,** 1276–1278.
82. N. Bashir and T. L. Gilchrist, *J. Chem. Soc., Perkin I,* **1973,** 868–872.
83. L. R. Bates and J. M. Jenkins, *Technical Note No. 25,* Ministry of Technology, Explosives Research and Development Establishment, Aug. 1970.
84. H. Bauer et al., *J. Chem. Soc., Perkin II,* **1972,** 662–667.
85. H. Bauer, G. R. Bedford, and A. R. Katritzky, *J. Chem. Soc.,* **1964,** 751–755.
86. W. Beckh and J. Tafel, *Ber.,* **1894, 27,** 2315–2316.
87. H. G. O. Becker and H.-J. Timpe, *J. Prakt. Chem.,* **1969,** *311,* 9–14.
88. M. Begtrup, *Acta Chem. Scand.,* **1971,** *25,* 249–251.
89. *Ibid.,* **1972,** *26,* 715–718.
90. *Ibid.,* **1975,** *B29,* 141–144.
91. M. Begtrup, K. Hansen, and C. Pedersen, *Acta Chem. Scand.,* **1967,** *21,* 1234–1238.
92. M. Begtrup and J. Holm, *J. Chem. Soc., Perkin I,* **1981,** 503–513.
93. M. Begtrup and P. A. Kristensen, *Acta Chem. Scand.,* **1969,** *23,* 2733–2740.
94. M. Begtrup and C. Pedersen, *Acta Chem. Scand.,* **1967,** *21,* 633–640.
95. M. Begtrup and K. V. Poulsen, *Acta Chem. Scand.,* **1971,** *25,* 2087–2098.
96. H. Behringer and H. J. Fischer, *Chem. Ber.,* **1961,** *94,* 1572–1584.
97. *Ibid.,* 2562–2566.
98. *Ibid.,* **1962,** *95,* 2546–2556.
99. H. Behringer and K. Kohl, *Chem Ber.,* **1956,** *89,* 2648–2653.
100. H. Behringer and M. Matner, *Tetrahedron Lett.,* **1966,** 1663–1669.
101. B. A. Belinka, Jr., A. Hassner, and J. M. Hendler, *J. Org. Chem.,* **1981,** *46,* 631–632.
102. F. R. Benson, *Heterocycl. Compds.,* **1967,** *8,* 1–104.
103. F. R. Benson, L. W. Hartzel, and E. A. Otten, *J. Am. Chem. Soc.,* **1954,** *76,* 1858–1862.
104. F. R. Benson, L. W. Hartzel, and W. L. Savell, *J. Am. Chem. Soc.,* **1950,** *72,* 1816–1818.
105. E. Bergmann and H. A. Wolff, *Ber.,* **1930,** *63,* 1176–1180.
106. J. K. Berlin and M. D. Coburn, *J. Heterocycl. Chem.,* **1975,** *12,* 235–237.
107. K. D. Berlin and L. A. Wilson, *Chem. Commun.,* **1965,** 280–281.
108. K. D. Berlin, L. A. Wilson, and L. M. Raff, *Tetrahedron,* **1967,** *23,* 965–984.
109. C. Bernard and L. Ghosez, *J. Chem. Soc. Chem. Commun.,* **1980,** 940–941.

110. A. Bertho, *Ber.*, **1924,** *57,* 1138–1142.

111. *Ibid.*, **1925,** *58,* 859–864.

112. A. Bertho, *J. Prakt. Chem.*, **1927,** *116,* 101–117.

113. G. Bianchetti, D. Pocar, and P. Dalla Croce, *Gazz. Chim. Ital.*, **1964,** *94,* 340–355.

114. G. Bianchi, C. DeMicheli, and R. Gandolfi, *Angew. Chem. Int. Ed. Engl.*, **1979,** *18,* 721–738.

115. L. Bieber, Dissertation, University of Würzburg, 1975.

116. H. Biltz and R. Weiss, *Ber.*, **1902,** *35,* 3519–3524.

117. J. M. Birchall, R. N. Haszeldine, and J. E. G. Kemp, *J. Chem. Soc. (C)*, **1970,** 1519–1523.

118. J. M. Birchall, R. N. Hazeldine, and A. R. Parkinson, *J. Chem. Soc.*, **1962,** 4966–4976.

119. R. Bird and A. J. Power, *Rep.-Aust. Mater. Res. Lab.*, **1978,** MRL-R-710; Chem. Abstr. 1978,89, 131860r.

120. L. Birkofer, A. Ritter, and P. Richter, *Chem. Ber.*, **1963,** *96,* 2750–2757.

121. L. Birkofer and P. Wegner, *Chem. Ber.*, **1966,** *99,* 2512–2517.

122. *Ibid.*, **1967,** *100,* 3485–3494.

123. P. Bitterli and H. Erlenmeyer, *Helv. Chim. Acta,* **1951,** *34,* 835–840.

124. M. Bleha and D. Lim, *Collect. Czech. Chem. Commun.*, **1973,** *38,* 46–49.

125. A. V. Blom, *Helv. Chim. Acta,* **1921,** *4,* 1036–1039.

126. H. Bock and K.-L. Kompa, *Z. Anorg. Allgem. Chem.*, **1964,** *332,* 238–246.

127. H. Bock and W. Wiegräbe, *Chem. Ber.*, **1966,** *99,* 1068–1076.

128. A. B. Boese, Jr., L. W. Jones, and R. T. Major, *J. Am. Chem. Soc.*, **1931,** *53,* 3530–3541.

129. W. Borsche and M. Feise, *Ber.*, **1907,** *40,* 378–386.

130. W. Borsche and D. Rantscheff, *Ann.*, **1911,** *379,* 152–158.

131. J. Bourgois, M. Bourgois, and F. Texier, *Bull. Soc. Chim. Fr.*, **1978,** Pt. 2, 485–527.

132. G. V. Boyd et al., *J. Chem. Soc., Perkin I,* **1977,** 1612–1615.

133. G. V. Boyd, T. Norris, and P. F. Lindley, *J. Chem. Soc., Perkin I,* **1976,** 1673–1688.

134. *Ibid.*, **1977,** 965–971.

135. J. H. Boyer, *Heterocycl. Compds.*, **1961,** *7,* 384–425.

136. J. H. Boyer, R. Borgers, and L. T. Wolford, *J. Am. Chem. Soc.*, **1957,** *79,* 678–680.

137. J. H. Boyer, M. S. Chang, and R. F. Reinish, *J. Org. Chem.*, **1960,** *25,* 286–287.

138. J. H. Boyer, J. Dunn, and J. Kooi, *J. Chem. Soc., Perkin I,* **1975,** 1743–1747.

139. J. H. Boyer and J. A. Hernandez, *J. Am. Chem. Soc.*, **1956,** *78,* 5124–5125.

140. J. H. Boyer and H. W. Hyde, *J. Org. Chem.*, **1960,** *25,* 458–459.

141. J. H. Boyer, W. E. Kreuger, and G. J. Mikol, *J. Am. Chem. Soc.*, **1967,** *89,* 5504–5505.

142. J. H. Boyer, W. E. Krueger, and R. Modler, *Tetrahedron Lett.*, **1968,** 5979–5982.

143. J. H. Boyer and E. J. Miller, Jr., *J. Am. Chem. Soc.*, **1959,** *81,* 4671–4673.

144. J. H. Boyer and D. Straw, *J. Am. Chem. Soc.*, **1952,** *74,* 4506–4508.

145. *Ibid.*, **1953,** *75,* 1642–1644.

146. F. T. Boyle and R. A. Y. Jones, *J. Chem. Soc., Perkin II,* **1973,** 160–164.

147. J. v. Braun and W. Rudolph, *Ber.*, **1941,** *74,* 264–272.

148. H. Bredereck et al., *Angew. Chem.*, **1963,** *75,* 825–830.

149. H. Bredereck, B. Föhlish, and K. Walz, *Ann.*, **1965,** *688,* 93–98.

150. E. Brehm, A. Haag, and G. Hesse, *Ann.*, **1970,** *737,* 80–86.

151. O. Bremer, *Ann.*, **1934,** *514,* 279–291.

152. *Ibid.*, **1935,** *518,* 274–289.

153. D. S. Breslow et al., *J. Am. Chem. Soc.*, **1967,** *89,* 2384–2390.

154. R. F. Bridger, *J. Am. Chem. Soc.*, **1972,** *94,* 3124–3132.

155. R. F. Bridger, *J. Org. Chem.*, **1970,** *35,* 1746–1750.

156. F. E. Brinckman, H. S. Haiss, and R. A. Robb, *Inorg. Chem.*, **1965,** *4,* 936–942.

156a. T. C. Britton and E. R. Wagner (Dow), U.S. Pat. 4,286,090, Aug. 25, 1981.

157. W. Broeckx et al., *Tetrahedron,* **1971,** *27,* 3527–3534.

158. L. G. S. Brooker, (Eastman Kodak), U.S. Pat. 2,689,849, Sept. 21, 1954.

159. B. R. Brown and D. L. Hammick, *J. Chem. Soc.*, **1947,** 1384–1386.

160. B. R. Brown, D. L. Hammick, and S. G. Heritage, *J. Chem. Soc.*, **1953,** 3820–3825.

161. H. C. Brown and R. J. Kassal, *J. Org. Chem.*, **1967,** *32,* 1871–1873.

162. R. F. C. Brown, F. Irvine, and R. J. Smith, *Aust. J. Chem.*, **1973,** *26,* 2213–2219.

163. A. W. Browne et al., *J. Am. Chem. Soc.*, **1923,** *45,* 2541–2550.

164. A. W. Browne and G. E. F. Lundell, *J. Am. Chem. Soc.*, **1909,** *31,* 435–448.

165. T. H. Brownlee (American Cyanamid), U.S. Pat. 3,405,114, Oct. 8, 1968.

166. W. Buder and A. Schmidt, *Chem. Ber.*, **1973,** *106,* 3812–3816.

167. W. Buder and A. Schmidt, *Z. Anorg. Allg. Chem.*, **1975,** *415,* 263–267.

168. W. E. Bull, J. A. Seaton, and L. F. Audrieth, *J. Am. Chem. Soc.*, **1958,** *80,* 2516–2518.

169. C. Bülow, *Ber.*, **1909,** *42,* 4429–4438.

170. K. Burger, J. Fehn, and A. Gieren, *Ann.*, **1972,** *757,* 9–14.

171. K. Burger, F. Manz, and A. Braun, *Synthesis,* **1975,** 250–251.

172. K. Burger, S. Tremmel, and H. Schickaneder, *J. Fluorine Chem.*, *1976,* **7,** 471–480.

173. E. M. Burgess and J. P. Sanchez, *J. Org. Chem.*, **1973,** *38,* 176–178.

174. *Ibid.*, **1974,** *39,* 940–948.

175. L. A. Burke et al., *J. Am. Chem. Soc.*, **1976,** *98,* 1685–1690.

176. *Ibid.*, **1978,** *100,* 3668–3674.

177. M. Busch, *Ber.*, **1892,** *25,* 445–452.

178. M. Busch, *J. Prakt. Chem.*, **1895,** *51,* 113–134.

179. M. Busch and G. Hefele, *J. Prakt. Chem.*, **1911,** *83,* 524–553.

180. M. Busch and R. Holbein, *Ber.*, **1907,** *40,* 2099–2102.

181. M. Busch, N. Patrascanu, and W. Weber, *J. Prakt. Chem.*, **1934,** *140,* 117–128.

182. M. Busch and H. Pfeiffer, *Ber.*, **1926,** *59,* 1162–1165.

183. M. Busch and W. Schmidt, *Ber.*, **1929,** *62,* 1449–1457.

184. R. N. Butler, *Adv. Heterocycl. Chem.*, **1977,** *21,* 323–435.

185. R. N. Butler et al., *J. Chem. Soc., Perkin I,* **1973,** 1357–1361.

186. *Ibid.*, **1979,** 2886–2890.

187. R. N. Butler and F. L. Scott, *J. Chem. Soc.* (*C*), **1968,** 1711–1716.

188. I. Butula, M. Antunovic, and I. Marusic, *Synthesis,* **1976,** 717–718.

189. J. I. G. Cadogan and J. B. Thomson, *Chem. Commun.*, **1969,** 770.

190. C. D. Campbell and C. W. Rees, *J. Chem. Soc.* (*C*), **1969,** 742–747.

191. H. C. Campbell and O. K. Rice, *J. Am. Chem. Soc.*, **1935,** *57,* 1044–1053.

192. T. W. Campbell and B. F. Day, *Chem. Rev.*, **1951,** *48,* 299–317.

193. R. A. Carboni et al., *J. Am. Chem. Soc.*, **1967,** *89,* 2618–2625.

194. *Ibid.*, 2626–2633.

195. S. Carboni et al., *J. Heterocycl. Chem.*, **1970,** *7,* 1037–1043.

196. W. R. Carpenter, *J. Org. Chem.*, **1962,** *27,* 2085–2087.

197. W. R. Carpenter, A. Haymaker, and D. W. Moore, *J. Org. Chem.*, **1966,** *31,* 789–792.

198. E. Cawkill et al., *J. Chem. Soc., Perkin I,* **1979,** 724–731.

199. A. Chakravorty, B. Behera, and P. S. Zacharias, *Inorg. Chem. Acta,* **1968,** *2,* 85–88.
200. E. A. Chandross and G. Smolinsky, *Tetrahedron Lett.,* **1960,** 19–22.
201. M. S. Chang and A. J. Matuszko, *J. Org. Chem.,* **1963,** *28,* 2260–2262.
202. G. Charrier and G. B. Crippa, *Gazz. Chim. Ital.,* **1923,** *53,* 462–469.
203. F. D. Chattaway, F. L. Gorton, and G. D. Parkes, *J. Chem. Soc.,* **1924,** 1980–1991.
204. M. R. Chaurasia and S. K. Sharma, *Heterocycles,* **1980,** *14,* 1759–1765.
205. G. W. H. Cheeseman and M. Rafiq, *J. Chem. Soc.* (*C*), **1971,** 452–454.
206. S. J. Childress, *J. Pharm. Sci.,* **1962,** *51,* 806–807.
207. C. Christophersen and A. Holm, *Acta Chem. Scand.,* **1971,** *25,* 2015–2022.
208. C. Christophersen and S. Treppendahl, *Acta Chem. Scand.,* **1971,** *25,* 625–630.
209. *Ibid.,* **1972,** *26,* 858–860.
210. L. Citerio, M. L. Saccarello, and P. Trimarco, *J. Heterocycl. Chem.,* **1979,** *16,* 289–292.
211. R. M. Claramunt et al., *Afinidad,* **1977,** *34,* 545–551.
212. R. M. Claramunt, R. Granadas, and E. Pedroso, *Bull. Soc. Chim. Fr.,* **1973,** 1854–1857.
213. R. E. Clark and R. D. Clark, *J. Org. Chem.,* **1977,** *42,* 1136–1139.
214. G. R. Clemo, W. H. Perkin, Jr., and R. Robinson, *J. Chem. Soc.,* **1924,** 1751–1804.
215. G. L. Closs and A. M. Harrison, *J. Org. Chem.,* **1972,** *37,* 1051–1052.
216. K. Clusius and E. Effenberger, *Helv. Chim. Acta,* **1955,** *38,* 1843–1847.
217. K. Clusius and H. Hürzeler, *Helv. Chim. Acta,* **1954,** *37,* 798–804.
218. K. Clusius and H. Schumacher, *Helv. Chim. Acta,* **1958,** *41,* 2264–2273.
219. K. Clusius and K. Schwarzenbach, *Helv. Chim. Acta,* **1959,** *42,* 739–748.
220. K. Clusius and M. Vecchi, *Helv. Chim. Acta,* **1956,** *39,* 1469–1483.
221. M. D. Coburn and J. K. Berlin, *Synthesis,* **1974,** 869.
222. P. Cohen-Fernandes and C. L. Habraken, *J. Org. Chem.,* **1971,** *36,* 3084–3086.
223. R. F. Coles and C. S. Hamilton, *J. Am. Chem. Soc.,* **1946,** *68,* 1799–1801.
223a. K. E. Cooper and E. H. Ingold, *J. Chem. Soc.,* **1926,** 1894–1896.
224. H. J. Cottrell, D. L. Pain, and R. Slack, *J. Chem. Soc.* **1954,** 2968–2969.
225. H. E. Crabtree, R. K. Smalley, and H. Suschitzky, *J. Chem. Soc.* (*C*), **1968,** 2730–2733.
226. J. M. Crossman, R. N. Haszeldine, and A. E. Tipping, *J. Chem. Soc., Dalton Trans.,* **1973,** 483–486.
227. W. D. Crow and H. McNab, *Aust. J. Chem.,* **1979,** *32,* 111–121.
228. W. D. Crow and M. N. Paddon-Row, *Tetrahedron Lett.,* **1972,** 2231–2234.
229. W. D. Crow and C. Wentrup, *Chem. Commun.,* **1968,** 1026–1027.
230. *Ibid.,* 1082–1083.
231. B. W. Cue, Jr., L. J. Czuba, and J. P. Dirlam, *J. Org. Chem.,* **1978,** *43,* 4125–4128.
232. D. Y. Curtin and N. E. Alexander, *Tetrahedron,* **1963,** *19,* 1697–1703.
233. D. Y. Curtin, R. J. Crawford, and D. K. Wedegaertner, *J. Org. Chem.,* **1962,** *27,* 4300–4305.
234. D. Y. Curtin and J. D. Druliner, *J. Org. Chem.,* **1967,** *32,* 1552–1557.
235. T. Curtius, *Ber.,* **1885,** *18,* 1283–1293.
236. *Ibid.,* **1900,** *33,* 2561–2562.
237. T. Curtius, *J. Prakt. Chem.,* **1900,** *62,* 83–126.
238. T. Curtius and A. Bertho, *Ber.,* **1926,** *59,* 565–589.
239. T. Curtius and J. Thompson, *Ber.,* **1906,** *39,* 3398–3409.
240. *Ibid.,* 4140–4144.
241. T. Curtius and K. Heidenreich, *Ber.,* **1894,** *27,* 2684–2685.

242. T. Curtius and F. Schmidt, *J. Prakt. Chem.*, **1923,** *105,* 177–198.

243. J. R. Dacey and D. M. Young, *J. Chem. Phys.*, **1955,** *23,* 1302–1304.

244. P. Dalla Croce et al., *J. Chem. Soc., Perkin I,* **1980,** 141–145.

245. D. Dal Monte Casoni, *Ann. Chim. (Rome),* **1958,** *48,* 783–787; Chem. Abstr., **1959,** *53,* 8123a.

246. R. E. Damschroder and W. D. Peterson, *Org. Synth.*, **1955,** *Coll. Vol. 3,* 106–108.

247. S. N. Danilov et al., *J. Gen. Chem. USSR, Engl. Ed.*, **1977,** *47,* 190–193.

248. A. B. Daruwala and U. Hornemann, *Org. Prep. Proc. Int.*, **1977,** *9,* 19–25.

249. B. F. Day, T. W. Campbell, and G. M. Coppinger, *J. Am. Chem. Soc.*, **1951,** *73,* 4687–4688.

250. K. L. Davies, R. C. Storr, and P. J. Whittle, *J. Chem. Soc., Chem. Commun.*, **1978,** 9–10.

251. J. Davoll and B. A. Lowy, *J. Am. Chem. Soc.*, **1951,** *73,* 1650–1655.

252. J. DeMendoza et al., *J. Chem. Res. (M)*, **1980,** 0514–0529.

253. G. C. Denault, P. C. Marx, and H. H. Takimoto, *J. Chem. Eng. Data,* **1968,** *13,* 514–516.

254. G. Denecker, G. Smets, and G. L'Abbé, *Tetrahedron,* **1975,** *31,* 765–775.

255. G. H. Denny et al., *J. Org. Chem.*, **1980,** *45,* 1662–1665.

256. O. Diels and P. Fritzche, *Ber.*, **1911,** *44,* 3018–3027.

257. O. Dimroth, *Ann.*, **1909,** *364,* 183–226.

258. *Ibid.*, **1909,** *364,* 183–226.

259. *Ibid.*, **1910,** *373,* 336–370.

260. O. Dimroth, *Ber.*, **1902,** *35,* 4041–4060.

261. *Ibid.*, **1903,** *36,* 909–913.

262. *Ibid.*, **1905,** *38,* 670–688.

263. *Ibid.*, **1906,** *39,* 3905–3912.

264. *Ibid.*, **1907,** *40,* 2376–2389.

265. O. Dimroth and G. de Montmollin, *Ber.*, **1910,** *43,* 2904–2915.

266. O. Dimroth and K. Pfister, *Ber.*, **1910,** *43,* 2757–2767.

267. O. Dimroth and W. Wislicenus, *Ber.*, **1905,** *38,* 1573–1576.

268. L. DiStefano and R. N. Castle, *J. Heterocycl. Chem.*, **1968,** *5,* 109–110.

269. F. M. D'Itri and A. I. Popov, *J. Am. Chem. Soc.*, **1968,** *90,* 6476–6481.

270. R. C. Dobbie and H. J. Emeléus, *J. Chem. Soc. (A)*, **1966,** 933–935.

271. B. A. Dolgoplosk et al., *Dokl. Akad. Nauk. SSSR, Engl. Ed.*, **1958,** *120,* 405–407.

272. B. A. Dolgoplosk, P. G. Urgryumov, and V. A. Krol, *Dokl. Akad. Nauk. SSSR,* **1954,** *96,* 757–760; *Chem. Abstr.*, **1955,** *49,* 9537b.

273. J. Donovan et al., *J. Chem. Soc., Perkin II,* **1972,** 1050–1056.

274. V. Dourtoglou, J.-C. Ziegler, and B. Gross, *Tetrahedron Lett.*, **1978,** 1269–1272.

275. M. P. Doyle, J. J. Maciejko, and S. C. Busman, *J. Am. Chem. Soc.*, **1973,** *95,* 952–953.

276. R. S. Drago and B. R. Karstetter, *J. Am. Chem. Soc.*, **1961,** *83,* 1819–1822.

277. R. S. Drago and F. E. Paulik, *J. Am. Chem. Soc.*, **1960,** *82,* 96–98.

278. J. D. Druliner, *J. Am. Chem. Soc.*, **1968,** *90,* 6879–6880.

279. G. F. Duffin, J. D. Kendall, and H. R. J. Waddington, *Chem. Ind.*, **1955,** 1355.

280. J. R. C. Duke, *Chem. Commun.*, **1971,** 2–3.

281. P. Dunn and D. Oldfield, *Aust. J. Chem.*, **1971,** *24,* 645–647.

282. J. A. Durden, H. A. Stansbury, and W. H. Catlette, *J. Chem. Eng. Data,* **1964,** *9,* 228–232.

283. F. P. Dwyer, *J. Am. Chem. Soc.*, **1941,** *63,* 78–81.

284. L. K. Dyall, *Aust. J. Chem.*, **1975,** *28,* 2147–2159.

285. L. K. Dyall and J. E. Kemp, *J. Chem. Soc. (B)*, **1968,** 976–979.

286. E. Dyer and P. A. Christie, *J. Polym. Sci., A-1,* **1968,** *6,* 729–742.

287. F. D. Eddy, K. Vaughan, and M. F. G. Stevens, *Can. J. Chem.,* **1978,** *56,* 1616–1620.

288. G. Ege and K. Gilbert, *Tetrahedron Lett.,* **1979,** 4253–4256.

289. F. Einberg, *J. Org. Chem.,* **1964,** *29,* 2021–2024.

290. F. J. Einberg, (United States of America as represented by the Secretary of the Army), U.S. Pat. 3,929,684, Dec. 30, 1975.

291. F. J. Einberg, (United States of America as represented by the Secretary of the Army), U.S. Pat. 4,013,596, March 22, 1977.

292. J. Elguero et al., *The Tautomerism of Heterocycles, Advances in Heterocyclic Chemistry,* Suppl. 1, Academic Press, New York, 1976.

293. J. Elguero et al., *Bull. Soc. Chim. Belg.,* **1975,** *84,* 1189–1196.

294. M. Elkins and L. Hunter, *J. Chem. Soc.,* **1938,** 1346–1350.

295. D. T. Elmore, *Chem. Ber.,* **1966,** *2,* 414.

296. F. Eloy, *J. Org. Chem.,* **1961,** *26,* 952–953.

297. A. V. El'tsov, A. Khanchmann, and N. I. Rtishchev, *Zh. Org. Khim., Engl. Ed.,* **1977,** *13,* 425–426.

298. J. K. Elwood and J. W. Gates, Jr., *J. Org. Chem.,* **1967,** *32,* 2956–2959.

299. J. Epsztajn et al., *J. Chem. Soc., Perkin I,* **1973,** 2622–2624.

300. J. Epsztajn, E. Lunt, and A. R. Katritzky, *Tetrahedron,* **1970,** *26,* 1665–1673.

301. J. Epsztajn and A. R. Katritzky, *Tetrahedron Lett.,* **1969,** 4739–4742.

302. H. Euler and A. Euler, *Ber.,* **1903,** *36,* 4253–4256; *Ber.,* **1904,** *37,* 47–49.

303. W. T. Evanochko and P. B. Shevlin, *J. Am. Chem. Soc.,* **1979,** *101,* 4668–4672.

304. N. A. Evans, R. B. Johns, and K. R. Markham, *Aust. J. Chem.,* **1967,** *20,* 713–722.

305. W. L. Evans, R. G. D. Moore, and J. E. Redding, *Anal. Chem.,* **1962,** *34,* 159–163.

306. E. Fahr, M. J. Richter, and N. Pelz, *Tetrahedron Lett.,* **1979,** 1741–1744.

307. E. Fanghänel et al., *J. Prakt. Chem.,* **1975,** *317,* 631–640.

308. E. Fanghänel et al., *J. Prakt. Chem.,* **1977,** *319,* 813–816.

309. E. Fanghänel, R. Hansel, and J. Hohlfeld, *J. Prakt. Chem.,* **1977,** *319,* 485–493.

310. W. V. Farrar, *J. Chem. Soc.,* **1964,** 906–907.

311. R. Faure et al., *Can. J. Chem.,* **1977,** *55,* 1728–1735.

312. R. Faure et al., *J. Heterocycl. Chem.,* **1977,** *14,* 1299–1304.

313. R. Faure et al., *Org. Mag. Resonance,* **1978,** *11,* 507–509.

314. B. T. Federoff and O. E. Sheffield, *Encyclopedia of Explosives and Related Items, PATR 2700,* Vol. 3, C 570 1966, Picatinny Arsenal; AD 653 029. Clearinghouse for Fed. Sci and Tech. Info., U.S. Dept. of Commerce.

315. H. Feuer and L. F. Spinialli, *J. Org. Chem.,* **1976,** *41,* 2981–2984.

316. R. Fields and J. P. Tomlinson, *J. Fluorine Chem.,* **1979,** *13,* 19–28.

317. L. F. Fieser and E. L. Martin, *J. Am. Chem. Soc.,* **1935,** *57,* 1835–1839.

318. A. Finch, B. Leach, and J. H. Morris, *Organomet. Chem. Rev.,* **1969,** 1–45.

319. K. T. Finley, *Triazoles: 1, 2, 3,* Wiley, New York, 1980.

320. W. G. Finnegan and R. A. Henry, *J. Org. Chem.,* **1959,** *24,* 1565–1567.

321. *Ibid.,* **1965,** *30,* 567–575.

322. W. G. Finnegan and R. A. Henry, (United States of America as represented by the Secretary of the Navy), U.S. Pat. 3,004,959, Oct. 17, 1961.

323. W. G. Finnegan, R. A. Henry, and E. Lieber, *J. Org. Chem.,* **1953,** *18,* 779–791.

324. W. G. Finnegan, R. A. Henry, and R. Lofquist, *J. Am. Chem. Soc.,* **1958,** *80,* 3908–3911.

325. E. Fischer, *Ber.*, **1898,** *31,* 2550–2574.
326. S. Fischer and C. Wentrup, *J. Chem. Soc., Chem. Commun.*, **1980,** 502–503.
327. W. Fischer and J.-P. Anselme, *J. Am. Chem. Soc.*, **1967,** *89,* 5284–5285.
328. I. J. Fletcher and A. R. Katritzky, *Chem. Commun.*, **1970,** 706–707.
329. L. Floch et al., *Collect. Czech. Chem. Commun.*, **1977,** *42,* 2945–2952.
330. L. Floch, A. Martovŏn, and M. Kŏsik, *J. Therm. Anal.*, **1977,** *12,* 407–411.
331. S. N. Foner and R. L. Hudson, *Adv. Chem.*, **1962,** *36,* 34–49.
332. S. N. Foner and R. L. Hudson, *J. Chem. Phys.*, **1958,** *29,* 442–443.
333. H. Fong and K. Vaughan, *Can. J. Chem.*, **1975,** *53,* 3714–3719.
334. R. G. Ford, *J. Mol. Spectrosc.*, **1977,** *65,* 273–279.
335. P. S. Forgione, G. S. Sprague, and H. J. Troffkin, *J. Am. Chem. Soc.*, **1966,** *88,* 1079–1080.
336. M. O. Forster and H. E. Fierz, *J. Chem. Soc.*, **1907,** *91,* 1942–1953.
337. *Ibid.*, **1908,** *93,* 1174–1179.
338. M. O. Forster, H. E. Fierz, and W. P. Joshua, *J. Chem. Soc.*, **1908,** *93,* 1070–1074.
339. N. I. Foster et al., *Synthesis,* **1980,** 572–573.
340. F. W. Fowler, *Adv. Heterocycl. Chem.*, **1971,** *13,* 45–76.
341. F. W. Fowler, A. Hassner, and L. A. Levy, *J. Am. Chem. Soc.*, **1967,** *89,* 2077–2082.
342. J. E. Francis et al., *Can. J. Chem.*, **1979,** *57,* 3320–3331.
343. D. Frank, G. Himbert, and M. Regitz, *Chem. Ber.*, **1978,** *111,* 183–188.
344. R. R. Fraser, Gurudata, and K. E. Haque, *J. Org. Chem.*, **1969,** *34,* 4118–4120.
345. M. Freund and T. Paradies, *Ber.*, **1901,** *34,* 3110–3122.
346. K. Fries and E. Roth, *Ann.*, **1912,** *389,* 318–344.
347. K. Freudenberg, *Ber.*, **1918,** *51,* 1668–1682.
348. P. Friedländer and S. v. Zakrzewski, *Ber.*, **1894,** *27,* 761–765.
349. K. Friedrich and H. K. Thieme, *Chem. Ber.*, **1970,** *103,* 1982–1991.
350. K. Fries, *Ann.*, **1927,** *454,* 131–172.
351. K. Fries and J. Empson, *Ann.*, **1912,** *389,* 345–367.
352. K. Fries, W. Franke, and W. Bruns, *Ann.*, **1934,** *511,* 241–267.
353. K. Fries, H. Guterbock, and H. Kuhn, *Ann.*, **1934,** *511,* 213–240.
354. K. Fujinuma, Y. Hashida, and K. Matsui, *Bull. Chem. Soc. Japan,* **1978,** *51,* 2394–2397.
355. R. Fusco, G. Bianchetti, and D. Pocar, *Gazz. Chim. Ital.*, **1961,** *91,* 933–957.
356. R. Fusco, L. Garanti, and G. Zecchi, *J. Org. Chem.*, **1975,** *40,* 1906–1909.
357. R. Fusco and R. Romani, *Gazz. Chim. Ital.*, **1946, 76,** 439–458.
358. S. M. Gadekar and J. L. Frederick, *J. Org. Chem.*, **1962,** *27,* 1383–1386.
359. G. J. Gainsford and A. D. Woolhouse, *Aust. J. Chem.*, **1980,** *33,* 2447–2454.
360. S. F. Gait et al., *J. Chem. Soc. Chem. Commun.*, **1972,** 982–983.
361. S. F. Gait et al., *J. Chem. Soc., Perkin I,* **1974,** 1248–1260.
362. S. F. Gait et al., *J. Chem. Soc., Perkin I,* **1975,** 19–25.
363. S. F. Gait, C. W. Rees, and R. C. Storr, *J. Chem. Soc., Chem. Commun.*, **1971,** 1545–1546.
364. T. C. Gallagher, M. J. Sasse, and R. C. Storr, *J. Chem. Soc., Chem. Commun.*, **1979,** 419–420.
365. G. Gaudiano et al., *Gazz. Chim. Ital.*, **1967,** *97,* 1411–1422.
366. G. Geiseler and W. König, *Z. Phys. Chem. (Leipzig)*, **1964,** *227,* 81–92.
367. P. Germeraad and H. W. Moore, *J. Org. Chem.*, **1974,** *39,* 774–780.
368. M. S. Gibson, *J. Chem. Soc.*, **1956,** 1076–1078.
369. M. S. Gibson, *J. Chem. Soc.*, **1963,** 3539–3541.

370. M. S. Gibson and A. W. Murray, *J. Org. Chem.*, **1962,** *27,* 4083–4084.

371. E. E. Gilbert and B. Veldhuis, *J. Heterocycl. Chem.*, **1969,** *6,* 779–782.

372. T. L. Gilchrist and G. E. Gymer, *Adv. Heterocycl. Chem.*, **1974,** *16,* 33–85.

373. T. L. Gilchrist, G. E. Gymer, and C. W. Rees, *J. Chem. Soc., Perkin I,* **1973,** 555–561.

374. T. L. Gilchrist, G. E. Gymer, and C. W. Rees, *J. Chem. Soc., Perkin I,* **1975,** 1–8.

375. T. L. Gilchrist, G. E. Gymer, and C. W. Rees, *J. Chem. Soc., Perkin I,* **1975,** 1747–1750.

376. T. L. Gilchrist, C. J. Moody, and C. W. Rees, *J. Chem. Soc., Perkin I,* **1979,** 1871–1878.

377. Y. Gilliams and G. Smets, *Makromol. Chem.*, **1968,** *117,* 1–11.

378. V. A. Ginsburg et al., *Dokl. Akad. Nauk. SSSR, Engl. Ed.*, **1962,** *142,* 4–7.

379. V. A. Ginsburg et al., *Dokl. Akad. Nauk. SSSR, Engl. Ed.*, **1962,** *142,* 354–357.

380. R. Gleiter, W. Rettig, and C. Wentrup, *Helv. Chim. Acta,* **1974,** *57,* 2111–2124.

381. E. E. Glover, private communication, Feb. 18, 1980.

382. E. E. Glover and K. T. Rowbottom, *J. Chem. Soc., Perkin I,* **1974,** 1792–1794.

383. E. E. Glover and M. Yorke, *J. Chem. Soc. (C)*, **1971,** 3280–3285.

384. J. Goerdeler and G. Gnad, *Chem. Ber.*, **1966,** *99,* 1618–1631.

385. H. Goldschmidt and V. Badl, *Ber.*, **1889,** *22,* 933–942.

386. S. Goldschmidt et al., *Ann.*, **1924,** *437,* 194–226.

387. S. Goldschmidt, *Ber.*, **1920,** *53,* 44–62.

388. S. Goldschmidt and K. Euler, *Ber.*, **1922,** *55,* 616–628.

389. S. Goldschmidt and K. Renn, *Ber.*, **1922,** *55,* 628–643.

390. R. Gompper and K. Schönafinger, *Chem. Ber.*, **1979,** *112,* 1514–1528.

391. R. Gompper and K. Schönafinger, *Chem. Ber.*, **1979,** *112,* 1529–1534.

392. R. Gompper and K. Schönafinger, *Chem. Ber.*, **1979,** *112,* 1535–1544.

393. A. Good and J. C. J. Thynne, *J. Chem. Soc. (B)*, **1967,** 684–685.

394. S. Götzky, *Ber.*, **1931,** *64,* 1555–1560.

395. C. Graebe and F. Ullmann, *Ann.*, **1896,** *291,* 16–17.

396. A. F. Graefe, *J. Org. Chem.*, **1958,** *23,* 1230–1231.

397. V. A. Grakauskas, doctoral dissertation, Illinois Institute of Technology, 1955.

398. P. Grammaticakis, *Compt Rend.*, **1957,** *245,* 1252–1254.

399. R. Granados, M. Rull, and J. Vilarrasa, *J. Heterocycl. Chem.*, **1976,** *13,* 281–284.

400. L. R. Grant (Rockwell International), U.S. Pat. 3,976,573, Aug. 24, 1976.

401. J. M. Grasshoff, J. L. Reid, and L. D. Taylor, *J. Polym. Sci., Polym. Chem. Ed.*, **1978,** *16,* 2403–2405.

402. J. M. Grasshoff and J. L. Reid (Polaroid Corp.), U.S. Pat. 3,936,401, Feb. 3, 1976.

403. M. L. Graziano and R. Scarpati, *J. Heterocycl. Chem.*, **1976,** *13,* 205–210.

404. N. N. Greenwood and J. H. Morris, *J. Chem. Soc.*, **1965,** 6205–6209.

405. P. A. Grieco and J. P. Mason, *J. Chem. Eng. Data,* **1967,** *12,* 623–624.

406. P. Griess, *Ber.*, **1888,** *21,* 1559–1566.

407. E. Griget, R. Pütter, and E. Muhlbauer, *Chem. Ber.*, **1965,** *98,* 3777–3784.

408. S. Gronowitz, C. Westerlund, and A.-B. Hörnfeldt, *Acta Chem. Scand. B.*, **1975,** *29,* 224–232.

409. P. Grünanger, P. V. Finzi, and C. Scotti, *Chem. Ber.*, **1965,** *98,* 623–628.

410. C. Grundmann and H.-D. Frommeld, *J. Org. Chem.*, **1966,** *31,* 157–162.

411. C. Grundmann and H. Ulrich, *J. Org. Chem.*, **1959,** *24,* 272–274.

412. Z. Grzonka and B. Liberek, *Rocz. Chem.*, **1971,** *45,* 967–980.

413. I. Hagedorn and H.-D. Winkelmann, *Chem. Ber.*, **1966,** *99,* 850–855.

413a. S. M. A. Hai, *Pakistan J. Sci. Ind. Res.*, **1974,** *17,* 59–64.

414. G. Hajos and A. Messmer, *J. Heterocycl. Chem.*, **1976,** *13,* 881–882.

415. J. H. Hall, *J. Org. Chem.*, **1968,** *33,* 2954–2956.

416. J. H. Hall, F. E. Behr, and R. L. Reed, *J. Am. Chem. Soc.*, **1972,** *94,* 4952–4958.

417. J. H. Hall and F. W. Dolan, *J. Org. Chem.*, **1978,** *43,* 4608–4612.

418. J. H. Hall and E. Patterson, *J. Am. Chem. Soc.*, **1967,** *89,* 5856–5861.

419. M. Hamana and S. Kumadaki, *Chem. Pharm. Bull.*, **1974,** *22,* 1506–1518.

420. R. N. Hanley et al., *J. Chem. Soc., Perkin I,* **1979,** 744–746.

421. R. N. Hanley, W. D. Ollis, and C. A. Ramsden, *J. Chem. Soc., Perkin I,* **1979,** 732–735.

422. *Ibid.*, 736–740.

423. *Ibid.*, 741–743.

424. R. M. Hann and C. S. Hudson, *J. Am. Chem. Soc.*, **1944,** *66,* 735–738.

425. H. Hansen, S. Hünig, and K. Kishi, *Chem. Ber.*, **1979,** *112,* 445–461.

426. P. E. Hansen and K. Undheim, *J. Chem. Soc., Perkin I,* **1975,** 305–308.

427. A. Hantzsch, *Ber.*, **1903,** *36,* 2056–2058.

428. R. J. Harder, R. A. Carboni, and J. E. Castle, *J. Am. Chem. Soc.*, **1967,** *89,* 2643–2647.

429. R. J. Harder and J. C. Kauer (DuPont), U.S. Pat. 3,262,944, July 26, 1966.

430. R. E. Harman et al., *Chem. Ind.*, **1970,** 1021–1022.

431. R. E. Harman et al., *J. Org. Chem.*, **1970,** *35,* 3444–3448.

432. W. W. Hartman and J. B. Dickey, *Org. Synth.*, **1943,** *Coll. Vol. II,* 163–164.

433. L. W. Hartzel and F. R. Benson, *J. Am. Chem. Soc.*, **1954,** *76,* 667–670.

434. F. E. Harvey, *Trans. Faraday Soc.*, **1933,** *29,* 653–658.

435. G. R. Harvey, *J. Org. Chem.*, **1966,** *31,* 1587–1590.

436. E. K. Harvill et al., *J. Org. Chem.*, **1950,** *15,* 662–670.

437. E. K. Harvill, C. W. Roberts, and R. M. Herbst, *J. Org. Chem.*, **1950,** *15,* 58–67.

438. M. Hashimoto and M. Ohta, *Bull. Chem. Soc. Japan,* **1962,** *35,* 766–767.

439. A. Hassner, D. Tang, and J. Keogh, *J. Org. Chem.*, **1976,** *41,* 2102–2104.

440. R. N. Haszeldine and A. E. Tipping, *J. Chem. Soc.* (*C*), **1966,** 1236–1241.

441. K. Hattori, E. Lieber, and J. P. Horwitz, *J. Am. Chem. Soc.*, **1956,** *78,* 411–415.

442. H. Hauptmann and A. Cid M. Périssé, *Chem. Ber.*, **1956,** *89,* 1081–1094.

443. M. Hauser (United States of America as represented by the Secretary of the Navy), U.S. Pat. 3,431,251, March 4, 1969.

444. K. H. Hausser, *Z. Naturforsch.*, **1956,** *11a,* 20–32.

445. Y. Hayashi, T. Watanabe, and R. Oda, *Tetrahedron Lett.*, **1970,** 605–608.

446. E. Hayon and M. Simic, *J. Am. Chem. Soc.*, **1972,** *94,* 42–47.

447. R. Hazard and A. Tallec, *Bull. Soc. Chem. Fr.*, **1976,** 433–438.

448. E. Hedaya, M. E. Kent, and D. W. McNeil, *Tetrahedron Lett.*, **1968,** 3415–3470.

449. M. Hedayutullah, *Bull. Soc. Chem. Fr.*, **1967,** 422–427.

450. A. F. Hegarty and E. P. Ahern, *J. Org. Chem.*, **1981,** *46,* 1342–1346.

451. A. F. Hegarty, J. B. Aylward, and F. L. Scott, *J. Chem. Soc.* (*C*), **1967,** 2587–2593.

452. A. F. Hegarty, K. Brady, and M. Mullane, *J. Chem. Soc., Perkin II,* **1980,** 535–560.

453. H. W. Heine and D. A. Tomalia, *J. Am. Chem. Soc.*, **1962,** *84,* 993–995.

454. B. T. Heitke and C. G. McCarty, *J. Org. Chem.*, **1974,** **39,** 1522–1526.

455. B. Helferich and H. Schirp, *Chem. Ber.*, **1953,** *86,* 547–556.

456. G. Heller, W. Dietrich, and G. Reichardt, *J. Prakt. Chem.*, **1928,** *118,* 138–148.

457. B. Hellrung and H. Balli, *Helv. Chim. Acta,* **1975,** *58,* 605–609.

458. H. Hemetsberger, D. Knittel, and H. Weidmann, *Monatsh. Chem.*, **1969,** *100,* 1599–1603.

458a. K. Henkel and F. Weygand, *Ber.*, **1943,** *76,* 812–816.

459. C. Hennart, *Bull. Soc. Chem. Fr.*, **1967,** 4691.

460. M. Henriet et al., *Tetrahedron Lett.*, **1980,** 223–226.

461. R. A. Henry, *J. Am. Chem. Soc.*, **1951,** *73,* 4470.

462. R. A. Henry (United States of America as represented by the Secretary of the Navy), U.S. Pat. 3,096,312, July 2, 1963.

463. R. A. Henry and W. M. Dehn, *J. Am. Chem. Soc.*, **1943,** *65,* 479–480.

464. R. A. Henry and W. G. Finnegan, *J. Am. Chem. Soc.*, **1954,** *76,* 923–926.

465. R. A. Henry, W. G. Finnegan, and E. Lieber, *J. Am. Chem. Soc.*, **1954,** 76, 88–93.

466. *Ibid.*, 2894–2898.

467. *Ibid.*, **1955,** *77,* 2264–2270.

468. R. M. Herbst, "Tetrazoles as Carboxylic Acid Analogs," in S. Graff, Ed., *Essays in Biochemistry,* Wiley, New York, 1955, pp. 141–155.

469. R. M. Herbst and W. L. Garbrecht, *J. Org. Chem.*, **1953,** *18,* 1283–1291.

470. R. M. Herbst and J. E. Klingbeil, *J. Org. Chem.*, **1958,** *23,* 1912–1916.

471. H. Herlinger et al. (Farkenfabriken Bayer), U.S. Pat. 3,163,646, Dec. 29, 1964.

472. M. E. Hermes and F. D. Marsh, *J. Am. Chem. Soc.*, **1967,** *89,* 4760–4764.

473. B. Hessett et al., *J. Chem. Soc., Dalton Trans.*, **1972,** 131–134.

474. F. Heusler, *Ann.*, **1890,** *260,* 227–250.

475. D. H. Hey, C. W. Rees, and A. R. Todd, *Chem. Ind.*, **1962,** 1332.

476. D. H. Hey, C. W. Rees, and A. R. Todd, *J. Chem. Soc. (C)*, **1968,** 1028–1029.

477. G. Himbert, D. Frank, and M. Regitz, *Chem. Ber.*, **1976,** *109,* 370–394.

478. G. Himbert and M. Regitz, *Ann.*, **1973,** 1505–1529.

479. G. Himbert and M. Regitz, *Chem. Ber.*, **1972,** *105,* 2975–2984.

480. *Ibid.*, **1974,** *107,* 2513–2516.

481. K. Hirakawa and Y. Tanabiki, *J. Org. Chem.*, **1982,** *47,* 280–284.

482. K. Hirota et al., *Heterocycles,* **1981,** *15,* 285–288.

483. J. Hjortás, *Acta Cryst.*, **1973,** *B29,* 1916–1922.

484. Hoechst Akt., Brit. Pat. 1,429,339, March 24, 1976.

485. L. Hoesch, *Helv. Chim. Acta,* **1981,** *64,* 38–48.

486. L. Hoesch et al., *Helv. Chim. Acta,* **1977,** *60,* 816–830.

487. L. Hoesch and A. S. Dreiding, *Helv. Chim. Acta,* **1975,** *58,* 980–1001.

488. L. Hoesch, N. Egger, and A. S. Dreiding, *Helv. Chim. Acta,* **1978,** *61,* 795–814.

489. L. Hoesch and B. Köppel, *Helv. Chim. Acta,* **1981,** *64,* 864–869.

490. R. W. Hoffmann et al., *J. Chem. Soc. (C)*, **1969,** 769–772.

491. K. A. Hofmann, H. Hock, and R. Roth, *Ber.*, **1910,** *43,* 1087–1095.

492. K. A. Hofmann and H. Hock, *Ber.*, **1911,** *44,* 2946–2956.

493. K. A. Hofmann and R. Roth, *Ber.*, **1910,** *43,* 682–688.

494. K. Hohenlohe-Oehringen, *Monatsh.*, **1958,** *89,* 557–561.

495. *Ibid.*, 562–569.

496. *Ibid.*, 588–596.

497. J. Hollaender and W. P. Neumann, *Angew. Chem. Int. Ed. Eng.*, **1970,** *9,* 804–805.

498. J. Hollaender, W. P. Neumann, and G. Alester, *Chem. Ber.*, **1972,** *105,* 1540–1552.

499. A. Holm, *Adv. Heterocycl. Chem.*, **1976,** *20,* 145–174.

500. A. Holm et al., *J. Org. Chem.*, **1975,** *40,* 431–436.

501. A. Holm et al., *Tetrahedron,* **1975,** *31,* 1783–1786.
502. A. Holm, L. Carlsen, and E. Larsen, *J. Org. Chem.,* **1978,** *43,* 4816–4822.
503. S.-Y. Hong and J. E. Baldwin, *Tetrahedron,* **1968,** *24,* 3787–3794.
504. W. D. Hooper, *Rev. Pure Appl. Chem.,* **1969,** *19,* 221–241.
505. J. R. E. Hoover and A. R. Day, *J. Am. Chem. Soc.,* **1956,** *78,* 5832–5836.
506. L. Horner and A. Gross, *Ann.,* **1955,** *591,* 117–134.
507. J. P. Horwitz and V. A. Grakauskas, *J. Am. Chem. Soc.,* **1957,** *79,* 1249–1253.
508. *Ibid.,* **1958,** *80,* 926–930.
509. P. G. Houghton, D. F. Pipe, and C. W. Rees, *J. Chem. Soc. Chem. Commun.,* **1979,** 771–772.
510. T. J. Houser and D. K. Humbert, *Macromolecules,* **1973,** *6,* 786–787.
511. B. Houston and T. B. Johnson, *J. Am. Chem. Soc.,* **1925,** *47,* 3011–3018.
512. H. Huber, *Angew. Chem., Int. Ed. Engl.,* **1982,** *21,* 64–65.
513. A. J. Hubert and G. Anthoine, *Bull. Soc. Chim., Belg.,* **1969,** *78,* 553–560.
514. L. Huff et al., *J. Org. Chem.,* **1970,** *35,* 2074–2076.
515. L. Huff and R. A. Henry, *J. Med. Chem.,* **1970,** *13,* 777–779.
516. R. Huisgen, *Angew. Chem. Int. Ed. Engl.,* **1963,** *2,* 565–598.
517. *Ibid.,* **1980,** *19,* 947–973.
518. R. Huisgen et al., *Ann.,* **1962,** *653,* 105–113.
519. R. Huisgen et al., *Chem. Ber.,* **1960,** *93,* 2106–2124.
520. *Ibid.,* **1965,** *98,* 3992–4013.
521. *Ibid.,* 4014–4021.
522. R. Huisgen, L. Möbius, and G. Szeimies, *Chem. Ber.,* **1965,** *98,* 1138–1152.
523. R. Huisgen, J. Sauer, and M. Seidel, *Ann.,* **1962,** *654,* 146–160.
524. R. Huisgen, J. Sauer, and M. Seidel, *Chem. Ber.,* **1962,** *94,* 2503–2509.
524a. R. Huisgen and G. Szeimies, *Chem. Ber.,* **1965,** *98,* 1153–1158.
525. R. Huisgen, G. Szeimies, and L. Möbius, *Chem. Ber.,* **1966,** *99,* 475–490.
526. R. Huisgen and I. Ugi, *Angew. Chem.,* **1956,** *68,* 705–706.
527. R. Huisgen and I. Ugi, *Ann.,* **1957,** *610,* 57–66.
528. R. Huisgen and I. Ugi, *Chem. Ber.,* **1957,** *90,* 2914–2927.
528a. W. E. Hull, M. Kunstlinger, and E. Breitmaier, *Angew. Chem. Int. Ed. Engl.,* **1980,** *19,* 924–926.
529. S. Hünig and K. H. Oette, *Ann.,* **1961,** *641,* 94–103.
530. S. Hünig and H. Quast, *Ann.,* **1968,** *711,* 139–156.
531. C. D. Hurd, *The Pyrolysis of Carbon Compounds,* The Chemical Catalog Company, Inc., New York, 1929.
532. A. Q. Hussein and J. C. Jochims, *Chem. Ber.,* **1979,** *112,* 1956–1972.
533. R. Hüttel and F. Büchele, *Chem. Ber.,* **1955,** *88,* 1586–1590.
534. R. Hüttel and A. Gebhardt, *Ann.,* **1947,** *558,* 34–47.
535. R. Hüttel and G. Welzel, *Ann.,* **1955,** *593,* 207–218.
536. A. T. Hutton and H. M. N. H. Irving, *J. Chem. Soc., Perkin II,* **1980,** 139–145.
537. W. Hutzenlaub, R. L. Tolman, and R. K. Robins, *J. Med. Chem.,* **1972,** *15,* 879–883.
538. H. Igeta, T. Tsuchiya, and T. Nakai, *Tetrahedron Lett.,* **1971,** 3117–3120.
539. T. Ignasiak, J. Suszko, and B. Ignasiak, *J. Chem. Soc., Perkin I,* **1975,** 2122–2129.
540. K. Ikeda et al., *J. Polym. Sci.,* **1973,** *11,* 1167–1176.
541. J. D. Illige, *Lawrence Berkeley Lab Report,* **1976,** LBL-5196; *Chem. Abstr.,* **1977,** *86,* 132628w.

542. T. Isida et al., *Bull. Chem. Soc. Jap.*, **1972,** *45,* 1471–1473.
543. *Ibid.*, **1973,** *46,* 1250–1253.
544. T. Isida et al., *J. Org. Chem.*, **1971,** *36,* 3807–3810.
545. K. Isomura, S. Kobayashi, and H. Taniguchi, *Tetrahedron Lett.*, **1968,** 3499–3502.
546. K. Isomura, M. Okada, and H. Taniguchi, *Tetrahedron Lett.*, **1969,** 4073–4076.
547. K. Isomura, M. Okada, and H. Taniguchi, *Chem. Lett.*, **1972,** 629–632.
548. T. Itai and G. Ito, *Chem. Pharm. Bull.*, **1962,** *10,* 1141–1145.
549. T. Itai and S. Kamiya, *Chem. Pharm. Bull.*, **1961,** *9,* 87–91.
550. *Ibid.*, **1963,** *11,* 348–355.
551. S. Ito, Y. Tanaka, and A. Kakehi, *Bull. Chem. Soc. Jap.*, **1976,** *49,* 762–766.
552. S. Ito, Y. Tanaka, and A. Kakehi, *Chem. Lett.*, **1982,** 59–60.
553. J. W. A. M. Janssen et al., *J. Org. Chem.*, **1973,** *38,* 1777–1782.
554. J. W. A. M. Janssen and C. L. Habraken, *J. Org. Chem.*, **1971,** *36,* 3081–3084.
555. J. M. Jenkins, *Chem. Ber.*, **1970,** *6,* 401.
556. K. A. Jensen et al., *Acta Chem. Scand. Ser. B.*, **1974,** *28,* 97–101.
557. K. A. Jensen, S. Burmester, and T. A. Bak, *Acta Chem. Scand.*, **1967,** *21,* 2792–2796.
558. K. A. Jensen, M. Due, and A. Hohm, *Acta Chem. Scand.*, **1965,** *19,* 438–442.
559. K. A. Jensen, A. Holm, and J. Wolff-Jensen, *Acta Chem. Scand.*, **1969,** *23,* 1567–1572.
560. K. A. Jensen and P. H. Nielsen, *Acta Chem. Scand.*, **1964,** *18,* 1–10.
561. K. A. Jensen and C. Pedersen, *Acta Chem. Scand.*, **1961,** *15,* 991–1002.
562. K. A. Jensen and C. Pedersen, *Adv. Heterocycl. Chem.*, **1964,** *3,* 263–284.
563. D. Jerchel and W. Edler, *Chem. Ber.*, **1955,** *88,* 1284–1295.
564. D. Jerchel and H. Fischer, *Ann.*, **1954,** *590,* 216–231.
565. S. Johne et al., *Pharmazie,* **1979,** 790–794.
566. K. E. Johnson et al., *J. Polym. Sci. Part B.*, **1966,** *4,* 997–999.
567. K. M. Johnston, G. H. Williams, and H. J. Williams, *J. Chem. Soc.* (*B*), **1966,** 1114–1118.
568. A. Jonas and H. v. Pechmann, *Ann.*, **1891,** *262,* 277–302.
569. D. W. Jones, *J. Chem. Soc., Perkin I,* **1976,** 1150–1155.
570. M. Julliard, G. Vernin, and J. Metzger, *Helv. Chim. Acta,* **1980,** *63,* 467–471.
571. P. K. Kadaba, *J. Org. Chem.*, **1976,** *41,* 1073–1075.
572. P. K. Kadaba and J. O. Edwards, *J. Org. Chem.*, **1961,** *26,* 2331–2335.
573. P. K. Kadaba, B. Stanovnik, and M. Tisler, *Tetrahedron Lett.*, **1974,** 3715–3718.
574. M. Kanaoka, *Chem. Pharm. Bull.*, **1958,** *6,* 382–386.
575. T. Kappe, W. Golser, and W. Stadlbauer, *Chem. Ber.*, **1978,** *111,* 2173–2178.
576. A. R. Katritzky, *Chem. Ind.*, **1965,** 331–335.
577. A. R. Katritzky and J. W. Mitchell, *J. Chem. Soc., Perkin I,* **1973,** 2624–2626.
578. J. C. Kauer and R. A. Carboni, *J. Am. Chem. Soc.*, **1967,** *89,* 2633–2637.
579. J. C. Kauer and W. A. Sheppard, *J. Org. Chem.*, **1967,** *32,* 3580–3592.
580. R. Kayama et al., *Bull. Chem. Soc. Jap.*, **1974,** *47,* 2825–2829.
581. M. Keating et al., *J. Chem. Soc., Perkin I,* **1972,** 1315–1316.
582. G. Keller et al., *Bull. Chim. Soc. Fr.*, **1975,** 1219–1222.
583. G. W. Kenner and R. J. Stedman, *J. Chem. Soc.*, **1952,** 2089–2094.
584. R. C. Kerber, *J. Org. Chem.*, **1972,** *37,* 1587–1592.
585. R. C. Kerber and P. J. Heffren, *J. Org. Chem.*, **1972,** *37,* 1592–1597.
586. T. J. King et al., *J. Chem. Soc., Perkin II,* **1979,** 1751–1757.

587. A. I. Kiprianov, V. Y. Pochinok, and A. I. Tolmachev, *Ukr. Khim. Zh.*, **1954,** *20,* 204–210; *Chem. Abstr.*, **1955,** *49,* 12834–12835.

588. W. Kirmse, *Chem. Ber.*, **1960,** *93,* 2353–2356.

589. W. Kirmse and H. Schütte, *Chem. Ber.*, **1968,** *101,* 1674–1688.

590. S. Kishimoto, S. Noguchi, and K. Masuda, *Chem. Pharm. Bull,* **1976,** *24,* 3001–3010.

591. D. Knittel et al., *Tetrahedron Lett.*, **1970,** 1459–1462.

592. D. Knittel, H. Hemetsberger, and H. Weidmann, *Monatsh. Chem.*, **1970,** *101,* 157–160.

593. L. Knorr and P. Duden, *Ber.*, **1893,** *26,* 103–108.

594. Y. Kobayashi, T. Kutsuma, and K. Morinaya, *Chem. Pharm. Bull.*, **1971,** *19,* 2106–2115.

595. J. Kobe, B. Stanovnik, and M. Tišler, *Monatsh. Chem.*, **1970,** *101,* 724–735.

595a. R. J. Kobylecki and A. McKillop, *Adv. Heterocycl. Chem.*, **1976,** *19,* 215–278.

596. M. Kočevar, B. Stanovnik, and M. Tišler, *J. Heterocycl. Chem.*, **1978,** *15,* 1175–1184.

597. K. H. Koch and E. Fahr, *Angew. Chem. Int. Ed.*, **1970,** *9,* 634–635.

598. H. Koga, M. Hirobe, and T. Okamoto, *Chem. Pharm. Bull.*, **1976,** *24,* 2267–2269.

599. H. Koga, M. Hirobe, and T. Okamoto, *Hukusokan Kagaku Toronkai Koen Yoshishu, 8th,* **1975,** 229–233; *Chem. Abstr.*, **1976,** *84,* 163800 j.

600. Y. D. Kondrashev, *Kristallografiya, Engl. Ed.*, **1962,** *6,* 413–418.

601. A. Könnecke, P. Lepom, and E. Lippmann, *Z. Chem.*, **1978,** *18,* 214–215.

602. A. Könnecke and E. Lippmann, *Z. Chem.*, **1977,** *17,* 261–262.

603. A. Könnecke, E. Lippmann, and E. Kleinpeter, *Tetrahedron Lett.*, **1976,** 533–536.

604. A. Könnecke, E. Lippmann, and E. Kleinpeter, *Z. Chem.*, **1975,** *10,* 402.

605. B. Koren, B. Stanovnik, and M. Tišler, *J. Heterocycl. Chem.*, **1977,** *14,* 621–625.

606. A. Kovačič, B. Stanovnik, and M. Tišler, *J. Heterocycl. Chem.*, **1968,** *5,* 351–354.

607. S. Kozima et al., *J. Organomet. Chem.*, **1972,** *44,* 117–126.

608. L. Krbechek and H. Takimoto, *J. Org. Chem.*, **1964,** *29,* 1150–1152.

609. R. Kreher and H. Wissmann, *Chem. Ber.*, **1973,** *106,* 3097–3099.

610. M. Kröger and F. Cramer, *Ann.*, **1977,** 1668–1675.

611. F. Krollpfeiffer, H. Pötz, and A. Rosenberg, *Ber.*, **1938,** *71,* 596–603.

612. F. Krollpfeiffer, A. Rosenberg, and C. Mühlhausen, *Ann.*, **1935,** *515,* 113–130.

613. J. J. Krutak et al., *J. Org. Chem.*, **1979,** *44,* 3847–3858.

614. J. J. Krutak, private communication, April 1, 1981.

615. K. Kubo, T. Nonaka, and K. Odo, *Bull. Chem. Soc. Jap.*, **1976,** *49,* 1339–1341.

616. M. Kuhn and R. Mecke, *Z. Anal. Chem.*, **1961,** *181,* 487–493.

617. R. Kuhn and D. Jerchel, *Ann.*, **1952,** *578,* 1–5.

618. R. Kuhn and D. Jerchel, *Ber.*, **1941,** *74,* 941–948.

619. R. Kuhn and H. Kainer, *Angew. Chem.*, **1953,** *65,* 442–446.

620. R. Kuhn and W. Münzing, *Chem. Ber.*, **1953,** *86,* 858–862.

621. R. Kuhn and O. Westphal, *Ber.*, **1940,** *73,* 1109–1113.

622. M. A. Kuznetsov and B. V. Ioffe, *Zh. Obshch. Khim., Eng. Ed.*, **1977,** *47,* 1492–1497.

623. E. P. Kyba and D. C. Alexander, *Tetrahedron Lett.*, **1976,** 4563–4564.

624. G. L'abbé, *Angew. Chem. Int. Ed.*, **1975,** *14,* 775–830.

625. G. L'abbé, *Chem. Rev.*, **1969,** *69,* 345–363.

626. G. L'abbe et al., *J. Am. Chem. Soc.*, **1974,** *96,* 3973–3978.

627. G. L'abbe et al., *J. Heterocycl. Chem.*, **1977,** *14,* 1417–1418.

628. G. L'abbe et al., *J. Org. Chem.*, **1974,** *39,* 3770–3772.

629. *Ibid.*, **1976,** *41,* 1875–1876.
630. G. L'abbe and G. Verhelst, *Angew. Chem., Int. Ed. Engl.*, **1976,** *15,* 489–490.
631. G. L'abbe, G. Verhelst, and S. Toppet, *J. Org. Chem.*, **1977,** *42,* 1159–1163.
632. G. L'abbe and G. Vermeulen, *Bull. Soc. Chim. Belg.*, **1982,** *91,* 97–98.
633. M. Lacan, I. Tabakovic, and Z. Cekovic, *Tetrahedron,* **1974, 30,** 2911–2917.
634. A. Ladenberg, *Ber.*, **1892,** *25,* 2768–2775.
635. G. Landen and H. W. Moore, *Tetrahedron Lett.*, **1976,** 2513–2516.
636. G. A. Lanovaya and V. F. Mishchenko, *Zh. Org. Khim., Engl. Ed.*, **1979,** *15,* 1996–1999.
637. L. A. Lazukina and V. P. Kukhar, *Zh. Org. Khim., Engl. Ed.*, **1979,** *15,* 2009–2010.
638. R. J. LeBlanc and K. Vaughan, *Can. J. Chem.*, **1972,** *50,* 2544–2551.
639. R. J. W. LeFevre and T. H. Liddicott, *J. Chem. Soc.*, **1951,** 2743–2748.
640. J. B. Leach and J. H. Morris, *J. Organomet. Chem.*, **1968,** *13,* 313–321.
641. J. B. Leach, J. H. Morris, and P. G. Perkins, *J. Chem. Soc. (A)*, **1970,** 1077–1081.
642. L. A. Lee and J. W. Wheeler, *J. Org. Chem.*, **1972,** *37,* 348–351.
643. J. A. Leermakers, *J. Am. Chem. Soc.*, **1933,** *55,* 3098–3105.
644. J. E. Leffler and S. K. Liu, *J. Am. Chem. Soc.*, **1956,** *78,* 1949–1952.
645. J. E. Leffler and R. D. Temple, *J. Am. Chem. Soc.*, **1967,** *89,* 5235–5246.
646. J. E. Leffler and L. J. Todd, *Chem. Ind.*, **1961,** 512.
647. J. E. Leffler and Y. Tsuno, *J. Org. Chem.*, **1963,** *28,* 902–906.
648. D. M. Lemal, T. W. Rave, and S. D. Mc Gregor, *J. Am. Chem. Soc.*, **1963,** *85,* 1944–1948.
649. K. Lempert and K. Zauer, *Acta. Chim., Acad. Sci. Hung.*, **1976,** *88,* 81–85.
650. N. J. Leonard and K. Golankiewicz, *J. Org. Chem.*, **1969,** *34,* 359–365.
651. N. J. Leonard and A. M. Hyson, *J. Am. Chem. Soc.*, **1949,** 71, 1961–1964.
652. C. Leuenberger et al., *Helv. Chim. Acta,* **1977,** *60,* 831–843.
653. C. Leuenberger, L. Hoesch, and A. S. Dreiding, *Helv. Chim. Acta,* **1982,** *65,* 217–228.
654. C. Leuenberger, L. Hoesch, and A. S. Dreiding, *J. Chem. Soc., Chem. Commun.*, **1980,** 1197–1198; L. Hoesch et al., *Helv. Chim. Acta,* **1982,** *65,* 682–696.
655. E. Lieber et al., *Chem. Ind.*, **1958,** 893–894.
656. E. Lieber et al., *Inorg. Synth.*, **1960,** *VI,* 42–46.
657. E. Lieber et al., *J. Am. Chem. Soc.*, **1951,** *73,* 2327–2329.
658. E. Lieber et al., *J. Org. Chem.*, **1957,** *22,* 1750–1751.
659. E. Lieber, T. S. Chao, and C. N. R. Rao, *J. Org. Chem.*, **1957,** *22,* 238–240.
660. *Ibid.*, 654–662.
661. E. Lieber, T. S. Chao, and C. N. R. Rao, *Org. Synth.*, **1963,** *Coll. Vol. IV,* pp. 380–382.
662. E. Lieber and T. Enkoji, *J. Org. Chem.*, **1961,** *26,* 4472–4479.
663. E. Lieber, C. B. Lawyer, and J. P. Trivedi, *J. Org. Chem.*, **1961,** 26, 1644–1646.
664. E. Lieber and D. R. Levering, *J. Am. Chem. Soc.*, **1951,** *73,* 1313–1317.
665. E. Lieber and R. L. Minnis, Jr., *Chem. Rev.*, **1965,** *65,* 377–384.
666. E. Lieber, E. Oftedahl, and C. N. R. Rao, *J. Org. Chem.*, **1963,** 28, 194–199.
667. E. Lieber, C. N. Pillai, and R. D. Hites, *Can. J. Chem.*, **1957,** *35,* 832–842.
668. E. Lieber and J. Ramachandran, *Can. J. Chem.*, **1959,** *37,* 101–109.
669. E. Lieber, C. N. R. Rao, and T. S. Chao, *J. Am. Chem. Soc.*, **1957,** *79,* 5962–5967.
670. E. Lieber, C. N. R. Rao, and R. C. Orlowski, *Can. J. Chem.*, **1963,** *41,* 926–931.
671. E. Lieber and G. B. L. Smith, *Chem. Rev.*, **1939,** *25,* 213–271.
672. H. Lind and H. Kristinsson, *Synthesis,* **1974,** 198–199.

673. J. M. Lindley et al., *J. Chem. Soc., Perkin I,* **1980,** 982–994.
674. R. O. Lindsay and F. F. H. Allen, *Org. Synth.,* **1955,** *Coll. Vol. 3,* 710–711.
675. K. H. Linke and H. J. Göhaussen, *Chem. Ber.,* **1971,** *104,* 301–306.
676. E. Lippmann, *Z. Chem.,* **1974,** *14,* 16–17.
677. E. Lippmann and A. Koennecke, *Z. Chem.,* **1975,** *15,* 226–227.
678. E. Lippmann, D. Reifegerste, and E. Kleinpeter, *Z. Chem.,* **1973,** *13,* 134–135.
679. E. Lippmann, R. Widera, and E. Kleinpeter, *Z. Chem.,* **1973,** *13,* 429–430.
680. H. L. Lochte, W. A. Noyes, and J. R. Bailey, *J. Am. Chem. Soc.,* **1922,** *44,* 2556–2567.
681. A. L. Logothetis, *J. Am. Chem. Soc.,* **1965,** *87,* 749–754.
682. R. A. Long, J. F. Gerster, and L. B. Townsend, *J. Heterocycl. Chem.,* **1970,** *7,* 863–869.
683. R. Longhi and R. S. Drago, *Inorg. Chem.,* **1963,** *2,* 85–88.
684. J. H. Looker and J. W. Carpenter, *Can. J. Chem.,* **1967,** 45, 1727–1734.
685. W. Lossen, *Ann.,* **1891,** *263,* 73–108.
686. H. W. Lucien, *J. Am. Chem. Soc.,* **1958,** *80,* 4458–4460.
687. L. Lunazzi et al., *J. Chem. Soc., Perkin II,* 1978, 686–691.
688. W. Lwowski, *Angew. Chem., Int. Ed. Engl.,* **1967,** *6,* 897–1012.
689. W. Lwowski, *J. Org. Chem.,* **1975,** *40,* 2608–2612.
690. W. Lwowski and T. W. Mattingly, Jr., *J. Am. Chem. Soc.,* **1965,** *87,* 1947–1958.
691. J. Lykkeberg and N. A. Klitgaard, *Acta. Chem. Scand.,* **1972,** *26,* 266–274.
692. *Ibid.,* 2687–2694.
693. B. M. Lynch and T.-L. Chan, *Can. J. Chem.,* **1963,** *41,* 274–277.
694. B. M. Lynch and Yuk-Yung Hung, *J. Heterocycl. Chem.,* **1965,** *2,* 218–219.
695. R. N. Macdonald et al., *J. Polym. Sci., Polym. Chem. Ed.,* **1974,** *12,* 663–668.
696. D. MacKay and D. D. McIntyre, *Can. J. Chem.,* **1982,** *60,* 990–999.
697. D. MacKay, D. D. McIntryre, and N. J. Taylor, *J. Org. Chem.,* **1982,** *47,* 532–535.
698. D. MacKay and W. A. Walters, *J. Chem. Soc. (C),* **1966,** 813–816.
699. S. M. Mackenzie and M. F. G. Stevens, *J. Chem. Soc. (C),* **1970,** 2298–2308.
700. E. Maguire (Crown City Plating Co.), U.S. Pat. 3,793,038, Feb. 19, 1974.
701. A. C. Mair and M. F. G. Stevens, *J. Chem. Soc. (C),* **1971,** 2317–2324.
702. S. P. Makarov et al., *J. Gen. Chem., USSR, Engl. Ed.,* **1968,** *38,* 685–690.
703. Y. Maki, K. Izuta, and M. Suzuki, *Tetrahedron Lett.,* **1972,** 1973–1976.
704. E. Manhart and K. v. Werner, *Synthesis,* **1978,** 705–706.
705. S. Marburg and P. A. Grieco, *Tetrahedron Lett.,* **1966,** 1305–1309.
706. P. Margaretha, S. Solar, and O. E. Polansky, *Angew. Chem., Int. Ed. Engl.,* **1971,** *10,* 412–413.
707. T. J. Maricich and V. L. Hoffman, *J. Am. Chem. Soc.,* **1974,** *96,* 7770–7781.
708. J. H. Markgraf et al., *J. Org. Chem.,* **1964,** *29,* 2629–2632.
709. J. H. Markgraf, W. T. Bachmann, and D. P. Hollis, *J. Org. Chem.,* **1965,** *30,* 3472–3474.
710. F. D. Marsh, *J. Org. Chem.,* **1972,** *37,* 2966–2969.
711. F. D. Marsh and M. E. Hermes, *J. Am. Chem. Soc.,* **1964,** *86,* 4506–4507.
712. *Ibid.,* **1965,** *87,* 1819–1820.
713. F. D. Marsh and H. E. Simmons, *J. Am. Chem. Soc.,* **1965,** *87,* 3329–3330.
714. D. Martin, *Chem. Ber.,* **1964,** *97,* 2689–2694.
715. D. Martin and A. Weise, *Chem. Ber.,* **1966,** *99,* 317–327.
716. A. Martvoň et al., *Collect. Czech. Chem. Commun.,* **1977,** *42,* 1557–1561.

717. K. Masuda, Y. Imashiro, and T. Kaneko, *Chem. Pharm. Bull.*, **1970,** *18,* 128–132.

718. K. Masuda, T. Kamiya, and K. Kashiwa, *Chem. Pharm. Bull.*, **1971,** *19,* 559–563.

719. A. Matsumoto, M. Yoshida, and O. Simamura, *Bull. Chem. Soc. Japan,* 1974, *47,* 1493–1495.

720. S. Mataka and J.-P. Anselme, *Chem. Lett.*, **1973,** 51–54.

721. M. A. Mathur, H. H. Sisler, and L. N. Morgenthaler, *Inorg. Chem.*, **1979,** *18,* 2350–2354.

722. K. Matsuda and L. T. Morin, *J. Org. Chem.*, **1961,** *26,* 3783–3787.

723. W. C. McCrone, D. Grabar, and E. Lieber, *Anal. Chem.*, **1951,** *23,* 543.

724. A. W. McGee, M. P. Servé, and P. G. Seybold, *J. Heterocycl. Chem.*, **1978,** 15, 1043–1045.

725. R. S. McDaniel and A. C. Oehlschlager, *Tetrahedron,* **1969,** *25,* 1381–91.

726. A. McKillop and R. J. Kobylecki, *J. Org. Chem.*, **1974,** *39,* 2710–2714.

727. J. M. McManus and R. M. Herbst, *J. Org. Chem.*, **1959,** *24,* 1462–1464.

728. *Ibid.*, 1643–1649.

729. J. S. Meek and J. S. Fowler, *J. Am. Chem. Soc.*, **1967,** *89,* 1967.

730. J. S. Meek and J. S. Fowler, *J. Org. Chem.*, **1968,** *33,* 985–991.

731. *Ibid.*, 3418–3421.

732. R. Meier, *Chem. Ber.*, **1953,** *86,* 1483–1492.

733. J. Meisenheimer, O. Senn, and P. Zimmermann, *Ber.*, **1927,** *60,* 1736–1748.

734. A. A. Mel'nikov et al., *Zh. Org. Khim., Engl. Ed.*, **1979,** *15,* 1677–1683.

735. G. Manecke and C.-S. Rühl, *Makromol. Chem.*, **1979,** *180,* 103–115.

736. A. Messmer and A. Gelleri, *Angew. Chem.*, **1965,** *77,* 171.

737. *Ibid., Int. Ed. Engl.*, **1967,** *6,* 261–262.

738. A. Michael, F. Luehn, and H. H. Higbee, *Am. Chem. J.*, **1898,** *20,* 377–395.

739. C. J. Michejda, S. R. Koepke, and D. H. Campbell, *J. Am. Chem. Soc.*, **1978,** *100,* 5978–5979.

740. J. S. Mihina and R. M. Herbst, *J. Org. Chem.*, **1950,** *15,* 1082–1092.

740a. T. W. Milligan and B. C. Minor, *J. Org. Chem.*, **1962,** *27,* 4663–4665.

741. H. Minato, M. Ibata, and H. Iwai, *Kogyo Kagaku Zasshi,* **1966,** *69,* 966–969; *Chem. Abstr.*, **1968,** *69,* 105656w.

742. H. Minato and M. Oku, *Bull. Chem. Soc. Jap.*, **1965,** *38,* 1529–1533.

743. H. Minato, M. Oku, and S. H.-P. Chan, *Bull. Chem. Soc. Jap.*, **1966,** *39,* 1049–1054.

744. S. Mineo, S. Kawamura, and K. Nakagawa, *Synth. Commun.*, **1976,** *6,* 69–74.

745. T. Mitsuhashi, Y. Osamura, and O. Simamura, *Tetrahedron Lett.*, **1965,** 2593–2596.

746. T. Mitsuhashi and O. Sinamura, *J. Chem. Soc.* (*B*), **1970,** 705–711.

747. D. Moderhack, *Ann.*, **1972,** *758,* 29–42.

748. D. Moderhack, *Chem. Ber.*, **1975,** *108,* 887–896.

749. E. Mohr, *Ber.*, **1898,** *31,* 2495–2496.

750. H. Moll and R. Vuille, *Chimia,* **1969,** *23,* 511–512.

751. H. Moll and R. Vuille, *Helv. Chim. Acta,* **1969,** *52,* 665–673.

752. J. A. Montgomery and H. J. Thomas, *J. Med. Chem.*, **1972,** *15,* 182–187.

753. H. W. Moore, *Acc. Chem. Res.*, **1979,** *12,* 125–132.

754. H. W. Moore, *Chem. Soc. Rev.*, **1973,** *2,* 415–455.

755. H. W. Moore et al., *J. Am. Chem. Soc.*, **1970,** *92,* 1675–1681.

756. H. W. Moore and D. S. Pearce, *Tetrahedron Lett.*, **1971,** 1621–1624.

757. H. W. Moore, H. R. Shelden, and D. F. Shellhamer, *J. Org. Chem.*, **1969,** *34,* 1999–2001.

758. H. W. Moore and W. Weyler, Jr., *J. Am. Chem. Soc.*, **1970,** *92,* 4132–4133.

759. H. W. Moore and D. S. Wilbur, *J. Am. Chem. Soc.*, **1978,** *100,* 6523–6525.

760. G. T. Morgan and L. P. Walls, *J. Chem. Soc.*, **1930,** 1502–1509.

761. R. M. Moriarty, J. Kliegman, and C. Shovlin, *J. Am. Chem. Soc.*, **1967,** *89,* 5958–5959.

762. R. M. Moriarty and P. Serridge, *J. Am. Chem. Soc.*, **1971,** *93,* 1534–1535.

763. J. H. Morris and P. G. Perkins, *J. Chem. Soc. (A)*, **1966,** 580–582.

764. W. L. Mosby and M. L. Silva, *J. Chem. Soc.*, **1965,** 1003–1012.

765. A. V. Moskvin et al., *Zh. Org. Chem., Engl. Ed.*, **1978,** *14,* 2248–2251.

766. E. Muckermann, *Ber.*, **1909,** *42,* 3449–3460.

767. E. Muckermann, *J. Prakt. Chem.*, **1911,** *84,* 278–292.

768. E. Müller and H. Haiss, *Chem. Ber.*, **1962,** *95,* 1255–1263.

769. E. Müller and W. Rundel, *Chem. Ber.*, **1956,** *89,* 1065–1071.

770. J. Müller and K. Dehnicke, *J. Organomet. Chem.*, **1968,** *12,* 37–47.

771. U. Müller and K. Dehnicke, *Angew. Chem., Int. Ed. Engl.*, **1966,** *5,* 841–842.

772. J. Munch-Petersen, *Org. Syn.*, **1963,** *Coll. Vol. IV.*, 715–717.

773. D. B. Murphy and J. P. Picard, *J. Org. Chem.*, **1954,** *19,* 1807–1814.

774. A. W. Murray and K. Vaughan, *Chem. Comm.*, **1967,** 1282–1283.

775. A. W. Murray and K. Vaughan, *J. Chem. Soc. (C)*, **1970,** 2070–2074.

776. A. Mustafa et al., *J. Am. Chem. Soc.*, **1956,** *78,* 145–149.

777. K. A. Muszkat, L. Hill, and B. Kirson, *Isr. J. Chem.*, **1963,** *1,* 27–31.

778. F. Muzik and Z. J. Allan, *Chem. List.*, **1952,** *46,* 487–488.

779. *Ibid.*, **1958,** *52,* 486–493.

780. K. Nagarajan and P. Rajagopalan, *Tetrahedron Lett.*, **1966,** 5525–5530.

781. H. K. Nagy, A. J. Tomson, and J. P. Horwitz, *J. Am. Chem. Soc.*, **1960,** *82,* 1609–1613.

782. R. Neidlein and J. Tauber, *Arch. Pharm. (Weinheim)*, **1971,** *304,* 687–695.

783. S. F. Nelsen et al., *J. Am. Chem. Soc.*, **1972,** *94,* 1610–1614.

784. S. F. Nelsen and R. Fibiger, *J. Am. Chem. Soc.*, **1972,** *94,* 8497–8501.

785. S. F. Nelsen and D. H. Heath, *J. Am. Chem. Soc.*, **1969,** *91,* 6452–6456.

786. S. F. Nelsen and R. T. Landis II, *J. Am. Chem. Soc.*, **1973,** *95,* 8707–8713.

787. S. F. Nelsen, R. T. Landis II, and J. C. Calabrese, *J. Org. Chem.*, **1977,** *42,* 4192–4197.

788. V. M. Neplyuev and P. S. Pel'kis, *Zh. Org. Khim., Engl. Ed.*, **1974,** 10, 1740–1743.

789. V. M. Neplyuev, *Zhr. Org. Khim., Engl. Ed.*, **1979,** *15,* 500–503.

790. E. P. Nesynov and M. M. Besprozvannaya, *Ukr. Khim. Zh., Engl. Ed.*, **1974,** 40, 67–69.

791. F. A. Neugebauer, *Angew. Chem., Int. Ed. Engl.*, **1973,** *12,* 455–464.

792. F. A. Neugebauer, *Tetrahedron,* **1970,** *26,* 4843–4851.

793. F. A. Neugebauer and H. Fischer, *Chem. Ber.*, **1973,** *106,* 1589–1594.

794. F. A. Neugebauer and G. A. Russell, *J. Org. Chem.*, **1968,** *33,* 2744–2746.

795. F. A. Neugebauer and H. Trischmann, *Ann.*, **1967,** *706,* 107–111.

796. H. Neunhoeffer, H.-D. Vötter, and H. Ohl, *Chem. Ber.*, **1972,** *105,* 3695–3703.

797. P. N. Neuman, *J. Heterocycl. Chem.*, **1970,** *7,* 1159–1160.

798. *Ibid.*, **1971,** *8,* 51–56.

799. H. Neunhoeffer, *Chemistry of 1,2,3-Triazines and 1,2,4-Triazines, Tetrazines and Pentazines,* Wiley, New York, 1979, p. 7.

800. H. Neunhoeffer, H.-D. Vötter, and M. Gais-Mutterer, *Tetrahedron Lett.*, **1973,** 219–222.

801. M. S. Newman, S. H. Lee, Jr., and A. B. Garrett, *J. Am. Chem. Soc.*, **1947,** *69,* 113–116.

802. E. Niecke and R. Kröher, *Angew. Chem., Int. Ed. Engl.*, **1976,** *15,* 692–693.

803. E. Niecke and R. Kröher, *Z. Naturforsch,* **1979,** *34B,* 837–842.

804. E. Niecke and H.-G. Schäfer, *Angew. Chem., Int. Ed. Engl.*, **1977,** *16,* 783.

805. S. v. Niementowski, *Ber.*, **1898,** *31,* 314–323.

806. R. Nietzki and N. Prinz, *Ber.*, **1893,** *26,* 2956–2960.

807. A. W. Nineham, *Chem. Rev.*, **1955,** *55,* 356–483.

808. V. L. Nirenburg and I. Y. Postovskii, *Khim. Geterotsikl. Soedin., Akad. Nauk Latv. SSR,* **1965,** 946–947; *Chem. Abstr.*, **1966,** *64,* 14184c.

809. E. Noelting, E. Grandmougin, and O. Michel, *Ber.*, **1892,** *25,* 3328–3342.

810. Y. Nomura et al., *Bull. Chem. Soc. Jap.*, **1981,** *54,* 261-266.

811. *Ibid.*, 2779–2785.

812. Y. Nomura et al., *Chem. Lett.*, **1979,** 187–190.

813. W. P. Norris, *J. Org. Chem.*, **1962,** *27,* 3248–3251.

814. W. P. Norris and R. A. Henry, *J. Org. Chem.*, **1964,** *29,* 650–660.

815. G. Norwitz, M. E. Everett, and M. T. Gurbarg, *Thermochim. Acta,* **1975,** *12,* 227–232.

816. W. Oberhummer, *Monatsh.*, **1933,** *63,* 285–300.

817. T. Oda, T. Maeshima, and K. Sugiyama, *J. Macromal. Sci. Chem.*, **1979,** *A13,* 715–725.

818. *Ibid.*, **1978,** *179,* 2331–2336.

819. M. S. O'Dell and B. deB. Darwent, *Can. J. Chem.*, **1970,** *48,* 1140–1147.

820. H. Oelschlaeger and U. Bremer, *Arch. Pharm.*, (*Weinheim*), **1971,** *304,* 717–722.

821. A. Ohsawa et al., *J. Chem. Soc., Chem. Commun.*, **1980,** 1182–1183.

822. *Ibid.*, **1981,** 1174.

823. M. Okawara, H. Yamazaki, and E. Imoto, *J. Chem. Soc. Jap., Ind. Chem. Sect.*, **1955,** *58,* 991-994; *Chem. Abstr.*, **1956,** *50,* 12878f.

824. G. A. Olah, S. C. Narang, and A. P. Fung, *J. Org. Chem.*, **1981,** *46,* 2706–2709.

825. E. Oliveri-Mandalá and T. Passalacqua, *Gazz. Chim. Ital.*, **1913,** *43, II,* 465–474.

826. W. D. Ollis and C. A. Ramsden, *Adv. Heterocyclic Chem.*, **1976,** *19,* 1–122.

827. C. E. Olsen and C. Pedersen, *Acta. Chem. Scand.*, **1973,** *27,* 2271–2278.

828. *Ibid.*, 2279–2286.

829. C. E. Olsen, *Acta Chem. Scand.*, **1973,** *27,* 2983–2988.

830. G. OmReddy et al., *Thermochim. Acta,* **1981,** *43,* 61–73.

831. T. A. Ondrus, E. E. Knaus, and C. S. Giam, *Can. J. Chem.*, **1979,** *57,* 2342–2349.

832. W. R. Orndorff and B. J. Ray, *Am. Chem. J.*, **1910,** *44,* 1–41.

833. T. Ottersen, *Acta. Chem. Scand.*, **1976,** *A30,* 351–357.

834. W. Otting, *Chem. Ber.*, **1956,** *89,* 2887–2896.

835. W. Otting and F. A. Neugebauer, *Chem. Ber.*, **1969,** *102,* 2520–2529.

836. K. I. Paciorek and R. Kratzer, *Inorg. Chem.*, **1964,** *3,* 594–595.

837. A. Padwa et al., *J. Org. Chem.*, **1978,** *43,* 66–72.

838. P. I. Paetzold, *Fortschr. Chem. Forsch.*, **1967,** *8,* 437–469.

839. P. I. Paetzold and P. P. Habereder, *J. Organmet. Chem.*, **1967,** *7,* 61–70.

840. P. I. Paetzold, P. P. Harbereder, and R. Müllbauer, *J. Organomet. Chem.*, **1967,** *7,* 45–50.

841. P. I. Paetzold and G. Stohr, *Chem. Ber.*, **1968,** *101,* 2874–2880.

842. S. H. Pantinkin, J. P. Horwitz, and E. Lieber, *J. Am. Chem. Soc.*, **1955,** *77,* 562–567.

843. A. J. Papa, *J. Org. Chem.*, **1966,** *31,* 1426–1430.

844. V. Papesch and R. M. Dodson, *J. Org. Chem.*, **1963,** *28,* 1329–1331.

845. L. A. Paquette, private communication, Jan. 13, 1981.

846. L. A. Paquette and R. J. Haluska, *J. Am. Chem. Soc.*, **1972,** *94,* 534–538.

847. E. W. Parnell, *J. Chem. Soc.*, **1961,** 4930–4933.

848. T. M. Patterson, R. K. Smalley, and H. Suschitzky, *Synthesis,* **1975,** 187–189.

849. *Ibid.*, 709–710.

850. H. v. Pechmann, *Ber.*, **1888,** *21,* 2751–2762.

851. H. v. Pechmann and W. Bauer, *Ber.*, **1909,** *42,* 659–674.

852. H. v. Pechmann and L. Frobenius, *Ber.*, **1894,** *27,* 703–706.

853. *Ibid.*, 898–900.

854. *Ibid.*, **1895,** *28,* 170–176.

855. H. v. Pechmann and P. Runge, *Ber.*, **1894,** *27,* 2920–2930.

856. C. Pedersen, *Acta Chem. Scand.*, **1958,** *12,* 1236–1240.

857. *Ibid.*, **1959,** *13,* 888–892.

858. *Ibid.*, **1964,** *18,* 2199–2200.

859. N. P. Peet and S. Sunder, *J. Heterocycl. Chem.*, **1977,** *14,* 561–566.

860. A. Peratoner and G. Oddo, *Gazz. Chim. Ital.*, **1895,** *25,* II, 13–21.

861. D. F. Percival and R. M. Herbst, *J. Org. Chem.*, **1957,** *22,* 925–933.

862. M. J. Perkins, *J. Chem. Soc.*, **1964,** 3005–3008.

863. V. A. Petrow and E. L. Rewald, *J. Chem. Soc.*, **1945,** 313–315.

864. J. C. Pezzullo and E. R. Boyko, *J. Org. Chem.*, **1973,** *38,* 168–169.

865. R. Pfleger, E. Garthe, and K. Rauer, *Chem. Ber.*, **1963,** *96,* 1827–1839.

866. R. Pfleger and H. G. Hahn, *Chem. Ber.*, **1957,** *90,* 2411–2416.

867. K. Pilgram, F. Görgen, and G. Pollard, *J. Heterocycl. Chem.*, **1971,** *8,* 951–959.

868. A. Pinner, *Ber.*, **1889,** *22,* 1600–1612.

869. J. Pinnow and E. Koch, *Ber.*, **1897,** *30,* 2850–2860.

870. V. Pirc et al., *J. Heterocycl. Chem.*, **1970,** *7,* 639–645.

871. W. H. Pirkle and P. L. Gravel, *J. Org. Chem.*, **1978,** *43,* 808–815.

872. J. Plenkiewicz, *Pol. J. Chem.*, **1979,** *53,* 521–524.

873. J. Plenkiewicz, *Tetrahedron,* **1978,** *34,* 2961–2966.

874. J. Plenkiewicz, *Tetrahedron Lett.*, **1975,** 341–342.

875. *Ibid.*, **1978,** 399–400.

876. J. Plenkiewicz and E. Osuchowska, *Pol. J. Chem.*, **1978,** *52,* 1597–1600.

877. D. Pocar et al., *Gazz. Chim. Ital.*, **1981,** *111,* 325–332.

878. V. Y. Pochinok et al., *Russ. Chem. Rev.*, **1975,** 44, 481–492.

879. S. Pohl, E. Niecke, and H.-G. Schäfer, *Angew. Chem. Int. Ed. Engl.*, **1978,** *17,* 136.

880. S. Polanc et al., *J. Org. Chem.*, **1974,** *39,* 2143–2147.

881. G. Ponzio, *Gazz. Chim. Ital.*, **1900,** *30, II,* 459–465.

882. G. Ponzio, *J. Prakt. Chem.*, **1898,** *57,* 160–172.

883. I. Y. Postovskii and V. L. Nirenburg, *Zh. Obschch. Khim.*, **1964,** *34,* 2517–2521.

884. K. T. Potts and S. Husain, *J. Org. Chem.*, **1970,** *35,* 3451–3456.

885. *Ibid.*, **1972,** *37,* 2049–2050.

886. F. Povazanec, J. Kovac, and A. Krutosikova, *Collect. Czech. Chem. Commun.*, **1976,** *41,* 1692–1697.

887. G. Powell, *J. Am. Chem. Soc.*, **1929,** *51,* 2436–2439.

888. H. Prakash and H. H. Sisler, *Indian J. Chem. Sect. A,* **1980,** *19A,* 825–828.

889. B. Prentice, *Ann.*, **1896,** *292,* 272–295.

890. P. N. Preston, K. K. Tiwari, and K. Turnbull, *J. Chem. Soc., Chem. Commun.*, **1976,** 343–344.
891. P. N. Preston and K. Turnbull, *J. Chem. Soc., Perkin I,* **1977,** 1229–1233.
892. R. W. G. Preston, S. H. Tucker, and J. M. L. Cameron, *J. Chem. Soc.*, **1942,** 500–504.
893. D. N. Purohit, *Talanta,* **1967,** *14,* 353–359.
894. H. Quast and L. Bieber, *Angew. Chem., Int. Ed. Engl.*, **1975,** *14,* 428–429.
895. H. Quast and S. Hünig, *Ann.*, **1968,** *711,* 157–173.
896. A. Quilico and C. Musante, *Gazz. Chim. Ital.*, **1941,** *71,* 327–342.
897. M. S. Raasch, *J. Org. Chem.*, **1970,** *35,* 3470–3483.
898. *Ibid.*, **1978,** *43,* 2500–2508.
899. R. Raap, *Can. J. Chem.*, **1969,** *47,* 3677–3681.
900. C. A. Ramsden, *Tetrahedron,* **1977,** *33,* 3203–3232.
901. H. C. Ramsperger, *J. Am. Chem. Soc.*, **1929,** *51,* 2134–2143.
902. H. C. Ramsperger and J. A. Leersmakers, *J. Am. Chem. Soc.*, **1931,** *53,* 2061–2067.
903. U. Rapp, H. A. Staab, and C. Wünsche, *Tetrahedron,* **1971,** *27,* 2679–2689.
904. M. A. Ratcliff, Jr. and J. K. Kochi, *J. Org. Chem.*, **1972,** *37,* 3268–3274.
905. J. O. Reed and W. Lwowski, *J. Org. Chem.*, **1971,** *36,* 2864–2869.
906. C. W. Rees, D. L. R. Reeves, and R. C. Storr, *J. Chem. Soc., Perkin I,* **1976,** 2178–2182.
907. C. W. Rees and A. A. Sale, *J. Chem. Soc., Perkin I,* **1973,** 545–550.
908. C. W. Rees, R. W. Stephenson, and R. C. Storr, *J. Chem. Soc., Chem. Commun.*, **1974,** 941–942.
909. C. W. Rees and R. C. Storr, *J. Chem. Soc., (C)*, **1969,** 756–760.
910. *Ibid.*, 1474–1477.
911. C. W. Rees, R. C. Storr, and P. J. Whittle, *Tetrahedron Lett.*, **1976,** 4647–4650.
912. M. Regitz, *Chem. Ber.*, **1966,** *99,* 2918–2930.
913. M. Regitz, *Newer Methods of Preparative Organic Chemistry,* **1971,** *VI,* 81–126.
914. M. Regitz and H. J. Geelhaar, *Chem. Ber.*, **1969,** *102,* 1743–1754.
915. M. Regitz and H. Schwall, *Ann.*, **1969,** *728,* 99–107.
916. W. T. Reichle, *Inorg. Chem.*, **1964,** *3,* 402–406.
917. J. W. van Reijendam and F. Baardman, *Synthesis,* **1973,** 413–414.
918. H. Reimlinger, *Chem. Ber.*, **1970,** *103,* 1900–1907.
919. H. Reimlinger, *Chem. Ind.*, **1972,** 294–295.
920. H. Reimlinger, W. R. F. Lingier, and J. J. M. Vandewalle, *Chem. Ber.*, **1975,** *108,* 3780–3786.
921. M. Remes et al., *Chem. Prum.*, **1973,** *23,* 133–134.
922. G. A. Reynolds and J. A. Van Allan, *J. Org. Chem.*, **1959,** *24,* 1478–1486.
923. F. O. Rice and C. J. Grelecki, *J. Am. Chem. Soc.*, **1957,** *79,* 2679–2680.
924. F. O. Rice and T. A. Luckenbach, *J. Am. Chem. Soc.*, **1960,** *82,* 2681–2682.
925. C. D. Ritchie and D. J. Wright, *J. Am. Chem. Soc.*, **1971,** *93,* 2429–2432.
926. M. Robba, M. Cugon de Sévricourt, and A.-M. Godard, *J. Heterocycl. Chem.*, **1978,** *15,* 1387–1391.
927. R. O. Roblin, Jr., et al., *J. Am. Chem. Soc.*, **1945,** *67,* 290–294.
928. V. V. Rode et al., *Vysokomol. Soedin., Ser. A, Engl. Ed.*, **1969,** *11,* 935–944.
929. F. Roelants and A. Bruglants, *Tetrahedron,* **1978,** *34,* 2229–2232.
930. H. Roesky and O. Glemser, *Chem. Ber.*, **1964,** *97,* 1710–1712.
931. H. de Rogier, N. D. An, and A. Martinet, *Compt. Rend., Ser. C,* **1976,** 282, 967–969.

932. N. W. Rokke, J. J. Worman, and W. S. Wadsworth, *J. Heterocycl. Chem.*, **1975,** *12,* 1031–1033.

933. L. M. Romanov, B. A. Dolgoplosk, and B. L. Erusalimsky, *Proc. Acad. Sci., USSR, Phys. Chem. Sect., Engl. Ed.*, **1957,** *112,* 105–108.

934. C. S. Rondestvedt and P. K. Chang, *J. Am. Chem. Soc.*, **1955,** *77,* 6532–6540.

935. C. S. Rondestvedt and S. J. Davis, *J. Org. Chem.*, **1957,** *22,* 200–203.

936. K. Rüfenacht, *Helv. Chim. Acta,* **1975,** *58,* 1521–1527.

937. M. Rull and J. Vilarrasa, *J. Heterocycl. Chem.*, **1977,** *14,* 33–36.

938. M. Rull and J. Vilarrasa, *Tetrahedron Lett.*, **1976,** 4175–4176.

939. H. Rutner and P. E. Spoerri, *J. Heterocycl. Chem.*, **1966,** *3,* 435–439.

940. L. Ruzicka, M. W. Goldberg, and M. Hurbin, *Helv. Chim. Acta,* **1933,** *16,* 1335–1339.

941. S. V. Rykov et al., *Vestn. Musk. Univ. Khim., Engl. Ed.*, **1972,** *13,* 97–99.

942. F. Sachs and M. Steiner, *Ber.*, **1909,** *42,* 3674–3683.

943. K. Sakai and J.-P. Anselme, *J. Org. Chem.*, **1972,** *37,* 2351–2352.

944. T. Sakamoto, Y. Terao, and M. Sekiya, *Chem. Pharm. Bull.*, **1977,** *25,* 731–739.

945. T. Sasaki, K. Kanematsu, and M. Murata, *J. Org. Chem.*, **1971,** *36,* 446–449.

946. T. Sasaki, K. Kanematsu, and M. Murata, *Tetrahedron,* **1971,** *27,* 5121–5129.

947. M. J. Sasse and R. C. Storr, *J. Chem. Soc., Perkin I,* **1978,** 909–912.

948. W. H. Saunders, Jr. and J. C. Ware, *J. Am. Chem. Soc.*, **1958,** *80,* 3328–3332.

949. Y. M. Saunier et al., *Tetrahedron,* **1976,** *32,* 1995–1999.

950. G. N. Sausen, *J. Org. Chem.*, **1968,** *33,* 2336–2338.

951. R. Scarpati and M. L. Graziano, *J. Heterocycl. Chem.*, **1972,** *9,* 1087–1091.

952. R. Scarpati and D. Sica, *Gazz. Chim. Ital.*, **1963,** *93,* 942–948.

953. P. Scheiner, *J. Org. Chem.*, **1965,** *30,* 7–10.

954. *Ibid.*, **1967,** *32,* 2022–2023.

955. P. Scheiner, *Tetrahedron,* **1967,** *24,* 349–356.

956. *Ibid.*, **1968,** *24,* 2757–2766.

957. P. Scheiner, "Triazoline Decomposition," in B. S. Thyagaragan, Ed., *Selective Organic Transformations,* Wiley-Interscience, New York, 1970, pp. 327–362.

958. P. Scheiner et al., *J. Am. Chem. Soc.*, **1965,** *87,* 306–311.

959. G. O. Schenck and H. Formanek, *Angew. Chem.*, **1958,** *70,* 505.

960. C. Schiele, *Ann.*, **1965,** *689,* 197–201.

961. C. Schiele, *Chimia,* **1967,** *21,* 413–420.

962. C. Schiele et al., *Z. Naturforsch. Part b,* **1966,** *21,* 400–409.

963. H. Schlesinger, *Angew. Chem.*, **1960,** *72,* 563.

964. A. Schmidt, *Chem. Ber.*, **1966,** *99,* 2976–2983.

965. *Ibid.*, **1967,** *100,* 3725–3735.

966. *Ibid.*, **1968,** *101,* 4015–4021.

967. *Ibid.*, **1970,** *103,* 3923–3927.

968. M. P. Schmidt and A. Hagenböcker, *Ber.*, **1921,** *54,* 2201–2207.

969. E. Schmitz, H. Fechner-Simon, and S. Schramm, *Ann.*, **1969,** *725,* 1–10.

970. S. W. Schneller and F. W. Clough, *Heterocycles,* **1975,** *3,* 135–138.

971. K. Schofield and R. S. Theobold, *J. Chem. Soc.*, **1949,** 2404–2408.

972. E. Schrader, *J. Prakt. Chem.*, **1917,** *95,* 312–326.

973. G. Schroeter, *Ber.*, **1909,** *42,* 2336–2339.

974. H. Schultheiss and E. Fluck, *Z. Naturforsch,* **1977,** *32b,* 257–264.

975. J. Schweng and E. Zbiral, *Ann.*, **1978,** 1089–1095.

976. F. L. Scott, R. N. Butler, and D. A. Cronin, *Angew. Chem., Int. Ed. Engl.*, **1965,** *4,* 950–951.

977. D. Seebach et al., *Helv. Chem. Acta,* **1978,** *61,* 1622–1647.

978. R. Selvarjan and J. H. Boyer, *J. Heterocycl. Chem.*, **1972,** *9,* 87–90.

979. M. P. Servé et al., *J. Heterocycl. Chem.*, **1975,** *12,* 811–812.

980. G. Seybold, U. Jersak, and R. Gompper, *Angew. Chem., Int. Ed. Engl.*, **1973,** *12,* 847–848.

981. M. A. E. Shaban and M. A. M. Nassr, *Bull. Chem. Soc. Jap.*, **1976,** *49,* 2606–2608.

982. A. S. Shawali, A. A. Fahmi, and N. F. Eweiss, *J. Heterocycl. Chem.*, **1979,** *16,* 123–128.

983. V. P. Shchipanov, *Zh. Org. Khim., Engl. Ed.*, **1966,** *2,* 347–350.

984. V. P. Shchipanov et al., *J. Org. Chem. USSR, Engl. Ed.*, **1965,** *1,* 2277–2283.

985. V. P. Shchipanov and I. Y. Postovskii, *Spektrosk., Tr. Sib. Soveshch., 4th,* **1965,** 144–146; *Chem. Abstr.*, **1971,** *74,* 41715h.

986. Y. F. Shealy et al., *J. Org. Chem.*, **1961,** *26,* 2396–2401.

987. Y. F. Shealy et al., *J. Pharm. Sci.*, **1968,** *57,* 83–86.

988. Y. F. Shealy and C. A. O'Dell, *J. Heterocycl. Chem.*, **1973,** *10,* 839–842.

988a. P. B. Shevlin, *J. Am. Chem. Soc.*, **1972,** *94,* 1379–1380.

989. P. B. Shevlin and S. Kammula, *J. Am. Chem. Soc.*, **1977,** *99,* 2627–2631.

990. M. S. S. Siddiqui and M. F. G. Stevens, *J. Chem. Soc., Perkin I,* **1974,** 2482–2486.

991. D. H. Sieh and C. J. Michejda, *J. Am. Chem. Soc.*, **1981,** *103,* 442–445.

992. D. H. Sieh, D. J. Wilbur, and C. J. Michejda, *J. Am. Chem. Soc.*, **1980,** *102,* 3883–3887.

993. H. Sieper, *Chem. Ber.*, **1967,** *100,* 1646–1654.

994. H. Sieper and P. Tavs, *Ann.*, **1967,** *704,* 161–165.

995. K. Sisido et al., *J. Organomet. Chem.*, **1971,** *33,* 337–346.

996. H. H. Sisler, R. M. Kren, and K. Utvary, *Inorg. Chem.*, **1969,** *8,* 2007–2008.

997. L. I. Skripnik et al., *Khim. Geterotsikl. Soedin., Engl. Ed.*, **1980,** 721–724.

998. R. K. Smalley, R. H. Smith, and H. Suschitzky, *Tetrahedron Lett.*, **1978,** 2309–2312.

999. R.K. Smalley and H. Suschitzky, *Tetrahedron Lett.*, **1966,** 3465–3469.

1000. D. Smith and P. J. Taylor, *J. Chem. Soc., Perkin II,* **1979,** 1376–1386; Suppl. 22552, pp. 1–17.

1001. G. B. L. Smith, *Inorg. Synth.*, **1939,** *I,* 81–84.

1002. J. R. L. Smith and J. S. Sadd, *J. Chem. Soc., Perkin I,* **1975,** 1181–1184.

1003. P. A. S. Smith, *Open Chain Nitrogen Compounds,* Vol. 2, W. A. Benjamin, New York, **1966,** pp. 211–268.

1004. *Ibid.*, pp. 333–343.

1005. P. A. S. Smith, *Organic Reactions,* **1946,** *III,* 337–449.

1006. P. A. S. Smith et al., *J. Org. Chem.*, **1970,** *35,* 2215–2221.

1007. P. A. S. Smith and J. H. Boyer, *Org. Synth.*, **1963,** *Coll. Vol. IV,* 75–78.

1008. P. A. S. Smith and B. B. Brown, *J. Am. Chem. Soc.*, **1951,** *73,* 2438–2441.

1009. P. A. S. Smith, J. M. Clegg, and J. H. Hall, *J. Org. Chem.*, **1958,** *23,* 524–529.

1010. P. A. S. Smith and H. Dounchis, *J. Org. Chem.*, **1973,** *38,* 2958–2963.

1011. P. A. S. Smith and J. H. Hall, *J. Am. Chem. Soc.*, **1962,** *84,* 480–485.

1012. P. A. S. Smith, J. H. Hall, and R. O. Kan, *J. Am. Chem. Soc.*, **1962,** *84,* 485–489.

1013. P. A. S. Smith and D. H. Kenny, *J. Org. Chem.*, **1961,** *26,* 5221–5224.

1014. P. A. S. Smith, L. O. Krbechek, and W. Resemann, *J. Am. Chem. Soc.*, **1964,** *86,* 2025–2033.

1015. P. A. S. Smith and E. Leon, *J. Am. Chem. Soc.*, **1958,** *80,* 4647–4654.

1016. G. Smolinsky, *J. Org. Chem.*, **1961,** *26,* 4108–4110.

1017. *Ibid.*, **1962,** *27,* 3557–3559.

1018. G. Smolinsky and C. A. Pryde, *J. Org. Chem.*, **1968,** *33,* 2411–2416.

1019. M. S. Solanski and J. P. Trivedi, *J. Indian Chem. Soc.*, **1971,** *48,* 843–846.

1020. M. Somei et al., *Chem. Pharm. Bull.*, **1978,** *26,* 2522–2534.

1021. S. Sommer and U. Schubert, *Angew. Chem., Int. Ed. Engl.*, **1979,** *18,* 696–697.

1022. A. H. Sommers and J. D. Barnes (Abbott Laboratories), U.S. Pat. 2,769,819, Nov. 6, 1956.

1023. P. Spagnolo, A. Tundo, and P. Zanirato, *J. Org. Chem.*, **1977,** *42,* 292–296.

1024. *Ibid.*, **1978,** *43,* 2508–2513.

1025. H. A. Staab, *Chem. Ber.*, **1956,** *89,* 1927–1940.

1026. B. Stanovnik, *J. Org. Chem.*, **1970,** *35,* 1138–1141.

1027. B. Stanovnik and M. Tišler, *J. Heterocycl. Chem.*, **1971,** *8,* 785–787.

1028. H. Staudinger and E. Hauser, *Helv. Chim. Acta,* **1921,** *4,* 861–886.

1029. J. Stawinski, T. Hozumi, and S. A. Narang, *Can. J. Chem.*, **1976,** *54,* 670–672.

1030. G. Stedman, *J. Chem. Soc.*, **1959,** 2943–2949.

1031. H. N. E. Stevens and M. F. G. Stevens, *J. Chem. Soc. (C)*, **1970,** 765–771.

1032. *Ibid.*, 2289–2298.

1033. M. A. Stevens, H. W. Smith, and G. B. Brown, *J. Am. Chem. Soc.*, **1960,** *82,* 3189–3192.

1034. M. F. G. Stevens, *Prog. Med. Chem.*, **1976,** *13,* 205–269.

1035. J. K. Stille and A. T. Chen, *Macromolecules,* **1972,** *5,* 377–384.

1036. J. K. Stille and L. D. Gotter, *Macromolecules,* **1969,** *2,* 465–468.

1037. *Ibid.*, 468–474.

1038. R. Stoermer, *Ber.*, **1909,** *42,* 3133–3134.

1039. E. Stöldt and R. Kreher, *Angew. Chem.*, **1978,** *90,* 215–216.

1040. R. Stollé, *Ber.*, **1924,** *57,* 1063–1065.

1041. *Ibid.*, **1926,** *59,* 1742–1747.

1042. *Ibid.*, **1929,** *62,* 1118–1126.

1043. R. Stollé, *J. Prakt. Chem.*, **1933,** *138,* 1–17.

1044. R. Stollé and G. Adam, *Ber.*, **1924,** *57,* 1656–1659.

1045. R. Stollé and F. Helworth, *Ber.*, **1914,** *47,* 1132–1141.

1046. R. Stollé and F. Henke-Stark, *J. Prakt. Chem.*, **1930,** *124,* 261–300.

1047. R. Stollé and A. Netz, *Ber.*, **1922,** *55,* 1297–1305.

1048. R. Stollé and H. Perrey, *Ber.*, **1930,** *63,* 670–677.

1049. O. Subba Rao and W. Lwowski, *J. Heterocycl. Chem.*, **1980,** *17,* 187–188.

1050. K. Sugiyama, T. Nakaya, and M. Imoto, *Bull. Chem. Soc. Jap.*, **1975,** *48,* 941–943.

1051. K. B. Sukumaran, C. S. Angadiyavar, and M. V. George, *Tetrahedron,* **1972,** *28,* 3987–3998.

1052. K. B. Sukumaran, S. Satish, and M. V. George, *Tetrahedron,* **1974,** *30,* 445–450.

1053. C. Süling, "Aromatische Triazene und höhere Azahomologe," in E. Müller, Ed., *Methoden der Organischen Chemie,* Vol. X/3, Georg. Thieme Verlag, Stuttgart, **1965,** pp. 695–743.

1054. R. J. Sundberg et al., *J. Org. Chem.*, **1972,** *37,* 719–724.

1055. W. Sundermeyer, *Angew. Chem.*, **1962,** *74,* 875.

1056. H. Suschitzky and R. E. Walrond, *J. Chem. Soc., Perkin I,* **1977,** 47–52.

1057. D. R. Sutherland and G. Tennant, *J. Chem. Soc., (C)*, **1971,** 706–713.

1058. *Ibid.*, 2156–2162.

1059. D. R. Sutherland, G. Tennant, and R. J. S. Vevers, *J. Chem. Soc., Perkin I,* **1973,** 943–949.

1060. J. W. Sutherland, *J. Phys. Chem.*, **1979,** *83,* 789–795.

1061. J. Světlik, J. Leško, and A. Martvon, *Monatsh. Chem.*, **1980,** *111,* 635–642.

1062. G. Szeimies and R. Huisgen, *Chem. Ber.*, **1966,** *99,* 491–503.

1063. G. Szeimies, K. Mannhardt, and W. Mickler, *Chem. Ber.*, **1977,** *110,* 2922–2938.

1064. G. Szeimies, U. Siefken, and R. Rinck, *Angew. Chem., Int. Ed. Engl.*, **1973,** *12,* 161–162.

1065. H. Taguchi, *Tetrahedron Lett.*, **1973,** 1137–1138.

1066. H. Takaku and M. Yoshida, *J. Org. Chem.*, **1981,** *46,* 589–593.

1067. A. Takamizawa and S. Hayashi, *Yakugaku Zasshi,* **1963,** *83,* 373–377; *Chem. Abstr.*, **1963,** *59,* 10022g.

1068. S. K. Talapatra, P. Chaudhuri, and B. Talapatra, *Heterocycles,* **1980,** *14,* 1279–1282.

1069. P. B. Talukdar et al., *Indian J. Chem.*, **1973,** *11,* 611–612.

1070. Y. Tanaka and S. I. Miller, *J. Org. Chem.*, **1973,** *38,* 2708–2712.

1071. C. A. Taylor and W. H. Rinkenbach, *J. Franklin Inst.*, **1923,** *196,* 551.

1072. E. C. Taylor and I. J. Turchi, *Chem. Rev.*, **1979,** *79,* 181–231.

1073. C. Temple, Jr. et al., *J. Org. Chem.*, **1966,** *31,* 935–938.

1074. C. Temple, Jr., C. L. Kussner, and J. A. Montgomery, *J. Org. Chem.*, **1967,** *32,* 2241–2245.

1075. C. Temple, Jr., R. L. McKee, and J. A. Montgomery, *J. Org. Chem.*, **1965,** *30,* 829–834.

1076. C. Temple, Jr. and J. A. Montgomery, *J. Org. Chem.*, **1965,** *30,* 826–829.

1077. G. Tennant and R. J. S. Vevers, *J. Chem. Soc., Chem. Commun.*, **1974,** 671–672.

1078. F. Texier and J. Bourgois, *J. Heterocycl. Chem.*, **1975,** *12,* 505–508.

1079. F. Texier and R. Carrie, *Bull. Soc. Chem. Fr.*, **1972,** 258–263.

1080. J. S. Thayer, *Organometal. Chem. Rev.*, **1966,** *1,* 157–178.

1081. W. Theilacker and E.-C. Fintelmann, *Chem. Ber.*, **1958,** *91,* 1597–1600.

1082. J. Thiele, *Ann.*, **1892,** *270,* 1–63.

1083. *Ibid.*, **1898,** *303,* 57–75.

1084. *Ibid.*, **1910,** *376,* 239–268.

1085. J. Thiele and J. T. Marais, *Ann.*, **1893,** *273,* 144–160.

1086. J. Thiele and K. Schleussner, *Ann.*, **1897,** *295,* 129–172.

1087. T. L. Thomas et al., *J. Med. Chem.*, **1970,** *13,* 196–203.

1088. W. E. Thun and W. R. McBride, *J. Org. Chem.*, **1969,** *34,* 2997–3002.

1089. W. E. Thun, D. W. Morris, and W. R. McBride, *J. Org. Chem.*, **1966,** *31,* 923–925.

1090. T. C. Thurber and L. B. Townsend, *J. Heterocycl. Chem.*, **1972,** *9,* 629–636.

1091. E. Tiede, *Ber.*, **1916,** *49,* 1742–1745.

1092. H. J. Timpe, *Adv. Heterocycl. Chem.*, **1974,** *17,* 213–253.

1093. H. J. Timpe, *Z. Chem.*, **1971,** *11,* 340–341.

1094. M. Tišler, *Synthesis,* **1973,** 123–136.

1095. W. Tochtermann, H. Gustmann, and C. Wolff, *Chem. Ber.*, **1978,** *111,* 566–573.

1096. D. Todd, *J. Am. Chem. Soc.*, **1949,** *71,* 1353–1355.

1097. W. M. Tolles, D. W. Moore, and W. E. Thun, *J. Am. Chem. Soc.*, **1966,** *88,* 3476–3479.

1098. N. H. Toubro and A. Holm, *J. Chem. Soc., Perkin I,* **1978,** 1440–1445.

1099. L. B. Townsend and T. C. Thurber, U.S. Pat. 4,092,305, May 30, 1978.

1100. M. Tremblay, *Can. J. Chem.*, **1965,** *43,* 1154–1157.

1101. O. Tsuge and H. Samura, *Chem. Lett.*, **1973,** 175–180.

1102. O. Tsuge and H. Samura, *Heterocycles,* **1974,** 27–32.

1103. O. Tsuge and H. Samura, *J. Heterocycl. Chem.*, **1971,** *8,* 707–710.

1104. O. Tsuge and H. Samura, *Org. Prep. Proc. Int.*, **1972,** *4,* 273–281.

1105. Y. Tsuno, *Mem. Inst. Sci. Ind. Res., Osaka Univ.*, **1958,** *15,* 183–189.

1106. R. C. Tuites, T. E. Whitsley, and L. M. Minsk (Eastman Kodak), U.S. Pat. 3,576,638, April 27, 1971.

1107. I. Ugi, *Adv. Heterocycl. Chem.*, **1964,** *3,* 373–383.

1108. I. Ugi, *Tetrahedron,* **1963,** *19,* 1801–1803.

1109. I. Ugi and R. Huisgen, *Chem. Ber.*, **1958,** *91,* 531–537.

1110. I. Ugi, H. Perlinger, and L. Behringer, *Chem. Ber.*, **1958,** *91,* 2324–2329.

1111. *Ibid.*, 2330–2336.

1112. *Ibid.*, **1959,** *92,* 1864–1866.

1113. F. Ullmann, *Ann.*, **1904,** *332,* 82–104.

1114. F. Ullmann and M. Fukui, *Ber.*, **1908,** *41,* 624–626.

1115. F. Ullmann and C. Gross, *Ber.*, **1910,** *43,* 2694–2704.

1116. F. Umland and C. Schleyerbach, *Angew. Chem.*, **1965,** *77,* 169.

1117. T. Urbanyi, M. Pagliery, and C. R. Rehm, *Anal. Chem.*, **1967,** *39,* 1656–1659.

1118. K. Utvary, *Inorg. Nucl. Chem. Lett.*, **1965,** *I,* 77–78.

1119. K. Utvary, *Monatsh.*, **1968,** *99,* 1483–1488.

1120. K. Utvary and H. H. Sisler, *Inorg. Chem.*, **1966,** *5,* 1835–1836.

1121. *Ibid.*, **1968,** *7,* 698–701.

1122. K. Utvary and G. Vitovee, *Monatsh.*, **1971,** *102,* 767–769.

1123. J. A. Van Allan et al., *J. Org. Chem.*, **1968,** *33,* 1100–1102.

1124. J.-M. Vandensavel, G. Smets, and G. L'Abbé, *J. Org. Chem.*, **1973,** *38,* 675–678.

1125. J. Vaughan and P. A. S. Smith, *J. Org. Chem.*, **1958,** *23,* 1909–1912.

1126. K. Vaughan and M. F. G. Stevens, *Chem. Soc. Rev.*, **1978,** *7,* 377–397.

1127. L. I. Vereschagin et al., *Zh. Org. Khim., Engl. Ed.*, **1979,** *15,* 1551–1557.

1128. G. Vernin and J. Metzger, *Synthesis,* **1978,** 921–924.

1129. L. Vignon and A. Simonet, *Bull. Soc. Chim. Fr., III,* **1905,** *33,* 655–659.

1130. L. Vignon and A. Simonet, *Compt. Rend.*, **1904,** *138,* 1104–1105.

1131. J. Vilarrasa, private communication, Aug. 31, 1981.

1132. A. Vogler, R. E. Wright, and H. Kunkely, *Angew. Chem., Int. Ed. Engl.*, **1980,** *19,* 717–718.

1133. A. Vollmar and A. Hassner, *J. Heterocycl. Chem.*, **1974,** *11,* 491–496.

1134. H. Vorbrüggen and K. Krolkiewicz, *Synthesis,* **1979,** 35.

1135. H. Voswinckel, *Ber.*, **1901,** *34,* 2349–2354.

1136. L. Wacker, *Ber.*, **1902,** *35,* 3920–3982.

1137. H. R. Waddington, G. F. Duffin, and J. D. Kendall (Ilford Ltd.), Brit. Pat. 785,334, Oct. 23, 1957.

1138. I. Wadsö, *Acta Chem. Scand.*, **1960,** *14,* 903–908.

1139. W. S. Wadsworth, Jr., *J. Org. Chem.*, **1969,** *34,* 2994–2997.

1140. G. Wagner and H. Gentzsch, *Arch. Pharm.* (*Weinheim*), **1968,** *301,* 923–933.

1141. P. Walker and W. A. Waters, *J. Chem. Soc.*, **1962,** 1632–1638.

1142. O. Wallach, *Ann.*, **1886,** *235,* 233–255.

1143. A. Walser, W. J. Zally, and R. I. Fryer, *J. Heterocycl. Chem.*, **1974,** *11,* 863–866.

1144. J. S. Watson and A. J. Waring, *Can. J. Chem.*, **1960,** *38,* 298–299.

1145. G. W. Watt and W. C. Fernelius, *Z. Anorg. Allgem. Chem.*, **1934,** *221,* 187–192.

1146. J. S. Webb and A. S. Tomcufcik (American Cyanamid), U.S. Pat. 2,714,110, July 26, 1955.

1147. E. Wedekind, *Ber.*, **1897,** *30,* 2993–2999.

1148. *Ibid.*, **1898,** *31,* 942–948.
1149. R. Weiss, R. K. Robins, and C. W. Noell, *J. Org. Chem.*, **1960,** *25,* 765–770.
1150. C. Wentrup, *Helv. Chim. Acta,* **1972,** *55,* 565–569.
1151. *Ibid.*, **1978,** *61,* 1755–1764.
1152. C. Wentrup, *Tetrahedron,* **1970,** *26,* 4969–4983.
1153. *Ibid.*, **1971,** *27,* 1281–1286.
1154. *Ibid.*, **1974,** *30,* 1301–1311.
1155. C. Wentrup and W. D. Crow, *Tetrahedron,* **1970,** *26,* 3965–3981.
1156. C. Wentrup, A. Damerius, and W. Reichen, *J. Org. Chem.*, **1978,** *43,* 2037–2041.
1157. C. Wentrup, C. Thétaz, and R. Gleiter, *Helv. Chem. Acta,* **1972,** *55,* 2633–2636.
1158. C. Wentrup and H. W. Winter, *J. Am. Chem. Soc.*, **1980,** *102,* 6159–6161.
1159. W. Weyler, Jr., D. S. Pearce, and H. W. Moore, *J. Am. Chem. Soc.*, **1973,** *95,* 2603–2610.
1160. E. Wiberg and H. Michaud, *Z. Naturforsch.*, **1954,** *9b,* 495–496.
1161. *Ibid.*, 496–497.
1162. *Ibid.*, 497.
1163. *Ibid.*, 497–499.
1164. *Ibid.*, 500.
1165. N. Wiberg et al., *Chem. Ber.*, **1979,** *112,* 2718–2729.
1166. *Ibid.*, **1980,** *113,* 2916–2927.
1167. N. Wiberg, H. Bayer, and H. Bachhuber, *Angew. Chem., Int. Ed. Engl.*, **1975,** *14,* 177–178.
1168. N. Wiberg, H. Bayer, and G. Ziegleder, *Z. Anorg. Allg. Chem.*, **1979,** *459,* 201–207.
1169. N. Wiberg and A. Gieren, *Angew. Chem., Int. Ed. Engl.*, **1962,** *1,* 664.
1170. N. Wiberg and W.-C. Joo, and H. Henke, *Inorg. Nucl. Chem. Lett.*, **1967,** *3,* 267–269.
1171. N. Wiberg and W. C. Joo, *Z. Naturforsch.*, **1966,** *21b,* 1234.
1172. N. Wiberg and H. J. Pracht, *Chem. Ber.*, **1972,** *105,* 1377–1387.
1173. *Ibid.*, 1392–1398.
1174. N. Wiberg and H. J. Pracht, *J. Organomet. Chem.*, **1972,** *40,* 289–299.
1175. N. Wiberg and W. Uhlenbrock, *Chem. Ber.*, **1972,** *105,* 63–66.
1176. N. Wiberg and G. Ziegleder, *Chem. Ber.*, **1978,** *111,* 2123–2129.
1177. O. Widman, *Ber.*, **1916,** *49,* 2778–2782.
1178. H. Wieland and H. Fressel, *Ann.*, **1912,** *392,* 133–151.
1179. H. Wieland and A. Süsser, *Ann.*, **1912,** *392,* 169–185.
1180. H. E. Wijers, L. Brandsma, and J. F. Arens, *Rec. Trav. Chim.*, **1967,** *86,* 670–672.
1181. P. F. Wiley, in H. Neunhoeffer and P. F. Wiley, Eds., *Chemistry of 1,2,3-Triazines, 1,2,4-Triazines, Tetrazines, and Pentazines,* Wiley, New York, 1979, pp. 1287–1295.
1182. R. H. Wiley, K. F. Hussung, and J. Moffat, *J. Org. Chem.*, **1956,** *21,* 190–192.
1183. R. H. Wiley and J. Moffat, *J. Org. Chem.*, **1957,** *22,* 995–996.
1184. F. G. Willey, *Angew. Chem., Int. Ed. Engl.*, **1964,** *3,* 138.
1185. C. Willgerodt, *Ber.*, **1892,** *25,* 2661–2664.
1186. C. Willgerodt and B. Hermann, *J. Prakt. Chem.*, **1890,** *42,* 126–133.
1187. M. M. Williams, W. S. McEwan, and R. A. Henry, *J. Phys. Chem.*, **1957,** *61,* 261–267.
1188. W. K. Wilmarth and N. Schwartz, *J. Am. Chem. Soc.*, **1955,** *77,* 4543–4551.
1189. E. Winkelmann, *Tetrahedron,* **1969,** *25,* 2427–2454.
1190. G. Wittig and H.-L. Dorsch, *Ann.*, **1968,** *711,* 46–54.
1191. G. Wittig and A. Krebs, *Chem. Ber.*, **1961,** *94,* 3260–3275.

1192. H. Wittmann et al., *Monats. Chem.*, **1969,** *100,* 602–609.
1193. A. Wohl and H. Schiff, *Ber.*, **1900,** *33,* 2741–2758.
1194. *Ibid.*, **1902,** *35,* 1900–1904.
1195. L. Wolff, *Ann.*, **1902,** *325,* 152–169.
1196. *Ibid.*, **1912,** *394,* 23–85.
1196a. R. B. Woodward and C. Wintner, *Tetrahedron Lett.*, **1969,** 2697–2700.
1197. V. P. Wystrach, *Heterocycl. Comp.*, **1967,** *8,* 105–161.
1198. P. Ykman, G. L'Abbé, and G. Smets, *J. Indian Chem. Soc.*, **1972,** *49,* 1245–1250.
1199. A. D. Yoffe, *Proc. Roy. Soc.*, **1951,** *A208,* 188–199.
1200. M. Yoshida, A. Matsumoto, and O. Simamura, *Bull. Chem. Soc. Jap.*, **1970,** *43,* 3587–3590.
1201. Y. Yukawa and Y. Tsuno, *J. Am. Chem. Soc.*, **1957,** *79,* 5530–5534.
1202. H. Zahn, B. Wollemann, and O. Waschka, *Hoppe-Seyler's Z. Physiol. Chem.*, **1954,** *294,* 100–107.
1202a. L. L. Zaika and M. M. Joullie, *J. Heterocycl. Chem.*, **1966,** *3,* 289–298.
1203. L. H. Zalkow and R. H. Hill, *Tetrahedron,* **1975,** *31,* 831–840.
1204. A. J. Zambito, E. Hendrzak, and R. D. Babson, *Nucleic Acid Chem., Part I,* **1979,** 3–12.
1205. E. Zbiral and J. Stroh, *Ann.*, **1969,** *725,* 29–36.
1206. K. Y. Zee-Cheng and C. C. Cheng, *J. Med. Chem.*, **1970,** *13,* 264–268.
1207. D. M. Zimmerman and R. A. Olofson, *Tetrahedron Lett.*, **1970,** 3453–3456.
1208. T. Zincke, *Ann.*, **1896,** *291,* 313–341.
1209. *Ibid.*, **1900,** *311,* 276–329.
1210. *Ibid.*, *313,* 251–298.
1211. *Ibid.*, **1909,** *370,* 297–314.
1212. T. Zincke and A. T. Lawson, *Ann.*, **1887,** *240,* 110–133.
1213. T. Zincke and P. Schwarz, *Ann.*, **1900,** *311,* 329–340.
1214. H. Zinner et al., *Ann.*, **1959,** *622,* 133–149.
1215. M. Zupan, B. Stanovnik, and M. Tišler, *Tetrahedron Lett.*, **1972,** 4179–4182.

3

PHYSICAL AND PHYSICAL-CHEMICAL PROPERTIES

The thermal responses of chemicals with three or more linked nitrogen atoms are one dimension of their existence. To expand the description of the high nitrogen compounds and to provide a basis for seeking structure-property relations, other physical attributes are discussed in this chapter.

Perhaps the most fundamental properties considered are molecular geometry, energy content, and degree of electrical asymmetry. Next are summarized properties of particular value for determination of structure—the absorption of ultraviolet, visible, and infrared radiation, absorption of radio-frequency radiation by nuclei in nitrogen NMR, and fragmentation by electron beams in mass spectrometry. Discussion then follows on acid-base behavior and, to the extent possible, the location of hydrogen atoms on nitrogen chains. Lastly, explosive properties are examined because they are an inherent characteristic of many compounds with linked nitrogens and indicate the probable degree of explosive hazard of the various high nitrogen classes.

3.1. MOLECULAR DIMENSIONS

Determination of atomic locations in molecules by measurement of bond distances and angles establishes structures, enables categorization of individual bonds and larger chemical groupings, allows geometric comparisons between compounds, and by extrapolation, facilitates estimates of energetics. Most measured molecular dimensions of high nitrogen compounds have been by X-ray crystallography, but some determinations have been made by gas electron diffraction and microwave spectra.

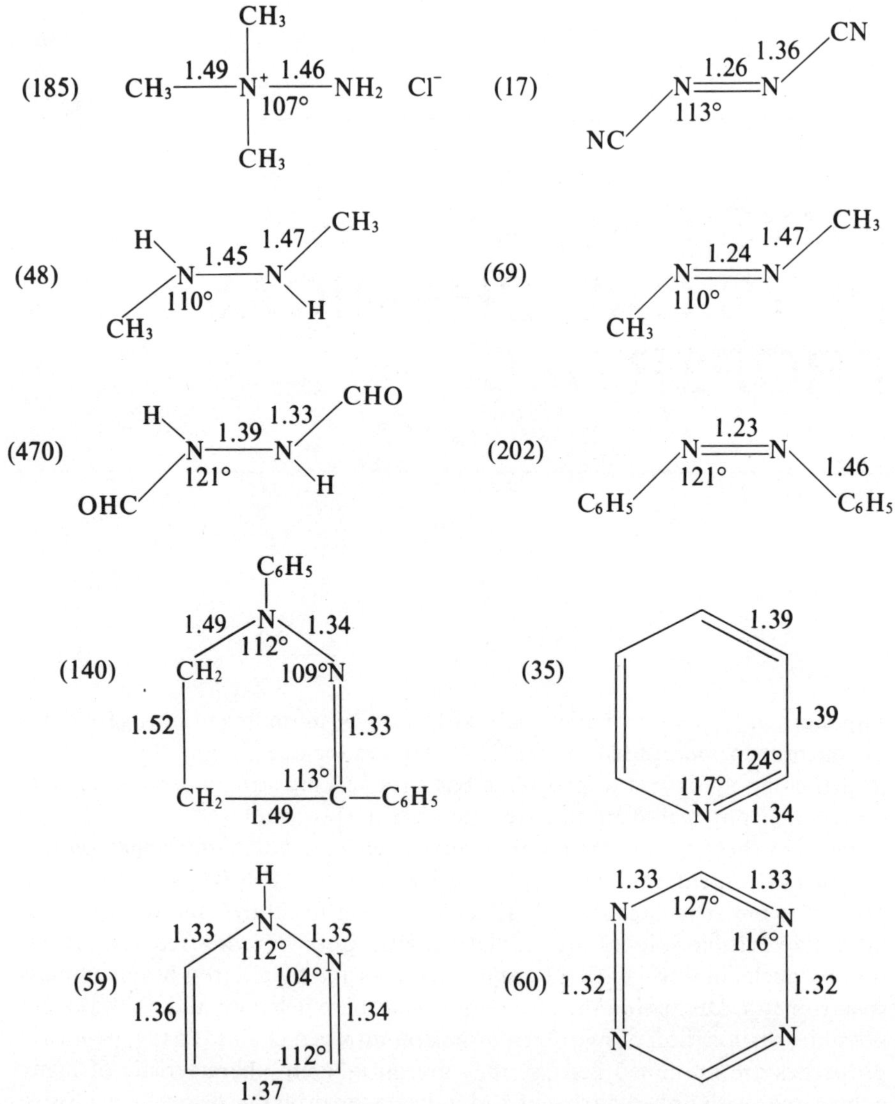

FIGURE 3.1. Measured dimensions of reference compounds.

For comparison with substances having three or more contiguous nitrogen atoms, bond distances and angles of pyridine and several compounds with two adjacent nitrogen atoms are listed (Figure 3.1).* In the acyclic reference com-

* Bond distances are in Ångstrom units and bond angles are in degrees. In both reference and high nitrogen compounds, bond distances are given to two decimal places so that values for different compounds will be compared at a similar level of accuracy. It is recognized that some bond distances have been measured with standard deviations as small as 0.001 Å, but this is not the case with the majority of compounds of present interest. Similarly, bond angles are reported to whole degrees.

pounds, nitrogen to nitrogen single bond distances vary from 1.46 Å in the trimethyl hydrazinium ion and 1.45 Å in 1,2-dimethylhydrazine (substituted by the electron-releasing methyl group) to 1.39 Å (substituted by the electron-withdrawing formyl group). Conversely, the cyano group slightly increases the azo bond length to 1.26 Å from 1.23 Å when substituted by phenyl. Linkages from phenyl and methyl to hydrazine and azo nitrogen are of normal length, but are shorter from cyano and formyl to nitrogen. In the partially reduced 1,3-diphenyl-2-pyrazoline, as well as the relatively aromatic pyrazole and *s*-tetrazine, all ring nitrogen-nitrogen bonds are shorter than any acyclic nitrogen-nitrogen single bonds, but are not as attenuated as acyclic nitrogen-nitrogen double bonds. The ring C—N distances in *s*-tetrazine and pyrazole do not deviate greatly from the 1.34 Å value in pyridine. Two different ring carbon-nitrogen bonds occur in 1,3-diphenyl-2-pyrazoline: a 1.49 Å single bond of usual length, and a 1.33 Å double bond, somewhat longer than an acyclic C═N of normal length, but comparable to that in pyrazole.

Molecular dimensions have been measured for several acyclic chains of three and four nitrogen atoms. The only triazane distances determined are for 2,2-dimethyltriazanium chloride, in which the N—N and C—N bonds are very close to those of 1,1,1-trimethylhydrazinium chloride (185). The NNC, CNC (111°), and NNN (110°) angles are all essentially tetrahedral.

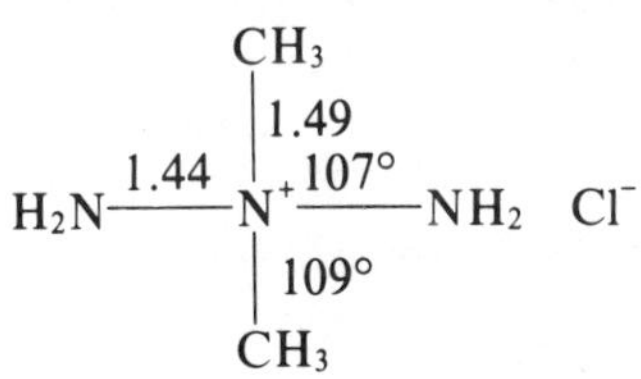

Introduction of a double bond into the three-nitrogen chain, as in the triazenes, results in shorter nitrogen-nitrogen single bond distances than in any of

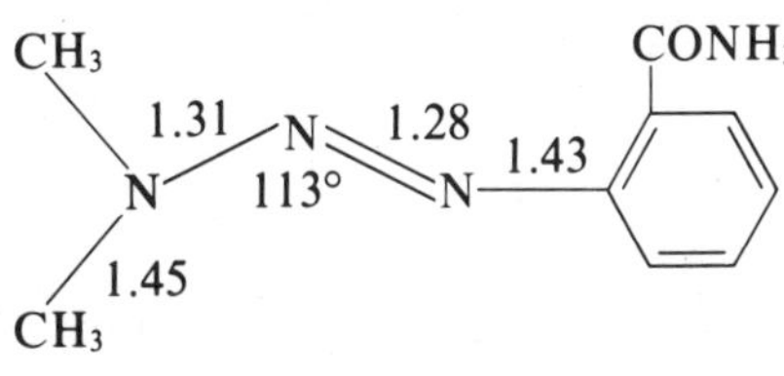

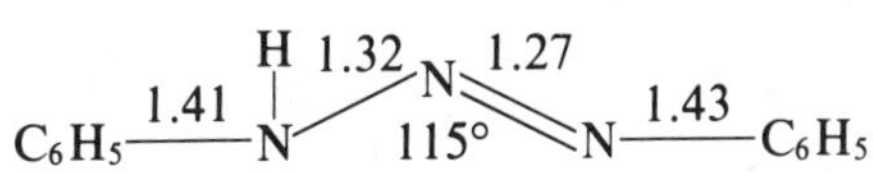

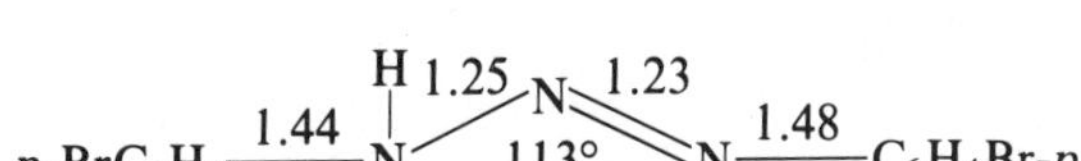

the acyclic reference compounds. Double bond lengths usually are slightly longer, indicating an interaction along the three-nitrogen chain. These altered bond lengths are shown by the measured values for 2-(3,3-dimethyl-1-triazeno)-phenyl-1-carboxamide (147), diazoaminobenzene (186), and *p,p′*-dibromodiazoaminobenzene (273). In the latter compound, both nitrogen-nitrogen bonds are at the azo bond length, and indeed, in each of these triazenes, the differences between the nominal single and double bonds are small. Carbon-nitrogen bonds to substituents in the 1,3-positions are slightly shorter than in related two-nitrogen compounds. Configurations of the substituents at the azo group of these triazenes in the solid state are trans. Hydrogen bonds appear to exist between the N-1 of one molecule and N-3 of an adjacent molecule in symmetrical 1,3-diaryltriazenes. In the unsymmetrical 1-*p*-nitrophenyl-3-phenyltriazene (274), the hydrogen is stabilized in the 1-position, closest to the nitrophenyl, and forms a hydrogen bond to an oxygen of a nitro group in a neighboring molecule. Nitro-

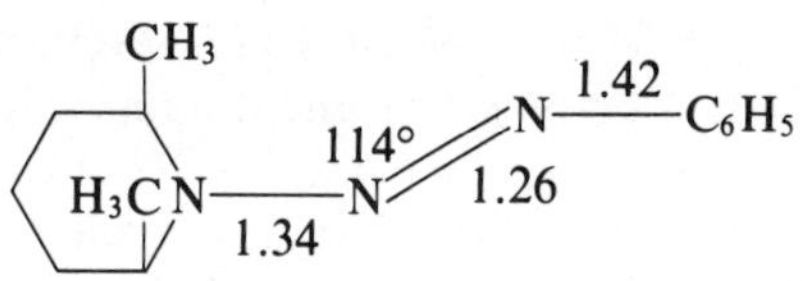

gen-nitrogen linkages are clearly differentiated into a single and a double bond. A sharp distinction also is found between the single and double nitrogen-nitrogen bond distances in the mixed acyclic-cyclic triazene, *N*-phenylazo-2,6-dimethylpiperidine (319).

In the 2-tetrazenes (362, 430), the azo bond length is normal and the single

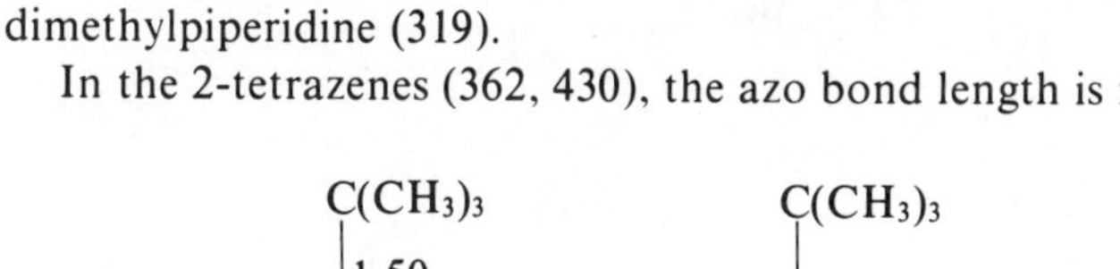

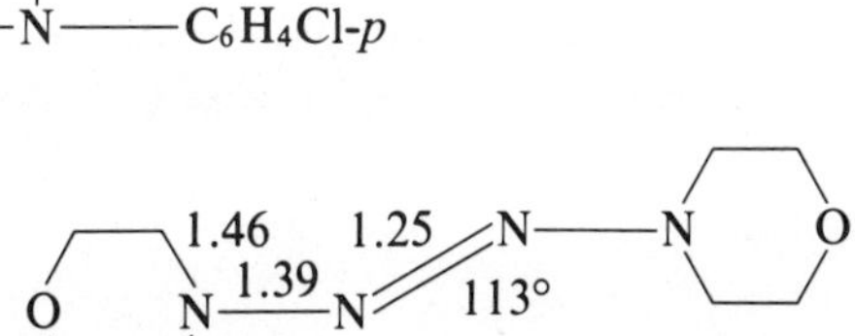

nitrogen-nitrogen bonds are similar to that of diformyl hydrazine. Carbon-nitrogen single bond distances fall in the normal range. Although the azo and

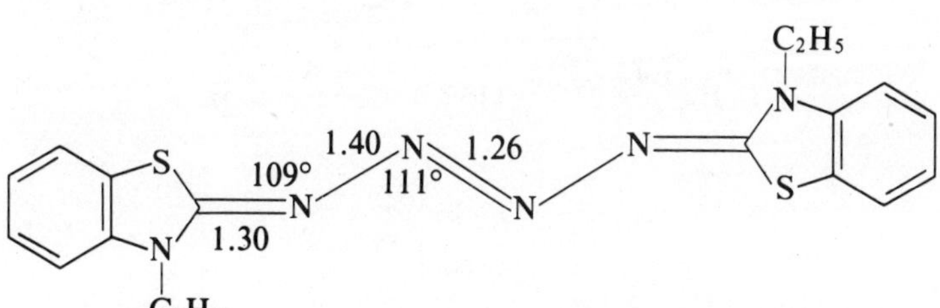

carbon-nitrogen double bond distances in 1,4-bis(*N*-ethyl-1,2-dihydrobenzothiazole-2-ylidene)-2-tetrazene (16), are nearly normal, the nitrogen-nitrogen single bond, shortened to the diformyl hydrazine length, implies some conjugation from the benzothiazole rings through the four-nitrogen, alternately unsaturated chain.

The six-nitrogen compound, 1,6-bis(*p*-chlorophenyl)-3,4-diacetyl-1,5-hexazdiene, consists of two planar halves connected through the two middle nitrogen atoms at a dihedral angle of 84.37° (320). Single nitrogen-nitrogen bond

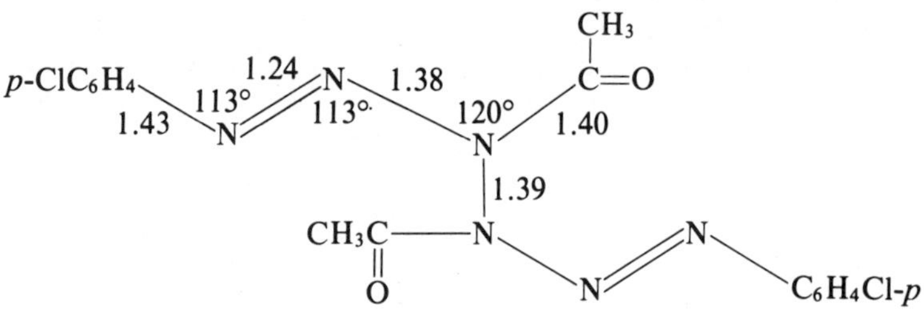

lengths are close to those of the tetrazenes and diformylhydrazine rather than as short as the distance in triazenes; the azo bond distance resembles that in azomethane and the tetrazenes. The C—N bond from acetyl to nitrogen is appreciably longer than other amide bonds, enabling the prediction of strong electrophilic character for the acetyl groups in this compound. (See Section 6.3, Reactions with Nucleophiles.)

Small nitrogen-nitrogen bond distances also are found in the nitrosohydrazine unit of 1-(4-chlorobenzyl)-1-nitroso-2-(imidazolidene-2-yl)-hydrazine mono-

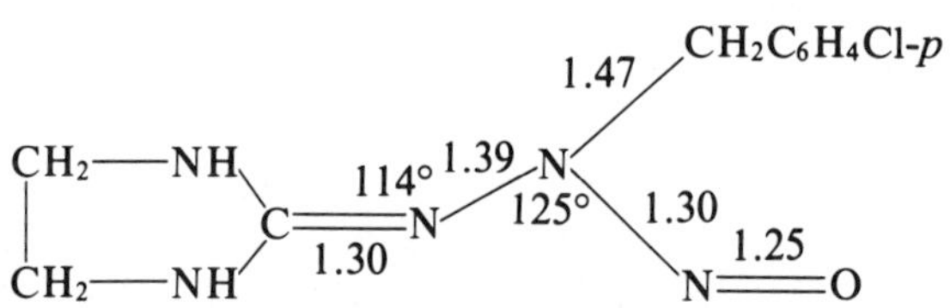

hydrate (378), reflecting the effects of unsaturation at either end of the three-nitrogen chain. Although the N=O bond distance is comparable to a bond order of about 1.5, the C=N link and the benzyl-nitrogen bond are substantially normal.

X-Ray crystal analyses confirm the open-chain structures of the amine ni-

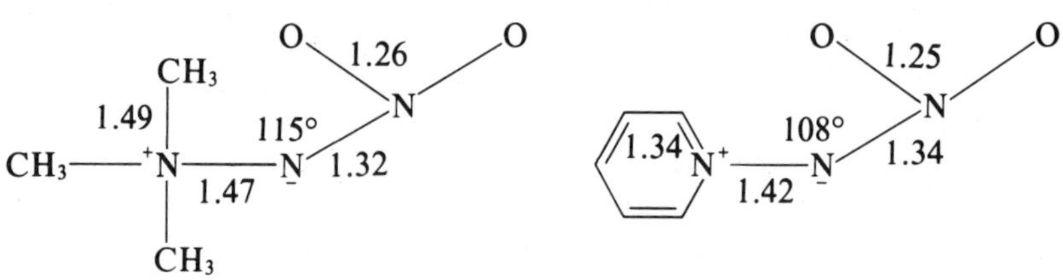

troimides and acyclic azimines. In trimethylamine nitroimide (93), the bond between the quaternary and imide nitrogens is not lessened from single bond length. Limited shortening of this bond in pyridine 1-nitroimide (25) indicates some conjugation with the pyridine ring. The bond between the imide and nitro nitrogens is essentially equal to the nitrogen-nitrogen bond in *s*-tetrazines and the N=O bonds of the nitro group also are the same length as the nitroso group in the nitrosohydrazine described previously. Accordingly, the negative charge, designated at the imide nitrogen, must be partially delocalized to the nitro oxygen atoms. Similar shortening occurs in the azimine moieties of azobenzene *N*-[4,5-bis(methoxycarbonyl)thiazol-2-yl]imide (76) and the four-nitrogen chain compound 2,3-dimethyl-1-phthalimide-azimine (347), with azimine nitrogen-

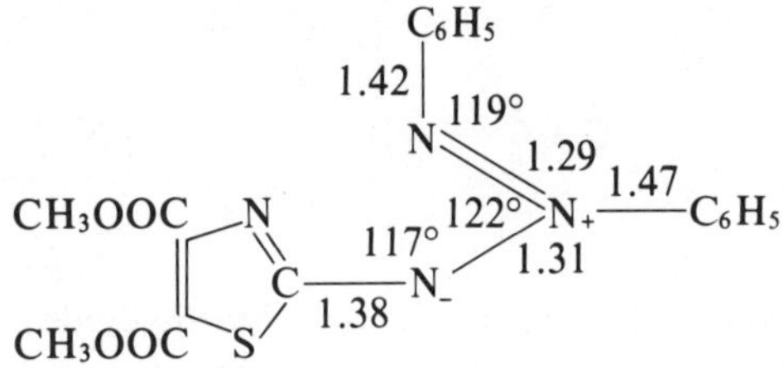

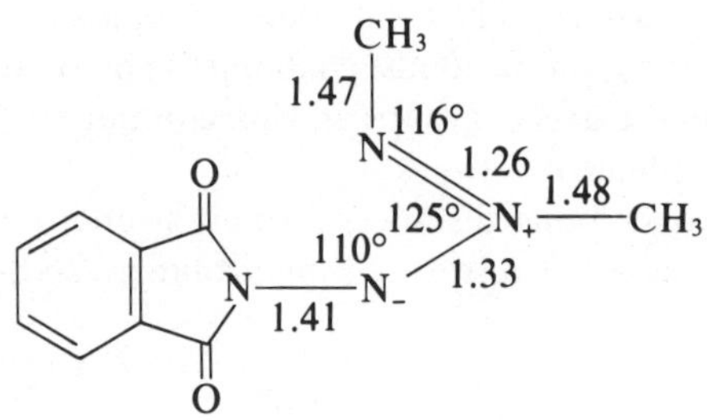

nitrogen bonds similar in length to the triazenes. Smaller single bond lengths of the exocyclic C—N bond in the thiazole azimine and the N—N bond to the phthalimide nitrogen in the phthalimide azimine indicate partial charge delocalization into these terminal rings. In the thiazole azimine, the shortened terminal bond from the end nitrogen to phenyl indicates conjugation between them, but there is no conjugation between the second nitrogen and its attached phenyl.

In the mixed acyclic-cyclic aminoazimine shown (486), the alternating nitrogen-nitrogen bond distances, as well as the shortened N—N and lengthened N=N, indicate conjugation. The considerable shortening of the bond between the end nitrogen of the aminoazimine chain and its triazoline ring carbon in comparison with the distance from the third nitrogen and its ring carbon suggests distribution of negative charge over the end nitrogen and adjacent carbonyl oxygen.

Of all the high nitrogen compounds, the shortest nitrogen-nitrogen bonds occur in the azides. Thus the terminal bond has a large triple bond character and approaches the elemental nitrogen distance of 1.10 Å. Indeed, a mean value

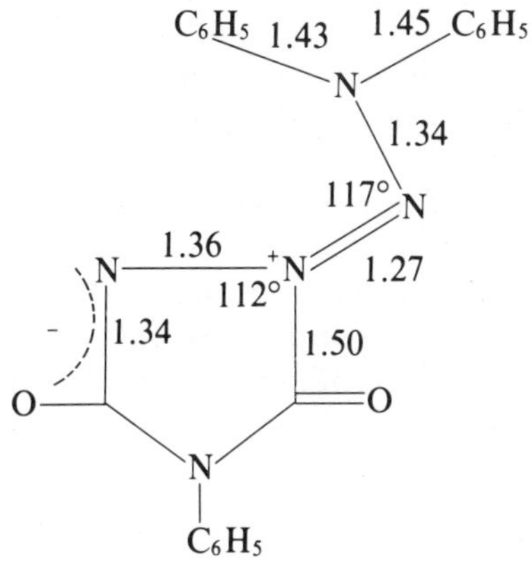

of 1.05 Å is found for the terminal nitrogen-nitrogen bond in the triazidocarbonium ion (353), compensated, however, by the relatively long inner nitrogen-nitrogen bond of 1.39 Å. Other inner nitrogen-nitrogen bonds, ranging from 1.24 to 1.27 Å, are very close to the reference azo distances. In methyl azide, the methyl-nitrogen distances has a normal C—N single bond length (315, 383). In contrast, the carbon-azide links to cyano (17), *p*-nitrophenyl (354), 4-phenyl-

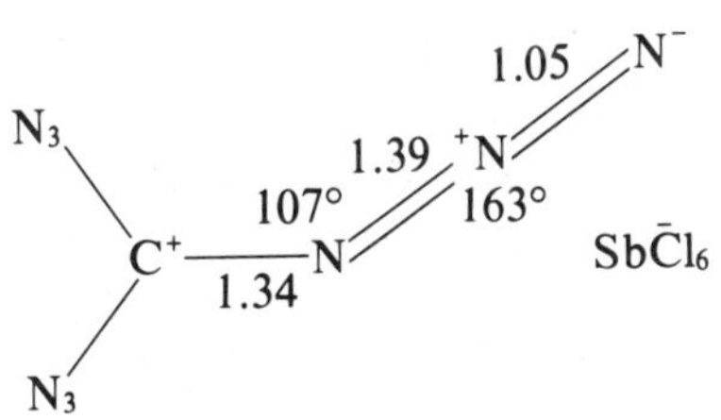

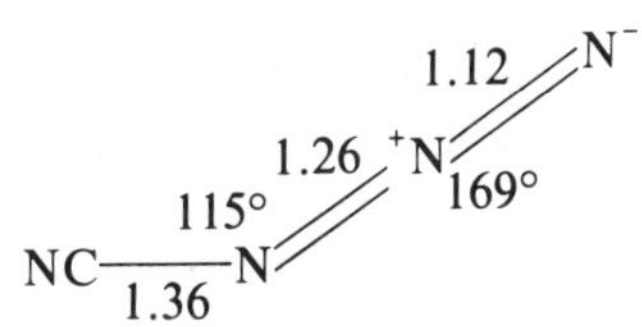

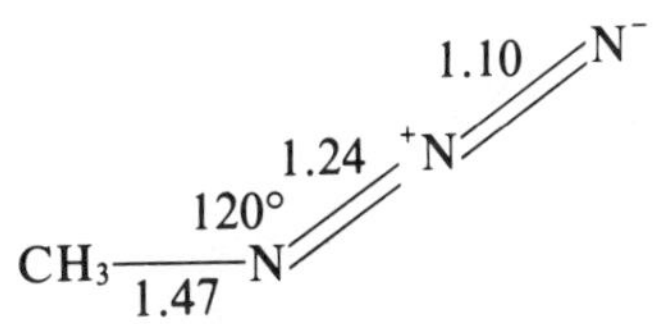

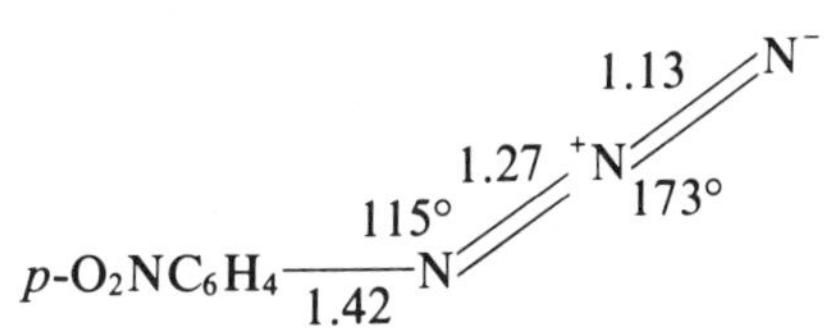

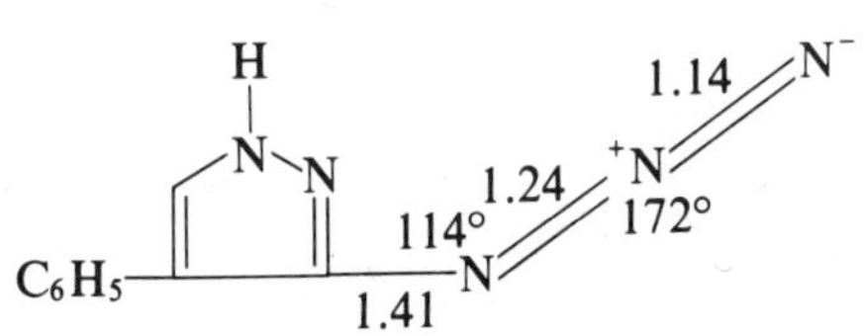

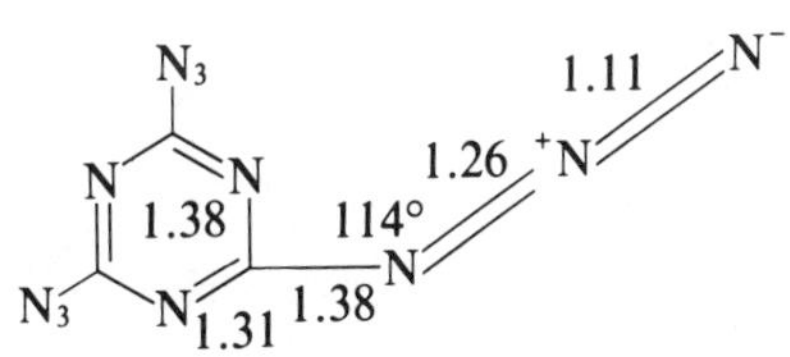

pyrazol-3-yl (133), diazidoguanidinium, and diazido triazinyl (270), all electron-withdrawing, are shortened, indicating double bond character and interaction between azide and the attached group. Although the CNN angle in the triazidocarbonium ion is only 107°, and that of methyl azide is 120°, CNN angles of the azides bound to electron-attracting groups are 114–115°. The azide ion is linear (316) and the presumption often is made that the NNN angle in hydrogen azide (427) and organic azides also is 180°. In Knaggs' classic demonstration in 1935 of the essential linearity of the azide group in cyanuric azide (270), however, deviation of three or four degrees from linearity was indicated as possible. Examination of organic azide angles measured subsequently shows these azide groups are not exactly linear, but have angles from 163° to 173°.

In reduced cyclic compounds with contiguous nitrogen atoms, the dimensions of partially unsaturated nitrogen chains are similar to those of open-chain analogs except for N—N distances. Thus the nitrogen-nitrogen double bonds of

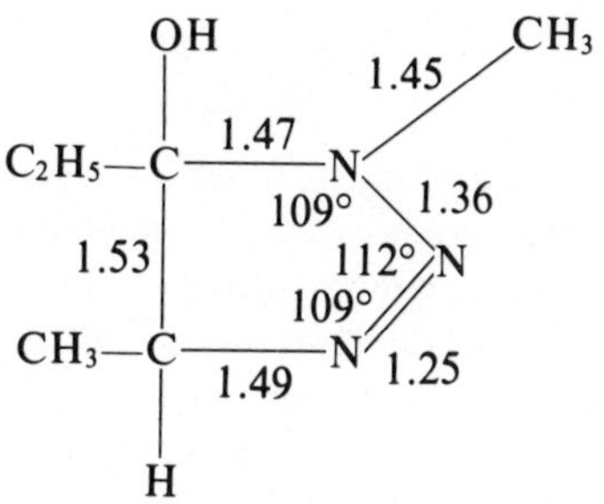

the triazolines are of normal length, but the single N—N bonds, ranging from 1.36 Å in 1,4-dimethyl-5-ethyl-5-hydroxy-Δ^2-1,2,3-triazoline (250) to 1.41 Å in the bistriazoline (allyl azide dimer) (388), are longer than the related distances in acyclic triazenes. The angles at N-2 of the triazolines (111° and 112°) correspond to the values of 111° to 115° found for open-chain triazenes. In the triazolines whose molecular dimensions have been measured, the three nitrogen

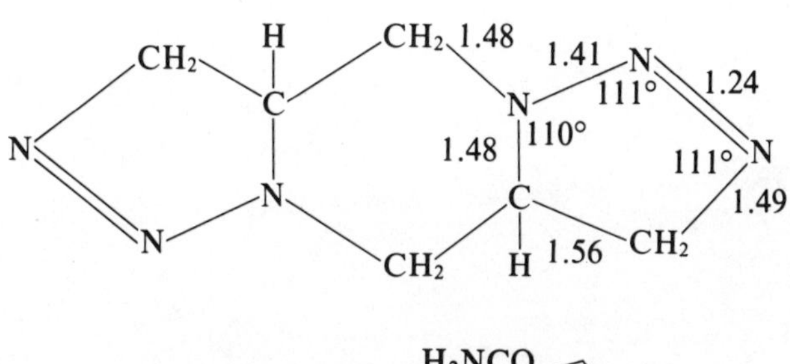

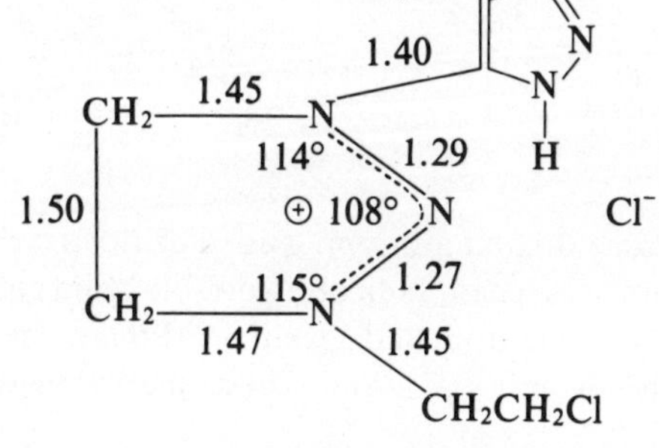

atoms and one ring carbon are planar but the remaining ring carbon is out of this plane (183, 250, 388).

The *v*-triazolinium salts, illustrated by 1-(2-chloroethyl)-3-(4-carbamoylpyrazol-3-yl)-Δ^2-1,2,3-triazolinium chloride (489), have nearly identical nitrogen-nitrogen distances, intermediate between *v*-triazoline nitrogen-nitrogen single and double bonds. The positive charge, accordingly, is delocalized over the three triazolinium nitrogen atoms. Carbon-nitrogen bonds in the ring and to the chloroethyl group are essentially of normal length, but the somewhat shortened bond to the aromatic pyrazole ring indicates interaction.

In the dimethyl ester of 3,5,6,7,8,8*a*-hexahydro-3-phenyl-1,2-benzotetrazinedicarboxylic acid, the 1- and 2-nitrogen atoms, bearing the methoxycarbonyl groups, are respectively above and below the plane of the tetrazine ring

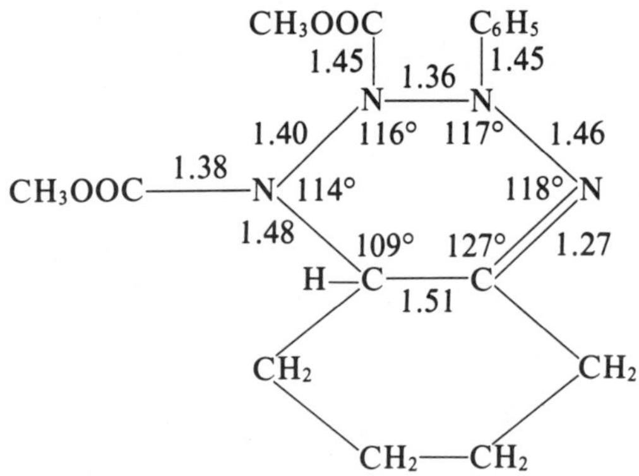

carbons and the 3- and 4-nitrogen atoms, forming a slightly distorted half chair conformation (445). The 1,2 N—N bond is almost the same length as in diformylhydrazine, but the 2,3 N—N bond is even shorter, being the same length as in pyrazole. The 3,4 N—N bond, however, is as long as that in trimethylhydrazinium chloride, and is adjacent to a very short carbon-nitrogen double bond. Conjugation of N-1 with methoxycarbonyl is indicated by the coplanarity of N-1 with the atoms to which it is attached, and the shorter exocyclic N—C bond to methoxycarbonyl than between methoxycarbonyl and N-2.

Single and double nitrogen-nitrogen bonds in 1,4-dimethyl-1,4,5,6-tetrahydro-*v*-tetrazine 2-*N*-oxide (430) are very similar to those in acyclic 2-tetra-

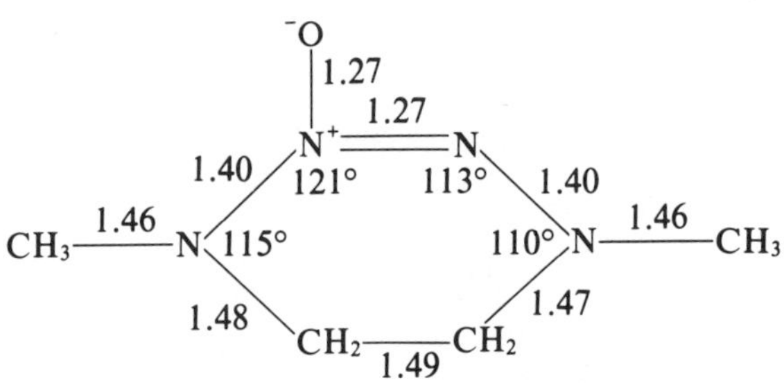

zenes. The *N*-oxide bond is essentially the same length as in other *N*-oxides and nitro and nitroso compounds, and the C—N bonds also are normal. Electronic repulsion in the four nitrogen chain is reflected in the molecule's twisted boat shape with nonplanar terminal nitrogen atoms.

Most five-membered, doubly unsaturated heterocycles with three or four sequential nitrogen atoms have the aromatic property of planarity, with shortened single and lengthened double bonds and are somewhat irregular pentagons. Nitrogen-nitrogen single bond distances in these rings usually are close to the 1.35 Å of pyrazole and the double bonds cluster around 1.32 Å, the nitrogen-nitrogen length in *s*-tetrazine.

In *v*-triazoles with a substituent or a hydrogen in the 1-position, there are alternate long and short ring bonds in the sequence long N—N and short N=N and long N—C. These characteristic bond distances are illustrated by 5-amino-4-carboxamido-*v*-triazole with ring hydrogen located at N-1 (253), 1-benzotriazolyl acetic acid (184), and 8-azaguanine monohydrate with its triazole ring hydrogen located at position 3 (441). Alternation also exists in 1-(α-benzoyl-

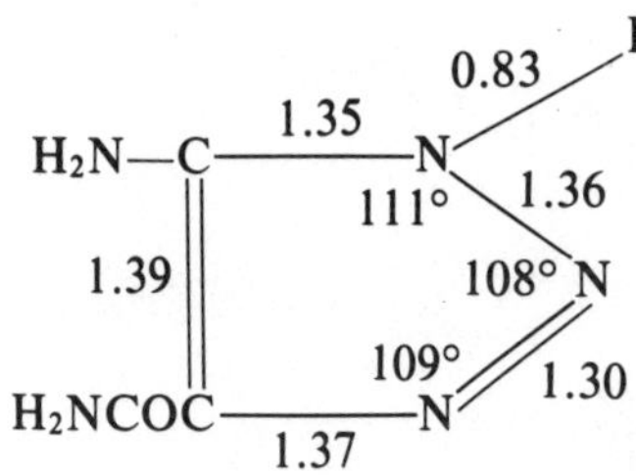

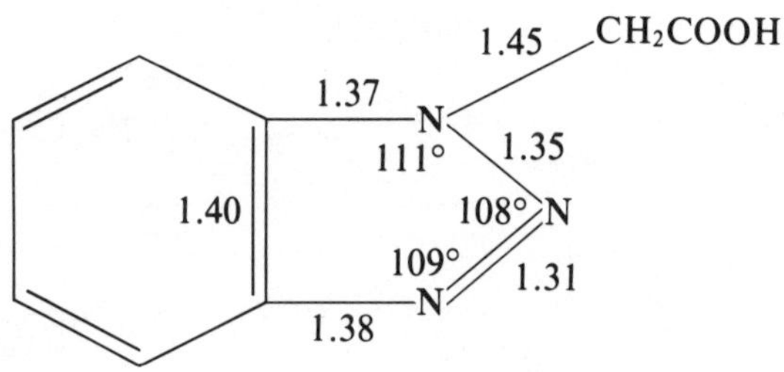

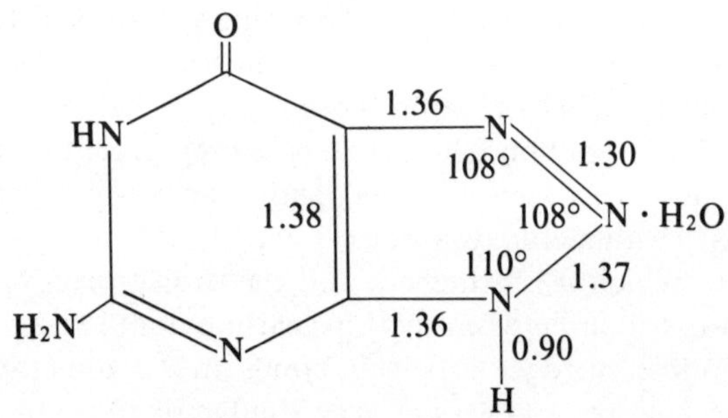

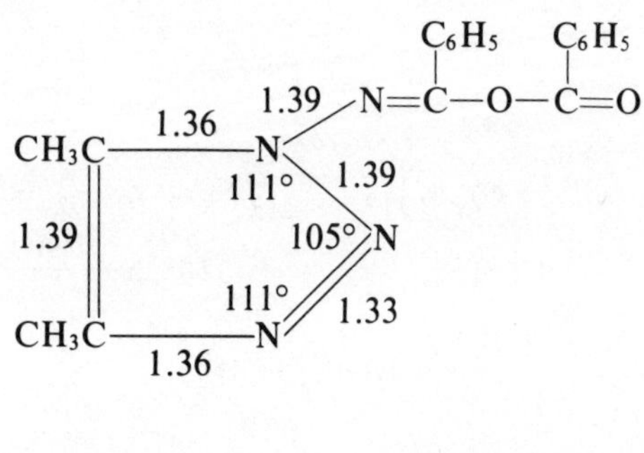

oxybenzylideneamino)-4,5-dimethyl-*v*-triazole with its four-nitrogen chain including an exocyclic nitrogen atom (47). The exocyclic N—N bond is essentially

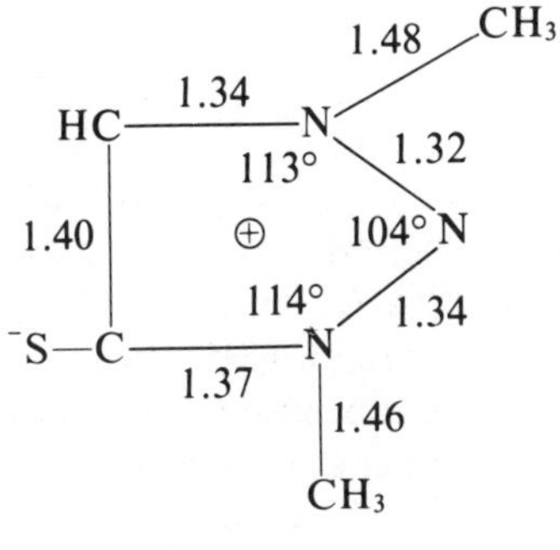

the same length as terminal N—N bonds in the open-chain 2-tetrazenes, implying some conjugation with the ring. Bond shortening, but not double bond fixation, is found in the mesoionic *v*-triazolium sulfides, represented by 1,3-dimethyl-*v*-triazolium-4-mercaptide (368). The 4,5 C—C bonds in these rings are substantially the same length as in benzene and pyridine (1.39 Å).

Formally, the N—N bonds of 2-substituted *v*-triazoles are equivalent, and

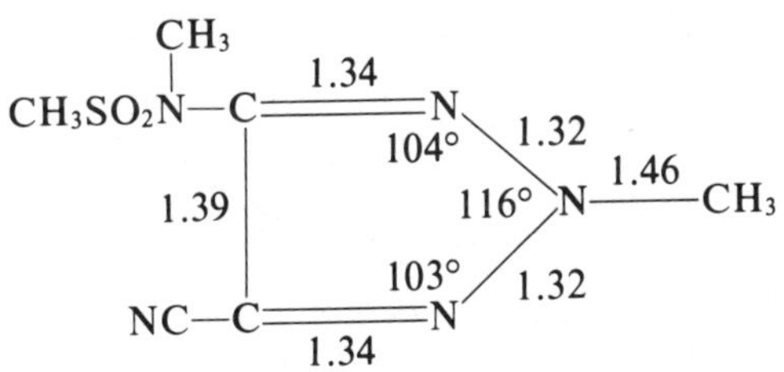

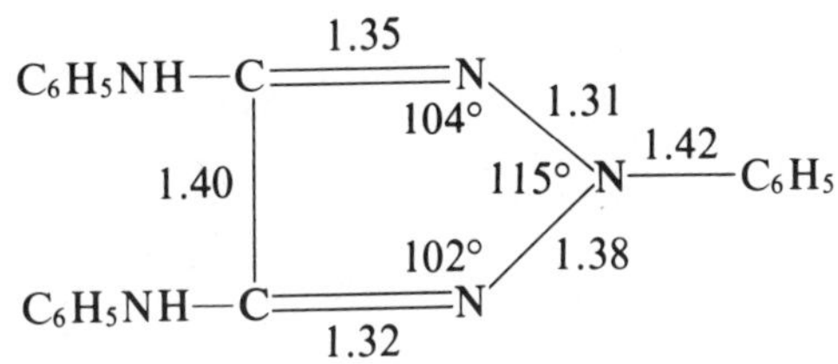

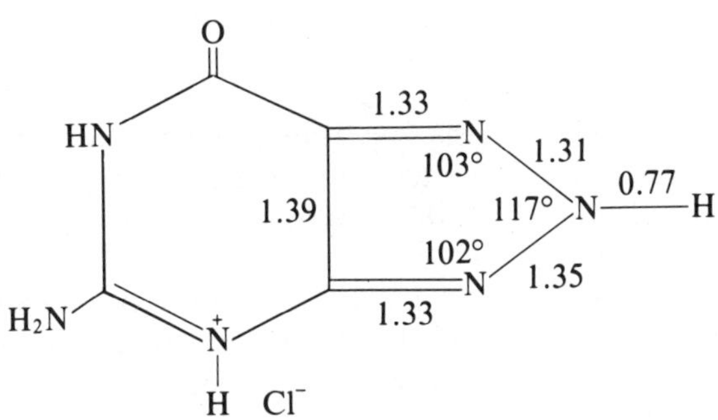

this is the case in a few instances such as 2-methyl-5-[methyl(mesyl)amino]-*v*-triazole-4-carbonitrile (254). In fact, N—N bonds of most 2-substituted *v*-triazoles or those with hydrogen at the 2-position are unequal. In 2-phenyl-4,5-dianilino-*v*-triazole the bond distortions seem to be the result of a different orientation of each anilino substituent (215) and in 8-azaguanine hydrochloride may be because of extensive intermolecular hydrogen bonding (278).

Bond lengths alternate in many tetrazoles, following the succession of a long N—N, short N═N, long N—N, and short N═C, illustrated by tetrazole (476), 1-(3-pyrazolyl)tetrazole (405), and tetrazolo[5,1-*i*]purine, shown by X-ray analysis to be tricyclic rather than an azidopurine (189). The hydrogen on nitrogen in crystalline tetrazole is at position 1, and is involved in hydrogen bonding to an adjacent molecule. In the pyrazolyl tetrazole, N—N and C═N are substantially the same in both the pyrazole and tetrazole rings, as well as in

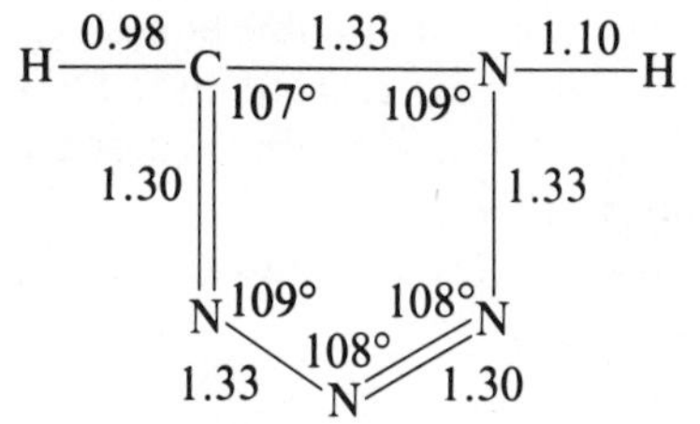

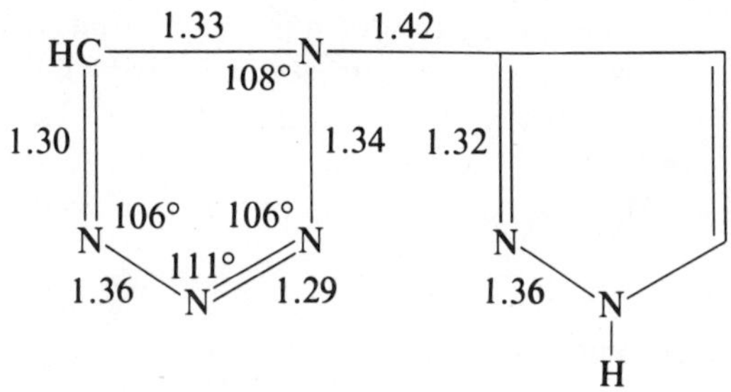

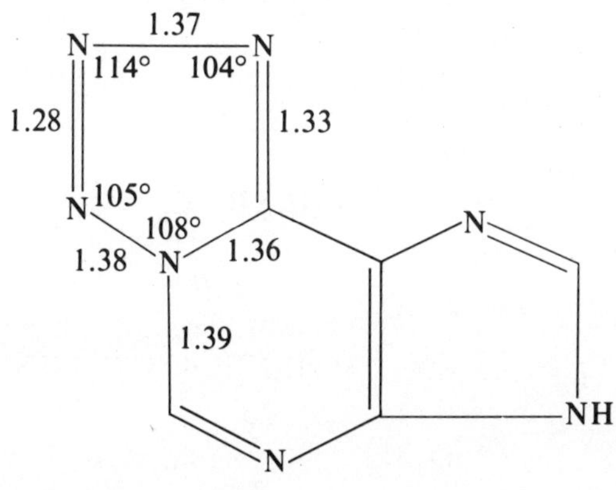

pyrazole; conjugation between the rings is indicated by the shortened tetrazole-pyrazole N—C bond. In addition to 5-aminotetrazole (81), bond alternation is exhibited by its hydrazinium salt (87) and by sodium tetrazolate (377), suggesting some charge localization in these anions.

The structure of tetracene as the hydrate of a dipolar 3-amidinium-1-(tetrazolate-5-yl)-1-tetrazene was established by X-ray analysis (141, 142). This formulation provides a rational location for the associated water molecule in tet-

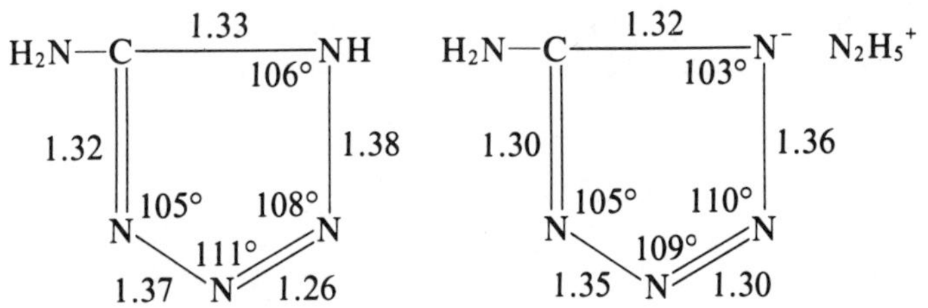

racene which is hydrogen bonded to an amidinium hydrogen and a tetrazole nitrogen. Bond distances in the tetrazolate ring of tetracene are too close to

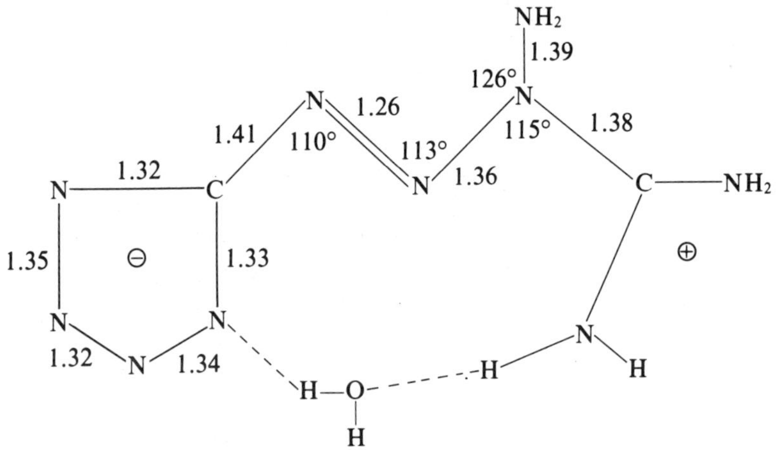

constitute evidence for bond alternation. Nitrogen-nitrogen bond lengths in the acyclic 1-tetrazene chain are similar to those in the 2-tetrazenes.

Only small differences in nitrogen-nitrogen bond lengths are found in cyclopolymethylenetetrazoles such as pentamethylenetetrazole (481), but bond alternation appears in the complex with iodine chloride, shown to be located at

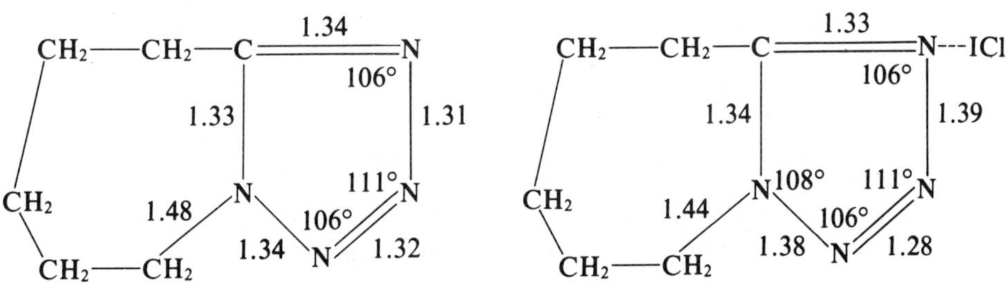

the nitrogen adjacent to the bridgehead carbon (32). Tetrazole bond angles are the same in the complex and the free pentamethylenetetrazole.

Both the mesoionic salt, 1,3-dimethyl-5-aminotetrazolium chloride (85), and

5-amino-2-methyltetrazole (86) have roughly similar bond distances except for the double bond character of the exocyclic carbon-nitrogen bond of the 1,3-

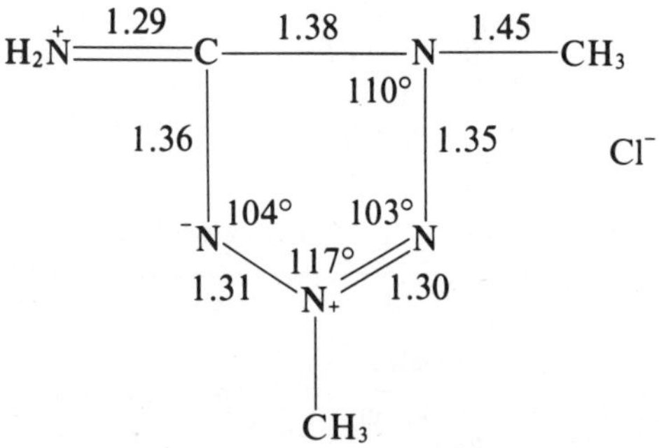

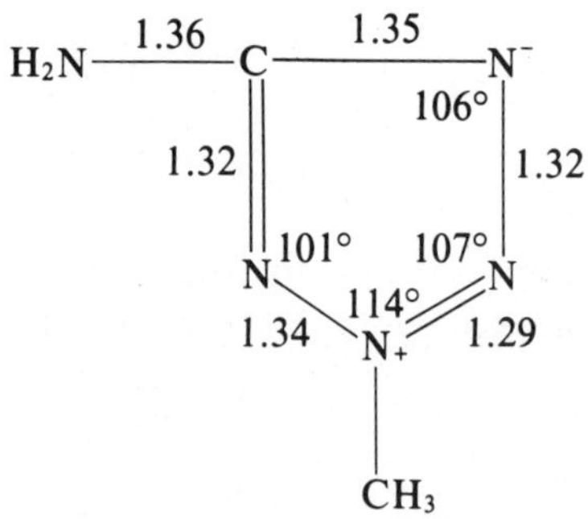

tetrazolium salt, suggesting the importance in each of dipolar forms such as those shown. Hydrogen bonds of the salt are from the aminium group to chloride, but in 5-amino-2-methyltetrazole, hydrogen bonding involves the amino group and ring nitrogen atoms of adjacent tetrazole rings.

In 3-phenyltetrazolo[4,5-*a*]pyridinium fluoborate (419), a 1,2-disubstituted tetrazolium salt, and 1,3-diphenyltetrazolium thiolate (266), some nitrogen-ni-

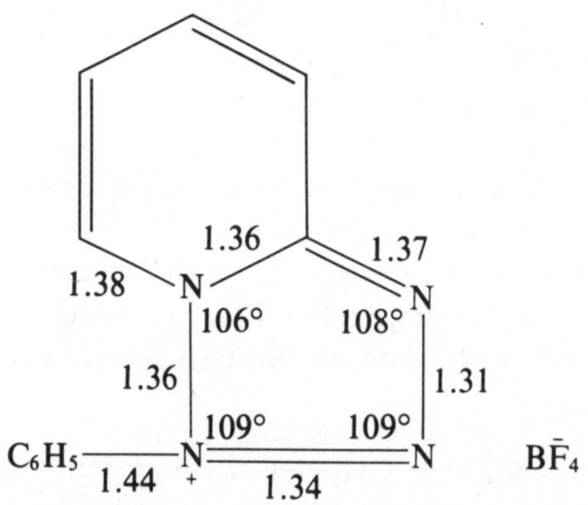

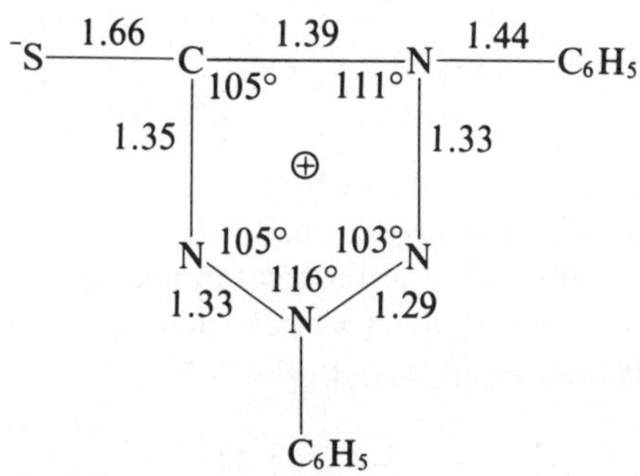

trogen bond alternation occurs, but is not significant in 2,3-disubstituted tetrazolium compounds, such as dehydrodithizone (287) or 2,3-diphenyl-5-(α-D-lyxofuranosyl)tetrazolium bromide (312). Nitrogen-nitrogen bond distances in the mesoionic 1,3 and 2,3-diphenyltetrazolium thiolates are close to each other. The similar carbon-sulfur distances in these isomers indicate a similar degree of negative charge on sulfur in each tetrazolium thiolate.

The planes of the substituent phenyl groups in these four tetrazolium compounds are inclined to the tetrazolium planes at dihedral angles of 45° (287) to 61° (312), and their carbon-nitrogen bonds are single bond lengths. Accord-

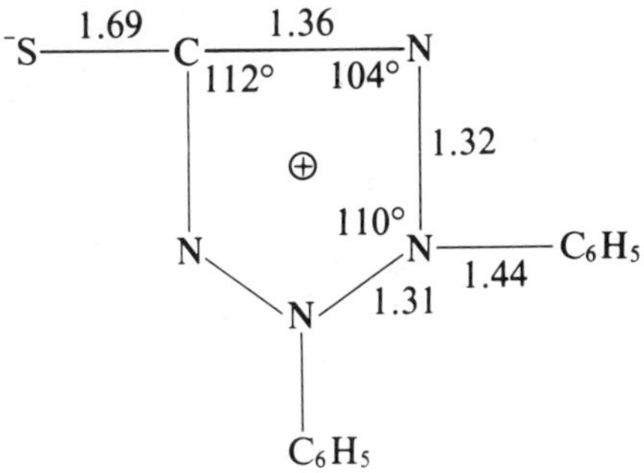

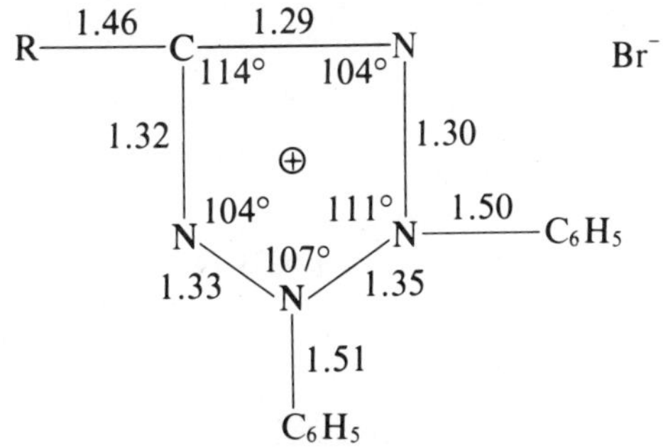

R = α-D-lyxofuranosyl

ingly, there is little or no conjugation between these phenyl groups and the tetrazolium rings.

Other planar five-membered heterocyclic compounds with three or four contiguous nitrogen atoms, having alternating shortened and lengthened bonds, indicating resonance stabilization, include 5-phenylthiatriazole (376), 2-ethyl-5-phenyl-1,2,3,4-thiatriazolium fluoborate (376), the mesoionic 3-phenyl-1,2,3,4-oxatriazolium-5-olate (375), and 1,4-dimethyl-Δ^2-tetrazaboroline (101).

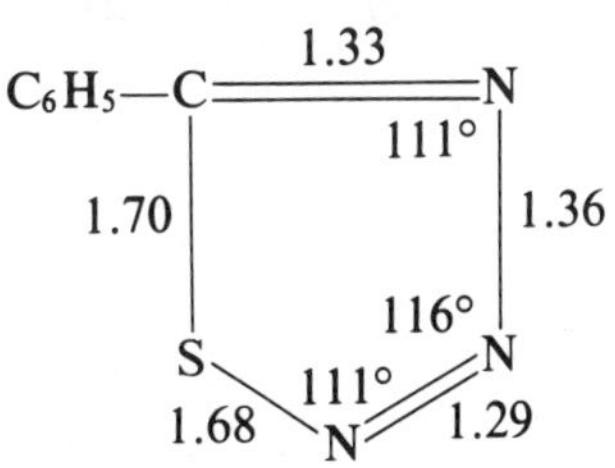

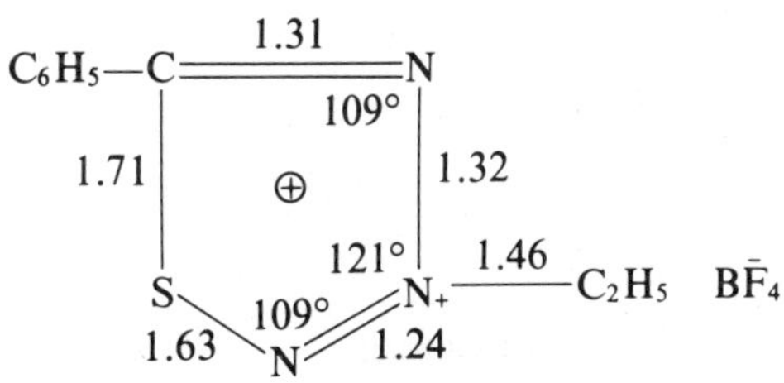

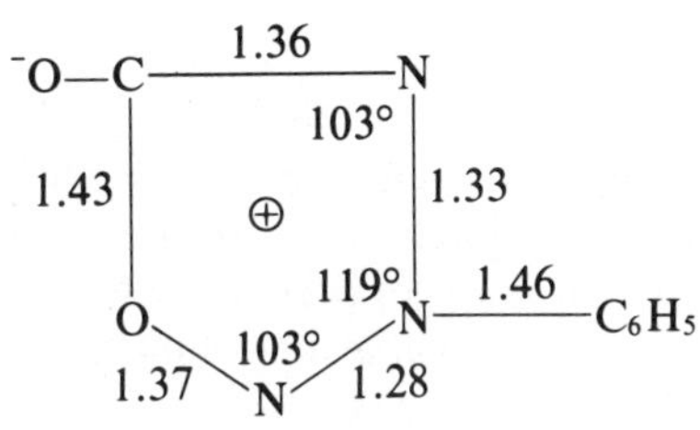

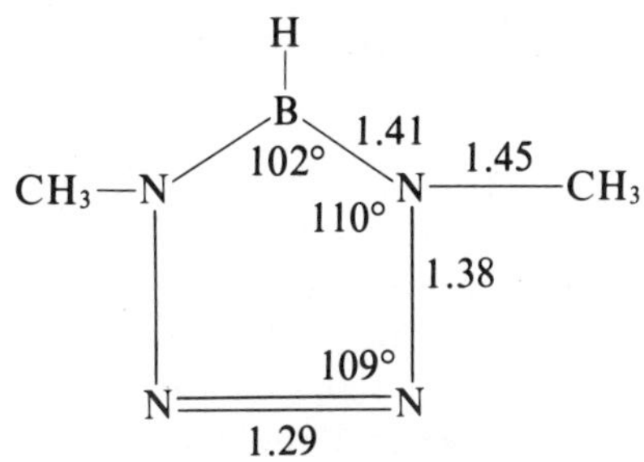

The six-membered *v*-triazine ring in 4,5,6-tri(*p*-anisyl)-1,2,3-triazine also has aromatic character, since it is planar and the contiguous nitrogen-nitrogen bonds are nearly identical with each other and with those in *s*-tetrazine (370). The ring carbon-nitrogen distances are slightly longer than in pyridine or *s*-tetrazine, but the carbon-carbon distances are close to those in pyridine.

In 2-azaadenosine, the six-membered *v*-triazine ring also is planar, its two essentially equal nitrogen-nitrogen bond distances are only slightly longer than

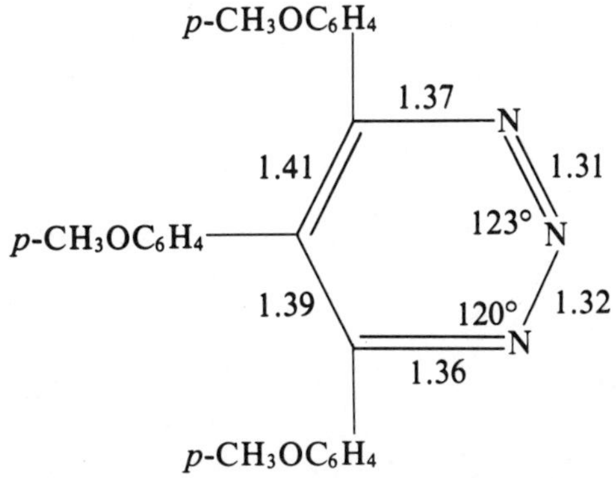

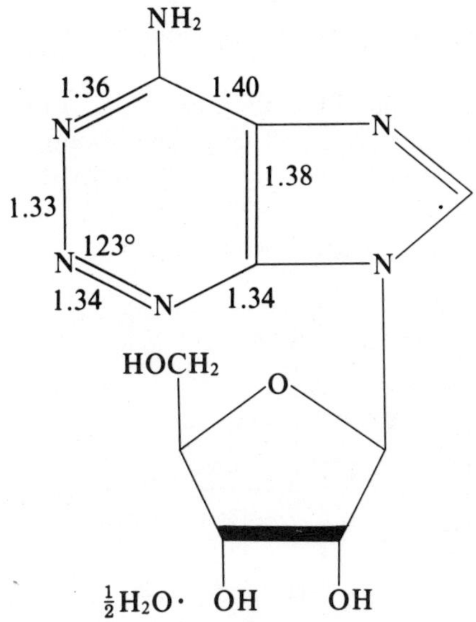

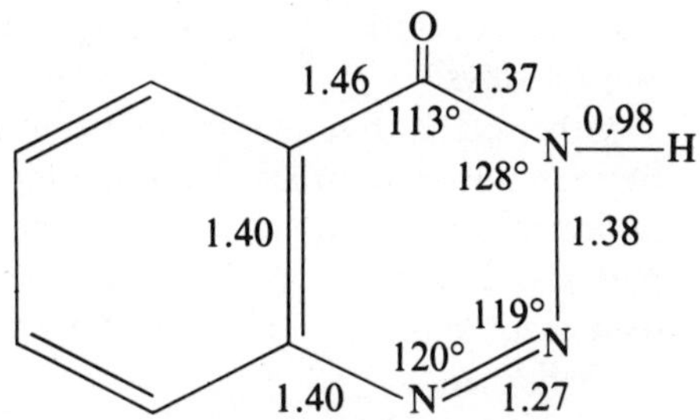

those in tri(p-anisyl)-1,2,3-triazine, its carbon-nitrogen and carbon-carbon bonds are close to those in v-triazoles, and, accordingly, it is aromatic (440). The five- and six-membered rings are inclined at a dihedral angle of 1.2°. The v-triazine ring in 1,2,3-benzotriazin-4-one, however, has a slight boat conformation with N-1 and the carbonyl carbon less than 0.02 Å above the plane of the remaining four ring atoms (227). The N═N is typical of open-chain triazenes, but the N—N is near that of pyrazole. Ring carbon-nitrogen distances, however, are closer to the ring C—N distances in monocyclic v-triazoles and in 1-benzotriazolyl acetic acid than to C—N distances in triazolines, indicating interaction of N-1 with the fused benzene ring and of N-3 with the adjacent carbonyl. Triazine ring hydrogen is located at N-3 and joins to other molecules through N—H····O bonds in polymers around screw axes.

Compounds with embedded nitrogen chains which have been measured are all planar and have bond distances similar to those in related aromatic hetero-

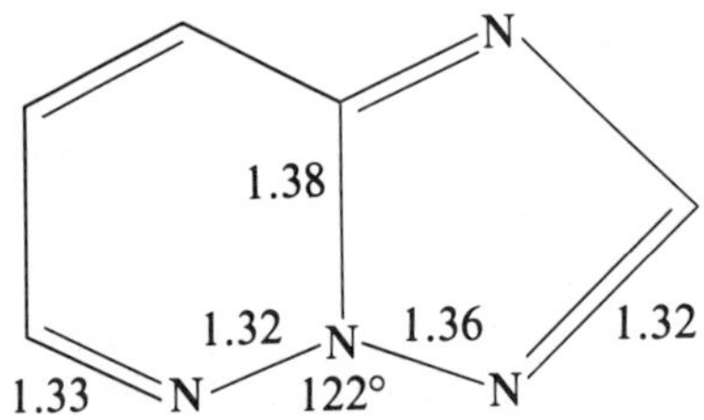

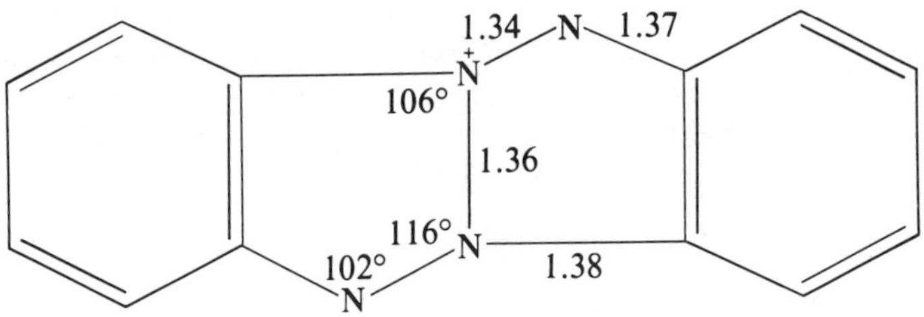

cyclic rings. In *s*-triazolo[1,5-*b*]pyridazine (190), with an embedded three-nitrogen chain, the N—N bond lengths are close to those of *v*-triazole. Nitrogen-nitrogen bonds in the four-nitrogen chain in dibenzo-1,3a,4,6a-tetrazapentalene (289) are close to each other in length and do not show the relatively large differences found between the end and middle nitrogen-nitrogen bonds in the nitrogen chain in acyclic tetrazenes or most tetrazoles. Bond distances and angles in embedded five-nitrogen chains have been measured in 1,6-dimethyl-7-ethoxycarbonylpyrazolo[1,5-*d*]tetrazole (14) and tetrazolo[1,5-*b*]pyridazine (190). Alternation is found in the bond lengths of the first three nitrogen-nitrogen bonds

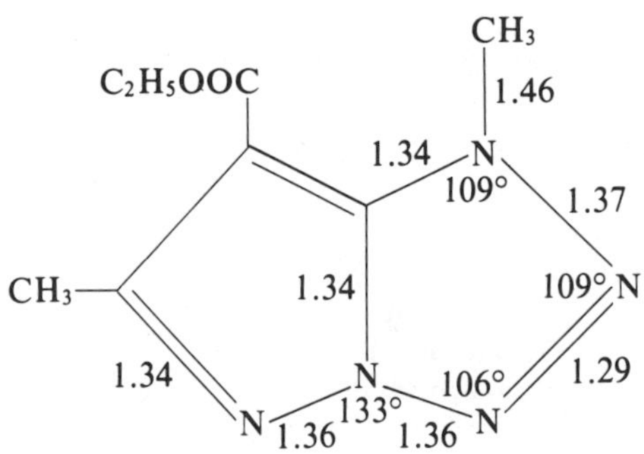

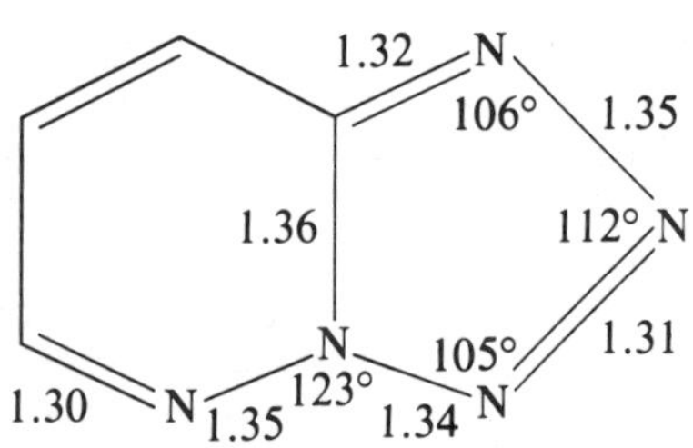

in the tetrazole moieties. In each ring system the bond length of the fourth N—N bond is similar to that in pyrazole. The external angles (122–133°) at the bridgehead nitrogens in these five-nitrogen chain heterocycles as well as that in *s*-triazolo[1,5-*b*]pyridazine, determined by the geometry of the bicyclic ring systems, are greater than in triazenes or triazolines (111–115°).

3.2. THERMOCHEMICAL PROPERTIES

Energy contents of a relatively small number of high nitrogen compounds and classes have been measured; most determinations are heats of combustion from which heats of formation have been derived. In part, the paucity of data is because many compounds with contiguous nitrogen atoms do not burn cleanly or completely. Nevertheless, reproducible combustion results have been attained with high nitrogen compounds by use of igniters and by adjusting oxygen pressure (103, 125, 499).

Compounds in which the high nitrogen moiety is a significant fraction are endoergic with respect to the elements in their standard states.

Only one modern heat of combustion determination is available for diazoaminobenzene. (Thermochemical values are in kcal/mole at 298°K for the liq-

$$C_6H_5N{=}NNHC_6H_5$$

$$\Delta H_f + 79.03 \ (155)$$

uid or solid state existing at this temperature.) The estimated enthalpy of formation of the parent triazene, $HN{=}NNH_2$, is +65 from hydrogenolysis of diazoaminobenzene to benzene and triazene, using the heat of hydrogenolysis of aniline to benzene and ammonia as a basis for calculation.

Hydrogen azide has a higher energy content than either of its electronic isosteres, diazomethane or allene (Table 3.1). Azides attached to aryl, heteroaryl, or cyano are more endothermic than hydrogen azide and those attached to aliphatic groups are less so, as expected from the energy content of these moieties (Table 3.2). The weaker azide bond is between the first and second nitrogen, $RN{-}N_2$, with a dissociation energy of 7 kcal in cyanogen azide (372) and 9 kcal in hydrogen azide (169).

Azides, consisting of three closely spaced nitrogen atoms, are stabilized by loss of considerable energy. Empirical resonance energies for hydrogen azide are estimated as 41 for $H\bar{N}{-}\overset{+}{N}{\equiv}N$ and 30.5 for $H{-}N{=}\overset{+}{N}{=}\bar{N}$ (438), and are 41 and 45 for cyclopentyl and cyclohexyl azides, respectively, calculated from the cumulative bond canonical form (156).

Table 3.1. Enthalpy of Formation of Hydrogen Azide, Diazomethane, and Allene

Isostere	ΔH_f	
HN_3	+71.66(g)[a]	(153, 195)
	+64.37(l)	
CH_2N_2	+46(g)	(290)
$CH_2{=}C{=}CH_2$	+45.9(g)	(413)

[a] Based on the heat of formation of aqueous azide ion (195).

Table 3.2. Enthalpy of Formation of Organic Azides

Azide	ΔH_f	
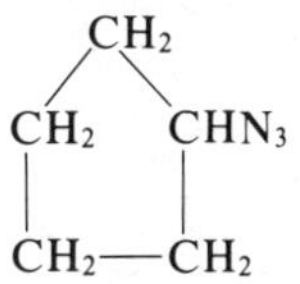	+42.79(l)	(156)
$HOCH_2CH_2N_3$	+22.47(l)	(155)
$C_6H_5N_3$	+82.3(l)	(153, 414)
NCN_3	$+108 \pm 5$(g)[a]	(372)
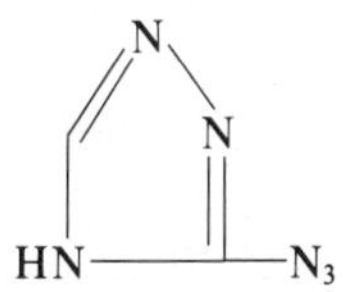	+105.5(c)	(125, 134)

[a] From photodissociation of cyanogen azide.

The aromatic fused *v*-triazoles are appreciably more endothermal than their structural analogs, indole and indene. (Thermochemical values for reference compounds are taken from Lange's *Handbook of Chemistry,* 12th ed., unless

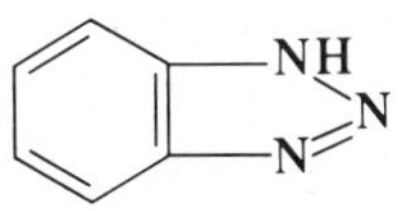

ΔH_f +59.74 (134, 155)

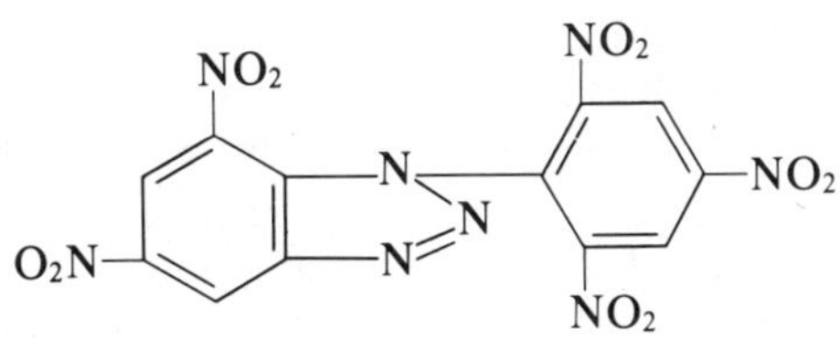

ΔH_f +71.7 (416)

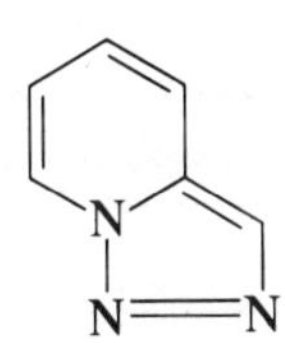

$\Delta H_f + 89.5 \pm 8$ (488)
(estimated from kinetic data)

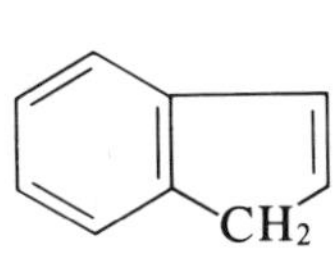

ΔH_f + 26.39

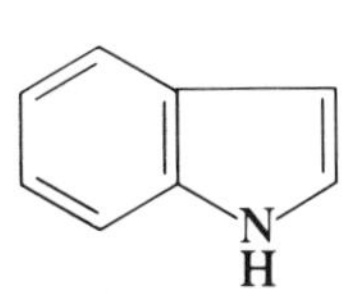

ΔH_f + 29.8

otherwise noted.) Benzotriazole is highly stabilized thermochemically, as indicated by its resonance energy of 74.7 compared to 41.8 for indole (510), calculated according to Coates and Sutton (111). Linear naphthotriazole is slightly more energetic than its angular isomer, as indicated by their heats of combustion (173).

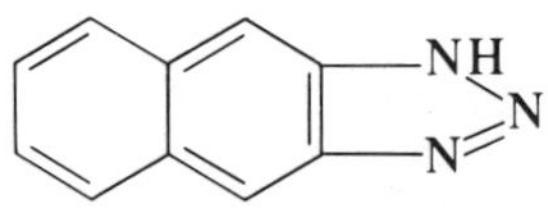

ΔH_c 1246.0

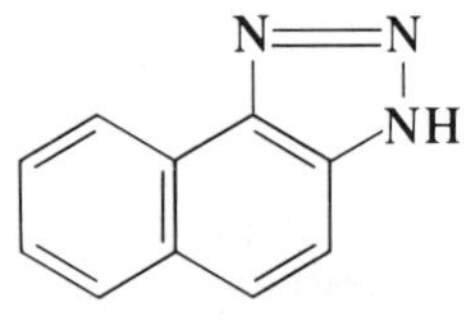

ΔH_c 1242.6

The heat of formation of 1,4-dimethyltetrazolinone is below the values for the aromatic *v*-triazoles and tetrazoles.

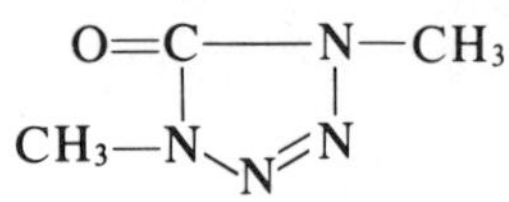

ΔH_f −6.6 (134, 499)

Tetrazole is the most endothermic of the five-membered aromatic carbon and nitrogen heterocycles, and is stabilized thermochemically to the greatest extent, indicated by increases of both enthalpy of formation and resonance energy as nitrogen is progressively substituted for carbon (Table 3.3). Theoretical calculations for five-membered aromatic nitrogen heterocycles also indicate increasing heats of formation as nitrogen content increases (128).

Modification of tetrazole by substitution of an electron-releasing group at position 5 renders the resulting molecule more stable thermochemically, but an electron-withdrawing group increases its endothermic character (Table 3.4).

Table 3.3. Enthalpy of Formation and Resonance Energy of Five-Membered Aromatic Nitrogen Heterocycles

Heterocycle	ΔH_f	Resonance Energy (510)
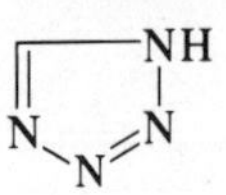	+56.7 (134, 339)	63.1
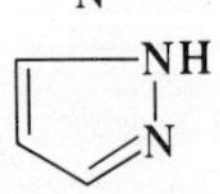	+28.3	32.7
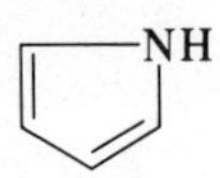	+15.1	17.4

Table 3.4. Enthalpy of Formation of 5-Monosubstituted Tetrazoles

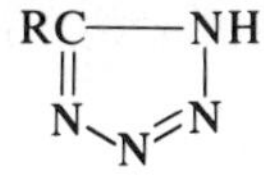

R	ΔH_f	Ref.
OH	+1.5	134, 499
H_2N	+49.7	134, 499
$(CH_3)_2N$	+43.6	134, 499
NC	+96.1	134, 499
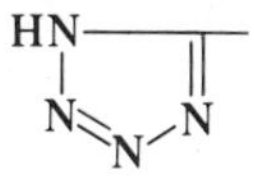	+127.1	44

The small heat of formation of the nominal 5-hydroxytetrazole is consistent with its structure as 5-tetrazolinone and is close to that of 1,4-dimethyl-tetrazolinone.

Cyclization of guanyl azide nitrate is exothermic, consistent with its rapid

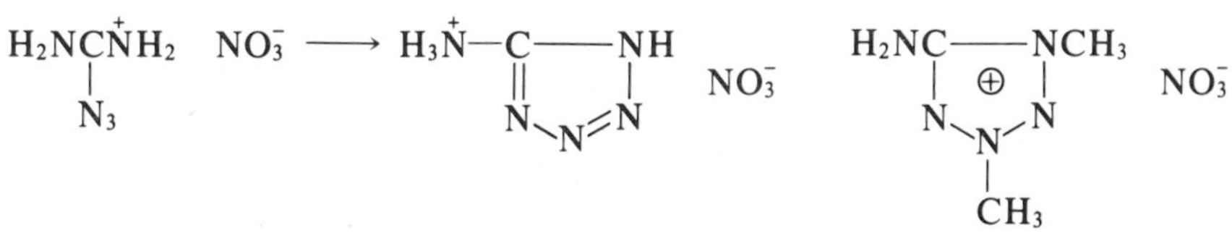

ΔH_f +3.8 $\quad$ ΔH_f −6.6 $\quad$ ΔH_f −1.3

ring closure (339). Salt formation decreases heat content, illustrated by 5-aminotetrazole nitrate and 1,3-dimethyltetrazolium nitrate (134).

Enthalpies of formation of isomeric 1,5- and 2,5-tetrazoles (134) are similar but show no structural regularity (Table 3.5).

Table 3.5. Enthalpy of Formation of 1,5- and 2,5-Disubstituted Tetrazoles

		ΔH_f	
R_5	R	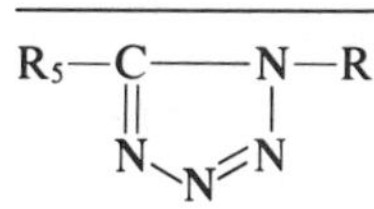	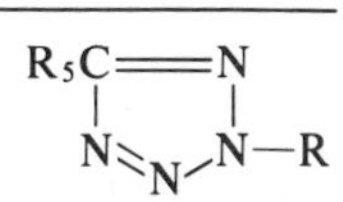
H_2N	CH_3	+46.25	+50.4
H_2N	$CH_2{=}CHCH_2$	+63.4	+67.6
CH_3	C_6H_5	+69.9	+65.7

Heats of formation of isomeric mono- and dibenzotetraazapentalenes indicate the compounds with the linear nitrogen chain to be slightly more endothermic than their branched isomers (103). The resonance energy contribution

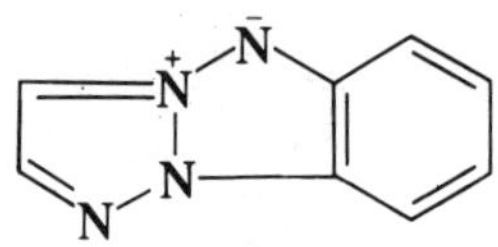

ΔH_f +118.5

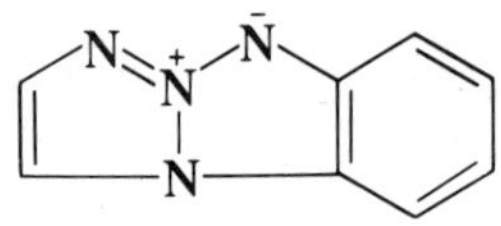

ΔH_f +113.0

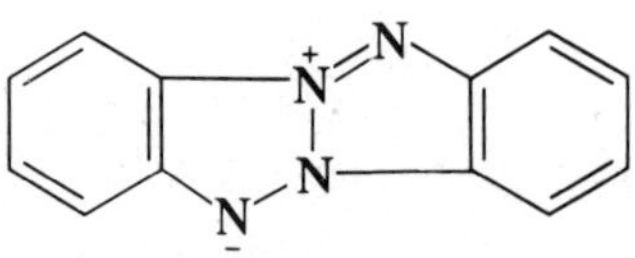

ΔH_f +126.0

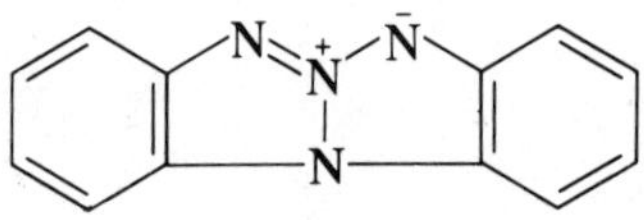

ΔH_f +122.0

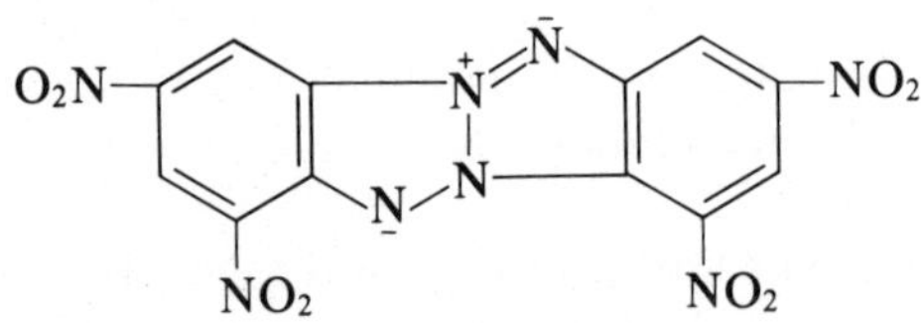

ΔH_f +110.2 (416)

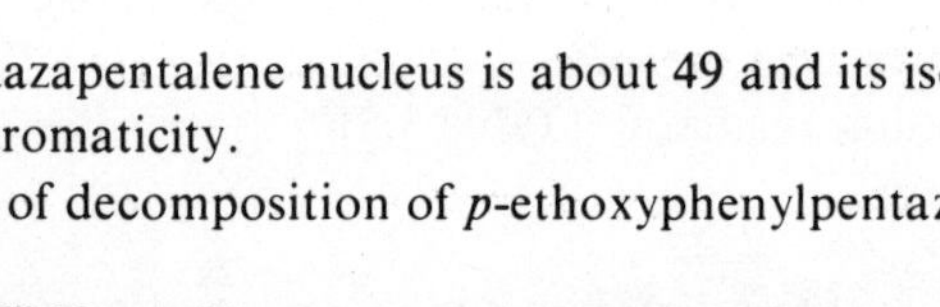

ΔH_f +109.8 (416)

of the linear tetraazapentalene nucleus is about 49 and its isomer's is about 57, indicating their aromaticity.

From the heat of decomposition of *p*-ethoxyphenylpentazole it may be con-

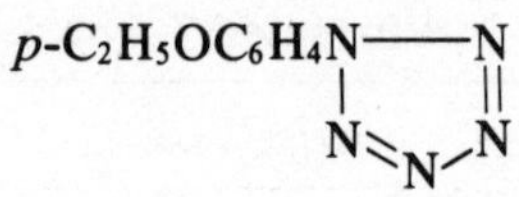

$$\longrightarrow p\text{-}C_2H_5OC_6H_4N_3 + N_2 \qquad \Delta H = 5.4$$

cluded that pentazole has a high resonance energy (473). *Ab initio* LCAO-SCF-MO calculations indicate pentazole to be 29 kcal more stable than $HN_3 + N_2$ and from 68 to 81 kcal more stable than either the *E* or *Z* open-chain diazoazide (418).

The calculated heat of formation of cyclic N_6, 323 kcal/mole, shows it to be unstable, decomposing without activation to $3N_2$ (128).

3.3. DIPOLE MOMENTS

Compounds with chains of three or more nitrogen atoms are almost always electrically asymmetric, in part because of moments originating in the non-bonding electrons on each nitrogen. Most acyclic high nitrogen compounds such as triazenes and azides have moments less than 4.0 Debye units, but many classes with cyclic nitrogen chains such as *v*-triazoles and tetrazoles have larger moments, indicating extensive electron displacement. A useful reference point is 4.8 D, the value for the movement of an electron 1 Å (455), a distance about $\frac{3}{4}$ of a nitrogen-nitrogen single bond length.

Although *trans*-azobenzene has no electric moment, substitution of anilino for phenyl to form diazoaminobenzene creates asymmetry; the contribution of the NH group is shown by the dipole moment. (Dipole moments cited are de-

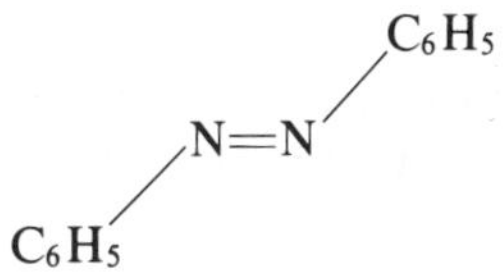

zero D

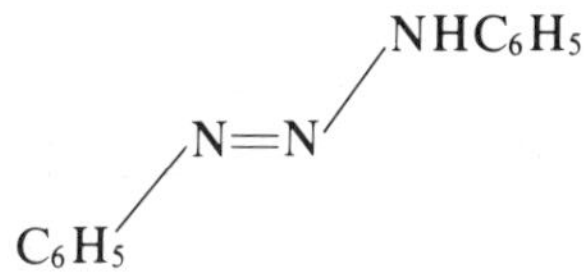

0.9 D (295)

termined in benzene unless stated otherwise.) Polarization of diazoaminobenzene decreases slightly as its concentration increases, suggesting association to less polar dimeric or polymeric hydrogen bonded forms (240, 295).

The small dipole moment of tetrakis(trimethylsilyl)-2-tetrazene indicates a trans configuration (496).

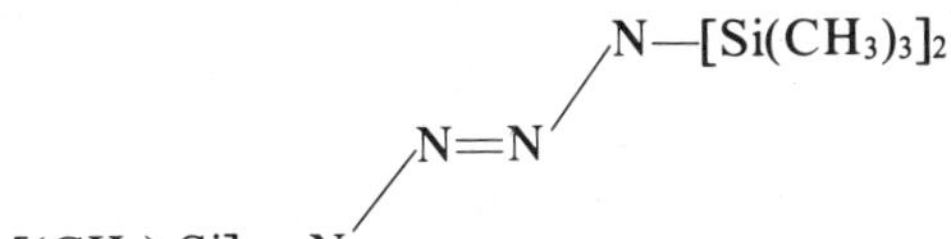

0.45 D

In the 1,3-diphenyltriazenes, electron-withdrawing groups increase dipole moments on substitution in one or both phenyl groups (Table 3.6) (295). Introduction of methyl in the aryl rings has little effect, but asymmetry increases when methyl replaces hydrogen on the nitrogen chain.

In 1,1-dimethyl-3-phenyltriazene, the negative end of the dipole is directed toward the azo portion of the triazene, as indicated by the effect of *p*-methyl in decreasing and *p*-chloro in increasing their moments (294).

$C_6H_5N{=}NN(CH_3)_2$	*p*-$CH_3C_6H_4N{=}NN(CH_3)_2$	*p*-$ClC_6H_4N{=}NN(CH_3)_2$
2.3 D	1.9 D	4.0 D

Table 3.6. Dipole Moments of 1,3-Diaryltriazenes

p-RC$_6$H$_4$N=N—N(R″)C$_6$H$_4$R′-p

R	R′	R″	μ, 25° (D)
Br	H	H	2.0
Br	Br	H	1.9
CH_3	H	H	1.05 (122)
CH_3	CH_3	H	0.9
H	H	CH_3	1.5
Br	Br	CH_3	2.5

Two resonance forms contribute to the structure of the triazenes and to the total moment. Mesomeric moments, derived from the vector difference between

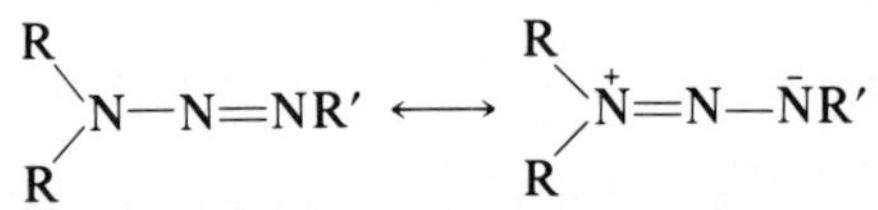

the moment of a conjugated system and that of an aliphatic analog, measure the interaction between a π-electron system and its substituent (279, 455, 480). Applied to a linear three-atom chain such as —N=N—N , the mesomeric moment is the difference between the measured or calculated moment of the unsaturated group and that derived from bond and group moments. For 1-phenyl and 1-p-halophenyl-3,3-dimethyltriazenes, the mesomeric moment of —N=N—N< is about 2.1 D, in comparison with 1.9 for —N=C—N< and 1.8 for >C=N—N< (389).

The 1,4-tetrazolines, which may be looked on as cyclic 2-tetrazenes, exhibit small degrees of polarity (260).

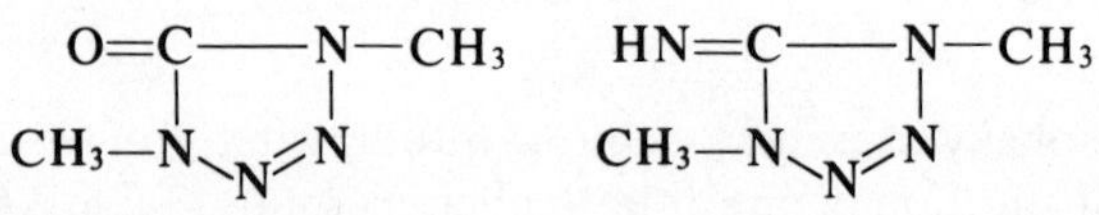

1.14 D 1.65 D

The component of the dipole moment of hydrogen azide along the azide axis is 0.847 D (19) in contrast to 0.2 D for its carbon isostere, allene (482), indicating the greater polarizability of nitrogen. Dipole moments of azides are close to those of chlorides and bromides* in both the aliphatic and aromatic series,

* Dipole moments of reference compounds are taken from Lange's *Handbook of Chemistry*, 12th ed., unless otherwise indicated.

showing the negative character of the azide group. Virtual cancellation of the moment in *p*-azidobromobenzene further illustrates the similarity.

$C_2H_5N_3$	C_2H_5Br
2.12 D (446a)	2.03 D

$C_6H_5N_3$	C_6H_5Br	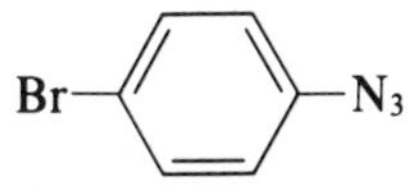
1.44 D (307)	1.55 D	0.02 D (446a)

In comparison with ethyl azide, the smaller moment of phenyl azide indicates resonance between azide and the aromatic ring, and is consistent with approximately equal contributions of the limiting resonance forms (307). Resonance also occurs in vinyl azide, as shown by low moments relative to ethyl

$$C_6H_5—\bar{N}—\overset{+}{N}{\equiv}N \longleftrightarrow C_6H_5—N{=}\overset{+}{N}{=}\bar{N}$$

azide both in solution and as a gas. The presence of two geometric forms of vinyl azide in the vapor phase is deduced from its microwave spectrum (166).

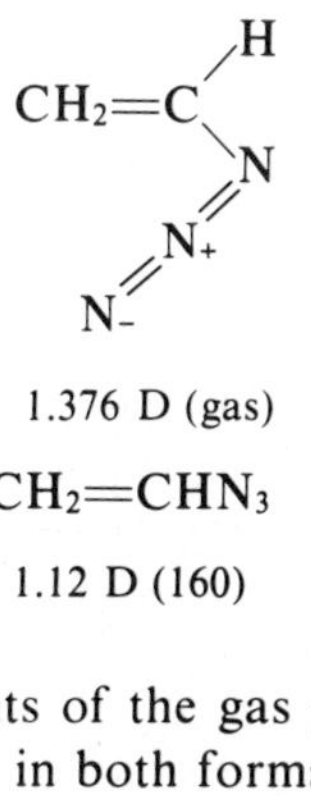	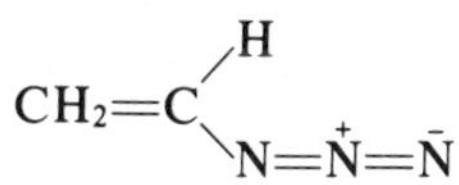
1.376 D (gas)	1.465 D (gas)
$CH_2{=}CHN_3$	NCN_3
1.12 D (160)	2.99 D (70)

Measured components of the gas phase moments, and *ab initio* calculations show that the dipole in both forms is nearly parallel to the C—N bond. The electron-withdrawing cyano in cyanogen azide group strongly polarizes the molecule resulting in a relatively high dipole moment.

In 2-azido imidazole and benzimidazole, comparison of calculated and determined dipole moments indicated that the azide is in the *Z* conformation,

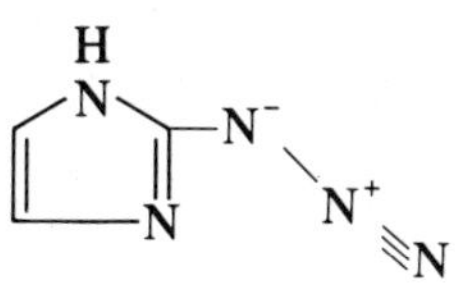	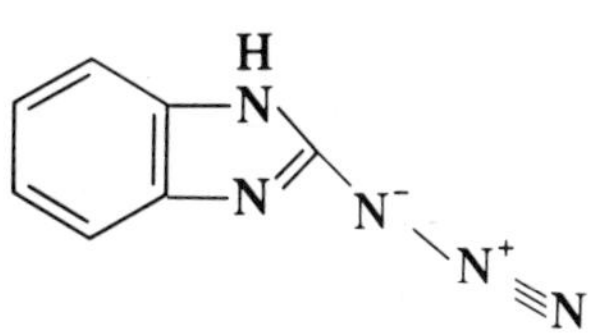
3.70 D	3.55 D

turned toward the pyridine rather than the pyrrole nitrogen (161). The conformation of 3-azido-*s*-triazole cannot be deduced from dipole moments, because of alternate prototropic forms, but can be narrowed to **I** or **II**, in which the azide is oriented away from the triazole ring. A methyl at N-1 or C-5 of the *s*-

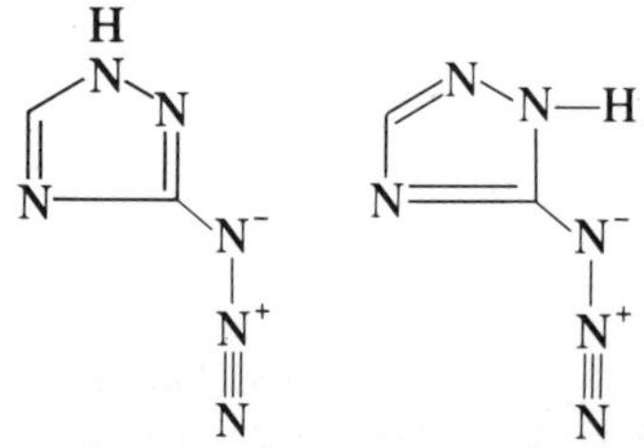

4.18 D

(I) **(II)**

triazole increases the dipole moment of the C-3 azides (387) but an electron-withdrawing group at C-5 counteracts the influence of the azide (387).

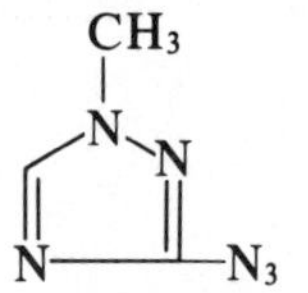

4.40 D (dioxane)

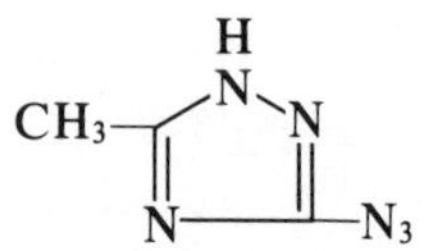

4.58 D (dioxane)

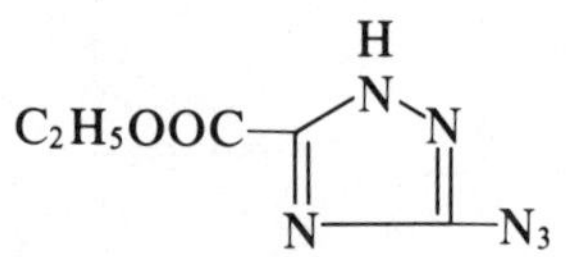

3.87 D (dioxane)

In the cyclic-cyclic three- and four-nitrogen chains, introducing two additional, nonconsecutive nitrogens into the ring contributing a single nitrogen to the chain increases the size of the moment (124). These compounds behave as *N*-halo nitrogen heterocycles; accordingly, the single nitrogen atom in the pyrrole or *s*-triazole ring is the positive end of the dipole.

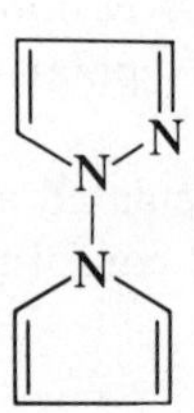

2.04 D (dioxane)

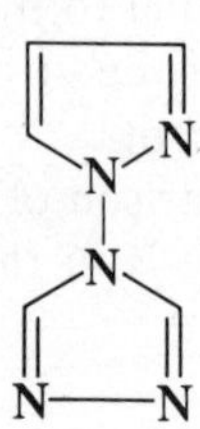

5.33 D (dioxane)

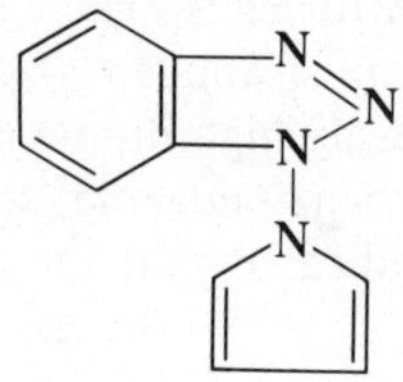

3.68 D (dioxane)

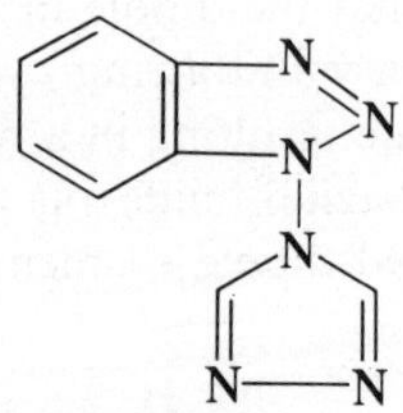

5.22 D (dioxane)

Dipole moments of corresponding 1- and 2-substituted *v*-triazoles reflect the geometry of the two series. The moment of *v*-triazole, in relation to the values

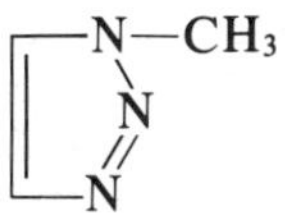

4.46 D (335)

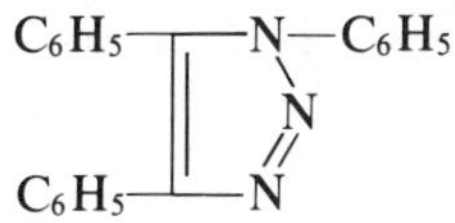

5.2 D (63)

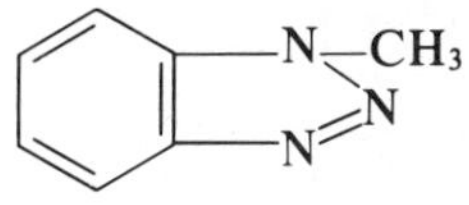

3.95 D (336)

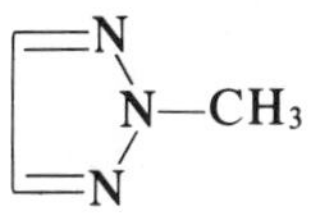

0.37 D (335)

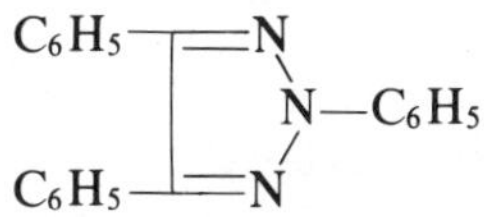

0.4 D (63)

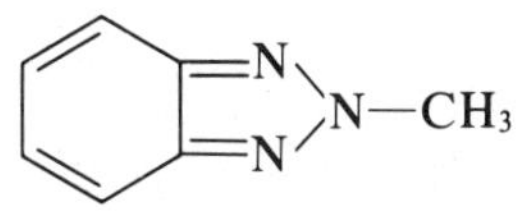

0.49 D (336)

for 1- and 2-methyl-*v*-triazole, indicates about 80% of the isomer with hydrogen on the 2-position at 25°. At least under the conditions of dipole moment meas-

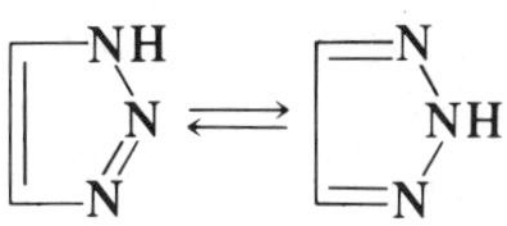

1.77 D (245)
1.85 D (335)

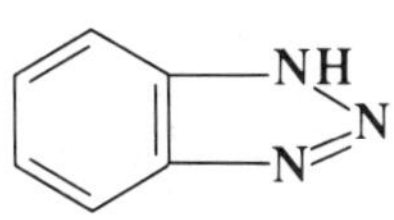

4.15 D (336)

urement in benzene solution, the hydrogen on nitrogen in benzotriazole is at position 1.

The dipole moment of 1,1′-bibenzotriazole indicates more polarity than

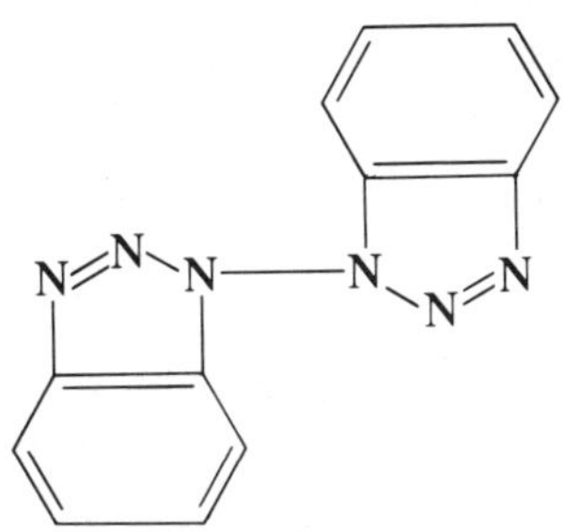

5.16 D (213)

benzotriazole or its 1-alkyl derivatives, but gives no information about the distribution of charge over the six-nitrogen chain.

In comparison to the *v*-triazoles, substitution of a ring carbon by another nitrogen, forming tetrazoles, leads to greater electrical asymmetry. Comparison of the dipole moments of tetrazole with those of 1- and 2-ethyltetrazole indicates the major form to be with hydrogen at position 1.

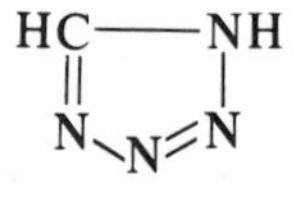

5.11 D (dioxane) (245)

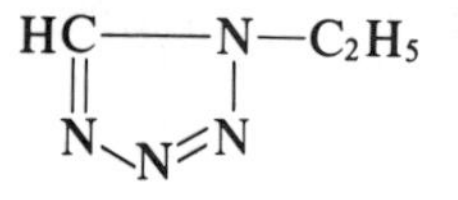

5.46 D (260)

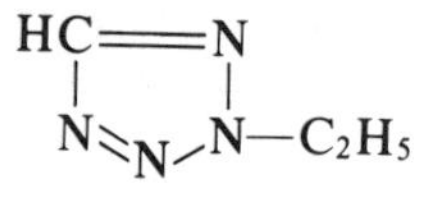

2.65 D (260)

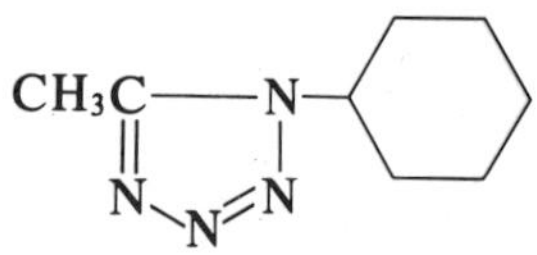

6.00 D (394)

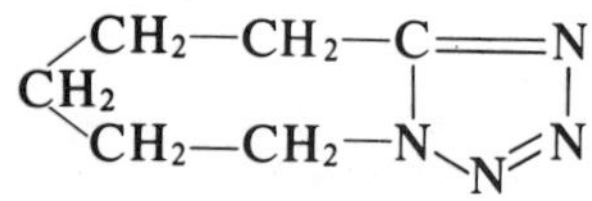

6.14 D (394)

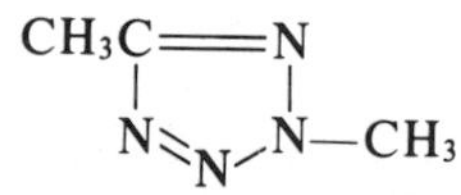

2.42 D (328)

Consideration of the charged resonance forms of tetrazole, illustrated for the 1*H* isomer, leads to the suggestion that the positive end of the dipole in tetrazole

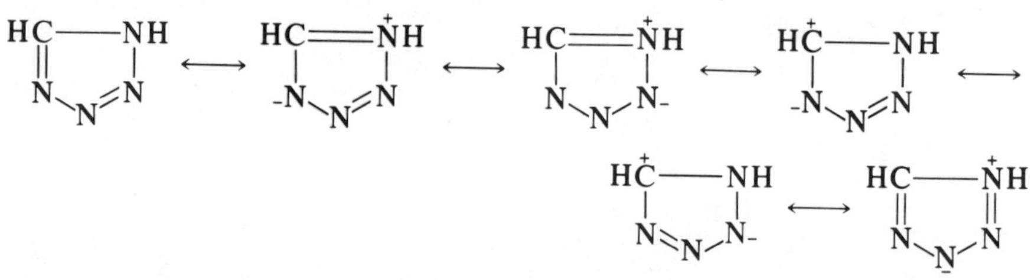

and 1-alkyltetrazoles is located near carbon and that the negative end is directed toward N-2 and N-3 (260). An electron-withdrawing substituent at posi-

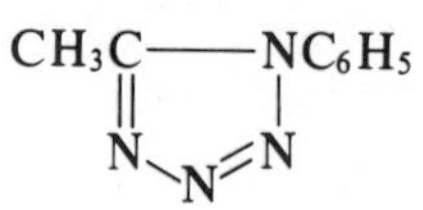

5.64 D (260)

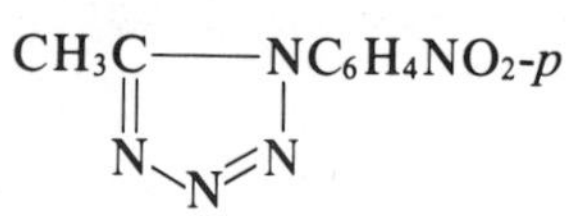

3.20 D (261)

tion 1 counteracts the effect of the electron-withdrawing tetrazole, as shown by the moments of 1-phenyl and 1-*p*-nitrophenyl-5-methyltetrazoles.

Resonance forms for 2-substituted tetrazoles, in contrast to the 1-isomers,

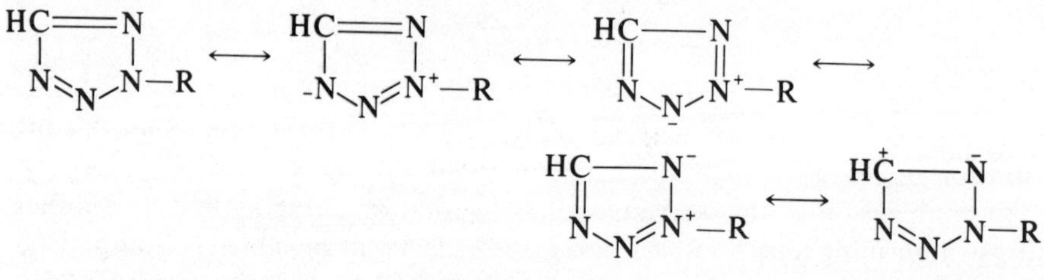

appear to have the positive end of the dipole near N-2 and the negative end between N-4 and the carbon (260).

A similar polarity difference to those between 1- and 2-alkyl tetrazoles is also found when a 5-alkylamino group is present (260). Both 5-aminotetrazole and

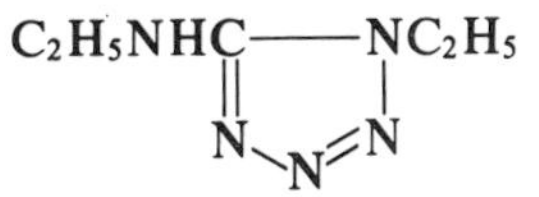

7.36 D

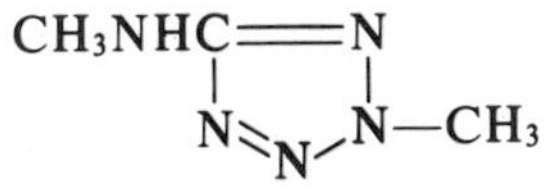

2.55 D

5-aminothiatriazole have similar dipole moments, showing like polar effects of NH and S.

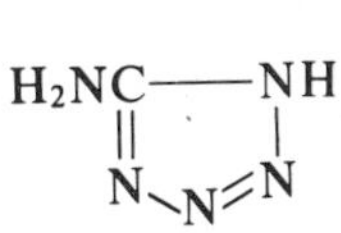

5.71 D (245) (dioxane)

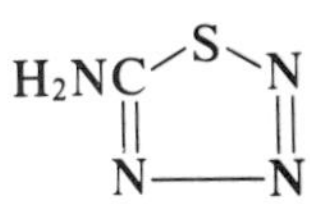

5.77 D (304) (dioxane)

Dipole moments for imidoyl azide and tetrazole isomers, calculated by the CNDO/S method, together with experimental data for the tautomeric systems, have been used to estimate equilibrium constants in the thiazole, benzothiazole, and benzoxazole series. Equilibrium values are qualitatively similar to those obtained from NMR measurements (Table 3.7) (159, 162).

Electrical symmetry is shown by dibenzo-1,3a,4,6a-tetraazapentalene, but

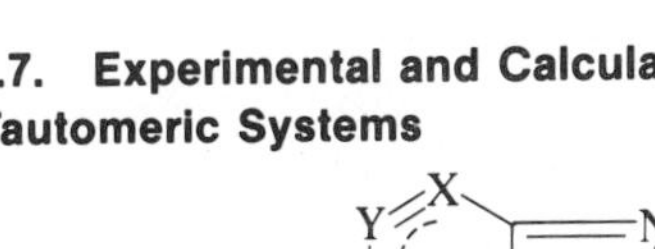

zero D

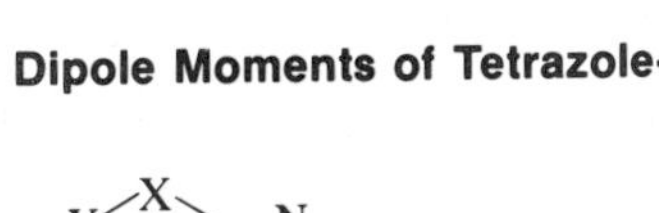

4.36 D

Table 3.7. Experimental and Calculated Dipole Moments of Tetrazole-Imidoyl Azide Tautomeric Systems

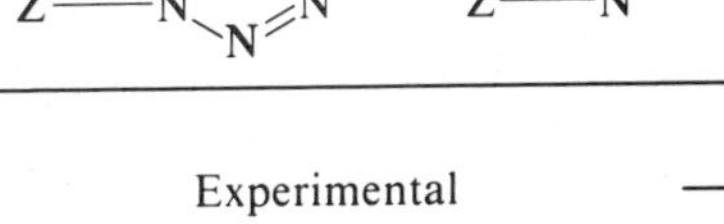

X	Y	Z	Experimental Dipole Moment (D)	$K_{A/T}$ (25°, dioxane) From μ	$K_{A/T}$ (25°, dioxane) From NMR
S	CH	CH	3.33	2.35	2.2
S	benzo		4.12	1.30	0.8
O	benzo		2.09	10.0	≥20

Table 3.8. Dipole Moments of 3-Aryl-1,2,3,4-Oxatriazolium Compounds

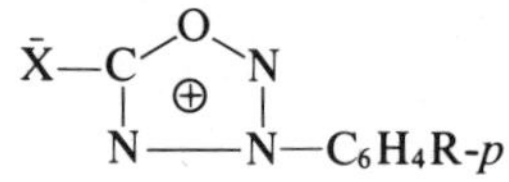

X	R	μ (D)
O	H	6.14
S	H	6.83
S	Cl	5.03
S	CH_3O	7.62
C_6H_5N	H	5.42
C_6H_5N	Cl	4.17
p-ClC_6H_4N	H	6.29

the unsymmetrical arrangement of nitrogen atoms in dibenzo-1,3a,6,6a-tetra-azapentalene is reflected in its dipole moment (96).

The structure of mesoionic nitrogen heterocycles involving a positively charged ring and a negatively charged exocyclic atom is confirmed by their dipole moments. Thus in the 3-phenyl-1,2,3,4-oxatriazole-5-thiones (Table 3.8), electron-withdrawing groups on phenyl decrease and electron-releasing groups increase the moments (203). Similarly in the 3-phenyl-5-phenylimino-1,2,3,4-oxatriazoles, an electron-attracting group on the ring phenyl decreases the compound's polarity, but when it is located on the exocyclic phenylimino, polarity is increased.

An exocyclic imino group on these mesoionic heterocycles lowers the dipole moment in comparison to sulfur or oxygen. Except for compounds with an exo-

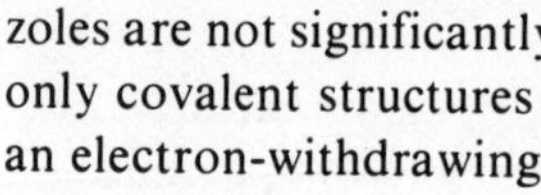

3.71 D

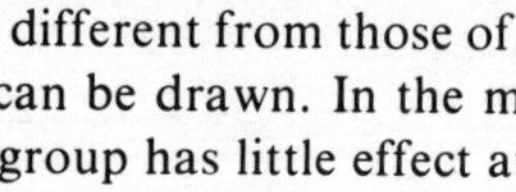

4.39 D

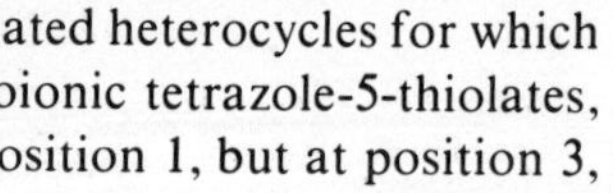

8.84 D

cyclic dicyanomethyl, dipole moments of mesoionic thiatriazoles and tetrazoles are not significantly different from those of related heterocycles for which only covalent structures can be drawn. In the mesoionic tetrazole-5-thiolates, an electron-withdrawing group has little effect at position 1, but at position 3, decreases the moment (Table 3.9) (203).

3.4. ELECTRONIC AND VIBRATIONAL SPECTRA

Absorption of ultraviolet, visible, and infrared radiation has been determined for representatives of most high nitrogen classes. The principal constitutive in-

Table 3.9. Dipole Moments of 1,3-Disubstituted Tetrazolium Compounds

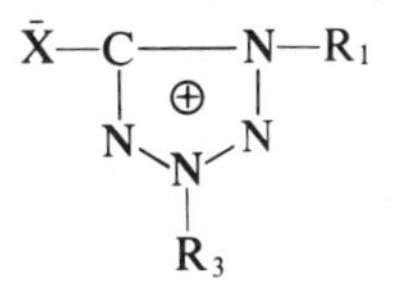

X	R_1	R_3	μ (D)
HN	CH_3	CH_3	4.02 (260)
O	C_6H_5	C_6H_5	5.28
S	C_6H_5	C_6H_5	6.52
S	*p*-ClC_6H_4	C_6H_5	6.47
S	C_6H_5	*p*-ClC_6H_4	5.61
$(NC)_2C$	C_6H_5	C_6H_5	9.54

formation derived from ultraviolet and visible spectra is the extent of electron delocalization in the molecules and accordingly, as far as the spectral data permit, participation of nitrogen chains in conjugation is identified. Infrared absorptions, however, are related to individual chemical bonds or assemblages of atoms and those discussed are specific to bonds of nitrogen to itself and other elements, and to nitrogen ring systems occurring in compounds with contiguous nitrogen atoms.

3.4.1. Ultraviolet and Visible Spectra

Relatively short wavelength maxima of 1,2-di(ethoxycarbonyl)-3,3-dimethyltriazane and 1,4-diphenyl-1,2,3,4-tetra(ethoxycarbonyl)tetrazane indicate the absence of conjugated chromophores in these reduced nitrogen chains, probably

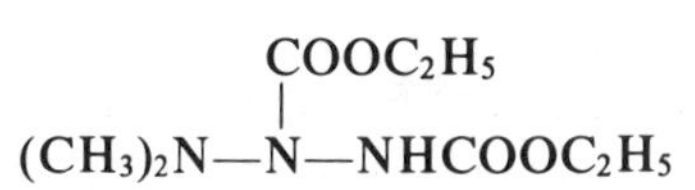

<220 (263)

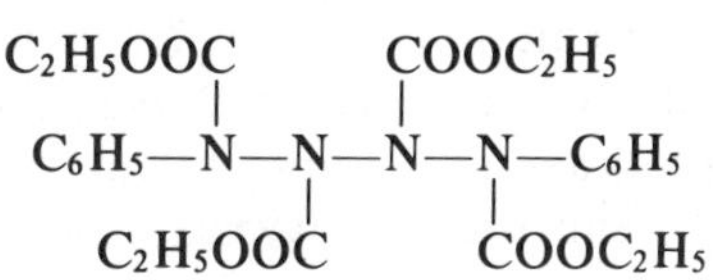

207(3.98) (33)

owing to hindered rotation of the ethoxycarbonyl groups. [Ultraviolet and visible maxima, λ_{max}, are given in nanometers, and extent of absorption as (log ϵ).]

Several heterocycles with embedded nitrogen chains, formally at the triazane, tetrazane, and pentazane oxidation level, however, absorb at longer wavelengths, including in a few cases a portion in the visible region. Participation is thereby indicated of some of the nitrogen atoms in conjugation through their nonbonding electrons. Thus to account for the electronic absorption of **I**, the tetra(methoxycarbonyl)pyridazine unit must interact with the benzene ring via at least one bridgehead nitrogen, and probably all the nitrogen atoms. Analogously, in **II** the skeleton of acetanilide (λ_{max} 235) presumably is conjugated with the remaining carbonyl through part of the tetrazane chain, and in **III** the

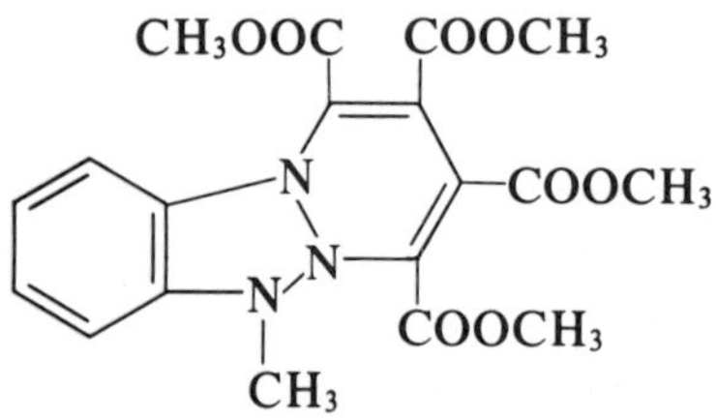

(I) red

214(4.20) (1)
280(3.97)
499(4.03)

embedded pentazane chain must be involved in conjugation between the acetyltoluidine residue and the ethoxycarbonyl groups.

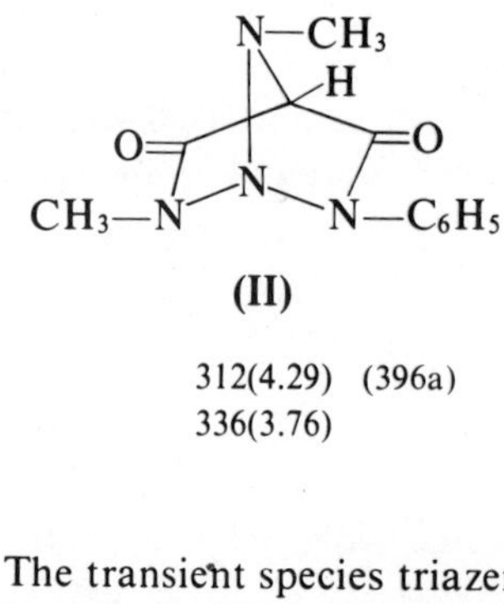

(II)

312(4.29) (396a)
336(3.76)

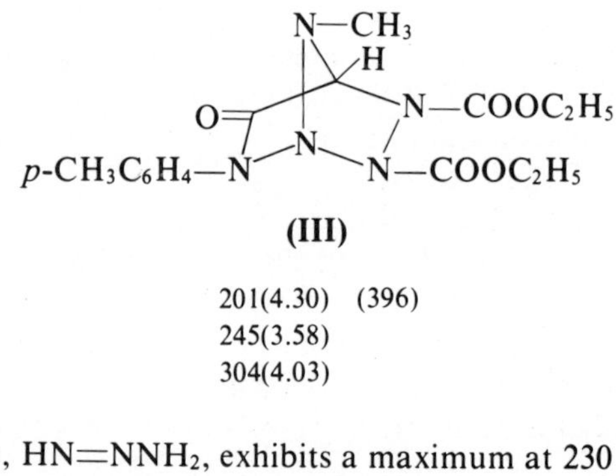

(III)

201(4.30) (396)
245(3.58)
304(4.03)

The transient species triazene, HN=NNH$_2$, exhibits a maximum at 230 nm in aqueous solution from pH 7.2 to 8.7 with log ϵ no higher than 3.7 (454). Its protonated form at pH 1.5–3.0 has a maximum at about 220 nm. There is no evidence of azo group absorption at 338 nm, characteristic of azomethane.

In the substituted triazenes absorption occurs at longer wavelengths. In comparison to the trialkyl derivative further bathochromic shifts result on replacement of an alkyl by cyano or phenyl, extending the conjugated system which also involves the free pair of electrons on amino nitrogen. Conjugation, indicated by the ultraviolet spectra of the triazenes (Table 3.10), is consistent with shortening of the N—N from 1.45 to a range of 1.34–1.25 Å and lengthening of the N=N bonds from 1.24 to 1.27 Å (Section 3.1). Introduction of a second phenyl shifts the maximum further to longer wavelengths. 1,3-Diaryltriazenes with hydrogen or alkyl on nitrogen absorb from 340 to 365 nm because of the aryl —N=N chromophore (123, 382); absorptions at shorter wavelengths are characteristic of aromatic amines.

Table 3.10. Ultraviolet Maxima of Triazenes

Triazene	$\lambda_{max}(\log \epsilon)$	Ref.
$n\text{-}C_4H_9—N{=}N—N(CH_3)—C_4H_9\text{-}n$	263	439
$NC—N{=}N—N(CH_3)_2$	268(4.14)	451
$C_6H_5—N{=}N—N(CH_3)_2$	225(3.98)	294
yellow		
	285(4.14)	
	308(4.11)	
$C_6H_5—N{=}N—N(CH_3)—C_6H_5$	235(4.28)	123
yellow		
	294(4.13)	
	340(4.25)	
$C_6H_5—N{=}N—NHCH_3$	276(4.1)	399
$C_6H_5—N{=}N—NHC_6H_5$	238(4.26)	123
yellow		
	287(3.89)	
	293(3.89)	
	346(4.26)	
$p\text{-}CH_3OC_6H_4—N{=}N—NH—C_6H_4OCH_3\text{-}p$	239(4.08)	123
yellow		
	301(4.07)	
	362(4.26)	
$p\text{-}ClC_6H_4—N{=}N—NHC_6H_4Cl\text{-}p$	238(4.24)	123
yellow		
	298(4.01)	
	356(4.35)	

Addition of oxide to the triazene chain causes a small bathochromic shift.

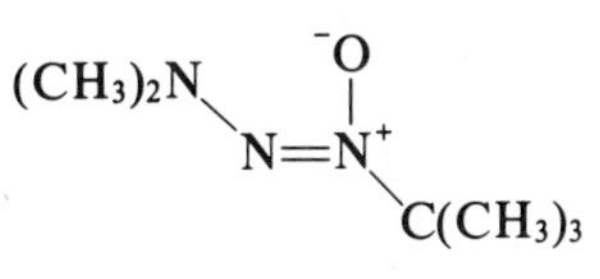

281(3.65) (230)

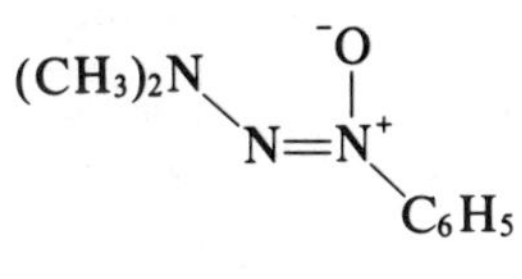

yellow
227(3.91) (230)
326(3.94)

Many diaryltriazenes are yellow because absorption of the 340–365 nm band extends into the blue region. Steric inhibition of planarity reduces the length of the conjugated system. When two of the four ortho positions in polyhalodiaryltriazenes are occupied by halogen atoms, the triazene is colored, but if all four ortho positions contain halogen, the product is colorless, suggesting that the two phenyl rings cannot achieve coplanarity (318).

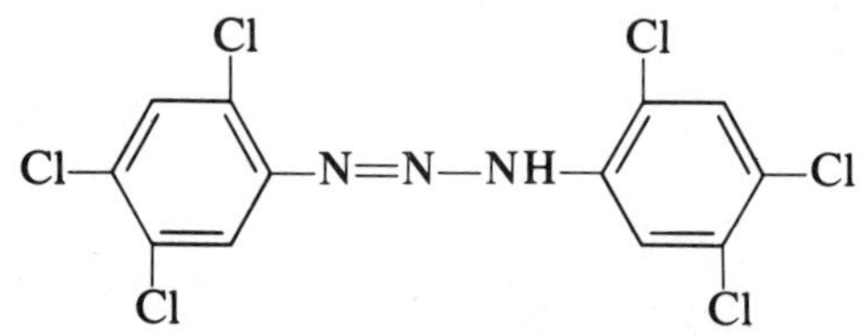

yellow
365(4.37)

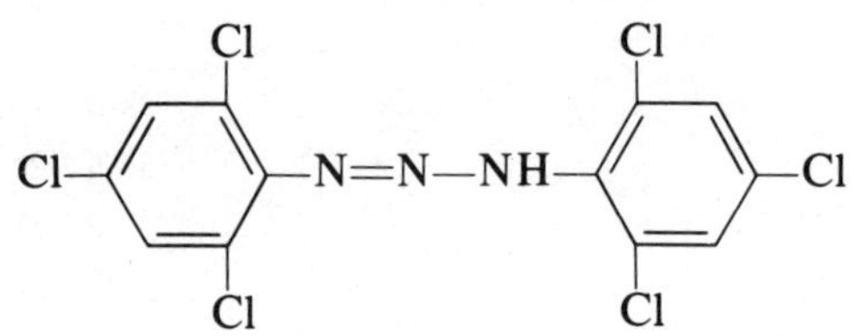

colorless
275–280(4.01)

The ultraviolet spectra of triazolines are similar to those of acyclic triazenes. Thus the norbornene triazoline with benzyl on nitrogen exhibits a maximum close to that of the trialkyltriazenes, and a 1-phenyl causes the expected batho-

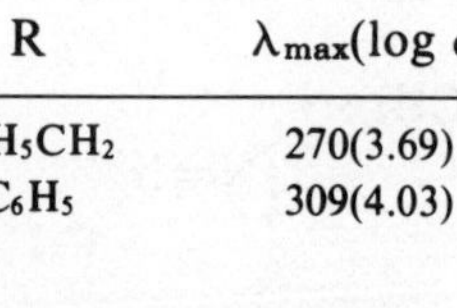

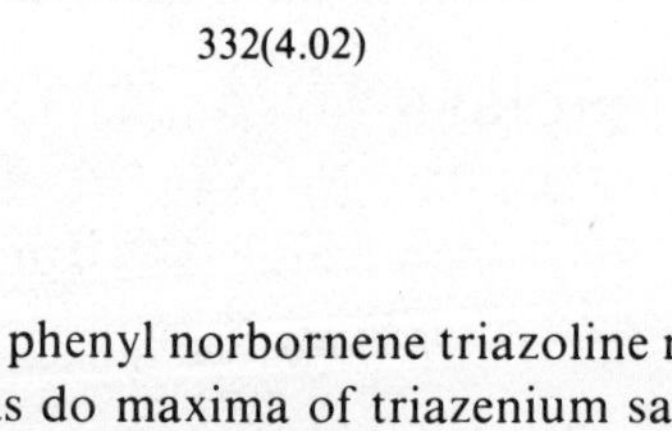

R	$\lambda_{max}(\log \epsilon)$
$C_6H_5CH_2$	270(3.69)
C_6H_5	309(4.03)

332(4.02)

chromic shift (421). Methylation of the phenyl norbornene triazoline moves the maximum to still longer wavelengths as do maxima of triazenium salts (Table 3.11) on substitution of aryl for alkyl on nitrogen (209).

The main conjugation in dibenzo[*d,f*][1,2,3]triazepine between the benzene rings involves the triazene chain rather than the junction of the phenyl groups, which are noncoplanar (175).

In the 2-tetrazenes, reduction of the nitrogen-nitrogen single bond to

Table 3.11. Ultraviolet Maxima of Triazenium Salts

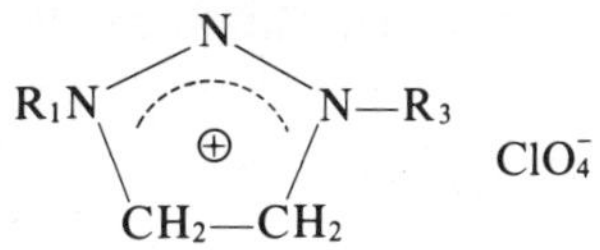

R_1	R_3	Color	$\lambda_{max}(\log \epsilon)$
$(CH_3)_3C$	CH_3	colorless	272(3.88)
C_6H_5	CH_3	colorless	326(4.05)
C_6H_5	C_6H_5	yellow	385(4.24)

1.38–1.39 Å (Section 3.1), although less than shortening in the triazenes, indicates conjugation along the four-nitrogen chain including nonbonded electrons on the terminal nitrogen atoms. As aryl replaces alkyl in the 2-tetrazenes, the maxima appear at longer wavelengths (Table 3.12). Steric inhibition of conjuga-

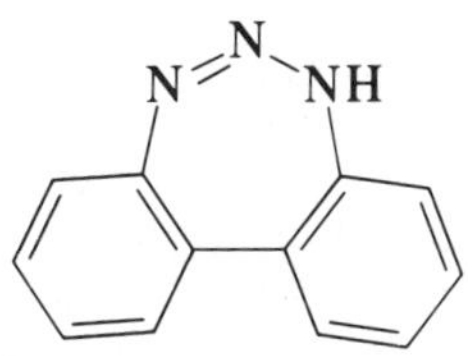

yellow
239(4.44)
265(3.99)
294(3.77)

Table 3.12. Ultraviolet Maxima of 2-Tetrazenes

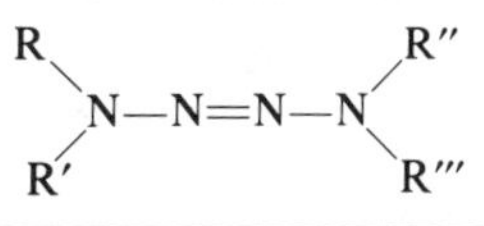

R	R′	R″	R‴	$\lambda_{max}(\log \epsilon)$	Ref.
CH_3	CH_3	CH_3	CH_3	277(3.92)	89
CH_3	CH_3OOC	CH_3	CH_3OOC	255sh(4.15) 270(4.26)	229
C_2H_5	CH_3CO	C_2H_5	CH_3CO	285(4.38)	466
C_2H_5	C_6H_5	C_2H_5	C_6H_5	353(4.50)	66
$(CH_3)_3C$	m-ClC_6H_4	$(CH_3)_3C$	m-ClC_6H_4	250(3.93) 311(3.75)	361
C_6H_5	C_6H_5	C_6H_5	C_6H_5	360(4.31)	66
$(CH_3)_3Si$	$(CH_3)_3Si$	$(CH_3)_3Si$	$(CH_3)_3Si$	254(3.99) 293(3.74)	494

tion is shown by the lower maximum of 1,4-di-*m*-chlorophenyl-1,4-di-*t*-butyl-2-tetrazene, in which the aryl groups are out of the plane of the nitrogen chain, in

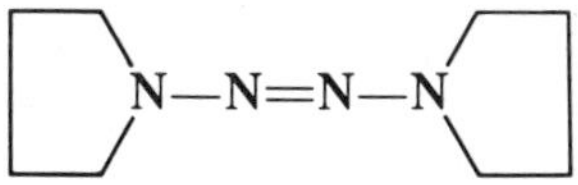

282(3.99) (297)

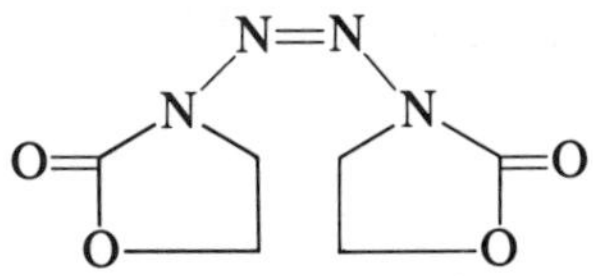

217(3.78) (167)
290(3.58)

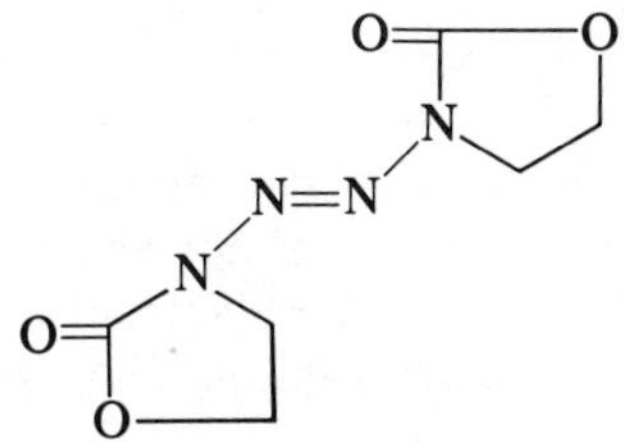

252(4.00) (167)
272(4.25)

comparison with that of 1,4-diethyl-1,4-diphenyl-2-tetrazene. Addition of hydrochloric acid to an ethanol solution of tetraethyltetrazene effects a hypsochromic shift from 284 nm to about 240 nm by removing a free pair of electrons from participation in conjugation and identifies this absorption as an $n \rightarrow \pi^*$ band (89). 2-Tetrazenes in which the end groups are part of a saturated ring absorb in the same region as the tetraalkyl analogs. The long wavelength maximum of *cis*-3,3′-azobis(2-oxazolidinone) is bathochromically shifted in comparison to the trans isomer, whose absorption is very similar to that of 1,3-di(methoxycarbonyl)-1,3-dimethyl-2-tetrazene.

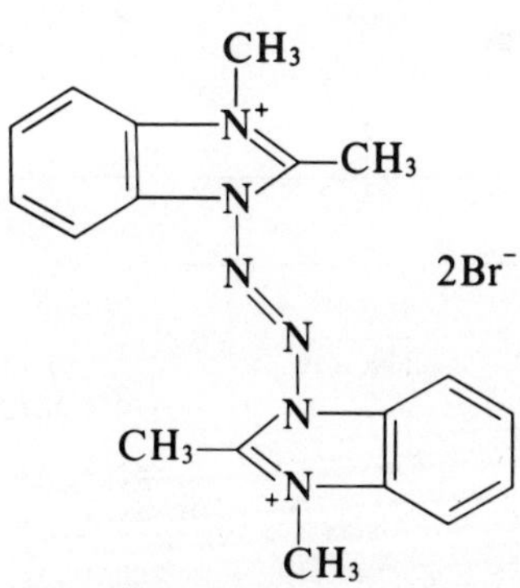

235sh(4.13) (187)
257sh(4.3)
264(4.42)
277(3.85)
344(4.12)

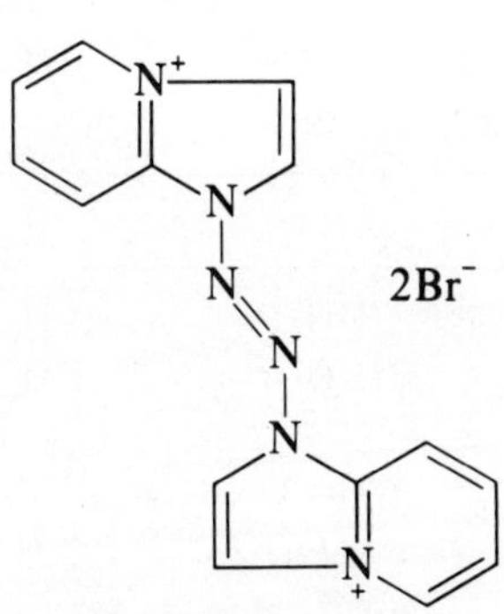

215sh(4.42) (188)
274(3.41)
366(4.64)

Bis-quaternary tetrazenes, such as 3,3′-dimethyl-1,1′-azobisbenzimidazolium and 1,1′-azobisimidazo[1,2-*a*]pyridinium salts, have characteristic long wavelength maxima near 350 nm.

Maxima of both the 1,4-dimethyl and diphenyl-1,4,5,6-tetrahydrotetrazines are hypsochromically shifted with respect to the related acyclic 2-tetrazenes.

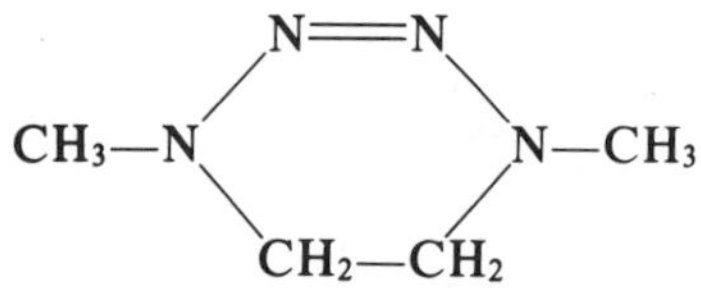

225(3.47) (430)
268(3.75)

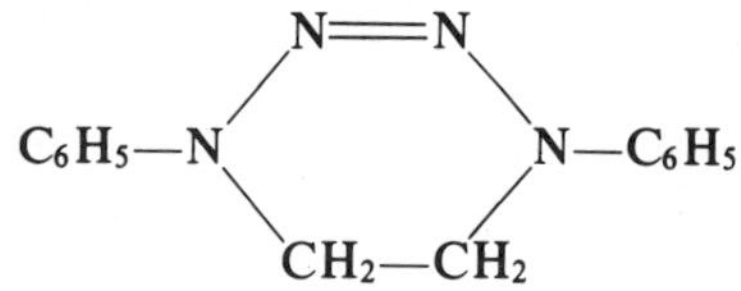

249(3.92) (280)
323–325(4.33)

The long wavelength maximum of 1,5-dimethyl-4-phenyltetrazoline, the equivalent of a monoaryl-trialkyl-2-tetrazene, lies between those of 1,4-diphenyl-

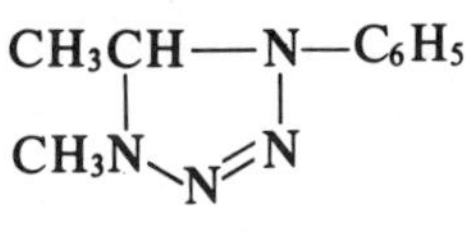

222(3.47) (241)
304(3.59)

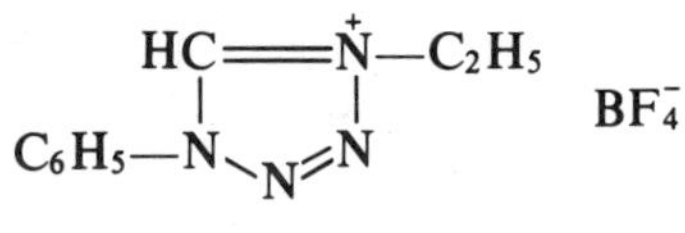

239(3.85) (277)

1,4-dialkyl and tetraalkyl-2-tetrazenes, and closely resembles that of 1,1-dimethyl-3-phenyltriazene. In contrast, the maximum of the aromatic

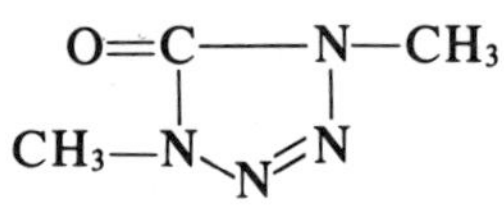

225(3.79) (479)

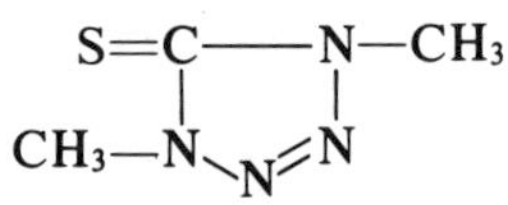

206(3.77) (469)
246(4.13)
275(3.19)

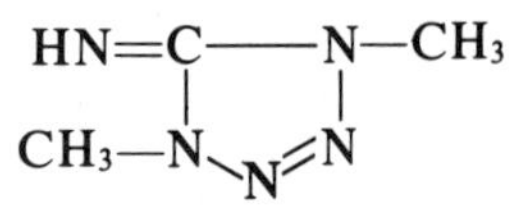

260(3.24) (358)

1-ethyl-4-phenyltetrazolium cation is appreciably below that of the uncharged 1,4-tetrazolines. The maximum reported for 1,4-dimethyltetrazolinone is below those of 1,4-diethyl-1,4-diacetyl-2-tetrazene and the corresponding thiono- and

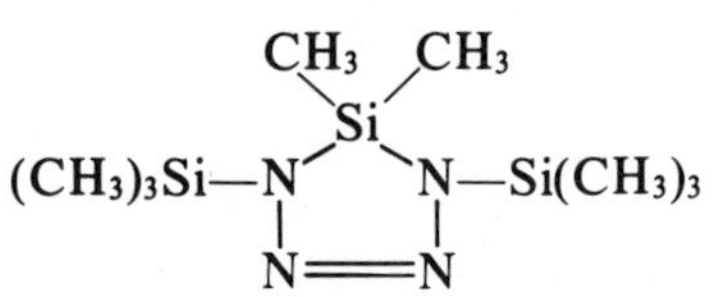

(IV)

246(3.77) (497)

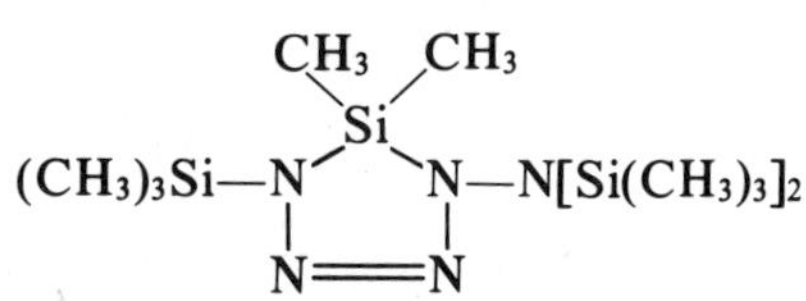

257(3.63) (497)

Table 3.13. Ultraviolet Maxima of Tetraazaborolines

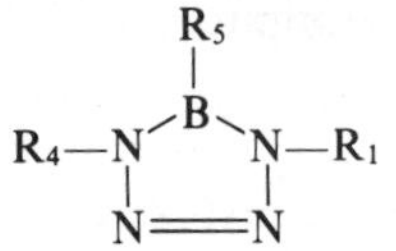

R_1	R_4	R_5	$\lambda_{max}(\log \epsilon)$	Ref.
CH_3	CH_3	H	198(3.80)	349
CH_3	CH_3	CH_3	200(3.96)	292
CH_3	C_6H_5	H	242(4.10)	349
C_6H_5	C_6H_5	CH_3	239(4.38)	292

iminotetrazolines. Replacement of ring carbon with boron, forming the tetraazaborolines (Table 3.13), shifts the wavelength of the maximum below that of the 1,4-tetrazolines. The cyclic silatetrazoline **(IV)** absorbs at longer wavelengths than the tetraazaborolines, but with less intensity. Replacing the trimethylsilyl group on **IV** with bis(trimethylsilyl)amino effects a small bathochromic shift.

Pentazdienes are colorless to bright yellow. In a series with phenyl, *p*-tolyl, and *p*-anisyl at the 1- and 5-positions, and benzyl or allyl at N-3, the long wavelength absorption occurs at 370–380 nm (log ϵ about 4.5) (345), longer than

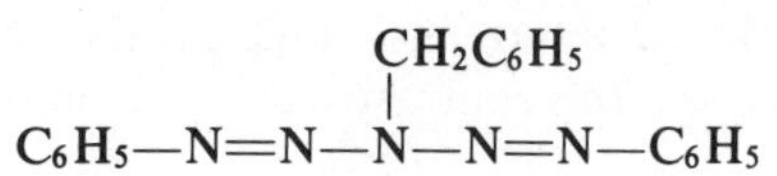

yellow

diarylmonoalkyl triazenes or the 2-tetrazenes. Nonbonding electrons on the middle nitrogen probably are involved in the conjugated system. The ultraviolet spectra of 1,6-bisaryl-3,4-diacetyl-1,5-hexazdienes are nearly superimposable

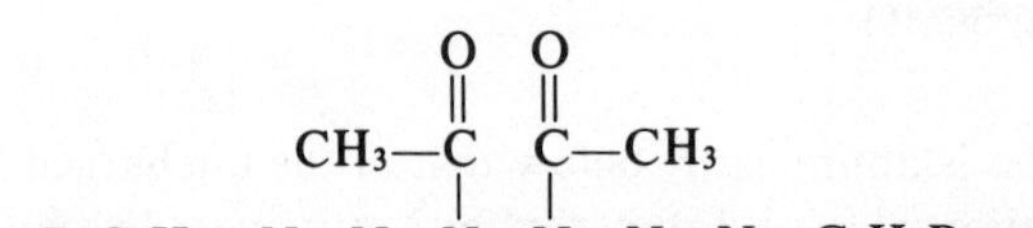

227
290

$CH_3CONHN{=}N{-}C_6H_4Br$-*p*

227
290

on those of the related 1-acetyl-3-aryltriazenes, suggesting little or no electronic interaction between the two three-nitrogen segments of the hexazdiene (194), consistent with the large dihedral angle between the two halves of 1,6-bis(*p*-chlorophenyl)-3,4-diacetyl-1,5-hexazdiene (320) (Section 3.1).

Disubstituted triazenes without hydrogen on the chain are at a higher oxidation level than triazenes with hydrogen or trisubstituted triazenes, and, combined as part of extended conjugated systems, absorb at longer wavelengths.

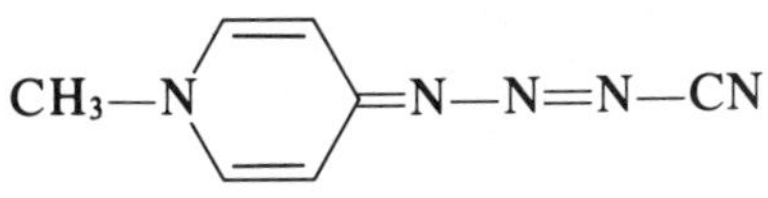

vivid yellow (255)
230(3.80) (461)
245(3.88)
376(4.59)

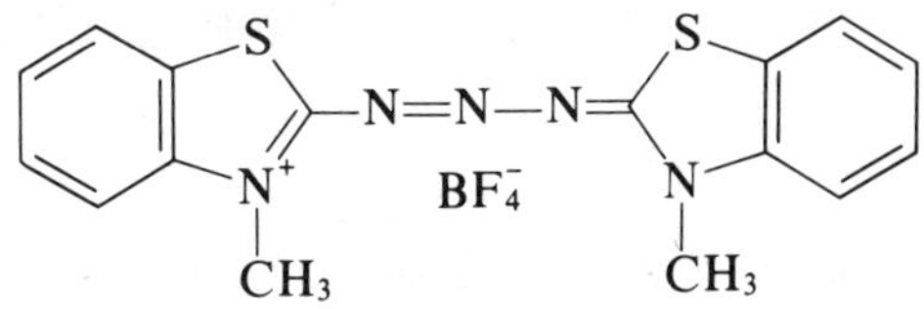

(V)

bright red
375(3.76) (398)
484(4.62)

The three-nitrogen segment of 6-phenoxy-2,2,2-tris(piperidino)2,2-dihydro-1,3,4,5,2-thiatriazaphosphorine is also at the same formal oxidation level. Although absorbing at a higher wavelength than 2-tetrazenes, *trans*-1,4-bis(3-ethylbenzothiazolinylidene)tetrazene, oxidized at both terminal nitrogens, has a

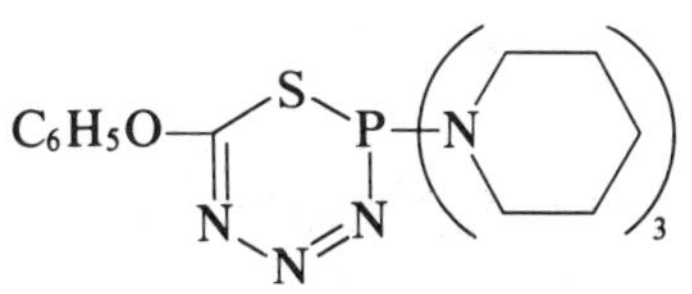

orange red
259(4.01) (391)
350(4.28)

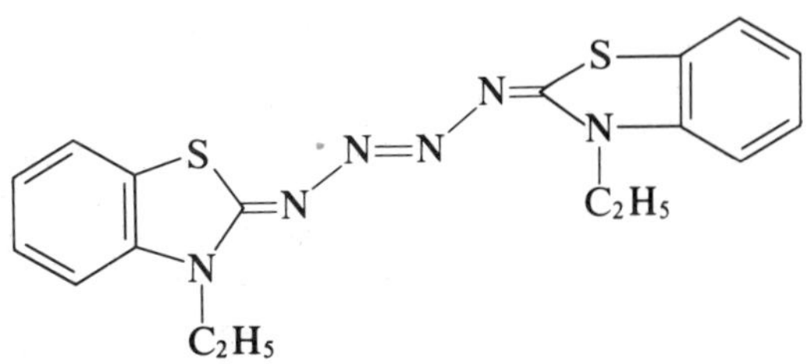

yellow
304(3.65) (40)
396(4.72)

lower maximum than the bis(3-methylbenzothiazolidine)triazene salt **(V)** suggesting less effective transmission of conjugation through the thiazoline rings. In comparison to the pentazdienes, pentaazapentamethine cyanines absorb at a still higher wavelength and with greater intensity, owing to their extended electron delocalized systems analogous to the bathochromic shift of triazatrimethine cyanines relative to triazenes.

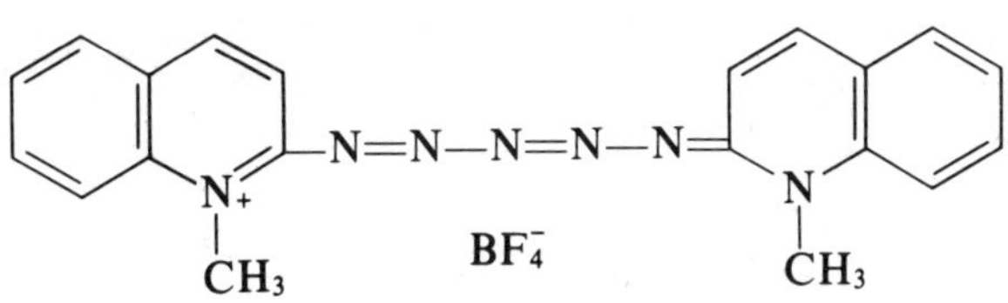

dark violet
513(4.85) (239)

Monoalkyl and monophenylnitrosohydrazines exhibit an $n \rightarrow \pi^*$ transition, probably involving unshared electrons on nitrogen. Nitrosohydrazines with

246(3.9) (119)

pale yellow
276(3.9) (119)

bright yellow (464)

yellow
371(1.6) (220)
387(1.7)
405(1.7)

more electron-withdrawing groups are colored, but with low absorption intensity. Although 1-nitroso-3-phenyl-Δ^2-pyrazoline is colorless (51, 359), *N*-nitrosopyrazoles are yellow. Nitrosotrimethyl tetrazene absorbs at a longer wavelength than the nitroso-monosubstituted hydrazines.

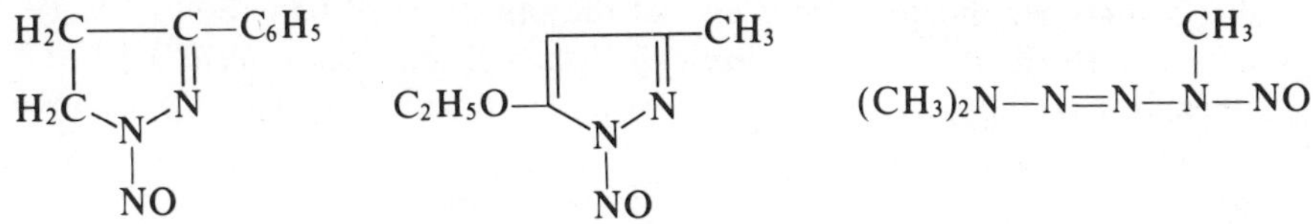

298(4.41) (359)

bright yellow (503)

slightly yellow
310(4.11) (466)

The amine nitroimides are colorless. On the basis of a CNDO2/S calculation, absorption of pyridine *N*-nitroimide at 300 nm is believed to be due to NO_2, but a comparable maximum is not found in the *s*-triazole nitroimide.

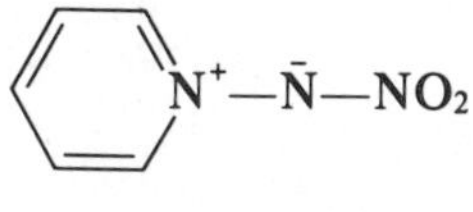

224(3.85) (24)
260(3.64)
300(3.26)

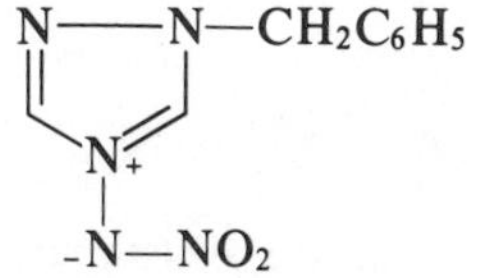

254(3.91) (468)

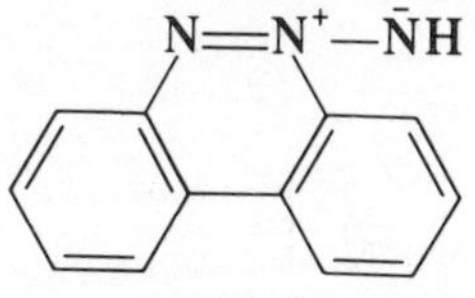

yellow
254(4.58) (175, 176)
297(3.95)
310(3.89)
323(3.84)
367(3.96)

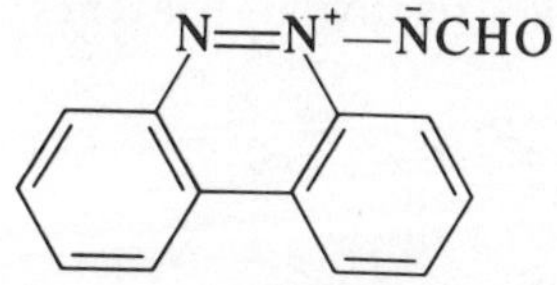

yellow
249(4.45) (366)
298(3.76)
425(3.98)

The benzo[*c*]cinnoline *N*-imides are colored, resembling the *N*-oxides. Incorporation of the three-nitrogen chain in a ring, eliminating the dipole and the azimine unsaturation, however, does not grossly change the absorption. Conjugation, involving unshared electrons on the partially embedded nitrogen atoms appears to be involved in the spectrum of the benzo[*c*]-*v*-triazolo[1,2*a*]cinnoline. The *N*-imides of benz[*cd*]indazole **(VI)** are intensely colored.

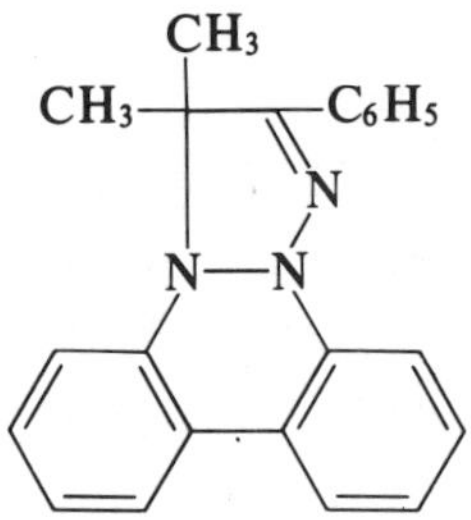

orange
254(4.51) (45)
260(4.49)
388(3.89)

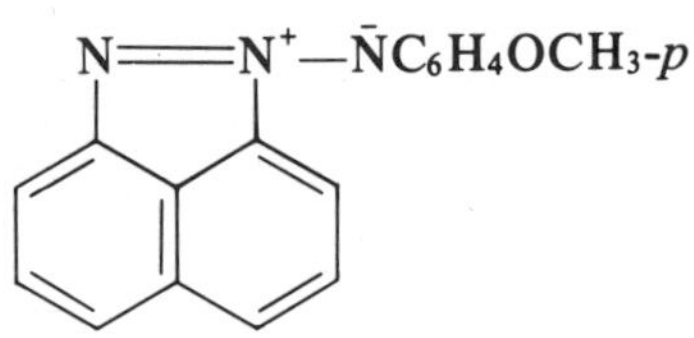

(VI)

dark red violet
293(4.32) (446)
302(4.25)
352(3.98)
531(4.39)
554(4.35)

Azimines in four-nitrogen chains absorb near or at higher wavelengths than the bis-quaternary tetrazenes. Ultraviolet spectra of the *cis*- and *trans*-2,3-diaryl-1-phthalimido azimines enabled assignment of configurations by comparison with spectra of corresponding azoxy aryl compounds; the absorption intensity near 340 nm is larger in the trans derivative. *trans*-2,3-Dialkyl-1-phthalimido-

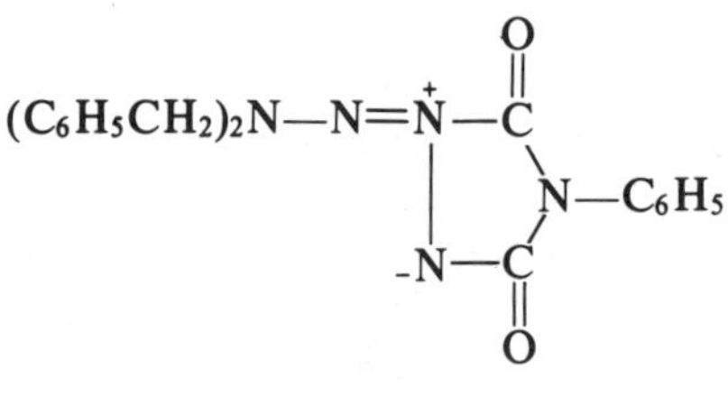

yellow
389(4.12) (9, 271)

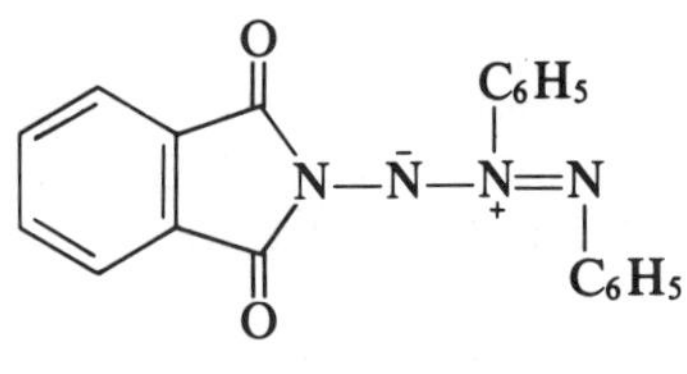

yellow
338(4.21) (228)

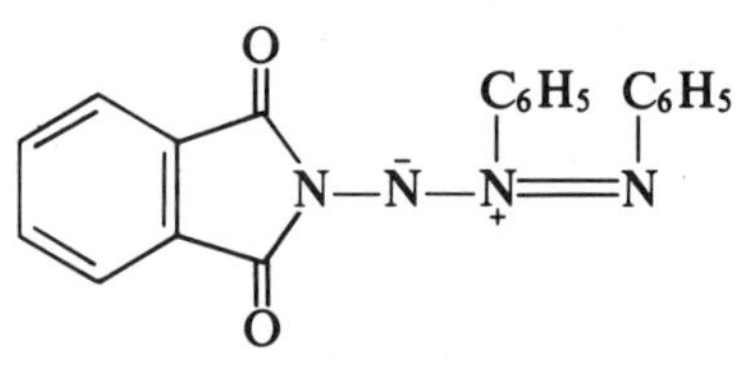

orange
342(3.83) (228)

Table 3.14. Ultraviolet Maxima of Organic Azides

Azide	$\lambda_{max}(\log \epsilon)$	Ref.
n-$C_4H_9N_3$	216(2.7)	110
	286(1.4)	
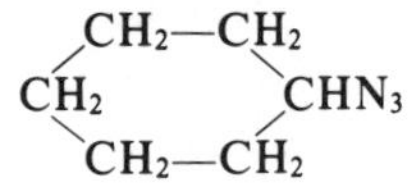	217(1.6)	110
	287(1.4)	
$(n$-$C_5H_{11})_3CN_3$	285(1.4)	2
$C_6H_5N_3$	246(4.37)	4
	276(3.60)	
p-$H_2NC_6H_4N_3$	260(4.26)	443
	307(3.49)	
p-$BrC_6H_4N_3$	255(4.25)	443
	285(3.48)	
	297(3.32)	

azimines are characterized by a shoulder at 258 nm (log ϵ ~4.15) and the cis configuration by a maximum at 270–278 nm (log ϵ ~4.00) (298a).

Hydrogen azide absorbs in the gas phase at 190 (log ϵ ~2.9) and at 264 nm (log ϵ ~1.3) (338). Alkyl azides absorb near 215 and 285 nm with low extinction coefficients, both considered to be $n \rightarrow \pi^*$ transitions (471). Maxima of aryl azides are much more intense than those of the aliphatic azides (Table 3.14). By comparison with the spectra of aromatic hydrocarbons, the main bands of aryl azides are assigned as $\pi \rightarrow \pi^*$ transitions.

Parent aromatic high nitrogen heterocycles absorb at lower wavelengths than substituted triazenes or azides. *v*-Triazole absorbs at the same wavelength as pyrrole. A small bathochromic shift occurs on substitution at position 4 by alkyl and a larger shift with phenyl (Table 3.15); 4-phenyl-*v*-triazole absorbs at the same value as diphenyl, emphasizing the aromatic character of the *v*-triazole ring.

Conversions of *v*-triazoles with hydrogen on ring nitrogen to their anions shifts the maxima to higher wavelengths (460).

1-Alkyl and aryl *v*-triazoles absorb at about the same wavelengths as *v*-triazoles unsubstituted on ring nitrogen, but the maxima of 2-substituted deriva-

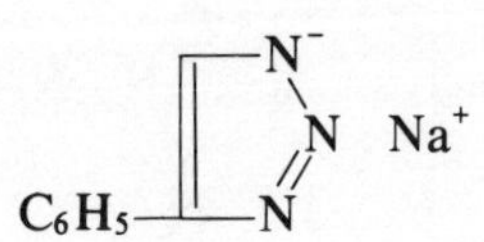

261(4.22)

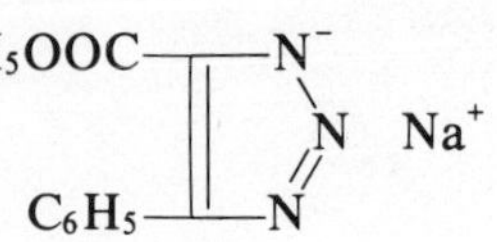

262(4.00)

Table 3.15. Ultraviolet Maxima of *v*-Triazoles Unsubstituted on Nitrogen

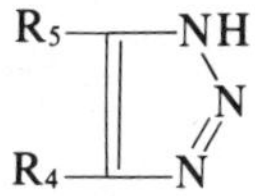

R_4	R_5	$\lambda_{max}(\log \epsilon)$	Ref.
H	H	210(3.6)	218
n-C_7H_{15}	H	215(3.7)	218
C_6H_5	H	245(4.2)	218
H_2NCO	H_2N	226(3.97)	201
		260(3.95)	
C_6H_5	C_2H_5OOC	240(3.99)	460

tives are at longer wavelengths (Table 3.16). Maxima of both cationic salts and mesoionic 1-methyl-3-phenyl-*v*-triazolium derivatives are higher than those of 1-phenyl-*v*-triazole.

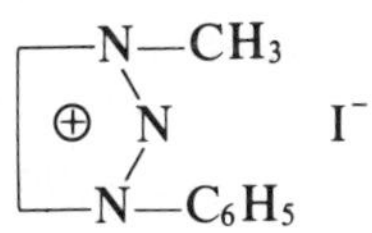

251.6(3.92) (411)

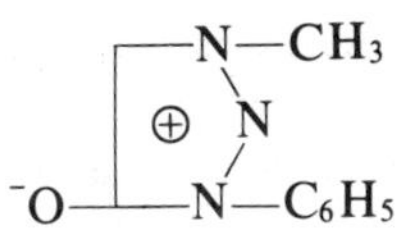

203(3.88) (396)
237(3.62)
298(3.82)

Table 3.16. Ultraviolet Maxima of 1- and 2-Substituted *v*-Triazoles

			$\lambda_{max}(\log \epsilon)$ (Ref.)			
R	R_4	R_5	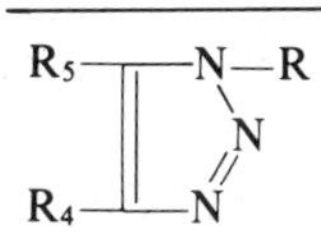		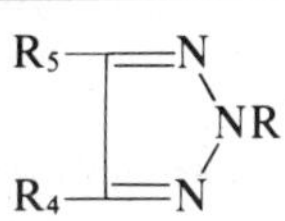	
$CH_3CH_2OCH_2CH_2$	H	H	210(3.63)	(291)		
CH_3	H	CH_3S	213(3.59)	(231)		
			221(3.54)		233(3.55)	(121)
			250(3.53)		256(3.67)	
CH_3	H_2NCO	NH_2	227(3.99)	(11)	217infl(3.81)	(11)
			260(3.92)		273(3.80)	
C_6H_5	H	H	243(4.01)	(197)	262(4.14)	(150)
C_6H_5	C_6H_5	H			295(4.27)	(329)

Table 3.17. Ultraviolet Maxima of 1- and 2-Substituted Benzotriazoles

	$\lambda_{max}(\log \epsilon)$	
R	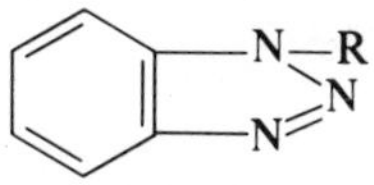	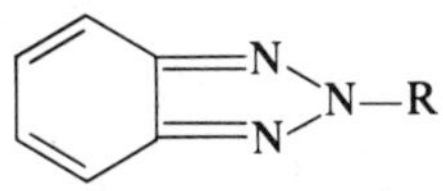
H	259(3.75)	
	275(3.71)	
CH_3	256(3.81)	274(3.96)
	262(3.81)	280(3.98)
	279(3.68)	285(3.97)
C_6H_5	236(4.10)	
	266(3.91)	221(4.25)
	292(3.82)	307(4.40)

Spectra of benzotriazole and its 1-methyl derivative are very similar, but 2-methyl and particularly 2-phenylbenzotriazole absorb at higher wavelenths (Table 3.17) (117). Both 2- and 3-methylbenzotriazole 1-oxide absorb at wavelengths over 40 nm longer than 1-methoxybenzotriazole, the spectrum of which differs only slightly from that of 1-methylbenzotriazole. The spectrum of 8-azapurine is similar to benzotriazole in neutral solution, but its maximum is

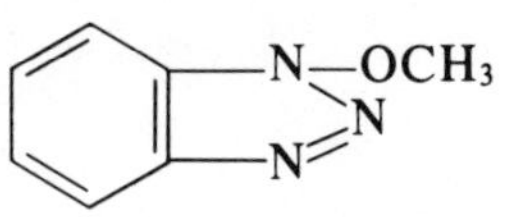

263(3.73) (431)
283(3.60)

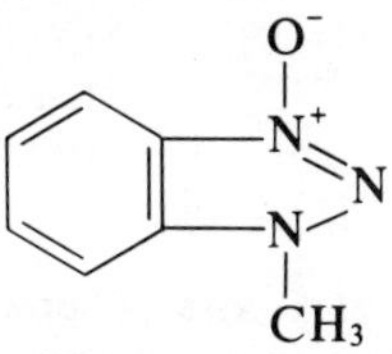

273(3.3) (78)
280(3.3)
325(3.9)

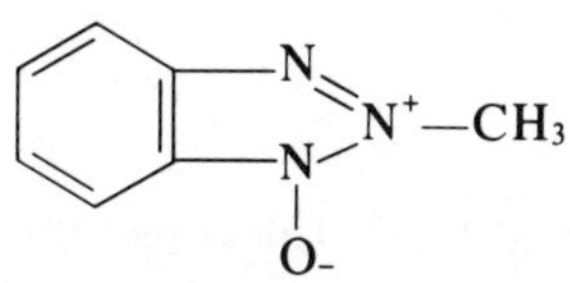

326(3.7) (78)

hypsochromically shifted in a strongly acid medium. At slightly acid pH values, neither 8-azaadenine nor 8-azaguanine differ greatly from 8-azapurine. In the 5-

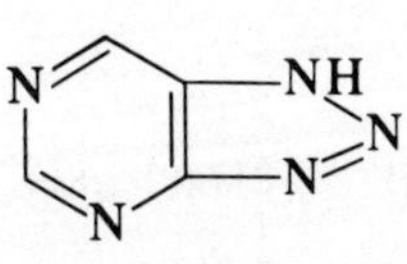

pH	
7.0	268(3.89) (10)
3.4	263(3.87)
0	248(3.91)

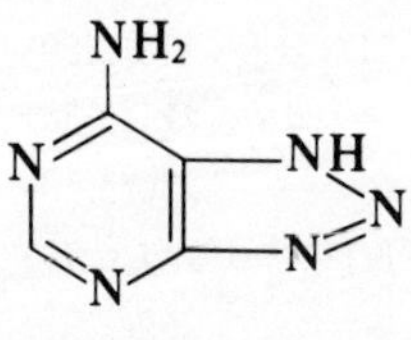

pH 4.5: 273(4.01) (13)

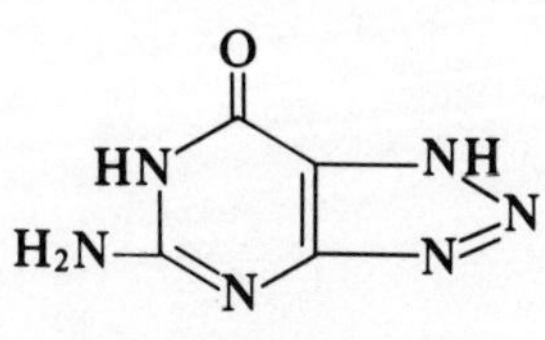

pH 3.8: 247(4.05) (10)
266(3.83)

phenyltriazolo[4,5-*d*]triazoles, the 2-methyl derivatives, for which only a dipolar structure can be written, exhibit maxima very slightly higher than the 1-isomer (508).

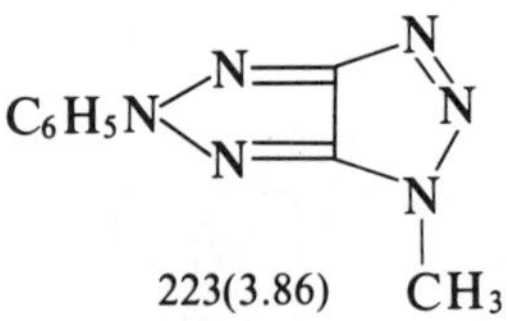

223(3.86)
299(4.33)

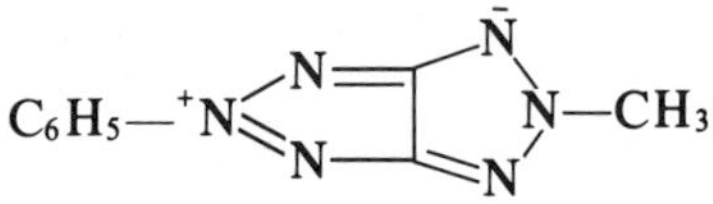

226(3.79)
301(4.40)

Neutral monocyclic *v*-triazines are colorless except for the tribromo compound. On protonation or alkylation, the *v*-triazines become colored, as are the

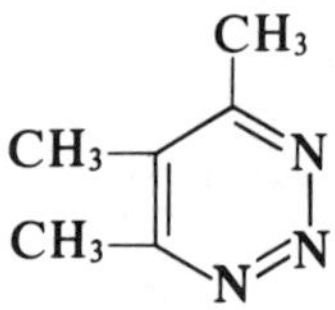

217(3.63) (109)
278(2.79)

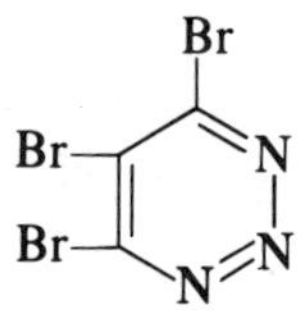

golden yellow
226(4.28) (191)
255sh
291sh

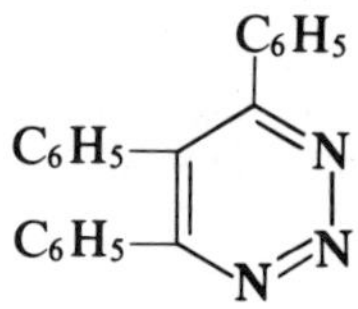

262(4.33) (100)
278sh(4.19)
297sh(3.91)

reduced, cross-conjugated *v*-triazine **(VII)** and the related 2,4,6-triphenyl-1,3,4,5-oxatriazine.

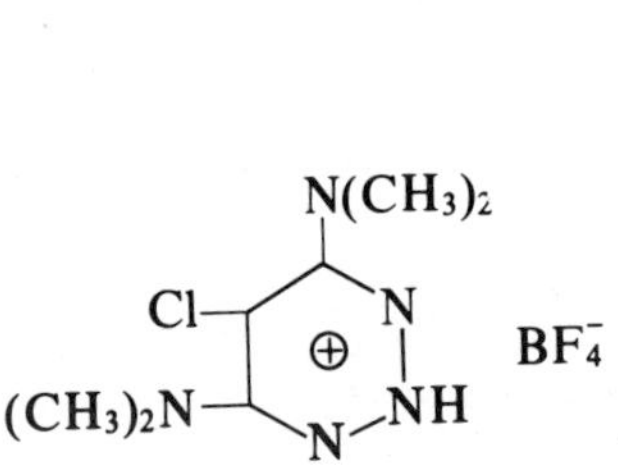

yellow
369(3.32) (192)

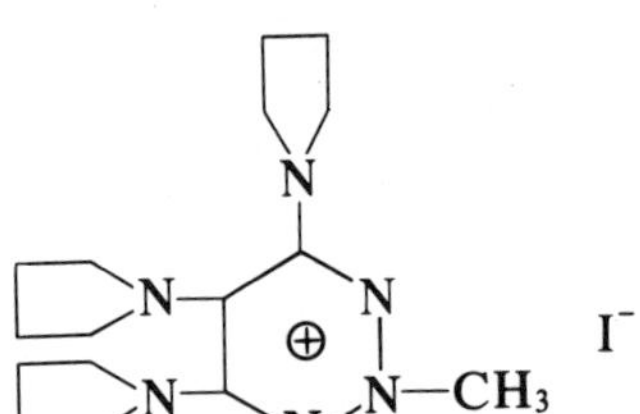

orange
456(3.77) (192)

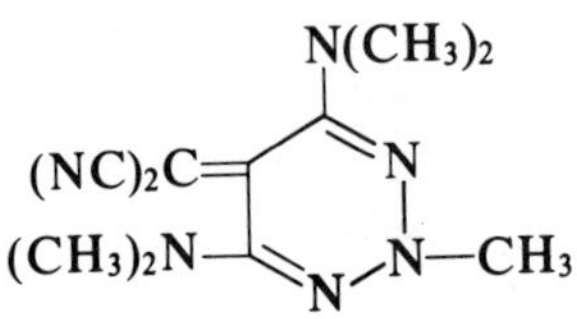

(VII)

red
484(4.20) (192)

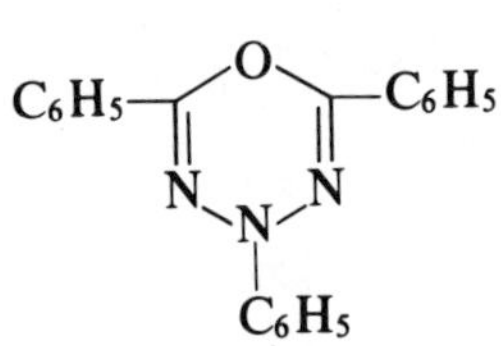

scarlet (174)
240(4.39)
315(4.10)
401(3.53)

The ultraviolet spectrum of 4-methylbenzotriazine does not differ greatly from that of trimethyl-*v*-triazine; a similar pattern is found with benzotriazinone, and a small bathochromic shift with 3-*p*-tolylbenzotriazinone. Naphtho[1,8-*de*]triazine and its 1-alkyl derivatives are red, but substitution at

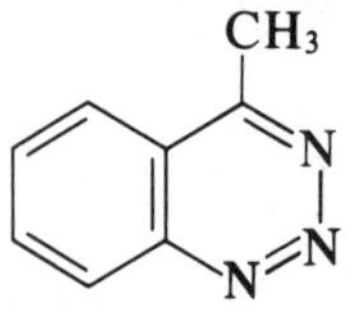

207(3.58) (5)
227(4.0)
275(2.83)

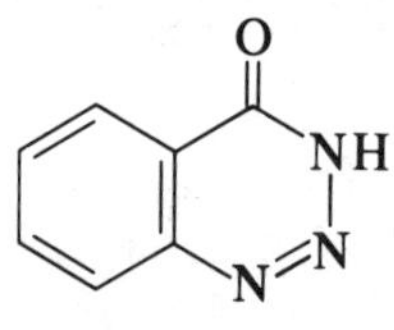

223(3.28) (88)
278(2.77)

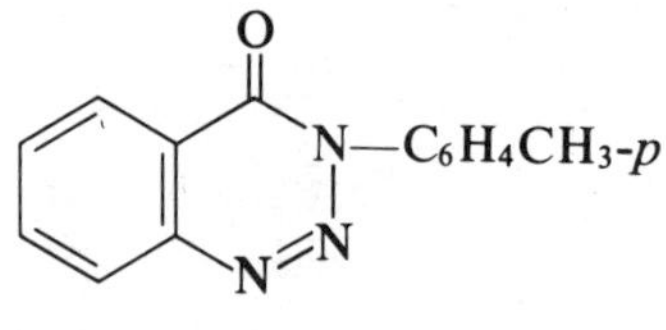

212(4.30) (340)
227(4.48)
292(3.84)

the 2-position shifts the long wavelength maxima to much longer wavelengths and the compounds are blue (386). Triazolo[1,5-*a*]pyridine **(VIII)** and, particu-

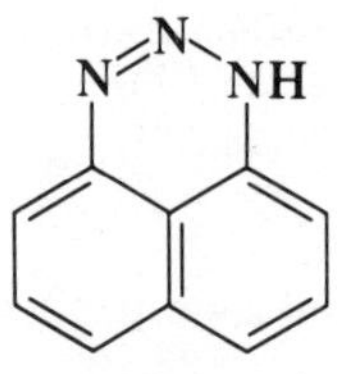

232.5(4.49)
338.5(4.00)
452(2.88)

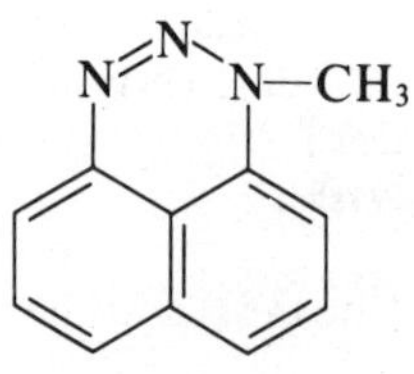

231.5(4.51)
338(4.01)
451(2.98)

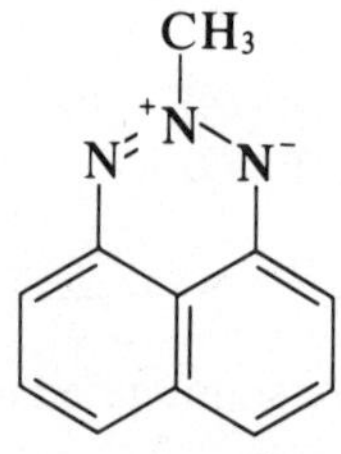

231.5(4.52)
355(4.11)
559(2.91)
603(2.91)
655(2.76)

larly pyrazolo[2,3-*b*]pyridazine absorb at longer wavelengths than their isomer, benzotriazole. In acid solution the maximum of **VIII** undergoes a hypsochromic shift to 267 nm, indicating the transitions to be $n \rightarrow \pi^*$.

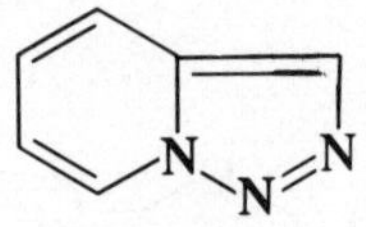

(VIII)

280(3.88) (77)

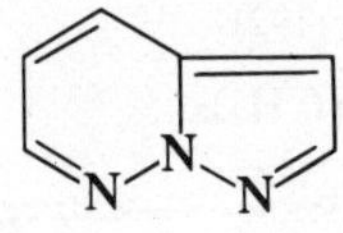

224(4.57) (403)
276(3.08)
283sh(3.03)
317(3.15)

Replacement of ring C—H in *v*-triazoles with sulfur, forming thiatriazoles, causes a bathochromic shift. 3-Phenyl thiatriazolium and oxatriazolium salts

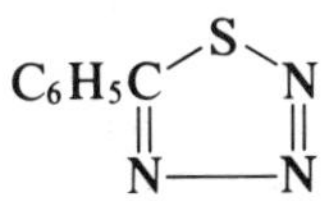

280(4.03) (267)

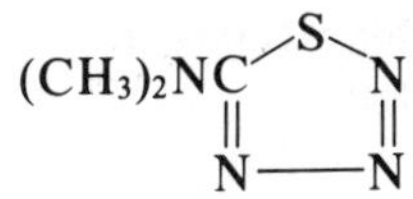

255sh(3.40) (304)
279(3.79)

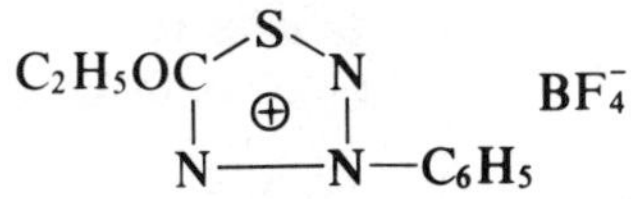

229infl(3.71) (205)
278infl(3.83)
313(4.02)

without other chromophores are colorless, but extension of conjugation in related mesoionic compounds moves the absorption into the visible.

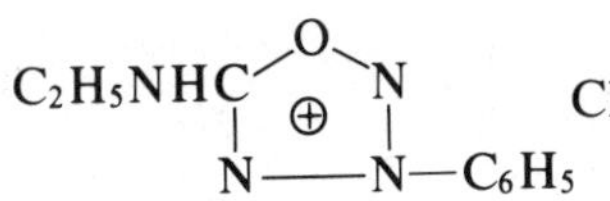

282(4.00) (106)

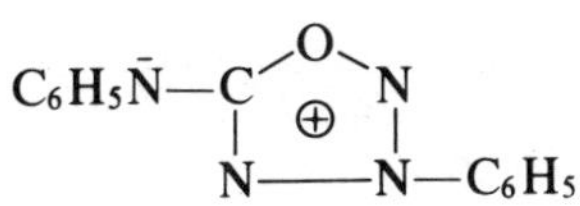

dark red
265(4.49) (206)
410(3.40)

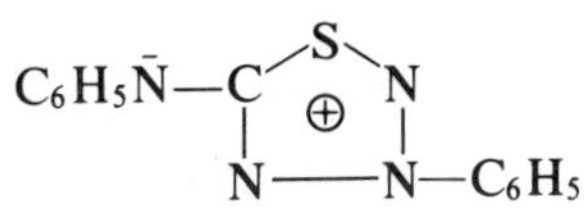

dark red (207)
279(4.47)
411(3.26)

Lengthening the nitrogen chain to four contiguous nitrogen atoms by substituting an amino group on monocyclic and fused ring *v*-triazoles and in naphtho[1,8-*de*]triazines has substantially the same effect on the spectra as introduction of alkyl.

Tetrazole and 5-alkyltetrazoles exhibit only low level end adsorption near and below 210 nm (151, 157). Maxima of 5-aryl, acyl, amino, and nitroamino tetrazoles are shown in Table 3.18. The position of maxima of 5-aryltetrazoles are almost unaffected by electron-withdrawing or electron-releasing substituents on the aryl group, but the more strongly electron-attracting benzoyl group effects a bathochromic shift. The maximum wavelength and absorption intensity of 5-phenyltetrazole and 4-phenyl-*v*-triazole are essentially the same. Since

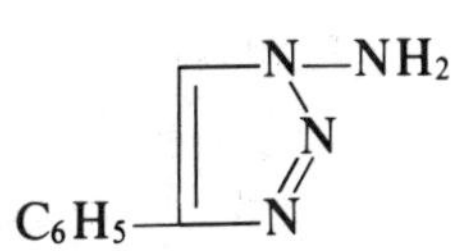

244(4.19) (221)

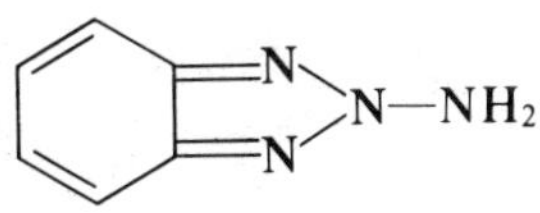

216(4.12) (94)
285(4.08)

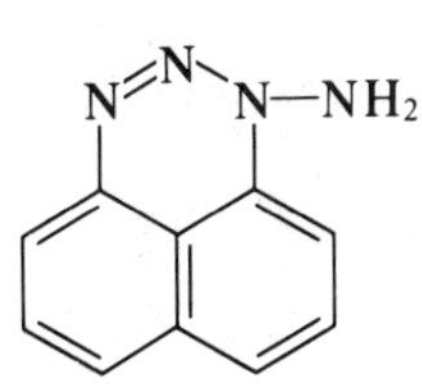

red
232.5(4.42) (404)
339(4.01)

Table 3.18. Ultraviolet Maxima of 5-Substituted Tetrazoles

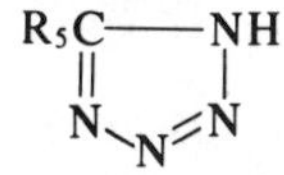

R_5	$\lambda_{max}(\log \epsilon)$	Ref.
C_6H_5	241(4.20)	157
p-ClC_6H_4	247(4.31)	157
p-$CH_3C_6H_4$	246(4.22)	157
C_6H_5CO	261(4.10)	163a
H_2N	218(3.49)	358
O_2NNH	277(4.11)	182
$O_2NN(CH_3)$	246(3.67)	182

the maxima of the potassium salts of both 5-nitraminotetrazole and 5-methylnitraminotetrazole are the same as the parent nitramino compounds, a high concentration of anion must be present in solutions of the parent compounds (182).

As with the *v*-triazoles, 2-substituted tetrazoles absorb at higher wavelengths than their 1-isomers (Table 3.19). Maxima for 1-aryltetrazoles are slightly hypsochromic with respect to the corresponding 5-aryltetrazoles and slightly less intense. Dipolar 5-benzyldimethylammonium tetrazolate has an upper maximum close to that of 2-benzyl-5-dimethylaminotetrazole, but a smaller extinction

Table 3.19. Ultraviolet Maxima of 1- and 2-Substituted Tetrazoles

		$\lambda_{max}(\log \epsilon)$ (Ref.)			
R	R_5	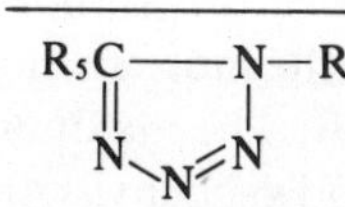		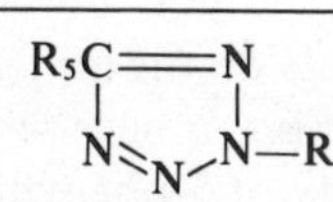	
C_6H_5	H	236(3.97)	(157)	280(4.2)	(237)
p-$H_2NC_6H_4$	H	278(3.97)	(314)	304(4.16)	(314)
p-$O_2NC_6H_4$	H	272.5(4.16)	(314)	279.5(4.24)	(314)
p-ClC_6H_4	H	242(4.15)	(157)		
CH_3	C_6H_5	232(4.00)	(151)	240(4.03)	(151)
CH_3	CH_3NH	232(3.50)	(224)	256(3.31)	(224)
$C_6H_5CH_2$	$(CH_3)_2N$	217(3.64) 239(3.54)	(236)	218(3.64) 272(3.40)	(236)

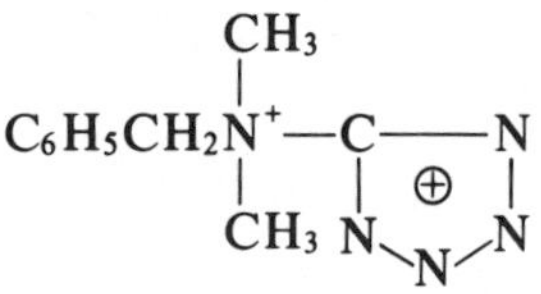

217(3.65) (236)
258(2.52)
263(2.66)
270(2.61)

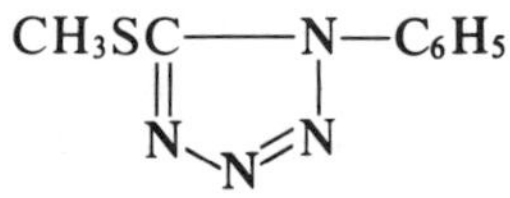

228(3.89) (46)
242sh(3.85)

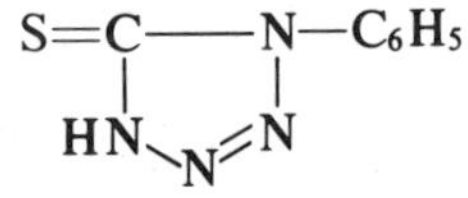

217.5(4.19) (304)
262(3.85)
280sh(3.81)

coefficient. In comparison to 5-methylthio-1-phenyltetrazole, the spectrum of 1-phenyl-5-thionotetrazoline is strongly shifted bathochromically and resembles that of 1,4-dimethyl-5-thionotetrazoline.

1-Ethyl-3-phenyltetrazolium fluoborate absorbs at a wavelength over 32 nm higher than its 1,4-isomer (277). In common with other high nitrogen compounds, as phenyl groups replace alkyl on ring nitrogen, bathochromic shifts and more intense absorption result. As with the mesoionic oxa- and thiatriazolium compounds, 1,3-diphenyltetrazolium-5-olate has a higher long wavelength maximum than the 1,3-tetrazolium salts because of a larger delocalized system. A similar relationship holds for the 2,3-disubstituted tetrazolium salts and related mesoionic compounds. As noted in Section 3.1, because the planes of the

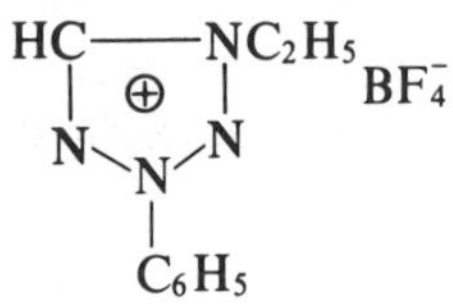

271.5(3.98) (277)

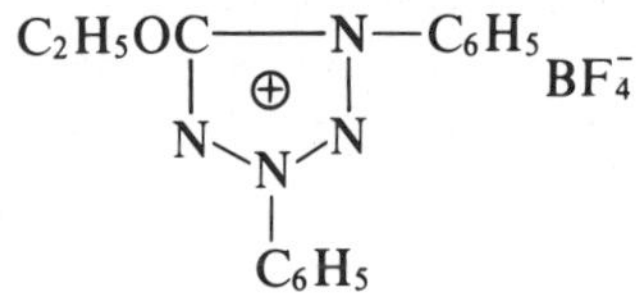

215sh(3.76) (206)
233infl(3.61)
296(4.02)

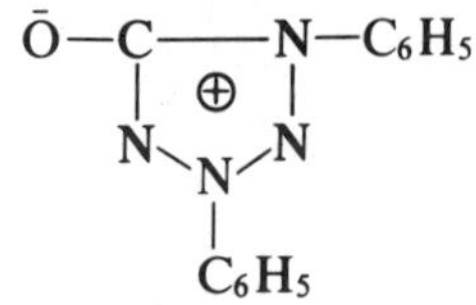

225(4.55) (206)
263(4.40)
326(4.51)

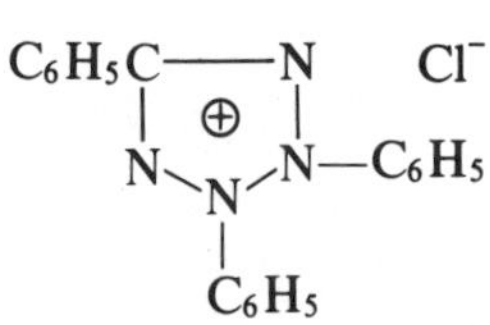

247(4.45) (18)

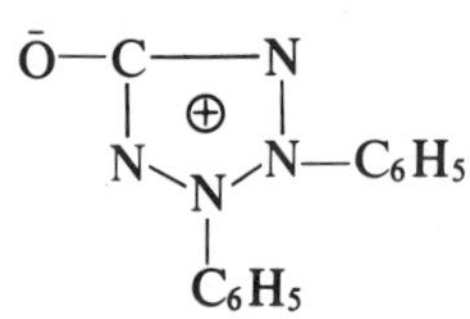

233(4.05) (204)
331(3.70)

phenyl groups are joined to the tetrazolium plane at an angle, conjugation between them is limited. 2,5-Diphenyl-3-tetrazolyltetrazolium betaine absorbs somewhat higher than the triphenyltetrazolium cation. Introduction of an electron-donating group into the 2-phenyl substituent at the ortho or para position leads to a strongly colored material, possibly owing to quinoid type conjugation as shown between the tetrazolium and the dimethylquinone imine moiety, which would be coplanar in this form.

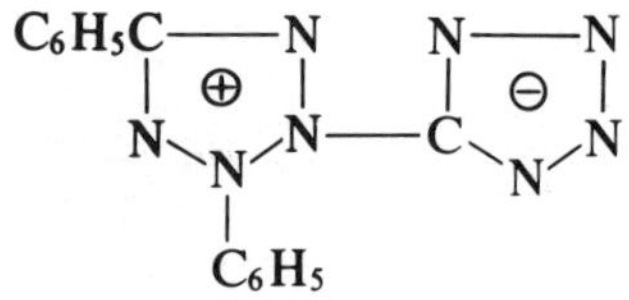

250 (286)
281

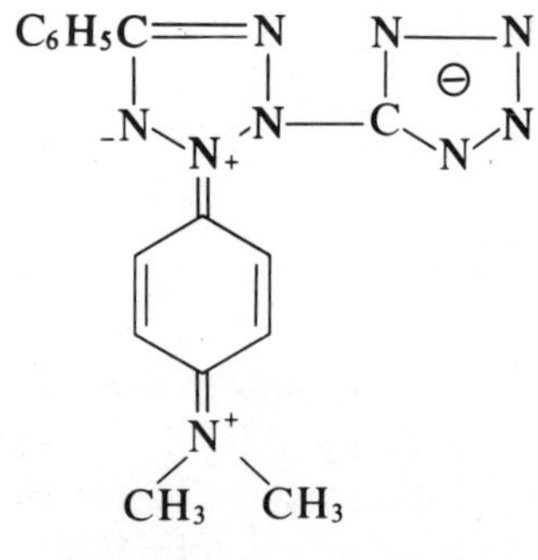

bright red (432, 434)
245(4.37)
460(4.10)

In comparison with the related tetrazolium salt, maxima of the radical, 5-*t*-butyl-2,3-bis(4-nitrophenyl) tetrazolinyl, are appreciably shifted to higher wavelengths (365). The tetracyclic phototetrazolinyl with its long conjugated system absorbs even higher.

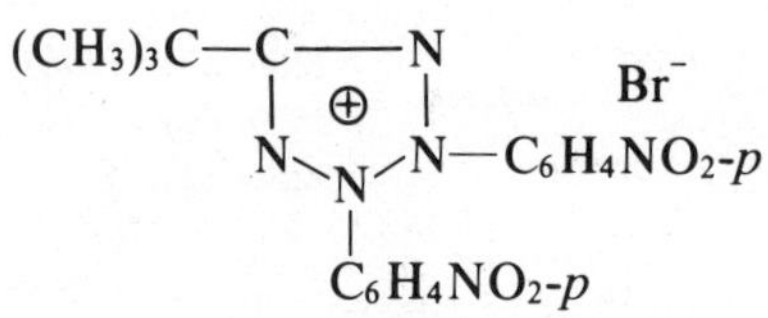

247(4.33)
285(4.11)

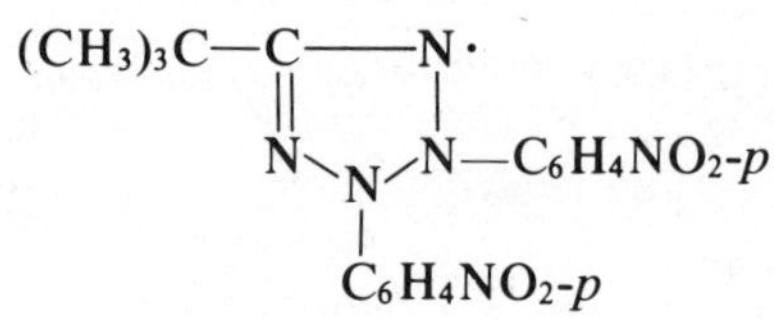

green black
276(4.18) 432(4.05)
385(4.01) 590(3.59)

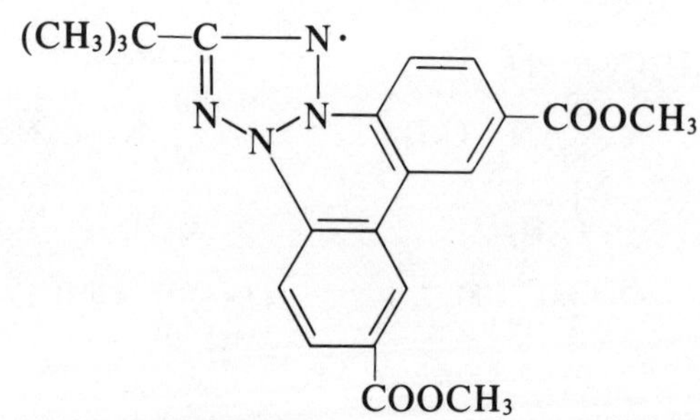

green black
257(4.58) 368(4.29)
272(4.55) 415sh(3.91)
287(4.24) 498(3.82)
297(4.27) 697(3.54)

Maxima of conjugated fused ring tetrazoles are shifted to much longer wavelengths than are 1- or 5-aryltetrazoles. In the 5,2′-polymethylene-1-phenyltetrazoles, as the number of methylene groups increases, forcing the phenyl out of

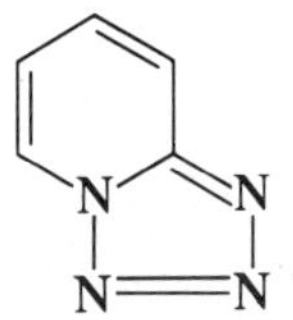

264.5(3.62) (75)
280.5(3.47)
293(3.16)

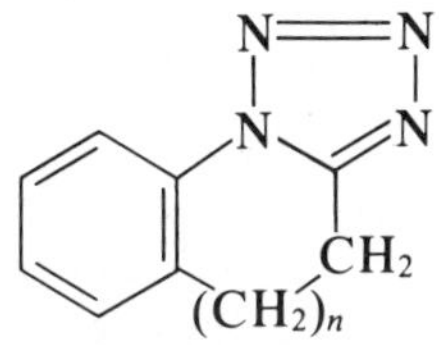

n	
1	250(4.1) (238)
2	235(3.9)
3	230sh(3.8)
4	end absorption

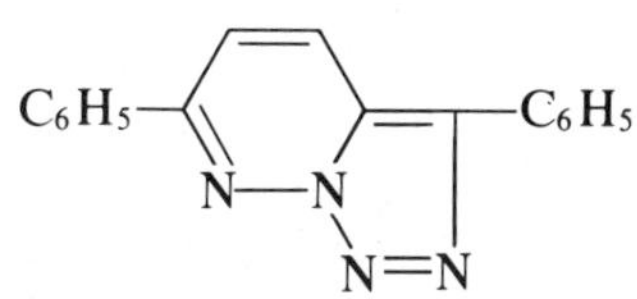

(IX)

yellow
242(4.1) (154)
275(4.4)
355(3.3)

coplanarity with the tetrazole and reducing conjugation, the maximum shifts to shorter wavelengths and the extent of absorption decreases. In 3,6-diphenylpyridazo[3,2-*c*]-1,2,3-triazole **(IX)** the location of the long wavelength maximum probably results both from the bathochromic effect of the phenyl groups and also from involvement of the nonbonding electrons of the bridgehead nitrogen in the delocalized system. In **IX** as well as in 1-methylpyrazolo[1,5-*d*]tetrazole, probable participation of the bridgehead nitrogen in the conjugated system is indicated by the shortening of the normal N—N bond length in the N—3

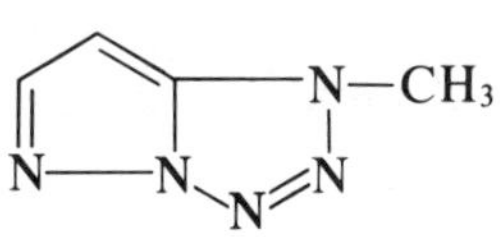

276(3.73) (14)
320sh(3.28)

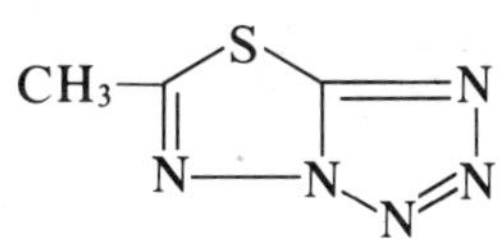

(X)

246(3.60) (256)

to N—4 and N—4 to N—5 bonds in the related 6-methyl-7-ethoxycarbonyl derivative and in tetrazolo[1,5-*b*]pyridazine (Section 3.1). The maximum of 2-methyltetrazolo[4,5-*a*]1,3,4-thiadiazole **(X)**, however, is at a lower wavelength than those of tetrazolo[1,5-*a*]pyridine, suggesting that the effective conjugated system of **(X)** is shorter.

Both mono- and dibenzo-1,3a,4,6a-tetraazapentalene **(XI** and **XIII)** have longer wavelength maxima than their isomers **XII** and **XIV** and, indeed, **XIII** is colored. Carboni et al. (96) noted similar spectral features of the dibenzotetraazapentalenes to those of their structural analog, chrysene; the long wavelength absorptions of **XIII** and **XIV** are much more intense than the corresponding bands of chrysene.

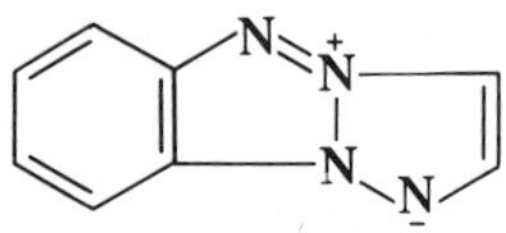

(XI)

236(4.44) (259)
278(3.48)
343(4.34)

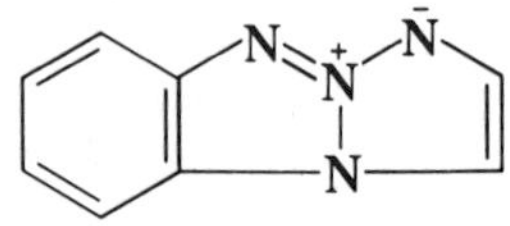

(XII)

232(4.46) (259)
285(3.58)
293(3.61)
326(4.18)
335(4.21)

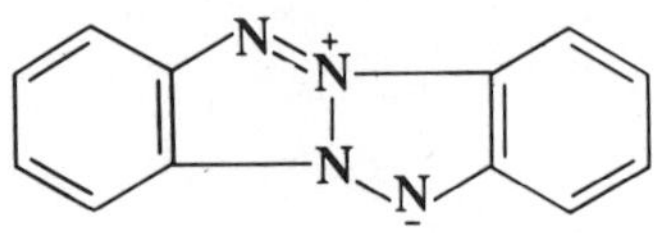

(XIII)

yellow
255(4.80) (96)
308(3.45)
323(3.61)
364(3.89)
382(4.37)
402(4.58)

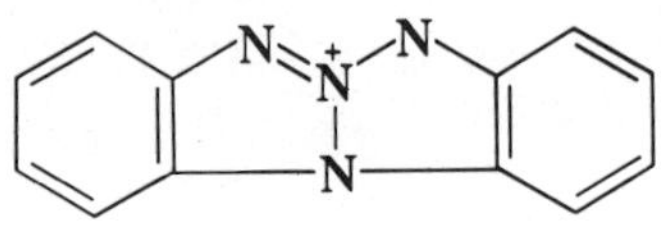

(XIV)

234(4.40) (96)
271(3.77)
380(3.92)
343(4.51)
356(4.60)

The effect of replacing the final carbon in the 1-aryltetrazoles by nitrogen, forming the aryl pentazoles, is to extend the maxima to higher wavelengths (473).

	$\lambda_{max}(\log \epsilon)$	
R	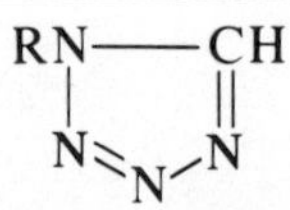	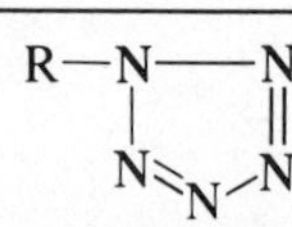
C_6H_5	236(4.0)	249(3.8)
p-ROC_6H_4	(p-CH_3O) 256(4.0)	(p-C_2H_5O) 272(4.1)
p-$(CH_3)_2NC_6H_4$	295(4.2)	329(4.2)

Similarity of the ultraviolet spectrum of 1,1′-bibenzotriazole to that of 1-methylbenzotriazole provides partial evidence for the 1,1′ linkage, and the spectrum of 1,2′-bibenzotriazole supports its assigned structure (213).

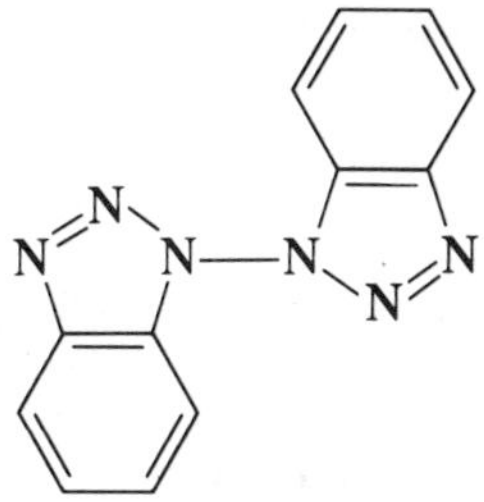

252(4.16)
288(3.89)

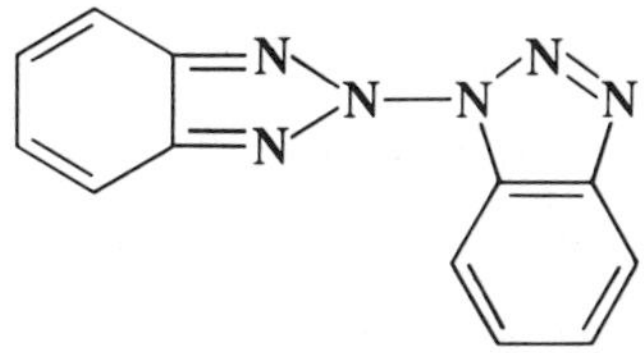

245sh(3.92)
282(4.24)
290sh(4.22)

The constitution of 1,1′-bi(5-phenyltetrazole) is indicated by absence of the azide band shown by its bi-imidoyl azide isomer (55). Since the maximum of the

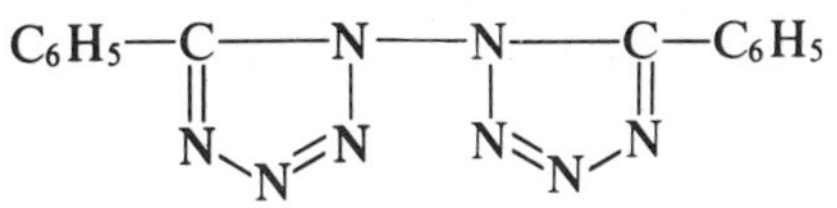

250(4.35)

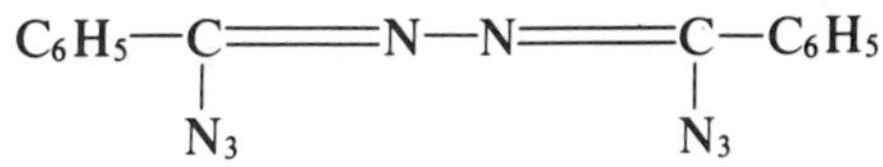

273(4.09)
361(4.04)

bitetrazole is at a somewhat longer wavelength than that of 1-methyl-5-phenyltetrazole (232 nm) some conjugation along the eight-nitrogen chain is indicated, although not as much as through the C=N—N=C chain of the biimidoyl azide, shown by its long wavelength maximum.

3.4.2. Infrared Spectra

Absorptions in the infrared associated with bonds in nitrogen chains and to some attached atoms are considered in this unit. Frequencies coupled with structural formulas are in reciprocal centimeters, and when absorption intensity is reported, it is indicated as strong (s), medium (m), weak (w), or broad (br). Spectra reported usually have been determined in mineral oil mulls, potassium bromide pellets, or liquid films. Sample media and solvents are mentioned only if relevant to the discussion.

A symmetrical stretching vibration is identified for the nitrogen-nitrogen single bond in 1,3-disubstituted triazenes. In salts of amine-nitric oxide adducts the N—N stretching band is assigned to absorption near 1130 cm^{-1} (316a), and

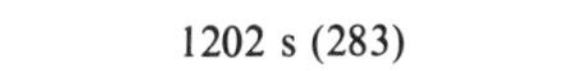

1202 s (283)

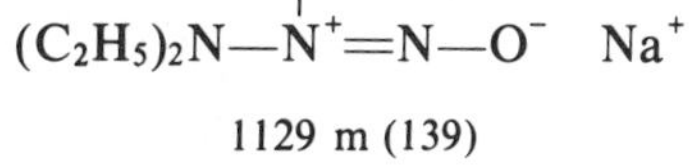

1129 m (139)

to bands in the 1045–1025 cm^{-1} region in triazanes, in 2-tetrazenes substituted with trimethylsilyl, germanyl, and stannyl radicals, and in 1-nitroso-1,4,4-

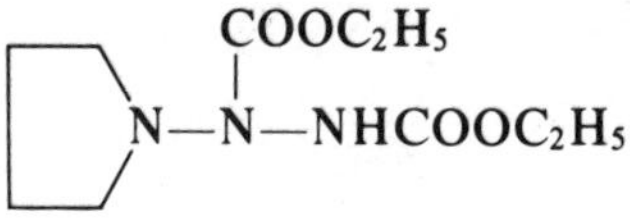

1025 w (313)

$[(CH_3)_3Si]_2N—N=N—N[Si(CH_3)_3]_2$

1036 vs (491)

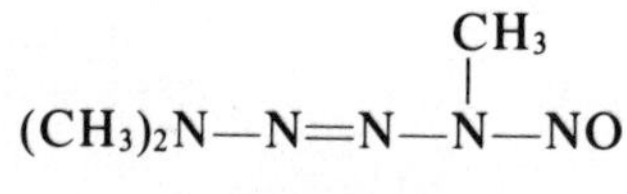

1045 (466)

trimethyl-2-tetrazene; the N—N bond in dimethylnitrosamine absorbs at 1049 cm^{-1} (219). Strong stretching vibrations from 840 to 819 cm^{-1} attributed to N—N are found for tetramethyltetrazene and for 2,3-diphenyl-5-(2-thenoyl)tetra-

$(CH_3)_2N—N=N—N(CH_3)_2$

819 s (216)

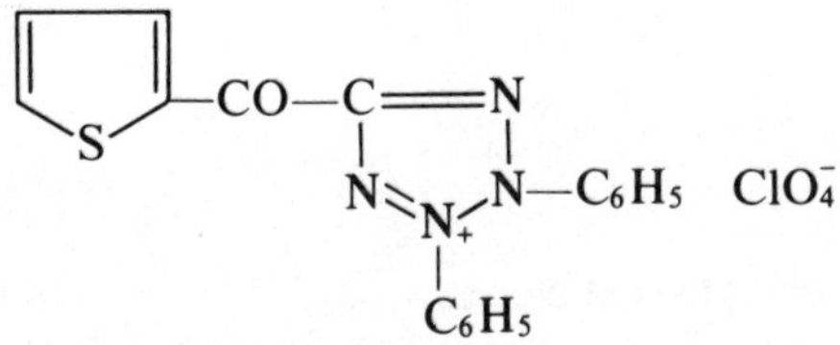

840 s (262)

zolium perchlorate, suggesting some localization of bonds in the tetrazolium salt.

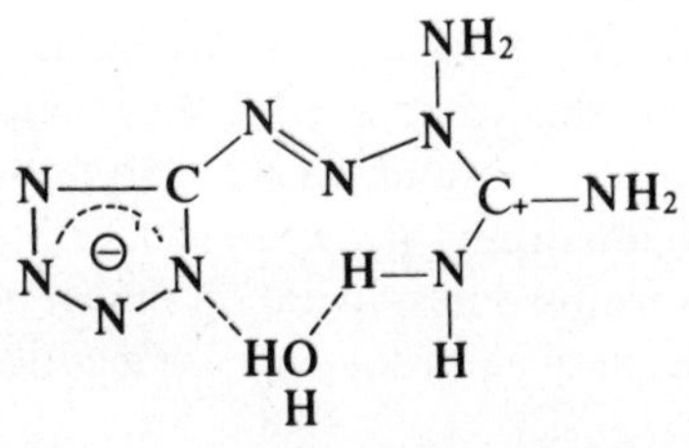

1626 (310)

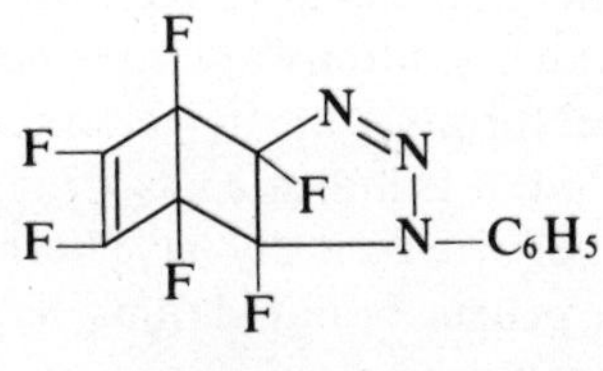

1600 (s) (43)

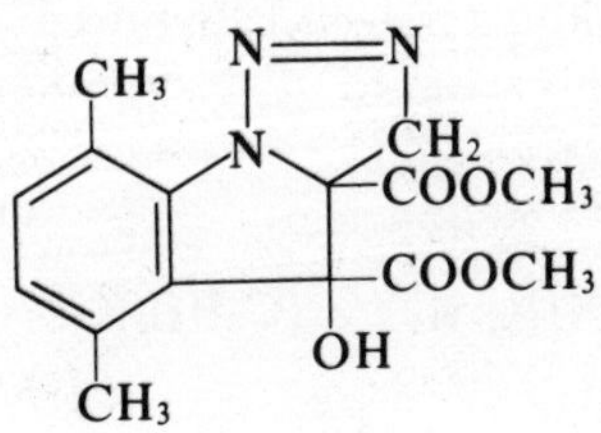

1525 (409)

Nitrogen-nitrogen double bonds in compounds with contiguous nitrogen atoms are associated with infrared bands from 1626 to 920 cm^{-1}. Tetracene and triazolines absorb at the upper end of this region, 1630–1575 cm^{-1}, where stretching vibrations of the azo group are found (56). A series of fused ring triazolines absorb in the 1530–1515 cm^{-1} range, assigned to the cyclic azo group (409).

Both triazolium salts and mesoionic triazolium compounds also absorb at 1620–1520 cm^{-1}; although these bands may be related to the azo linkage, it has been suggested that they are associated with the triazolium ring (411).

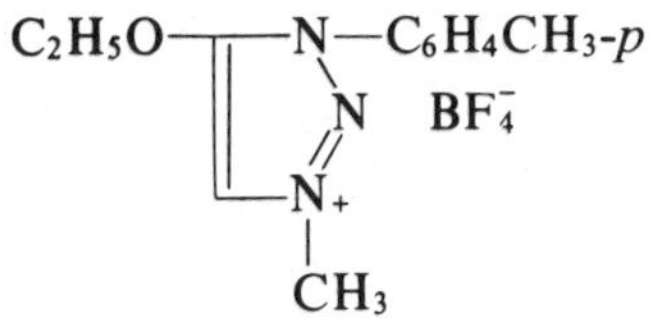

1620 s (396)

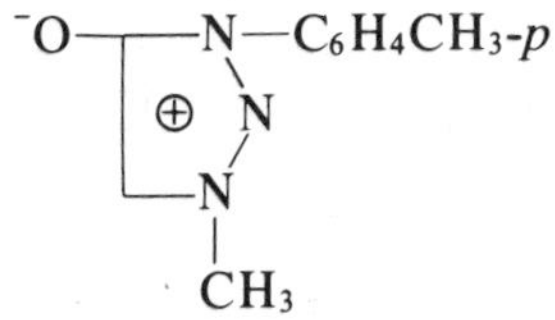

1520 m (396)

Only one N=N infrared absorption is reported for 1,3-dialkyltriazenes, but a doublet is shown by many triazenes substituted with an alkyl and an aryl, in which two azo locations are possible. Assignment of the 1495–1480 cm^{-1} stretching vibration to the isomer with azo adjacent to alkyl, and the 1415–1410 cm^{-1} band to the isomer with N=N conjugated with aryl is based in part on the

$n\text{-}C_4H_9N{=}NNHC_2H_5$

1385 (120)

$p\text{-}NCC_6H_4NHN{=}NCH_3 \rightleftharpoons p\text{-}NCC_6H_4N{=}NNHCH_3$

1482 s   1415 w (477)

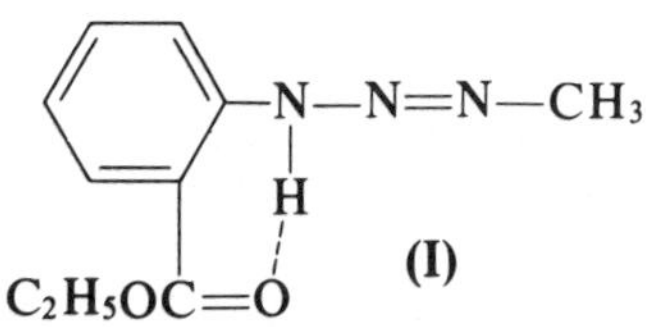

(I)

1495 s (477)
1410 w

$C_6H_5N{=}NNHC_6H_5$

1416 s (283)

$p\text{-}ClC_6H_4N{=}N{-}\overset{+}{N}{\equiv}\overset{-}{C}$

1440 w (240a)

higher absorption intensity of the intramolecularly hydrogen bonded isomer of 1-methyl-3-*o*-ethoxycarbonylphenyltriazene **(I)**. The weak azo absorption of the aryl diazoisocyanide falls closer to that of triazenes with the nitrogen-nitrogen double bond conjugated with aryl. The strong N=N absorption of diazoaminobenzene, in which azo is always conjugated with aryl, is consistent with this assignment.

For the symmetrical *trans*-tetramethyltetrazene, some investigations, as expected, found only Raman absorption for N=N at 1480 cm^{-1} (167) and 1475

cm^{-1} (216), but one instance of the forbidden infrared absorption (1480 br) has been reported (360).

A variety of four-nitrogen heterocycles exhibit bands in the 1450–1360 cm^{-1}

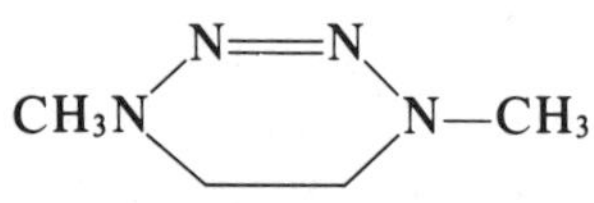

1450 br (360)

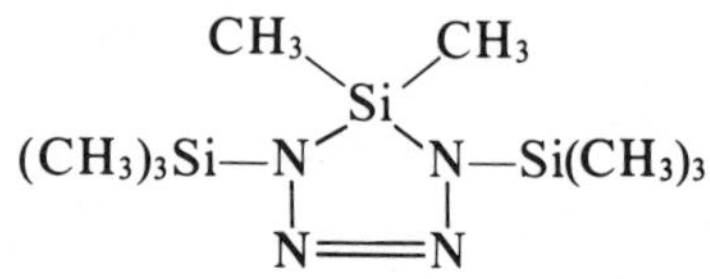

1426 (497)

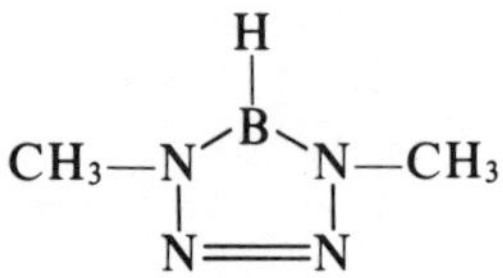

1363 s (136)

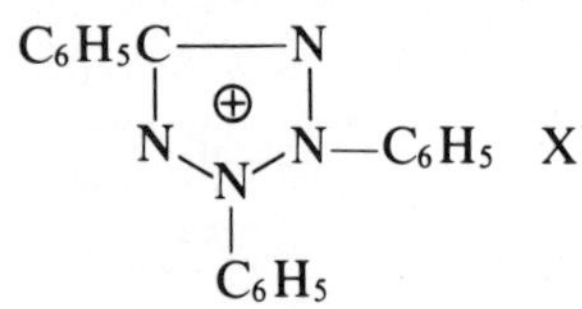

931 (23)

region, which must be associated with N=N. Absorption from 932 to 920 cm^{-1} in 2,3-tetrazolium salts is ascribed to C—N=N—C, illustrated by the triphenyl derivative.

Bands from 1190 to 1140 cm^{-1} are assigned to —N=N—N< in trisubstituted

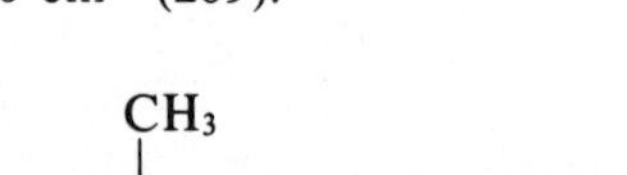

1185

1140

triazenes (209). The N=–=N=–=N unit in triazenium salts absorbs from 1310 to 1260 cm^{-1} (209).

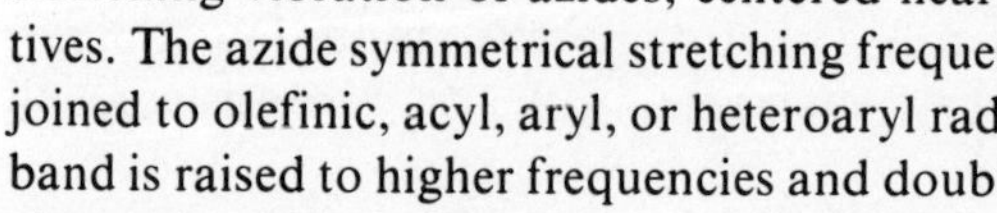

1309

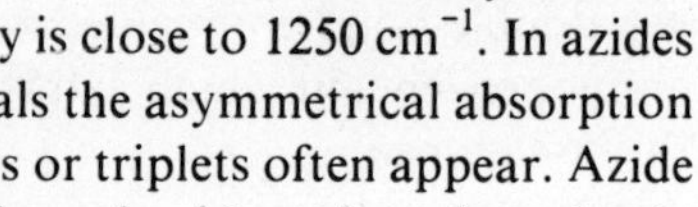

1260

Probably the infrared absorption most frequently sought in the investigation of high nitrogen compounds is the presence or absence of the asymmetrical stretching vibration of azides, centered near 2100 cm^{-1} for most alkyl derivatives. The azide symmetrical stretching frequency is close to 1250 cm^{-1}. In azides joined to olefinic, acyl, aryl, or heteroaryl radicals the asymmetrical absorption band is raised to higher frequencies and doublets or triplets often appear. Azide absorption also is found not only in nitryl azide and azidoamines, but also in diazonium amine salts, which may be viewed as protonated or alkylated azides.

A plot of infrared intensities of *p*-substituted benzene azides near 1600 cm^{-1}

$n\text{-}C_4H_9N_3$	$CF_3CFICF_2N_3$	$C_6H_5C(N_3){=}CH_2$
2083 (301)	2169 (41)	2202 (306) 2142
$C_6H_5N_3$	$p\text{-}CH_3C_6H_4N_3$	$p\text{-}O_2NC_6H_4N_3$
2130 (447) 2095	2140 (447) 2107 2055	2122 (447) 2104 2089
$C_6H_5CON_3$	NCN_3	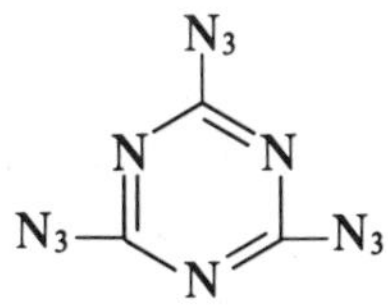
2174 (437) 2137	2143 s (330) 2090 s	2160 s (436) 2148 w

(aromatic band) against a measure of donor or acceptor character of the substituents results in a straight line, indicating that azide acts as either an electron

O_2NN_3	$(CH_3)_2NN_3$	$[H{-}N(H){-}\overset{+}{N}{\equiv}N]\ \overset{-}{Sb}Cl_6$
2185 s (138) 2165 s	2110 vs (67)	2290 s (424)

donor or acceptor when the para substituent is, respectively, an acceptor or donor (257). Azide is a donor in *m*-phenyl derivatives.

Characteristic ring stretching and ring breathing vibrations are found in monocyclic *v*-triazoles, thiatriazoles, and tetrazoles. The most distinguishing *v*-triazole absorptions (Table 3.20) are the C—H in-plane and ring breathing bands from 1235 to 1000 cm^{-1} and ring bending in-plane bands from 990 to 945 cm^{-1} (71, 79, 218, 303, 456).

Benzotriazole, its 1-phenyl derivative (374), and 1,5,6-trimethylbenzotriazole

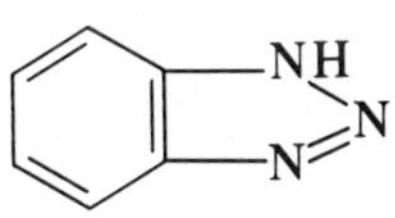

1600 (374)

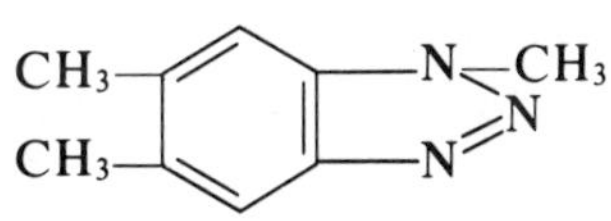

1600–1580 (423)

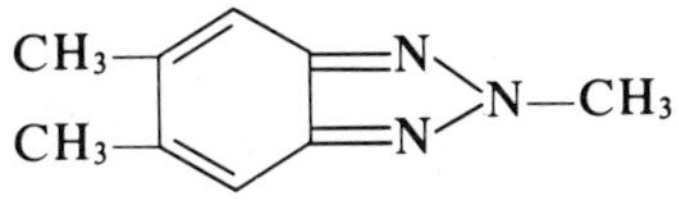

1570–1550 (423)

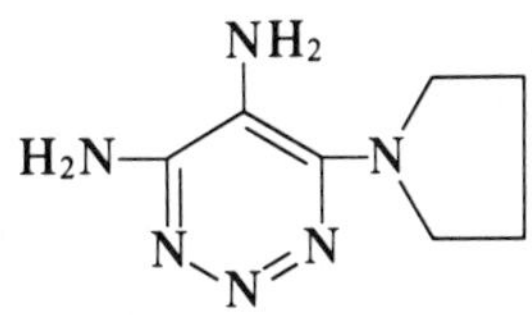

1575 (346)

Table 3.20. Infrared Absorptions of *v*-Triazoles

v-Triazole	Ring Stretching	C—H In-plane Ring Breathing and Ring Bending	Ref.
	1533 m, 1440 m, 1413 m, 1383 w	1164 w, 1075 s 976 s, 955 s	73
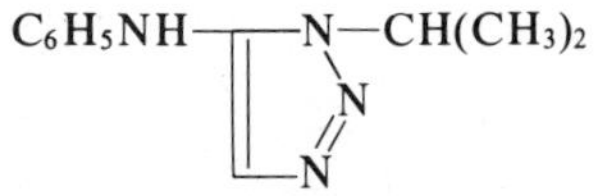	1577	1111, 1081, 976	456
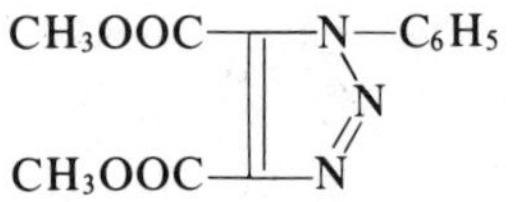	1530, 1451	1235, 1105, 1010	265
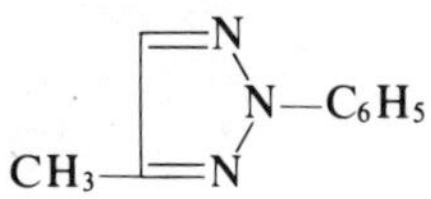	1524 m, 1401 m	1042 m, 964 vs	71

all absorb near 1600 cm^{-1}, attributed to N=N, and suggesting that the hydrogen on ring nitrogen is located at position 1. Absorption of 2,5,6-trimethylbenzotriazole is shifted to a lower wave number. A ring stretching vibration is shown by 4,5-diamino-6-pyrrolidino-*v*-triazine.

Ring breathing in thiatriazoles appears to extend from 1120 to 920 cm^{-1}. Modification of some thiatriazole ring skeletal stretching vibrations, originally proposed (302, 400) for the vicinity of 1600 cm^{-1}, became necessary because of the absence of absorption at 1610–1560 cm^{-1} by 5-methylthiothiatriazole (105). Ring stretching vibrations, however, from 1550 to 1300 cm^{-1} seem probable. In-plane vibration frequencies calculated for 5-aminothiatriazole are close to observed absorptions (284).

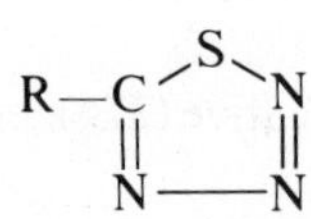

R	Ring Stretching	C—H In-plane Ring Breathing and Ring Bending	Ref.
H_2N	1525, 1377, 1315	1120, 940	284
$C_2H_5OOCCH_2NH$	1551 s, 1540 s	1086 m, 1021 m, 920 w	164
$(CH_3)_3CO$	1480 s, 1450 vs	1060 m, 920 m	243
$(C_6H_5)_3CS$	1493 s, 1445 s, 1319 m, 1306 m	1082 s, 1035 m, 1002 m	105

Table 3.21. Infrared Absorption of Tetrazoles

	Ring Stretching	C—H In-plane Ring Breathing and Ring Bending	Other Absorptions	Ref.
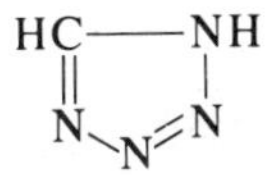		1089, 1060, 1018	763, 741	310
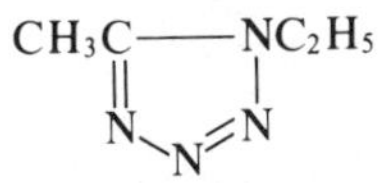	1520 s, 1404 s, 1377 m	1089 s, 1050 m, 1004 m, 969 m	712	293
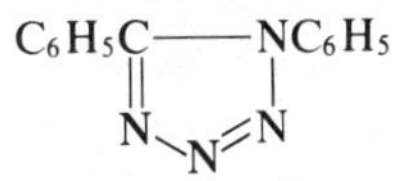	1500 s, 1450 s, 1320	1105 m	745 s	15
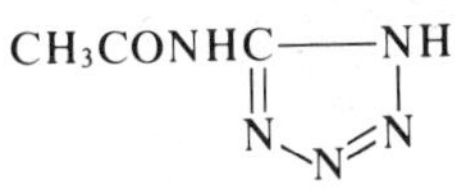	1629, 1593	1061, 1040, 1000	759, 740	433

Most monocyclic tetrazoles exhibit ring stretching vibrations from 1640 to 1335 cm^{-1} and ring breathing bands from 1095 to 900 cm^{-1} (400); some tetrazoles also absorb in the 765–710 cm^{-1} range (Table 3.21). Tetrazoles with 1-methyl-5-aryl substituents absorb at higher wave numbers than their isomers with aryl at position 1 (407).

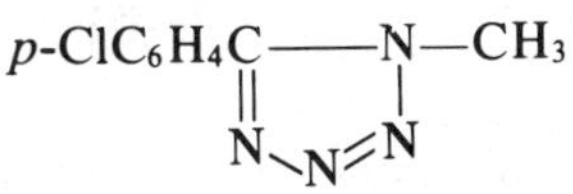

1200, 1093, 976

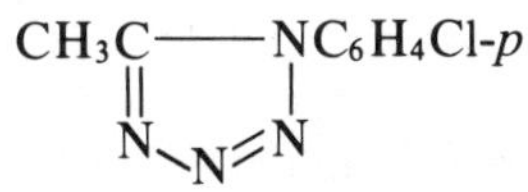

1094, 983, 712

Absorption attributed to ring C=N shifts to lower frequencies in the sequence of the isomeric 5-substituted amino **(II)**, 5-amino 1-substituted **(III)** and 5-amino 2-substituted tetrazoles (429). A band that is found near 1260 cm^{-1} for **II** and **III**, but that is absent in the 2-isomer, may be indicative of cyclic N=N—N. Frequencies assigned to cyclic N=N—N in many tetrazoles (457), thiatriazoles (302), and *v*-triazoles (303) are usually somewhat higher, 1335–1270 cm^{-1}.

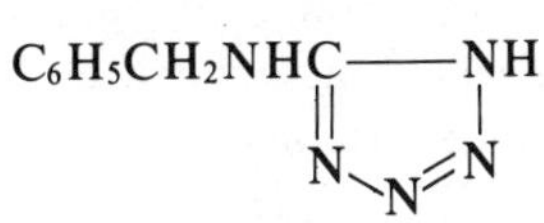

(II)

1634
1256

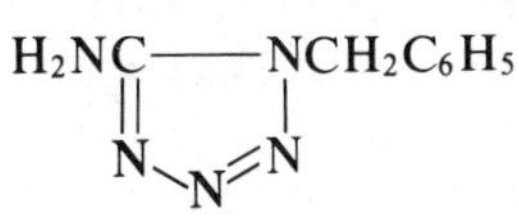

(III)

1600
1266

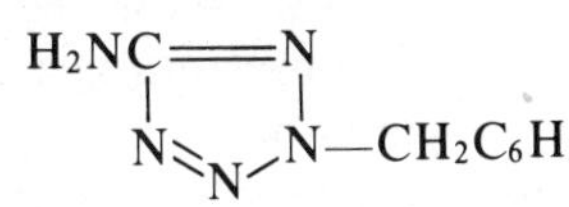

1560

On formation of a tetrazolate anion, a characteristic absorption appears near 1125 cm^{-1} and ring breathing bands shift to lower frequencies (199). In

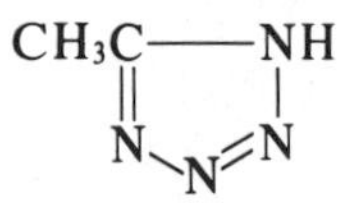

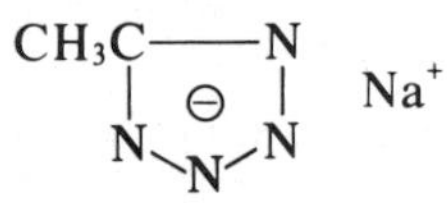

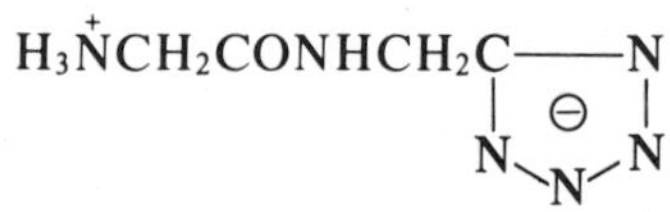

1112 s	1126 s	1130 s
1089 w	1103 s	1115 m
1052 vs	1069 w	1075 m
	1029 w	1032 m

tetrazole analogs of peptides, the tetrazolate anion absorbs in the 1130–1113 cm^{-1} region (200). Absorptions observed and those calculated by vibrational analysis for sodium tetrazolate monohydrate differ somewhat from the values found for sodium 5-methyltetrazolate and tetrazolate peptides (178).

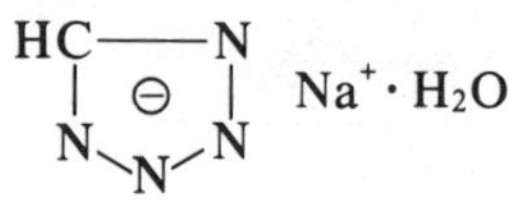

	ν, obs.	ν, calc.
sym ring breathing	1161 vs	1138
sym ring deformation	1065 vs	962
asym ring deformation	1023 vs	1063
asym ring deformation	1015 vs	1013
asym ring deformation	702 vs	730

A characteristic doublet in the 746–727 cm^{-1} range is attributed to the tetraazapentalenes (39).

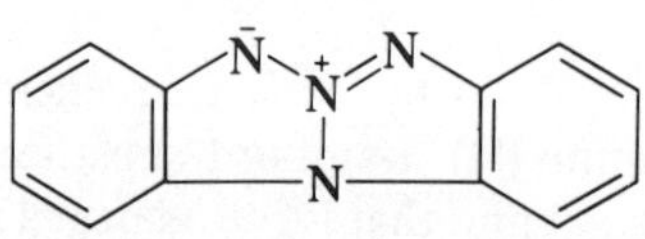

730 (96)

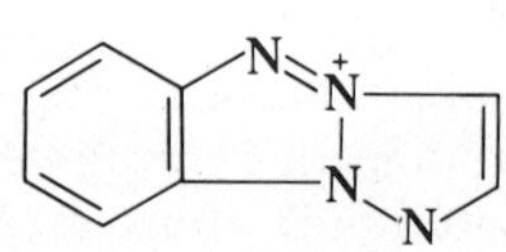

731 vs (97)

The stretching frequency of NH in contiguous nitrogen chains varies from nearly 3500 to below 2800 cm^{-1}, generally in the same range found for imino groups in lower nitrogen compounds. Diazoaminobenzene is associated as a hydrogen-bonded dimer in most solvents, but forms an intermolecular bond with pyridine, while the absorption shifts to 3155 cm^{-1} (146). The NH moiety in the methylaminodiazonium ion is involved in intermolecular hydrogen bonding to other cations, accounting for the low wave number of its absorption. As expected, two close NH bands appear for amino groups in high nitrogen compounds.

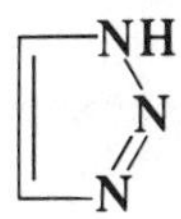

3470 s (73)

$C_6H_5N{=}NNHC_6H_5$

3326 (CCl_4) (146)

$\bar{N}{=}\overset{+}{N}{=}NH$

3320 s (137)

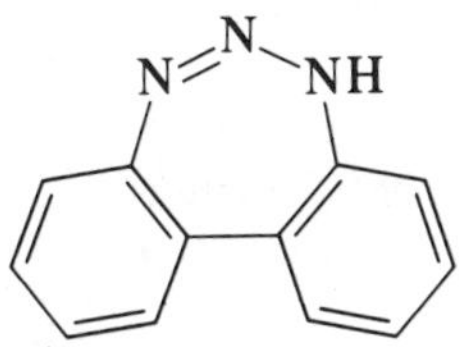

3271 (175)

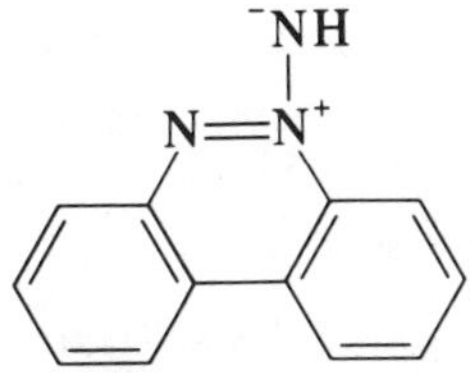

3150 (177)

$[N{\equiv}\overset{+}{N}{-}\overset{CH_3}{\overset{|}{N}}H]\ SbCl_6^-$

2790 m (424)

Disubstituted triazene oxides contain an intramolecular hydrogen bond, whose NH absorption shows the expected lower frequency in comparison to

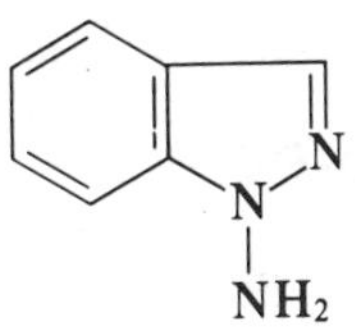

3330, 3200 (5)

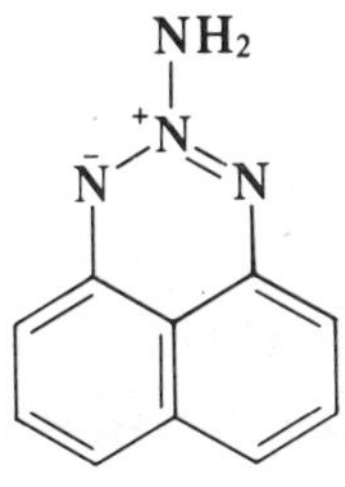

3300, 3140 (404)

$[(CH_3)_3Si]_2N{-}N{=}N{-}NH_2$

3395 s, 3260 vs (492)

other triazenes. Three bands in the 1530–1430 cm^{-1} range appear to be characteristic of triazene oxides. *v*-Triazole oxides lack the usual absorption of *v*-triazoles from 1110 to 1050 cm^{-1}, but the bands from 965 to 945 cm^{-1} shift to slightly higher frequencies with decreased intensity (72).

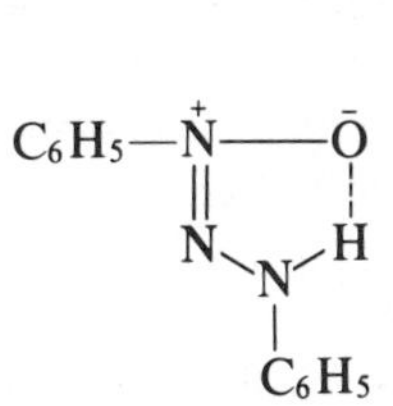

3160 s (143)
1510 s, 1475 br,
1440 s

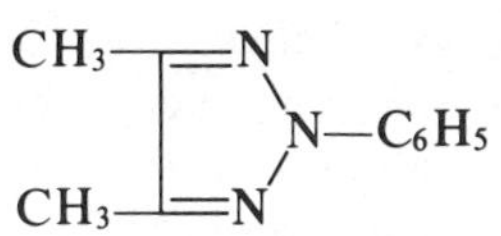

950 ($\epsilon = 192$)

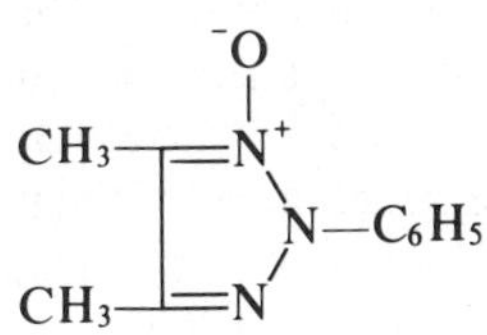

960 ($\epsilon = 38$)

Nitroso absorption is observed at several locations in the infrared spectrum. Nitro groups attached to nitrogen show a characteristic symmetrical stretching

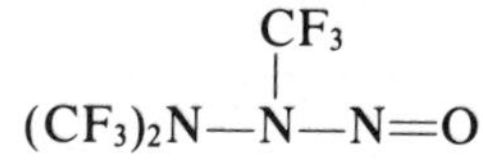

1621 s (220)

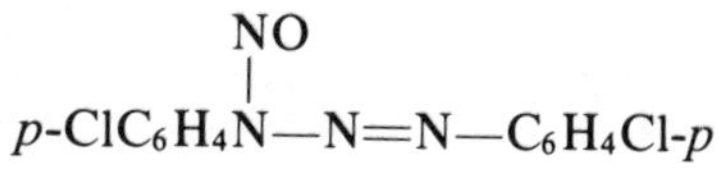

1470 (356)

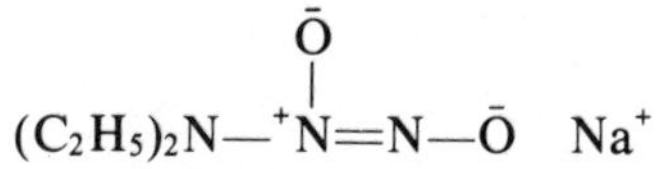

1225 s, 1210 s (139)
1167 sh

vibration in the 1295–1270 cm^{-1} region. Similarly, the amine nitroimides absorb from 1300 to 1260 cm^{-1} and 1420 to 1380 cm^{-1}.

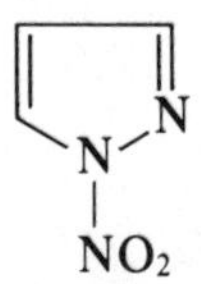

1287 (242)

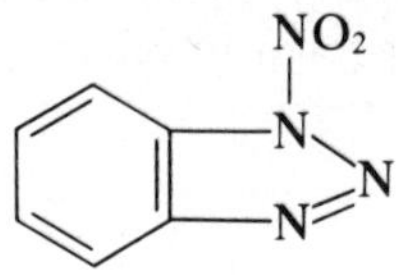

1280 (113)

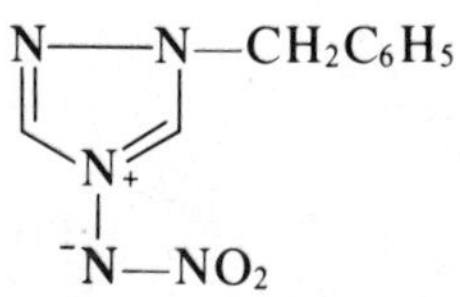

1420 vs, 1300 vs (468)

3.5. NITROGEN NUCLEAR MAGNETIC RESONANCE

Both ^{14}N and ^{15}N NMR spectra enable distinction between differently located and differently bonded nitrogen atoms, including those in compounds with nitrogen chains (299, 300). ^{14}N NMR is limited by large line widths and uncertainties because of its quadrupole moment. Both ^{14}N and ^{15}N NMR spectra, however, have large chemical shifts, negligibly divergent from each other, which facilitate identification of nitrogen atoms in varying chemical environments through comparison with spectra of structurally related nitrogen compounds. The signals of ^{15}N NMR spectra are sharp and $^{15}N—H$, $^{15}N—^{13}C$, and $^{15}N—^{15}N$ coupling constants are easily measured; presently it is impractical to determine ^{14}N coupling constants. Because of the low natural abundance of ^{15}N, its principal use at this time is with labeled compounds.

In 1-aryl-3,3-dimethyltriazenes labeled with ^{15}N, increased electron-withdrawing character of the *p*-substituent on phenyl deshields N-3 and shields N-1, consistent with a significant contribution from the dipolar form with a negative charge at N-1 (Table 3.22) (31). The shifts at N-2 are relatively constant. Many series of contiguous nitrogen compounds have a small range of nitrogen-nitrogen coupling constants, as exemplified by the triazenes.

As with the ^{15}N NMR spectra of the triazenes, the ^{14}N line for azo nitrogen in

Table 3.22. ^{15}N Chemical Shifts and Nitrogen-Nitrogen Coupling Constants in 1-Aryl-3,3-Dimethyltriazenes

$$\underset{}{p\text{-X—C}_6\text{H}_4\text{—}\overset{}{\underset{1}{^{15}\text{N}}}\text{=}\underset{2}{^{15}\text{N}}\text{—}\underset{3}{^{15}\text{N}}(\text{CH}_3)_2} \rightleftarrows p\text{-X—C}_6\text{H}_4\text{—}\underset{1}{^{15}\text{N}^-}\text{—}\underset{2}{^{15}\text{N}}\text{=}\underset{3}{^{15}\text{N}^+}(\text{CH}_3)_2$$

X	^{15}N Chemical Shift, δ (ppm)[a]			Nitrogen-Nitrogen Coupling Constant (Hz)	
	N-1	N-2	N-3	$^1J(N_1N_2)$	$^1J(N_2N_3)$
H	332.1	424.1	127.3	12.8	14.0
O_2N	320.0	428.0	138.0	12.8	13.4
CH_3O		421.1	124.4	12.2	14.0

[a] Relative to $^{15}NH_4Cl$.

2-tetrazenes is further downfield than for the end, three-coordinate nitrogen (Table 3.23). Chemical shifts of double-bonded nitrogen in a series of azo compounds are almost linearly related to ultraviolet $n \rightarrow \pi^*$ transition energies (282, 333), and the values for the 2-tetrazenes qualitatively support this correlation.

Different ^{15}N NMR spectra are found for cis and trans isomers of 1-phenyl-3-[3-methylbenzothiazolinylidene-2]triazene labeled with ^{15}N in the triaza chain (158). Although coupling of N-1 and N-3 is through the intervening bonds of the nitrogen chain, location of the electron pairs in the all-trans structure must facilitate coupling, whereas in the cis isomer, this is not possible. Similarity of chemical shifts and coupling constants of 1-ethyl-1-*p*-nitrophenyl-3-[3-methylbenzothiazolinylidene-2]-triazenium cation and the corresponding unalkylated triazene in trifluoracetic acid indicates preferred protonation of the triazene in the 1-position (158).

Table 3.23. ^{14}N Chemical Shifts in 2-Tetrazenes

2-Tetrazene	Reference Substance	^{14}N Chemical Shift, δ (ppm): =N—	^{14}N Chemical Shift, δ (ppm): >N—	Ref.
$(CH_3)_2N—N=N—N(CH_3)_2$	NH_4^+ aq.	383(331)[a]	116(496)	333
	$NaNO_3$ aq.	29(280)	−275(750)	282
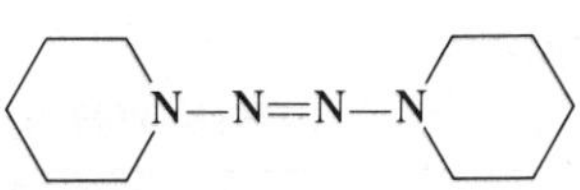	$NaNO_3$ aq.	41	−308	282
$[(CH_3)_3Si]_2N—N=N—N—[Si(CH_3)_3]_2$	$NaNO_3$ aq.	45(355)	−246	282

[a] Values in parentheses are line widths at half-heights.

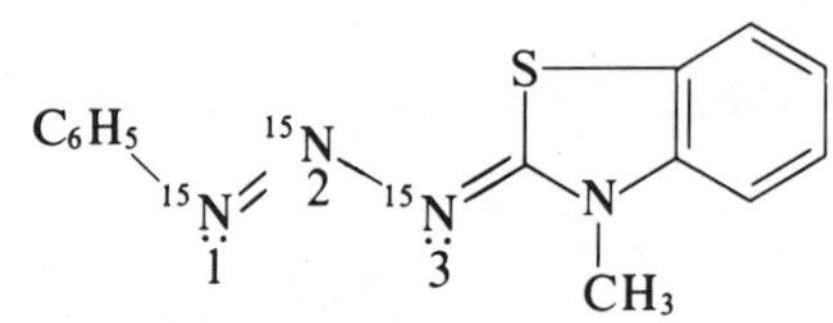

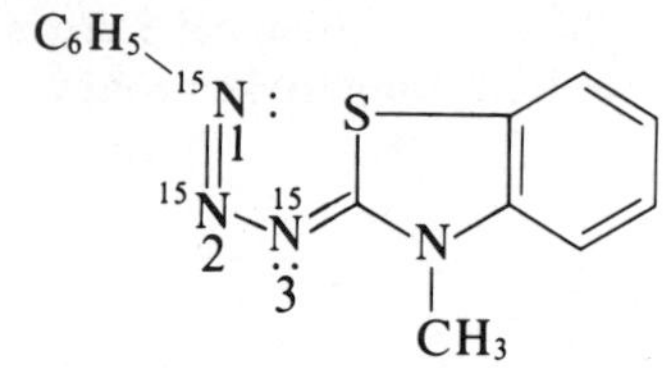

all trans				cis-trans			
	δ (ppm)[a]		J (Hz)		δ (ppm)		J (Hz)
N_1	404.9	(N_1N_2)	13.2	N_1	388.9	(N_1N_2)	18.9
N_2	466.8	(N_2N_3)	17.7	N_2	439.0	(N_2N_3)	15.0
N_3	270.2	(N_1N_3)	10.7	N_3	270.2	(N_1N_3)	0

[a]Relative to NH_4^+Cl.

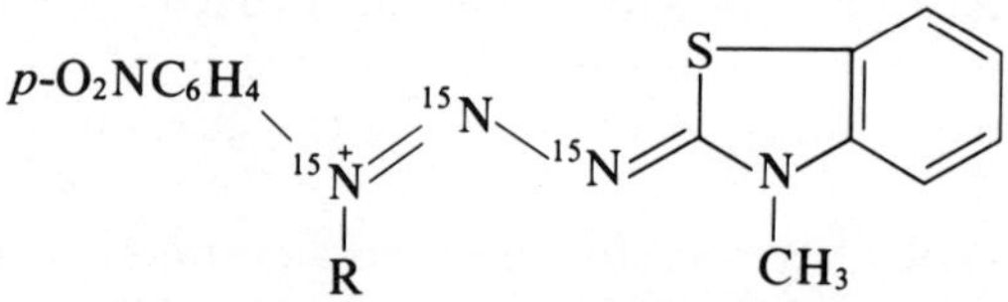

Coupling Constants (Hz)

	$R = C_2H_5$	$R = H$
$J(N_1N_2)$	13.5	11.8
$J(N_2N_3)$	11.8	11.8
$J(N_1N_3)$	5.1	5.9

Chemical shifts of trimethylamine nitroimide and of the nitric oxide adduct of diethylamine are roughly similar.

$(CH_3)_3\overset{+}{N}—\overset{-}{N}—NO_2$
2 1
^{14}N δ (ppm) (332)
(relative to NH_4^+)

N-1	343
N-2	242

$(C_2H_5)_2N—{}^{15}\underset{|}{\overset{^-O}{}}N^+={}^{15}N—\bar{O}\quad (C_2H_5)_2\overset{+}{N}H_2$
2 1
^{15}N δ (ppm) (428)
(relative to NH_4^+)

		$J(N_1N_2)$ (Hz)
N-1	386.5	16.9
N-2	215	

Evidence for the structures of *cis*- and *trans*-2,3-diphenyl-1-phthalimido-2,3-di[^{15}N]-azimine comes, in part, from ^{15}N and ^{13}C NMR spectra (298). That the

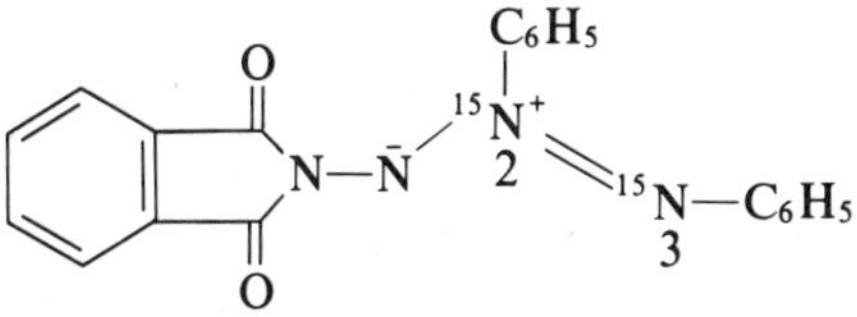

^{15}N δ (ppm) (relative to CH_3NO_2)		$J(N_2N_3)$ (Hz)
N-3	−59.6	15.0
N-2	−63.7	

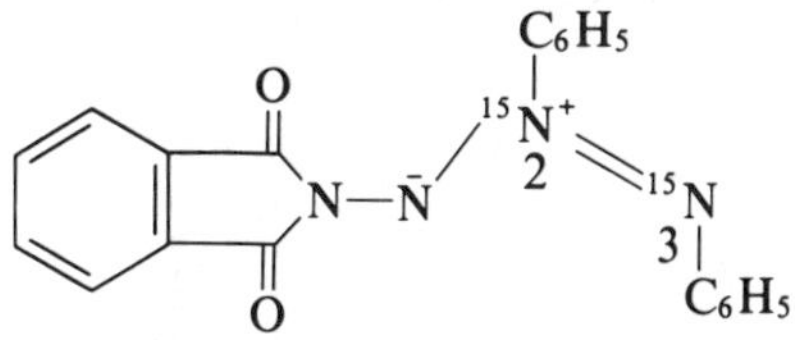

^{15}N δ (ppm) (relative to CH_3NO_2)		$J(N_2N_3)$ (Hz)
N-3	−60.5	13.4
N-2	−64.7	

^{15}N labeled nitrogen atoms are nonequivalent in each isomer is shown by their different chemical shifts, the relatively large values of the N_2N_3 coupling constants indicates that these nitrogen atoms are adjacent, and comparison of their ^{15}N and ^{13}C NMR spectra with those of *cis*- and *trans*-di(^{15}N)azoxybenzene shows a similarity, also indicated by comparison of ultraviolet spectra.

In the azide ion, the two terminal atoms have the same chemical shift, showing their equivalence, but the signal of the middle nitrogen differs (49). In covalent azides, including hydrogen azide, the shifts are distinct for each of the three nitrogen atoms (Table 3.24). As expected, chemical shifts in ^{14}N and ^{15}N NMR spectra for the same azide are often close, as illustrated by the values for methyl azide. Shifts of the central azide nitrogen fall within a small range of about 20 ppm and do not appear to depend on the nature of the attached organic group. Compared to methyl azide, the ^{15}N chemical shifts of the aryl azides indicate greater magnetic shielding at the nitrogen attached to the aromatic group and at the end nitrogen, further indicating the ability of azide to participate in resonance with an attached conjugated system. Nitrogen atoms in trimethylsilyl azide are less shielded than those of *t*-butyl azide. The coupling constant of an inner nitrogen-nitrogen pair is larger than that of a terminal pair, additionally emphasizing the difference between the three nitrogen atoms in covalent azides.

In the ^{14}N NMR spectra of five-membered heterocycles such as *v*-triazoles and tetrazoles, trivalent pyrrole-type nitrogen atoms undergo larger chemical shifts than do pyridine-type nitrogen atoms (501). Shifts of N-1 in 1-methyl *v*-triazole, benzotriazole, and tetrazole are nearly the same, and the size of the N-2 shifts in the 2-substituted series, though smaller, leaves no doubt that the nitrogen atoms carrying methyl are similar in magnetic shielding. Thus these shifts may be used to identify azole ring systems.

Table 3.24. Nitrogen NMR Spectra of Azides

$R{-}N_a{=}N_b^+{=}N_c^- \rightleftarrows RN^-{-}N^+{\equiv}N$

Azide	^{14}N δ (ppm)[a]			^{15}N δ (ppm)[a]			^{15}N J (Hz)		Ref.
	a	*b*	*c*	*a*	*b*	*c*	N_aN_b	N_bN_c	
NaN_3	277(55)	128(22)	277						49
HN_3	300(100)	129(24)	165(100)						49
CH_3N_3	320(101)	128(17)	170(19)	318.25	126.75	168.00	14.4	8.2	49, 352, 357
$(CH_3)_3CN_3$	282(180)	130.5(30)	159(40)						357
$C_6H_5N_3$				284.95	133.16	143.88	13.4	7.8	352
p-$O_2NC_6H_4N_3$				278.53	136.52	140.57	13.8	7.4	352
$C_6H_5CON_3$	322(430)	140(300)	237(73)						49
NCN_3				311.76	146.21	143.98	16.0	6.0	352
$(CH_3)_3SiN_3$	316.5(123)	137.8(15.7)	203.5(12.3)						357

[a] Shifts relative to NO_3^-.

^{14}N δ (ppm)
(relative to CH_3NO_2)

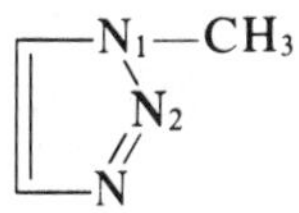

N-1 144(125)
N-2 28(480)

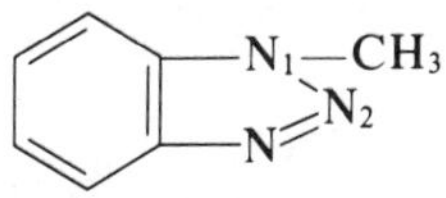

N-1 148(300)
N-2 40(600)

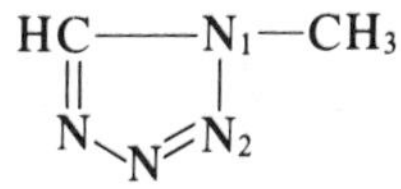

N-1 150(160)
N-2 17(600)

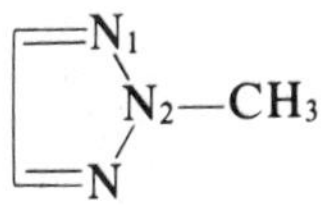

N-1 53(250)
N-2 132(105)

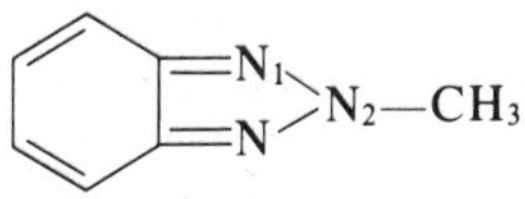

N-1 50(500)
N-2 118(250)

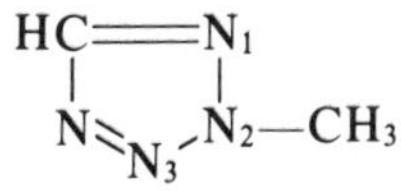

N-1 55(200)
N-2 103(105)
N-3 10(300)

By comparison of the ^{14}N NMR spectra of *v*-triazole and tetrazole with those of the 1- and 2-methyl derivatives, a high proportion of the isomer with hydrogen at position 2 is indicated (501).

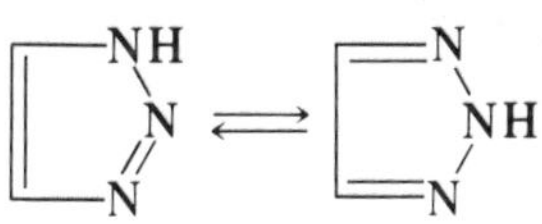

N-1 60(250)
N-2 128(200)

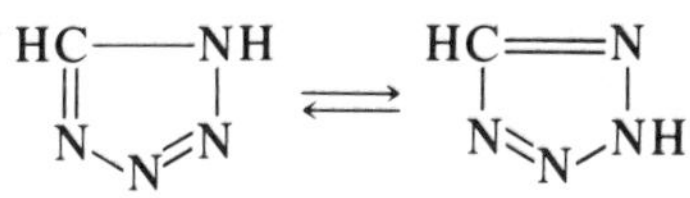

N-1 25(480)
N-2 106(350)

Nitrogen-14 chemical shifts of unmeasured azoles can be estimated by empirical additivity rules involving increments in shifts between known azoles (448). Calculated π-charge densities in azoles correlate with ^{14}N shifts for both pyrrole and pyridine types of nitrogen atoms (501).

Nitrogen-14 chemical shifts and relatively sharp lines are obtained for pyrrole-type nitrogen atoms at the junction of five- and six-membered rings in a series of azolopyridines with bridgehead nitrogens, using the differential saturation method which involves selective variation of the intensity of the radio-frequency radiation (502). The data for *v*-triazolo[1,5-*a*]pyridine and tetrazolopyridine (relative to nitromethane) illustrate the signals found for the indolizine and other types of nitrogen atoms. Shifts for each nitrogen in trimethylene tetra-

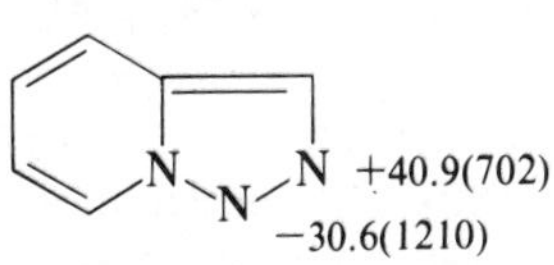

+119.9(156)

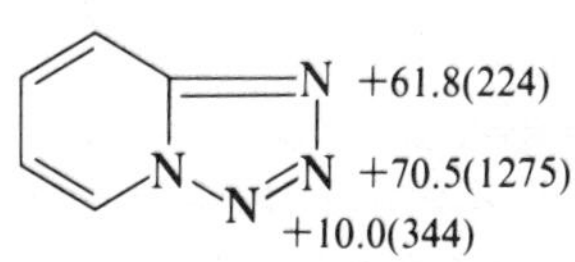

+129.2(68)

zole are assigned by comparison of the ^{14}N NMR spectra of pyrazole and 1,2,4-triazole derivatives (37). As the size of the cyclopolymethylene ring increases,

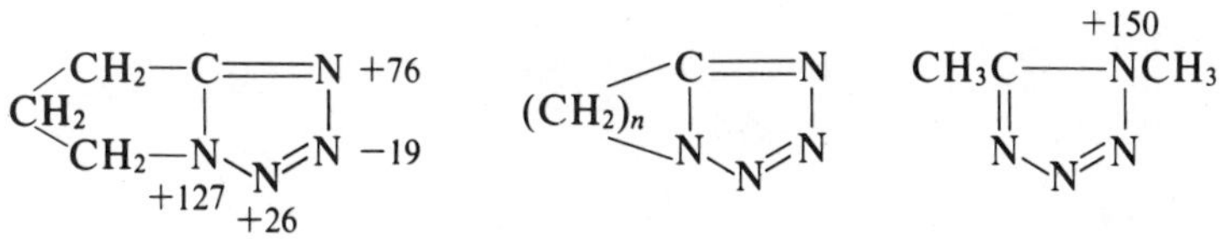

n	^{14}N δ (ppm) (relative to NO_3^-)
3	+127
5	+137
6	+140

the shift of the bridgehead nitrogen approaches that of N-1 of 1,5-dimethyltetrazole.

5-Phenyltetrazole labeled with ^{15}N at each nitrogen exhibits resonance peaks at 84.85 and 26.04 ppm upfield from nitrate, but when labeled only at N-1 and

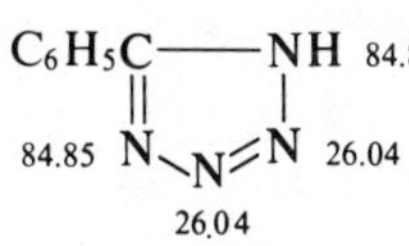

N-4 shows only the peak at 84.85 ppm (422). Absence of transposition of ring nitrogen during ultraviolet irradiation of 1,4-labeled 5-phenyltetrazole follows from finding only the 84.85 ppm signal in recovered irradiated material.

Because of the high instability of *p*-dimethylaminophenylpentazole, to obtain a ^{14}N NMR spectrum requires a solvent in which decomposition is minimized. Appearance of a signal +70 upfield from nitromethane is direct evidence

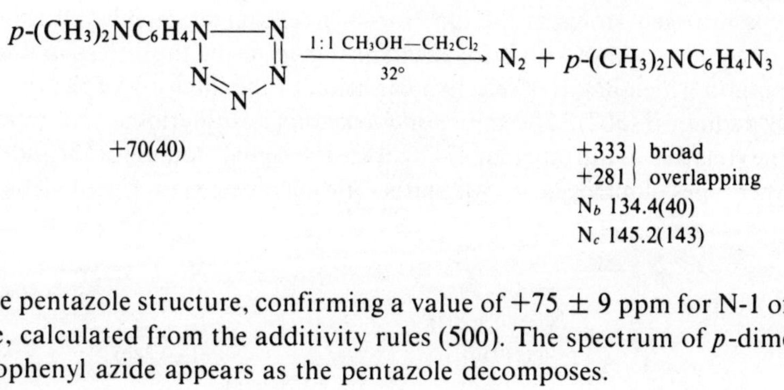

for the pentazole structure, confirming a value of +75 ± 9 ppm for N-1 of pentazole, calculated from the additivity rules (500). The spectrum of *p*-dimethylaminophenyl azide appears as the pentazole decomposes.

3.6. MASS SPECTROMETRY

Under electron impact, most high nitrogen compounds form molecular ions, thus specifying molecular weight. The several types of nitrogen chains fragment in characteristic ways, often close to those of thermal decomposition. For each high nitrogen example, *m/z* (*m/e*) ratios are given of the ionized nitrogenous and other relevant fragments formed, with probable constitutive interpretations. When available, intensities are listed parenthetically as a percent of the base peak. Main fragmentations are indicated on each structural formula.

In high nitrogen chains at the most reduced level, under conditions of mass spectrometry, fragments with multiple linked nitrogen atoms tend to persist; frequently substituents on the nitrogen chains are eliminated before nitrogen-nitrogen bonds are broken. Thus ethoxycarbonyl units are lost initially from 1-[1,2-bis(ethoxycarbonyl)hydrazino]piperidine and fragments with three-nitrogen sequences are present in moderate to large intensity. Although stable

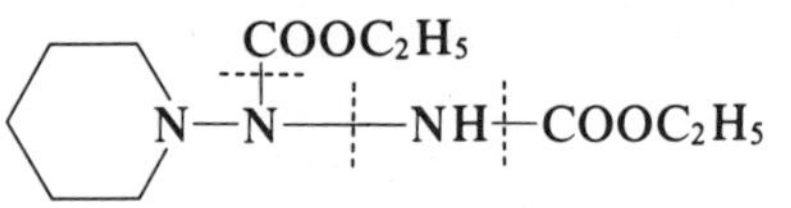

(313)

$M^{+\cdot}$	259(13.4)
$(M—COOC_2H_5)^+$	186(100)
$(M—NHCOOC_2H_5)^+$	171(15.8)
$(CH_2)_5NNHNH^+$	114(33)
$(CH_2)_5NNH^+$	99(5.0)
$(CH_2)_5N^+$	84(92)

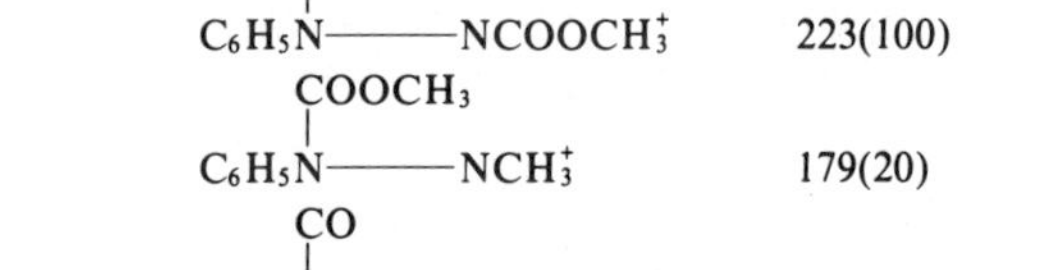

(33)

$M^{+\cdot}$	446(10)
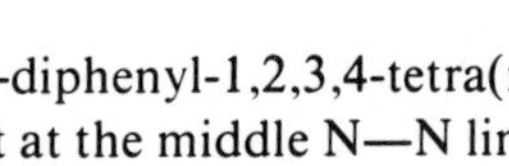	223(100)
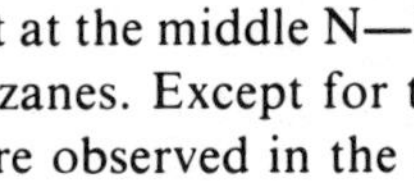	179(20)
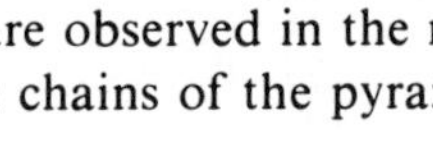	147(8)

thermally to 200°, 1,4-diphenyl-1,2,3,4-tetra(methoxycarbonyl)tetrazane breaks under electron impact at the middle N—N link, in analogy to the thermal dissociation of other tetrazanes. Except for the molecular ion, only ions with two-nitrogen fragments are observed in the mass spectrum.

The four-nitrogen chains of the pyrazolidine-dione and urazole dimers, do not survive electron impact and only two-nitrogen ions are detected in their mass spectra (393). Embedded, reduced three, five, and six nitrogen chains, however, maintain their integrity to a large extent during mass spectrometry. Indeed the mass spectrum of tetra(methoxycarbonyl)pyridazino[1,2-*a*]benzotriazole **(I)** shows that methoxycarbonyl groups are split off; all peaks below

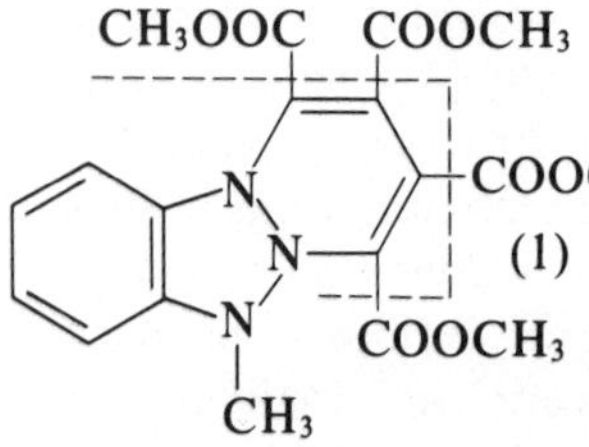

(I)

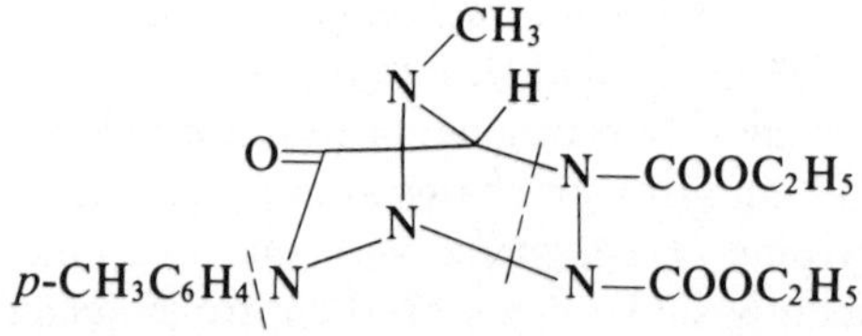

(396)

(II)

$M^{\ddagger}$	417(0.2)
$(M—COOCH_3)^+$	358(100)
$[M\text{-}4(COOCH_3) + H]^+$	182(4)

$M^{\ddagger}$	363(15)
$(M—COOC_2H_5)^+$	290(19)
$(M—COOC_2H_5—COOC_2H_4)^+$	218(33)
$[M—(N—COOC_2H_5)_2]^{\ddagger}$	189(27)
$CH_3C_6H_4^{\ddagger}$	91(100)

m/z 182 have abundances less than 1%, indicating little fragmentation of the nitrogen chain. The pentaazabicyclo[2.2.1]heptane **(II)** similarly loses ethoxycarbonyl, as well as an azodicarboxylic ester unit (converse of synthesis) and the tolyl substituent (base peak), none of which involve destruction of the *v*-triazolinone moiety. Intermediate decomposition structures of the bi(oxoindazolo-[1,2-*a*]benzotriazolyl) **(III)** cannot be exactly specified, but carbon monoxide is lost and the N—N bond joining the two main parts of the molecule is broken.

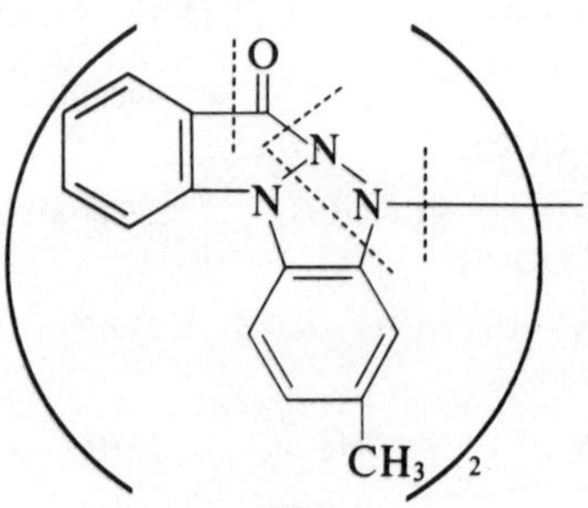

(472)

(III)

$M^{\ddagger}$	472
$(M—CO)^{\ddagger}$	444(4)
$(M—2CO)^{\ddagger}$	416(11)
$(M/2)^+$	236(90)
$(M/2—CO)^+$	208(92)
$(M/2—CO—N_2)^+$	180(100)

The halved molecule also loses carbon monoxide, and after the *v*-triazole ring is exposed, a molecule of nitrogen is lost.

At the triazene oxidation level, in comparison with the more reduced nitrogen sequences, greater fragmentation of the nitrogen chain occurs and two- and one-nitrogen fragments are abundant. Each tautomer of the 1-methyl-3-phenyltriazenes undergoes its own decomposition; the anilinium radical ion arises by migration of a hydrogen from methyl in the anilino tautomer.

$C_6H_5NH{-}N{=}NCH_3$

$\rightleftarrows$ (272)

$C_6H_5N{=}N{-}NHCH_3$

$M^{\ddagger}$	135(10)
$(M{-}NCH_3)^{\ddagger}(\underline{C_6H_5N_2H^{\ddagger}})$	106(15.0)
$(M{-}NHCH_3)^{+}(\underline{C_6H_5N_2^{+}})$	105(18.3)
$(M{-}N_2CH_2)^{\ddagger}(\underline{C_6H_5NH_2^{\ddagger}})$	93(100)
$C_6H_5^{+}$	77(53.3)

$C_6H_5N{=}N{-}N(CH_3)_2$ (272)

$M^{\ddagger}$	149(11.3)
$(M{-}N(CH_3)_2)^{+}(\underline{C_6H_5N_2^{+}})$	105(35.8)
$C_6H_5^{+}$	77(100)

From both monomethyl and dimethylphenyltriazene, a methyl amine is eliminated first, followed by loss of nitrogen from the resulting benzenediazonium ion. Release of nitrogen from 1-aryl-3-methyltriazenes is facilitated by *p*-cyano, acetyl, or methoxycarbonyl, but not nitro substituents on aryl (8).

The main fragmentations of 1-aryltriazolines result from elimination of nitrogen or by reversal of synthesis with loss of aryl azide. A minor decomposition of

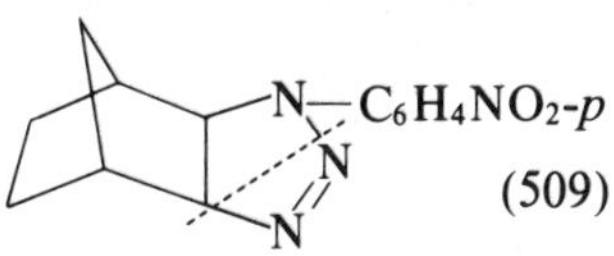

(509)

$(M{-}N_2)^{\ddagger}$ 230 (23)

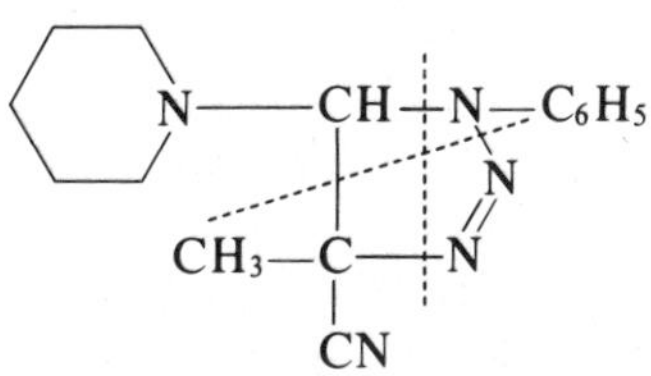

(463)

(IV)

$M^{\ddagger}$	269(1)
$(M{-}CH_3CH(CN)N_2)^{+}$	187(20)
$(M{-}C_6H_5N_3)^{\ddagger}$	150(100)
$(M{-}C_6H_5N_3{-}CH_3)^{+}$	135(64)

IV, similar to those of the acyclic triazenes, is the loss of the 4-triazoline carbon and its substituents including a diazo group.

Fragmentation of both the phthalimidodiazene oxides and the nonacylated trisubstituted triazene oxides occurs between the unsaturated nitrogens (230).

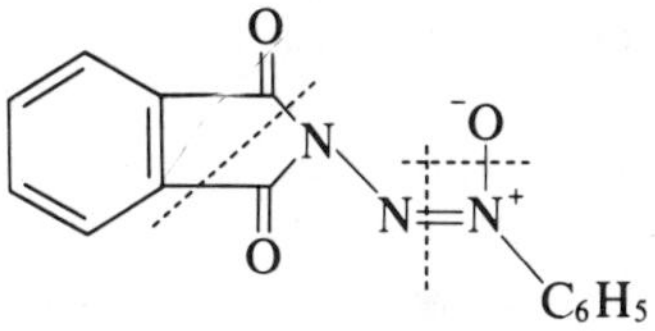

Ion	m/e
$M^{+\cdot}$	267(2)
$C_6H_4(CO)_2NH^{+\cdot}$	147(6)
$C_6H_5NO^{+\cdot}$	107(4)
$C_6H_5CO^+$	105(10)
$C_6H_4CO^{+\cdot}$	104(100)
$C_6H_5N^{+\cdot}$	91(43)
$C_6H_5^+$	77(18)
$C_6H_4^{+\cdot}$	76(41)

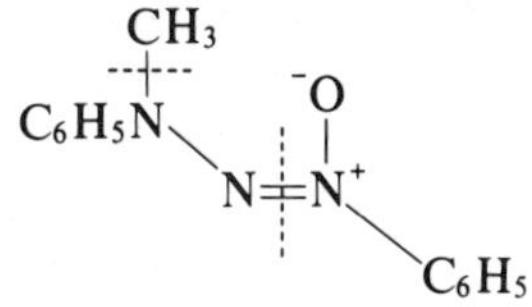

Ion	m/e
$M^{+\cdot}$	227(16)
$(M-C_6H_5NO)^{+\cdot}$	120(9)
$\underline{(C_6H_5N(CH_3)N)^{+\cdot}}$	
$C_6H_5N_2^+$	105(25)
$C_6H_5N^{+\cdot}$	91(5)
$C_6H_5^+$	77(100)

Small quantities of the nitrosobenzene unit are involved in the decomposition of both triazene oxides, either by loss of a neutral fragment or as a radical ion.

Determination of the molecular weight of 2-tetrazene, N_4H_4, by mass spectrometry contributed to establishment of its constitution (493, 495). The spectra

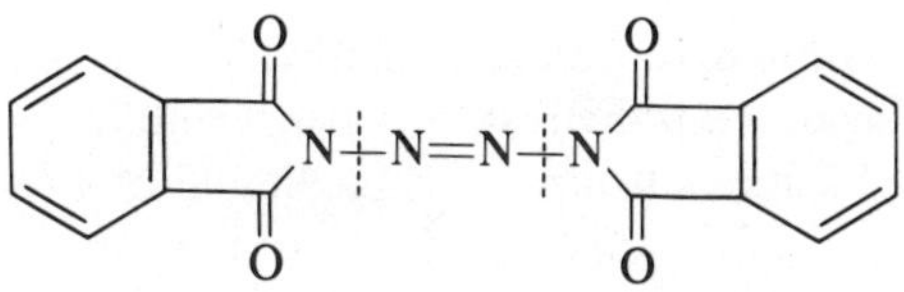

(229)

Ion	m/e
$M^{+\cdot}$	320(31)
$(M-N_2)^{+\cdot}$	292(8)
$(M-N_2-C_6H_4NCO)^+$	174(25)
$C_6H_4CONCO^+$	146(78)
$C_6H_4CO^{+\cdot}$	104(98)
$C_6H_4^{+\cdot}$	76(100)

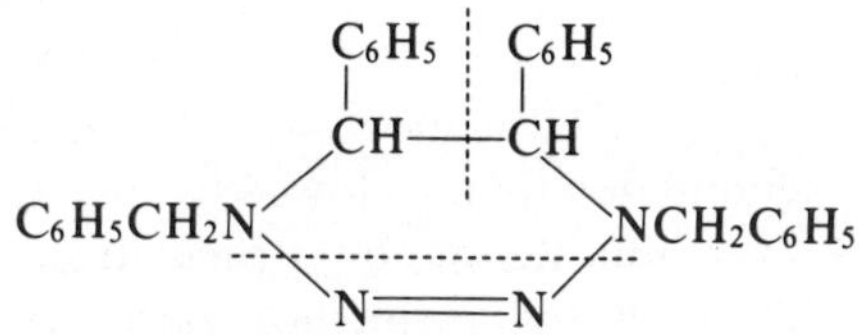

(334)

Ion	m/e
$M^{+\cdot}$	418
$(M-N_2)/2^{+\cdot}$	195
$\underline{(C_6H_5CH{=}NCH_2C_6H_5)^{+\cdot}}$	
$C_6H_5CH_2NH^+$	106
$C_6H_5CH_2^+$	91

of 1,1,4,4-tetrasubstituted 2-tetrazenes indicates loss of nitrogen and fragmentation of the substituents, illustrated by *trans*-1,4-bisphthaloyl-2-tetrazene. The *v*-tetrazines and tetraazaborolines, both cyclic tetrazenes, also eliminate nitrogen on electron impact; the *v*-tetrazines characteristically split between the two ring carbon atoms (280, 360). In contrast, the 1-aryltetrazolines do not release nitrogen, but undergo cycloreversions to actual or potential starting 1,3-dipoles. Thus fragmentation in both directions occur with 4-methyl-1-phenyl-5-methyliminotetrazoline **(V)** and 1-phenyl-4-vinyltetrazolinone **(VI)** forming, respectively, two carbodiimides and two isocyanates along with aryl and alkyl azides. Elimination of alkyl azide predominates, shown for **V** by the relative abundances of each carbodiimide, and for **VI** by high resolution mass spectrometry which indicates a 4/7 ratio of $CH_2{=}CHNCO^{+\cdot}/CH_2{=}CHN_3^{+\cdot}$. Cycloreversion is less important for tetrazolinones substituted only with alkyl groups, where alkyl fragmentation predominates (126).

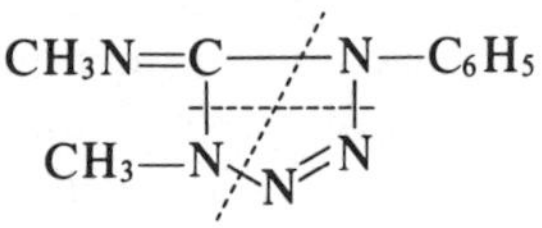

(410)

(V)

Ion	m/e (%)
$M^{\dagger}$	189(15)
$(M-CH_3N_3)^{\dagger}$ $\underline{(CH_3N{=}C{=}NC_6H_5)^{\dagger}}$	132(100)
$C_6H_5N^{\dagger}$	91(87)
$C_6H_5^{\dagger}$	77(81)
$(M-C_6H_5N_3)^{\dagger}$ $\underline{(CH_3N{=}C{=}NCH_3)^{\dagger}}$	70(19)
$CH_3N{=}C^{\dagger}$	41(20)

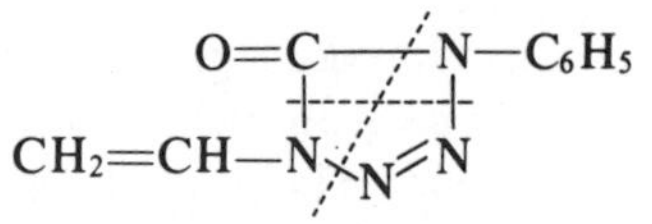

(126)

(VI)

Ion	m/e (%)
$M^{\dagger}$	188(54)
$(M-CH_2{=}CHN_3,CH_2{=}CHNCO)^{\dagger}$ $\underline{(C_6H_5NCO,\ C_6H_5N_3)^{\dagger}}$	119(100)
$CH_2{=}CHNCO^{\dagger}$	69(<2)

Pentazdienes do not form molecular ions (34, 369) but sequentially release two molecules of nitrogen. The base peak of 1,5-di-(*p*-nitrophenyl)-3-*n*-butylpentazdiene, probably the radical ion of *p*-nitroaniline, could arise by transfer of hydrogen from the butyl group to the *p*-nitrophenyl-nitrogen moiety in the pentazdiene or the intermediate triazene.

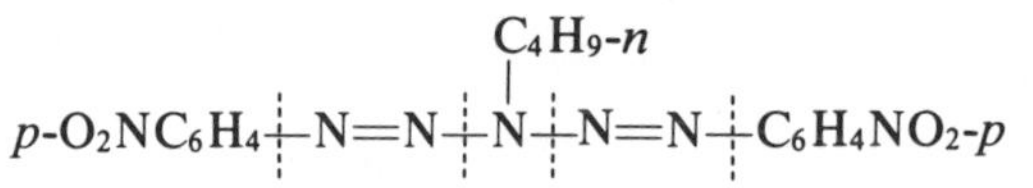

(369)

Ion	m/e (%)
$(M-N_2-C_6H_4NO_2)^{+}$ $\underline{(O_2NC_6H_4N{=}NNC_4H_9)^{+}}$	221(64)
$O_2NC_6H_4N{=}N-N{=}CHC_3H_7^{\dagger}$	220(70)
$O_2NC_6H_4N_2^{+}$	150(55)
$O_2NC_6H_4NH_2^{\dagger}$	138(100)
$(O_2NC_6H_4N_2^{+}-N_2)^{+}$ $\underline{(O_2NC_6H_4)^{+}}$	122(24)

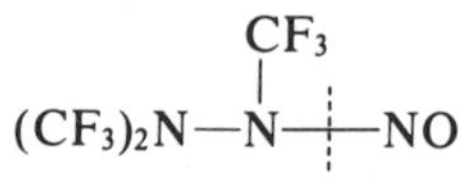

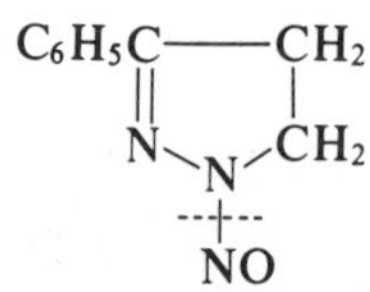

Nitroso tris(trifluoromethyl)hydrazine (132, 220) and 1-nitroso-3-phenyl-Δ^2-pyrazoline (214) lose nitric oxide initially in the mass spectrometer but nitrosotrimethyl-2-tetrazene eliminates its nitroso group as methyl nitrosamine ion.

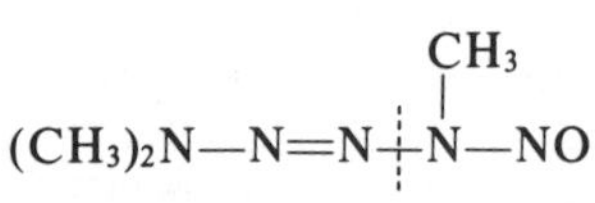

(466)

Ion	m/e
$M^{\dagger}$	131
CH_3NNO^{+}	59
$CH_3N{=}CH_2^{\dagger}$	43
$N_2^{\dagger}$	28

During mass spectrometry, substituted imine fragments of azimines migrate to the principal carbon skeleton. Thus nitrogen is released from the *N*-imide (175) and *N*-methylimide (176) of benzocinnoline on electron impact, but is accompanied by transfer of the imide or methylimide group and attachment to the diphenyl skeleton before further fragmentation. The four-nitrogen chain of

2,3-diphenyl-1-phthalimidoazimine is lost in part as nitrogen, requiring a migration of the azobenzene moiety and its combination with the phthaloyl unit to account for the *m/z* 314 peak. In another decomposition path, the four-nitrogen chain is severed, forming 2-phenylbenzotriazole and phthalimide. Essentially no difference is found between the fragmentation of cis and trans isomers.

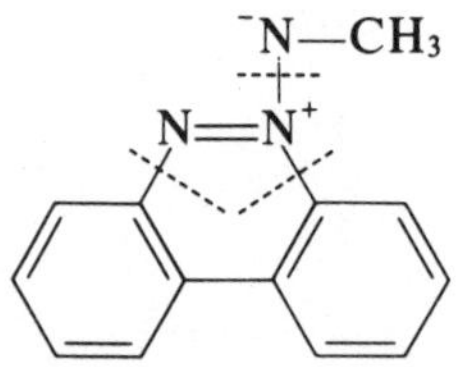

(176)

$M^{\ddagger}$	209
$(M—N_2)^{\ddagger}$	181
$(M—N_3)^+$	167
$(M—N_2—CH_3N)^{\ddagger}$	152

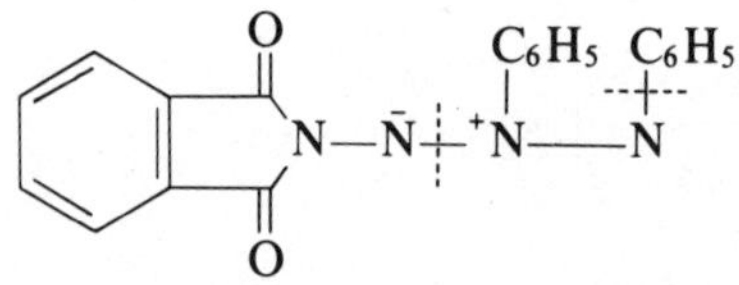

(298)

$M^{\ddagger}$	342(2)
$(M—N_2)^{\ddagger}$	314(12)
$C_6H_4N_3C_6H_5^{\ddagger}$	195(12)
$C_6H_4(CO)_2NH^{\ddagger}$	147(6)
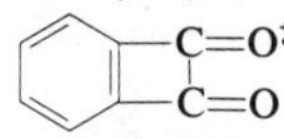	132(8)
$C_6H_5N_2^+$	105(10)
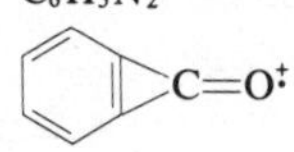	104(100)
$C_6H_5^+$	77(83)

Molecular ions are formed in the mass spectrometer from hydrogen azide, methyl azide, and aryl azides, but under usual conditions are absent or present in low abundance from secondary and tertiary alkyl azides. These molecular ions lose nitrogen readily. The major pathway for decomposition of secondary and tertiary azides is by elimination of N_3 or HN_3; small quantities of N_3^+ ions are formed from hydrogen azide and methyl azide. After aryl azides lose N_2, the

$H\bar{N}—\overset{+}{N}\equiv N$ (169)

$M^{\ddagger}$	43(100)
$(M—H)^+(\underline{N_3^+})$	42(5.8)
$(M—N)^+(\underline{N_2H})^+$	29(8.3)
$(M—NH)^{\ddagger}(\underline{N_2})^{\ddagger}$	28(7.3)
$(M—N_2)^{\ddagger}$	15(16.8)

$CH_3\bar{N}—\overset{+}{N}\equiv N$ (169)

$M^{\ddagger}$	57(48.3)
$(M—CH_3)^+(\underline{N_3})^+$	42(1.9)
$(M—N_2—H)^+(\underline{CH_2N})^+$	28(100)
$(M—N_2—H_2)^{\ddagger}(\underline{HCN})^{\ddagger}$	27(32)

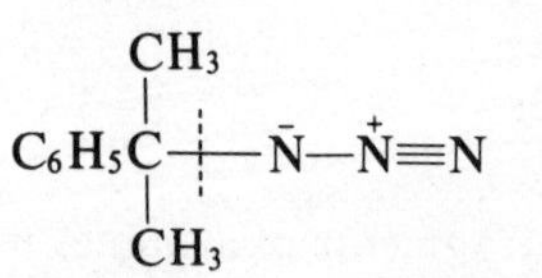

(2)

$M^{\ddagger}$	161(0.5)
$(M—N_2)^{\ddagger}$	133(3)
$(M—N_3)^+$	119(100)
$(M—HN_3)^{\ddagger}$	118(76)

$C_6H_5\bar{N}—\overset{+}{N}\equiv N$ (171)

$(M+1)^+$	120(3)
$M^{\ddagger}$	119(36)
$(M—N_2)^{\ddagger}(\underline{C_6H_5N})^{\ddagger}$	91(100)
$(M—N_2—HCN)^{\ddagger}(\underline{C_5H_4})^{\ddagger}$	64(78)

third nitrogen is removed as HCN, with fracture of the aromatic ring. Acyl azides such as benzoyl azide (171) and cyanogen azide (330) also lose nitrogen from their molecular ions. Analysis by mass spectrometry of the decomposition products of hydrogen azide by electrodeless electrical discharge indicates, besides nitrogen, ammonia, and the blue product obtained by freezing the products on a surface cooled by liquid nitrogen, the presence of a mass 30 peak (2%), considered to be diimide, HN=NH (165).

Fragmentation under electron impact of aromatic heterocycles with three or more contiguous nitrogen atoms principally involves loss of elemental nitrogen,

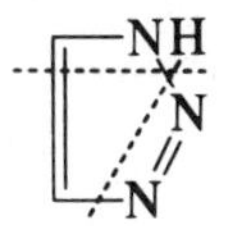 (324)

$(M + 1)^+$	70(3.7)
$M^{+\cdot}$	69(100)
$(M—HCN)^{+\cdot}$	42(42.3)
$(M—N_2)^{+\cdot}$	41(12.3)
$(M—N_2—H)^+$	40(14.6)
$(M—HCN—N)^+$	28(10.8)

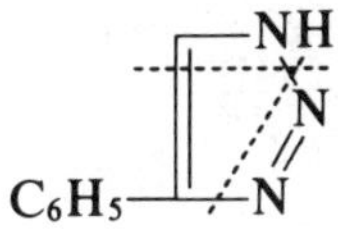 (343a)

$(M + 1)^+$	146(13)
$M^{+\cdot}$	145(100)
$(M—HCN)^{+\cdot}$	118(21)
$(M—N_2—H)^+$	116(7)
$(M—N_3)^+$	103(5)
$(M—N_2—CN)^+$	91(16)

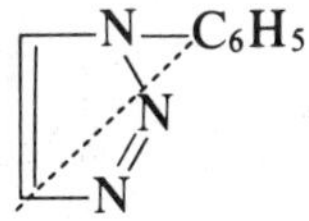 (27)

$(M + 1)^+$	146(2)
$M^{+\cdot}$	145(24)
$(M—N_2)^{+\cdot}$	117(47)
$(M—N_2—HCN)^+$	90(9)
$(C_6H_5)^+$	77(100)

sometimes simultaneously with other accompanying atoms or groups such as hydrogen, hydrogen cyanide, carbon monoxide, or sulfur. Elimination of hydrogen cyanide predominates in the decomposition of monocyclic *v*-triazoles unsubstituted on nitrogen, but other *v*-triazoles, including benzotriazoles, decompose mainly by release of nitrogen. Most *v*-triazoles exhibit a characteristic M + 1 ion. Cation radicals, $C_6H_5N^{+\cdot}$, *m/z* 91, are formed from benzotriazole,

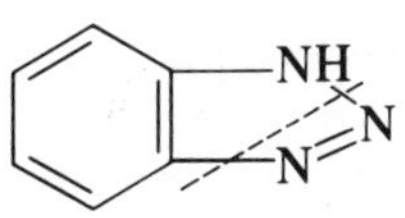 (325)

$(M + 1)^+$	120(6)
$M^{+\cdot}$	119(100)
$M—N_2)^{+\cdot}$	91(68)
$M—N_2—HCN)^+$	64(40)

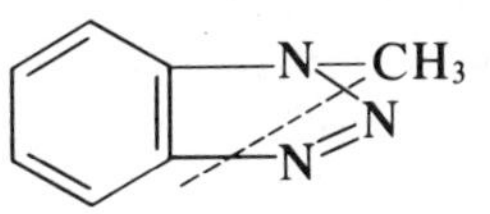 (325)

$(M + 1)^+$	134(8)
$M^{+\cdot}$	133(100)
$(M—N_2)^{+\cdot}$	105(83)
$(M—N_2—CH_3)^+(\underline{C_6H_4N})^+$	90(45)
$(C_6H_4N—HCN)^+$	63(15)

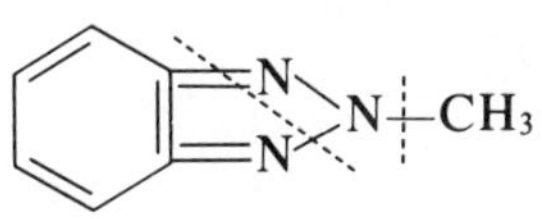 (325)

$(M + 1)^+$	134(8)
$M^{+\cdot}$	133(100)
$(M—CH_3)^+$	118(0.5)
$(M—N_2)^{+\cdot}$	105(5)
$(M—N_2—CH_3)^+(\underline{C_6H_4N})^+$	90(28)
$(C_6H_4N—HCN)^+$	63(12)

phenyl azide, *v*-triazolo[1,5-*a*]pyridine, and many other precursors (326); mass analyzed kinetic energy spectrometry shows that all the metastable $C_6H_5N^{+\cdot}$ radical ions from different sources are isomerized to a common structure before further fragmentation.

v-Triazine loses nitrogen, hydrogen cyanide, and acetylene. Benzotriazinones eliminate carbon monoxide after loss of nitrogen, with subsequent isomerization of the remaining radical ion. The decomposition of 5-phenylthiatriazole resembles its pyrolysis in the elimination of nitrogen and sulfur.

(371)

Ion	m/e (%)
$M^{+\cdot}$	81(47)
$(M-N_2)^{+\cdot}$	53(69)
$HCN^{+\cdot}$	27(13)
$C_2H_2^{+\cdot}$	26(100)

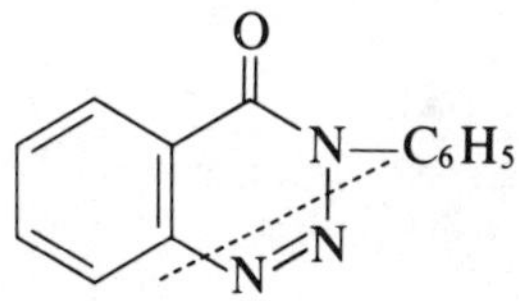

(249)

Ion	m/e (%)
$(M+1)^+$	224(5.5)
$M^{+\cdot}$	223(31)
$(M-N_2)^{+\cdot}$	195(100)
$(M-N_2-CO)^{+\cdot}$ (carbazole)$^{+\cdot}$	167(65)
$C_6H_5NCO^+$	119(25)

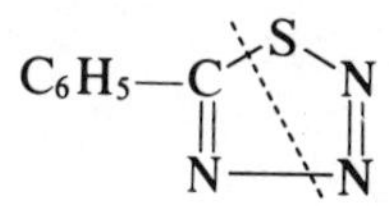

(244)

Ion	m/e (%)
$M^{+\cdot}$	163(2)
$(M-N_2)^{+\cdot}$	135(10)
$(M-N_2-S)^{+\cdot}$	103(100)

Nitrogen is evolved from all tetrazoles on electron impact, and is often the main fragment, but some azide or hydrogen azide is usually lost from those

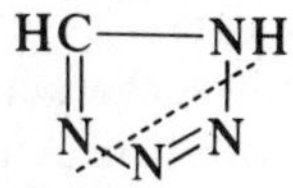

(168)

Ion	m/e (%)
$M^{+\cdot}$	70(13)
$(M-HCN)^{+\cdot}(HN_3)^{+\cdot}$	43(14)
$(M-N_2)^{+\cdot}$	42(100)
$(H\overset{+}{N}_3^{\cdot}-N^{\cdot})^+(HN_2)^+$	29(22)
$(M-CH_2N_2,N_3)^+$ $(N_2^{+\cdot},HCNH^+)$	28(60)
$(M-HN_3)^{+\cdot}$	27(23)

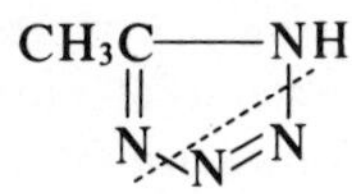

(168)

Ion	m/e (%)
$M^{+\cdot}$	84(11)
$(M-N_2)^{+\cdot}(CH_3CN_2H)^{+\cdot}$	56(100)
$(M-HN_2)^+(CH_2CN_2H)^+$	55(23)
$(M-N_3)^+(CH_3CNH)^+$	42(21)
$(CH_3CN_2H^+-HN_2)^+(C_2H_3)^+$	29(35)
$(CH_3CN_2H^+-C_2H_3)(HN_2)^+$	27(76)

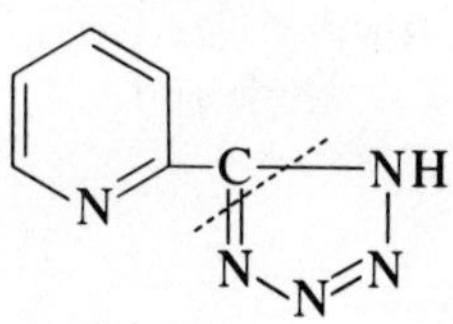

(487)

Ion	m/e (%)
$M^{+\cdot}$	147(3.8)
$(M-N_2)^{+\cdot}$	119(7.8)
$(M-N_3)^+$	105(8.0)
$(M-HN_3)^{+\cdot}$	104(2.2)
$(M-2N_2)^{+\cdot}$ $(C_5H_4NCH)^{+\cdot}$	91(100)

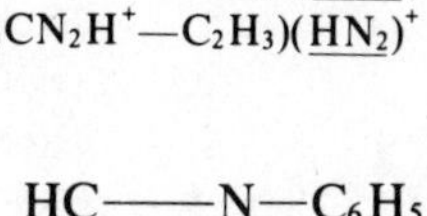

(21)

Ion	m/e (%)
$M^{+\cdot}$	146(2)
$(M-N_2)^{+\cdot}(C_6H_5N_2CH)^{+\cdot}$	118(100)
$(C_6H_5N_2CH^{+\cdot}-HCN)^{+\cdot}$ $(C_6H_5N)^{+\cdot}$	91(75)
$(M-CHN_4)^+(C_6H_5)^+$	77(40)

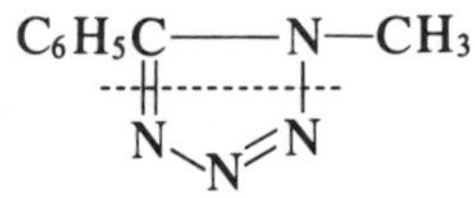

(170)

$(M + 1)^+$	161(2.5)
$M^{\ddagger}$	160(20.5)
$(M—N_2)^{\ddagger}$	132(5.7)
$(M—CH_2N_2)^{\ddagger}$	
$(C_6H_5CN_2H)^{\ddagger}$	118(21.2)
$(C_6H_5CN_2H—N_2)^{\ddagger}$	
$(\underline{C_7H_6})^{\ddagger}$	90(26.2)
$C_6H_5^{\ddagger}$	77(30.6)
$HN_3^{\ddagger}$	43(100)

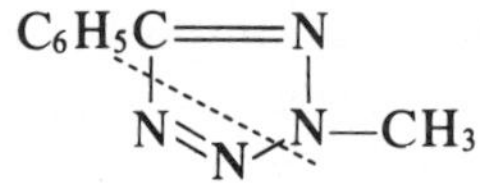

(170)

$(M + 1)^+$	161(1.6)
$M^{\ddagger}$	160(11.5)
$(M—N_2)^{\ddagger}$	132(100)
$(M—N_2—H)^+$	
$(C_6H_5CN_2CH_2)^+$	131(42.2)
$(C_6H_5CN_2CH_2^{\ddagger}—HCO)^+$	104(73.3)
$C_6H_5^{\ddagger}$	77(54.4)
$CH_3N_2^{\ddagger}$	43(37.8)

with hydrogen on ring nitrogen and the 1-substituted derivatives. Loss of the entire tetrazole ring also is observed (21, 22). The 2-aryltetrazoles lose more N_2 and HCN than do their 1-isomers, enabling distinction between them (276).

Mass spectrometry confirms the structure of **VII** as 1-phenyltetrazolinone rather than the isomeric 5-hydroxy-1-phenyltetrazole, because of loss of hydrogen azide and formation of phenyl isocyanate, in common with the fragmentation of tetrazolinones, and the absence of primary nitrogen formation. 1,4-Diphenyl-5-hydroxy-*v*-triazole, in contrast, loses nitrogen.

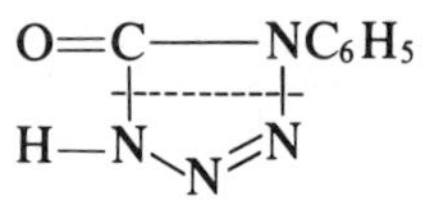

(410)

(VII)

$M^{\ddagger}$	162(6)
$(M—HN_3)^{\ddagger}(\underline{C_6H_5NCO})^{\ddagger}$	119(100)
$(C_6H_5NCO^+—CO)^{\ddagger}(\underline{C_6H_5N})^{\ddagger}$	91(72)
$(C_6H_5N^+—HCN)^{\ddagger}(\underline{C_5H_4})^{\ddagger}$	64(71)

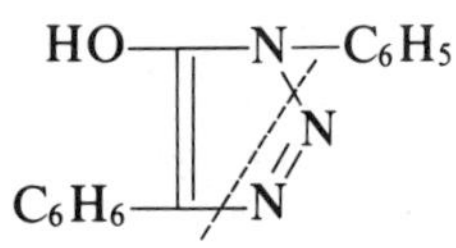

(92)

$M^{\ddagger}$	237
$(M—N_2)^{\ddagger}$	209

The nitrogen chain in compounds with single, exocyclic substituted amino groups attached to cyclic nitrogen in pyrazole fragments differently from compounds with substituted amino joined to benzotriazole or tetrazole. Initial fragmentations of 1-dimethylamino-3-phenylpyrazole occur both by loss of a methyl group followed by migration of the second methyl to N-2, and by deamination with transfer of hydrogen from a methyl to N-1 (268). In the first pathway, acetylene and nitrogen are expelled, leaving the *N*-methylbenzonitrile positive ion. The four- and five-nitrogen chains of the 1-phenylmethyleneamine derivatives of benzotriazole **VIII** and tetrazole **IX**, however, each lose nitrogen in two steps. Rearrangement of the carbon skeleton of **VIII** during nitrogen loss leads to a fluorenyl cation (401), and hydrogen cyanide splits out along with the first molecule of nitrogen released from **IX** (355).

(268)

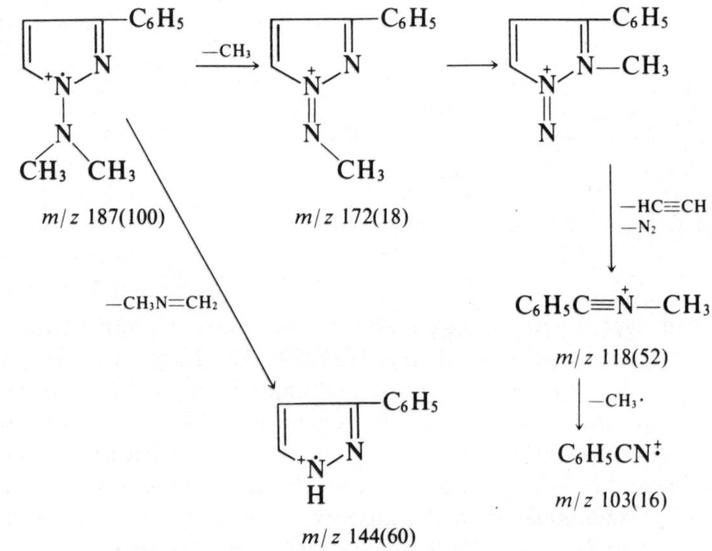

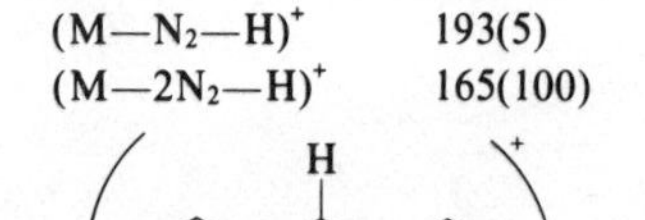

(VIII)

$M^{\ddagger}$	222(40)
$(M—N_2—H)^+$	193(5)
$(M—2N_2—H)^+$	165(100)

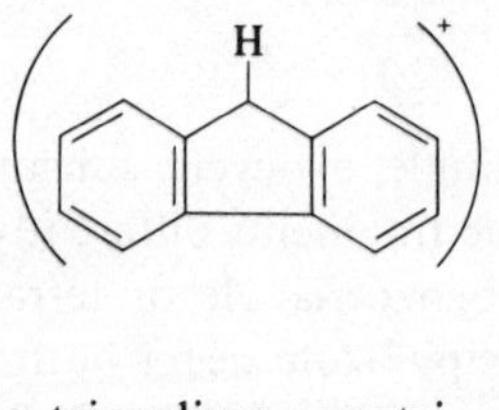

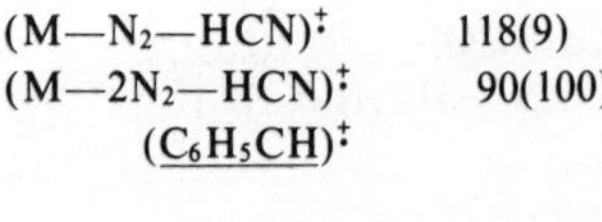

(IX)

$M^{\ddagger}$	173(2)
$(M—N_2—HCN)^{\ddagger}$	118(9)
$(M—2N_2—HCN)^{\ddagger}$ $(C_6H_5CH)^{\ddagger}$	90(100)

Mesoionic triazolium, oxatriazolium, thiatriazolium, and 1,3-tetrazolium compounds fragment by loss of two adjacent ring atoms and their attached radicals, if any. Thus $C_6H_5C{\equiv}\overset{+}{N}C_6H_5$ is eliminated from the diphenyltriazolium **(X)**, NO from the oxatriazolium, NS from the thiatriazolium, and benzenediazonium from the 1,3-diphenyltetrazolium **(XI)** compounds.

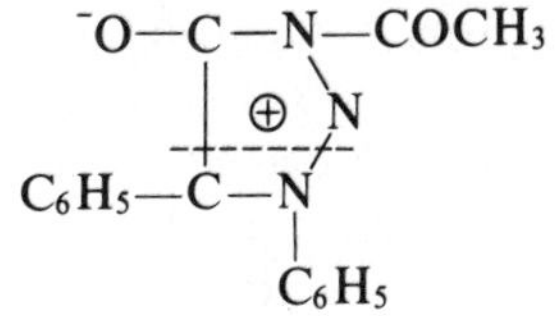

(104)

(X)

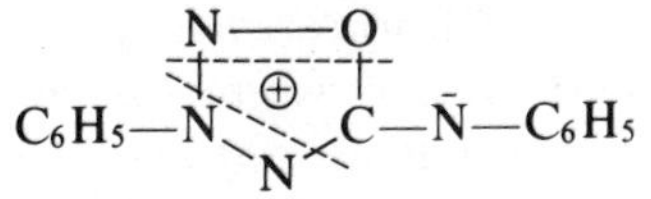

(208)

$M^{\ddagger}$	279(4)
$(M-COCH_3)^+$	236(78)
$C_6H_5C\equiv NC_6H_5^{\ddagger}$	180(100)

$M^{\ddagger}$	238(18)
$(M-NO)^+$	208(10)
$C_6H_5C=NC_6H_5^{\ddagger}$	180(11)
$C_6H_5NCO^{\ddagger}$	119(32)
$C_6H_5N_2^+$	105(49)
$C_6H_5^{\ddagger}$	77(100)

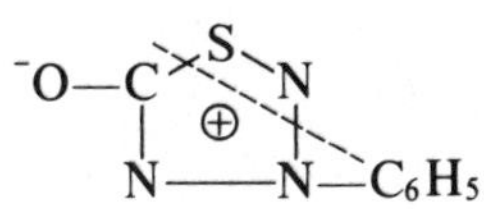

(208)

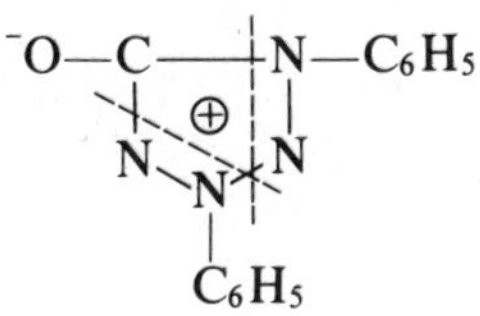

(208)

(XI)

$M^{\ddagger}$	179(27)
$(M-NS)^+$	133(23)
$C_6H_5N_2^+$	105(12)
C_6H_5	77(100)

M^+	238(10)
$C_6H_5NCO^{\ddagger}$	119(<1)
$C_6H_5N_2^+$	105(48)
$C_6H_5^{\ddagger}$	77(100)

2,3-Diphenyltetrazolium-5-thiolate releases nitrogen under electron impact, which must involve a rearrangement or loss of phenyl (252). First formed in mass spectrometry of 2,3-diphenyl tetrazolium salts are the tetrazolium cation (T^+) and a $(T + 1)^{\ddagger}$ radical ion (458). The base peak results from elimination of a phenyl and attached nitrogen and the dehydrogenated phenylmethylene phenylhydrazone radical ion by fission of the tetrazolium ring. Rearrangement and acquisition of hydrogen from phenyl must be involved in the loss of N_2H.

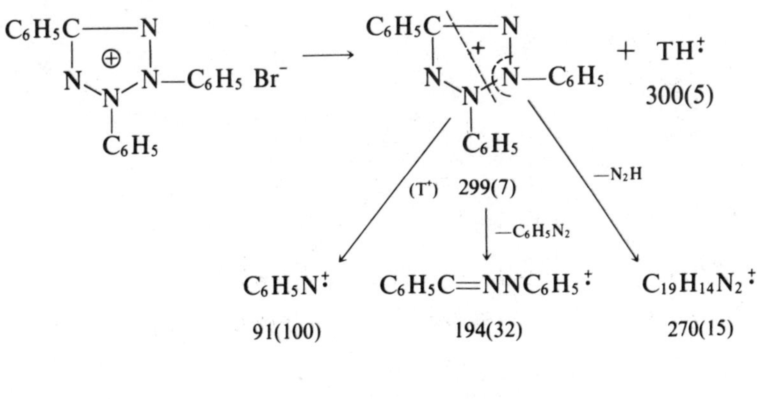

Elemental nitrogen is released from embedded nitrogen chains in aromatic heterocycles, even if prior rearrangement is required, as with *s*-triazolo[1,5-*b*]-pyridazine **(XII)**. Except for molecular ions, fragments with extended contiguous nitrogen chains do not appear to exist after electron impact. Hydrogen cyanide or cyanide radical usually is also lost during fragmentation of these

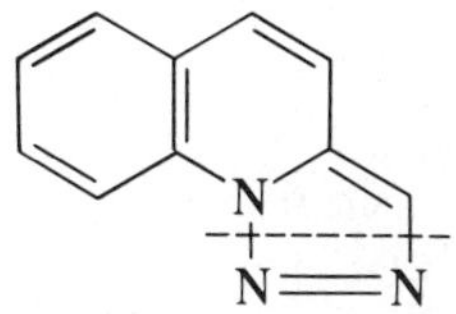

(3)

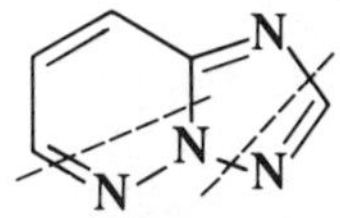

(459)

(XII)

Ion	m/z (%)
$M^{+\cdot}$	169(42)
$(M-N_2)^{+\cdot}$	141(100)
$(M-N_2-H)^{+}$	140(68)
$(M-N_2-HCN)^{+\cdot}$	114(53)

Ion	m/z (%)
$M^{+\cdot}$	120(100)
$(M-N_2)^{+\cdot}$	92(14)
$(M-N_2-HCN)^{+\cdot}$	65(25)

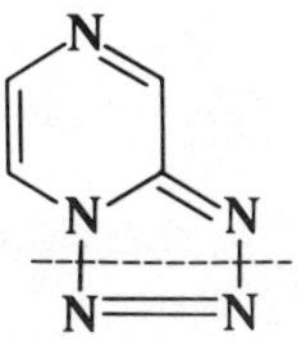

(487)

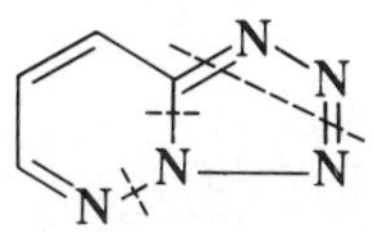

(392)

Ion	m/z (%)
$(M+1)^{+}$	122(1)
$M^{+\cdot}$	121(5)
$(M-HCN)^{+\cdot}$	94(2.6)
$(M-N_2)^{+\cdot}$	93(44)
$(M-N_2-HCN)^{+\cdot}$	66(100)

Ion	m/z (%)
$M^{+\cdot}$	121(6)
$(M-CHN_2)^{+}$	80(59)
$(M-2N_2)^{+\cdot}$	65(85)
$(M-CHN_2-N_2)^{+}$	52(41)
$(M-2N_2-HCN)^{+\cdot}$	38(100)

embedded nitrogen chain compounds. Since the mass spectrum of 6-azidotetrazolo[1,5-*b*]pyridazine **(XIII)** shows a very prominent peak corresponding to the loss of six nitrogen atoms, it is concluded that the molecular ion has the bis-tetrazolo structure (392).

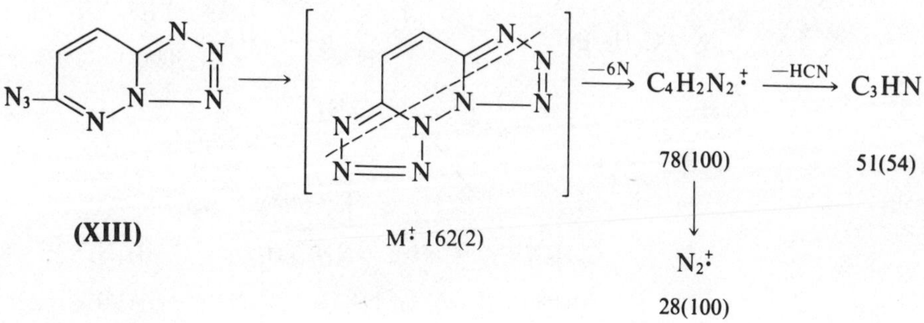

(XIII) $M^{+\cdot}$ 162(2)

3.7. ACID-BASE PROPERTIES

The degree of release or acceptance of protons on chains of three or more nitrogen atoms varies as much as or more than in compounds with one or two nitrogen atoms. Quantitative measures of acidity are reported as pK_a values, usually for aqueous solution on the pH-H_0 acidity scale; basicity is indicated as the pK_a of the conjugate protonated cation, frequently measured in a solvent rich in sulfuric or other strong acid. For many classes of high nitrogen compounds, however, acid or base properties are known only through the formation of salts. Tautomers are possible with 1,3-disubstituted triazenes, as well as *v*-triazoles and tetrazoles unsubstituted on ring nitrogen, and probable locations of hydrogen on the nitrogen chain or equilibrium compositions are given where they have been determined.

Parent high nitrogen compounds with ionizable hydrogen on nitrogen range in acidity from weaker than phenol (pK_a 9.99) to about the strength of acetic acid (pK_a 4.76). (Acidity constants for reference compounds are taken from Lange's *Handbook of Chemistry,* 12th ed., unless otherwise noted.) Structural modification of hydrazoic acid is not possible, but the acidity of triazene, *v*-triazole, and tetrazole can be decreased or increased by introduction of electron-releasing or -withdrawing groups at other hydrogen positions.

$HN{=}NNH_2$	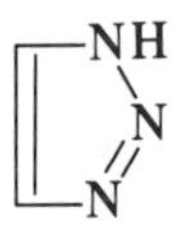	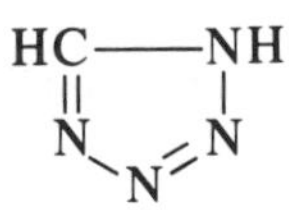	HN_3
pK_a	pK_a	pK_a	pK_a
11.37 (454)	9.26 (210)	4.90 (211)	4.62 (90)
	9.42 (12)	4.79 (343)	

Acidity constants of 1,3-disubstituted triazenes do not appear to have been determined, but their acidic character is illustrated by formation of sodium (50), potassium (144), and heavy metal (483) salts of diazoaminobenzene. Total hydrolysis of the sodium salt by water shows its pK_a to be greater than 14. This low degree of acidity of diazoaminobenzene is consistent with its slight degree of hydrogen bonded association to a cyclic dimer in benzene ($K_{12} = 0.52$) (490). In contrast, 1,3-di-(*p*-nitrophenyl)triazene, which can exist in both a normal and an *aci* form, is strong enough to liberate carbon dioxide from sodium carbonate (145), indicating a pK_a of less than 6.35; it is completely ionized in dime-

p-$O_2NC_6H_4$—N=NNH—$C_6H_4NO_2$-*p*

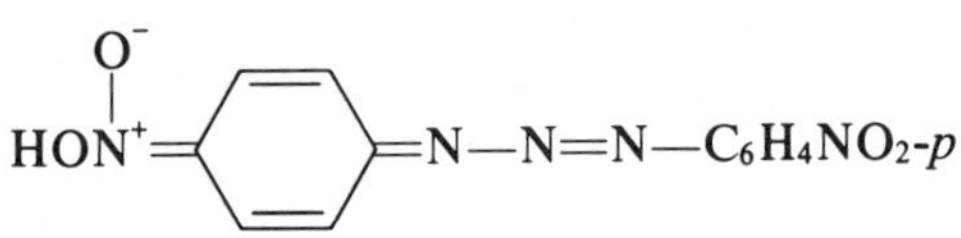

thylformamide, acetone, and methanol (420). Acidities of the cyclic acyltriazenes, benzotriazonone, and its pyrido analog reflect the electron-withdrawing

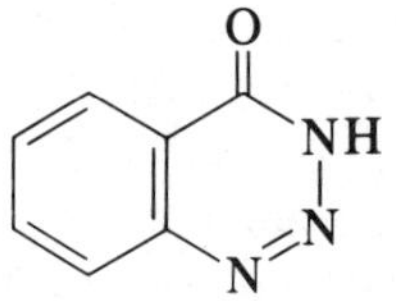

pK_a 9.37(417)
(80% aqueous Methyl Cellosolve)

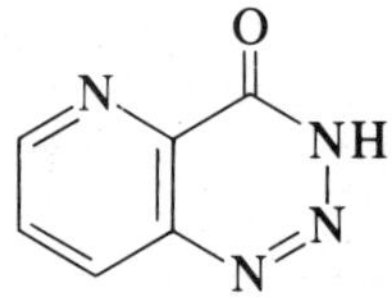

pK_a 8.40(417)
(80% aqueous Methyl Cellosolve)

properties of the carbonyl group. NMR studies of 3-methyl-1-*p*-tolyltriazene indicate that in concentrated $CDCl_3$ hydrogen on the nitrogen chain is located adjacent to methyl and the azo linkage is conjugated with the substituted phenyl (477). Triazene hydrogen in arylmethyltriazenes with strongly electron-

p-$CH_3C_6H_4$—N=N—$NHCH_3$

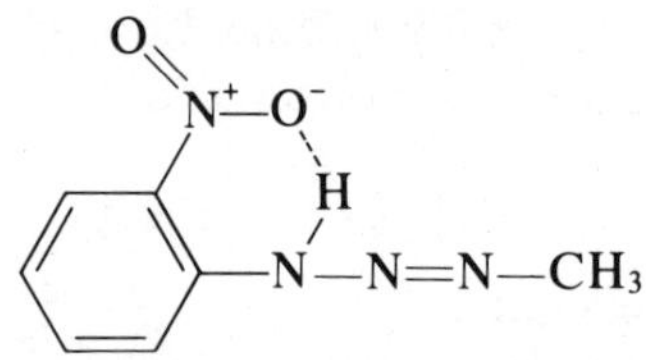

withdrawing ortho substituents is adjacent to aryl and participates in intramolecular hydrogen bonding, causing nonconjugation of azo with aryl. In dilute $CDCl_3$ solution, both tautomers of 3-methyl-1-*p*-tolyltriazene are present, as is also the case in hexadeuterodimethyl sulfoxide; in the latter solvent the tautomeric equilibrium is considerably slowed, presumably because of intermolecular hydrogen bonding of the NH protons to the strong acceptor DMSO (234). With para-substituted electron-withdrawing groups on aryl the nonconjugated tautomer is present in $CDCl_3$ in the largest proportion (118, 477). In deuterated DMSO, the ratio of conjugated to unconjugated tautomer 1-*p*-nitrophenyl-3-methyltriazene is 49:51 (234). Quantitative measurement of a series of unsymmetrical diaryltriazenes (Table 3.25) singly labeled with ^{15}N, indicate the tautomer conjugated with substituted aryl to be predominant except with a *p*-nitro substituent (344).

Triazene oxides exhibit both intra- and intermolecular hydrogen bonding (54); their acidities (397) approximate the low value of the parent triazene. The

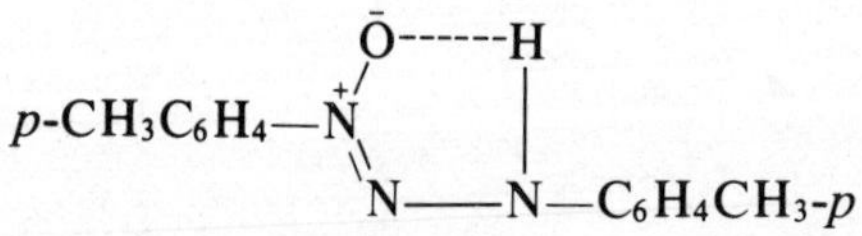

pK_a 12.17

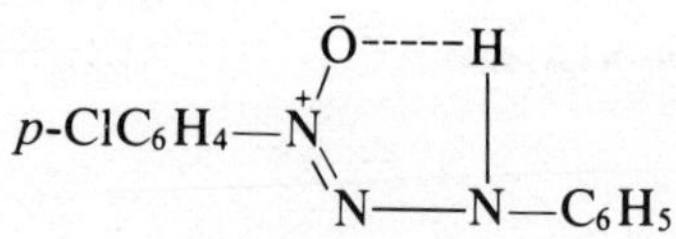

pK_a 10.64

Table 3.25. Tautomeric Equilibria in Unsymmetrical Diaryltriazenes

$$XC_6H_4\text{—N=N—NHC}_6H_5 \underset{CCl_4}{\rightleftarrows} XC_6H_4\text{NHN=NC}_6H_5$$

X	K (Conjugated/Unconjugated) (484)
p-O_2N	0.47
m-O_2N	1.74
p-Cl	1.46
m-Cl	1.33
p-CH_3	1.91
m-CH_3	1.19

hydrogen on N-3 in the 1,5-disubstituted pentazdienes is acidic. Thus 1,5-diphenyl, di-*p*-tolyl, and di-*p*-nitrophenylpentazdienes dissolve in alkaline solution, forming colored anions, from which the pentazdienes are reprecipitated on acidification (384, 385).

$$p\text{-}CH_3C_6H_4\text{—N=N—}\overset{\displaystyle H}{\overset{|}{N}}\text{—N=N—}C_6H_4CH_3\text{-}p$$

v-Triazoles with simple aryl substitution or fusion on carbon do not differ significantly in acidity from the parent compound, but derivatives with electron-withdrawing substituents are much stronger. The pK_a values of substituted benzotriazoles are inversely linearly related to σ_m of substituents in the 4 and 7 positions and to σ_p in the 5 and 6 positions in a Hammett relationship (26).

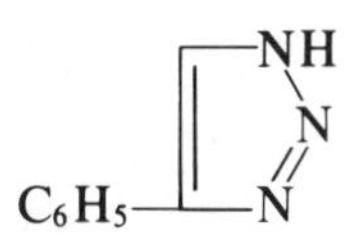

pK_a 9.19(460) (50% ethanol)

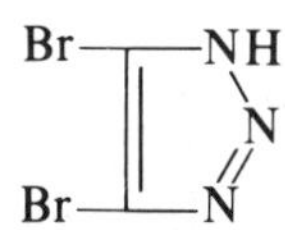

pK_a 5.37(210)

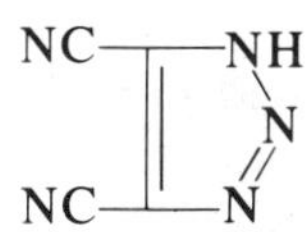

pK_a 1.47(462)

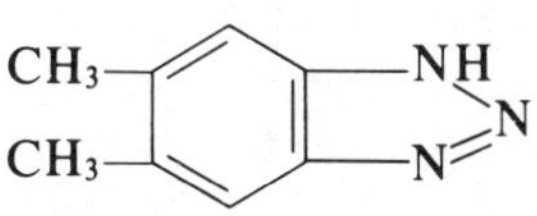

pK_a 9.98(26)

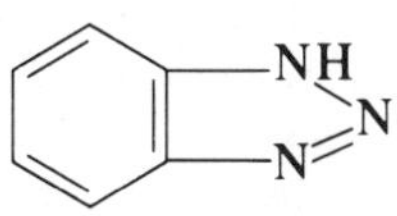

pK_a 8.38(210) 9.05(26)

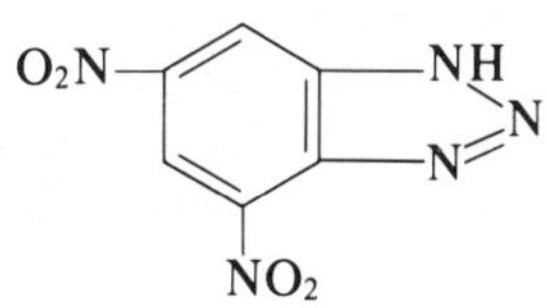

pK_a 4.10(26)

Location of hydrogen on nitrogen in *v*-triazole is not definite. Comparison of the dipole moments in benzene (335), ^{14}N NMR chemical shifts in methanol and gas phase photoelectron spectra (379) of *v*-triazole with those of its 1- and

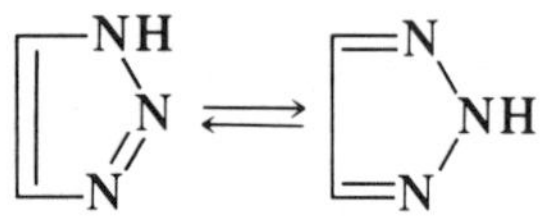

2-methyl derivatives indicates a major proportion of the 2-tautomer in the parent heterocycle. Proton NMR spectra in deuterated acetone, however, show approximately equal amounts of the tautomers (415), and ^{13}C—H coupling constants in $CDCl_3$ imply a 3:2 proportion of 1*H* to 2*H* tautomers (52). Microwave spectra of *v*-triazole in the gas phase have been interpreted to demonstrate preponderance of the 1*H* isomer (64, 450), but further investigation indicates the 2*H*:1*H* ratio to be approximately 100:1 (50).

Both dipole moment (336) and ultraviolet spectra (117) show that benzotriazole exists principally as the 1*H* isomer and proton NMR (408) confirms this assignment. At −100° only the 1*H* tautomer is found, undergoing rapid exchange with the equivalent 3-position (367).

Most physical properties of tetrazole indicate that the 1*H* isomer predominates. Based on dipole moments of tetrazole in dioxane (245) and of 1- and 2-ethyltetrazoles in benzene (260), between 80 and 95% of the 1*H* form is present (148). The same conclusion is reached from ^{1}H NMR of tetrazole and its 1- and

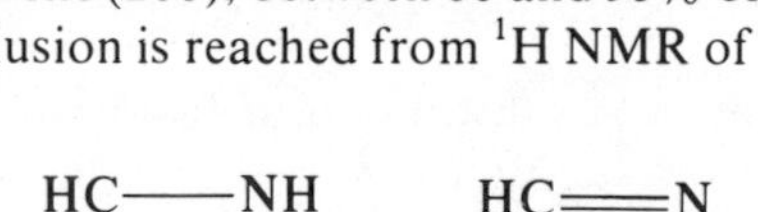

2-methyl derivatives in hexadeuteroacetone, where the chemical shifts indicate 85% of the 1*H* and 15% of the 2*H* tautomers (415). Carbon-13 coupling constants of these tetrazoles in hexadeuteroacetone also indicate preponderance of the 1*H* tautomer (53), confirmed by ^{13}C chemical shifts in a series of solvents including water (149). Nitrogen-14 NMR data (501) and gas phase photoelectron spectra (379), however, favor the existence of more 2-isomer.

Analogous acidic character of corresponding 5-substituted tetrazoles (RCN_4H) and carboxylic acids (RCO_2H) was perceived and elaborated by Herbst (225). 5-Alkyl tetrazoles are somewhat weaker, but 5-phenyltetrazole and its *m*- and *p*-substituted derivatives are slightly stronger than the related carboxylic acids (Table 3.26) because of increased resonance in the 5-aryltetrazolates in comparison to the benzoates. Indeed, 5-azidotetrazole is a strong acid because of resonance stabilization between azide and tetrazole in the anion (311). Acidity constants of the 5-*m*- and *p*-phenyltetrazoles correlate with σ_m and σ_p (251). As expected, acidities are greater in tetramethylguanidine in which

Table 3.26. Acidity Constants of Corresponding 5-Substituted Tetrazoles and Carboxylic Acids

	pK_a	
R	RCN_4H	RCO_2H
CF_3	1.7 (211)	0.25
m-BrC_6H_4	4.38[a]	5.22[a]
p-BrC_6H_4	4.69[a]	5.35[a]
C_6H_5	4.98[a] 4.61[b] (99)	5.73[a]
p-$CH_3C_6H_4$	5.17[a]	5.94[a]
CH_3	5.63 (211) 4.60[b] (99)	4.76

[a] 50% ethanol; tetrazoles (251), carboxylic acids (82).
[b] In 1,1,3,3-tetramethylguanidine.

a linear correlation is found between the acid constants of 5-alkyltetrazoles and Taft σ^* constants (99).

The similarity of tetrazole to carboxyl is emphasized by the closeness of the pK values of the cations of the amino acids glycine and D,L-phenylalanine and their tetrazole analogs (Table 3.27) (341).

Table 3.27. Acidity Constants of Corresponding Amino Acids and Amino Tetrazoles

Amino Acid/Amino Tetrazole	pK_1	pK_2
$H_3\overset{+}{N}CH_2COOH$	2.34	9.60
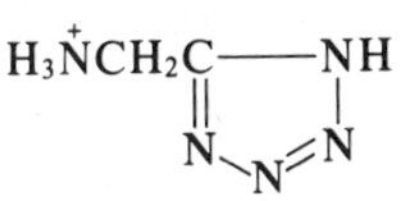	2.62	8.54
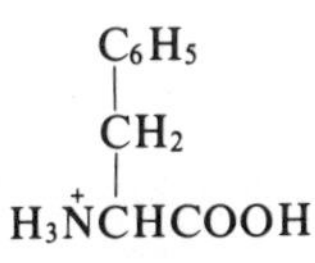	1.83	9.13
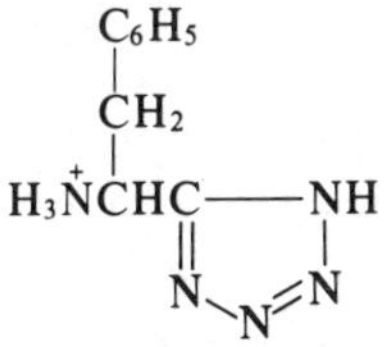	1.93	8.18

Since the ring hydrogen of 1-substituted 5-tetrazolinones and tetrazoline thiones exists primarily on nitrogen rather than on oxygen or sulfur its ionization also must be from nitrogen. The lower pK_a value of 5-phenoxytetrazole in comparison with 1-phenyl-5-tetrazolinone shows the greater resonance stability of the tetrazole anion over that of the tetrazolinone. Such a large difference is not observed, however, between 1-substituted tetrazolinethiones and the isomeric 5-substituted mercaptotetrazoles, consistent with little extra resonance stabilization of the anion of 5-methylmercaptotetrazole indicated by ultraviolet spectra (304).

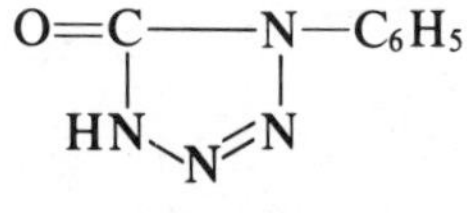

pK_a 5.53(235)
(50% ethanol)

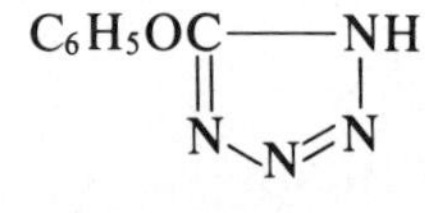

pK_a 3.8(331)
(50% methanol)

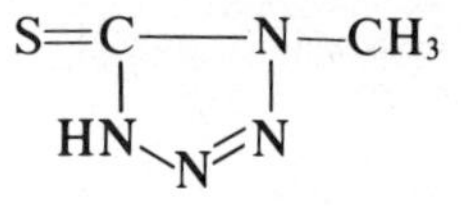

pK_a 3.86(304)
(50% ethanol)

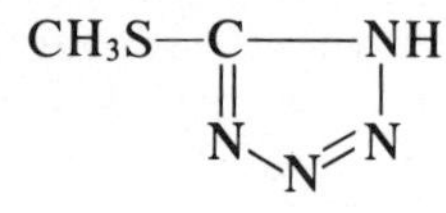

pK_a 4.07(304)
(50% ethanol)

Similarity between 5-amino and substituted aminotetrazoles and the analogous carboxylic acids cannot be determined because free carbamic acids do not exist. 5-Amino and alkylaminotetrazoles are weaker than 5-alkyltetrazoles, but substituents on amino such as aryl or acyl, which can conjugate with the tetrazole ring through the nitrogen atom, increase acidity of the tetrazole hydrogen (Table 3.28). 5-Acetylaminotetrazole is a monobasic acid in water, but both the ring and amide hydrogens are acidic in ethylenediamine (322).

Table 3.28. Acidity Constants of 5-Aminotetrazoles

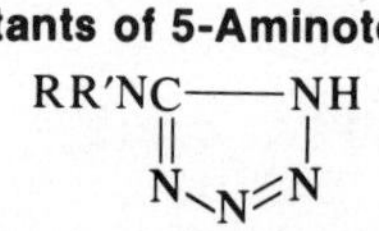

R	R′	pK_a	Ref.
H H	H	5.93 6.44 (50% methanol)	179
CH_3	H	6.06	180
CH_3	CH_3	5.92	179
C_6H_5	H	5.49 (50% methanol)	181
p-$O_2NC_6H_4$	H	4.34 (50% methanol)	181
CH_3CO	H	4.53	226
n-C_4H_9NHCO	H	4.85 (30% ethanol)	127

Except for ammonium, quaternary, and amine oxide type nitrogen, in principle all nitrogen atoms in nitrogen chains should exhibit some degree of basicity. In reality, however, high nitrogen compounds are protonated or alkylated on only one or at most two nitrogen atoms in the chain, probably reflecting reduced availability of electron pairs because of resonance interaction along the nitrogen chain, steric restrictions, or repulsion owing to potential proximity of positive charges. Note is made that hydrazine forms a diprotonated cation only in the solid state, which completely hydrolyzes to the monoprotonated cation in water (29).

In aqueous solution, basicities of parent high nitrogen compounds which have been measured range from triazene, similar to aniline (anilinium $pK_a = 4.60$), to hydrazoic acid, weaker as a base than nearly all organic nitrogen compounds. Tetrazane, during its momentary existence, appears to form two cations, $N_4H_8^{2+}$ and $N_4H_7^+$, with pK_a values between 5 and 9, similar to those of the protonated cations of ethylenediamine, 6.85 and 9.93 (222).

$HN{=}N\overset{+}{N}H_3$ — pK_a 4.95(454)

pK_a 1.17(12)

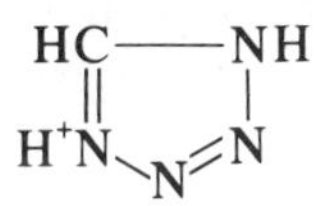

pK_a −3.0(444)

$H_2N_3^+$ — pK_a −6.21(36), −7.19(7)

Substituted triazenes also are weak bases, illustrated by the formation in benzene of hydrochlorides of diazoaminobenzene (342) and 1,3,3-triphenyltriazene (478), which are entirely dissociated in water. Instability of triazenes in acid media renders measurement of their basicity difficult.

Nucleophilicity of 1,1,4,4-tetramethyltetrazene is shown by its strong protonation and its alkylation to form quaternary salts (see Section 2.4.2). Replacement of one methyl by formyl significantly reduces basicity, and replacement of

$(CH_3)_2\overset{+}{N}HN{=}NN(CH_3)_2$ — pK_a 6.20(337)

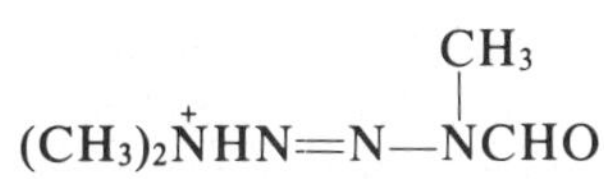

pK_a −0.57(466)

two methyls by formyl reduces the pK_a to approximately −4 (466). Existence of salts of azimines with acids (Section 2.2.2) and of salts and quaternaries of tetraazapentalenes (Section 2.6.6) demonstrates their basicity.

Basic character of organic azides is indicated by formation at room temperature of a 1 : 1 adduct of methyl azide with antimony pentachloride (193) and the low-temperature conversion of hydrazoic acid to diazonium amine salts by the combined action of antimony pentachloride and either hydrogen chloride or *t*-butyl chloride (424).

Basicity of 1-methyl-*v*-triazole is close to that of *v*-triazole, but the 2-methyl isomer is weaker, as is benzotriazole. Schofield relates the difference between

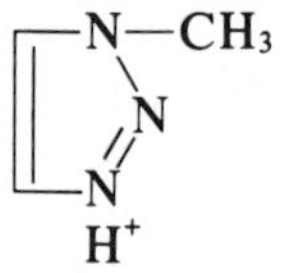

pK_a 1.25(12)

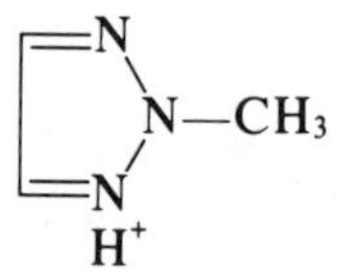

pK_a <1(12)

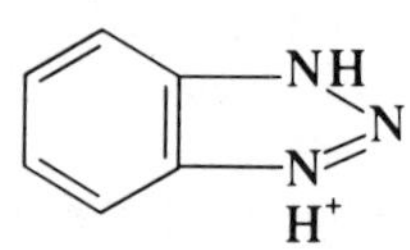

pK_a 0.40(26)

the 1- and 2-methyl-*v*-triazoles to the greater basicity of imidazole in comparison to pyrazole; 1-methyl-*v*-triazole protonates to an imidazolium-type cation, but the 2-methyl isomer can only form a pyrazolium-like cation (426).

Similarly the unsubstituted and 1-alkyl substituted tetrazoles are slightly stronger bases than tetrazole, but 2-substituted tetrazoles are weaker. Introduction of phenyl at the 1-tetrazole position reduces basicity. From correlation of pK_a values of aryl tetrazoles with inductive σ constants (6, 351), and from ^{1}H and ^{13}C NMR data and calculated electron densities at each nitrogen (277), it is

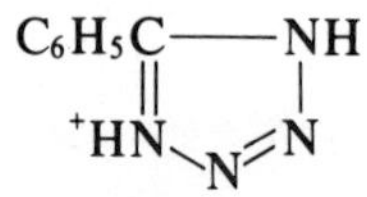

pK_a −2.45(453)

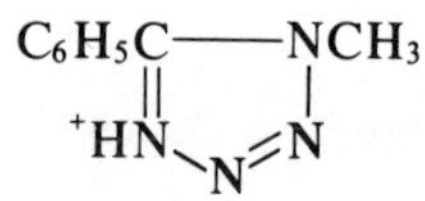

pK_a −2.50(351)

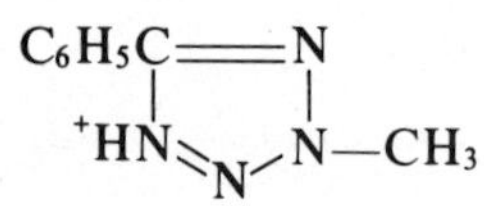

pK_a −3.27(351)

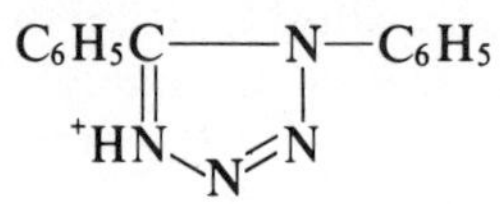

pK_a −2.96(373)

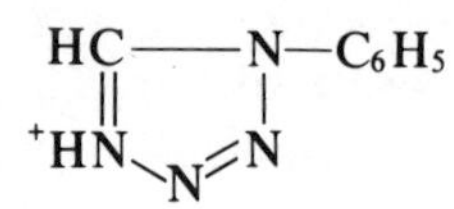

pK_a −3.41(373)

concluded that protonation of these tetrazoles occurs at N-4. Conductance measurements of tetrazole, 5-methyl-, 5-phenyl-, and cyclopolymethylene tetrazoles in anhydrous formic acid indicate them to be monoprotic bases in this solvent (152, 485). Pentamethylenetetrazole and related tetrazoles can be titrated in formic acid with *p*-toluenesulfonic acid (395). The basic character of tetrazole and 5-phenyltetrazole also is shown by depression of the freezing point of absolute sulfuric acid, approaching the formation of two ions per mole of solute (212). Only one nitrogen per mole of tetrazole compound is involved in this ionization.

$$C_6H_5CN_4H + H_2SO_4 \longrightarrow C_6H_5CN_4H_2^+ + HSO_4^-$$

High nitrogen compounds containing an additional nitrogen not in the chain or oxygen or sulfur atoms often protonate on these atoms. Thus the strongly basic characters of 5-imino-3-phenyl-1,2,3,4-oxatriazole and of 5-imino-1,3- and

1,4-tetrazolium compounds reflect their structures as substituted amidines and guanidines rather than the basicity of their nitrogen chains.

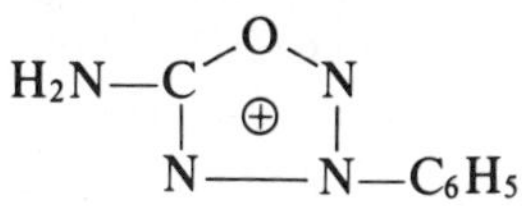

pK_a ~6(107)

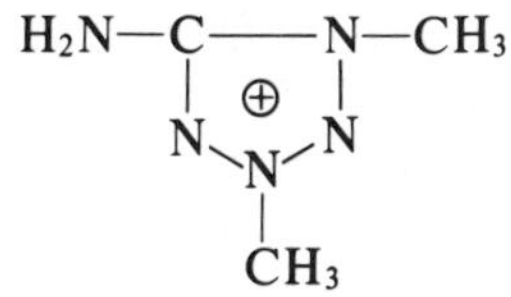

pK_a 11.5(224)

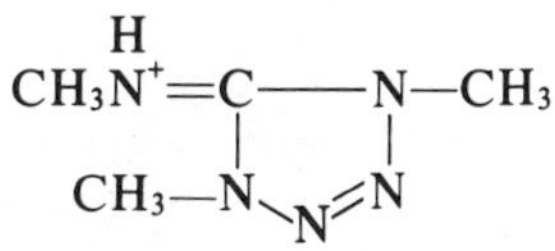

pK_a 9.57(61)

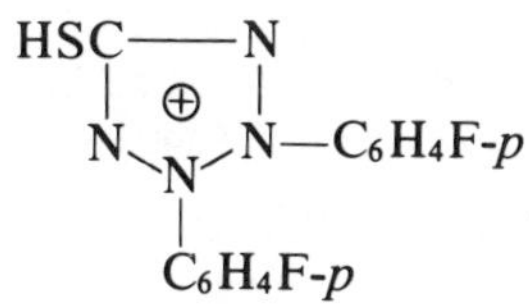

pK_a −1.51(269)

Amine nitroimides are weak bases as shown by the acidity constants for the trimethylamine and pyridine nitroimide cations; based on calculated and experimental spectra, protonation of these cations is on oxygen rather than nitrogen (24).

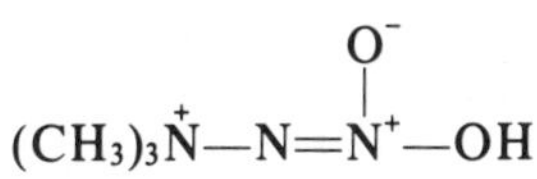

pK_a −4.2

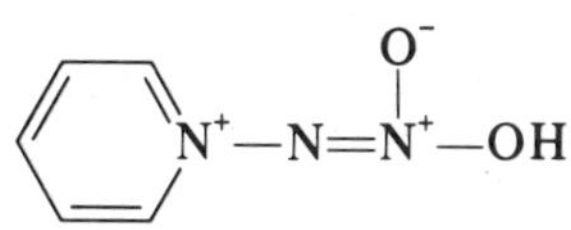

pK_a −4.6

In sulfuric acid solutions of bases containing both nitrogen and oxygen, correlation of pK_a values of cations with H_0 indicates nitrogen protonation, but

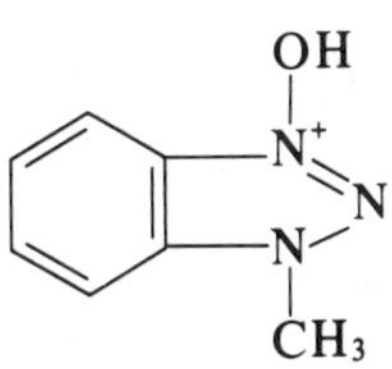

pK_a −0.98

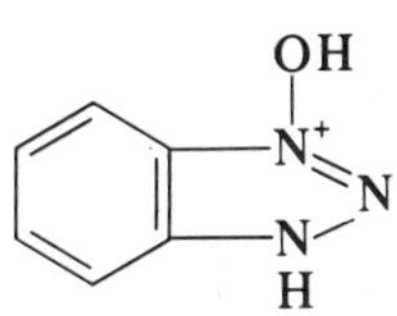

pK_a −0.95

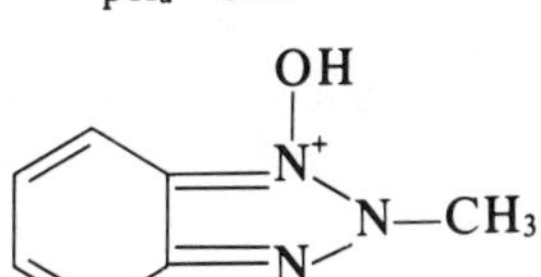

pK_a −1.86

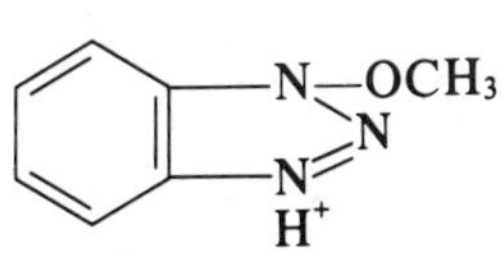

pK_a −0.69

correlation with H_A, based on amide indicators, indicates oxygen protonation (247, 507). Applied to benzotriazole oxides and related model compounds, and supported by ultraviolet spectra data, the correlations indicate protonation of the oxides is on oxygen and that of 1-methoxybenzotriazole is on nitrogen (78). Closeness of the pK_a values for the cations of the 3-methyl 1-oxide and of the 1-oxide, and the divergence from the value for the 2-methyl 1-oxide, supports the 1,3-structure of the 1-oxide.

3.8. EXPLOSIVE PROPERTIES

Mechanical, thermal, or electrical stimulation causes explosive response in compounds of several different high nitrogen classes. Standard quantitative tests of sensitivity and shattering effect have been applied to only a small number of these substances, largely azides, *v*-triazoles, and tetrazoles. Most indications of explosiveness have been obtained during handling and characterization of compounds with linked nitrogen atoms and are qualitative in nature.

Standard tests of sensitivity to impact involve dropping a weight onto a quantity of test substance held in place and determining either the minimum height causing an explosion or the height at which explosion occurs in a given percentage of trials. The work (in kg-cm) to cause explosion (or its absence) under the conditions of each test, given for each substance with quantitative data, allows an approximate ranking of sensitivity. Impact sensitivities of reference compounds in common types of apparatus are as follows:

Compound	Bureau of Mines Apparatus (475), 2 kg wt (cm) (*kg-cm*)	Picatinny Arsenal Apparatus (475), 2 kg wt (in.) (*kg-cm*)	Los Alamos Scientific Laboratory Machine (363), 2.5 kg wt, 50% Fire level (cm) (*kg-cm*)
Mercury fulminate	5(10)	2(10.2)	
Lead azide, undesensitized	10(20)	3(15.2)	
Cyclonite (RDX)	32(64)	8(40.6)	23(57.5)

Qualitative observations indicate sensitivity of 1-substituted triazenes and a few disubstituted triazenes to mechanical provocation. Dimroth remarked that the tendency of 1-phenyltriazene **(I)** to explode on handling is so great that extraordinary patience is required to weigh it. Some triazenes with cyano substitution either on the nitrogen chain or further removed explode from a blow (3-cyano-1-phenyltriazene) **(II)**, or on powdering (*o*-cyanophenylazocyanoguanidine) **(III)**.

$C_6H_5N{=}NNH_2$	$C_6H_5N{=}NNHCN$	$o\text{-}NCC_6H_4{-}N{=}NNH{-}C(NH_2){=}NCN$
(I)	**(II)**	**(III)**
(129)	(504)	(321)

Azides are traditionally associated with explosive sensitivity. *t*-Butyl azidoformate **(IV)**, propargyl azide **(V)**, hydrazinodicarboxylic azide **(VI)**, and cyanuric azide **(VII)** are more sensitive to impact than mercury fulminate in the various quantitative tests. Guanyl azide nitrate **(VIII)** and phenylphosphoric diazide **(IX)** are both less sensitive than lead azide.

Conclusions about sensitivity to impact also can be deduced from qualitative information (Table 3.29), compared, as appropriate, with quantitative data. Unsaturation adjacent to azide, as in vinyl azide, or one carbon removed, as in propargyl azide **(V)**, leads to sensitivity to mechanical stimulation, but in 4-azido-1-butyne **(X)** separation by two methylene groups confers insensitivity. Although structural factors can enhance explosive sensitivity in the azides, the main determinant is proportion of azide. Thus cyanuric azide **(VII)** (61.8% azide N) and 1,8-diazido naphthalene **(XIII)** (40% N) are both sensitive to im-

Quantitative Impact Sensitivity

AZIDES

$(CH_3)_3COC(=O)N_3$	$HC{\equiv}CCH_2N_3$	$N_3C(=O)NHNHC(=O)N_3$	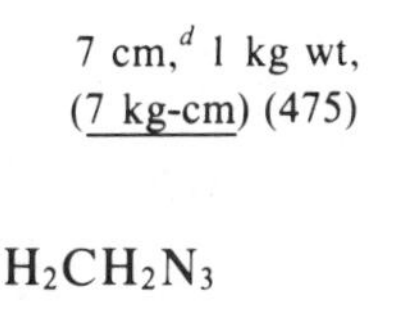
(IV)	**(V)**	**(VI)**	**(VII)**
2.8 kg-cm[a] (163)	1.9 kg-in 33% fire level[b] (4.8 kg-cm) (248)	70 cm, 100 g wt[c] (7 kg-cm) (264)	7 cm,[d] 1 kg wt, (7 kg-cm) (475)

$(H_2N)_2C^+N_3NO_3^-$	$C_6H_5PO(N_3)_2$	$HC{\equiv}CCH_2CH_2N_3$
(VIII)	**(IX)**	**(X)**
20 cm[d] (40 kg-cm) (57)	20 cm 2 kg wt (40 kg-cm) (38)	insensitive at 38 kg-in.[a] (>96.5 kg-cm) (248)

[a] Nitroglycerin 1.6 kg-cm.
[b] Cyclonite 10 kg-in. (74.5 kg-cm) 50% fire level.
[c] PbN_6 80–85 cm.
[d] Bureau of Mines Apparatus, 2 kg wt, unless otherwise stated.

Table 3.29. Qualitative Sensitivity of Azides to Impact or Friction

Sensitive	Insensitive
$CH_2{=}CHN_3$ (498)	$CF_3CF_2CF_2CH_2N_3$ (98)
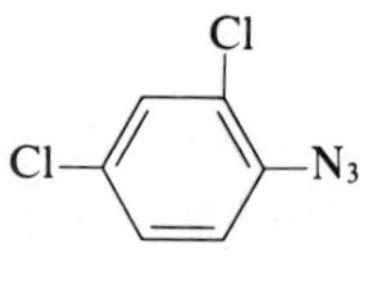 (102) **(XII)**	**(XI)** 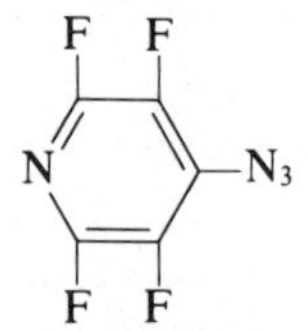 (42)
$(C_6H_5)_2C(N_3)_2$ (348a)	
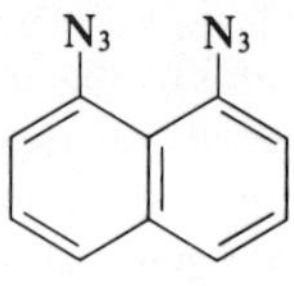 (232) **(XIII)**	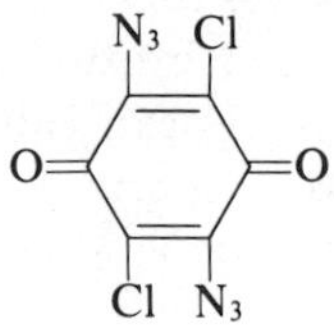 (172)
$O_2NNHC({=}NH)N_3$ (305)	$F_2NC({=}NF)N_3$ (84)
$O{=}C(N_3)_2$ (114, 115)	$H_2NC({=}O)NHC({=}O)N_3$ (28)
$P(N_3)_4^+SbCl_6^-$ (425)	$(CH_3)_3SiOSO_2N_3$ (281)

pact. Smith indicates for azides containing carbon, oxygen, and nitrogen that explosiveness ends not far from a ratio of $(C + O)/N = \frac{3}{1}$ (442) and the principle is generally borne out when extended to azides containing other elements.

$(CH_3)_2NN_3$

highly sensitive (67, 68)

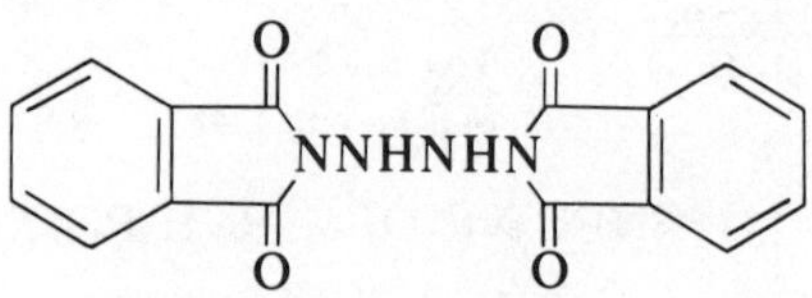

explodes on scratching (20)

$C_6H_5N{=}NN(NO)C_6H_5$ (356)

explodes on handling with a wooden spatula

$(CH_3)_2NN{=}NN^+(CH_3){=}CHC(NO_2){=}NO_2^-$

insensitive to impact (467)

Dilution of azide with halogen in 1-azido-heptafluorobutane **(XI)** (18.7% N) is sufficient for insensitivity; in contrast, 1-azido-2,4-dichlorobenzene **(XII)** (22.3% N) with a similar proportion of azide explodes when struck. Pure cyanogen azide detonates on mild mechanical shock, but its 30% solution in acetonitrile does not explode under the impact of a 6 kg weight dropped 80 cm (330).

No indications of impact sensitivity are available for simple tetrazenes. The four-nitrogen amine azides, bisphthaloyltetrazane and diaryl nitrosotriazenes, are very sensitive to mechanical stimulation. Dipolar 1,1,4-trimethyl-4-β,β-dinitrovinyl-2-tetrazene and its homologs, however, are insensitive to impact.

Impact sensitivity data are available only for v-triazoles with nitro and picryl groups, and accordingly the contribution of the v-triazole ring cannot be precisely assessed. Since 1-picryl-v-triazole **(XIV)** is more impact sensitive than picramide [23 in., Picatinny Arsenal Machine (475)], the 1-substituted v-triazole moiety must enhance susceptibility to mechanical shock. The appreciable difference in impact sensitivity between 1- **(XIV)** and 2-picryl-v-triazole **(XV)** is remarkable. Although sensitivity values of specific compounds are not reported, 1-picrylbenzotriazole, its nitro derivatives, and several dipicrylbenzobistriazoles are indicated as comparable to cyclonite (58, 112).

Quantitative Impact Sensitivity

v-TRIAZOLES[a]

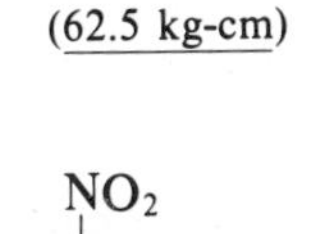

(364) 25 cm
(62.5 kg-cm)

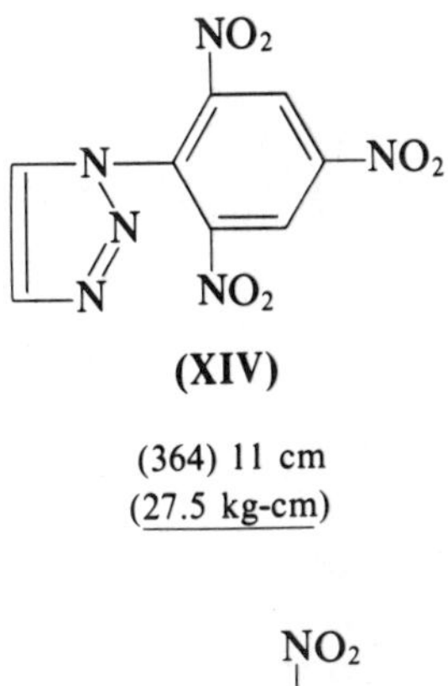

(XIV)

(364) 11 cm
(27.5 kg-cm)

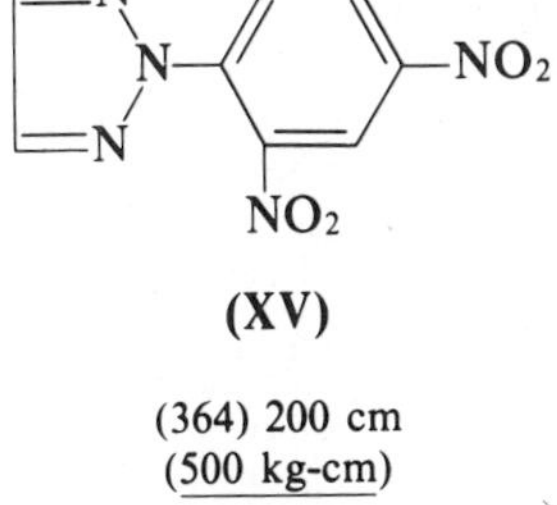

(XV)

(364) 200 cm
(500 kg-cm)

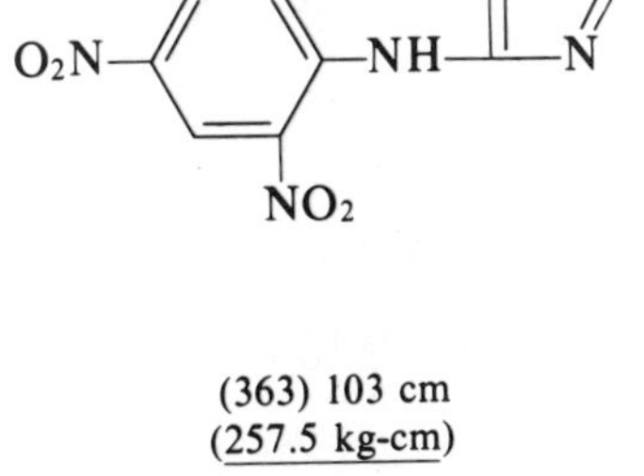

(363) 103 cm
(257.5 kg-cm)

[a] LASL Machine, 2.5 kg wt, 40% fire level.

Thiatriazoles with sulfur at the 5-position, such as **XVI** and **XVII**, are sensitive to impact, even under water. Impact sensitivity has not been observed in

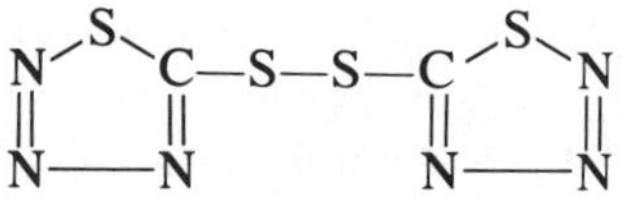

(XVI)

(83)

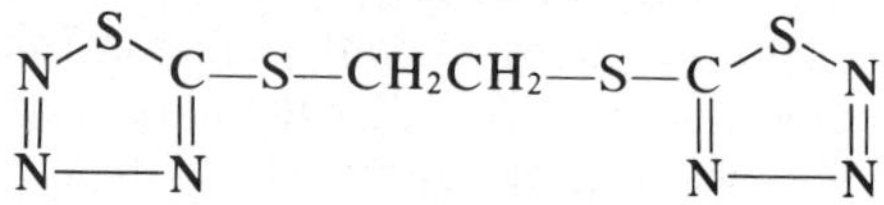

(XVII)

(390)

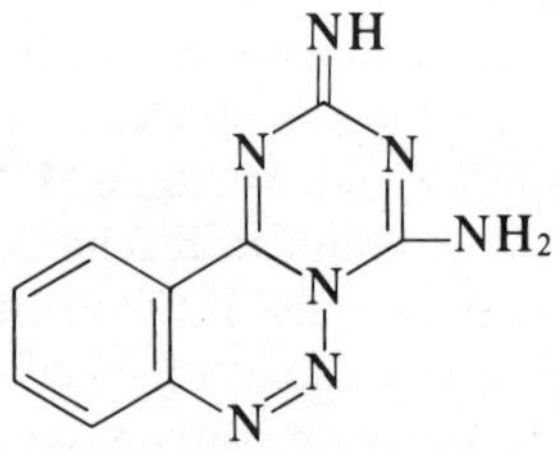

(XVIII)

(321)

Quantitative Impact Sensitivity

Tetrazoles

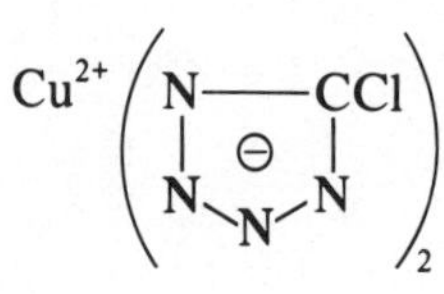

(XIX)

1 in.[a] (5.1 kg-cm)
(475)

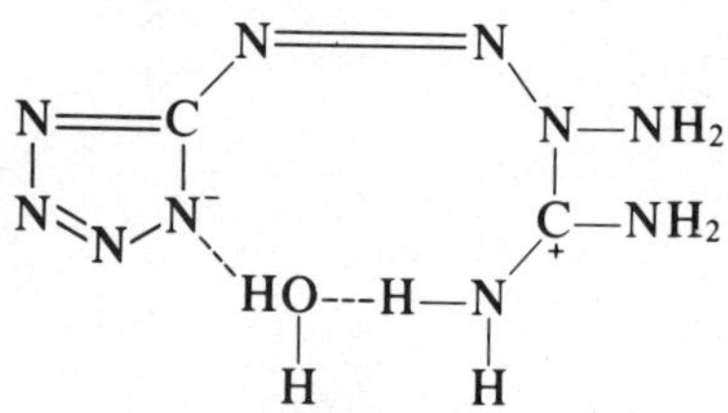

(XX)

7 cm^b (14 kg-cm)
2 in.[a] (10.2 kg-cm)
(475)

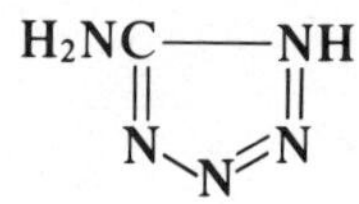

(XXI)

80 cm^b (160 kg-cm)
(57)

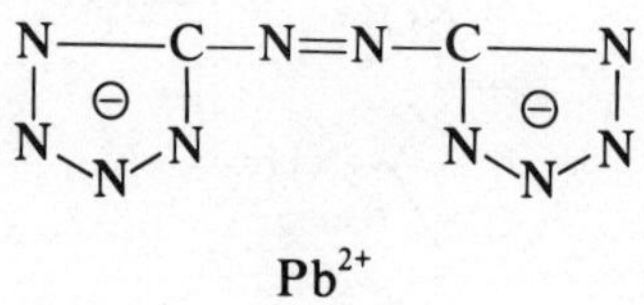

Pb^{2+}

(XXII)

12 cm^b (24 kg-cm)
(350)

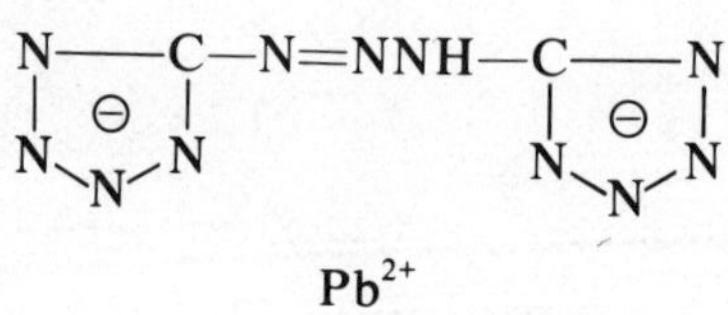

Pb^{2+}

(XXIII)

28 cm^b (56 kg-cm)
(350)

[a] Picatinny Arsenal Apparatus, 2 kg wt.
[b] Bureau of Mines Apparatus, 2 kg wt.

many *v*-triazines, but the triazinobenzotriazine shown **(XVIII)** explodes on being powdered in a glass mortar.

Copper chlorotetrazole **(XIX)** detonates from impact more easily than mercury fulminate. Absence of a report of impact sensitivity of 5-chlorotetrazole (451a) suggests that salt formation with a heavy metal increases sensitivity to mechanical stimulation. Although 5-aminotetrazole **(XXI)** is relatively insensitive, that it explodes at all under impact indicates a contribution of the tetrazole ring to sensitivity. Impact sensitivity is enhanced by introduction of an azo **(XXII)**, triazene **(XXIII)**, or tetrazene **(XX)** moiety as well as salt formation with lead. Qualitative observations indicate that silver tetrazole, 5-azidotetrazole,

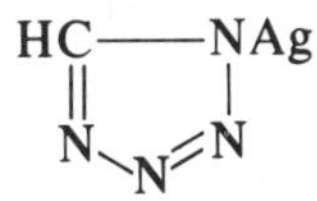

(296)

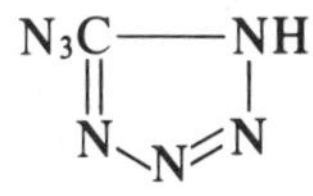

(465)

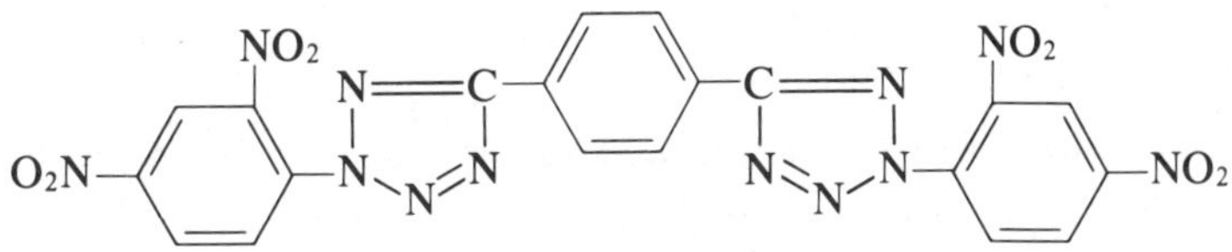

(275)

and bis[2(2,4-dinitrophenyl)-tetrazolyl-5]-1,4-benzene all explode under impact. The potassium (309) and sodium (330) salts of 5-azidotetrazole are even more sensitive to friction than 5-azidotetrazole. Tetranitro-1,3a,4,6a-dibenzotetraazapentalene (Z-TACOT) **(XXIV)** is slightly more sensitive to impact than cyclonite. 2,5-Diphenyltetraazaborole **(XXV)** is insensitive to mechanical shock.

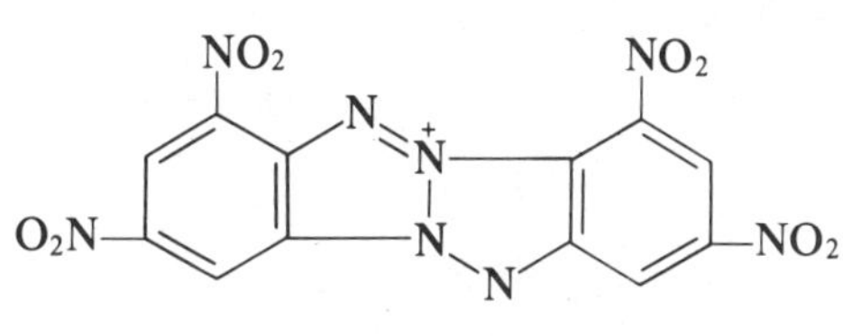

(XXIV)

12 in. (61 kg-cm) (30)
Cyclonite 14.75 in. (75 kg-cm)
Picatinny Arsenal Apparatus,
2 kg wt, 50% fire

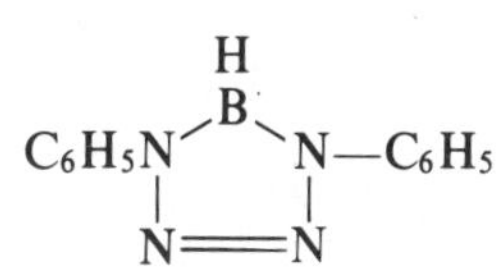

(XXV)

(196)

A large number of compounds with five contiguous nitrogen atoms are sensitive to impact or friction. Thus 1-nitroso-1,4,4-trimethyltetrazene **(XXVI)** is as sensitive as cyanuric azide, several pentazdienes **(XXVII)** explode on striking or rubbing, the nitroimides of benzotriazole **(XXVIII)** are very sensitive, and phenylpentazole **(XXIX)** explodes above −10° at the lightest touch.

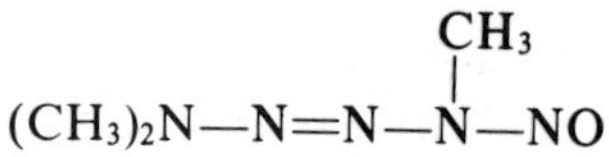

(XXVI)

expl. at 7 cm (1 kg wt)
(7 kg-cm) (466)

$p\text{-}CH_3C_6H_4N{=}NNHN{=}NC_6H_4CH_3\text{-}p$

(XXVII)

(384)

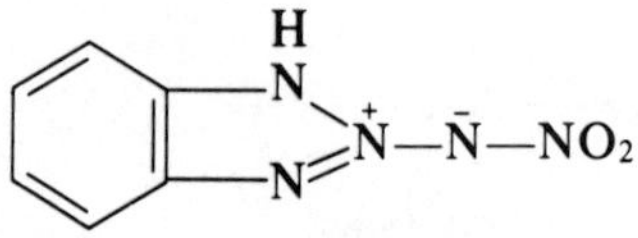

(XXVIII)

(258)

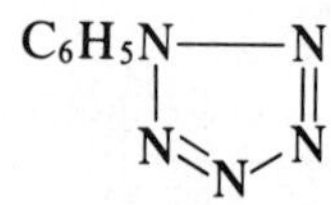

(XXIX)

(473)

Most compounds with six, seven, eight, or ten contiguous nitrogen atoms are highly sensitive to impact or friction.

(233)

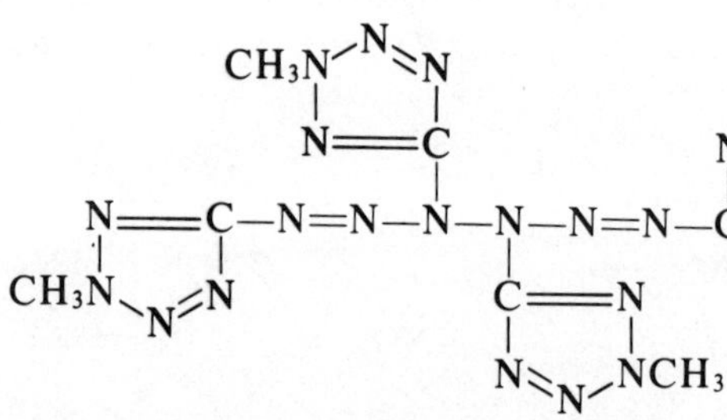

(91)

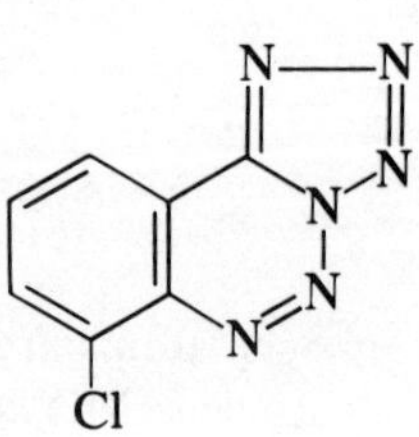

(80)

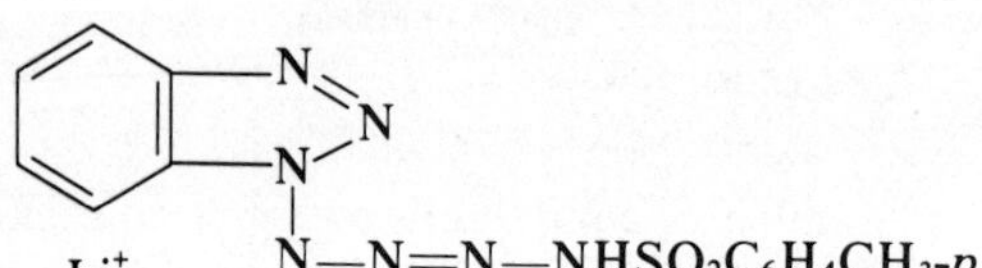

(262)

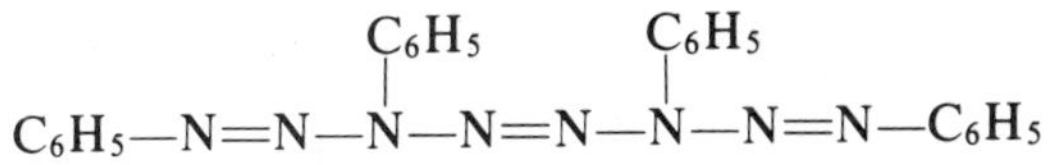

(505)

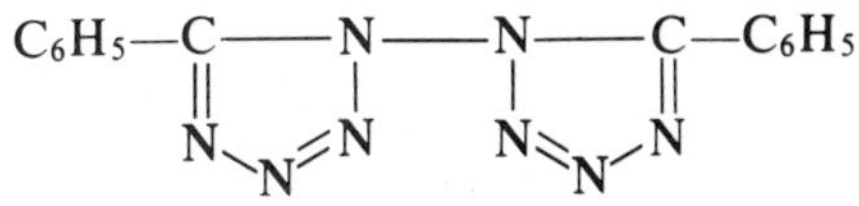

(55)

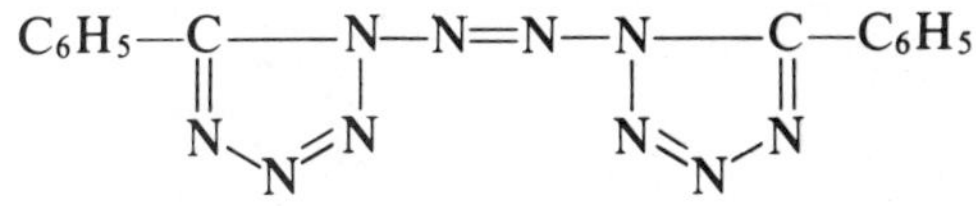

(452)

Many high nitrogen compounds are susceptible to explosion from both mechanical and thermal stimulation. It must be remembered, however, that other members of a class with one or more thermally or impact sensitive compounds are not necessarily dangerous. Nevertheless, caution is urged in handling high nitrogen compounds structurally related to chemicals that explode on being heated or struck.

Initiation of explosion either mechanically or thermally follows a similar course. Impact on an explosive creates hot spots; if their temperature is sufficiently high, decomposition reaches a rate such that the reaction system is adiabatic, followed by acceleration until explosion. Similarly, under thermal stimulation at a temperature sufficient for decomposition to begin, hot spots also are formed, balance between generation of heat and its dissipation is attained and then exceeded, and decomposition of the sample accelerates to explosion (74). In Chapter 2, temperatures at which many high nitrogen compounds explode are reported. These explosion temperatures, usually not determined under standard test conditions, are summarized for major classes, together with examples (Table 3.30).

The three-nitrogen compounds with explosion temperatures up to about 70° are the monoaryltriazenes, triazenes with an aziridine as part of the nitrogen chain, acyl azides such as benzene sulfinyl azide or acetyl azide, and 5-alkoxythiatriazoles. A sample of *t*-butyl azidoformate has exploded in a freezer at −20°, although by-product carbonyldiazide may have been responsible (506), and there is a record of violent explosion of a cooled methanol solution of about 19% methyl azide, evidently resulting from contact with hot glass (198). Explosion temperatures of the heavy metal azides (Table 3.31) are higher than those of organic azides (317). 5-Tetrazolyldiazonium chloride, solutions of which explode at 0° or lower, is the most thermally sensitive high nitrogen compound (Section 2.6.2). Potassium tetrazolyl azide, but not tetrazolyl azide, explodes when placed on a melting point bar at 60° or above (309). The five-nitrogen benzotriazole nitroimides explode in the 74–81° range (258). Not all

Table 3.30. Summary of Thermal Initiation of Explosion of High Nitrogen Classes

High Nitrogen Class	Explosion Temp. Range (°C)	Examples		
Triazenes				
Monosubstituted triazenes	Room temp.—68	*m*-$BrC_6H_4N{=}NNH_2$ expl. ~ room temp. (130)	*p*-$C_2H_5OOCC_6H_4N{=}NNH_2$ expl. 68°, mp (130)	
Di and trisubstituted triazenes	70–233	$C_6H_5N{=}NNHCN$ mp 72° expl. (504)	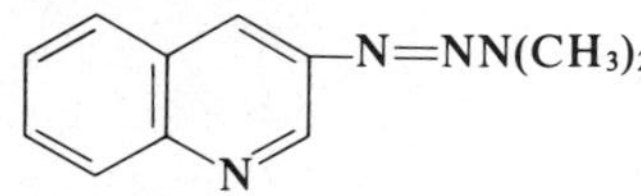 mp 131.5° expl. (412)	
		$C_6H_5N{=}NN(C_6H_5)_2$ expl. 175–180° (478)		
Azides				
Monoazides	11–200	$C_6H_5SON_3$ expl. 11° (327)	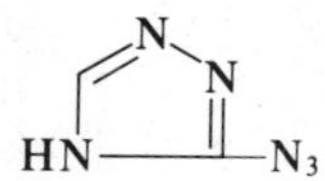 expl. > 121–122°, mp (323)	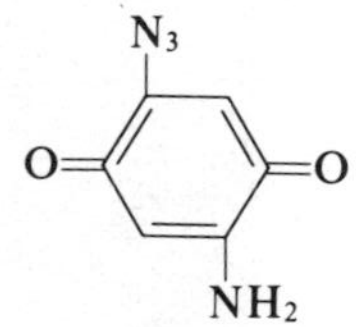 expl. 200° (348)
Diazides	83–205	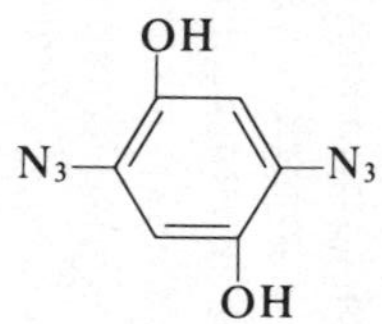 expl. 100° (348)	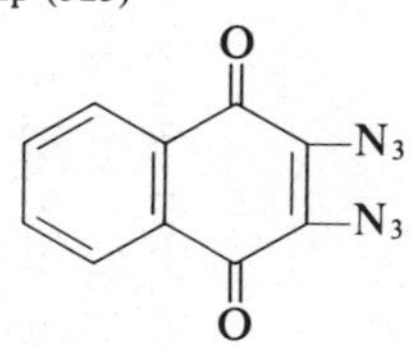 expl. > 136°, mp (172)	 expl. > 190° (65)

Acyclic-cyclic Three-nitrogen chains	Room temp.—172	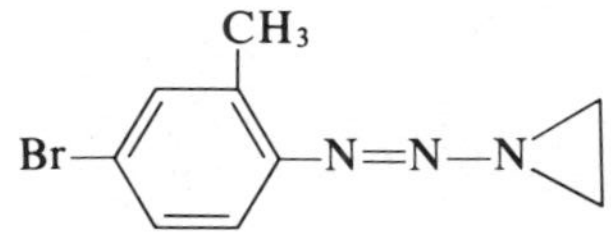 expl. room temp. in 20–30 min (223)	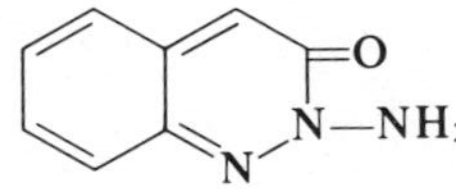 expl. 135° (402)	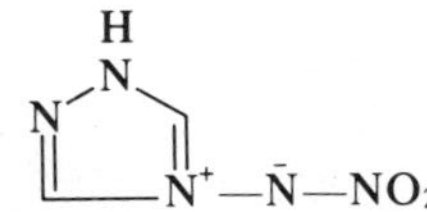 expl. 172° (258)
v-Triazoles Monocyclic	163–233	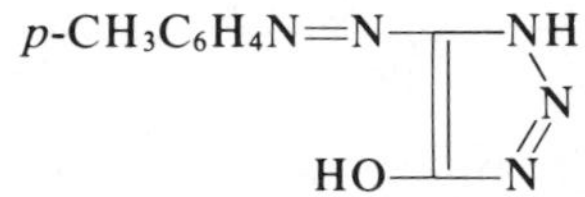 expl. 163° (116)	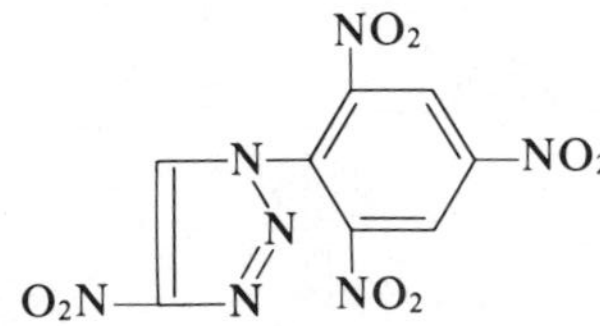 expl. 233° (364)	
Fused ring	144–430	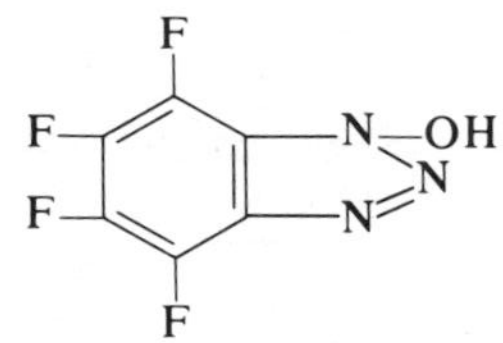 expl. 144° (62)	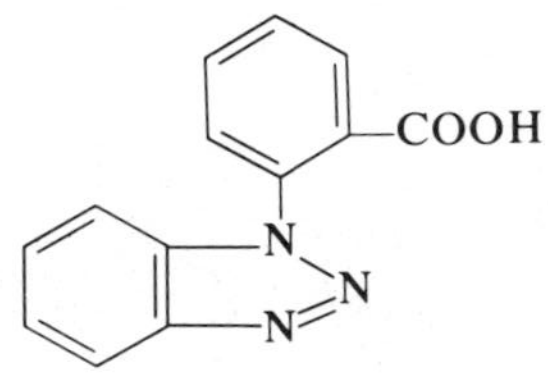 expl. 217° (108)	
		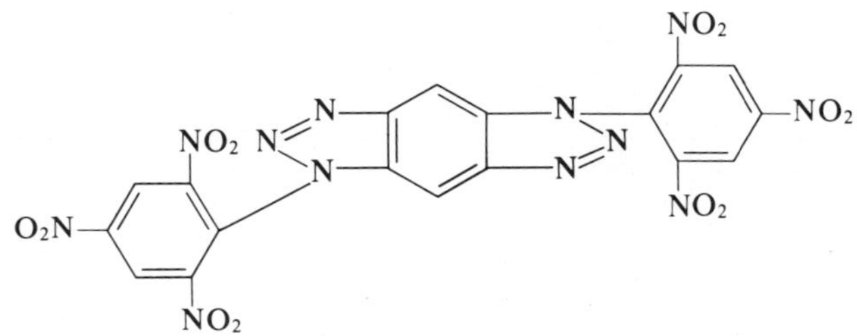 expl. 328° (58)		

Table 3.30. *(Continued)*

High Nitrogen Class	Explosion Temp. Range (°C)	Examples
v-Triazines	199–320	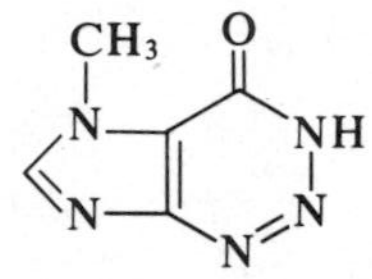 expl. 199–200° (380) 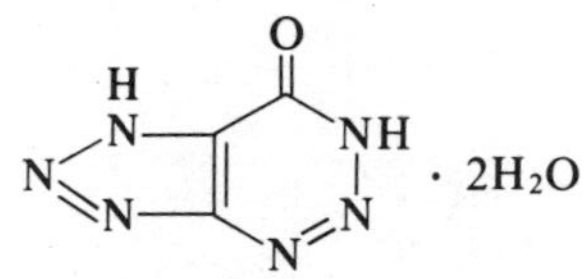 expl. 270° (435)
Thiatriazoles	Room temp.—130	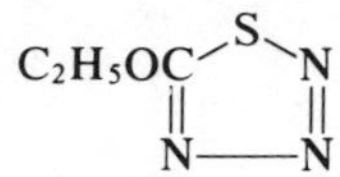 Deflagrates 44–45° (246) 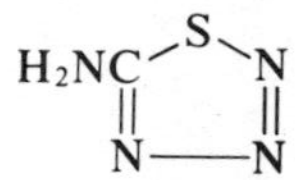 expl. 130° (285)
Tetrazenes	94–170	$p\text{-}O_2NC_6H_4N{=}N(CH_2C_6H_5)N{-}NH_2$ expl. > 94–96°, mp (505) $(CH_3)_2NN{=}NN(CH_3)_2$ expl. > 130°, bp atm press. (406) 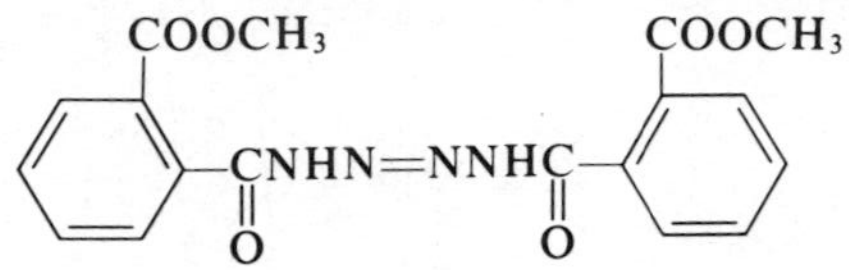 expl. 165° (229)

Tetrazoles	60–309	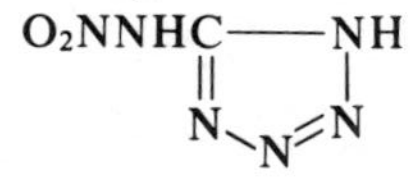 expl. 140° (305)  expl. 239° (350) 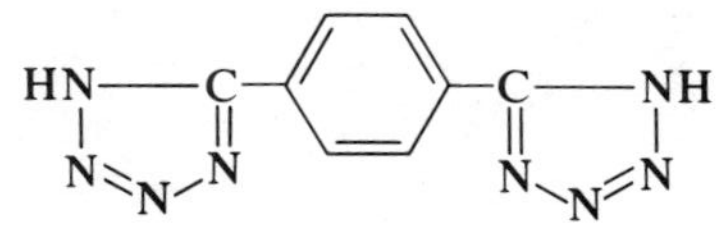 expl. 309°, mp (381)

Table 3.31. Explosion Temperatures of Heavy Metal Azides

Azide	Explosion Temp. (°C)
AgN_3	250
$Hg(N_3)_2$	270
$Pb(N_3)_2$	350

longer-chain nitrogen compounds, however, are as susceptible to thermally induced explosion (see Section 2.7.1 *et seq.*). Those pentazdienes and hexazdienes which explode on heating do so, respectively, from 80 to 166° and from 90 to 131°. Octaztrienes explode from 51 to 64° and the 10-nitrogen azo-1,1′-bistetrazoles from 122 to 141°.

Most compounds containing embedded three-, four-, and five-nitrogen chains are not sensitive to thermal initiation of explosion or, as with Z-TACOT **(XXIV)**, require a high temperature. Tetrazolo(4,5-*c*)-1,2,3-benzotriazine **(XXX)** with one nitrogen embedded in a six-nitrogen chain, however, explodes on heating.

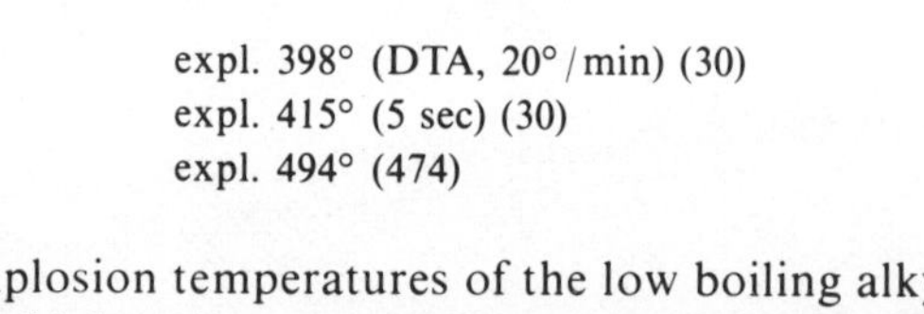

(XXIV)

expl. 398° (DTA, 20°/min) (30)
expl. 415° (5 sec) (30)
expl. 494° (474)

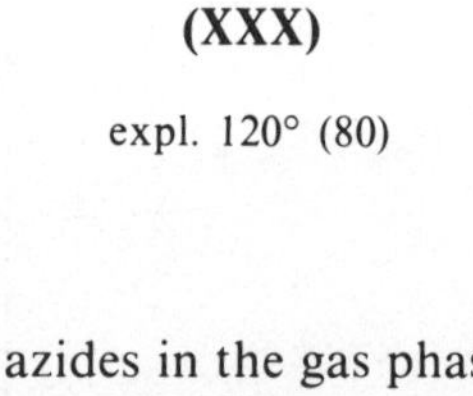

(XXX)

expl. 120° (80)

Explosion temperatures of the low boiling alkyl azides in the gas phase are much higher than those of liquid and solid azides. Methyl azide explodes at over 500° (131) and trifluoromethyl azide at 330° (322a), and the explosion limits of ethyl azide range from 258.5° and 105.2 mm to 320.2° and 3.1 mm (95).

Dust explosibility, expressed as an explosibility index, is measured by the product of the factors electric spark ignition sensitivity and explosion severity, and is compared to that of Pittsburgh coal with an index of 1. Explosibility indices of diazoaminobenzene and of benzotriazole are both over 10, each with severe ignition sensitivity and explosion severity. The index of 4,4′-diphenyldi-(sulfonyl azide) is 3.8 with strong ignition sensitivity and severe explosion severity (135).

Table 3.32. Brisance of High Nitrogen and Reference Explosives

Explosive	Sand Test (g Crushed)[a]	Sensitivity to Initiation (g)	
		$Hg(ONC)_2$	PbN_6
Cyanuric azide	32.2		0.20
Guanylazide nitrate (57)	55.2	0.23	
Copper chlorotetrazole	27.4		0.30
5-Aminotetrazole (57)	4.7	0.25 + 0.25 g tetryl	
Tetracene	28.0	0.40	
Lead azide	19.0		
Mercury fulminate	23.4		
TNT	48.0	0.24	0.27
Cyclonite	60.2	0.19	0.05

[a] Quantity of 200 g of sand, initially retained on a 30 mesh sieve, crushed by explosion of 0.4 g explosive to pass a 30 mesh sieve. Data are from Ref. 475 except as noted.

Measurements of brisance (shattering power) of high nitrogen compounds, and those of some reference explosives are listed in Table 3.32, together with the quantities of initiating compound required for detonation. Cyanuric azide, copper chlorotetrazole, and tetracene are slightly more brisant than the primary explosives lead azide and mercury fulminate, but do not approach the values for the high explosives TNT and cyclonite. Release of nitrogen from the azide, tetrazole, and tetrazene groups in these compounds and their endothermic

Table 3.33. Detonation Rates of High Nitrogen and Reference Explosives

Explosive	Density (g/ml)	Detonation Rate[a] (km/sec)
Mercury fulminate	2.0	3.50
	4.0	5.00
Lead azide	2.0	4.07
	4.0	5.18
Cyanuric azide	1.15	5.55–5.60
TNT	1.56	6.64
	1.654	6.98 (416)
Z-TACOT	1.85	7.25 (416)
Cyclonite	1.65	8.18

[a] Detonation rates from reference 475 except as noted.

energy content results in their pulverizing effect. In comparison with the brisance of copper chlorotetrazole, the low value of 5-aminotetrazole indicates a small effect for the tetrazole ring, although even initiation with mercury fulminate and tetryl may have been inadequate. The large brisance of guanyl azide nitrate comes largely from the combination of nitrate oxygen with hydrogen and carbon and only in small part from the azide group.

Detonation rates (Table 3.33) of lead azide and cyanuric azide are similar to those of other initiators such as mercury fulminate. Z-TACOT propagates its explosion at a velocity characteristic of high explosives, between the rates of TNT and cyclonite.

REFERENCES

1. P. J. Abbott et al., *J. Chem. Soc., Perkin I,* **1972,** 2182–2190.
2. R. A. Abramovitch and E. P. Kyba, *J. Am. Chem. Soc.,* **1974,** *96,* 480–488.
3. R. A. Abramovitch and T. Takaya, *J. Org. Chem.,* **1972,** *37,* 2022–2029.
4. R. Abu-Eittah and A. El-Shahawy, *Appl. Spectrosc.,* **1971,** *25,* 90–93.
5. B. M. Adger et al., *J. Chem. Soc., Perkin I,* **1975,** 31–40.
6. N. D. Agibalova et al., *Zh. Org. Khim., Engl. Ed.,* **1972,** *8,* 2460–2463.
7. *Ibid.,* **1973,** *9,* 1605–1609.
8. T. P. Ahern, H. Fong, and K. Vaughan, *Can. J. Chem.,* **1977,** *55,* 1701–1709.
9. R. Ahmed and J.-P. Anselme, *Can. J. Chem.,* **1972,** *50,* 1778–1780.
10. A. Albert, *J. Chem. Soc. (B),* **1966,** 427–433.
11. *Ibid., (C),* **1968,** 2076–2083.
12. A. Albert, *Phys. Methods Heterocycl. Chem.,* **1963,** *1,* 1–108.
13. A. Albert and K. Tratt, *J. Chem. Soc. (C),* **1968,** 344–347.
14. E. Alcalde et al., *J. Heterocyclic Chem.,* **1978,** *15,* 395–399.
15. A. Alemagna, *Compt. Rend.,* **1963,** *256,* 1258–1260.
16. R. Allmann, *Acta Crystallogr.,* **1967,** *22,* 246–251.
17. A. Almenningen et al., *Acta. Chem. Scand.,* **1973,** *27,* 1531–1540.
18. F. P. Altman, *Prog. Histochem. Cytochem.,* **1976,** *9,* 1–56.
19. E. Amble and B. P. Dailey, *J. Chem. Phys.,* **1950,** *18,* 1422.
20. D. J. Anderson, T. L. Gilchrist, and C. W. Rees, *Chem. Commun.,* **1971,** 800.
21. A. Antonowa et al., *J. Prakt. Chem.,* **1973,** *315,* 313–322.
22. A. Antonowa, R. Herzschuh, and S. Hauptmann, *Z. Chem.,* **1977,** *17,* 65–67.
23. G. Arnold and C. Schiele, *Spectrochim. Acta,* **1969,** *25A,* 671–673.
24. J. Arriau et al., *J. Chem. Soc., Perkin II,* **1976,** 1245–1248.
25. J. Arriau et al., *Tetrahedron Lett.,* **1974,** 3865–3868.
26. W. C. Aten and K. H. Büchel, *Z. Naturforsch,* **1970,** *25b,* 961–965.
27. J. L. Aubagnac, P. Campion, and P. Guenot, *Org. Mass Spectrom.,* **1978,** *13,* 571–574.
28. L. F. Audrieth and P. G. Gordon, *J. Org. Chem.,* **1955,** *20,* 244–249.
29. L. F. Audrieth and B. A. Ogg, *The Chemistry of Hydrazines,* Wiley, New York, **1951,** pp. 167–180.
30. L. Avrami, H. J. Jackson, and M. S. Kirshenbaum, *U.S. Natl. Tech. Inf. Serv.,* AD Rep. 1973, No. 775371/8GA.

31. T. Axenrod, P. Mangiaracina, and P. S. Pregosin, *Helv. Chim. Acta*, **1976,** *59,* 1655–1660.
32. N. C. Baenziger et al., *J. Am. Chem. Soc.*, **1967,** *89,* 6463–6465.
33. B. D. Baigrie, J. I. G. Cadogan, and J. T. Sharp, *J. Chem. Soc., Perkin I,* **1975,** 1065–1068.
34. K. M. Baines et al., *J. Org. Chem.*, **1981,** *46,* 856–859.
35. B. Bak, L. Hansen-Nygaard, and J. Rastrup-Andersen, *J. Mol. Spectrosc.*, **1958,** *2,* 361–368.
36. T. A. Bak and E. L. Praestgaard, *Acta Chem. Scand.*, **1957,** *11,* 901–902.
37. E. B. Baker and A. I. Popov, *J. Phys. Chem.*, **1972,** *76,* 2403–2404.
38. R. A. Baldwin, *J. Org. Chem.*, **1965,** *30,* 3866–3871.
39. E. Balicka and Z. Eckstein, *Rocz. Chem.*, **1975,** 1927–1933.
40. H. Balli, *Angew. Chem., Int. Ed. Engl.*, **1966,** *5,* 132.
41. R. E. Banks and M. J. McGlinchey, *J. Chem. Soc.* (*C*), **1971,** 3971.
42. R. E. Banks and G. R. Sparkes, *J. Chem. Soc., Perkin I,* **1972,** 2964–2970.
43. M. G. Barlow et al., *J. Chem. Soc., Perkin I,* **1973,** 1798–1802.
44. E. E. Baroody and G. A. Carpenter, *J. Chem. Eng. Data,* **1979,** *24,* 3–6.
45. J. J. Barr and R. C. Storr, *J. Chem. Soc., Perkin I,* **1979,** 192–197.
46. J. R. Bartels-Keith, M. T. Burgess, and J. M. Stevenson, *J. Org. Chem.*, **1977,** *42,* 3725–3731.
47. H. Bauer et al., *J. Chem. Soc., Perkin II,* **1972,** 662–667.
48. W. H. Beamer, *J. Am. Chem. Soc.*, **1948,** *70,* 2979–2982.
49. W. Beck et al., *J. Chem. Soc., Dalton* (*2*), **1972,** 245–247.
50. W. Beckh and J. Tafel, *Ber.*, **1894,** *27,* 2315–2316.
51. S. G. Beech, J. H. Turnbull, and W. Wilson, *J. Chem. Soc.*, **1952,** 1686–1690.
52. M. Begtrup, *J. Chem. Soc. Chem. Commun.*, **1974,** 702–703.
53. M. Begtrup, *J. Chem. Soc., Perkin II,* **1976,** 736–741.
54. B. Behera and P. S. Zacharias, *Spectrochem. Acta,* **1971,** 27A, 2273–2275.
55. H. Behringer and H. J. Fischer, *Chem. Ber.*, **1962,** *95,* 2546–2556.
56. L. J. Bellamy, *The Infra-red Spectra of Complex Molecules,* 3rd ed., Chapman and Hall, London, **1975,** p. 295.
57. F. R. Benson, *Picatinny Arsenal Tech. Rep.*, PATR 1409, *Compounds of High Nitrogen Content in Propellent Powders,* U.S. Natl. Tech. Inf. Serv. No. AD-493 263.
58. J. K. Berlin and M. D. Coburn, *J. Heterocycl. Chem.*, **1975,** *12,* 235–237.
59. J. Berthou, J. Elguero, and C. Rérat, *Acta Crystallogr., Sect. B.*, **1970,** *26,* 1880.
60. F. Bertinotti, G. Giacomello, and A. M. Liquori, *Acta Cryst.*, **1956,** *9,* 510–514.
61. G. Bianchi et al., *J. Chem. Soc., B.*, **1971,** 2355–2358.
62. J. M. Birchall, R. N. Haszeldine, and J. E. G. Kemp, *J. Chem. Soc.* (*C*), **1970,** 1519–1523.
63. L. Birkofer, A. Ritter, and H. Uhlenbrauck, *Chem. Ber.*, **1963,** *96,* 3280–3288.
64. G. L. Blackman et al., *J. Molec. Spectrosc.*, **1977,** *65,* 313–318.
65. A. H. Blatt, "Compilation of Data on Organic Explosives," National Defense Research Committee, Office of Scientific Research and Development, OSRD 2014, 1944, P.B. 130, 159.
66. M. Bleha and D. Lim, *Collect. Czech. Chem. Commun.*, **1973,** *38,* 46–49.
67. H. Bock and K.-L. Kompa, *Angew. Chem. Int. Ed. Engl.*, **1962,** *1,* 264.
68. H. Bock and K.-L. Kompa, *Z. Anorg. Allgem. Chem.*, **1964,** *332,* 238–246.
69. H. Boersch, *Sitzungsber. Akad. Wiss. Wien,* **1935,** *144,* 1–27.
70. K. Bolton, R. D. Brown, and F. R. Burden, *Chem. Phys. Lett.*, **1972,** *15,* 79–80.
71. E. Borello and A. Zecchina, *Spectrochim. Acta,* **1963,** *19,* 1703–1715.
72. E. Borello, A. Zecchina, and E. Guglielminotti, *J. Chem. Soc.* (*B*), **1966,** 1243–1245.

73. *Ibid.*, **1969,** 307–311.
74. F. P. Bowden and A. D. Yoffe, *Endeavour,* **1962,** *21,* 125–136.
75. J. D. Bower, *J. Chem. Soc.,* **1957,** 4510–4513.
76. G. V. Boyd et al., *J. Chem. Soc., Perkin I,* **1977,** 1612–1615.
77. J. H. Boyer and L. T. Wolford, *J. Am. Chem. Soc.,* **1958,** *80,* 2741–2743.
78. F. T. Boyle and R. A. Y. Jones, *J. Chem. Soc., Perkin II,* **1973,** 160–164.
79. O. Bremer, *Ann.,* **1935,** *518,* 274–289.
80. T. C. Britton and E. R. Wagner (Dow), U.S. Pat. 4,286,090, Aug. 25, 1981.
81. K. Britts and I. L. Karle, *Acta Cryst.,* **1967,** *22,* 308–312.
82. H. C. Brown, D. H. McDaniel, and O. Häfliger, "Dissociation Constants," in E. A. Braude and F. C. Nached, Eds. *Determination of Organic Structures by Physical Methods,* Academic Press, New York, 1955, pp. 567–662.
83. A. W. Browne et al., *J. Am. Chem. Soc.,* **1923,** *45,* 2541–2550.
84. T. H. Brownlee (American Cyanamid), U.S. Pat. 3,405,114, Oct. 8, 1968.
85. J. H. Bryden, *Acta Cryst.,* **1955,** *8,* 211–216.
86. *Ibid.,* **1956,** *9,* 874–878.
87. *Ibid.,* **1958,** *11,* 31–37.
88. D. Buckley and M. S. Gibson, *J. Chem. Soc.,* **1956,** 3242–3243.
89. W. E. Bull, J. A. Seaton, and L. F. Audrieth, *J. Am. Chem. Soc.,* **1958,** *80,* 2516–2518.
90. E. A. Burns and F. D. Chang, *J. Phys. Chem.,* **1959,** *63,* 1314–1317.
91. R. N. Butler and V. C. Garvin, *J. Chem. Res., Synop.,* **1982,** 183.
92. N. T. Buu and J. T. Edward, *Can. J. Chem.,* **1972,** *50,* 3719–3729.
93. A. F. Cameron, N. J. Hair, and D. G. Morris, *J. Chem. Soc., Perkin II,* **1972,** 1071–1076.
94. C. D. Campbell and C. W. Rees, *J. Chem. Soc. (C),* **1969,** 742–744.
95. H. C. Campbell and O. K. Rice, *J. Am. Chem. Soc.,* **1935,** *57,* 1044–1053.
96. R. A. Carboni et al., *J. Am. Chem. Soc.,* **1967,** *89,* 2618–2625.
97. *Ibid.,* 2626–2633.
98. W. R. Carpenter, *J. Org. Chem.,* **1962,** *27,* 2085–2087.
99. J. A. Caruso, P. G. Sears, and A. I. Popov, *J. Phys. Chem.,* **1967,** *71,* 1756–1760.
100. E. A. Chandross and G. Smolinsky, *Tetrahedron Lett.,* **1960,** 19–22.
101. C. H. Chang, R. F. Porter, and S. H. Bauer, *Inorg. Chem.,* **1969,** *8,* 1677–1683.
102. F. D. Chattaway, F. L. Garton, and G. D. Parkes, *J. Chem. Soc.,* **1924,** 1980–1991.
103. Y. T. Chia and H. E. Simmons, *J. Am. Chem. Soc.,* **1967,** *89,* 2638–2643.
104. A. Chinone and M. Ohta, *Chem. Lett.,* **1972,** 969–970.
105. C. Christophersen and A. Holm, *Acta Chem. Scand.,* **1971,** *25,* 2015–2022.
106. C. Christophersen and S. Treppendahl, *Acta. Chem. Scand.,* **1971,** *25,* 625–630.
107. *Ibid.,* **1972,** *26,* 858–860.
108. G. R. Clemo, W. H. Perkin, Jr., and R. Robinson, *J. Chem. Soc.,* **1924,** 1751–1804.
109. G. L. Closs and A. M. Harrison, *J. Org. Chem.,* **1972,** *37,* 1051–1052.
110. W. D. Closson and H. B. Gray, *J. Am. Chem. Soc.,* **1963,** *85,* 290–294.
111. G. E. Coates and L. E. Sutton, *J. Chem. Soc.,* **1948,** 1187–1196.
112. M. D. Coburn, *J. Heterocycl. Chem.,* **1973,** *10,* 743–746.
113. P. Cohen–Fernandes and C. L. Habraken, *J. Org. Chem.,* **1971,** *36,* 3084–3086.
114. T. Curtius and A. Bertho, *Ber.,* **1926,** *59,* 565–589.
115. T. Curtius and K. Heidenreich, *Ber.,* **1894,** *27,* 2684–2685.

116. T. Curtius and J. Thompson, *Ber.*, **1906,** *39,* 4140–4144.
117. D. Dal Monte Casoni et al., *Gazz. Chim. Ital.*, **1958,** *88,* 977–1034; *Chem. Abstr.*, **1959,** *53,* 18945.
118. T. A. Daniels, S. Sidi, and K. Vaughan, *Can. J. Chem.*, **1977,** *55,* 3751–3754.
119. S. N. Danilov et al., *J. Gen. Chem. USSR, Engl. Ed.*, **1977,** *47,* 190–193.
120. S. N. Danilov, L. N. Yastrebov, and L. N. Burova, *Zh. Obschch. Khim., Engl. Ed.*, **1970,** *40,* 2235–2236.
121. J. P. Daris and J. M. Muchowski, *J. Heterocycl. Chem.*, **1975,** *12,* 761–762.
122. H. R. Davidson, *J. Chem. Educ.*, **1950,** *27,* 598–603.
123. B. F. Day, T. W. Campbell, and G. M. Coppinger, *J. Am. Chem. Soc.*, **1951,** *73,* 4687–4688.
124. J. DeMendoza et al., *J. Chem. Res.* (*M*), **1980,** 0514–0529.
125. G. C. Denault, P. C. Marx, and H. H. Takimoto, *J. Chem. Eng. Data,* **1968,** *13,* 514–516.
126. G. Denecker, G. Smets, and G. L'Abbé, *Tetrahedron,* **1975,** *31,* 765–775.
127. G. H. Denny et al., *J. Org. Chem.*, **1980,** *45,* 1662–1665.
128. M. J. S. Dewar, *Pure Appl. Chem.*, **1975,** *44,* 767–782.
129. O. Dimroth, *Ber.*, **1907,** *40,* 2376–2389.
130. O. Dimroth and K. Pfister, *Ber.*, **1910,** *43,* 2757–2767.
131. O. Dimroth and W. Wislicenus, *Ber.*, **1905,** *38,* 1573–1576.
132. R. C. Dobbie and H. J. Emeléus, *J. Chem. Soc.* (*A*), **1966,** 933–935.
133. P. Domiano and A. Musatti, *Cryst. Struct. Comm.*, **1974,** *3,* 713–715.
134. E. S. Domalski, *J. Phys. Chem. Ref. Data,* **1972,** *1,* 221–277.
135. H. G. Dorsett and J. Nagy, *Dust Explosibility of Chemicals, Drugs, Dyes and Pesticides,* Bureau of Mines Report of Investigation, 7132, U.S. Dept. of the Interior, 1968.
136. A. J. Downs and J. H. Morris, *Spectrochim. Acta,* **1966,** *22,* 957–967.
137. D. A. Dows and G. C. Pimentel, *J. Chem. Phys.*, **1955,** *23,* 1258–1262.
138. M. P. Doyle, J. J. Maciejko, and S. C. Busman, *J. Am. Chem. Soc.*, **1973,** *95,* 952–953.
139. R. S. Drago and F. E. Paulik, *J. Am. Chem. Soc.*, **1960,** *82,* 96–98.
140. B. Duffin, *Acta Cryst.*, **1968,** *B24,* 1256–1261.
141. J. R. C. Duke, *Chem. Commun.*, **1971,** 2–3.
142. J. R. C. Duke, private communication, Aug. 12, 1980.
143. R. L. Dutta and R. Sharma, *J. Indian Chem. Soc.*, **1980,** *57,* 550–552.
144. F. P. Dwyer, *Aust. Chem. Inst., J. Proc.*, **1939,** *6,* 348–361.
145. F. P. Dwyer, *J. Soc. Chem. Ind.*, **1938,** *57,* 351–357.
146. L. K. Dyall, *Aust. J. Chem.*, **1960,** *13,* 230–235.
147. S. L. Edwards et al., *Acta Cryst.*, **1977,** *B33,* 276–278.
148. J. Elguero et al., *The Tautomerism of Heterocycles, Advances in Heterocyclic Chemistry,* Suppl. 1, Academic Press, New York, 1976.
149. J. Elguero, C. Marzin, and J. D. Roberts, *J. Org. Chem.*, **1974,** *39,* 357–363.
150. H. El Khadem and Z. M. El-Shafei, *J. Chem. Soc.*, **1958,** 3117–3119.
151. B. Elpern and F. C. Nachod, *J. Am. Chem. Soc.*, **1950,** *72,* 3379–3382.
152. R. H. Erlich and A. I. Popov, *J. Phys. Chem.*, **1970,** *74,* 338–340.
153. B. L. Evans, A. D. Yoffe, and P. Gray, *Chem. Rev.*, **1959,** *59,* 515–568.
154. N. A. Evans, R. B. Johns, and K. R. Markham, *Aust. J. Chem.*, **1967,** *20,* 713–722.
155. T. F. Fagley, E. Klein, and J. F. Albrecht, *J. Am. Chem. Soc.*, **1953,** *75,* 3104–3106.
156. T. F. Fagley and H. W. Myers, *J. Am. Chem. Soc.*, **1954,** *76,* 6001–6003.

157. F. G. Fallon and R. M. Herbst, *J. Org. Chem.*, **1957,** *22,* 933–936.
158. E. Fanghänel et al., *J. Prakt. Chem.*, **1978,** *320,* 618–624.
159. R. Faure et al., *Can. J. Chem.*, **1977,** *55,* 1728–1735.
160. G. Favini and I. R. Bellobono, *Gazz. Chim. Ital.*, **1962,** *92,* 468–478.
161. J.-P. Fayet et al., *Bull. Soc. Chim. Belg.*, **1978,** *87,* 189–193.
162. J.-P. Fayet et al., *Compt. Rend., Ser. C,* **1976,** 157–158.
163. W. J. Fenlon, *Chem. Eng. News,* May 24, 1976, p. 3.
163a. B. F. Fisher, A. J. Tomson, and J. P. Horwitz, *J. Org. Chem.*, **1959,** *24,* 1650–1654.
164. L. Floch et al., *Collect. Czech. Chem. Commun.*, **1977,** *42,* 2945–2952.
165. S. N. Foner and R. L. Hudson, *J. Chem. Phys.*, **1958,** *28,* 719–720.
166. R. G. Ford, *J. Mol. Spectrosc.*, **1977,** *65,* 273–279.
167. P. S. Forgione, G. S. Sprague, and H. J. Troffkin, *J. Am. Chem. Soc.*, **1966,** *88,* 1079–1080.
168. D. M. Forkey and W. R. Carpenter, *Org. Mass Spectrom.*, **1969,** *2,* 433–445.
169. J. F. Franklin et al., *J. Am. Chem. Soc.*, **1958,** *80,* 298–302.
170. R. R. Fraser and K. E. Haque, *Can. J. Chem.*, **1968,** *46,* 2855–2859.
171. R. T. M. Fraser, N. C. Paul, and M. J. Bagley, *Org. Mass Spectrom.*, **1973,** *7,* 83–88.
172. K. Fries and P. Ochwat, *Ber.*, **1923,** *56,* 1291–1304.
173. K. Fries, R. Walter, and K. Schilling, *Ann.*, **1935,** *516,* 248–285.
174. G. J. Gainsford and A. D. Woolhouse, *Aust. J. Chem.*, **1980,** *33,* 2447–2454.
175. S. F. Gait et al., *J. Chem. Soc., Perkin I,* **1974,** 1248–1260.
176. *Ibid.*, **1975,** 19–25.
177. S. F. Gait, C. W. Rees, and R. C. Storr, *J. Chem. Soc., Chem. Commun.*, **1971,** 1545–1546.
178. L. L. Garber, L. B. Sims, and C. H. Brubaker, Jr., *J. Am. Chem. Soc.*, **1968,** *90,* 2518–2523.
179. W. L. Garbrecht and R. M. Herbst, *J. Org. Chem.*, **1953,** *18,* 1003–1013.
180. *Ibid.*, 1022–1029.
181. *Ibid.*, 1269–1281.
182. J. A. Garrison and R. M. Herbst, *J. Org. Chem.*, **1957,** *22,* 278–283.
183. A. Gieren, K. Burger, and J. Fehn, *Angew. Chem., Int. Ed. Engl.*, **1972,** *11,* 223–224.
184. F. Giordano and A. Zagari, *Acta Cryst.*, **1977,** *B33,* 1288–1290.
185. T. J. Giordano, G. J. Palenik, and H. H. Sisler, *Inorg. Chem.*, **1976,** *4,* 751–754.
186. V. F. Gladkova and Y. D. Kondrashev, *Kristallogr., Engl. Ed.*, **1972,** *17,* 23–28.
187. E. E. Glover, K. T. Rowbottom, and D. C. Bishop, *J. Chem. Soc., Perkin I,* **1973,** 842–845.
188. E. E. Glover and M. Yorke, *J. Chem. Soc. (C)*, **1971,** 3280–3285.
189. J. P. Glusker et al., *Acta Crystallogr., Sect B.*, **1968,** *24,* 359–366.
190. L. Golič et al., *Acta Cryst.*, **1978,** *B34,* 1136–1140.
191. R. Gompper and K. Schönafinger, *Chem. Ber.*, **1979,** *112,* 1529–1534.
192. *Ibid.*, 1535–1544.
193. J. Goubeau, E. Allenstein, and A. Schmidt, *Chem. Ber.*, **1964,** *97,* 884–890.
194. V. A. Grakauskas, doctoral dissertation, Illinois Institute of Technology, 1955.
195. P. Gray and T. C. Waddington, *Proc. Roy. Soc., London,* **1956,** *A235,* 106–119.
196. N. N. Greenwood and J. H. Morris, *J. Chem. Soc.*, **1965,** 6205–6209.
197. P. Grünanger, P. V. Finzi, and E. Fabbri, *Gazz. Chim. Ital.*, **1960,** *90,* 413–425.
198. C. Grundmann and H. Haldenwanger, *Angew. Chem.*, **1950,** *62A,* 410.
199. Z. Grzonka, B. Lieberek, and Z. Palacz, *Zesz. Nauk. Wyds. Mat., Fiz. Chem., Uniw. Gdanski Chem.*, **1971,** *1,* 201–211.

200. *Ibid.*, **1972,** *2,* 91–94.
201. J. H. Hall, F. E. Behr, and R. L. Reed, *J. Am. Chem. Soc.*, **1972,** *94,* 4952–4958.
202. G. C. Hampson and J. M. Robertson, *J. Chem. Soc.*, **1941,** 409–413.
203. R. N. Hanley et al., *J. Chem. Soc., Perkin I,* **1978,** 600–604.
204. *Ibid.*, **1979,** 744–746.
205. R. N. Hanley, W. D. Ollis, and C. A. Ramsden, *J. Chem. Soc., Perkin I,* **1979,** 732–735.
206. *Ibid.*, 736–740.
207. *Ibid.*, 741–743.
208. *Ibid.*, 747–751.
209. H. Hansen, S. Hünig, and K. C. Kishi, *Chem. Ber.*, **1979,** *112,* 445–461.
210. L. D. Hansen et al., *J. Am. Chem. Soc.*, **1968,** *90,* 6588–6592.
211. L. D. Hansen, E. J. Baca, and P. Scheiner, *J. Heterocycl. Chem.*, **1970,** *7,* 991–996.
212. A. Hantzsch, *Ber.*, **1930,** *63,* 1782–1786.
213. R. J. Harder, R. A. Carboni, and J. E. Castle, *J. Am. Chem. Soc.*, **1967,** *89,* 2643–2647.
214. R. L. Hardie and R. H. Thomson, *J. Chem. Soc.*, **1958,** 1286–1290.
215. R. L. Harlow et al., *Acta Cryst.*, **1977,** *B33,* 3423–3428.
216. W. C. Harris et al., *Inorganic Chem.*, **1974,** *13,* 2297–2301.
217. G. S. Hartley and R. J. W. LeFèvre, *J. Chem. Soc.*, **1939,** 531–535.
218. L. W. Hartzel and F. R. Benson, *J. Am. Chem. Soc.*, **1954,** *76,* 667–670.
219. R. N. Haszeldine and B. J. H. Mattinson, *J. Chem. Soc.*, **1955,** 4172–4185.
220. R. N. Haszeldine and A. E. Tipping, *J. Chem. Soc. (C)*, **1966,** 1236–1241.
221. S. Hauptmann, H. Wilde, and K. Moser, *J. Prakt. Chem.*, **1971,** *313,* 882–888.
222. E. Hayon and M. Simic, *J. Am. Chem. Soc.*, **1972,** *94,* 42–47.
223. H. W. Heine and D. A. Tomalia, *J. Am. Chem. Soc.*, **1962,** *84,* 993–995.
224. R. A. Henry, W. G. Finnegan, and E. Lieber, *J. Am. Chem. Soc.*, **1954,** *76,* 2894–2898.
225. R. M. Herbst, "Tetrazoles as Carboxylic Acid Analogs," in S. Graff, Ed., *Essays in Biochemistry,* Wiley, New York, 1955, pp. 141–155.
226. R. M. Herbst and W. L. Garbrecht, *J. Org. Chem.*, **1953,** *18,* 1283–1291.
227. J. Hjortás, *Acta Cryst.*, **1973,** *B29,* 1916–1922.
228. L. Hoesch et al., *Helv. Chim. Acta,* **1977,** *60,* 816–830.
229. L. Hoesch and A. S. Dreiding, *Helv. Chim. Acta,* **1975,** *58,* 980–1001.
230. L. Hoesch and B. Köppel, *Helv. Chim. Acta,* **1981,** *64,* 864–869.
231. S. Hoff and A. P. Blok, *Rev. Trav. Chim. Pays-Bas,* **1974,** *93,* 317–319.
232. R. W. Hoffmann et al., *J. Chem. Soc. (C)*, **1969,** 769–772.
233. K. A. Hofmann and H. Hock, *Ber.*, **1911,** *44,* 2946–2956.
234. D. L. Hooper and K. Vaughan, *J. Chem. Soc., Perkin II,* **1981,** 1161–1165.
235. J. P. Horwitz, B. E. Fisher, and A. J. Tomasewski, *J. Am. Chem. Soc.*, **1959,** *81,* 3076–3079.
236. L. Huff et al., *J. Org. Chem.*, **1970,** *35,* 2074–2076.
237. R. Huisgen and H.-J. Koch, *Ann.*, **1955,** *591,* 200–231.
238. R. Huisgen and I. Ugi, *Ann.*, **1957,** *610,* 57–66.
239. S. Hünig and H. Quast, *Ann.*, **1968,** *711,* 139–156.
240. L. Hunter, *J. Chem. Soc.*, **1937,** 320–324.
240a. T. Ignasiak, J. Suszko, and B. Ignasiak, *J. Chem. Soc., Perkin I,* **1975,** 2122–2129.
241. T. Isida et al., *Bull. Chem. Soc. Jap.*, **1973,** *46,* 1250–1253.
242. J. W. A. M. Janssen and C. L. Habraken, *J. Org. Chem.*, **1971,** *36,* 3081–3084.

243. K. A. Jensen, A. Holm, and J. Wolff-Jensen, *Acta. Chem. Scand.*, **1969,** *23,* 1567–1572.

244. K. A. Jensen et al., *Acta Chem. Scand., Ser. B,* **1974,** *28,* 97–101.

245. K. A. Jensen and A. Friediger, *Kgl. Danske Videnskab., Selskab, Mat-fys. Medd.,* **1943,** *20,* 1–54; *Chem. Abstr.,* **1945,** *39,* 2068.

246. K. A. Jensen, A. Holm, and B. Thorkilsen, *Acta. Chem. Scand.,* **1964,** *18,* 825.

247. C. D. Johnson, A. R. Katritzky, and N. Shakir, *J. Chem. Soc. (B),* **1967,** 1235–1237.

248. K. E. Johnson et al., *J. Polym. Sci., Part B,* **1966,** *4,* 997–999.

249. R. A. W. Johnstone et al., *J. Chem. Soc., (C),* **1970,** 1238–1241.

250. K. Kaas, *Acta Cryst.,* **1973,** *B29,* 1458–1463.

251. J. Kaczmarek, H. Smagowski, and Z. Grzonka, *J. Chem. Soc., Perkin II,* **1979,** 1670–1674.

252. H. Kagami and S. Motoki, *Bull. Chem. Soc. Jap.,* **1979,** *52,* 3463–3464.

253. A. Kálmán, *J. Chem. Soc., Perkin II,* **1974,** 1849–1852.

254. A. Kálmán, L. Párkányi, and J. Schawartz, *Acta Cryst.,* **1977,** *B33,* 3097–3102.

255. S. Kamiya, private communication, Sept. 30, 1981.

256. M. Kanaoka, *Chem. Pharm. Bull.,* **1958,** *6,* 382–386.

257. A. R. Katritzky et al., *J. Am. Chem. Soc.,* **1970,** *92,* 6855–6860.

258. A. R. Katritzky and J. W. Mitchell, *J. Chem. Soc., Perkin I,* **1973,** 2624–2626.

259. J. C. Kauer and R. A. Carboni, *J. Am. Chem. Soc.,* **1967,** *89,* 2632–2637.

260. M. H. Kaufman, F. M. Ernsberger, and W. S. McEwen, *J. Am. Chem. Soc.,* **1956,** *78,* 4197–4201.

261. M. H. Kaufman and A. L. Woodman, *J. Phys. Chem.,* **1958,** *62,* 508–509.

262. M. Keating et al., *J. Chem. Soc., Perkin I,* **1972,** 1315–1316.

263. G. W. Kenner and R. J. Stedman, *J. Chem. Soc.,* **1952,** 2089–2094.

264. W. Kesting, *Ber.,* **1924,** *57,* 1321–1324.

265. S. K. Khetan and M. V. George, *Can. J. Chem.,* **1967,** *45,* 1993–1995.

266. T. J. King et al., *J. Chem. Soc., Perkin II,* **1979,** 1751–1757.

267. W. Kirmse, *Chem. Ber.,* **1960,** *93,* 2353–2356.

268. S. Kishimoto, S. Noguchi, and K. Masuda, *Chem. Pharm. Bull.,* **1976,** *24,* 3001–3010.

269. A. M. Kiwan, A. Y. Kassim, and H. M. Maarafi, *CIM Spec. Vol. Ser. 21, Proc. Int. Solvent Extr. Conf. 1977,* **1979,** 765–768.

270. I. E. Knaggs, *Proc. Roy. Soc., Ser. A.,* **1935,** *150,* 576–601.

271. K. H. Koch and E. Fahr, *Angew Chem., Int. Ed. Engl.,* **1970,** *9,* 634–635.

272. G. F. Kolar, *Mass Spectrom. Biochem. Med., Symp.,* **1974,** 267–275; *Chem. Abstr.,* **1975,** *82,* 15796f.

273. Y. D. Kondrashev, *Kristallogr., Engl. Ed.,* **1962,** *6,* 413–418.

274. Y. D. Kondrashev, *Zh. Strukt. Khim., Engl. Ed.,* **1974,** *15,* 441–446; *Chem. Abstr.,* **1974,** *81,* 69695v.

275. A. Könnecke, P. Lepom, and E. Lippmann, *Z. Chem.,* **1978,** *18,* 214–215.

276. A. Könnecke and E. Lippmann, *Org. Mass Spectrom.,* **1976,** *11,* 167–170.

277. A. Könnecke, E. Lippmann, and E. Kleinpeter, *Tetrahedron,* **1977,** *33,* 1399–1403.

278. D. L. Kozlowski, P. Singh, and D. J. Hodgson, *Acta. Crystallogr.,* **1974,** *B30,* 2806–2811.

279. J. Kraft and S. Walker, *Phys. Methods Heterocycl. Chem.,* **1971,** *IV,* 237–264.

280. R. Kreher and H. Wissmann, *Chem. Ber.,* **1973,** *106,* 3097–3099.

281. H. R. Kricheldorf and E. Leppert, *Synthesis,* **1975,** 49–50.

282. J. Kroner et al., *J. Chem. Soc., Faraday II,* **1978,** *74,* 1909–1919.

283. R. Kübler, W. Lüttke, and S. Weckherlin, *Z. Electrochem.,* **1960,** *64,* 650–658.

284. S. A. Kudchadker and C. N. R. Rao, *Proc. Indian Acad. Sci., Sect. A,* **1977,** *86A,* 265–272.

285. M. Kuhn and R. Mecke, *Z. Anal. Chem.,* **1961,** *181,* 487–493.

286. R. Kuhn and H. Kainer, *Angew. Chem.,* **1953,** *65,* 442–446.

287. Y. Kushi and Q. Fernando, *J. Am. Chem. Soc.,* **1970,** *92,* 1965–1968.

288. M. Lacan, I. Tabakovic, and Z. Cekovic, *Tetrahedron,* **1974,** *30,* 2911–2917.

289. M. B. Laing et al., *Acta Cryst.,* **1976,** *B32,* 2518–2520.

290. A. Langer, J. A. Hipple, and D. P. Stevenson, *J. Chem. Phys.,* **1954,** *22,* 1836–1844.

291. G. A. Lanovaya and V. F. Mischenko, *Zh. Org. Khim., Engl. Ed.,* **1979,** *15,* 1996–1999.

292. J. B. Leach and J. H. Morris, *J. Organomet. Chem.,* **1968,** *13,* 313–321.

293. L. A. Lee, R. Evans, and J. W. Wheeler, *J. Org. Chem.,* **1972,** *37,* 343–347.

294. R. J. W. LeFèvre and T. H. Liddicoet, *J. Chem. Soc.,* **1951,** 2743–2748.

295. R. J. W. LeFèvre and H. Vine, *J. Chem. Soc.,* **1937,** 1805–1809.

296. M. J. Leleu, *Cah. Notes Doc.,* **1978,** *92,* 445–449; *Chem. Abstr.,* **1979,** *90,* 11461k.

297. D. M. Lemal, T. W. Rave, and S. D. McGregor, *J. Am. Chem. Soc.,* **1963,** *85,* 1944–1948.

298. C. Leuenberger et al., *Helv. Chim. Acta,* **1977,** *60,* 831–843.

298a. C. Leuenberger, L. Hoesch, and A. S. Dreiding, *Helv. Chim. Acta,* **1981,** *64,* 1219–1233.

299. G. C. Levy and R. L. Lichter, *Nitrogen-15 Nuclear Magnetic Resonance Spectroscopy,* Wiley-Interscience, New York, 1979.

300. R. L. Lichter, *Determination of Organic Structures by Physical Methods,* **1971,** *4,* 195–232.

301. E. Lieber et al., *Anal. Chem.,* **1957,** *29,* 916–918.

302. E. Lieber et al., *Can. J. Chem.,* **1958,** *36,* 801–809.

303. *Ibid.,* 1441–1443.

304. *Ibid.,* **1959,** *37,* 563–574.

305. E. Lieber et al., *J. Am. Chem. Soc.,* **1951,** *73,* 2327–2329.

306. E. Lieber et al., *Spectrochim. Acta,* **1963,** *19,* 1135–1144.

307. E. Lieber, J. S. Curtice, and C. N. R. Rao, *Chem. Ind.,* **1966,** 586–591.

308. E. Lieber and T. Enkoji, *J. Org. Chem.,* **1961,** *26,* 4472–4479.

309. E. Lieber and D. R. Levering, *J. Am. Chem. Soc.,* **1951,** *73,* 1313–1317.

310. E. Lieber, D. R. Levering, and L. J. Patterson, *Anal. Chem.,* **1951,** *23,* 1594–1604.

311. E. Lieber, S. H. Patinkin, and H. H. Tao, *J. Am. Chem. Soc.,* **1951,** *73,* 1792–1795.

312. K. B. Lindberg et al., *Acta Cryst.,* **1978,** *B34,* 3484–3486.

313. K.-H. Linke and H. J. Göhausen, *Chem. Ber.,* **1971,** *104,* 301–306.

314. E. Lippmann and A. Koennecke, *Z. Chem.,* **1975,** *15,* 226–227.

315. R. L. Livingston and C. N. R. Rao, *J. Phys. Chem.,* **1960,** *64,* 756–759.

316. F. J. Llewellyn and F. E. Whitmore, *J. Chem. Soc.,* **1947,** 881–884.

316a. R. Longhi and R. S. Drago, *Inorg. Chem.,* **1963,** *2,* 85–88.

317. J. K. Luchs, *Photogr. Sci. Eng.,* **1966,** *10,* 334–337.

318. V. O. Lukashevich and E. S. Lisitsyna, *Dokl. Akad. Nauk. SSSR, Engl. Ed.,* **1965,** *160,* 31–34.

319. L. Lunazzi et al., *J. Chem. Soc., Perkin II,* **1978,** 686–691.

320. D. MacKay, D. D. McIntyre, and N. J. Taylor, *J. Org. Chem.,* **1982,** *47,* 532–535.

321. S. M. MacKenzie and M. F. G. Stevens, *J. Chem. Soc. (C),* **1970,** 2298–2308.

322. T. P. Maher and G. R. Yohe, *J. Org. Chem.,* **1958,** *23,* 1082–1083.

322a. S. P. Makarov et al., *J. Gen. Chem. USSR, Engl. Ed.,* **1968,** *38,* 685–690.

323. W. Manchot and R. Noll, *Ann.,* **1905,** *343,* 1–27.

324. A. Maquestiau et al., *Org. Mass Spectrom.,* **1973,** *7,* 271–276.

325. *Ibid.*, 1267–1277.

326. *Ibid.*, **1978,** *13,* 518–526.

327. T. J. Maricich and V. L. Hoffmann, *J. Am. Chem. Soc.*, **1974,** *96,* 7770–7781.

328. J. H. Markgraf, W. T. Bachmann, and D. P. Hollis, *J. Org. Chem.*, **1965,** *30,* 3472–3474.

329. G. Märkl and C. Martin, *Tetrahedron Lett.*, **1973,** 4503–4506.

330. F. D. Marsh, *J. Org. Chem.*, **1972,** *37,* 2966–2969.

331. D. Martin and A. Weise, *Chem. Ber.*, **1966,** *99,* 317–327.

332. J. Mason, W. v. Bronswijk, and J. G. Vinter, *J. Chem. Soc., Perkin II,* **1977,** 469–473.

333. J. Mason and J. G. Vinter, *J. Chem. Soc., Dalton Trans.*, **1975,** 2522–2526.

334. S. Mataka and J.-P. Anselme, *Chem. Lett.*, **1973,** 51–54.

335. P. Mauret et al., *J. Chim. Phys.*, **1973,** *70,* 1483–1485.

336. P. Mauret, J.-P. Fayet, and M. Fabre, *Bull. Soc. Chim. Fr.*, **1975,** 1675–1678.

337. W. R. McBride and H. W. Kruse, *J. Am. Chem. Soc.*, **1957,** *79,* 572–576.

338. J. R. McDonald, J. W. Rabalais, and S. P. McGlynn, *J. Chem. Phys.*, **1970,** *52,* 1332–1340.

339. W. S. McEwan and M. W. Rigg, *J. Am. Chem. Soc.*, **1951,** *73,* 4725–4727.

340. A. McKillop and R. J. Kobylecki, *J. Org. Chem.*, **1974,** *39,* 2710–2714.

341. J. M. McManus and R. M. Herbst, *J. Org. Chem.*, **1959,** *24,* 1643–1649.

342. L. Mennier, *Compt. Rend.*, **1900,** *131,* 50–53.

343. J. S. Mihina and R. M. Herbst, *J. Org. Chem.*, **1950,** *15,* 1082–1092.

343a. S. I. Miller, R.-R. Lii, and Y. Tanaka, *J. Chem. Soc., Perkin I,* **1979,** 15–19.

344. T. Mitsuhashi and O. Simamura, *Chem. Ind.*, **1964,** 578–579.

345. H. Moll and R. Vuille, *Chimia,* **1969,** *23,* 511–512.

346. J. A. Montgomery and H. J. Thomas, *J. Med. Chem.*, **1972,** *15,* 182–187.

347. R. Moor et al., *Helv. Chim. Acta,* **1979,** *62,* 1202–1209.

348. H. W. Moore, H. R. Sheldon, and D. F. Shellhamer, *J. Org. Chem.*, **1969,** *34,* 1999–2001.

348a. R. M. Moriarty and J. M. Kliegman, *J. Am. Chem. Soc.*, **1967,** *89,* 5959–5960.

349. J. H. Morris and P. G. Perkins, *J. Chem. Soc.* (*A*), **1966,** 580–582.

350. H. Morrisson, *Util. Elem. Pyrotechniques Explos. Syst. Spatiaux, Collg. Int.*, **1968,** 111–120.

351. A. V. Moskvin et al., *Zh. Org. Chem., Engl. Ed.*, **1978,** *14,* 2248–2251.

352. J. Mueller, *Z. Naturforsch., B. Anorg. Chem., Org. Chem.*, **1979,** *34B,* 437–441.

353. U. Mueller and H. Baernighausen, *Acta Crystallogr., Sect. B.*, **1970,** *26,* 1671–1679.

354. A. Mugnoli, C. Mariani, and M. Simonetta, *Acta Cryst.*, **1965,** *19,* 367–372.

355. E. Müller, R. Beutler, and B. Zeeh, *Ann.*, **1968,** *719,* 72–79.

356. E. Müller and H. Haiss, *Chem. Ber.*, **1962,** *95,* 1255–1263.

357. J. Müller, *J. Organometal. Chem.*, **1973,** *51,* 119–123.

358. D. B. Murphy and J. P. Picard, *J. Org. Chem.*, **1954,** *19,* 1807–1814.

359. K. Nagarajan and P. Rajagopalan, *Tetrahedron Lett.*, **1966,** 5525–5530.

360. S. F. Nelsen and R. Fibiger, *J. Am. Chem. Soc.*, **1972,** *94,* 8497–8501.

361. S. F. Nelsen and R. T. Landis II, *J. Am. Chem. Soc.*, **1973,** *95,* 8707–8713.

362. S. F. Nelsen, R. T. Landis II, and J. C. Calabrese, *J. Org. Chem.*, **1977,** *42,* 4192–4197.

363. P. N. Neuman, *J. Heterocycl. Chem.*, **1970,** *7,* 1159–1160.

364. *Ibid.*, **1971,** *8,* 51–56.

365. F. A. Neugebauer, *Tetrahedron,* **1970,** *26,* 4843–4851.

366. F. A. Neugebauer and H. Fischer, *Chem. Ber.*, **1973,** *106,* 1589–1594.

367. A. N. Nesmeyanov et al., *Tetrahedron,* **1969,** *25,* 4667–4670.

368. K. Nielsen, *Acta Chem. Scand.*, **1975,** *A29,* 647–650.
369. H. Oelschläger and H. Blume, *Arzneim.-Forsch,* **1976,** *26,* 303–311.
370. E. Oeser and L. Schiele, *Chem. Ber.*, **1972,** *105,* 3704–3708.
371. A. Ohsawa et al., *J. Chem. Soc., Chem. Comm.*, **1981,** 1174.
372. H. Okabe and A. Mele, *J. Chem. Phys.*, **1969,** *51,* 2100–2106.
373. V. A. Ostrovskii, *Zh. Org. Khim., Engl. Ed.*, **1977,** *13,* 2251–2254.
374. D. G. O'Sullivan, *J. Chem. Soc.*, **1960,** 3653–3658.
375. T. Ottersen, *Acta Chem. Scand.*, **1975,** *A29,* 799–801.
376. *Ibid.*, **1976,** *A30,* 351–357.
377. G. J. Palenik, *Acta Cryst.*, **1963,** *16,* 596–600.
378. *Ibid.*, **1965,** *19,* 47–56.
379. M. H. Palmer, I. Simpson, and J. R. Wheeler, *Z. Naturforsch,* **1981,** *36A,* 1246–1252.
380. R. P. Panzica and L. B. Townsend, *J. Heterocycl. Chem.*, **1972,** *9,* 623–628.
381. A. J. Papa, *J. Org. Chem.*, **1966,** *31,* 1426–1430.
382. C. Párkányi et al., *Helv. Chim. Acta,* **1981,** *64,* 171–175.
383. L. Pauling and L. O. Brockway, *J. Am. Chem. Soc.*, **1937,** *59,* 13–20.
384. H. v. Pechmann and L. Frobenius, *Ber.*, **1894,** *27,* 898–900.
385. *Ibid.*, **1895,** *28,* 170–176.
386. M. J. Perkins, *J. Chem. Soc.*, **1964,** 3005–3008.
387. M. A. Pervozvanskaya et al., *Zh. Org. Khim., Engl. Ed.*, **1975,** *11,* 1995–1997.
388. J. C. Pezzullo and E. R. Boyko, *J. Org. Chem.*, **1973,** *38,* 168–169.
389. C. Pigenet and H. Lumbroso, *Bull. Soc. Chim. Fr.*, **1972,** 3743–3750.
390. K. Pilgram and F. Görgen, *J. Heterocycl. Chem.*, **1971,** *8,* 899–902.
391. K. Pilgram, F. Görgen, and G. Pollard, *J. Heterocycl. Chem.*, **1971,** *8,* 951–959.
392. V. Pirc et al., *J. Heterocycl. Chem.*, **1970,** *7,* 639–645.
393. W. H. Pirkle and P. L. Gravel, *J. Org. Chem.*, **1978,** *43,* 808–815.
394. A. I. Popov and R. D. Holm, *J. Phys. Chem.*, **1962,** *66,* 158–160.
395. A. I. Popov and J. C. Marshall, *J. Inorg. Nucl. Chem.*, **1961,** *19,* 340–347.
396. K. T. Potts and S. Husain, *J. Org. Chem.*, **1970,** *35,* 3451–3456.
396a. *Ibid.*, **1972,** *37,* 2049–2050.
397. D. N. Purohit and N. L. Sogani, *J. Chem. Soc.*, **1964,** 2820–2822.
398. H. Quast and S. Hünig, *Ann.*, **1968,** *711,* 157–173.
399. P. Ramart-Lucas, J. Hoch, and M. Grumez, *Bull. Chim. Fr.*, **1949,** 447–454.
400. C. N. R. Rao and R. Venkataraghavan, *Can. J. Chem.*, **1964,** *42,* 43–49.
401. U. Rapp, H. A. Staab, and C. Wünsche, *Tetrahedron,* **1971,** *27,* 2679–2689.
402. C. W. Rees and A. A. Sale, *J. Chem. Soc., Perkin I,* **1973,** 545–550.
403. C. W. Rees, R. W. Stephenson, and R. C. Storr, *J. Chem. Soc., Chem. Commun.*, **1974,** 941–942.
404. C. W. Rees and R. C. Storr, *J. Chem. Soc.*, (*C*), **1969,** 756–760.
405. H. Reimlinger, G. S. D. King, and M. A. Peiren, *Chem. Ber.*, **1970,** *103,* 2821–2827.
406. E. Renouf, *Ber.*, **1880,** *13,* 2169–2174.
407. C. W. Roberts, G. F. Fanta, and J. D. Martin, *J. Org. Chem.*, **1959,** *24,* 654–657.
408. N. K. Roberts, *J. Chem. Soc.*, **1963,** 5556–5558.
409. H. deRogier, N. D. An, and A. Martinet, *Compt. Rend., Ser. C,* **1976,** *282,* 967–969.
410. N. W. Rokke, J. J. Worman, and W. S. Wadsworth, *J. Heterocycl. Chem.*, **1975,** *12,* 1031–1033.

411. C. S. Rondestvedt and P. K. Chang, *J. Am. Chem. Soc.*, **1955,** *77,* 6532–6540.

412. C. S. Rondestvedt and S. J. Davis, *J. Org. Chem.*, **1957,** *22,* 200–203.

413. F. D. Rossini et al., *Selected Values of Physical and Thermodynamic Properties of Hydrocarbons and Related Compounds,* American Petroleum Institute Research Project 44, 1953, p. 477.

414. W. A. Roth and F. Müller, *Ber.*, **1929,** *62,* 1188–1194.

415. M. L. Roumestant et al., *Tetrahedron Lett.*, **1969,** 495–498.

416. P. E. Rouse, Jr., *J. Chem. Eng. Data,* **1976,** *21,* 16–20.

417. K. Rüfenacht, *Helv. Chim. Acta,* **1975,** *58,* 1521–1527.

418. M. Sana et al., *Nouv. J. Chim.*, **1979,** *3,* 607–621.

419. K. Sasvari et al., *Acta Cryst.*, **1979,** *B35,* 2145–2149.

420. E. Sawicki, T. R. Hauser, and T. W. Stanley, *Anal. Chem.*, **1959,** *31,* 2063–2065.

421. P. Scheiner, *Tetrahedron,* **1968,** *24,* 2757–2766.

422. P. Scheiner and W. M. Litchman, *J. Chem. Soc., Chem. Commun.*, **1972,** 781–782.

423. C. W. Shellhammer, J. Schroeder, and N. Joop, *Tetrahedron,* **1970,** *26,* 497–510.

424. A. Schmidt, *Chem. Ber.*, **1966,** *99,* 2976–2983.

425. *Ibid.*, **1970,** *103,* 3923–3927.

426. K. Schofield, M. R. Grimmett, and B. R. T. Keene, *The Azoles,* Cambridge University Press, Cambridge, 1976, p. 24.

427. V. Schomaker and K. Spurr, *J. Am. Chem. Soc.*, **1942,** *64,* 1184–1187.

428. H. Schultheiss and E. Fluck, *Z. Naturforsch,* **1977,** *32b,* 257–264.

429. F. L. Scott and J. C. Tobin, *J. Chem. Soc. (C)*, **1971,** 703–706.

430. D. Seebach et al., *Helv. Chim. Acta,* **1978,** *61,* 1622–1647.

431. M. P. Servé et al., *J. Heterocyclic Chem.*, **1976,** *13,* 509–512.

432. V. P. Shchipanov, *Zh. Org. Khim., Engl. Ed.*, **1979,** *15,* 1793–1798.

433. V. P. Shchipanov et al., *J. Org. Chem. USSR, Engl. Ed.*, **1965,** *1,* 2277–2283.

434. V. P. Shchipanov and G. F. Grigor'eva, *Khim. Geterotsikl. Soedin. Engl. Ed.*, **1974,** 235–238.

435. Y. F. Shealy et al., *J. Org. Chem.*, **1961,** *26,* 2396–2401.

436. S. J. Shearer et al., *J. Chem. Phys.*, **1968,** *48,* 1138–1144.

437. Y. N. Sheinker, L. B. Senyavina, and V. N. Zheltova, *Dokl. Akad. Nauk. SSSR, Engl. Ed.*, **1965,** *160,* 215–218.

438. N. V. Sidgwick, *The Chemical Elements and their Compounds,* Vol. I, Clarendon Press, Oxford, 1950, pp. 715–718.

439. D. H. Sieh and C. J. Michejda, *J. Am. Chem. Soc.*, **1981,** *103,* 442–445.

440. P. Singh and D. J. Hodgson, *Acta Crystallogr. Sect. B,* **1979,** *B35,* 973–976.

441. J. Sletten, E. Sletten, and L. H. Jensen, *Acta Cryst.*, **1968,** *B24,* 1692–1698.

442. P. A. S. Smith, *Open Chain Nitrogen Compounds,* Vol. II, W. A. Benjamin, New York, 1966, p. 214.

443. P. A. S. Smith, J. H. Hall, and R. O. Kan, *J. Am. Chem. Soc.*, **1962,** *84,* 485–489.

444. M. M. Sokolova et al., *Zh. Org. Khim.*, **1974,** *10,* 1085–1088.

445. S. Sommer and V. Schubert, *Angew. Chem., Int. Ed. Engl.*, **1979,** *18,* 696–697.

446. P. Spagnolo, A. Tundo, and P. Zanirato, *J. Org. Chem.*, **1978,** *43,* 2508–2513.

446a. H. O. Spauschus and J. M. Scott, *J. Am. Chem. Soc.*, **1951,** *73,* 210–212.

447. S. Stankovsky and S. Kovac, *Chem. Zvesti,* **1974,** *28,* 243–246.

448. L. Stefaniak, *Bull. Acad. Pol. Sci., Ser. Sci. Chem.*, **1978,** *26,* 291–302.

449. M. F. G. Stevens, *Prog. Med. Chem.*, **1976,** *13,* 205–269.

450. O. L. Stiefvater, H. Jones, and J. Sheridan, *Spectrochim. Acta,* **1970,** *26A,* 825–833.

451. E. Stöldt and R. Kreher, *Angew. Chem.,* **1978,** *90,* 215–216.

451a. R. Stollé, *Ber.,* **1929,** *62,* 1118–1126.

452. R. Stollé, *J. Prakt. Chem.,* **1933,** *138,* 1–17.

453. V. N. Strel'tsova et al., *Zh. Org. Khim., Engl. Ed.,* **1974,** *10,* 1093–1096.

454. J. W. Sutherland, *J. Phys. Chem.,* **1979,** *83,* 789–795.

455. L. E. Sutton, *Determination Org. Struct. Phys. Methods,* **1955,** *1,* 373–425.

456. J. Světlik, J. Leško, and A. Martvon, *Monatsh. Chem.,* **1980,** *111,* 635–642.

457. J. Světlik, A. Martvon, and J. Leško, *Chem. Zvesti,* **1979,** *33,* 521–527.

458. J. Tamas, J. Hegedüs-Vajda, and A. Messmer, *Acta Chim. Acad. Sci., Sci. Hung.,* **1979,** *99,* 193–200.

459. Y. Tamura, J. H. Kim, and M. Ikeda, *J. Heterocycl. Chem.,* **1975,** *12,* 107–110.

460. Y. Tanaka, S. R. Velen, and S. I. Miller, *Tetrahedron,* **1973,** *29,* 3271–3283.

461. M. Tanno and S. Kamiya, *Chem. Pharm. Bull.,* **1979,** *27,* 1824–1829.

462. E. G. Taylor, *Can. J. Res.,* **1942,** *20B,* 161–167.

463. F. Texier and J. Bourgois, *J. Heterocycl. Chem.,* **1975,** *12,* 505–508.

464. J. Thiele, *Ann.,* **1910,** *376,* 239–268.

465. J. Thiele and H. Ingle, *Ann.,* **1895,** *287,* 233–265.

466. W. E. Thun and W. R. McBride, *J. Org. Chem.,* **1969,** *34,* 2997–3002.

467. W. E. Thun, D. W. Moore, and W. R. McBride, *J. Org. Chem.,* **1966,** *31,* 923–925.

468. H.-J. Timpe, *Z. Chem.,* **1971,** *11,* 340–341.

469. W. Tochtermann, H. Gustmann, and C. Wolff, *Chem. Ber.,* **1978,** *111,* 566–573.

470. Y. Tomiie, C. H. Koo, and I. Nitta, *Acta Cryst.,* **1958,** *11,* 774–781.

471. A. Trenin, "General and Theoretical Aspects," in S. Patai, Ed.; *The Chemistry of the Azido Group,* Interscience, New York, 1971, pp. 1–55.

472. O. Tsuge and H. Samura, *Org. Prep. Proc. Int.,* **1972,** *4,* 273–281.

473. I. Ugi, H. Perlinger, and L. Behringer, *Chem. Ber.,* **1958,** *91,* 2324–2329.

474. T. Urbanski and S. K. Vasudeva, *J. Sci. Ind. Res.,* **1978,** *37,* 250–255.

475. U.S. Army Material Command, *Engineering Design Handbook, Explosive Series, Properties of Explosives of Military Interest,* AMCP No. 706-177, Jan. 1971.

476. N. Van der Putten, D. Heijdenrijk, and H. Schenk, *Cryst. Struct. Commun.,* **1974,** *3,* 321–322.

477. K. Vaughan, *J. Chem. Soc., Perkin II,* **1977,** 17–20.

478. L. Vignon and A. Simonet, *Compt. Rend.,* **1904,** *138,* 1104–1105.

479. W. S. Wadsworth, Jr., *J. Org. Chem.,* **1969,** *34,* 2994–2997.

480. S. Walker, *Physical Methods in Heterocyclic Chemistry,* **1963,** *1,* 189–215.

481. D. L. Ward et al., *Acta Cryst.,* **1979,** *B35,* 1413–1419.

482. H. E. Watson and K. L. Ramaswamy, *Proc. Roy. Soc. (London),* **1936,** *156,* 130–143.

483. G. W. Watt and W. C. Fernelius, *Z. Anorg. Allg. Chem.,* **1934,** *221,* 187–192.

484. S. Weckerlin and W. Lüttke, *Tetrahedron Lett.,* **1964,** 1711–1717.

485. T. C. Wehman and A. I. Popov, *J. Phys. Chem.,* **1968,** *72,* 4031–4036.

486. J. E. Weidenborner et al., *Angew. Chem., Int. Ed. Engl.,* **1973,** *12,* 236–237.

487. C. Wentrup, *Helv. Chem. Acta,* **1978,** *61,* 1755–1764.

488. C. Wentrup, *Tetrahedron,* **1974,** *30,* 1301–1311.

489. J. E. Whinnery and W. H. Watson, *Acta Cryst.,* **1972,** *B28,* 635–640.

490. N. E. White and M. Kilpatrick, *J. Phys. Chem.,* **1955,** *59,* 1044–1053.

491. N. Wiberg et al., *Chem. Ber.*, **1979,** *112,* 2718–2729.
492. *Ibid.*, **1980,** *113,* 2916–2927.
493. N. Wiberg, H. Bayer, and H. Bachhuber, *Angew. Chem., Int. Ed. Engl.*, **1975,** *14,* 177–178.
494. N. Wiberg, H. Bayer, and G. Ziegleder, *Z. Anorg. Allg. Chem.*, **1979,** *459,* 201–207.
495. N. Wiberg, H.-W. Häring, and S. K. Vasisht, *Z. Naturforsch. B,* **1979,** *34B,* 356–357.
496. N. Wiberg and W. Uhlenbrock, *Angew. Chem., Int. Ed. Engl.*, **1970,** *9,* 70–71.
497. N. Wiberg and G. Ziegleder, *Chem. Ber.*, **1978,** *111,* 2123–2129.
498. R. H. Wiley and J. Moffat, *J. Org. Chem.*, **1957,** *22,* 995–996.
499. M. M. Williams, W. S. McEwan, and R. A. Henry, *J. Phys. Chem.*, **1957,** *61,* 261–267.
500. M. Witanowski et al., *J. Cryst. Mol. Struct.*, **1975,** *5,* 137–140.
501. M. Witanowski et al., *Tetrahedron,* **1972,** *28,* 637–653.
502. M. Witanowski et al., *J. Magn. Resonance,* **1976,** *21,* 185–195.
503. L. Wolff, *Ber.*, **1904,** *37,* 2827–2836.
504. L. Wolff and H. Lindenhayn, *Ber.*, **1904,** *37,* 2374–2381.
505. A. Wohl and H. Schiff, *Ber.*, **1900,** *33,* 2741–2758.
506. H. Yajima, H. Kawatani, and Y. Kiso, *Chem. Pharm. Bull,* **1970,** *18,* 850–851.
507. K. Yates, J. B. Stevens, and A. R. Katritzky, *Can. J. Chem.*, **1964,** *42,* 1957–1970.
508. M. Yoshida, A. Matsumoto, and O. Simamura, *Bull. Chem. Soc. Japan,* **1970,** *43,* 3587–3590.
509. L. H. Zalkow and R. H. Hill, *Tetrahedron,* **1975,** *31,* 831–840.
510. H. Zimmermann and H. Geisenfelder, *Z. Electrochem.*, **1961,** *65,* 368–371.

4

NONBIOLOGICAL USES

There are 205 different chemicals with three or more contiguous nitrogen atoms, including intermediates, manufactured in or imported into the United States for commercial purposes other than agricultural or pharmaceutical. The *v*-triazoles are the most numerous class, followed by the tetrazoles, organic azides, and triazenes (Table 4.1). These data are derived from the chemical substances inventory (102) and premanufacturing notifications of the Toxic Substances Control Act and are complete through 1981.

About 165 of these high nitrogen compounds are involved in end uses rather than as synthetic intermediates or starting materials. To the extent that information is available, application areas and functions served by these chemicals are discussed in this chapter.*

4.1. CORROSION INHIBITION

Benzotriazole, its *C*-methyl derivative "tolyltriazole," and other derivatives are extensively used as corrosion inhibitors for copper and copper alloys and to a

* With the advent of the inventory of the Toxic Substances Control Act (TSCA), listing over 55,000 substances, it is possible for the first time to know which chemicals are in actual commercial use in the United States, except for the 1.2% whose identity was claimed confidential. Further, from the on-line TSCA-Plant and Production Data Base, a part of the NIH/EPA Chemical Information System (104), nonconfidential production or import volumes, company identity, and related data can be found. In the past, patents have been a source of information to determine chemicals used commercially but information so obtained is frequently misleading, both because not all patents are practiced and because their disclosures of chemical compounds often are sufficiently numerous or diffuse so that substances actually used in commerce cannot be identified. Information specifying commercial chemicals in reference books, encyclopedias, journal articles, and promotional literature is usually incomplete, particularly for substances with minor uses, or whose identity has not been disclosed or coupled with a use. Since these same information sources are used to learn of commercial applications, a similar degree of uncertainty exists in assessing both use areas and the way a chemical functions in a particular use.

Table 4.1. Distribution of Commercial High Nitrogen Compounds in the United States by Chemical Class, 1981

Triazenes	25	*v*-Triazines	4
Inorganic azides	8	Tetrazoles	34
Organic azides	32	Tetrazolium salts	7
v-Triazoles	93	Tetrazapentalenes	2

lesser extent for steel, cast iron, silver, and cadmium, usually in combination with phosphate or silicate inhibitors or organic anticorrosion agents such as amines or mercaptobenzothiazole.

The film produced on copper surfaces treated with benzotriazoles is polymeric, chemically bonded to the metal or metal oxide surface, and is less than 50 Å thick (25). Together with the cuprous oxide film normally present, the copper benzotriazole complex resists washing with water and the action of organic solvents. Besides being a physical barrier, the copper benzotriazole film plays an electrochemical role, acting as a cathodic inhibitor of the oxygen reduction corrosion reaction (32). Benzotriazole also is strongly adsorbed on iron; its concentration in the electrolyte and its surface coverage fit a Langmuir adsorption isotherm (20).

Some specific applications of benzotriazole and *C*-methylbenzotriazole include protection of copper wire and sheet (24) and the surfaces of copper and copper alloys before soldering (18). By protecting brass-coated steel tire cord against corrosion, adhesion to rubber is improved (91). Together with antioxidants, the benzotriazoles inhibit the effect of copper in promoting oxidative degradation of polypropylene used for wire insulation (47, 48). Benzotriazole is added to ethylene glycol antifreeze compositions both to protect copper parts and to extend the life of the glycol (71, 75), to the water coolant for alternator stators (111), and to lubricants (33, 77) to inhibit copper and copper alloy corrosion.

Modified benzotriazoles serve as corrosion inhibitors in applications requiring both more hydrophilic and more lipophilic derivatives. Alkanolamine salts of benzotriazole in aqueous solution inhibit rusting of iron and also show biocidal activity (90). Corrosion of iron and steel is reduced by treatment with solutions of the water-soluble sodium **(I)**, ammonium, or ethanolamine salts of 1-hydroxybenzotriazole (98). The Mannich base made from benzotriazole, formaldehyde, and diethanolamine **(II)** is a metal deactivator and corrosion in-

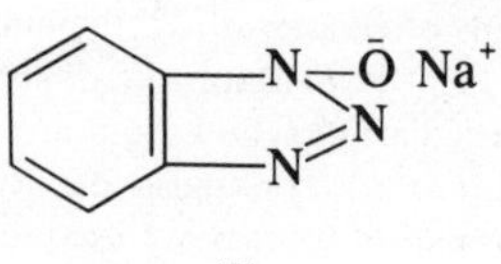

(I)

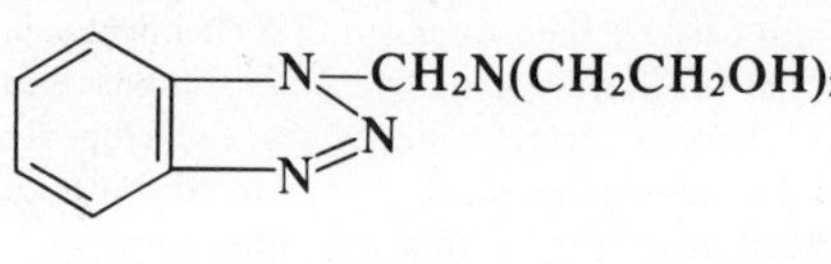

(II)

hibitor in hydraulic fluids and is particularly effective in antifreeze solutions in protecting brass (81). Reaction products of benzotriazole and oleic acid are extensively used in lubricating oil and greases to combat copper corrosion (5, 6, 70). The products probably consist of *N*-oleoyl derivatives **(III)** and benzotriazolium oleates. Tetrahydrobenzotriazole is more soluble in mineral oil than

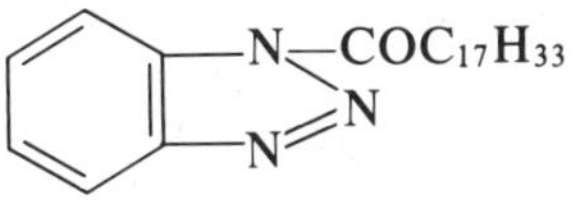

(III)

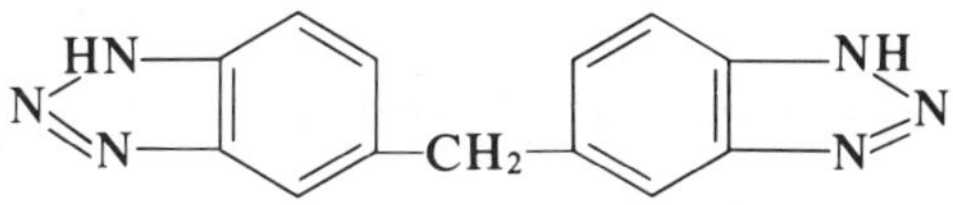

(IV)

benzotriazole and inhibits corrosion of copper, silver, and steel in this medium more effectively (83). 5,5′-Methylenebisbenzotriazole **(IV)** is a copper passivator in a polyester lubricant for gas turbine engines (19), protects copper foil against hydrogen sulfide, and is a tarnish inhibitor for cadmium (84).

4.2. PLASTICS

Cellular polymers can be produced by heating mixtures of chemical blowing agents and resins in defined temperature ranges where gas is liberated. By heating a composition containing unvulcanized rubber and *p*-toluenesulfonyl azide at 153° where the rubber vulcanizes and the azide decomposes, foam rubber is obtained (4). At temperatures of 250–300°, mixtures of 5-phenyltetrazole **(V)**

p-$CH_3C_6H_4SO_2N_3$

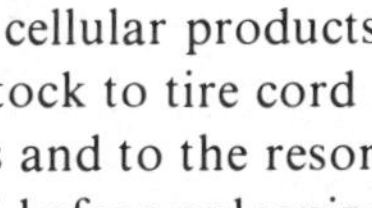

(V)

and thermoplastic materials such as polypropylene (72), polycarbonates (23), and polyethylene terephthalate (34) form cellular products.

Adhesion of vulcanizable rubber tire stock to tire cord is improved by substances which can bond both to the fibers and to the resorcinol-formaldehyde resin latex with which tire cord is treated before vulcanization. The bisazidoformate of β-hydroxyethyl isophthalate **(VI)**, applied to polyester or aramid fibers, grafts to the surface by heating in two stages for short periods at 107 and 171° (49, 50, 51, 56). Treated tire cord is thus primed for adhesion to the resorcinol-formaldehyde resin latex in a vulcanization treatment at 153°. The fiber adheres to rubber so firmly that cohesive failure of the rubber occurs rather than the rubber-fiber bond. Coupling of siliceous fillers with plastics is achieved by trimethoxysilyl sulfonylazides, illustrated by the ethylbenzene derivative **(VII)** (55, 100, 101). Bonding to a filler such as clay or glass occurs by hydrolysis of the methoxysilane to silanol groups, which condense with surface hydroxyl groups on the silicate. The treated filler attaches to polypropylene or other polymer by decomposition of the sulfonyl azide during melt compounding at 210°.

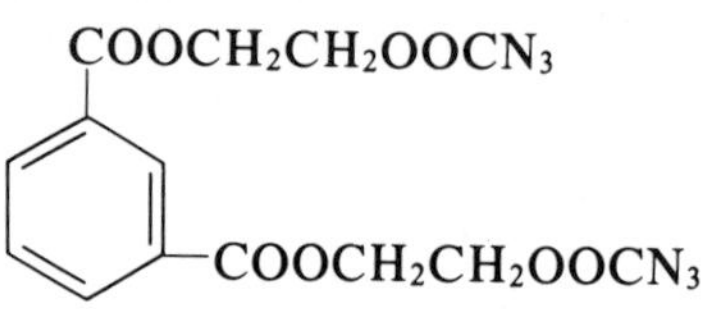

(VI)

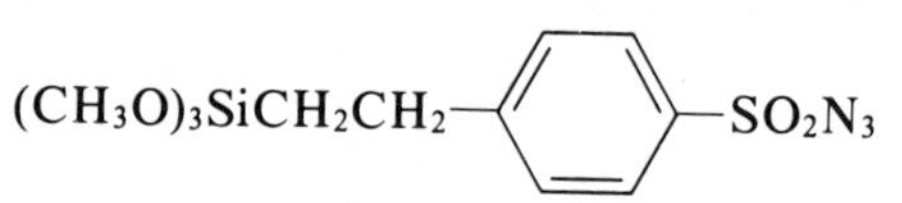

(VII)

These uses of azides depend on thermal elimination of nitrogen, formation of nitrenes, and their subsequent combination with components of the plastic system. Understanding of the reactions involved comes from photochemical investigations of nitrenes from simpler azides (paralleling thermal behavior), which indicate that both singlet and triplet states are present (2, 14). Insertion into

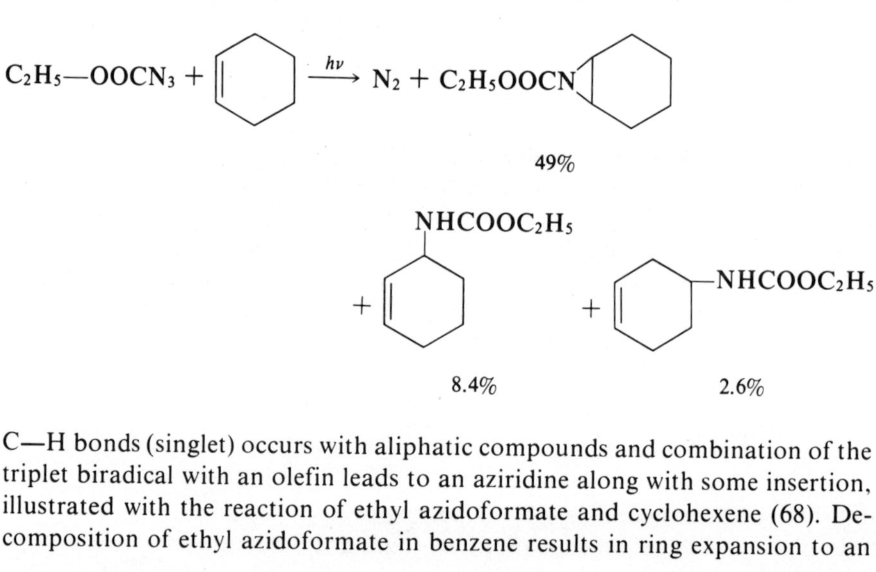

C—H bonds (singlet) occurs with aliphatic compounds and combination of the triplet biradical with an olefin leads to an aziridine along with some insertion, illustrated with the reaction of ethyl azidoformate and cyclohexene (68). Decomposition of ethyl azidoformate in benzene results in ring expansion to an

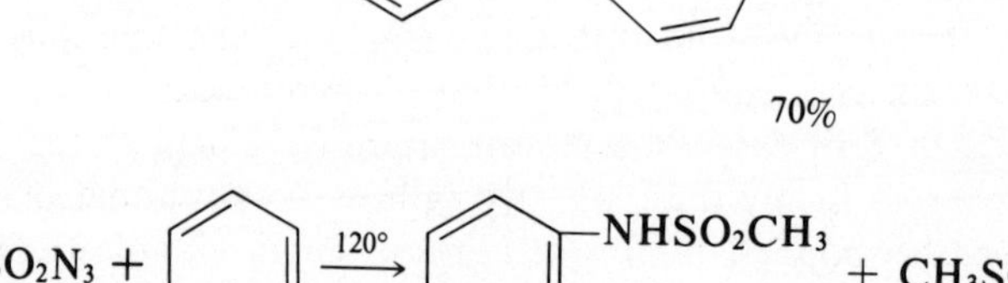

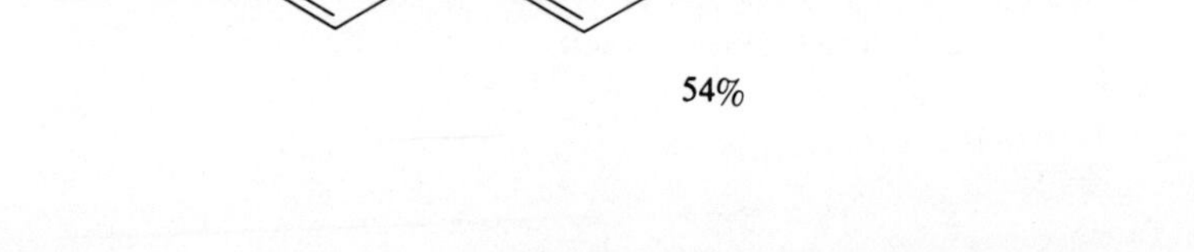

azepine (45); at 120° methanesulfonyl azide forms the sulfonanilide and methanesulfonamide through abstraction of hydrogen by the nitrene (3).

One of the larger uses of high nitrogen compounds is that of 2-(2′-hydroxyphenyl)benzotriazoles as ultraviolet absorbers in plastics. Ultraviolet absorbers prevent or slow deterioration of many polymers by converting ultraviolet radiation in sunlight to heat. The 2-(2′-hydroxyphenyl)benzotriazoles have high absorptivities from below 290 nm, the lower wavelength of sunlight at the earth's surface, to nearly 400 nm, the beginning of visible light, but are essentially

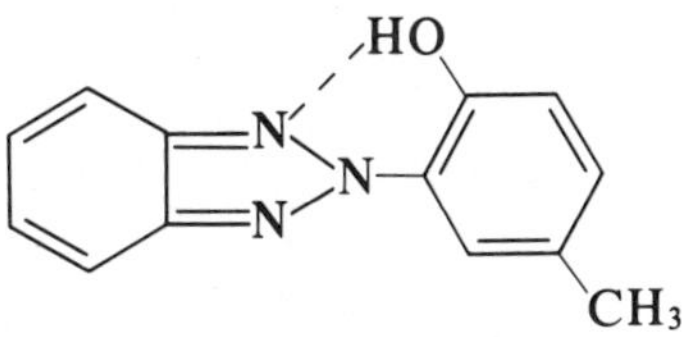

(VIII)

(21, 52, 53)
acrylics, polystyrene, cellulose esters, polyvinyl chloride, polyesters, polyvinylidene chloride

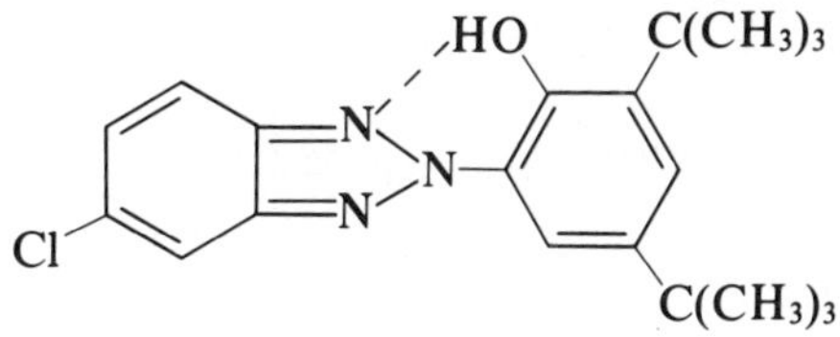

(28, 40, 52, 53)
polyethylene, polypropylene, polyesters

without color (21, 40, 41). A bathochromic shift of the maximum of 2-(2′-hydroxy-5′-methylphenyl)benzotriazole **(VIII)** to 340 nm from that of unsubstituted 2-phenylbenzotriazole, 307 nm (26), results from the 2-hydroxy group, which is hydrogen bonded to triazole nitrogen. Absorption maxima of the 2-(2′-hydroxyphenyl)benzotriazoles are in the region of the solar spectrum causing

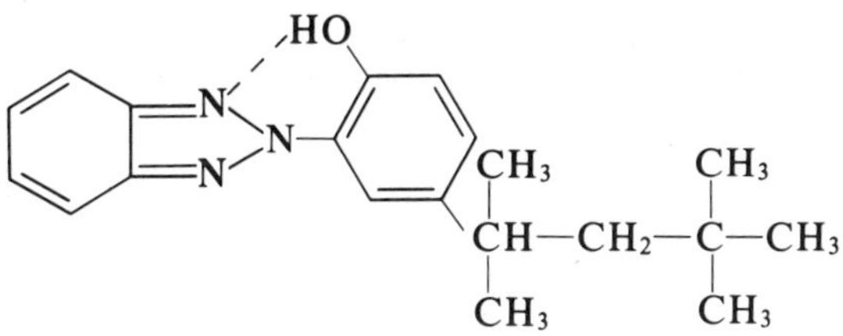

(13)
polyethylene, polypropylene

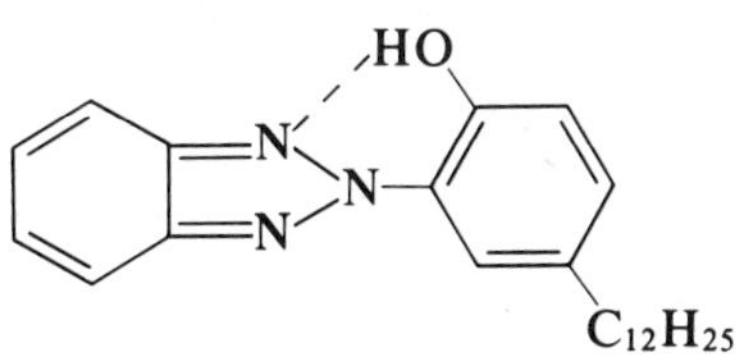

(35)
polyethylene, polypropylene

photodegradation of most plastics. Examples of plastics for which these ultraviolet absorbers are recommended are shown with each structure. The largest use for ultraviolet stabilizers is in polyolefins (8); accordingly, hydroxyphenylbenzotriazoles with lipophilic substituents such as bis-*t*-butyl, *t*-octyl, or dodecyl are required to achieve compatibility with polyethylene and polypropylene. Hydroxyphenylbenzotriazoles also are used as ultraviolet absorbers in polyurethane coatings on wood (88), in colorless lacquers for nonferrous metals (114), in photographic film to decrease sensitivity to room light (93), or to protect dye images (65).

4.3. TEXTILES AND PAPER

Embossed areas of nylon carpet can be produced by application of benzotriazole and acetic acid. Fibers are shrunk and the pile height is reduced, possibly through breaking of hydrogen bonds between polymer chains, allowing linear contraction of the fibers (78). The shrinking agent may be incorporated in a dye or pigment printing paste so that color appears in register with the embossed area.

C.I. Mordant Green 9 is a bluish green dye for dyeing and printing wool, bathochromically shifted by the arylazo substituent from the red of the parent

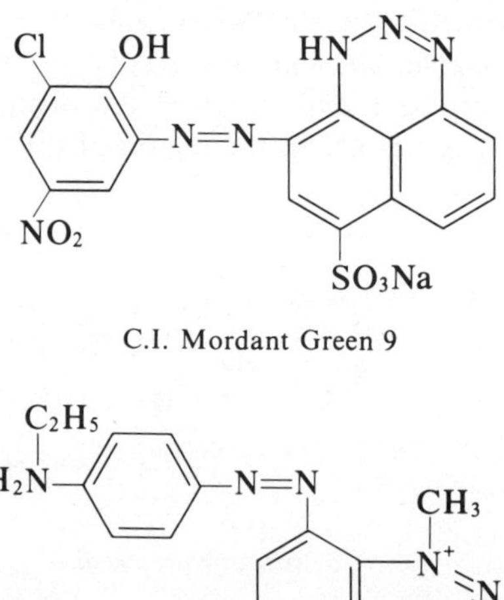

C.I. Mordant Green 9

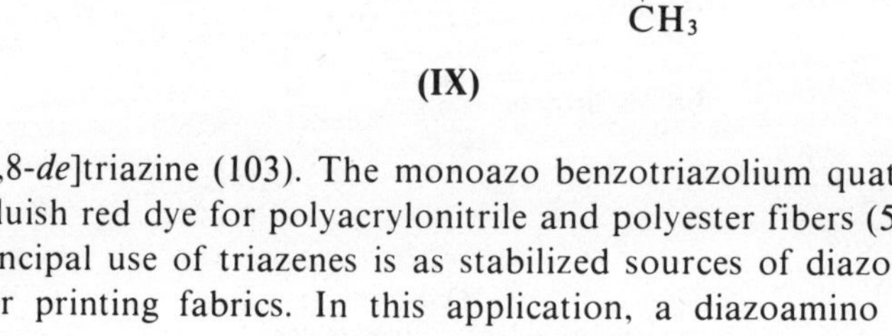

(IX)

naphtho[1,8-*de*]triazine (103). The monoazo benzotriazolium quaternary salt **(IX)** is a bluish red dye for polyacrylonitrile and polyester fibers (57).

The principal use of triazenes is as stabilized sources of diazonium compounds for printing fabrics. In this application, a diazoamino compound formed from a diazonium salt and salts of amino acids such as sarcosine **(X)**, *N*-methyltaurine, or 2-ethylamino-5-sulfobenzoic acid is mixed with a naphthol or other coupling agent in a paste. After printing on the cloth, steaming with acetic

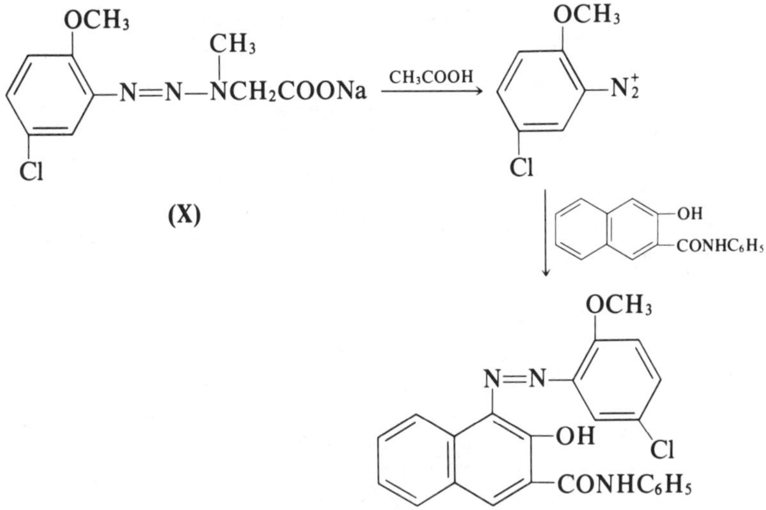

(X)

or formic acid decomposes the triazene, liberating the diazonium compound which couples immediately, forming an azo dye on the fiber (109). Other extensively used triazenes are shown.

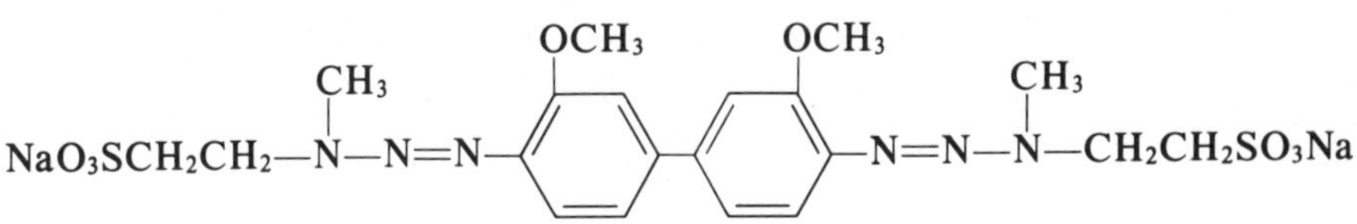

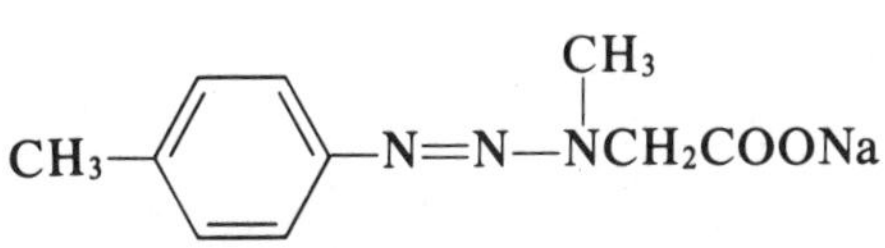

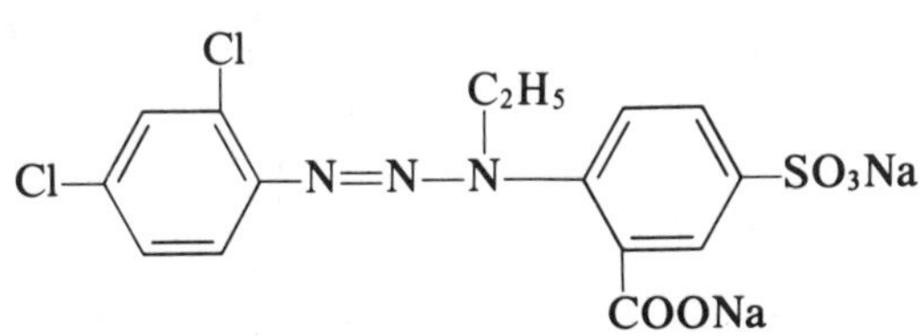

A major use in the high nitrogen group is that of 2-substituted *v*-triazoles as fluorescent brighteners. In this application the colorless fluorescent brighteners absorb ultraviolet light and emit blue-violet or blue-green fluorescence (43, 117). Applied in the same manner as dyes to substrates such as textiles, paper, or plastics, they create the impression of whiteness by optical compensation of a yellow cast resulting from absorption of blue-violet light by the substrate. A major method of application is in detergents for laundering. The most prominent structural types of triazole brighteners are naphtho[1,2-*d*]triazoles, monocyclic *v*-triazoles, and a quaternized *v*-triazolo[4,5-*b*]pyridine **(XI)**, illustrated together with their applications areas.

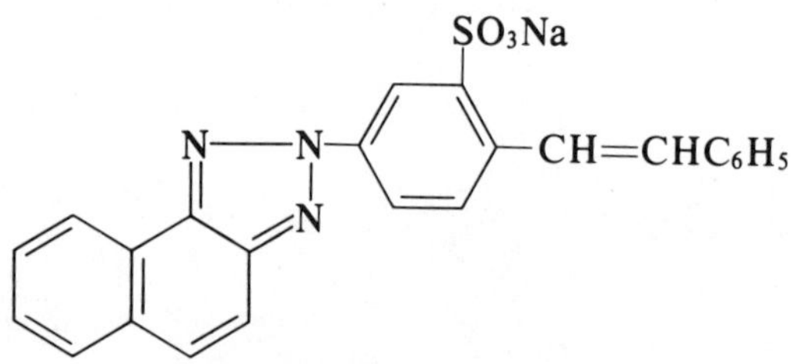

C.I. Fluorescent Brightening Agent No. 46
(62, 118)
polyester, polyacrylonitrile, polyvinyl chloride, polystyrene, detergents

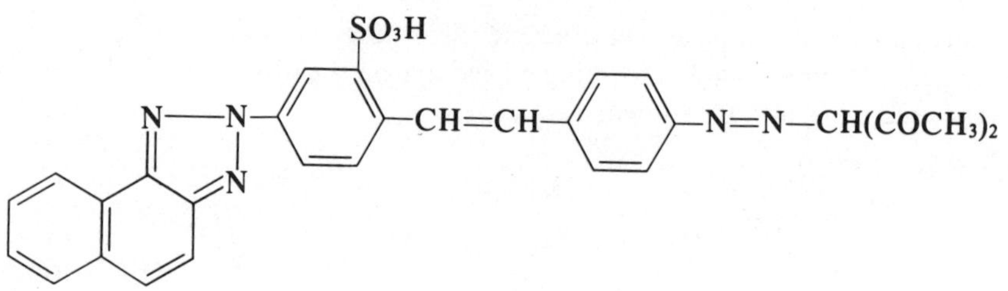

(112)
paper

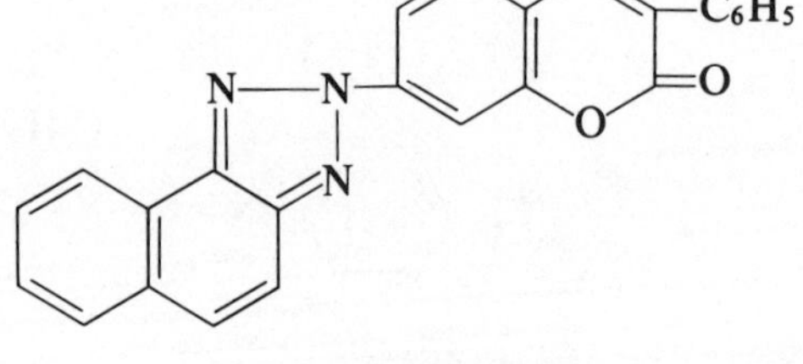

(38)
cellulose acetate, nylon, polyester, polyvinyl chloride, polystyrene, poly(acrylonitrile-butadiene-styrene)

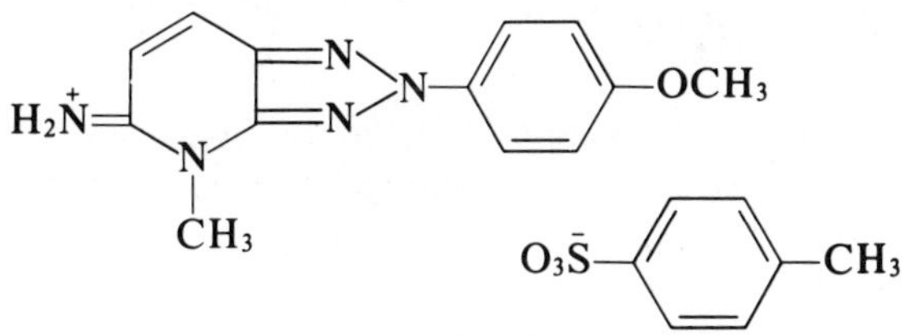

(XI)

(16, 17)
polyacrylonitrile

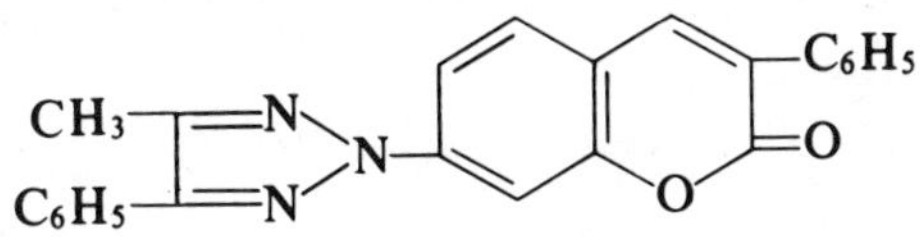

(74)
polyester

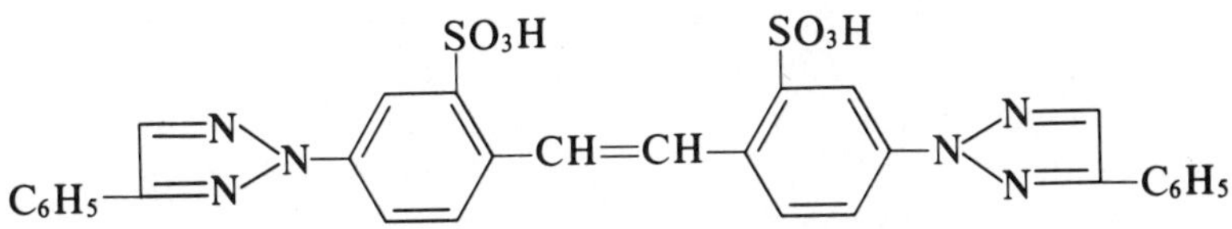

(29, 30)
cotton, rayon, nylon, detergents

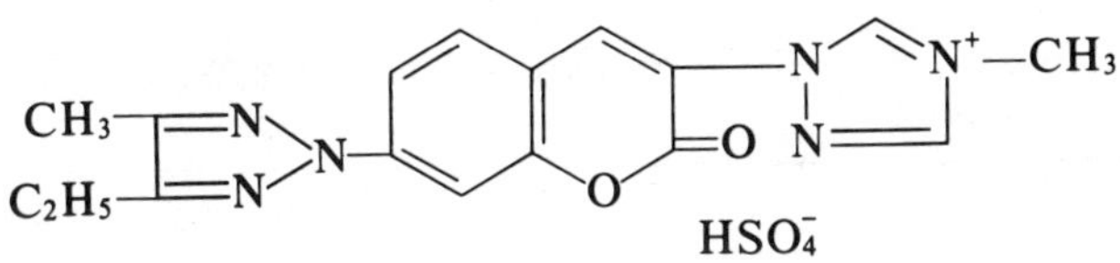

(36)
polyester

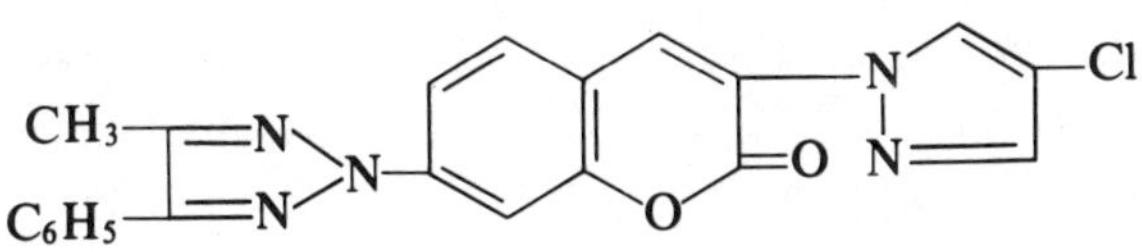

(22)
polyester

4.4. PHOTOGRAPHY

Benzotriazoles, 5-thionotetrazolines, related thiotetrazoles, and tetrazolium salts are useful in photographic development, in controlling fog, and in assisting the formation of direct positive and color prints.

Fog in photographic images can result from the action of a developer in producing silver in areas which are not part of the image-forming exposure. One of the most extensively used organic antifoggants is 1-phenyl-2-tetrazoline-5-thione **(XII)**, more frequently called phenylmercaptotetrazole in the photographic literature (59). Other antifoggants include 1-methyl-2-tetrazoline-5-thione **(XIII)** (31) benzotriazole and its 5-chloro **(XIV)** (99), nitro (73), and methyl (69, 76) derivatives. They are usually applied in the developing bath but also may be incorporated in a nonemulsion layer of the photographic element. Reduction of

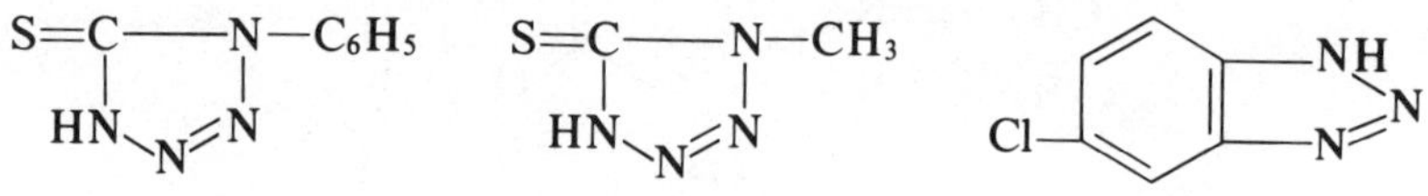

(XII) **(XIII)** **(XIV)**

silver ions is inhibited by these antifoggants by selective adsorption on silver halide as well as at silver and silver sulfide nuclei which are present (59). Infrared spectra indicate that both 1-phenyl-2-tetrazoline-5-thione and benzotriazole exist on the surface of silver bromide as silver salts of the stabilizers (97). Solubility data for the silver antifoggants in comparison with those of silver chloride and bromide (Table 4.2) show that metathesis occurs with these compounds displacing chloride and bromide ions. 5-Nitrobenzotriazole reacts only with silver chloride but 1-phenyl-2-tetrazoline-5-thione reacts with silver iodide

Table 4.2. Solubility Products of Silver Halides and Silver Antifoggants

Silver Salt	Solubility Product in water, pK_{sp}, 25° (82)
Chloride	9.8
Bromide	12.3
Iodide	16.1
1-Phenyl-2-tetrazoline-5-thione	16.2
Benzotriazole	13.5
5-Methylbenzotriazole	13.6
5-Chlorobenzotriazole	12.8
5-Nitrobenzotriazole	11.8

Table 4.3. Heats of Adsorption of Antifoggants

Antifoggant	Heat of Adsorption ($mcal/m^2$) (115) AgBr	Ag^0	Ag_2S
Benzotriazole	2	11	41
1-Phenyl-2-tetrazoline-5-thione	43	53	93
1-Methyl-2-tetrazoline-5-thione	179	60	104

as well as the other silver halides. Strong adsorption of the tetrazoline thiones, in comparison with benzotriazole, on silver bromide, silver, and silver sulfide is shown by the heats of adsorption (Table 4.3).

Positive photographs are produced by a reversal process in which, after conventional development of the negative image, the film is fogged by uniform exposure to light; the remaining silver halide is then reduced in a developer containing 1-phenyl-2-tetrazoline-5-thione, followed by simultaneous bleaching and fixing (39, 54). This process depends on the strong adsorption of the tetrazolinethione to the developed silver and resistance of the coated silver to subsequent bleaching. Benzotriazole and methylbenzotriazole are similarly used (95).

Bleaching of silver in optical motion picture sound track in color films is inhibited by a 1-(3-alkylamidophenyl)-2-tetrazoline-5-thione such as the caproamide derivative **(XV)**, applied after initial development and before bleaching (11, 12). Both the acetamide and caproamide derivatives, applied to an

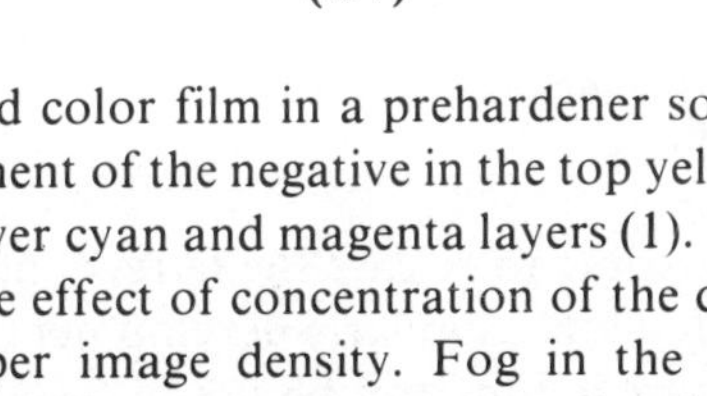

(XV)

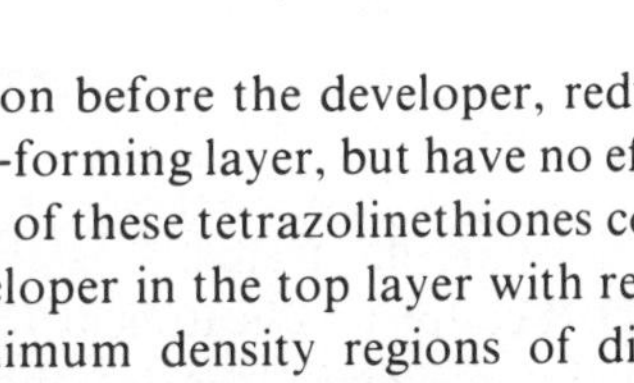

(XVI)

exposed color film in a prehardener solution before the developer, reduce development of the negative in the top yellow-forming layer, but have no effect on the lower cyan and magenta layers (1). Use of these tetrazolinethiones counteracts the effect of concentration of the developer in the top layer with resultant improper image density. Fog in the minimum density regions of diffusion transfer color photographic materials is suppressed, giving clear white areas, by inclusion of 1-(3,4-dichlorophenyl)-2-tetrazoline-5-thione **(XVI)** in nonsensitive layers of the emulsion (80).

Development-inhibitor-releasing (DIR) couplers in subtractive three-color photography release a development inhibitor during coupling with oxidized *p*-

phenylenediamine color developers when the cyan, magenta, or yellow image dyes are formed (9). The DIR couplers control development of the emulsion layers in proportion to the silver or dye image formed, thus avoiding general restraint and loss of speed which can occur when a free development inhibitor such as 1-phenyl-2-tetrazoline-5-thione **(XII)** is added to the developer solution. The DIR compound is incorporated either in an emulsion layer or between layers of a color film. Structures of DIR couplers which release 1-phenyl-2-tetrazo-

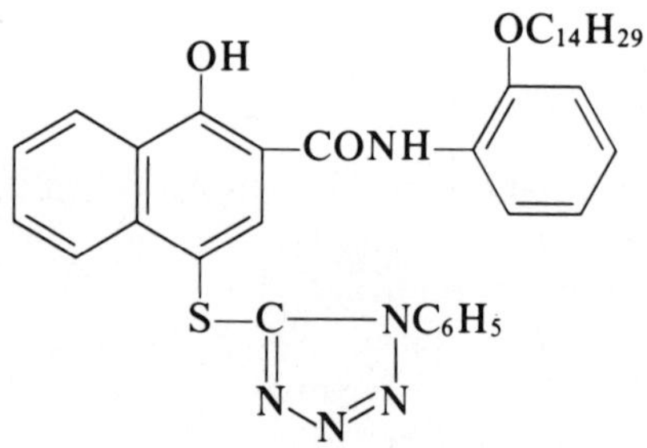

cyan DIR coupler
(10, 58, 61)

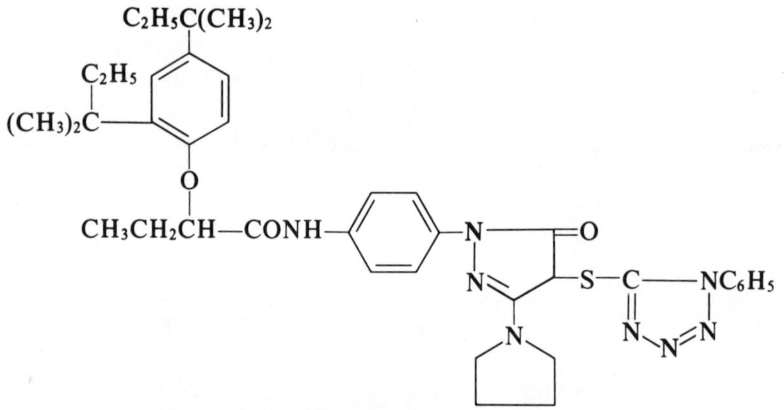

magenta DIR coupler
(60)

line-5-thione are illustrated. It is also possible to release the tetrazolinethione from a DIR coupler by heating at 75–100° (116). Similarly development can be controlled in an image transfer photographic film unit by use of 5-(β-cyano-

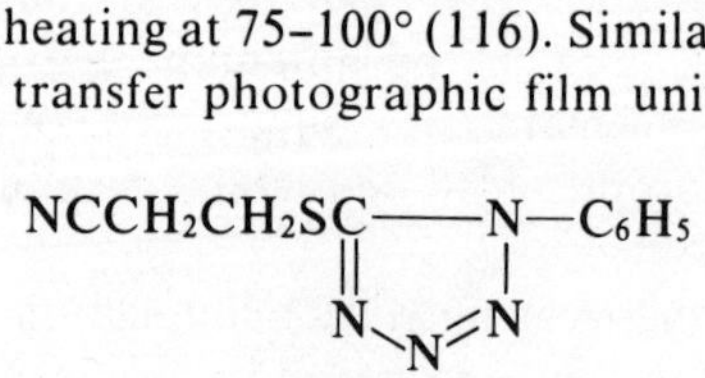

(XVII)

ethylthio)-1-phenyltetrazole **(XVII)** in an emulsion layer, an adjacent layer, or in an image-receiving layer (46). This development restrainer liberates the tetrazolinethione on processing with an alkaline solution, substantially inhibiting further development.

Several color photography processes involve the reduction of colorless tetrazolium salts to colored formazans. In a silver-dye-bleach process (113) a tetra-

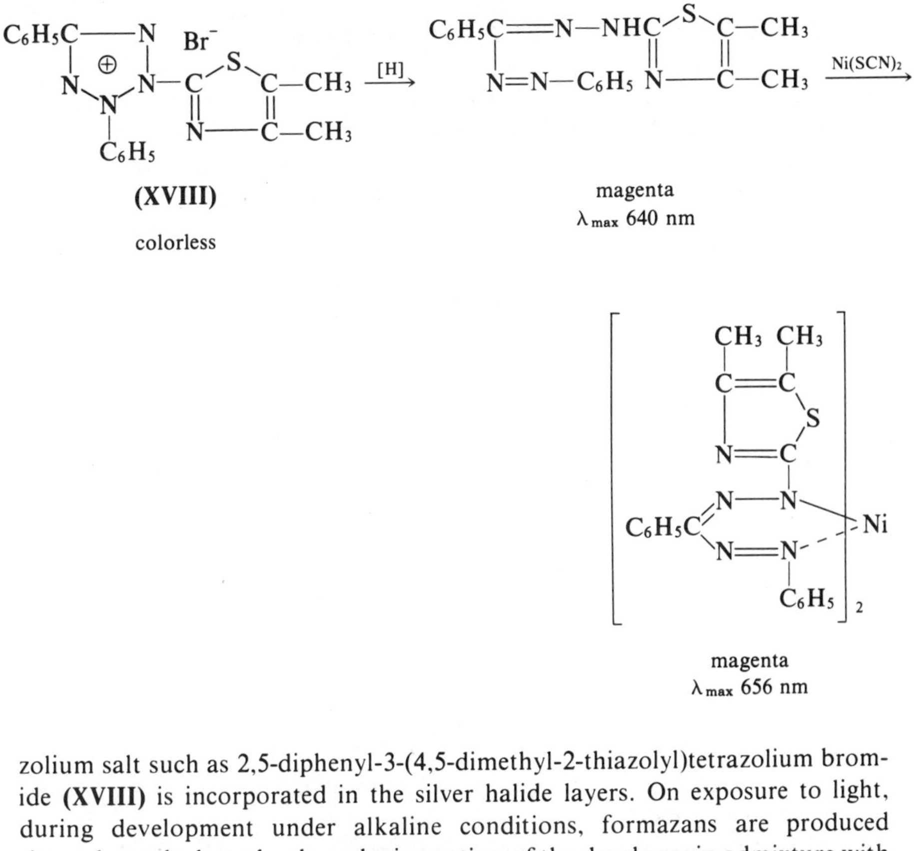

(XVIII)

colorless

magenta
λ_{max} 640 nm

magenta
λ_{max} 656 nm

zolium salt such as 2,5-diphenyl-3-(4,5-dimethyl-2-thiazolyl)tetrazolium bromide **(XVIII)** is incorporated in the silver halide layers. On exposure to light, during development under alkaline conditions, formazans are produced throughout the layer by the reducing action of the developer in admixture with metallic silver. In the bleach step, formazan and silver are bleached imagewise and after washing a positive formazan image remains. The formazan image is rendered more light stable and its adsorption may be shifted bathochromically by chelation with metal ions.

Tetrazolium salts are light sensitive (79) and photographic systems have been proposed utilizing this property, but the images formed are usually weak, and latent images, as formed by irradiation of silver halides, are not produced. Direct printout systems are described in which depth is enhanced by adding light-sensitive materials such as ferric or vanadic salts (42), 2-nitrobenzyl alcohol, or β-naphthyl disulfide (15) to the tetrazolium layer.

In a physical development process, consisting of formation of nuclei of group IB or VIIIB metals by exposure of their salts to light, followed by contacting with a bath containing tetrazolium salt and a reducing agent (67, 44), large amplifications of formazan formation are obtained. The metal nuclei function catalytically, and with palladium, 10^6–10^7 formazan molecules result for each palladium atom.

4.5. REPROGRAPHICS AND PHOTORESISTS

Photolithographic plates are prepared by coating with 2,6-[di(4′-azidophenylmethylene)]-4-methylcyclohexanone **(XIX)** or a related diazide such as 4,4′-diazidodiphenylamine **(XX)** and a polymer such as styrene-butadiene (89), dialkyl isophthalate prepolymer (85), or cyclized rubber (105). Areas of the plates exposed to light are hardened by cross-linking and become receptive to greasy

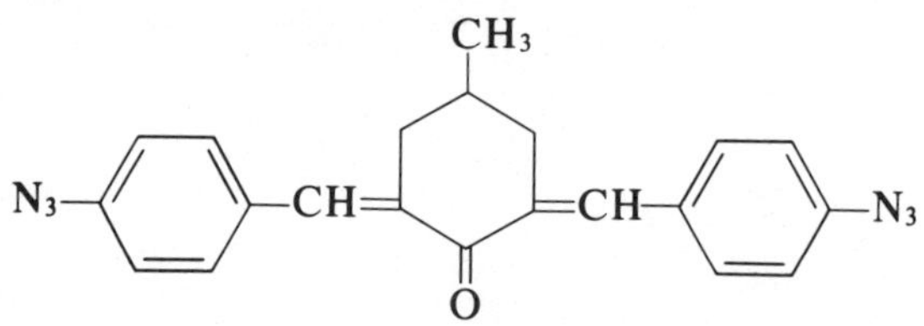

(XIX)

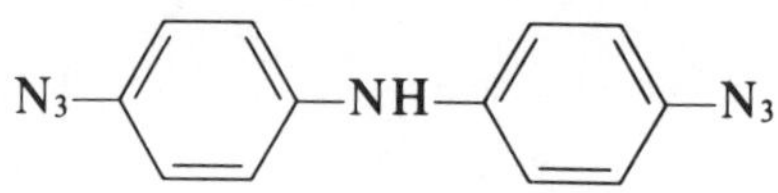

(XX)

inks; unexposed polymer is removed by solvent treatment. In photoresists, **XIX** is used extensively to insolubilize cyclized rubber, after which unexposed areas are removed by etching (89). The principal application of **XIX** is in the manufacture of microelectronic parts such as silicon or germanium semiconductors or thin metal films on ceramics (27). Both **XIX** and **XX** are sensitive from 250 to over 450 nm, and the region of spectral sensitivity can be extended to 530 nm by addition of 1-nitropyrene or similar compounds (105). Spectral sensitization appears to be by triplet-triplet energy transfer. Insolubilization of a linear polymer with hydroxyl groups, such as the polyether of bisphenol-A and epichlorohydrin, in a photolithographic printing plate is achieved by photochemical decomposition of isophthaloyl diazide **(XXI)** (66). The probable mechanism involves conversion of the dicarboxylic acid azide to diisocyanate, cross-linking the polymer by urethane formation. To determine if proper exposure of a presensitized photolithographic plate has been achieved before development or

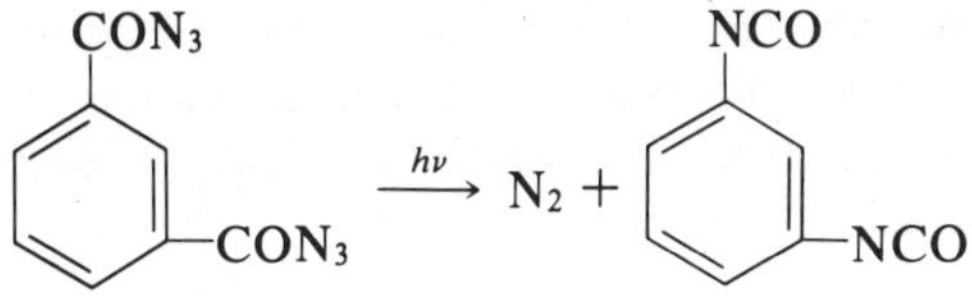

(XXI)

printing, a colorless leuco dye such as that of crystal violet and an azide such as 2-azido-1-carbobutoxymethylcarbamylbenzimidazole **(XXII)** are incorporated in the lithographic coating (86). The nitrene resulting from the azide dehydrogenates the leuco compound to the triphenylmethane dye.

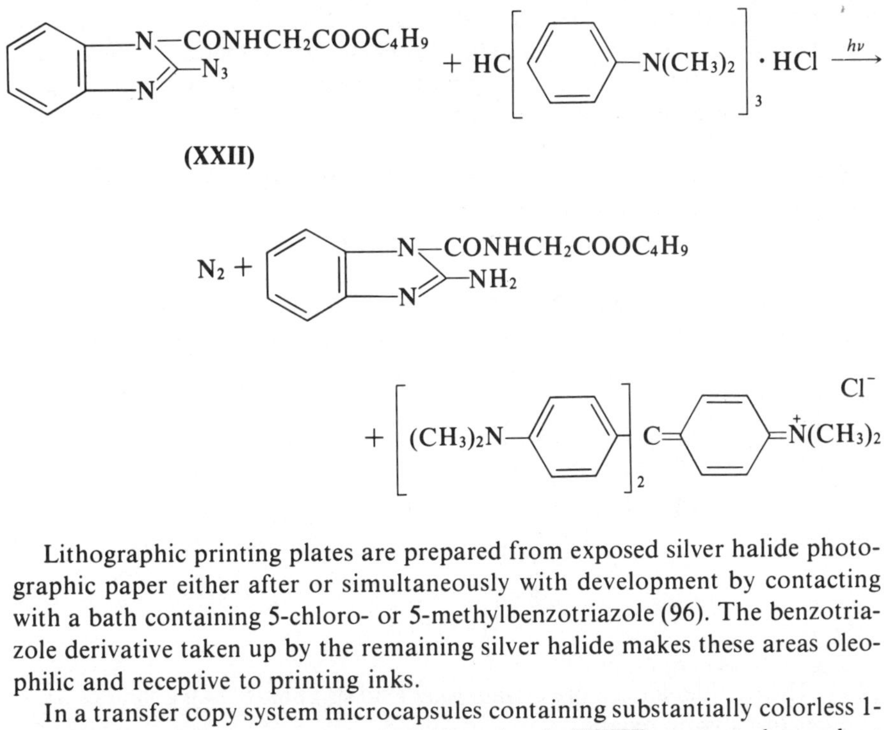

Lithographic printing plates are prepared from exposed silver halide photographic paper either after or simultaneously with development by contacting with a bath containing 5-chloro- or 5-methylbenzotriazole (96). The benzotriazole derivative taken up by the remaining silver halide makes these areas oleophilic and receptive to printing inks.

In a transfer copy system microcapsules containing substantially colorless 1-bis-(*p*-dimethylaminophenyl)methylbenzotriazole **(XXIII)** are coated on a sheet of colorless carbon paper. When this coated paper is contacted by pressure or impact (e.g., by typewriter type) with a copy paper containing an electron acceptor, such as kaolin or bentonite, a deep blue coloration results (64, 108).

Triphenylmethyl azide, $(C_6H_5)_3CN_3$, is a radiation-sensitive promotor in

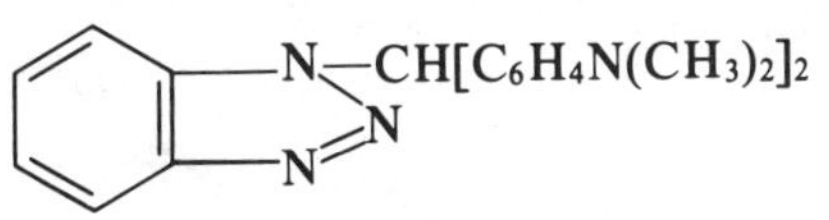

(XXIII)

diazo paper coatings, increasing the speed of the diazotype process, through its fission to nitrene, which initiates a chain decomposition of the diazonium compounds in the formulation when exposed to a light source (87).

4.6. EXPLOSIVES

Lead azide is a component of explosive trains in detonators, primers, and ignition-delay units (110). Because its sensitivity to impact and friction can be controlled by addition of other materials such as dextrin or polyvinyl alcohol, or by inclusion of lead carbonate during manufacture, and because it is stable to over 100° and has useful explosive characteristics, it is the primary explosive employed most extensively. It is used in detonators, both alone and in primer mixtures with other materials such as lead styphnate, barium nitrate, or tetracene. In a detonator, after initiation by the primer, the lead azide in turn initiates detonation in a secondary explosive such as cyclonite, tetryl, or pentaerythritol tetranitrate.

Tetracene, although appreciably less stable thermally than lead azide and other primer components, is used to improve ignition by the action of a firing pin (increases stab sensitivity) (110). It is not efficient in initiating high explosives (107).

The double salt of lead nitroaminotetrazole and lead styphnate **(XXIV)** is a component of primer compositions for ignition of propellant powders (92, 94). It is more stable thermally than lead styphnate alone, and withstands 210° for

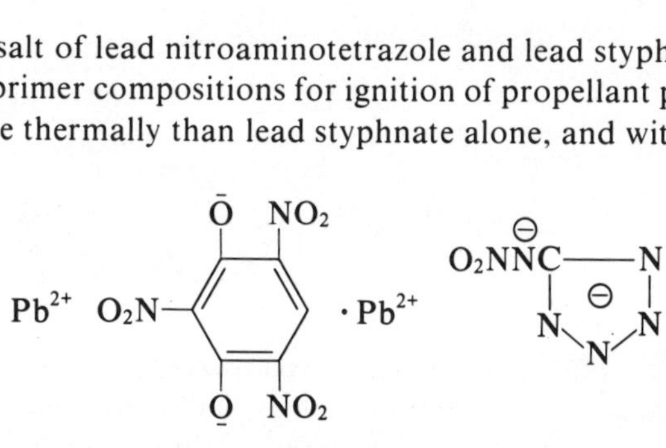

(XXIV)

24 hr. The double salt is more brisant than lead styphnate (24.0 g sand vs. 16.1) and performs more reliably.

TACOT is the most thermally resistant explosive known, withstanding temperatures to 375° (7, 106). It can be used for applications requiring heat resistance such as explosive charges for high temperature oil wells, as a tapping device for open hearth furnaces, or in operations in supersonic aircraft skins where temperatures reach 300–340° at Mach 2.

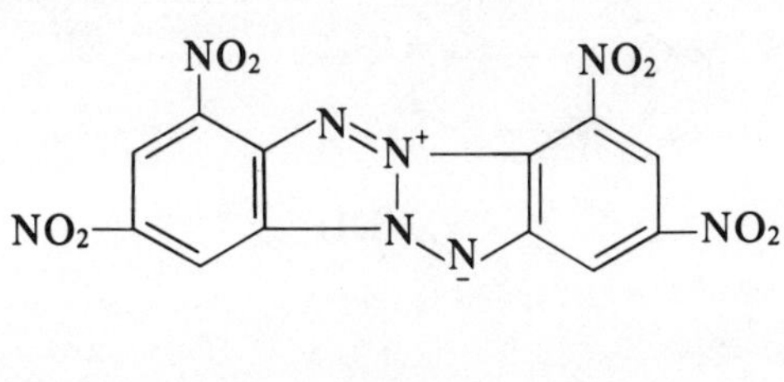

4.7. LAMPS AND VACUUM TUBES

Thermal decomposition of the azides of barium and cesium in lamps and

$$Ba(N_3)_2 \longrightarrow Ba^0 + 3N_2$$

vacuum tubes yields the corresponding metals which serve as getters to eliminate oxygen and improve tube performance (37, 63).

REFERENCES

1. J. R. Abbott and I. F. Salminen (Eastman Kodak), U.S. Pat. 3,295,976, Jan. 3, 1967.
2. R. A. Abramovitch and E. P. Kyba, "Decomposition of Organic Azides," in S. Patai, Ed., *The Chemistry of the Azide Group,* Wiley, New York, 1971, pp. 221–329.
3. R. A. Abramovitch and V. Uma, *Chem. Commun.,* **1968,** 797–798.
4. F. H. Adams (American Cyanamid), U.S. Pat. 2,830,029, April 8, 1958.
5. H. J. Andress, Jr. (Mobil Oil), U.S. Pat. 3,788,993, Jan. 29, 1974.
6. H. J. Andress, Jr. (Mobil Oil), U.S. Pat. 3,969,237, July 13, 1976.
7. Anonymous, *Chem. Eng. News,* Nov. 5, **1962,** 49–50.
8. Anonymous, *Mod. Plast.,* Sept. **1981,** 71–74.
9. C. R. Barr, J. R. Thirtle, and P. W. Vittum, *Photogr. Sci. Eng.,* **1969,** *13,* 74–80; 214–217.
10. C. R. Barr, J. Williams, and K. E. Whitmore (Eastman Kodak), U.S. Pat. 3,227,554, Jan. 4, 1966.
11. H. J. Bello and C. F. Holtz (Eastman Kodak), U.S. Pat. 3,705,799, Dec. 12, 1972.
12. M. J. Bevis, H. J. Bello, and C. F. Holtz (Eastman Kodak), U.S. Pat. 3,705,803, Dec. 12, 1972.
13. R. J. Boyle (American Cyanamid), U.S. Pat. 3,230,194, Jan. 18, 1966.
14. D. S. Breslow, W. D. Wills, and L. O. Amberg, *Rubber Chem. Technol.,* **1970,** *43,* 605–612.
15. R. T. Brongo, J. C. Fleming, and J. W. Manthey (Eastman Kodak), U.S. Pat. 3,887,374, June 3, 1975.
16. B. G. Buell and R. S. Long (American Cyanamid), U.S. Pat. 3,049, 438, Aug. 14, 1962.
17. B. G. Buell and R. S. Long (American Cyanamid), U.S. Pat. 3,058,989, Oct. 16, 1962.
18. J. H. Burkhart and E. C. Slick (GTE Sylvania), U.S. Pat. 3,716,421, Feb. 13, 1973.
19. D. C. Byford and P. M. Blanchard (British Petroleum), U.S. Pat. 3,790,481, Feb. 5, 1974.
20. R. J. Chin and K. Nobe, *J. Electrochem. Soc.,* **1971,** *118,* 545–548.
21. Ciba-Geigy Product Information Bulletin, *Tinuvin® P. Ultraviolet Light Absorber.*
22. U. Claussen (Bayer), U.S. Pat. 3,947,412, March 30, 1976.
23. K. T. Collington, *Eur. J. Cell. Plast.,* **1978,** 107–113.
24. J. B. Cotton, P. A. Mack, and R. Price (ICI), U.S. Pat. 3,837,964, Sept. 24, 1974.
25. J. B. Cotton and I. R. Scholes, *Brit. Corros. J.,* **1967,** *2,* 1–5.
26. D. Dal Monte Casoni et al., *Gazz. Chim. Ital.,* **1958,** *88,* 977–1034; *Chem. Abstr.,* **1959,** *53,* 18945.
27. W. S. DeForest, *Photoresist,* McGraw-Hill, **1975,** pp. 214–215.
28. A. D. DiBattista and J. D. Spivack (Geigy), Ger. Pat. 2,042,807, March 11, 1971.
29. A. Dolars and O. Neuner (Farbenfabriken Bayer), U.S. Pat. 3,666,758, May 30, 1972.

30. A. Dolars, O. Neuner and R. Putter (Farbenfabriken Bayer), U.S. Pat. 3,485,831, Dec. 23, 1969.
31. D. R. Douglas, U.S. Pat. 3,832,180, Aug. 27, 1974.
32. I. Dugdale and J. B. Cotton, *Corros. Sci.*, **1963,** *3,* 69–74.
33. A. M. Durr, Jr. and R. A. Krenowicz (Continental Oil), U.S. Pat. 3,923,672, Dec. 2, 1975.
34. Dynamit Nobel Akt., Brit. Pat. 1,570,320, June 25, 1980.
35. Eastman Kodak Co., Ger. Pat. 1,166,623, March 26, 1964.
36. Farbenfabriken Bayer Akt., Fr. Pat. 2,004,074, Nov. 21, 1969.
37. H. Ficheroulle and A. Kovache, *Mem. Poudres,* **1951,** *33,* 7–19.
38. F. Fleck, H. Balzer and H. Aebli (Sandoz), U.S. Pat. 3,288,801, Nov. 29, 1966.
39. J. H. Gallet and R. F. Garcia (Itek), U.S. Pat. 3,640,717, Feb. 8, 1972.
40. Geigy Industrial Chemicals Product Information Bulletin, *Tinuvin® 327 Ultraviolet Absorber.*
41. Geigy Product Information Bulletin, *Tinuvin® 328 Ultraviolet Absorber.*
42. Gevaert Photo Producten N. V., Brit. Pat. 670,883, April 30, 1952.
43. H. Gold, "Fluorescent Brightening Agents," in K. Venkataraman, Ed., *The Chemistry of Synthetic Dyes,* Vol. 5, Academic Press, New York, 1971, pp. 535–679.
44. H. J. Gysling and M. Lelental (Eastman Kodak), U.S. Pat. 4,042,392, Aug. 16, 1977.
45. K. Hafner and C. König, *Angew. Chem.,* **1963,** *75,* 89–90.
46. H. A. Hammond, W. J. Humphlett, and I. F. Salminen (Eastman Kodak), U.S. Pat. 4,009,029, Feb. 22, 1977.
47. R. H. Hansen (Bell Telephone Labs.), U.S. Pat. 3,367,907, Feb. 6, 1968.
48. R. H. Hansen, T. DeBenedictis, and W. M. Martin, *Poly. Eng. Sci.,* **1965,** *5,* 223–226.
49. J. N. Haynes (Hercules), U.S. Pat. 3,686,231, Aug. 22, 1972.
50. J. N. Haynes (Hercules), U.S. Pat. 3,814,657, June 4, 1974.
51. J. N. Haynes (Hercules), U.S. Pat. 3,946,051, March 23, 1976.
52. H. Heller et al. (Geigy), U.S. Pat. 3,004,896, Oct. 17, 1961.
53. H. Heller et al., (Geigy), U.S. Pat. 3,189,615, June 15, 1965.
54. R. W. Henn and N. H. King, *Photogr. Sci. Eng.,* **1967,** *11,* 363–364.
55. Hercules Incorporated, Bulletin OPD-24D.
56. Hercules Incorporated, Bulletin ORS-302.
57. Hoechst Aktiengesellschaft, Brit. Pat. 1,404,097, Aug. 28, 1975.
58. T. Iijima, T. Obikawa, and M. Tanaka (Konishiroku Photo Industry), U.S. Pat. 4,153,460, May 8, 1979.
59. T. H. James, "The Mechanism of Development," in T. H. James, Ed., *The Theory of the Photographic Process,* 4th ed., Macmillan, New York, **1977,** pp. 373–403.
60. K. Kato, S. Saito, and H. Kawaguchi (Konishiroku Photo Industry), U.S. Pat. 4,171,975, Oct. 23, 1979.
61. T. Kato et al. (Fuji Photo Film), U.S. Pat. 4,145,219, March 20, 1979.
62. E. Keller, R. Zweidler, and H. Hausermann (Geigy), U.S. Pat. 2,784,183, March 5, 1957.
63. B. Kopelman (Sylvania Electric Products), U.S. Pat. 3,514,660, May 26, 1970.
64. F. H. Kranz (Allied Chemical), U.S. Pat. 2,981,738, April 25, 1961.
65. M. Kreuder et al. (Agfa-Gavaert), U.S. Pat. 3,687,671, Aug, 29, 1972.
66. U. L. Laridon, A. J. Conix, and G. A. Delzenne (Gevaert), U.S. Pat. 3,278,305, Oct. 11, 1966.
67. M. Lelental and H. J. Gysling, *J. Photogr. Sci.,* **1978,** *26,* 135–143.
68. W. Lwowski and T. W. Mattingly, Jr., *J. Am. Chem. Soc.,* **1965,** *87,* 1947–1958.
69. O. Matthies and B. Wendt (Agfa Ansco), U.S. Pat. 1,758,577, May 13, 1930.
70. Mobil Oil Corp., Brit. Pat. 1,387,134, March 12, 1975.

71. R. F. Monroe and A. J. Maciejewski (Dow), U.S. Pat. 3,291,741, Dec. 13, 1966.

72. L. D. Moore and J. J. Randall (Borg-Warner), U.S. Pat. 3,442,829, May 6, 1969

73. W. Neugebauer and M. K. Reichel (Keuffel & Esser), U.S. Pat. 2,656,271, Oct. 20, 1953.

74. O. Neuner, A. Dorlars, and O. Berendes (Farbenfabriken Bayer), U.S. Pat. 3,646,052, Feb. 29, 1972.

75. A. W. Oberhofer, J. J. Benko, and J. C. Drozd (Nalco), U.S. Pat. 3,959,166, May 25, 1976.

76. K. Ohkubo, J. Noguchi, and K. Ohmura (Fugi Photo Film), Ger. Pat. 2,041,019, March 4, 1971.

77. A. O. M. Okorodudu (Mobil Oil), U.S. Pat. 3,986,967, Oct. 19, 1976.

78. L. B. Palmer and R. P. Conger (Congoleum Industries), U.S. Pat. 3,849,158, Nov. 19, 1974.

79. H. v. Pechmann and P. Runge, *Ber.*, **1894,** *27,* 2920–2930.

80. M. Peters and A. v. Koenig (Agfa-Gavaert), Ger. Pat. 2,529,488, Jan. 27, 1977.

81. A. Popplewell and D. R. Clark (Ciba-Geigy), Ger. Pat. 2,601,719, July 29, 1976.

82. J. Pouradier, A. Pailliotet, and C. R. Berry, "Properties of Silver Halides I. Structure and Thermodynamic Properties," in T. H. James, Ed., *The Theory of the Photographic Process,* 4th ed., Macmillan, New York, **1977,** pp. 1–12.

83. D. R. Randall, E. A. Cox, and A. Atkinson (Geigy), U.S. Pat. 3,597,353, Aug. 3, 1971.

84. D. R. Randall and D. K. Howard (Geigy), U.S. Pat. 3,531,414, Sept. 29, 1970.

85. F. J. Rauner (Eastman Kodak), U.S. Pat. 3,475,176, Oct. 28, 1969.

86. F. J. Rauner, M. P. Cunningham, and R. C. Van Hanehem (Eastman Kodak), U.S. Pat. 4,139,390, Feb. 13, 1979.

87. J. J. A. Robillard (U.S. Phillips), U.S. Pat. 3,515,554, June 2, 1970.

88. E. C. Rothstein, *Am. Chem. Soc., Div. Org. Coatings, Plast. Chem., Prepr.,* **1967,** *27,* 406–419.

89. J. J. Sagura and J. A. Van Allan (Eastman Kodak), U.S. Pat. 2,940,853, June 14, 1960.

90. J. W. Schick and R. H. Davis (Mobil Oil), U.S. Pat. 4,107,060, Aug. 15, 1978.

91. R. M. Shemenski, Sr. (Goodyear Tire & Rubber), U.S. Pat. 4,269,645, May 26, 1981; U.S. Pat. 4,269,877, May 26, 1981; U.S. Pat. 4,283,460, Aug. 11, 1981.

92. J. W. Silva and E. A. Staba (Olin Mathieson), U.S. Pat. 3,463,086, Aug. 26, 1969.

93. J. G. Smith and T. I. Abbott (Eastman Kodak), U.S. Pat. 3,350,204, Oct. 31, 1967.

94. E. A. Staba (Olin Mathieson), U.S. Pat. 3,310,569, Mar. 21, 1967.

95. R. E. Stauffer (Eastman Kodak), U.S. Pat. 2,497,917, Feb. 21, 1950.

96. J. G. Strem, J. S. Juskey, and E. P. Moschovis (APECO), U.S. Pat. 3,764,323, Oct. 9, 1973.

97. M. Tamura et al., *J. Phys. Chem.,* **1962,** *66,* 559–560.

98. T. Tanaka, K. Tanikawa, and K. Arakawa (Nippon Steel), U.S. Pat. 3,895,170, July 15, 1975.

99. T. Tani (Fuji Photo Film), U.S. Pat. 4,030,927, June 21, 1977.

100. J. B. Thompson (Hercules), U.S. Pat. 3,697,551, Oct. 10, 1972.

101. J. B. Thompson (Hercules), U.S. Pat. 3,706,592, Dec. 19, 1972.

102. Toxic Substances Control Act, Chemical Substance Inventory, U.S. Environmental Protection Agency, Initial Inventory, May, 1979, Cumulative Supplement, June 1980.

103. The Society of Dyers and Colourists, *Colour Index,* 3rd ed., Vol. 4, 1971, p. 4136.

104. TSCA-Plant and Production Data Base, NIH/EPA Chemical Information System, Fein-Marquart Associates, Inc., Data Base Update Aug. 15, 1981.

105. T. Tsunoda, Y. Yamaoka, and G. Nagamatsu, *Photogr. Sci. Eng.,* **1973,** *17,* 390–393.

106. T. Urbanski and S. K. Vasudeva, *J. Sci. Ind. Res.,* **1978,** *37,* 250–255.

107. U.S. Army Material Command, *Engineering Design Handbook,* Explosives Series, Properties of Explosives of Military Interest, AMCP 706-177, January, 1971.

108. A. E. Vassiliades (Champion Intl.), U.S. Pat. 3,993,831, Nov. 23, 1976.

109. K. Venkataraman, *The Chemistry of Synthetic Dyes*, Vol. 1, Academic Press, New York, **1952,** pp. 228–30, 691–693.

110. W. Voreck, N. Slagg, and L. Avrami, "The Role of Azides in Explosive Trains," in H. D. Fair and R. F. Walker, Eds., *Energetic Materials*, Vol. 2, Plenum Press, New York, 1977, pp. 249–291.

111. K. H. Wall and I. Davies, *J. Appl. Chem.*, **1965,** *15*, 389–392.

112. W. E. Wallace, U.S. Pat. 3,578,653, May 11, 1971.

113. R. B. Wilson and J. V. Williams (Eastman Kodak), U.S. Pat. 3,503,741, March 31, 1970.

114. J. Winchester, *Farbe Lack*, **1967,** *73*, 424–426.

115. H. W. Wood, *J. Photogr. Sci.*, **1966,** *14*, 72–78.

116. M. J. Youngquist (Eastman Kodak), U.S. Pat. 3,700,457, Oct. 24, 1972.

117. R. Zweidler and H. Hefti, *Encyclopedia of Chemical Technology*, 3rd. ed., Vol. 4, Wiley, New York, pp. 213–226.

118. R. Zweidler and E. Keller (Geigy), U.S. Pat. 2,713,057, July 12, 1955.

5

BIOLOGICAL PROPERTIES

In comparison to those high nitrogen compounds used for nonbiological purposes, relatively few of these substances thus far have attained commercial status as agricultural chemicals or pharmaceuticals in the United States. Only agricultural agents and drugs sold in 1982 are discussed below; candidates and obsolete products are omitted. A large number of compounds with three or more linked nitrogen atoms, however, affect living systems, from single cells to whole organisms. Studies of biological activity have involved principally the triazenes, azides, *v*-triazoles, and tetrazoles; many other high nitrogen classes remain uninvestigated. The biological property of toxicity is considered both because of its importance as a characteristic of all chemicals, and because of its relevance to both nonbiological and biological applications.

5.1. AGRICULTURAL USES

The benzotriazinone insecticide, azinphos-methyl, applied to fruits, vegetables, nuts, alfalfa, clover, wheat, cotton, and ornamental, forest, and shade trees (20,

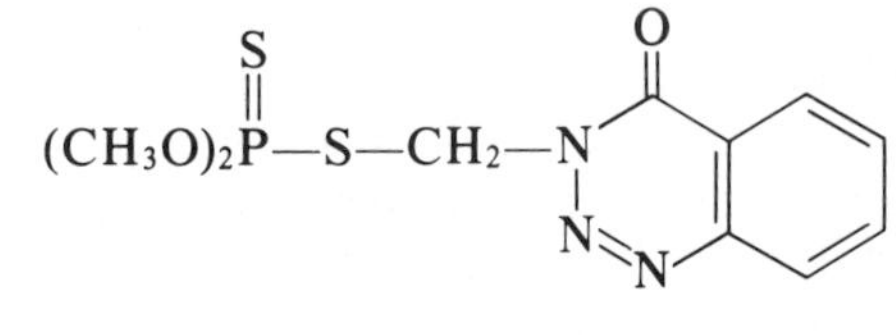

azinphos-methyl

21, 24, 76, 108), is a persistent contact and stomach poison against aphids, armyworms, boll weevils, grasshoppers, scales, and other insects (102, 105).

Tetrazolium salts are used to determine seed variability (32, 44) and to detect living microorganisms in soil (112) and activated sludge (35) through enzymatic reduction to colored formazans.

5.2. PHARMACEUTICAL USES

The triazene 5-(3,3-dimethyl-1-triazeno)imidazole-4-carboxamide **(I)**, dacarbazine, also called DTIC, is used in the treatment of malignant melanoma, and, in combination with other antineoplastic agents, of soft tissue sarcomas and

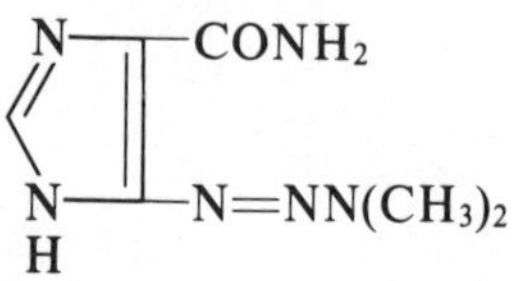

(I)

dacarbazine

Hodgkins' disease (10, 71). Overall response in malignant melanoma is about 20% (114). Dacarbazine functions as an alkylating agent, inhibiting the synthesis of RNA and protein, probably through the action of its metabolite, 5-(3-methyl-1-triazeno)imidazole-4-carboxamide **(III)** (10, 36), resulting from the initially formed 3-hydroxymethyl derivative **(II)** (39). Metabolism of triazenes is discussed further in Section 5.3, Toxicity.

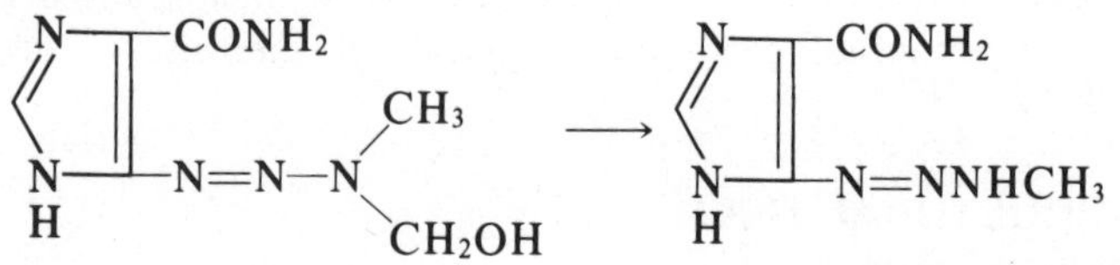

(II) → **(III)**

The modified cephalosporins, cefazolin **(IV)** and cefamandole **(V)**, and the related 1-oxacephalosporin, moxalactam **(VI)**, each with tetrazole moieties, are broad-spectrum antibiotics, active against many gram-positive and gram-negative organisms (23, 29, 88, 116). In their bactericidal action the enzyme transpeptidase is inactivated, preventing cross-linking of peptidoglycan chains, thus interfering with synthesis of the bacterial cell wall (57). They are active against both penicillin-resistant and penicillin-sensitive strains of *Staphlococcus aureus,* beta-hemolytic streptococci, *S. pneumoniae, E. coli, Klebsiella* species, *Haemophilus influenzae,* and other bacteria (29, 30, 34, 91) and are used against infections of the respiratory tract, genitourinary tract, skin, bone, and joint, and in septicemia (69, 70).

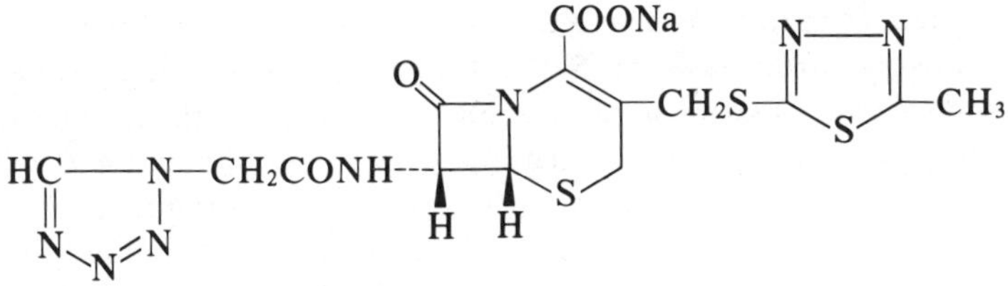

(IV)

cefazolin sodium

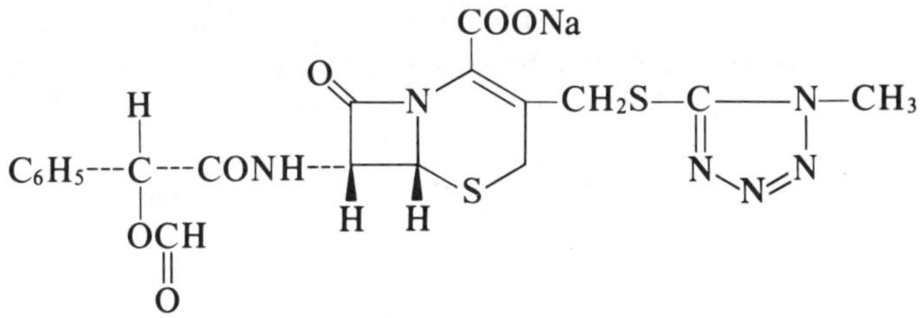

(V)

cefamandole nafate

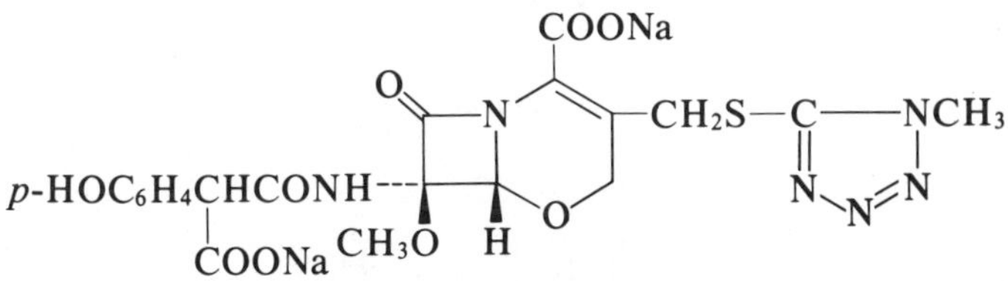

(VI)

moxalactam disodium

Pentylenetetrazole **(VII)** is a central nervous system stimulant administered to enhance mental and physical activity of elderly patients (68). It stimulates all

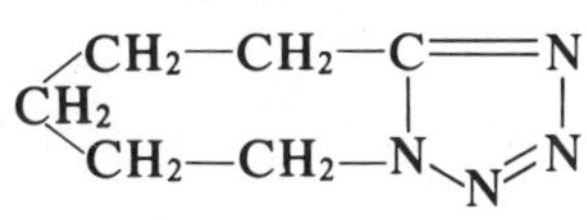

(VII)

pentylenetetrazole

levels of the spinal cord and brain (26). Pentylenetetrazole may act by changing permeability of the central nervous system to potassium ion, increasing extraneuronal potassium and thereby partially depolarizing neuronal membranes and increasing excitability.

Reduction of tetrazolium salts to formazans is used to assay oxidases such as serum lactate dehydrogenase (11, 59) and succinic dehydrogenase (60), usually in the presence of an intermediate electron carrier such as phenazine methosulfate, and for histochemical detection and localization of oxidative enzymes (2). In clinical bacteriological laboratory operation, measurement of bacterial growth by tetrazolium reduction is involved in rapid determination of minimum inhibitory concentrations of antibiotics (4, 107).

Dehydrogenases are localized in cells by deposits of formazans; the red formazan resulting from triphenyltetrazolium chloride is difficult to distinguish from the natural color of mammalian tissue. Some other tetrazolium salts with readily observed color and which reduce more easily than the triphenyltetrazolium salts are shown, identified by common initials and trivial name, if any, together with colors, as seen in sections of rat liver (2).

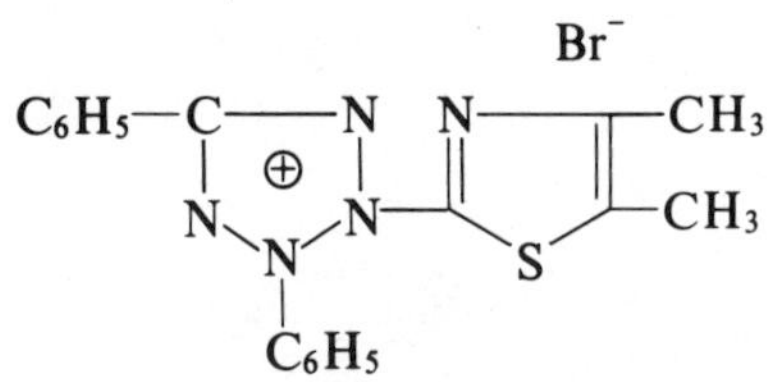

MTT
purple, blue, and gray

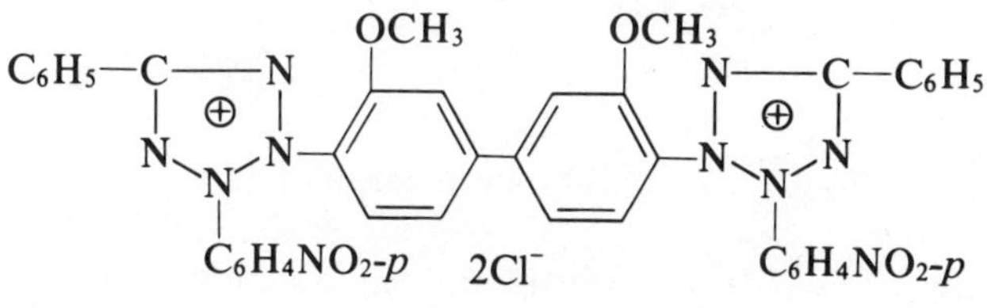

NBT nitroblue tetrazolium
deep blue-red

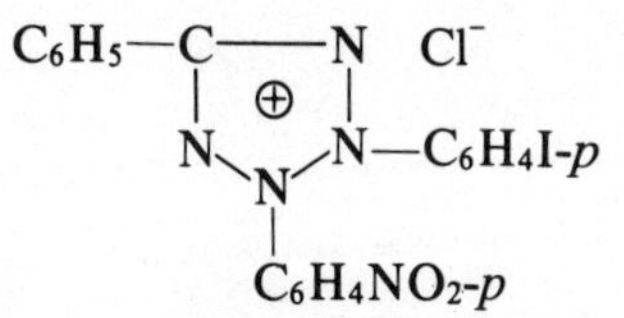

INT
orange-red

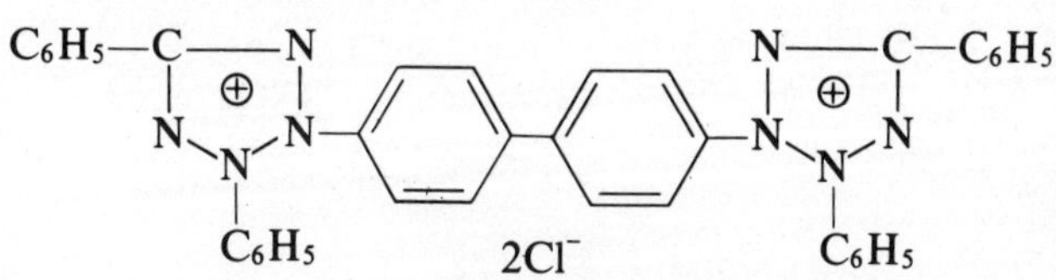

NT neotetrazolium
purple-red

5.3. TOXICITY

Reported toxicity values of the triazenes extend from extremely toxic to practically nontoxic,* illustrated by the quantities lethal to 50% of the mice tested by the oral route for triazenes **I** and **II**.

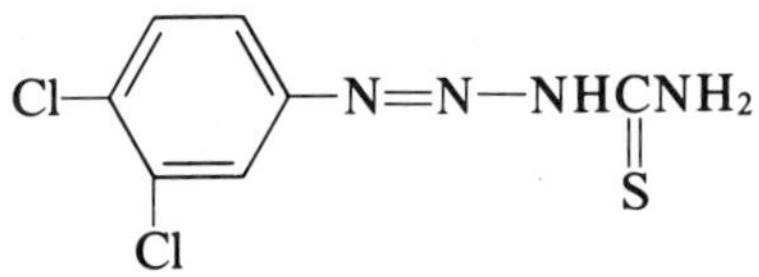

(I)

LD_{50} (female) 0.28 mg/kg
(13)

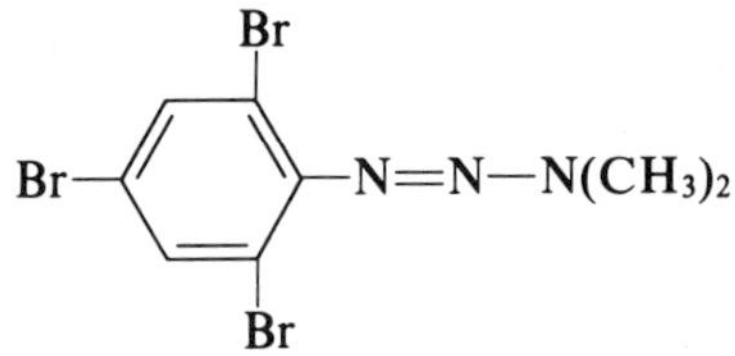

(II)

$LD_{50} > 5000$ mg/kg
(38)

$$p\text{-}XC_6H_4N{=}N\underset{}{\overset{CH_3}{\overset{|}{N}}}{-}R$$

(III)

The concentration (C) corresponding to an LD_{50} in a series of 3-alkyl-3-methyl 1-*p*-substituted phenyl triazenes **(III)** is related to the electronic character of the group on phenyl (σ^+), and to a lesser extent to the lipophilic character (P) of the series members, by the equation shown with a coefficient of correlation (r) of 0.913 and a standard deviation (s) of 0.110 (31). Administration of 3,3-di-

$$\log \frac{1}{C} = -0.0241\ (\log P)^2 - 0.264\ \sigma^+ + 3.490$$

methyl-1-phenyl triazene to mice, either intraperitoneally or orally, impairs righting reflexes and causes ataxia, dyspnea, and depression; in rats, pathological changes of the kidney, spleen, and bone marrow result (93). Acute toxicities and carcinogenic characteristics of some triazenes are tabulated (Table 5.1).

The 1-aryl-3,3-dialkyltriazenes induce tumors at locations remote from the injection site, causing malignancies in the brain, peripheral nervous system, and kidneys (17, 73, 74). Two pathways lead to carcinogenesis by the aryldimethyl triazenes: local activity results from hydrolysis to aryl diazonium ion, and systemic activity arises by hydroxylation of and loss of one *N*-methyl, followed by liberation of an aromatic amine and methyl diazonium ion (17, 38, 73). In these

* Qualitative designations of toxicity are given using the terminology of Hodge and Sterner (33).

Table 5.1. Acute Toxicity and Carcinogenicity of Triazenes

RN=N—NR′R″

R	R′	R″	Acute Toxicity, Rat,[a,b] LD_{50} (mg/kg) (Ref.)	Carcinogenicity: Species; Route[b]; Dose (mg/kg); Frequency, Duration of Administration (Ref.)
C_6H_5	H	CH_3	420 (52)	9/9 mice dead from tumors; sc; 45; 1/week, 4 weeks (92)
C_6H_5	H	C_6H_5	224 (84)	3 carcinomas, 3 papillomas/20 mice; equivocal tumorigenic agent, no controls; dermal; 5% solution; 3/week, 545 days (37)
C_6H_5	CH_3	CH_3	310 (73) sc 310	11/12 rats dead from tumors; oral; 50; 1/week, 330 days (73)
p-$CH_3OC_6H_4$	CH_3	CH_3	375 (73) sc 450	19/21 rats dead from tumors; sc; 50; 1/week, 316 days (73)
p-HOC_6H_4	CH_3	CH_3	25 (38) sc 7	
p-ClC_6H_4	CH_3	CH_3	sc 330 (73)	5/13 rats dead from tumors; sc; 50; 1/week, 490 days (73)

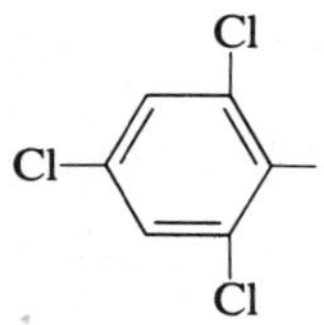	CH_3	CH_3	1500 (38)	
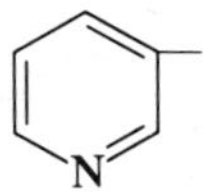	CH_3	CH_3	160 (73) sc 130	8/8 rats dead from tumors; sc; 26; 1/week, 265 days (73)
	$C_6H_5NHN{=}\overset{+}{N}(O^-)CH_3$		sc 550 (73)	2/2 rats dead from tumors; sc; 400–470, single dose, 510–608 days (73)

[a] Acute toxicities are by the oral route unless otherwise indicated.
[b] Intraperitoneal ip; intravenous iv; subcutaneous sc.

mechanisms, the aryl and methyl cations are the active carcinogenic agents. 1-Aryl-3,3-dialkyltriazenes with electron-releasing substituents on aryl have half-lives of less than 200 min at pH 7 and 37°C and predominantly produce aryl diazonium ions (40); triazenes with electron-withdrawing groups on aryl have

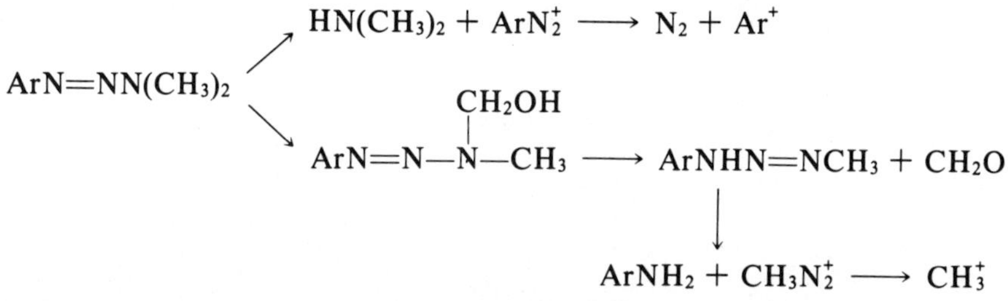

longer half-lives and yield monoalkyltriazenes which are alkylating agents, principally producing systemic effects. Intraperitoneal injection in rats of 3,3-dimethyl-1-phenyltriazene labeled with ^{13}C in one methyl group results in the formation of labeled 7-methylguanine, isolated from liver, brain, and kidney RNA and liver DNA, showing that the dimethyltriazene is metabolized in vivo to form an alkylating agent (43). Similarly, 3-methyl-1-phenyltriazene, administered subcutaneously to normal or pregnant rats produces 7-methylguanine in RNA and DNA as well as minor amounts of 3-methylcytosine (49).

Many aryl dialkyltriazenes are both antineoplastic and tumorigenic (25, 38). Thus the antitumor agent 5-(3,3-dimethyl-1-triazeno)imidazole-4-carboxamide

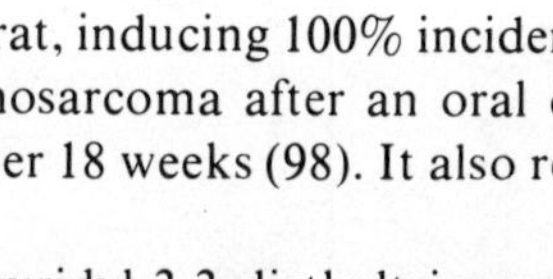

(IV)

(IV) is carcinogenic in the rat, inducing 100% incidence of mammary adenocarcinoma and thymic lymphosarcoma after an oral dose of 7.3 mg/day for 14 weeks, with 21/24 dead after 18 weeks (98). It also reduces primary immune responses in mice (75).

Both 1-phenyl- and 1-pyridyl-3,3-diethyltriazene given to pregnant rats at the 15th day of gestation elicit tumors in the nervous system of their offspring; 1-phenyl-3,3-dimethyltriazene, however, does not induce malignant tumors in the offspring until the 23rd (last) day of gestation, indicating that different enzymes, available at different times during ontogenesis, effect oxidative dealkylation of the ethyl and methyl derivatives (17).

Teratogenic effects are produced in the skeleton of chick embryos by 3,3-dimethyl-1-phenyltriazene and a series of related aryl dialkyltriazenes except for 3,3-dimethyl-1-*p*-nitrophenyltriazene (15). Both 3,3-dimethyl-1-phenyltriazene and the corresponding monomethyl compound, given as single subcutaneous injections to pregnant rats on the 15th day of gestation cause cleft palates in the offspring (18).

Mutagenic effects in *Salmonella typhimurium* (G46 strain), as measured by increase in the number of revertant colonies, appear on treatment with 3,3-dimethyl-1-phenyltriazene, or the corresponding 1-*p*-chlorophenyl, *p*-bromophenyl, 2,4,6-trichlorophenyl, 2,4,6-tribromophenyl, 3-nitro, and 3-carboxyphenyl derivatives when activated by microsomal fractions from the livers of phenobarbitone-treated mice (48). The microsomal fraction apparently furnishes enzymes which mediate the oxidation of an alkyl group, leading to its removal. No mutagenic or toxic effects were induced with the 1-(*o*- or *p*-carboxyphenyl) or the *p*-methyl, *p*-methoxy-, or *p*-acetamidophenyl-3,3-dimethyltriazenes. In metaphase during cell divisions in Chinese hamster cells treated with 3,3-dimethyl-1-phenyltriazene and particularly 3,3-dimethyl-1-(3-pyridyl)triazene, mutagenicity is exhibited by sister chromatid exchange, which involves transfer of chromosome material between DNA helices (6, 97).

With a different *S. typhimurium* strain (TA 92), also activated by rat liver microsomes, a quantitative relation in mutagenic potency was found with 1-substituted phenyl 3,3-dialkyltriazenes (109). As in the toxicity relationship

$$\log \frac{1}{C} = 1.09 \log P - 1.63\, \sigma^{+} + 5.58$$

$$r = 0.974 \qquad s = 0.315$$

with some of the same triazenes, increasing electron-releasing character of the aryl group enhances mutagenicity.

Trialkyltriazenes (96) such as 3-benzyl-1,3-dimethyltriazene **(V)** are mutagenic in *S. typhimurium* without activation, as also is the arylmonoalkyltriazene, 1-methyl-3-*p*-tolyltriazene, and the cyclic triazene, 1-ethyl-Δ^2-triazoline **(VI)** (101). Each of these compounds should readily form carbonium ions, which would be expected to alkylate nucleic acids.

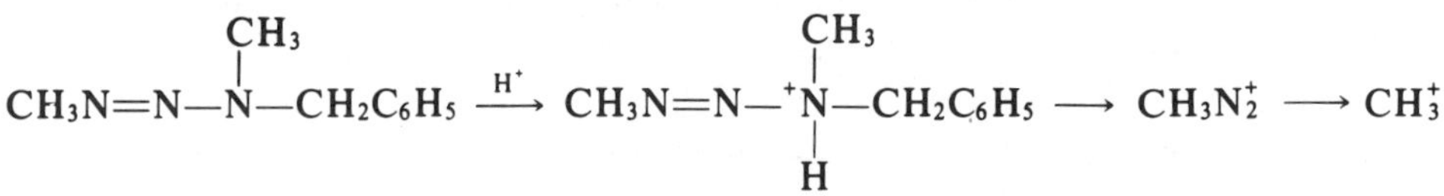

(V)

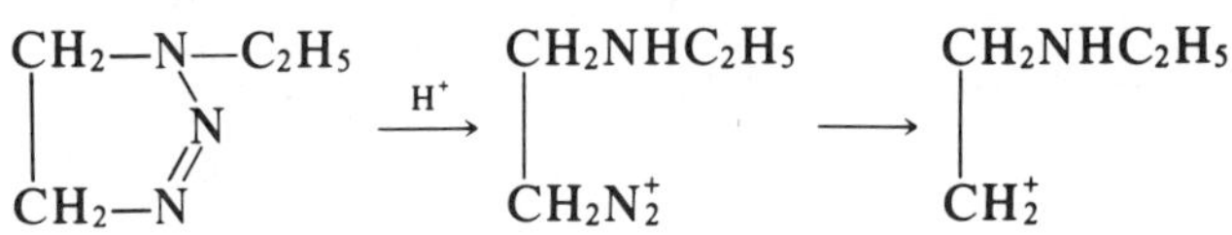

(VI)

Tetramethyltetrazene **(VII)** is much less toxic than most triazenes, and 1,5-diphenyl-3-methylpentazdiene **(VIII)** is moderately toxic. In both its acute toxicity and carcinogenicity, *N*-nitrosotrimethylhydrazine **(IX)** is similar to nitrosodimethylamine (19, 113).

$$(CH_3)_2N{-}N{=}N{-}N(CH_3)_2$$

(VII)

LD$_{50}$ (oral, rat) 1.78 ml/kg
(approximately 1500 mg/kg)
(99)

$$C_6H_5N{=}N{-}N(CH_3){-}N{=}NC_6H_5$$

(VIII)

LD$_{50}$ (oral, rat) 212 mg/kg
(50, 79)

$$(CH_3)_2N{-}N(CH_3){-}N{=}O$$

(IX)

LD$_{50}$ (oral, rat) 95 mg/kg
(19)

Hydrazoic acid and the azide salts are highly toxic, inhibiting heme iron enzymes, especially cytochrome oxidase, catalase, and peroxidase, and uncoupling oxidative phosphorylation (7, 63, 117). A concentration of 1100 ppm of hydrogen azide in air is lethal to rats in 1 hr (78), and headache, reduced blood pressure, and acceleration of heart rate were observed in humans exposed to 0.3–3.9 ppm (28). Rats administered 3 mg/kg of sodium azide intraperitoneally four times a day for 4 days became akinetic and exhibited tremor of the forelimbs; histological examination revealed lesions in the nervous system, engorgement of capillaries in the corpus striatum, some demyelination of the optic tract, and edema of the lung (56). Intravenous injection of 12 mg/kg of sodium azide in rhesus monkeys causes a rapid drop in blood pressure, followed by convulsions, hypokinesis, unconsciousness, and death; extensive neural damage is found, including almost complete removal of the cerebellar lining (54, 55). Nonfatal nervous system effects have been observed after ingestion by humans of 5–10 mg and 60 mg of sodium azide (89). Sodium azide induced mutations in *S. typhimurium* (22), but was noncarcinogenic (104) in rats after administration for 18 months.

HN_3	NaN_3	KN_3
LD$_{50}$ (ip, mouse) 22 mg/kg (28)	LD$_{50}$ (oral, rat) 27 mg/kg (65); (ip, mouse) 28 mg/kg (28)	LD$_{50}$ (oral, rat) 60–80 mg/kg (65)

Acyl azides, such as *t*-butoxycarbonyl azide **(X)**, which hydrolyze relatively easily, approach ionic azides in toxicity. Ingestion of cyanogen azide in acetonitrile solution by rats produces cyanosis and death at a dose of less than 72 mg/kg (51). Most other organic azides, however, are less toxic. 2,4,6-Tribromophenylazide is the only member of a series of substituted phenyl azides to cause mutagenic changes in *S. typhimurium,* but is not carcinogenic (5).

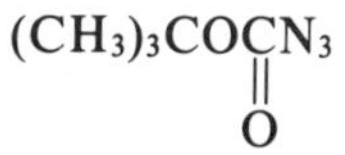

(X)

LD$_{50}$ (oral, rat) 57 mg/kg (42)

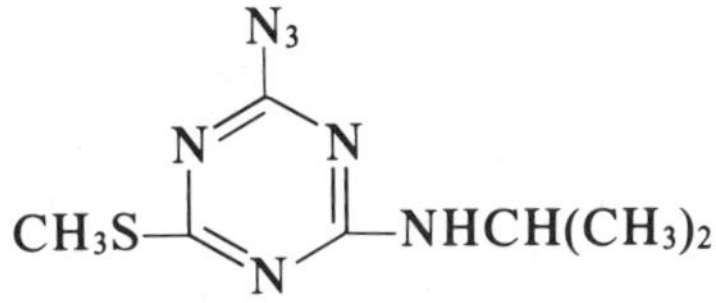

LD$_{50}$ (oral, rat) 3600–5800 mg/kg (21)
(dermal, rat) >3000 mg/kg

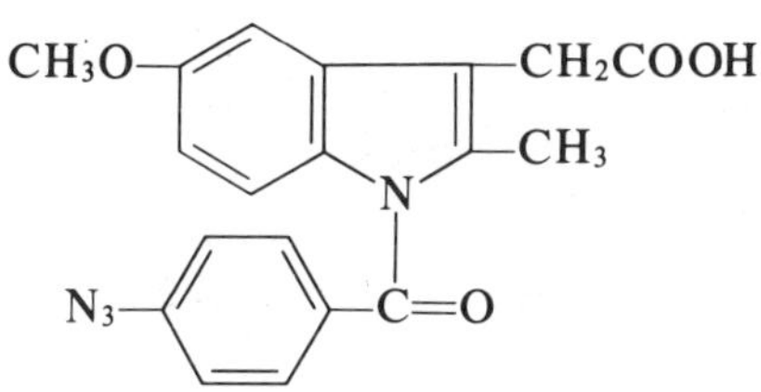

LD$_{50}$ (oral, mouse) 250 mg/kg (103)

Acute toxicities of the monocyclic and fused ring *v*-triazoles range from moderately toxic to relatively harmless. 1-(4-Aminophenyl)-4-diethylcarboxamido-5-methyl-*v*-triazole hydrochloride **(XI)** induces loss of righting reflex in rats at a median oral dose of 30 mg/kg (3). The *v*-triazole fluorescent whiteners, **XII, XIII,** and **XVI**, cause dermal toxicity at doses from 500 to 2000 mg/kg, the maximum technically capable of application. No skin irritation results from **XII** and **XIII**, and application of **XVI** causes slight irritation. Eye irritation was moderate in the case of **XII** and slight in that of **XIII**, and none was caused by **XVI** (100). Righting reflex in mice is lost on intravenous administration of 55 mg/kg benzotriazole **(XIV)** and 50 mg/kg 1-ethylbenzotriazole **(XV)** (16). No teratogenic effects or toxicity to the mothers in rats or rabbits are caused by **XII** (46). The ED$_{50}$ of **XVII** (ip) causing impairment of motor activity in mice is 168.4 mg/kg (1).

Mice and rats, respectively receiving up to 2.35 and 1.21% benzotriazole in their feed for 104 and 78 weeks, had lower mean body weights than controls, indicating some chronic toxicity (62). Survival, however, did not decrease and was over 60% in both dosed and control groups. Some increased incidence of alveolar/bronchiolar carcinomas in mice and brain tumors in rats occurred in the treated animals, but there was no clear evidence demonstrating carcino-

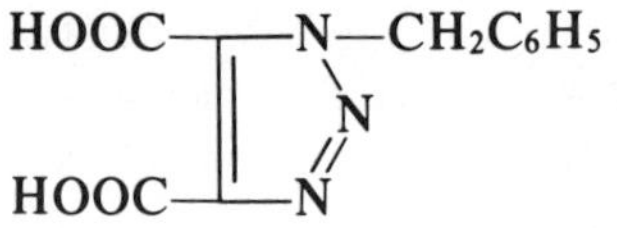

LD_{50} (iv, mouse)
180 mg/kg (85)

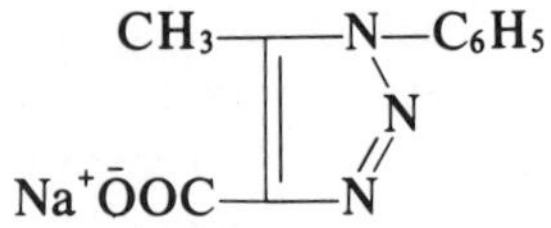

LD_{50} (oral, rat)
approx. 1500 mg/kg (14)

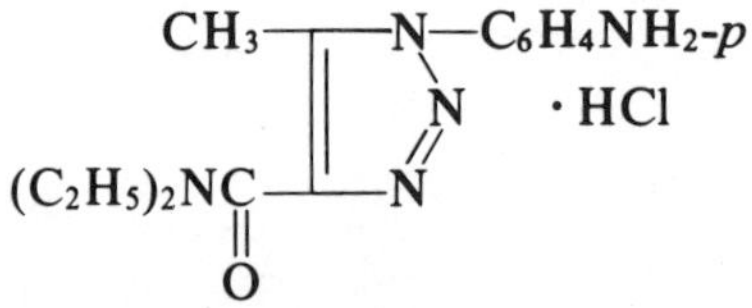

(XI)

LD_{50} (oral, rat) 494 mg/kg (3)

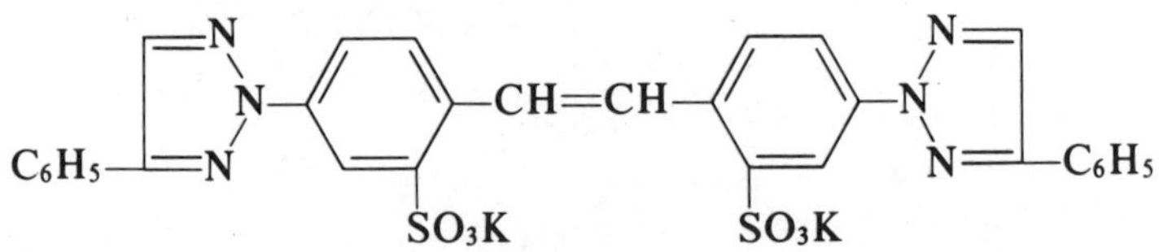

(XII)

LD_{50} (oral, rat) >15,000 mg/kg (100)

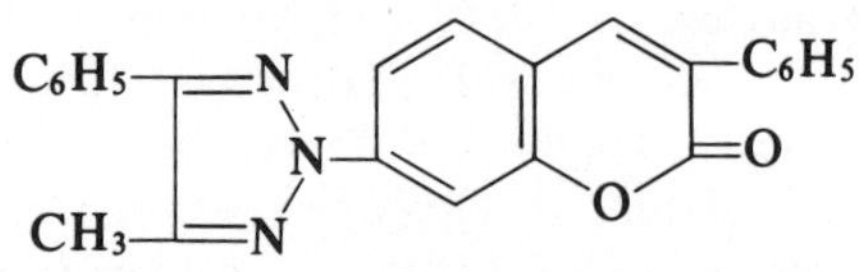

(XIII)

LD_{50} (oral, rat) >2500 mg/kg (100)

genicity in these species. It was concluded that benzotriazole probably poses no carcinogenic risk to man.

Acute toxicities of the 8-azapurines indicate them to be moderately toxic. 8-Azaguanine **(XVIII)** is more toxic to young than adult animals. Studies with ^{14}C-labeled material show rapid excretion, largely as its metabolite, 8-azaxanthine. The retained 8-azaguanine is incorporated into tissue RNA, substituting for guanine, with little or no incorporation in DNA (66). Teratogenic effects in the skeletal system of mice are caused by 8-azaguanine **(XVIII)** (64).

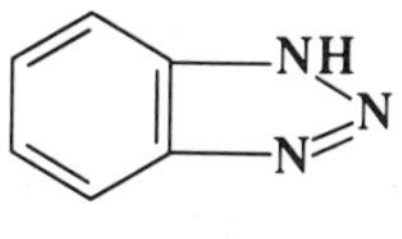

(XIV)

LD_{50} (iv, mouse) 238 mg/kg (16)

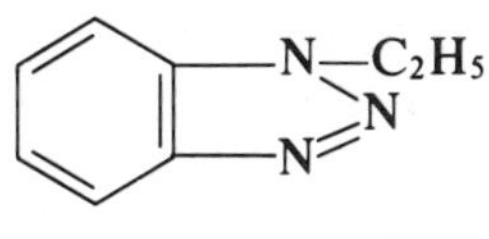

(XV)

LD_{50} (iv, mouse) 125 mg/kg (16)

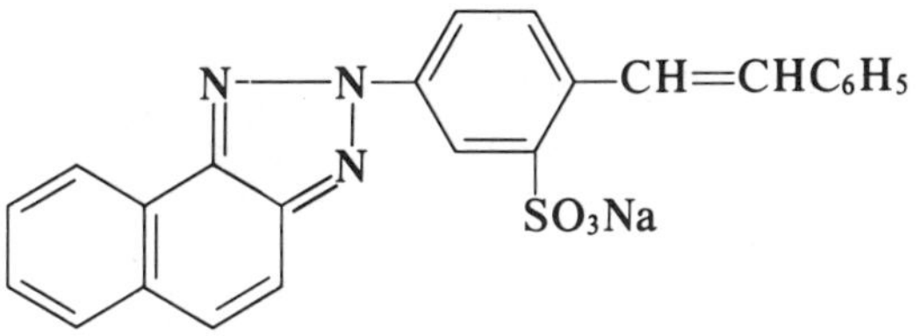

(XVI)

LD_{50} (oral, rat) >10,250 mg/kg (100)

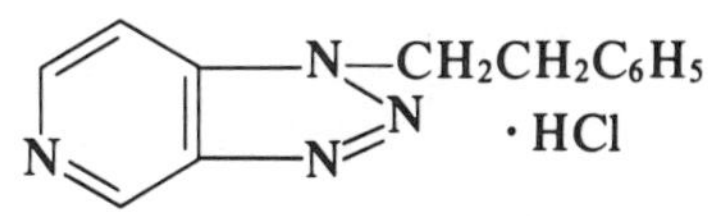

(XVII)

LD_{50} (oral, mouse) 850 mg/kg
(ip, mouse) 414 mg/kg (1)

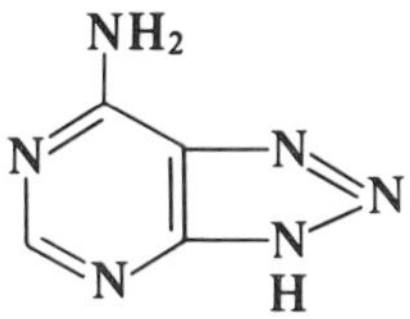

LD_{50} (ip, mouse)
315 mg/kg (86)

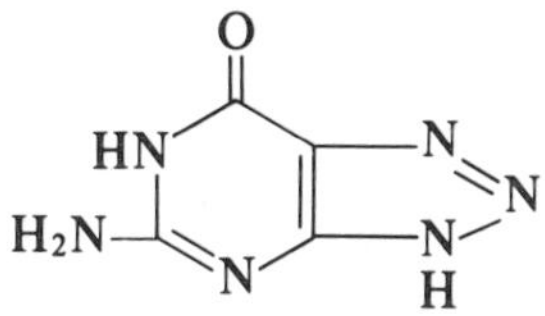

(XVIII)

LD_{50} (ip, mouse)
120 mg/kg (66)

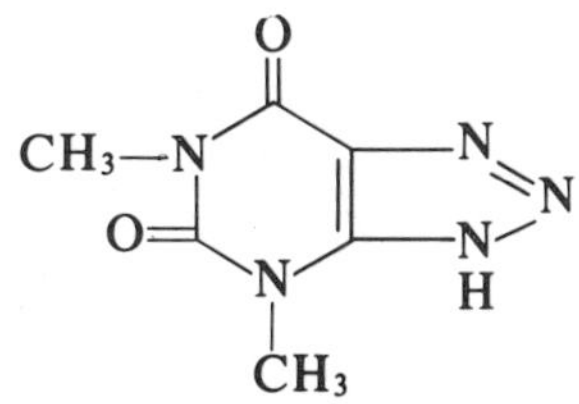

LD_{50} (ip, mouse)
515 mg/kg (87)

Introduction of a nitrogen atom at position 8 into an oncogenic 3-hydroxypurine reduces but does not eliminate its tumorigenic activity. Thus subcutaneous administration of 3-hydroxy-8-azaxanthine **(XIX)** induces sarcomas in 50% of the rats after 14 months, but is much less potent than 3-hydroxyanthine **(XX)**,

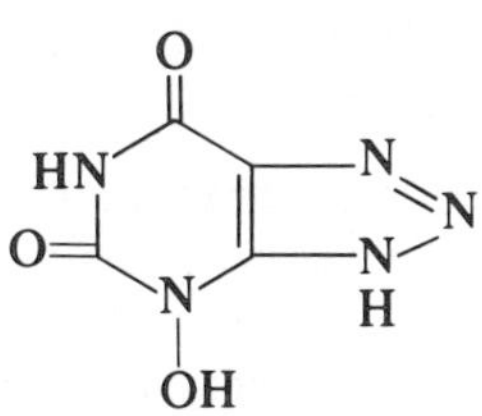

(XIX)

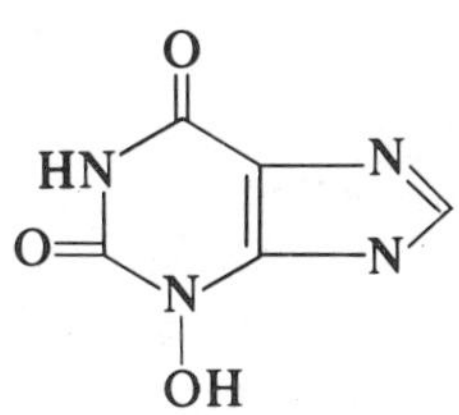

(XX)

which causes 90% incidence of tumors in 7 months (9). 3-Acetoxy-8-azaxanthine is a very weak mutagen in *Bacillus subtilis* (53).

Single, intraperitoneal doses of 0.5–2.0 mg of 2-azadenine **(XXI)** are lethal to mice, but are counteracted by simultaneously administered adenine and, less effectively, by hypoxanthine (115). Benzotriazinone **(XXII)** and 7-methyl-7*H*-pyrazolo(3,4-*d*)-*v*-triazin-4-ol **(XXIII)** are highly and moderately toxic, respectively.

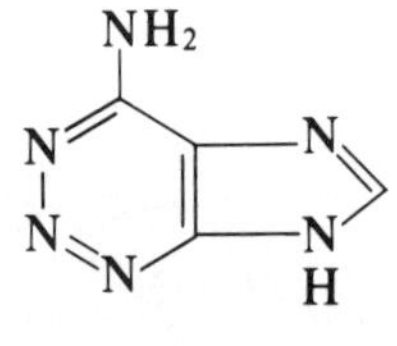

(XXI)

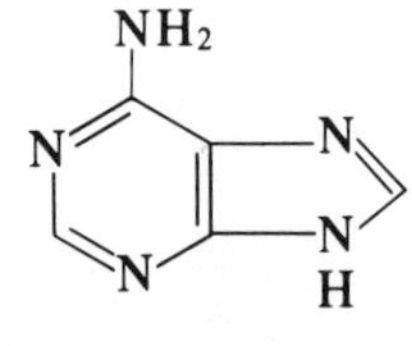

adenine

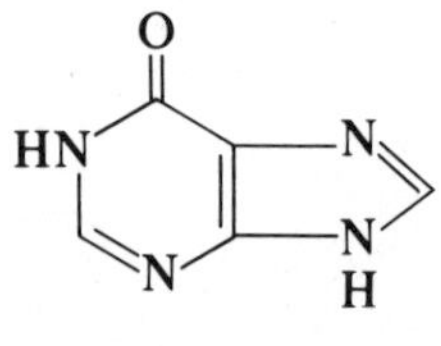

hypoxanthine

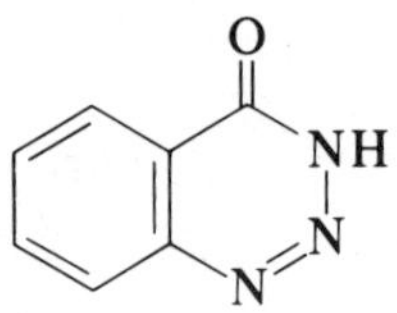

(XXII)

LD_{50} (ip, mouse)
50 mg/kg (77)

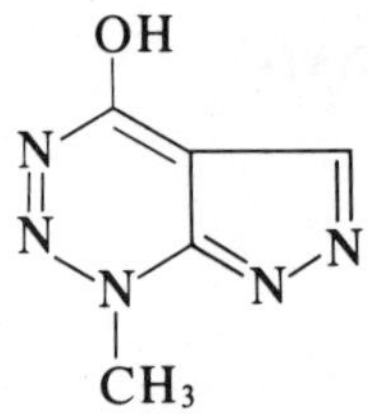

(XXIII)

LD_{50} (iv, mouse)
180 mg/kg (80)

The acute toxicity of azinphos-methyl, containing a dithiophosphorate moiety in addition to the benzotriazinone ring, is categorized as highly toxic. Its no-effect level in the rat and dog is 0.125 mg/kg/day. A maximum acceptable daily intake for man of 0.0025 mg/kg was recommended by the Joint FAO-WHO

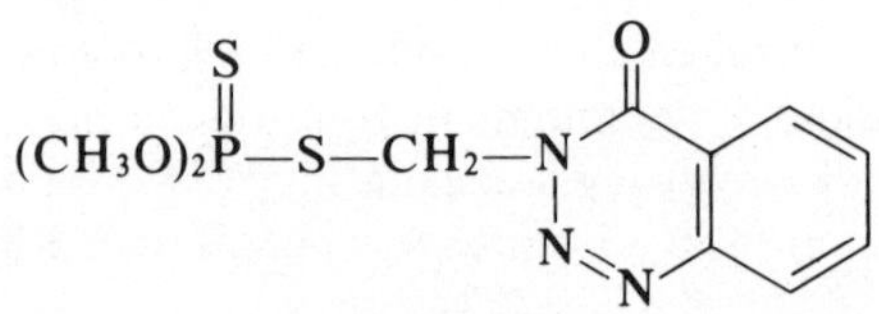

LD_{50} (oral, rat)
25 mg/kg (67)

Meeting on Pesticide Residues (111). Cholinesterase levels in a group of 21 agricultural field workers exposed to foliage-borne azinphos-methyl residues decreased 15% during 5 normal work days, but no clinical signs of organophosphate intoxication were evident (41). Azinphos-methyl does not appear to be teratogenic to mice or rats (94), but a slight decrease in length and lowered cholinesterase activity is found in chick embryos from treated eggs (106). Although increased numbers of tumors of the pancreatic islets and follicular cells of the thyroid occurred in male rats receiving 156 ppm azinphos-methyl in the diet for 80 weeks, it was concluded that these incidences cannot be clearly implicated as a chemically induced effect. The insecticide is not carcinogenic in female rats or mice of either sex (61). Azinphos-methyl has no mutagenic effect in *Salmonella typhimurium* or in *Streptomyces coelicolor* (12).

Acute toxicities of the tetrazoles investigated range from highly toxic to practically nontoxic. Pentylenetetrazole **(XXIV)** is a powerful convulsant. At

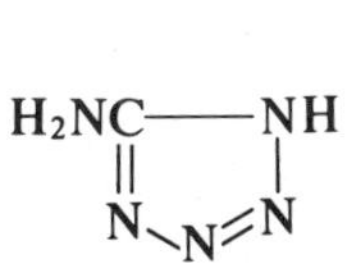

LD_{50} (ip, mouse) 2500 mg/kg (72)

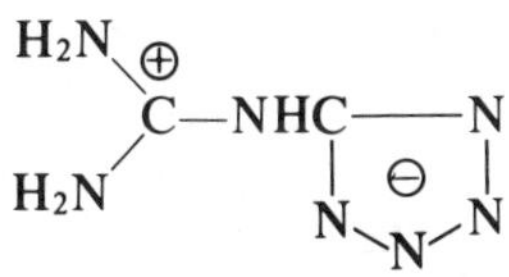

LD_{50} (ip, mouse) 180 mg/kg (81)

LD_{50} (ip, mouse) 178 mg/kg (58)

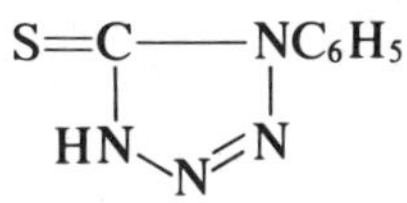

LD_{50} (ip, mouse) 250 mg/kg (82)

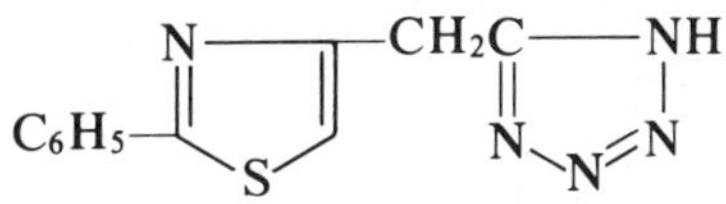

LD_{50} (oral, mouse) 740 mg/kg (47)

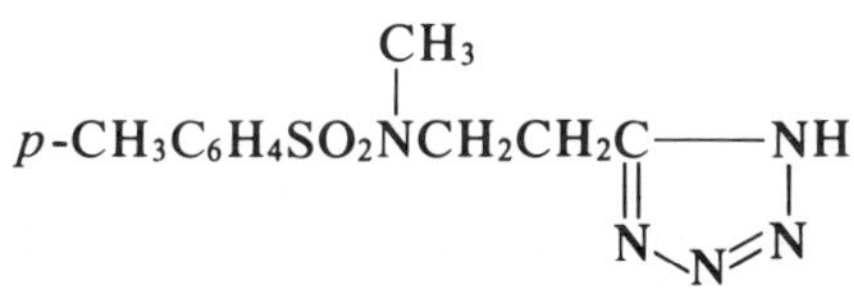

LD_{50} (ip, mouse) 825 mg/kg (95)

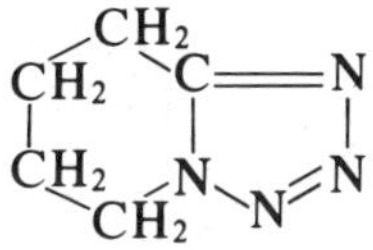

LD_{50} (iv, mouse) 56 mg/kg (83)

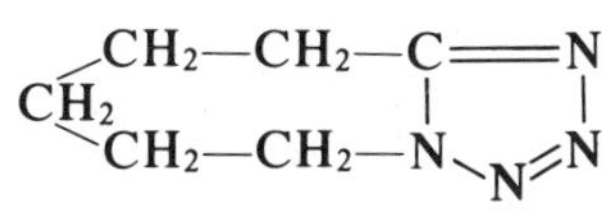

(XXIV)

LD_{50} (oral, rat) 212 mg/kg (rectal, rat) 107 (8)

the LD_{50} dose orally in rats, the mean times to mild and severe convulsions are 7.2 and 26.7 min, respectively, followed by mortality in an average of 65 min (8). Death results from anoxia, medullary depression, and respiratory arrest (27). There is no evidence of carcinogenicity from pentylenetetrazole given 12–18 mg/kg intravenously to rabbits every 2–3 days for 9 weeks (45).

Acute toxicity values for tetrazolium salts indicate them to be extremely toxic. Mice receiving sublethal doses of the tetrazolium salts were hyperirritable, hyperactive, and exhibited muscular twitching. Triphenyltetrazolium chlo-

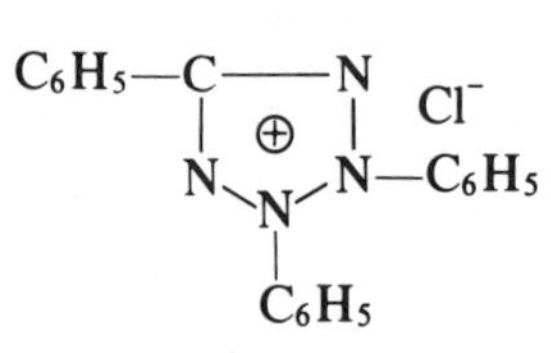

(XXV)

LD_{50} (ip, mouse) 0.40 mg/kg
(90)

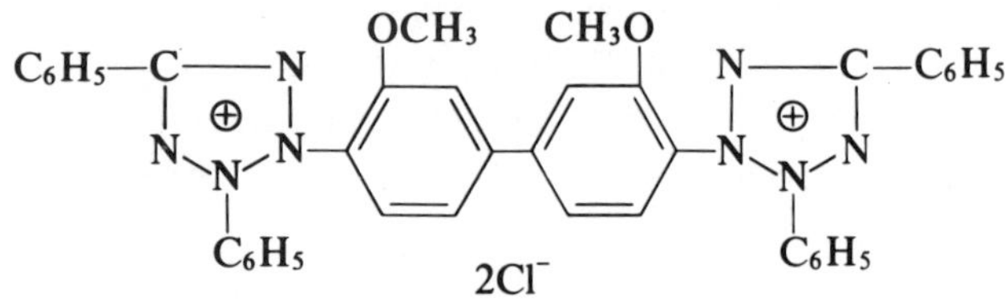

LD_{50} (ip, mouse) 0.15 mg/kg
(90)

ride **(XXV)** is not mutagenic in *E. coli* or *S. typhimurium* in the presence or absence of a microsomal preparation from the livers of Aroclor-induced rats (110). Direct mutations without activation occur in both bacteria, however,

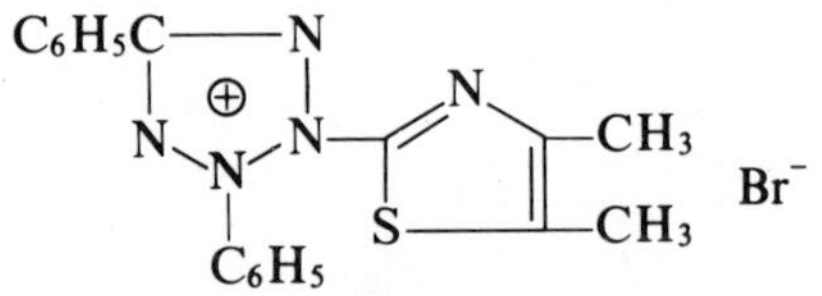

(XXVI)

when exposed to 3-(4,5-dimethylthiazolyl-2)2,5-diphenyltetrazolium bromide **(XXVI)** (110).

REFERENCES

1. A. Ahmad, M. M. Vohra, and G. Achari, *Jap. J. Pharmacol.*, **1967,** *17,* 622–630.
2. F. P. Altman, *Prog. Histochem. Cytochem.*, **1976,** *9,* 1–56.
3. R. Aston and H. Cullumbine, *Toxicol. Appl. Pharmacol.*, **1959,** *1,* 150–155.
4. R. C. Bartlett and M. F. Mazens, *Antimicrob. Agents Chemother.*, **1979,** *15,* 769–774.
5. H. Bartsch et al., *Mutat. Res.*, **1980,** *76,* 1–50.
6. T. Bauknecht et al., *Human Genet.*, **1977,** *35,* 229–307.
7. K. Bogucka and L. Wojtczak, *Biochem. Biophys. Acta,* **1966,** *122,* 381–392.
8. E. M. Boyd and J. Singh, *Anesth. Analg. Curr. Res.*, **1967,** *46,* 395–400.
9. G. B. Brown et al., *Cancer Res.*, **1973,** *33,* 1113–1118.
10. P. Calabresi and R. E. Parks Jr., in A. G. Gilman, L. S. Goodman, and A. Gilman, Eds., *The Pharmacological Basis of Therapeutics,* 6th ed., 1980, Macmillan, New York, Chapter 55, pp. 1256–1313.
11. R. D. Capps et al., *Clin. Chem.*, **1966,** *12,* 406–413.
12. A. Carere et al., *Chem.-Biol. Interactions,* **1978,** *22,* 297–308.

13. K. W. Cochran and K. P. DuBois, *Fed. Proc.*, **1949,** *8,* 282–283.
14. R. W. Cunningham, E. J. Fellows, and A. E. Livingston, *J. Pharmacol.*, **1941,** *73,* 312–318.
15. C. P. Dagg et al., *Proc. Soc. Expl. Biol. Med.*, **1955,** *90,* 489–494.
16. E. F. Domino, K. R. Unna, and J. Kerwin, *J. Pharmacol. Expt. Therap.*, **1952,** *105,* 486–497.
17. H. Druckrey, *Xenobiotica,* **1973,** *3,* 271–303.
18. H. Druckrey et al., *Experientia,* **1967,** *23,* 1042–1043.
19. H. Druckrey et al., *Z. Krebsforsch,* **1967,** *69,* 103–201.
20. *Farm Chemicals Handbook,* Meister Publishing Co., Willoughby, Oh., 1982, p. C145.
21. *Ibid.*, p. C25.
22. *Fed. Regist.*, **1981,** *46,* 14552.
23. *Ibid.*, 61068-61071.
24. R. E. Fye, R. L. Walker, and A. R. Hopkins, *J. Econ. Entomol.*, **1957,** *50,* 700–701.
25. A. Gescher et al., *Biochem. Pharmacol.*, **1981,** *30,* 89–93.
26. A. G. Gilman, L. S. Goodman, and A. Gilman, *The Pharmacological Basis of Therapeutics,* 6th ed., 1980, Macmillan, New York, p. 588.
27. R. E. Gosselin et al., *Clinical Toxicology of Commercial Products,* 4th ed., 1976, Williams and Wilkins, Baltimore, p. 241.
28. J. D. P. Graham, J. M. Rogan, and D. G. Robertson, *J. Ind. Hyg. Toxicol.*, **1948,** *30,* 98–102.
29. W. H. Hall et al., *Curr. Ther. Res. Clin. Exp.*, **1977,** *21,* 374–389.
30. W. H. Hall, B. J. Opfer, and D. N. Gerding, *Antimicrob. Agents Chemother.*, **1980,** *17,* 273–279.
31. C. Hansch et al., *J. Med. Chem.*, **1978,** *21,* 574–577.
32. R. L. Harty, L. G. Paleg, and D. Aspinall, *Aust. J. Exp. Agric. Anim. Husb.*, **1972,** *12,* 517–522.
33. H. C. Hodge and J. H. Sterner, *Am. Ind. Hyg. Assoc. Q.*, **1949,** *10,* 93–96.
34. G. R. Hodges and S. Saslaw, *Am. J. Med. Sci.*, **1973,** *265,* 23–32.
35. P. H. Jones and D. Prasad, *J. Water Pollut. Control Fed.*, **1969,** *41,* R441–R449.
36. M. Julliard and G. Vernin, *Bull. Soc. Chim. Fr.*, **1981,** II, 150–159.
37. A. H. M. Kirby, *Br. J. Cancer,* **1948,** *2,* 290–294.
38. G. F. Kolar, *Adv. in Antimicrobial and Antineoplastic Chemotherapy,* Vol II, University Press, Baltimore, 1972, pp. 121–123.
39. G. F. Kolar, M. Maurer, and M. Wildschätt, *Cancer Lett.*, **1980,** *10,* 235–241.
40. G. F. Kolar and R. Preussmann, *Z. Naturforsch B.*, **1971,** *26,* 950–953.
41. J. F. Kraus et al., *Arch. Environ. Contemp. Toxicol.*, **1977,** *5,* 471–485.
42. W. Krause et al., *Acta. Biol. Med. Ger.*, **1970,** *25,* 875–881.
43. F. W. Krüger, R. Preussmann, and N. Niepelt, *Biochem. Pharmacol.*, **1971,** *20,* 529–533.
44. G. Lakon, *Ber. Dtsch. Bot. Ges.*, **1939,** *57,* 191–203; **1942,** *60,* 299–305.
45. W. Laubender and F. Barth, *Arch. Exp. Pathol. Pharmakol.*, **1941,** *198,* 5–11.
46. P. Lorke and L. Machemer, *Environ. Qual. Saf. Suppl.*, **1975,** *4,* 223–229.
47. R. Madroñero et al., *Eur. J. Med. Chem.*, **1974,** *9,* 445–448.
48. C. Malaveille, G. F. Kolar, and H. Bartsch, *Mutat. Res.*, **1976,** *36,* 1–10.
49. G. P. Margison, A. J. Likhachev, and G. F. Kolar, *Chem. Biol. Interactions,* **1979,** *25,* 345–353.
50. J. Marhold, private communication, July 10, 1982.
51. F. D. Marsh, *J. Org. Chem.*, **1972,** *37,* 2966–2969.
52. M. Matrka, V. Rambousek, and V. Zverina, *Neoplasma,* **1978,** *25,* 153–156.

53. R. W. McCuen, G. Stöhrer, and F. M. Sirotnak, *Cancer Res.*, **1974,** *34,* 378–384.
54. F. A. Mettler, *Fed. Proc., Fed. Am. Soc. Exp. Biol.*, **1972,** *31,* 1504–1507.
55. F. A. Mettler and D. S. Sax, *Brain,* **1972,** *95* (Pt. 3), 505–516.
56. K. Miyoshi, *Acta. Neuropathol.*, **1967,** *9,* 199–216.
57. *Modern Drug Encyclopedia and Therapeutic Index,* 15th ed., Yorke Medical Books, New York, 1979, pp. 153–158.
58. R. B. Moffett, A. Robert, and L. L. Skaletzky, *J. Med. Chem.*, **1971,** *14,* 963–968.
59. M. M. Nachlas et al., *Anal. Biochem.*, **1960,** *1,* 317–326.
60. M. M. Nachlas, S. I. Margulies, and A. M. Seligman, *J. Biol. Chem.*, **1960,** *235,* 499–503.
61. National Cancer Institute, *Bioassay of Azinphosmethyl for Possible Carcinogenicity,* Technical Report Series, No. 69, 1978.
62. National Cancer Institute, *Bioassasy of 1H-Benzotriazole for Possible Carcinogenicity,* Carcinogenesis Technical Report Series, No. 88, 1978.
63. P. Nicholls, *Biochem J.*, **1964,** *90,* 331–343.
64. H. Nishimura and H. Nimura, *J. Embryol. Exp. Morphol.*, **1958,** *6,* 593–595.
65. R. P. Ouellette and J. A. King, *Chemical Week Pesticides Register,* McGraw-Hill, New York, 1977, pp. 256, 275.
66. R. E. Parks and K. C. Agarwal, *Handb. Exp. Pharmakol.*, **1975,** *38,* Pt. 2, Antineoplast. Immunosuppr. Agents, 458–467.
67. J. Pasquet et al., *Toxicol. Appl. Pharmacol.*, **1976,** *37,* 85–92.
68. *Physicians Desk Reference,* 36th ed., 1982, Medical Economics Co., Oradell, N.J., p. 1013.
69. *Ibid.*, pp. 1118–1120.
70. *Ibid.*, p. 1120.
71. *Ibid.*, p. 1326.
72. A. V. Piskarev, V. S. Nesterenko, and A. N. Kost, *Russ. Pharmacol. Toxicol., Engl. Ed.*, **1978,** *41,* 249–250.
73. R. Preussmann et al., *Z. Krebsforsch. Klin. Onkol.*, **1974,** *81,* 285–310.
74. R. Preussmann et al., *Ann. N.Y. Acad. Sci.*, **1969,** *163,* 697–716.
75. P. Puccetti, A. Giampietri, and M. C. Fioretti, *Experientia,* **1978,** *34,* 799–800.
76. C. F. Rainwater, *Agric. Chem.*, **1956, 2,** 32–33.
77. *Registry of Toxic Effects of Chemical Substances,* 1979, National Institute for Occupational Safety and Health, DM 0800000, Vol. 1, p. 277.
78. *Ibid.*, MW 2800000, Vol. 1, p. 761.
79. *Ibid.*, SB 1300000, Vol. 2, p. 206.
80. *Ibid.*, UR 0825000, Vol. 2, p. 427.
81. *Ibid.*, XF 7472000, Vol. 2, p. 587.
82. *Ibid.*, XF 7700000, Vol. 2, p. 587.
83. *Ibid.*, XF 8240000, Vol. 2, p. 588.
84. *Ibid.*, XY 2625000, Vol. 2, p. 635.
85. *Ibid.*, XZ 4580000, Vol. 2, p. 642.
86. *Ibid.*, XZ 5685000, Vol. 2, p. 644.
87. *Ibid.*, XZ 5957000, Vol. 2, p. 644.
88. L. B. Reller et al., *Antimicrob. Agents Chemother.*, **1973,** 488–497.
89. S. G. N. Richardson, C. Giles, and C. H. J. Swan, *J. Clin. Pathol.*, **1975,** *28,* 350–351.
90. A. M. Rutenburg, R. Gofstein, and A. M. Seligman, *Cancer Res.*, **1950,** *10,* 113–121.
91. R. J. Scherertz, R. Dacey, and M. A. Sande, *J. Antimicrob. Chemother.*, **1976,** *2,* 159–165.

92. F. A. Schmidt and D. J. Hutchinson, *Cancer Res.*, **1974,** *34,* 1671–1675.
93. J. Scholler, F. S. Philips, and S. S. Sternberg, *Proc. Soc. Exp. Biol. Med.*, **1957,** *96,* 491–496.
94. R. D. Short et al., *Report 1978, EPA/600/1-78/056,* Order No. PB-288457.
95. J. S. Shukla and R. Rastogi, *Indian J. Pharm. Sci.*, **1978,** *40,* 212–214.
96. D. H. Sieh, A. W. Andrews, and C. J. Michejda, *Mutat. Res.*, **1980,** *73,* 227–235.
97. S. R. Sirianni and C. C. Huang, *Proc. Soc. Exp. Biol. Med.*, **1978,** *158,* 269–274.
98. J. L. Skibba, E. Erturk, and G. T. Bryan, *Cancer,* **1970,** *26,* 1000–1005.
99. H. F. Smyth, Jr. et al., *Am. Ind. Hyg. Assoc. J.*, **1969,** *30,* 470–476.
100. P. Thomann and L. Krueger, *Environ. Qual. Saf. Suppl.*, **1975,** *4,* 193–198.
101. H. F. Thomas et al., *Mutat. Res.*, **1979,** *60,* 25–32.
102. W. T. Thomson, *Agricultural Chemicals, Book 1, Insecticides, Acaricides and Ovicides,* 1979–1980 Revision, Thomson Publications, Fresno, p. 171–172.
103. S. Tricerri et al., *Eur. J. Med. Chem.*, **1979,** 181–183.
104. B. Ulland, E. K. Weisburger, and J. H. Weisburger, *Toxicol. Appl. Pharmacol.*, **1973,** *25,* 446.
105. G. Unterstenhöfer, *Mededel. Landbouwhogeschool en Opzoekingsstas. Staat Gent.*, **1958,** *23,* 770–779; *Chem. Abstr.*, **1960,** *54,* 7049h.
106. D. G. Upshall, J. C. Roger, and J. E. Pasida, *Biochem. Pharmacol.*, **1968,** *17,* 1529–1542.
107. T. Urban and C. Jarstrand, *J. Antimicrob. Chemother.*, **1981,** *8,* 363–369.
108. U.S. Environmental Protection Agency, *Pesticide Product Information,* 6th ed., April, 1980.
109. B. H. Venger et al., *J. Med. Chem.*, **1979,** *22,* 473–476.
110. S. Venitt and C. Crofton-Sleigh, *Mutat. Res.*, **1979,** *68,* 107–116.
111. G. Vettorazzi, *Residue Rev.*, **1976,** *63,* 1–44.
112. H. Wada, M. Sarto, and Y. Takai, *Soil Sci. Plant Nutr.* (*Tokyo*), **1978,** 349–356.
113. J. S. Wishnok and M. C. Archer, *Br. J. Cancer,* **1976,** *33,* 307–311.
114. R. E. Wittes, J. T. Wittes, and R. B. Golbey, *Cancer,* **1978,** *41,* 415–421.
115. D. W. Woolley and E. Shaw, *J. Biol. Chem.*, **1951,** *189,* 401–410.
116. V. L. Yu, R. M. Vickers, and J. J. Zuravleff, *Antimicrob. Agents Chemother.*, **1980,** *17,* 96–98.
117. R. A. Zvyagil'skaya, K. Bogucka, and L. Wojtczak, *Acta Biochim. Pol.*, **1969,** *16,* 163–173.

6

REACTIONS

This chapter is concerned with reactions of high nitrogen compounds with common types of chemical agents. First considered are the responses of polynitrogen chains to reagents which form carbon-nitrogen bonds, conditions under which the nitrogen chains are stable to these reagents, and those causing rupture. Next is a section on the behavior of nitrogen chains with electrophiles other than those forming carbon-nitrogen bonds, followed by a summary of reactions with nucleophiles, excepting those forming C—N bonds. Finally, reactions of nitrogen chains with oxidants and with reductants are discussed.

6.1. ALKYLATION, ARYLATION, AND ACYLATION

The terminal amino groups of 2,2-dimethyltriazanium chloride resist further alkylation to trialkyltriazanium salts with methyl and ethyl bromides, and the starting triazanium salt is recovered unchanged (185).

Compounds with N—H groups as part of a nitrogen chain usually are readily alkylated and acylated. Diazoaminobenzene in alkaline medium is converted to 3-benzyl-1,3-diphenyltriazene (168); a yield of 90% can be attained in the same

$$C_6H_5N{=}NNHC_6H_5 \xrightarrow[\substack{C_2H_5OH \\ \text{reflux}}]{\substack{C_6H_5CH_2Cl \\ NaOC_2H_5}} C_6H_5N{=}NN(CH_2C_6H_5){-}C_6H_5 \quad 69\%$$

reaction if conducted in the presence of benzyltriethylammonium chloride as a phase transfer catalyst (538). Unsymmetrically disubstituted triazenes in principle should form two isomers on alkylation or acylation, but in the majority of such substitution reactions, only one product is found. Examples of alkylation in which two isomers have been isolated are shown. Although much effort has been expended in studies of alkylation or acylation of the imino nitrogen in triazenes to establish the location of hydrogen on nitrogen in disubstituted tria-

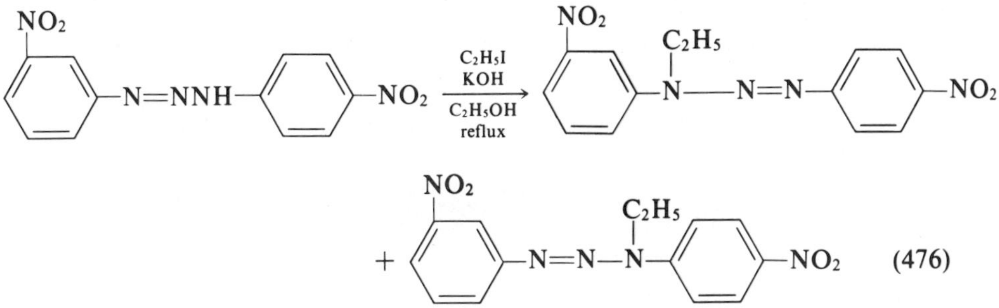

(476)

$$C_6H_5CH_2—N{=}NNH—C_4H_9\text{-}n \xrightarrow[\substack{t\text{-butanol} \\ \text{room temp.}}]{\substack{CH_3I \\ KOC(CH_3)_3}}$$

$$C_6H_5CH_2—\underset{}{N}(CH_3)—N{=}N—C_4H_9\text{-}n + C_6H_5CH_2N{=}N—N(CH_3)—C_4H_9\text{-}n \qquad (474)$$

39% 60%

zenes, the rapid tautomeric equilibrium precludes the use of this type of chemical evidence (103) (see Section 3.7, Acid-Base Properties).

Nitrogen chains of trisubstituted triazenes are broken on reaction with powerful alkylating agents. Trimethyloxonium fluoborate splits triazenes to dia-

$$p\text{-}CH_3C_6H_4—N{=}N—N(CH_3)—C_6H_4CH_3\text{-}p \xrightarrow[\text{ice cooling}]{\substack{(CH_3)_3O^+\ BF_4^- \\ CH_2Cl_2}} p\text{-}CH_3C_6H_4N_2^+\ BF_4^-$$

99%

$$+\ (CH_3)_3\overset{+}{N}C_6H_4CH_3\text{-}p\ BF_4^-$$

74%

zonium and quaternary salts (208) and tetracyanoethylene breaks the triazene chain of 3-benzyl-1,3-di-*p*-chlorophenyltriazene (538).

$$p\text{-}ClC_6H_4—N{=}N—N(CH_2C_6H_5)—C_6H_4Cl\text{-}p \xrightarrow[50\text{–}60^\circ]{\substack{(NC)_2C{=}C(CN)_2 \\ CH_3OH}} p\text{-}ClC_6H_4NHN{=}C(CN)_2$$

50%

$$+\ p\text{-}ClC_6H_4NHCH_2C_6H_5$$

Although 3-benzenesulfonyl-1-phenyltriazene is alkylated by acrylonitrile at the nitrogen adjacent to phenyl, 3-acetyl-1-phenyltriazene methylates next to acetyl. The cyclic triazene 4,4-diphenyl-5-triazolinone also alkylates adjacent to

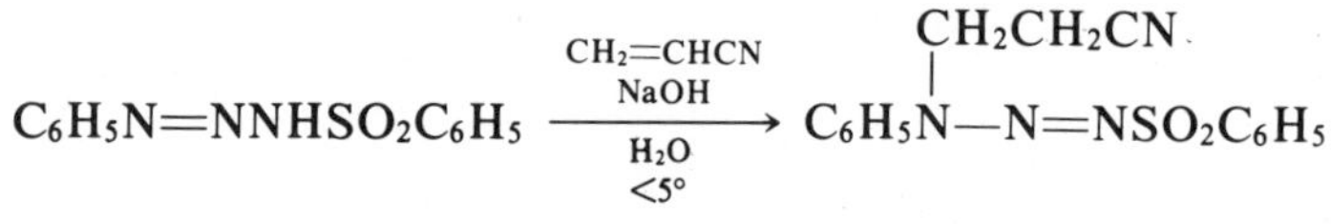
(210)

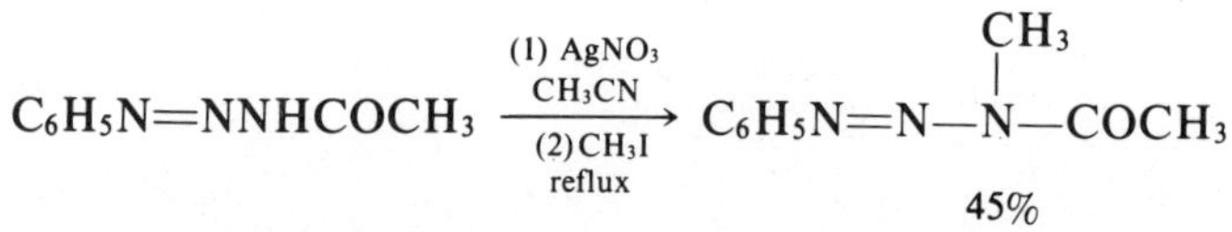
(339)

carbonyl. *N*-Aminopyrazoles and benzocinnoline imides alkylate at the nitrogen atoms outside the rings.

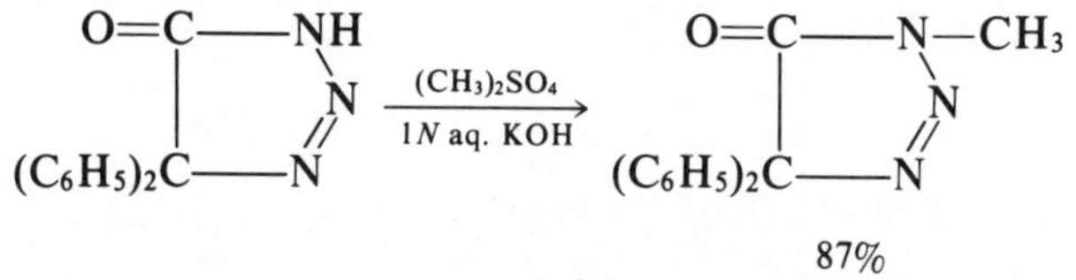
(243)

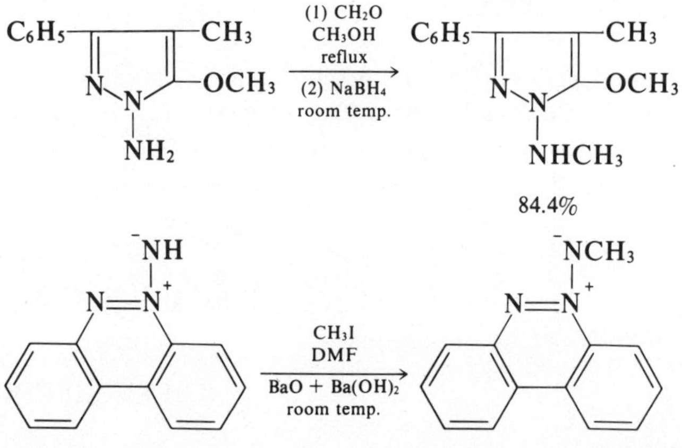
(4)

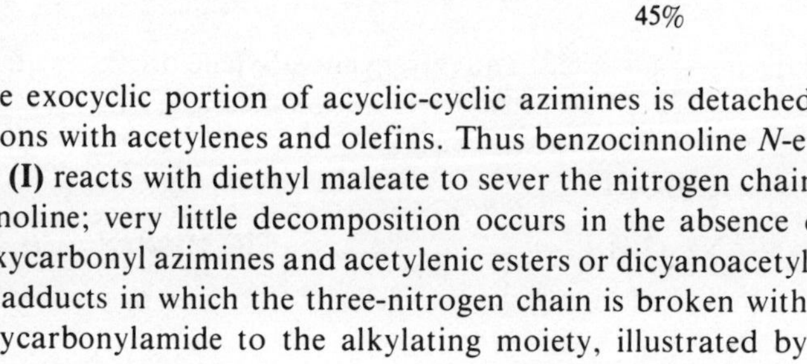
(384)

The exocyclic portion of acyclic-cyclic azimines is detached in exothermic reactions with acetylenes and olefins. Thus benzocinnoline *N*-ethoxycarbonylimine **(I)** reacts with diethyl maleate to sever the nitrogen chain, forming benzocinnoline; very little decomposition occurs in the absence of olefin (111). Alkoxycarbonyl azimines and acetylenic esters or dicyanoacetylene combine to form adducts in which the three-nitrogen chain is broken with transfer of the alkoxycarbonylamide to the alkylating moiety, illustrated by reactions of **I**

(111) and the 2-methoxycarbonylimide of 3-methyl-4-ethylcinnoline **(II)** (431). The expected fused triazoline intermediates are not isolated.

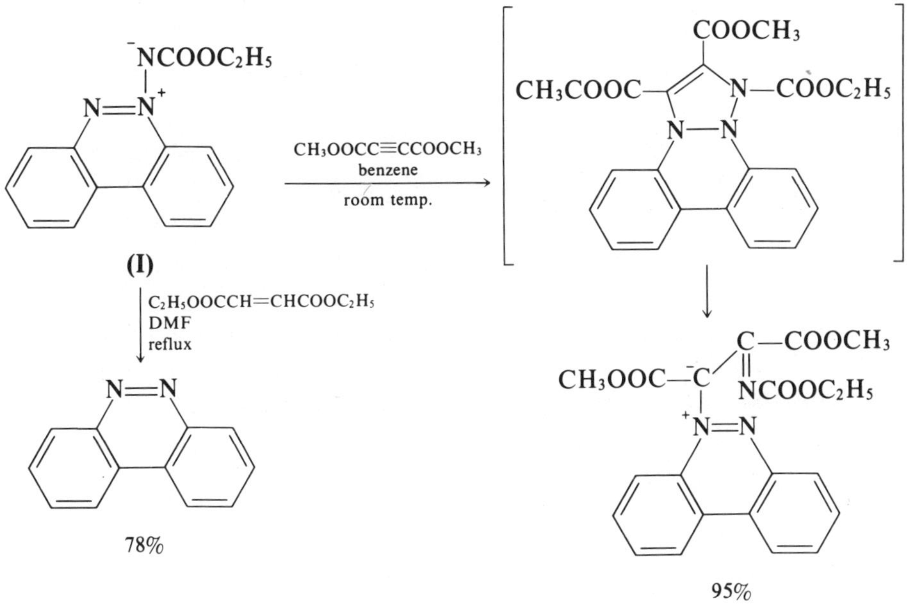

(I) 78% 95%

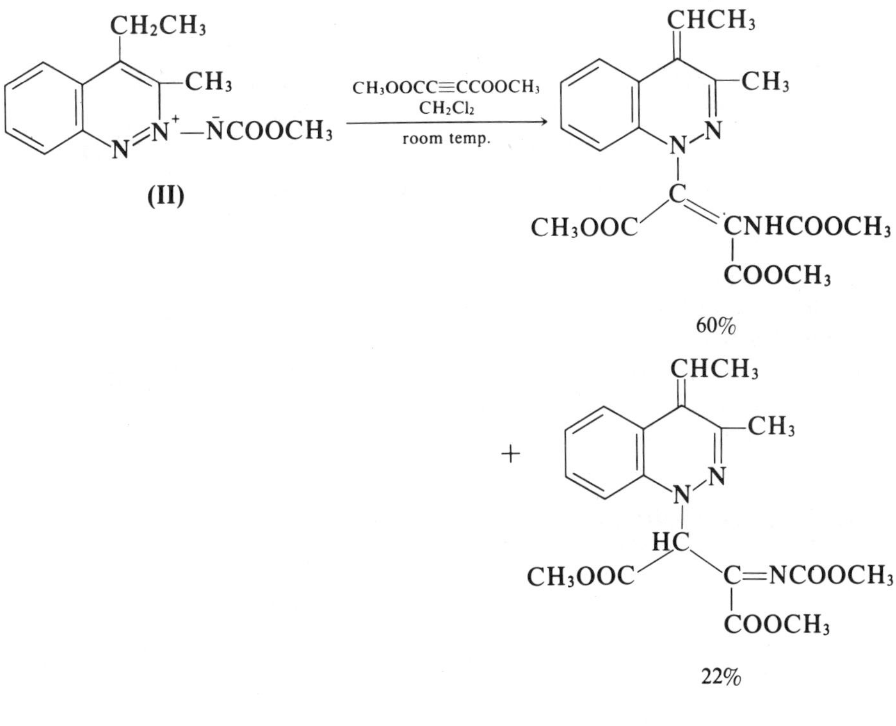

(II)

The nitrogen chain in *N*-methylbenzocinnoline imide also is broken on reaction with dimethyl acetylenedicarboxylate resulting in a dibenzoimidazolinodi-

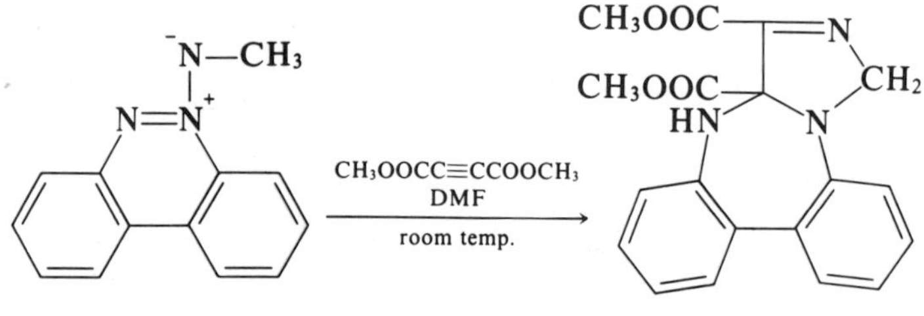

80%

hydrodiazepine (427). The lithium salt of dibenzotriazepine methylates on nitrogen under mild conditions (172), but the cyclic three-nitrogen chain in dibenzotriazepine is opened by diethyl malonate forming a hydrazone.

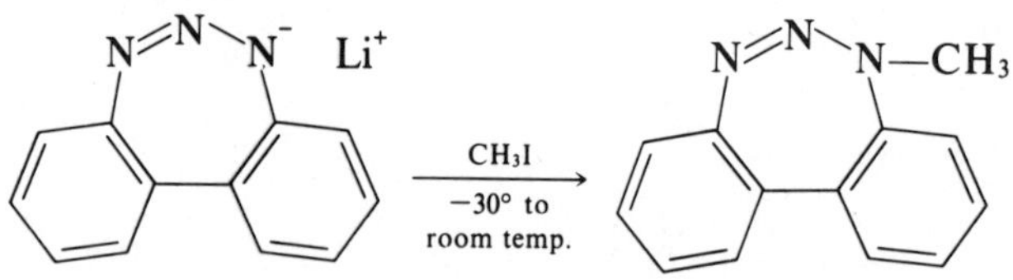

67%

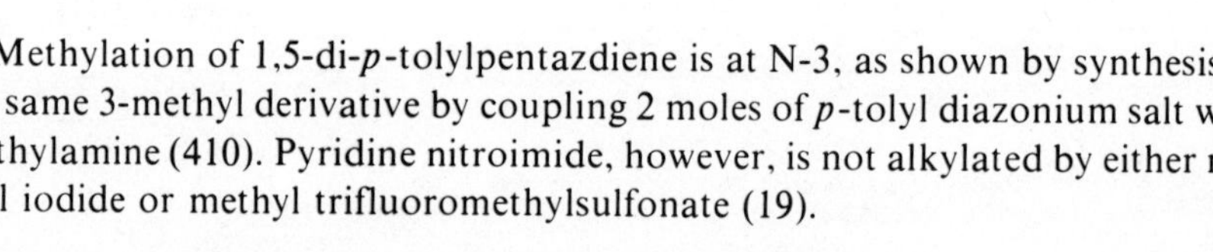

(172)

60%

Methylation of 1,5-di-*p*-tolylpentazdiene is at N-3, as shown by synthesis of the same 3-methyl derivative by coupling 2 moles of *p*-tolyl diazonium salt with methylamine (410). Pyridine nitroimide, however, is not alkylated by either methyl iodide or methyl trifluoromethylsulfonate (19).

$$p\text{-}CH_3C_6H_4\text{—}N\text{=}NNHN\text{=}N\text{—}C_6H_4CH_3\text{-}p \xrightarrow[\substack{CH_3OH \\ \text{room temp.}}]{\substack{CH_3I \\ NaOCH_3}}$$

$$p\text{-}CH_3C_6H_4\text{—}N\text{=}N\text{—}\underset{}{\overset{\displaystyle CH_3}{\overset{|}{N}}}\text{—}N\text{=}N\text{—}C_6H_4CH_3\text{-}p$$

The nitrogen chain in 1-formyl-2-phenyl-3-phenylmethylenetriazane survives further acylation (560). Acylation of triazenes usually requires less strenu-

$$C_6H_5CH{=}N{-}N(C_6H_5){-}NHCHO \xrightarrow[\text{heat}]{CH_3COCl,\ (CH_3CO)_2O} C_6H_5CH{=}N{-}N(C_6H_5){-}N(COCH_3)(CHO)$$

ous conditions than alkylation, illustrated by the preparation of 3-acetyl-1,3-diphenyltriazene (237). Monoaryl triazenes react easily with isocyanates at the

$$C_6H_5N{=}NNHC_6H_5 \xrightarrow[\text{room temp.}]{(CH_3CO)_2O,\ \text{ether}} C_6H_5N{=}N{-}N(C({=}O){-}CH_3){-}C_6H_5 \quad 74\%$$

terminal nitrogen to form acyl aryl triazenes (132); success of this reaction constitutes additional proof of the existence of monosubstituted triazenes.

$$p\text{-}C_2H_5OOCC_6H_4N{=}NNH_2 \xrightarrow[\text{cooling}]{C_6H_5NCO,\ \text{ether}} p\text{-}C_2H_5OOCC_6H_4N{=}NNHCONHC_6H_5$$

Goldschmidt made an extensive study of the acylation of disubstituted triazenes with phenyl isocyanate, identifying the resulting single product trisubstituted triazenes by their decomposition to symmetrically disubstituted ureas (191). Location of the phenylcarbamoyl group on the triazene chain does not seem to depend on the electronic character of the triazene substituents, since products are obtained with acylation adjacent to both electron-releasing and electron-withdrawing groups.

$$p\text{-}CH_3C_6H_4{-}N{=}NNHC_6H_5 \xrightarrow{C_6H_5NCO} p\text{-}CH_3C_6H_4{-}N(C_6H_5NHC{=}O){-}N{=}N{-}C_6H_5$$

$$p\text{-}BrC_6H_4{-}N{=}NNH{-}C_6H_5 \xrightarrow{C_6H_5NCO} p\text{-}BrC_6H_4{-}N(C_6H_5NHC{=}O){-}N{=}N{-}C_6H_5$$

$$\beta\text{-}C_{10}H_7{-}N{=}NNH{-}C_6H_5 \xrightarrow{C_6H_5NCO} \beta\text{-}C_{10}H_7{-}N(C_6H_5NHC{=}O){-}N{=}N{-}C_6H_5$$

$$\beta\text{-}C_{10}H_7{-}N{=}NNHC_6H_4Br\text{-}p \xrightarrow{C_6H_5NCO} \beta\text{-}C_{10}H_7{-}N{=}NN(O{=}CNHC_6H_5){-}C_6H_4Br\text{-}p$$

$$C_6H_5N{=}NNHCH_3 \xrightarrow{C_6H_5NCO} C_6H_5N{=}N{-}N(O{=}CNHC_6H_5){-}CH_3 \qquad (130)$$

Triazenes with an acylating substituent two or three atoms removed from N-3 undergo intramolecular acylation to five- or six-membered rings, illustrated by formation of 3-methyl-1-*p*-tolyl-*v*-triazolium-5-olate (420) and 3-methylbenzotriazinone (317).

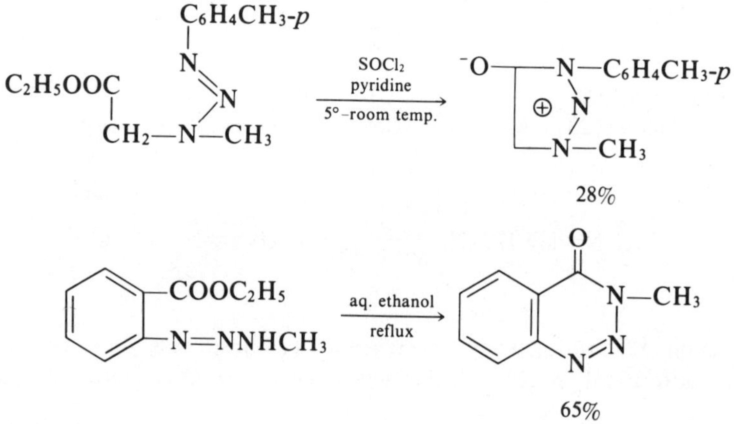

1-Aminopyridazinium chloride acylates readily to a dipolar compound (289), and *N*-aminopyridazines with adjacent amino acylate to *s*-triazolo[1,5-*b*]-pyridazines with an embedded nitrogen (509).

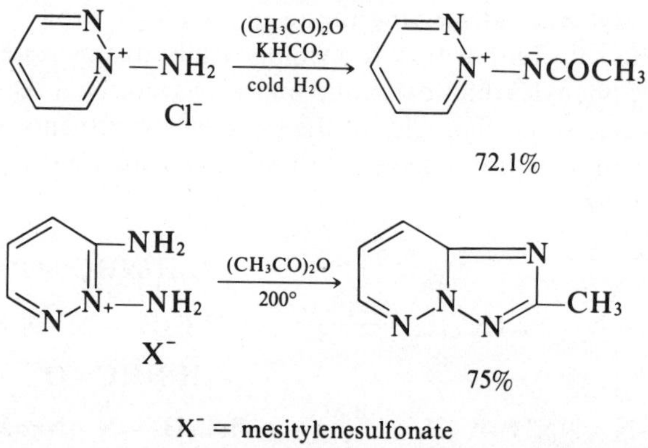

Phenyl isocyanide adds to disubstituted triazenes in the presence of cuprous chloride, forming *N*-arylazoformamidines (528).

Acylation of 1-alkyl-3-aryltriazene oxides causes rearrangement of the oxygen to the α-carbon of the alkyl group, its esterification, and acylation at the N-1 of the triazene oxide chain (162).

The nitrogen chain of triazene oxide **(III)** remains intact on treatment with acetic anhydride, but oxygen migrates to adjacent carbon. Under the influence of trifluoroacetic anhydride at low temperature, however, the triazene segment is converted to azide (162).

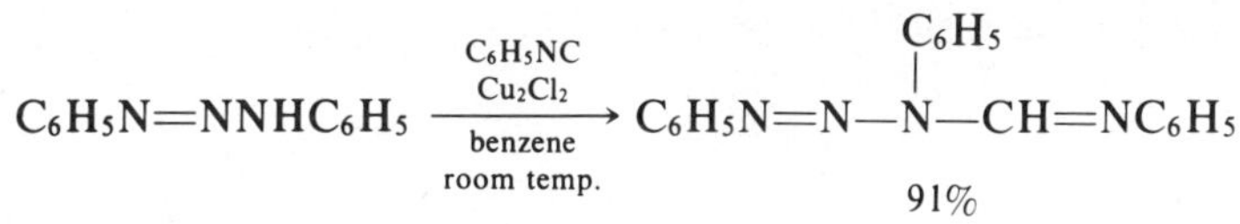

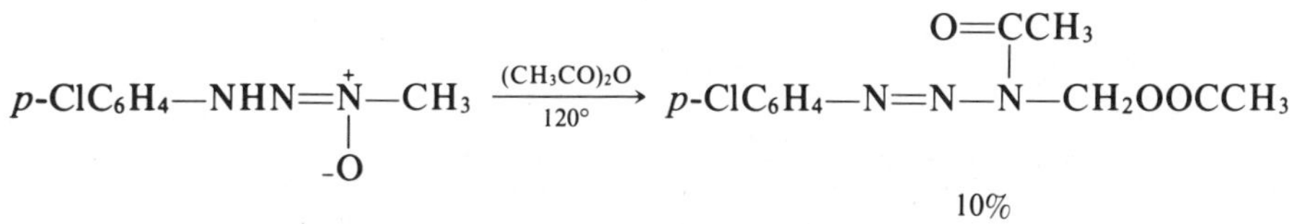

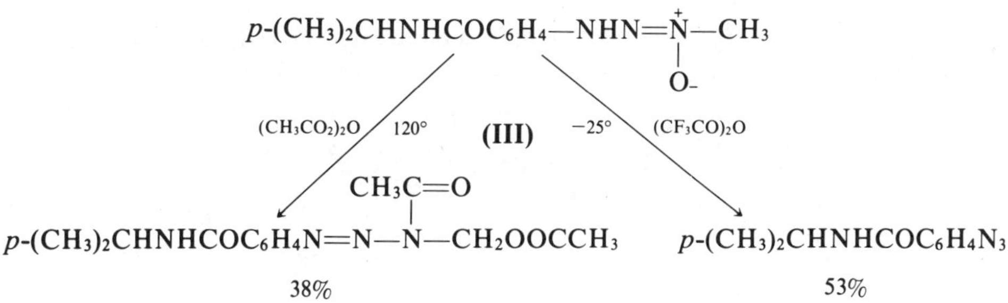

Benzoylation of 4,4-diphenyl-5-triazolinone causes the elimination of nitrogen and formation of the corresponding oxazoline (243), and in a similar dis-

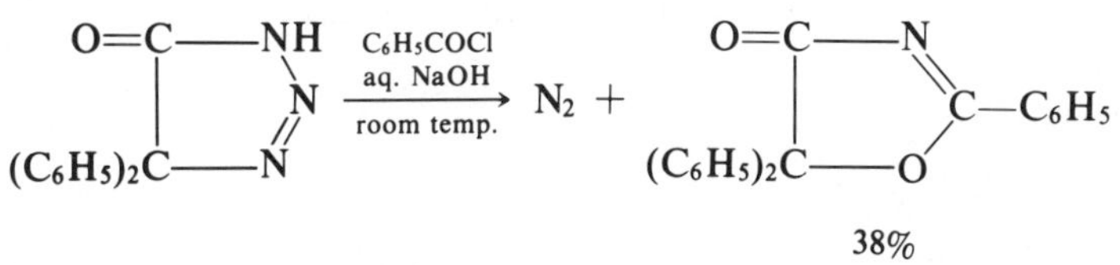

placement, 1-*p*-nitrophenyl-4-methylene-5-dimethylaminotriazoline is converted by trimethylbenzonitrile oxide to an oxazine (135a). The nitrogen chain in di-

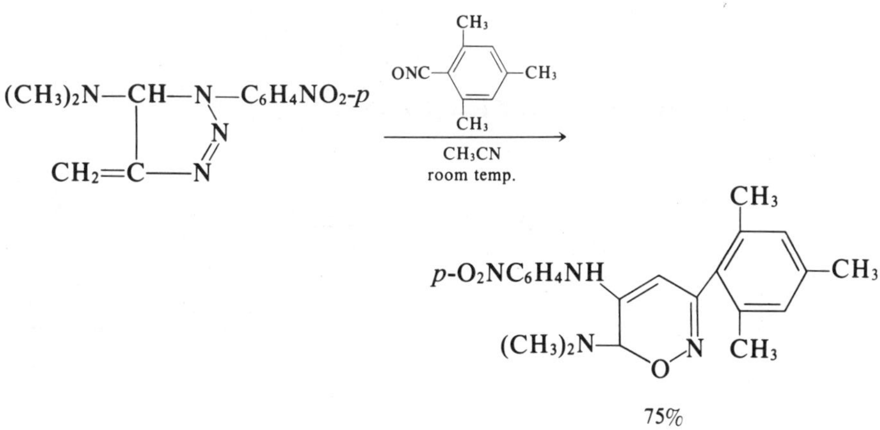

benzotriazepine is easily disrupted by benzoyl chloride replacing the masked diazonium group by chloride (172).

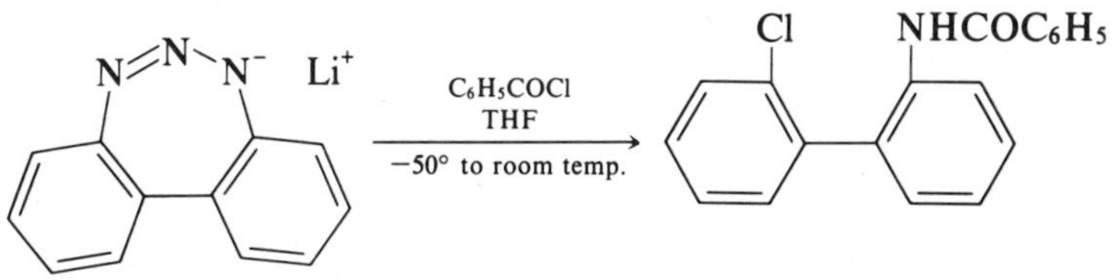

70%

The side chain of pyridine nitroimide is eliminated by refluxing with acetic anhydride (19).

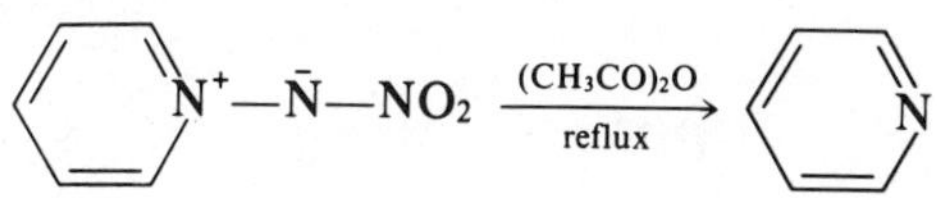

50%

Alkylidene groups are linked to the ends of nitrogen chains by the characteristic reaction of amino with carbonyl, exemplified by the reactions of aldehydes with derivatives of triazane, triazene, 1,3-tetrazene, and 1-aminotetrazole. Isolation of 1-formyl-2-phenyltriazane from the reduction of nitrosoformylphenyl

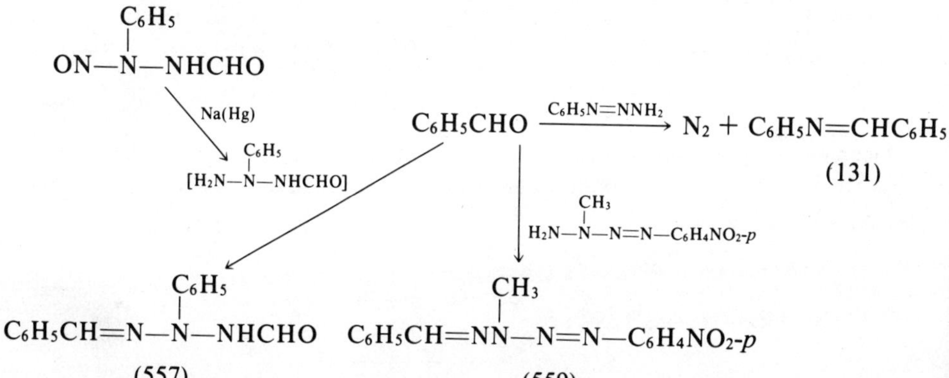

(131)

(557)

(559)

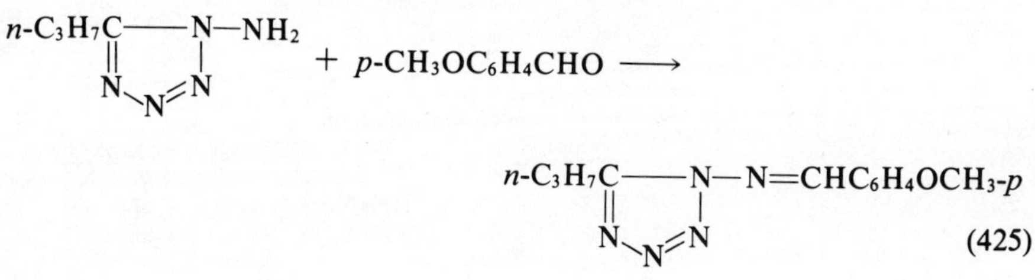

(425)

hydrazine is not successful, but its presence is indicated by formation of the phenylmethylene derivative when benzaldehyde is added to the reaction mixture after completion of the reduction. Monosubstituted triazene chains, however, are broken in the presence of aldehydes, and only the Schiff's base of the substituted amine is obtained.

Reaction of the exocyclic amino group of *N*-aminoheterocycles such as *N*-aminopyrazole with 2,5-diethoxytetrahydrofuran and *N,N*-dimethylformamide azine leads, respectively, to 1-pyrrolyl and 4-*s*-triazolyl cyclic-cyclic high nitrogen compounds (125).

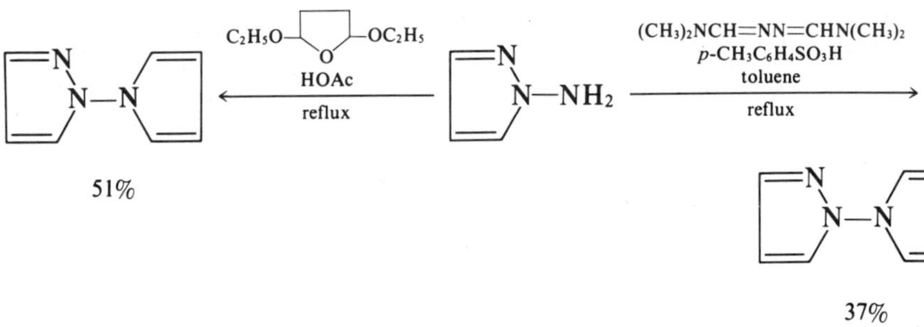

Quaternary salts are formed by alkylation of compounds with three to six contiguous nitrogen atoms. Addition occurs only at the end of the nitrogen chain in triazenes, triazolines, tetraazapentalenes, and 1,1′-bibenzotriazoles. The small quantity of quaternary obtained from the open-chain triazene contrasts with the higher yields of triazolium salts from intramolecular cyclization of trisubstituted triazenes carrying ω-hydroxy- or bromoalkyl groups, or the

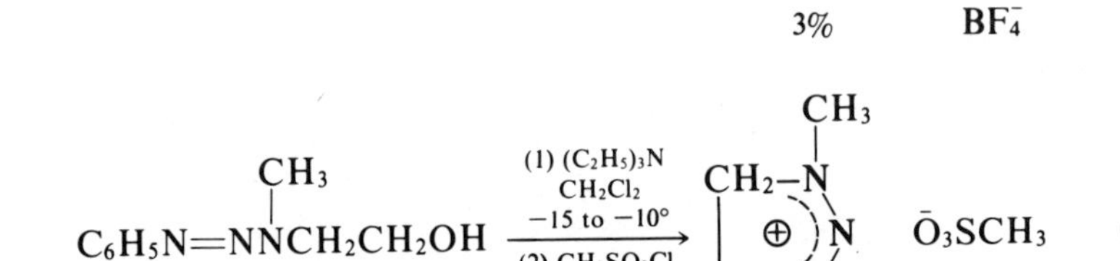

(208)

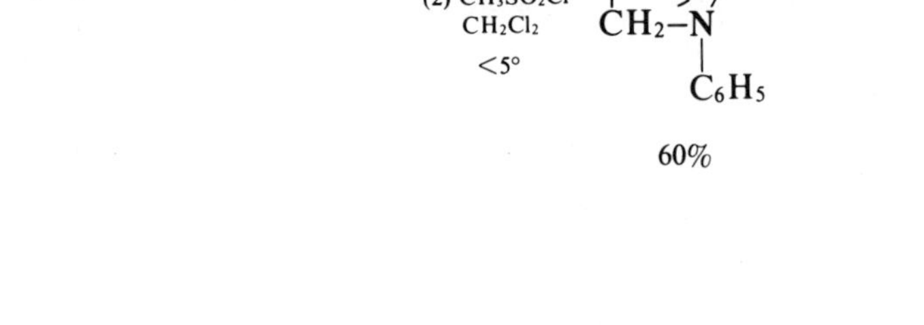

(208)

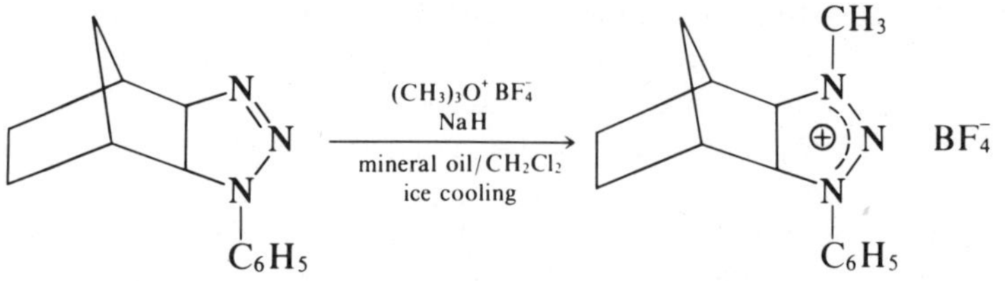

(208)

100%

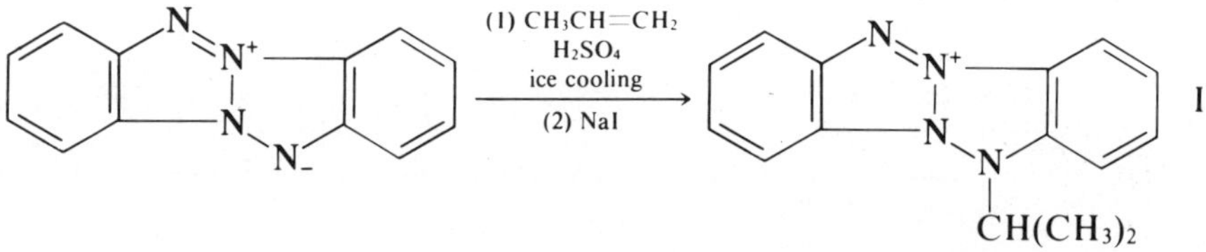

88%

(106)

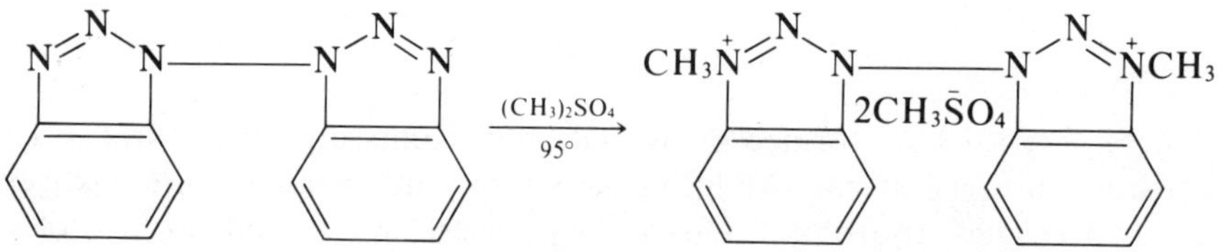

86%

(209)

alkylation of triazolines. Unstable quaternary salts are obtained from tetraalkyl-2-tetrazenes by alkylation at low temperatures, but the tetrazene chain is broken at higher temperatures with formation of quaternary ammonium salts.

$$(CH_3)_2N{-}N{=}N{-}N(CH_3)_2 \xrightarrow[0^\circ]{CH_3Br \atop ether} (CH_3)_2N{-}N{=}N{-}\overset{+}{N}(CH_3)_3Br^- \qquad (343)$$

$$(C_6H_5CH_2)_2N{-}N{=}N{-}N(CH_2C_6H_5)_2 \xrightarrow[reflux]{CH_3I \atop benzene} N_2 + (C_6H_5CH_2)_2\overset{+}{N}(CH_3)_2I^- \qquad (551)$$

The nitrogen chain of 1-*t*-butyltriazoline is stable to trimethyloxonium fluoborate, which effects quaternization, but is destroyed by methyl iodide (208). Treatment of tetramethyltetrazene with acetic anhydride (382, 383) or phthalic or other anhydrides (362) leads initially to the tetrazene radical cation, detected by ESR, followed by loss of nitrogen and formation of the radical cation of tetramethylhydrazine. For each mole of tetramethyltetrazene decomposed one

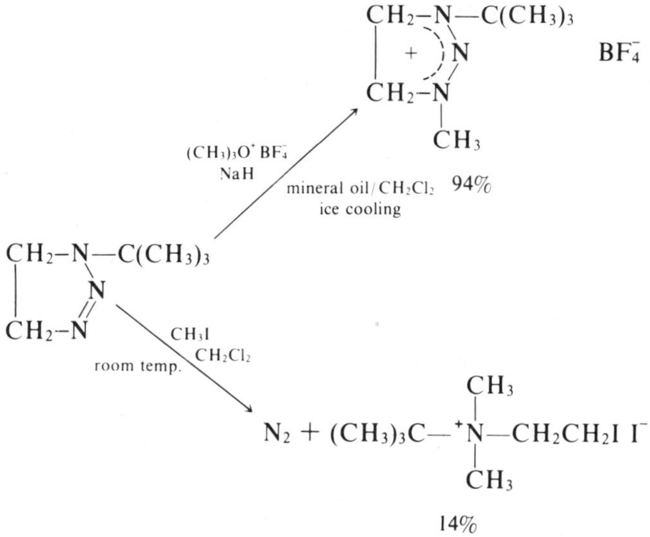

mole of dimethylacetamide is formed (383). Extensive fragmentation of tetra(trimethylsilyl)tetrazene results from interaction with benzenesulfonyl isocyanate (547).

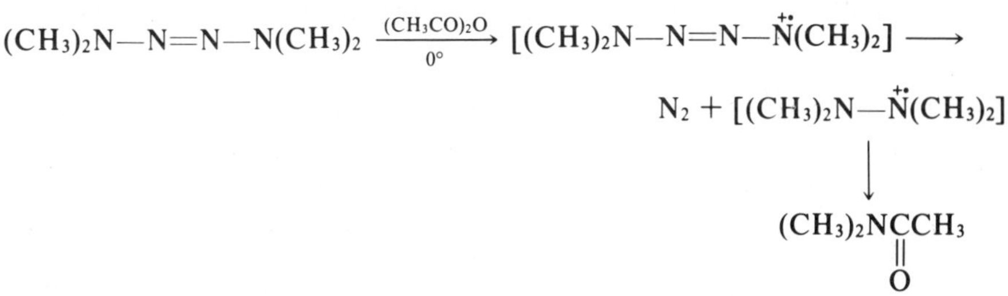

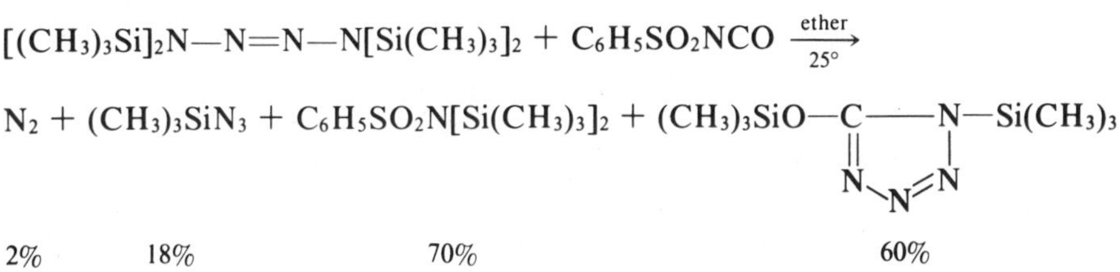

Under the influence of alkylating, arylating, or acylating agents the three-nitrogen azide chain may remain together or transfer two or one nitrogen atoms to the reagent. (Reactions in which nitrenes are thermally generated have been considered in Section 2.1.7. Azides.)

Azide ion remains intact in its nucleophilic substitution of functional groups on carbon such as halide. This is the principal method for formation of organic azides. Alkyl halides are converted to azides in good yields by heating with sodium azide in monoalkyl ethers of ethylene glycols (Carbitols) (327). Azidium salts are obtained easily from halogen-containing quaternary salts (30) and two chlorine atoms on the same carbon are readily replaced by silver azide (198).

$$n\text{-}C_4H_9Br \xrightarrow[95^\circ]{\substack{NaN_3 \\ CH_3OCH_2CH_2OCH_2CH_2OH}} n\text{-}C_4H_9N_3$$

78.3%

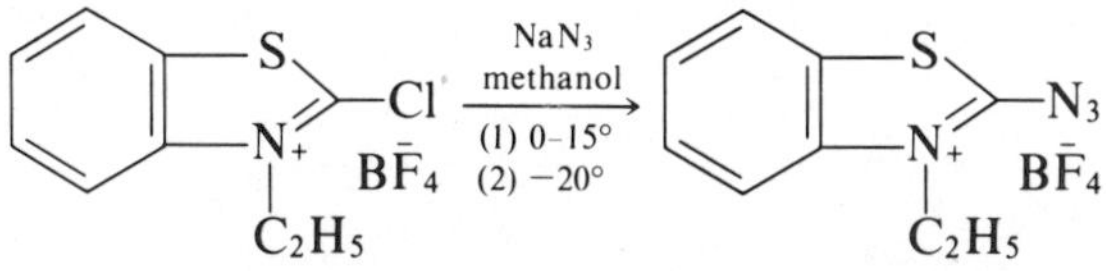

75%

$$(C_6H_5)_2CCl_2 \xrightarrow[ether]{AgN_3} (C_6H_5)_2C(N_3)_2$$

88%

Mild conditions can be used with alkyl halides if the reaction occurs in a solvent system in which azide ion is unsolvated and its counterion is solvated. Thus potassium azide in benzene containing the dimethyl ether of octaethylene glycol, which has an affinity for potassium ion, reacts with benzyl bromide at

$$C_6H_5CH_2Br \xrightarrow[\substack{CH_3O(CH_2CH_2O)_8CH_3 \\ \text{room temp.}}]{\substack{KN_3 \\ benzene}} C_6H_5CH_2N_3$$

100%

room temperature to form benzyl azide in quantitative yield (322). In the absence of the polyoxyethylene glycol ether, the yield of benzyl azide is 14%. Although halide is the usual displaced group in nucleophilic substitutions by azide ion, other leaving groups include *p*-toluenesulfonates and nitrates; even nitro

$$ClCH_2CH_2OSO_2C_6H_4CH_3\text{-}p \xrightarrow[reflux]{\substack{NaN_3 \\ methanol/H_2O}} ClCH_2CH_2N_3$$

65% (553)

$$(CH_3)_3CONO_2 \xrightarrow[25^\circ]{\substack{LiN_3 \\ DMF}} (CH_3)_3CN_3$$

75% (348)

can participate in the reaction, illustrated in the comparative formations of *p*-azidonitrobenzene from *p*-chloronitrobenzene and *p*-dinitrobenzene (202).

$$p\text{-}O_2NC_6H_4Cl \xrightarrow[60^\circ]{NaN_3,\ DMSO} p\text{-}O_2NC_6H_4N_3 \xleftarrow[\text{room temp. to } 40^\circ]{NaN_3,\ DMSO} p\text{-}O_2NC_6H_4NO_2$$

75% 91%

Azide ion also opens the epoxide ring in oxiranes forming azido alcohols. Styrene oxide is attacked at the more positive carbon next to phenyl but with propylene oxide, azide substitutes largely at the terminal carbon, consistent with the electron-releasing effect of methyl on the adjacent carbon.

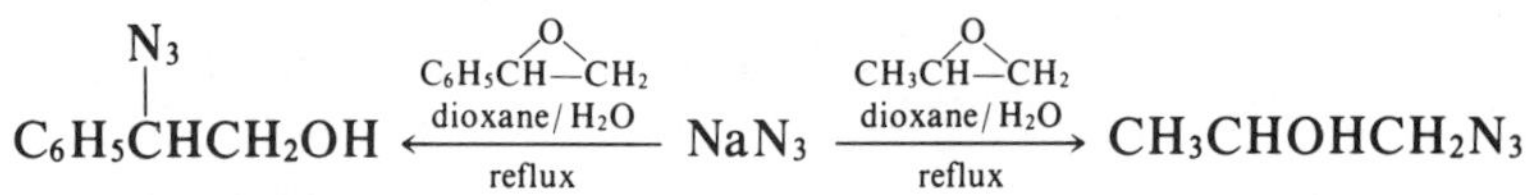

63% (356) 60–80% (533)

In an S_N1 reaction, alcohols forming carbonium ions easily, such as triphenylcarbinol, react with hydrazoic acid in the presence of strong acids.

$$(C_6H_5)_3COH \xrightarrow[\text{ice cooling to } 25^\circ]{HN_3/CHCl_3,\ H_2SO_4} (C_6H_5)_3CN_3$$

70% (18a)

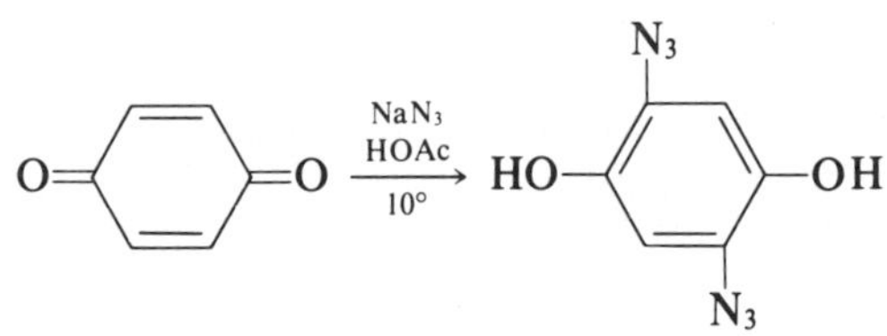

33% (368)

Two moles of hydrogen azide add to 1,4-benzoquinone to form 2,5-diazido-1,4-benzohydroquinone.

Iodine azide adds stereospecifically, usually trans, to carbon-carbon unsaturation, illustrated by combination with *cis*- and *trans*-2-butene, considered to be electrophilic processes (160). The action of base on such β-iodoazides is a general synthesis of vinyl azides which also occurs regiospecifically (215). Thus erythro-1-azido-2-iodo-1-phenylpropane, prepared from *trans*-1-phenyl-1-propene, loses hydrogen iodide to form *cis*-1-azido-1-phenyl-1-propene. Other olefins undergo more complicated reactions with iodine azide, including rearrangement and combination of initially formed azides with the acetonitrile solvent resulting in tetrazoles. Further, both electrophilic and radical pathways can be followed, illustrated by the tabulation of products from reaction of norbornene under oxygen (favoring an electrophilic process) and under nitrogen (favoring a radical process) (Table 6.1) (100). Rearrangement and trans addition in one di-

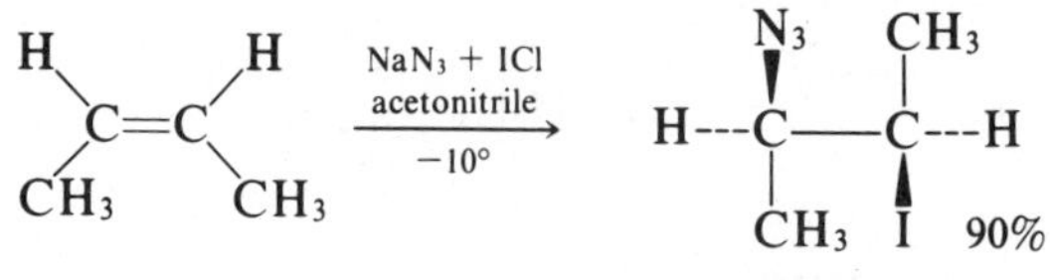

threo-2-azido-3-iodobutane

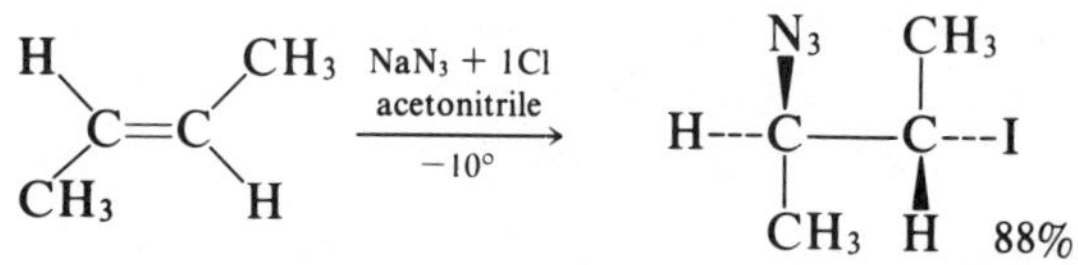

erythro-2-azido-3-iodobutane

rection occur in the electrophilic process; these same reactions as well as trans addition in the opposite sense and cis addition take place in the radical process.

Acylation of azide ion by acyl halides proceeds easily in high yields, as

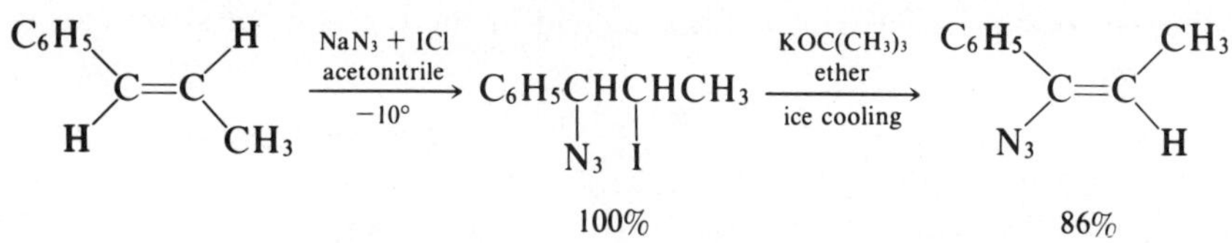

Table 6.1. Norbornene-Iodine Azide Reaction Products (Acetonitrile, −10 to −15°) (% Yield)

	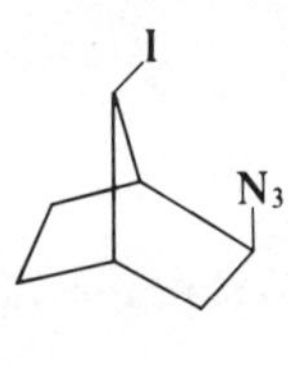	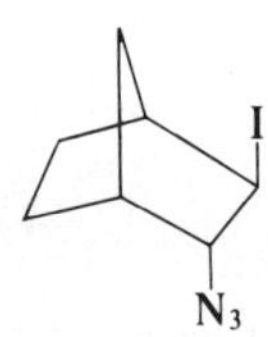	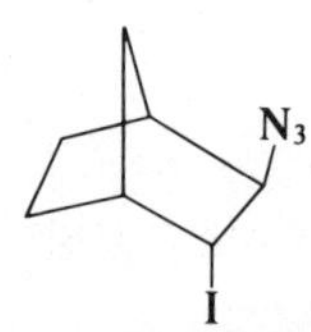	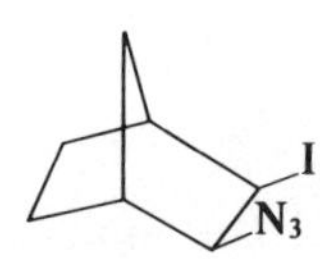
Under oxygen	49	8	—	—
Under nitrogen	26	2	15	33

	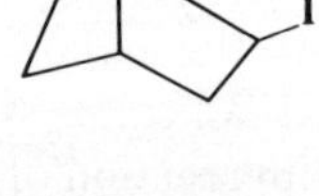	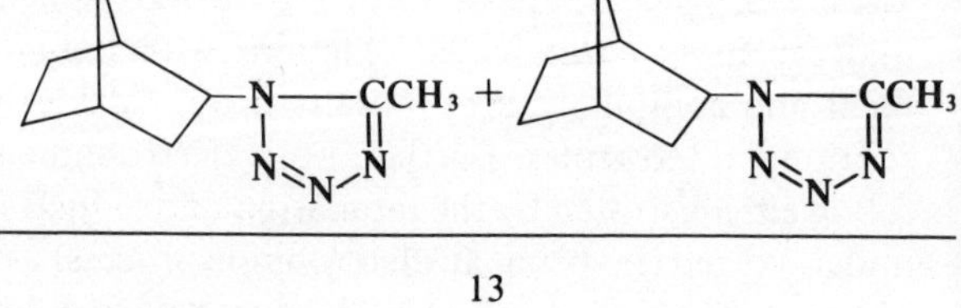
Under oxygen	5	13
Under nitrogen	3	11

shown by the syntheses of *m*-nitrobenzoyl azide (372), the elastomer cross-linking agent, bis[2-(azidocarbonyl)oxy]ethyl-1,3-benzene dicarboxylate (219, 220), and *p*-toluenesulfonylazide (254). Cyanogen azide is obtained in nearly

$$m\text{-}O_2NC_6H_4COCl \xrightarrow[20\text{–}25^\circ]{NaN_3,\ \text{acetone}/H_2O} m\text{-}O_2NC_6H_4CON_3 \quad 80\text{–}90\%$$

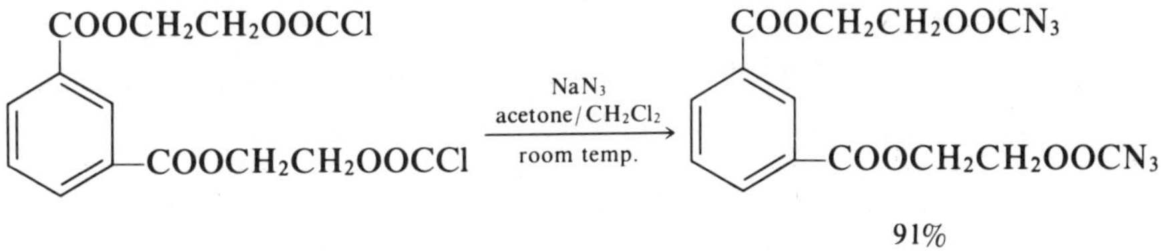

$$p\text{-}CH_3C_6H_4SO_2Cl \xrightarrow[0^\circ]{NaN_3,\ \text{acetone}/H_2O} p\text{-}CH_3C_6H_4SO_2N_3 \quad 90\%$$

quantitative yield in solution in acetonitrile or other anhydrous solvent from the reaction of sodium azide and cyanogen chloride (350). In aqueous solution, cyanogen bromide combines with 2 moles of azide to form either sodium 5-

$$ClCN \xrightarrow[0\text{–}12^\circ]{NaN_3,\ \text{acetonitrile}} NaCl + N_3CN \quad 98\%$$

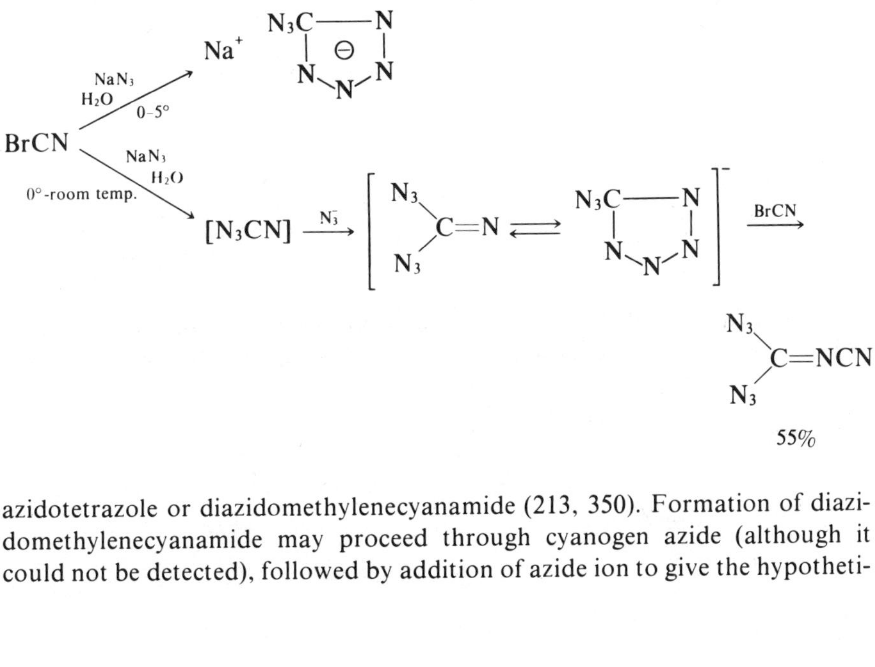

azidotetrazole or diazidomethylenecyanamide (213, 350). Formation of diazidomethylenecyanamide may proceed through cyanogen azide (although it could not be detected), followed by addition of azide ion to give the hypothetical *gem*-diazide, possibly in equilibrium with azidotetrazolate anion; reaction of this intermediate with cyanogen bromide leads to diazidomethylenecyanamide. Note is made that tetrazoles convert readily to imidoyl azides under the influence of acylating agents including cyanogen bromide (p. 473).

The dimer of azidotetrachloroantimony effects substitution of azide for chloride in carbon tetrachloride, chloroformimidium salts, and phosphorus trichloride.

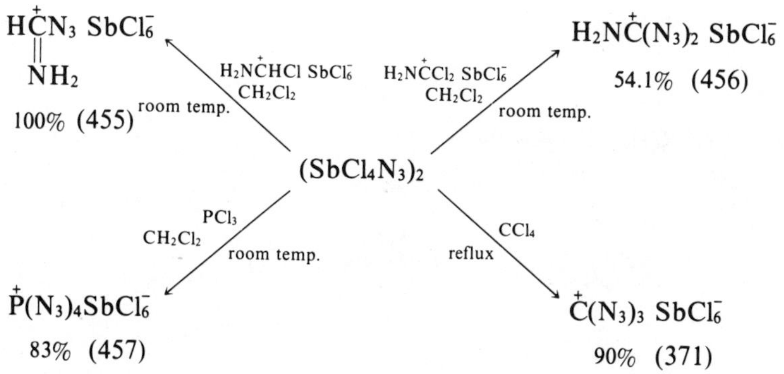

Replacement of diazonium by azide ion is the main method for its arylation, illustrated by the preparation of 2-azidobiphenyl and *p*-azidobenzaldehyde, an intermediate in the manufacture of photolithographic and photoresist azides. Diazonium compounds from weaker aromatic bases, conveniently diazotized

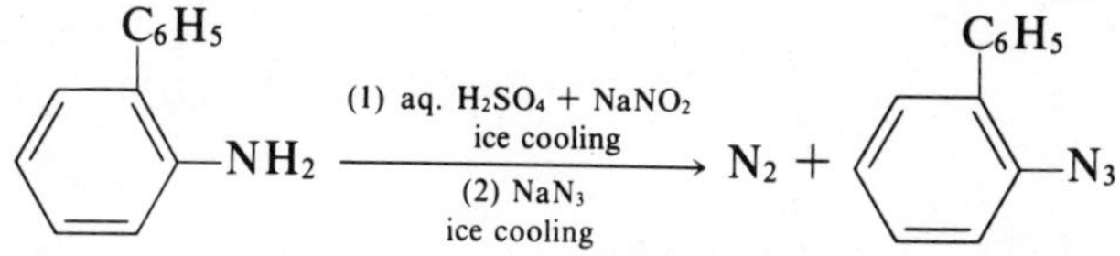

88% (478)

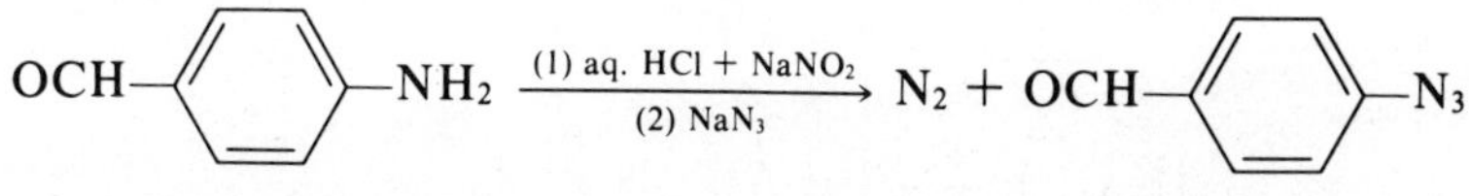

53% (158)

by amyl nitrite in acid medium, also react to form azides. Superficially, substitution of diazonium by azide appears similar to the S_N2 reaction with chloride,

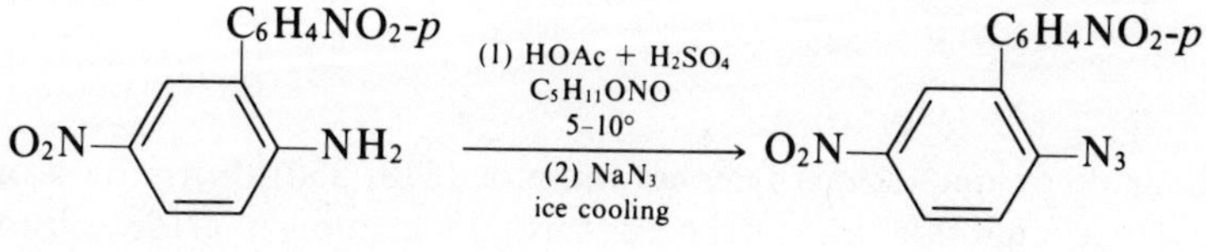

78% (478)

bromide, or thiocyanate, but the rate of the process with azide is appreciably faster (324) and involves coupling to form intermediate diazoazides and pentazoles, discussed in Section 2.9, Pentazoles. Benzotriazinones, reacting as masked diazonium compounds, also arylate azide ions (180).

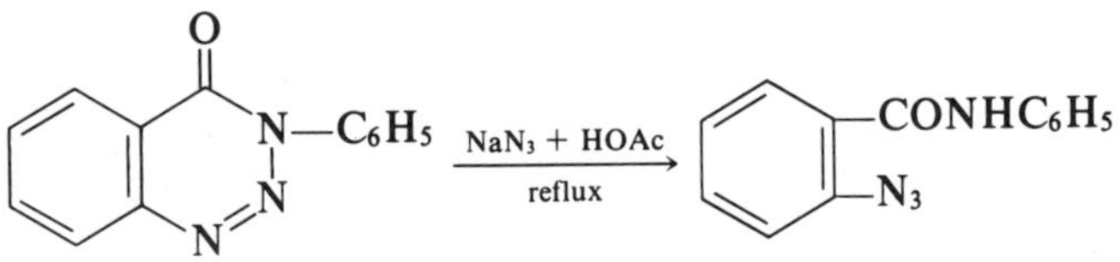

95% (180)

Alkyl and aryl groups in Grignard reagents and other metal organic compounds combine with the end nitrogen of alkyl, aryl, and acyl azides forming 1,3-disubstituted triazenes after hydrolysis of the intermediate metal adduct. This synthesis enables the preparation of disubstituted triazenes carrying two alkyl groups. Yields from acyl azides are poor because of side reactions consuming Grignard reagent. Cyanide also adds to terminal azide nitrogen forming cyanotriazenes.

$$C_6H_5N_3 \xrightarrow[\text{ether}, <0^\circ]{CH_3MgI} \left[C_6H_5N{=}N{-}N(CH_3){-}MgI \right] \xrightarrow{\text{aq. } NH_4Cl,\ NH_4OH} C_6H_5N{=}NNHCH_3 \qquad (130)$$

75%

$$C_6H_5CH_2N_3 \xrightarrow[\text{ice cooling}]{LiC_4H_9\text{-}n,\ \text{pentane}} \left[C_6H_5CH_2N{=}NN(C_4H_9\text{-}n){-}Li \right] \xrightarrow{\text{aq. } NH_4Cl,\ NH_4OH} C_6H_5CH_2N{=}NNHC_4H_9\text{-}n \qquad (464)$$

92%

$$C_6H_5CON_3 \xrightarrow[\text{ether, ice cooling}]{C_6H_5MgBr} C_6H_5CONHN{=}NC_6H_5 \qquad (64)$$

8%

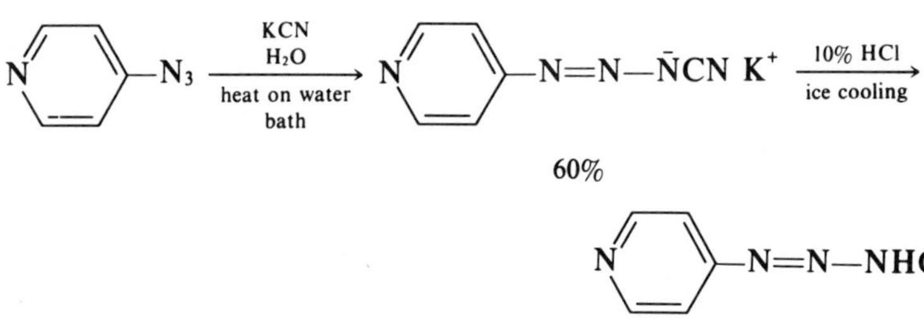

(510a)

Sulfonyl azides can be used in the same type of procedure to transfer the azide group, via an intermediate triazene, to form alkyl, aryl, or heteroaryl azides. Thus tosyl azide transfers azide to the norbornane anion and heating the

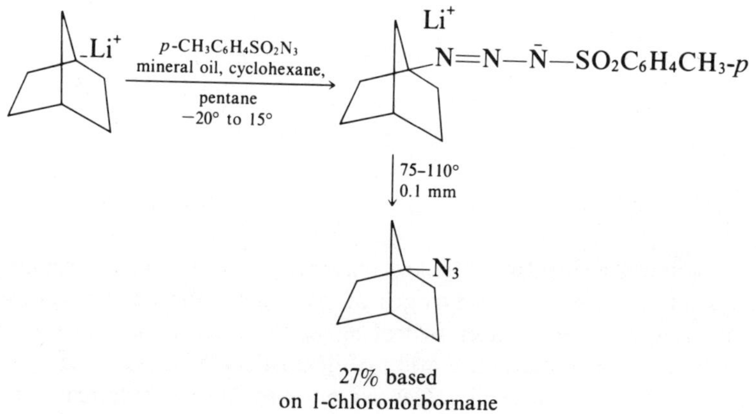

intermediate triazene liberates 1-azidonorbornane (429). Alternatively treatment of a sulfonyl triazene salt with alkali hydroxides or sodium pyrophosphate liberates the azide without the danger of a heating step, shown in the

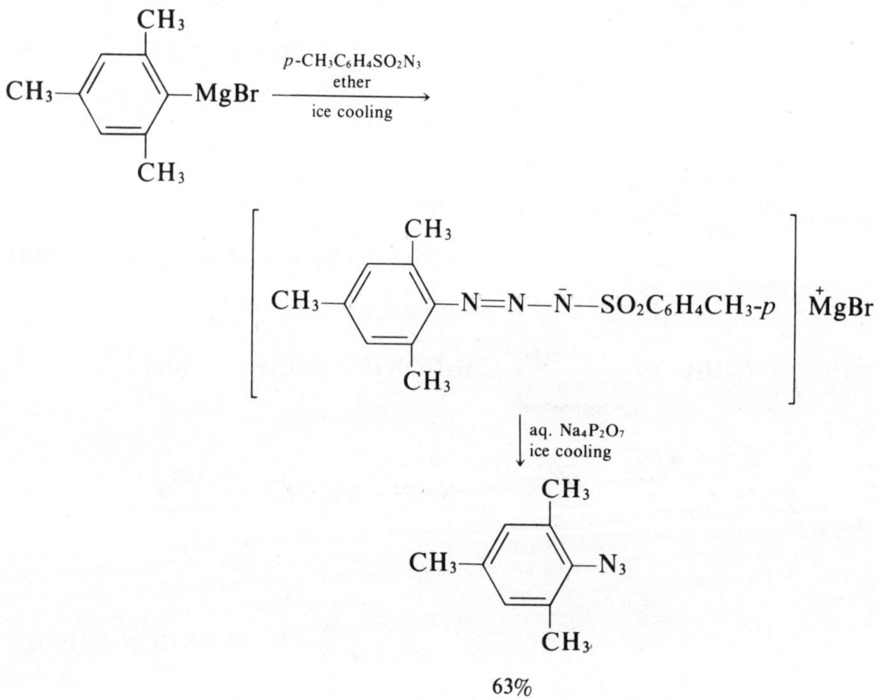

preparation of mesityl azide (481). No treatment to free the azide product is necessary in the synthesis of activated tertiary aliphatic azides (544).

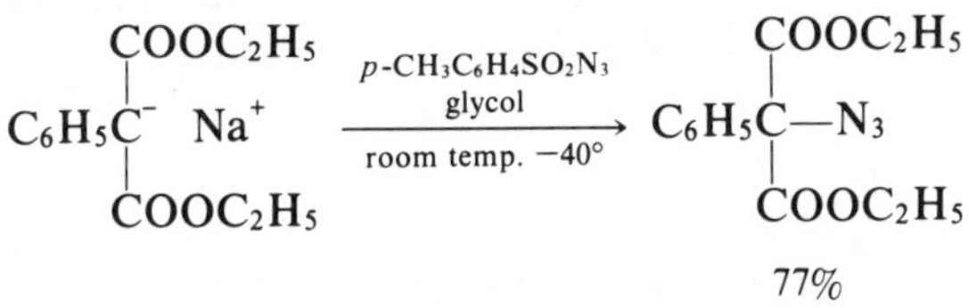

A different nucleophilic alkylation is involved in the reaction of acyl azides with 2 moles of carbonyl stabilized sulfonium ylids forming vinyl triazenes (535). The product with phenacylidenemethylphenylsulfurane **(IV)** and ethoxy-

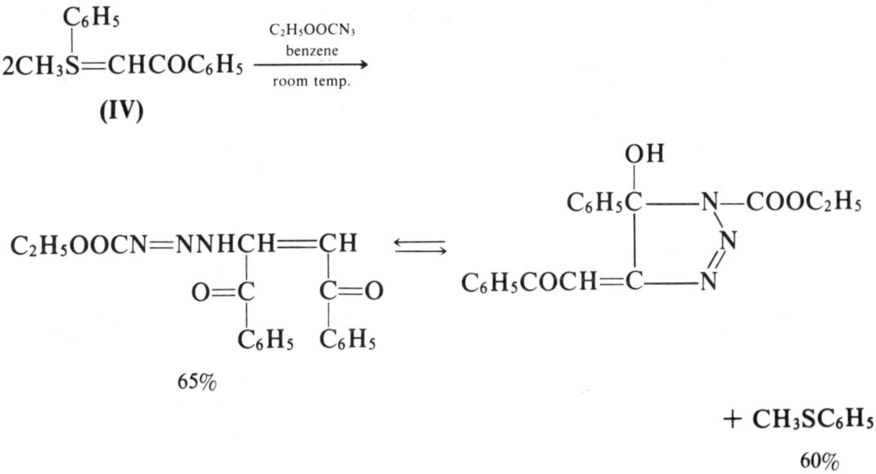

(IV)

carbonyl azide is a ketovinyltriazene in the solid state, but exists as a 5-hydroxytriazoline in benzene solution (see Section 2.3.2, Triazolines). With 3 moles of sulfonium ylid a triazenocyclopropane is formed. Use of a benzoyl azide substituted with an electron-withdrawing group for reaction with ethoxycarbonyl-

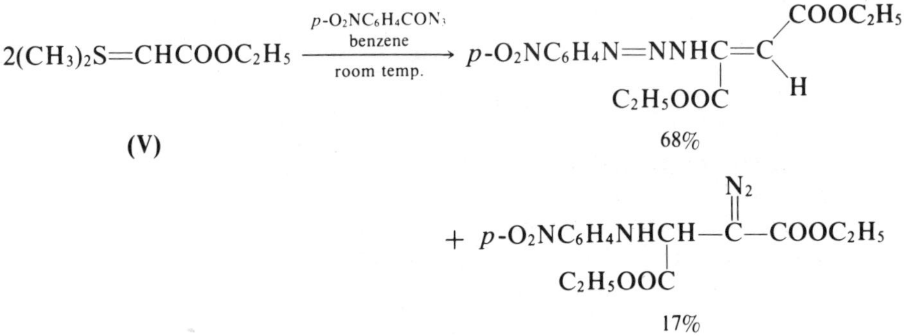

(V)

methylene dimethylsulfurane **(V)** produces both triazene and also a small amount of diazo compound (534).

The unstable diazonium amine salts are obtained from hydrogen azide or methyl azide adducts with antimony pentachloride (199) by alkylation at low temperature (454).

$$CH_3N_3 + SbCl_5 \xrightarrow[35\text{–}40°]{CH_2Cl_2} Cl_5\bar{S}b\text{—}\underset{}{\overset{CH_3}{\overset{|}{N}}}\text{—}\overset{+}{N}{\equiv}N \xrightarrow[-78°]{(CH_3)_3CCl,\ CH_2Cl_2} (CH_3)_3C\overset{CH_3}{\overset{|}{N}}\text{—}\overset{+}{N}{\equiv}N\ SbCl_6^-$$

Sulfonyl azides, azidium salts, and other heterocyclic azides interact with active methylene compounds and hydroxyaromatics in the presence of equimolar quantities of base to form diazo compounds and sulfonamides or heterocyclic amines. Thus a diazo group is transferred from tosyl azide to diacylmethanes,

$$(C_6H_5CO)_2CH_2 \xrightarrow[(2)\ 6N\ HCl]{(1)\ p\text{-}CH_3C_6H_4SO_2N_3,\ \text{piperidine},\ CH_2Cl_2,\ 0\text{–}3°} p\text{-}CH_3C_6H_4SO_2NH_2 + \underset{70\%}{(C_6H_5CO)_2C{=}N_2} \quad (438)$$

β-ketoesters, or malonic esters with the use of piperidine, triethylamine, or alkali ethoxide; diazocyclopentadiene results from lithium cyclopentadiene. When the reaction with cyclopentadiene is carried out with tosyl azide labeled

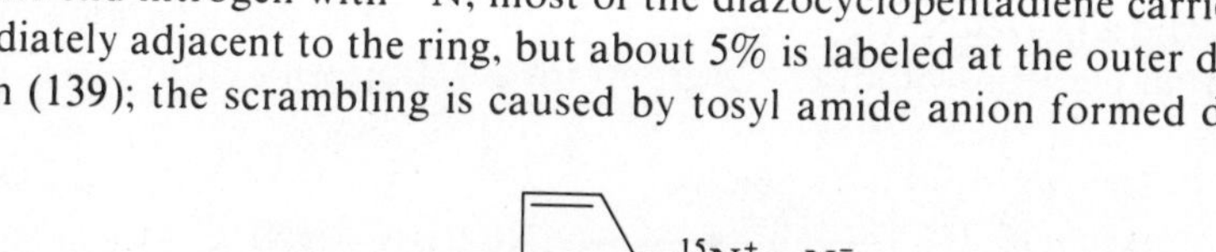

(136)

35%

at the end nitrogen with ^{15}N, most of the diazocyclopentadiene carries ^{15}N immediately adjacent to the ring, but about 5% is labeled at the outer diazo position (139); the scrambling is caused by tosyl amide anion formed during the

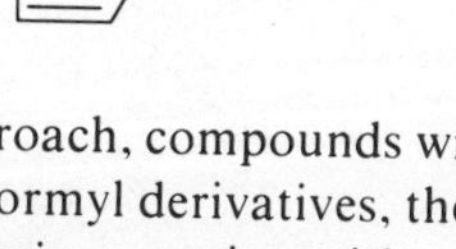

reaction (109). In a related approach, compounds with a single carbonyl next to methylene are converted to α-formyl derivatives, thereby enhancing the activation of the remaining C—H; during reaction with tosyl azide, the formyl is displaced with formation of a diazo compound (439).

Azidium salts effect diazo transfer usually in comparable or better yields

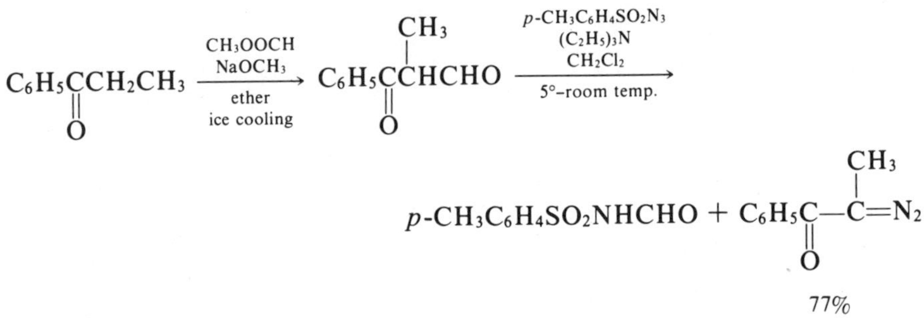

than does tosyl azide. Similarly, heterocyclic azides also are reduced to amines during the diazo transfer reaction with acetylacetone or other 1,3-dicarbonyl

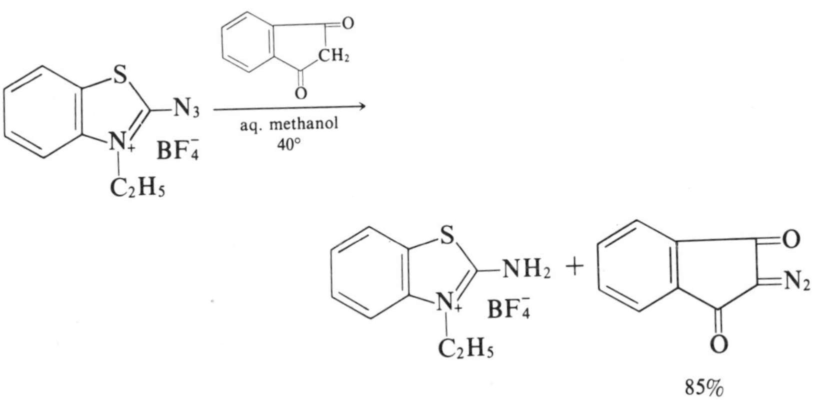

(28)

compounds (490), with formation of the diazodiketone in practically quantitative yield as determined by ^{1}H NMR (475).

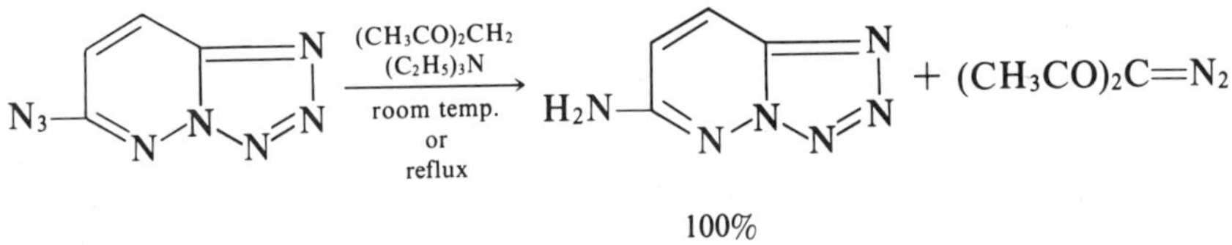

Azo compounds result from the transfer of two nitrogens from acyl azides or azidium salts to phenols (513), heterocycles (23, 412), and *o*-alkylaryl Grignard reagents (481).

Various interactions are known in which azides furnish one nitrogen atom to a reactive one-carbon unit. Reactions with dichlorocarbene, carbon monoxide, and isocyanides lead, respectively, to isocyanide dichlorides, isocyanates, and carbodiimides. The reaction with carbon monoxide probably proceeds by a ni-

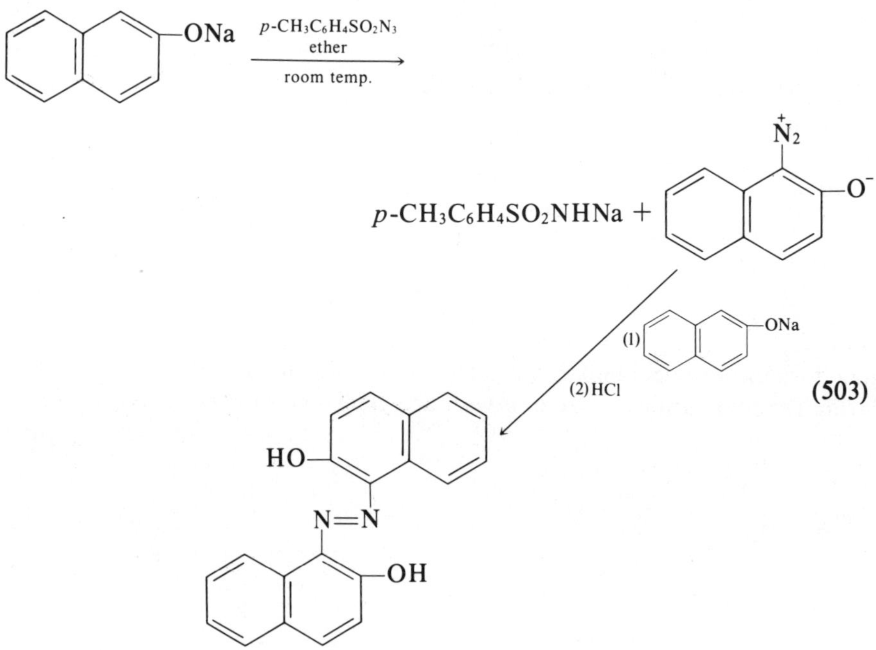

(503)

39%

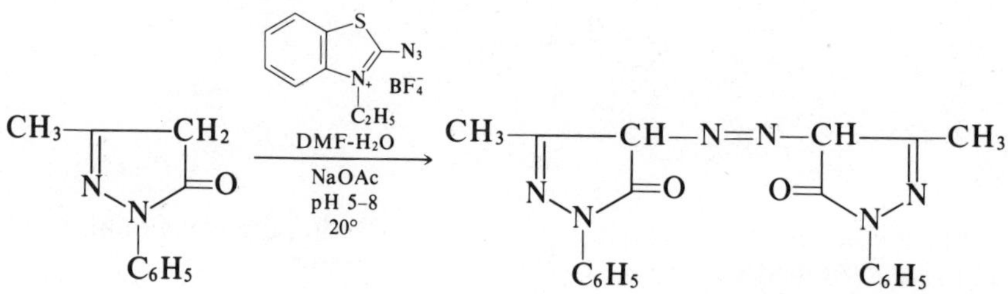

(29)

97%

$$CHCl_3 \xrightarrow{KOC(CH_3)_3} [CCl_2]$$

$$n\text{-}C_8H_{17}N_3 + [CCl_2] \xrightarrow[0^\circ]{\text{pentane}} N_2 + n\text{-}C_8H_{17}N{=}CCl_2 \qquad (25)$$

89%

$$C_6H_5N_3 + CO \xrightarrow[\substack{200\text{–}300 \\ \text{atm} \\ CCl_2FCF_2Cl}]{160\text{–}180^\circ} N_2 + C_6H_5NCO \qquad (57, 58)$$

47%

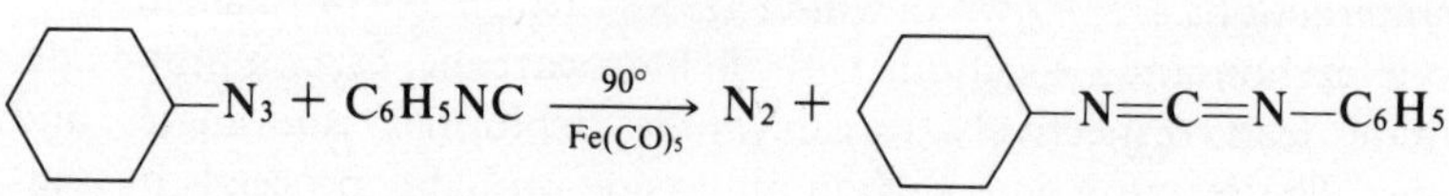

(449)

51%

trene intermediate, but this is unlikely for the reaction with dichlorocarbene, because of the mild conditions used.

On boiling *N*-(*o*-azidophenyl)pyrrolidine in acetic anhydride, the azide nitrogen remaining after loss of elemental nitrogen undergoes internal alkylation as well as acetylation (361).

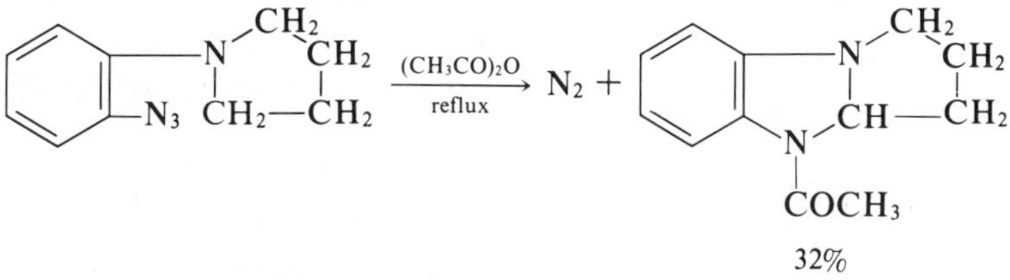

Methylene cycloalkanes react with aromatic sulfonyl azides, probably through a triazoline, to form the next higher arene sulfonimidocycloalkene as shown by the reaction of *o*-nitrobenzenesulfonyl azide with methylene cyclopentane. The same product is obtained from cyclohexene. Yields are nearly quantitative.

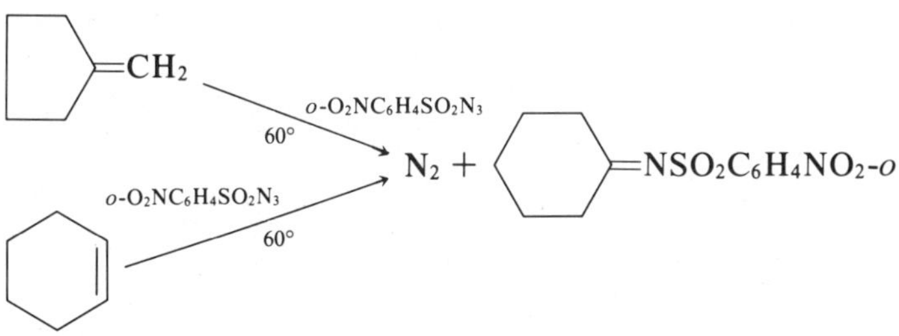

(561)

Benzoyl azide reacts with the α-carbon of imines and enamines to form *N*-benzoyl amidines, identified by hydrolysis to secondary amides (90). During the

$$(CH_3)_2CHCH{=}NCH_3 \xrightarrow[\text{room temp.}]{C_6H_5CON_3,\ CH_3CN} N_2 + C_6H_5CONH\overset{\overset{\displaystyle NCH_3}{\|}}{C}CH(CH_3)_2 \xrightarrow[H_2O]{H^+} C_6H_5CONHCOCH(CH_3)_2 \quad 69\%$$

reaction of 4-(1-cyclohexene-1-yl)morpholine, the six-membered ring contracts to cyclopentyl. Acyclic vinyl ethers react exothermally with benzene sulfonylazide, probably through a triazoline, to form a 1:1 polymer adduct and a dimer as the principal products (161). Reaction of the azide with dihydropyran is less vigorous, but gives *N*-phenylsulfonyltetrahydropyranone-2-imine in quantita-

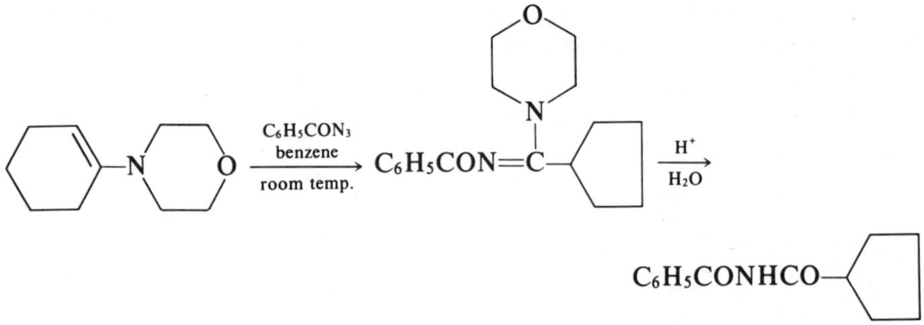

82%

tive yield. Methanesulfonyl azide furnishes one nitrogen for a diaziridine ring on reaction with ketenimines (314).

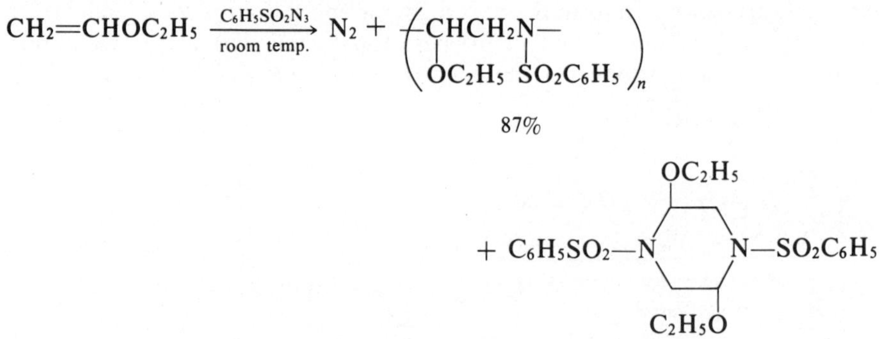

87%

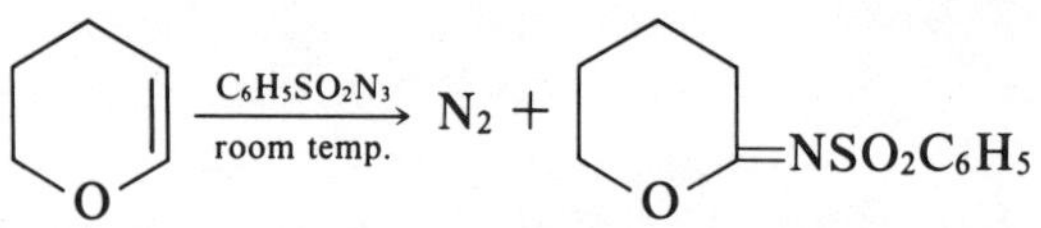

100%

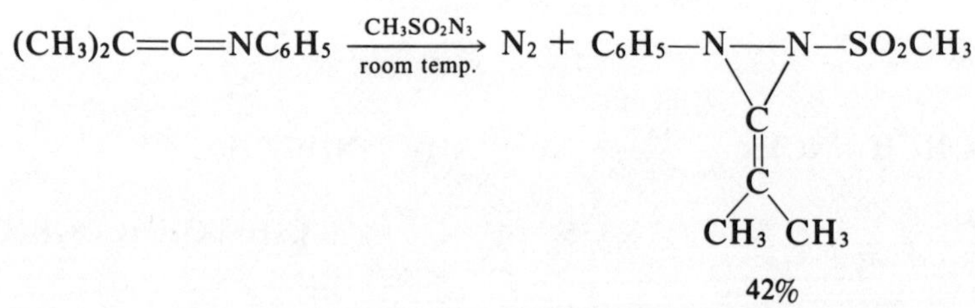

42%

Interaction of phenyl azide and a mixture of acetyl chloride, acetic anhydride, and aluminum chloride leads to *p*-chloroacetanilide, probably through preliminary formation of *N*-chloroacetanilide (73). Aromatic azides with an

$$C_6H_5N_3 \xrightarrow[AlCl_3]{CH_3COCl + (CH_3CO)_2O} N_2 + [C_6H_5NClCOCH_3] \longrightarrow p\text{-}ClC_6H_4NHCOCH_3$$

electron-withdrawing para substituent and a vacant ortho position form oxazoles on heating with a carboxylic acid and polyphosphoric acid (176). In alum-

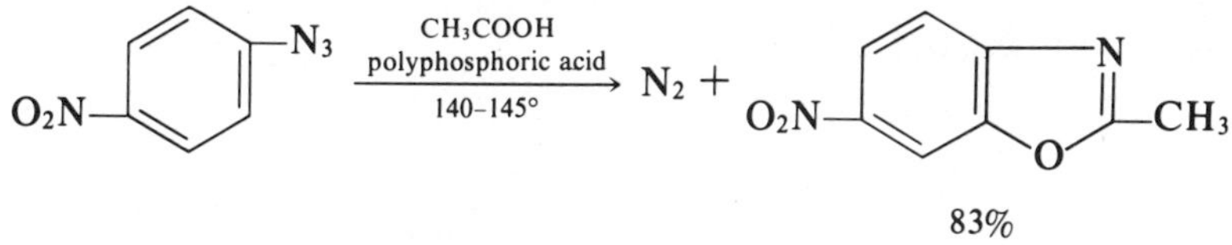

inum chloride-catalyzed reactions, azidoacetone is arylated by reaction with benzene and benzhydryl azide undergoes rearrangement to form phenylmethylene anil.

$$C_6H_6 + CH_3COCH_2N_3 \xrightarrow{AlCl_3} N_2 + \underset{35\%}{CH_3COCH_2NHC_6H_5} \qquad (301)$$

$$C_6H_6 + (C_6H_5)_2CHN_3 \xrightarrow[50^\circ]{AlCl_3} N_2 + \underset{44\%}{C_6H_5CH{=}NC_6H_5} + \underset{8\%}{(C_6H_5)_2CHC_6H_5} \qquad (302)$$

Aromatic thioketones, but not ketones, react with phenyl azide forming ketimines (461).

$$C_6H_5N_3 + S{=}C(C_6H_5)_2 \xrightarrow{110^\circ} N_2 + C_6H_5N{=}C(C_6H_5)_2$$

In a slow reaction, alkyl azides combine with aryl isothiocyanates to produce 1:1 adducts, 1-alkyl-4-aryltetrazoline-5-thiones, and also 1:2 adducts, principally thiadiazolidines and dithiazolidines. Products isolated from the combination of benzyl azide and phenylisothiocyanate after 42 days are shown.

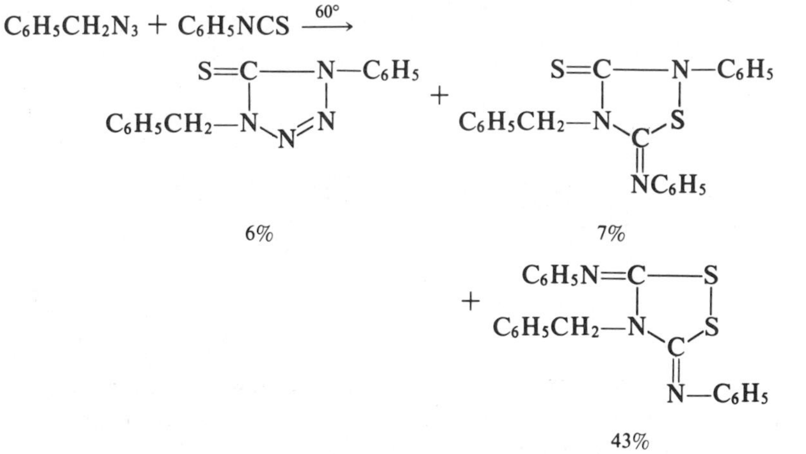

(311)

The combination of aroylisothiocyanates with alkyl azides produces 1,2,4-thiadiazoles in largest quantity with smaller quantities of 1,2,4-dithiazolidines.

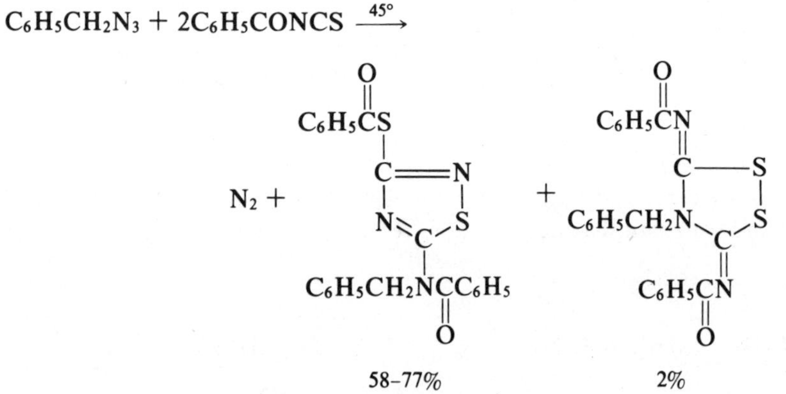

(308)

Benzotriazinones can be alkylated at several locations on the nitrogen chain, as well as on the 4-oxygen. In many studies only the N-3 isomer has been found, shown in the reaction of 3-dimethylaminopropyl chloride with 7-chlorobenzotriazinone, and in the formation of 3-hydroxymethylbenzotriazinone, the first step in the manufacture of azinphos-methyl (335, 475). In a more extended

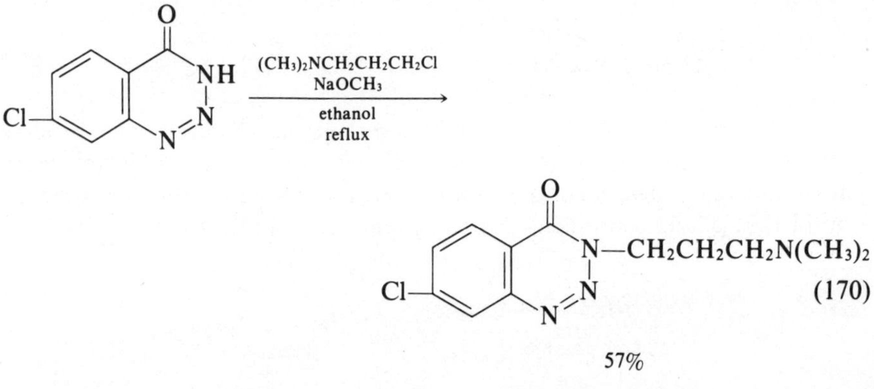

(170)

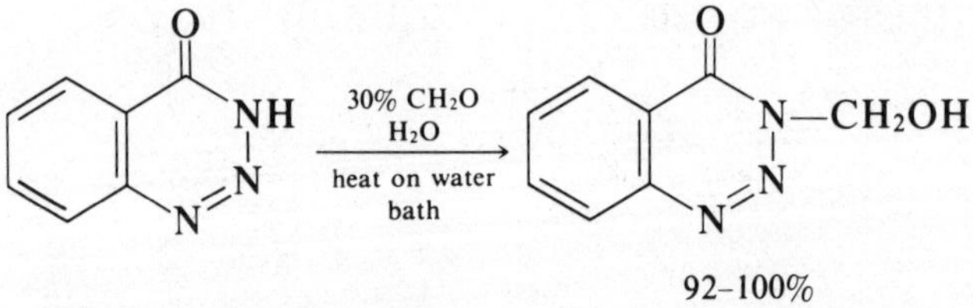

study with 2,3,4,6-tetraacetyl-α-D-glycopyranosyl bromide, alkylation was found at N-3, N-2 and on oxygen (542). The product substituted at N-2 is a betaine. With triethyloxonium fluoborate, benzotriazinone is converted to a mixture of 3-N ethyl and *O*-ethyl derivatives in a ratio of about 3:1, with no evi-

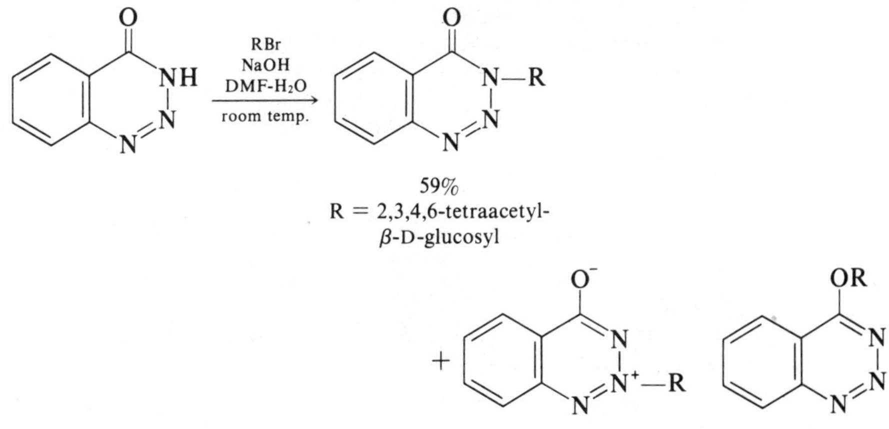

59%
R = 2,3,4,6-tetraacetyl-β-D-glucosyl

21% 17%

dence of a 2-substituted betaine (489). In contrast, benzotriazine-4-thione is alkylated only on sulfur (489). During methylation of 4-imino-3-phenyl-3,4-dihydrobenzotriazine, the 3-4 nitrogen-carbon bond is broken and both possible methyl triazenes are formed (494).

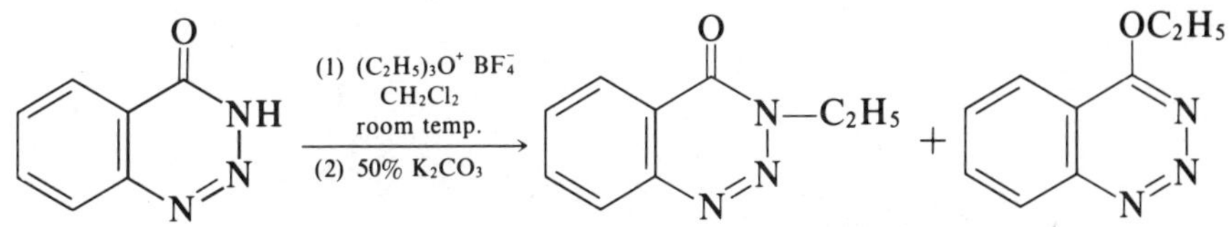

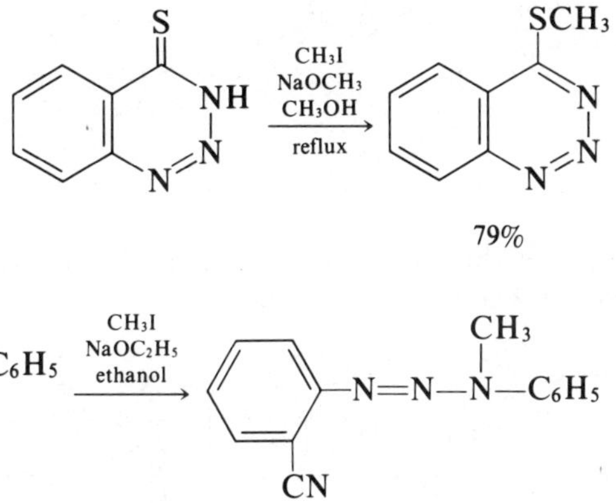

79%

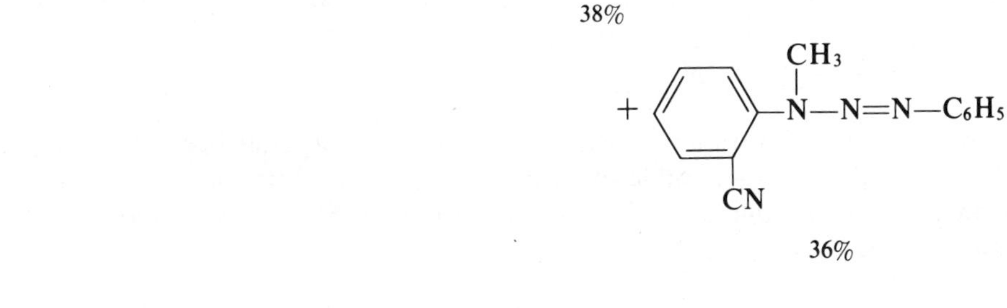

38%

36%

Table 6.2. Alkylation of Naphtho[1,8-*de*]-triazine

R	Substituted Naphtho[1,8-*de*]triazine, Yield (%) 1-Isomer	2-Isomer
C_2H_5	34	35
$C_{12}H_{25}$	40	59
$C_6H_5CH_2$	36.6	24.1
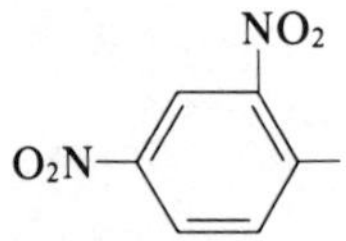	14	54

Substitution of naphtho[1,8-*de*]triazine leads to both 1- and 2-substituted isomers, illustrated by yields of a series of alkyl and aryl derivatives (Table 6.2) (512).

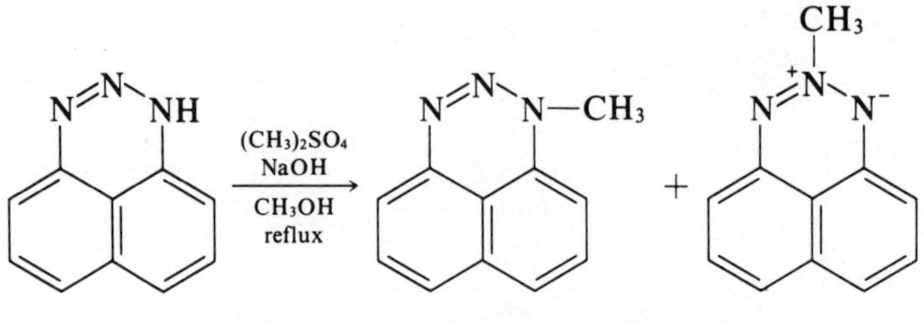

37% 31%

(414)

The nitrogen chain of the condensed *v*-triazine **(VI)** is broken on reaction with diethylmalonate, forming a hydrazone.

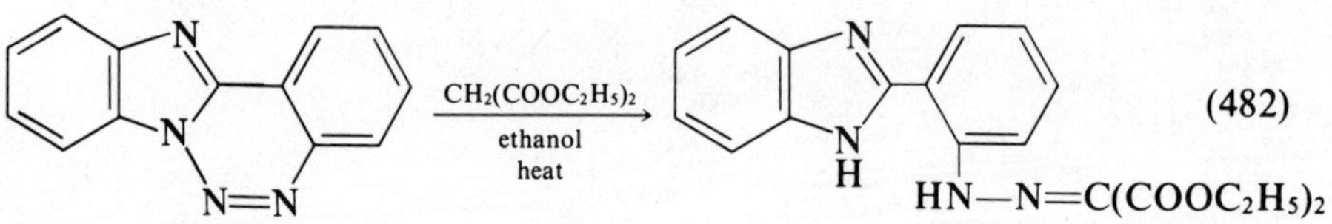

(VI) 88%

(482)

Monocyclic 4,5,6-triamino-*v*-triazines acylate at the 2-nitrogen, rendering the 5-amino group susceptible to hydrolysis with the formation of 2-acyl-4,6-diamino-5-oxo-2,5-dihydro-*v*-triazine derivatives (196). Benzotriazinone acylates at N-3.

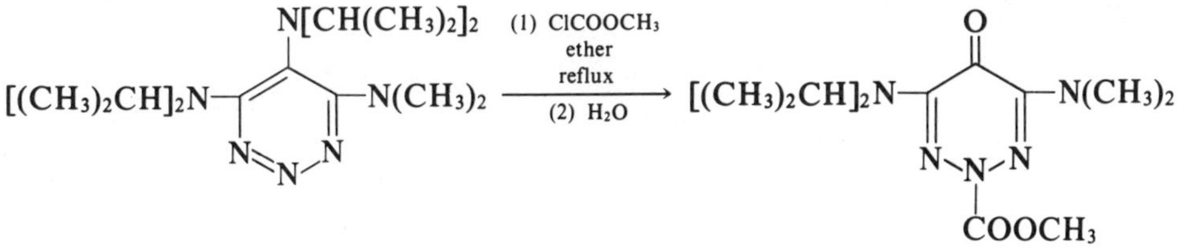

86%

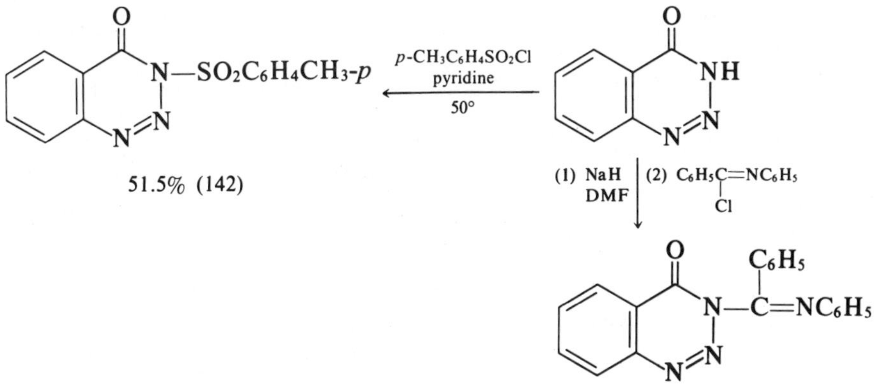

51.5% (142)

80% (409)

Nitrogen is displaced from fused ring *v*-triazines by isocyanates (236) and isothiocyanates (482).

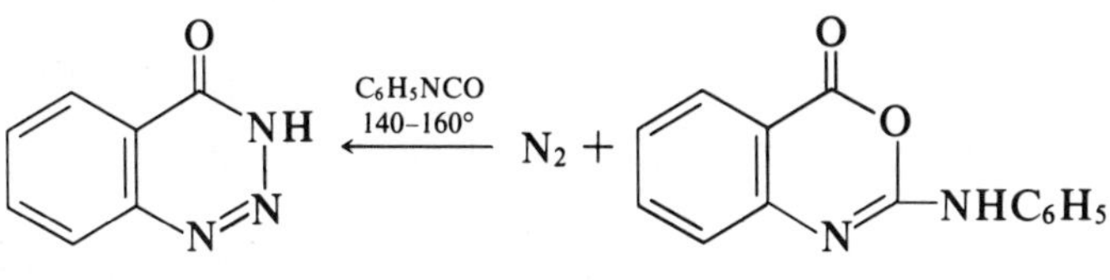

80%

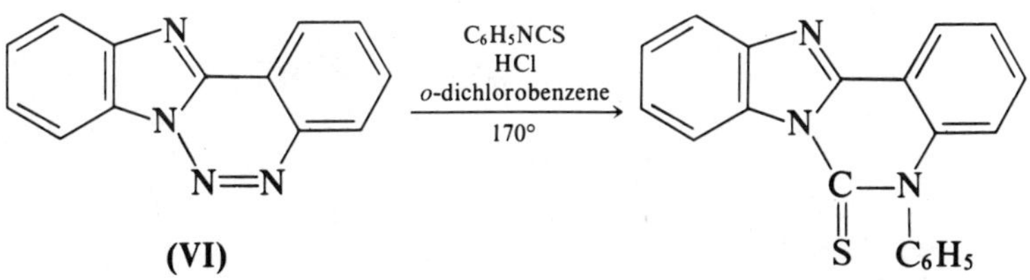

(VI)

36.6%

The principal site for alkylation of *v*-triazines to quaternary salts is at the middle nitrogen, as illustrated with monocyclic *v*-triazines and 4-substituted aminobenzotriazines. Treatment of a 4-arylaminobenzotriazinium iodide with base produces a betaine. Alkylation and arylation of benzotriazinone leads to

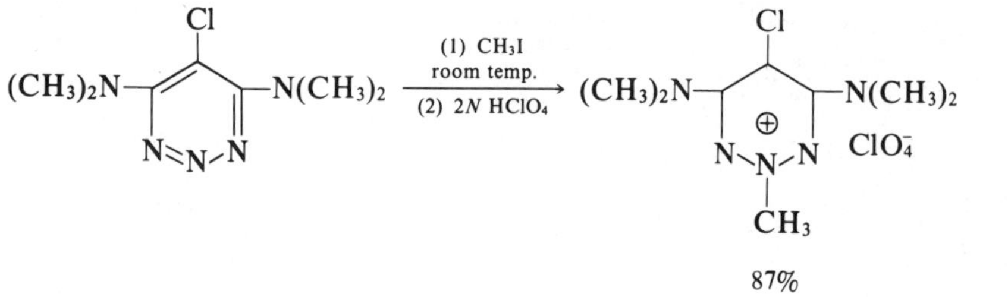

(196)

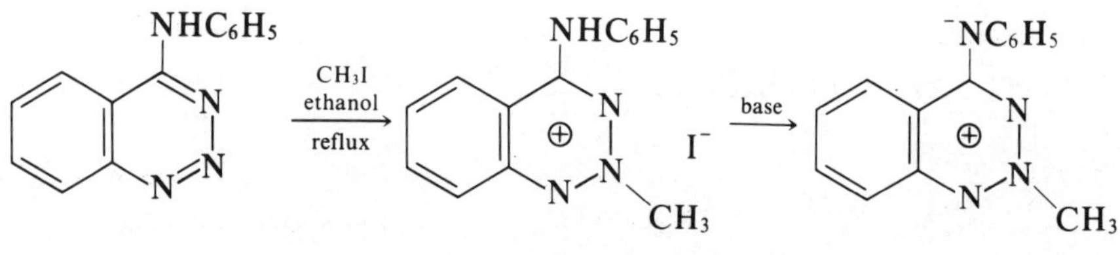

86% (495)

betaines also substituted at N-2 (359). Quaternization of 1-methylnaphtho-[1,8-*de*]-*v*-triazine, however, occurs at the 3-position (42, 512).

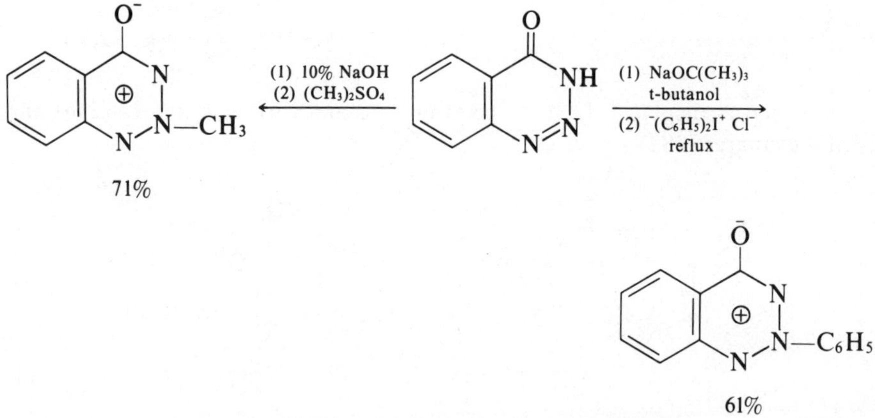

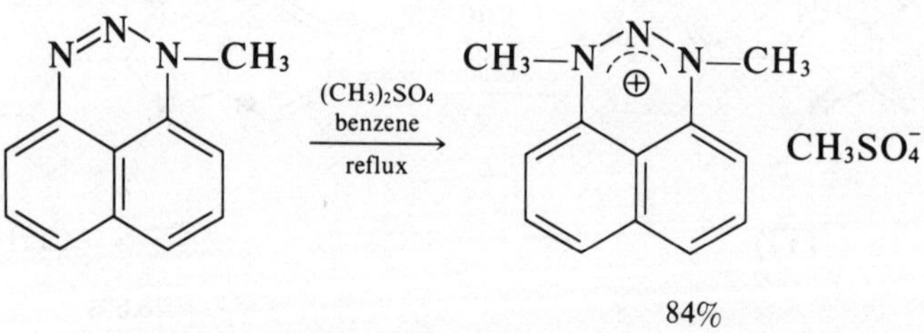

Both 5-phenyl and 5-alkylthiothiatriazoles are alkylated by triethyloxonium fluoborate with formation of 3-substituted thiatriazolium salts (244). Alkylation of 5-monosubstituted aminothiatriazoles, however, is at the 4-nitrogen,

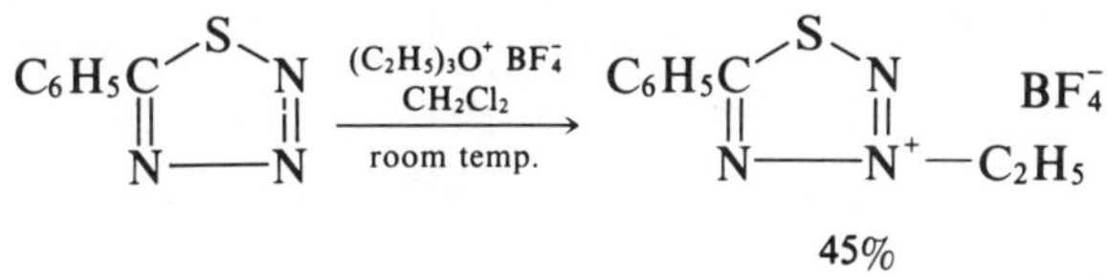

forming 5-substituted ammonium salts (527). Mesoionic 3-phenyloxatriazolium-5-thiolate alkylates at exocyclic sulfur (206).

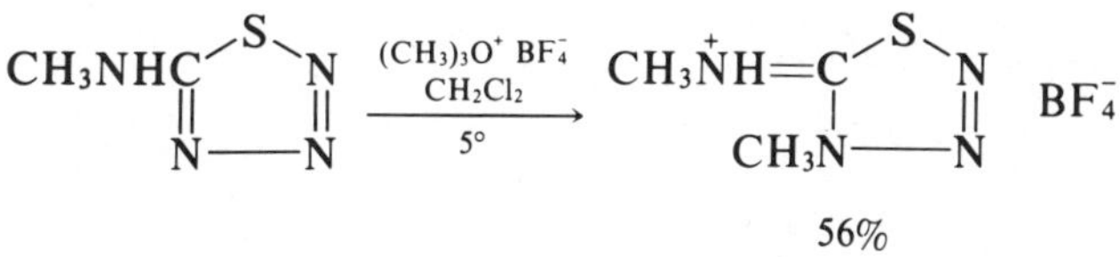

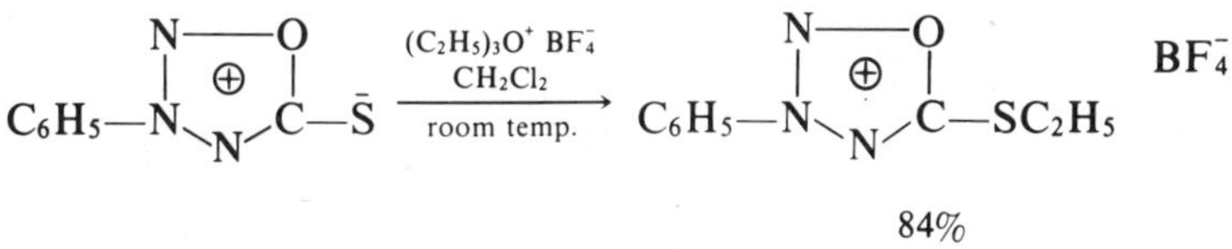

The thiatriazole ring in 5-aminothiatriazole is destroyed by the action of benzoyl chloride, forming formamidine disulfide and 3,5-di(benzoylamino)-1,2,4-thiadiazole (313). Reaction of 5-anilinothiatriazole with isocyanates also

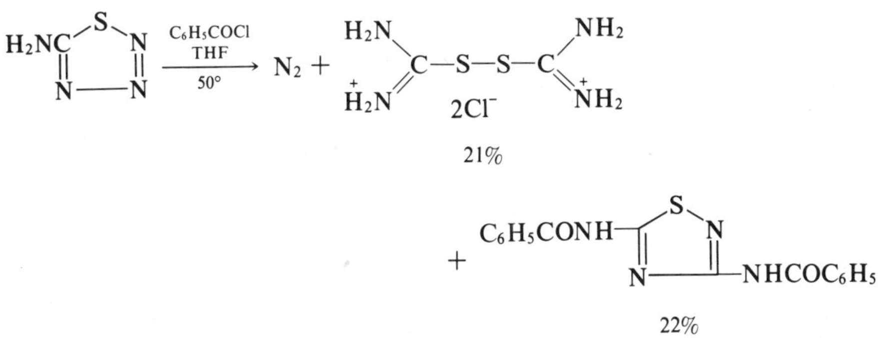

leads to a 1,2,4-thiadiazole replacing two ring nitrogens with the NC atoms of the isocyanate (281). Acylation of 5-aminothiatriazole with benzoyl chloride or

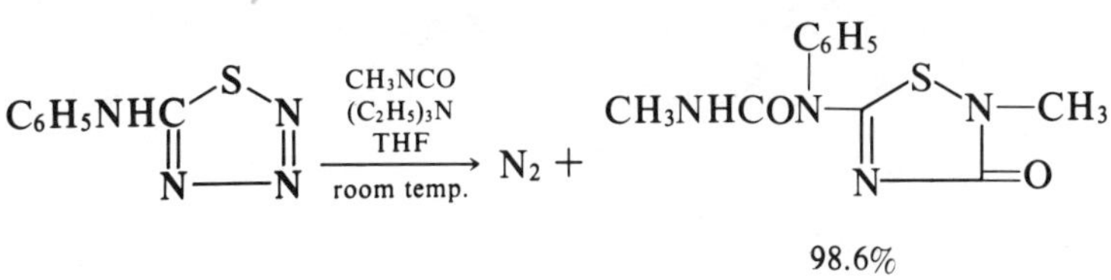

with imidoyl chlorides in the presence of base leads, with displacement of nitrogen, to a 6*a*-λ^4-thia-1,6-heteropentalene, evidently through an intermediate 1,2,4-thiadiazole or an open-chain dipolar isomer. Thiadiazole can be isolated in some instances.

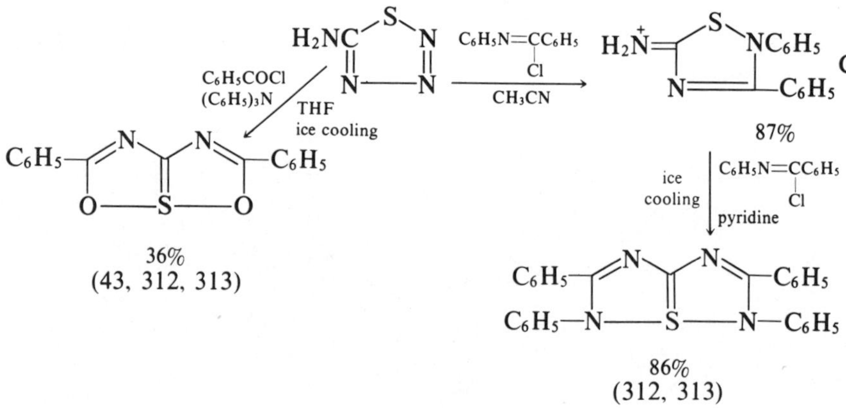

Nitrogen is lost from thiatriazolines by the action of reagents with cumulative double bonds.

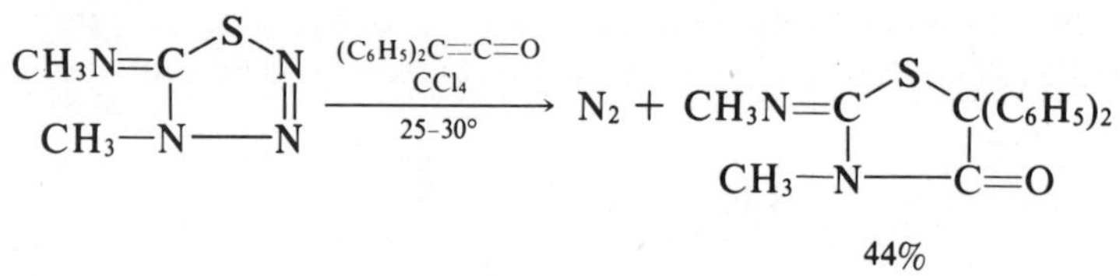

(527)

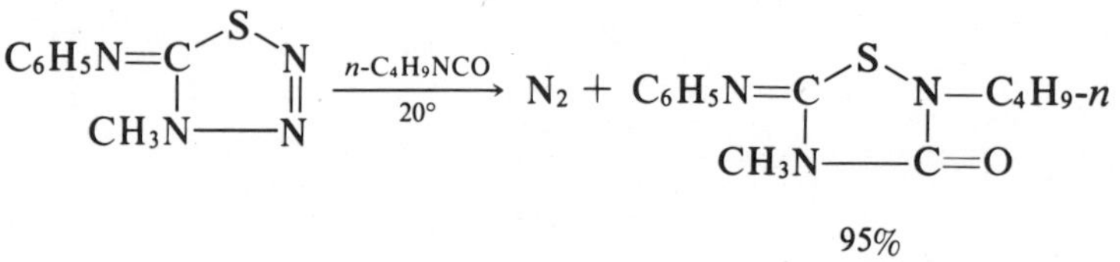

(310)

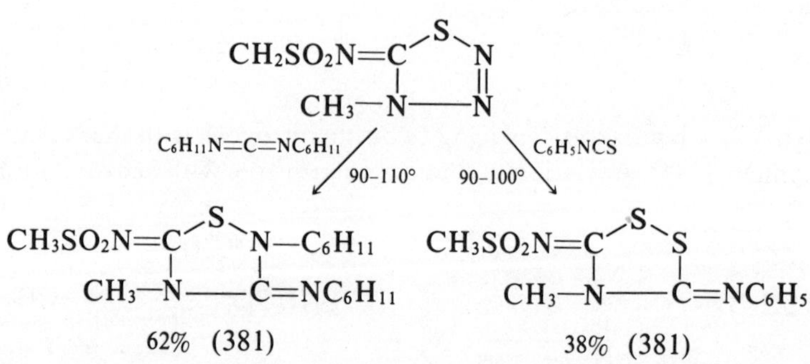

Both *v*-triazoles and tetrazoles with hydrogen on ring nitrogen can be alkylated or arylated under suitable conditions at each ring nitrogen position. The relative proportions of alkylation at each position are partially determined by reaction conditions, by the reagent and by substituents on ring carbon, but as yet are not completely predictable. Tanaka and Miller in a limited generaliza-

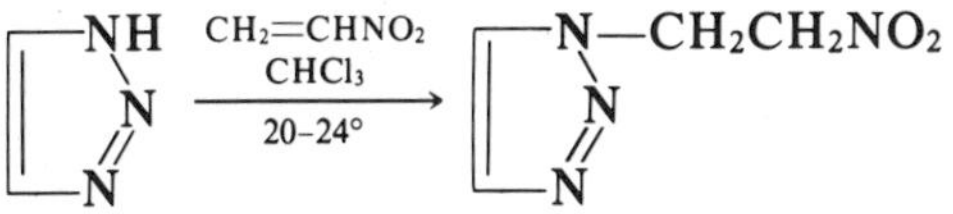

(189)

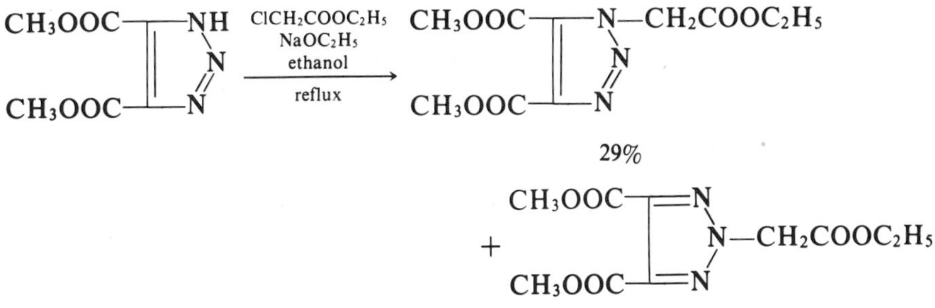

52% (511)

tion (511) note that neutral, un-ionized *v*-triazoles usually alkylate with alkyl halides and unsaturated compounds at N-1; that *v*-triazole anions are attacked by unsaturates at N-2; but that alkyl halides attack at both N-1 and N-2. Results of alkylation of some monocyclic *v*-triazoles are tabulated in Table 6.3, showing reasonable concordance with the generalization.

Diazomethane alkylates *v*-triazole and 4,5-dibromo-*v*-triazole at both N-1 and N-2, but *v*-triazoles substituted with benzoyloxy at position 4 are alkylated only at N-2.

The 4-aryl-5-triphenylphosphonium-*v*-triazole ylides are attacked only at N-2 of the *v*-triazole ring (510).

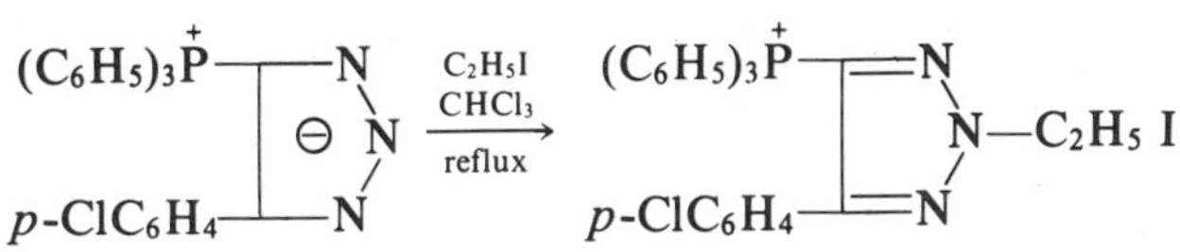

71%

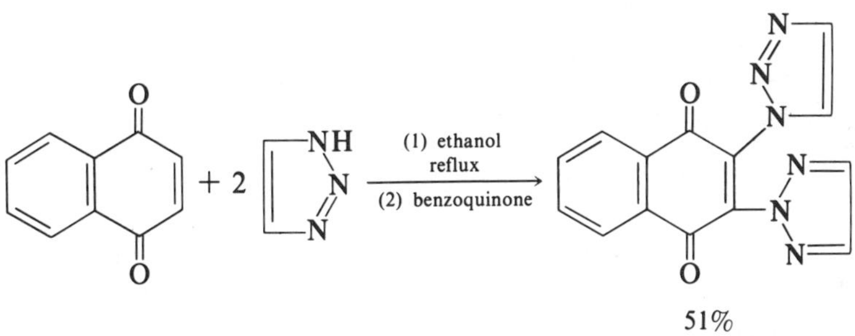

51%

Table 6.3. Alkylation of *v*-Triazoles

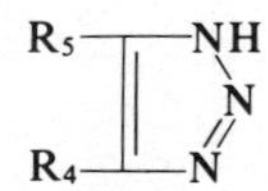

R_4	R_5	Alkylating or Arylating Agent	Conditions	Yield (%) 1-Isomer	Yield (%) 2-Isomer	Ref.
H	H	$CH_2{=}CHCH_2Br$	Room temp.	26.6	2.1	189
H	H	CH_2N_2	Ether, 0°	42	46	411
Br	Br	CH_2N_2	Ether, room temp.	35	46	51
C_6H_5COO	H	CH_2N_2	Ether, 0°		42	44
C_6H_5COO	C_6H_5	CH_2N_2	Ether, 0°		71	44
H	H	n-C_3H_7Br	$NaOCH_3$, CH_3OH, room temp.	50	16	189
H	H	o-$FC_6H_4NO_2$	Na_2CO_3, DMF, reflux	20	62	106
H	H	$CH_2{=}CHCN$	Trimethyl benzyl ammonium hydroxide, dioxane, 25–30°		99	189
H	H	$HC{\equiv}CCOC_6H_5$	2 drops $(C_2H_5)_3N$, CH_3OH, reflux	67.3	5	536
Br	Br	$(CH_3)_2SO_4$	1 *N* NaOH, room temp.	29	29	401
C_6H_5	H	$ClCH_2CH_2COOC_2H_5$	$NaOCH_3$, DMF, reflux	26	17	511

Both 1- and 2-triazole substitutions are found in the product resulting from addition of 2 moles of *v*-triazole to 1,4-naphthoquinone, followed by reoxidation to a quinone (178).

Condensation of 4-amino-5-phenyl-*v*-triazole with acetylacetone under basic conditions leads to 5,7-dimethyl-3-phenyl-*v*-triazolo[3,4-*a*]pyrimidine, but in acid medium the triazole ring is decomposed, as discussed in the section on stability toward electrophiles (504).

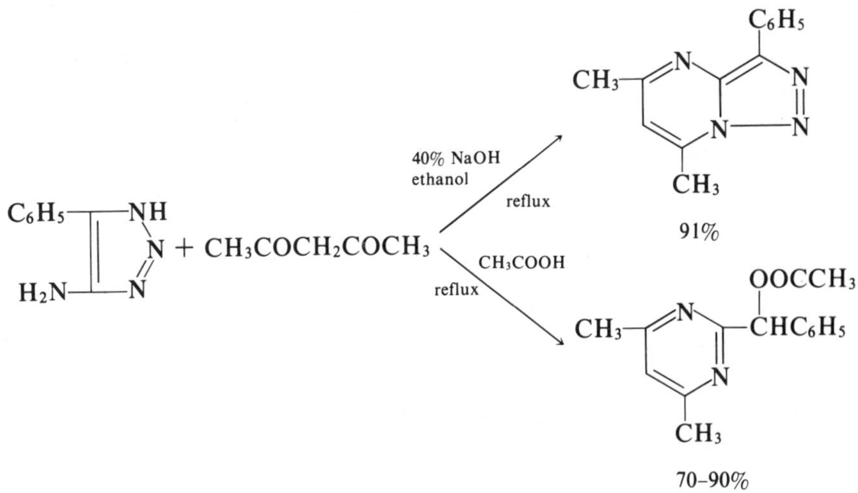

A *v*-triazolo[3,4-*c*]-*as*-triazine results by combined alkylation and coupling of 4-carboxamido-5-diazo-*v*-triazole and nonane-4,6-dione (395).

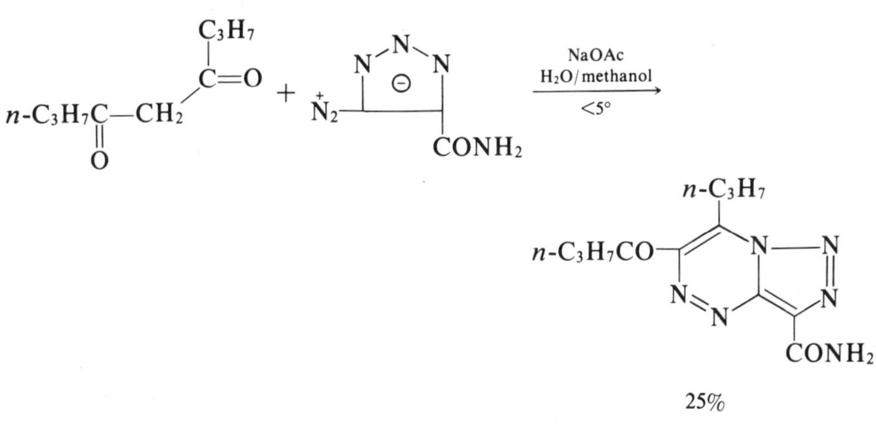

Results of methylation of benzotriazole are consistent with the conclusions from monocyclic *v*-triazoles. Methyl iodide under neutral conditions substitutes only at N-1, but under alkaline conditions both isomers are formed (Table 6.4).

Table 6.4. Methylation of Benzotriazole

Methylating Agent	Conditions	Methylation Ratios (408)	
		1-Isomer	2-Isomer
CH_3I	KOH, CH_3OH, reflux	61	39
$(CH_3)_2SO_4$	NaOH, CH_3OH, reflux	65	35
CH_2N_2	C_2H_5OH	35	65

Alkylation of benzotriazole anion under phase transfer conditions leads to 1-substitution in 52-80% and 2-substitution in 18-37% yield (72). Although basic

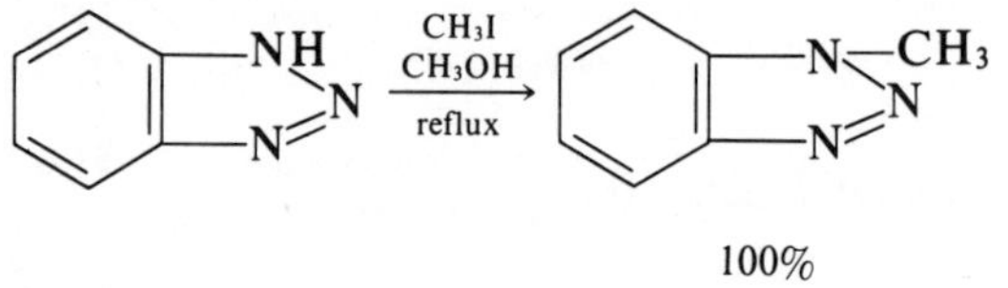

(408)

cyclohexyl tosylate forms only 2-cyclohexyl benzotriazole, neutral cyclohexyl phosphate forms only the 1-isomer (468). All three isomers are formed in the alkylation of 5-substituted benzotriazoles (453). Yields of each isomer are almost the same whether the 5-substituent is methyl, phenyl, or nitro. Michler's

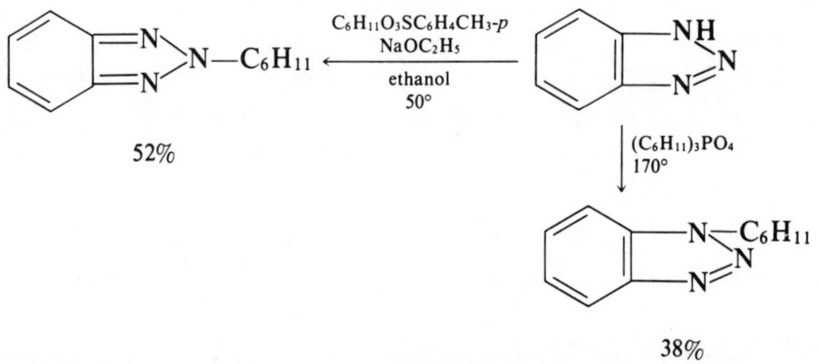

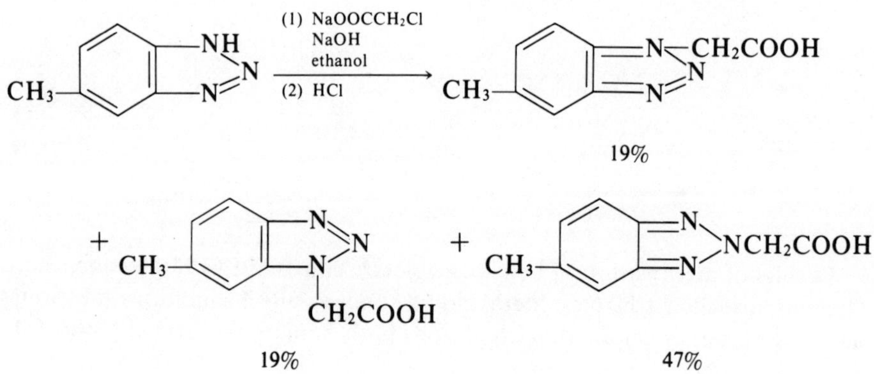

hydrol, a secondary alcohol, alkylates benzotriazole in the presence of ammonium hydroxide to form the transfer copy material, 1-bis(*p*-dimethylaminophenyl)methylbenzotriazole (297).

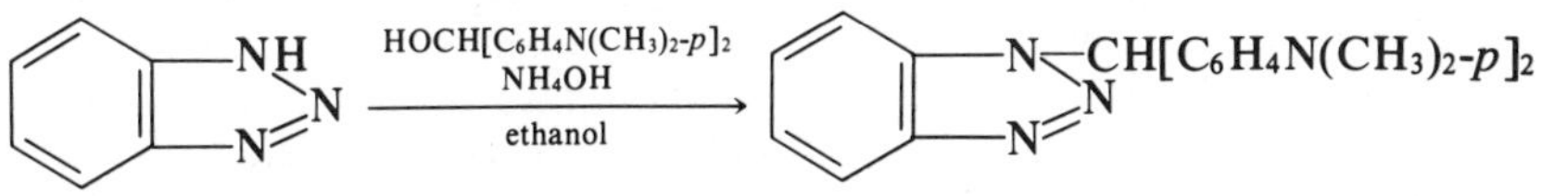

Mannich type reactions with *v*-triazoles carrying hydrogen on ring nitrogen lead to 1-substituted derivatives.

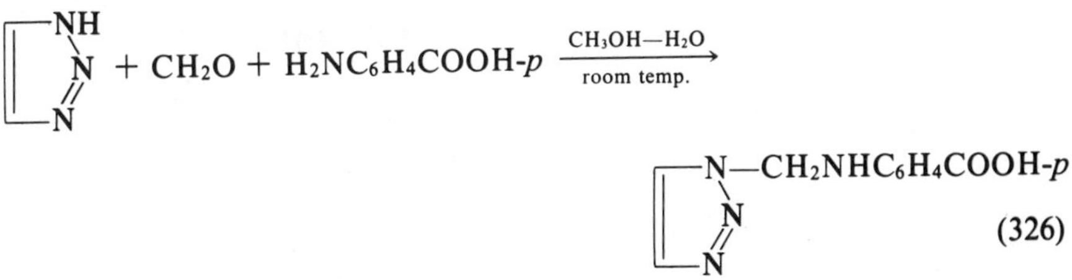

(326)

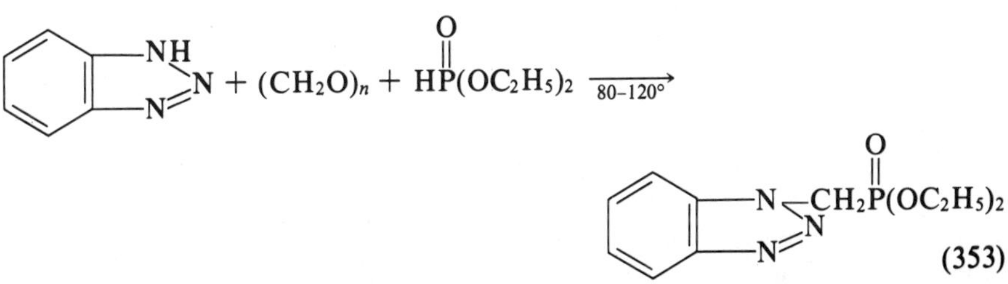

(353)

77%

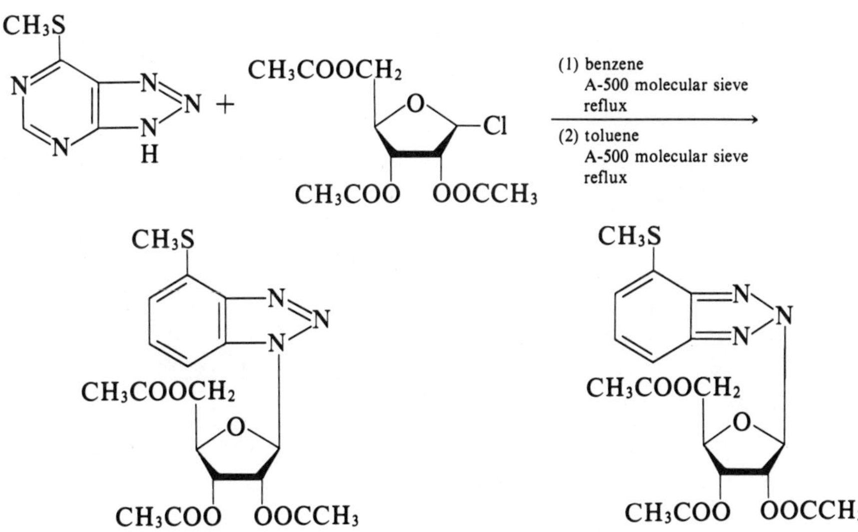

61%

12%

Isomeric 3- and 2-*O*-acetylated nucleosides of 7-methylthio-*v*-triazolo[4,5-*d*]-pyrimidine are obtained from reaction with 2,3,5-tri-*O*-acetyl-D-ribofuranosyl chloride (365); a molecular sieve converts an initial 2:1 mixture of 3- to 2-substituted *v*-triazolo[4,5-*d*]pyrimidine to a much higher proportion of 3-isomer. Subsequent ammoniation of the methylthio-3-nucleoside replaces the methylthio group with amino and effects deacetylation, forming 8-azaadenosine.

Addition of 1-chlorobenzotriazole to olefins leads to mixtures of 1- and 2-substituted chloroalkyl benzotriazoles with the 2-isomer predominating (436). In the reaction with styrene a minor quantity of styrene dichloride also is formed.

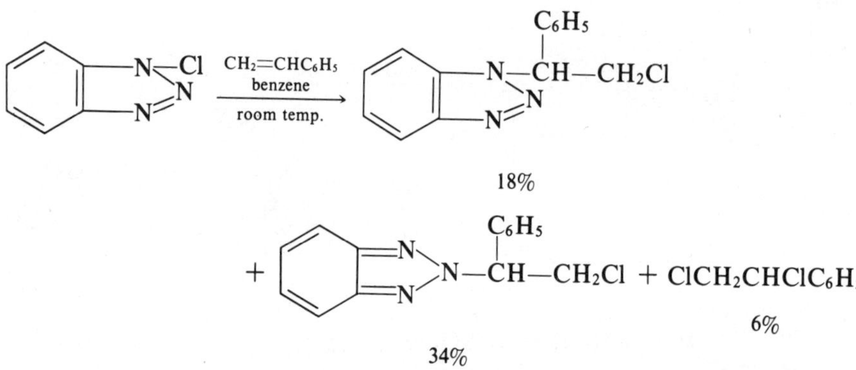

Alkylation of 1-substituted *v*-triazoles to form quaternary salts occurs at the 3-position. The same salt is obtained by methylating 1-benzyl-*v*-triazole and by benzylating 1-methyl-*v*-triazoles (552), and ethylation of 1-benzyl-4-phenyl-*v*-triazole leads to the 3-ethyl quaternary (195). Ethylation of 1-benzyl-5-phenyl-*v*-triazole displaces the benzyl group, forming the 1,3-diethyl-4-phenyl-*v*-triazolium

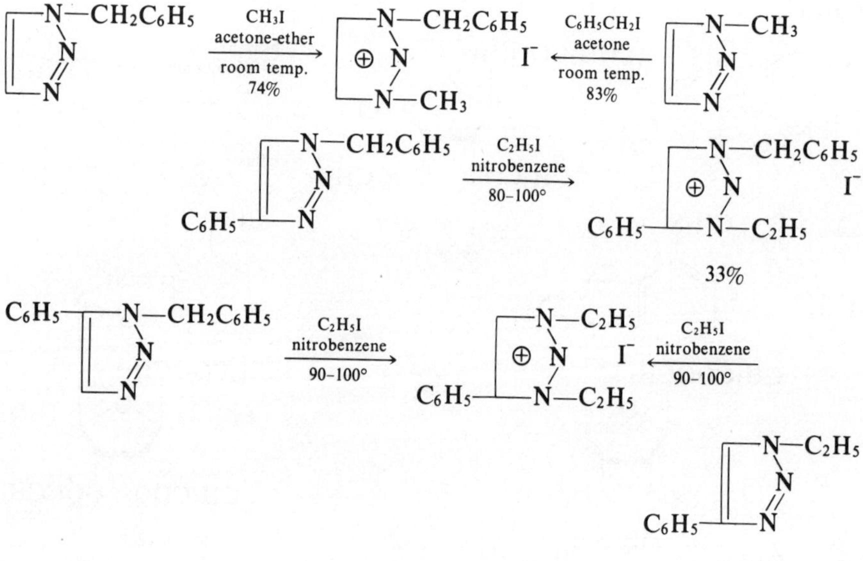

cation, also obtained from 1-ethyl-4-phenyl-*v*-triazole (195). Methylation of 4-hydroxy-1-methyl-*v*-triazole with diazomethane produces both the mesoionic triazolium compound and 4-methoxy-1-methyl-*v*-triazole (49). In a slow rear-

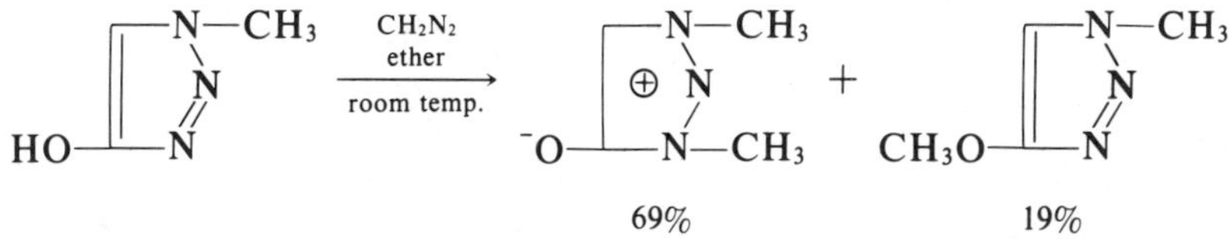

rangement, methyl migrates from methoxy in 1-benzyl-4-methoxy-*v*-triazole to N-3, forming 1-benzyl-3-methyl-*v*-triazolium 4-oxide (52). Alkylation of 1-

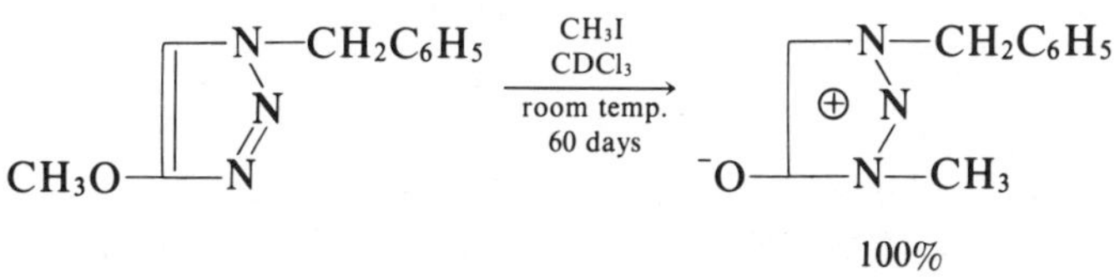

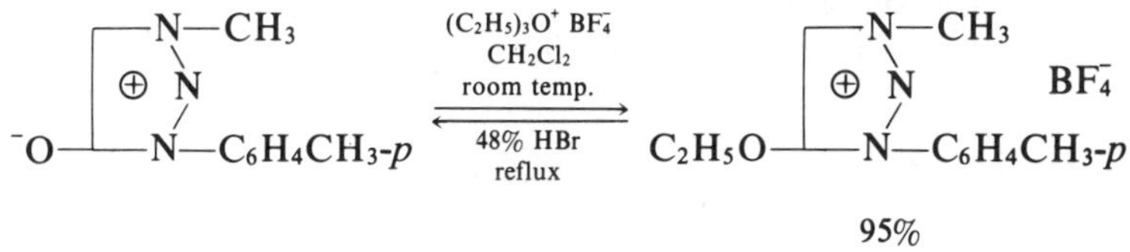

methyl-3-*p*-tolyl-*v*-triazole 4-oxide occurs on the exocyclic oxygen; refluxing with hydrobromic acid regenerates the starting betaine (420).

Greater difficulty is encountered in alkylating 2-substituted *v*-triazoles in comparison with 1-substituted isomers, consistent with the lower basicity of the 2-isomers. With strong alkylating agents such as methylfluorosulfonate, however, 2-alkyl and aryl-*v*-triazoles can be converted to 1,2-disubstituted *v*-triazol-

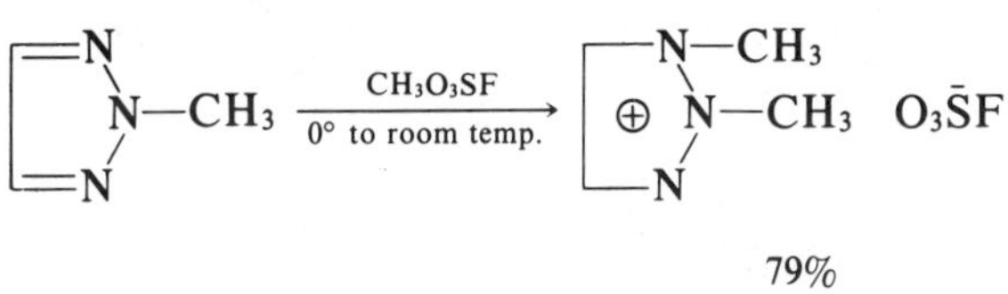

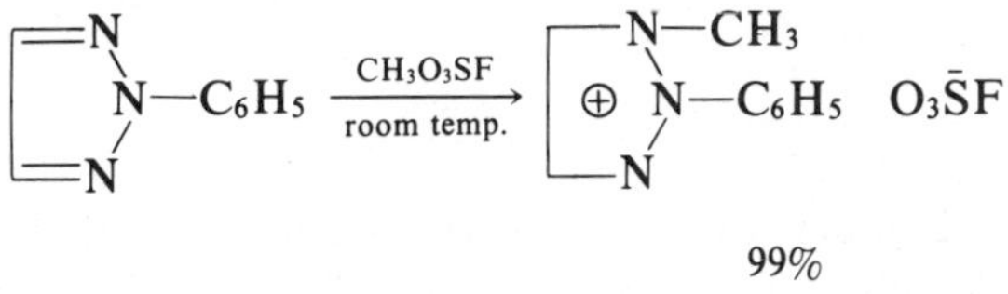

ium salts (53). Treatment of 1,2-disubstituted *v*-triazolium salts with *N*-bromoacetamide and sodium methoxide leads to the 1,2-disubstituted *v*-triazolone

(VII) through the 5-bromo compound, which need not be isolated (53). On heating **VII** with benzoyl chloride, the 1-alkyl group is removed with formation of 4-benzoyloxy-2-phenyl-*v*-triazole (46).

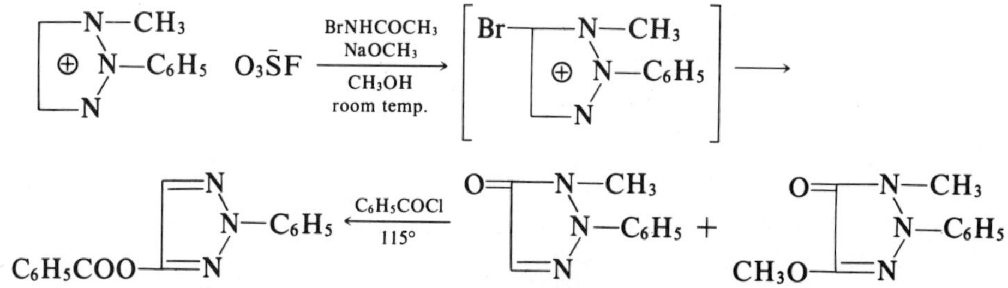

(VII)

55% 91% 4%

Both 1,3- and 1,2-benzotriazolium salts result from alkylation, respectively, of 1- and 2-substituted benzotriazoles (304). Destructive distillation of both 1,2- and 1,3-benzotriazolium iodides at elevated temperatures forms 1-methyl- and 2-methylbenzotriazole and methyl iodide.

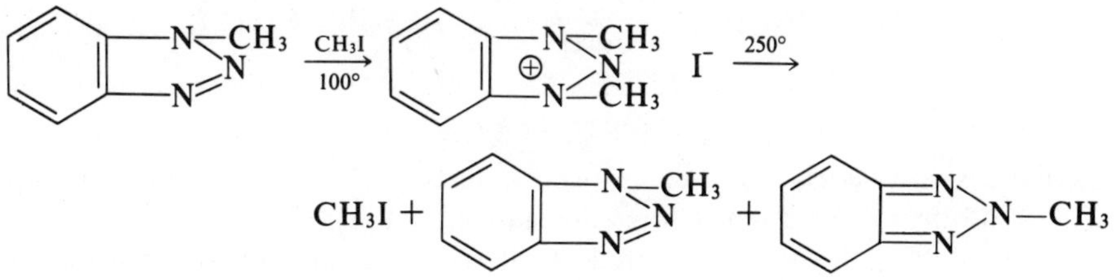

82.7% 11.6%

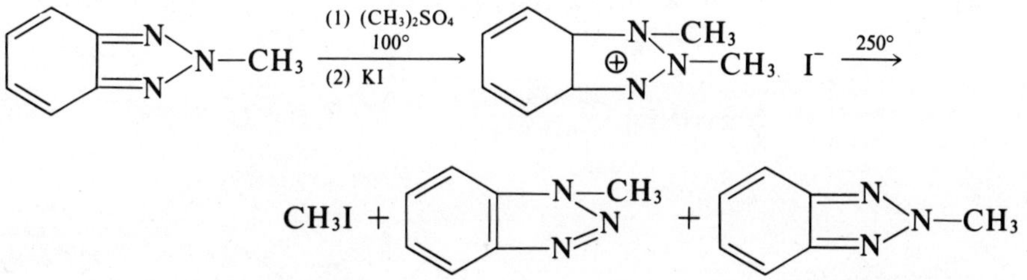

75.6% 12.4%

Treatment of 2-(*o*-benzoylphenyl)benzotriazole with triethyl phosphite effects internal alkylation of N-1 by the carbonyl carbon, forming a dibenzotriazapentalene betaine (318).

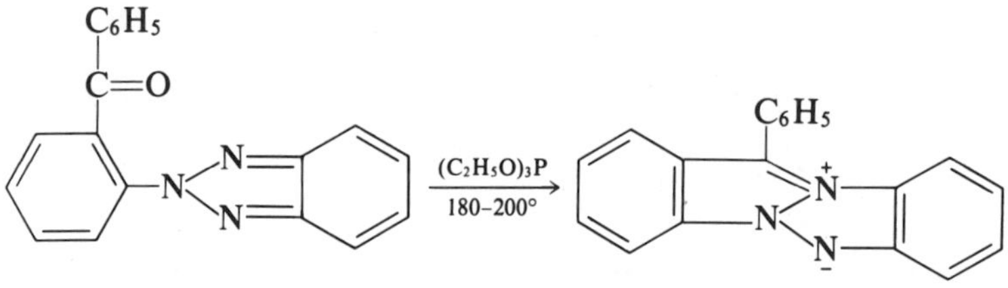

Alkylation of 5-monosubstituted tetrazoles occurs predominately at N-2 with the netural molecule and usually, but not always, with tetrazole anions. Other tetrazole alkylations are listed in Table 6.5.

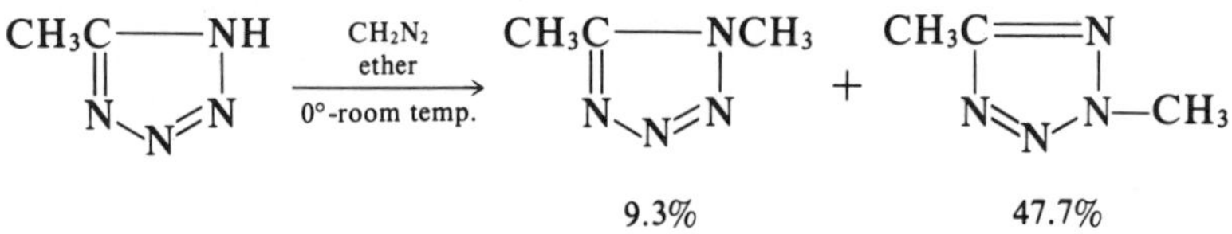

(349)

The major alkylation site with dicyclohexylcarbodiimide and *t*-butanol is at N-2, when the 5-substituent is phenyl, amino, methylthio, and chloro (230). In

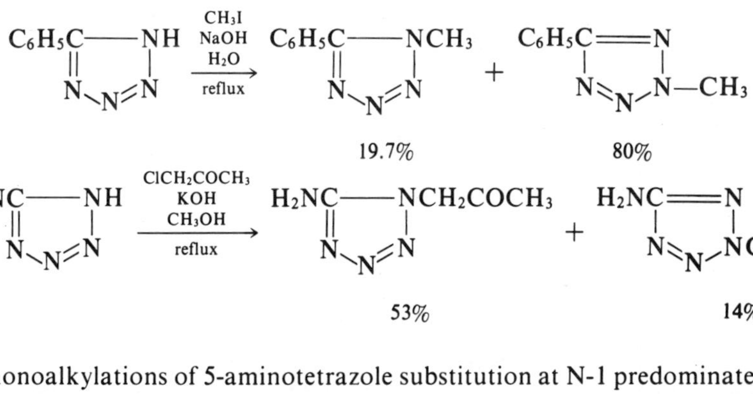

(229)

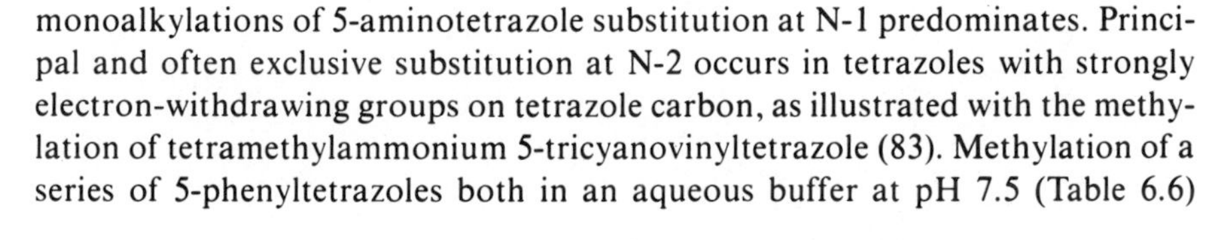

(144)

monoalkylations of 5-aminotetrazole substitution at N-1 predominates. Principal and often exclusive substitution at N-2 occurs in tetrazoles with strongly electron-withdrawing groups on tetrazole carbon, as illustrated with the methylation of tetramethylammonium 5-tricyanovinyltetrazole (83). Methylation of a series of 5-phenyltetrazoles both in an aqueous buffer at pH 7.5 (Table 6.6)

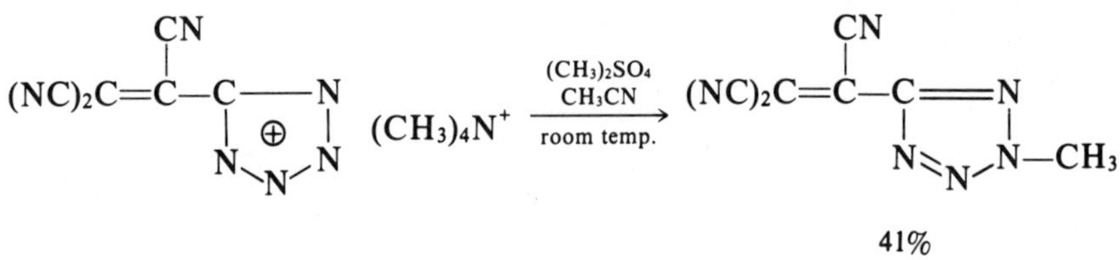

Table 6.5. Alkylation of Tetrazoles

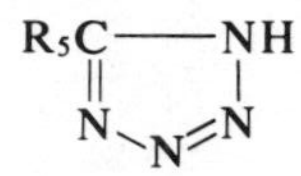

R_5	Alkylating Agent	Conditions	Yield (%) 1-Isomer	Yield (%) 2-Isomer	Ref.
Br	CH_2N_2	Ether, cold	13	68	216
C_6H_5	t-C_4H_9OH	Dicyclohexyl-carbodiimide, Cu_2Cl_2, room temp.	15	59	230
H	$HC{\equiv}CCOC_6H_5$	2–3 drops $(C_2H_5)_3N$, ethanol, reflux	55.6		536
CH_3	$BrCH_2COOC_2H_5$	$(C_2H_5)_3N$, acetone, 0°–room temp.	49	22	426
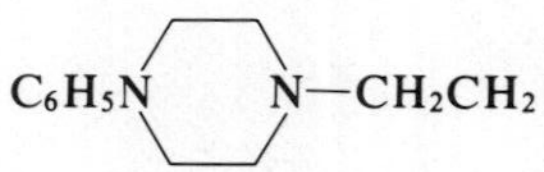	C_2H_5I	Na tetrazolate, ethanol, reflux	43	57	251
H_2N	$(CH_3)_2SO_4$	NaOH, H_2O, 92–95°	36–53	23–32	231
CF_3	n-$C_8H_{17}Br$	Na tetrazolate, CH_3CN, reflux	4	71	108

Table 6.6. Methylation of 5-Phenyltetrazole at pH 7.5

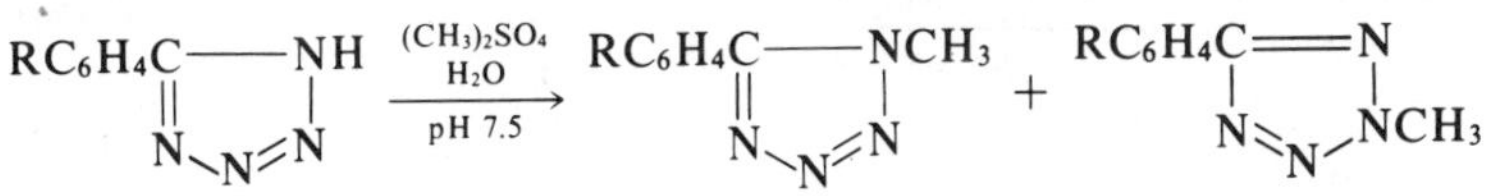

R	Ratios	
	1-Isomer	2-Isomer
p-CH_3	30	70
H	28	72
p-Cl	26	74
m-NO_2	20	80

(470) and in acetonitrile (406) leads principally to 2-substituted derivatives. The proportion of 2-methyl derivative increases as the electron-withdrawing character of the phenyl substituent increases. In a comparison of methylation of anions of 5-*o*- and *p*-substituted phenyltetrazoles, the *o*-substituted derivatives are substituted approximately equally at both N-1 and N-2, but methylation at N-2 predominates with *p*-substituted derivatives, illustrated with the 5-tolyl and chlorophenyltetrazoles (Table 6.7) (94). This difference is related to inhibition of interannular conjugation in the case of the *o*-substituted phenyl compounds in which the phenyl and tetrazole rings are twisted out of planar configuration. A linear correlation is found for the plot of Hammett σ values against the log of the ratio of yields of 2- to 1-derivatives for *o*-substitutents; the ratio of isomer yields for *p*-substituted phenyltetrazoles roughly parallels the ortho plot but does not fit the correlation.

Di-*p*-tolyliodonium bromide arylates sodium 5-phenyltetrazolate in *t*-butyl alcohol at both ring positions and also the solvent. The products are mixtures of *m*- and *p*-tolyl derivatives, suggesting a benzyne intermediate.

Table 6.7. Methylation of 5-*o*- and *p*-Tolyl and Chlorophenyltetrazoles

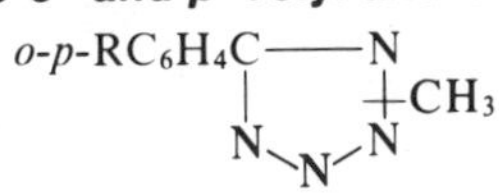

R	Yields			
	o-Substituted		*p*-Substituted	
	1-Isomer	2-Isomer	1-Isomer	2-Isomer
CH_3	38.7	48.3	30	62.7
Cl	43.1	42.0	23.7	64.2

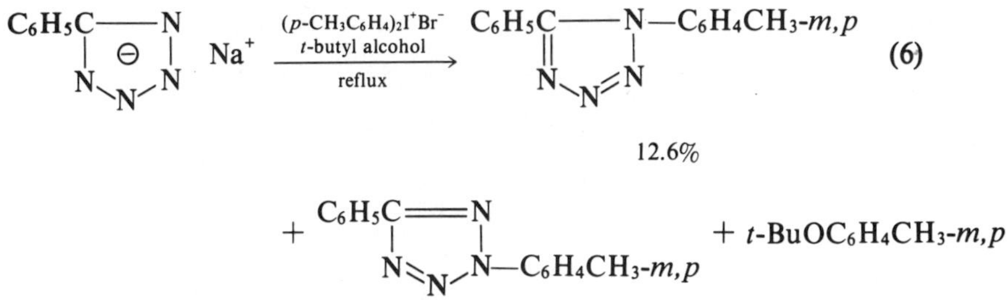

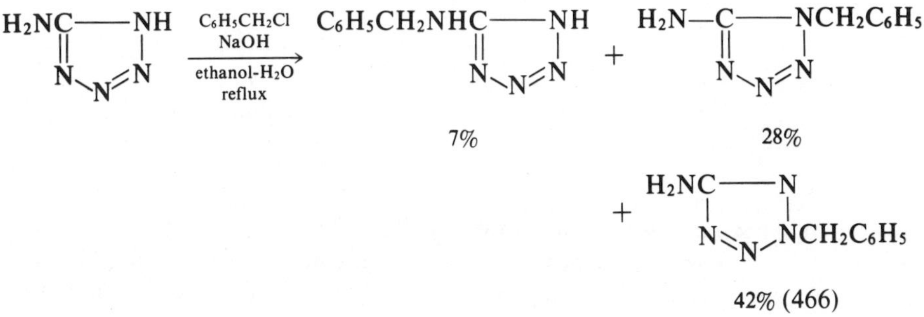

When groups at the 5 position on tetrazole can be substituted, both products with ring and exocyclic substitution can be obtained on alkylation, illustrated by tetrazoles carrying amino, hydrazone, and carbonyl oxygen at C-5. A dipolar compound results from benzylation of 5-dimethylaminotetrazole **(VIII)**. Alkylation of 1-methyl-5-aminotetrazole under neutral conditions leads principally to the formation of 1,4-dimethyl-5-iminotetrazoline obtained in part as a salt and also mesoionic 1,3-dimethyl-5-iminotetrazoline.

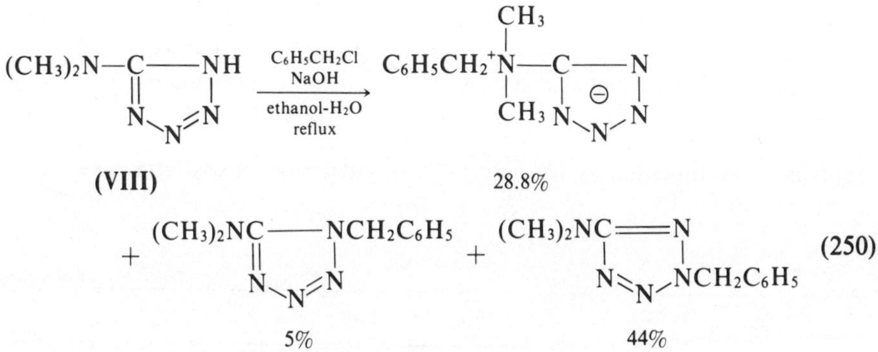

Methylation of tetrazolinone with diazomethane takes place on the ring at N-1, N-4, and N-2 as well as on oxygen (216). 1-Methyltetrazolinone is converted to 1,4-dimethyltetrazolinone, 5-methoxy-1-methyltetrazole, and a small amount of the mesoionic 1,3-dimethyltetrazolium-5-oxide **(IX)**. 2-Methyltetrazolinone is converted to 5-methoxy-2-methyltetrazole and the mesoionic tetrazolium compound **(IX)**.

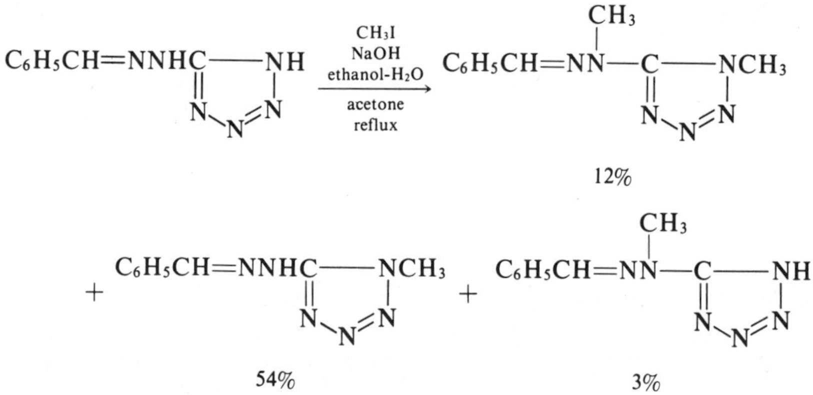

(97)

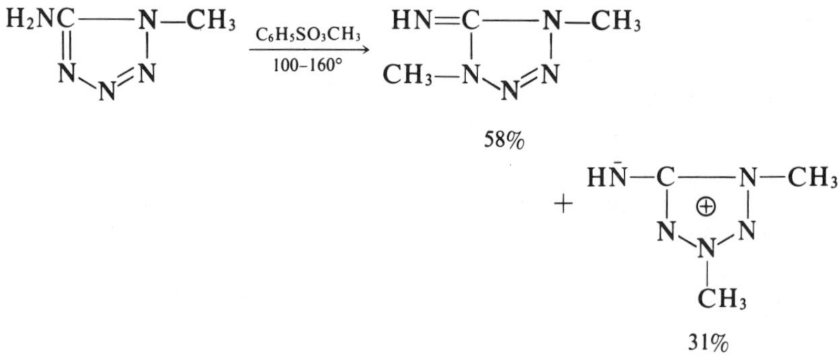

(232)

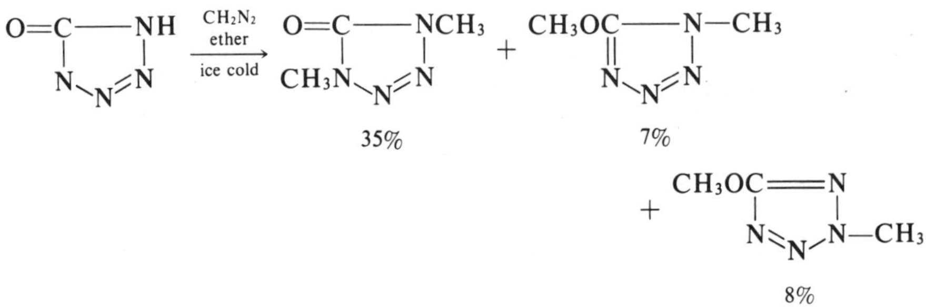

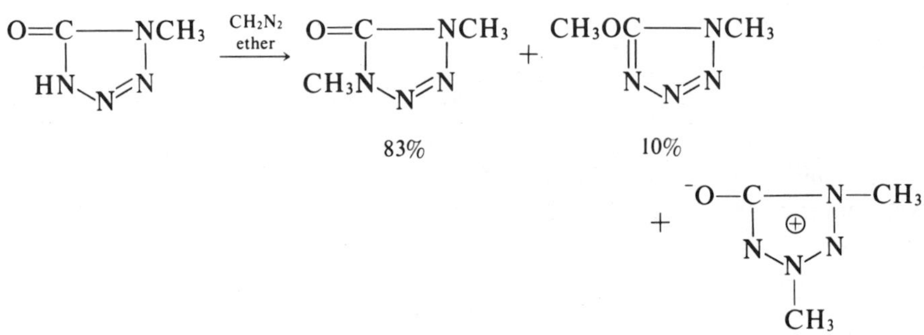

(IX)

1%

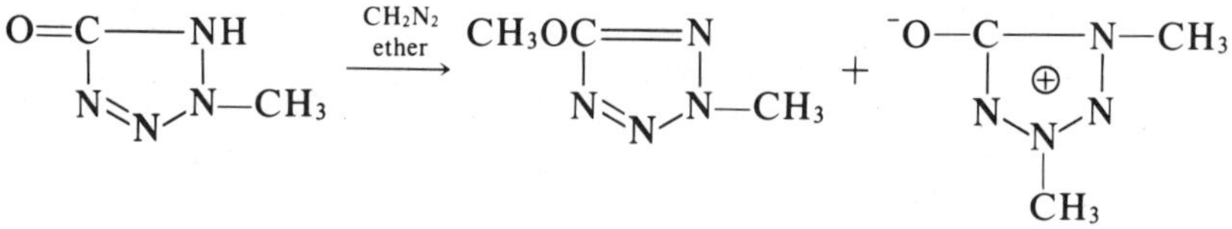

(IX)

17% 40%

Ring closure of imidoyl azides with electron-withdrawing substituents to tetrazoles occurs from the action of diazomethane on both cyanoguanyl azide and *N*-cyano-*N*′-methylguanyl azide, forming 1,4-dimethyl-5-cyaniminotetrazoline (393).

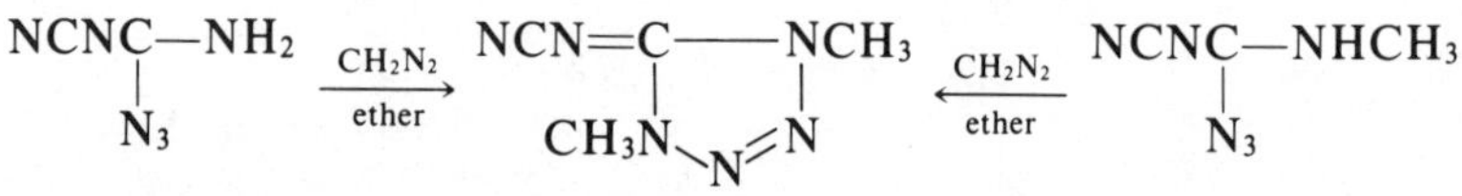

5-Aminotetrazole condenses with 1,3-dicarbonyl compounds to form bicyclic derivatives by substitution at both ring nitrogen and the amino group. Thus acetylacetone and ethyl acetoacetate react to form tetrazolo[1,5-*a*]pyrimidines (79, 88).

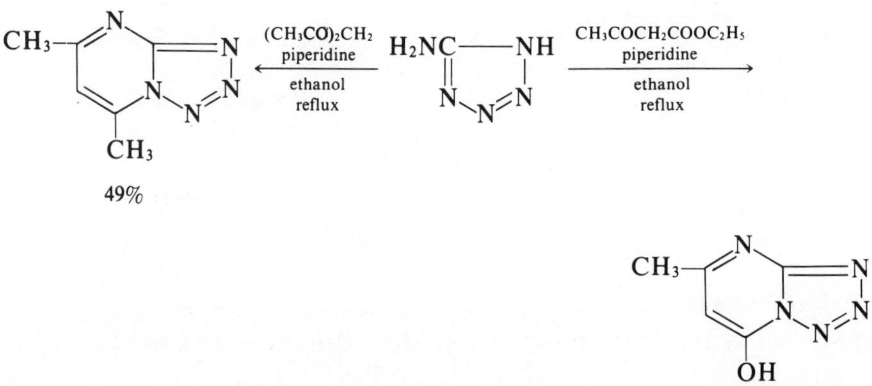

49%

40%

On heating in the presence of sodium iodide, alkyl in 5-alkoxy-1-aryltetrazoles migrates from oxygen to N-4 (539).

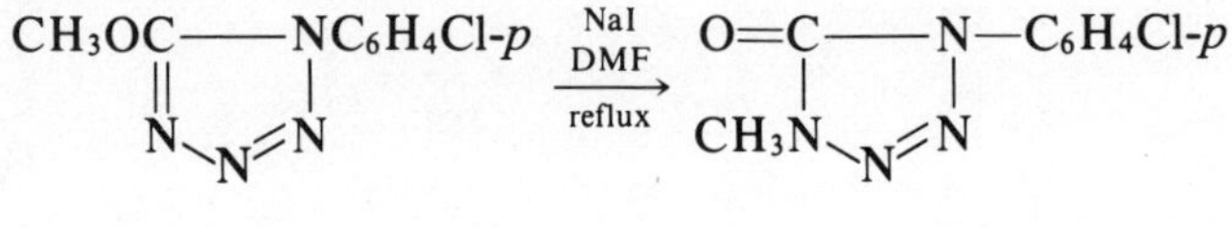

84%

Dimethylacetylene dicarboxylate reacts with 2,3-diphenyltetrazolium-5-thiolate, alkylating N-1 and sulfur, opening the tetrazolium ring between N-1 and N-2, and forming the acyclic azimine, azobenzene *N*-[4,5-bis(methoxycarbonyl)-thiazol-2-yl]imide (74).

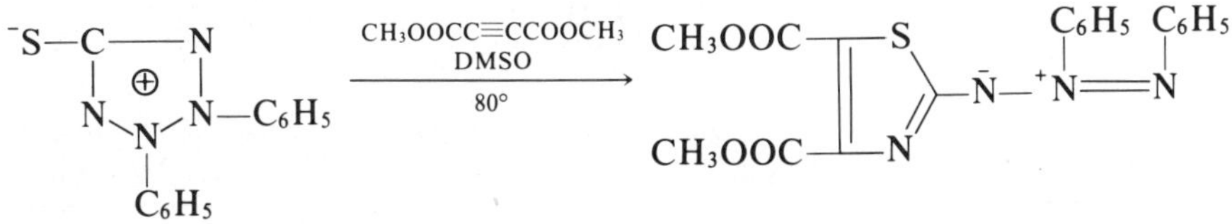

23%

Alkylation of 1-substituted tetrazoles can lead to either or both 1,4- or 1,3-disubstituted tetrazolium salts depending in part on the nature of the 5-substituent. 2-Substituted tetrazoles alkylate to form 1,3-ditetrazolium salts. Only a single 1,4-tetrazolium isomer is found in the addition of methyl groups to 1-vinyltetrazole, tetrazolopyridine, or the five-nitrogen 1-diethylaminotetrazole **(X)**. It is notable that **X** alkylates on the tetrazole ring rather than at the exocyclic nitrogen. Location of alkylation of 1-mono- and 1,5-disubstituted tetra-

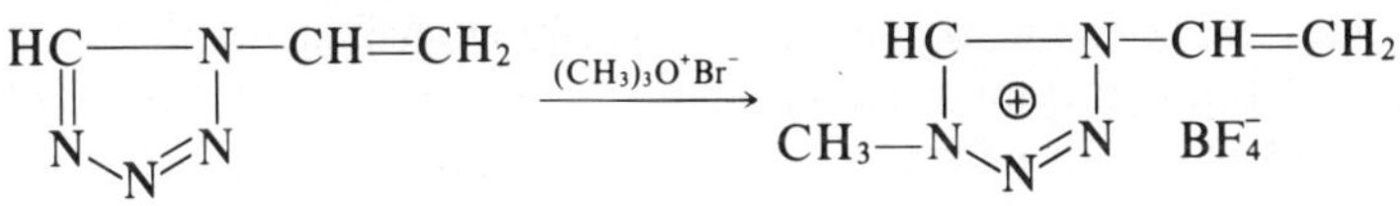

(568)

83%

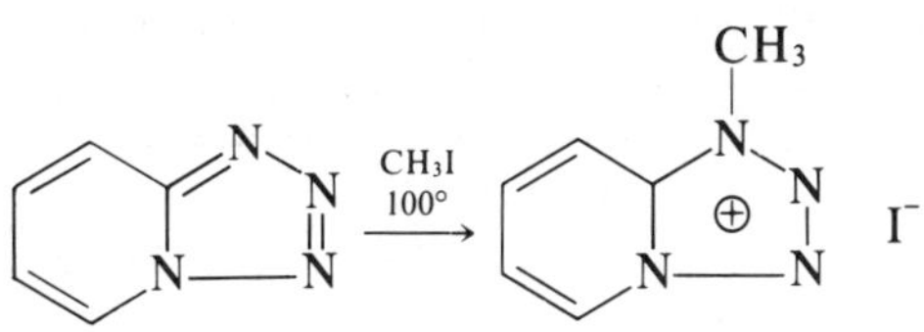

(16)

60%

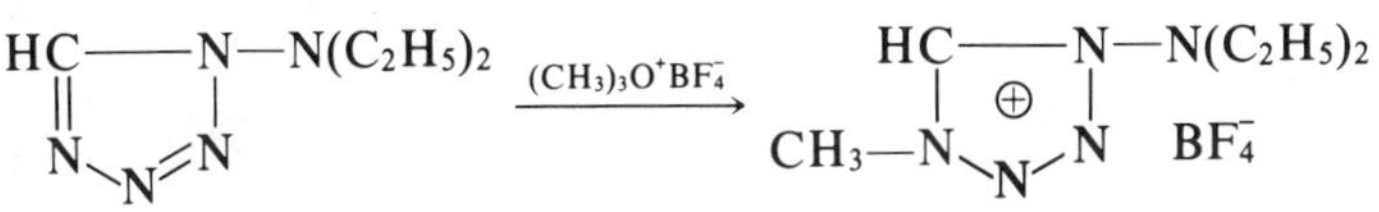

(568)

(X) 87%

zoles was assigned first to the 4-position (138); specific search for the 1,3-isomer often shows its presence in small proportion. Thus ethylation of 1-phenyltetrazole gives a 96% combined yield of 1,4- and 1,3-quaternary salts in the propor-

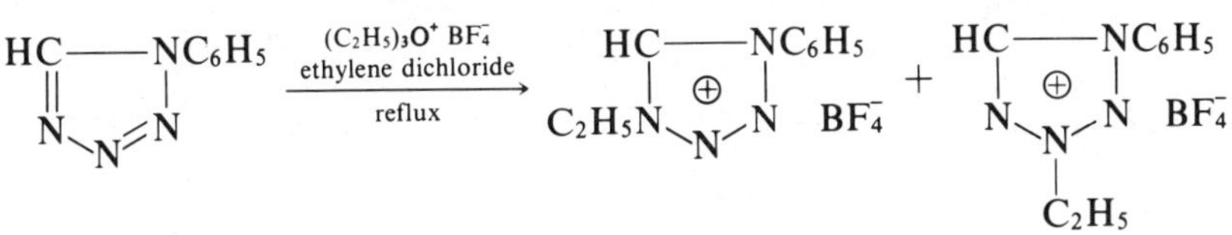

Table 6.8. Quaternization of 1-Methyl-5-Methylthiotetrazole

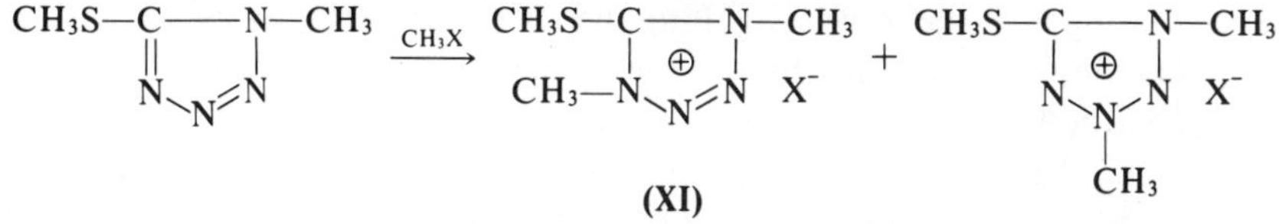

(XI)

Alkylating Agent	Solvent	Reaction Temp. (°C)	Ratio of 1,4 to 1,3 Tetrazolium
$(CH_3)_2SO_4$	—	100–120	49:51
$(CH_3)_3O^+ BF_4^-$	$ClCH_2CH_2Cl$	25	30:70
CH_3O_3SF	CH_2Cl_2	−28 to +25	17:83

tion of 84:16 (295). The products of alkylation of 5-alkylthio 1-substituted tetrazoles are mixtures of 1,4- **(XI)** and 1,3-tetrazolium salts, the proportions of which depend on the nature of the alkylating agent (423). Although use of dimethyl sulfate with 1-methyl-5-methylthiotetrazole yields nearly equal proportions of tetrazolium isomers, the more powerful reagents trimethyloxonium salt and methyl fluorosulfonate form higher proportions of 1,3-tetrazolium compound (Table 6.8). In contrast, 1-*t*-butyl-5-chlorotetrazole **(XII)**, with a

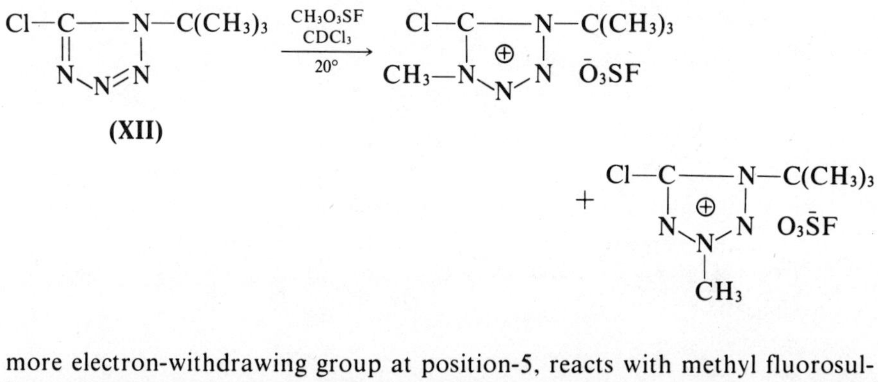

(XII)

more electron-withdrawing group at position-5, reacts with methyl fluorosulfonate to form 1,4- and 1,3-quaternary salts in the ratio of 2:1 (423). No alkylation of **XII** occurs with dimethyl sulfate. Only 3-ethyl-5-methylthio-1-phenyl-

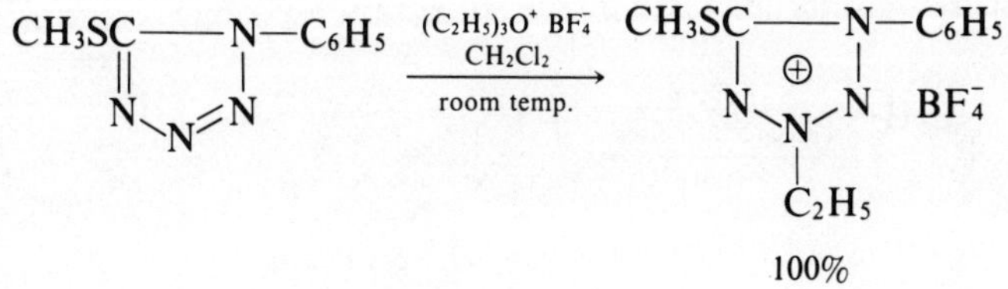

tetrazolium cation is formed on ethylation of 5-methylthio-1-phenyl-tetrazole (39). Further alkylation of 1,4-dimethyl-5-thionotetrazoline is on sulfur, forming the same 1,4-tetrazolium salt **(XI)** obtained from 1-methyl-5-methylthiotetrazole (423).

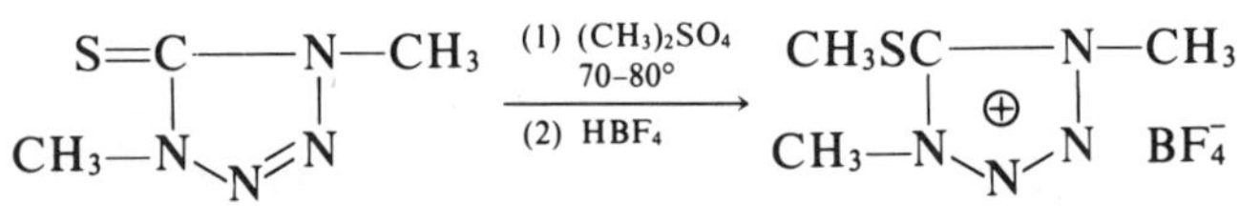

(XI)

69%

High yields of 1,3-tetrazolium salts are obtained from alkylation of 2-methyl- and 2-phenyltetrazoles. The mesoionic 1,3-dimethyl-tetrazolium-5-imide re-

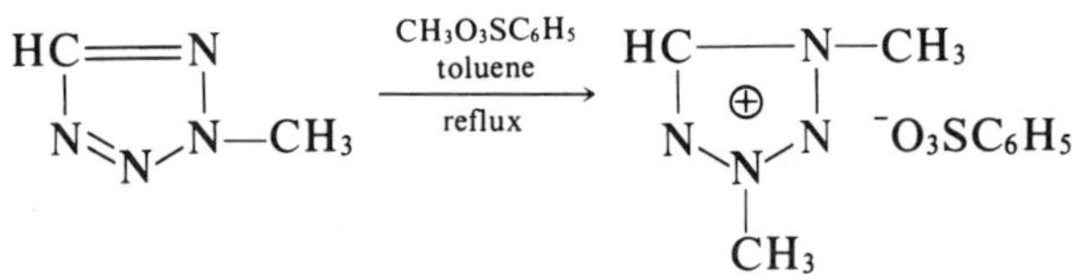

(394)

100%

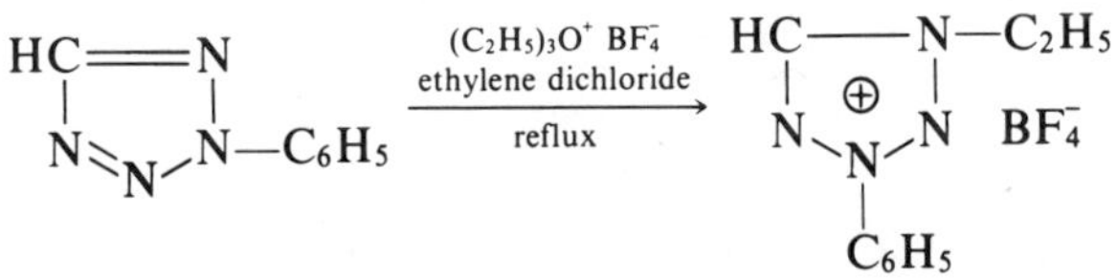

(295)

94%

sults from alkylation of 5-amino-2-methyltetrazole followed by neutralization of the salt (232).

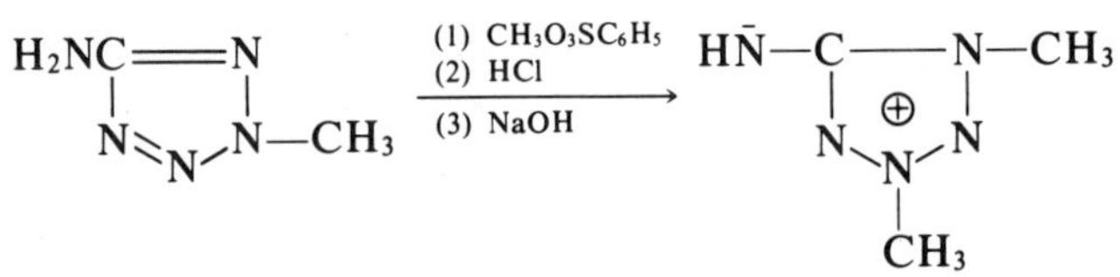

35%

Acetylation of *v*-triazole with acetyl chloride at low temperatures leads principally to 1-acetyl-*v*-triazole with small amounts of the 2-isomer; at higher temperature acetic anhydride leads only to the thermochemically more stable 2-acetyl derivative (67). (See Section 2.3.3, Monocyclic *v*-Triazoles.) 1-Acetyl-*v*-triazole also can be obtained through nucleophilic attack on 2-trimethylsilyl-*v*-triazole by acetyl chloride. Acylation at N-1 occurs with 4,5-diphenyl-*v*-triazole

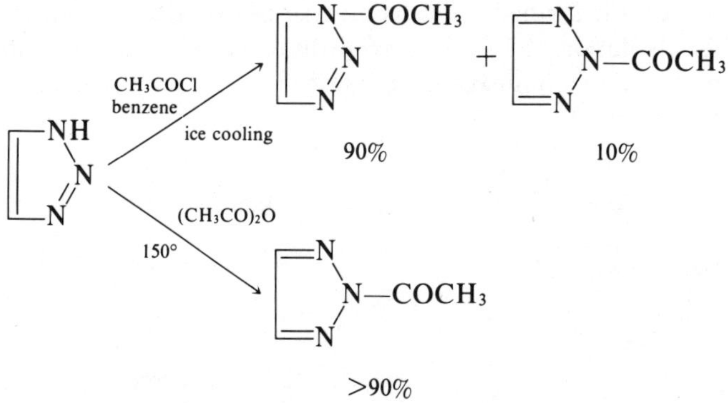

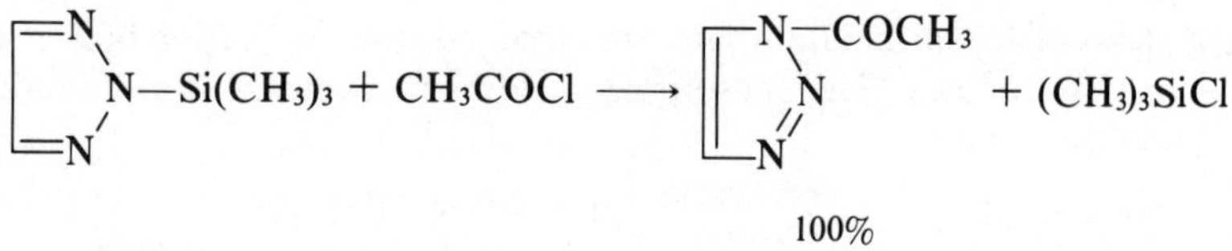

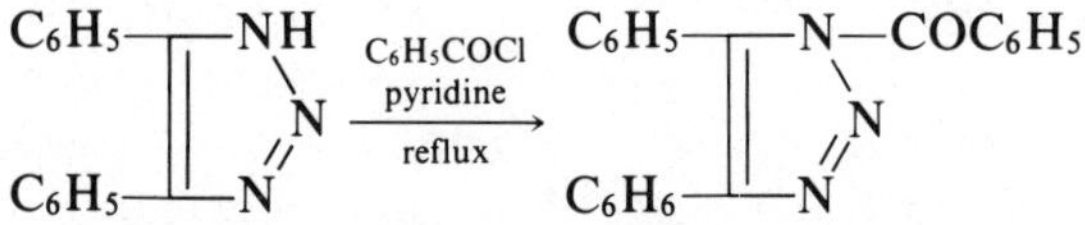

(258)

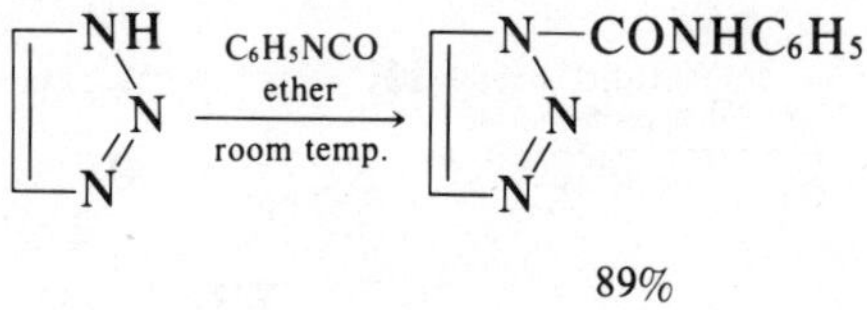

(484)

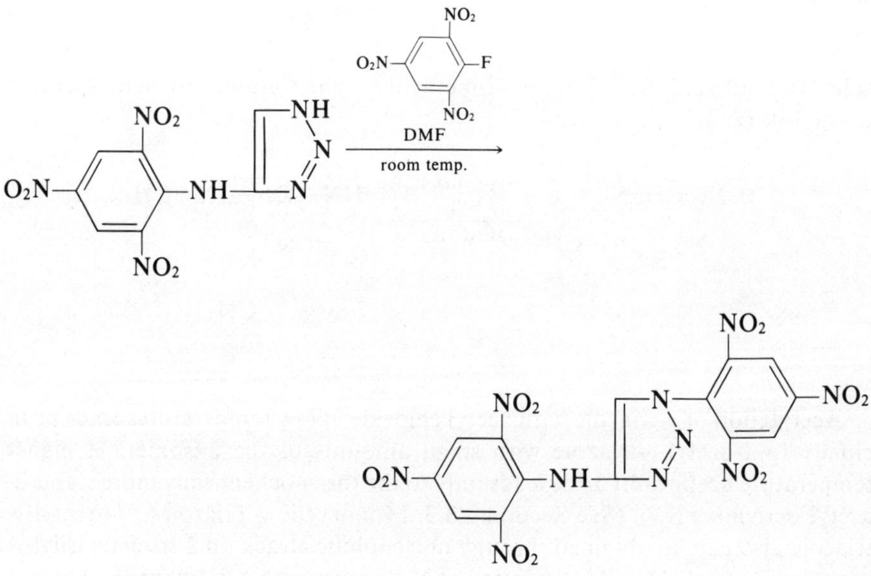

15% (385)

and benzoyl chloride, *v*-triazole and phenyl isocyanate, and 4-picrylamino-*v*-triazole and picryl fluoride. Benzoylation of 4-hydroxy-*v*-triazole, with a relatively electron-releasing substituent, occurs at N-2, as well as at hydroxyl, and the anionic *v*-triazole ring of 4-aryl-5-triphenylphosphonium *v*-triazole acylates at N-2, as in its alkylation.

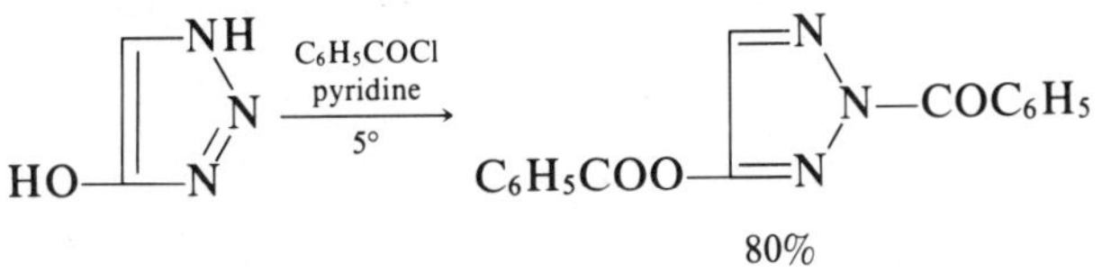

(44)

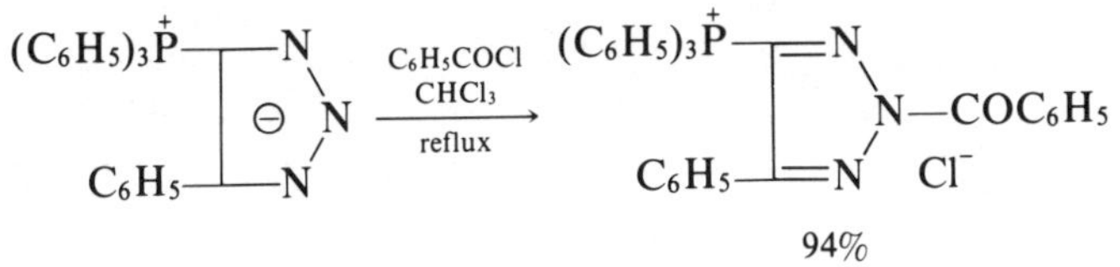

(510)

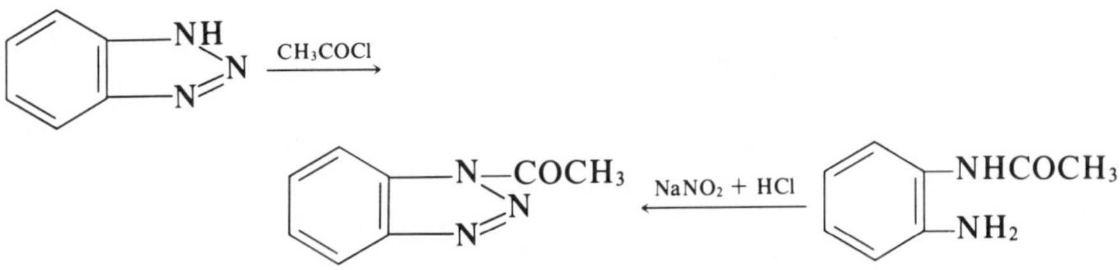

The same 1-acetylbenzotriazole is obtained both by acetylating benzotriazole and by nitrosating monoacetyl-*o*-phenylenediamine, establishing the acetyl

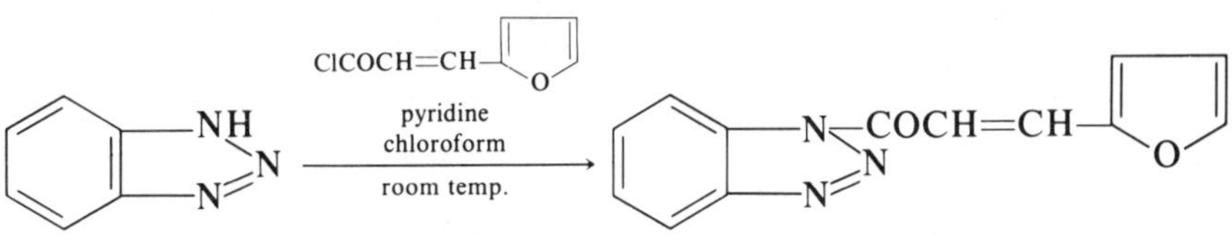

58% (303)

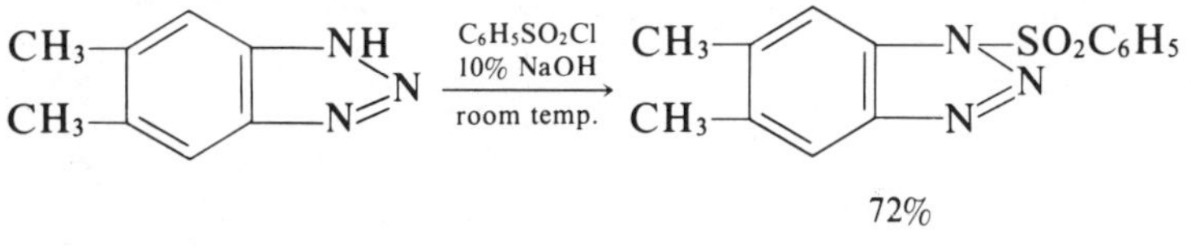

(59)

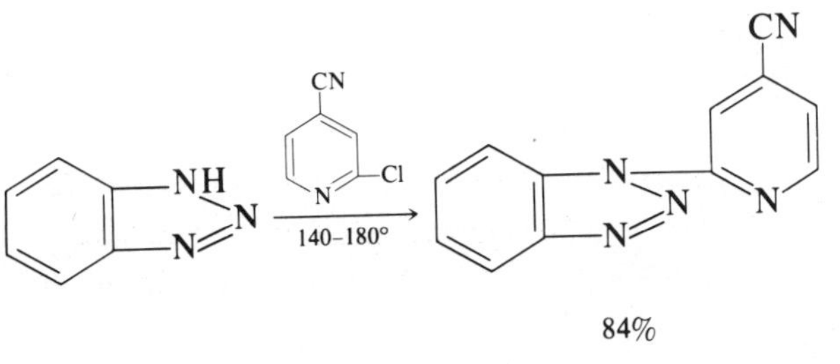

(377)

location (55). Only 1-acyl derivatives are obtained from benzotriazoles even when acylated under basic conditions. Fusion of 2-chloro-4-cyanopyridine with benzotriazole leads to 1-(4-cyanopyridyl-2)benzotriazole (377). Although picryl fluoride reacts with benzotriazole to form only 1-picryl benzotriazole, and only 1- and 3-substitution of beno[1,2-*d*:4,5-*d*]bistriazole results, reaction with benzo-[1,2-*d*:4,5-*d'*]bistriazole-4,8-dione forms only the bis-2-substituted quinone (62).

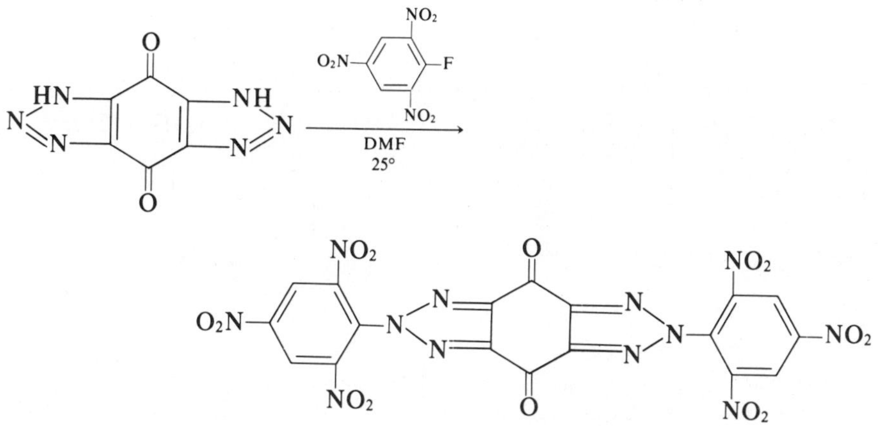

96%

On heating 1-benzoyl-4,5-diphenyl-*v*-triazole, the triazole ring is broken with loss of nitrogen and formation of triphenyloxazole. A much higher temperature is required with the *v*-triazole derivative than in the analogous reaction with 2-acyltetrazoles, but which probably occurs through an intermediate similar to

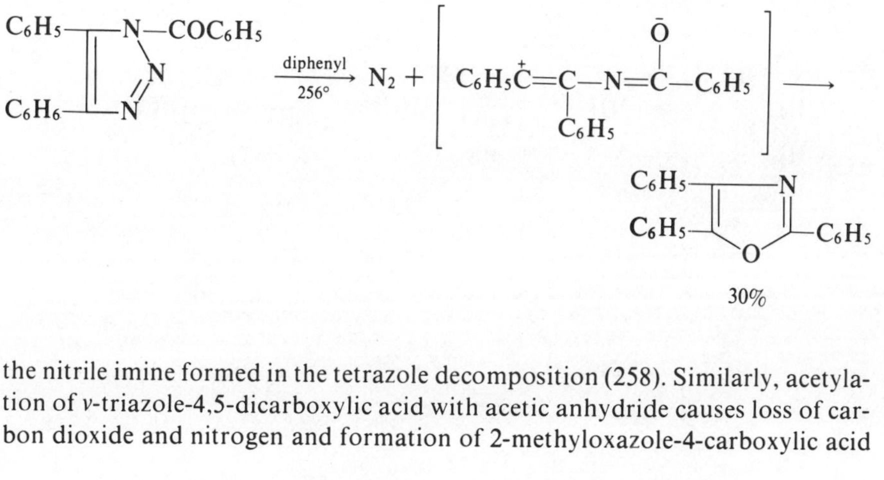

30%

the nitrile imine formed in the tetrazole decomposition (258). Similarly, acetylation of *v*-triazole-4,5-dicarboxylic acid with acetic anhydride causes loss of carbon dioxide and nitrogen and formation of 2-methyloxazole-4-carboxylic acid

(15). The triazole ring of benzotriazole is destroyed by heating with the cyclic imide chloride, 2-chloro-4-methyl-5-nitropyridine (378, 379).

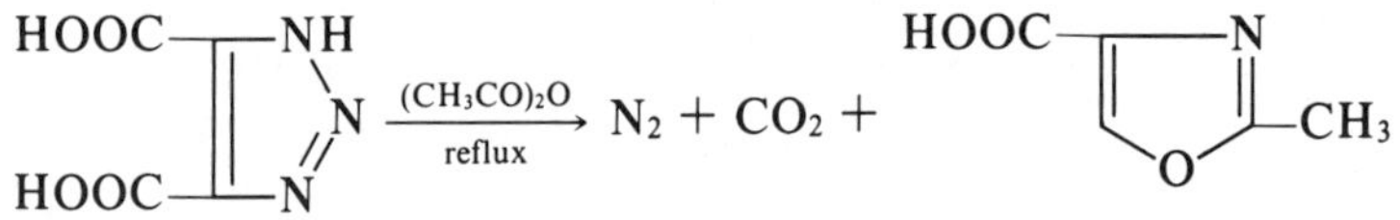

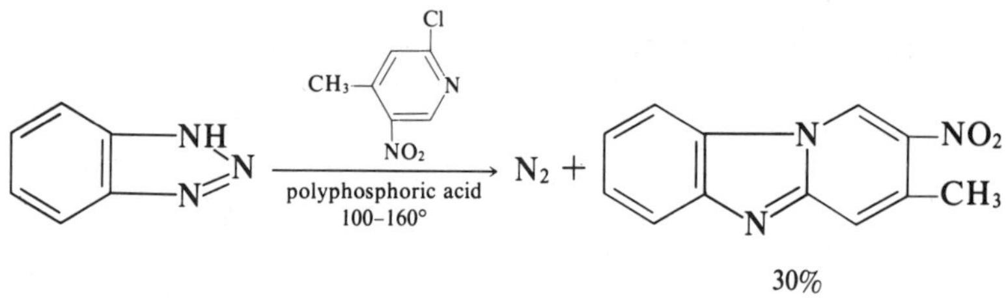

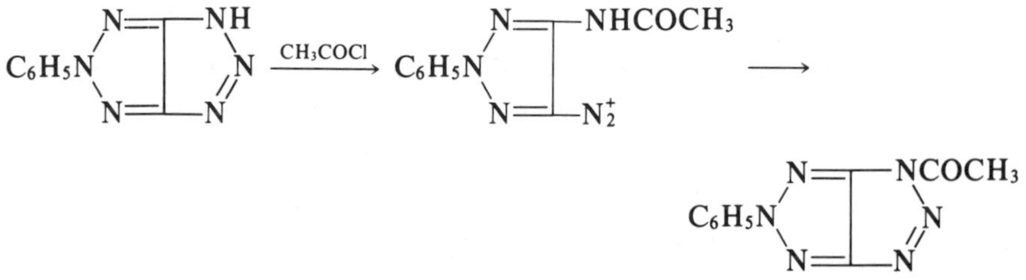

Treatment of 5-phenyl-1*H*-triazolo[4,5-*d*]triazole with acetyl chloride opens the unsubstituted triazole ring to the 4-acetylamino-5-diazonium compound (520); in the absence of acid, the triazole ring is reformed. The ring system remains intact, however, on methylation of the silver salt (566).

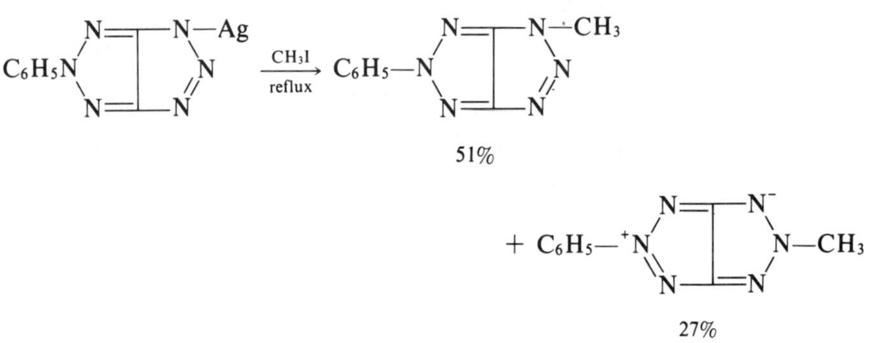

Under mild conditions tetrazole is acetylated on the ring, forming an *N*-acetyl derivative (253, 294, 483, 540) substituted at N-1. Nitrile oxides react with 5-alkyl and acyl tetrazoles to form 1- and 2-hydroxamoyltetrazoles (417).

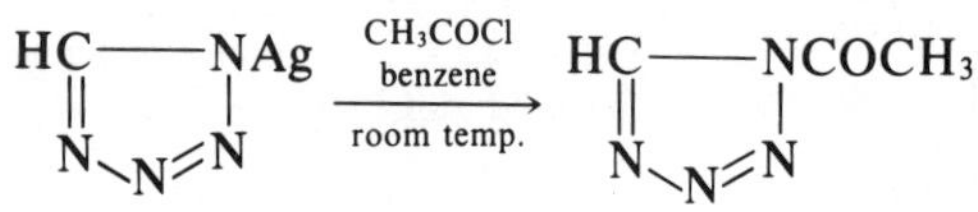

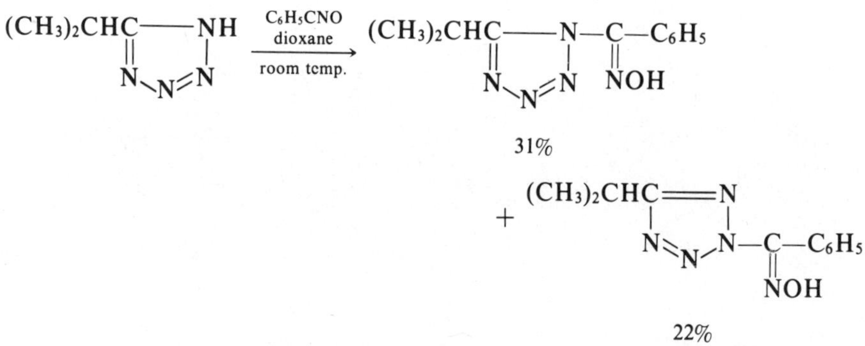

1-Phenyl-2-tetrazoline-5-thione combines with isocyanides to form a mixture of *N*- and *S*-acyltetrazoles which activate *N*-protected amino acids for forma-

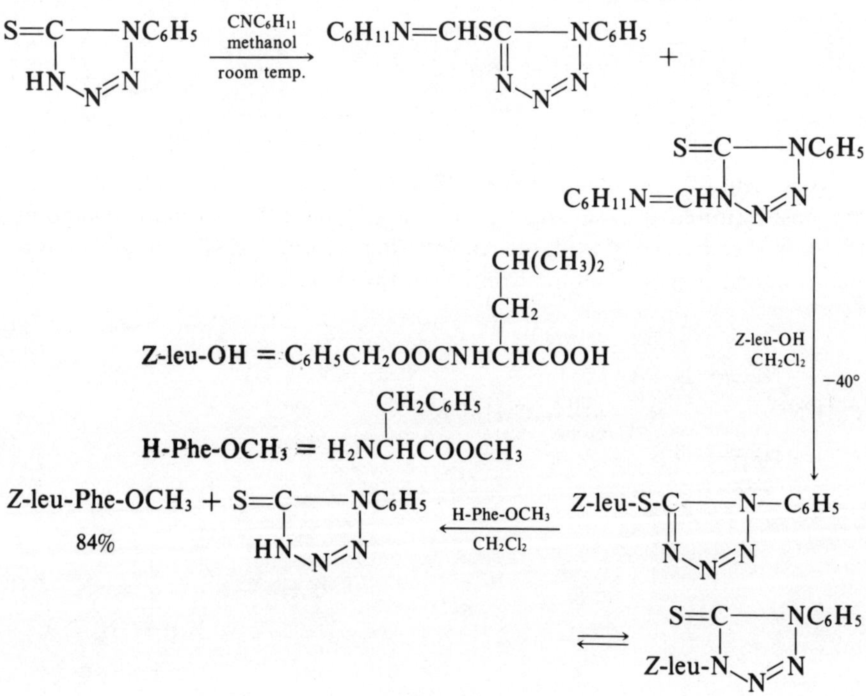

tion of peptides by reaction with amino acid esters (459). Phenyltetrazolinethione-isocyanide adducts also effect cyclization of ω-hydroxycarboxylic acids to macrocyclic lactones (458).

5-Acylamino-1-alkyl or aryltetrazoles are obtained by acylation of the 5-alkylamino or arylaminotetrazole. Indeed, the same product is obtained by acylating both the 1-substituted 5-aminotetrazole and the 5-substituted aminotetra-

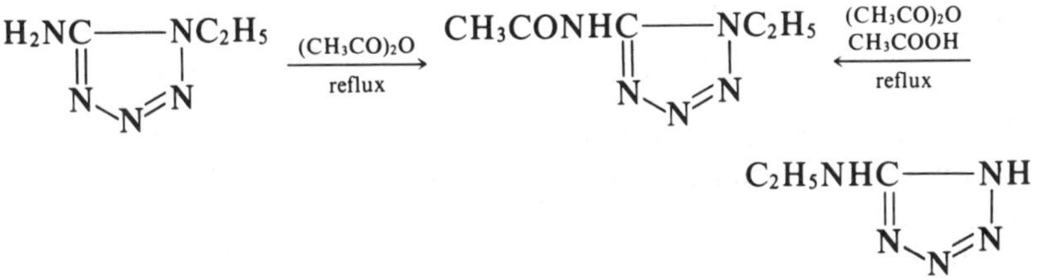
(234)

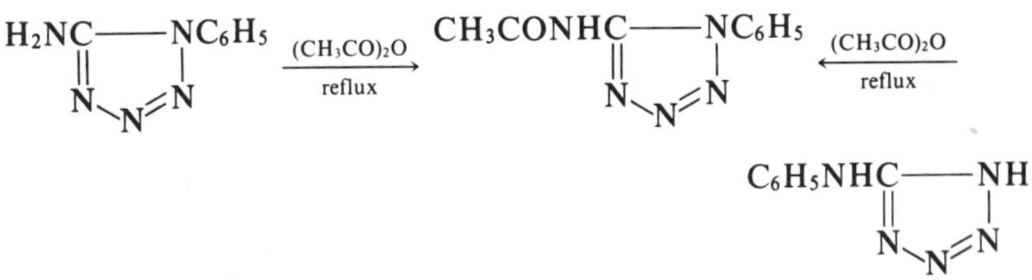
(235)

zole. Conversion of the 5-substituted aminotetrazoles requires a rearrangement probably through a guanyl azide intermediate of the same type involved in the thermal isomerization of 1-aryl-5-amino and 5-alkylaminotetrazoles (Section 2.6.2). No equilibrium appears to be involved and acylation fixes the aryl derivatives in the 1-aryl-5-acylamino form rather than in the 5-arylamino form favored in the thermal isomerization equilibria of 1-aryl-5-aminotetrazoles; 1-acetyl-5-anilinotetrazole, which can be obtained by acetylation in the cold, rearranges on heating.

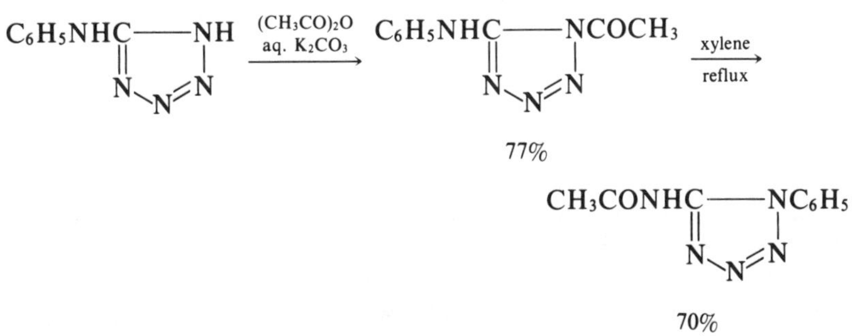
(235)

Reaction of 5-aminotetrazole with arenesulfonyl chlorides causes ring opening to sulfonylguanyl azides which can be converted to tetrazoles by base (272). Cyanogen bromide also reacts with 5-amino-, methylamino-, and dimethylaminotetrazole to form cyanoguanyl azides (393). (See Section 2.16.2, Imidoyl Azide-Tetrazole Isomerization.) The absence of reaction of 1-methyl-5-aminotetrazole with tosyl chloride suggests that the attack of the sulfonyl chloride is

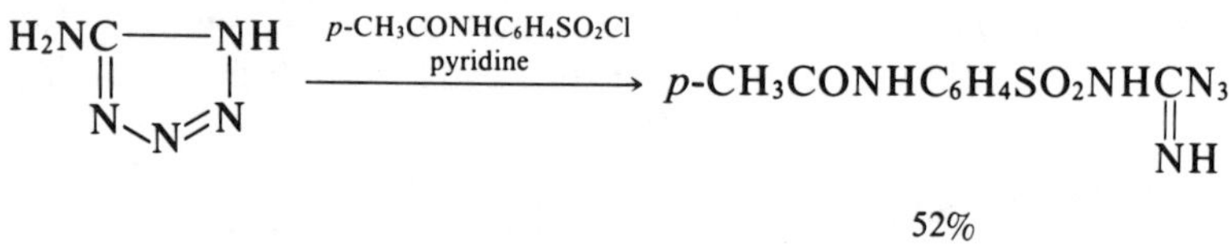

on ring NH rather than amino. Sulfonyl chlorides react with 5-dimethylaminotetrazole in pyridine to form a pyridinium intermediate, which decomposes on

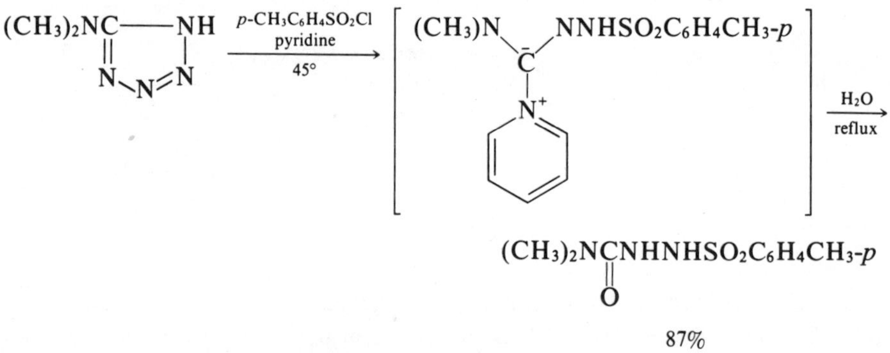

treatment with water to 1-sulfonylsemicarbazides (271). 5-Phenoxytetrazole, however, is substituted by cyano at the 2-position without destruction of the tetrazole ring (351).

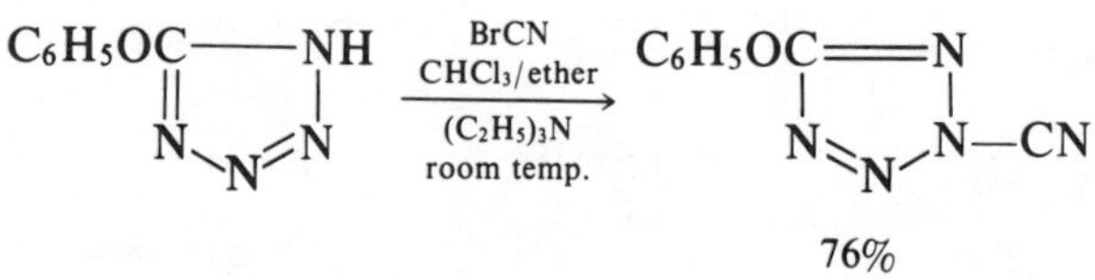

Internal acylation occurs on heating 1- and 2-methyl-5-tetrazolyl hydrazidic halides, forming 1,2,4-triazolo[4,3-*d*]tetrazoles, but the tetrazole ring in unsubstituted tetrazolyl hydrazidic halides is opened to a 3-azido-5-aryl-1,2,4-triazole.

The amino group in 5-aminotetrazole and its ring-substituted derivatives can be acylated without affecting the tetrazole ring by the action of acetic acid (79) or *N,N*-dimethylacetamide or by heating with acetic anhydride for a short time. Continuing to heat the 5-acetylaminotetrazole reaction system for a longer time causes complete destruction of the tetrazole ring and formation of 2-acetamido-5-methyl-1,3,4-oxadiazole.

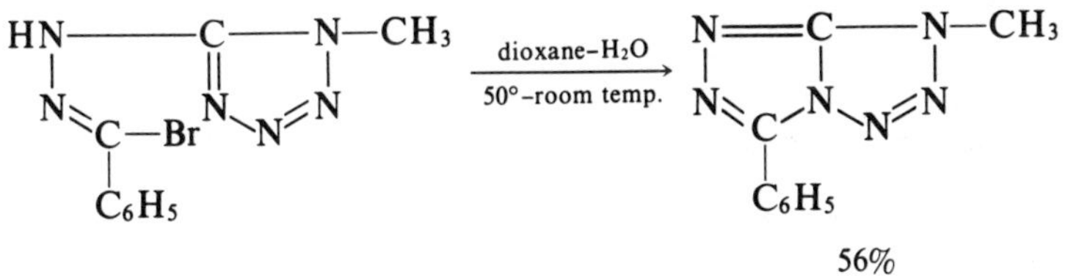

(95)

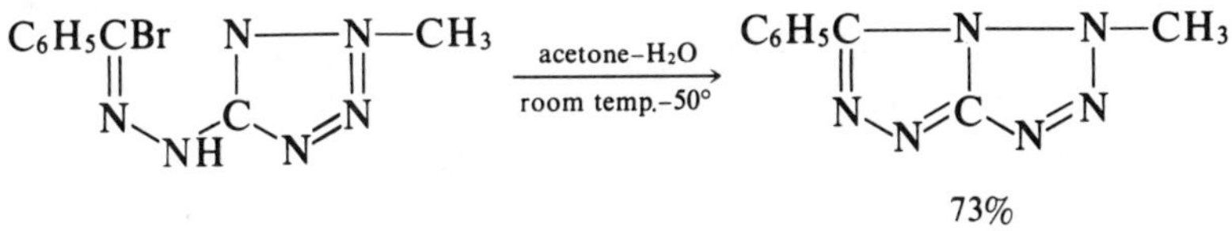

(96)

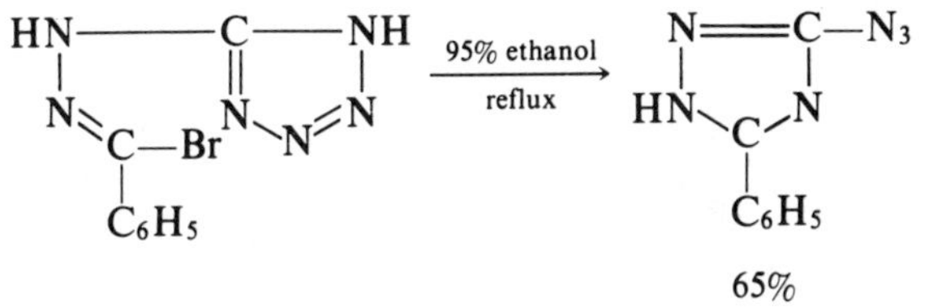

(465)

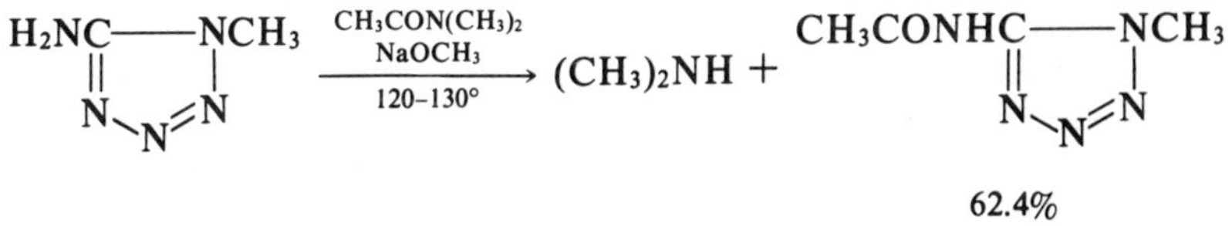

(143)

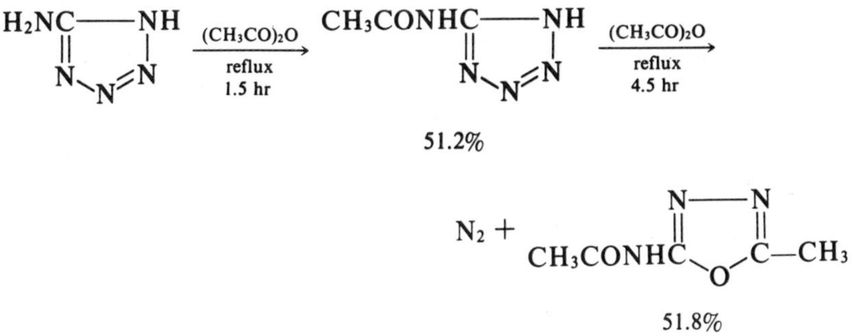

(143)

As indicated in Chapter 2, 2-acyltetrazoles decompose on heating to nitrogen and acyl nitrile imine. The intermediate *N*-acyltetrazoles need not be isolated and usually the 5-substituted tetrazole is heated with acyl chloride, anhydrides, or isocyanates to form oxadiazoles directly. In like manner isothiocyanates and imide chlorides attack the tetrazole ring forming, respectively, 1,3,4-thiadiazoles and 1,2,4-triazoles. Fused 1,2,4-triazolo heterocycles result from the reaction of 2-chloroazines with the triethylamine salts of 5-aryltetrazoles.

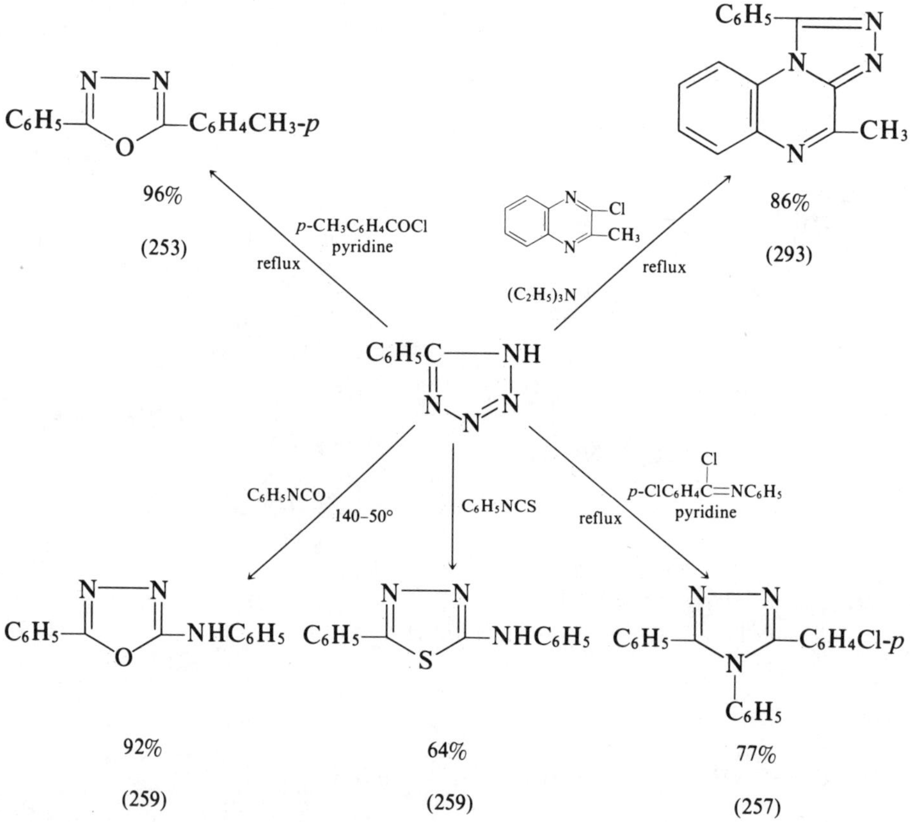

Alternatively dicyclohexylcarbodiimide and carboxylic acids convert 5-substituted tetrazoles to oxadiazoles (421). By use of formic acid in this method, 2-

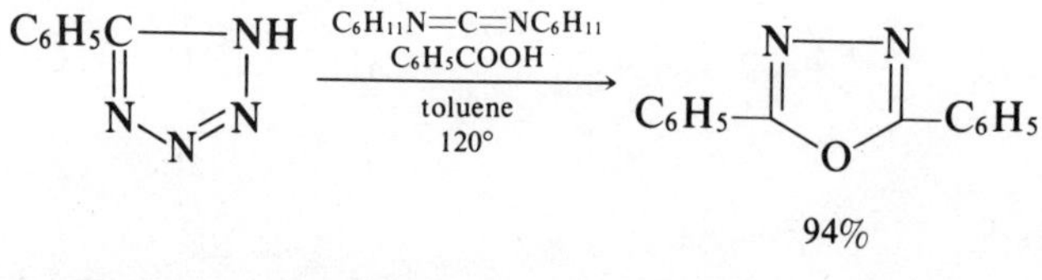

monosubstituted oxadiazoles may be prepared from tetrazoles; they are not available by the acid chloride procedure because of the nonexistence of formyl chloride.

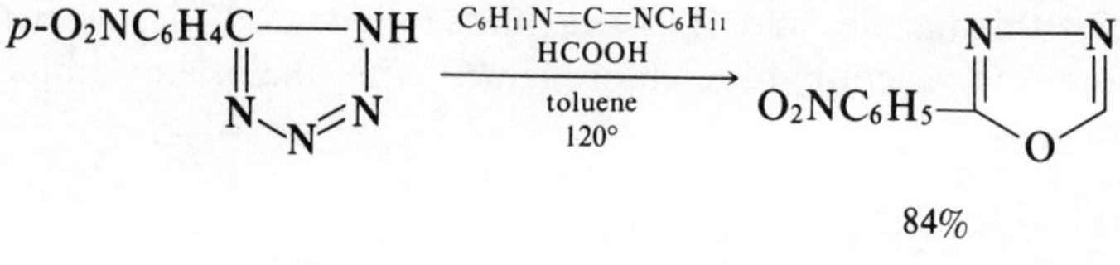

6.2. REACTIONS WITH ELECTROPHILES

In this section stability and cleavage of high nitrogen compounds under the influence of electrophiles are examined, except for alkylating and acylating agents, discussed in the preceding section. Effects on nitrogen chains are considered of protonic and Lewis acids and of reagents capable of nitrosation, nitration, and halogenation.

Terminal amino and substituted amino groups are severed from triazanes and tetrazanes by acids forming one-nitrogen compounds with nitrogen at a re-

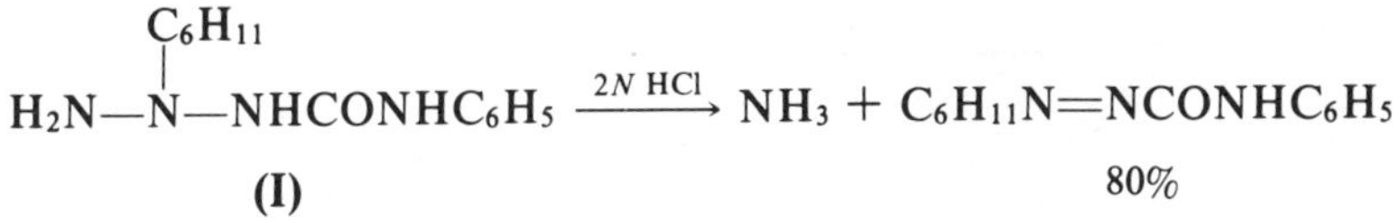

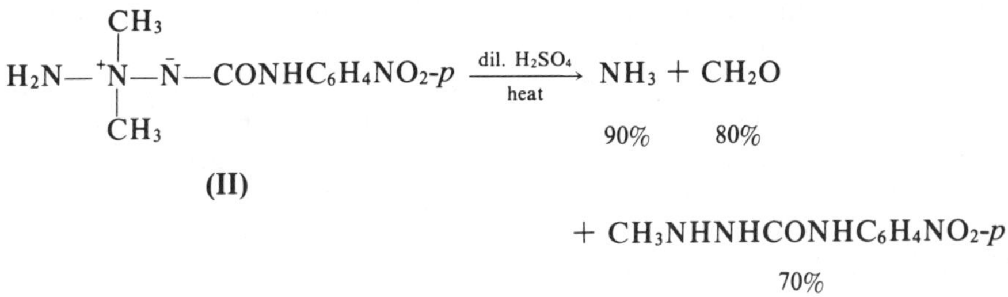

duced level, usually accompanied by dismutation products. Thus acid hydrolysis of 2-cyclohexyltriazane-1-carbanilide **(I)** forms ammonia and cyclohexylazocarbanilide, and triazane betaines such as **II** decompose rapidly in acid solution leading to ammonia, formaldehyde, and a semicarbazide (460). Acetanilide is removed from the ends of 1,4-diacetyl-1,2,3,4-tetraphenyltetrazane

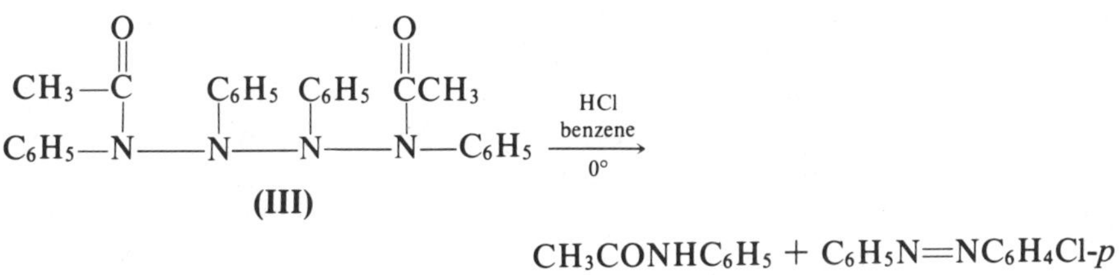

(III) by the action of hydrogen chloride in benzene; simultaneous oxidation is indicated both by the azo compound produced and by the substitution of one

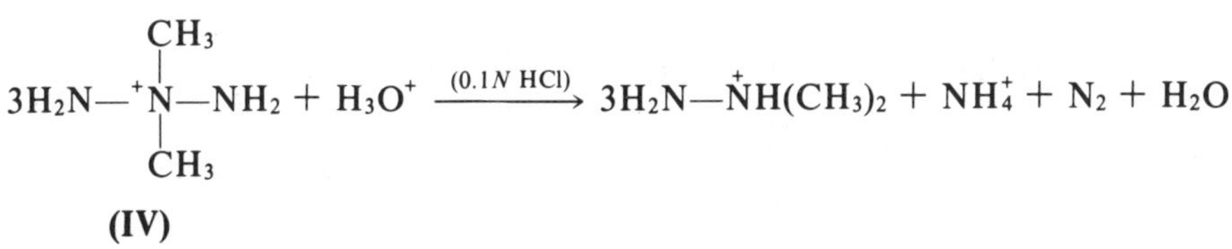

Table 6.9. Stability of Triazanium Salts in Cold Aqueous Solution

Triazanium Salt	Half-life (min)
$N_3H_7^{2+}\ SO_4^{2-}$	2040
$N_3H_6^+\ ClO_4^-$	1260
$N_3H_6^+\ Cl^-$	1020
$N_3H_6^+\ O\bar{O}CCH_3$	380
$N_3H_6^+\ CN^-$	65

benzene ring with chlorine (193). Although 2,2-dimethyltriazanium chloride **(IV)** is stable in neutral aqueous solution, it hydrolyzes rapidly in dilute acid, forming dimethylhydrazinium and ammonium ions together with elemental nitrogen (532). Stabilities of unsubstituted triazanium salts in ice-cold solution, prepared from the sulfate by methathesis with barium salts, depend on their anions, varying linearly with the acidity constant of the acid form of the anion, illustrated by half-lives of a series of triazanium salts in 0.1*N* aqueous solution (Table 6.9) (332).

Triazenes are very sensitive to acids, decomposing in aqueous media to nitrogen, amines, and alcohols or phenols. Indeed, 3-methyl-1-phenyltriazene suspended in water dissolves with decomposition on saturation with carbon dioxide (130), and a suspension at pH 1 and 20° is entirely destroyed in less than 1 hr (14); silica gel also catalyzes its decomposition (282). 3,3-Dimethyl-1-phenyltriazene decomposes below pH 5 (283). Triazenes and pentazdienes can be analyzed by heating at 150° in concentrated sulfuric acid, liberating double bonded nitrogen as elemental nitrogen (277). An ionic mechanism is involved in which a diazonium ion results first from breaking the bond between N-2 and N-3, followed by loss of nitrogen and formation of a carbonium ion, if the resulting group can stabilize a positive charge. Alkyl carbocations beyond ethyl rearrange as indicated by the formation of both *n*-butyl alcohol and *sec*-butyl alcohol from acid treatment of 1,3-di-*n*-butyl-3-methyltriazene **(V)** (473). Decompo-

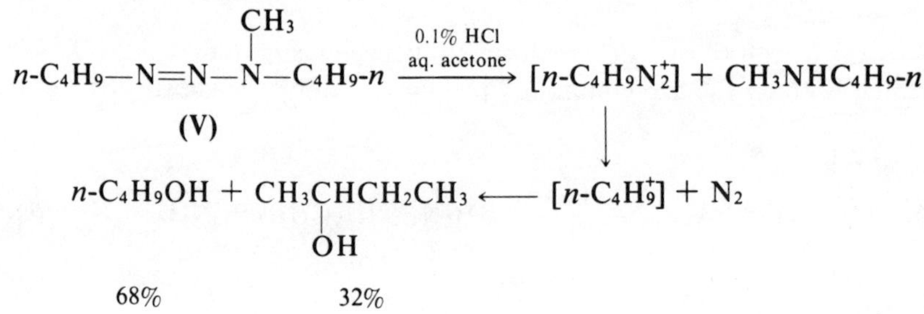

sition of 1-alkyl-3-aryltriazenes by carboxylic acids enables esterification of these acids by alkyl under gentle conditions (545). The sequence, consisting of

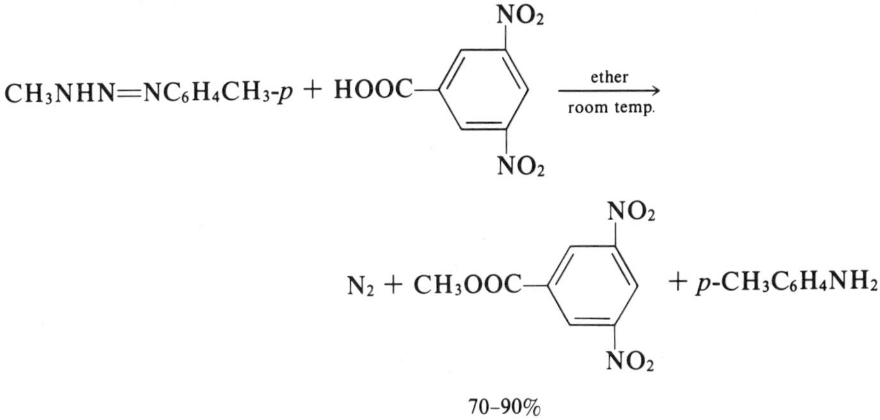

formation of a triazene and its subsequent decomposition by acid, is a method for deaminating alkyl amines (546).

As expected, boron trifluoride, aluminum chloride (299, 300), and cuprous chloride (334) cause the decomposition of triazenes. Products formed are principally the result of combination of radicals from the decomposition of liberated diazonium salts and their reaction with the benzene solvent.

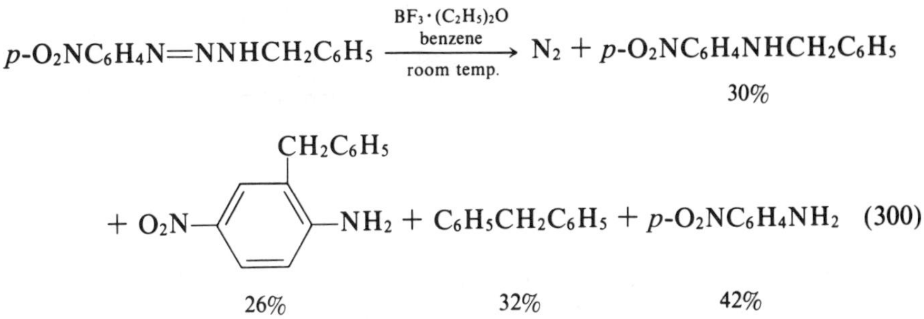

$$C_6H_5N{=}NNHCOOC_2H_5 \xrightarrow[30\text{–}40^\circ]{AlCl_3,\ \text{benzene}} \underset{}{N_2} + \underset{87\%}{C_6H_5C_6H_5} + \underset{32\%}{C_2H_5C_6H_5} + \underset{5\%}{C_6H_5NH_2} \quad (299)$$

Decomposition of aryl triazenes by acids, to diazo components for reaction with coupling agents to form dyes *in situ* on textiles is discussed in Chapter 4. Such triazenes are surrogate, or masked, diazonium compounds. In the same manner, acid treatment of 3-cyano-1-phenyltriazene **(VI)** in the presence of 2-

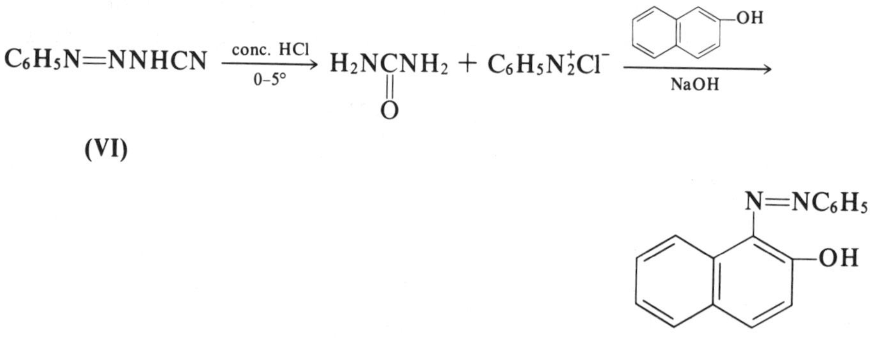

(VI)

96%

naphthol leads to an azo derivative (564), and the intermediate benzenediazonium ion resulting from acid decomposition of *N*-phenylazopyrrolidine **(VII)**

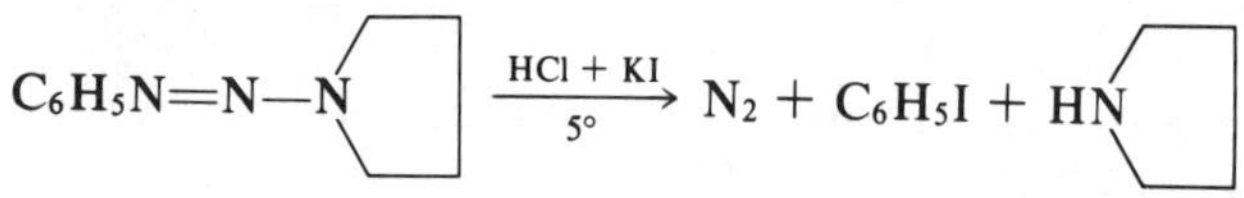

(VII)

is converted to iodobenzene in the presence of potassium iodide (159). Similarly, aromatic fluorides are obtained from aryl triazenes by treatment with hydrogen fluoride in pyridine (445).

$$p\text{-}C_2H_5OOCC_6H_4N{=}NN(CH_3)_2 \xrightarrow[18^\circ]{70\%\ HF,\ \text{pyridine}} N_2 + p\text{-}C_2H_5OOCC_6H_4F$$

90%

Triazenes with 1,3-diaryl substituents undergo conversion to aminoazo compounds, usually in the presence of a salt of an amine component of the triazene. From diazoaminobenzene in aniline as solvent and aniline hydrochloride as acid, the major product is *p*-aminoazobenzene with a small quantity of ortho isomer (554). This rearrangement, the object of much study and discussion

$$C_6H_5N{=}NNHC_6H_5 \xrightarrow[C_6H_5NH_2]{C_6H_5NH_2\cdot HCl} C_6H_5N{=}NC_6H_4NH_2\text{-}p + C_6H_5N{=}NC_6H_4NH_2\text{-}o$$

93% 4%

(283), seems to be an intermolecular process, as originally proposed by Friswell and Green (169), in which the triazene is protonated and undergoes fission to diazonium ion and amine, followed by coupling to the benzene ring carrying the amino group to form the aminoazo compound. Thus diazoaminobenzene in dimethylaniline containing dimethylaniline hydrochloride produces *p*-dimethylaminoazobenzene (446). The most compelling evidence, however, for the intermolecular nature of the rearrangement involves systems of diazoaminobenzene, aniline, and aniline hydrochloride, labeled variously with ^{15}N. In all cases, when the middle nitrogen is labeled, it is found in the nitrogen of the azo group attached to the benzene ring carrying amino (113, 543), showing a quantitative

$$C_6H_5N{=}NNHC_6H_5 \xrightarrow[\substack{C_6H_5N(CH_3)_2 \\ \text{room temp.}}]{C_6H_5N(CH_3)_2 \cdot HCl} p\text{-}(CH_3)_2NC_6H_4N{=}NC_6H_5$$

$$C_6H_5N{=}^{15}N{-}NHC_6H_5 \xrightarrow[30\text{–}40^\circ]{\substack{C_6H_5NH_2 \cdot HCl \\ C_6H_5NH_2}} C_6H_5N{=}^{15}N{-}C_6H_4NH_2$$

exchange between the C_6H_5NH of the triazene and the benzene ring of the aniline salt. The possibility that ortho coupling may be partially intramolecular is raised by experiments in which the ortho/para ratio of triazene rearrangement products is increased by an increase in solvent viscosity (397).

Metal salts of disubstituted triazenes undergo reaction with nitrosyl chloride to form unstable nitrosotriazenes (370). Condensation of 1,3-disubstituted triazenes with diazonium compounds to form pentazdienes is discussed in Chapter 7, Synthesis.

$$C_6H_5N(Ag){-}N{=}NC_6H_5 \xrightarrow[-70^\circ]{\substack{ONCl \\ \text{ether}}} C_6H_5N(NO){-}N{=}NC_6H_5$$

Acids effect breakdown of 1,3-disubstituted 1-tetrazenes to diazonium salts or their decomposition products and hydrazines or derivatives, sometimes with rearrangement. Thus the benzaldehyde condensation product of 1-*p*-nitrophenyl-3-phenyltetrazene **(VIII)** on treatment with cold hydrochloric acid forms the arylhydrazone derived from migration of *p*-nitrophenylazo to the phenylmethylene group (217). If different aryl groups are substituted at N-1 and

$$C_6H_5CH{=}N{-}N(C_6H_5){-}N{=}N{-}C_6H_4NO_2\text{-}p \xrightarrow[<0^\circ]{\substack{HCl \\ \text{aq. ethanol}}}$$

(VIII)

$$C_6H_5N_2^+Cl^- + \underset{99\%}{C_6H_5CH{=}NNHC_6H_4NO_2\text{-}p}$$

N-3, the arylazo with the most electron-withdrawing character is found in the resulting aryl hydrazone. Under more vigorous conditions, acids convert 1,3-disubstituted 1-tetrazenes to phenols and hydrazines, as in the decomposition of 3-phenyl-1-*p*-tolyl-1-tetrazene to nitrogen, *p*-cresol, phenylhydrazine, and a trace of *p*-tolylazide (548).

$$p\text{-}CH_3C_6H_4N{=}N{-}\underset{}{\overset{C_6H_5}{\overset{|}{N}}}{-}NH_2 \xrightarrow[\text{reflux}]{\text{dil. }H_2SO_4} N_2 + p\text{-}CH_3C_6H_4OH + C_6H_5NHNH_2\cdot H_2SO_4 + CH_3C_6H_4N_3$$

Acid cleavage of 2-tetrazenes liberates nitrogen and oxidized and reduced species, derived for the most part from the 1- and 4-nitrogen atoms and their substituents. Tetraethyltetrazene, which is unchanged in cold, dilute acid, decomposes on heating to nitrogen, mono- and diethylamine, and acetaldehyde (551). Radicals are detected by ESR during the acid decomposition of tetra-

$$(C_2H_5)_2N{-}N{=}N{-}N(C_2H_5)_2 \xrightarrow[70^\circ]{7\%\text{ HCl}} N_2 + C_2H_5NH_2 + (C_2H_5)_2NH + CH_3CHO$$

methyltetrazene, suggesting initial participants in its breakdown. Thus in carbon tetrachloride containing 0.2–2.0 *M* acetic acid, a transient dimethylamino-solvent radical is formed as well as cation radicals of tetramethyltetrazene and tetramethylhydrazine (383). Tetramethyltetrazene cation radical also is formed in concentrated sulfuric acid, decaying slowly over several days; dimethyl-

$$(CH_3)_2N{-}N{=}N{-}N(CH_3)_2 \xrightarrow{\text{conc. }H_2SO_4} (CH_3)_2N{-}N{=}N{-}\overset{+\bullet}{N}(CH_3)_2 + (CH_3)_2\overset{+}{N}H_2 + CH_3\overset{+}{N}H{=}CH_2 + (CH_3)_2\overset{\overset{+}{H}}{N}{-}\overset{\overset{+}{H}}{N}(CH_3)_2$$

aminium and methylene methylaminium cations and a small amount of tetramethylhydrazine dication are detected by ^{1}H and ^{13}C NMR (345).

Tetramethyltetrazene combines with aluminum trimethyl to form an adduct whose ^{1}H NMR spectrum shows it to be symmetrically bonded in a cyclic structure (149, 150). Decomposition of the complex at 90° leads to nitrogen, tetra-

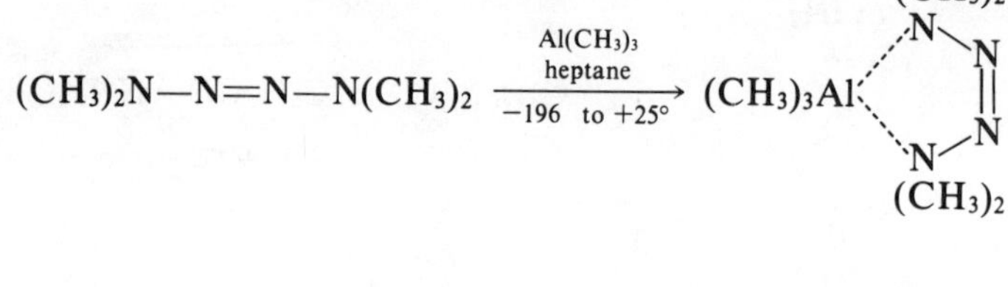

methyltetrazene, $(CH_3)_2AlN(CH_3)_2$, a polymer, and traces of methane and methyl azide.

Tetraaryl tetrazenes decompose on treatment with acid to diaryl amines and oxidized products such as *N,N'*-diphenylbenzidine from tetraphenyltetrazene (205) and *N*-chloro-di-*p*-tolylamine, di-*p*-tolylamine, and ditolyldimethylphenazine from tetra-*p*-tolyltetrazene and hydrochloric acid (550).

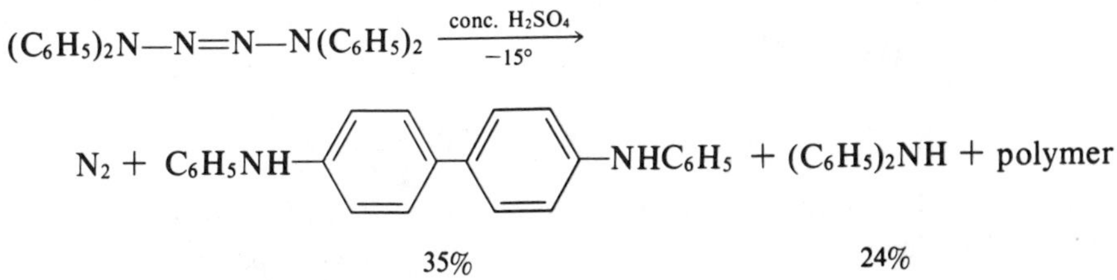

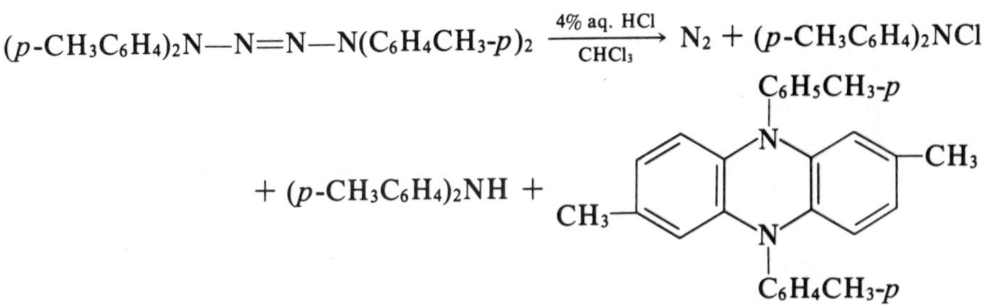

The azo group in 2,2',3,3'-tetramethyl-1,1'-azobenzimidazolium bromide similarly is removed on treatment with picric acid (187). On heating 1,1'-azobis(5-*p*-tolyltetrazole) in acetic acid, 5-*p*-tolyltetrazole is formed (499).

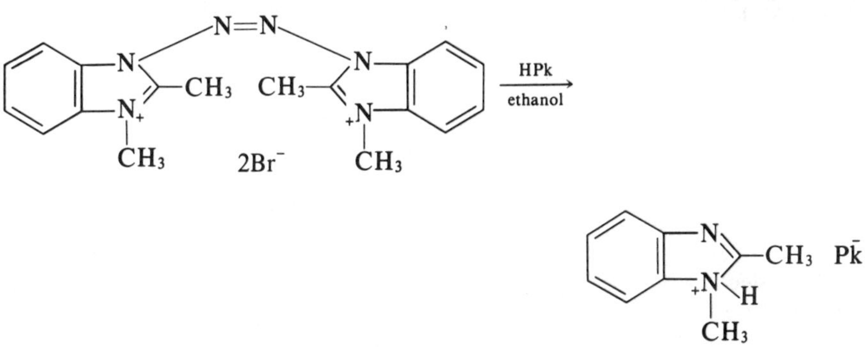

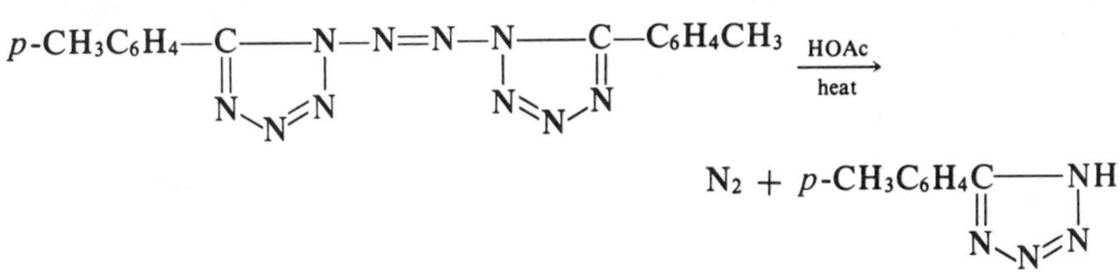

Fission of the 2-tetrazene chain to form products with one and three nitrogen atoms, but not elemental nitrogen, also is known. α,α'-Dimethoxytetraethyl tetrazene **(IX)** is hydrolyzed by water to ethylazide and ethylamine (119).

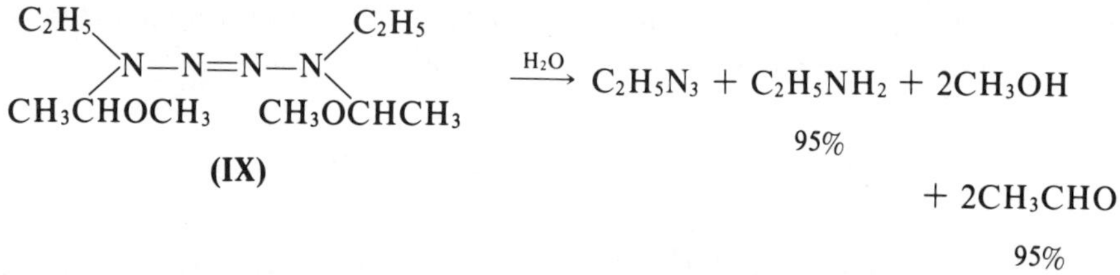

(IX)

Treatment of 1,4-dimethyl-1,4-di(2-pyridyl)-2-tetrazene **(X)** with hot hydrobromic acid leads to methyltetrazolopyridinium bromide, containing three of the tetrazene nitrogen atoms, and 2-methylaminopyridine (16).

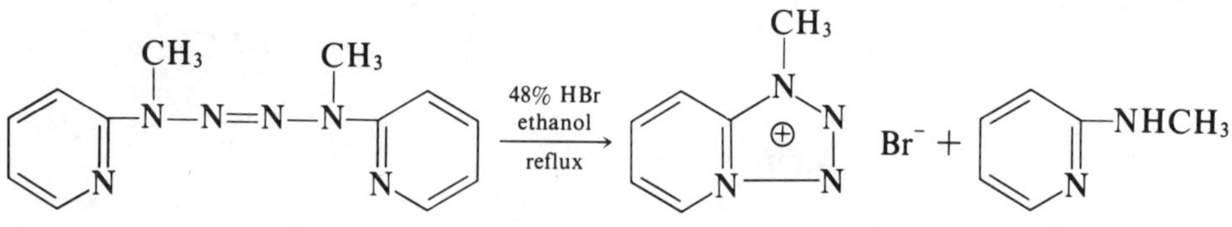

(X) 84%

Stereoisomerization of 2,3-disubstituted phthalimidoazimines occurs on treatment under mildly acid conditions. Thus (2*E*)-2,3-diisopropyl-1-phthalimidoazimine is converted to the 2*Z* form on exposure to silica, methyl iodide, zinc chloride, or trifluoracetic acid, giving a *Z/E* mixture in the proportion of 95:5 (323).

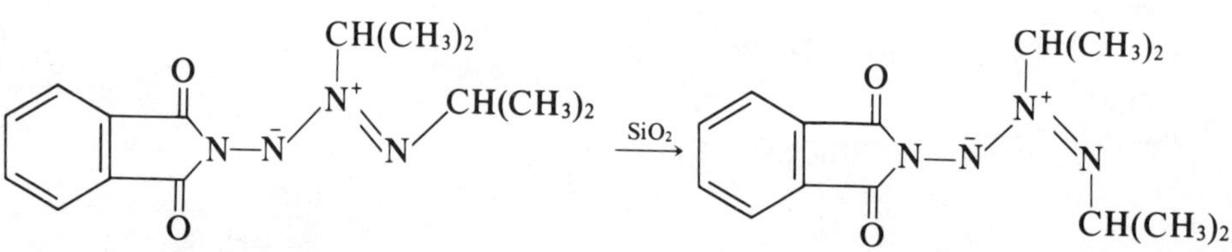

Acid decomposition of pentazdienes appears to involve initial protonation followed by decomposition to disubstituted triazene and aryldiazonium ion. The triazene is rapidly decomposed further and the final products are those expected from these intermediates. Based on the kinetics of the destruction of 1,5-

$$C_6H_5N{=}N{-}N(CH_3){-}N{=}NC_6H_5 \xrightarrow[\text{heat}]{\text{dil. } H_2SO_4} N_2 + CH_3NH_2 + CH_3OH + C_6H_5NH_2 + C_6H_5OH$$

(XI)

diaryl-3-methylpentazdienes by perchloric acid in 50% ethanol, the reaction is pseudomonomolecular; its rate is directly related to the acidity of the medium and increases as the electron-withdrawing character of aryl substituents increases (443). 1,5-Diphenyl-3-methylpentazdiene **(XI)**, which is stable in cold dilute sulfuric acid, decomposes on heating to nitrogen, methylamine, methyl alcohol, aniline, and phenol (190). The action of hydrochloric acid on bis(benzeneazo)glycine **(XII)** liberates nitrogen and carbon dioxide, and leads to diazoaminobenzene (92).

$$C_6H_5{-}N{=}N{-}\underset{\displaystyle (CH_2COOH)}{N}{-}N{=}N{-}C_6H_5 \xrightarrow[\text{room temp.}]{\text{conc. HCl, ethanol}} N_2 + CO_2 + C_6H_5N{=}NNHC_6H_5$$

(XII)

Hexazdienes slowly dissolve in cold sulfuric acid forming diazonium salts (247).

$$p\text{-}ClC_6H_4N{=}N{-}\underset{CH_3C{=}O}{N}{-}\underset{O{=}CCH_3}{N}{-}N{=}NC_6H_4Cl\text{-}p \xrightarrow[\text{cool}]{\text{conc. } H_2SO_4} p\text{-}ClC_6H_4N_2^+\ H\bar{SO_4}$$

Nitroso groups are removed by acids from nitrogen chains by loss as nitrous acid or nitric oxide, by formation of azide, or in some *N*-nitrosopyrazolinones and pyrazolidinones through rearrangement to a ring carbon.

Nitrous acid is the product of acid hydrolysis of 3,4-dimethyl-5-ethoxy-1-nitrosopyrazole **(XIII)** (562) and 1-nitroso-1,4,4-trimethyl-2-tetrazene **(XIV)** (522). Hydrolysis to nitrous acid suggests a high degree of polarity of the ni-

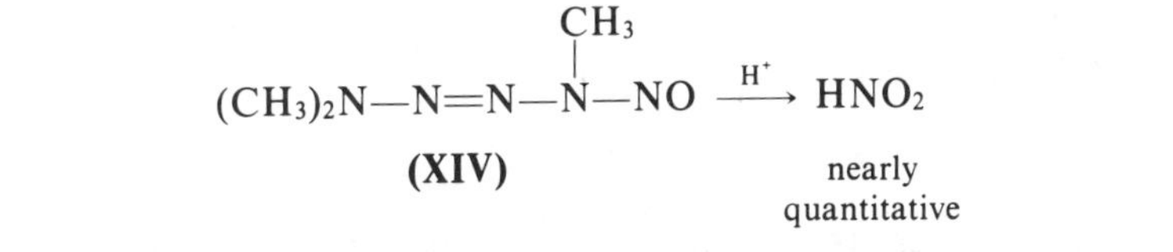

(XIII)

$$(CH_3)_2N{-}N{=}N{-}\underset{\displaystyle (CH_3)}{N}{-}NO \xrightarrow{H^+} HNO_2$$

(XIV) nearly quantitative

troso bond to nitrogen, so that hydroxyl preferentially attaches to incipiently positive nitrosyl. Although 3-methyl-2-nitroso-1-phenyl-5-pyrazolidone **(XV)** can be recrystallized unchanged from glacial acetic acid after heating at 40–50° (288), the cold acid splits nitric oxide from 1-nitroso-5-phenyl-3-pyrazolidone **(XVI)** (369). The nitric oxide-diethylamine reaction product **(XVII)** liberates nitric oxide on treatment with hydrochloric acid (135). In these decompositions, it can be assumed that the nitrogen to which the nitroso group is attached retains sufficient basicity so that a proton can displace nitric oxide.

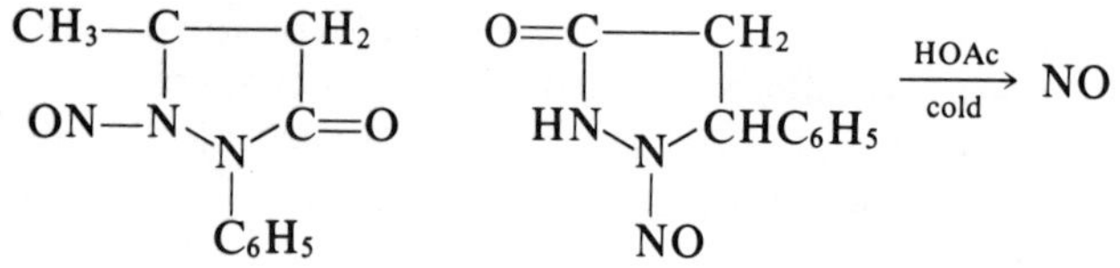

(XV) (XVI)

$$(C_2H_5)_2NN_2\bar{O}_2(C_2H_5)_2N\overset{+}{H}_2 \xrightarrow{HCl} NO + (C_2H_5)_2N\overset{+}{H}_2\,Cl^-$$

(XVII)

Monoalkyl or aryl nitrosohydrazines on treatment with acids lose water and rearrange to azides, illustrated by the formation of benzyl azide (120, 558) and phenyl azide (56). Electron-withdrawing substituents on phenyl accelerate the formation of aryl azides.

$$C_6H_5CH_2N(NO){-}NH_2 \xrightarrow[\text{warm}]{\text{dil. }H_2SO_4} C_6H_5CH_2N_3$$

$$C_6H_5N(NO){-}NH_2 \xrightarrow[\text{ethanol, }15^\circ]{HSO_3NH_2} C_6H_5N_3$$

Nitrosation of ethyl nitrosohydrazinoacetate **(XVIII)** leads to ethyl diazoacetate and the evolution of nitrous oxide (124).

$$C_2H_5OOCCH_2N(NO){-}NH_2 \xrightarrow[\text{ice cooling}]{NaNO_2 + H_2SO_4,\ H_2O} C_2H_5OOCCHN_2 + N_2O$$

(XVIII)

On acid treatment, the nitroso group of 1-nitroso-5-*m*-nitrophenyl-3-pyrazolinone **(XIX)** migrates to the unoccupied 4-position, forming 4-isonitroso-3-*m*-

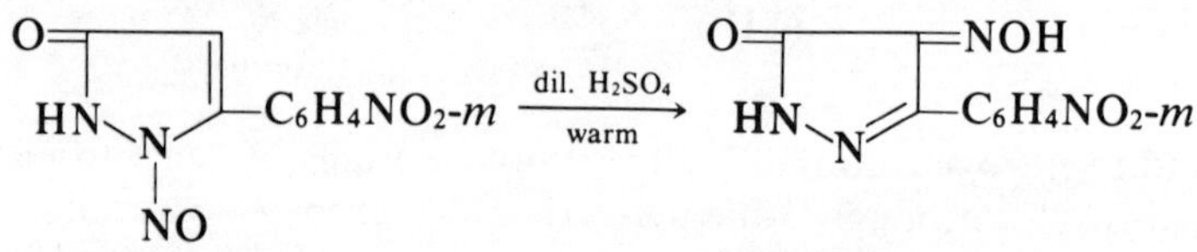

(XIX)

nitrophenyl-5-pyrazolinone (122). In a closely similar reaction, *N*-nitropyrazole **(XX),** which is stable in boiling water, rearranges to 4-nitropyrazole in

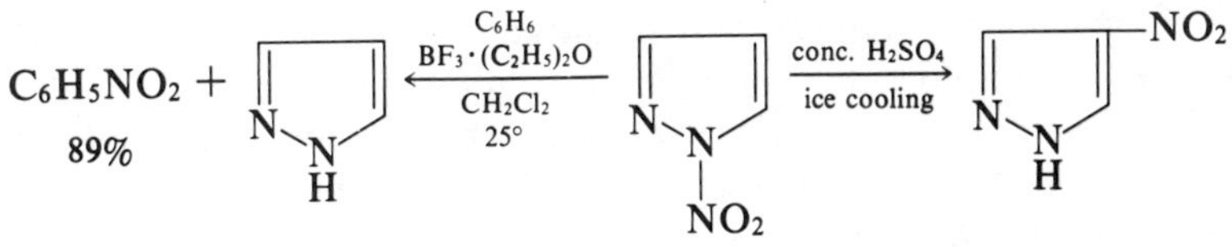

89% (XX) 80%

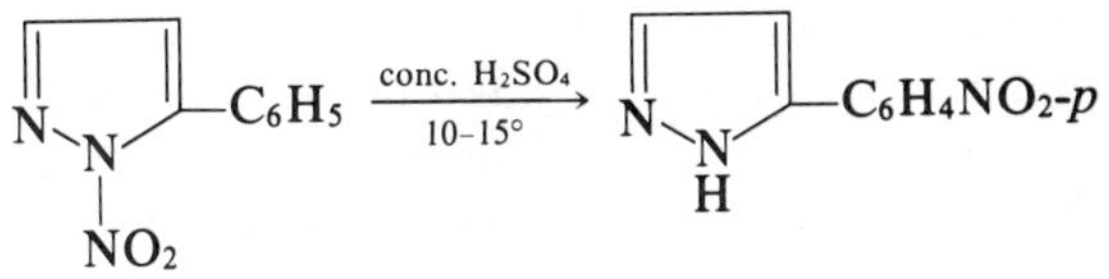

(XXI)

sulfuric acid (263). That the reaction involves dissociation of nitronium ion from the pyrazole nitrogen is indicated by the conversion of added anisole or phenols to nitroanisoles or nitrophenols (274); such a dissociation is not observed in the thermal isomerization of *N*-nitropyrazoles. Indeed *N*-nitropyrazole, in the presence of protonic or Lewis acids, effectively nitrates aromatic compounds (400), and the phenyl group in 1-nitro-5-phenylpyrazole **(XXI)** undergoes nitration when dissolved in sulfuric acid (123).

The amine nitroimides such as pyridine-1-nitroimide **(XXII)** are inert toward concentrated sulfuric acid and the nitrating agents, acetyl nitrate, or nitronium

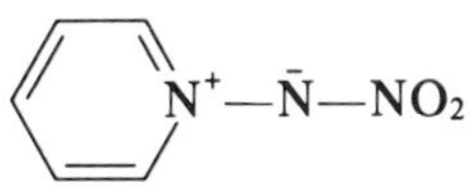

(XXII)

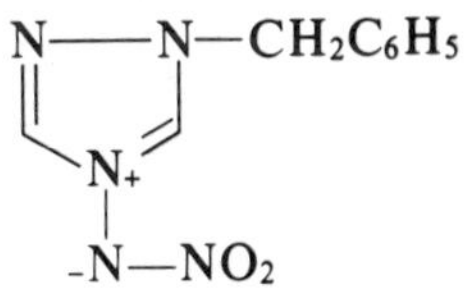

(XXIII)

fluoborate (19, 145). 1-Benzyl-4-nitroimino-1,2,4-triazolium ylid **(XXIII)** is recrystallized unchanged from dilute hydrochloric acid (524).

Electrophiles act on most azides to eliminate nitrogen, frequently accompan-

$$CH_3CH_2N_3 \xrightarrow[80°]{H_2SO_4-SO_3} \begin{cases} N_2 + [CH_3CH{=}NH] \longrightarrow CH_3CHO + NH_3 \quad 86\% \\ N_2 + [CH_2{=}N{-}CH_3] \longrightarrow CH_2O + CH_3NH_2 \quad 14\% \end{cases}$$

$$(CH_3)_2CHN_3 \xrightarrow[\text{room temp.}]{\text{HCl-SbCl}_5,\ CH_2Cl_2}$$

$$N_2 + \underset{68.5\%}{(CH_3)_2C{=}\overset{+}{N}H_2\ \overset{-}{S}bCl_6} + \underset{31.5\%}{CH_3CH{=}\overset{+}{N}HCH_3\ \overset{-}{S}bCl_6} \quad (199, 399a)$$

ied by rearrangement of the organic substituent. In aliphatic azides, besides loss of nitrogen, both shift of hydrogen and migration of alkyl occur, initially forming imines or iminium ions which can yield carbonyl compounds and ammonia or amines on hydrolysis. Ethyl azide is converted mainly to acetaldehyde and ammonia, with formation of small quantities of formaldehyde and methylamine (469). Isopropyl azide is converted by antimony pentachloride and hydrogen chloride to 2-propylideniminium and a mixture of *cis*- and *trans*-*N*-methylethylideniminium ions.

Rearrangement from the action of acids is more extensive in alicyclic azides, illustrated by the decomposition of cyclopentyl azide **(XXIV)** principally to Δ'-piperideine and a minor quantity of cyclopentylidene imine (75). An even more

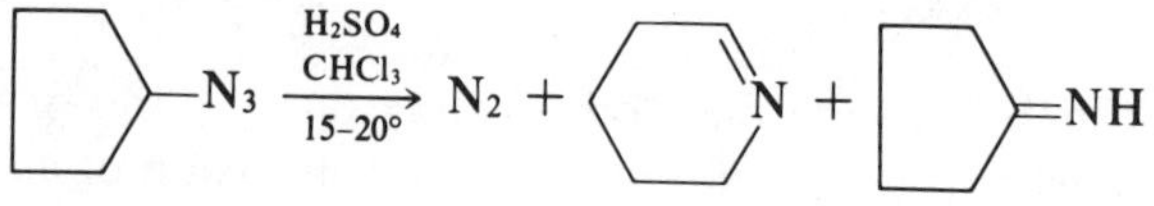

(XXIV) 79.8% 2.2%

profound acid-induced rearrangement occurs with azidoquinones such as 2-azido-6-isopropyl-3-methylquinone **(XXV)** to cyanoalkylidene γ-lactones (367).

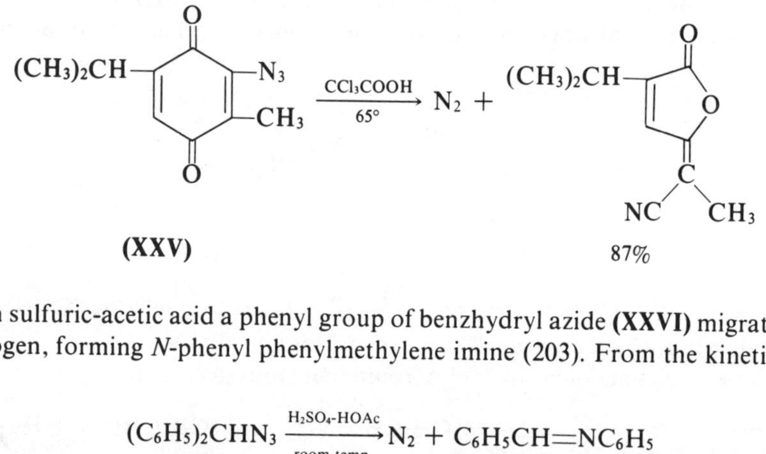

(XXV) 87%

In sulfuric-acetic acid a phenyl group of benzhydryl azide **(XXVI)** migrates to nitrogen, forming *N*-phenyl phenylmethylene imine (203). From the kinetics of

$$(C_6H_5)_2CHN_3 \xrightarrow[\text{room temp.}]{H_2SO_4\text{-HOAc}} N_2 + C_6H_5CH{=}NC_6H_5$$

(XXVI)

the reaction, it is concluded that loss of nitrogen occurs first and is not synchronous with the migration step. Both hydrolysis and rearrangement of diazidodiphenylmethane **(XXVII)** occur on treatment with sulfuric acid (198).

$$(C_6H_5)_2C(N_3)_2 \xrightarrow[\text{room temp.}]{75\%\ H_2SO_4} N_2 + HN_3 + C_6H_5NHCOC_6H_5$$

(XXVII) 98%

Nitrosation also causes destruction of organic azides, liberating nitrogen and forming nitrous oxide. Thus triphenylmethyl azide **(XXVIII)** is converted by nitrosonium fluoborate quantitatively to triphenylmethyl fluoborate (137).

$$\underset{\textbf{(XXVIII)}}{(C_6H_5)_3CN_3} \xrightarrow[\text{room temp.}]{\substack{NO^+\,BF_4^-\\ \text{acetonitrile}}} N_2 + N_2O + (C_6H_5)_3C^+\ BF_4^-$$

Although electrophiles effect decomposition of most azides, conditions exist, particularly with aryl azides, whereby electrophilic substitutions are effected elsewhere in the molecule with only partial attack on the azide group. Phenyl azide, for example, is brominated (479); under the influence of nitric acid alone, *m*-nitrophenyl azide is further nitrated (87); and azido-nitronaphthalenes result from both 1- and 2-azidonaphthalene (157). In contrast, the monoazidonaphthalenes decompose vigorously and 2,3-diazido-1,4-naphthoquinone decomposes explosively on addition of concentrated sulfuric acid (167). A bromoamino diphenyl results from the action of hydrogen bromide on *o*-azidodiphenyl (478), and sulfonation of phenyl azide forms, after hydrolysis, *p*-aminophenol sulfonic acid (469), showing compensating oxidation and reduction also indicated by the formation of bromoanilines in the bromination of phenyl azide.

$$C_6H_5N_3 \xrightarrow[0\text{–}10^\circ]{\substack{Br_2\\ CCl_4}} \underset{54\%}{p\text{-}BrC_6H_4N_3} + \underset{23\%}{x\text{-}BrC_6H_4NH_2}$$

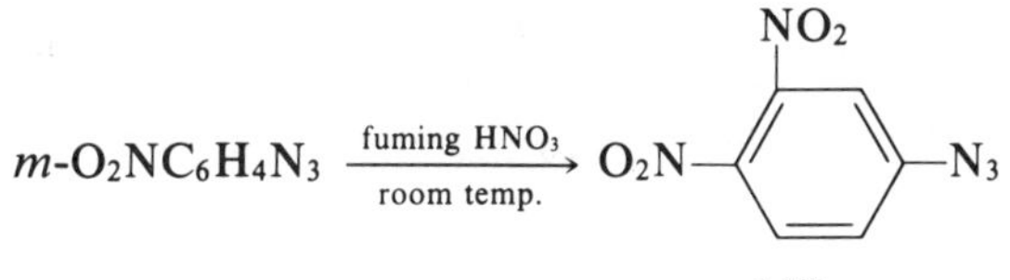

94%

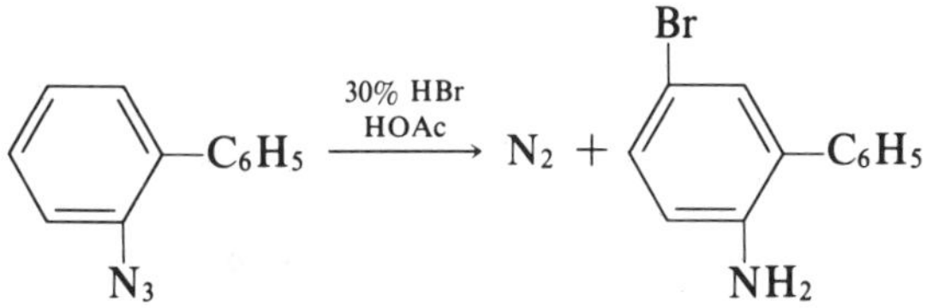

82%

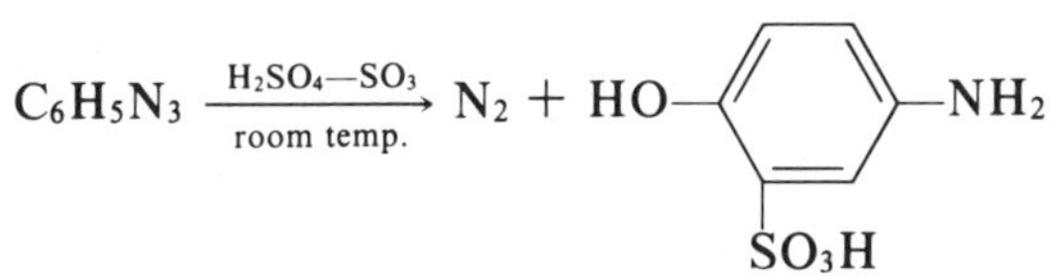

58%

Although ring expansion of aryl azides to azepines usually takes place under the influence of nucleophiles, ring expansion of phenyl azide occurs in acetic acid solution, along with oxidation-reduction substitution (508). A small yield of azobenzene also results from treating phenyl azide with aluminum chloride (73).

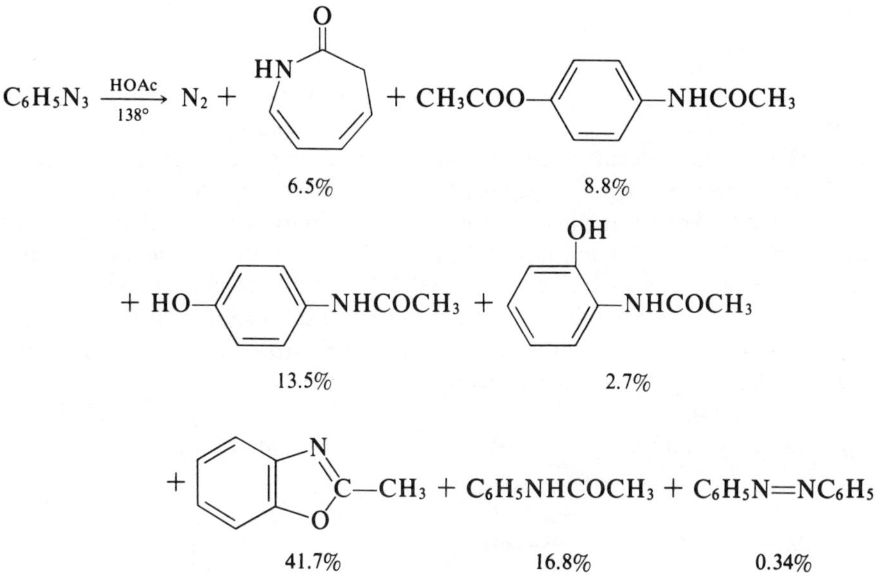

An adduct forms between acid azides and boron trifluoride or boron trichloride as low as −60°; nitrogen is lost on warming and the Curtius rearrangement occurs (146). Carboxylic acid azides decompose under the influence of both

$$p\text{-}O_2NC_6H_4CON_3 \xrightarrow[-60^\circ]{BF_3,\ \text{toluene}} p\text{-}O_2NC_6H_4\underset{\underset{\displaystyle O{:}BF_3}{\|}}{C}\text{—}N_3 \xrightarrow{-20^\circ \text{ to } 0^\circ} N_2 + p\text{-}O_2NC_6H_4NCO{:}BF_3$$

Table 6.10. Decomposition of Benzoyl Azide by Lewis Acids

Catalyst	Temp. (°C)	Catalytic Rate Constant in Nitrobenzene, $\times 10^4$
None	54.9	9.8
$AlCl_3$	10.5	340
$TiCl_4$	10.5	370
$SbCl_5$	10.5	5470

protonic and Lewis acids at much faster rates than their thermal rearrangement to isocyanates, illustrated by the catalytic rate constants for benzoyl azide with several Lewis acids (Table 6.10) (117).

In a series of qualitative experiments at 75°, Newman and Gildenhorn (387) determined that the time to collect half the volume of gas from benzoyl azide under the influence of different acids decreases as acid strength increases, and is faster than with a nonacidic solvent (Table 6.11).

In sulfuric-acetic acid, hydrazidic azides rearrange to semicarbazides (222).

$$C_6H_5C(N_3)\!=\!NNHC_6H_4NO_2\text{-}p \xrightarrow[\text{room temp.}]{H_2SO_4\text{-}HOAc} N_2 + \underset{84\%}{C_6H_5NHCONHNHC_6H_4NO_2\text{-}p}$$

A nondestructive conversion of a heterocyclic azide occurs in the action of ethanolic hydrochloric acid at the carbonyl group of 2-azidopyrido[1,2-*a*]pyrimid-4-one **(XXIX)** (296). Both opening of the pyrimidine ring and cyclization of azide with adjacent nitrogen take place, forming 1-(2′-pyridyl)-5-carboxymethyltetrazole and its ethyl ester.

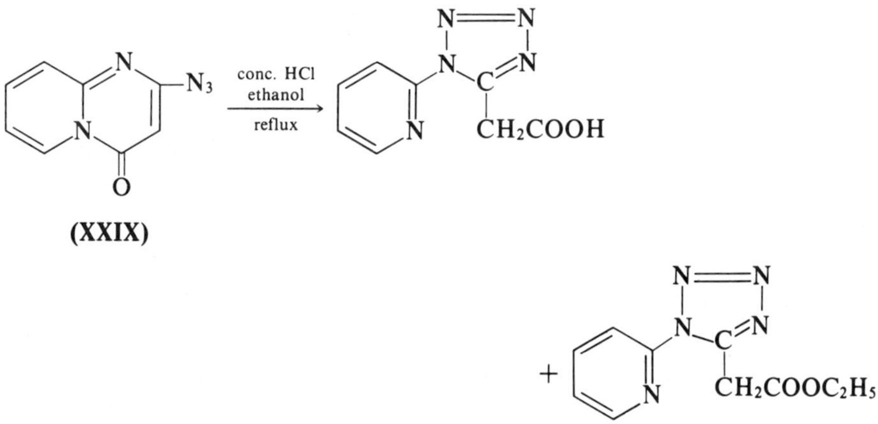

(XXIX)

Table 6.11. Decomposition of Benzoyl Azide in Acidic Media

Acid	Solvent	Time to Evolve $\frac{1}{2}$ N_2 from Benzoyl Azide, 75° (minutes)
—	Dioxane	60
	100% HOAc	20
50% $ClCH_2COOH$	HOAc	13
50% CCl_3COOH	HOAc	7
18% BF_3	Dioxane	7
3.5% HCl	Dioxane	6

The three-nitrogen chain of triazolines remains intact under the influence of acids, if the elements of a base can be removed from the 4- and 5-positions, to form a *v*-triazole. For example, removal of morpholine from 5-morpholino-1-*p*-nitrophenyl-5-phenyltriazoline **(XXX)** on acid treatment leads to 1-*p*-nitro-

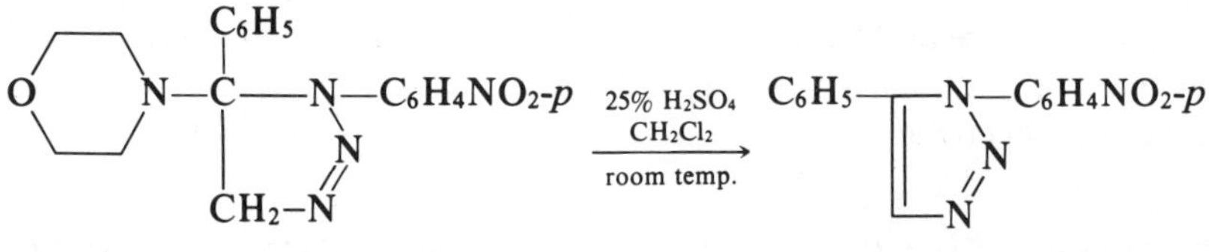

(XXX) 88%

phenyl-5-phenyl-*v*-triazole (256). In like manner, pyrazolidinotriazolines aromatize on heating with acetic acid by opening of the pyrazolidine ring (269).

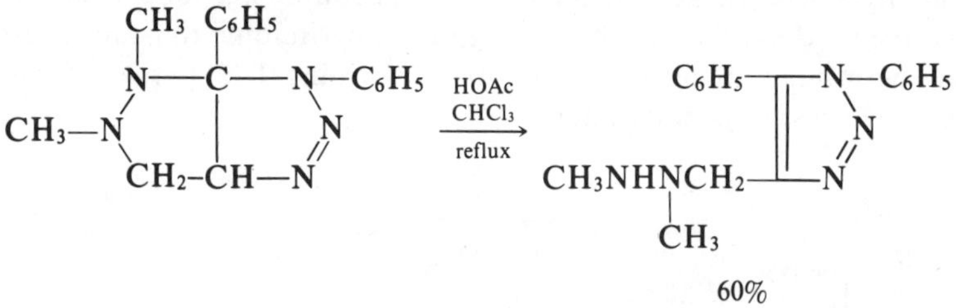

Indeed, aromatization of 1-(1′-phenylvinyl)-5-pyrrolidino-4,5-tetramethylenetriazoline **(XXXI)** occurs on passing a solution through a column of silica gel (392).

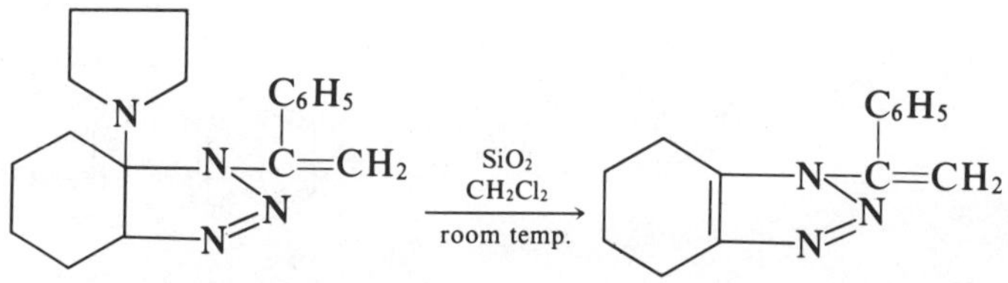

(XXXI) 40%

If there is no hydrogen available on triazoline carbon to participate in removal of a base, acid treatment of triazoline causes breaking of the ring and elimination of nitrogen, probably through an intermediate aliphatic diazonium ion, illustrated by the decomposition of 4,4-dimethyl-1,5-diphenyl-5-(1-piperidino)triazoline **(XXXII)** (373). The action of acid on the triazoline adduct of santene and phenyl azide **(XXXIII)**, which cannot form a ketone without breaking a carbon-carbon bond, loses nitrogen with formation of an amino alcohol (10). Treatment of 1-phenyltriazoline **(XXXIV)** similarly opens the three-nitrogen chain (223) and 4,4-diphenyl-5-iminotriazoline **(XXXV)**, without a substituent at N-1, also loses nitrogen easily, forming benzilic acid amidine (241).

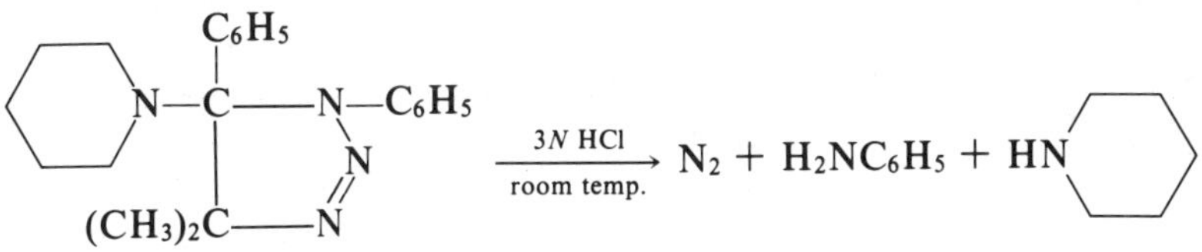

(XXXII)

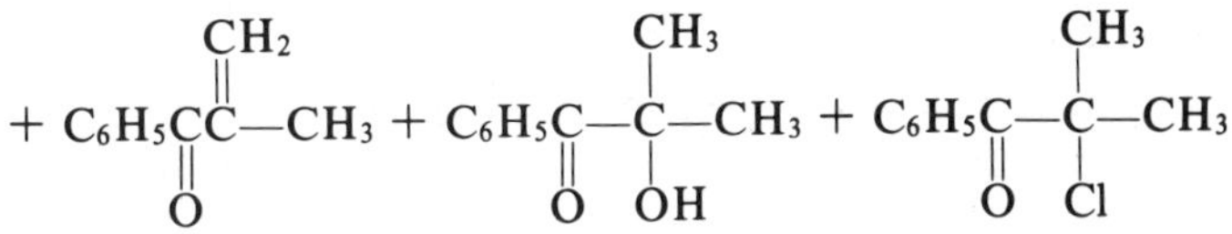

49% 19% 15%

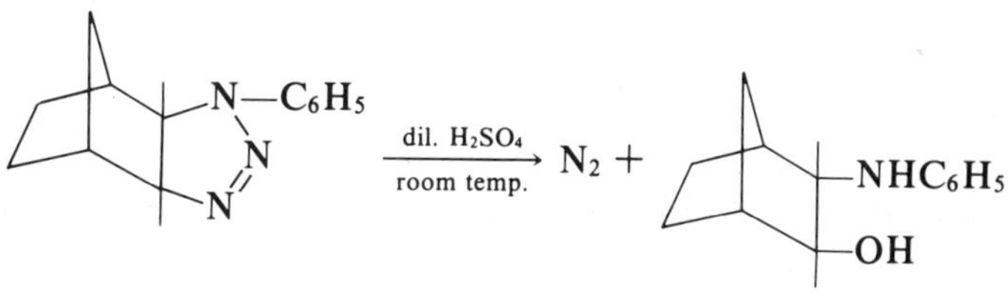

(XXXIII)

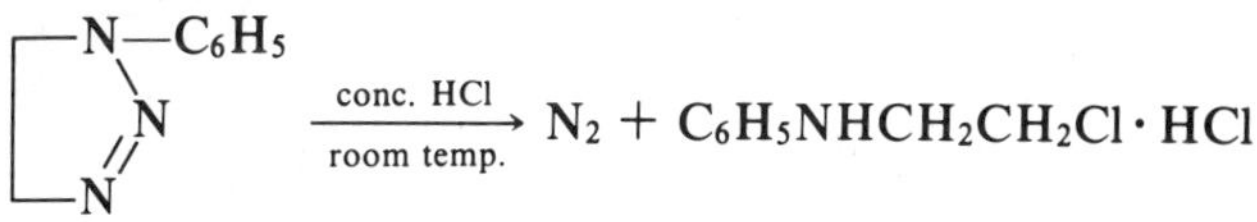

(XXXIV) 87%

Ring expansion occurs on acid treatment of spiro[fluorenetriazolines] such as **XXXVI**, with formation of 9-arylaminophenanthrenes in higher yield than by pyrolysis or photolysis (238).

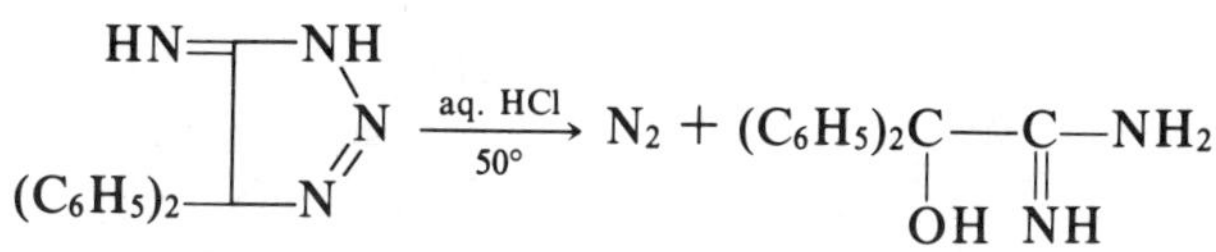

(XXXV) 83%

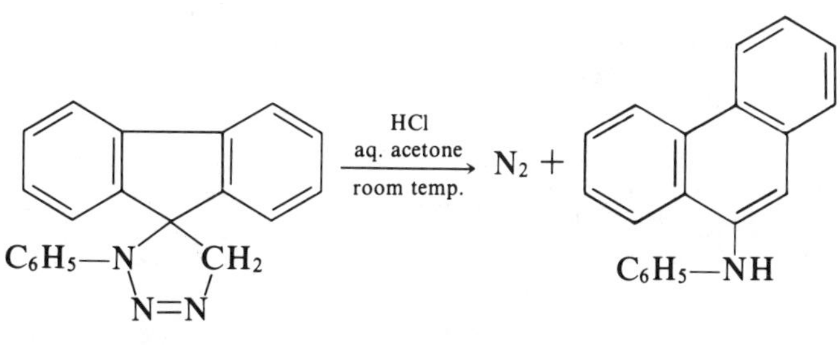

(XXXVI) 85%

5-Hydroxy-5-alkyl-1-substituted triazolines **(XXXVII)** are exceedingly acid labile, losing nitrogen with migration of the 5-alkyl group to the adjacent diazonium carbon (404).

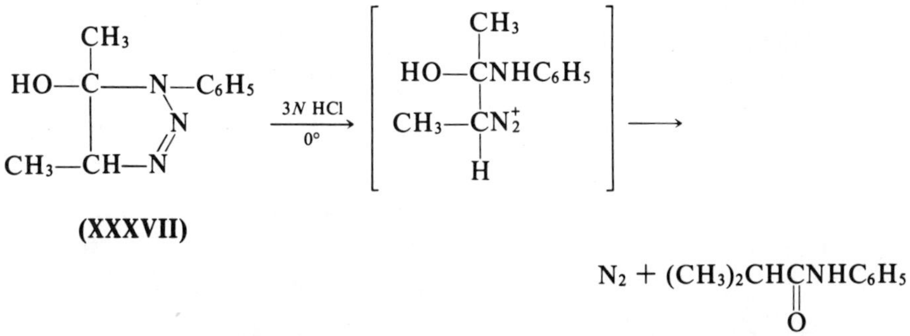

(XXXVII)

93%

v-Triazoles without hydroxyl groups on carbon usually resist destruction by acids to about 100° and in many cases are not destroyed under more strenuous conditions. Thus 4-amino-v-triazole **(XXXVIII)** (507) and 1-benzyl-5,6-dimethyl-v-triazolo[4,5-*b*]pyrazine **(XXXIX)** (338) are recovered unchanged after refluxing in aqueous hydrochloric acid. During hydrolysis of the acetyl groups of 1,7-diacetylbenzo[1,2-*d*:4,5-*d*′]bistriazole **(XL)** by refluxing in 50% ethanol-sulfuric acid no destruction of the bistriazole ring system occurs (114).

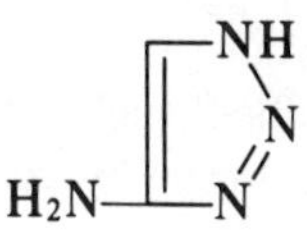

(XXXVIII)

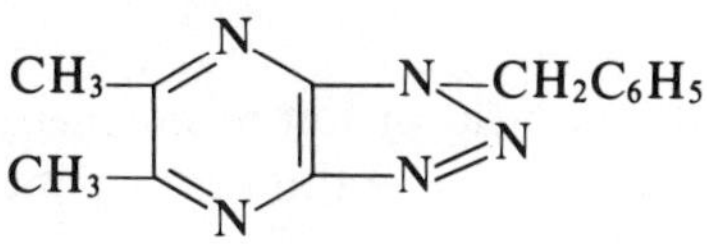

(XXXIX)

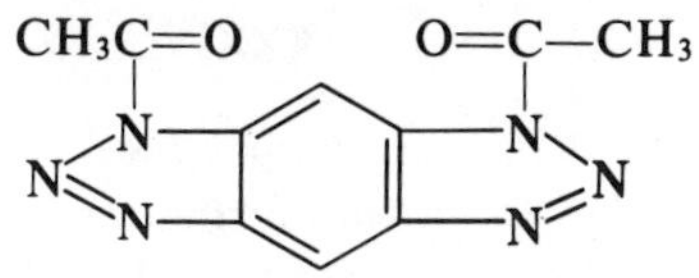

(XL)

N-Bromo and chloro derivatives result from the action of hypohalites on 4,5-dimethyl-v-triazole (265). Substitution with chlorine at N-1 of benzotriazole is

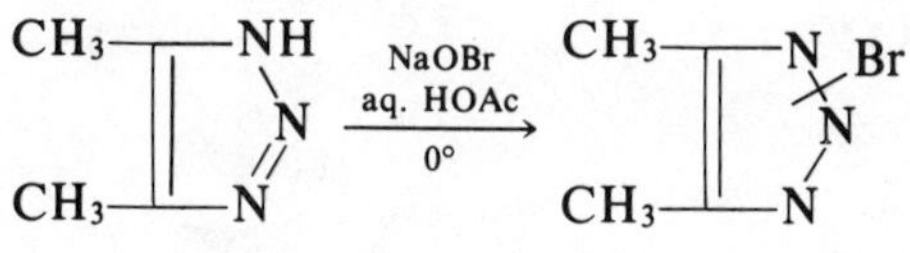

60%

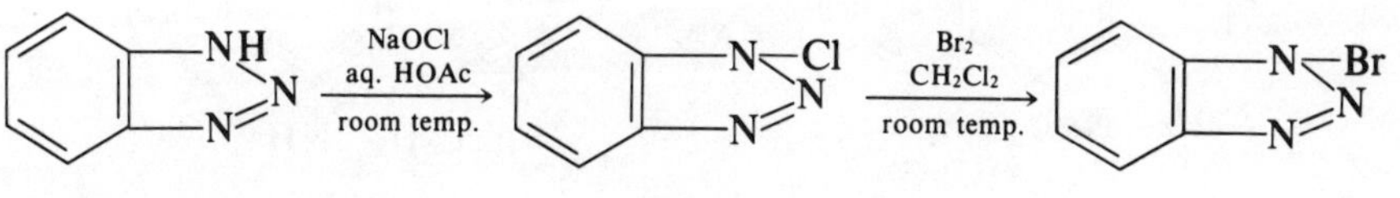

90% 80%

easily accomplished with hypochlorite (435). Although direct substitution on nitrogen by bromine or iodine is not successful, these halogens react with 1-chlorobenzotriazole to form the corresponding 1-halo derivatives (452). Benzotriazole nitrosates (98) and nitrates (116) at the 1-position; amination is at both N-1 and N-2 (101).

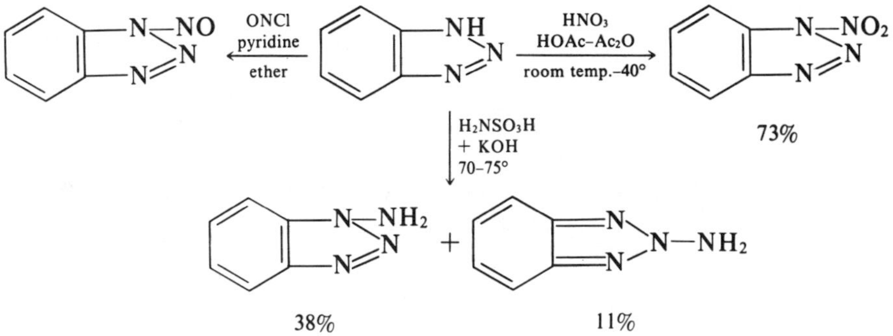

Electrophilic substitution more frequently takes place at *v*-triazole carbon. Thus *v*-triazole is converted into 4,5-dibromo-*v*-triazole by reaction with bro-

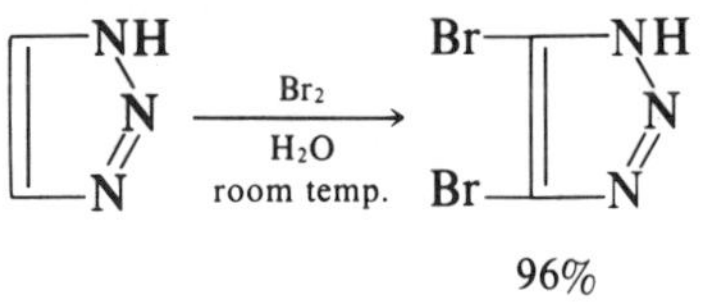

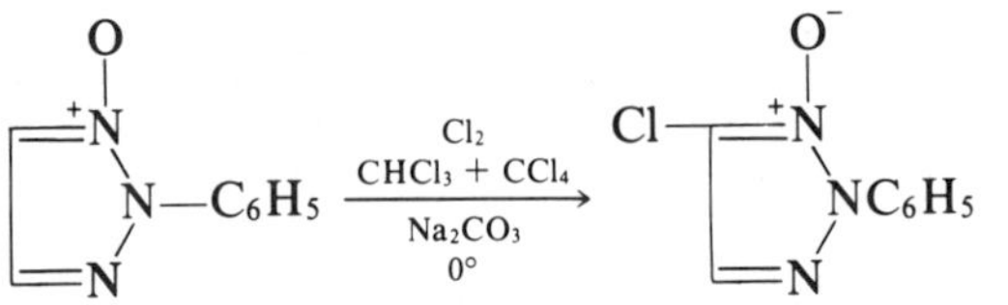

(XLI) 75%

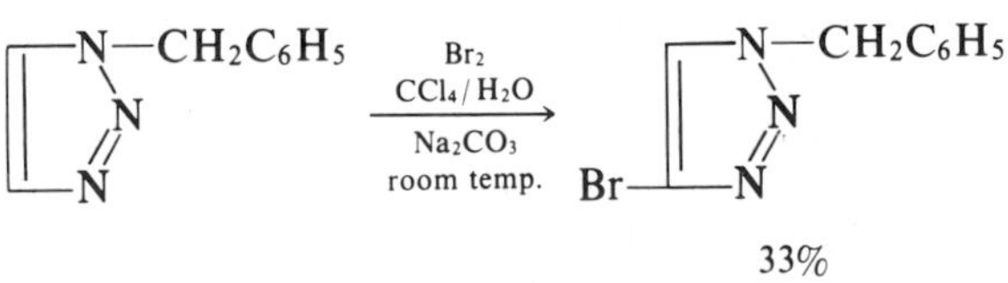

mine (264), 2-phenyl-*v*-triazole-1-oxide **(XLI)** chlorinates or brominates at position 5 (50), and 1-benzyl-*v*-triazole brominates at position 4 (45). Initial nitration of 2-phenyl-*v*-triazole **(XLII)** is in the benzene ring at the para position (444), but the next nitro group enters at *v*-triazole carbon (341). Diazonium salts couple with 4- or 5-hydroxy-*v*-triazoles unsubstituted at the other *v*-triazole carbon (126). Nitrosation of 5-hydroxy-1-phenyl-*v*-triazole-4-carboxylic

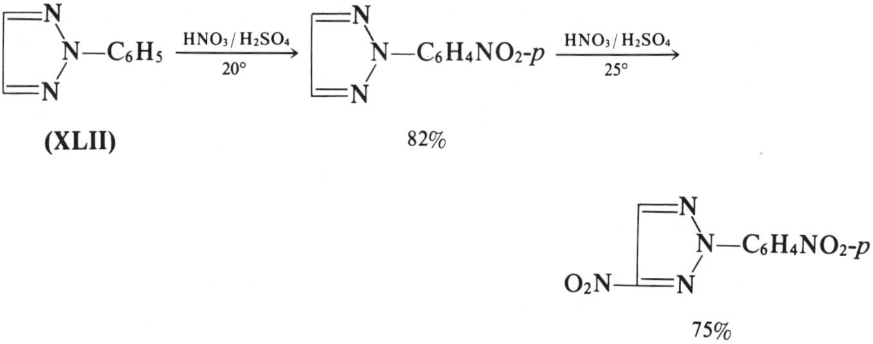

(XLII)

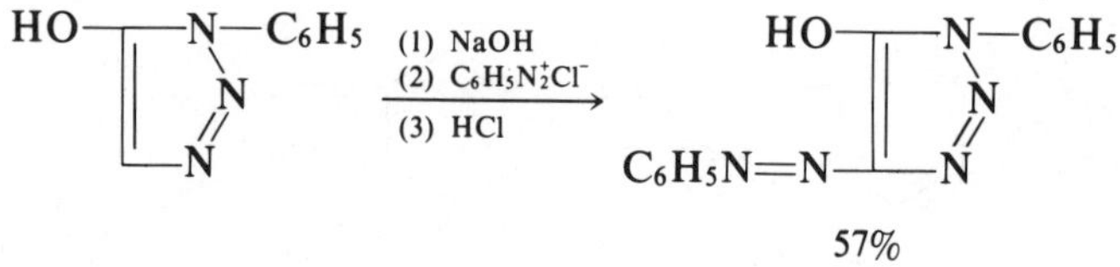

acid **(XLIII)** displaces carboxyl; the triazole ring of the hydroxy nitroso-1-phenyl-*v*-triazole is broken on treatment with acid (133). Both **XLIII** (129) and the corresponding amino compound (128) are decomposed by acid.

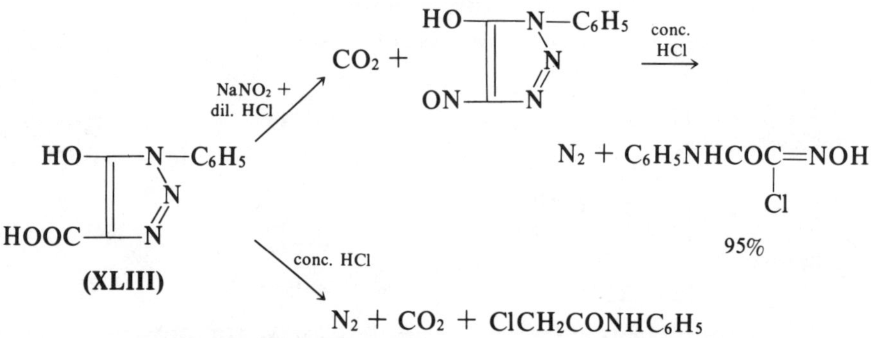

(XLIII)

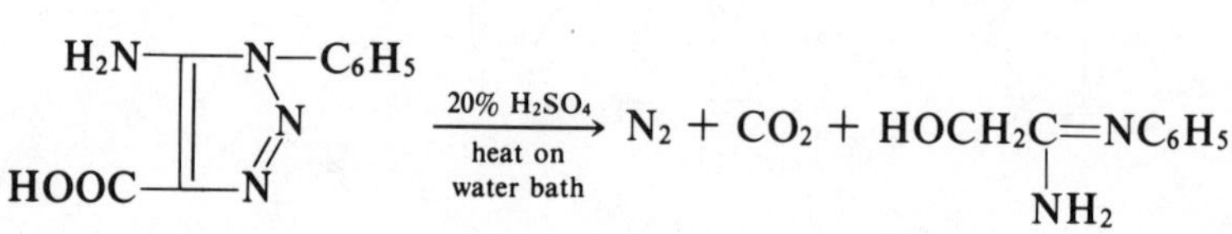

Derivatives of 1-(2′-pyridyl)benzotriazole form carbolines and pyrido[1,2-*a*]-benzimidazoles in low yield on heating in polyphosphoric acid (376); the temperature required is much lower than in the purely thermal Graebe-Ullmann reaction.

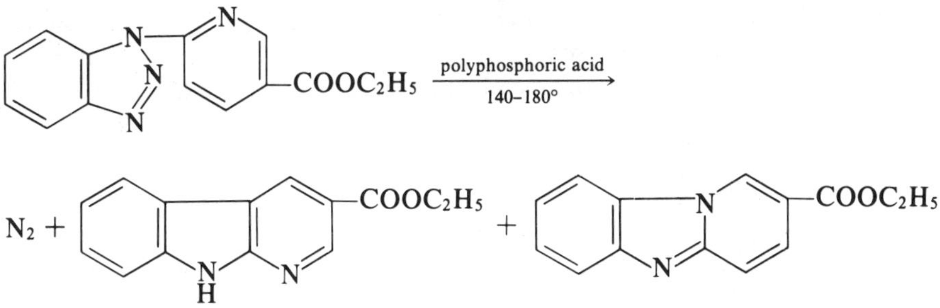

15% <1%

Pyrido[1,2-*e*]triazole **(XLIV)** or its 1-phenyl derivative alkylate carboxylic acids and phenols with concomitant loss of nitrogen, forming 2-pyridyl methanol esters and ethers (78). Tetrazolopyridine is stable under these conditions. Nitration of **XLIV** does not destroy the ring system, but chlorination or bromination does (276).

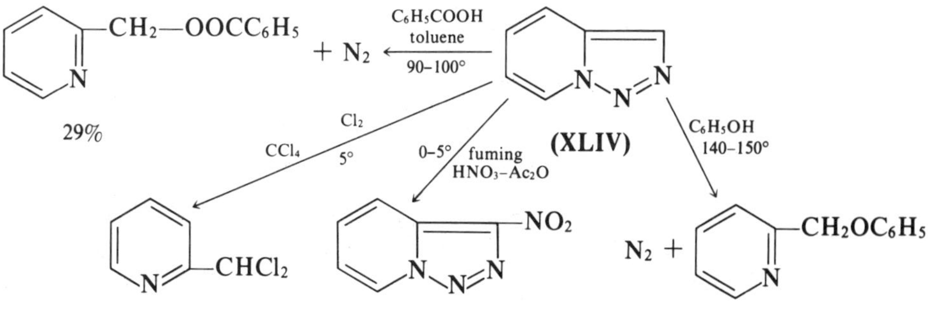

29%

(XLIV)

67% 20–25% 22%

Fused ring *v*-triazoles with one embedded nitrogen attached to cyclic azo are as susceptible to disruption by electrophiles or more so than monocyclic *v*-triazoles or *v*-triazoles annelated at the carbon positions. Thus the triazole ring of 5,7-dimethyl-3-phenyl-*v*-triazolo[3,4-*a*]pyrimidine opens with loss of nitrogen on heating in glacial acetic acid (504). A particularly illustrative example is the

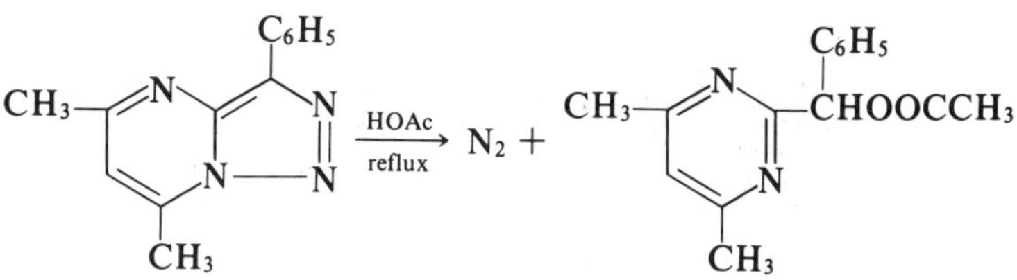

reaction of hot acetic acid with the tricyclic compound **(XLV)** containing two fused *v*-triazole rings, one fused at the 4,5-positions, which is stable under these conditions, and the other with an embedded nitrogen and an exposed azo which

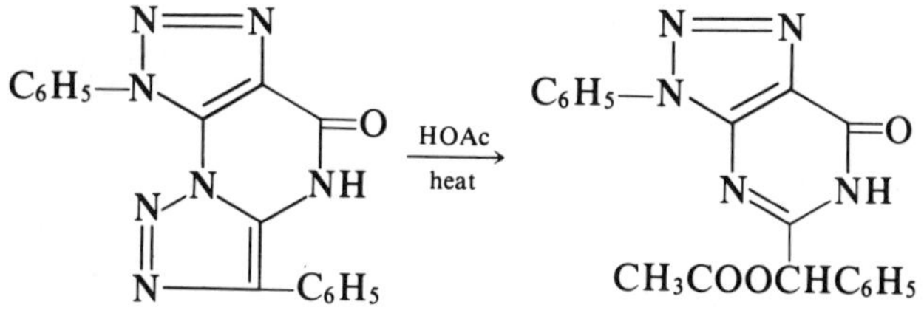

(XLV) 91%

is lost as nitrogen (503). The closely related *v*-triazolo[1,5-*b*]pyridazine with a four-nitrogen chain also loses nitrogen from the triazolo portion on heating with acetic acid, forming pyridazine-3-methyl acetate (355) and the three-nitro-

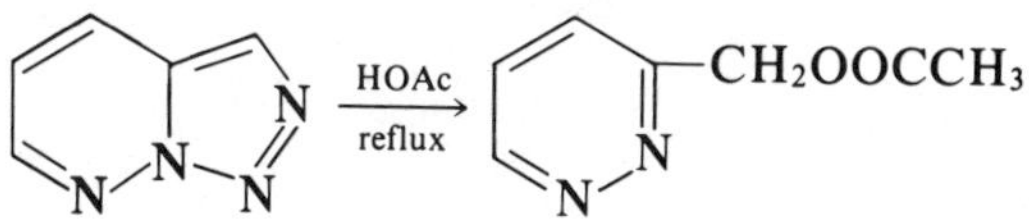

80%

gen chain of 6-chloro-*s*-triazolo[1,5-*b*]pyridazine 3-oxide with the middle nitrogen embedded also is broken by acid, splitting off an HCNO unit (80).

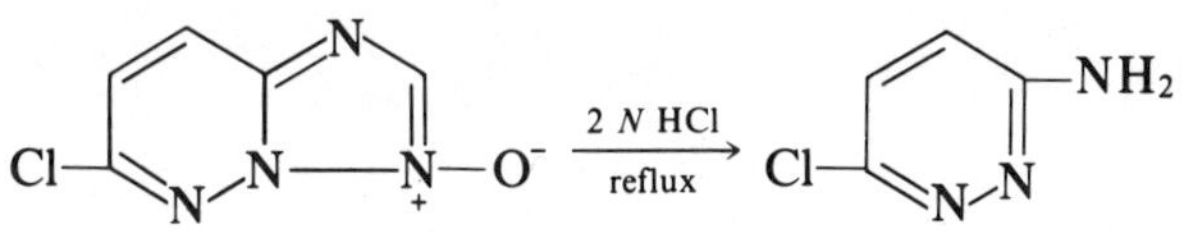

3,4,5-Triphenyl-*v*-triazine is hydrolyzed to 1,2,3-triphenylpropane-1,3-dione by hydrochloric acid in acetic acid (112), and in 4-amino-6-cyanamino-1,2,3,5-

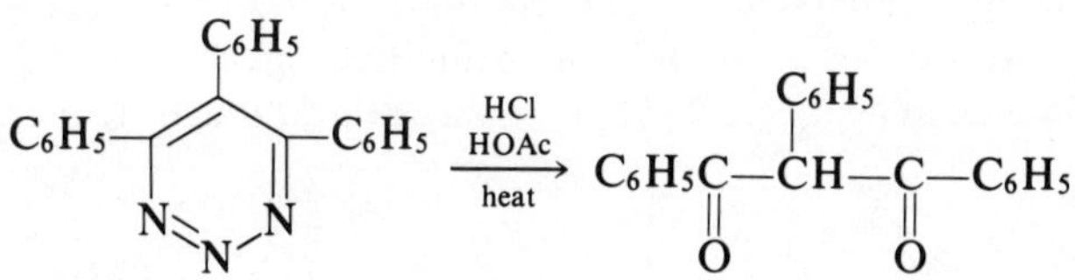

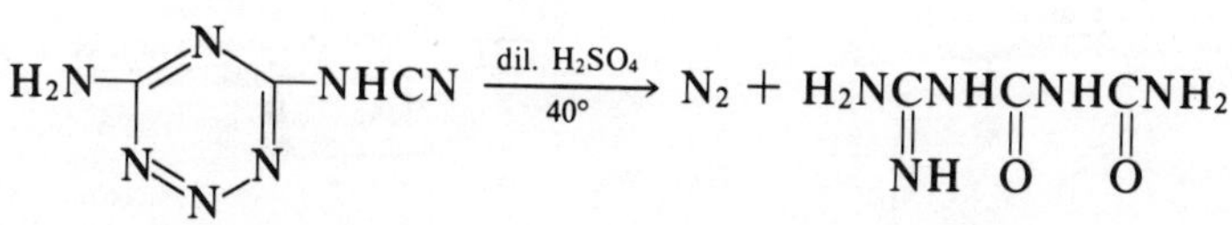

(XLVI) 80% 73%

tetrazine **(XLVI),** the three-nitrogen sequence is broken on warming with dilute sulfuric acid (305).

Benzotriazine and 4-substituted benzotriazines hydrolyze rapidly and extensively in the presence of acids to *o*-aminobenzene carbonyl derivatives (5).

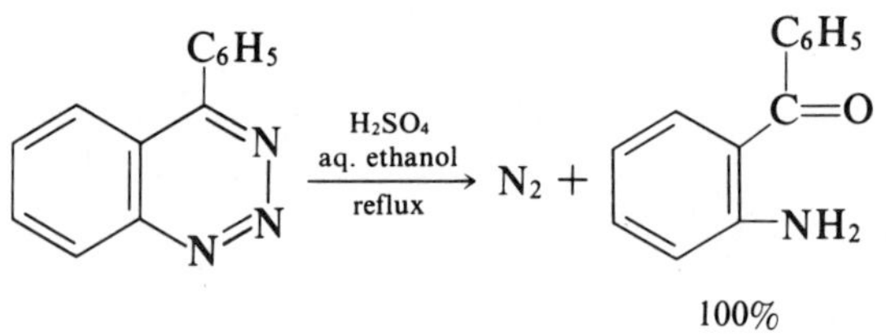

Heating benzotriazinones in acid unmasks the diazonium group inherent in the triazine chain, and in the presence of a phenol or aromatic amine, coupling occurs to form azo dyes, even in concentrated acid (35, 85). Instead of forming 4-chlorobenzotriazine by the action of phosphorus pentachloride and phosphorus oxychloride on benzotriazinone, the masked diazo group is replaced by

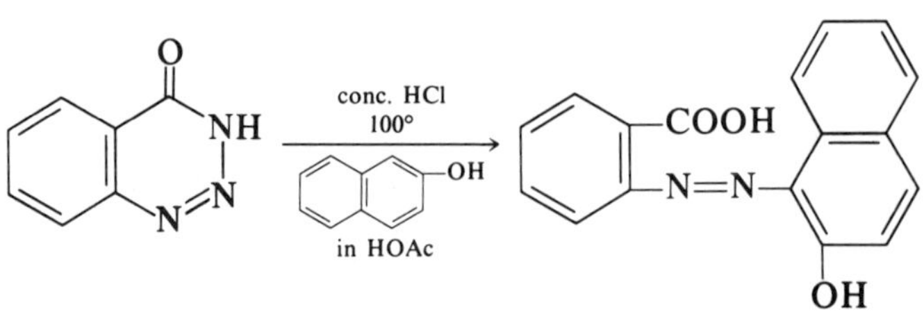

chlorine and the amide linkage is converted to cyanide (85). Acid hydrolysis of 3-methylbenzotriazinone yields salicylic acid and *N*-methylsalicylamide (347).

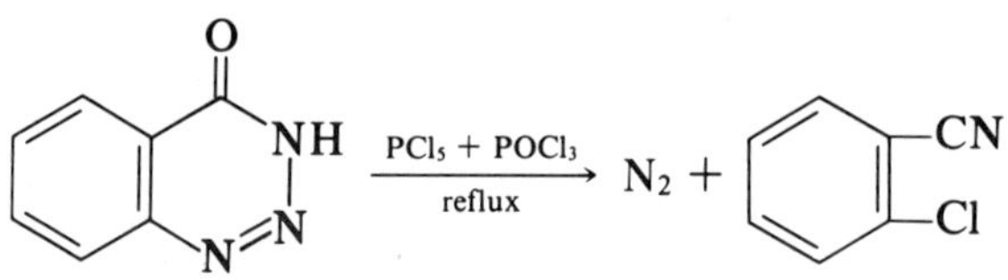

Benzotriazinones with larger substituents at N-3 such as phenethyl, however, are recovered nearly quantitatively from boiling 10 *N* hydrochloric acid.

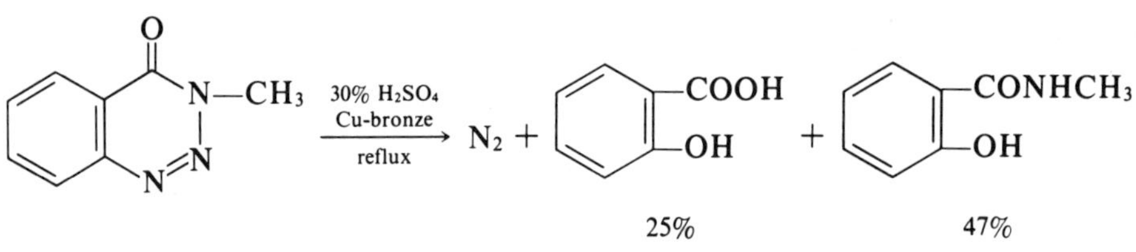

It is possible to nitrate 3-phenylbenzotriazinone **(XLVII)** on phenyl without destroying the *v*-triazine ring, but heating in phosphoric acid causes cyclization to phenanthridine with loss of nitrogen (183).

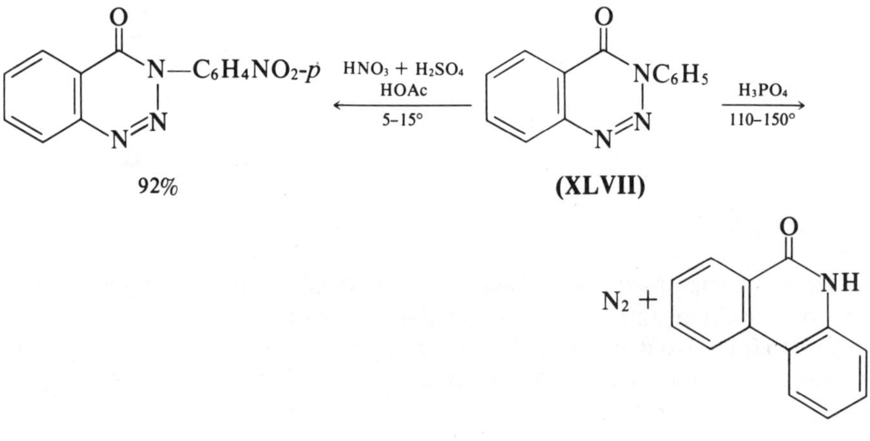

92%

(XLVII)

51%

The three-nitrogen chain is retained in 4-methylbenzotriazine 1-oxide **(XLVIII)** on treatment with dilute sulfuric acid with ring opening to *o*-azidoacetophenone; the triazine oxide **(XLVIII)**, however, can be dissolved in

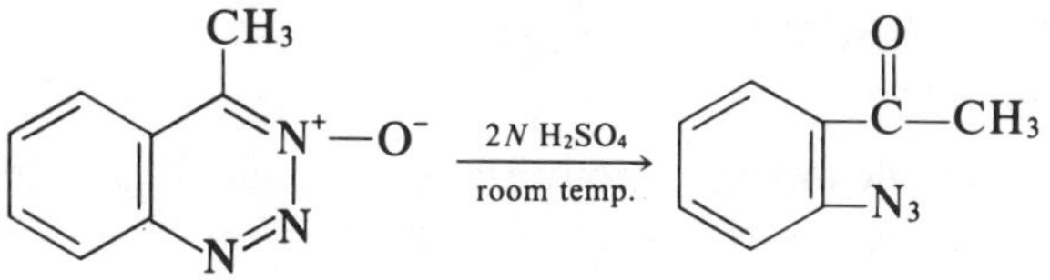

(XLVIII) 95%

concentrated hydrochloric or sulfuric acid and reprecipitated unchanged on dilution with ice (360).

3,4-Dihydro-4-imino-3-*p*-tolylbenzotriazine **(XLIX)** undergoes the Dimroth rearrangement to 4-*p*-toluidinobenzotriazine on heating in acetic acid (471, 493).

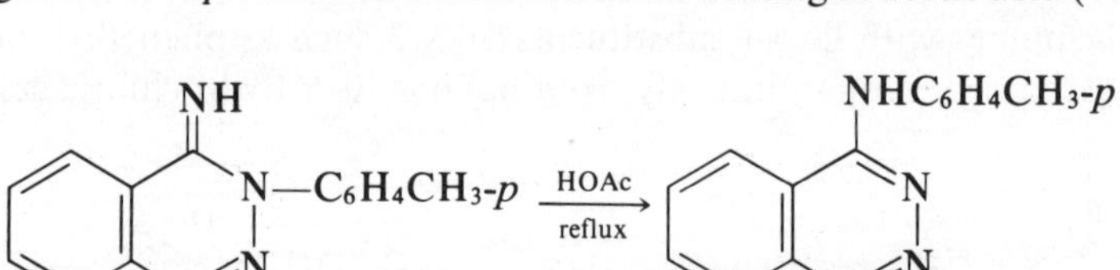

(XLIX) 96%

Although 4-arylimino-3-substituted-3,4-dihydrobenzotriazines such as **L** undergo ring fusion and form azonaphthol derivatives on heating in acetic acid containing 2-naphthol, they are recovered unchanged after refluxing in acetic acid alone (471).

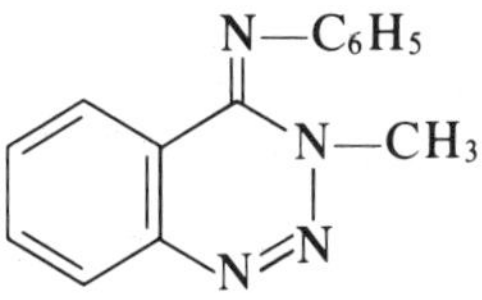

(L)

The action of acetic and sulfuric acids on 3,3′-bibenzotriazinone destroys the six-nitrogen chain, elminating nitrogen and forming bis(salicyloyl hydrazide) (225).

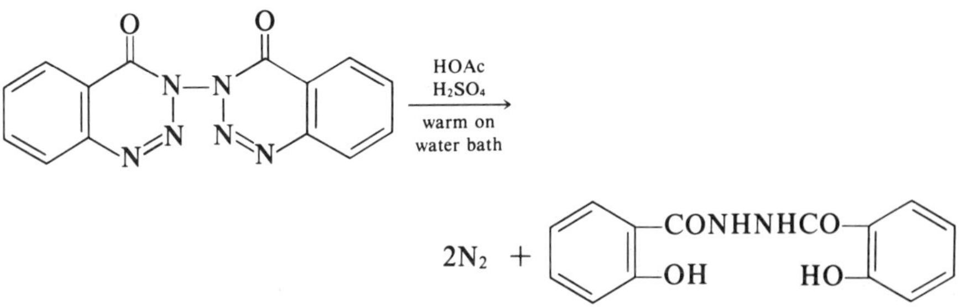

Hydroxylamine-*O*-sulfonic acid reacts with naphtho[1,8-*de*]triazine both to aminate the nitrogen chain and to break the ring, forming 1-amino-8-azidonaphthalene, probably through rearrangement of the initially formed 2-aminonaphtho[1,8-*de*]triazine (432). Acid treatment of both 1- and 2-amino-

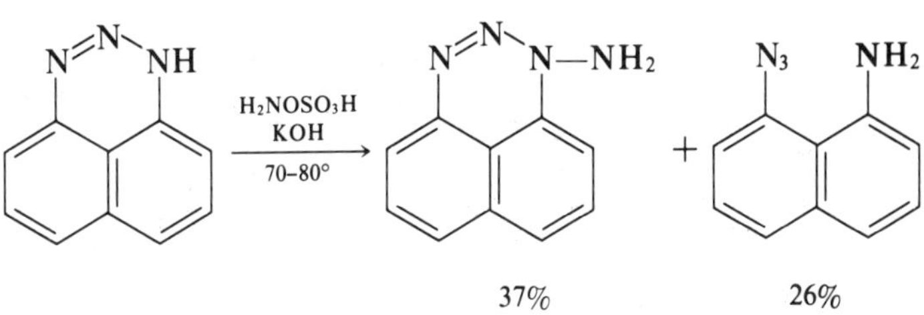

naphtho[1,8-*de*]triazine causes rearrangement to 1-amino-8-azidonaphthalene, but more vigorous conditions are required to convert the 1-amino derivative.

Presence of a potential diazonium group is shown by the behavior of dibenzotriazepine **(LI)** in acid. It undergoes reaction with hydrogen iodide to replace the diazonium equivalent with iodide and is recovered unchanged after treatment with concentrated hydrochloric acid, probably because of equilibrium formation of a stable dication (171, 172). In dilute hydrochloric or 50% acetic

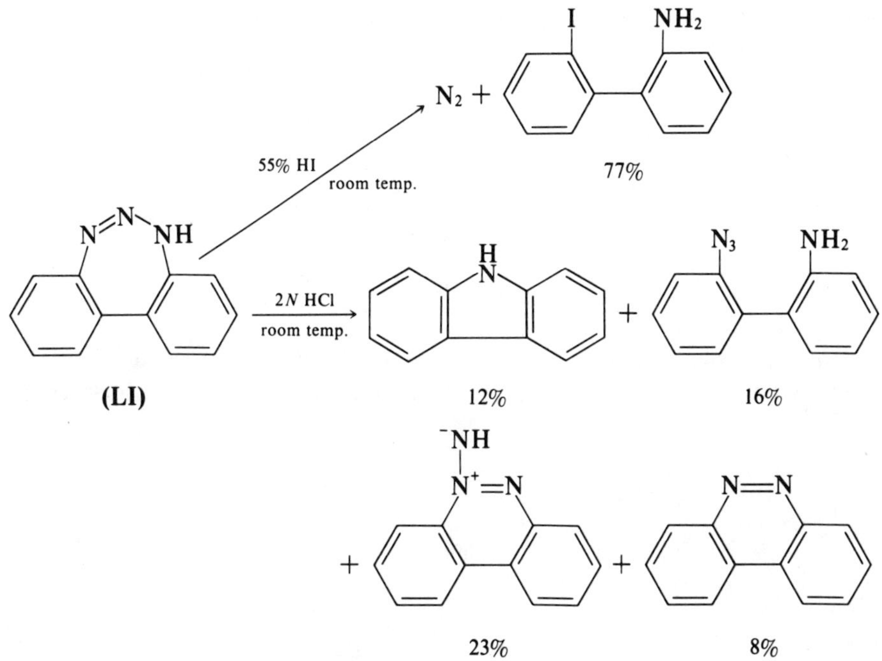

acid, however, extensive fragmentation occurs as well as isomerization to benzo-[*c*]cinnoline *N*-imide.

The presence of one embedded nitrogen in a three-nitrogen chain has no effect on the diazonium reactivity of benzimidazo[1,2-*c*][1,2,3]benzotriazine as shown by the Schiemann reaction with fluoboric acid (482) and hydrolysis to a phenol (567).

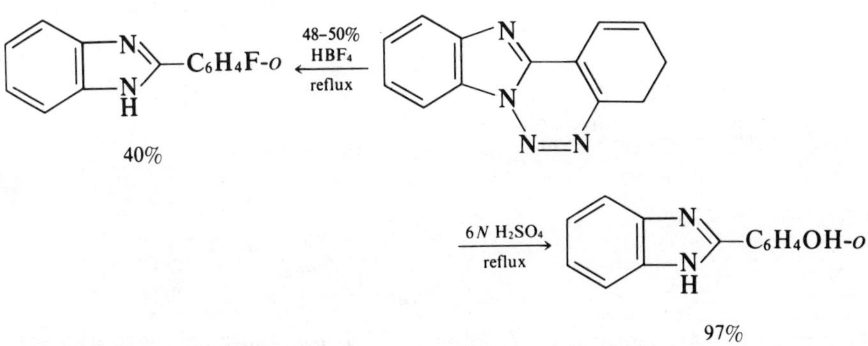

The 3-alkyl oxatriazolium-5-olates are stable to boiling water (152) and dilute mineral acids below 40°C (267) but decompose at higher temperatures in the presence of more concentrated acid. Thus 3-cyclohexyloxatriazolium-5-olate **(LII)** is hydrolyzed to cyclohexanol, hydrazoic acid, and carbon dioxide (77). Similarly, the acid decomposition of the 3-isopropyl derivative **(LIII)**, which

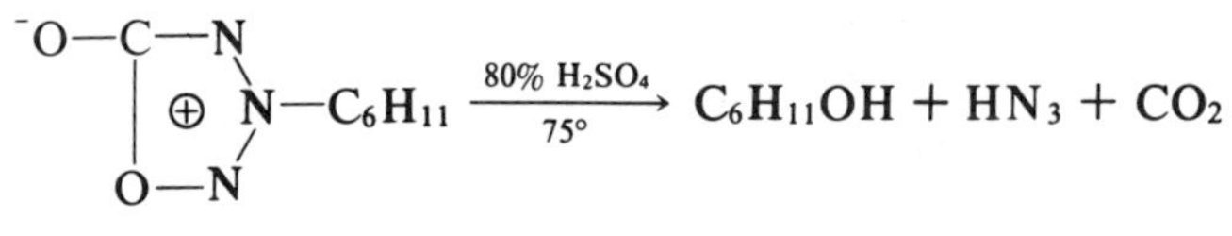

(LII)

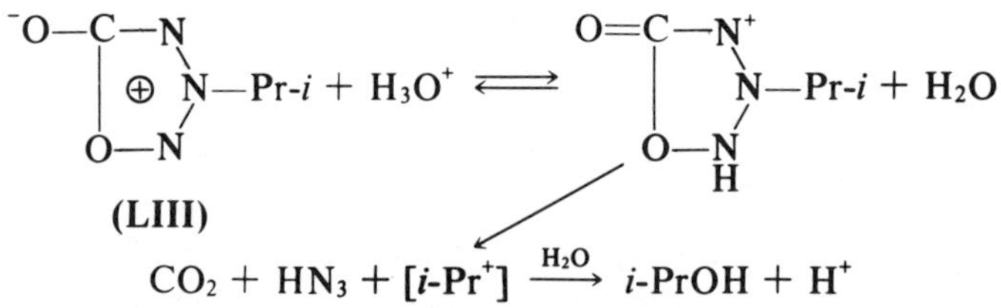

(LIII)

appears to first involve protonation of the oxatriazole ring, occurs at an appreciable rate only at a concentration above 8 *M* perchloric acid; kinetics are first order in H_0 and effectiveness is $HClO_4 > H_2SO_4 > HCl$ (267).

Thiatriazoles are very sensitive to acids, usually depositing sulfur and liberating nitrogen, illustrated by the decomposition of 5-phenylthiatriazole (480).

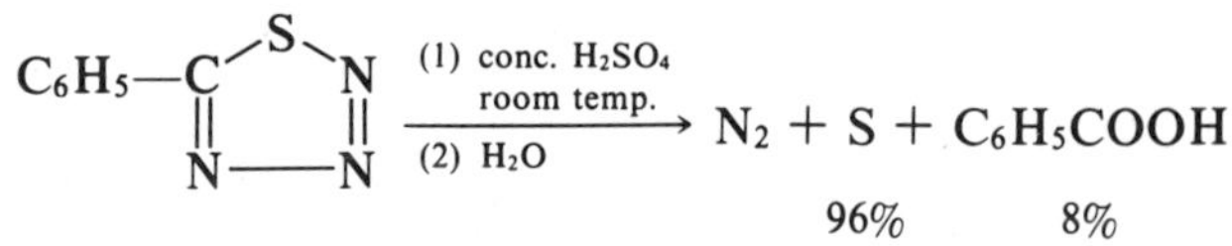

Hydrolysis of 5-alkylaminothiatriazoles with hydrochloric acid produces formamidine disulfide, alkylamine hydrochloride, hydrogen sulfide, hydrazoic acid, nitrogen, and carbon dioxide; no sulfur or thiocyanic acid is found (270). The initial step is probably protonation of the thiatriazole ring, followed by loss

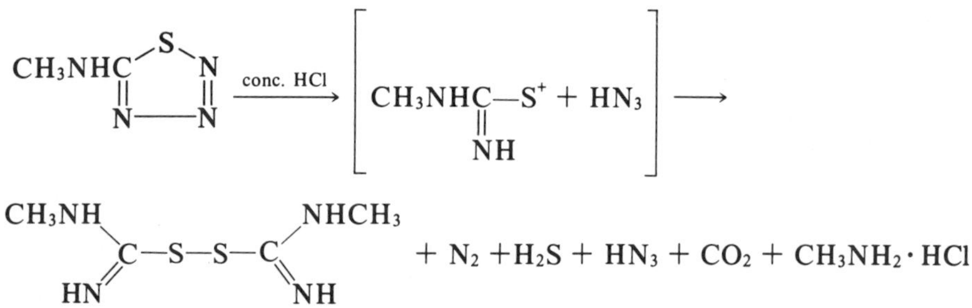

of nitrogen, reduction of the positive ion by hydrogen azide, and combination of the fragments to give formamidine disulfide. Acid hydrolysis of 5-*p*-tolylaminothiatriazole leads to an aminobenzothiazole and a small amount of sulfur (463). 5-Phenylthiatriazole is not attacked by chlorine and 5-*p*-methoxyphenylthiatriazole is inert to a mixture of glacial acetic, sulfuric acid, and yellow fuming nitric acid (480).

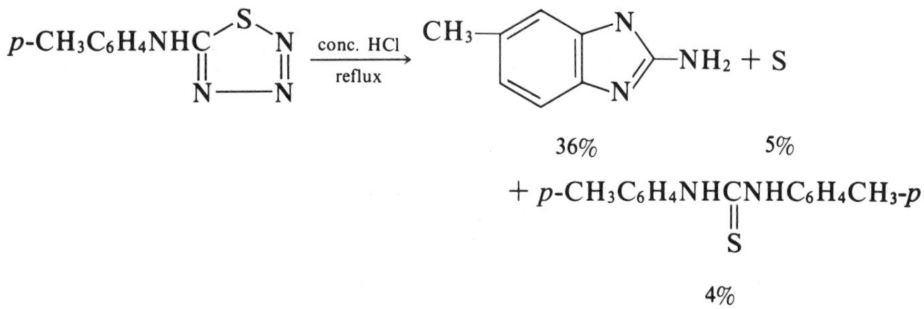

Hydrolysis of tetrazole with sulfuric acid yields nitrogen, ammonia, and carbon dioxide, but no formic acid (519). Many 5-substituted tetrazoles decom-

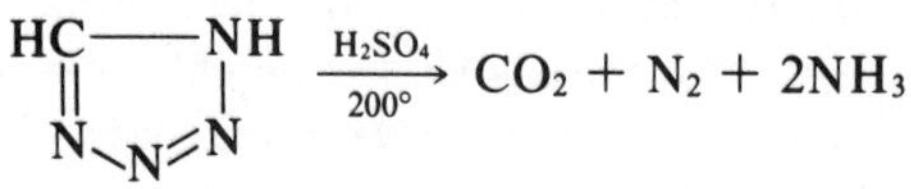

pose under the influence of strong acid with migration of the substituent at tetrazole carbon to adjacent nitrogen, forming a product with an amino group attached to the former 5-substituent, ammonia or amine from the other adjacent nitrogen and carbon dioxide from tetrazole carbon. Thus 5-phenyltetrazole decomposes with concentrated acid at 220°, but not at 180°, to form aniline and ammonia (336), 5-aminotetrazole forms hydrazine and ammonia (519), and 5-(1-phenanthryl)1-phenyltetrazole **(LIV)** forms 1-phenanthrylamine and aniline (477).

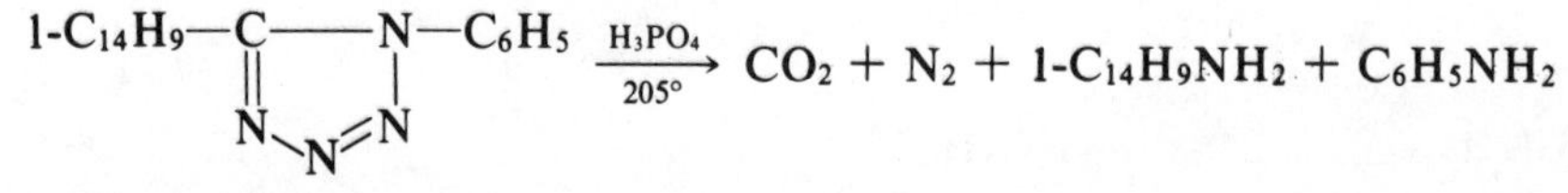

(LIV)

5,5′-Azobis(tetrazole) exists only in the form of its salts; acidification of the sodium salt destroys one tetrazole ring, liberating 5-hydrazinotetrazole, nitrogen, and formic acid (518). Hydrazotetrazole dissolves in concentrated hydrochloric acid and precipitates unchanged on dilution.

Long heating of 4,5-dihydrotetrazoloquinazolones with aqueous acids opens the tetrazole ring to an azide which is hydrolyzed to a dihydroquinazol-2,4-dione (60).

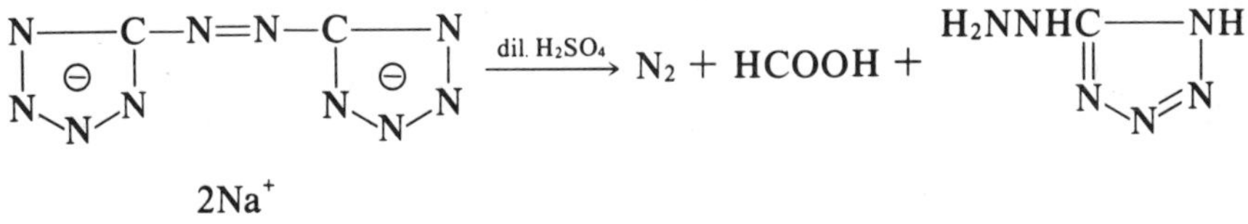

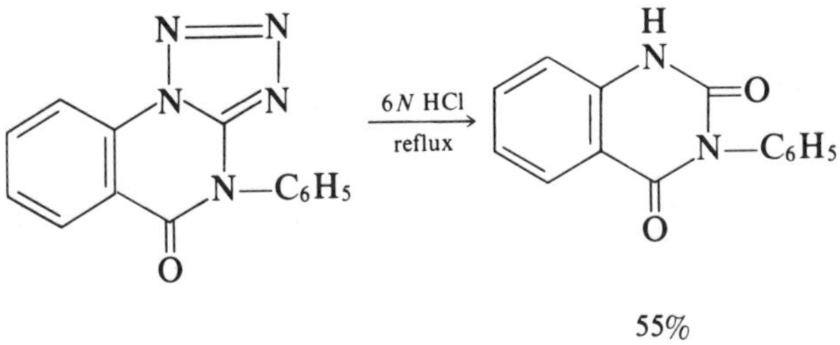

The action of phosphorus oxychloride and dimethylformamide on 6-ethyl-3-(5-tetrazolyl)chromone **(LV)** causes immediate evolution of nitrogen and after addition of water, *N,N*-dimethylformamide-N^2-(6-ethylchromone-3)carbohydrazone is obtained (390).

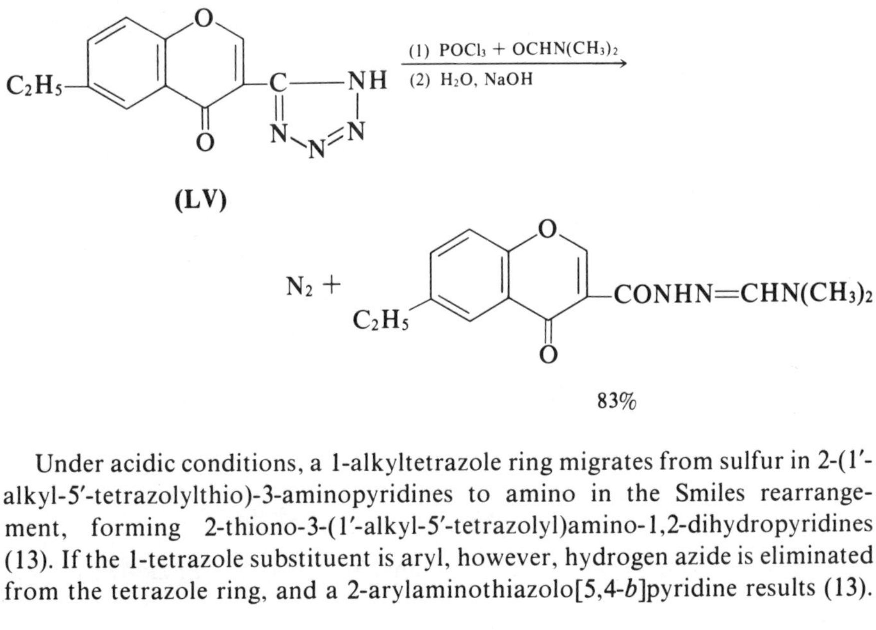

Under acidic conditions, a 1-alkyltetrazole ring migrates from sulfur in 2-(1′-alkyl-5′-tetrazolylthio)-3-aminopyridines to amino in the Smiles rearrangement, forming 2-thiono-3-(1′-alkyl-5′-tetrazolyl)amino-1,2-dihydropyridines (13). If the 1-tetrazole substituent is aryl, however, hydrogen azide is eliminated from the tetrazole ring, and a 2-arylaminothiazolo[5,4-*b*]pyridine results (13).

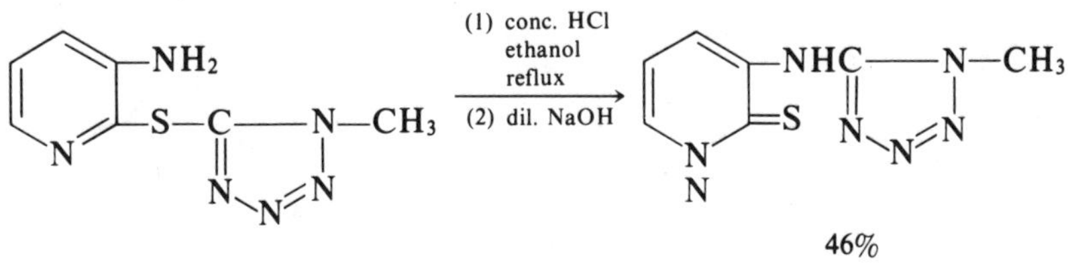

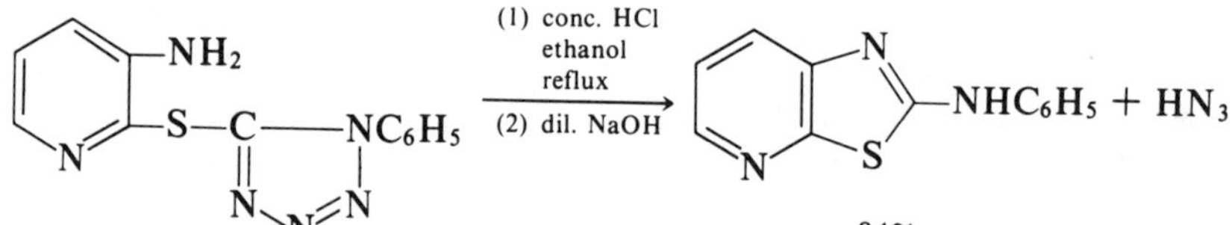

84%

Bromine reacts slowly with 1-phenyltetrazole displacing the hydrogen at tetrazole carbon to form 5-bromo-1-phenyltetrazole (500), but the two-step procedure consisting of the reaction of mercuric acetate with 1-phenyltetrazole to form a 5-mercury acetate derivative, followed by bromination, is more efficient (501).

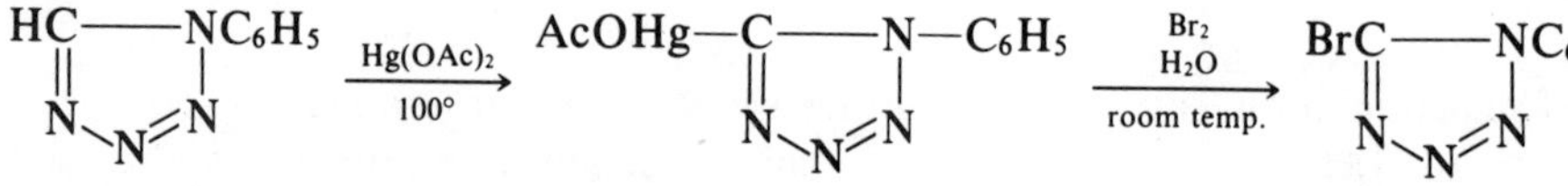

95% quantitative

The tetrazole ring in 1,5-diphenyltetrazole remains intact on sulfonation or nitration, with substitution principally in the para positions of each phenyl (81).

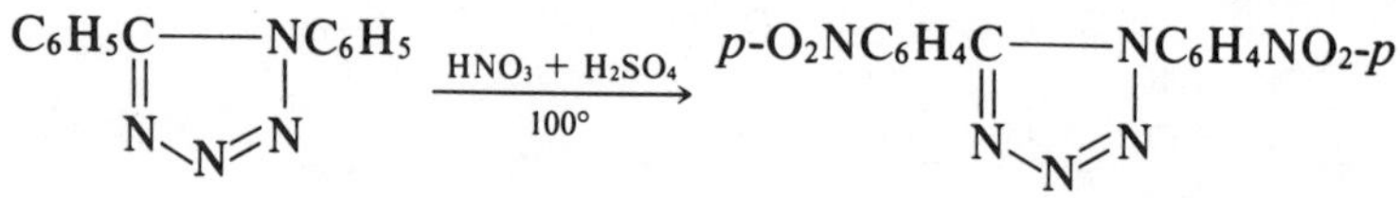

Nitration of 5-amino- and 5-alkylaminotetrazoles occurs at the exocyclic nitrogen rather than on the tetrazole ring. Thus the nitrate salts of 5-methylaminotetrazole is dehydrated by sulfuric acid to the nitramine (177).

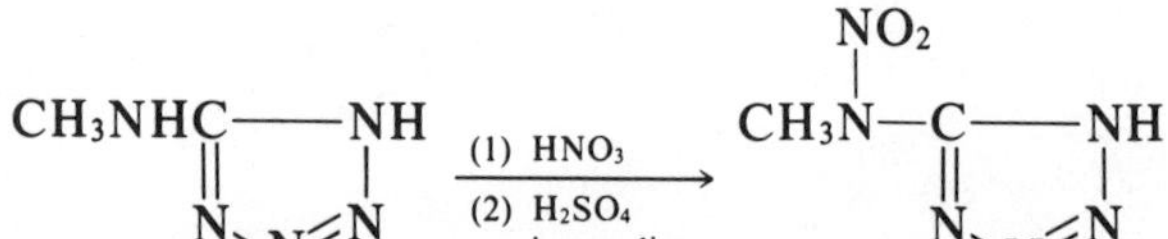

35%

The tetrazole ring, however, can be aminated by hydroxylamine-*O*-sulfonic acid (425).

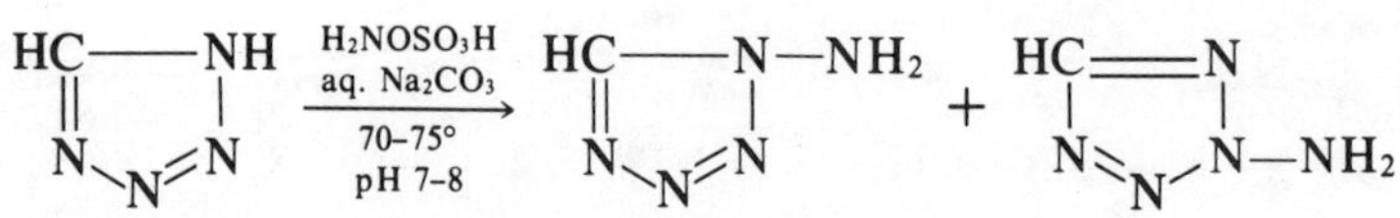

25% 13%

Table 6.12. Decomposition of Phenylpentazole in Acidic and Nonacidic Media

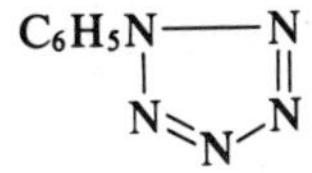

Solvent	$t_{1/2}$ (min)
CCl_4	3.4
CH_3OH	11.8
75% HOAc–25% H_2O	19.8
HCOOH	50

Acetic and formic acids decrease the decomposition rate of phenylpentazole in comparison to methanol and nonacidic solvents, as indicated by half-lives at 0° (Table 6.12) (529).

2,5-Diphenyltetraazaboroline is hydrolyzed to aniline and boric acid with evolution of hydrogen and nitrogen by hydrochloric acid (201). By analogy with the low-temperature hydrolysis of tetra(trimethylsilyl)-2-tetrazene to tetra-

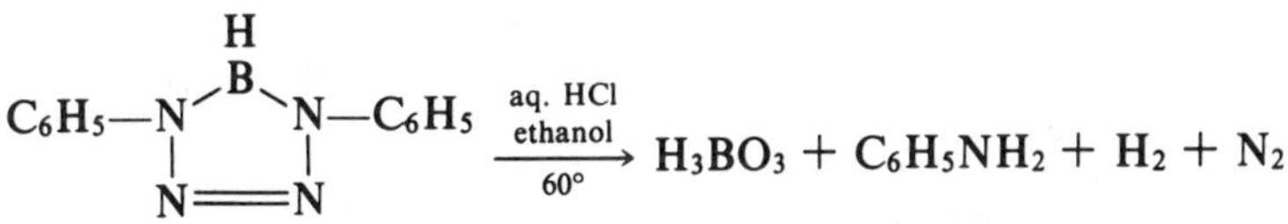

zene (548), search was made for pentazene from the hydrolysis of 1-[bis (trimethylsilyl)amino]5,5-dimethyl-4-trimethylsilyl-4,5-dihydro-5-silatetrazole **(LVI)**; the only nitrogenous products even at −78°, however, are nitrogen, hydrogen azide, hydrazine, and ammonia (549).

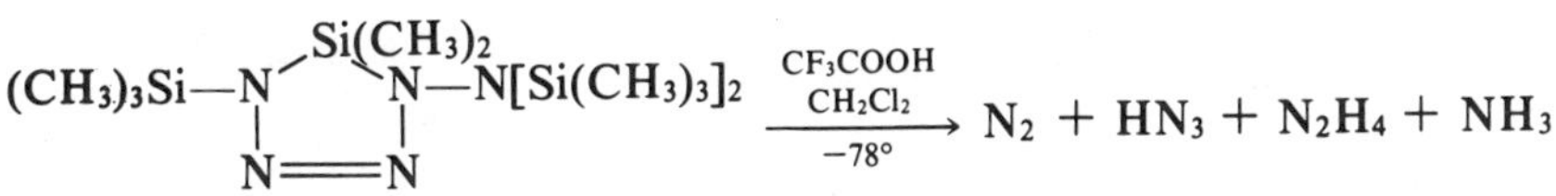

(LVI) 85% 17% 58% 79%

The four-nitrogen acyclic-cyclic nitrogen chain in 1-(*N*-ureido)-*v*-triazoles remains intact during acid hydrolysis to 1-amino-*v*-triazoles (11) and 1-anilino-5-methyl-*v*-triazole **(LVII)** dissolves in concentrated hydrochloric acid, but is precipitated unchanged on addition of water (563).

Nitrosyl chloride effects removal of the exocyclic nitrogen from 1-benzoylamino-5-phenyltetrazole **(LVIII)** by decomposition to nitrogen, benzoic acid, and 5-phenyltetrazole (54). Amino groups attached to nitrogen in ring systems are replaced by hydrogen through nitrosation, illustrated by the deaminations

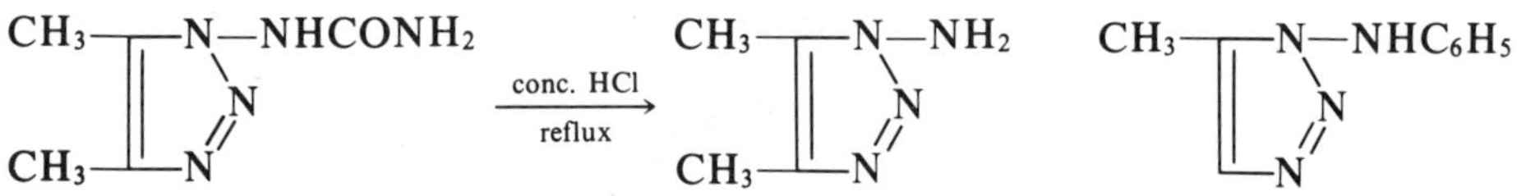

65% (LVII)

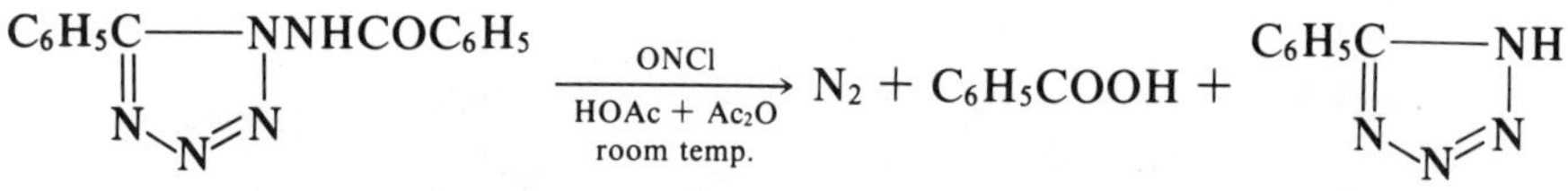

(LVIII)

of 2-aminocinnolin-3-one **(LIX)** (430), 1-amino-4-phenyl-*v*-triazole (218), and 1-amino-naphtho[1,8-*de*]triazine (432). In similar manner, the imino group of benzocinnoline-*N*-imide also is removed on nitrosation (173).

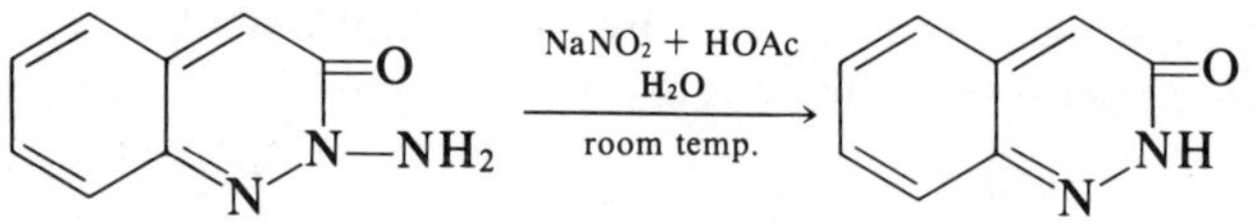

(LIX) 96%

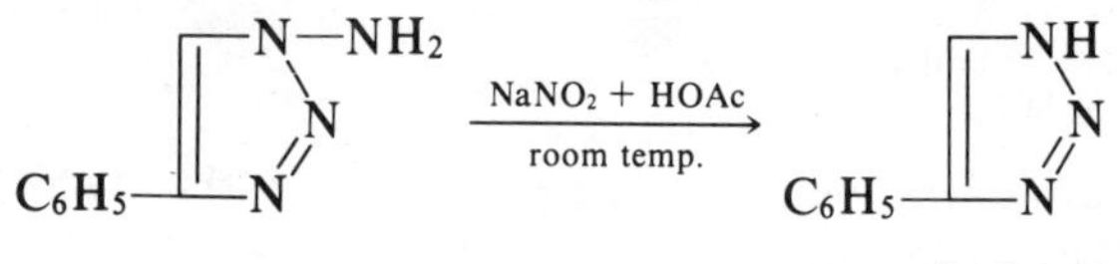

quantitative

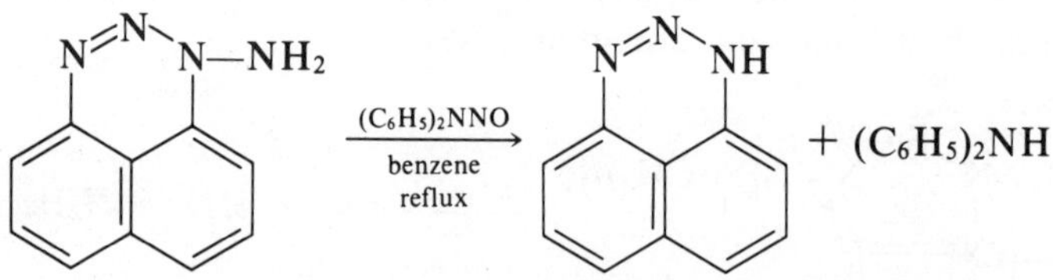

85% 85%

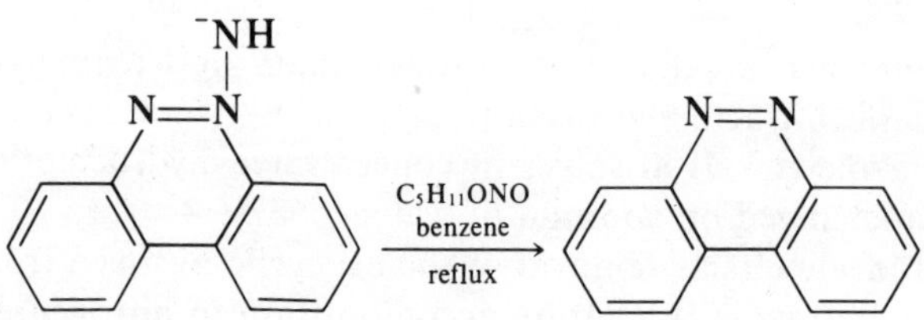

96%

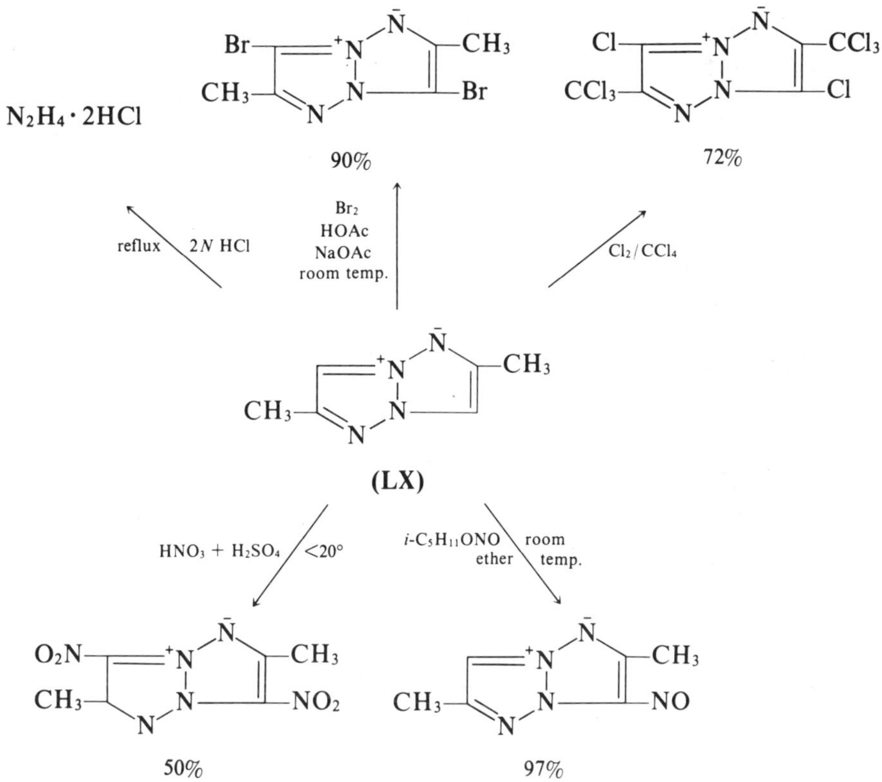

The nitrogen chain in 2,7-dimethyl-tetraazapentalene **(LX)** is broken by heating with mineral acid, forming hydrazine, but is stable to a variety of electrophilic agents at mild temperature (415). Mononitrosation, dinitration, and dibromination takes place at unsubstituted ring carbons of **LX**, but chlorine substitutes for all hydrogens without destroying the tetraazapentalene ring. Dibenzo-1,3a,4,6a-tetraazapentalene **(LXI)** is halogenated, nitrated, and sulfonated in the benzene rings with no disruption of the nitrogen chain (106).

6.3. REACTIONS WITH NUCLEOPHILES

Characteristic reactions of high nitrogen compounds with nucleophiles are described in this section, except for those included in Section 6.1. The principal nucleophiles which react with nitrogen chain compounds are alkali hydroxides and amines, but actions of trivalent phosphorus derivatives, sulfinates, ionic azides, and alkoxides also are discussed.

Although there are exceptions, most acyclic high nitrogen chains decompose on exposure to nucleophiles, usually at temperatures below those at which thermal degradation begins.

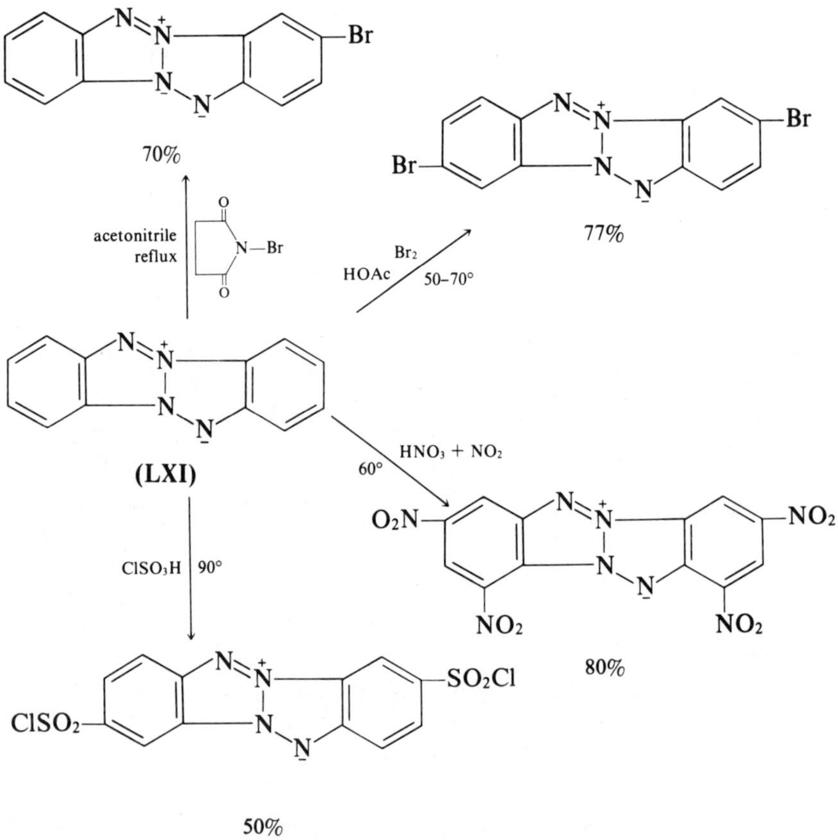

Triazanium salts are unstable in the presence of hydroxide ion; in aqueous solution both triazanium hydroxide (332) and 2,2-dimethyltriazanium hydroxide are over half decomposed in less than 30 min (532). Basic hydrolysis of 2,2-

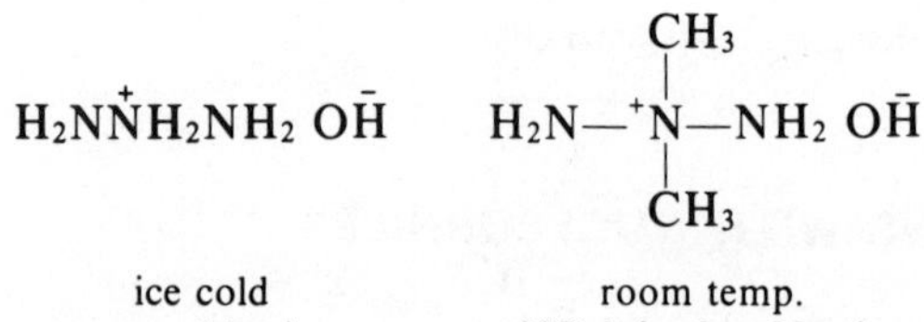

ice cold $t_{1/2} = 14$ min   room temp. 38% left after 35 min

dimethyltriazanium chloride **(I)** leads principally to formaldehyde dimethylhydrazone, following approximately first-order kinetics in the triazanium salt (354). Ammonia also is a hydrolysis product, but dimethylamine, 1,1-dimethylhydrazine, and tetramethyltetrazene are absent. **I** dissolves in liquid ammonia but does not decompose, and the solution can be used for preparation of the corresponding bromide and iodide from the potassium halides (532).

$$H_2N-\overset{+}{N}(CH_3)_2-NH_2\ Cl^- \xrightarrow[25^\circ]{1N\ KOH} (CH_3)_2N-N{=}CH_2 + NH_3$$

(I) 12–20%

1,3-Bis-*p*-nitro, chloro, and methylphenyltriazenes are all recovered unchanged after refluxing in morpholine (bp 129°) (181). Diazoaminobenzene, however, decomposes on heating γ-picoline (bp 143°) by a radical reaction in which nitrogen and α- and β-phenyl-γ-picolines are formed in a 45:55 ratio

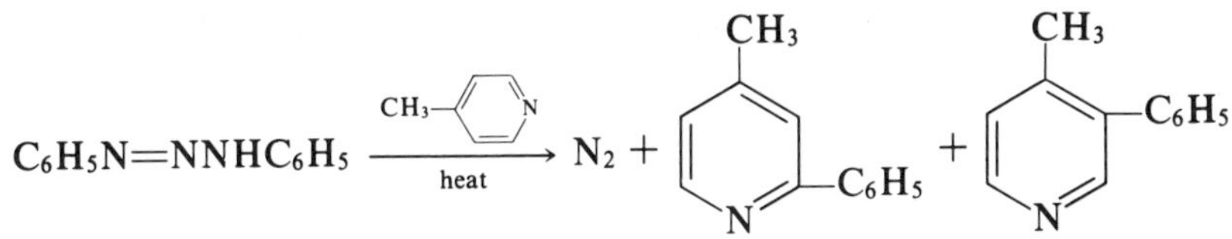

(537). The trisubstituted triazene unit in **II** is remarkably stable in concentrated alcoholic potassium hydroxide, surviving while the amino group on the quinazoline moiety is hydrolyzed (181). An increase in pH retards decomposition of 3-

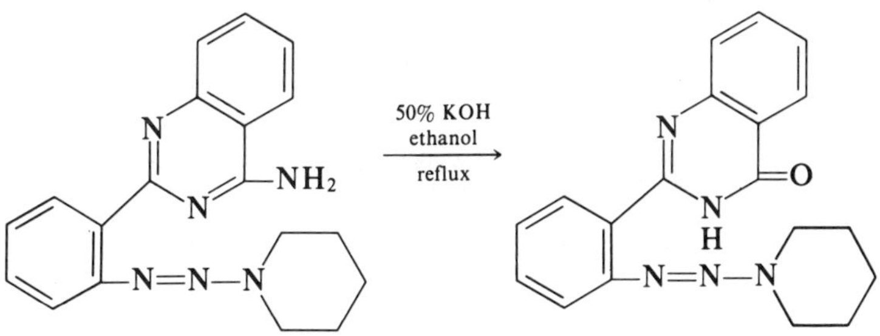

(II)

alkyl-1-phenyltriazenes; less than 5% of 3-methyl-1-phenyltriazene is decomposed at 20° after 12 hr at pH 10.96 (14). 1-(*p*-Nitrophenyl)triazenes substituted at N-3 with an alkyl carrying an electron-withdrawing group decompose readily in the presence of base, forming *p*-nitroaniline and α-diazo derivatives. Thus 1-*p*-nitrophenyl-3-ethoxycarbonylmethyltriazene **(III)** is converted quantitatively to ethyl diazoacetate by a catalytic quantity of potassium hydroxide in ethanol (357). Indeed, such triazenes may be formed *in situ* and be decomposed by

$$C_2H_5OOCCH_2NHN{=}NC_6H_4NO_2\text{-}p \xrightarrow[\text{ethanol}]{\text{cat. quant. KOH}} C_2H_5OOCHN_2 + H_2NC_6H_4NO_2\text{-}p$$

(III) 100%

triethylamine to diazo compounds at temperatures as low as −70°. Bis(3-ethylbenzothiazolyl-2)triazacarbocyanine fluoborate **(IV)** is readily hydrolyzed with loss of nitrogen (31, 33). Cyanotriazenes derived from azidium salts decompose

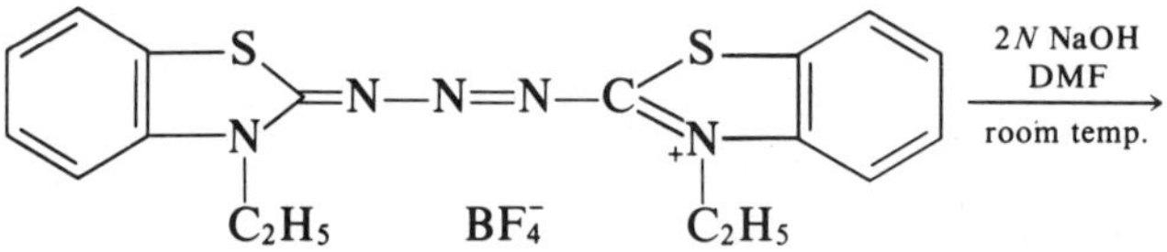

(IV)

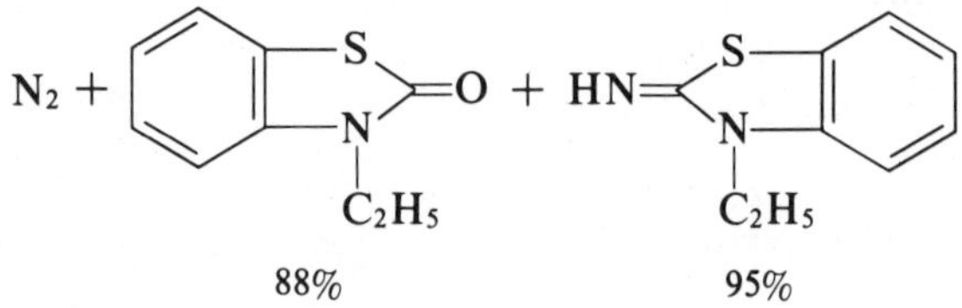

under alkaline conditions to nitrogen, a ketoheterocycle, and cyanamide in equimolar proportions (228).

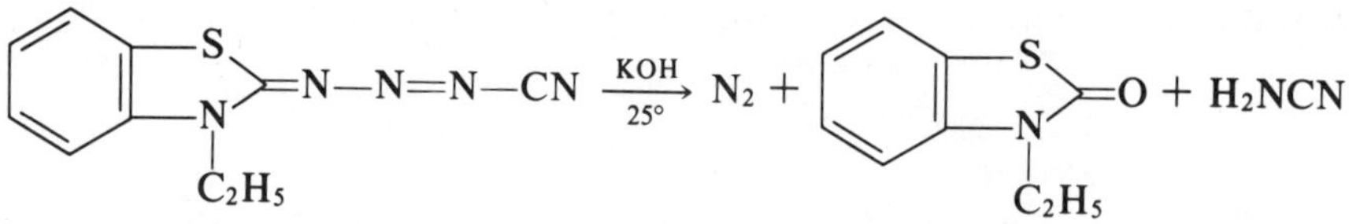

Rearrangement of aldehyde condensation products of 1,3-diaryltetrazenes to formazans is facilitated by alkoxide (93).

$$C_6H_5N{=}N{-}\underset{}{\overset{C_6H_5}{\overset{|}{N}}}{-}N{=}CHC_6H_4NO_2\text{-}m \xrightarrow[\text{ice cooling}]{NaOC_2H_5,\ \text{ethanol}} C_6H_5NHN{=}\underset{\underset{C_6H_5N{=}N}{|}}{C}{-}C_6H_4NO_2\text{-}m$$

Heating 1,4-dimethyl-1,4-diethoxycarbonyl-2-tetrazene **(V)** with piperidine causes elimination of an ethoxycarbonyl and cyclization to 1,4-dimethyltetra-

$$C_2H_5OOC{-}\overset{CH_3}{\overset{|}{N}}{-}N{=}N{-}\overset{CH_3}{\overset{|}{N}}{-}COOC_2H_5 \xrightarrow[\text{reflux}]{\text{piperidine}}$$

(V)

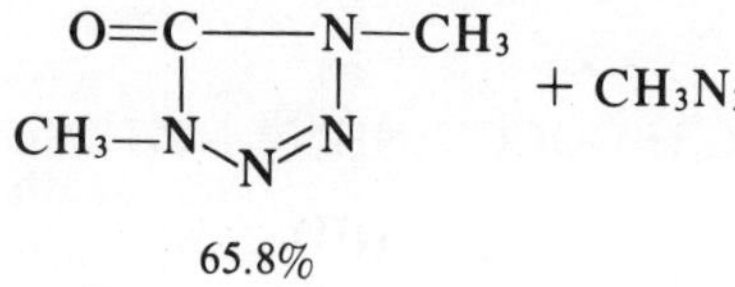

zolinone; a small amount of methyl azide is formed by an alternate route (541). No tetrazolinone results from decomposition of 1,4-dimethoxycarbonyl-1,4-diphenyl-2-tetrazene, but only methyl *N*-phenylcarbamate and polymer.

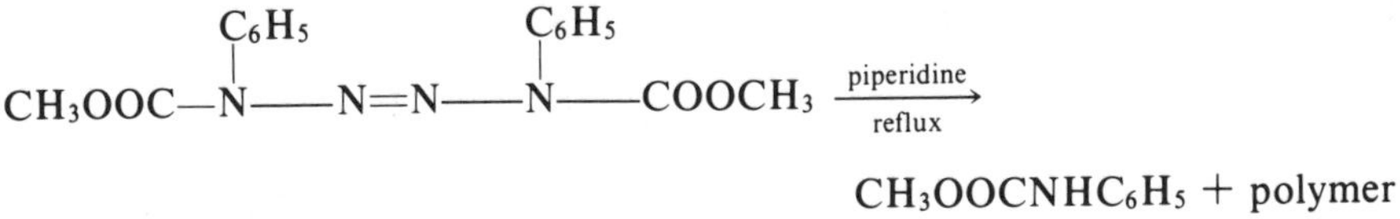

Pentaalkyltetrazenium salts react with nucleophiles such as arylsulfinate and triphenylphosphine by exchange with the trialkylamine group, forming, respectively, 3,3-dialkyl-1-arylsulfonyltriazenes and triazene phosphonium salts; the probable intermediate is an *N*-dialkylaminodiazonium ion (498).

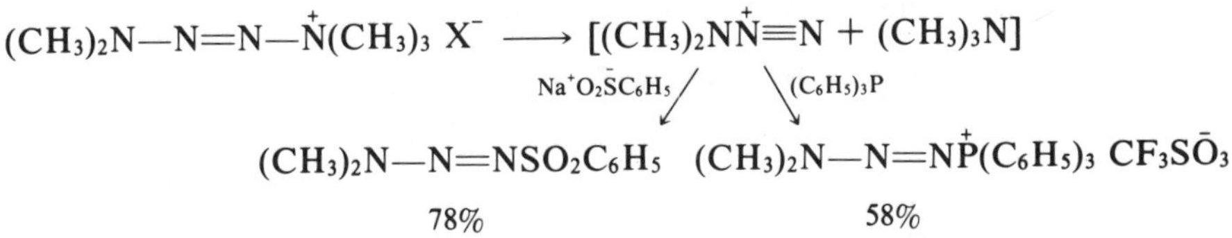

The five-nitrogen chain in bis(3-methylbenzothiazolyl-2)pentaazacarbocyanine decomposes on treatment with lithium benzenesulfinate to nitrogen and di- and triazacyanines (424).

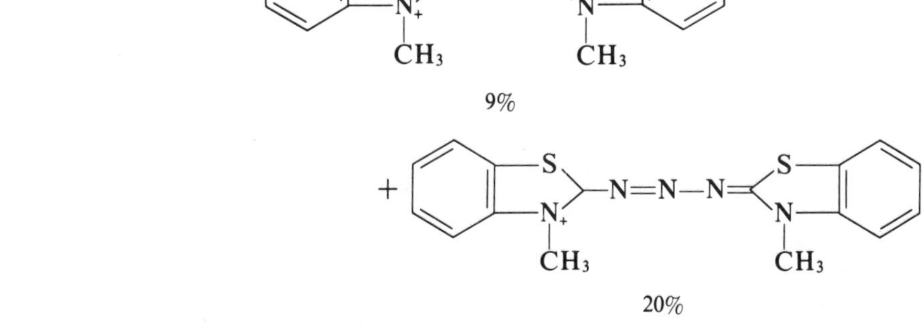

Cold concentrated sodium hydroxide acts on 1,6-bis(5′-tetrazolyl)-1,5-hexazdiene forming a sodium salt, which on dilution with water liberates nitrogen and forms 5-azidotetrazole and 5-aminotetrazole (240). No ammonia results.

Similarly, hot alcoholic potassium hydroxide cleaves 1,6-bis(*p*-chlorophenyl)-3,4-diacetyl-1,5-hexazdiene **(VI)** into nitrogen, *p*-chlorophenyl azide, and *p*-chloroaniline, accounting for virtually all the nitrogen in the chain (247). At

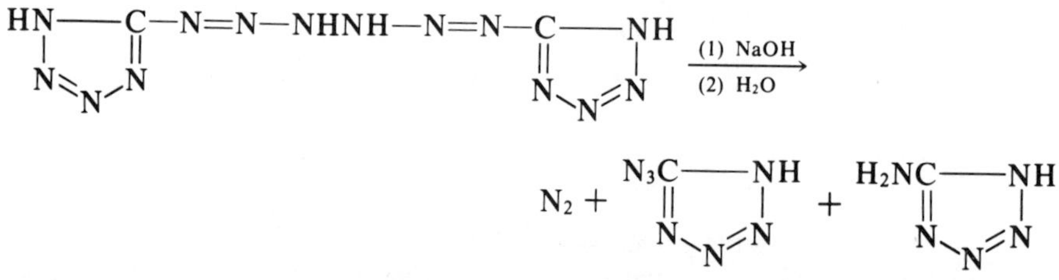

lower temperatures and with more dilute alkali the hexazdiene is split to 3-acetyl-1-*p*-chlorophenyltriazene in high yield. The acetyl groups of **VI** combine with amines, alkoxides, and phenoxides to form amides and esters (342).

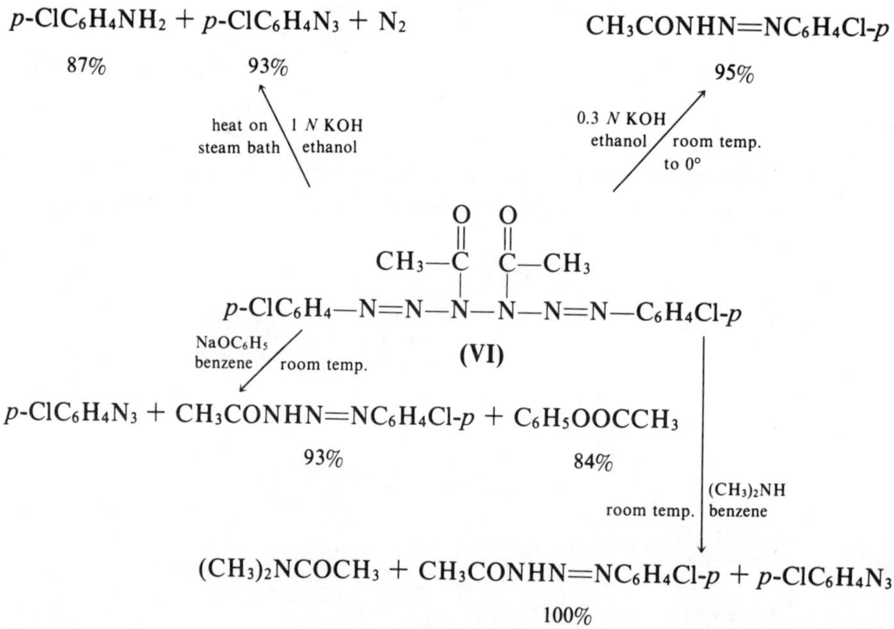

The N—N linkage in nitrosohydrazines between nitroso and hydrazine nitrogen is relatively resistant to alkali. Thus 1,2-dimethyl-2-nitroso-1-phenylhydrazine **(VII)** is changed very little on continuous heating with alkali (211), nitroso-1,2-diisopropylhydrazine reacts with sodium hydroxide to form a sodium salt **(VIII)** (333), and nitrosobenzyl hydrazine **(IX)** is unchanged on warming with 15% sodium hydroxide on the water bath (558). Ethyl nitrososemicarbazino acetate **(X)**, with other points of attack, evolves gas on heating with 10% potassium hydroxide (24). Although not affected by aqueous base at room

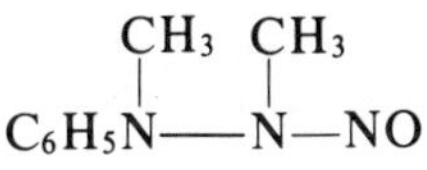

(VII)

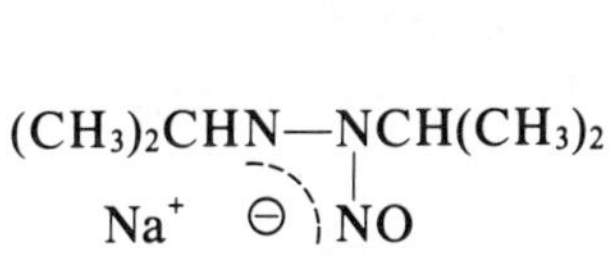

(VIII)

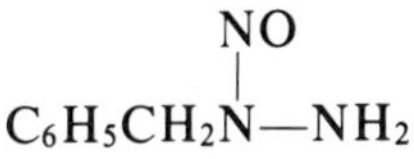

(IX)

(X)

temperature, the nitroso group of 1-nitroso-3-methyl-4-phenylazo-5-pyrazolone **(XI)** is removed by hot alkali (89). Nitrosotriazenes are denitrosated by low

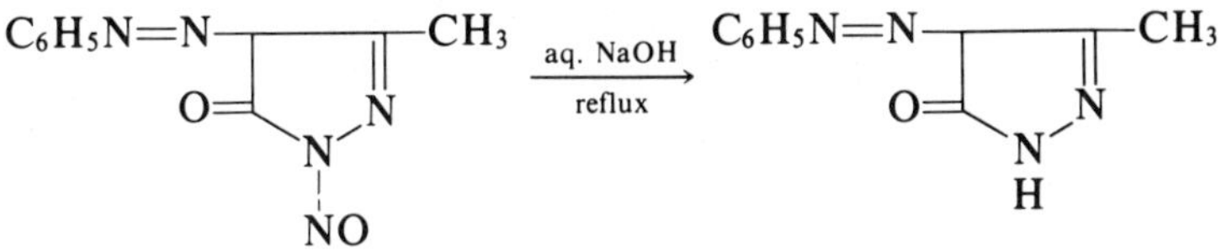

(XI)

temperature treatment with amines with concomitant oxidation of the amine (370).

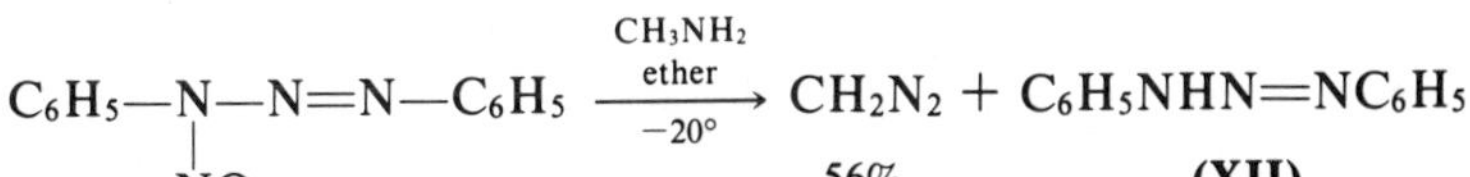

(XII)

Neutralization of 3-morpholinosydnonimine hydrochloride with sodium bicarbonate leads through an open-chain 1-cyanomethyl-1-nitrosohydrazine to nitric oxide and a denitrosated hydrazone (352).

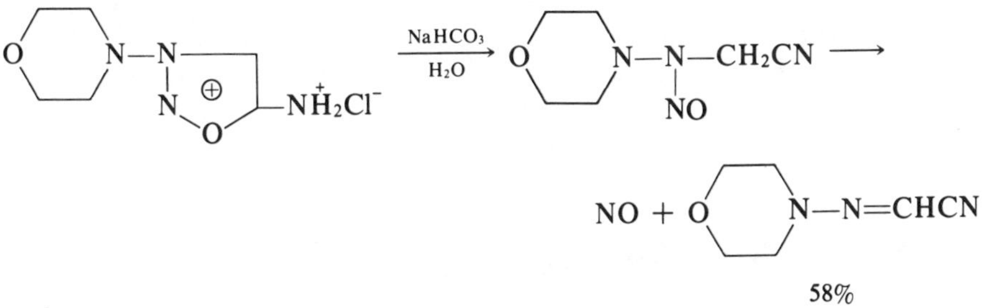

The *N*-nitro group of 1,3-dinitropyrazole **(XII)** is displaced by secondary amines such as morpholine forming *N*-nitromorpholine and a small amount of

N-nitrosomorpholine (204). In contrast, secondary amines substitute at the 5-position of 1,4-dinitropyrazole **(XIII)** with displacement of the *N*-nitro group as

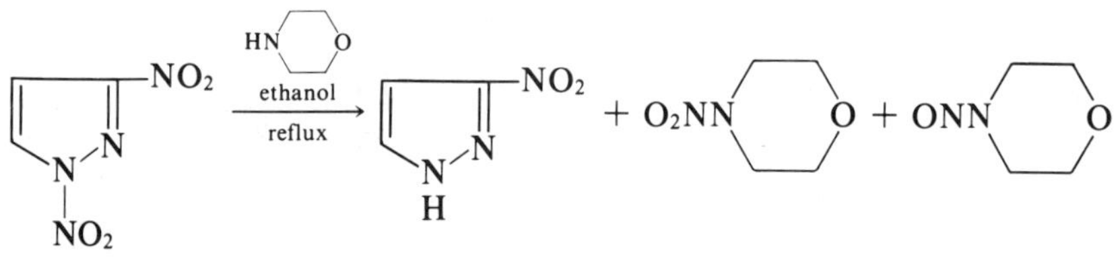

(XII) 84% 84% 14%

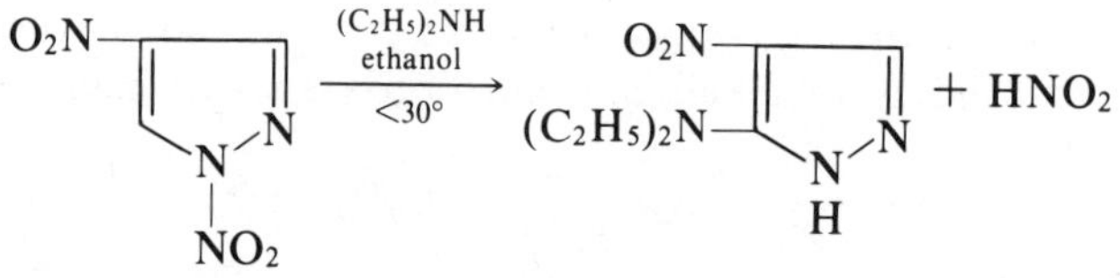

(XIII)

nitrous acid, but without formation of *N*-nitroamine. Pyrazole and *C*-substituted pyrazoles can act as the amine component in this synthesis, forming 4-nitro-3(5)(1′-pyrazolyl)pyrazoles in 80–95% yields (115).

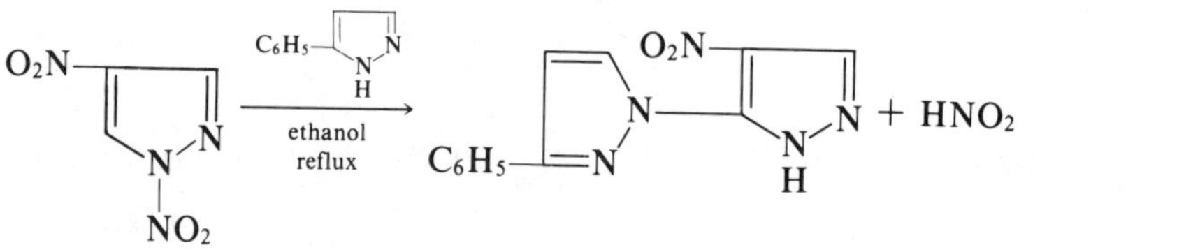

(XIII)

1-Aryl-3-methyl-3-nitrotriazenes dissolve in basic solvents such as pyridine or aqueous alkali; on dilution of a freshly prepared alkaline solution with water, the *N*-nitrotriazene precipitates, but after a few hours fails to do so and on acidification, diazo compound and nitrous acid are formed (38). Solutions of 1-aryl-3-methyl-3-nitrotriazenes in sodium methylate, after standing several hours, form arylamine, nitrite, formaldehyde, and nitrogen, illustrated with 3-methyl-3-nitro-1-(*o*-nitrophenyl)triazene.

$$o\text{-}O_2NC_6H_4N{=}N{-}\underset{\displaystyle NO_2}{\underset{|}{N}}{-}CH_3 \xrightarrow[\text{1.5–2.0 hr}]{\text{16.5\% NaOCH}_3\text{, methanol}}$$

$$N_2 + CH_2O + \bar{NO_2} + o\text{-}O_2NC_6H_4NH_2$$

54–59% 53.5% 43%

Trimethylamine nitroimide is stable on refluxing in aqueous alkali (145). Pyridine-1-nitroimide, however, undergoes ring opening on heating with sodium deuteroxide. Investigation by proton NMR reveals initial exchange of the 2,6-hydrogen atoms of pyridine with deuterium, followed by formation of glutaconic aldehyde derivatives (19).

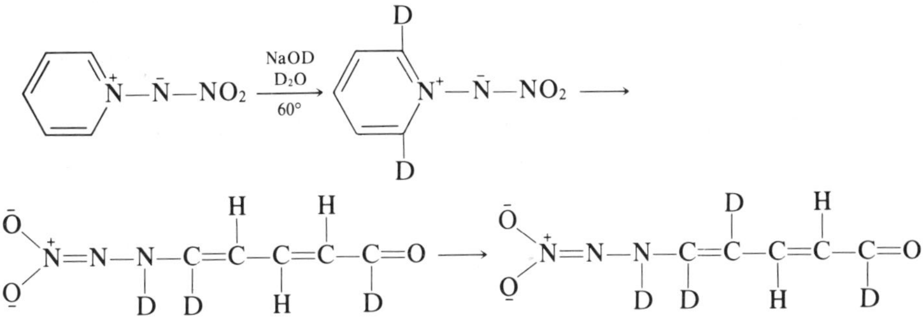

Alkyl and unactivated aryl azides are generally inert to attack by bases. Thus 1- and 2-azidonaphthalene and 2-azido-5- and 8-nitronaphthalene are unchanged on boiling in alcoholic potassium hydroxide (157). In the presence of an activating group ortho or para to azide, as in 2-azido-1-nitronaphthalene, azide is readily displaced in refluxing alcoholic potassium hydroxide. Similarly azide in azidium salts such as **XIV** can be replaced on treatment with alkali at room temperature. With dimethylamine an intermediate adduct forms which hydrolyzes with loss of azide ion (26, 27).

Some azides react with *p*-toluenesulfinate ion or triphenylphosphine to form adducts in which the nitrogen chain remains intact, illustrated with 1-ethyl-2-azidoquinolinium fluoborate **(XV)** (26) and tosyl azide (71), respectively. Formation of phosphazide adducts or their decomposition products, called the Staudinger reaction, has been extensively studied (194, 491, 492). The product

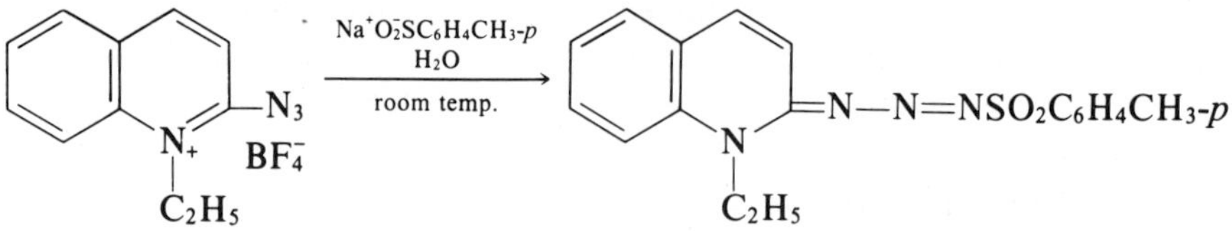

(XV) 75%

$$p\text{-}CH_4C_6H_4SO_2N_3 \xrightarrow{(C_6H_5)_3P} p\text{-}CH_3C_6H_4SO_2N{=}N{-}N{=}P(C_6H_5)_3$$

from triphenylmethyl azide and triphenylphosphine is stable in ether but dissociates in benzene (61). Phosphazides usually form readily, even at very low

$$(C_6H_5)_3C{-}N_3 + (C_6H_5)_3P \underset{\text{benzene}}{\overset{\text{ether}}{\rightleftharpoons}} (C_6H_5)_3C{-}N{=}N{-}N{=}P(C_6H_5)_3$$

temperatures, as shown with guanyl azide hydrochloride (86). All, however, lose nitrogen on heating, resulting in phosphine imines; in many instances only

$$(H_2N)_2\overset{+}{C}N_3\ Cl^- \xrightarrow[\substack{CH_2Cl_2 \\ -78°}]{\substack{(C_6H_5)_3P \\ \text{ethanol}}} (H_2N)_2\overset{+}{C}{-}N{=}N{-}N{=}P(C_6H_5)_3 \xrightarrow[\text{reflux}]{CH_2Cl_2}$$

$$N_2 + (H_2N)_2\overset{+}{C}{-}N{=}P(C_6H_5)_3\ Cl^-$$

quantitative

the phosphine imine is obtained. Thus reaction of triethylphosphine and benzoyl azide in ether solution at room temperature leads to triethylphosphine benzoylimine

$$N_3COC_6H_5 \xrightarrow[\text{room temp.}]{\substack{(C_2H_5)_3P \\ \text{ether}}} N_2 + (C_2H_5)_3P{=}NCOC_6H_5$$

82%

(491). Hydrogen azide reacts with trialkyl phosphites yielding dialkyl-*N*-alkylaminophosphates, presumably by isomerization of the trialkylimidophosphate first formed, as well as diethylphosphite (278). By use of tosyl azide, labeled with ^{15}N in the terminal or middle positions, the point of addition of triphenyl

$$HN_3 \xrightarrow[\text{room temp.}]{\substack{(C_2H_5O)_3P \\ \text{benzene}}} N_2 + [(C_2H_5O)_3P{=}NH] \longrightarrow$$

$$N_2 + (C_2H_5O)_2PHO + (C_2H_5O)_2\underset{\substack{\Vert \\ O}}{P}NHC_2H_5 + (C_2H_5O)_2\underset{\substack{\Vert \\ O}}{P}NH_2$$

14% 34% 10%

phosphine is established as the end nitrogen (70). In the further decomposition of the phosphazide formed from tosylazide, a four-membered cyclic transition state has been proposed (320).

$$p\text{-}CH_3C_6H_4SO_2\bar{N}{-}N^+{\equiv}N \xrightarrow[60^\circ]{(C_6H_5)_3P \text{ benzene}} p\text{-}CH_3C_6H_4SO_2N{=}N{-}N{=}P(C_6H_5)_3 \longrightarrow$$

$$\left[\begin{array}{c} p\text{-}CH_3C_6H_4SO_2{-}N{-}{-}{-}P(C_6H_5)_3 \\ | \quad\;\; | \\ N{=}N \end{array}\right] \longrightarrow N{\equiv}N + p\text{-}CH_3C_6H_4SO_2N{=}P(C_6H_5)_3$$

A variation in the attack of nucleophiles on azidium compounds is the reaction with azide ion. At low temperatures an imineazide results, which reacts further with a second molecule of azidium salt, losing two molecules of nitrogen to form a triazatrimethine cyanine (31, 32).

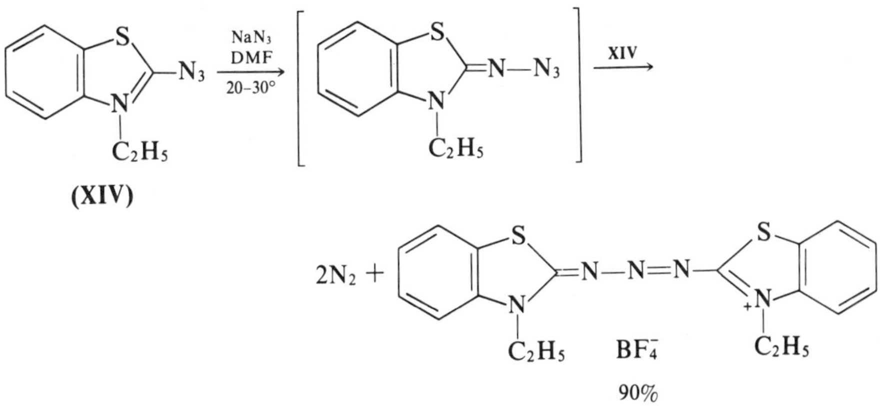

(XIV)

Analogous to the diazo transfer from tosyl azide to carbon is its reaction with anions of primary amines to produce alkyl and aryl azides, illustrated by the conversion of cyclohexylamine to cyclohexylazide (17, 154). Anions of hydrazones, such as benzophenone hydrazone, react with tosyl azide to form

$$p\text{-}CH_3C_6H_4SO_2N_3$$

$$C_6H_{11}NH_2 \xrightarrow[THF]{CH_3Li \text{ ether}} C_6H_{11}\bar{N}H \qquad (C_6H_5)_2C{=}N\bar{N}H \xleftarrow[THF]{CH_3MgCl} (C_6H_5)_2C{=}NNH_2$$

$$[C_6H_{11}NHN{=}NNSO_2C_6H_4CH_3\text{-}p] \qquad [(C_6H_5)_2C{=}NNHN{=}NNSO_2C_6H_4CH_3\text{-}p]$$

$$\downarrow \qquad\qquad\qquad \downarrow$$

$$\underset{35\%}{C_6H_{11}N_3} + H_2NSO_2C_6H_4CH_3\text{-}p \qquad \underset{50\%}{(C_6H_5)_2C{=}N_2} + N_2 + \underset{48\%}{H_2NSO_2C_6H_4CH_3\text{-}p}$$

diazoalkanes (155). In the reaction with amines a four-nitrogen adduct, and with hydrazones a five-nitrogen adduct, presumably are intermediates.

Similarly, reaction of the mono- and dianions of 1,1-dibenzylhydrazine with tosyl azide, possibly through dibenzylamine azide, leads to bibenzyl, benzaldehyde dibenzylhydrazone, and 3,3-dibenzyl-1-tosyltriazene (291). Some heterocyclic amines such as 3-aminopyridine and 1-aminoisoquinoline can receive a

$$(C_6H_5CH_2)_2NNH_2 \xrightarrow[\text{THF}]{\text{hexane},\ n\text{-}C_4H_9Li} (C_6H_5CH_2)_2N\bar{N}H$$

$$p\text{-}CH_3C_6H_4SO_2N_3 \xrightarrow{(C_6H_5CH_2)_2N\bar{N}H} [(C_6H_5CH_2)_2N{-}N_3] \longrightarrow$$

$$\underset{23\%}{C_6H_5CH_2CH_2C_6H_5} + \underset{30\%}{(C_6H_5CH_2)_2NN{=}CHC_6H_5}$$

$$+ \underset{37\%}{H_2NSO_2C_6H_4CH_3\text{-}p} + \underset{16\%}{(C_6H_5CH_2)_2NN{=}NSO_2C_6H_4CH_3\text{-}p}$$

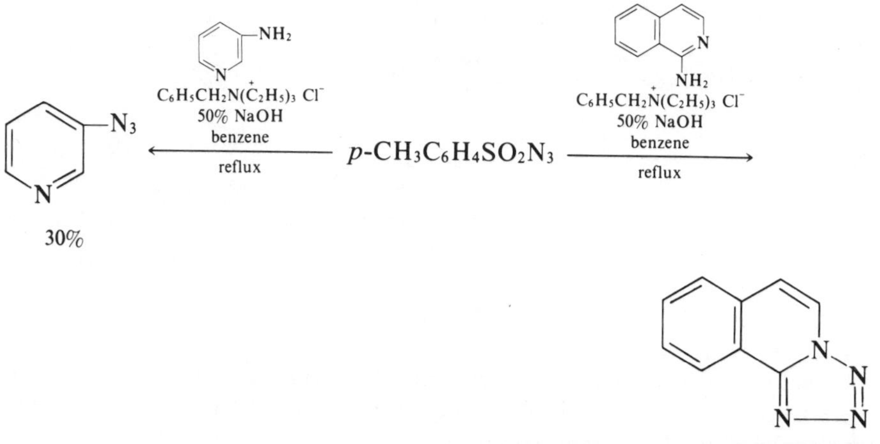

30%

74%

diazo moiety from tosyl azide under phase transfer conditions, forming the corresponding heterocyclic azide or tetrazoloazine (486).

On heating aryl azides with primary amines, nitrogen is lost with rearrangement to an azepine, shown in the reaction of phenyl azide and benzylamine (261). The probable intermediate is a phenyl nitrene (18). In the corresponding

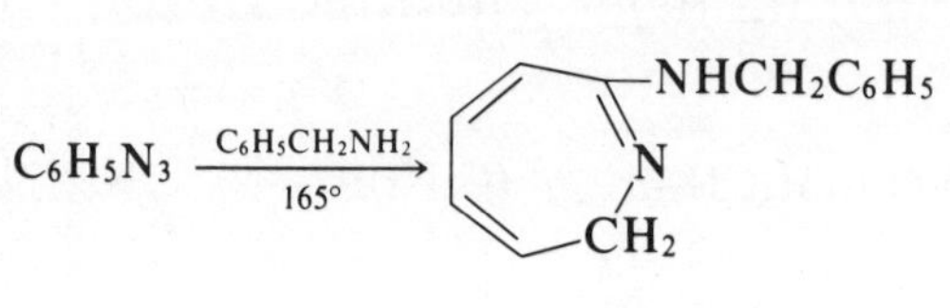

41%

reaction of aniline and phenylazide labeled with ^{14}C at the carbon attached to azide, essentially all the labeled carbon in the azepine is joined to anilino (255).

Alkyl azides illustrated by hexyl azide react with singlet ethoxycarbonyl nitrene, generated from *N*-(*p*-nitrobenzenesulfonyloxy)urethane, to form ethoxycarbonylhydrazones, probably by rearrangement of an initially formed, unisolated azo compound (182). Yields are those obtained when 20% of the hexyl

$$p\text{-}O_2NC_6H_4SO_2ONHCOOC_2H_5 \xrightarrow{(C_2H_5)_3N} [NCOOC_2H_5]$$

$$n\text{-}C_6H_{13}N_3 \xrightarrow[\substack{CH_3NO_2 \\ \text{room temp.}}]{[NCOOC_2H_5]} N_2 + [n\text{-}C_6H_{13}N{=}NCOOC_2H_5] \longrightarrow$$

$$\underset{11\%}{n\text{-}C_5H_{11}CH{=}NNHCOOC_2H_5} + \underset{14\%}{C_2H_5OOCNH_2} + \underset{3\%}{C_2H_5OOCNHNHCOOC_2H_5}$$

azide is decomposed. Phenyl azide, without a hydrogen for tautomerism, forms the expected azo compound.

$$C_6H_5N_3 \xrightarrow{[NCOOC_2H_5]} N_2 + C_6H_5N{=}NCOOC_2H_5$$

In heterocycles bearing azide adjacent to ring nitrogen, reactions with amines at other locations can induce ring closure to a tetrazole. Thus diethylamine

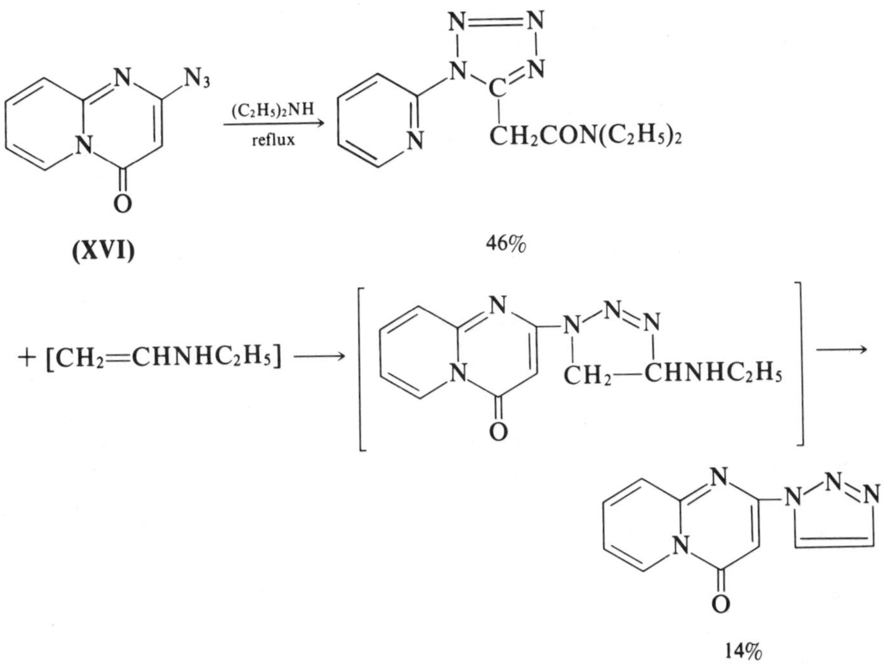

reacts with 2-azidopyrido[1,2-*a*]pyrimidin-4-one **(XVI)** to open the pyrimidone ring with formation of 1-(2-pyridyl)-5-(diethylacetamido)tetrazole (296, 419). At the same time 2-(1-*v*-triazolyl)pyrido[1,2-*a*]pyrimidine-4-one is obtained, probably by a sequence consisting of dehydrogenation of diethylamine to vinylethylamine, addition of **XVI** to the vinyl group to form a triazoline, and loss of ethylamine. Dehydrogenation of secondary aliphatic amines also appears to be involved in the conversion of 6-azidotetrazolo[1,5-*b*]pyridazine **(XVII)** to alkylated aminomethylene aminotetrazolo[1,5-*b*]pyridazine illustrated with diethylamine (418, 419). Loss of diazomethane from the intermediate triazoline leads to the ethylaminemethylene amino derivative.

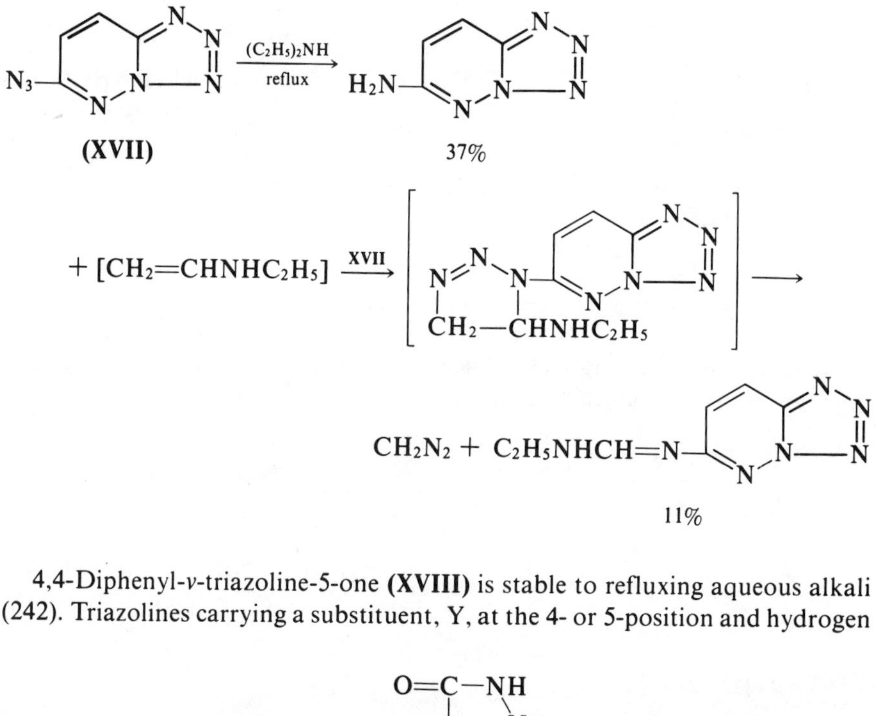

(XVII) 37% 11%

4,4-Diphenyl-*v*-triazoline-5-one **(XVIII)** is stable to refluxing aqueous alkali (242). Triazolines carrying a substituent, Y, at the 4- or 5-position and hydrogen

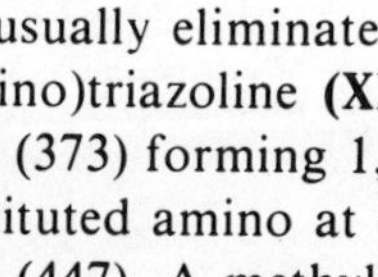

(XVIII)

on the other carbon position usually eliminate HY on treatment with base. Thus 1,5-diphenyl-5-(1-piperidino)triazoline **(XIX)** loses piperidine on treatment with an alkaline solution (373) forming 1,5-diphenyl-*v*-triazole, and the polycyclic triazoline with substituted amino at the fusion bridgehead **(XX)** is deaminated to a fused triazole (447). A methylsulfonyl group is removed by base from 1-methoxy-5,5-bis(methylsulfonyl)-*v*-triazoline **(XXI)** to form 1-methoxy-*v*-triazole-5-methylsulfone (21).

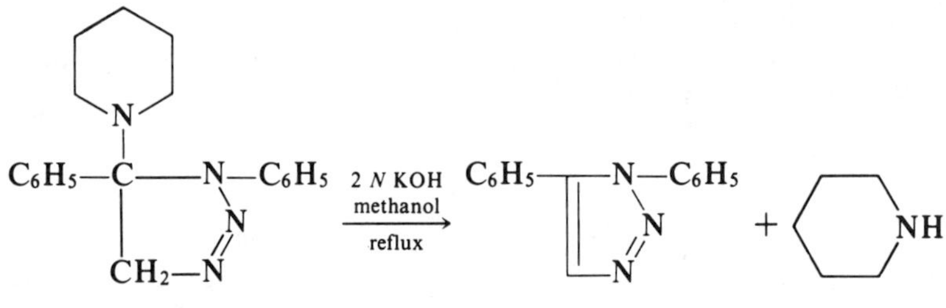

(XIX) 97%

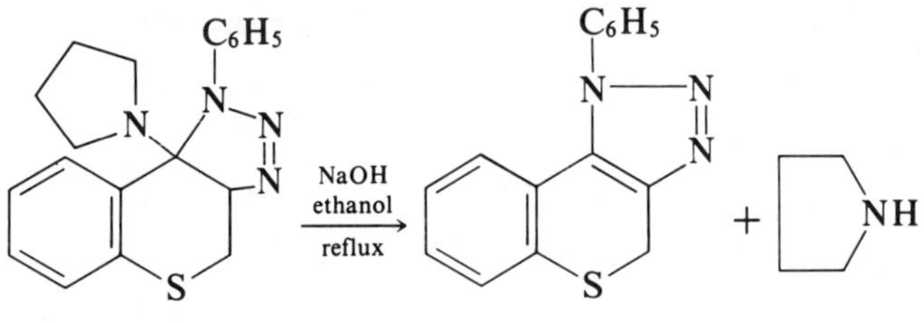

(XX) 80%

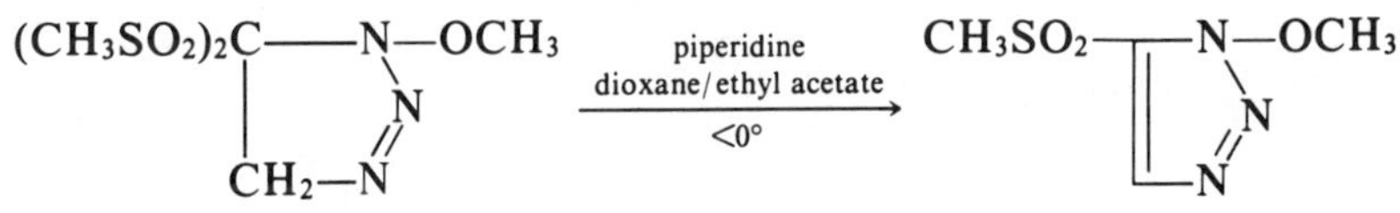

(XXI) 49.5%

Water is eliminated from 1-alkyl-5-hydroxy-*v*-triazolines to form *v*-triazoles in much higher yield than from the corresponding 1-aryl compounds (405). The initial effect of methanolic potassium hydroxide on 4,5-dimethyl-5-

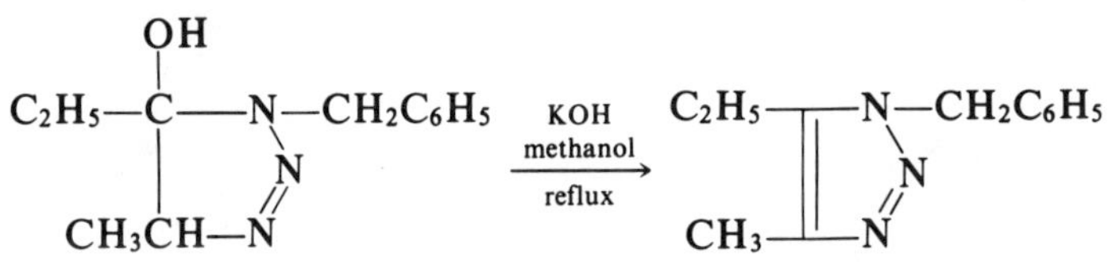

70%

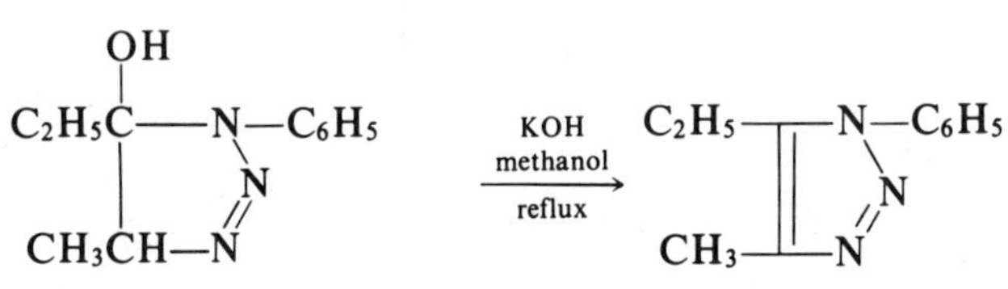

7%

hydroxy-1-phenyl-*v*-triazoline is to accelerate attainment of equilibrium between the diastereomeric hydroxytriazolines and the open-chain triazene ketone isomer, followed by elimination of water from the hydroxytriazolines to form the *v*-triazole and of aniline from the triazene ketone leading to diazobutanone (403). The cyclic nitrogen chain in 4-methoxycarbonyl-1-phenyltriazoline

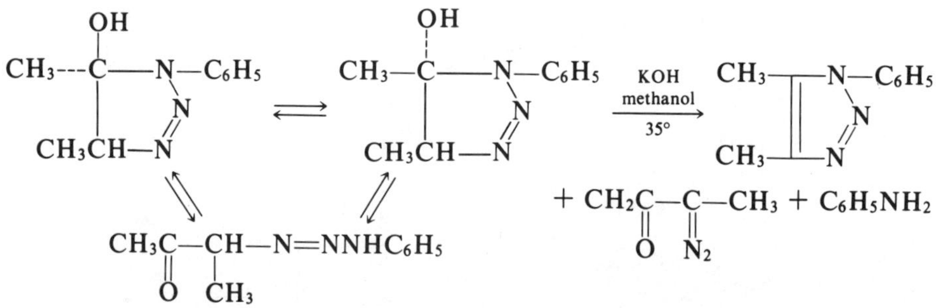

is opened by triethylamine to the acyclic isomer, methyl 3-anilino-2-diazopropionate (260, 505).

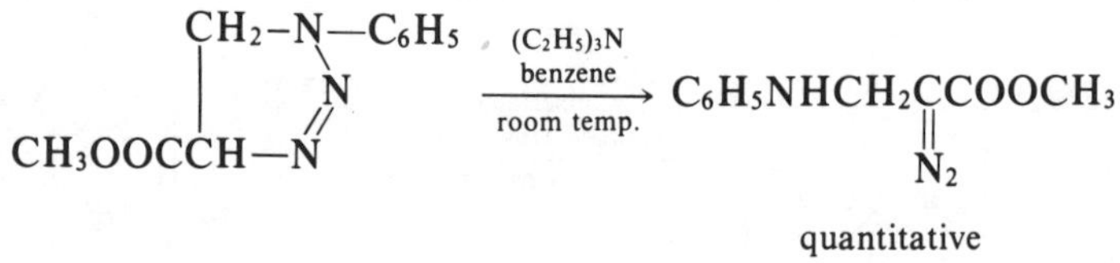

1,3-Disubstituted triazolinium salts are dehydrogenated to triazolium salts by sodium hydride with some ring opening (208).

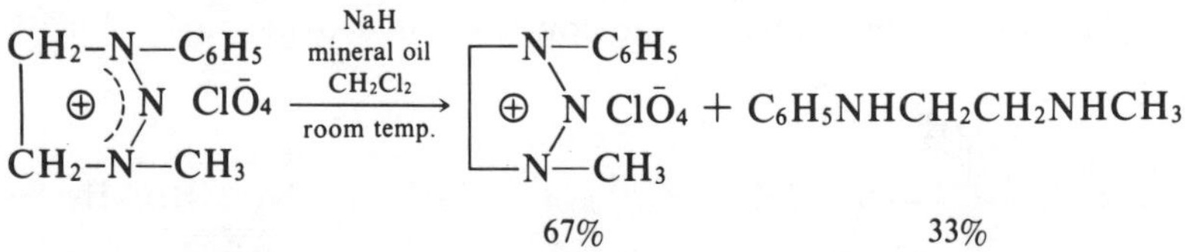

The influence of bases on 1-substituted 5-amino-*v*-triazoles is to facilitate the Dimroth rearrangement to 5-substituted amino-*v*-triazoles. Thus 5-amino-1-phenyl-*v*-triazole is converted by heating with pyridine to 4-anilino-*v*-triazole (127). Electron-withdrawing groups at N-1 enhance the rates of Dimroth rear-

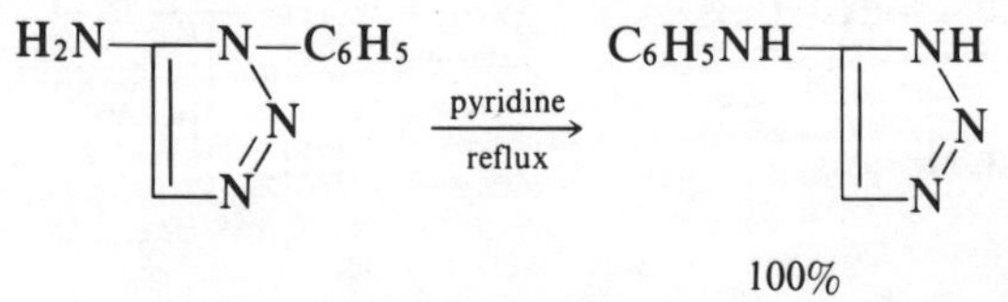

rangements (328). Equilibrium compositions are obtained in the presence of base starting with 1-benzyl-5-amino-*v*-triazole and related compounds (7).

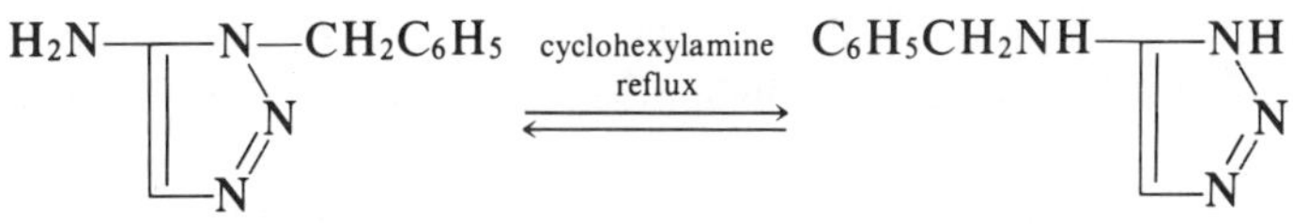

70%

Except for the 5-amino-*v*-triazoles, the *v*-triazole nitrogen sequence is usually stable to nucleophiles and their reactions with *v*-triazoles consist of hydrolysis or rearrangement of nitrogen substituents and displacement of groups on triazole carbon.

Strongly electron-withdrawing substituents on triazole nitrogen are hydrolyzed under the influence of base as in the conversion of 2-(2′,4′-dinitrophenyl)-4-nitro-*v*-triazole to 4-nitro-*v*-triazole (140). 1-Acyl groups on *v*-triazoles mi-

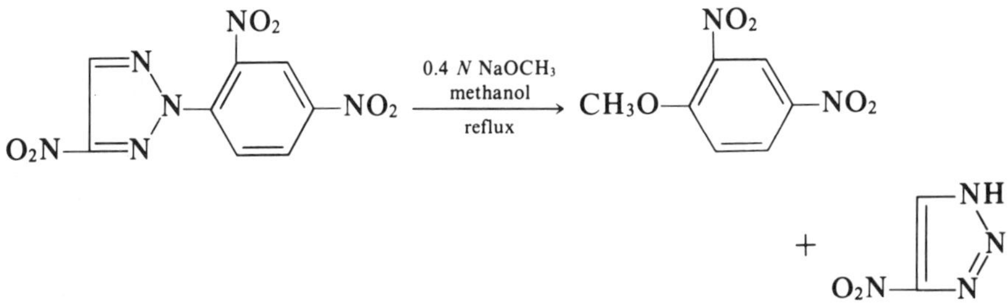

92%

grate intramolecularly to the 2-position on addition of amines; starting with a mixture of 1-ethoxycarbonyl-4-phenyl-*v*-triazole **(XXII)** and 1-ethoxycarbonyl-5-phenyl-*v*-triazole **(XXIII)** in a ratio of 53/47, in the presence of triethylenediamine after 220 min, the composition has shifted to a preponderance of 2-ethoxycarbonyl-4-phenyl-*v*-triazole **(XXIV)** (565).

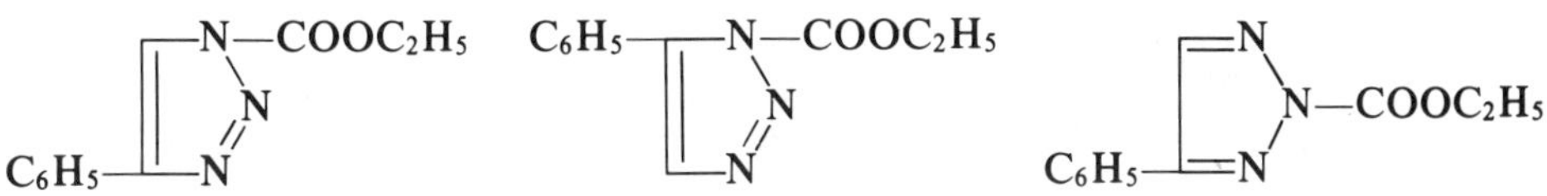

(XXII) 37.3% **(XXIII)** 19.7% **(XXIV)** 43.0%

The methylol group of 1-hydroxymethyl-5-hydroxybenzotriazole migrates to the benzene ring in alkaline solution (166).

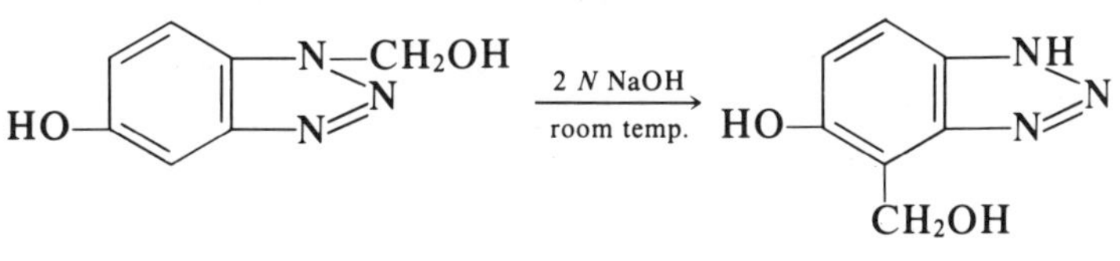

The 5-chloro group in 5-chloro-1-phenyl-*v*-triazole **(XXV)** readily undergoes nucleophilic displacement by methylamine (128). With 5-chloro-2-phenyl-*v*-tria-

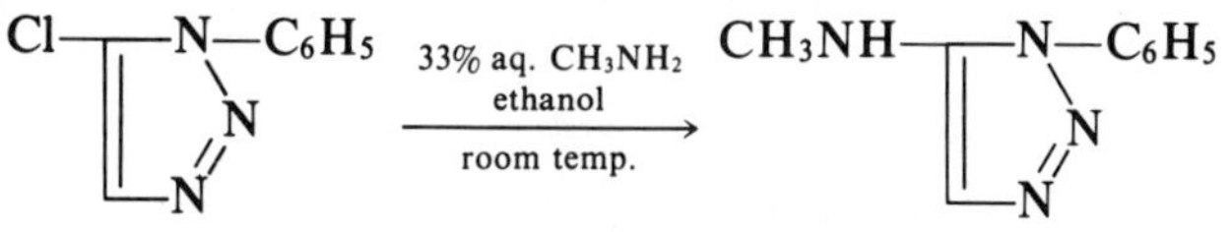

(XXV)

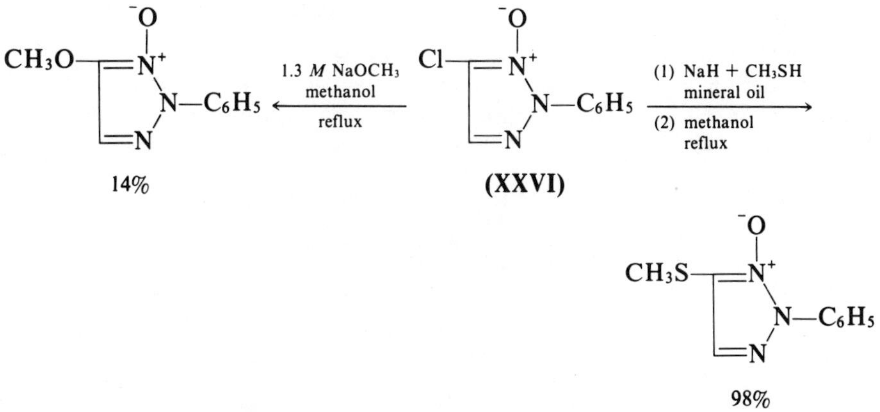

zole 1-oxide **(XXVI)**, methoxide is less reactive than the more nucleophilic methanethiolate (50). 4-Aryl-5-triphenylphosphonium-*v*-triazole ylides hydrolyze in hot aqueous base to 4-aryl-*v*-triazole and triphenylphosphine oxide (510).

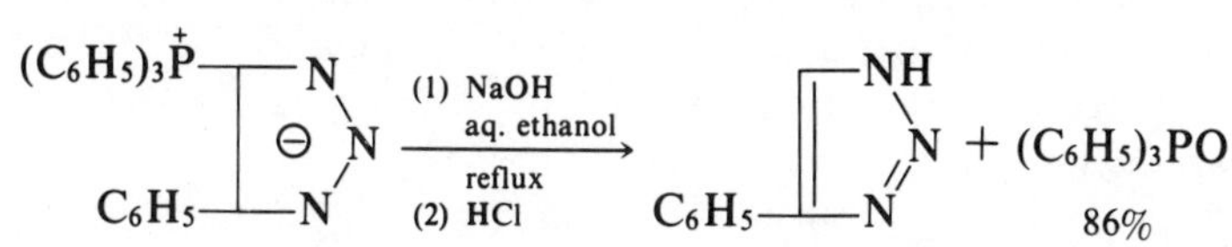

Solution of 1-benzyl-4-bromo-3-methyl-*v*-triazolium tosylate **(XXVII)** in aqueous potassium hydroxide at room temperature establishes an equilibrium

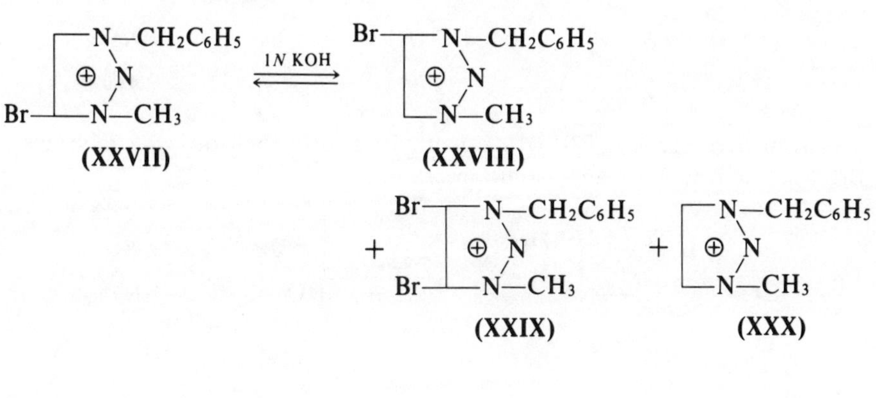

mixture consisting of the tosylates of the 5-bromo isomer **(XXVIII)**, the 4,5-dibromo **(XXIX)** and the unbrominated *v*-triazolium **(XXX)** derivative (45). The same equilibrium results when the 5-bromo salt **(XXVIII)** or a mixture of the dibromo **(XXIX)** and unsubstituted salt **(XXX)** is dissolved in aqueous alkali. No *v*-triazolium 4- or 5-oxides are formed, even on refluxing the bromotriazolium salts with sodium hydroxide. If, however, 1-methyl-3-phenyl-*v*-triazolium tosylate **(XXXI)** is treated with *N*-bromoacetamide and dilute sodium hydroxide, both possible *C*-bromo-*v*-triazolium-*C*-oxides are formed. Heating **XXVII** with

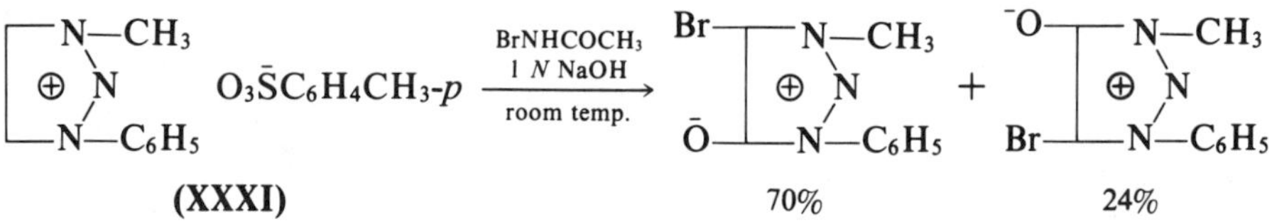

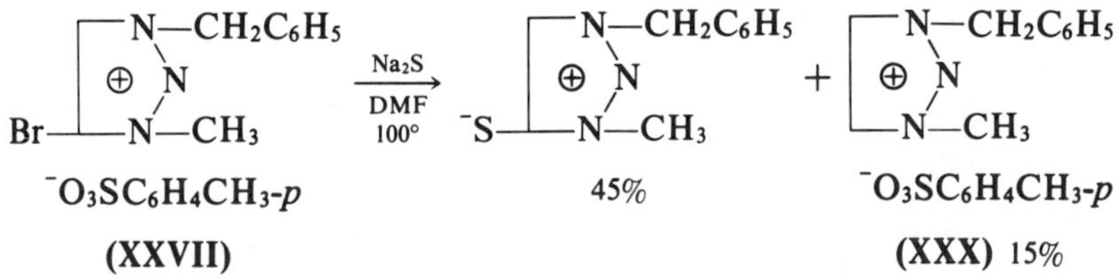

sodium sulfide leads to the *v*-triazolium sulfide and some unsubstituted salt **(XXX)** (47). Direct thiation of 1,3-disubstituted *v*-triazolium salts by elemental sulfur in the presence of sodium hydride effects replacement of hydrogen; halogen on *v*-triazolium carbon is not displaced (48).

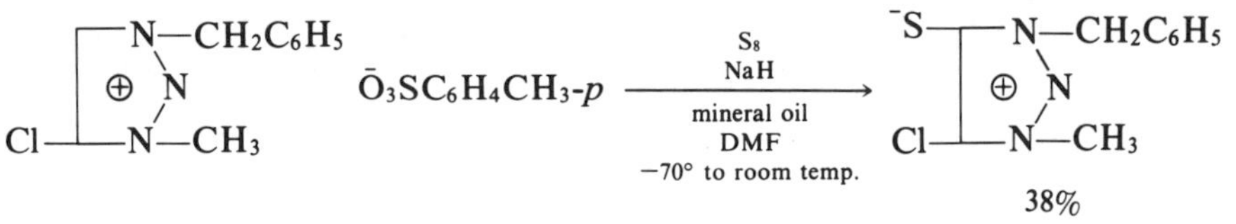

Heating 1,2-disubstituted 5-bromo-*v*-triazolium salts with aqueous sodium hydroxide leads to 1,2-disubstituted 5-*v*-triazolinones (53).

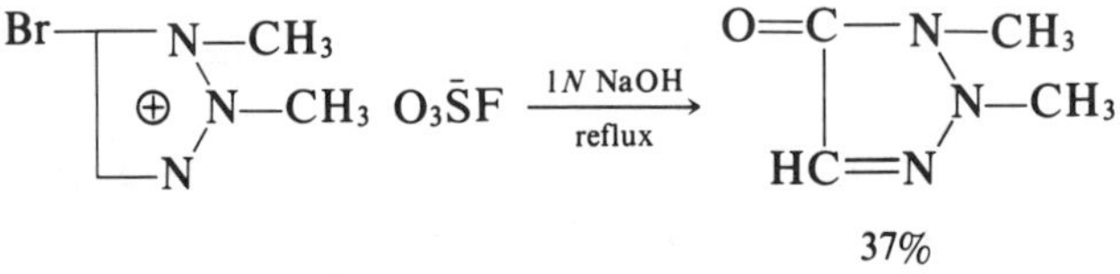

The 1-methoxy group of 1-methoxy-2-phenyl-*v*-triazolium fluoborate **(XXXIa)** activates the 5-position toward attack by nucleophiles with simultaneous loss of the methoxy group in a nucleophilic addition-elimination, illustrated by reactions at room temperature with hydroxide, fluoride, and azide ions and with ammonia (50). Triethylamine effects removal of the methyl group and formation of 2-phenyl-*v*-triazole 1-oxide.

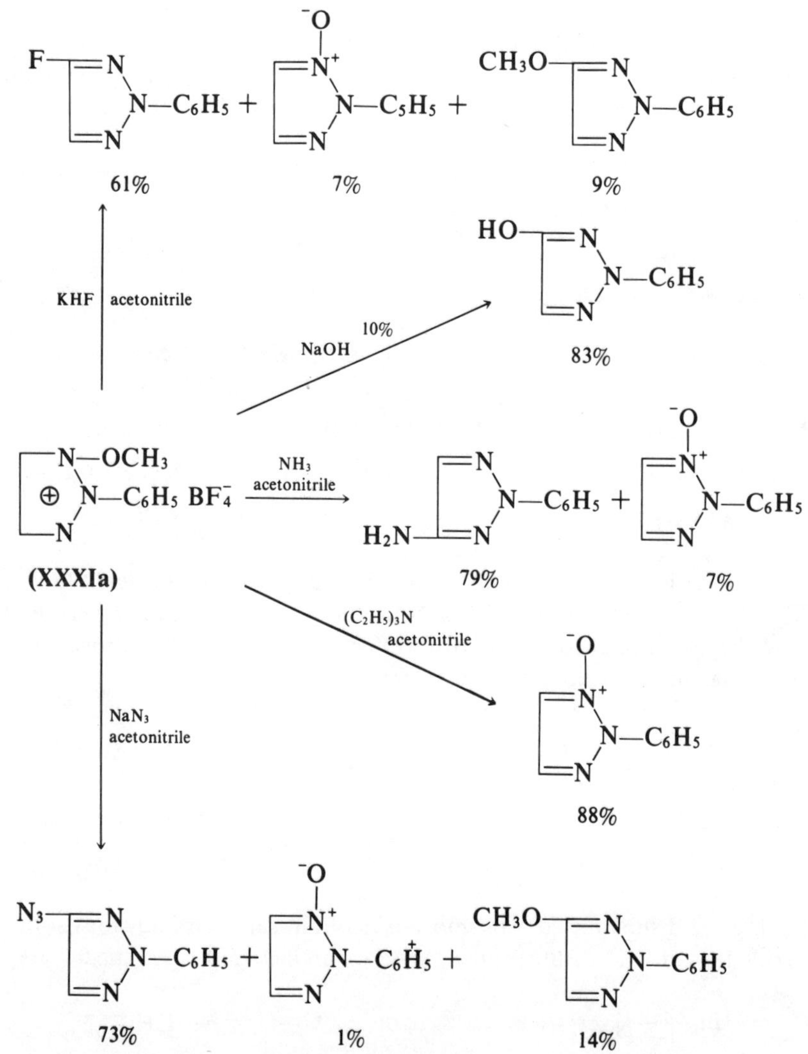

Benzotriazines, such as the 4-anilino derivative **(XXXII)** (497), benzotriazinone (374), and the parent compound (5) react with amines and hydrazines to displace nitrogen and form *o*-aminobenzene derivatives containing the nucleophilic moiety. A monosubstituted triazene may be an intermediate. The 4-aralkylaminobenzotriazines, however, are unaffected by refluxing piperidine or morpholine (497). The three-nitrogen chain is partially opened in 1,3-dimethylnaphtho[1,8-*de*]1,2,3-triazonium methylsulfate **(XXXIII)** by treatment with potassium carbonate and acetone (42). Hydrolytic destruction of both the imi-

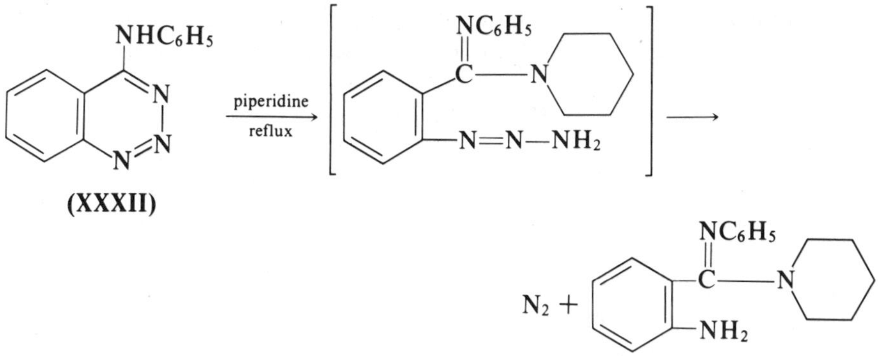

(XXXII)

72%

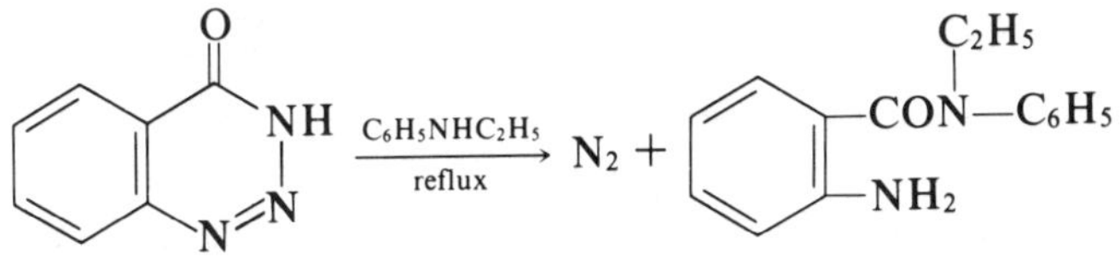

85%

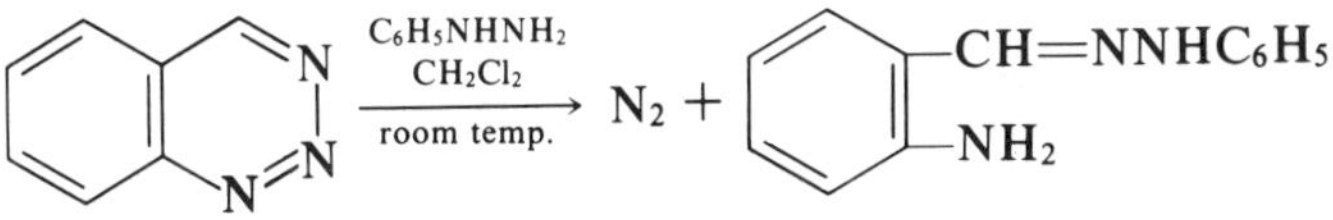

89%

dazole and *v*-triazine rings occurs on treatment of 2-aza-9-cyclopentyladenine **(XXXIV)** with dilute alkali (366).

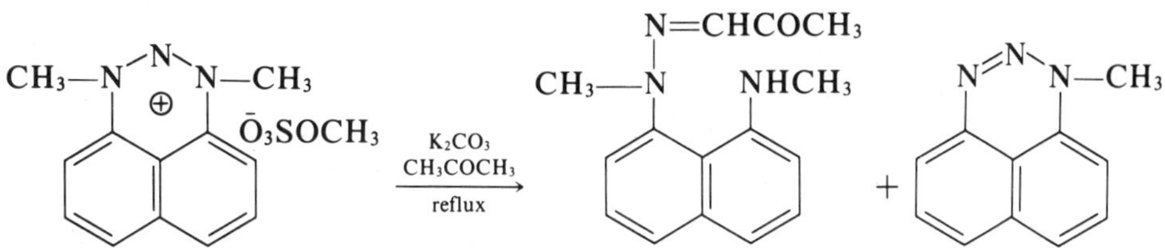

(XXXIII)

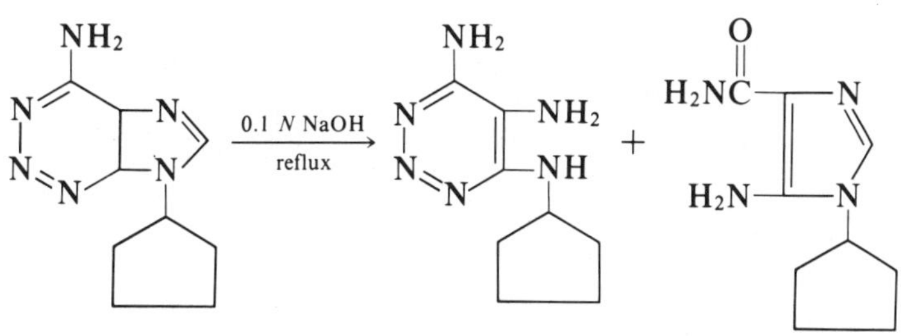

(XXXIV) 36% 41%

Alkaline reagents open 4-amino and 4-hydroxybenzotriazine 3-oxides to *o*-azidobenzoic derivatives (212, 472). An azide group also is formed from the ac-

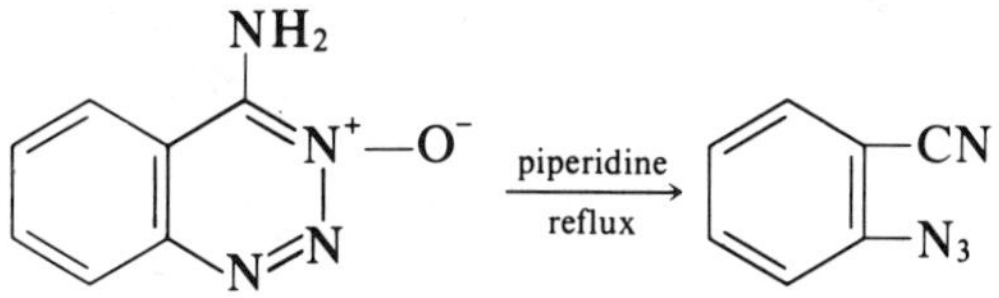

77%

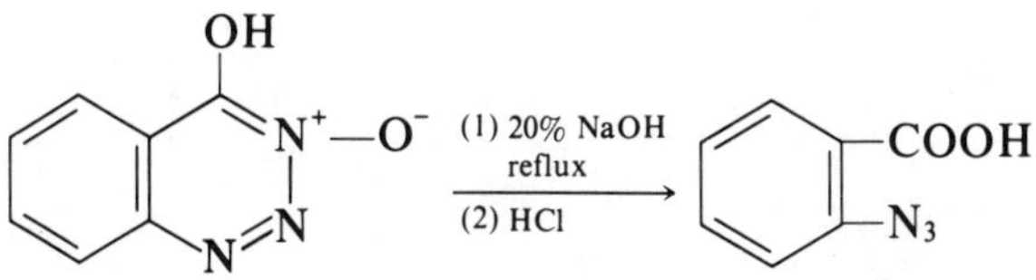

83%

tion of hydrazine on dibenzotriazepine, leading to 2-amino-2′-azidobiphenyl and a small quantity of carbazole (172).

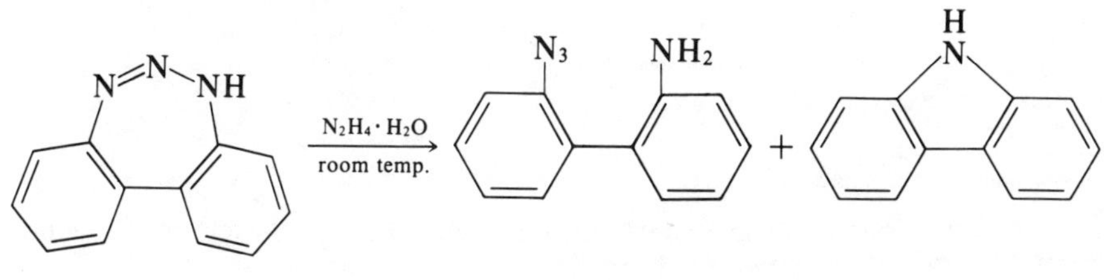

70% 10%

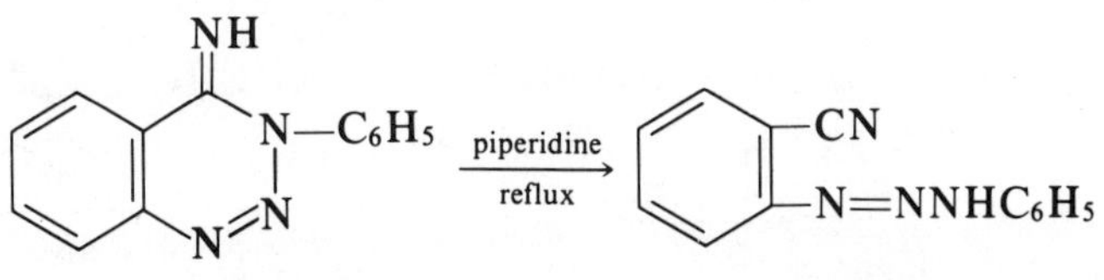

80%

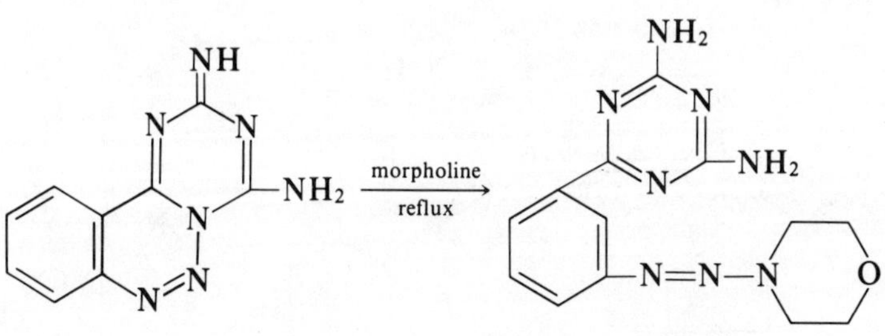

85%

The nitrogen chain is preserved as a triazene by the action of amines on 3,4-dihydro-4-iminobenzotriazines; a triazene also results from fused ring derivatives by combination of amine nitrogen with the triazine masked diazo group (472).

Under the influence of alkaline media, the *N*-acetyl and phenylmethylene derivatives of 3-aminobenzotriazinone decompose to liberate ammonia, forming *o*-azidobenzoic acid and the phenylmethylene derivative of *o*-hydrazinobenzoic acid, respectively (224, 226). On treatment with aqueous alkali, 2-amino-

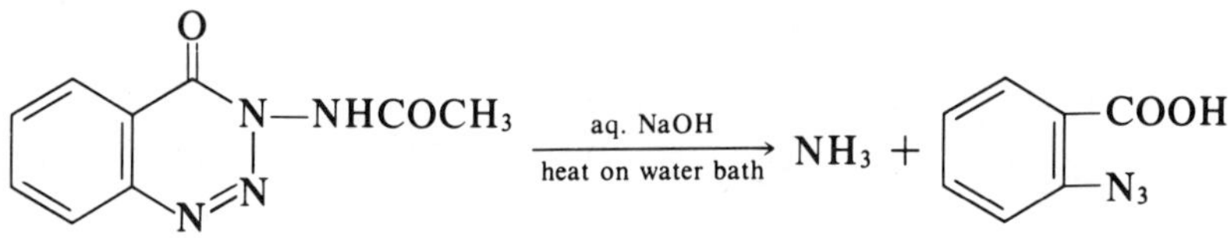

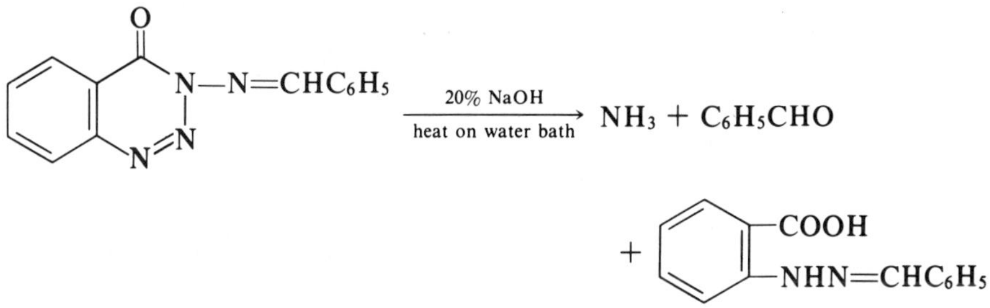

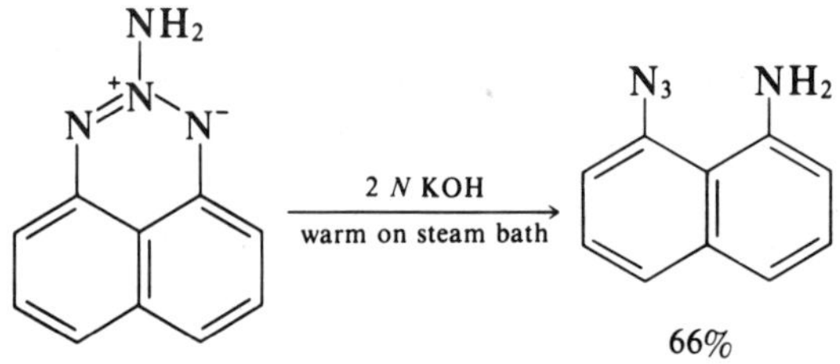

naphtho[1,8-*de*]triazine, but not the 1-isomer, rearranges to 1-amino-8-azidonaphthalene (432). Heating 3,3′-bibenzotriazinone with alcoholic potassium hydroxide liberates nitrogen (225).

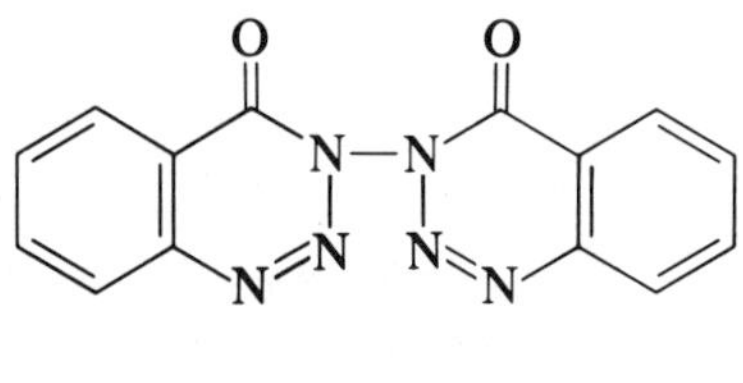

Treatment of 2-phenylbenzothiatriazine 1,1-dioxide **(XXXV)** with dilute sodium hydroxide in the presence of metallic copper induces loss of nitrogen and bond formation between the two benzene rings to produce a diphenylsultam, analogous to the synthesis of diphenyl derivatives from diazonium compounds and benzenoid hydrocarbons (531).

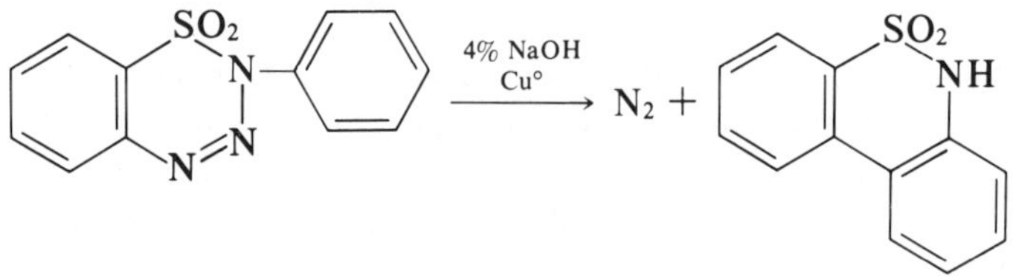

(XXXV) 96%

3-Benzyloxatriazolium-5-olate hydrolyzes to benzyl nitrosohydrazine on heating with ammonium hydroxide (152). Oxatriazolium-5-thiolates such as the

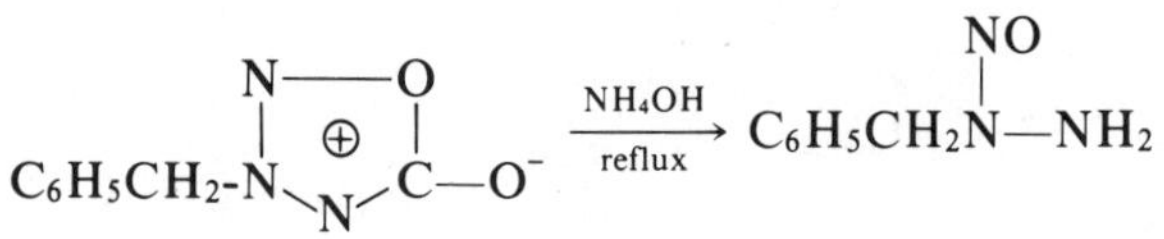

23%

3-phenyl derivative **(XXXVI)** isomerize to thiatriazolium-5-olates by heating with ethanolic ammonia (206).

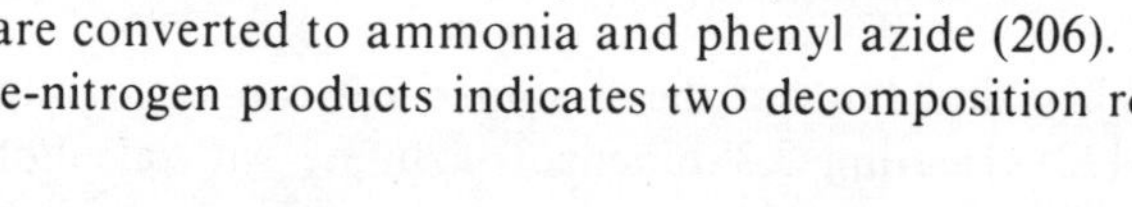

(XXXVI) 40%

On refluxing with aqueous sodium hydroxide **(XXXVI)** and its thiatriazolium olate isomer are converted to ammonia and phenyl azide (206). Formation of three- and one-nitrogen products indicates two decomposition routes.

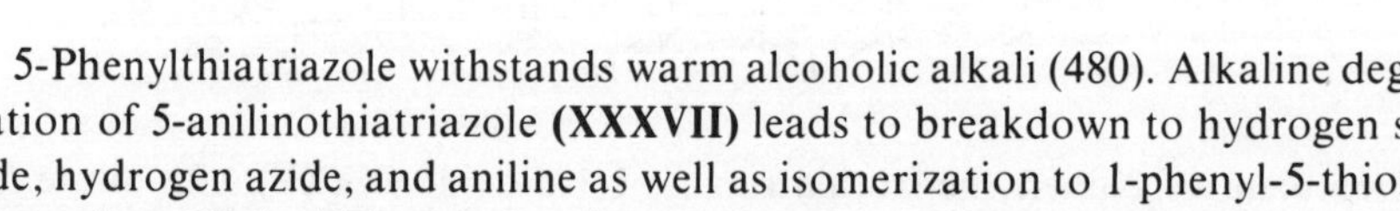

(XXXVI)

5-Phenylthiatriazole withstands warm alcoholic alkali (480). Alkaline degradation of 5-anilinothiatriazole **(XXXVII)** leads to breakdown to hydrogen sulfide, hydrogen azide, and aniline as well as isomerization to 1-phenyl-5-thionotetrazoline (330). Although heating 5-alkylaminothiatriazoles with base does

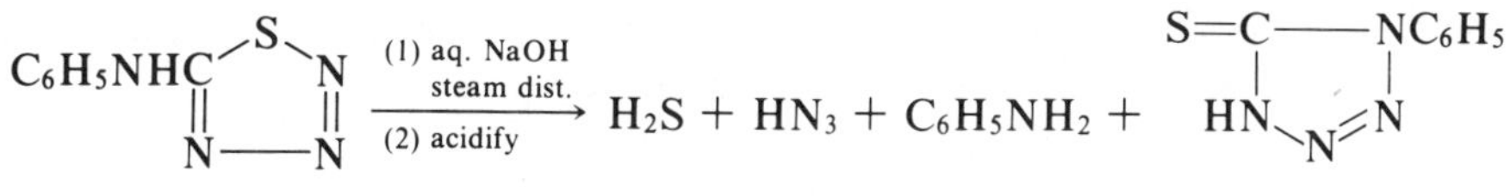

(XXXVII) 34%

not effect isomerization to 5-thionotetrazolines, reaction of 5-methylaminothiatriazole with sodium azide leads to a small yield of 1-methyl-5-thionotetrazoline (331). Many thiatriazolines and thiatriazoles decompose under basic conditions

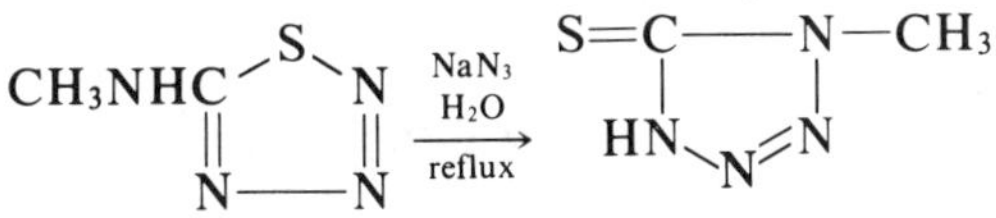

15%

to sulfur, nitrogen, and a compound containing the 5-substituent, 5-carbon, and sometimes an adjacent nitrogen. Thus 4-benzyl-5-tosyliminothiatriazoline **(XXXVIII)** is converted under the influence of sodium methylate to the elemen-

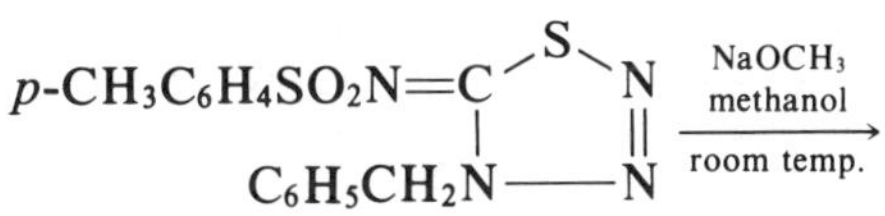

(XXXVIII)

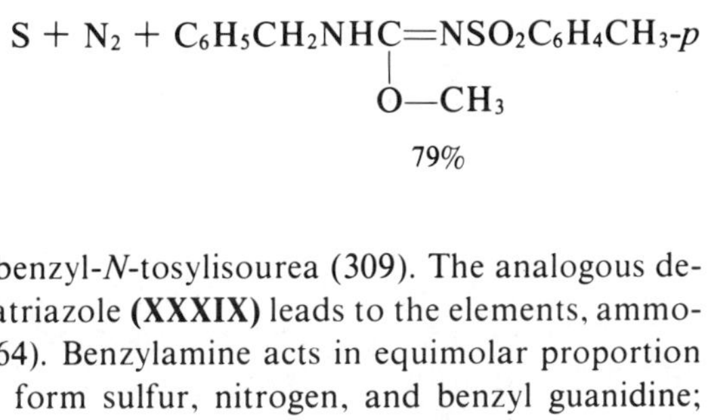

79%

tary products and *O*-methyl-*N*-benzyl-*N*-tosylisourea (309). The analogous decomposition of 5-allylaminothiatriazole **(XXXIX)** leads to the elements, ammonia, and allyl isothiocyanate (164). Benzylamine acts in equimolar proportion on 5-aminothiatriazole **(XL)** to form sulfur, nitrogen, and benzyl guanidine; with four equivalents of benzylamine the decomposition is that of the anion of

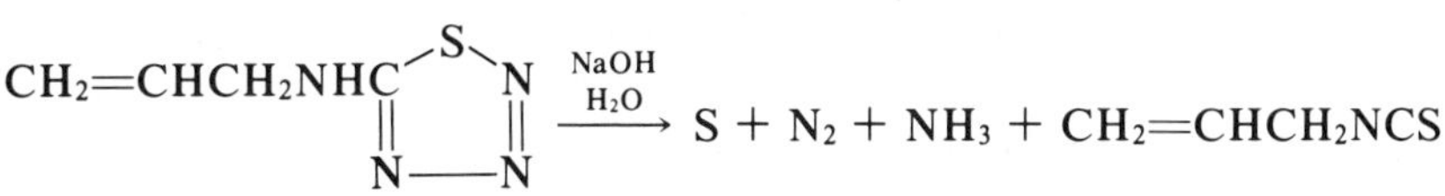

(XXXIX)

(XL) which ejects azide ion, forming benzylammonium azide together with benzylthiourea and *sym*-dibenzylthiourea (329).

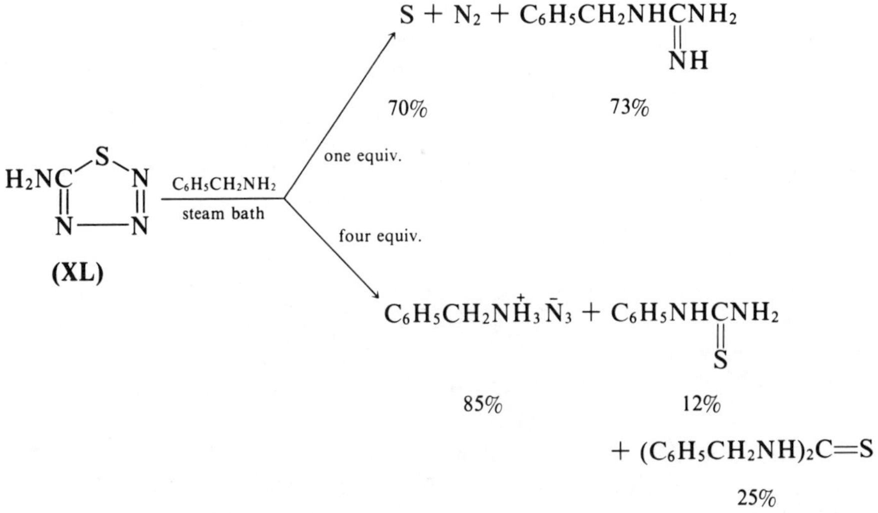

(XL)

The thiatriazole ring is destroyed in the reaction of 5-aryloxythiatriazoles with phosphorus triamides to form 1,3,4,5,2-thiatriazaphosphorines, discussed under Synthesis, Chapter 7. With triphenylphosphine, a weaker nucleophile, and 5-*p*-chlorophenoxythiatriazole, nitrogen, *p*-chlorophenylcyanate, and triphenylphosphine sulfide result (416).

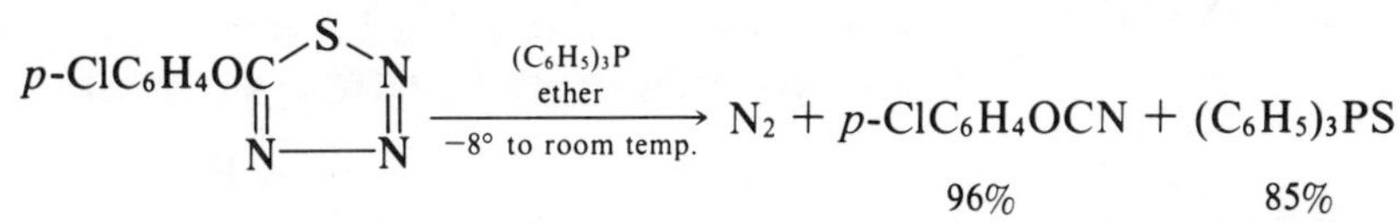

The stability of 5-monosubstituted tetrazole anions is reflected in the behavior of several such tetrazoles on heating with alkali. 5-Phenyltetrazole is stable to heating to dryness with potassium hydroxide to a final temperature of 250° (336, 337). 5-*p*-Tosylaminotetrazole is stable to refluxing in 1 *N* sodium hydroxide (272) and methyl-5-tetrazolyl sulfone is not decomposed by molten potas-

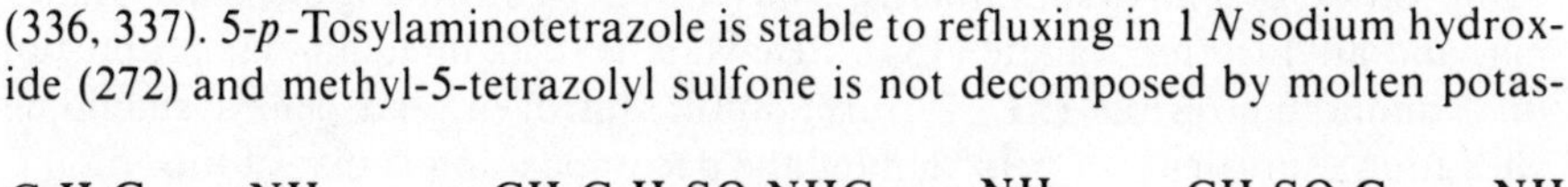

sium hydroxide (163). The tetrazole ring remains intact on alkaline fusion of potassium tetrazole-5-sulfonate to form tetrazolinone.

1-Phenyl-5-chlorotetrazole **(XLI)** undergoes nucleophilic displacement of

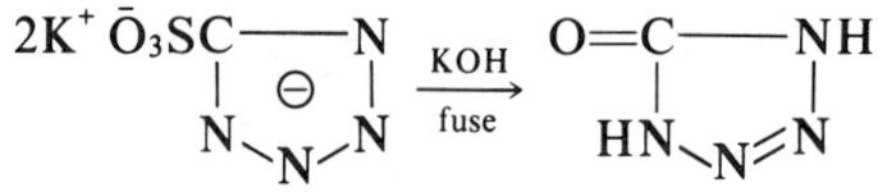

chloride with phenols (375) and azide (280). The tetrazole ring in **XLI**, however, is decomposed by butyllithium or magnesium, liberating nitrogen and forming

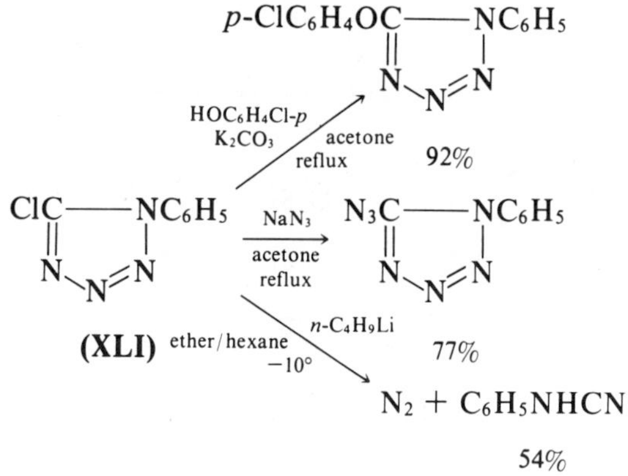

phenylcyanamide. Other 1-aryltetrazoles such as 1-*p*-nitrophenyltetrazoles (501) also are converted to cyanamides under alkaline conditions. 1-Methyltetrazole, however, hydrolyzes further to amines, ammonia, and carbon dioxide, but not formic acid (402). The hydrolysis of 5-amino-1-phenyltetrazole

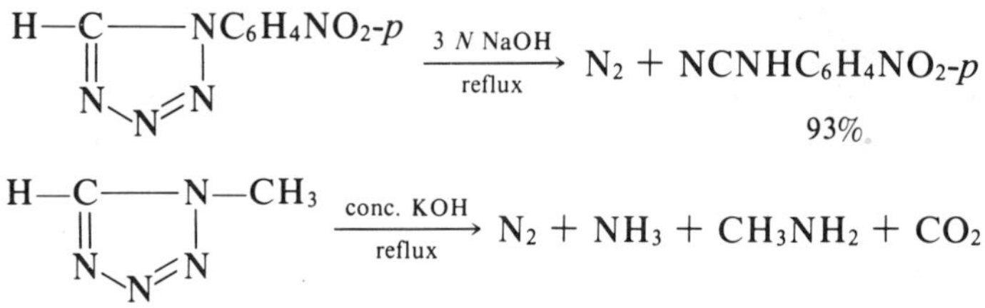

with sodium hydroxide to aniline, ammonia, sodium azide, and sodium carbonate suggests initial ring opening to *N*-phenylguanyl azide (233). 1-Methyl-4-

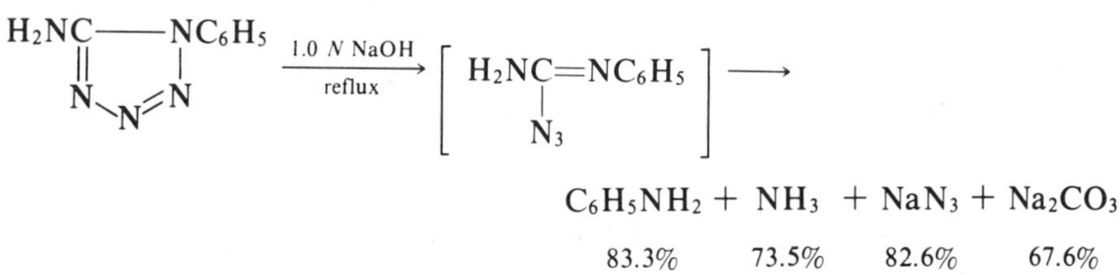

phenyltetrazolinone is split by concentrated alkali to phenyl azide and methylamine (462).

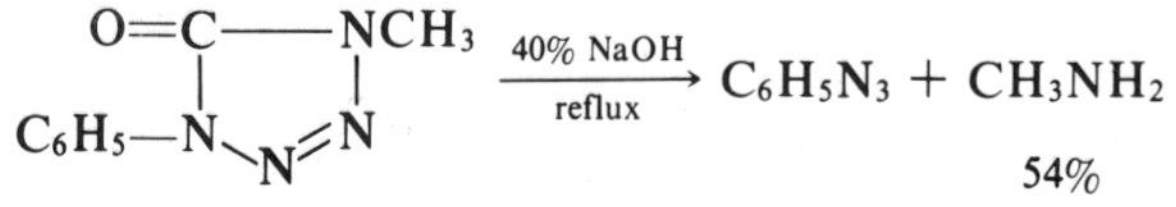

Heating 1,3-dimethyltetrazolium chloride **(XLII)** at 60° in deuterium oxide with potassium bicarbonate-carbonate buffer of pH 11.15 effects complete exchange of the 5-proton and partial exchange of the protons of the 3-methyl

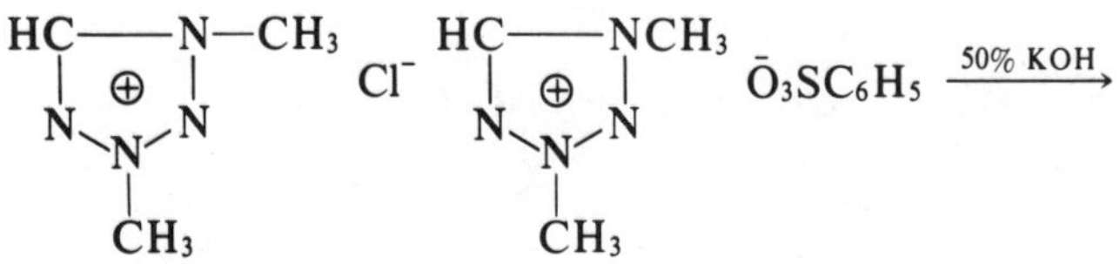

(XLII)

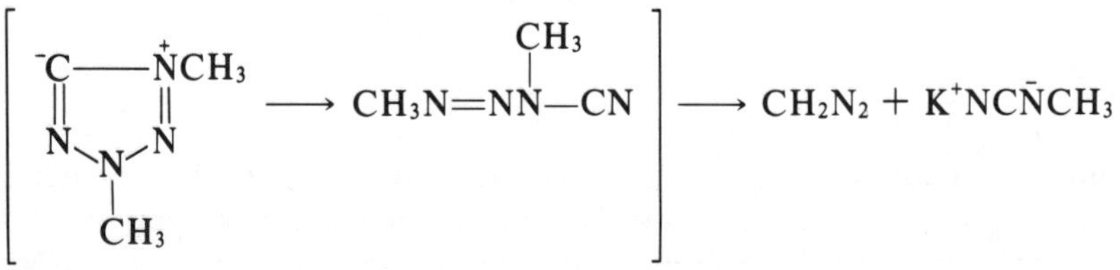

group (394). In concentrated potassium hydroxide the benzenesulfonate salt hydrolyzes to diazomethane and potassium methyl cyanamide, probably via a dipolar intermediate. Similarly the nitrogen chain of 1-methyl-3-phenyltetrazolium iodide cleaves to 3-cyano-3-methyl-1-phenyltriazene.

Under mild basic conditions, 1,4-disubstituted tetrazolium salts are converted to carbodiimides and nitrogen (568).

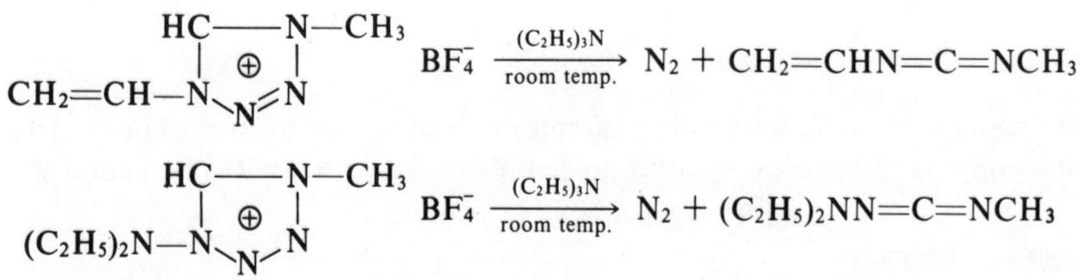

The fused *v*-triazole **(XLIII)** with an embedded nitrogen is recovered unchanged after heating under reflux with aqueous alkali (514), and the 5-nitrogen chain of 6-azidopyrazino[2,3-*d*]tetrazolo[4,5-*b*]pyridazine **(XLIV)** also with an embedded nitrogen, remains intact during replacement of the azide by methoxide (134). In 6-chloro-*s*-triazolo[1,5-*b*]pyridazine-3-oxide **(XLV)** the nitrogen

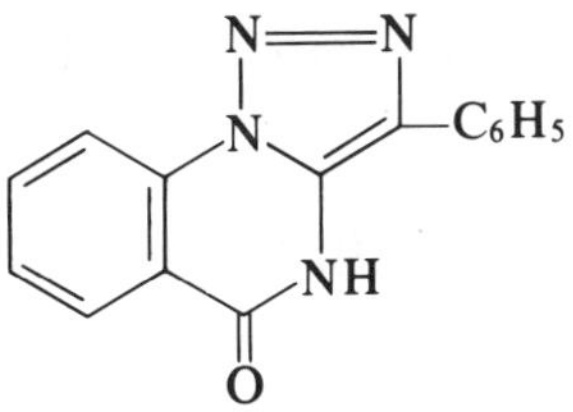

(XLIII)

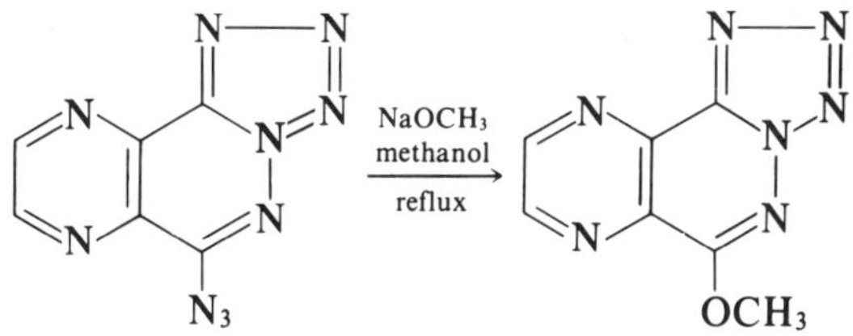

(XLIV) 75%

chain is at least partially stable toward hydrazine, but is broken by dilute alkali (80). Nitrogen is lost from benzimidazolo[1,2-*c*][1,2,3]benzotriazine on heating with alcoholic potassium hydroxide (482).

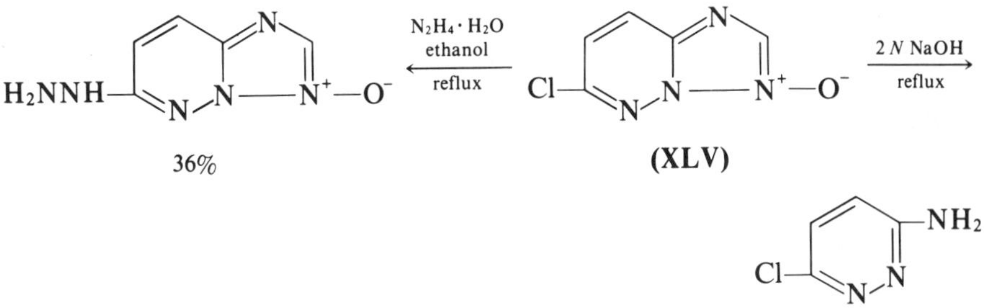

36% (XLV)

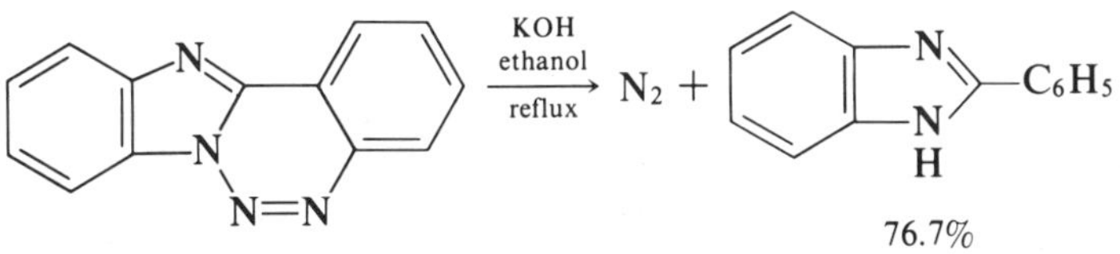

76.7%

6.4. OXIDATION

Oxidations in which nitrogen chains remain intact are considered first, followed by those in which the nitrogen sequence is broken without significantly affecting the attached radicals. Next oxidations are discussed in which organic and noncarbon substituents of high nitrogen compounds are consumed, with and without destruction of the high nitrogen moieties. Finally, reactions are summarized in which high nitrogen compounds function as oxidants.

Some reactions of high nitrogen compounds with oxidizing agents are known in which the nitrogen chain remains intact while being oxidized to a higher level, sometimes with accompanying rearrangement; attached organic groups are essentially unaffected. Phenyltriazene is oxidized quantitatively to phenyl azide by sodium hypobromite or alkaline ammoniacal silver ion (131).

$$C_6H_5N{=}NNH_2 \xrightarrow[-15^\circ]{NaOBr} C_6H_5N_3$$

N-Aminopyrazoles (399), indazoles (5), and pyrazolo heterocycles (184) rearrange during lead tetraacetate oxidation to 1,2,3-triazines.

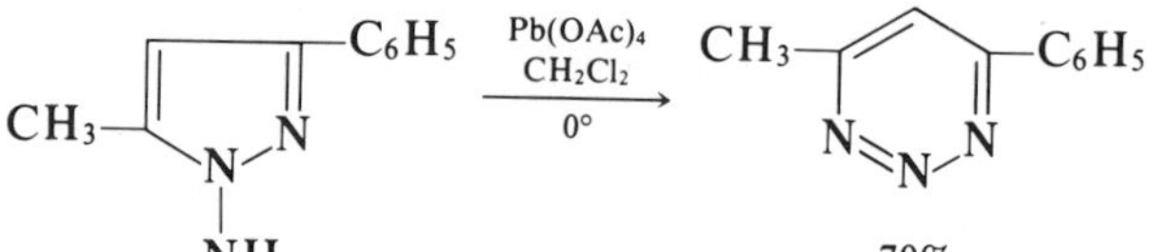

70%

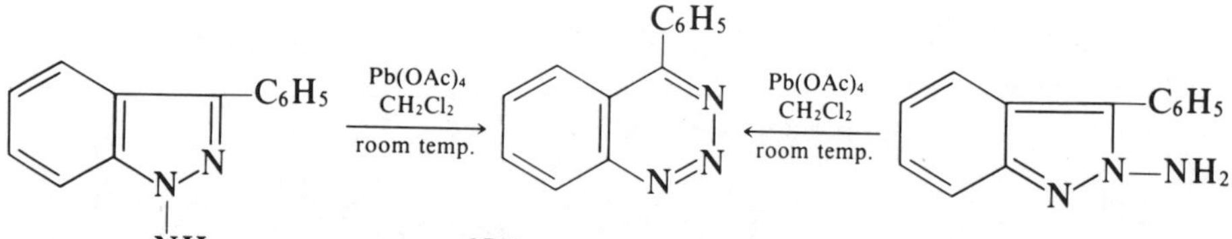

97% 90%

Dihydro-1,2,3-triazines are oxidized by air to 1,2,3-triazines (399).

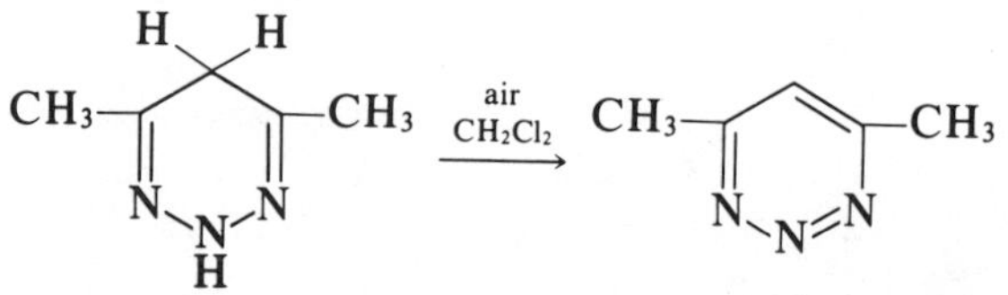

1,3-Diaryltriazene 1-oxides result from the action of perbenzoic acid on 1,3-diaryltriazenes (364). Trisubstituted triazenes do not react. If the polarities of the aryl substituents differ appreciably the isomer obtained is that with the oxide nearest the most electron-releasing group. Both N-1 and N-2 oxides of

$$C_6H_5N{=}NNHC_6H_4NO_2\text{-}p \xrightarrow[0^\circ]{C_6H_5CO_3H,\ \text{ether}} C_6H_5\overset{+}{N}(O^-){-}NNHC_6H_4NO_2\text{-}p$$

86%

monocyclic 1,2,3-triazines result from the action of *m*-chloroperbenzoic acid (399). Somewhat more vigorous conditions are required for peracetic acid to transfer an oxygen to 1-(2′-pyridyl)benzotriazole at N-3 (249).

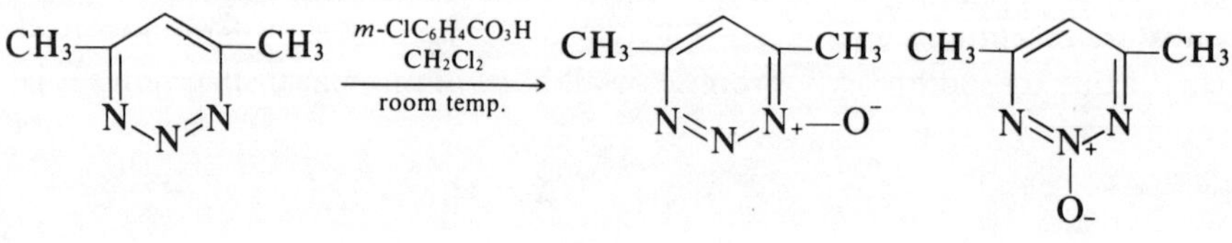

20% 48%

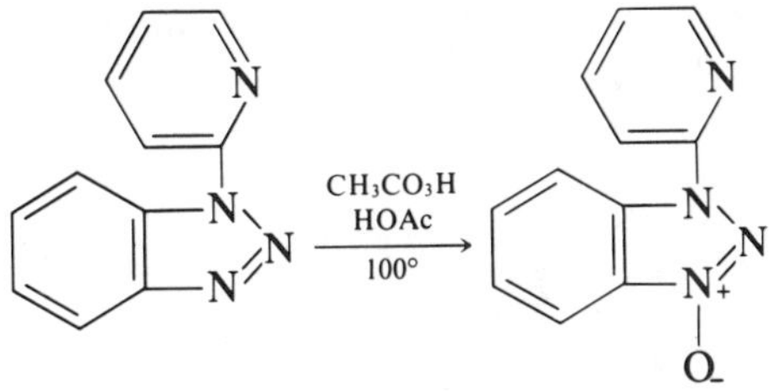

50%

Neither 2-alkyl nor 2-arylbenzotriazinium betaines are attacked by hydrogen peroxide, peracetic acid, or *m*-chloroperbenzoic acid (359).

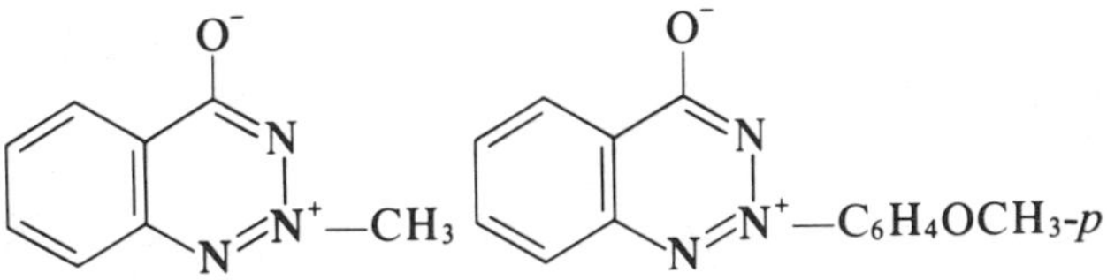

With other oxidants and conditions, nitrogen chains are broken. Nitrogen in triazenes and pentazdienes can be determined by oxidation with chromic-sulfuric acid; azo groups are liberated as elemental nitrogen and amines result from the remaining nitrogen (277). Electrochemical oxidation of diazoamino-benzene in acetonitrile is slow and diffusion controlled. It involves loss of one electron with formation of a cation radical which reacts with residual water forming a triazenium ion and oxygen, followed by rearrangement to *p*-amino-azobenzene (252). Lead tetraacetate converts 1,3-diaryltriazenes in aromatic

$$C_6H_5N{=}NNHC_6H_5 \xrightarrow{\text{electrolysis}} \bar{e} + [C_6H_4N{=}\dot{\overset{+}{N}}NHC_6H_5] \xrightarrow{H_2O}$$

$$O_2 + [C_6H_5N{=}N\overset{H}{\underset{H}{N^{+}}}C_6H_5] \longrightarrow C_6H_5N{=}NC_6H_4NH_2\text{-}p$$

solvents to azo compounds and biaryls (99). Hexazdienes are not formed, in contrast to the results with permanganate (517). Oxidative couplings such as the

$$C_6H_5N{=}NNHC_6H_5 \xrightarrow[40^\circ]{\text{Pb(OAc)}_4,\ \text{benzene}} N_2 + \underset{40.9\%}{C_6H_5N{=}NC_6H_5} + \underset{40\%}{C_6H_5C_6H_5}$$

conversions of triazenes to hexazdienes or tetrazenes to octazatrienes are discussed in Chapter 7, Synthesis. 1,3-Diphenyltriazene 1-oxide **(I)** is oxidized by silver oxide, 2,4,6-tri-*t*-butylphenoxyl, or other oxidizing agent through a triazene oxide radical, possibly located on oxygen, to nitrogen and nitrosobenzene, phenyl radical, and diphenyl nitrogen oxide radical, believed to have arisen by combination of phenyl radical with nitrosobenzene (1).

$$C_6H_5\overset{+}{N}(O^-){=}N{-}NHC_6H_5 \xrightarrow[\text{room temp.}]{Ag_2O,\ \text{benzene}} [C_6H_5{-}N(O^\bullet){-}N{=}NC_6H_5] \longrightarrow$$

(I)

$$N_2 + C_6H_5NO + [C_6H_5\cdot] \longrightarrow [(C_6H_5)_2NO\cdot]$$

$$C_6H_5NO \xleftarrow[H^+]{HgO} C_6H_5{-}N(NO){-}NH_2 \xrightarrow{AgNO_3} C_6H_5N_3$$

(II)

Nitrosobenzene is formed from nitrosophenylhydrazine **(II)** by mercuric oxide in acid solution; phenyl azide results from oxidation of **(II)** with silver nitrate, cupric acetate, and neutral mercuric oxide (36). Nitrogen is the only gaseous product from oxidation of phenyl, *n*-propyl, and benzylnitrosohydrazines with potassium permanganate (316). In neutral solution, ferric chloride forms both nitrogen and nitrous oxide, whereas iodine forms nitrogen, nitrous oxide,

$$\underset{47.0\%}{NO} + \underset{13.2\%}{N_2O} + \underset{39.8\%}{N_2} \xleftarrow[H_2O]{I_2} C_6H_5CH_2N(NO){-}NH_2 \xrightarrow[H_2O]{FeCl_3} \underset{65.6\%}{N_2} + \underset{34.5\%}{N_2O}$$

and nitric oxide. The nitroso group of 1-*p*-bromophenyl-1-nitroso-2-phenylmethylenehydrazine is oxidized by oxygen to nitro with simultaneous rearrangement from nitrogen to carbon (151).

$$C_6H_5CH{=}N{-}N(NO)C_6H_4Br\text{-}p \xrightarrow[\text{room temp.}]{O_2,\ \text{toluene}} C_6H_5C(NO_2){=}NNHC_6H_4Br\text{-}p$$

Phenyl azide in hot carbon tetrachloride decomposes rapidly in the presence of benzoyl peroxide, but hardly at all in its absence; nitrogen and *p*-chlorophenyliminomethylene dichloride are formed (319).

$$C_6H_5N_3 \xrightarrow[79.8^\circ]{(C_6H_5CO)_2O_2,\ CCl_4} \underset{104\%}{N_2} + \underset{67\%}{p\text{-}ClC_6H_4N{=}CCl_2}$$

Decomposition of a mixture of benzenesulfonyl azide and *t*-butylhydroperoxide in *p*-xylene produces benzenesulfonyl *p*-xylidide almost quantitatively

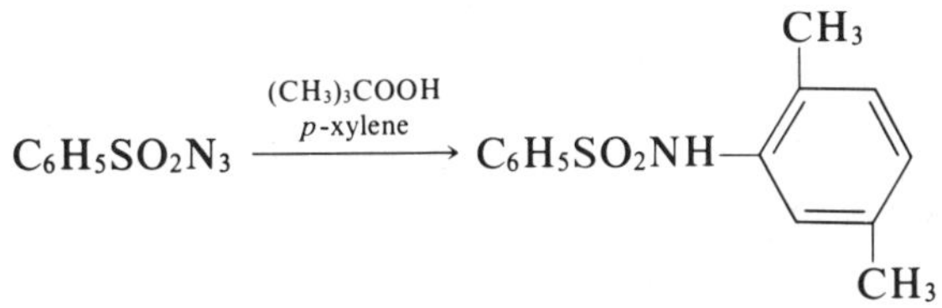

(321). Hydrogen peroxide converts 4-azido-2-picoline **(II)** to 4,4′-azoxybis-2-picoline 1,1′-dioxide (268).

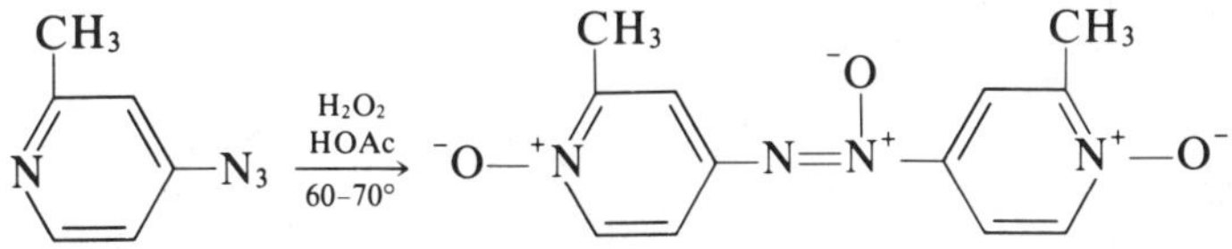

(II)

Oxidation of 2-aminoindazole with lead tetraacetate in benzene leads to indazole, and with mercuric oxide to 2,2′-biindazole (450).

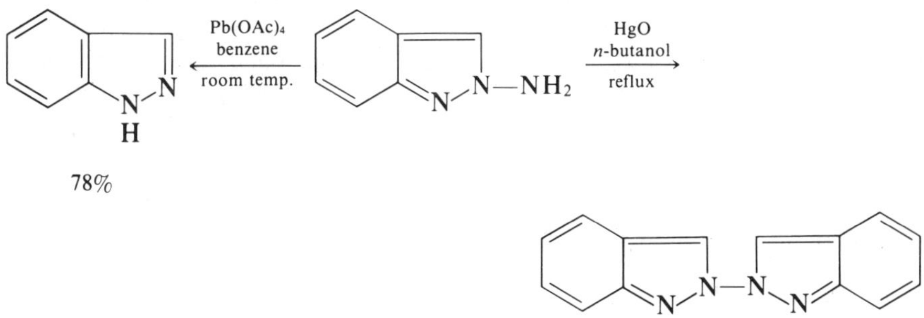

78%

34%

Tetrabenzyltetrazene is oxidized by lead tetraacetate to benzyl azide and benzaldehyde (290).

$$(C_6H_5CH_2)_2NN{=}NN(CH_2C_6H_5)_2 \xrightarrow[5\text{–}10^\circ]{Pb(OAc)_4,\ \text{benzene}} \underset{6.4\%}{C_6H_5CH_2N_3} + \underset{20\text{–}25\%}{C_6H_5CHO} + \underset{10\%}{(C_6H_5CH_2)_2NH} + \underset{13\%}{C_6H_5CH_2OH}$$

The nitrogen chain remains intact on oxidation of 1,4-dimethyl-1,4,5,6-tetrahydro-*v*-tetrazine 2-oxide **(III)** with potassium permanganate to 1,4-dimethyltetrazolinone 2-oxide or with manganese dioxide to 1,4-dimethyl-5-oxo-1,4,5,6-tetrahydro-*v*-tetrazine 2-oxide (467). Hydrogen peroxide, however, splits the nitrogen chain of **III** to *N,N′*-dimethyl-*N,N′*-dinitroso-1,2-diaminoethane.

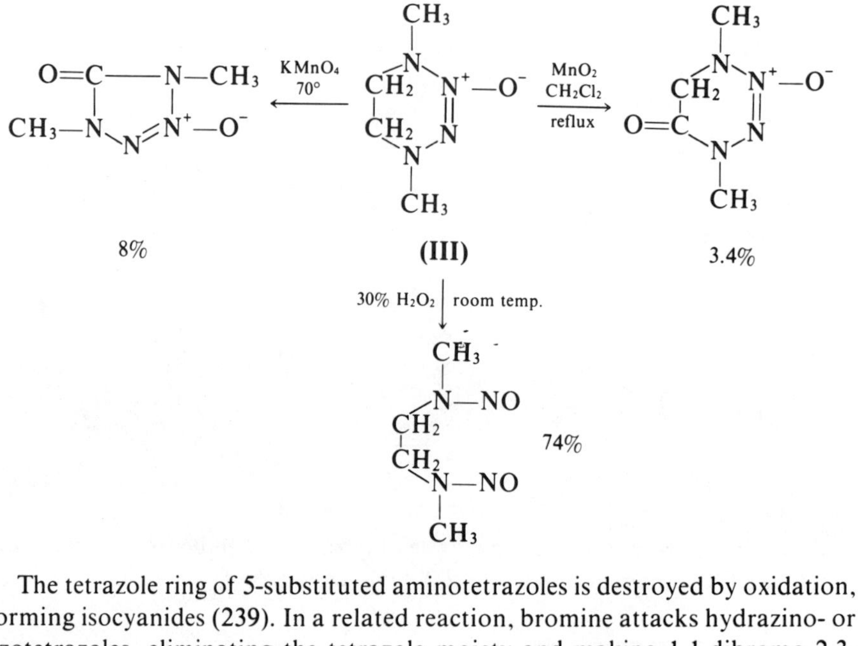

8% (III) 3.4% 74%

The tetrazole ring of 5-substituted aminotetrazoles is destroyed by oxidation, forming isocyanides (239). In a related reaction, bromine attacks hydrazino- or azotetrazoles, eliminating the tetrazole moiety and making 1,1-dibromo-2,3-diazabutadiene derivatives, shown with *p*-nitrophenylmethylene-5-tetrazolyl-hydrazone (110, 398, 464) and 5,5′-hydrazotetrazole **(IV)** (518). One or two tetra-

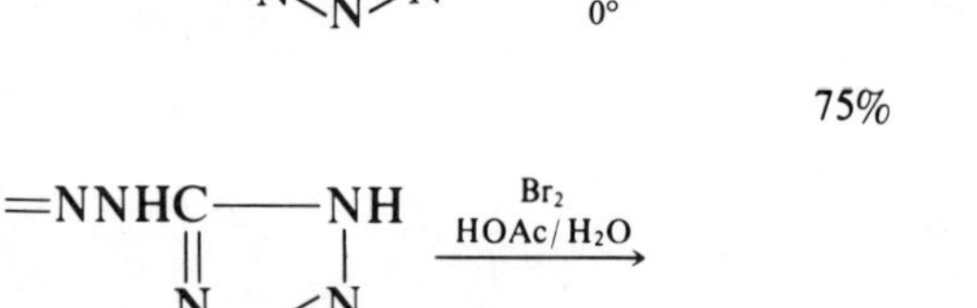

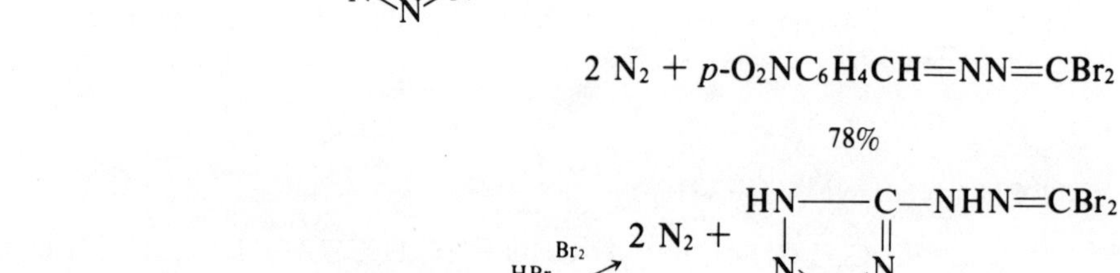

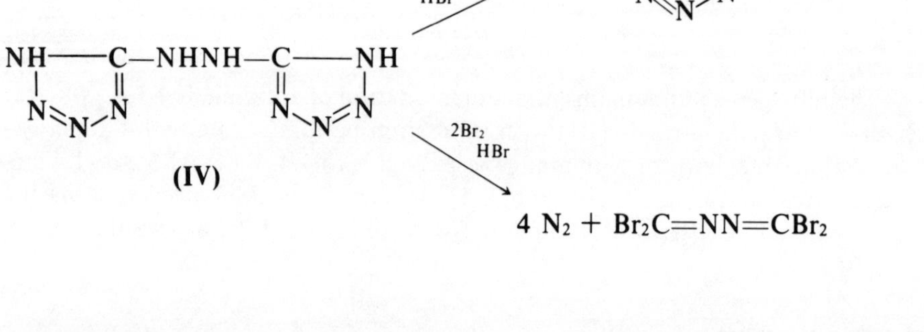

(IV)

zole rings are decomposed in **IV** depending on the quantity of bromine used. Chlorination of arylmethylene hydrazinotetrazoles yields trichlorodiazabutadienes, and tribromo derivatives also can be obtained by further bromination of tetrazolylhydrazidic bromides (398).

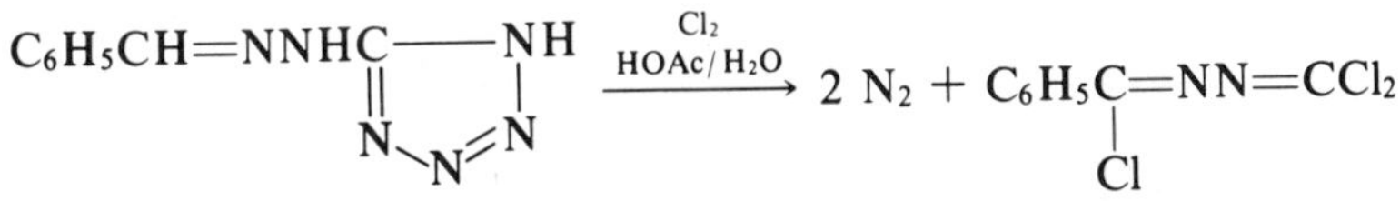

62%

Oxidation of fused ring 1-amino-*v*-triazoles is an elegant method for preparing arynes. Thus 1-aminobenzotriazole **(V)** is oxidized by lead tetraacetate, evolving 2 moles of nitrogen to form benzyne in nearly quantitative yield (101). When no trapping agent is present, biphenylene results. *N*-Bromosuccinimide and nickel peroxide also generate benzyne from **V**.

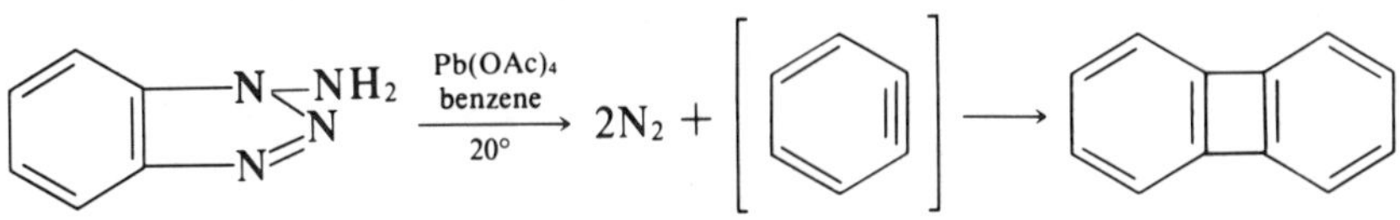

(V) 79%

Some other arynes prepared from 1-amino-*v*-triazoles by this reaction are benzynequinone (437), pyridyne (156), and 9,10-phenanthryne (40). Cycloalkynes such a cycloheptyne also can be prepared *in situ* from the corresponding

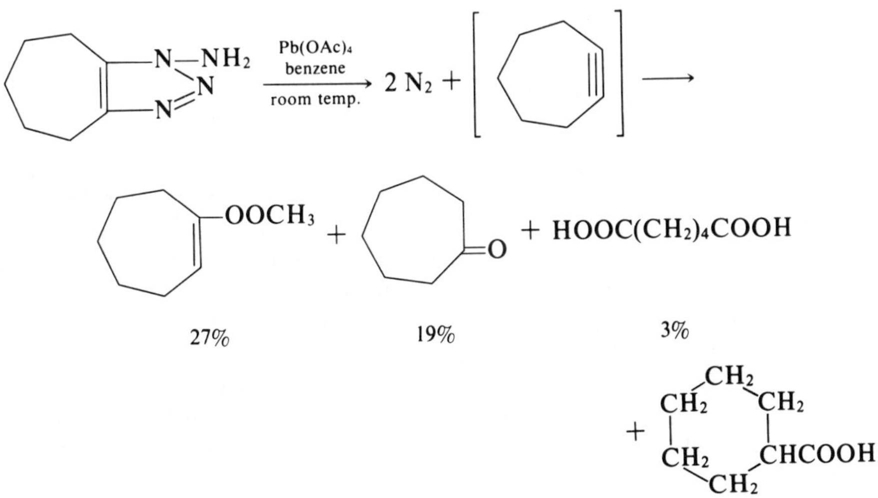

27% 19% 3% 27%

1-amino-*v*-triazoles (555, 556). Cycloheptyne is stable at −76° for a few hours. In a related oxidation, 1-aminonaphtho[1,8-*de*]triazine **(VI)** forms 1,8-dehydronaphthalene which undergoes 1,2-addition to olefins, acetylenes, and aromatic compounds (433, 434).

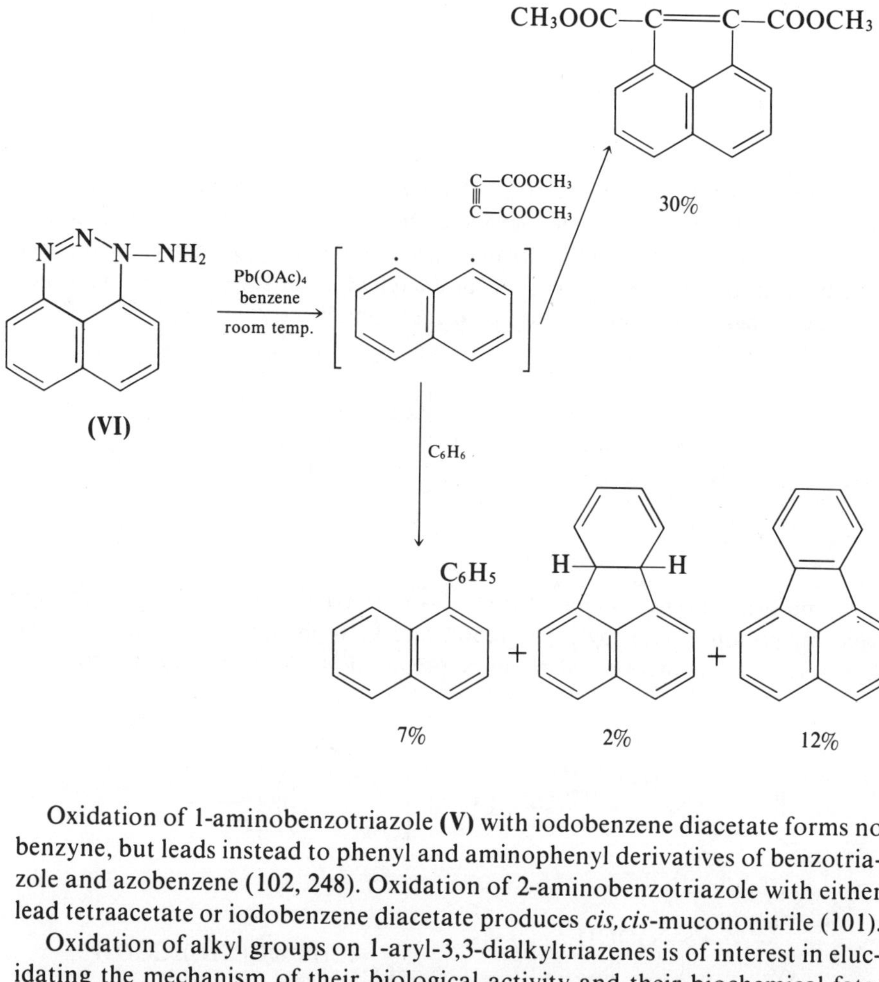

Oxidation of 1-aminobenzotriazole **(V)** with iodobenzene diacetate forms no benzyne, but leads instead to phenyl and aminophenyl derivatives of benzotriazole and azobenzene (102, 248). Oxidation of 2-aminobenzotriazole with either lead tetraacetate or iodobenzene diacetate produces *cis,cis*-muconontrile (101).

Oxidation of alkyl groups on 1-aryl-3,3-dialkyltriazenes is of interest in elucidating the mechanism of their biological activity and their biochemical fate. Under the influence of an hydroxylase, degradation is on the intact triazene, involving hydroxylation of alkyl (Section 5.3, Toxicity) and its elimination as aldehyde. Under the influence of a microsomal fraction from rat liver, together with oxygen and a nicotinamide adenine diphosphate (NADPH) generating system *in vitro,* formaldehyde is formed from triazenes carrying at least one methyl group and acetaldehyde is formed from 3,3-diethyl-1-(3′-pyridyl)triazene (422). Formation of formaldehyde by the *in vitro* oxidation of aryldimethyltriazene is inhibited by aryl diazonium ions produced by hydrolysis of the tri-

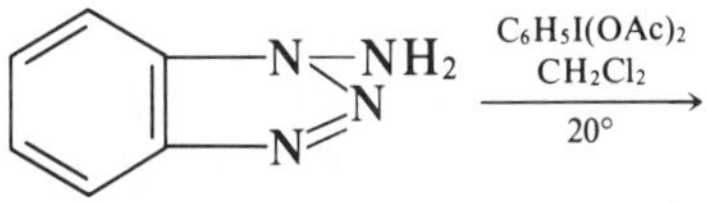

(V)

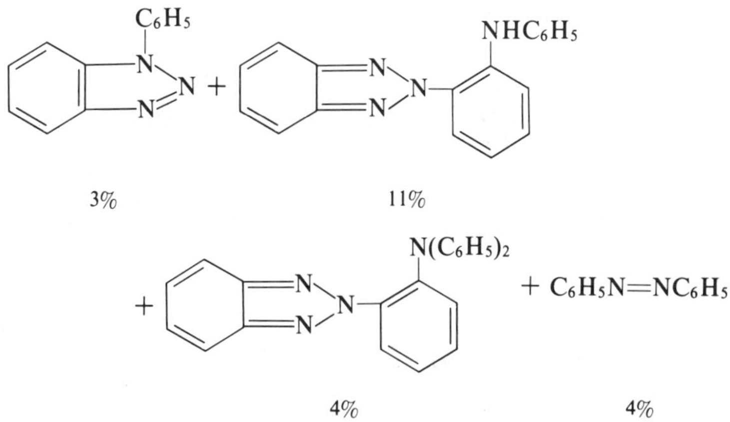

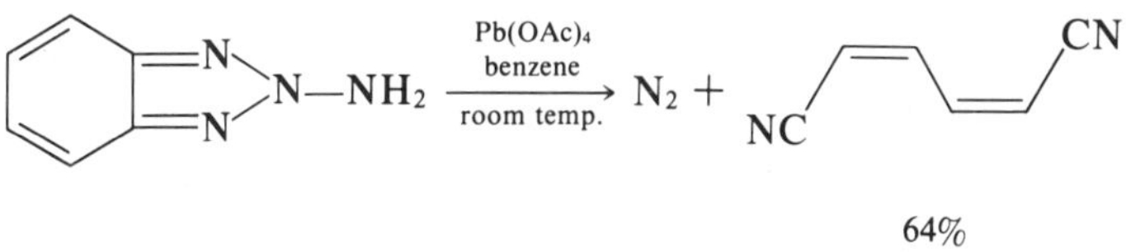

substituted triazene and by monomethyl aryltriazene produced by the oxidative demethylation, consistent with the suggested pathway (186).

$$C_6H_5N{=}NN(CH_3)_2 \xrightarrow[37^\circ]{O_2;\ \text{microsomal fraction, NADPH}} CH_2O + C_6H_5NH_2 + [C_6H_5N{=}NNHCH_3]$$

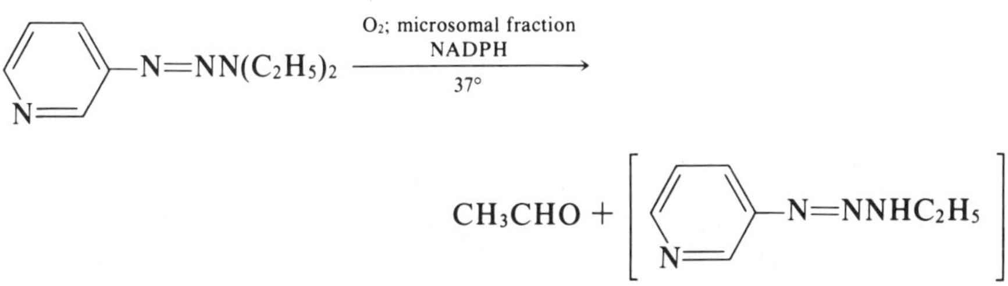

In a nonbiological system, molecular oxygen in the presence of ferrous ethylenediaminetetraacetic acid oxidizes 3,3-dimethyl-1-phenyltriazene forming aniline, probably through monomethylphenyltriazene (571).

Tetramethyltetrazene **(VII)** is oxidized to 1,1,4-trimethyl-4-formyltetrazene by one equivalent of potassium permangante and to 1,4-diformyl-1,4-dimethyltetrazene with two equivalents (522). Tetranitromethane converts **VII** to 1,1,4-trimethyl-4-β,β-dinitrovinyl-2-tetrazene, probably through the cation radical of **VII** (523, 526). Similarly azodicarboxylic esters condense quantitatively with a methyl group of **VII** (141). Nitrosation of **VII**, optimally at pH 2, effects demethylation, forming 1,1,4-trimethyl-4-nitroso-2-tetrazene and, with partial fission of the nitrogen chain, nitrosotrimethylhydrazine (363).

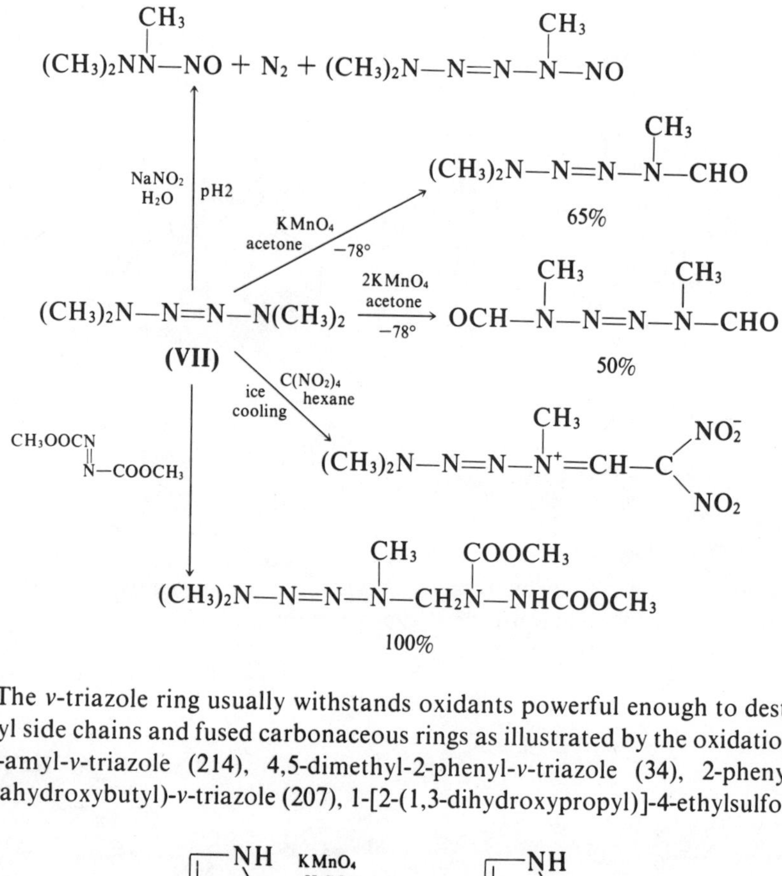

The *v*-triazole ring usually withstands oxidants powerful enough to destroy alkyl side chains and fused carbonaceous rings as illustrated by the oxidation of 4-*n*-amyl-*v*-triazole (214), 4,5-dimethyl-2-phenyl-*v*-triazole (34), 2-phenyl-4-tetrahydroxybutyl)-*v*-triazole (207), 1-[2-(1,3-dihydroxypropyl)]-4-ethylsulfonyl-

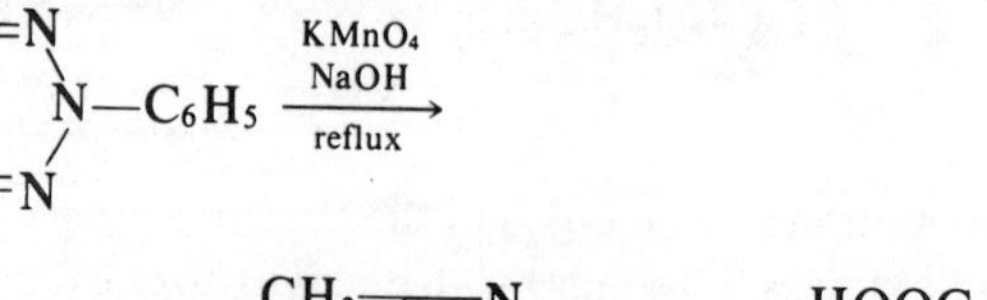

(214)

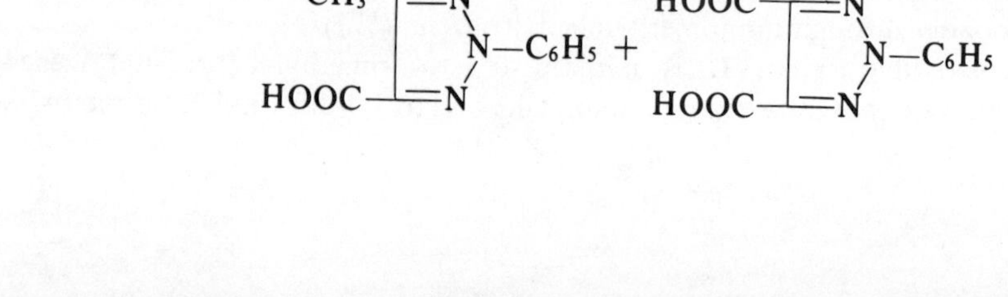

(34)

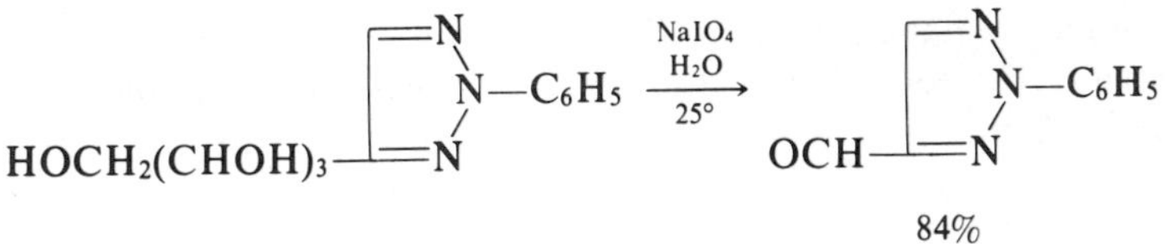

(207)

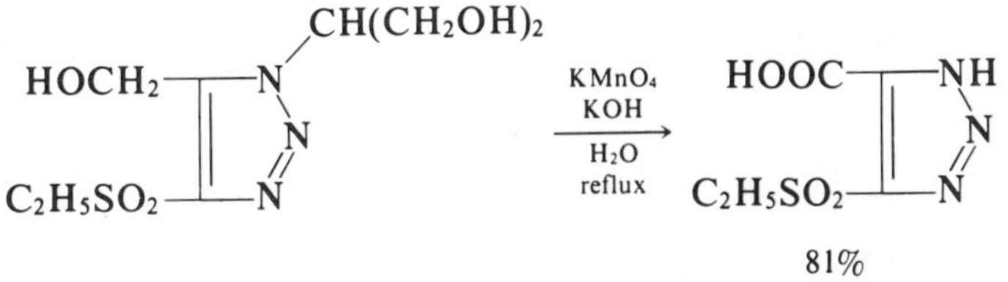

(148)

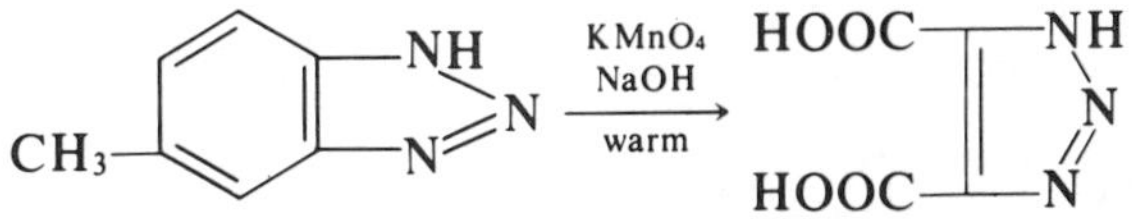

(69)

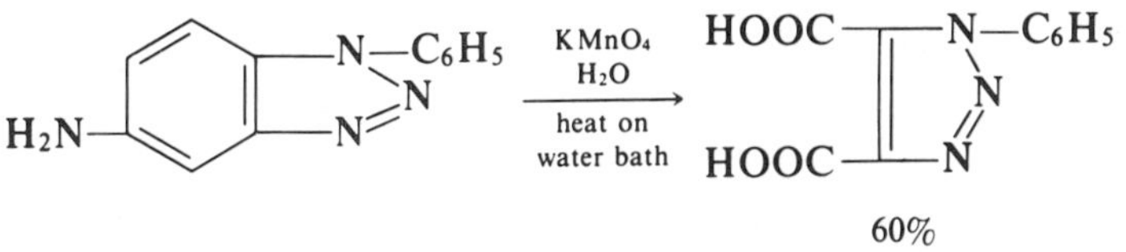

(165)

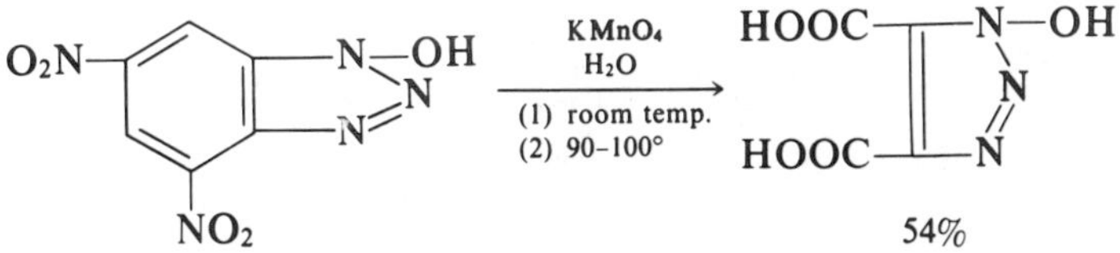

(262)

5-hydroxymethyl-*v*-triazole (148), 5-methylbenzotriazole (69), and 5-amino-1-phenylbenzotriazole (165). Even the 1-hydroxy-*v*-triazole moiety remains intact on permanganate oxidation of 1-hydroxy-4,6-dinitrobenzotriazole, despite the nitro groups on benzene which usually inhibit oxidative destruction (262). Unsubstituted phenyl attached to *v*-triazole nitrogen usually is not attacked by oxidizing agents short of complete oxidative consumption of the molecule. The presence of hydroxyl on *v*-triazole carbon renders the triazole nitrogen chain susceptible to destruction (129). Hot dilute sulfuric acid oxidizes 3-phenyloxatriazolium-5-thiolate forming nitrogen dioxide and phenol (206). In the first synthesis of tetrazole, Bladin oxidized 2-(aminophenyl)tetrazole-5-carboxylic

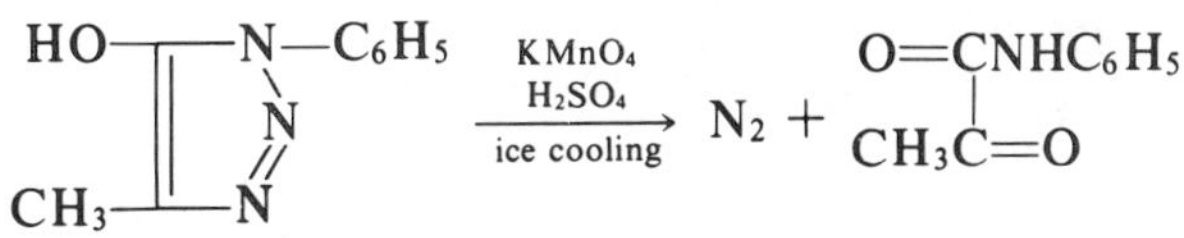

acid with alkaline permanganate (68). The carbon attached to tetrazole nitrogen in tetrazolo[5,1-*a*]isoquinoline also is oxidized by alkaline permanganate (441). Embedded nitrogen chains survive oxidation in the case of 3-

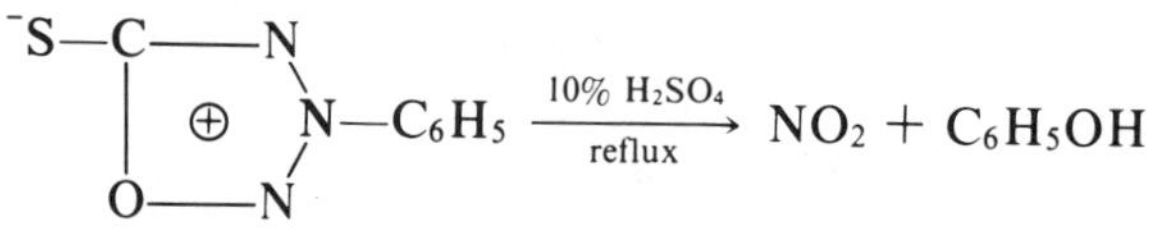

48%

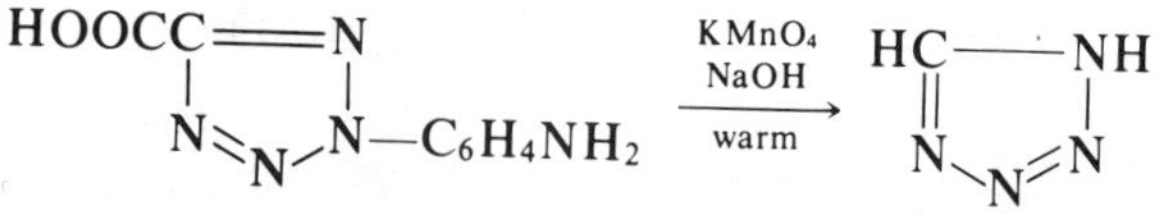

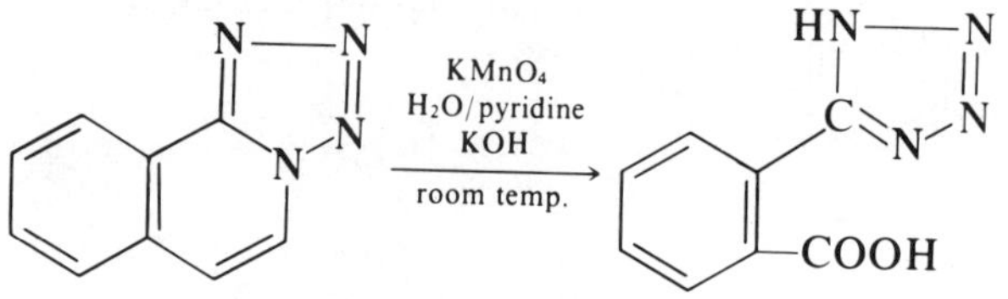

54%

phenylpyrido[1,2-*e*]triazole (440), the fused heteropentalene **(VIII)** (8), and 2,6-dimethyl-3,7-diformyltetraazapentalene (415).

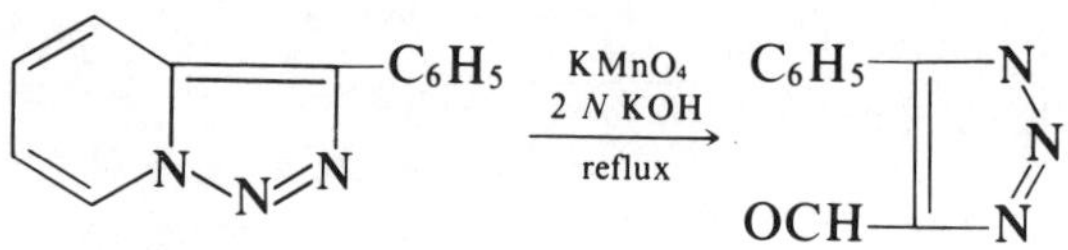

70%

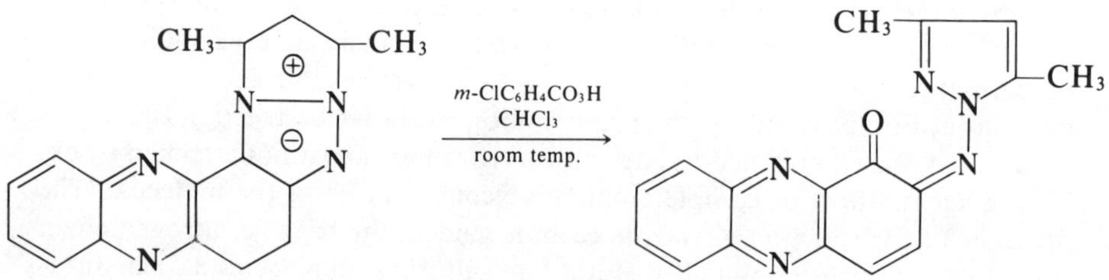

(VIII)

85%

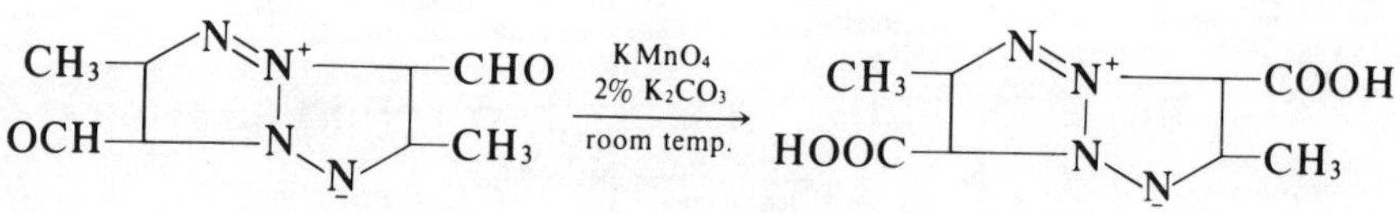

42%

The three-nitrogen chain of 2-phenyl-*s*-triazolo[2,3-*b*]pyridazine is broken during oxidation by permanganate (570). Although the four-nitrogen chain of dibenzo-1,3a,4,6a-tetraazapentalene is stable to permanganate, it is cleaved by peracetic acid to 2-(*o*-nitrosophenyl)benzotriazole (106).

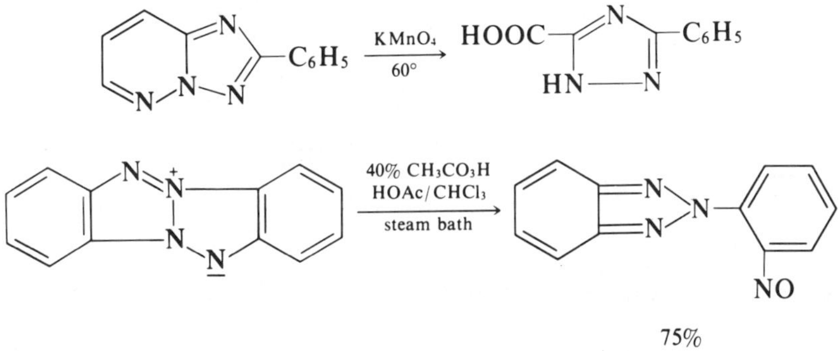

v-Triazolines are oxidized to *v*-triazoles by removal of hydrogen at the 4,5-positions (260). Good yields are obtained by use of phase transfer catalysts, as illustrated with 1-*p*-tolyl-5-(4′-pyridyl)-*v*-triazoline (279). In a related oxidation,

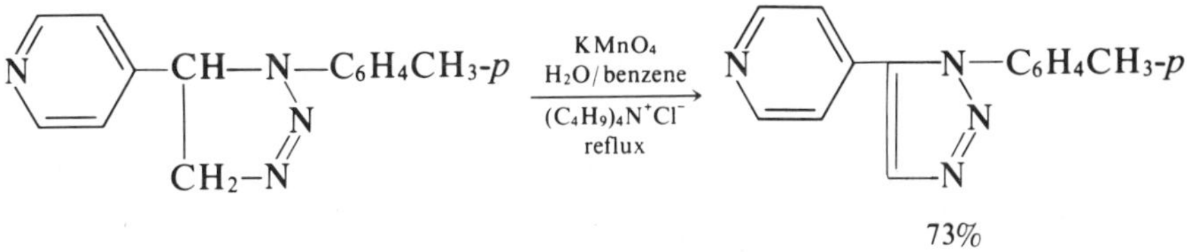

the triazoline-diazoamine system, predominantly in the acyclic form, is converted, presumably through a diazoimine, to the favored *v*-triazole (516).

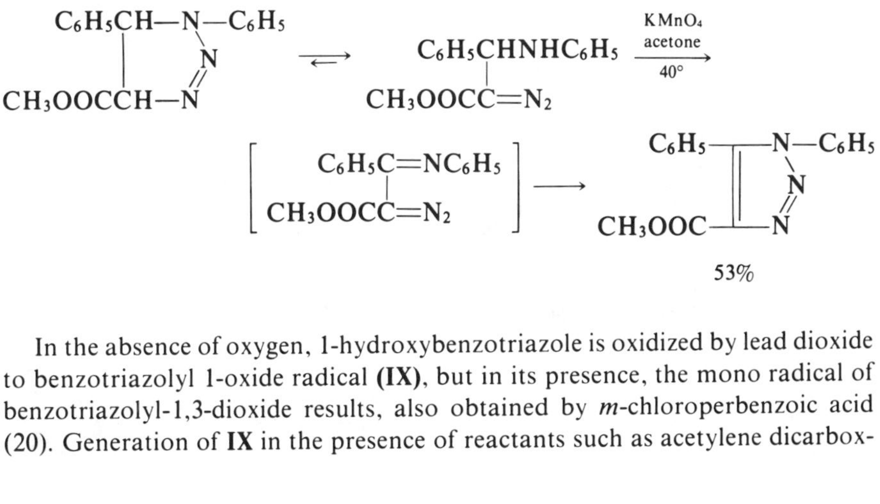

In the absence of oxygen, 1-hydroxybenzotriazole is oxidized by lead dioxide to benzotriazolyl 1-oxide radical **(IX)**, but in its presence, the mono radical of benzotriazolyl-1,3-dioxide results, also obtained by *m*-chloroperbenzoic acid (20). Generation of **IX** in the presence of reactants such as acetylene dicarbox-

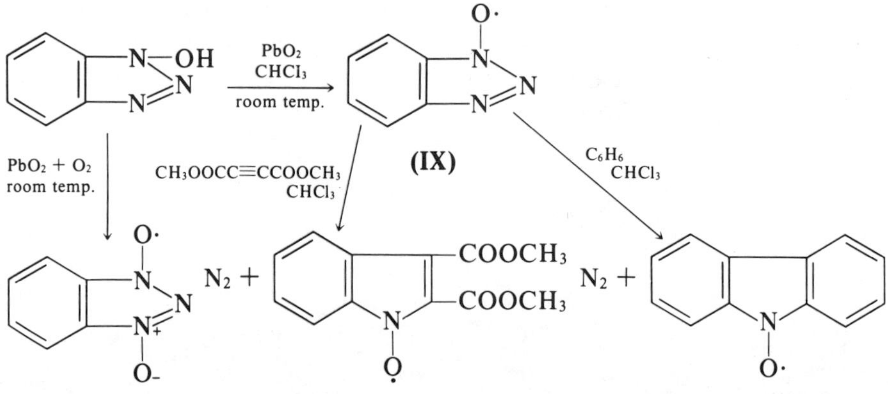

ylic ester or benzene leads to new radicals with loss of nitrogen from the benzotriazole ring.

5-Phenylthiatriazole is resistant to peroxide (480) and peracetic acid (287). Thiatriazole-5-thiolate ion **(X)** is oxidized by iodine in potassium iodide in a

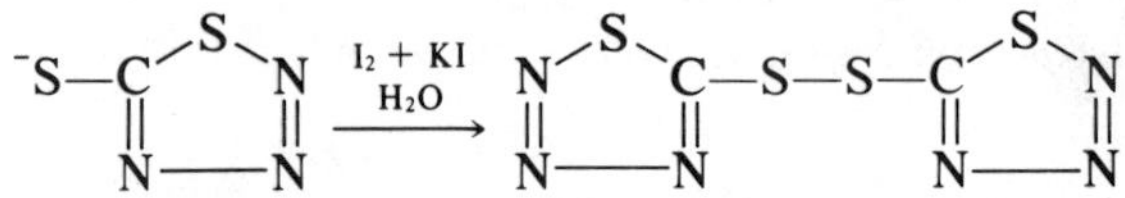

(X)

two-electron change to the disulfide (84). Potassium dichromate oxidizes **X** beyond the disulfide stage in an 8-electron reaction, acid permanganate effects a 12-electron, and alkaline permanganate a 17-electron reaction (386). None of these more extensive oxidations are of sufficient precision and reliability for use in analysis.

5-Tetrazoline thione is oxidized by dilute nitric acid to tetrazole and by per-

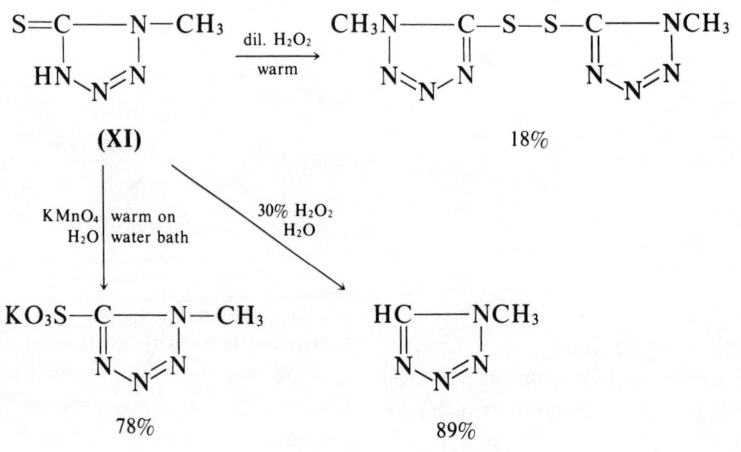

manganate to 5-tetrazole sulfonic acid (163). 1-Methyl-5-tetrazoline-thione **(XI)** loses its sulfur with 30% hydrogen peroxide, forms a disulfide with dilute peroxide, and is oxidized to a sulfonic acid by permanganate (501). Chromic acid also effects desulfurization of 5-tetrazoline thiones (280). 1-Phenyl-5-methylthiotetrazole is oxidized to the sulfone by permanganate (501).

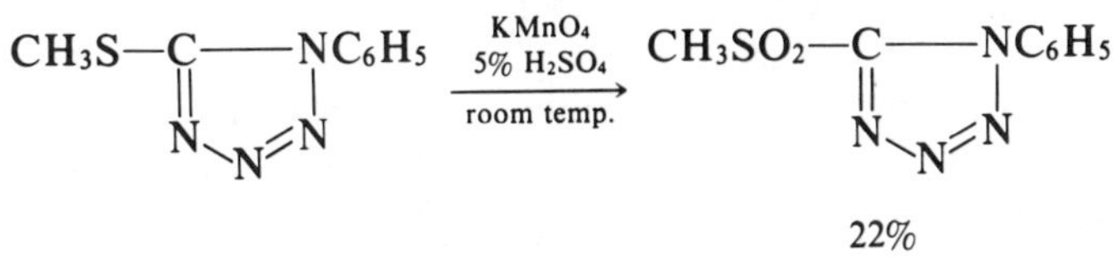

Alkaline permanganate converts 5-aminotetrazole to azotetrazole salts (518).

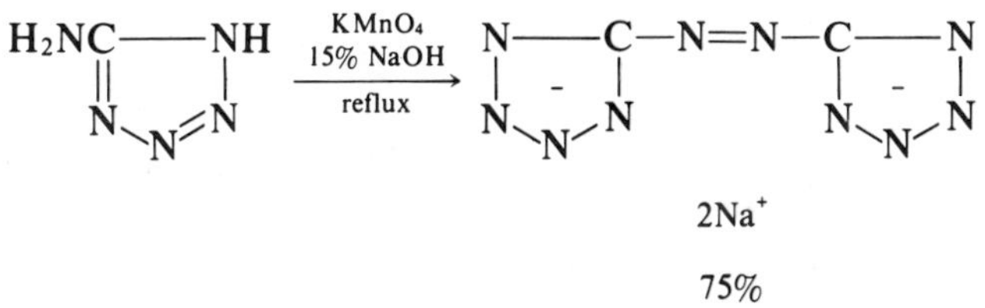

1-Chlorobenzotriazole **(XI)** oxidizes alcohols to aldehydes and ketones, hydrazo to azo compounds, and 1-amino-4,5-diphenyl-*v*-triazole to diphenylacetylene, simultaneously being converted to benzotriazole hydrochloride (435). 1-Bromobenzotriazole has similar oxidant properties (452).

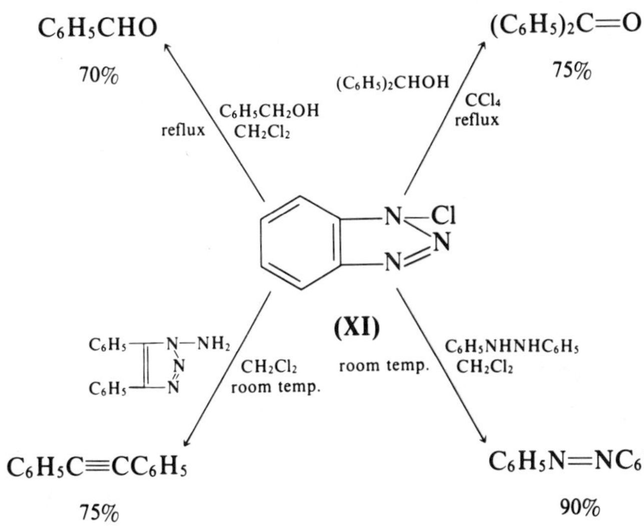

6.5. REDUCTION

The range of behavior of high nitrogen compounds toward reducing agents includes fragmentation to products with fewer nitrogen atoms, reduction of the nitrogen sequence to a lower oxidation level without breaking the chain, reduction of other groups in the molecule leaving the contiguous nitrogens unchanged, and total resistance to reductants.

The nitrogen chain of triazanes and triazanium salts is reduced by hydrogen iodide to ammonia and a hydrazine derivative, illustrated by the decomposition of 2,2-diethyltriazanium betaine-1-carbonylanilide (460). Titration of liberated iodine enables estimation of the triazane (197).

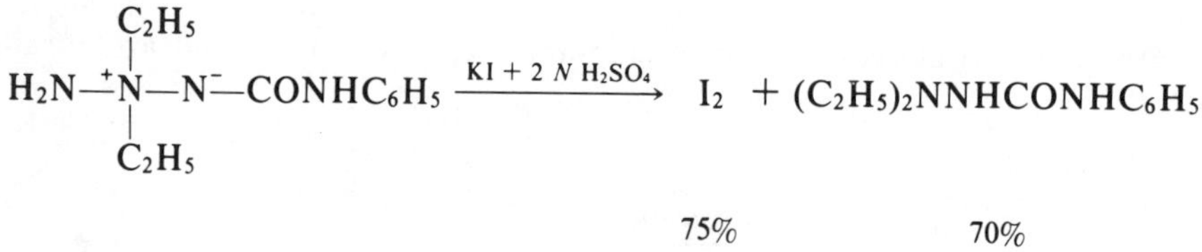

Triazenes reduce to an amine and a hydrazine. Disubstituted triazenes with the same group at positions 1 and 3 form only one amine and one hydrazine, as shown with the bis-*s*-triazolyltriazene **(I)** (179), whereas with an unsymmetrical

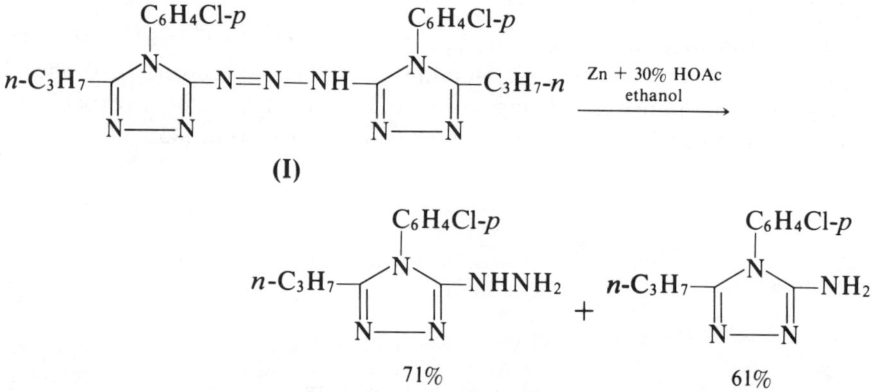

diazoamino compound such as 1-phenyl-3-*p*-tolyltriazene (391), two amines and two hydrazines are obtained. 3,3-Dimethyl-1-phenyltriazene is reduced po-

$$C_6H_5N{=}NNHC_6H_4CH_3\text{-}p \xrightarrow[\text{cool in ice water}]{SnCl_2 + \text{conc. HCl}}$$

$$C_6H_5NH_2 + p\text{-}CH_3C_6H_4NH_2 + C_6H_5NHNH_2 + p\text{-}CH_3C_6H_4NHNH_2$$

$$C_6H_5N{=}NN(CH_3)_2 \xrightarrow[\text{polarograph}]{[H]} C_6H_5NH_2 + H_2NN(CH_3)_2$$

100%

larographically to aniline and dimethylhydrazine with an uptake of four electrons (283). In the reduction of 1-(*o*-cyanophenyl)-3-methyl-3-phenyltriazene with acidic stannous chloride, hydrolysis to monomethylaniline and *o*-cyanobenzenediazonium ion probably precedes reductive cyclization to 3-aminoindazole (494). 1-*p*-Bromophenyl-3-methyl-3-phenyltriazene 1-oxide is reduced to

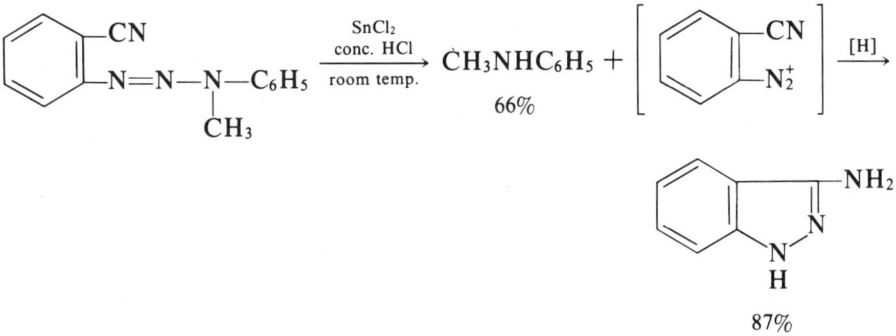

the corresponding triazene, *p,p'*-dibromoazobenzene, *p*-bromoaniline, methylaniline, and ammonia (37). The system of 1,1,4,4-tetraphenyl-2,3-dibenzoyltetra-

$$p\text{-}BrC_6H_4\overset{+}{N}(O^-){=}NN(CH_3){-}C_6H_5 \xrightarrow{Zn\,+\,HOAc}$$

$$p\text{-}BrC_6H_4N{=}NN(CH_3){-}C_6H_5 + p\text{-}BrC_6H_4N{=}NC_6H_4Br\text{-}p + p\text{-}BrC_6H_4NH_2 + CH_3NHC_6H_5 + NH_3$$

zane and its derived radical is converted by hydroquinone to 1,1-diphenyl-2-benzoylhydrazine (192). Aluminum amalgam splits the bond between N-2 and

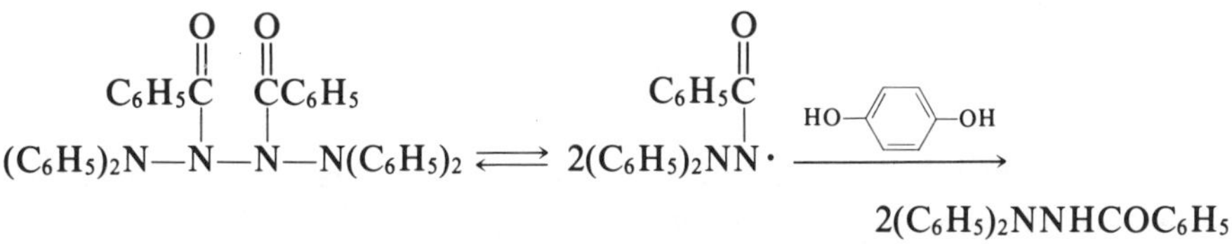

$$C_6H_5N{=}N{-}N(C_6H_5){-}NH_2 \xrightarrow{Al(Hg)} 2C_6H_5NHNH_2$$

$$(C_6H_5)_2N{-}N{=}N{-}N(C_6H_5)_2 \xrightarrow[\text{ethanol}]{H_2/Pd/CaSO_4} N_2 + \underset{93\%}{2(C_6H_5)_2NH}$$

N-3 in 1,3-diphenyltetrazene, forming phenylhydrazine (559). Tetraalkyl (66) and tetraaryl (407) 2-tetrazenes lose nitrogen and form secondary amines on reduction.

Treatment of 1,5-diphenyl-3-methylpentazdiene with zinc and acetic acid splits the molecule to phenylhydrazine and methylamine (190), and the benzothiazole pentaazapentamethine cyanine **(II)** is converted by titanium(III) chloride to the corresponding azidium salt and hydrazone (424).

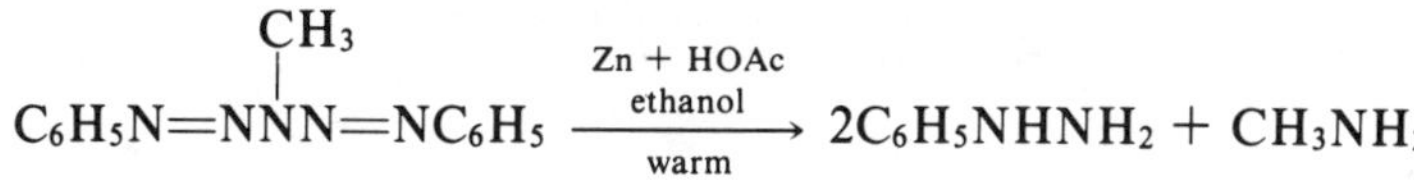

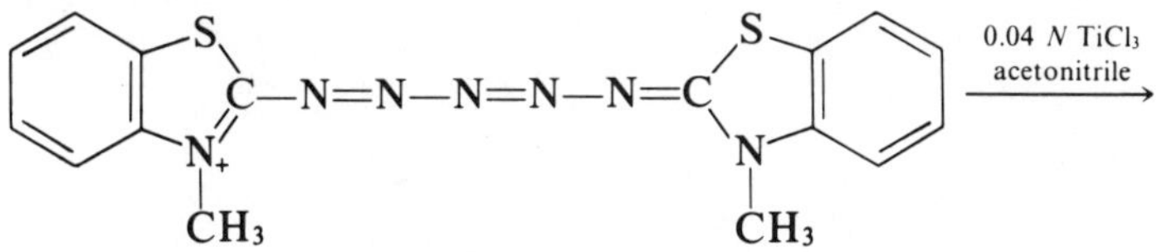

(II)

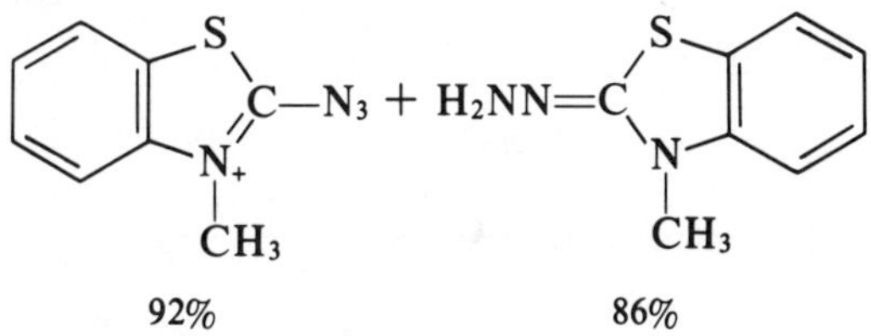

The nitrogen chain remains intact when 1,4-dimethyl-1,4,5,6-tetrahydro-*v*-tetrazine 2-oxide is deoxygenated by trimethylphosphite (50% yield) or lithium aluminum hydride (457).

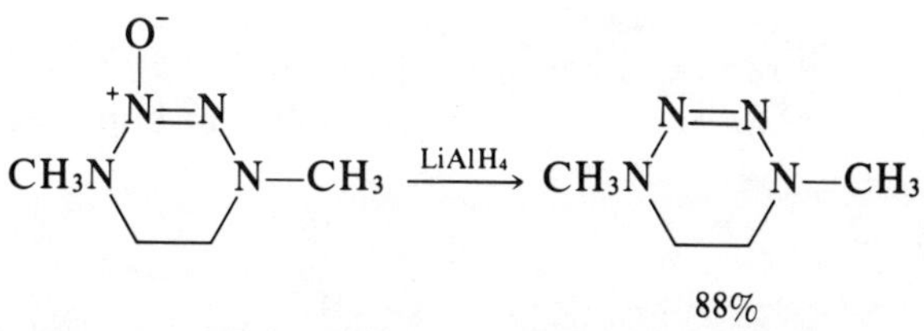

Reduction of 1,2-dimethyl-1-nitroso-2-phenylhydrazine splits the hydrazine bond and converts nitroso to amino, forming monomethylaniline and methylhydrazine (211). Nitroso also is reduced to amino in the preparation of triazanes, discussed in Chapter 7, Synthesis. The main reduction of nitrosohydrazines, however, consists of cleaving the nitroso-nitrogen bond, forming a hydrazine or products of further fission of nitrogen-nitrogen or nitrogen-car-

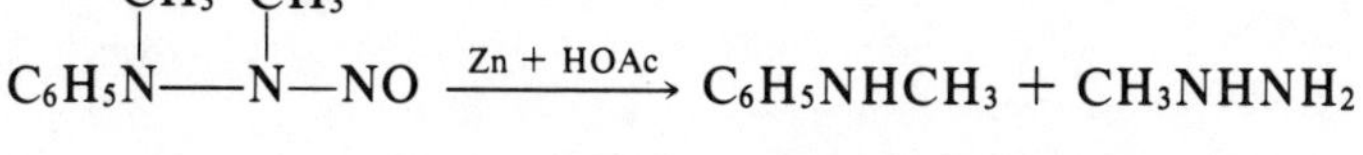

bon bonds, illustrated with benzoylnitrosophenylhydrazine (521) and nitrosotrimethylhydrazine **(III)** (200).

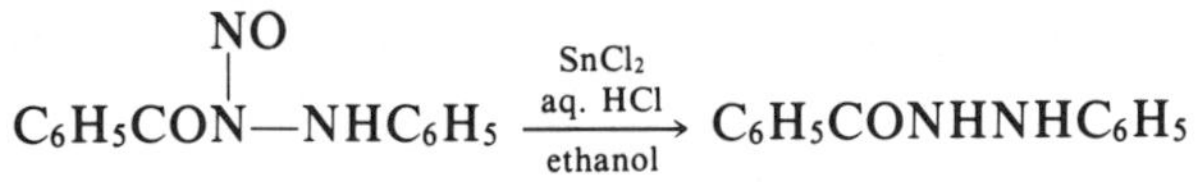

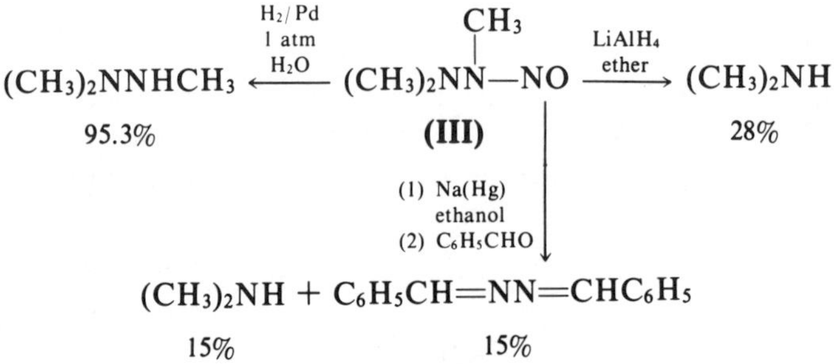

Catalytic hydrogenation of dibenzo[*d,f*][1,2,3]triazepine leads to 2-aminobiphenyl (172), but its isomer, *N*-iminobenzocinnoline, and *N*-imino derivatives lose the exocyclic nitrogen (173). The *N*-amino group in *N*-aminopyridazinium salts is removed similarly by hydrogenolysis (289).

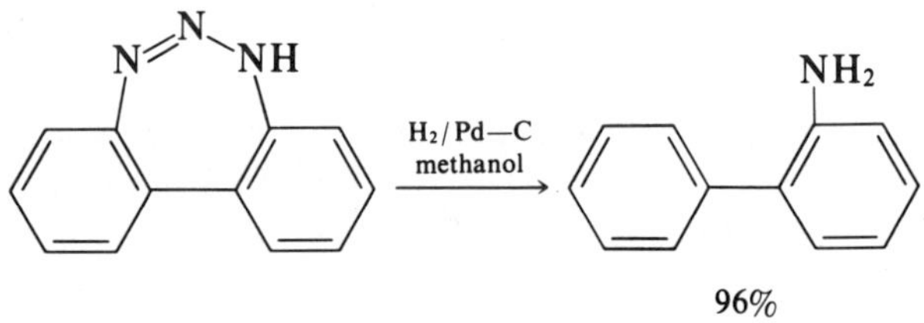

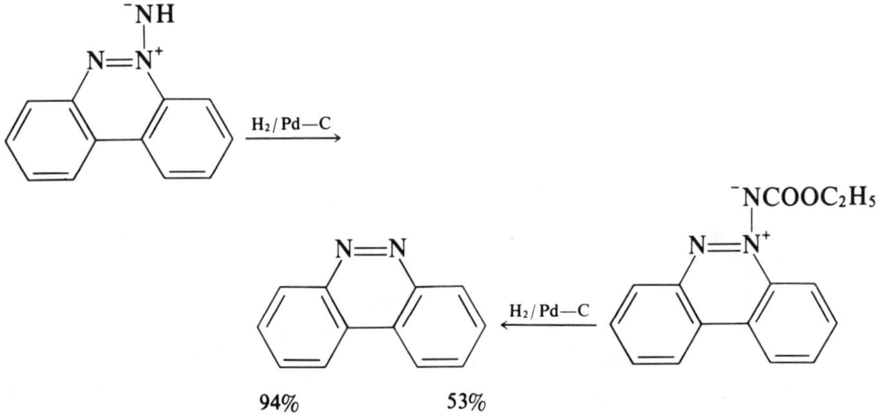

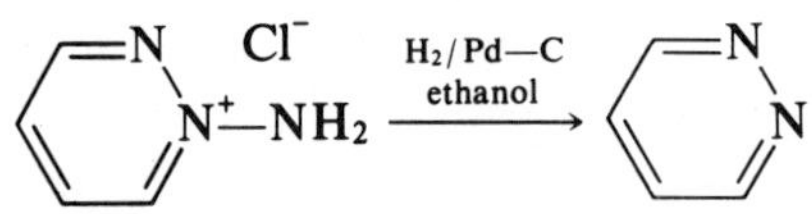

The cyclic triazene moiety in 1,5-diphenyl-4-nitro-*v*-triazoline remains unchanged during reduction of nitro to amino (442).

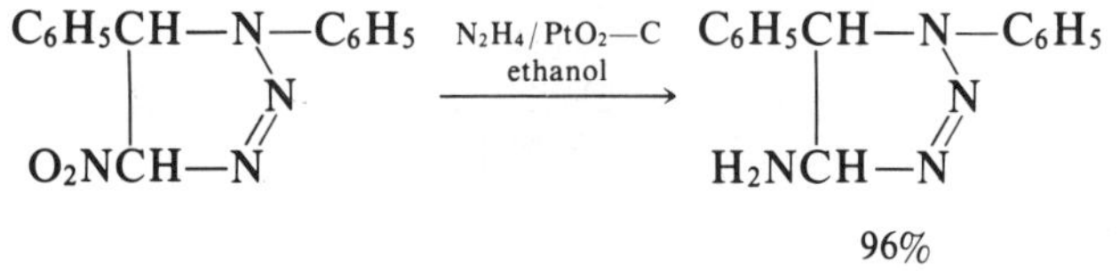

Catalytic hydrogenation of α-azidodiphenylacetonitrile leads in part to the aminonitrile, and also to the triazoline derivative, presumably through the monosubstituted triazene, the three-nitrogen chain remaining intact (241).

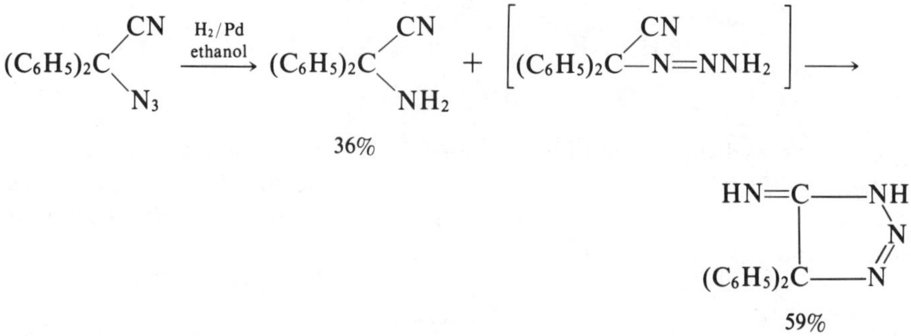

Reduction of organic azides to amines is accomplished by a variety of reagents, usually in good yields. Arsenite reduction of alkyl and aryl azides with measurement of liberated nitrogen enables quantitative analysis (530). Alkyl

$$C_6H_5CH_2N_3 \xrightarrow[\text{methanol} \atop 0\text{–}40^\circ]{AsO_3^{3-} \atop \text{KOH}} N_2 + C_6H_5CH_2NH_2 + AsO_4^{3-}$$

$$n\text{-}C_8H_{17}N_3 \xrightarrow{\text{NaI} \atop CCl_3COOH} N_2 + I_2 + n\text{-}C_8H_{17}NH_3^+ \ I^-$$

azides also are determined gasometrically with the nitrogen released by the action of iodide (107). Other combinations of reductants and azides are sodium amalgam with phenyl azide (121), sodium hydrosulfite with 2-azido-4-benzenesulfonamidophenol (3), sodium borohydride with 2,5-diazido-1,4-benzoquinone (368), and titanium(III) chloride with a variety of heterocyclic azides (487). 2-Azido-3-ethylbenzothiazolium fluoborate is converted quantitatively by sulfite ion to 2-imino-3-ethyl-benzothiazoline in the pH range 0–6 (227). Reduction of azides is frequently used in syntheses of amines, for example, with alicyclic (451) and aromatic (2) compounds, amino acids (82, 388), steroids (12), carbohydrates (380), nucleosides (246), and purines (502).

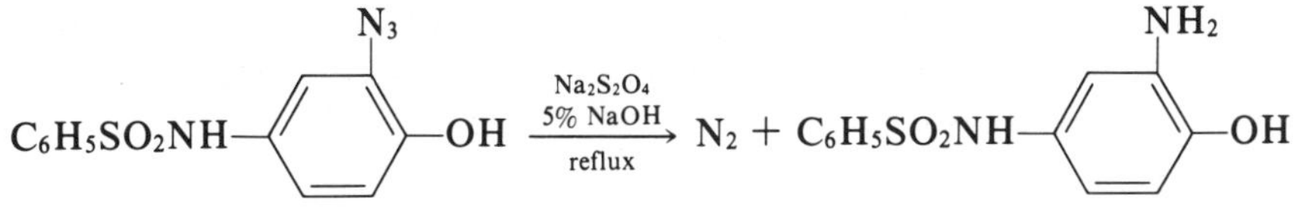

83.5%

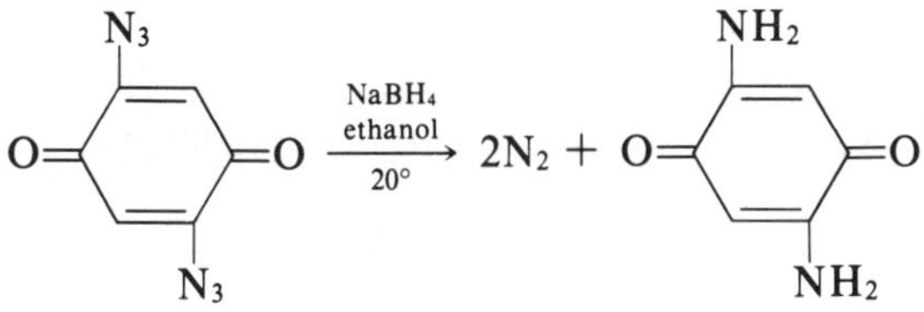

62%

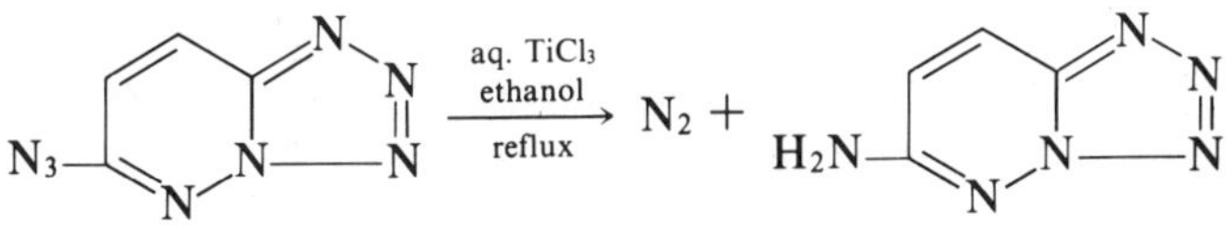

54%

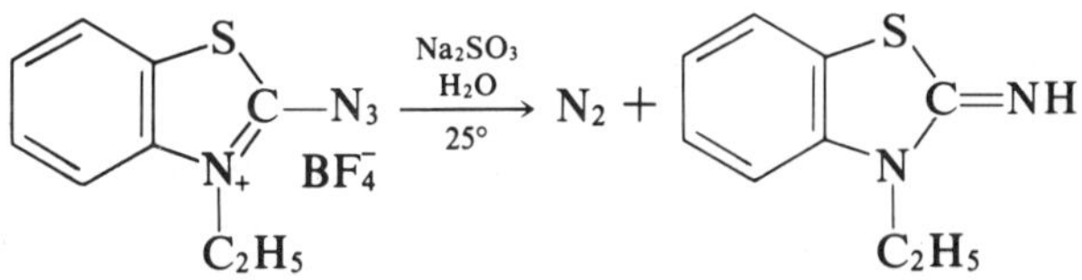

100%

In the reduction of carbamoyl azides, through the use of labeling with ^{15}N in the azide, the source of amide nitrogen is shown to be the nitrogen attached to the carbonyl group, with elimination of the two outer nitrogens. Similar loss of the middle and end nitrogen during reduction is found with phenyl, benzoyl, and tosyl azides. If the reaction with carbamoyl azides is conducted at 80° or

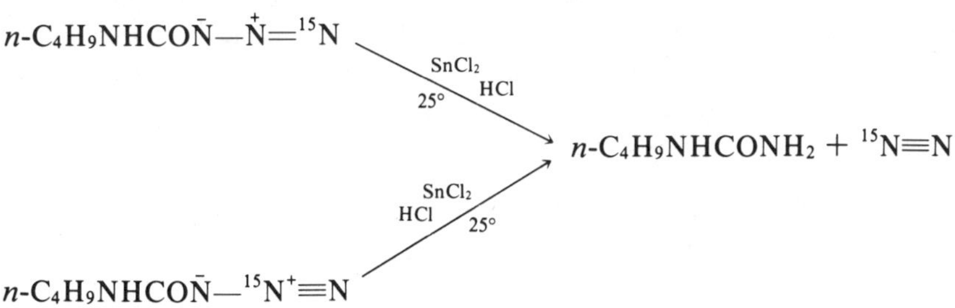

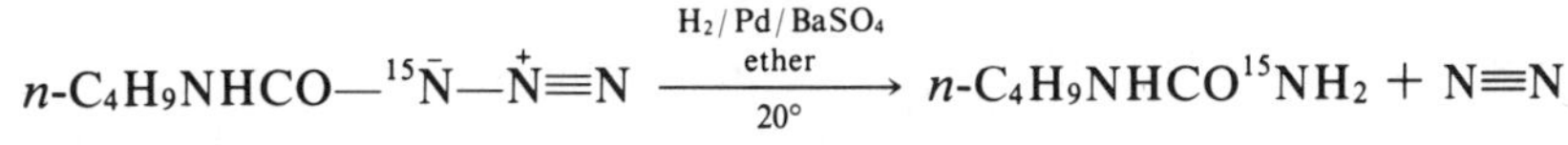

above, dissociation to isocyanate and hydrogen azide occurs, with recombination to carbamoyl azide with equal quantities of initially attached and terminal nitrogen joined to carbonyl (174).

$$n\text{-}C_4H_9NHCO\overset{-}{N}\text{—}\overset{+}{N}\equiv{}^{15}N \underset{}{\overset{>80^\circ}{\rightleftarrows}} n\text{-}C_4H_9NCO + H\overset{-}{N}\text{—}\overset{+}{N}\equiv{}^{15}N \rightleftarrows n\text{-}C_4H_9CO{}^{15}\overset{-}{N}\text{—}\overset{+}{N}\equiv N$$

Aryl azidosulfates are reduced to aryl sulfamate esters in methanol in the presence of copper powder with formation of formaldehyde (221).

$$C_6H_5OSO_2N_3 \xrightarrow[\text{reflux}]{CH_3OH,\ Cu^\circ} \underset{86\%}{C_6H_5OSO_2NH_2} + CH_2O + N_2$$

Reductive fission of the *v*-triazole ring in 2-phenyl-*v*-triazole-4-carboxylic acid occurs on treatment with sodium amalgam, forming hydrogen cyanide and

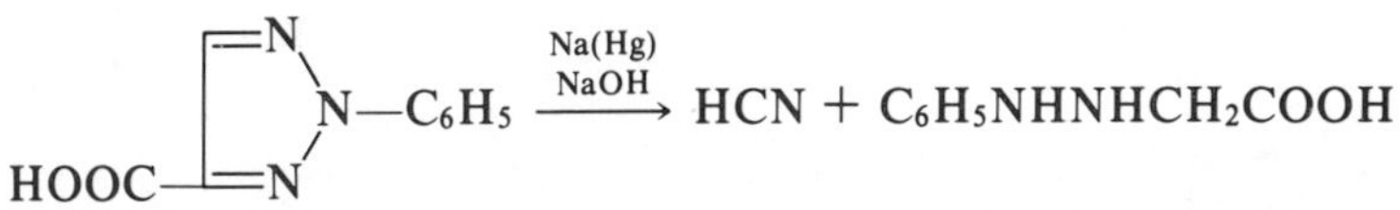

phenylhydrazinoacetic acid (275). Benzotriazole undergoes ring opening to *o*-aminophenylhydrazine on electrolytic reduction with consumption of four electrons (147). Polarography indicates a reversible reduction of 2-methylbenzotri-

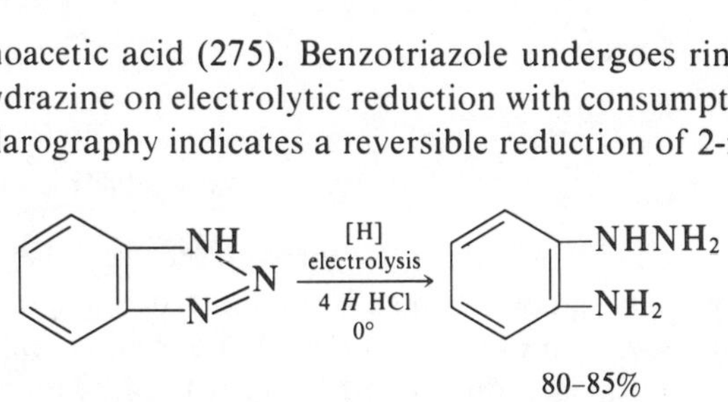

80–85%

azole to a dihydro derivative in alkaline solution (340). Catalytic hydrogenation of benzotriazole with Raney nickel at 100° and 100 atm pressure results in formation of *o*-phenylenediamine and ammonia (396). Electrochemical reduction at a copper amalgam cathode in sulfuric acid leads to 2,3-diaminophenazine

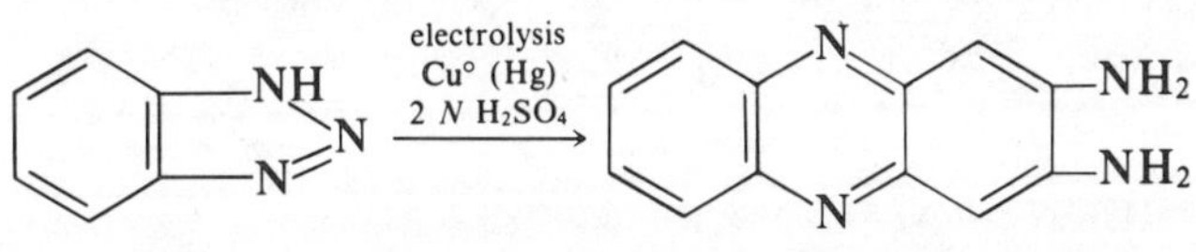

(448). Sodium in liquid ammonia also splits benzotriazole and its 1- and 2-methyl derivatives at the 1,2- and 2,3-bonds, forming *o*-phenylenediamine or 1-methyl *o*-phenylenediamine (104). 4,5,6,7-Tetrachloro-1-hydroxybenzotriazole is cleaved to tetrachloro-*o*-phenylenediamine by zinc and hydrochloric acid (91).

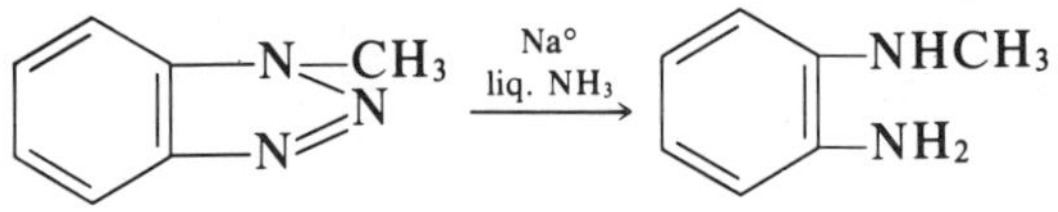

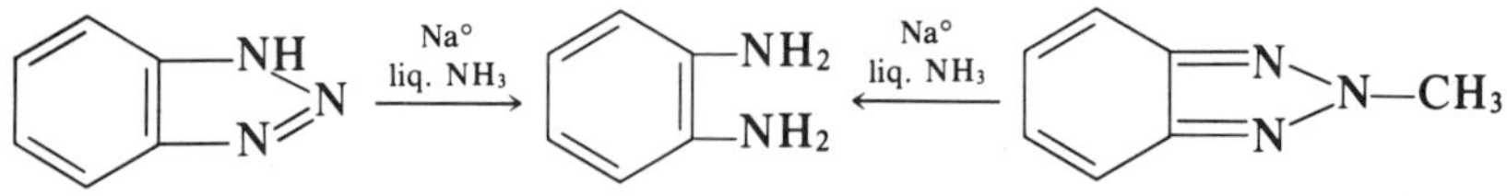

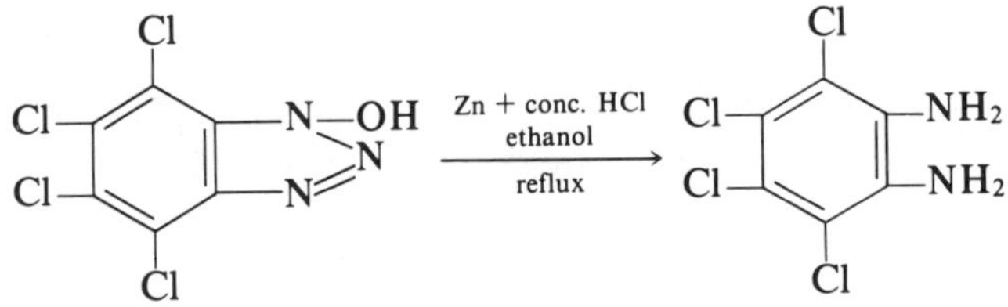

81%

Sodium dithionite opens the nitrogen chain in the bicyclic salt, 3-methyl-1-phenyl-*v*-triazolo[5,1-*a*]pyridinium chloride (307).

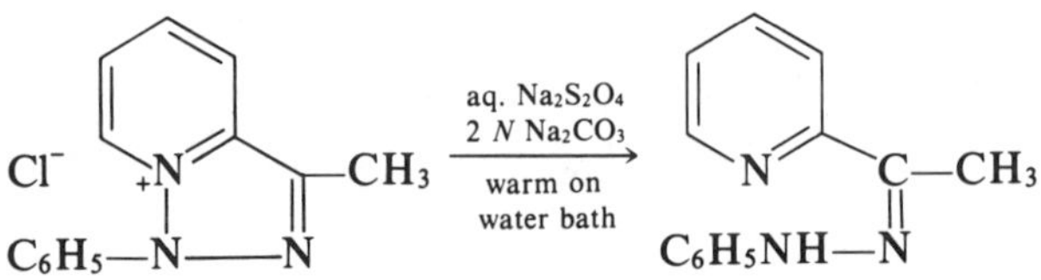

The triazole ring remains intact while reduction occurs elsewhere in the molecule, illustrated by reduction of the nitro group in 4,5-dimethyl-2-*p*-nitrophenyl-*v*-triazole (118), desulfurization of 5-methylthio-1-phenyl-*v*-triazole (286), hydrogenation of the fused benzene ring in benzotriazole (396, 428), and of one and two phenyl groups in 2,5-diphenyl-2*H*,5*H*-triazolo[4,5-*d*]triazole (566).

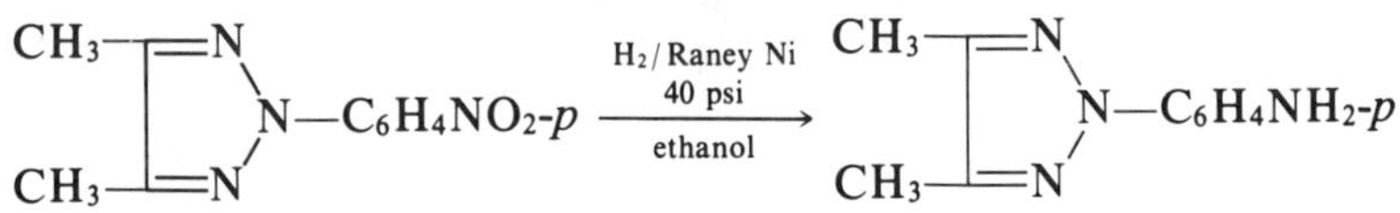

97%

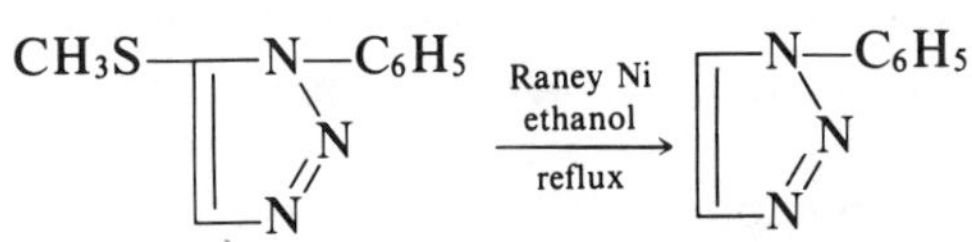

56%

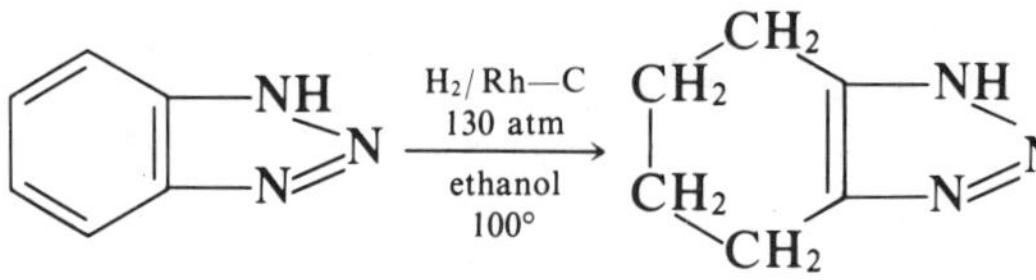

89%

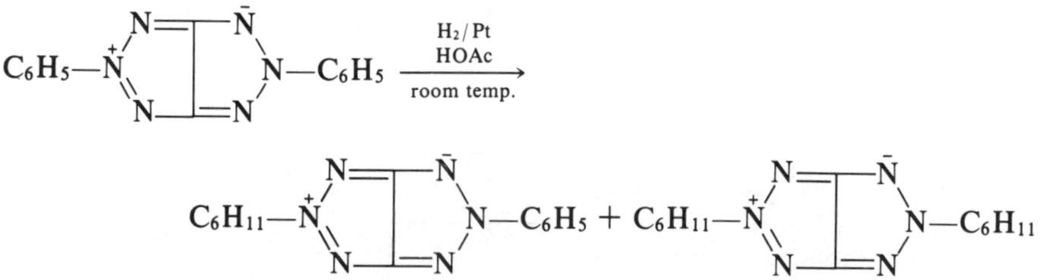

30% 15%

Benzyl groups are removed from *v*-triazole nitrogen by hydrogenolysis either by sodium and liquid ammonia (6a) or catalytically (7a). Similarly, hydroxyl groups on the *v*-triazole ring are replaced with hydrogen by reduction, and the oxygen of *N*-oxides can be removed by reducing or deoxygenating agents.

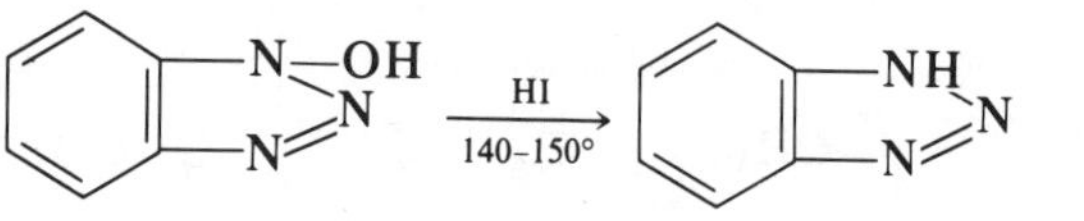

(569)

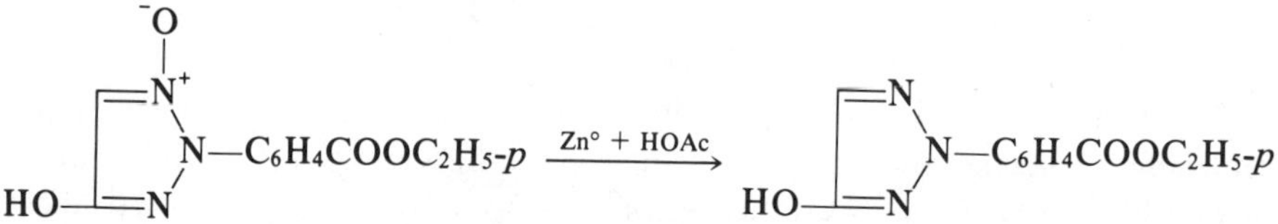

90%

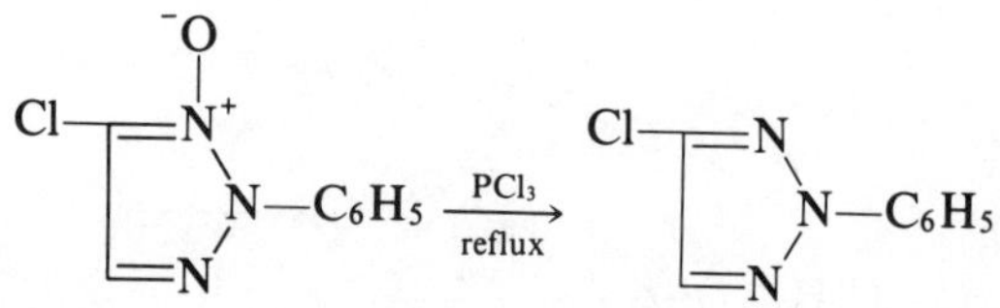

95%

Cyclic voltammetry shows reduction of 1,3-disubstituted-*v*-triazolium and *v*-triazolinium cations to radicals in acetonitrile to be reversible (63). The magnitude of the reduction potentials (Table 6.13) indicates that 1,3-*v*-triazolinium cations with aryl substituents are more easily reduced than those with alkyl substituents, that comparable *v*-triazolinium ions have lower reduction potentials than 3,4,5,6-tetrahydro-*v*-triazinium ions, and that 1,3-triazolium cations

Table 6.13. Electrolytic Reduction of 1,3-Disubstituted v-Triazolinium, Tetrahydro-v-triazinium, and v-Triazolium Cations

Cyclic Three-Nitrogen Cation Type	R_1	R_3	Reduction Potential at Pt Electrode vs. Ag/AgCl in Acetonitrile
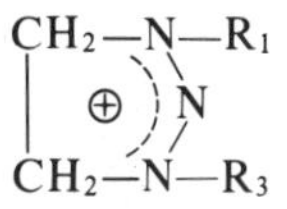	C_6H_5	p-$CH_3C_6H_4$	−0.615
	p-$CH_3OC_6H_4$	p-$CH_3OC_6H_4$	−0.715
	C_6H_5	CH_3	−0.780
	$(CH_3)_3C$	CH_3	−1.065
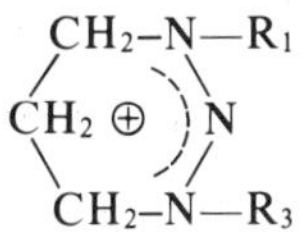	C_6H_5	p-$CH_3C_6H_4$	−0.788
	p-$CH_3OC_6H_4$	p-$CH_3OC_6H_4$	−1.180
	C_6H_5	CH_3	−1.325
	$(CH_3)_3C$	CH_3	−1.730

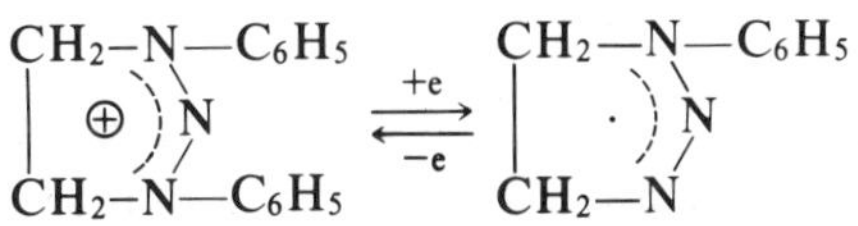

require a still greater potential for reduction. Within each series, a linear relationship is found between the reduction potential of the aryl derivatives and the Hammett sigma constants of the *m*- or *p*-substituents.

1,2-Disubstituted *v*-triazolium salts are reduced to Δ^3-triazolines (266).

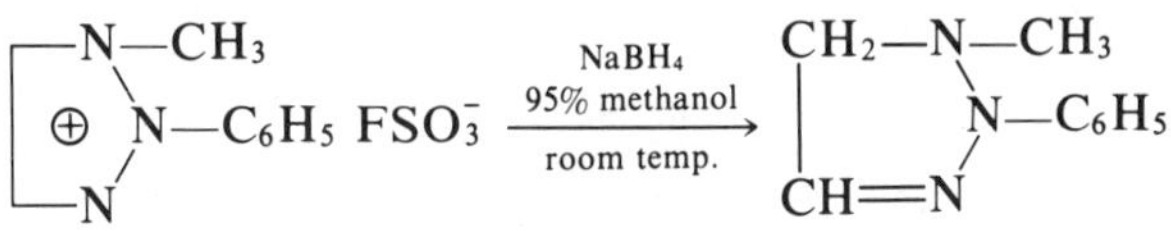

92.2%

Catalytic hydrogenation of *v*-triazines or their 1- or 2-oxides leads to 1,4-dihydro-*v*-triazines (399). The oxides also can be reduced to the dihydro compounds with sodium borohydride. Ring contraction occurs on hydrogenation

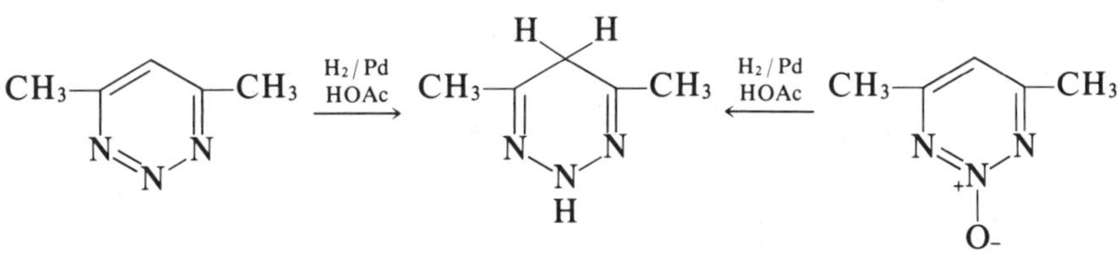

of 3,4,5-triphenyl-*v*-triazine (112). Similarly, electrochemical reduction of 3-phenylbenzotriazinone forms indazolinone, aniline, and benzanilide (245);

yields of indazolinone and aniline decrease and that of benzanilide increases as concentration of acid in the electrolyte increases. Other reductive ring openings

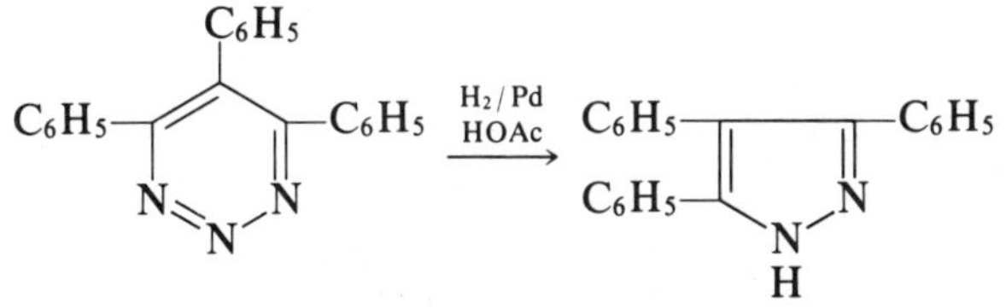

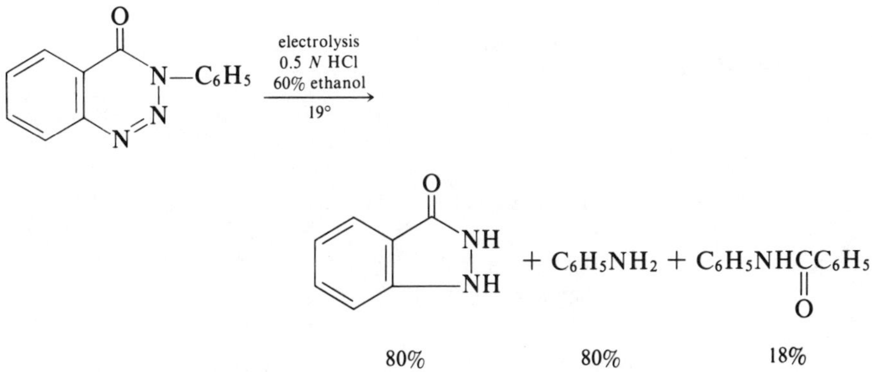

of benzotriazine derivatives are the formation of *N*-phenyl-*o*-aminobenzamidine from 4-anilinobenzotriazine **(IV)** in which all bonds between nitrogens in the triazine sequence are severed (496), and the formation of *o*-aminobenzoyl-*p*-bromophenylhydrazine from 2-*p*-bromophenylbenzotriazinone betaine **(V)** by breaking the 1,2-bond (284).

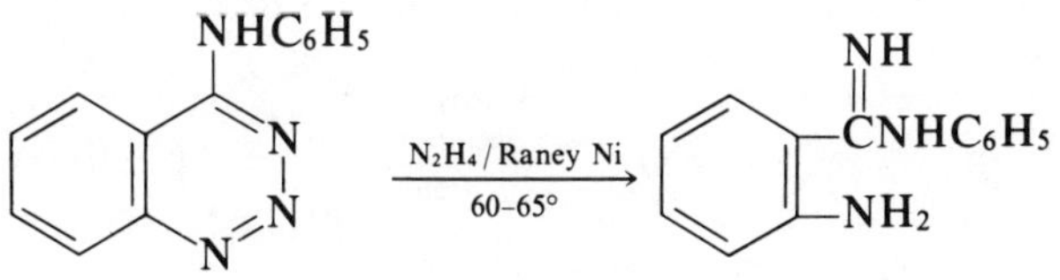

(IV) 31%

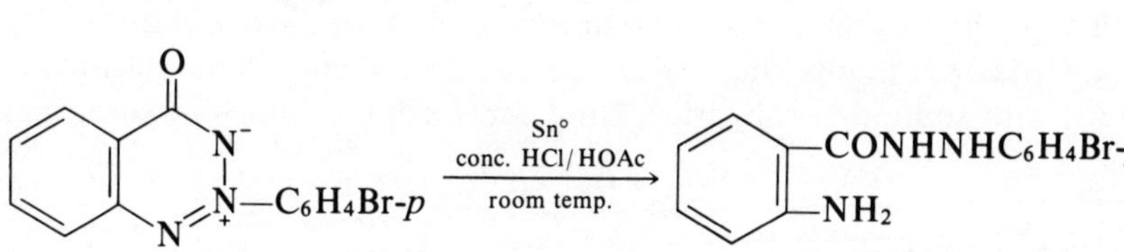

(V) 68%

The triazine chain remains unbroken on electrolytic removal of oxygen from 2-azaadenine 1-oxide (358).

Although 5-substituted aminothiatriazoles are not affected by hydrogen sulfide (330), hydriodic acid liberates iodine immediately from 5,5′-thiatriazole

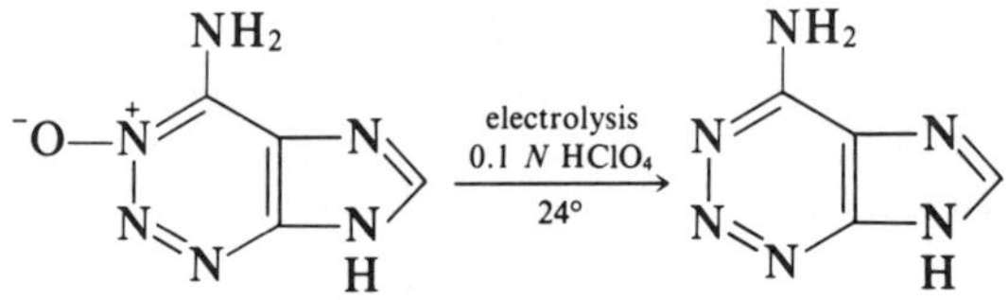

disulfide (84). 5-Phenylthiatriazole is reduced by lithium aluminum hydride, and following treatment with alkali gives dibenzoyl disulfide (480).

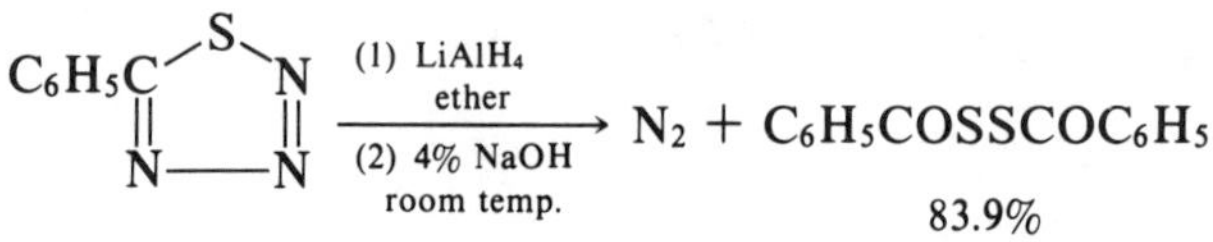

The nitrogen chain is entirely lost on deoxygenation of 1-nitrosobenzotriazole by ethyldiphenylphosphinate, probably to benzotriazole-1-nitrene, forming benzyne which is trapped by tetracyclone (98). Exocyclic nitrogen atoms are

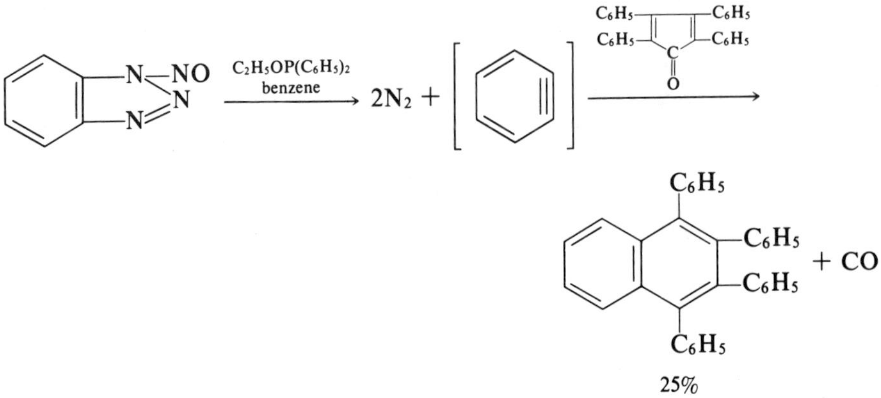

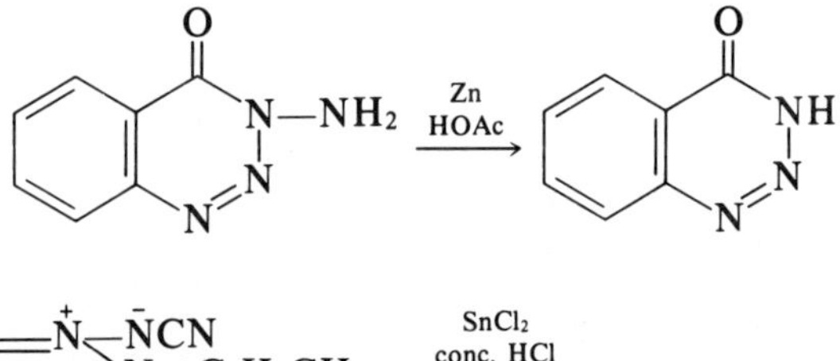

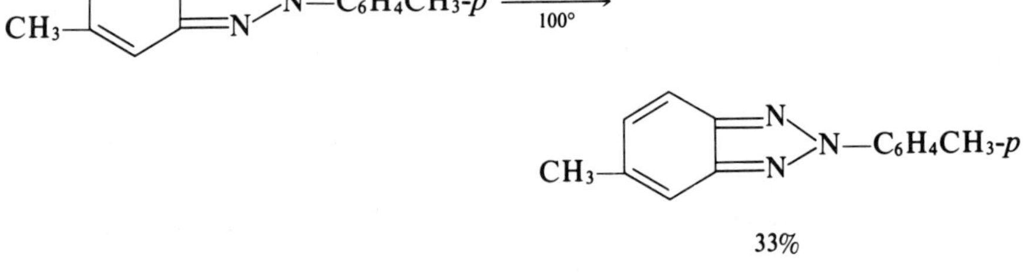

removed in the reductions of 3-aminobenzotriazinone (224) and 5-methyl-2-*p*-tolylbenzotriazole-1-cyanoimide (41). Catalytic hydrogenation of 1,1′-azobis(2-phenylimidazo[1,2-*a*]pyridinium)bromide **(VI)** causes the loss of nitrogen from the tetrazene chain and the reduction of the phenyl and fused pyridine rings (188).

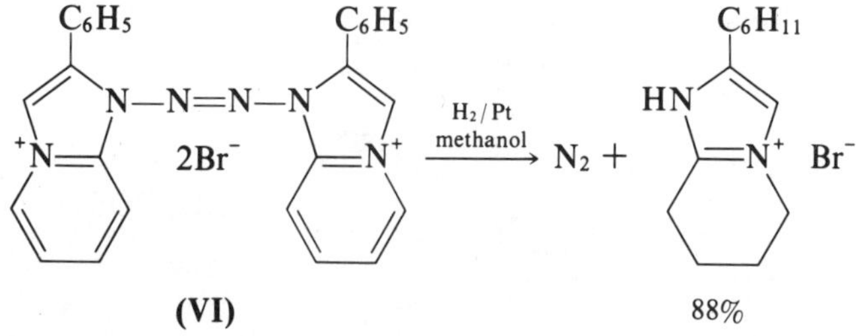

Benzyl groups in the 1- or 1,4-positions of 5-iminotetrazolines are removed on hydrogenolysis without affecting the four-nitrogen sequence (413). Reduc-

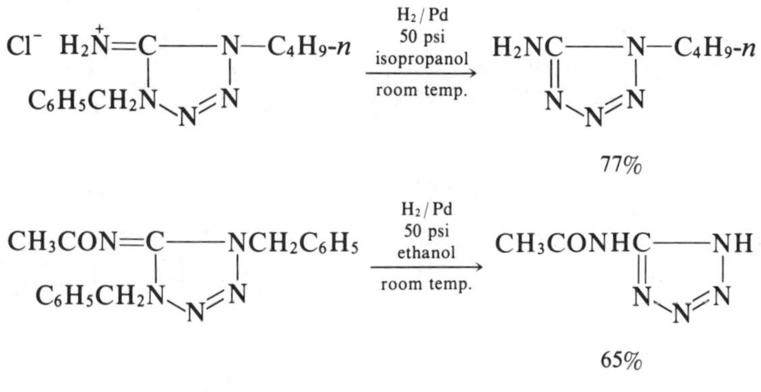

tion of 1,4,5-trisubstituted tetrazolium salts with sodium borohydride leads to Δ^2-tetrazolines, but 1,3-dimethyltetrazolium iodide is unchanged (266).

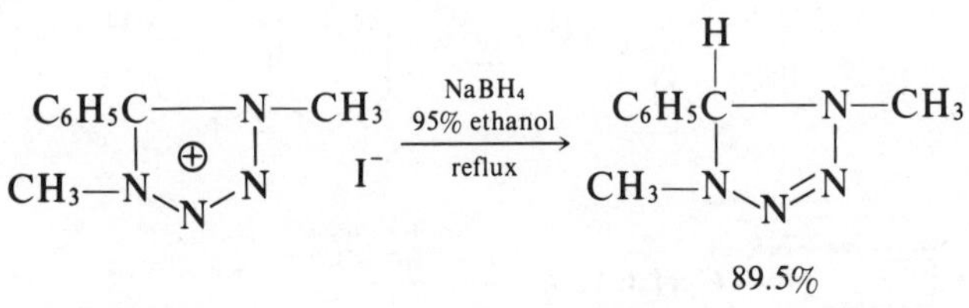

The tetrazole ring of 1,5-disubstituted tetrazoles is cleaved by lithium aluminum hydride to secondary amines (315). Thus cyclopentamethylenetetrazole is reduced to hexamethylenimine and 5-methyl-1-phenyltetrazole forms ethylphenylamine. Hydrogenolysis of 5-methylnitraminotetrazole (177) and 5-benzylmethylaminotetrazole (153) removes nitro and benzyl, respectively, without

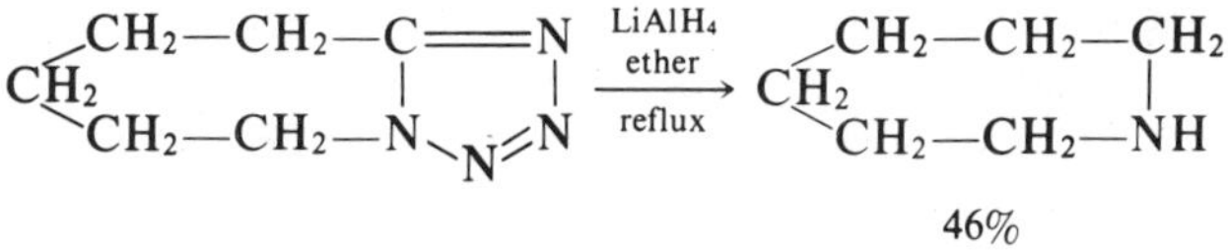

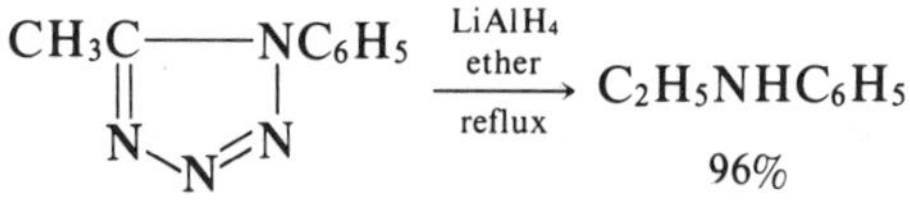

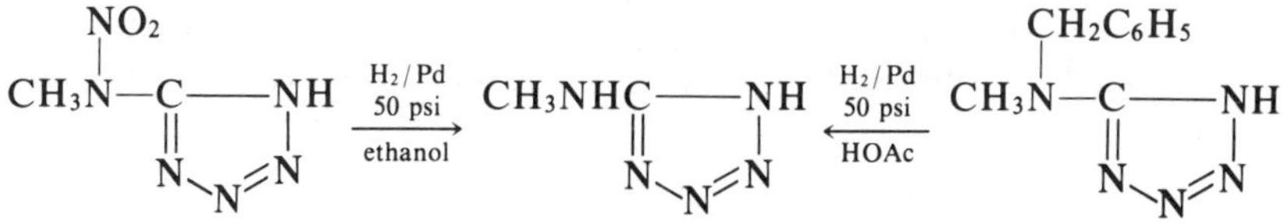

affecting the tetrazole ring. Sodium 5,5′-azobistetrazolate is reduced to hydrazotetrazole by stannous chloride (518). 5-Chloro-1-phenyltetrazole is unchanged on hydrogenation with platinum at 40 psi for 2 hr (280).

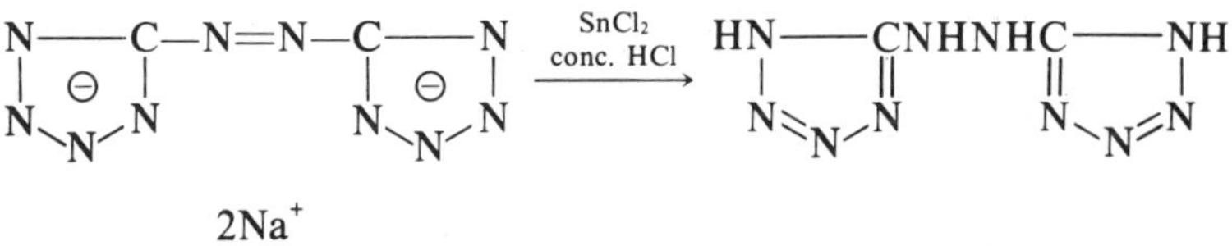

2,3-Disubstituted tetrazolium salts are reduced to formazans, breaking the 2,3 N—N bond, by such chemical agents as alkaline ascorbic acid and hydroxylamine as well as by more powerful reductants such as sodium hydrosulfite (389). Enzymatic reduction is discussed in Chapter 5, Biological Properties. Cat-

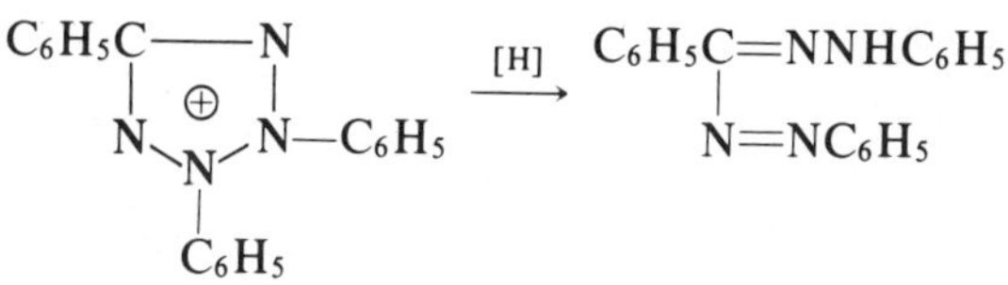

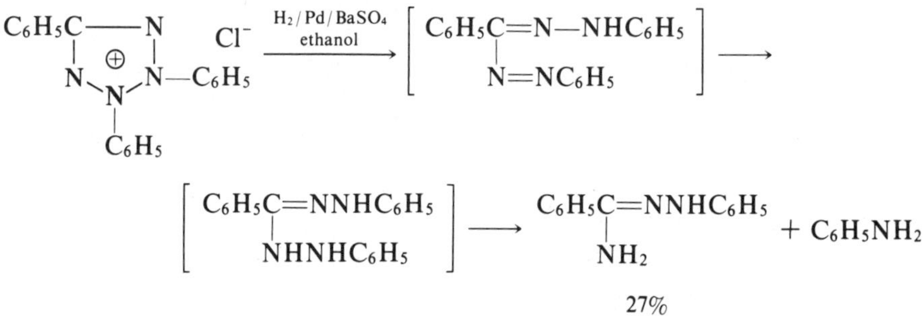

alytic hydrogenation of tetrazolium salts proceeds through the formazan and hydrazidine stages to the amidrazone; further reduction forms the hydrazone (273). Conversion of a tetrazolium cation to a formazan is a two-electron reduction with addition of one proton, going through a tetrazolinyl radical which is stable in acetonitrile in the time scale of the reduction (506). Electrochemical reduction of tetrazolium cation to tetrazolinyl is a reversible process, but the addition of a proton is irreversible. The enolate anion of propiophenone reduces tetrazolium salts to high concentrations of tetrazolinyl radicals; a concentration of 80% radical is obtained from 2,3-diphenyl-5-*p*-nitrophenyltetrazolium bromide in dimethyl sulfoxide containing potassium *t*-butoxide, *t*-butyl alcohol, and propiophenone (346). Careful reduction of 2,3-diphenylene-5-phenyltetrazolium chloride with sodium hydrosulfite forms an isolable radical (306).

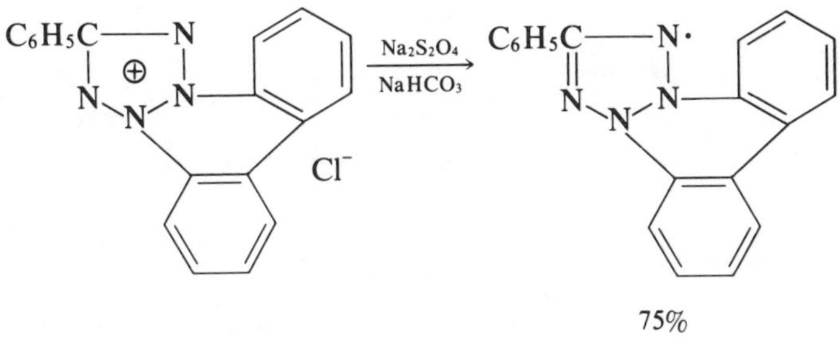

Reductive fission of cyclic-cyclic nitrogen-nitrogen bonds requires moderately stringent conditions as shown in reactions of biindazole (298) and the 1,1′-

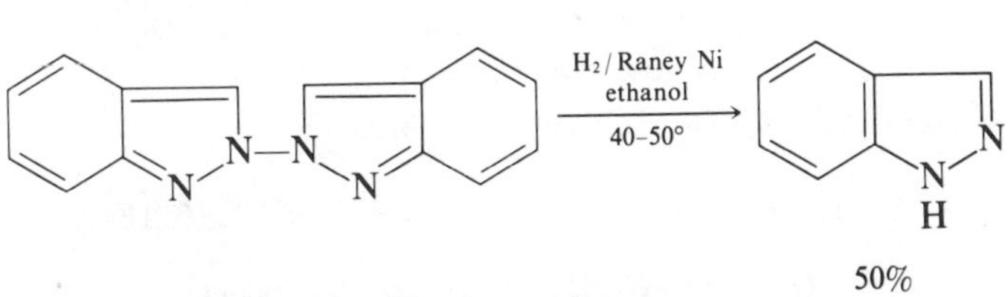

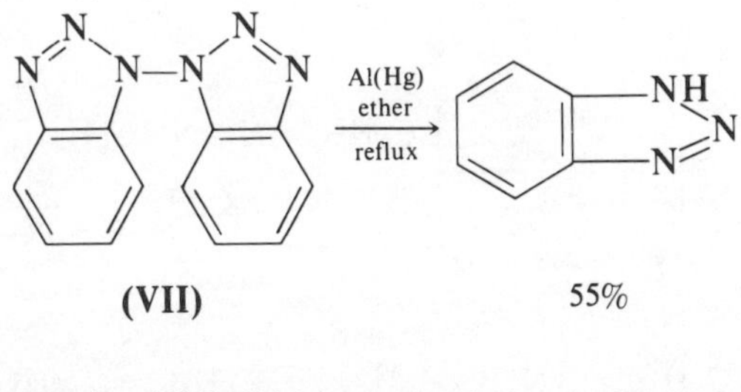

(VII)

and 1,2′-bibenzotriazoles (209). Lithium aluminum hydride does not affect **VII**. Hydrogenation of 1,1′-bi(5-phenyltetrazole) does not split the bond joining tetrazole rings, but breaks the nitrogen chain within one tetrazole ring, forming *N*-(5-phenyltetrazolyl-1)-benzamidine (54).

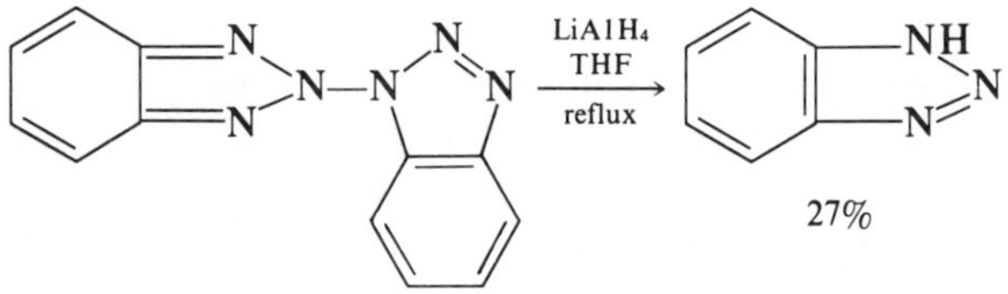

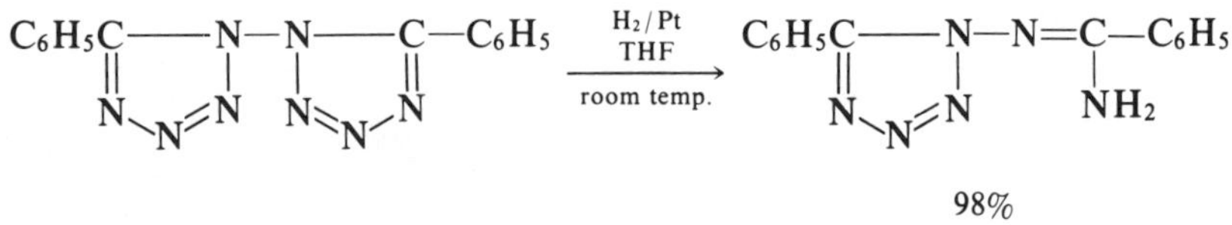

The nitrogen chain of heterocycles with an embedded nitrogen and an exposed azo group is decomposed under reducing conditions with loss of nitrogen. Thus the triazole ring in the *v*-triazolo-1,2,4-benzotriazine oxide **(VIII)** is split and the oxide eliminated by sodium hydrosulfite (515). Ethanol reduction

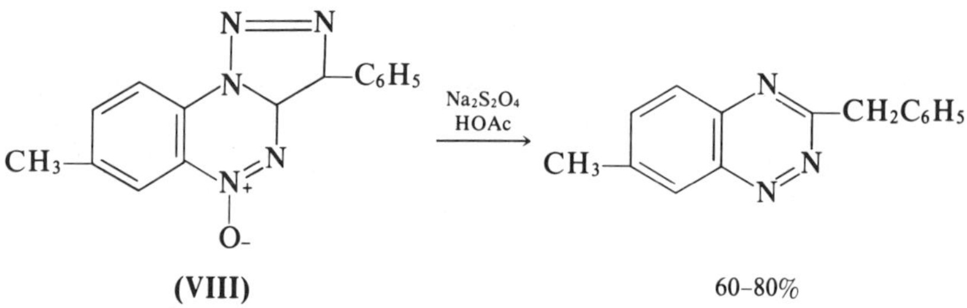

of *s*-triazinobenzotriazines such as **IX** removes the masked diazonium unit but hydrazine and Raney nickel forms an amine from the cyclic azo (344). Nitrogen

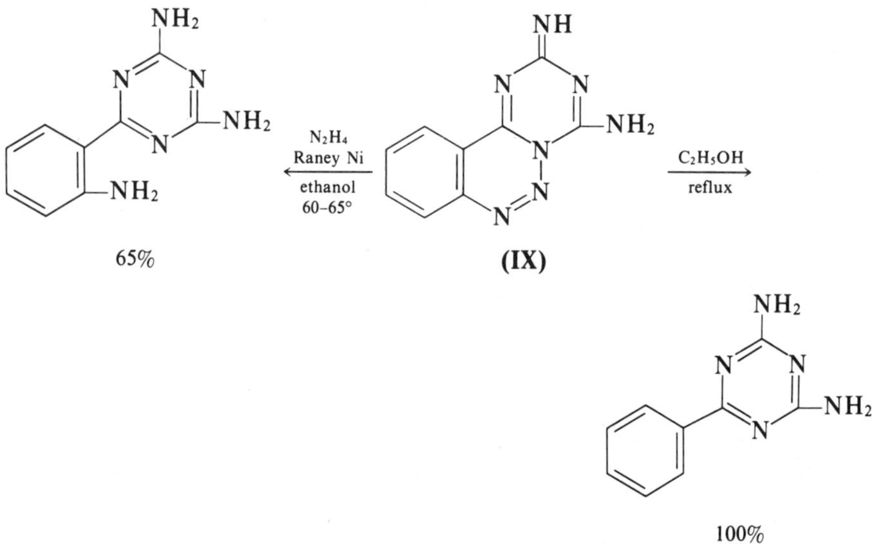

chains in compounds in which the middle nitrogen is shared between two rings are more resistant to reductive decomposition than are compounds in which an end nitrogen is shared. Thus the ring nitrogen sequence of 6-azido-*s*-triazolo-[1,5-*b*]pyridazine 3-oxide **(X)** remains intact during different reduction reactions. In methanol, hydrogenation reduces only the azide; in hydrochloric acid

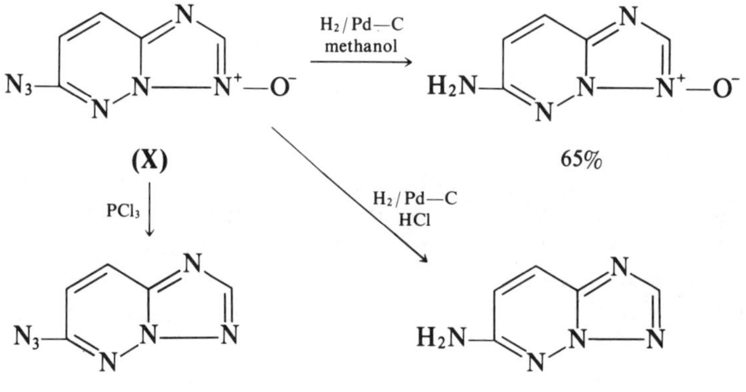

solution the oxide is removed as well, and deoxygenation without affecting the azide is effected with phosphorus trichloride (80). Sharing of two nitrogens between rings also results in stability toward reduction. The ring system and the nitrogen chain of 5*H*-pyrazolo[1′,2′:1,2]1,2,3-triazolo[5,4-*a*]phenazinyliumide **(XI)** are unaffected on hydrogenation, but one double bond in the phenazine

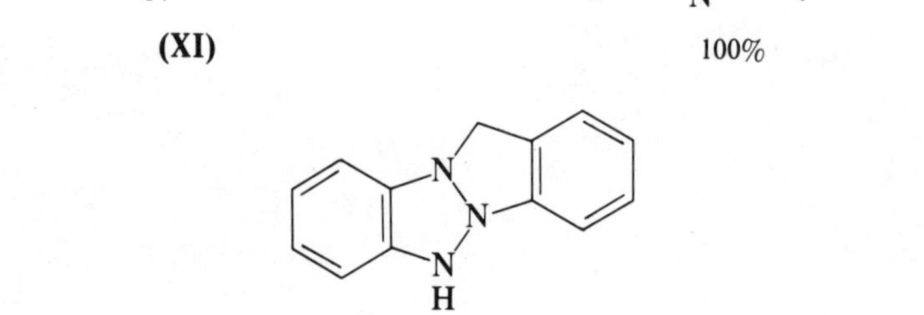

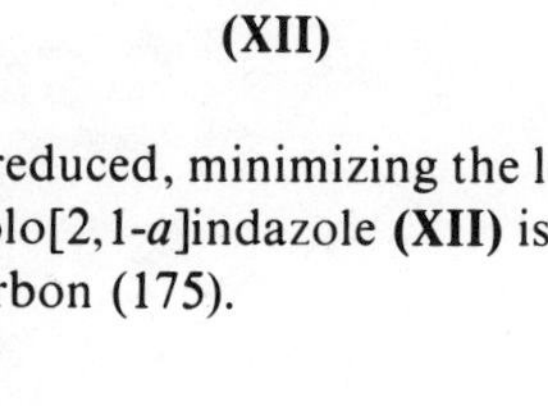

(XII)

portion of the molecule is reduced, minimizing the loss of aromaticity (8). Similarly, 5*H*,11*H*-benzotriazolo[2,1-*a*]indazole **(XII)** is unchanged on hydrogenation over palladium on carbon (175).

Stannous chloride breaks the four-nitrogen chain and converts the azo unit in pyrazolo[5,1-*c*]benzotriazine to a hydrazine group (9).

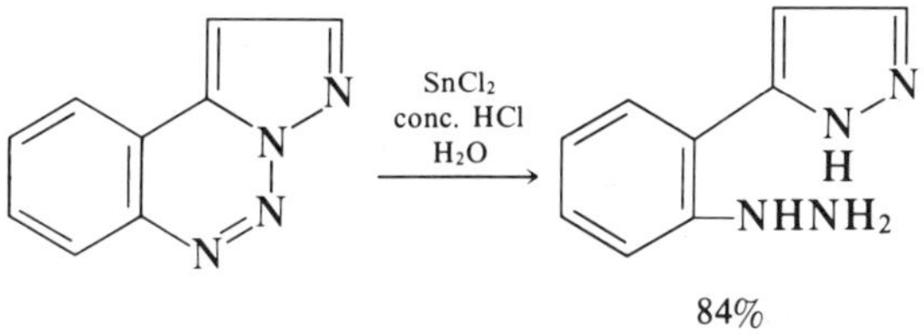

Tetrazolopyridine is reduced to 2-aminopyridine in alkaline solution (76). A nitrogen-nitrogen bond in dibenzo-1,3a,4,6a-tetraazapentalene is cleaved by

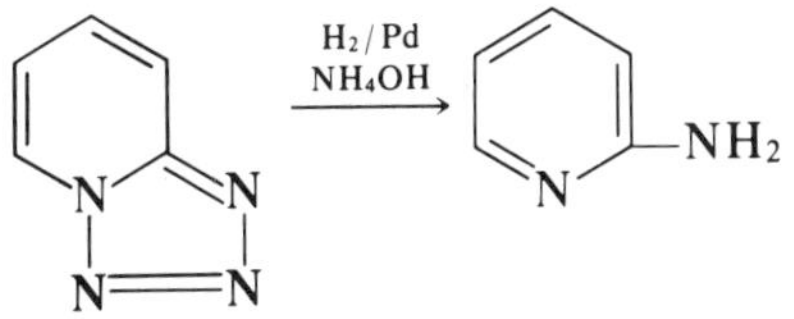

reduction with lithium aluminum hydride forming 2-(*o*-aminophenyl)benzotriazole (105).

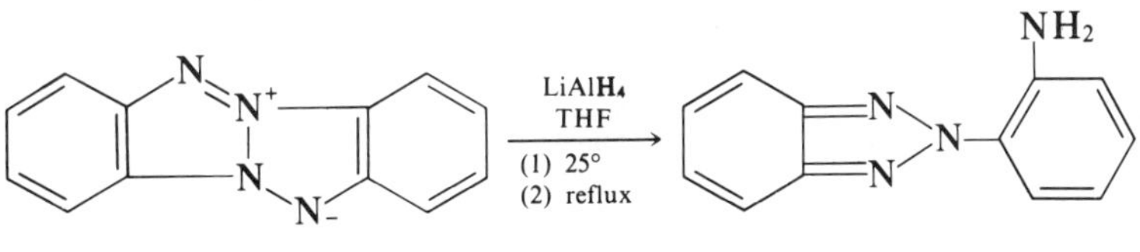

Only the chlorine is removed on hydrogenation of 7-amino-8-chlorotetrazolo[1,5-*b*]pyridazine without affecting the embedded five-nitrogen chain.

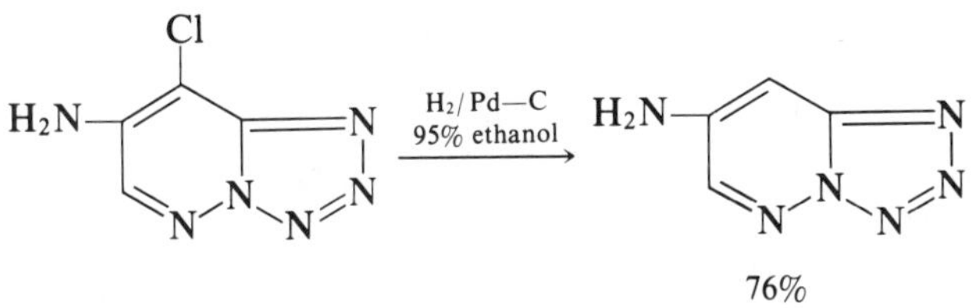

REFERENCES

1. G. A. Abakumov, V. K. Cherkasov, and G. A. Razuvaev, *Dokl. Akad. Nauk. SSSR, Engl. Ed.*, **1971,** *197,* 265–267.
2. R. Adams and D. C. Blomstrom, *J. Am. Chem. Soc.*, **1953,** *75,* 3405–3408.
3. R. Adams and L. Whitaker, *J. Am. Chem. Soc.*, **1956,** *78,* 658–663.

4. G. Adembri et al., *J. Chem. Soc., Perkin I,* **1977,** 971–979.

5. B. M. Adger et al., *J. Chem. Soc., Perkin I,* **1975,** 31–40.

6. T. Akiyama, Y. Imasaki, and M. Kawanisi, *Chem. Lett.,* **1974,** 229–230.

6a. A. Albert, *J. Chem. Soc. (C),* **1969,** 152–160.

7. *Ibid.,* **1970,** 230–235.

7a. A. Albert, *J. Chem. Soc., Perkin I,* **1978,** 513–519.

8. A. Albini, G. F. Bettinetti, and G. Minoli, *J. Chem. Soc., Perkin I,* **1981,** 1821–1825.

9. C. Alberti, *Gazz. Chim. Ital.,* **1957,** *87,* 772–780.

10. K. Alder and G. Stein, *Ann.,* **1933,** *501,* 1–48.

11. N. E. Alexandrou and S. Adamopoulos, *Synthesis,* **1976,** 482–483.

12. Y. Ali and A. C. Richardson, *Carbohydr. Res.,* **1967,** *5,* 441–448.

13. H. W. Altland, *J. Org. Chem.,* **1976,** *41,* 3395–3399.

14. V. Y. Andakushkin, B. A. Dolgoplosk, and I. I. Radchenko, *J. Gen. Chem. USSR, Engl. Ed.,* **1956,** *26,* 3307–3314.

15. D. J. Anderson et al., *J. Chem. Soc., Perkin I,* **1973,** 550–555.

16. S. Anderson, E. E. Glover, and K. D. Vaughan, *J. Chem. Soc., Perkin I,* **1975,** 1232–1234.

17. J.-P. Anselme and W. Fischer, *Tetrahedron,* **1969,** *25,* 855–859.

18. M. Appl and R. Huisgen, *Chem. Ber.,* **1959,** *92,* 2961–2967.

18a. C. L. Arcus and R. J. Mesley, *J. Chem. Soc.,* **1953,** 178–181.

19. J. Arriau et al., *J. Chem. Soc., Perkin II,* **1976,** 1245–1248.

20. H. G. Aurich and W. Weiss, *Chem. Ber.,* **1973,** *106,* 2408–2414.

21. H. J. Backer, *Rec. Trav. Chim.,* **1950,** *69,* 1223–1232.

22. A. S. Bailey, B. R. Brown, and M. C. Churn, *J. Chem. Soc. (C),* **1971,** 1590–1591.

23. A. S. Bailey and J. J. Merer, *J. Chem. Soc. (C),* **1966,** 1345–1348.

24. J. R. Bailey and W. T. Read, *J. Am. Chem. Soc.,* **1914,** *36,* 1747–1766.

25. J. E. Baldwin and J. E. Patrick, *Chem. Commun.,* **1968,** 968.

26. H. Balli, *Ann.,* **1961,** *647,* 11–18.

27. H. Balli, *Helv. Chim. Acta,* **1974,** *57,* 1912–1919.

28. H. Balli et al., *Helv. Chim. Acta,* **1978,** *61,* 97–103.

29. H. Balli and R. Gipp, *Ann.,* **1966,** *699,* 133–134.

30. H. Balli and F. Kersting, *Ann.,* **1961,** *647,* 1–10.

31. *Ibid.,* **1963,** *663,* 96–103.

32. *Ibid.,* 103–107.

33. H. Balli and R. Maul, *Helv. Chim. Acta,* **1976,** *59,* 148–155.

34. O. Baltzer and H. v. Pechmann, *Ann.,* **1891,** *262,* 302–313.

35. E. Bamberger and A. v. Goldberger, *Ber.,* **1898,** *31,* 2636–2640.

36. E. Bamberger and H. Hauser, *Ann.,* **1910,** *375,* 316–333.

37. E. Bamberger and A. Stiegelmann, *Ber.,* **1899,** *32,* 3554–3560.

38. N. M. Baranchik, I. V. Grachev, and D. Z. Zavel'skii, *Zh. Obsch. Khim., Engl. Ed.,* **1958,** *28,* 1400–1409.

39. J. R. Bartels-Keith, M. T. Burgess, and J. M. Stevenson, *J. Org. Chem.,* **1977,** *42,* 3725–3731.

40. J. W. Barton and A. R. Grinham, *J. Chem. Soc., Perkin I,* **1972,** 634–637.

41. H. Bauer, G. R. Bedford, and A. R. Katritzky, *J. Chem. Soc.,* **1964,** 751–755.

42. H. Beecken, *Angew. Chem., Int. Ed. Engl.,* **1967,** *6,* 360.

43. R. J. S. Beer and I. Hart, *J. Chem. Soc., Chem. Commun.,* **1977,** 143–144.

44. M. Begtrup, *Acta Chem. Scand.*, **1969,** *23,* 2025–2030.
45. *Ibid.*, **1971,** *25,* 249–259.
46. *Ibid.*, **1972,** *26,* 715–718.
47. *Ibid.*, 1243–1257.
48. *Ibid.*, **1975,** *B29,* 141–144.
49. M. Begtrup, K. Hansen, and C. Pedersen, *Acta Chim. Scand.*, **1967,** *21,* 1234–1238.
50. M. Begtrup and J. Holm, *J. Chem. Soc., Perkin I,* **1981,** 503–513.
51. M. Begtrup and P. A. Kristensen, *Acta Chem. Scand.*, **1969,** *23,* 2733–2740.
52. M. Begtrup and C. Pedersen, *Acta Chem. Scand.*, **1969,** *23,* 1091–1100.
53. M. Begtrup and K. V. Poulsen, *Acta Chem. Scand.*, **1971,** *25,* 2087–2098.
54. H. Behringer and H. J. Fischer, *Chem. Ber.*, **1962,** *95,* 2546–2556.
55. F. Bell and J. Kenyon, *J. Chem. Soc.*, **1926,** 954.
56. E. Y. Belyaev et al., *Zh. Org. Khim., Engl. Ed.*, **1976,** *12,* 405–408.
57. R. P. Bennett and W. B. Hardy, *J. Am. Chem. Soc.*, **1968,** *90,* 3295–3296.
58. R. P. Bennett and W. B. Hardy (American Cyanamid), U.S. Pat. 3,370,078, Feb. 20, 1968.
59. F. R. Benson, L. W. Hartzel, and W. L. Savell, *J. Am. Chem. Soc.*, **1952,** *74,* 4917–4920.
60. N. N. Bereshchagina and I. Y. Postovskii, *J. Gen. Chem. USSR, Engl. Ed.*, **1964,** *34,* 1758–1761.
61. E. Bergmann and H. A. Wolff, *Ber.*, **1930,** *63,* 1176–1180.
62. J. K. Berlin and M. D. Coburn, *J. Heterocycl. Chem.*, **1975,** *12,* 235–237.
63. H. Berneth, H. Hansen, and S. Hünig, *Ann.*, **1980,** 285–290.
64. A. Bertho, *J. Prakt. Chem.*, **1927,** *116,* 101–117.
65. S. N. Bhat, *Curr. Sci.*, **1973,** *42,* 198–199.
66. L. Birkofer, *Ber.*, **1942,** *75,* 429–441.
67. L. Birkofer and P. Wegner, *Chem. Ber.*, **1967,** *100,* 3485–3494.
68. J. A. Bladin, *Ber.*, **1892,** *25,* 1411–1413.
69. *Ibid.*, **1893,** *26,* 545–546.
70. H. Bock and M. Schnöller, *Chem. Ber.*, **1969,** *102,* 38–49.
71. H. Bock and W. Wiegräbe, *Angew. Chem.*, **1963,** 789–790.
72. R. Böhm, *Pharmazie,* **1978,** *33,* 83–84.
73. W. Borsche and H. Hahn, *Chem. Ber.*, **1949,** *82,* 260–263.
74. G. V. Boyd et al., *J. Chem. Soc., Perkin I,* **1977,** 1612–1615.
75. J. H. Boyer et al., *J. Am. Chem. Soc.*, **1956,** *78,* 325–327.
76. J. H. Boyer, M. S. Chang, and R. F. Reinish, *J. Org. Chem.*, **1960,** *25,* 286–287.
77. J. H. Boyer and J. A. Hernandez, *J. Am. Chem. Soc.*, **1956,** *78,* 5124–5125.
78. J. H. Boyer and L. T. Wolford, *J. Am. Chem. Soc.*, **1958,** *80,* 2741–2743.
79. L. E. Brady and R. M. Herbst, *J. Org. Chem.*, **1959,** *24,* 922–926.
80. J. Bartož-Stres et al., *Tetrahedron Lett.*, **1975,** 4429–4430.
81. J. v. Braun and W. Rudolph, *Ber.*, **1941,** *74,* 264–272.
82. H. Bretschneider, N. Karpitschka, and G. Piekavski, *Monats.*, **1953,** *84,* 1084–1090.
83. M. Brown and R. E. Benson, *J. Org. Chem.*, **1966,** *31,* 3849–3851.
84. A. W. Browne et al., *J. Am. Chem. Soc.*, **1923,** *45,* 2541–2550.
85. D. Buckley and M. S. Gibson, *J. Chem. Soc.*, **1956,** 3242–3243.
86. W. Buder and A. Schmidt, *Z. Naturforsch.*, **1975,** *B30,* 503–505.
87. W. E. Bull, J. A. Seaton, and L. F. Audrieth, *J. Am. Chem. Soc.*, **1958,** *80,* 2516–2518.

88. C. Bülow, *Ber.*, **1909,** *42,* 4429–4438.
89. C. Bülow and K. Haas, *Ber.*, **1910,** *43,* 2647–2662.
90. R. D. Burpitt and V. W. Goodlett, *J. Org. Chem.*, **1965,** *30,* 4308–4309.
91. D. E. Burton et al., *J. Chem. Soc.* (*C*), **1968,** 1268–1273.
92. M. Busch, N. Patrascanu, and M. Weber, *J. Prakt. Chem.*, **1934,** *140,* 117–128.
93. M. Busch and H. Pfeiffer, *Ber.*, **1926,** *59,* 1162–1165.
94. R. N. Butler and V. C. Garvin, *J. Chem. Soc., Perkin I,* **1981,** 390–393.
95. R. N. Butler and F. L. Scott, *J. Chem. Soc.* (*C*), **1967,** 239–244.
96. *Ibid.*, **1968,** 1711–1716.
97. R. N. Butler and F. L. Scott, *J. Org. Chem.*, **1966,** *31,* 3182–3187.
98. J. I. G. Cadogan and J. B. Thomson, *Chem. Commun.*, **1969,** 770.
99. C. M. Camaggi, M. Tiecco and A. Tundo, *J. Chem. Soc.* (*B*), **1968,** 680–683.
100. R. C. Cambie et al., *J. Chem. Soc., Perkin I,* **1982,** 961–965.
101. C. D. Campbell and C. W. Rees, *J. Chem. Soc.* (*C*), **1969,** 742–747.
102. *Ibid.*, 752–756.
103. T. W. Campbell and B. F. Day, *Chem. Rev.*, **1951,** *48,* 299–317.
104. N. O. Cappel and W. C. Fernelius, *J. Org. Chem.*, **1940,** *5,* 40–47.
105. R. A. Carboni et al., *J. Am. Chem. Soc.*, **1967,** *89,* 2618–2625.
106. *Ibid.*, 2626–2633.
107. W. R. Carpenter, *Anal. Chem.*, **1964,** *36,* 2352–2353.
108. W. R. Carpenter, *J. Org. Chem.*, **1962,** *27,* 2085–2087.
109. C. Casewit and J. D. Roberts, *J. Am. Chem. Soc.*, **1980,** *102,* 2364–2368.
110. E. Cawkill et al., *J. Chem. Soc., Perkin I,* **1979,** 724–731.
111. S. R. Challand et al., *J. Chem. Soc., Perkin I,* **1975,** 26–31.
112. E. A. Chandross and G. Smolinsky, *Tetrahedron Lett.*, **1960,** 19–22.
113. K. Clusius and H. R. Weisser, *Helv. Chim. Acta,* **1952,** *35,* 1524–1527.
114. M. D. Coburn and J. K. Berlin, *Synthesis,* **1974,** 869.
115. P. Cohen-Fernandes et al., *J. Org. Chem.*, **1979,** *44,* 4156–4160.
116. P. Cohen-Fernandes and C. L. Habraken, *J. Org. Chem.*, **1971,** *36,* 3084–3086.
117. R. A. Coleman, M. S. Newman, and A. B. Garrett, *J. Am. Chem. Soc.*, **1954,** *76,* 4534–4538.
118. R. F. Coles and C. S. Hamilton, *J. Am. Chem. Soc.*, **1946,** *68,* 1799–1801.
119. J. Cramer, H. Hansen, and S. Hünig, *J. Chem. Soc., Chem. Commun.*, **1974,** 264–265.
120. T. Curtius, *Ber.*, **1900,** *33,* 2561–2562.
121. T. Curtius, *J. Prakt. Chem.*, **1895,** *52,* 210–226.
122. T. Curtius and P. A. Bleicher, *J. Prakt. Chem.*, **1924,** *107,* 86–98.
123. D. Dal Monte Casoni, *Ann. Chim.* (*Rome*), **1958,** *48,* 783–787; *Chem. Abstr.*, **1959,** *53,* 8123a.
124. A. Darapsky and M. Prabhakar, *Ber.*, **1912,** *45,* 1654–1665.
125. J. DeMendoza et al., *J. Chem. Res.* (*M*), **1980,** 0514–0529.
126. O. Dimroth, *Ann.*, **1904,** *335,* 1–112.
127. *Ibid.*, **1909,** *364,* 183–226.
128. *Ibid.*, **1909,** *364,* 183–226.
129. O. Dimroth, *Ber.*, **1902,** *35,* 4041–4060.
130. *Ibid.*, **1905,** *38,* 670–688.
131. *Ibid.*, **1907,** *40,* 2376–2389.
132. O. Dimroth and K. Pfister, *Ber.*, **1910,** *43,* 2757–2767.

133. O. Dimroth and L. Taub, *Ber.*, **1906,** *39,* 3912–3920.

134. L. DiStefano and R. N. Castle, *J. Heterocycl. Chem.*, **1968,** *5,* 109–110.

135. W. v. E. Doering and C. H. De Puy, *J. Am. Chem. Soc.*, **1953,** *75,* 5955–5957.

135a. D. D'Oria, *J. Chem. Res.* (*S*), **1980,** 242–243.

136. M. P. Doyle and W. Wierenga, *J. Am. Chem. Soc.*, **1972,** *94,* 3896–3899.

137. R. S. Drago and F. E. Paulik, *J. Am. Chem. Soc.*, **1960,** *82,* 96–98.

138. G. F. Duffin, J. D. Kendall, and H. R. J. Waddington, *Chem. Ind.*, **1955,** 1355.

139. R. O. Duthaler, H. G. Förster, and J. D. Roberts, *J. Am. Chem. Soc.*, **1978,** *100,* 4974–4979.

140. T. E. Eagles, M. A. Khan, and B. M. Lynch, *Org. Prep. Proced.*, **1970,** *2,* 117–119.

141. F. Effenberger and P. Fischer, *Tetrahedron,* **1970,** *26,* 3029–3037.

142. G. Ege, *Chem. Ber.*, **1968,** *101,* 3079–3088.

143. F. Einberg, *J. Org. Chem.*, **1967,** *32,* 3687–3689.

144. *Ibid.*, **1970,** *35,* 3978–3980.

145. J. Epsztajn and A. R. Katritzky, *Tetrahedron Lett.*, **1969,** 4739–4742.

146. E. Fahr and L. Neumann, *Angew. Chem.*, **1965,** *77,* 591.

147. M. Falsig and P. E. Iversen, *Acta Chem. Scand.*, **1977,** *B31,* 15–20.

148. A. Farrington and L. Hough, *Carbohydr. Res.*, **1974,** *38,* 107–115.

149. N. R. Fetter and B. Bartocha, *Can. J. Chem.*, **1962,** *40,* 342–347.

150. N. R. Fetter, F. E. Brinckman, and D. W. Moore, *Can. J. Chem.*, **1962,** *40,* 2184–2188.

151. H. Feuer and L. F. Spinialli, *J. Org. Chem.*, **1976,** *41,* 2981–2984.

152. W. G. Finnegan and R. A. Henry, *J. Org. Chem.*, **1965,** *30,* 567–575.

153. W. G. Finnegan, R. A. Henry, and E. Lieber, *J. Org. Chem.*, **1953,** *18,* 779–791.

154. W. Fischer and J.-P. Anselme, *J. Am. Chem. Soc.*, **1967,** *89,* 5284–5285.

155. W. Fischer and J.-P. Anselme, *Tetrahedron Lett.*, **1968,** 877–879.

156. G. W. J. Fleet, I. Fleming, and D. Philippides, *J. Chem. Soc.* (*C*), **1971,** 3948–3950.

157. M. O. Forster and H. E. Fierz, *J. Chem. Soc.*, **1907,** *91,* 1942–1953.

158. M. O. Forster and H. M. Judd, *J. Chem. Soc.*, **1910,** *97,* 254–264.

159. N. I. Foster et al., *Synthesis,* 1980, 572–573.

160. F. W. Fowler, A. Hassner, and L. A. Levy, *J. Am. Chem. Soc.*, **1967,** *89,* 2077–2082.

161. J. E. Franz et al., *J. Org. Chem.*, **1966,** *31,* 2847–2853.

162. K. Freter, F. Hess, and K. Grozinger, *Ann.*, **1973,** 811–820.

163. M. Freund and T. Paradies, *Ber.*, **1901,** *34,* 3110–3122.

164. M. Freund and H. P. Schwarz, *Ber.*, **1896,** *29,* 2491–2499.

165. K. Fries, *Ann.*, **1927,** *454,* 131–172.

166. K. Fries, H. Güterbock, and H. Kühn, *Ann.*, **1934,** *511,* 213–240.

167. K. Fries and P. Ochwat, *Ber.*, **1923,** *56,* 1291–1304.

168. R. J. Friswell and A. G. Green, *Ber.*, **1886,** *19,* 2034–2037.

169. R. J. Friswell and A. G. Green, *J. Chem. Soc.*, **1885,** 917–924.

170. S. M. Gadekar and E. Ross, *J. Org. Chem.*, **1961,** *26,* 613–615.

171. S. F. Gait et al., *J. Chem. Soc., Chem. Comm.*, **1972,** 982–983.

172. *Ibid., Perkin I,* **1974,** 1248–1260.

173. *Ibid.*, **1975,** 19–25.

174. R. Galland and A. Heesing, *Chem. Ber.*, **1973,** *106,* 2580–2589.

175. I. Ganes, R. Taranu, and D. Postescu, *Stud. Univ. Babes-Bolyai, Ser. Chem.*, **1967,** *2,* 105–107; *Chem. Abstr.*, **1968,** *69,* 10401v.

176. R. Garner, E. B. Mullock, and H. Suschitzky, *J. Chem. Soc. (C)*, **1966**, 1980–1983.
177. J. A. Garrison and R. M. Herbst, *J. Org. Chem.*, **1957**, *22*, 278–283.
178. W. Gauss, H. Heitzer, and S. Petersen, *Ann.*, **1972**, *764*, 131–144.
179. H. Gehlen and J. Dost, *Ann.*, **1963**, *665*, 144–149.
180. A. Gescher, M. F. G. Stevens, and C. P. Turnbull, *J. Chem. Soc., Perkin I*, **1977**, 103–106.
181. *Ibid.*, 107–114.
182. H. H. Gibson, H. R. Gaddy, and C. S. Blankenship, *J. Org. Chem.*, **1977**, *42*, 2443–2448.
183. M. S. Gibson, *J. Chem. Soc.*, **1963**, 3539–3541.
184. T. L. Gilchrist, G. E. Gymer, and C. W. Rees, *J. Chem. Soc., Chem. Comm.*, **1973**, 819–820.
185. T. J. Giordano and H. H. Sisler, *Inorg. Chem.*, **1977**, *16*, 2043–2046.
186. T. Giraldi, C. Nisi, and G. Sava, *Biochem. Pharmacol.*, **1975**, *24*, 1793–1797.
187. E. E. Glover, K. T. Rowbottom, and D. C. Bishop, *J. Chem. Soc., Perkin I*, **1973**, 842–845.
188. E. E. Glover and M. Yorke, *J. Chem. Soc. (C)*, **1971**, 3280–3285.
189. H. Gold, *Ann.*, **1965**, *688*, 205–216.
190. H. Goldschmidt and V. Badl, *Ber.*, **1889**, *22*, 933–942.
191. H. Goldschmidt and E. Molinari, *Ber.*, **1888**, *21*, 2557–2579.
192. S. Goldschmidt et al., *Ann.*, **1924**, *437*, 194–226.
193. S. Goldschmidt and K. Euler, *Ber.*, **1922**, *55*, 616–628.
194. Y. G. Gololobov, I. N. Zhmurova, and L. F. Kasukhin, *Tetrahedron*, **1981**, *37*, 437–472.
195. R. Gompper, *Chem. Ber.*, **1957**, *90*, 382–386.
196. R. Gompper and K. Schönafinger, *Chem. Ber.*, **1979**, *112*, 1535–1544.
197. R. Gösl, *Angew. Chem.*, **1962**, *74*, 470.
198. S. Götzky, *Ber.*, **1931**, *64*, 1555–1560.
199. J. Goubeau, E. Allenstein, and A. Schmidt, *Chem. Ber.*, **1964**, *97*, 884–890.
200. A. F. Graefe, *J. Org. Chem.*, **1958**, *23*, 1230–1231.
201. N. N. Greenwood and J. H. Morris, *J. Chem. Soc.*, **1965**, 6205–6209.
202. P. A. Grieco and J. P. Mason, *J. Chem. Eng. Data*, **1967**, *12*, 623–624.
203. C. H. Gudmundsen and W. E. McEwan, *J. Am. Chem. Soc.*, **1957**, *79*, 329–334.
204. C. L. Habraken and E. K. Poels, *J. Org. Chem.*, **1977**, *42*, 2893–2895.
205. G. S. Hammond, B. Seidel, and R. E. Pincock, *J. Org. Chem.*, **1963**, *28*, 3275–3279.
206. R. N. Hanley, W. D. Ollis, and C. A. Ramsden, *J. Chem. Soc., Perkin I*, **1979**, 732–735.
207. R. M. Hann and C. S. Hudson, *J. Am. Chem. Soc.*, **1944**, *66*, 735–738.
208. H. Hansen, S. Hünig, and K. Kishi, *Chem. Ber.*, **1979**, *112*, 445–461.
209. R. J. Harder, R. A. Carboni, and J. E. Castle, *J. Am. Chem. Soc.*, **1967**, *89*, 2643–2647.
210. W. B. Hardy and F. H. Adams (American Cyanamid), U.S. Pat. 2,766,227, Oct. 9, 1956.
211. C. D. Harries, *Ber.*, **1894**, *27*, 696–702.
212. D. Harrison and A. C. B. Smith, *J. Chem. Soc.*, **1960**, 2157–2160.
213. C. V. Hart, *J. Am. Chem. Soc.*, **1928**, *50*, 1922–1930.
214. L. W. Hartzel and F. R. Benson, *J. Am. Chem. Soc.*, **1954**, *76*, 667–670.
215. A. Hassner and F. W. Fowler, *J. Org. Chem.*, **1968**, *33*, 2686–2691.
216. K. Hattori, E. Lieber, and J. P. Horwitz, *J. Am. Chem. Soc.*, **1956**, *78*, 411–415.
217. H. Hauptmann and A. Cid M. Périssé, *Chem. Ber.*, **1956**, *89*, 1081–1094.
218. S. Hauptmann, H. Wilde, and K. Moser, *J. Prakt. Chem.*, **1971**, *313*, 882–888.
219. J. N. Haynes (Hercules), U.S. Pat. 3,686,231, Aug. 22, 1972.
220. J. N. Haynes (Hercules), U.S. Pat. 3,814,657, June 4, 1974.

221. M. Hedayatullah and A. Guy, *Tetrahedron Lett.*, **1975,** 2455–2458.
222. A. F. Hegarty, J. B. Aylward, and F. L. Scott, *J. Chem. Soc.* (*C*), **1967,** 2587–2593.
223. H. W. Heine and D. A. Tomalia, *J. Am. Chem. Soc.*, **1962,** *84,* 993–995.
224. G. Heller, *J. Prakt. Chem.*, **1925,** *111,* 36–53.
225. G. Heller, W. Dietrich, and G. Reichardt, *J. Prakt. Chem.*, **1928,** *118,* 138–148.
226. G. Heller and A. Siller, *J. Prakt. Chem.*, **1927,** *116,* 9–16.
227. B. Hellrung and H. Balli, *Helv. Chim. Acta,* **1975,** *58,* 596–604.
228. *Ibid.*, 605–609.
229. R. A. Henry, *J. Am. Chem. Soc.*, **1951,** *73,* 4470.
230. R. A. Henry, *J. Heterocyclic Chem.*, **1976,** *13,* 391–392.
231. R. A. Henry and W. G. Finnegan, *J. Am. Chem. Soc.*, **1954,** *76,* 923–926.
232. R. A. Henry, W. G. Finnegan, and E. Lieber, *J. Am. Chem. Soc.*, **1954,** *76,* 2894–2898.
233. *Ibid.*, **1955,** *77,* 2264–2270.
234. R. M. Herbst and W. L. Garbrecht, *J. Org. Chem.*, **1953,** *18,* 1283–1291.
235. R. M. Herbst and J. E. Klingbeil, *J. Org. Chem.*, **1958,** *23,* 1912–1916.
236. H. Herlinger, *Angew Chem.*, **1964,** *76,* 437.
237. F. Heusler, *Ber.*, **1891,** *24,* 4156–4161.
238. K. Hirakawa and Y. Tanabiki, *J. Org. Chem.*, **1982,** *47,* 280–284.
239. G. Höfle and B. Lange, *Angew. Chem., Int. Ed. Engl.*, **1976,** *15,* 113–114.
240. K. A. Hofmann and H. Hock, *Ber.*, **1911,** *44,* 2946–2956.
241. K. Hohenlohe-Oehringen, *Monatsh.*, **1958,** *89,* 557–561.
242. *Ibid.*, 562–569.
243. *Ibid.*, 588–596.
244. A. Holm et al., *J. Org. Chem.*, **1975,** *40,* 431–436.
245. H. H. Holst and H. Lund, *Acta Chem. Scand.*, **1979,** *B33,* 233–234.
246. J. P. Horwitz et al., *J. Org. Chem.*, **1962,** *27,* 3045–3048.
247. J. P. Horwitz and V. A. Grakauskas, *J. Am. Chem. Soc.*, **1957,** *79,* 1249–1253.
248. P. G. Houghton and C. W. Rees, *J. Chem. Res.* (*S*), **1980,** 303.
249. A. J. Hubert and G. Anthoine, *Bull. Soc. Chim., Belg.*, **1969,** *78,* 553–560.
250. L. Huff et al., *J. Org. Chem.*, **1970,** *35,* 2074–2076.
251. L. Huff and R. A. Henry, *J. Med. Chem.*, **1970,** *13,* 777–779.
252. J. Huguet, M. Libert, and C. Caullet, *Bull. Soc. Chim. Fr.*, **1972,** 4860–4864.
253. R. Huisgen et al., *Chem. Ber.*, **1960,** *93,* 2106–2124.
254. *Ibid.*, **1965,** *98,* 3992–4013.
255. R. Huisgen and M. Appl, *Chem. Ber.*, **1958,** *91,* 12–21.
256. R. Huisgen, L. Möbius, and G. Szeimies, *Chem. Ber.*, **1965,** *98,* 1138–1152.
257. R. Huisgen, J. Sauer, and M. Seidel, *Chem. Ber.*, **1960,** *93,* 2885–2891.
258. R. Huisgen and M. Seidel, *Chem. Ber.*, **1961,** *94,* 2509–2511.
259. R. Huisgen, H. J. Sturm, and M. Seidel, *Chem. Ber.*, **1961,** *94,* 1555–1562.
260. R. Huisgen, G. Szeimies, and L. Möbius, *Chem. Ber.*, **1966,** *99,* 475–490.
261. R. Huisgen, D. Vossius, and M. Appl, *Chem. Ber.*, **1958,** *91,* 1–12.
262. R. Huisgen and V. Weberndörfer, *Chem. Ber.*, **1967,** *100,* 71–78.
263. R. Hüttel and F. Büchele, *Chem. Ber.*, **1955,** *88,* 1586–1590.
264. R. Hüttel and A. Gebhardt, *Ann.*, **1947,** *558,* 34–47.
265. R. Hüttel and G. Welzel, *Ann.*, **1955,** *593,* 207–218.

266. T. Isida et al., *Bull. Chem. Soc. Jap.*, **1973,** *46,* 1250–1253.
267. E. A. Isukul, R. Ranson, and J. G. Tillett, *J. Org. Chem.*, **1976,** *41,* 3040–3041.
268. T. Itai and S. Kamiya, *Chem. Pharm. Bull.*, **1961,** *9,* 87–91.
269. R. Jacquier et al., *Bull. Soc. Chim. Fr.*, **1971,** 4078–4084.
270. K. A. Jensen and A. Holm, *Acta Chem. Scand.*, **1964,** *18,* 570–572.
271. K. A. Jensen, A. Holm, and S. Rachlin, *Acta Chem. Scand.*, **1966,** *20,* 2795–2806.
272. K. A. Jensen and C. Pedersen, *Acta Chem. Scand.*, **1961,** *15,* 991–1002.
273. D. Jerchel and R. Kuhn, *Ann.*, **1950,** *568,* 185–192.
274. J. W. A. M. Janssen and C. L. Habraken, *J. Org. Chem.*, **1971,** *36,* 3081–3084.
275. A. Jonas and H. v. Pechmann, *Ann.*, **1891,** *262,* 277–302.
276. G. Jones et al., *J. Chem. Soc., Perkin I,* **1981,** 78–81.
277. M. Jureček, J. Šramkova, and P. Kozak, *Coll. Czech. Chem. Commun.*, **1976,** *41,* 2187–2197.
278. M. I. Kabachnik and V. A. Gilyarov, *Invest. Akad. Nauk. SSSR, Otdel. Khim. Nauk, Engl. Ed.*, **1961,** 755–757.
279. P. K. Kadaba, *Synthesis,* **1978,** 694–695.
280. J. C. Kauer and W. A. Sheppard, *J. Org. Chem.*, **1967,** *32,* 3580–3592.
281. G. Kaugars and V. L. Rizzo, *J. Org. Chem.*, **1979,** *44,* 3840–3843.
282. M. Kawanisi, I. Otani, and H. Nozaki, *Tetrahedron Lett.*, **1968,** 5575–5576.
283. G. Kazemifard, F. Moattar, and J. Reisch, *Acta Pharm. Jugosl.*, **1978,** *28,* 151–154.
284. R. C. Kerber, *J. Org. Chem.*, **1972,** *37,* 1587–1592.
285. H. V. Kidd, *J. Org. Chem.*, **1937,** *2,* 198–208.
286. T. Kindt-Larsen and C. Pedersen, *Acta Chem. Scand.*, **1962,** *16,* 1800–1801.
287. W. Kirmse, *Chem. Ber.*, **1960,** *93,* 2353–2356.
288. L. Knorr and P. Duden, *Ber.*, **1893,** *26,* 103–108.
289. Y. Kobayashi, T. Kutsuma, and K. Morinaya, *Chem. Pharm. Bull.*, **1971,** *19,* 2106–2115.
290. G. Koga, *J. Am. Chem. Soc.*, **1969,** *91,* 4323–4324.
291. G. Koga and J.-P. Anselme, *J. Org. Chem.*, **1970,** *35,* 960–964.
292. A. Könnecke, R. Dörre, and E. Lippmann, *Tetrahedron Lett.*, **1978,** 2071–2074.
293. A. Könnecke and E. Lippmann, *Z. Chem.*, **1978,** *18,* 175–177.
294. A. Könnecke, E. Lippmann, and E. Kleinpeter, *Tetrahedron Lett.*, **1976,** 533–536.
295. A. Könnecke, E. Lippmann, and E. Kleinpeter, *Tetrahedron,* **1977,** *33,* 1399–1403.
296. M. Kovacic et al., *J. Heterocycl. Chem.*, **1974,** *11,* 949–952.
297. F. H. Kranz (Allied Chemical), U.S. Pat. 2,981,738, April 25, 1961.
298. L. Krbechek and H. Takimoto, *J. Org. Chem.*, **1964,** *29,* 1150–1152.
299. R. Kreher and K. Goth, *Z. Naturforsch.*, **1976,** *31b,* 131–134.
300. *Ibid.*, 217–221.
301. R. Kreher and G. Jäger, *Angew. Chem.*, **1965,** *77,* 963.
302. R. Kreher and G. Jäger, *Z. Naturforsch.*, **1964,** *19b,* 657–658.
303. A. Kreutzberger and H. van der Goot, *J. Heterocycl. Chem.*, **1975,** *12,* 665–667.
304. F. Krollpfeiffer, A. Rosenberg, and C. Mühlhausen, *Ann.*, **1935,** *515,* 113–130.
305. K. Kubo, T. Nonaka, and K. Odo, *Bull. Chem. Soc. Jap.*, **1976,** *49,* 1339–1341.
306. R. Kuhn and D. Jerchel, *Ann.*, **1952,** *578,* 1–5.
307. R. Kuhn and W. Münzing, *Chem. Ber.*, **1952,** *85,* 29–37.
308. G. L'abbé et al., *Bull. Soc. Chim. Belg.*, **1979,** *88,* 245–251.
309. G. L'abbé et al., *J. Am. Chem. Soc.*, **1974,** *96,* 3973–3978.

310. G. L'abbé and G. Verhelst, *Angew. Chem., Int. Ed. Engl.*, **1976,** *15,* 489–490.
311. G. L'abbé, G. Verhelst, and S. Toppet, *J. Org. Chem.*, **1977,** *42,* 1159–1163.
312. G. L'abbé, G. Verhelst, and G. Vermeulen, *Angew. Chem., Int. Ed. Engl.*, **1977,** *16,* 403–404.
313. G. L'abbé and G. Vermeulen, *Bull. Soc. Chim. Belg.*, **1981,** *90,* 89–98.
314. G. L'abbé, C.-C. Yu, and S. Toppet, *Angew. Chem., Int. Ed. Engl.*, **1977,** *16,* 475–477.
315. R. A. La Forge, C. E. Cosgrove, and A. D'Adamo, *J. Org. Chem.*, **1956,** *21,* 988–992.
316. G. A. Lanovaya and L. A. Kalaida, *J. Gen. Chem. USSR, Engl. Ed.*, **1977,** *47,* 1478–1480.
317. R. J. Le Blanc and K. Vaughan, *Can. J. Chem.*, **1972,** *50,* 2544–2551.
318. J. H. Lee et al., *Chem. Lett.*, **1974,** 951–954.
319. J. E. Leffler and H. H. Gibson, *J. Am. Chem. Soc.*, **1968,** *90,* 4117–4121.
320. J. E. Leffler and R. D. Temple, *J. Am. Chem. Soc.*, **1967,** *89,* 5235–5246.
321. J. E. Leffler and Y. Tsuno, *J. Org. Chem.*, **1963,** *28,* 190–194.
322. H. Lehmkuhl, F. Rabet, and K. Hauschild, *Synthesis,* **1977,** 184–186.
323. C. Leuenberger, L. Hoesch, and A. S. Dreiding, *Helv. Chim. Acta,* **1981,** *64,* 1219–1233.
324. E. S. Lewis and M. D. Johnson, *J. Am. Chem. Soc.*, **1960,** *82,* 5408–5410.
325. D. D. Libman and R. Slack, *J. Chem. Soc.*, **1956,** 2253–2257.
326. J. J. Licari et al., *J. Am. Chem. Soc.*, **1955,** *77,* 5386–5387.
327. E. Lieber, T. S. Chao, and C. N. R. Rao, *J. Org. Chem.*, **1957,** *22,* 238–240.
328. *Ibid.*, 654–662.
329. E. Lieber and C. N. Pillai, *J. Org. Chem.*, **1957,** *22,* 1054–1056.
330. E. Lieber, C. N. Pillai, and R. D. Hites, *Can. J. Chem.*, **1957,** *35,* 832–842.
331. E. Lieber and J. Ramachandran, *Can. J. Chem.*, **1959,** *37,* 101–109.
332. K. H. Linke and R. Turley, *Z. Anorg. Allg. Chem.*, **1970,** *377,* 139–143.
333. H. L. Lochte, W. A. Noyes and J. R. Bailey, *J. Am. Chem. Soc.*, **1922,** *44,* 2556–2567.
334. L. J. Lohr, *Anal. Chem.*, **1953,** *25,* 1117–1118.
335. W. Lorenz (Farbenfabriken Bayer), U.S. Pat. 2,758,115, Aug. 7, 1965.
336. W. Lossen, *Ann.*, **1891,** *263,* 73–108.
337. *Ibid.*, **1897,** *298,* 54–116.
338. C. A. Lovelette and L. Long, Jr., *J. Org. Chem.*, **1972,** *37,* 4124–4128.
339. O. A. Luk'yanov, Y. B. Salamonov, and B. N. Khasapov, *Izv. Akad. Nauk. SSSR Ser. Khim.*, **1981,** 2404–2405.
340. H. Lund, *Disc. Faraday Soc.*, **1968,** *45,* 193–201.
341. B. M. Lynch and T.-L. Chan, *Can. J. Chem.*, **1963,** *41,* 274–277.
342. D. Mackay, D. D. McIntyre, and N. J. Taylor, *J. Org. Chem.*, **1982,** *47,* 532–535.
343. D. MacKay and W. A. Waters, *J. Chem. Soc.* (*C*), **1966,** 813–816.
344. S. M. Mackenzie and M. F. G. Stevens, *J. Chem. Soc.* (*C*), **1970,** 2298–2308.
345. L. J. Madgzinski et al., *Can. J. Chem.*, **1978,** *56,* 1657–1667.
346. O. W. Maender and G. A. Russell, *J. Org. Chem.*, **1966,** *31,* 442–446.
347. A. C. Mair and M. F. G. Stevens, *J. Chem. Soc.* (*C*), **1971,** 2317–2324.
348. P. Margaretha, S. Solar, and O. E. Polansky, *Angew. Chem., Int. Ed. Engl.*, **1971,** *10,* 412–413.
349. J. H. Markgraf, W. T. Bachmann, and D. P. Hollis, *J. Org. Chem.*, **1965,** *30,* 3472–3474.
350. F. D. Marsh, *J. Org. Chem.*, **1972,** *37,* 2966–2969.
351. D. Martin and A. Weise, *Chem. Ber.*, **1966,** *99,* 317–327.
352. K. Masuda, Y. Imashiro, and T. Kaneko, *Chem. Pharm. Bull.*, **1970,** *18,* 128–132.
353. G. L. Matevosyan et al., *Zh. Obsch. Khim., Engl. Ed.*, **1980,** *50,* 1217–1220.

354. M. A. Mathur, H. H. Sisler, and L. N. Morgenthaler, *Inorg. Chem.*, **1979,** *18,* 2350–2354.

355. G. Maury et al., *Bull. Soc. Chim. Belg.*, **1982,** *91,* 153–161.

356. W. E. McEwen, W. E. Conrad, and C. A. Vanderwerf, *J. Am. Chem. Soc.*, **1952,** *74,* 1168–1171.

357. J. F. McGarrity, *J. Chem. Soc. Chem. Comm.*, **1974,** 558–559.

358. F. A. McGinn and G. B. Brown, *J. Am. Chem. Soc.*, **1960,** *82,* 3193–3195.

359. A. McKillop and R. J. Kobylecki, *J. Org. Chem.*, **1974,** *39,* 2710–2714.

360. J. Meisenheimer, O. Senn, and P. Zimmermann, *Ber.*, **1927,** *60,* 1736–1748.

361. O. Meth-Cohn, R. K. Smalley, and H. Suschitzky, *J. Chem. Soc.*, **1963,** 1666–1669.

362. C. J. Michejda and D. Romans, *Tetrahedron Lett.*, **1969,** 4213–4216.

363. C. J. Michejda, T. J. Tipton, and D. H. Campbell, *IARC Sci. Publ.*, **1976,** *14,* Environ. N-Nitroso Compd. Anal. Form., Proc. Work Conf. 1975, 255–260; *Chem. Abstr.*, **1977,** *86,* 188746a.

364. T. Mitsuhashi and O. Simamura, *J. Chem. Soc. (B)*, **1970,** 705–711.

365. J. A. Montgomery and R. D. Elliott, *Nucl. Acid. Chem.*, **1978,** 687–691.

366. J. A. Montgomery and H. J. Thomas, *J. Med. Chem.*, **1972,** *15,* 182–187.

367. H. W. Moore and H. R. Shelden, *J. Org. Chem.*, **1968,** *33,* 4019–4024.

368. H. W. Moore, H. R. Shelden, and D. F. Shellhamer, *J. Org. Chem.*, **1969,** *34,* 1999–2001.

369. E. Muckermann, *Ber.*, **1909,** *42,* 3449–3460.

370. E. Müller and H. Haiss, *Chem. Ber.*, **1962,** *95,* 1255–1263.

371. U. Müller and K. Dehnicke, *Angew. Chem., Int. Ed. Engl.*, **1966,** *5,* 841–842.

372. J. Munch-Petersen, *Org. Syn.*, **1963,** *Coll. Vol. IV,* 715–717.

373. M. E. Munk and Y. K. Kim, *J. Am. Chem. Soc.*, **1964,** *86,* 2213–2217.

374. A. W. Murray and K. Vaughan, *J. Chem. Soc. (C)*, **1970,** 2070–2074.

375. W. J. Musliner and J. W. Gates, Jr., *J. Am. Chem. Soc.*, **1966,** *88,* 4271–4273.

376. P. Nantka-Namirski and J. Kalinowski, *Acta Pol. Pharm.*, **1973,** *30,* 1–8.

377. *Ibid.*, **1974,** *31,* 125–136.

378. *Ibid.*, 137–145.

379. P. Nantka-Namiriski and J. Kalinowski, Polish Pat. 68,831, Oct. 31, 1973; *Chem. Abstr.*, **1974,** *80,* 133436v.

380. U. G. Nayak and R. L. Whistler, *J. Org. Chem.*, **1968,** *33,* 3582–3585.

381. R. Neidlein and K. Salzmann, *Synthesis,* **1975,** 52–54.

382. S. F. Nelsen, *J. Am. Chem. Soc.*, **1966,** *88,* 5666–5667.

383. S. F. Nelsen, R. B. Metzler, and M. Iwamura, *J. Am. Chem. Soc.*, **1969,** *91,* 5103–5107.

384. F. A. Neugebauer and H. Fischer, *Chem. Ber.*, **1973,** *106,* 1589–1594.

385. P. N. Neuman, *J. Heterocycl. Chem.*, **1970,** *7,* 1159–1160.

386. E. A. Neves and D. W. Franco, *Talanta,* **1979,** *26,* 81–84.

387. M. S. Newman and H. L. Gildenhorn, *J. Am. Chem. Soc.*, **1948,** *70,* 317–319.

388. E. D. Nicolaides, R. D. Westland, and E. L. Whittle, *J. Am. Chem. Soc.*, **1954,** *76,* 2887–2891.

389. A. W. Nineham, *Chem. Rev.*, **1955,** *55,* 356–483.

390. A. Nohara, *Tetrahedron Lett.*, **1974,** 1187–1190.

391. E. Nölting and F. Binder, *Ber.*, **1887,** *20,* 3004–3018.

392. Y. Nomura et al., *Bull. Chem. Soc. Jap.*, **1981,** *54,* 261–266.

393. W. P. Norris and R. A. Henry, *J. Org. Chem.*, 1964, *29,* 650–660.

394. W. P. Norris and R. A. Henry, *Tetrahedron Lett.*, **1965,** 1213–1215.

395. T. Novinson et al., *J. Med. Chem.*, **1976,** *19,* 517–520.

396. H. Oelschlaeger and U. Bremer, *Arch. Pharm.* (*Weinheim*), **1971,** *304,* 717–722.

397. Y. Ogata, Y. Nakagawa, and M. Inaishi, *Bull. Chem. Soc. Jap.*, **1981,** *54,* 2853–2854.

398. J. K. O'Halloran et al., *J. Chem. Soc., Perkin I,* **1972,** 2214–2219.

399. A. Ohsawa et al., *J. Chem. Soc., Chem. Commun.*, **1980,** 1182–1183.

399a. G. A. Olah and D. J. Donovan, *J. Org. Chem.*, **1978,** *43,* 860–865.

400. G. A. Olah, S. C. Narang, and A. P. Fung, *J. Org. Chem.*, **1981,** *44,* 2706–2709.

401. J. E. Oliver and J. B. Stokes, *J. Heterocycl. Chem.*, **1970,** *7,* 961–965.

402. E. Oliveri-Mandala, *Atti Accad. Lincei,* **1910,** *19,* I, 228–231.

403. C. E. Olsen, *Acta Chem. Scand.*, **1973,** *27,* 2989–2996.

404. *Ibid.*, **1975,** *B29,* 953–962.

405. C. E. Olsen and C. Pedersen, *Tetrahedron Lett.*, **1968,** 3805–3809.

406. V. A. Ostrovskii, I. Y. Shirobokov, and G. I. Koldobskii, *Zh. Org. Khim., Engl. Ed.*, **1981,** *17,* 131–135.

407. C. Paal and W.-N. Yao, *Ber.*, **1930,** *63,* 57–66.

408. M. H. Palmer et al., *J. Chem. Soc., Perkin II,* **1975,** 1695–1700.

409. T. M. Patterson, R. K. Smalley, and H. Suschitzky, *Synthesis,* **1975,** 709–710.

410. H. v. Pechmann and L. Frobenius, *Ber.*, **1895,** *28,* 170–176.

411. C. Pedersen, *Acta Chem. Scand.*, **1959,** *13,* 888–892.

412. W. Pelz (Agfa), U.S. Pat. 2,950,273, Aug. 23, 1960.

413. D. F. Percival and R. M. Herbst, *J. Org. Chem.*, **1957,** *22,* 925–933.

414. M. J. Perkins, *J. Chem. Soc.*, **1964,** 3005–3008.

415. R. Pfleger, E. Garthe, and K. Rauer, *Chem. Ber.*, **1963,** *96,* 1827–1839.

416. K. Pilgram, F. Görgen, and G. Pollard, *J. Heterocycl. Chem.*, **1971,** *8,* 951–959.

417. J. Plenkiewicz and T. Zdrojewski, *Bull. Soc. Chim. Belg.*, **1981,** *90,* 193–194.

418. S. Polanc, B. Stanovnik, and M. Tišler, *J. Heterocycl. Chem.*, **1973,** *10,* 565–567.

419. S. Polanc, B. Stanovnik, and M. Tišler, *J. Org. Chem.*, **1976,** *41,* 3152–3155.

420. K. T. Potts and S. Husain, *J. Org. Chem.*, **1970,** *35,* 3451–3456.

421. F. Povazanec, J. Kovac, and J. Svoboda, *Collect. Czech. Chem. Commun.*, **1980,** *45,* 1299–1300.

422. R. Preussmann, A. v. Hodenberg, and H. Hengy, *Biochem. Pharm.*, **1969,** *18,* 1–13.

423. H. Quast and L. Bieber, *Chem. Ber.*, **1981,** *114,* 3253–3272.

424. H. Quast and S. Hünig, *Ann.*, **1968,** *711,* 157–173.

425. R. Raap, *Can. J. Chem.*, **1969,** *47,* 3677–3681.

426. R. Raap and J. Howard, *Can. J. Chem.*, **1969,** *47,* 813–819.

427. M. J. Rance et al., *J. Chem. Soc., Chem. Commun.*, **1974,** 658–659.

428. D. R. Randall, E. A. Cox, and A. Atkinson (Geigy), U.S. Pat. 3,597,353, Aug. 3, 1971.

429. J. O. Reed and W. Lwowski, *J. Org. Chem.*, **1971,** *36,* 2864–2869.

430. C. W. Rees and A. A. Sale, *J. Chem. Soc., Perkin I,* **1973,** 545–550.

431. C. W. Rees, R. W. Stephenson, and R. C. Storr, *J. Chem. Soc., Chem. Comm.*, **1974,** 941–942.

432. C. W. Rees and R. C. Storr, *J. Chem. Soc.* (*C*), **1969,** 756–760.

433. *Ibid.*, 760–764.

434. *Ibid.*, 765–769.

435. *Ibid.*, 1474–1477.

436. *Ibid.*, 1478–1483.

437. C. W. Rees and D. E. West, *J. Chem. Soc. (C)*, **1970,** 583–589.
438. M. Regitz, *Chem. Ber.*, **1966,** *99,* 3128–3147.
439. M. Regitz and F. Menz, *Chem. Ber.*, **1968,** *101,* 2622–2632.
440. H. Reimlinger, *Chem. Ber.*, **1964,** *97,* 3493–3502.
441. H. Reimlinger, W. R. F. Lingier, and J. J. M. Vandewalle, *Chem. Ber.*, **1975,** *108,* 3780–3786.
442. G. Rembarz, B. Kirchhoff, and G. Dongowski, *J. Prakt. Chem.*, **1966,** *33,* 199–205.
443. M. Remes et al., *Collect. Czech. Chem. Commun.*, **1975,** *40,* 2104–2108.
444. J. L. Riebsomer, *J. Org. Chem.*, **1948,** *13,* 815–821.
445. M. N. Rosenfeld and D. A. Widdowson, *J. Chem. Soc., Chem. Commun.*, **1979,** 914–916.
446. E. Rosenhauer and H. Unger, *Ber.*, **1928,** *61,* 392–398.
447. L. M. Rossi and P. Trimarco, *Synthesis,* **1978,** 465–467.
448. A. A. Rysakov et al., *Elektrokhimiya,* **1974,** *10,* 1888–1891; *Chem. Abstr.*, **1975,** *82,* 117799b.
449. T. Saegusa, Y. Ito, and T. Shimizu, *J. Org. Chem.*, **1970,** *35,* 3995–3996.
450. K. Sakai and J.-P. Anselme, *J. Org. Chem.*, **1972,** *37,* 2351–2352.
451. T. Sasaki et al., *J. Org. Chem.*, **1977,** *42,* 3741–3743.
452. M. J. Sasse and R. C. Storr, *J. Chem. Soc., Perkin I,* **1978,** 909–912.
453. C. W. Schellhammer, J. Schroeder, and N. Joop, *Tetrahedron,* **1970,** *26,* 497–510.
454. A. Schmidt, *Chem. Ber.*, **1966,** *99,* 2976–2983.
455. *Ibid.*, **1967,** *100,* 3319–3325.
456. *Ibid.*, 3725–3735.
457. *Ibid.*, **1970,** *103,* 3923–3927.
458. U. Schmidt and M. Dietsche, *Angew. Chem., Int. Ed. Engl.*, **1981,** *20,* 771–772.
459. *Ibid.*, **1982,** *21,* 143–144.
460. E. Schmitz, H. Fechner-Simon, and S. Schramm, *Ann.*, **1969,** *725,* 1–10.
461. A. Schönberg and W. Urban, *J. Chem. Soc.*, **1935,** 530–532.
462. G. Schroeter, *Ber.*, **1909,** *42,* 2336–2349.
463. F. L. Scott, *Experientia,* **1957,** *13,* 275–276.
464. F. L. Scott and D. A. Cronin, *Chem. Ind.*, **1964,** 1757.
465. F. L. Scott, D. A. Cronin, and J. K. O'Halloran, *J. Chem. Soc. (C)*, **1971,** 2769–2775.
466. F. L. Scott and J. C. Tobin, *J. Chem. Soc. (C)*, **1971,** 703–706.
467. D. Seebach et al., *Helv. Chim. Acta,* **1978,** *61,* 1622–1647.
468. M. P. Servé, P. G. Seybold, and H. P. Covault, *J. Heterocycl. Chem.*, **1978,** *15,* 127–128.
469. K. W. Sherk, A. G. Houpt, and A. W. Browne, *J. Am. Chem. Soc.*, **1940,** *62,* 329–331.
470. M. L. Shpak et al., *Zh. Org. Khim., Engl. Ed.*, **1978,** *14,* 2252–2254.
471. M. S. S. Siddiqui and M. F. G. Stevens, *J. Chem. Soc., Perkin I,* **1974,** 609–610.
472. *Ibid.*, 611–615.
473. D. H. Sieh and C. J. Michejda, *J. Am. Chem. Soc.*, **1981,** *103,* 442–445.
474. D. H. Sieh, D. J. Wilbur, and C. J. Michejda, *J. Am. Chem. Soc.*, **1980,** *102,* 3883–3887.
475. M. Sittig, *Pesticide Manufacturing and Toxic Materials Central Encyclopedia,* Noyes Data Corp., Park Ridge, N.J., 1980, pp. 69–73.
476. C. Smith and C. H. Watts, *J. Chem. Soc.*, **1910,** *97,* 562–571.
477. P. A. S. Smith, *J. Am. Chem. Soc.*, **1954,** *76,* 436–441.
478. P. A. S. Smith and B. B. Brown, *J. Am. Chem. Soc.*, **1951,** *73,* 2438–2441.
479. P. A. S. Smith, J. H. Hall, and R. O. Kan, *J. Am. Chem. Soc.*, **1962,** *84,* 485–489.
480. P. A. S. Smith and D. H. Kenny, *J. Org. Chem.*, **1961,** *26,* 5221–5224.
481. P. A. S. Smith, C. D. Rowe, and L. B. Bruner, *J. Org. Chem.*, **1969,** *34,* 3430–3433.

482. R. H. Spector and M. M. Joullié, *J. Heterocycl. Chem.*, **1969,** *6,* 605–612.
483. H. A. Staab, *Chem. Ber.*, **1956,** *89,* 1927–1940.
484. H. A. Staab and W. Benz, *Ann.*, **1961,** *648,* 72–82.
485. B. Stanovnik, private communication, July 5, 1982.
486. B. Stanovnik et al., *J. Heterocycl. Chem.*, **1979,** *16,* 1567–1570.
487. B. Stanovnik et al., *Synthesis,* **1978,** 65–66.
488. B. Stanovnik et al., *Tetrahedron Lett.*, **1978,** 3059–3062.
489. B. Stanovnik and M. Tišler, *J. Heterocycl. Chem.*, **1971,** *8,* 785–787.
490. B. Stanovnik et al., *Synthesis,* **1977,** 491–492.
491. H. Staudinger and E. Hauser, *Helv. Chim. Acta,* **1921,** *4,* 861–886.
492. H. Staudinger and J. Meyer, *Helv. Chim. Acta,* **1919,** *2,* 635–646.
493. H. N. E. Stevens and M. F. G. Stevens, *J. Chem. Soc.* (*C*), **1970,** 765–771.
494. *Ibid.*, 2284–2289.
495. *Ibid.*, 2289–2298.
496. *Ibid.*, 2308–2312.
497. M. F. G. Stevens, *J. Chem. Soc., Perkin I,* **1974,** 615–620.
498. E. Stöldt and R. Kreher, *Angew. Chem.*, **1978,** *90,* 215–216.
499. R. Stollé, *J. Prakt. Chem.*, **1933,** *138,* 1–17.
500. R. Stollé et al., *J. Prakt. Chem.*, **1932,** *134,* 282–309.
501. R. Stollé and F. Henke-Stark, *J. Prakt. Chem.*, **1930,** *124,* 261–300.
502. D. R. Sutherland and J. Pickard, *J. Heterocycl. Chem.*, **1974,** *11,* 457–458.
503. D. R. Sutherland and G. Tennant, *J. Chem. Soc.* (*C*), **1971,** 706–713.
504. *Ibid.*, 2156–2162.
505. G. Szeimies and R. Huisgen, *Chem. Ber.*, **1966,** *99,* 491–503.
506. I. Tabakovic, M. Trkovnik, and Z. Grujic, *J. Chem. Soc., Perkin II,* **1979,** 166–171.
507. H. Taguchi, *Tetrahedron Lett.*, **1973,** 1137–1138.
508. H. Takeuchi and K. Koyama, *J. Chem. Soc., Perkin I,* **1982,** 1269–1273.
509. Y. Tamura, J. H. Kim, and M. Ikeda, *J. Heterocycl. Chem.*, **1975,** *12,* 107–110.
510. Y. Tanaka and S. I. Miller, *J. Org. Chem.*, **1973,** *38,* 2708–2712.
511. Y. Tanaka and S. I. Miller, *Tetrahedron,* **1973,** *29,* 3285–3296.
511a. M. Tanno and S. Kamiya, *Chem. Pharm. Bull.*, **1979,** *27,* 1824–1829.
512. P. Tavs, H. Sieper, and H. Beecken, *Ann.*, **1967,** *704,* 150–160.
513. J. M. Tedder and B. Webster, *J. Chem. Soc.*, **1960,** 4417–4420.
514. G. Tennant, *J. Chem. Soc.* (*C*), **1966,** 2290–2295.
515. *Ibid.*, **1967,** 1279–1283.
516. F. Texier and R. Carrié, *Bull. Soc. Chim. Fr.*, **1971,** 3642–3648.
517. W. Theilacker and E.-C. Fintelmann, *Chem. Ber.*, **1958,** *91,* 1597–1600.
518. J. Thiele, *Ann.*, **1898,** *303,* 57–75.
519. J. Thiele and H. Ingle, *Ann.*, **1895,** *287,* 233–265.
520. J. Thiele and K. Schleussner, *Ann.*, **1897,** *295,* 129–172.
521. J. Thiele and K. Sieglitz, *Ann.*, **1910,** *375,* 334–335.
522. W. E. Thun and W. R. McBride, *J. Org. Chem.*, **1969,** *34,* 2997–3002.
523. W. E. Thun, D. W. Moore, and W. R. McBride, *J. Org. Chem.*, **1966,** *31,* 923–925.
524. H.-J. Timpe, *Z. Chem.*, **1971,** *11,* 340–341.
525. W. Tochtermann, H. Gustmann, and C. Wolff, *Chem. Ber.*, **1978,** *111,* 566–573.
526. W. M. Tolles, D. W. Moore, and W. E. Thun, *J. Am. Chem. Soc.*, **1966,** *88,* 3476–3479.

527. N. H. Toubro and A. Holm, *J. Chem. Soc., Perkin I,* **1978,** 1440–1445.

528. S. Treppendahl, P. Jakobsen, and G. Vernin, *Acta Chem. Scand.,* **1978,** *B32,* 777–778.

529. I. Ugi, H. Perlinger, and L. Behringer, *Chem. Ber.,* **1958,** *91,* 2324–2329.

530. *Ibid.,* 2330–2336.

531. F. Ullmann and C. Gross, *Ber.,* **1910,** *43,* 2694–2704.

532. K. Utvary, *Monatsh.,* **1968,** *99,* 1483–1488.

533. C. A. VanderWerf, R. Y. Heisler, and W. E. McEwen, *J. Am. Chem. Soc.,* **1954,** *76,* 1231–1235.

534. E. Van Loock, G. L'abbé, and G. Smets, *J. Org. Chem.,* **1971,** *36,* 2520–2524.

535. E. Van Loock, G. L'abbé, and G. Smets, *Tetrahedron,* **1972,** *28,* 3061–3067.

536. L. I. Vereschagin et al., *Zh. Org. Khim., Engl. Ed.,* **1979,** *15,* 1551–1557.

537. G. Vernin, A. K. El Shafei, and J. Metzger, *J. Chem. Res.* (*S*), **1980,** 150–151.

538. G. Vernin and J. Metzger, *Synthesis,* **1978,** 921–924.

539. A. Vollmar and A. Hassner, *J. Heterocycl. Chem.,* **1974,** *11,* 491–496.

540. I. Wadsö, *Acta Chem. Scand.,* **1960,** *14,* 903–908.

541. W. S. Wadsworth, Jr., *J. Org. Chem.,* **1969,** *34,* 2994–2997.

542. G. Wagner and H. Gentzsch, *Arch. Pharm.* (*Weinheim*), **1968,** *301,* 923–933.

543. S. Weckherlin and W. Lüttke, *Ann.,* **1966,** *700,* 59–64.

544. S. J. Weininger et al., *J. Org. Chem.,* **1974,** *39,* 1591–1592.

545. E. H. White, A. A. Baum, and D. E. Eitel, *Org. Synth.,* **1973,** *Coll. Vol. V,* 797–800.

546. E. H. White and H. Scherrer, *Tetrahedron Lett.,* **1961,** 758–762.

547. N. Wiberg et al., *Chem. Ber.,* **1980,** *113,* 2916–2927.

548. N. Wiberg, H. Bayer, and H. Bachhuber, *Angew. Chem., Int. Ed. Engl.,* **1975,** *14,* 177–178.

549. H. Wiberg and G. Ziegleder, *Chem. Ber.,* **1978,** *111,* 2123–2129.

550. H. Wieland, *Ber.,* **1908,** *41,* 3498–3506.

551. H. Wieland and H. Fressel, *Ann.,* **1912,** *392,* 133–151.

552. R. H. Wiley and J. Moffat, *J. Am. Chem. Soc.,* **1955,** *77,* 1703–1704.

553. R. H. Wiley and J. Moffat, *J. Org. Chem.,* **1957,** *22,* 995–996.

554. F. H. Witt, *Ber.,* **1913,** *46,* 2557–2559.

555. G. Wittig and A. Krebs, *Chem. Ber.,* **1961,** *94,* 3260–3275.

556. G. Wittig and J. Meske-Schüller, *Ann.,* **1968,** *711,* 65–75.

557. A. Wohl, *Ber.,* **1900,** *33,* 2759–2760.

558. A. Wohl and C. Oesterlin, *Ber.,* **1900,** *33,* 2736–2744.

559. A. Wohl and H. Schiff, *Ber.,* **1900,** *33,* 2741–2758.

560. *Ibid.,* **1902,** *35,* 1900–1904.

561. R. A. Wohl, *J. Org. Chem.,* **1973,** *38,* 3862–3864.

562. L. Wolff, *Ber.,* **1904,** *37,* 2827–2836.

563. L. Wolff, *Ann.,* **1902,** *325,* 152–169.

564. L. Wolff and H. Lindenhayn, *Ber.,* **1904,** *37,* 2374–2381.

565. P. Ykman, G. L'abbé, and G. Smets, *Tetrahedron Lett.,* **1970,** 5225–5228.

566. M. Yoshida, A. Matsumoto, and O. Simamura, *Bull. Chem. Soc. Jap.,* **1970,** *43,* 3587–3590.

567. L. L. Zaika and M. M. Joullié, *J. Heterocycl. Chem.,* **1966,** *3,* 444–449.

568. D. M. Zimmerman and R. A. Olofson, *Tetrahedron Lett.,* **1970,** 3453–3456.

569. T. Zincke and P. Schwarz, *Ann.,* **1900,** *311,* 329–340.

570. M. Zupan, B. Stanovnik, and M. Tišler, *Tetrahedron Lett.,* **1972,** 4179–4182.

571. V. Zverina et al., *Collect. Czech. Chem. Commun.,* **1971,** *36,* 1598–1603.

7

SYNTHESIS

Syntheses of classes with three or more linked nitrogens are outlined in this chapter. Discussion is first from the viewpoint of the combination of nitrogen atoms creating nitrogen-nitrogen bonds, followed by a summary of ring closures of intact nitrogen chains. Some syntheses of high nitrogen compounds involving reactions with preformed nitrogen chains have been described in Chapter 6, such as alkylation of azide ion to yield aliphatic azides, electrophilic amination of *v*-triazoles and tetrazoles to form *N*-amino derivatives, and oxidation of triazenes to produce triazene oxides.

7.1. FORMATION OF NITROGEN-NITROGEN BONDS

In this section, syntheses are described in which nitrogen-nitrogen bonds are formed, building nitrogen sequences in high nitrogen classes except for those involving cyclization of intact chains with three or more nitrogen atoms, considered in Section 7.2.

Oxidation and reduction are involved directly or indirectly in establishing all nitrogen-nitrogen bonds. Many such reactions consist of combinations of nitrogen species which attain valence levels between those of the nitrogen reactants, such as the interactions of ammonia with nitric acid to form nitrous oxide or of anilinium ion with nitrous acid to produce benzenediazonium ion (146). Most formations of nitrogen-nitrogen bonds in syntheses of high nitrogen compounds are the result of reactions between nitrogen atoms at different oxidation states; in other words, relatively electron-poor and electron-rich nitrogen atoms combine. In other syntheses of nitrogen-nitrogen linkages, external oxidants or reductants convert nitrogen atoms to electron configurations suitable for joining nitrogen at other oxidation levels. The few formations of nitrogen-nitrogen bonds from nitrogen atoms nominally at the same oxidation number are dismutations which result in nitrogen products in both relatively higher and lower oxidation states or in the formation of nitrogen chains, often dipolar, containing nitrogen atoms at different oxidation levels.

Triazanes are obtained by addition of amines to azo compounds substituted with electron-withdrawing groups. Aminoformyloxaziridines aminate mono-

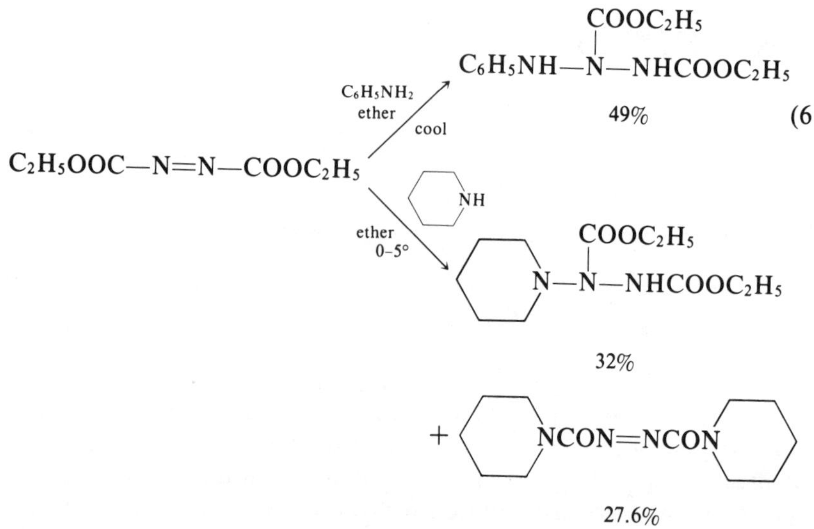

(61, 51)

(77)

and unsymmetrical disubstituted hydrazines forming aminoformyltriazanes **(I)** (264). Triazanium salts result from the amination of hydrazines. 2-Dialkylamino-

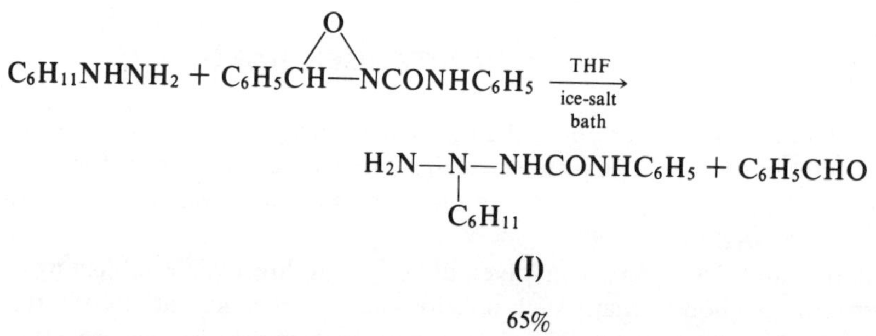

(264)

$$N_2H_4 + H_2NOSO_3H \xrightarrow[0^\circ]{\text{methanol}} H_2N\overset{+}{N}H_2NH_2\ HSO_4^- \qquad (79)$$

66%

$$(CH_3)_2NNH_2 + H_2NCl \xrightarrow[\text{ice cooling}]{\text{benzene}} H_2N{-}^{+}N(CH_3)_2{-}NH_2\ Cl^- \qquad (275)$$

30–50%

1,3,2-dioxaphospholanes react with chloramine and ammonia, forming dialkyl triazanium salts and a polymeric phosphorus compound.

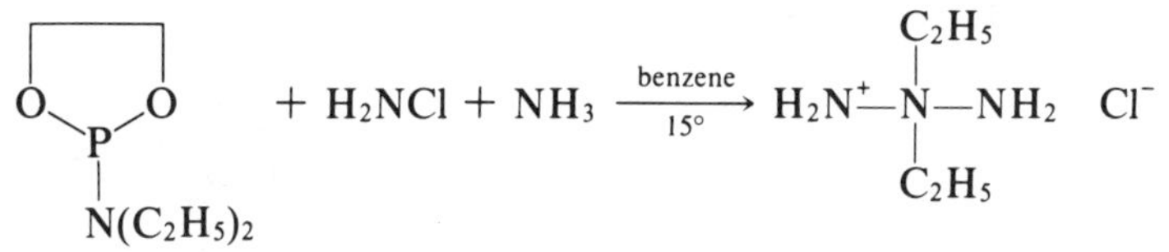

(306)

62%

The principal synthesis of triazenes is coupling of a diazonium salt with primary and secondary aromatic and aliphatic amines usually in slightly basic aqueous solution. Amidines and amides also couple with diazonium compounds. The triazene from *p*-chlorobenzenediazonium chloride and formamide

$$C_6H_5N_2^+Cl^- + C_6H_5NH_2 \xrightarrow[\text{ice cooling}]{\text{NaOAc, }H_2O} C_6H_5N{=}NNHC_6H_5 \qquad (113)$$

82–85%

$$C_6H_5N_2^+Cl^- + (CH_3)_2NH \xrightarrow[\text{cool}]{Na_2CO_3,\ H_2O} C_6H_5N{=}NN(CH_3)_2 \qquad (74)$$

93%

$$C_6H_5N_2^+Cl^- + H_2N\underset{\underset{NH}{\|}}{C}NHCN \xrightarrow[\text{ice cooling}]{\text{NaOAc, }H_2O} C_6H_5N{=}NNH\underset{\underset{NH}{\|}}{C}NHCN \qquad (313)$$

71%

can be dehydrated to form *p*-chlorobenzenediazoisocyanide **(II)** (147). Coupling also occurs in nonaqueous systems, illustrated by the combination of sodium acetanilide with benzenediazonium fluoride in tetrahydrofuran to pro-

$$p\text{-}ClC_6H_4N_2^+Cl^- + H_2N\underset{\underset{O}{\|}}{C}H \xrightarrow[\text{reflux}]{NaHCO_3,\ \text{ether}/H_2O} p\text{-}ClC_6H_4N{=}NNHCHO \xrightarrow[\text{pyridine, } -5^\circ]{SOCl_2}$$

70%

$$p\text{-}ClC_6H_4N{=}NNC$$

(II)

50%

duce 3-acetyl-1,3-diphenyltriazene **(III)** and the formation of *cis*-1-*p*-nitrophenyl-3-[3-methylbenzothiazolinylidene-2]triazene **(IV)**. Separate preparation of

$$C_6H_5N_2^+BF_4^- + CH_3CO\bar{N}C_6H_5 \xrightarrow[5^\circ]{THF} C_6H_5N{=}N\underset{}{N}(COCH_3){-}C_6H_5 \quad (55)$$

(where $CH_3CO\bar{N}C_6H_5$ is formed from $CH_3CONHC_6H_5$ with NaH)

(III)

43%

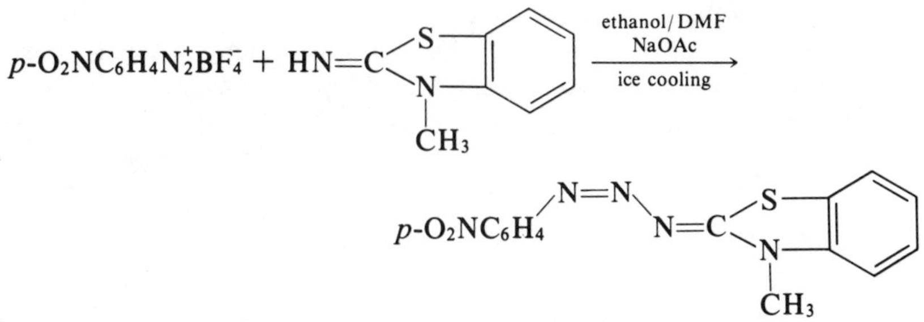

(76)

(IV)

70%

diazonium salts is not necessary, as illustrated by the conversion of aniline to diazoaminobenzene in hexane (308).

$$2C_6H_5NH_2 \xrightarrow[\text{room temp.}]{i\text{-}C_5H_{11}ONO,\ \text{hexane}} C_6H_5N{=}NNHC_6H_5$$

80%

Triazene oxides result from coupling diazonium compounds with hydroxylamine derivatives. Reaction of *O,N*-dimethylhydroxylamine leads to an *O*-

$$p\text{-}CH_3C_6H_4N_2^+Cl^- + HONHCH_2C_6H_5 \xrightarrow[0\text{–}5^\circ]{NaOAc,\ H_2O} p\text{-}CH_3C_6H_4N{=}N\overset{+}{N}(O^-){-}CH_2C_6H_5 \quad (14)$$

70%

methoxytriazene, but the product from *O*-methylhydroxylamine loses methanol to form an azide (34).

Diazonium salts couple with the anions of alkyl and aryl nitramines to form *N*-nitrotriazenes.

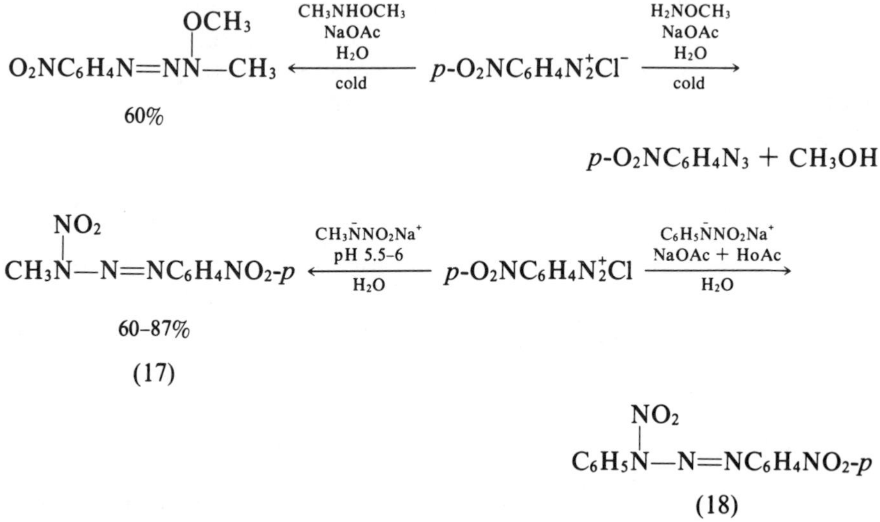

Nitrosation of 2,2′-diaminobiphenyl with aqueous nitrous acid leads to the cyclic triazene, dibenzo[*d,f*][1,2,3]triazepine **(V)**, but under anhydrous conditions, amyl nitrite forms the isomeric benzo[*c*]cinnoline *N*-imide **(VI)** and by-products (88).

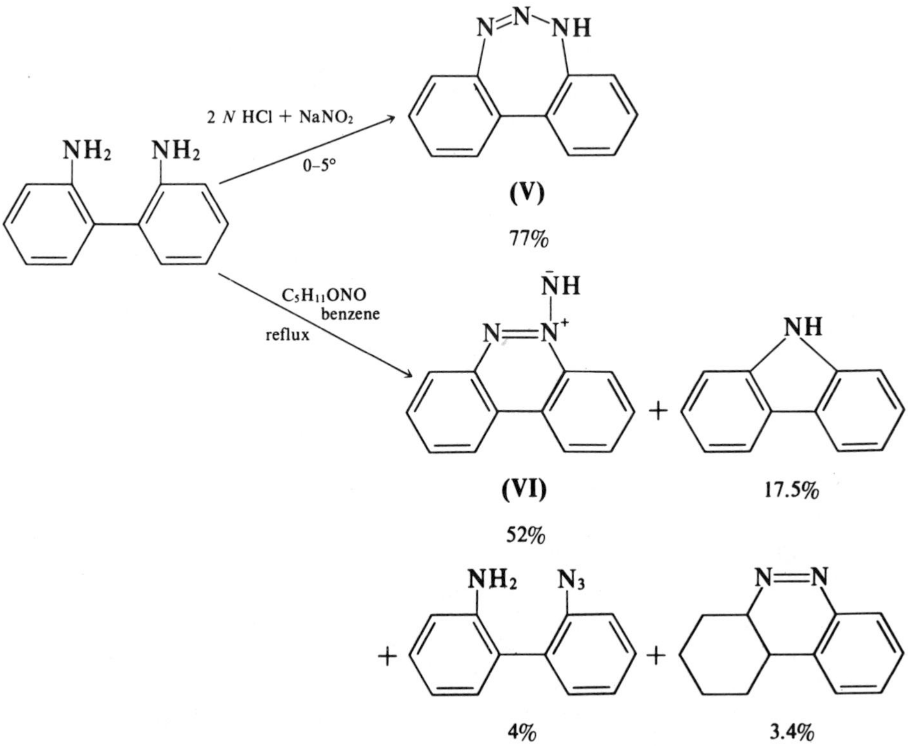

Hydrazines and derivatives containing an NH group condense with diazonium compounds to form 1-tetrazenes. Diazonium coupling with aldehyde phenylhydrazones occurs on hydrazine nitrogen in the pH range 4–8; at pH 3 and below, coupling is on the benzene ring (119).

$$p\text{-HOOCC}_6\text{H}_4\text{N}_2^+\text{Cl}^- \xrightarrow[\text{cool}]{\text{H}_2\text{NNHC}_6\text{H}_5,\ \text{NaOAc}} \text{HOOCC}_6\text{H}_4\text{N}{=}\text{NN}(\text{NH}_2)\text{—C}_6\text{H}_5 \quad (318)$$

80%

$$\text{C}_6\text{H}_5\text{N}_2^+\text{OAc} \xrightarrow[\text{pH 6.9},\ 0^\circ]{\text{C}_6\text{H}_5\text{CH}{=}\text{NNHC}_6\text{H}_5,\ \text{ethanol}} \text{C}_6\text{H}_5\text{CH}{=}\text{N—N}(\text{C}_6\text{H}_5)\text{—N}{=}\text{N—C}_6\text{H}_5$$

66%

Pentazdienes result from coupling 2 moles of diazonium compound with 1 mole of ammonia or primary amine. Pentazdienes, including derivatives with

$$2p\text{-CH}_3\text{C}_6\text{H}_4\text{N}_2^+\text{Cl}^- + \text{conc. NH}_3 \xrightarrow[0^\circ]{} p\text{-CH}_3\text{C}_6\text{H}_4\text{N}{=}\text{NNHN}{=}\text{NC}_6\text{H}_4\text{CH}_3\text{-}p \quad (226)$$

21%

$$2\text{C}_6\text{H}_5\text{N}_2^+\text{Cl}^- + \text{aq. CH}_3\text{NH}_2 \xrightarrow[0^\circ]{} \text{C}_6\text{H}_5\text{N}{=}\text{NN}(\text{CH}_3)\text{N}{=}\text{NC}_6\text{H}_5 \quad (3)$$

69%

different substituents on positions 1 and 5, can be prepared by diazo coupling of triazenes.

$$p\text{-CH}_3\text{C}_6\text{H}_4\text{N}{=}\text{NNHC}_6\text{H}_4\text{CH}_3\text{-}p + \text{C}_6\text{H}_5\text{N}_2^+\text{Cl}^- \xrightarrow[\text{cool}]{\text{NaOCH}_3,\ \text{methanol}} p\text{-CH}_3\text{C}_6\text{H}_4\text{N}{=}\text{NN}(\text{C}_6\text{H}_4\text{CH}_3\text{-}p)\text{N}{=}\text{NC}_6\text{H}_5 \quad (227)$$

68–82%

Hexazdienes are obtained by condensation of 2 moles of diazonium compounds with symmetrical diacylhydrazines (133).

v-Triazolines result from the addition of diazomethane to Schiff's bases (156). Yields are highest with electron-withdrawing groups in the *N*-aryl substituent. A cyclic-cyclic bitriazoline with a six-nitrogen chain is formed by combination of 2 moles of diazomethane with hexafluoroacetone azine (43).

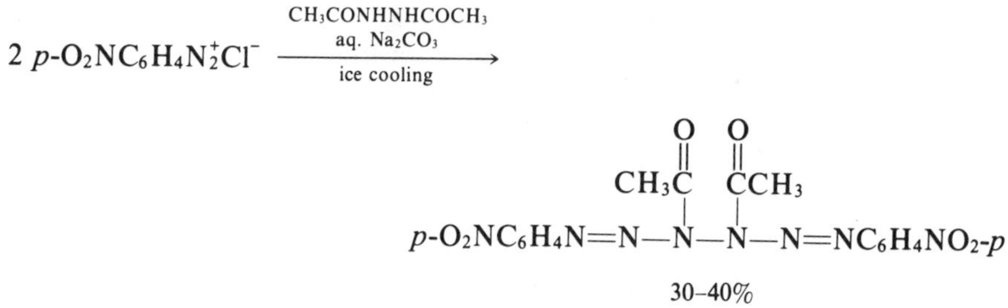

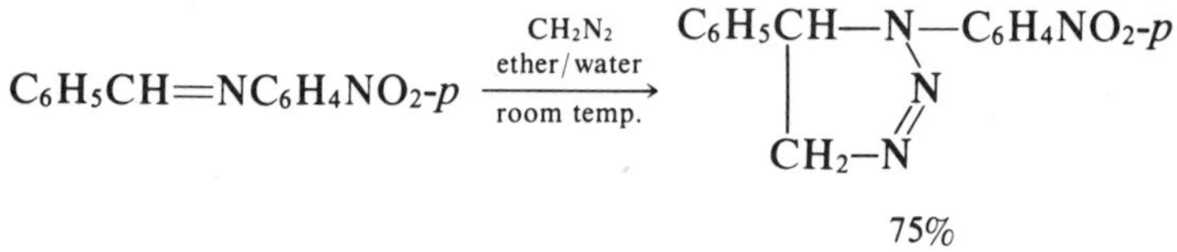

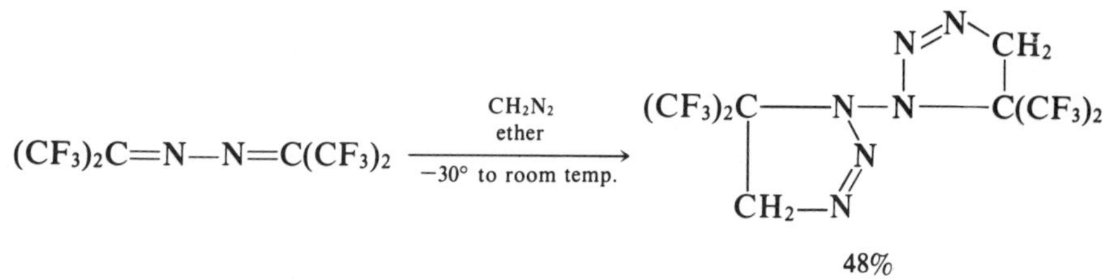

Triazenes or triazene oxides are formed by reaction of nitroso compounds with hydrazines. Normally, less than one-third of the nitroso derivative combines with the hydrazine to make a three-nitrogen chain, most of the nitroso

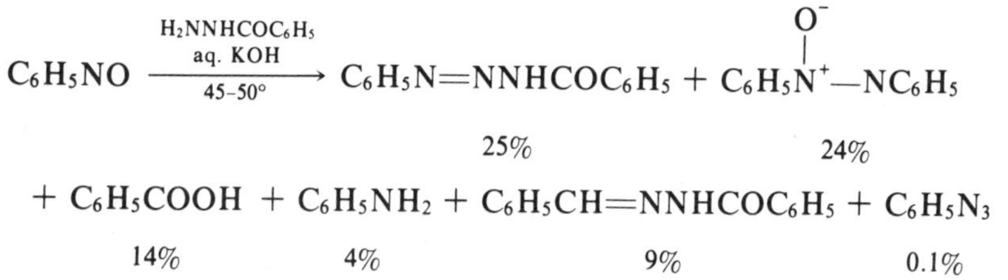

(150)

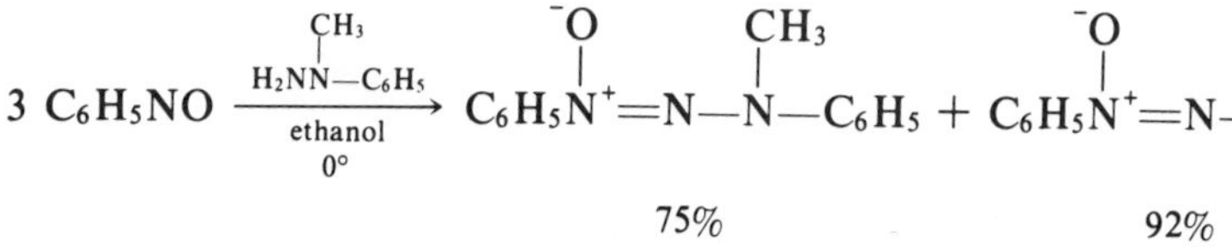

(15, 128a)

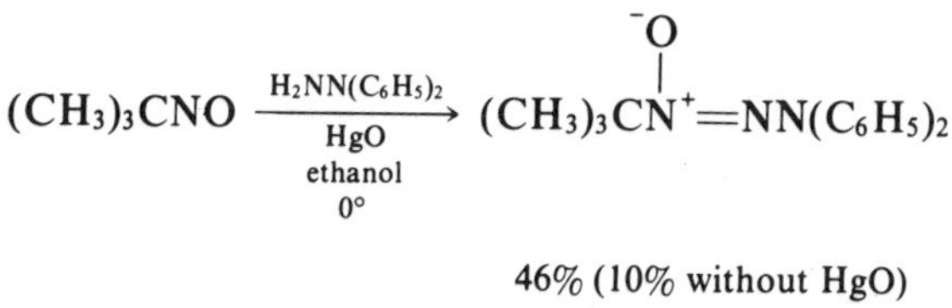

(128a)

compound acting as an oxidant forming an azoxy product. When mercuric oxide is included as a reactant, fair yields result from use of equimolar quantities of nitroso and hydrazine. Phthalimido nitrene, generated oxidatively from *N*-aminophthalimide, unites with nitroso compounds to form 3,3-phthaloyltriazene oxides. *N*-(*N*-Nitrosoarylamino)methyl succinimides **(VII)** react with sec-

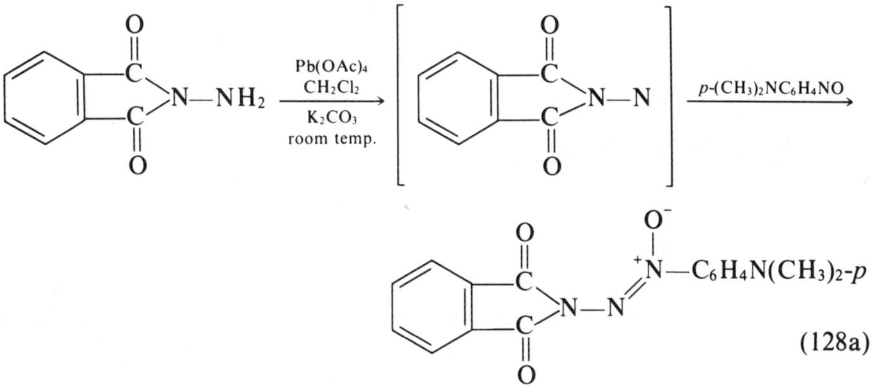

(128a)

92%

ondary amines to yield 1-aryl 3,3-disubstituted triazenes with accompanying loss of *N*-methylsuccinimide and water (258).

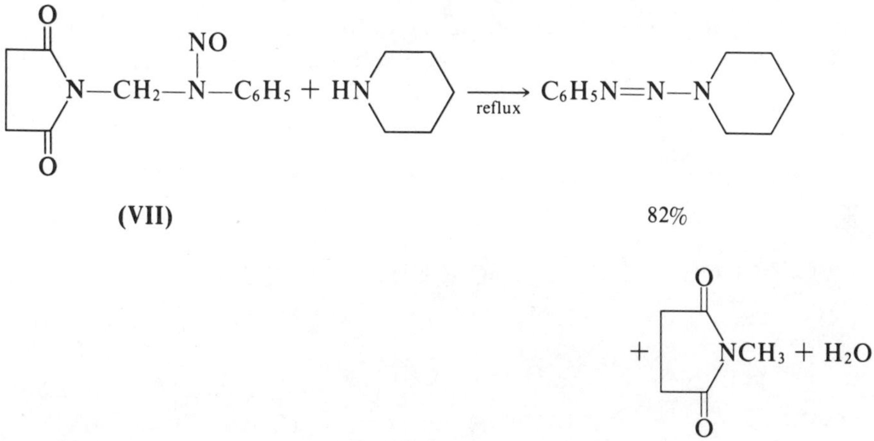

(VII) 82%

A sulfonyltriazene results from the transfer of a tosyl nitrogen unit from tosyl azide to lithium 1,1-dibenzylhydrazide (164).

$$(C_6H_5CH_2)_2N\bar{N}HLi^+ \xrightarrow[-60^\circ \text{ to } -10^\circ]{N_3SO_2C_6H_4CH_3\text{-}p \atop \text{THF}} N_2 + (C_6H_5CH_2)_2N{-}N{=}NSO_2C_6H_4CH_3\text{-}p$$

11%

Single nitrogen-nitrogen bonds are formed by the reaction of nitrous acid and other sources of nitrosyl with imino groups, illustrated by the preparation of acyclic and acyclic-cyclic nitrosohydrazines and acyclic nitrosotriazenes.

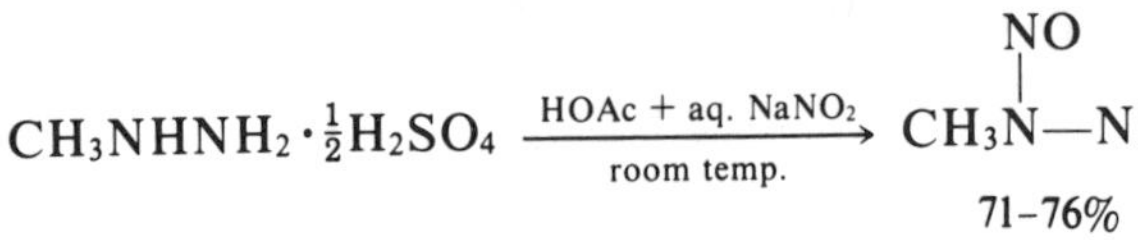

(298)

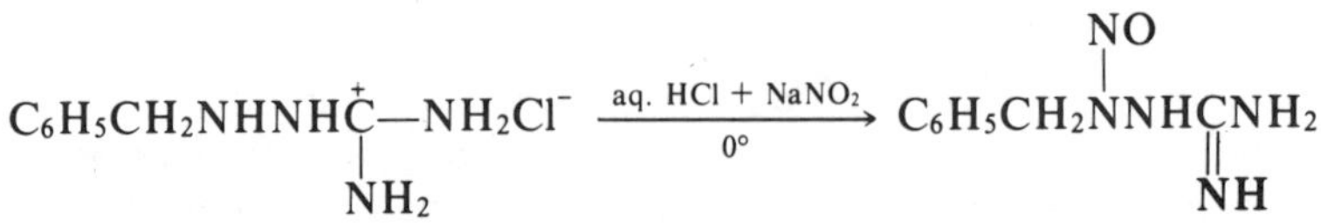

(80)

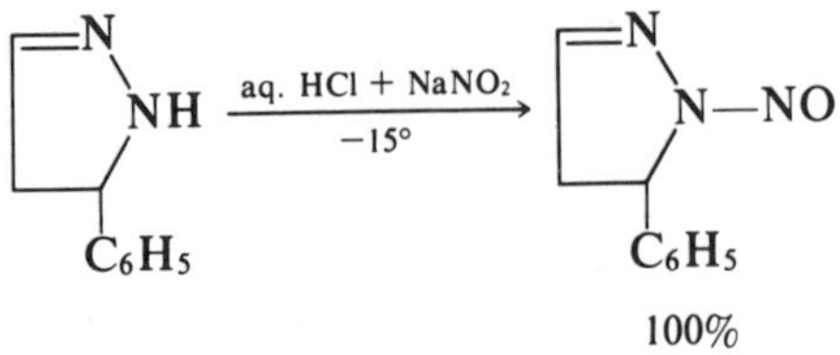

(8)

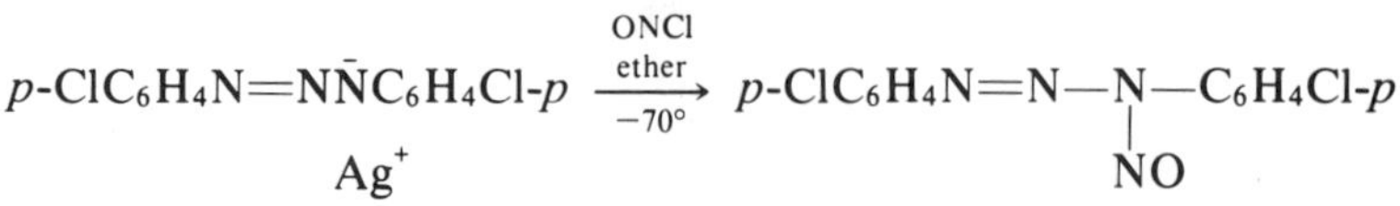

(207)

Nitric oxide combines with amines to form three-nitrogen anions (68) and with hydrazyl radicals to form nitrosohydrazines. Hydrazyl radicals are generated either by dissociation of tetrazanes or by oxidation of trisubstituted hydrazines.

$$(n\text{-}C_4H_9)_2NH \xrightarrow[-78^\circ]{\text{NO, ether}} \underset{80\%}{(n\text{-}C_4H_9)_2N\bar{N}_2O_3\ (n\text{-}C_4H_9)_2\overset{+}{N}H_2} \quad (67)$$

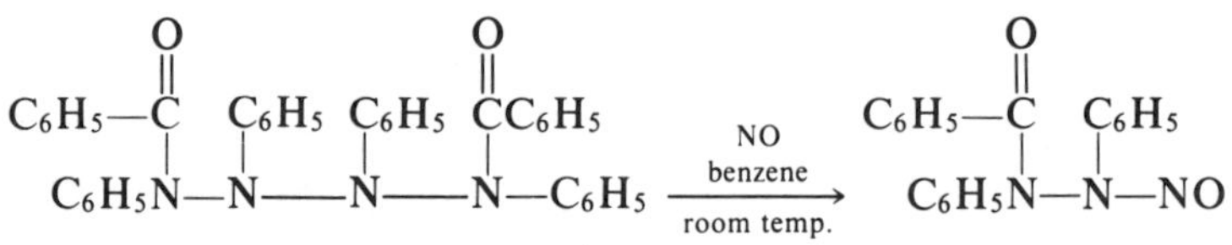

(100)

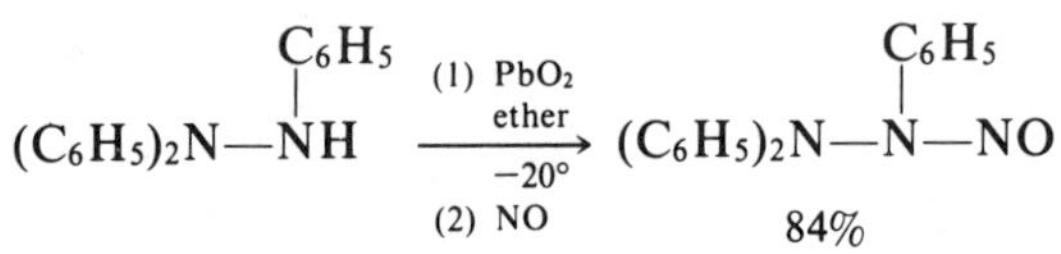

(98)

An acyclic-cyclic silapentazene results from the oxidation of bis[2-lithio-1,2-bis(trimethylsilyl)hydrazine]dimethylsilane with ethyl nitrite, which also furnishes a nitrogen for the five-nitrogen chain (316).

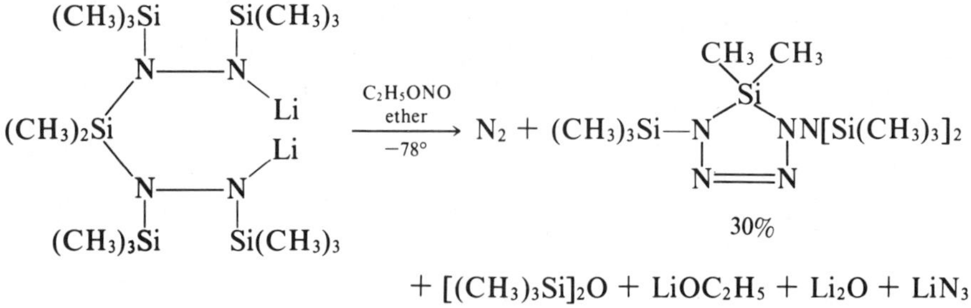

1-Nitropyrazoles result from nitration of pyrazoles with hydrogen on ring nitrogen (153a). Acyclic and acyclic-cyclic hydrazinium nitrate salts with an unsubstituted amino group are converted to amine nitroimides by anhydride reagents with or without additional nitric acid. The sulfonyl moiety in the 1-alkyl-4-benzenesulfonylimino-1,2,4-triazolium betaine is replaced by nitro.

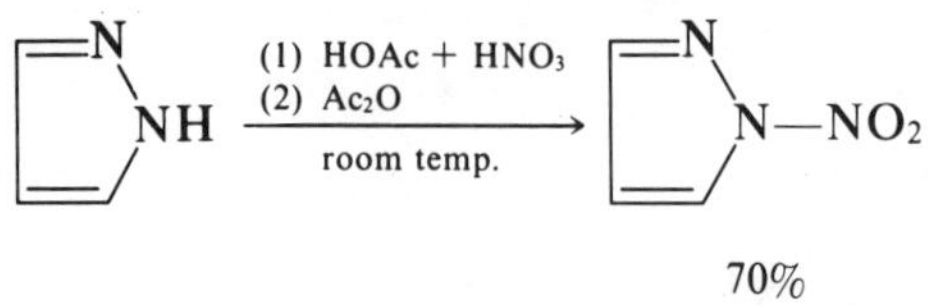 (145)

$$(CH_3)_3\overset{+}{N}NH_2\ NO_3^- \xrightarrow[0^\circ]{HNO_3 + Ac_2O} (CH_3)_3\overset{+}{N}\overset{-}{N}NO_2 \qquad (75)$$

72%

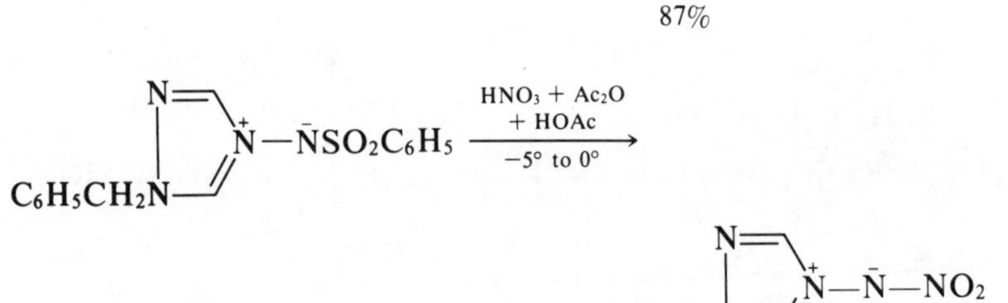 (75)

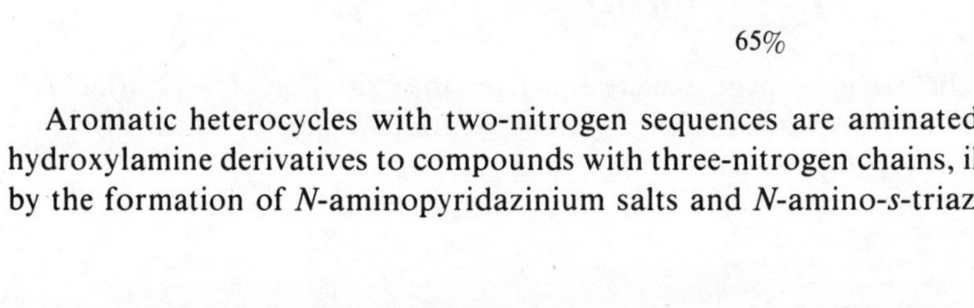 (300)

Aromatic heterocycles with two-nitrogen sequences are aminated with *O*-hydroxylamine derivatives to compounds with three-nitrogen chains, illustrated by the formation of *N*-aminopyridazinium salts and *N*-amino-*s*-triazoles. The

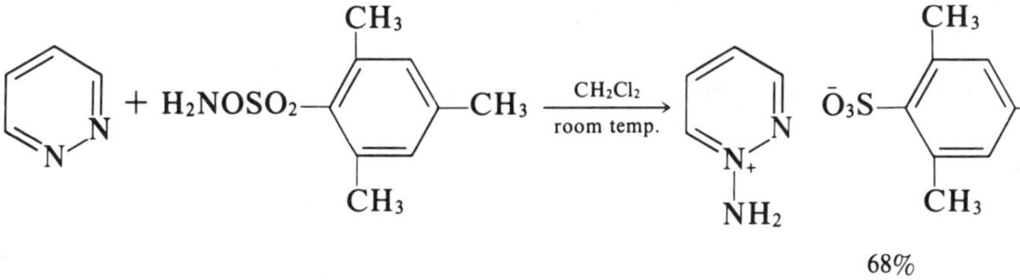

(293)

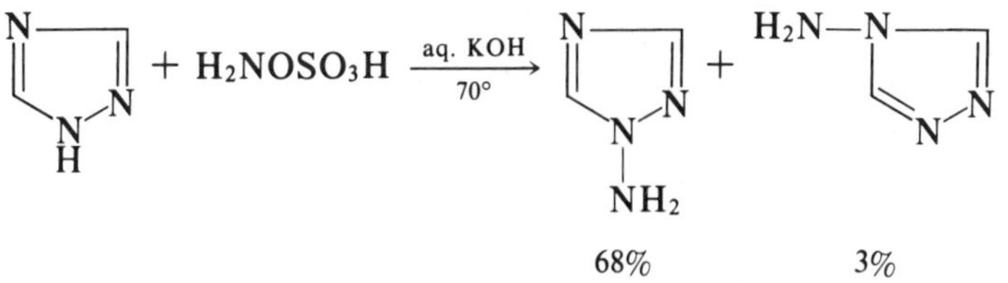

(59)

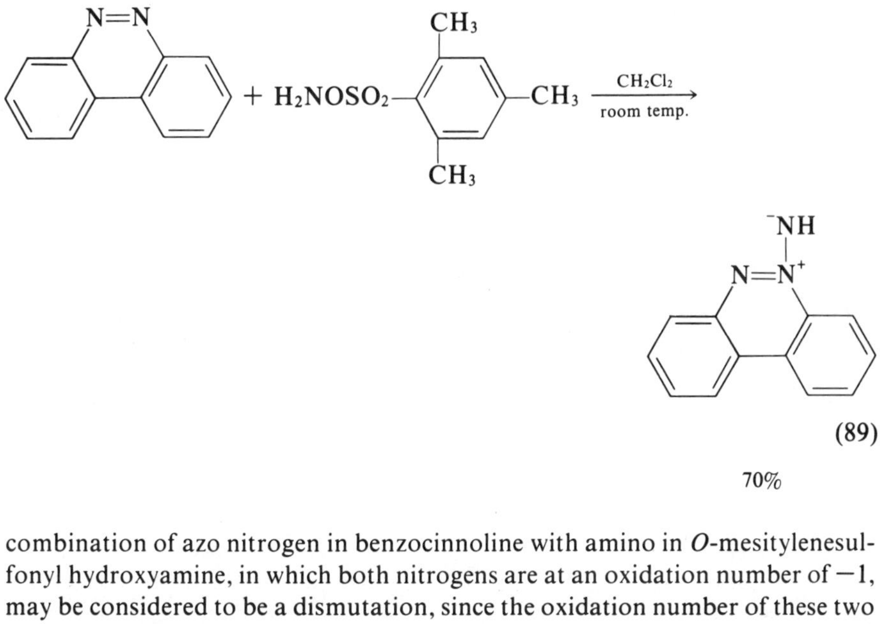

(89)

70%

combination of azo nitrogen in benzocinnoline with amino in *O*-mesitylenesulfonyl hydroxyamine, in which both nitrogens are at an oxidation number of −1, may be considered to be a dismutation, since the oxidation number of these two nitrogens in benzo[*c*]cinnoline *N*-imide are −2 for imide and zero for the quaternary nitrogen.

The acyclic-cyclic three-nitrogen chain in 1-aminopyrazolin-5-ones is formed by thermal rearrangement of 5-isoxazolylhydrazines (1).

Alkoxycarbonyl azimines are obtained by the interaction of aliphatic and aromatic azo compounds with alkoxy nitrenes generated from *N*[(*p*-nitrophenyl)sulfonyloxy]carbamate **(VIII)** (180). Similarly the *N*-aminophthal-

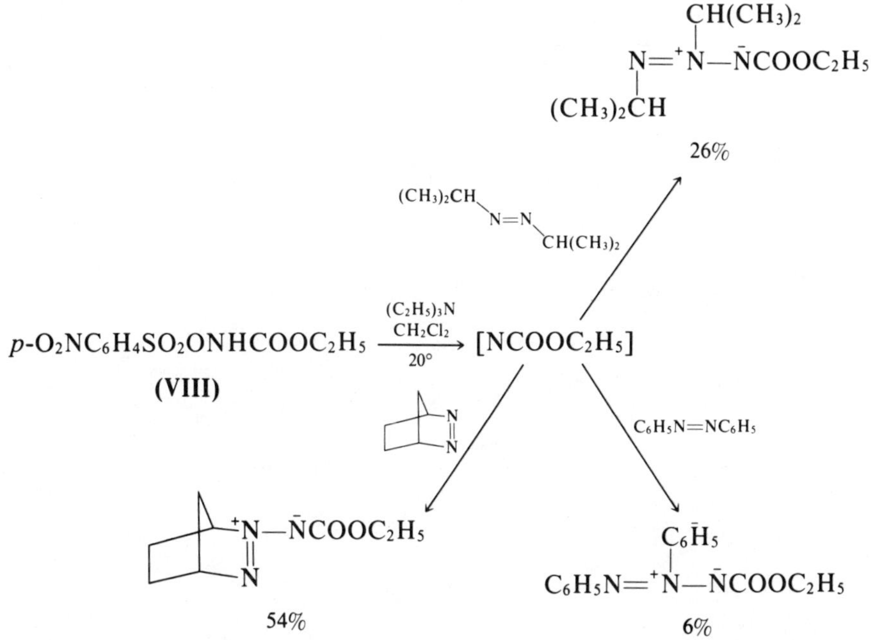

imide hydrazone of *o*-azidobenzaldehyde decomposes via the azide derived nitrene, which combines with hydrazone nitrogen to form 2(*N*-phthalimido)-indazole (257).

N-Acylbenzo[*c*]cinnoline *N*-imides result from oxidation of 2,2′-bis(2-acylhydrazino)biphenyls (215).

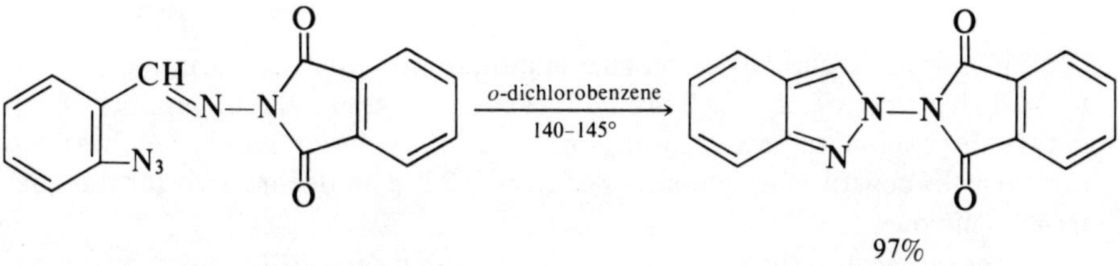

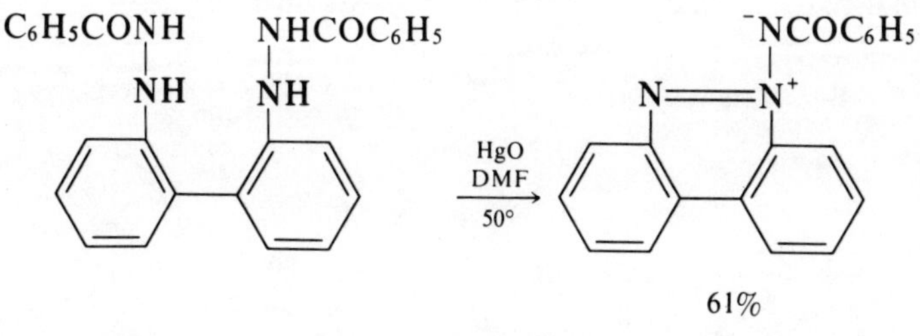

Sodium azide, the principal source of the azide moiety in high nitrogen compounds, is manufactured by interaction of nitrous oxide with sodium amide, obtained from sodium and ammonia, either as a gas-solid reaction or in liquid ammonia (251, 316a). Formation of both inorganic and organic azides occurs by reaction of

$$2NaNH_2 \xrightarrow[170\text{–}190^\circ]{N_2O} NaN_3 + NaOH + NH_3$$

organic nitrites or nitrous acid with hydrazine or monosubstituted hydrazines.

$$N_2H_4 \cdot H_2O \xrightarrow[\substack{\text{methanol/ether} \\ \text{ice cooling}}]{\substack{C_2H_5ONO \\ NaOCH_3}} NaN_3 + H_2O \qquad (284, 299)$$

$$C_6H_5NHNH_2 \xrightarrow[0\text{–}5^\circ]{\text{aq. HCl} + NaNO_2} \underset{65\text{–}68\%}{C_6H_5N_3} \qquad (187)$$

$$H_2NNHCOOC(CH_3)_3 \xrightarrow[10\text{–}15^\circ]{\text{aq. HOAc} + NaNO_2} \underset{64\text{–}82\%}{N_3COOC(CH_3)_3} \qquad (45)$$

Aryl azides result from the interaction of diazonium salts of oxidizing anions such as perbromide or hexachloroplumbate with ammonia. Sulfonyl azides transfer two nitrogens to alkyl and aryl amines, forming the corresponding azides. Phase transfer catalysts usually enhance the yield of azide. α-Azidocar-

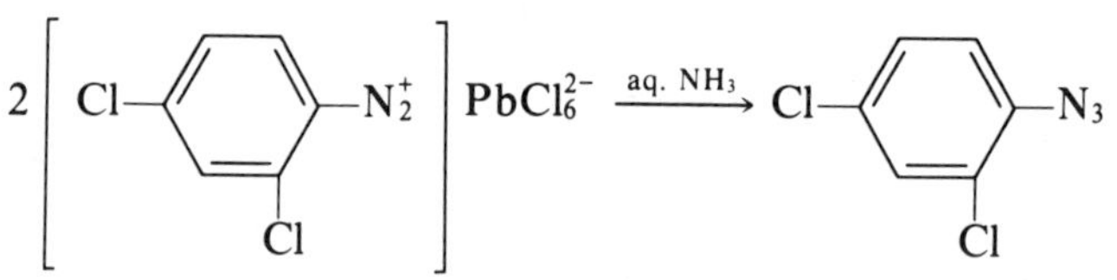

(49)

almost quant.

$$C_6H_5NH_2 + N_3SO_2C_6H_4CH_3\text{-}p \xrightarrow[\text{reflux}]{\substack{\text{aq. NaOH} \\ C_6H_5CH_2\overset{+}{N}(C_2H_5)_3\ Cl^- \\ \text{benzene}}} \underset{94\%}{C_6H_5N_3} \qquad (213)$$

$$\underset{\displaystyle NH_2}{CH_3CHCOOC_2H_5} + CF_3SO_2N_3 \xrightarrow[\text{room temp.}]{\substack{(C_8H_{17})_3\overset{+}{N}CH_3\ Cl^- \\ CH_2Cl_2}} \underset{\substack{\displaystyle N_3 \\ 62\%}}{CH_3CHCOOC_2H_5} \qquad (323)$$

boxylate salts are obtained from the action of trifluoromethylsulfonyl azide on α-amino acids under mild conditions with high retention of optical activity.

Tetra(ethoxycarbonyl)tetrazane is obtained by photochemical reduction of ethylazodicarboxylate with isopropanol (261).

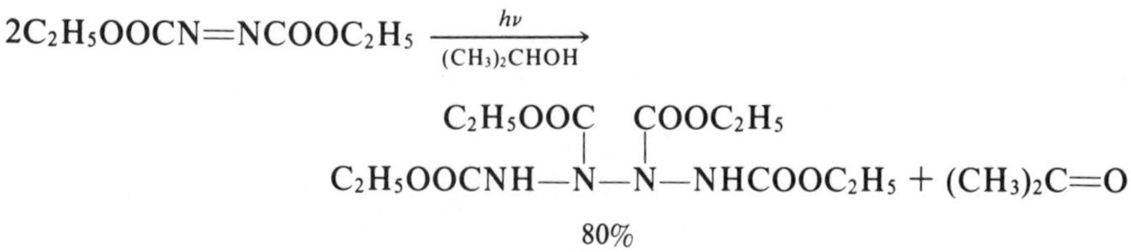

Oxidative coupling is extensively used for the synthesis of acyclic, acyclic-cyclic, and cyclic nitrogen chains. Tetrazanes, including pyrazolidine-dione dimers, are obtained by oxidation of trisubstituted hydrazines forming a nitrogen-nitrogen single bond. Similarly, oxidation of the imino group in disubstituted triazenes leads to hexazdienes such as the 1,3,4,6-tetra(p-chlorophenyl) derivative **(IX)** (297).

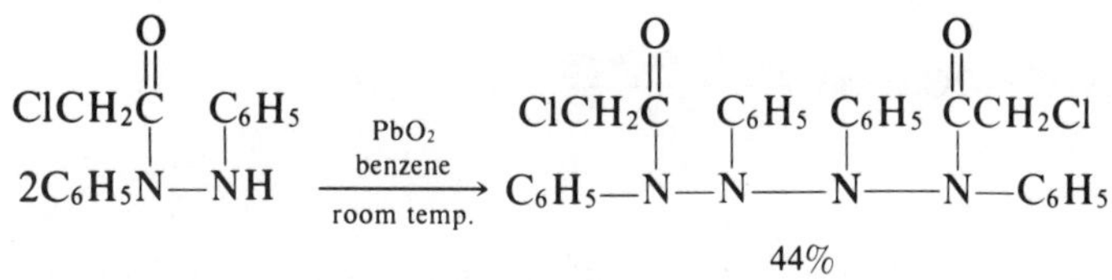

(99)

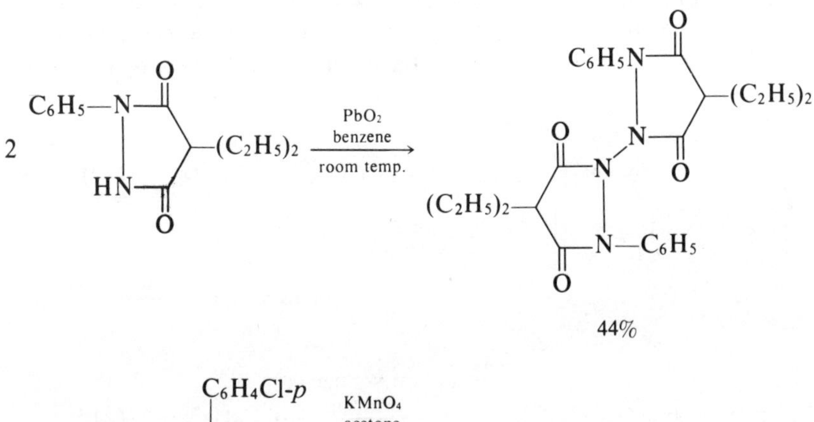

(234)

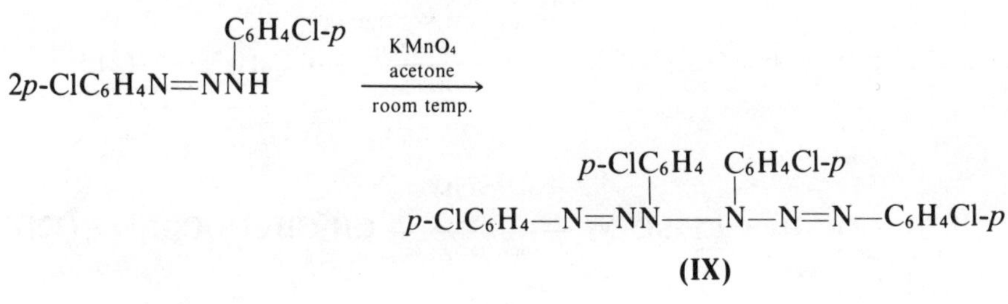

(IX)

20%

2-Tetrazenes result from oxidation of unsymmetrical disubstituted hydrazines with formation of an azo linkage. The mechanism of the reaction appears

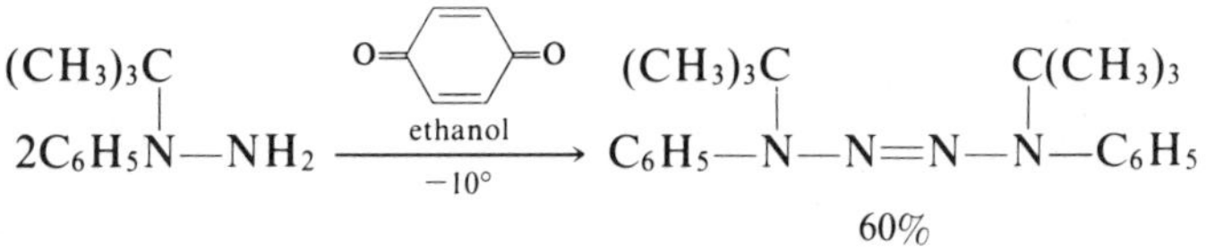

(214)

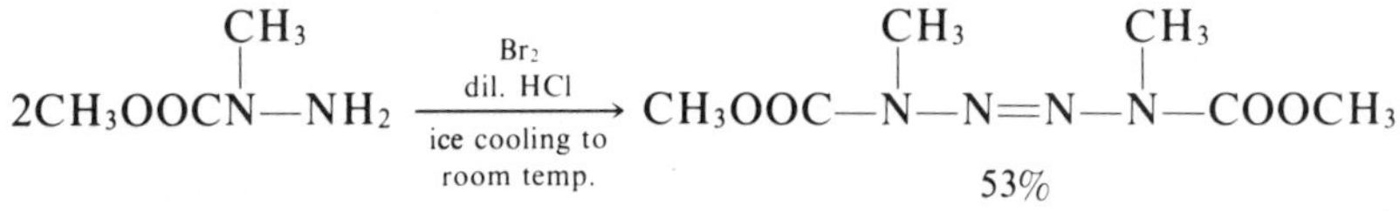

(311)

to consist of dehydrogenation of the hydrazine to a diazenium ion ($R_2\overset{+}{N}{=}NH$) followed by combination with its conjugate base, the diazene (191). This reaction course requires both diazene and diazenium ion to be present and is com-

$$R_2NNH_2 \xrightarrow{[O]} R_2N^+{=}NH$$

$$R_2N^+{=}NH + H_2O \rightleftharpoons R_2N^+{=}N^- + H_3O^+$$

$$R_2N^+{=}N^- + R_2\overset{+}{N}{=}NH \longrightarrow R_2\overset{+}{N}HN{=}NNR_2$$

patible with absence of tetrazene formation at acid concentrations greater than 1 *N* HCl and in strongly basic solutions. Further support comes from the results of mixing separately oxidized 1,1-dimethylhydrazine and 1,1-diethylhydrazine: tetramethyltetrazene, 1,1-diethyl-4,4-dimethyltetrazene, and tetraethyltetrazene are formed in the theoretical random ratio of 1:2:1, consistent with the combination of two different diazenium ions with the corresponding diazenes.

Tetrazenes also are obtained from diazenes formed by other routes. Sodium

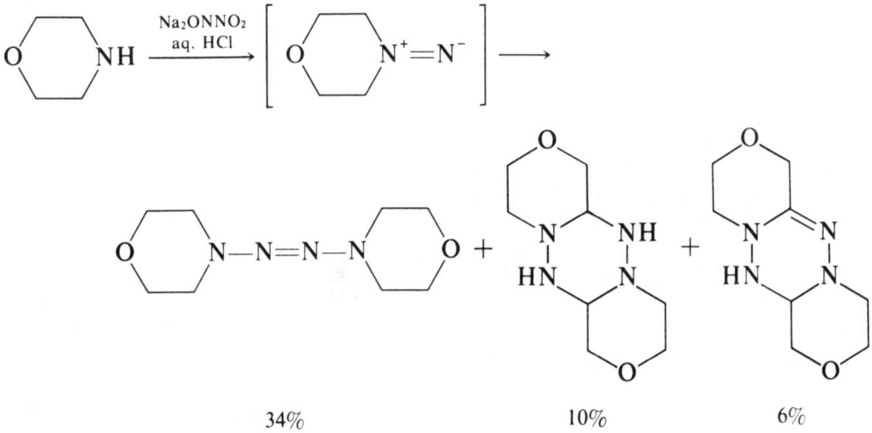

α-oxyhyponitrite (Angeli's salt) reacts with piperidine and morpholine to give acyclic-cyclic tetrazenes and tricyclic hydrazines, probably through an initially formed diazene (178). 2-Tetrazenes result from 1,1-dialkyl-2-benzenesulfonyl hydrazine salts by reaction in tetraglyme with diazene, formed by loss of sulfinate salt (177). In diethylene glycol, only a hydrazone rearrangement product of the diazene is formed, but not a tetrazene.

$$(CH_3)_2N\overset{Na^+}{\bar{N}}SO_2C_6H_5 \longrightarrow [(CH_3)_2\overset{+}{N}{=}\overset{-}{N}] \xrightarrow[110-120°]{\text{tetraglyme}} (CH_3)_2N{-}N{=}N{-}N(CH_3)_2 \quad 76\%$$

Methoxyalkyltetrazenes are formed by treatment of 1,1-diethyl-2-ethoxydiazenium fluoborate with methoxide. The ethoxy group is lost with creation of an azo group and substitution of methoxy on ethyl (52).

$$2(C_2H_5)_2N\overset{+}{N}OC_2H_5\ BF_4^- \xrightarrow[-40°]{KOCH_3,\ \text{methanol}} C_2H_5{-}N(CH(OCH_3)CH_3){-}N{=}N{-}N(CH(OCH_3)CH_3){-}C_2H_5 \quad 15\text{–}32\%$$

The nitrosamine of *p*-chloroaniline, obtained at low temperature, combines with *p*-chlorophenylhydrazine to form a tetrazene (208).

$$p\text{-}ClC_6H_4NH_2 \xrightarrow[-78°]{ONCl} p\text{-}ClC_6H_4NHNO \xrightarrow[-60°]{p\text{-}ClC_6H_4NHNH_2,\ \text{ether}} p\text{-}ClC_6H_4N{=}N{-}N(NH_2){-}C_6H_4Cl\text{-}p \quad 65\%$$

Tetra(trimethylsilyl)tetrazene **(X)** is formed from bis(trimethylsilyl) diimine in high yield at low temperature under the catalysis of silicon tetrafluoride in a dismutation reaction; products involving the catalyst also are found (314). Much lower yields of the silyltetrazene are obtained by thermolysis at 150–160° (14%) (314a) or combined thermolysis and hydrolysis at 75° (19%) (315).

$$(CH_3)_3SiN{=}NSi(CH_3)_3 \xrightarrow[-196°]{SiF_4} \underset{\textbf{(X)}\ 90\%}{[(CH_3)_3Si]_2NN{=}NN[Si(CH_3)_3]_2}$$

$$+ [(CH_2)_3Si]_2NN[Si(CH_3)_2]_2 + [(CH_3)_3Si]_3N_4SiF_3 + [(CH_3)_3Si]_3N_2SiF_3 + (CH_3)_3SiF$$

1,2-Diacyl-1,2,3,6-tetrahydrotetrazines result from addition of diacylazo compounds to conjugated azo olefins (279). This reaction is not a dismutation since the oxidation number of the nitrogens in the diacyl azo derivatives with a strong electron-withdrawing group is +1, whereas that of the nitrogens in the azoolefins is −1.

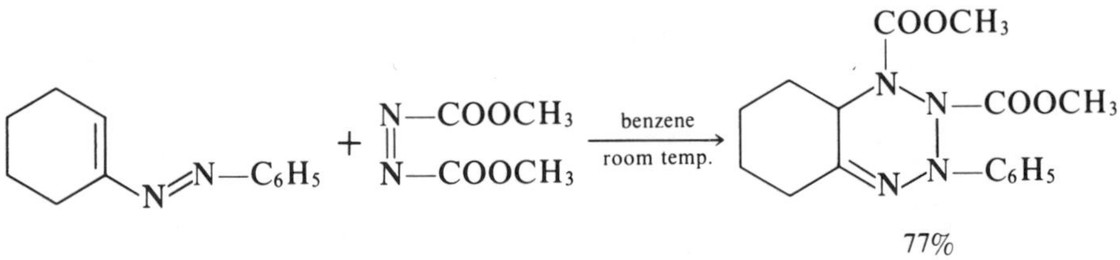

In addition to preparation of tetrazenes, oxidative coupling has been applied to the synthesis of 1,4-tetrazolines, tetrahydrotetrazines, cyclic bistetrazenes,

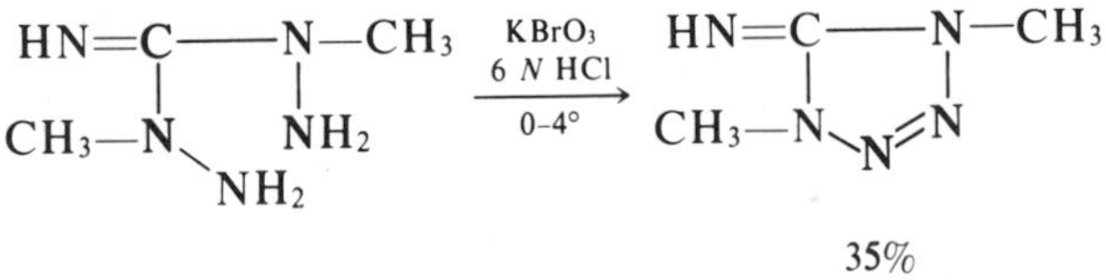

(192)

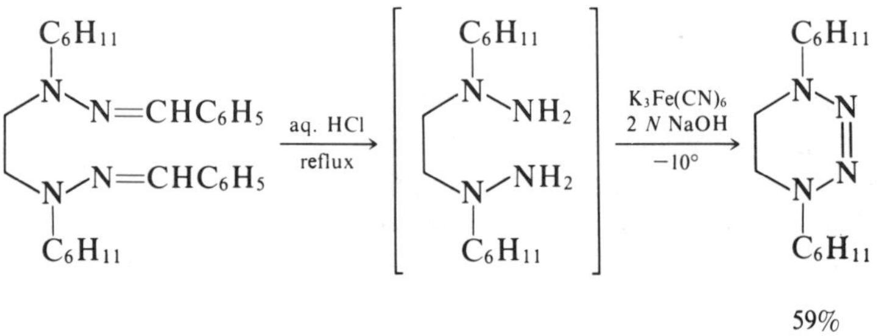

(166)

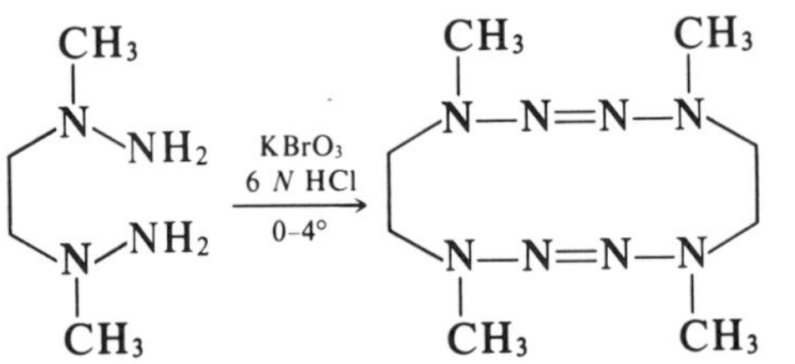

(153)

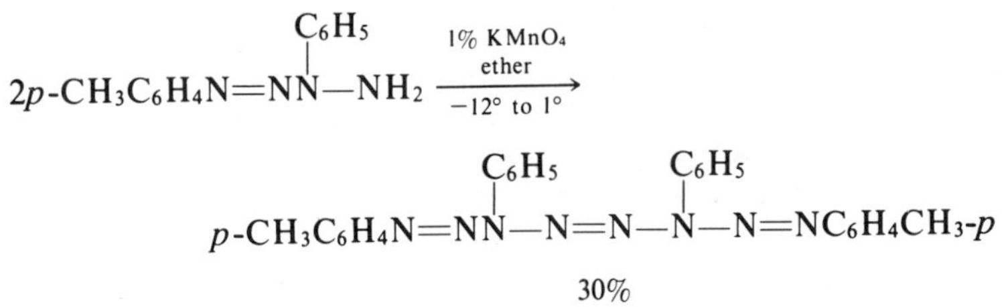

(201, 318)

octaztrienes, 1,1′-azobisbenzotriazolium salts, and 1,1′-azobistetrazoles. Some of the variety of oxidants used are indicated in the examples.

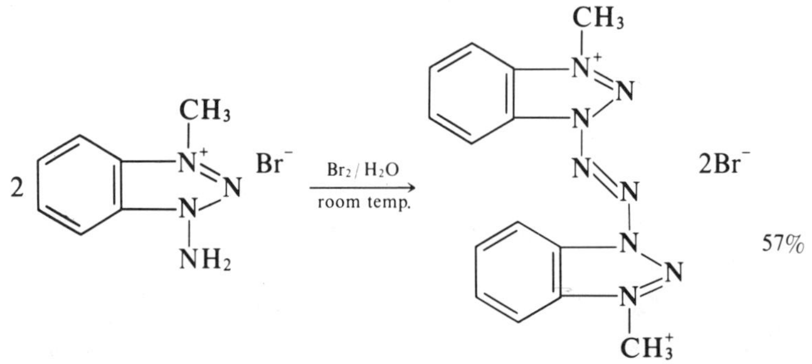

(96)

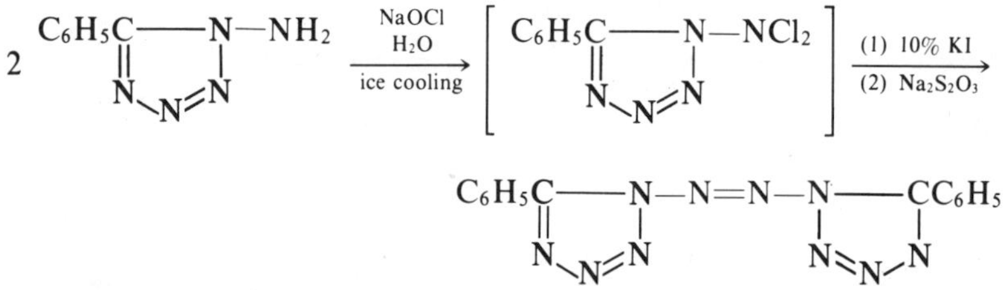

(285)

The chlorine-containing unstable intermediate first formed by hypochlorite treatment of 1-amino-5-phenyltetrazole is converted to the 1,1′-azobistetrazole by the action of potassium iodide.

Reaction of azidium salts with azide ion at 20–30° leads to triazacarbocyanines **(XI)** (9, 10), proceeding through a diazo-azo intermediate **(XII)**, stable at −60 to −50°; by addition of other azidium salts to solutions of **(XII)**, unsym-

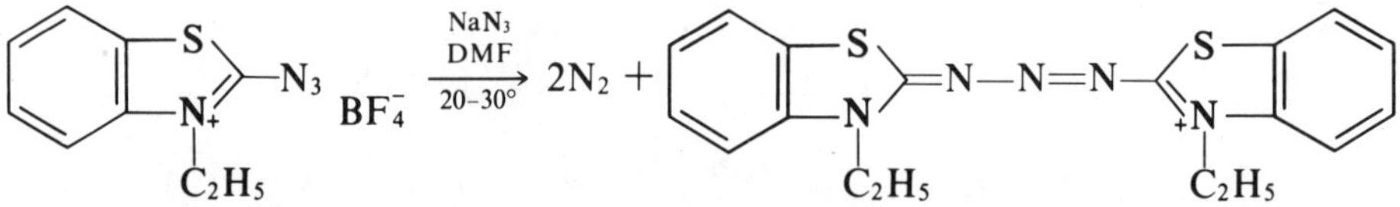

BF_4^-

(XI)

90%

metrical triazacarbocyanines are prepared. Loss of the four nitrogen chain leaves an ylide **(XIII)** which can undergo addition to the terminal nitrogen of azide in an azidium salt. Formation of **XIII** by other routes, such as decarboxylation of a 3-alkyl-benzothiazolium-2-carboxylic acid in the presence of an azidium salt, also leads to a triazacyanine (222). 2-Arylsulfonylazo-3-methylbenzothiazolium salts **(XIV)** react with heterocyclic imines to form triazacyanines, accompanied by related oxidized and reduced products (142, 242).

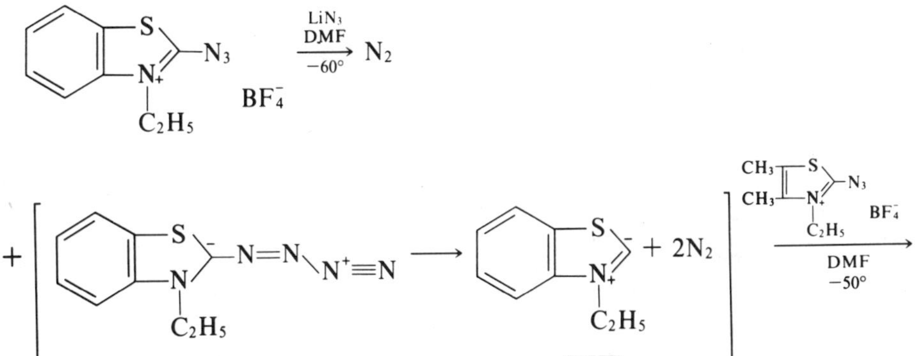

(XII) (XIII)

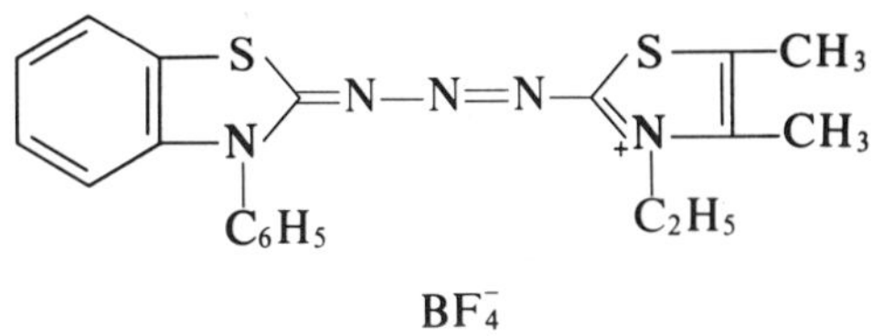

62%

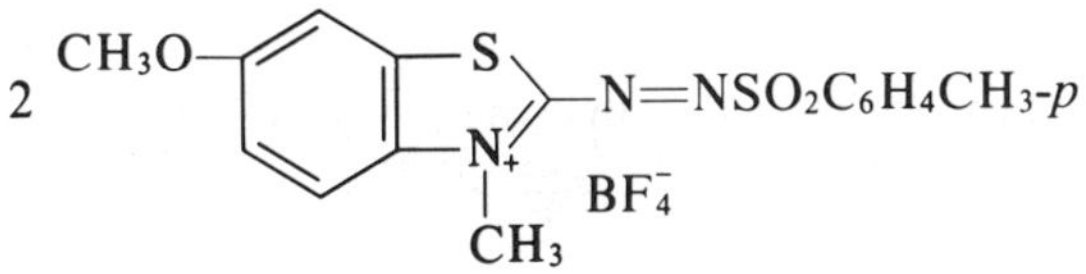

(XIV)

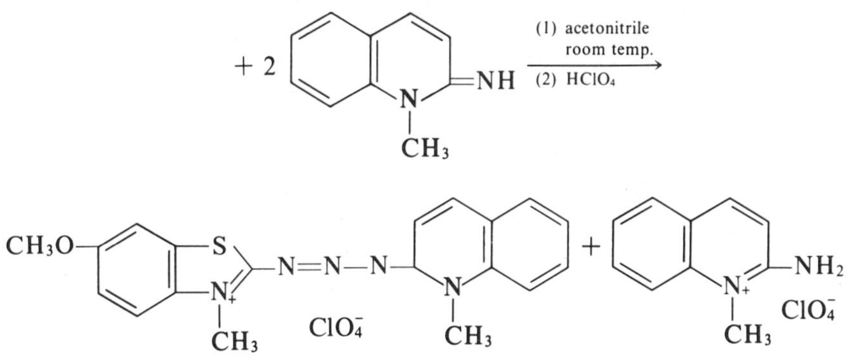

97%

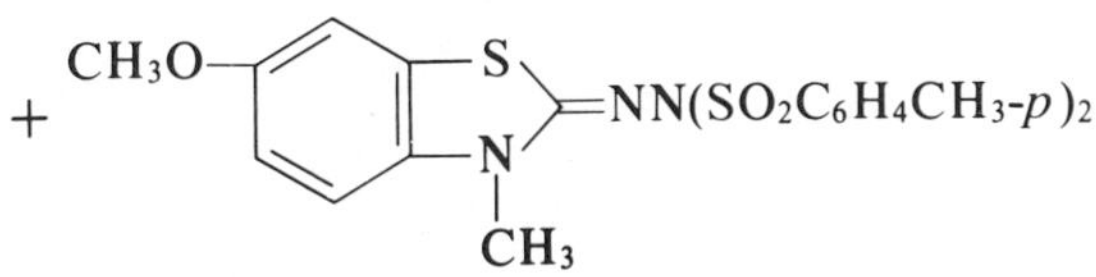

44%

The tetraazatetramethine **(XVI)** is obtained in low yield by oxidation of 3-methyl-2-benzothiazolinone hydrazone **(XV)**; a better preparation is oxidation of the azidium salt with azide ion (22).

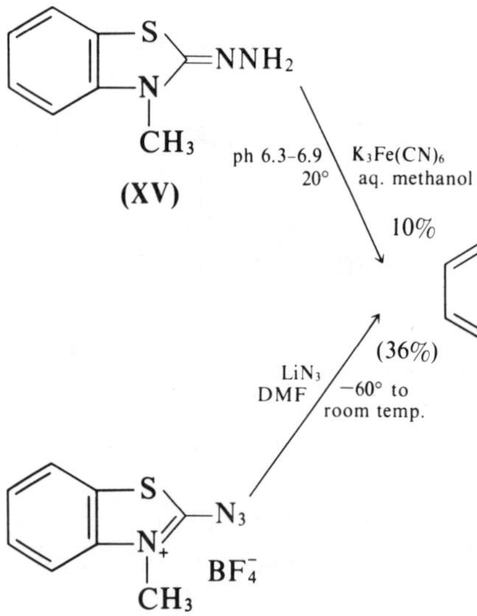

Pentaazapentacyanines form on acidic oxidation of heterocyclic hydrazones with azidium salts in dimethylformamide-formic acid solution, together with some tetraazapentamethinecyanine (144). The 3-methylbenzothiazole penta-

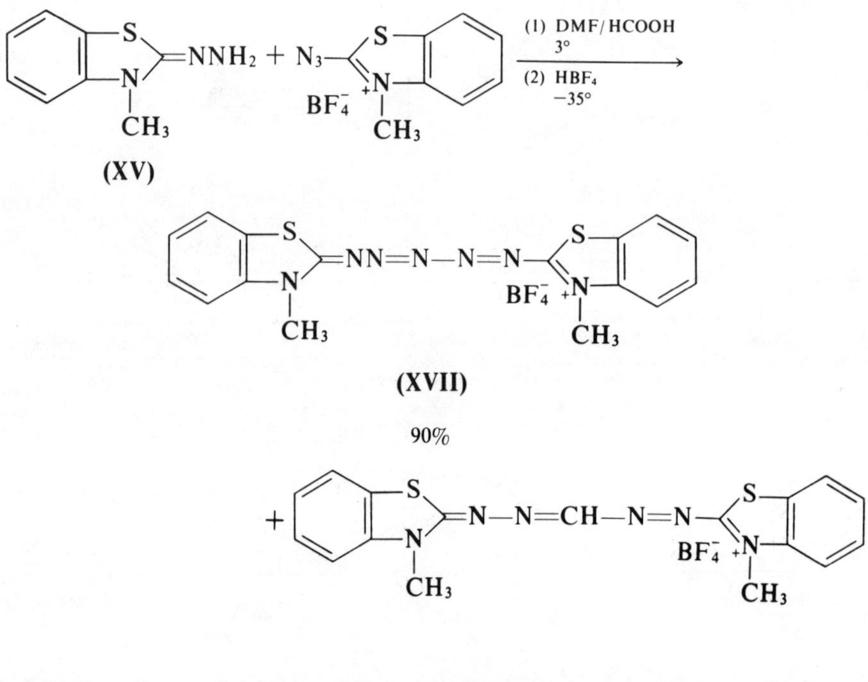

azapentacyanine **(XVII)** also results in 40–43% yield by oxidation of the hydrazone **(XV)** with lead tetraacetate (143, 144).

N-Azidodimethylamine is prepared by reaction of *N*-chlorodimethylamine

$$(CH_3)_2NCl \xrightarrow[\text{(1) room temp. (2) reflux}]{LiN_3,\ THF} (CH_3)_2NN_3 \quad 83\%$$

and alkali azides (18, 33). A four-nitrogen dipolar 1-tetrazene chain results from the reaction of pyridine and 2-azido-4,6-dimethoxytriazine (86).

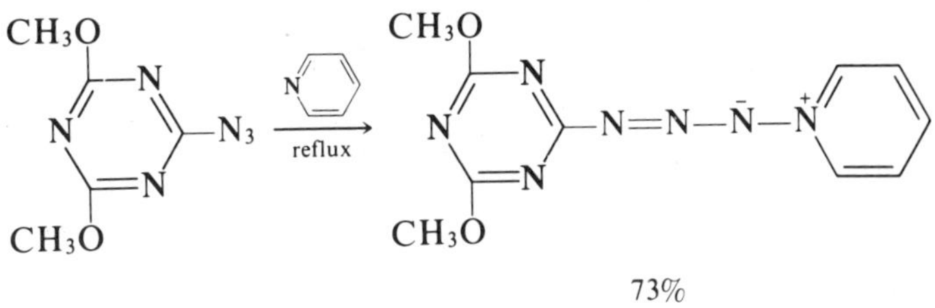

Four-nitrogen azimines are obtained by combination of azo compounds with diazenes. 1,1-Dibenzylhydrazine monoanion reacts with tosyl azide to form dibenzyl diazene **(XVIII)**, probably through the amine azide; reaction of **(XVIII)** with 4-phenyl-1,2,4-triazolidine-3,5-dione yields the acyclic-cyclic azimine

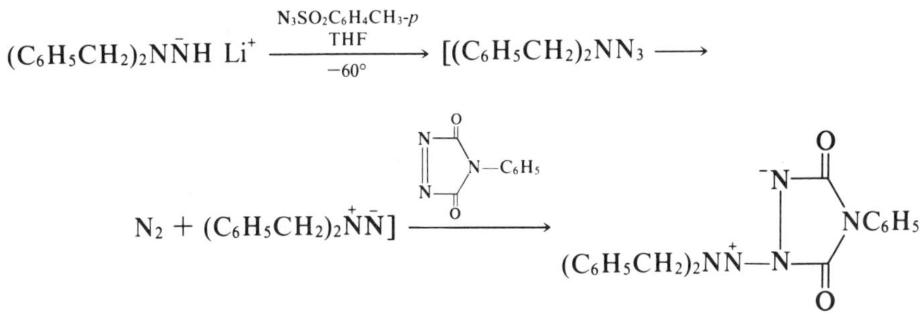

(XVIII) **(XIX)**

(XIX) (4). Phthalimido azimines arise by joining azo compounds with the diazene oxidatively generated from *N*-aminophthalimide. *cis*-Azobenzene forms the *cis*-azimine **(XX)** along with small quantities of the trans isomer and *trans*-azobenzene.

The lithium salt of the seven-nitrogen 1-(benzotriazolyl-1)-4-*p*-toluene-

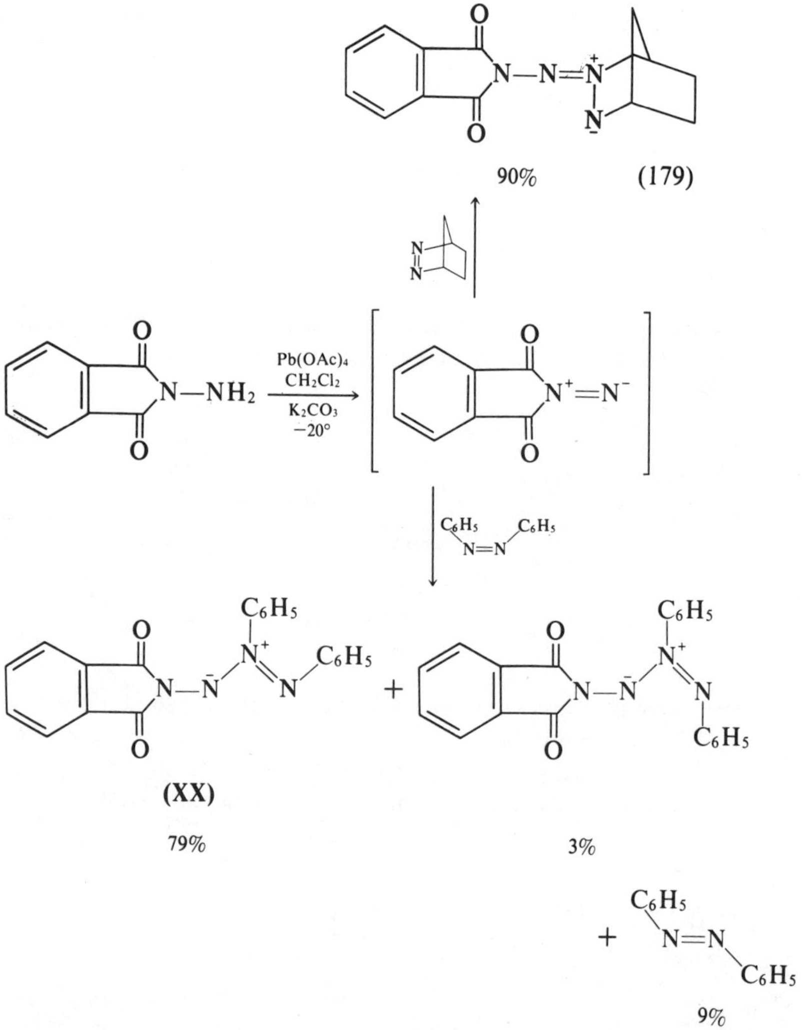

(128)

sulfonyltetrazene is prepared by reaction of 1-aminobenzotriazole with tosyl azide (161).

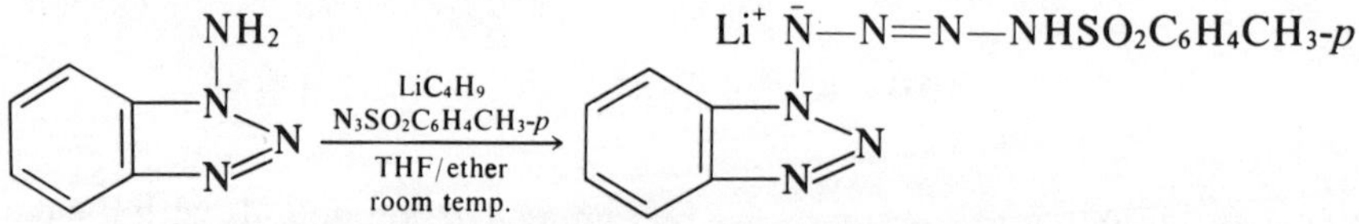

Tetrahydro-*v*-tetrazine *N*-oxides and lithium oxide result from an unusual dismutation of α-lithio nitrosamines. Electron spin resonance spectra indicate the presence of radicals, and the carbon-carbon and nitrogen-nitrogen bonds formed are proposed to arise by radical combination (267). Alternatively, after

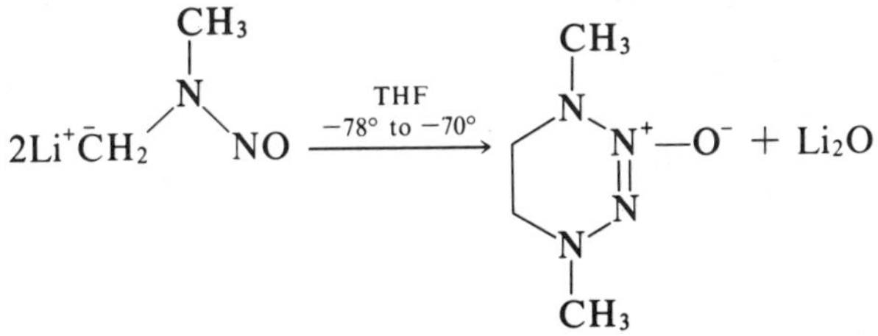

50%

formation of the carbon-carbon bond by linkage of radicals, loss of oxide ion from one nitroso group, producing lithium oxide, could result in a diazene which would combine with the remaining nitroso group, forming the cyclic tetrazene oxide.

The action of nitrous acid and equivalents on two amino groups in the same molecule located two and three atoms apart is a principal method for the synthesis of *v*-triazoles, tetrazoles, and *v*-triazines; three nitrogen-nitrogen bonds result. In the formation of *v*-triazoles, one amino group may be monosubstituted or its equivalent may be located within a ring. Heterocycles, such as 1-aminoquinolizinium salts **(XXI)**, with a bridgehead quaternary nitrogen and an amino group two atoms removed, rearrange to fused triazolo derivatives.

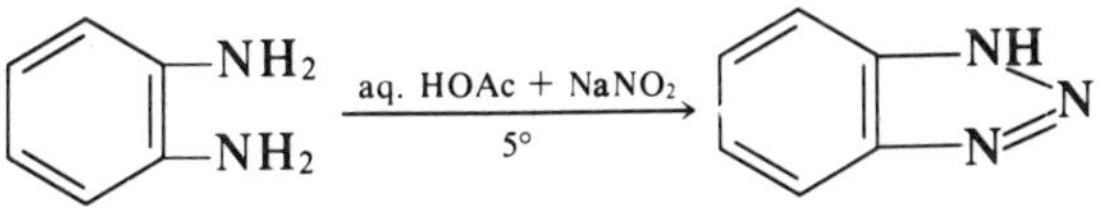

(57)

75–81%

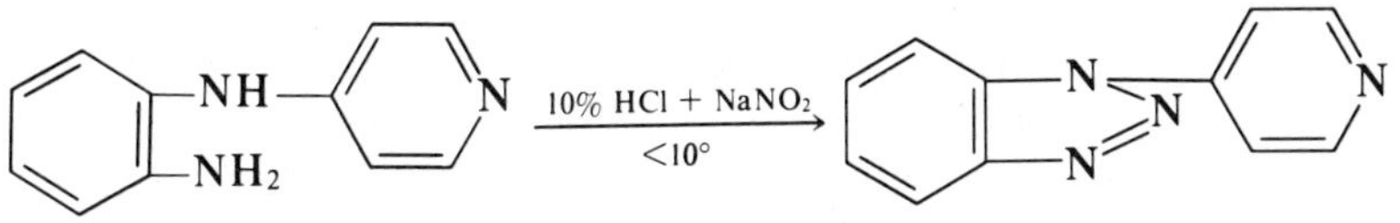

(255)

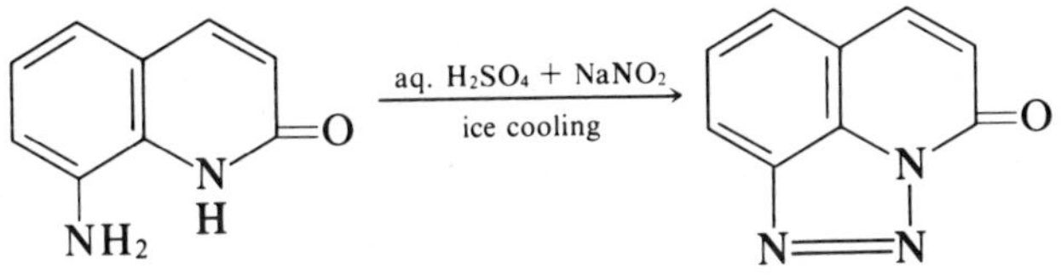

(106)

93.5%

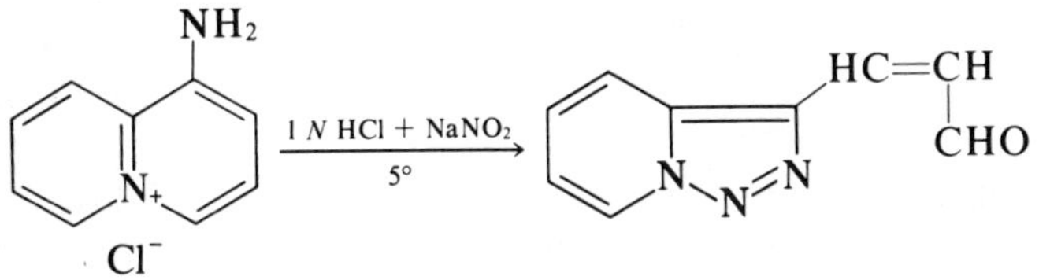

(58)

(XXI) 43%

The six-nitrogen chain in 1,1′-bibenzotriazole is formed by nitrosation of *o,o′*-diaminoazobenzene (111).

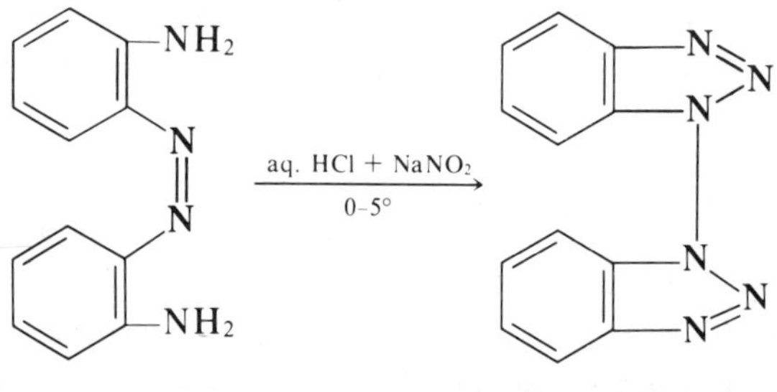

63%

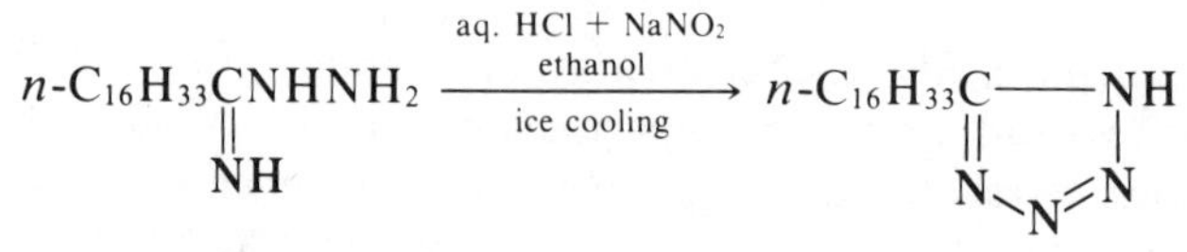

86%

(160)

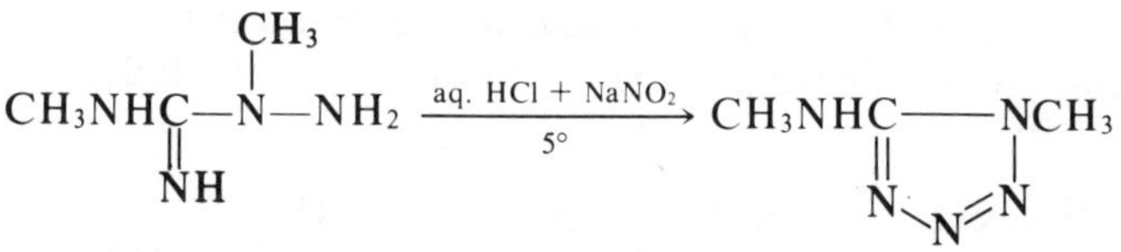

51.5%

(210)

Nitrosation of unsubstituted hydrazidines leads to 5-monosubstituted tetrazoles, hydrazidines monosubstituted on nitrogen adjacent to carbon form 1-substituted tetrazoles, and those substituted at the end of the hydrazine unit form 2-substituted tetrazoles. Diazotized 1-alkyl-2-aminoguanidines primarily cyclize on the alkyl-substituted nitrogen, but also form small quantities of the isomeric 5-alkylaminotetrazole. Nitrosation of hydrazidines unsubstituted in the hydrazine moiety is assumed to go through intermediate imidoyl azides.

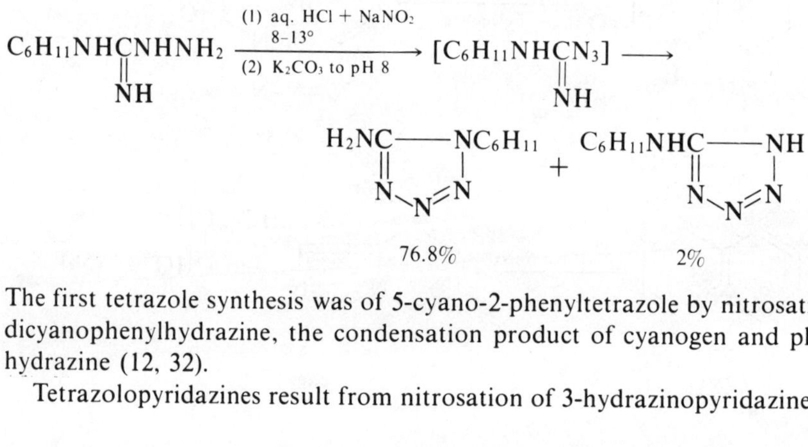

76.8% 2%

(81)

The first tetrazole synthesis was of 5-cyano-2-phenyltetrazole by nitrosation of dicyanophenylhydrazine, the condensation product of cyanogen and phenylhydrazine (12, 32).

Tetrazolopyridazines result from nitrosation of 3-hydrazinopyridazines and

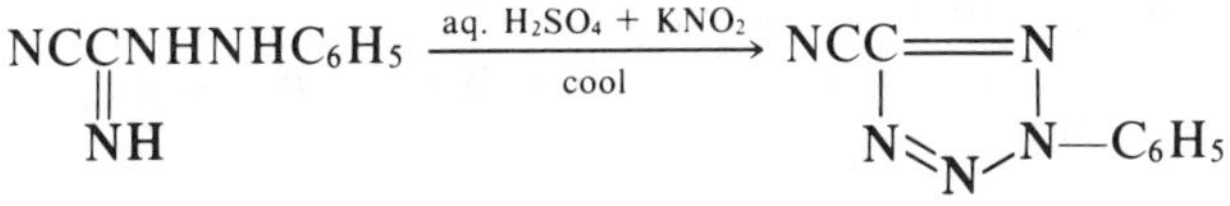

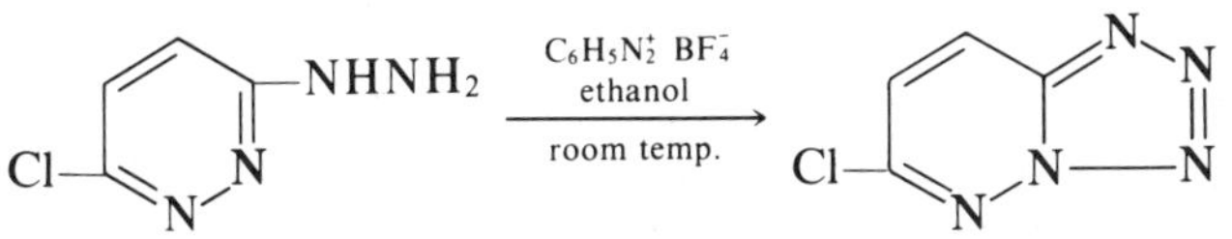

64% (60)

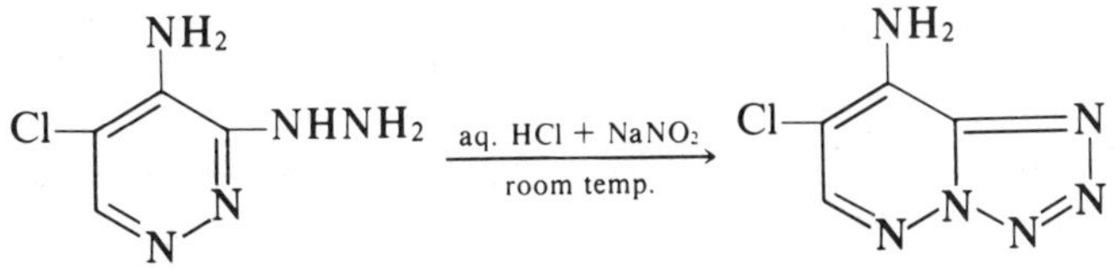

93% (168)

also by the action of benzenediazonium ion, possibly through an intermediate tetrazene. 1,1′-Bi-5-phenyltetrazole, with an eight-nitrogen chain, results from nitrosation of *N*-(5-phenyltetrazolyl-1)benzhydrazidine.

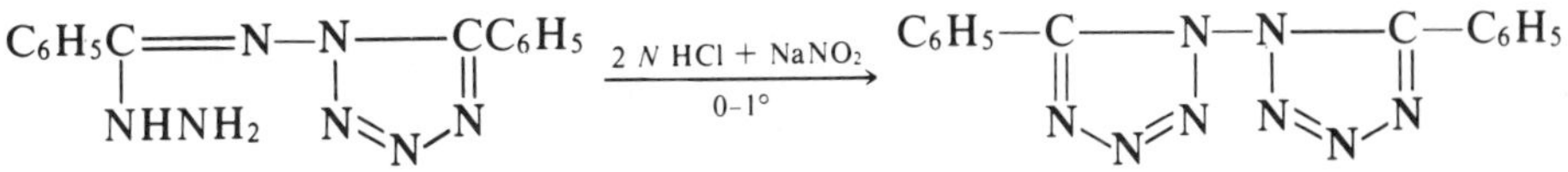

98% (28)

Benzotriazines and related fused ring triazines are formed by nitrosation of

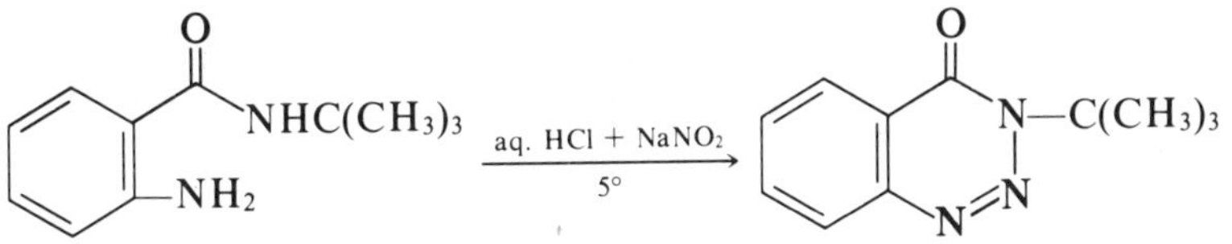

60% (23)

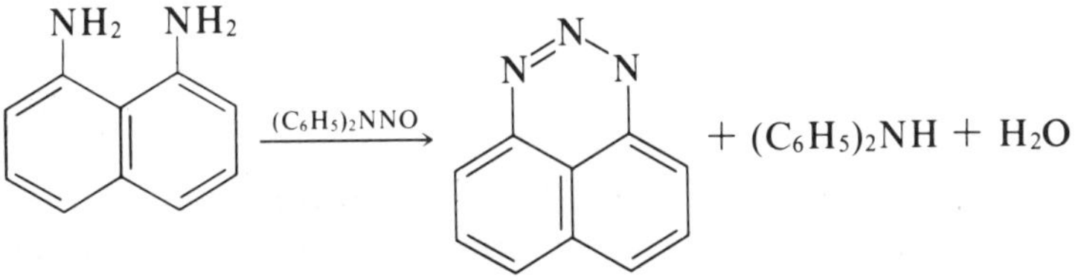

97% (272)

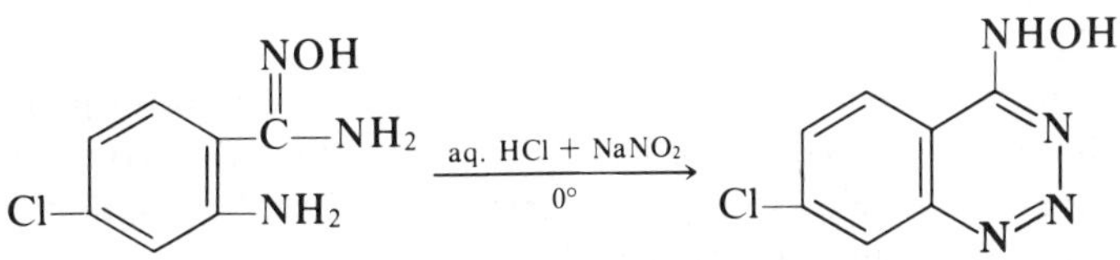

70% (104)

an aromatic amino compound having an exocyclic reduced nitrogen, four atoms removed. Benzotriazine 3-oxides result from diazotizing oximes of *o*-aminobenzene carbonyl compounds.

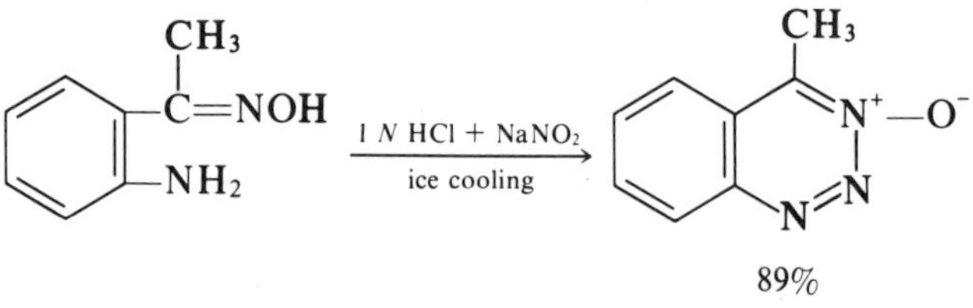

(197)

Six-nitrogen chains in cyclic-cyclic bitriazines and in benzylimidazo[4,5-*d*]-tetrazolo[1,5-*f*]-*v*-triazine **(XXII)** are made by the nitrosation reaction.

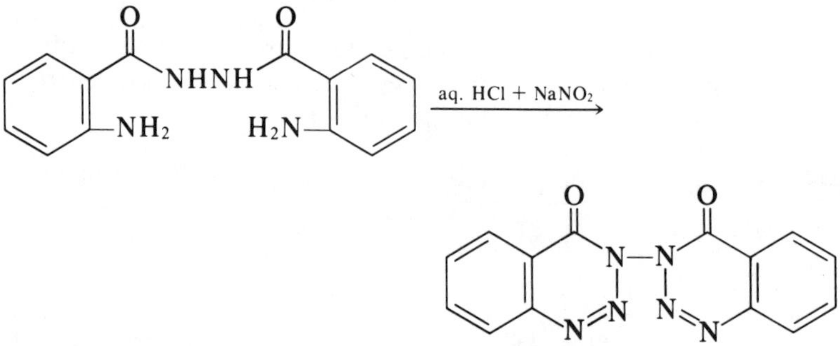

(122)

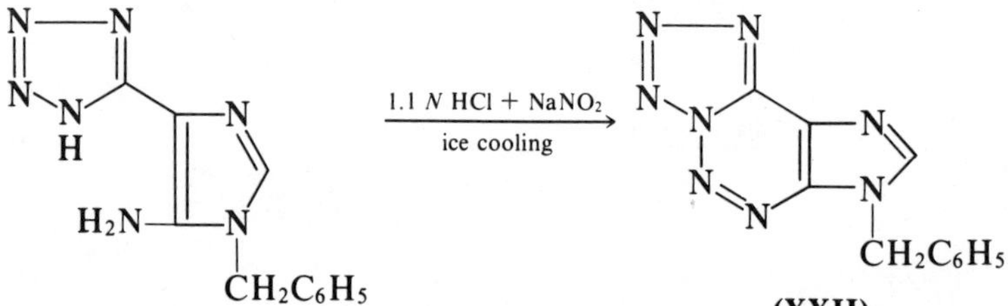

(295)

(XXII)

74%

Thiatriazoles result from diazotization of thiocarbonyl hydrazides. Oxatriazolium compounds are obtained by diazotization of hydrazines carrying on ω-carbon substituent bearing hydrolyzable groups, such as potassium phenylhydrazone methane disulfonate and the reaction product of phenylhydrazine and cyanogen bromide, made *in situ*.

Benzene azotrinitromethane undergoes hydrolysis and cyclization in acetic acid, forming 3-phenyloxatriazolium-5-olate (78, 244). The NO portion of the ring must arise from nitro, either through rearrangement or dissociation from the trinitromethyl moiety and nitrosation of azo.

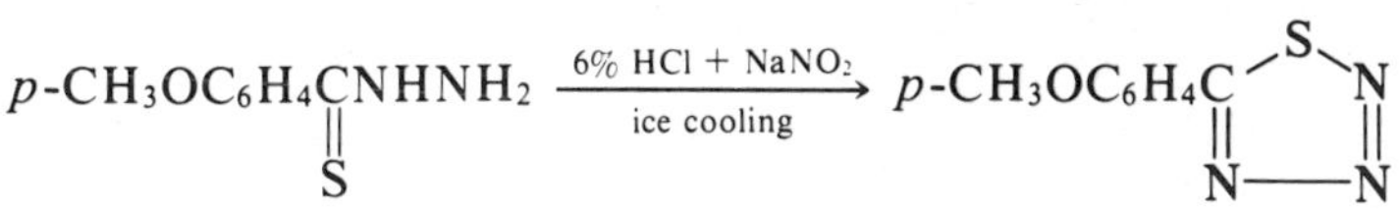
(278)

47.8%

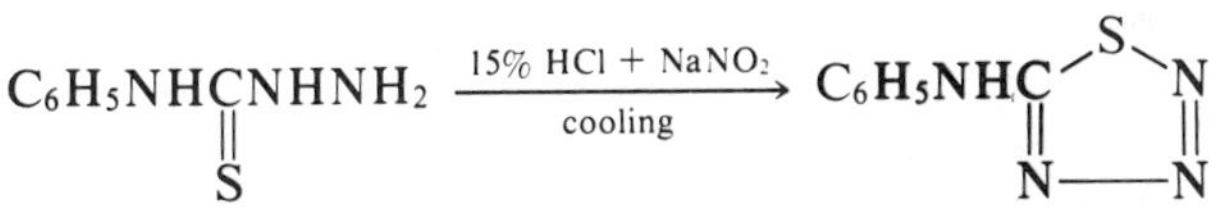
(181)

89%

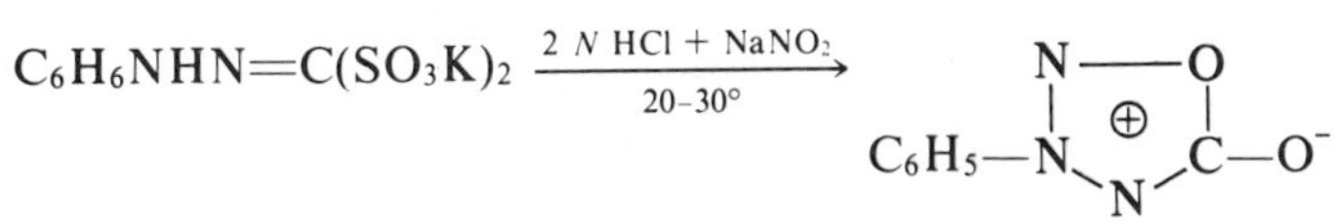
(78)

64%

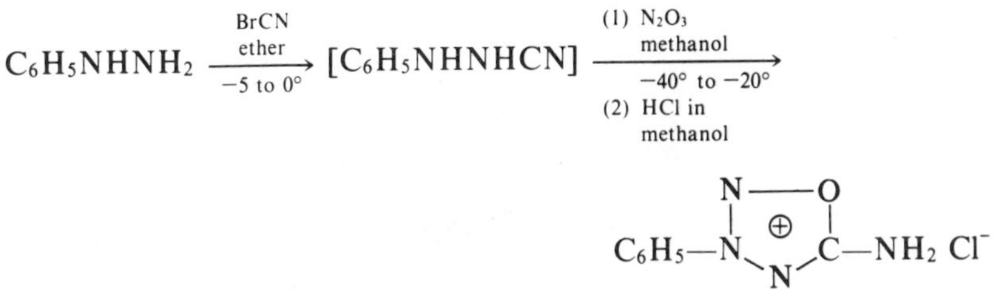
(190)

60.5%

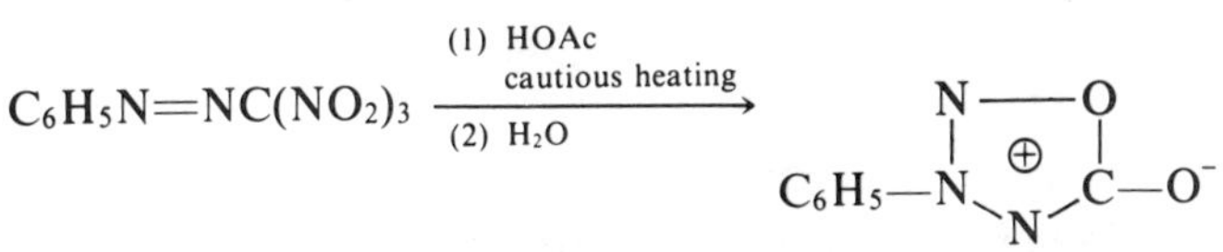

Diazoketone esters or diketones are cyclized to *v*-triazoles unsubstituted in the ring by the action of ammonia, illustrated with diazoacetoacetic ester (320). *v*-Triazoles form on addition of diazomethane to unsaturated groups contain-

ing at least one nitrogen, shown by its reaction with nitriles, cyanates, ketene-imines, carbodiimides, and isothiocyanates. Ring methylation and other secondary reactions also occur when N-unsubstituted *v*-triazoles are formed initially.

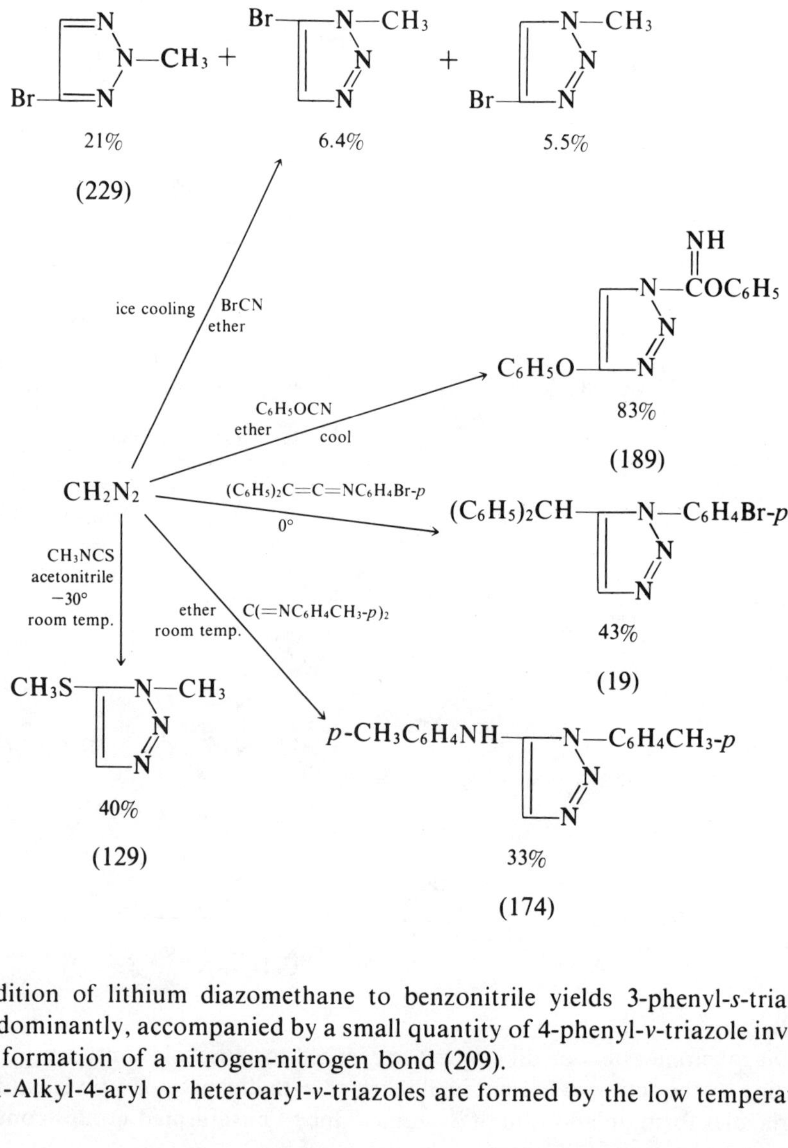

Addition of lithium diazomethane to benzonitrile yields 3-phenyl-*s*-triazole predominantly, accompanied by a small quantity of 4-phenyl-*v*-triazole involving formation of a nitrogen-nitrogen bond (209).

1-Alkyl-4-aryl or heteroaryl-*v*-triazoles are formed by the low temperature

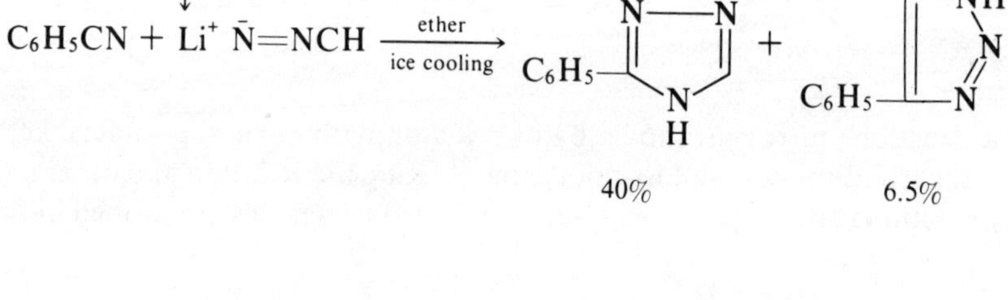

combination of lithium nitrosoamines with non-enolizable nitriles, illustrated by the reaction of lithium dimethylnitrosoamine and benzonitrile (266).

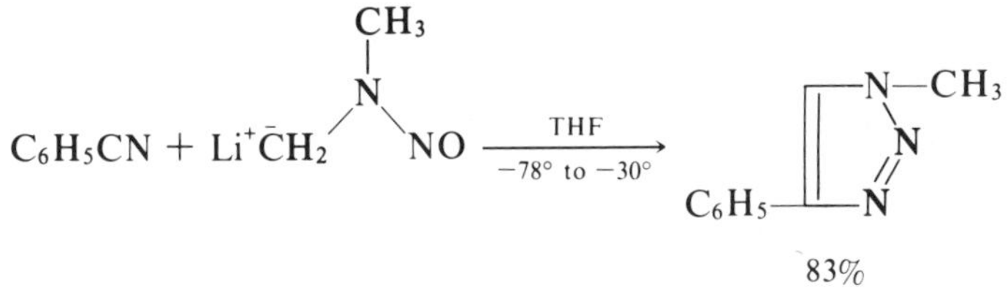

N-oxygen substituted benzotriazoles result from *o*-nitrobenzene azo or hydrazine derivatives or their precursors by combination of the terminal nitrogen with nitro. The corrosion inhibitor 1-hydroxybenzotriazole is made by reaction of hydrazine hydrate with *o*-nitrochlorobenzene (36) or through the action of potassium hydroxide on *o*-nitrophenylhydrazine (328). Perfluoro-1-aminobenzotriazole results by reaction of hydrazine with decafluoroazoxybenzene, probably through the unisolated *o*-hydrazino intermediate (31).

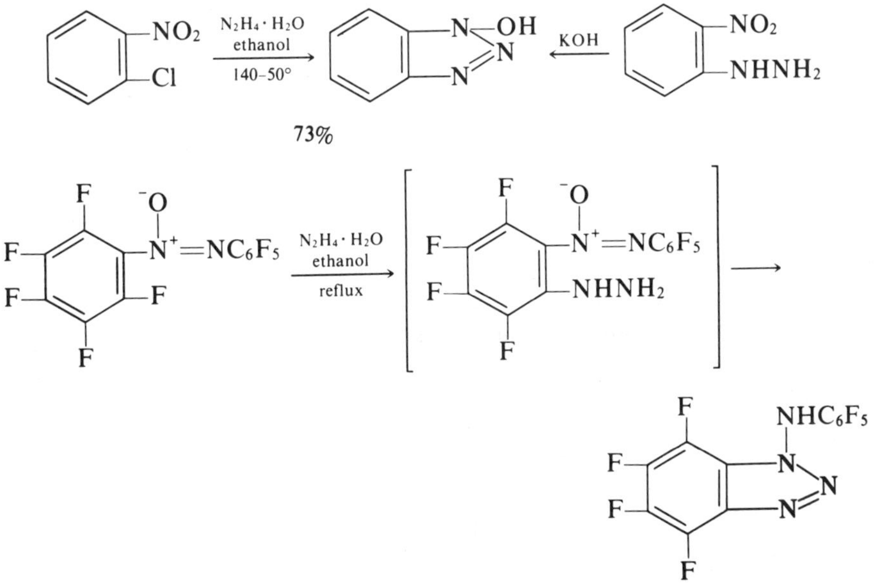

Sodium sulfide removes one oxygen from *o*-nitroazobenzene **(XXIII)** forming 2-phenylbenzotriazole oxide **(XXIV)** (13), but triethylphosphite converts **(XXIII)** to 2-phenylbenzotriazole, probably through its nitrene (44).

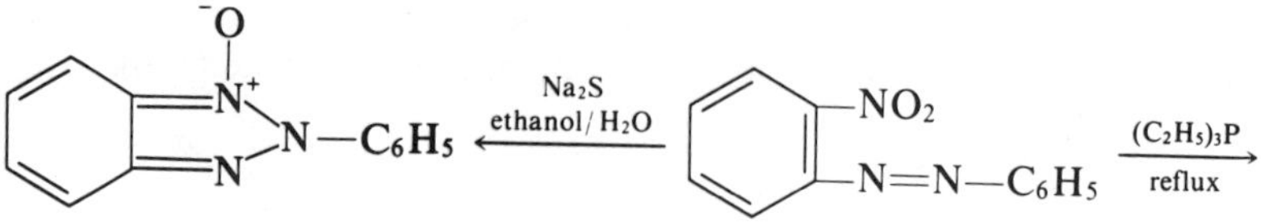

(XXIV) (XXIII)

84%

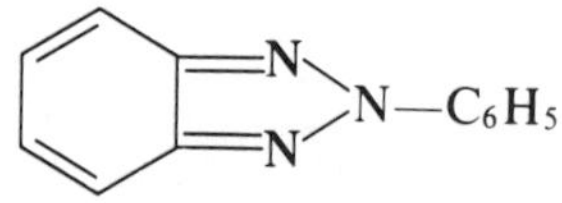

72.5%

Cyclization of *o*-nitrohydrazinobenzenes to 2-substituted benzotriazoles and benzotriazole oxides proceeds even when elimination of methyl from a dimethylhydrazino group is required.

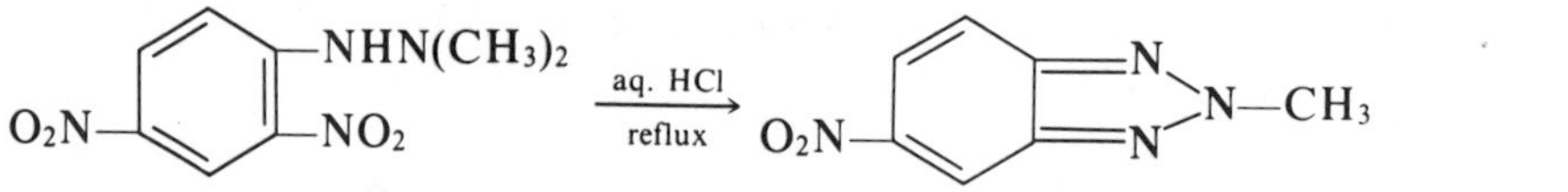

16%

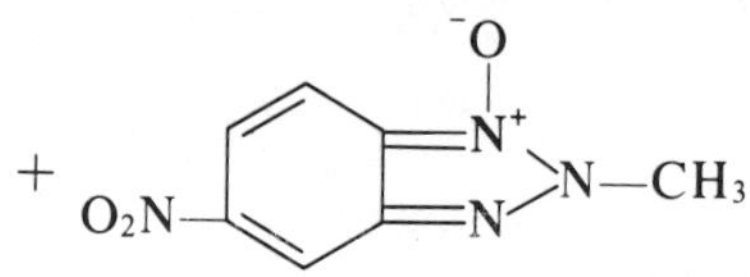

23%

Diazirine moieties adjacent to carbonyl react with hydrazines to form 1-amino-*v*-triazole derivatives. Similarly, diaziridines (121) with an *o*-nitrophenyl

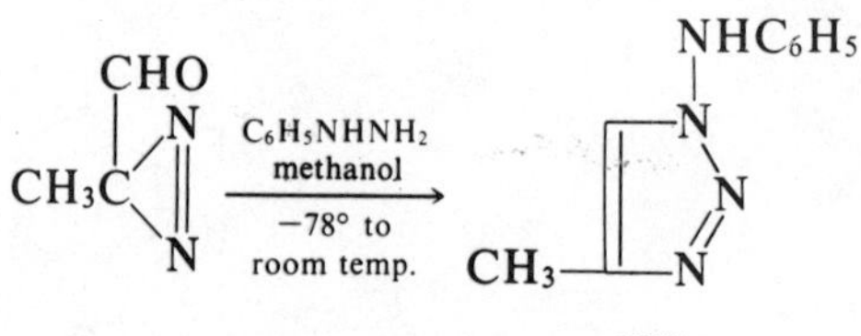

(265)

27%

moiety on N-1, alkyl on N-2, and a cyclohexylidine group on carbon rearrange to 2-alkylbenzotriazole 1-oxides such as the 2-methyl derivative **(XXV)** with simultaneous elimination of cyclohexanone. Acid treatment of 1-*o*-nitrophenyl-2-methylhydrazine also leads to **XXV**. Manufacture of 2-(2′-hydroxyphenyl)-

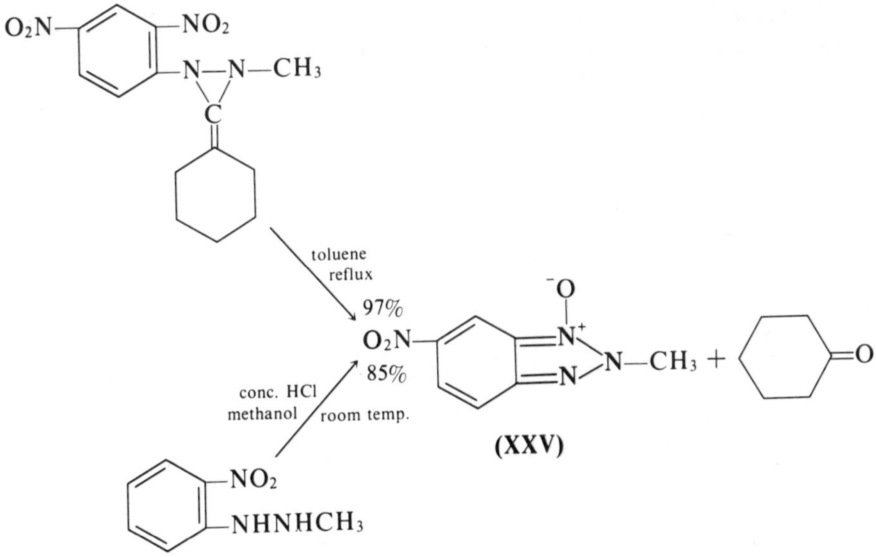

(XXV)

benzotriazole ultraviolet absorbers is by a similar route, illustrated by the synthesis of 2-(2′-hydroxy-5-*t*-octylphenyl)benzotriazole (39).

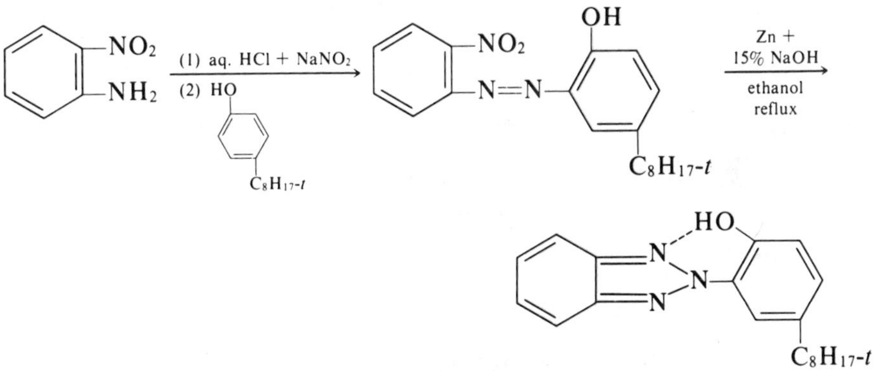

Diazomethane adds to the diazonium moiety of *p*-nitrobenzenediazonium chloride to form 1-*p*-nitrophenyltetrazole along with *p*-nitrophenyl cyanamide (139).

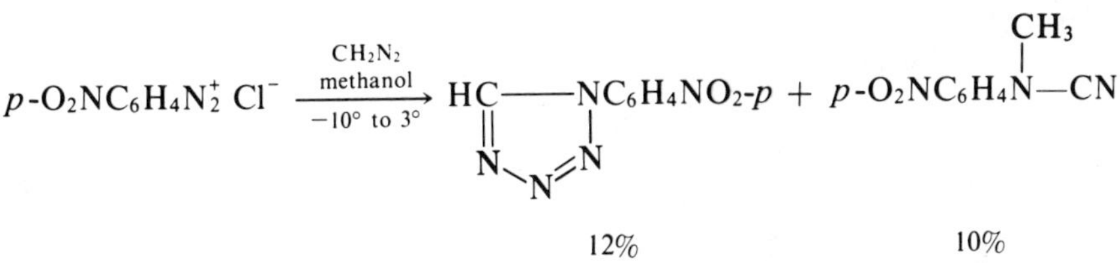

v-Triazoles are formed from osazones by a dismutation reaction with simultaneous loss of ammonia or amines (38). 1,2-Oxime hydrazones or their acyl derivatives are converted to 2-substituted *v*-triazoles by joining the oxime and hydrazone nitrogens with loss of water or carboxylic acid. Elements of meth-

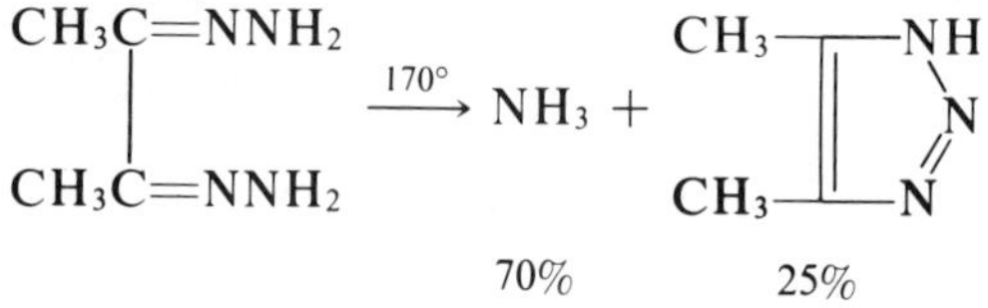

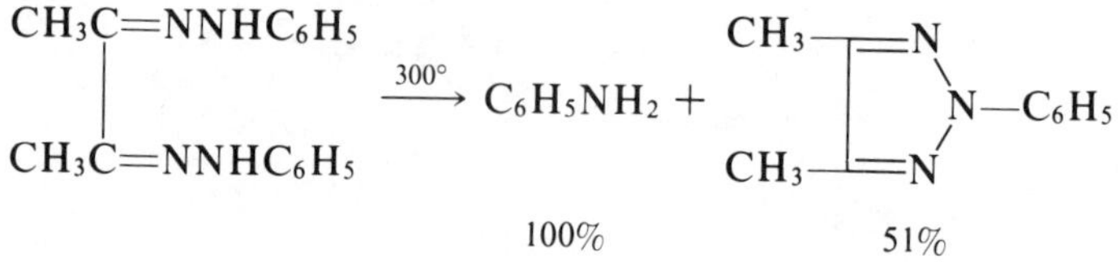

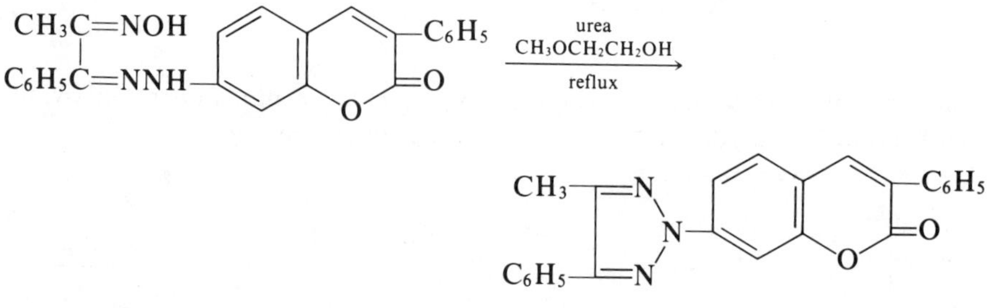

(25)

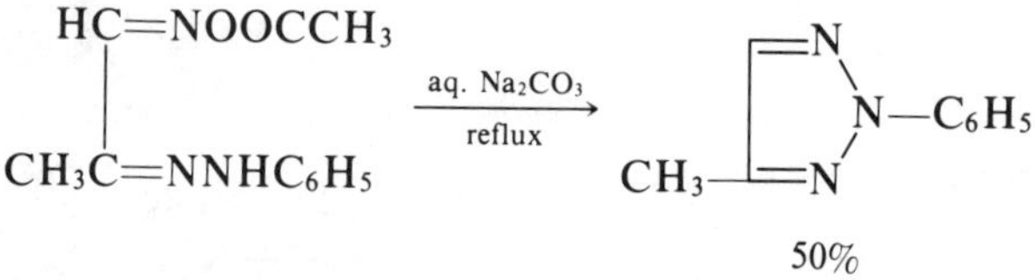

(154)

anol are eliminated from biacetyloxime methylphenylhydrazone with formation of 4,5-dimethyl-2-phenyl-*v*-triazole on heating with acetic anhydride (11).

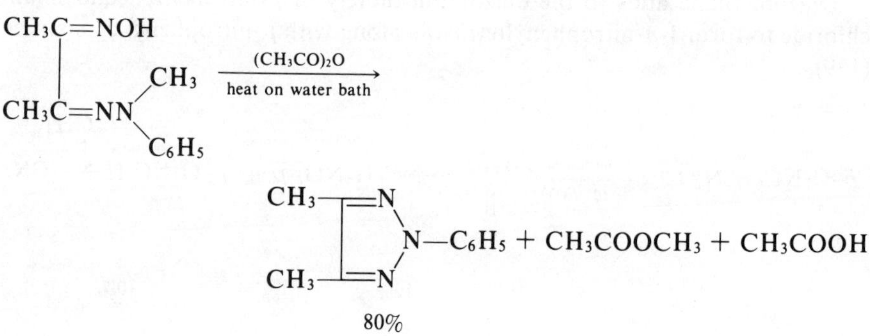

Osazones and bisazoethylenes, each with identical two-nitrogen units, are converted under nonoxidative conditions to 1-amino-*v*-triazoles. In each case a redistribution of oxidation states occurs among the nitrogen atoms in the four-nitrogen chains of the products, and the resonance energy of the system increases. An embedded four-nitrogen heterocycle results from cyclization of bis-

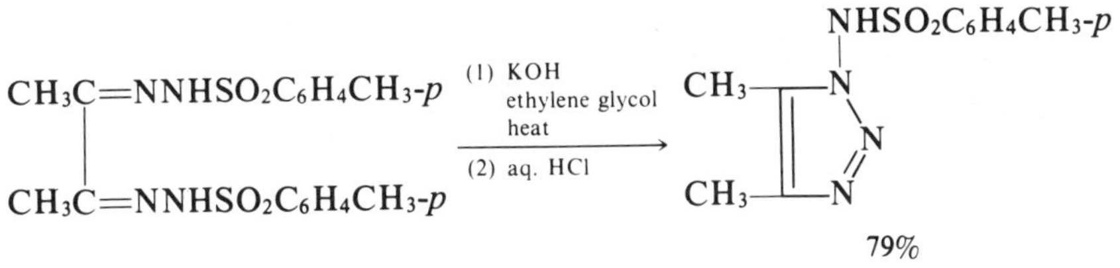
(16)

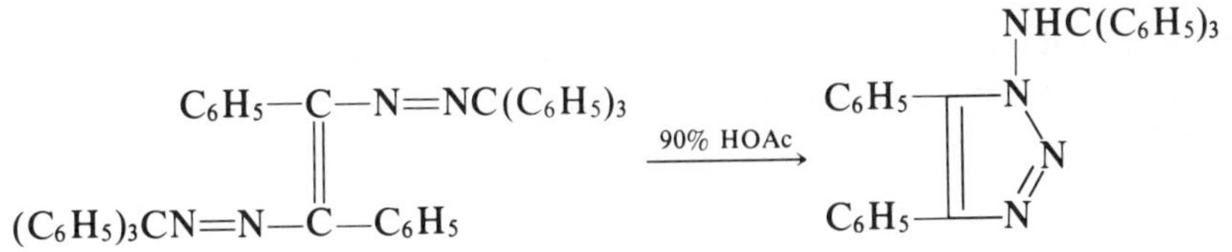
(54)

azoethylenes, to an unisolated, presumably dipolar, 1-aminotriazolium compound, and its condensation with acetylene dicarboxylic ester (287).

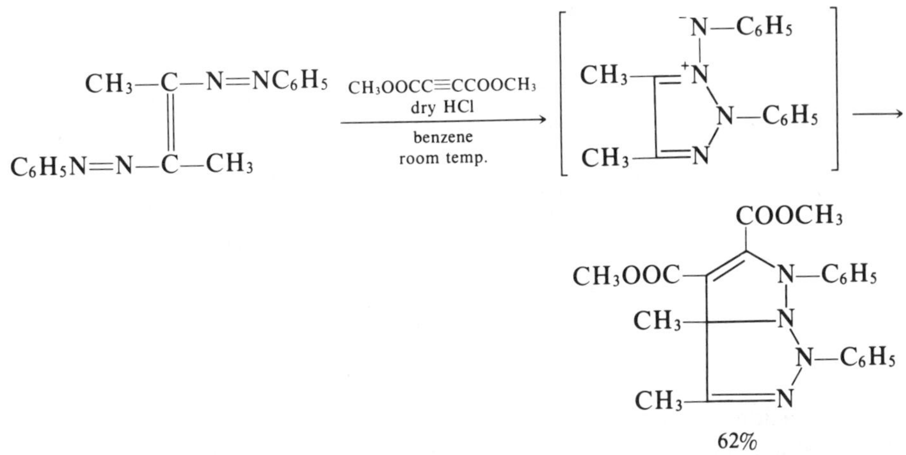

A 1-hydroxy-*v*-triazole forms on alkaline treatment of a 1,2-oxime tosylhydrazone through a diazene by elimination of *p*-toluenesulfinic acid (20).

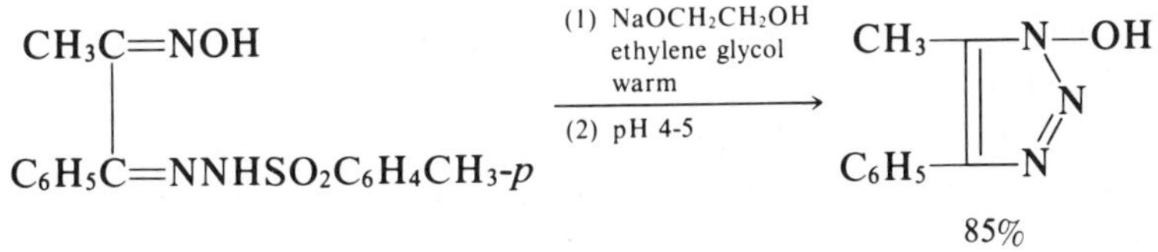

Oxidation of *N*-aminoimidazoles leads to both *v*-triazole and 1,2,4-triazine derivatives (127). Similarly, *v*-triazoles are obtained from the oxidation of 1,2,4-

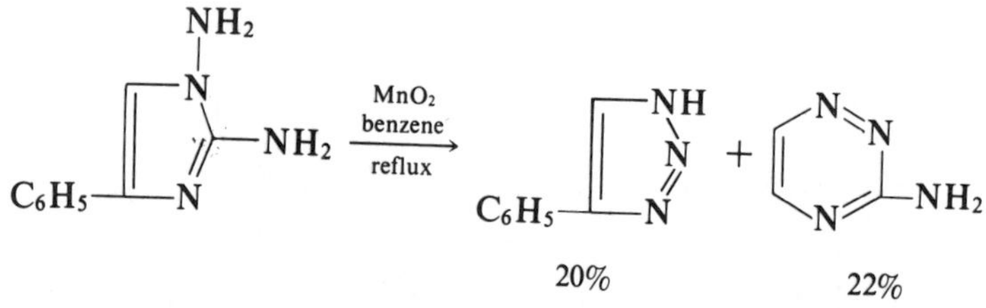

triazinones by chloramine (247); the 1-ethoxyethyl-*v*-triazole derivative arises from the action of chloramine on the ether solvent. The *N*-oxide of 3-methyl-6-phenyl-1,2,4-triazine 4-oxide **(XXVI)** serves as internal oxidant to form 4-phenyl-*v*-triazole (219).

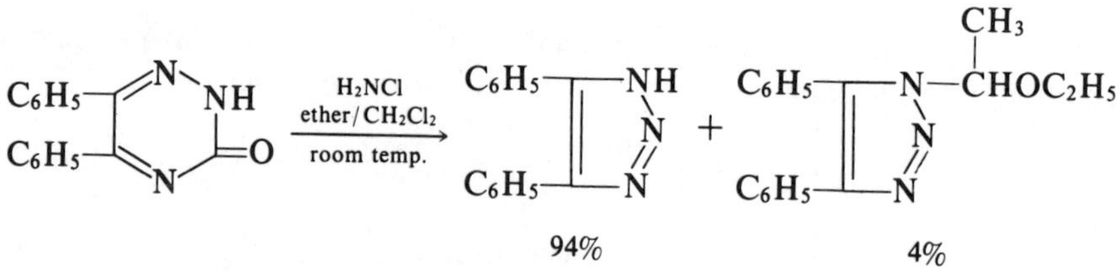

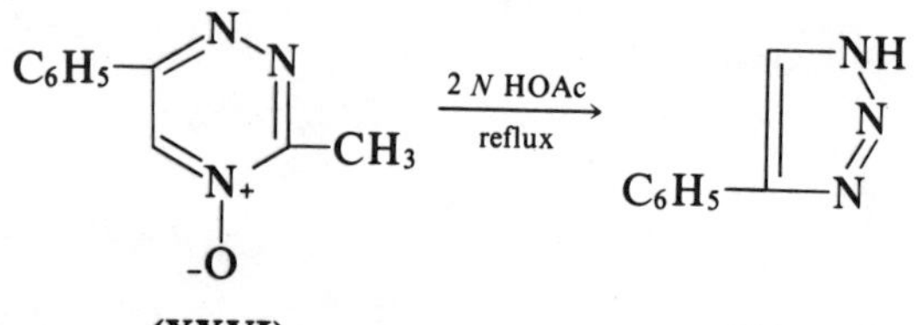

(XXVI)

2,3-Diazidonaphthoquinone undergoes a dismutation on treatment with triphenylphosphine, forming the triphenylphosphine derivatives of 2-aminonaphtho[2,3-*d*]triazole-4,9-dione **(XXVII)** and 2,3-diaminonaphtho-4,9-dione; hydrolysis liberates **XXVII** (205).

Oxidation of bisaryl hydrazones of 1,2-dicarbonyl compounds leads to 2-substituted *v*-triazoles with loss of one terminal hydrazone nitrogen and its substituent. The reaction has been extensively applied to osazones of carbohydrates (71, 118). Unsubstituted and acyl vicinal-hydrazones can be oxidized to yield 1-substituted *v*-triazoles in which all hydrazone nitrogen atoms are retained. Oxidation of 1,2-oxime hydrazones leads to 2-substituted *v*-triazole 1-oxides.

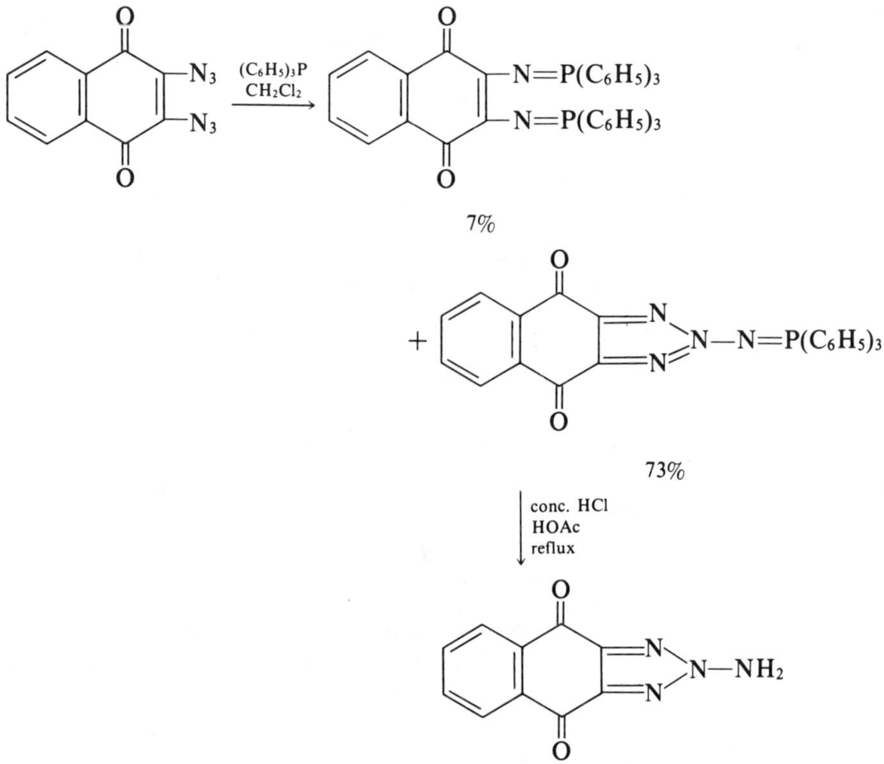

(XXVII)

91%

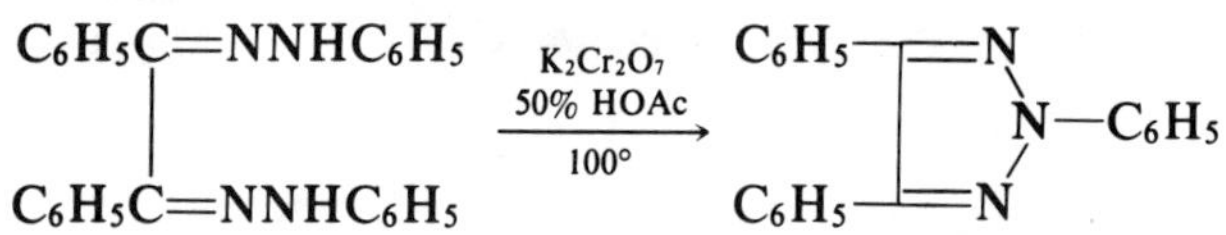

(72)

65%

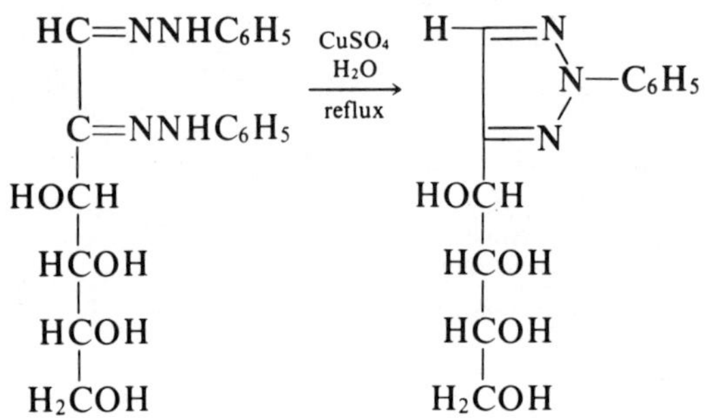

(109)

67%

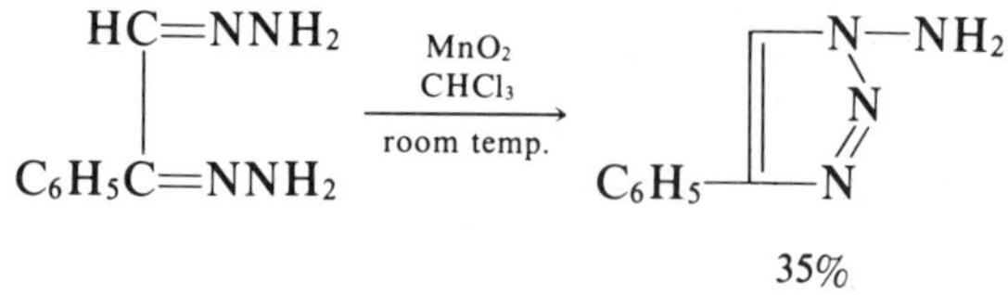

(120)

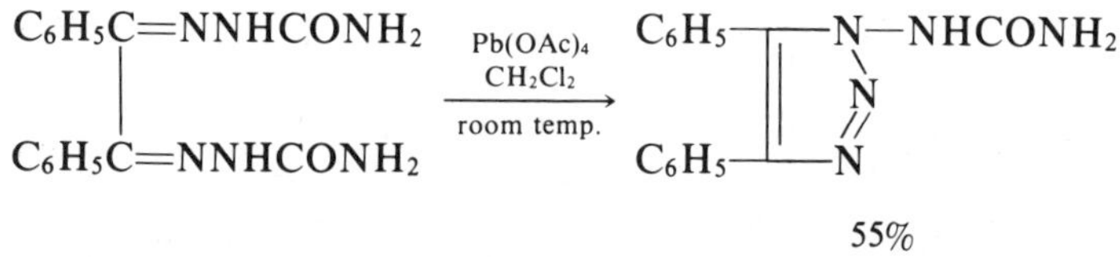

(6)

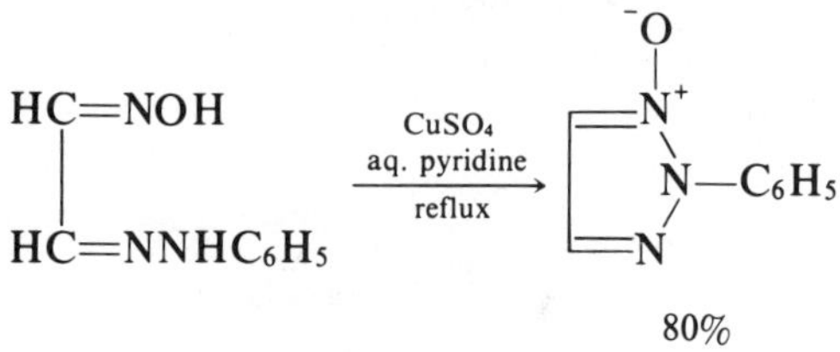

(27)

Oxidation of compounds with amino and azo groups on adjacent carbons produces 2-substituted *v*-triazoles. The process, which is called triazolation (97), is the main method for synthesis of *v*-triazole fluorescent brighteners, and can be applied to both monocyclic and fused ring *v*-triazoles. The principal oxidant for this ring closure is copper sulfate in pyridine or ammonium hydroxide, but sodium hypochloride (84, 162), sodium dichromate (212), and lead tetraacetate (69) also are effective. Dipolar benzenesulfonyl and cyano derivatives of 1-amino-2-phenylbenzotriazole are obtained by diazotization of *o*-aminoazobenzene, followed by coupling with sulfinate or cyanide salts, reactions involving formation of nitrogen-nitrogen bonds not only by nitrosation, but also between diazo and azo nitrogen atoms (24).

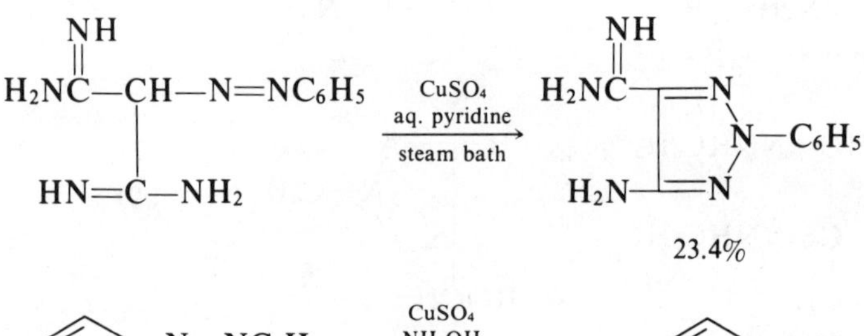

(253)

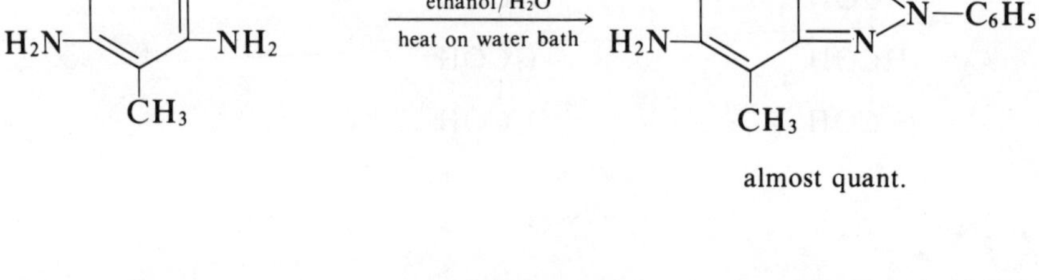

(263)

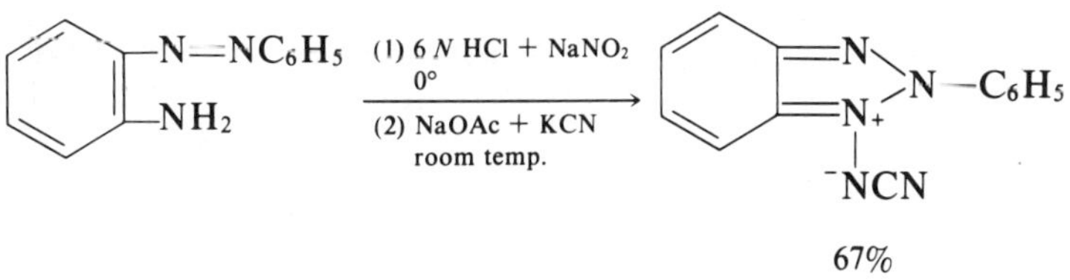

ν-Triazines and benzotriazines result from oxidation of *N*-aminopyrazoles and indazoles. In a similar reaction, the hydrazone of *o*-aminoacetophenone

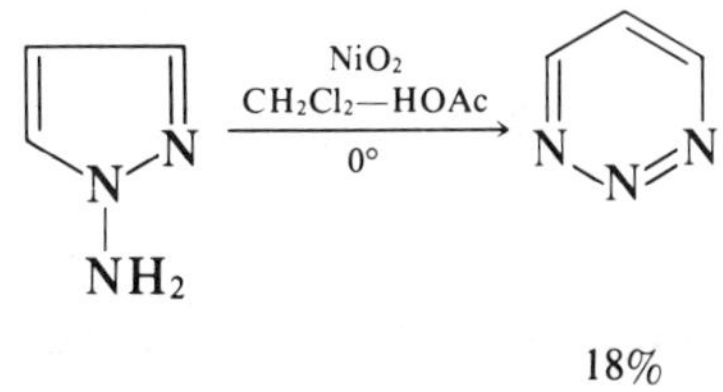

(220)

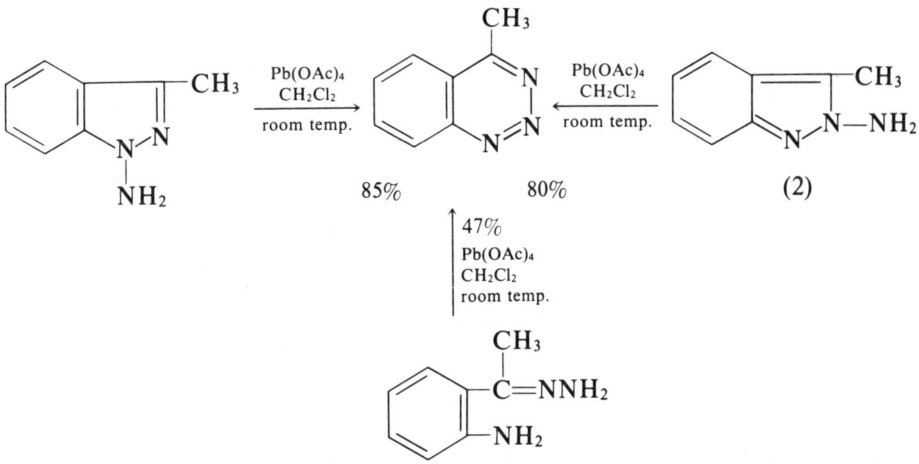

(2)

also is oxidized to a benzotriazine. 2-Methylbenzotriazinium betaine 1-oxide results by oxidizing *o*-nitrobenzaldehyde methylhydrazone.

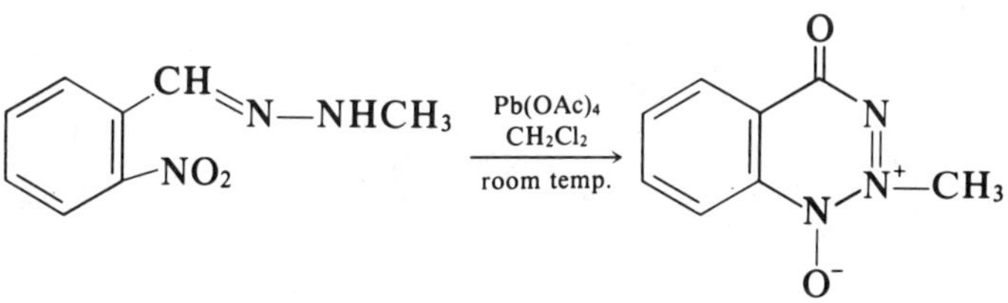

(193)

Electrolysis of cyanamide in alkaline solution results in several 1,2,3,5-tetrazines and cyanamide-related by-products (167).

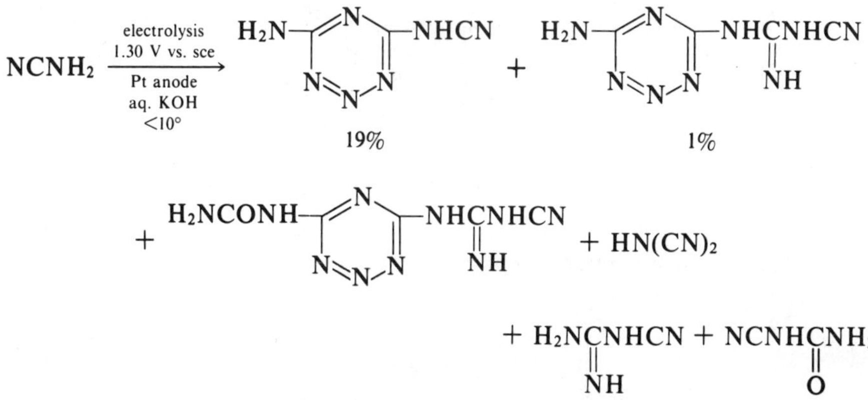

Dimroth rearrangement of imidazo[4,5-*e*]-2,3,4-thiadiazine leads to 4-thionoimidazo[4,5-*d*]-*v*-triazine. 1,2,3-Oxadiazoles and thiadiazoles with a nitrogen

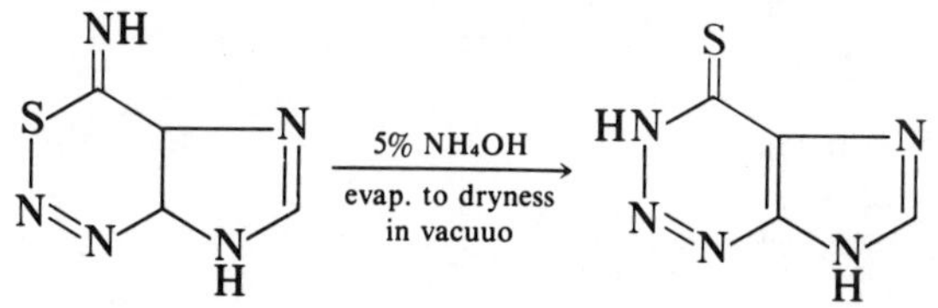
(238)

76%

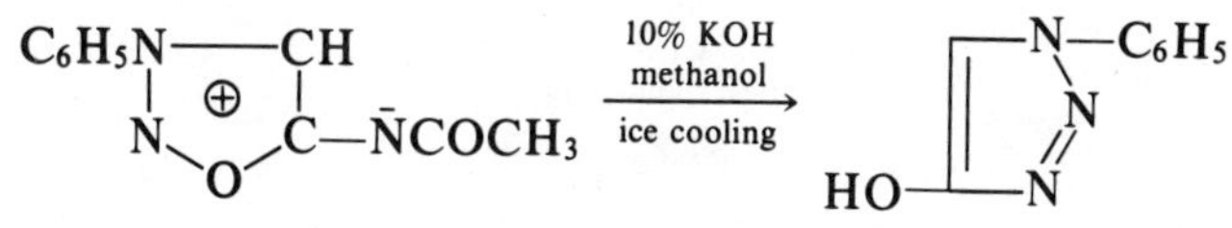
(56)

48%

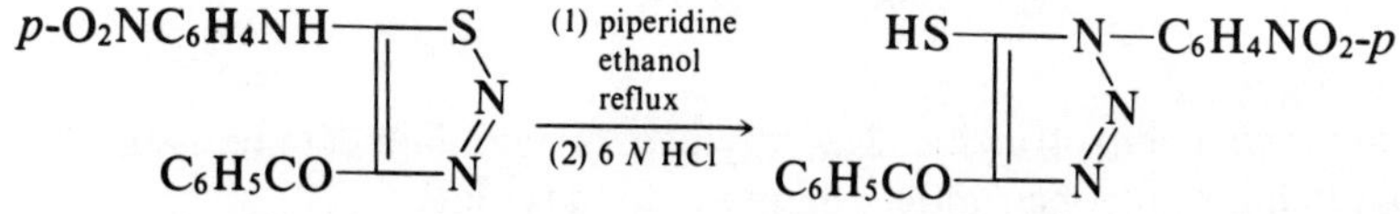
(56)

96%

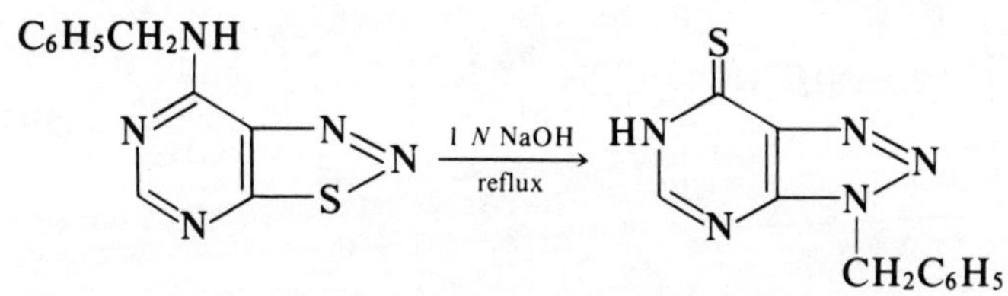
(4a)

90%

substituent on the ring carbon adjacent to ring oxygen or sulfur rearrange under alkaline conditions to *v*-triazoles. In a similar interconversion, amino-[1,2,3]thiadiazole[5,4-*d*]pyrimidines form thiono-8-azapurines.

Cyclic three-nitrogen chains result from the combination of the nitrogen of phenyl isocyanate with dipolar diazonium pyrazoles and *v*-triazoles. Com-

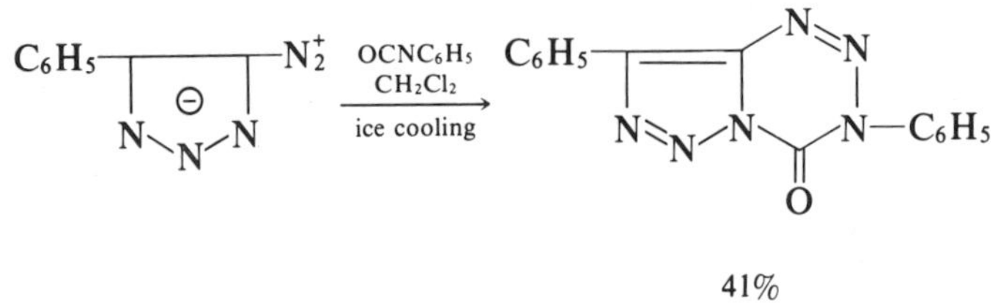

(70)

pounds with embedded four-nitrogen chains are obtained from the addition of phenyl isocyanate or isothiocyanate to the 2,5-positions of 1,3-dimethyl *v*-triazolium-4-olates **(XXVIII)** (237). Formation of an embedded five-nitrogen chain occurs by addition of azodicarboxylic ester to **XXVIII** (236, 237).

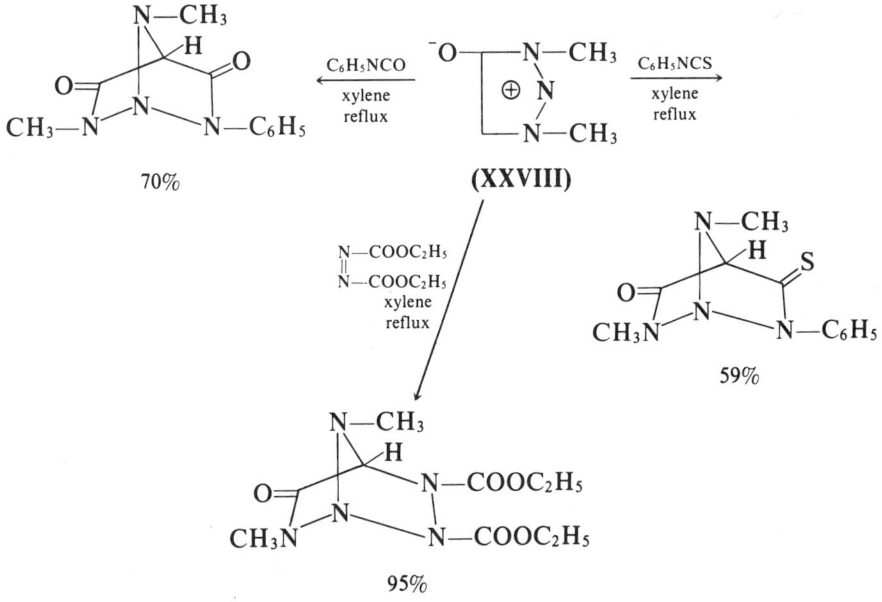

Naphtho[2,1-*e*]-*v*-triazolo[1,5-*b*]-*as*-triazine **(XXIX)** with an embedded four-nitrogen chain is formed in a thermal rearrangement of *v*-triazolyl[4-azo-1]-2-naphthol, in which two *v*-triazole nitrogens are joined to the azo group, probably through the unisolated intermediate shown (309).

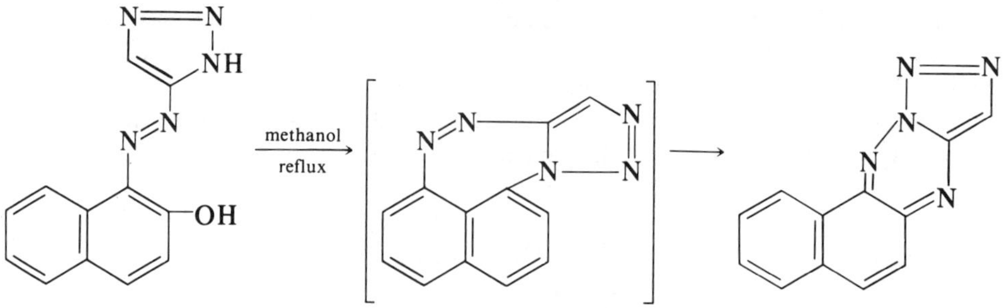

(XXIX)

80%

Aryl azides transfer two nitrogen atoms to aldehyde hydrazones while acquiring two hydrogens, forming 2-substituted tetrazoles and aryl amines.

v-Triazole and tetrazole rings result from the transfer of two nitrogen fragments from tosyl to 2-amino or phenacyl-substituted nitrogen heterocyles.

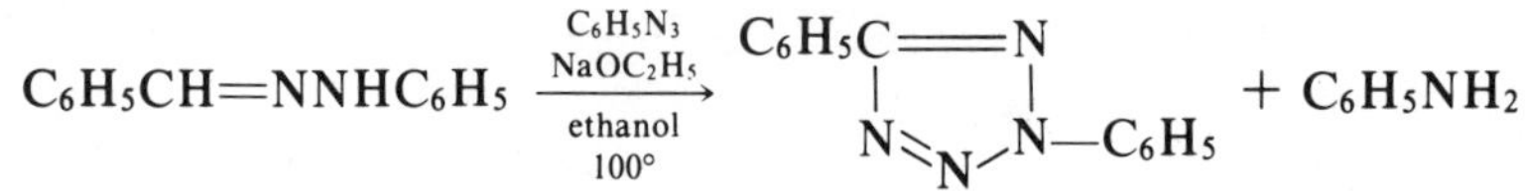

(64)

75%

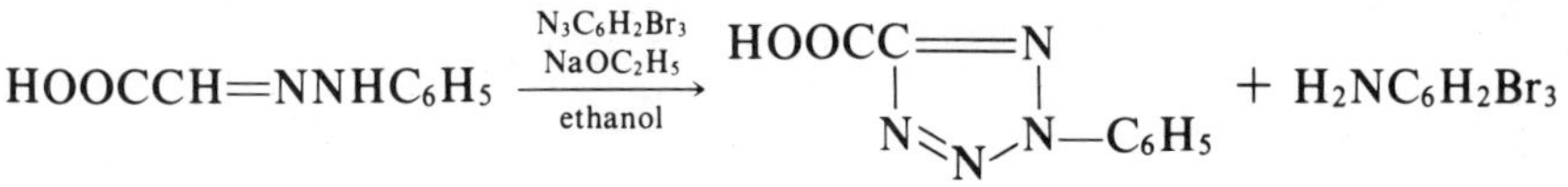

(65)

70%

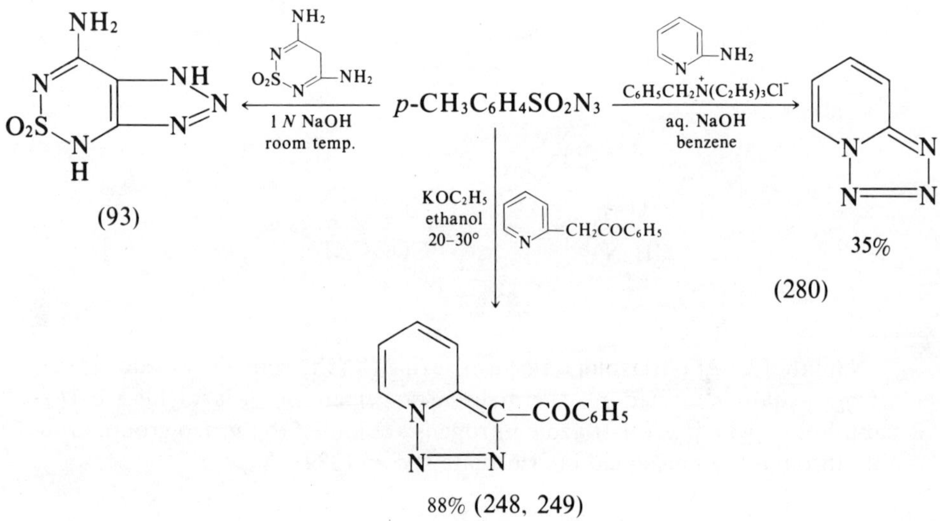

(93)

35%

(280)

88% (248, 249)

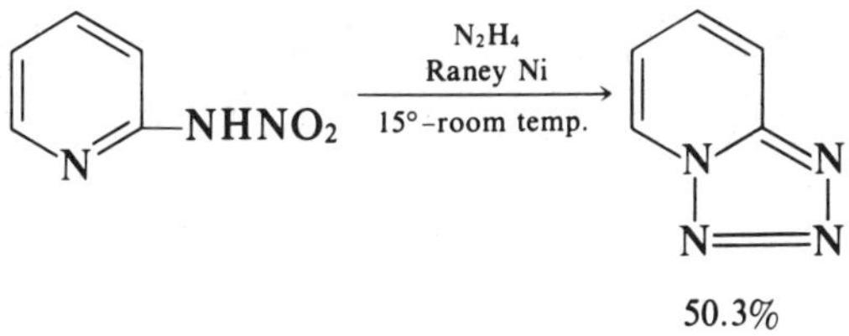

(224a)

Synthesis of tetrazolopyridine by the action of hydrazine on 2-nitraminopyridine may involve intermediate formation of 2-azidopyridine through oxidation-reduction reactions of the nitramino group with hydrazine.

Mesoionic 1,3-diphenyltetrazolium-5-olate is formed by coupling potassium phenylhydrazonomethanedisulfonate with diazonium salts, evidently through a 1-tetrazene which cyclizes with hydrolysis of the disulfonate.

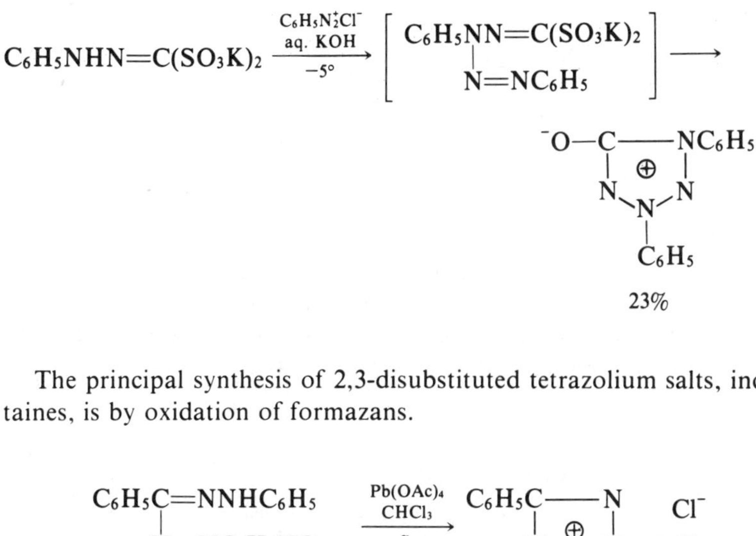

(78, 108)

The principal synthesis of 2,3-disubstituted tetrazolium salts, including betaines, is by oxidation of formazans.

(7)

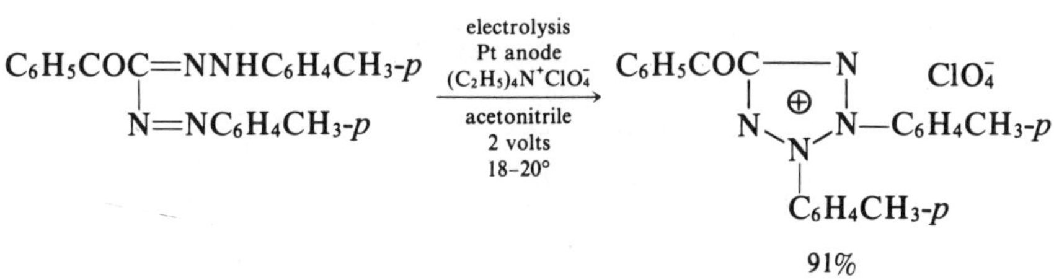

(172)

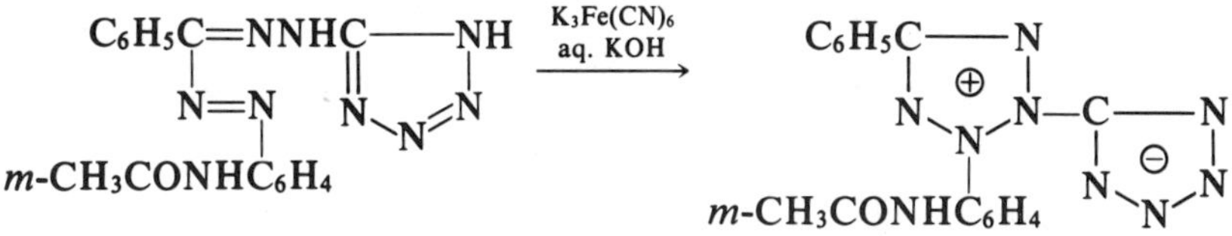

(269)

50%

Aldehyde benzenesulfonyl hydrazones react with diazonium salts to yield 2,5-disubstituted tetrazoles. The unisolated formazan loses benzenesulfinic acid forming an azodiazene intermediate.

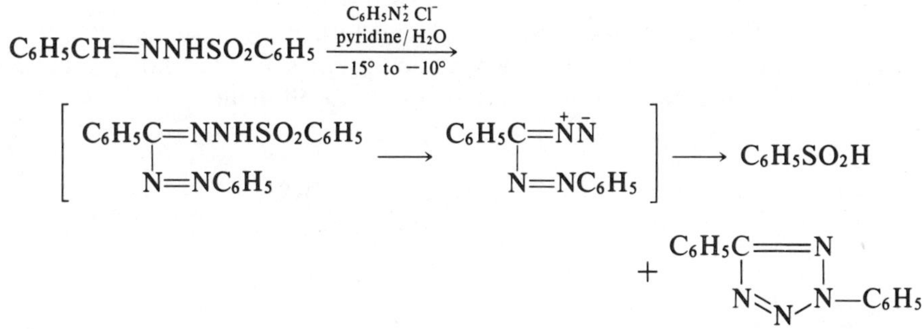

71% (148, 268)

Loss of benzenesulfinic acid also occurs in the conversion of hydrazidines prepared from *N*-phenylsulfonylbenzhydrazidoyl chloride **(XXX)** and benzoyl hydrazine or phenyl hydrazine, respectively, to 5-phenyl and 2,5-diphenyltetrazoles. The benzoyl group hydrolyzes during the formation of 5-phenyltetrazole. Air oxidation of the intermediate may be involved in this synthesis of 2,5-diphenyltetrazole.

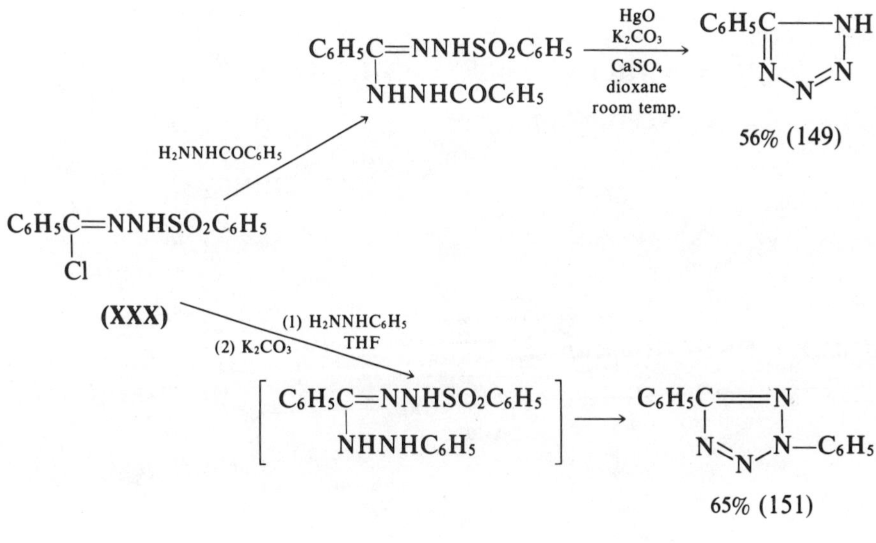

56% (149)

65% (151)

v-Triazolo[1,5-*a*]pyridines result from the action of tosyl hydrazine on α-pyridyl ketones whereby two nitrogens are transferred to complete the ring. Al-

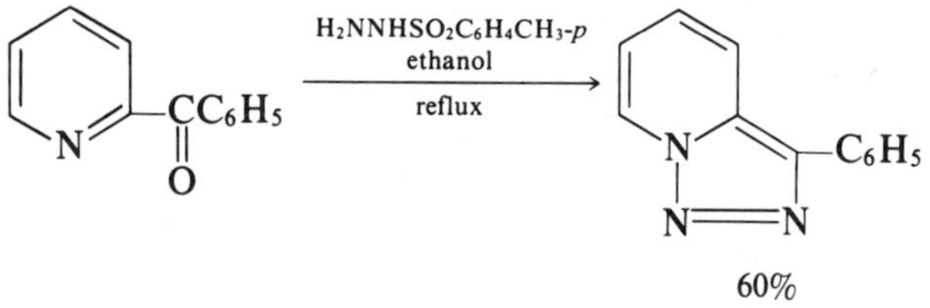

(250)

though the pyridyl ketone tosyl hydrazone is not an intermediate in this synthesis, tosyl hydrazones of 3-substituted quinoxaline 1-oxides are converted by alkaline treatment to *v*-triazolo[1,5-*a*]quinoxaline 1-oxides, probably through an intermediate diazene.

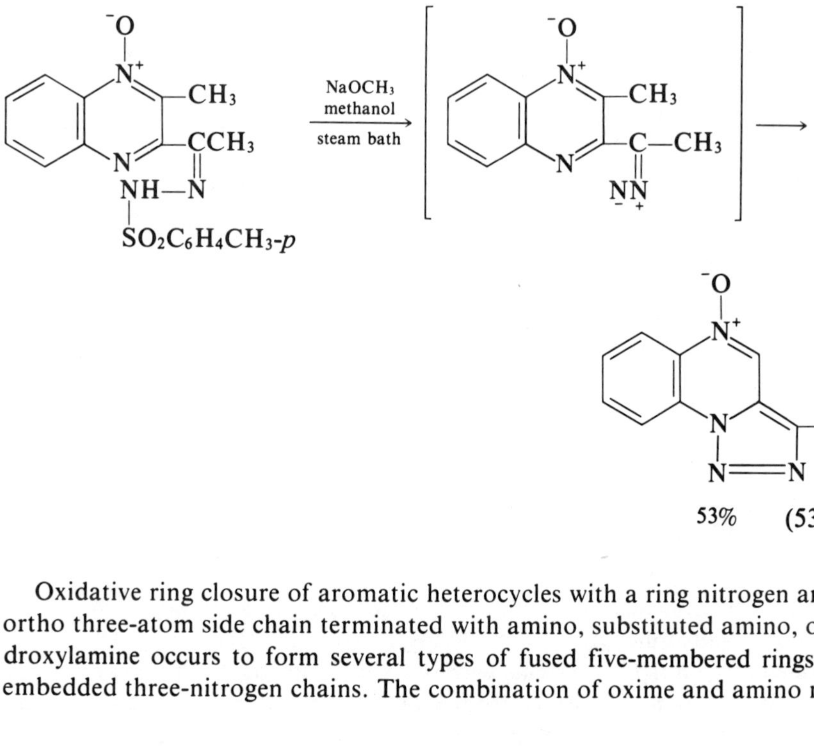

(53)

Oxidative ring closure of aromatic heterocycles with a ring nitrogen and an ortho three-atom side chain terminated with amino, substituted amino, or hydroxylamine occurs to form several types of fused five-membered rings with embedded three-nitrogen chains. The combination of oxime and amino nitro-

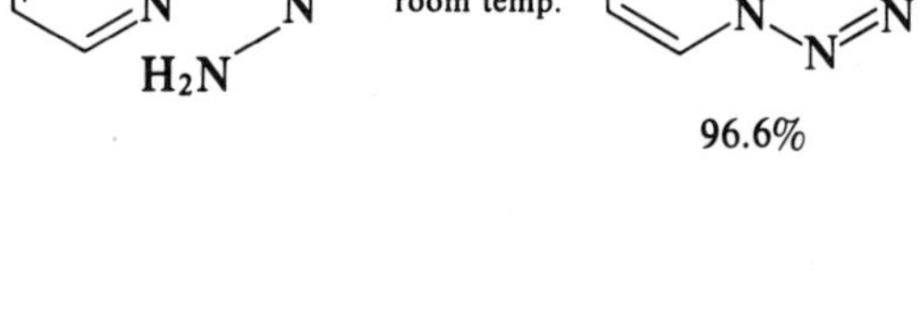

(202)

gens of *N*-amino-2-pyridinecarbonyloxime salts **(XXXI)** to form *v*-triazolo[1,5-*a*]-pyridines and the cyclization of pyridazines *N*-substituted at position 3 **(XXXII)** with a hydroxamic acid group, are similar reactions.

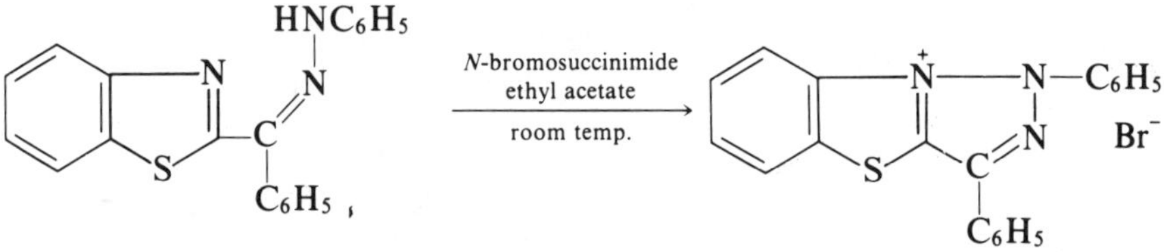

69% (199)

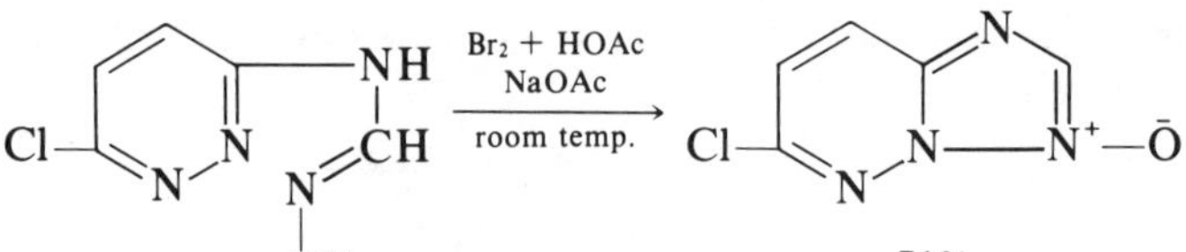

71% (40)

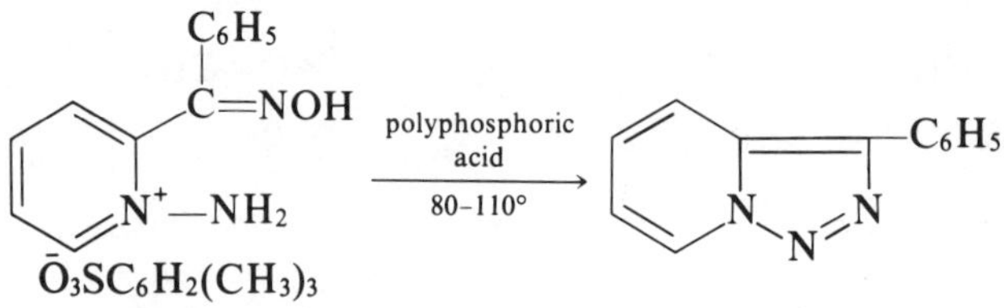

(XXXI) 76% (292)

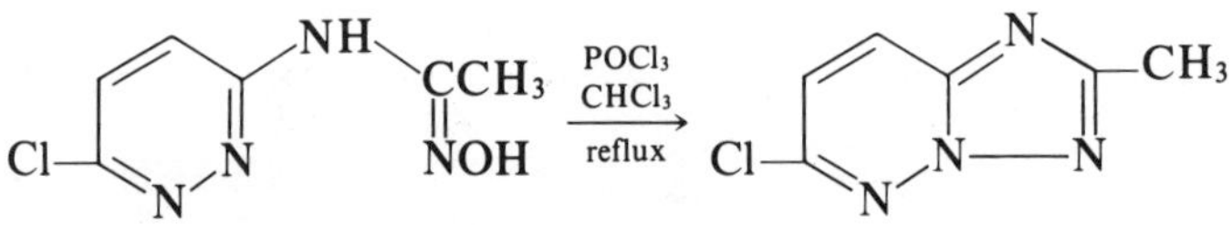

(XXXII) (235, 281)

N-pyrazolyl moieties adjacent to an azide or nitro on an aromatic ring, as in 1-(5-chloro-2-nitrophenyl)-3,5-dimethylpyrazole **(XXXIII)**, are cyclized to embedded three-nitrogen systems with two shared nitrogens through an ortho nitrene. By-products are formed from **XXXIII** by a triplet nitrene C—H insertion reaction and reduction of nitro to amino. The nitrene from 2-azido-1-(3,5-dimethylpyrazolyl)- phenazine, however, forms the dipolar pyrazolo derivative with little attack on the methyl groups. Deoxygenation of 1-(*o*-nitrophenyl)-indazole leads through the nitrene to intramolecular cyclization to the dipolar 6-dehydroindazolo-[1,2-*a*]benzotriazole and also to the hexaaza-*as*-indacene **(XXXIV)** with an embedded six-nitrogen chain (301).

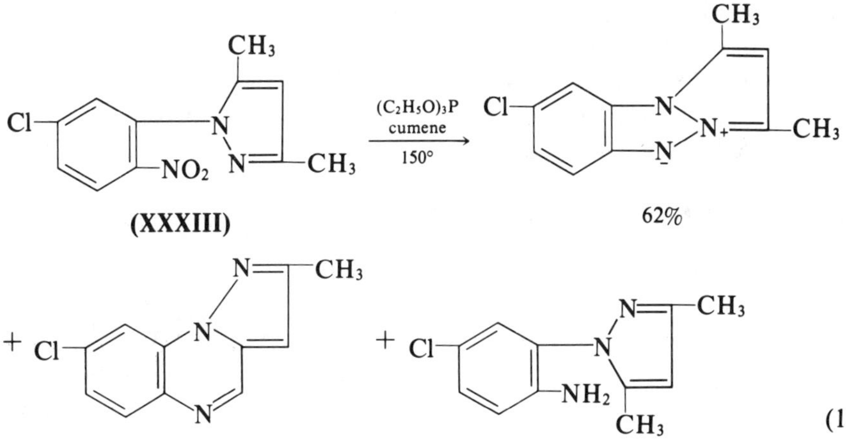

(XXXIII) 62%

16% 11% (194)

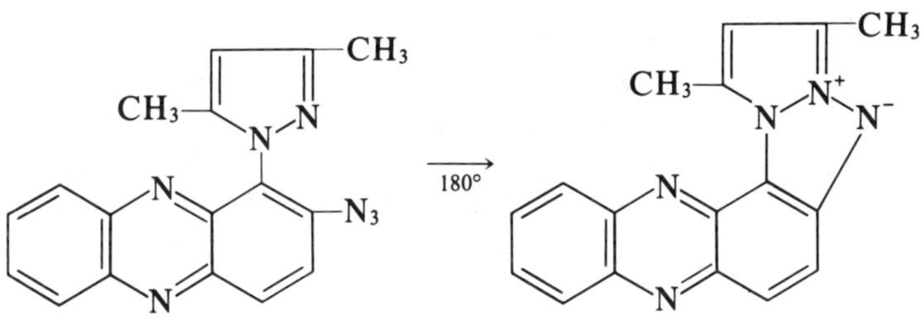

90% (5)

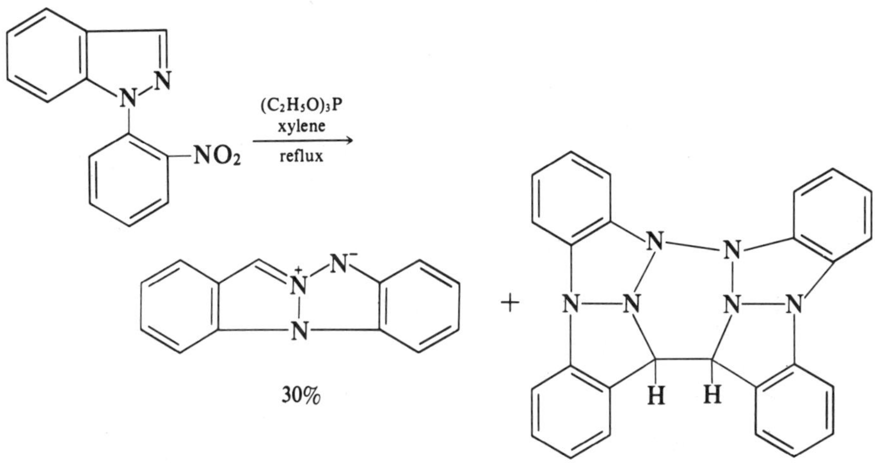

30%

(XXXIV)

30.6%

Symmetrical bicyclic 1,3a,4,6a-tetraazapentalene derivatives result from condensation of two molecules of α,β-diketobutyric ester α-monohydrazone

(232). In addition to formation of two carbon-nitrogen bonds, the inner hydrazone nitrogens unite, creating the embedded dipolar four-nitrogen chain with a redistribution of oxidized and reduced nitrogen atoms. The reaction is thermochemically favorable because of the considerable resonance energy in the tetraazapentalene ring system.

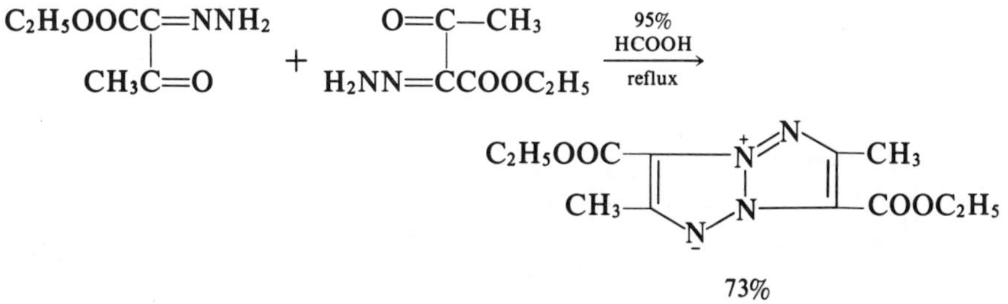

Benzo or dibenzotetraazapentalenes are formed by combination of nitrenes generated thermally from *o*-azidoaryl or by deoxygenation of *o*-nitroaryl groups with nitrogen atoms in *o*-azo or *v*-triazole substituents. Thus dibenzo-1,3a,4,6a-tetraazapentalene results from heating *o,o'*-diazidoazobenzene directly, stepwise through 2-*o*-azidophenylbenzotriazole (46) or from *o,o'*-dinitroazobenzene (44).

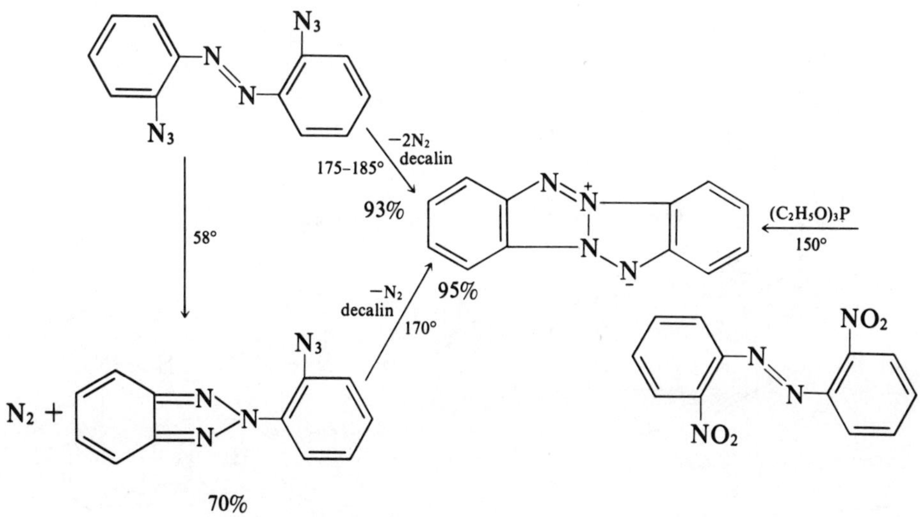

Tetraazapentalenes with a branched nitrogen chain are obtained from 1-substituted *o*-azido or *o*-nitroaryl *v*-triazoles. The related benzotriazolo[2,1-*a*]-naphtho[1,8-*de*]triazine is obtained from 2-*o*-nitrophenylnaphtho[1,8-*de*]triazine by deoxygenation (273).

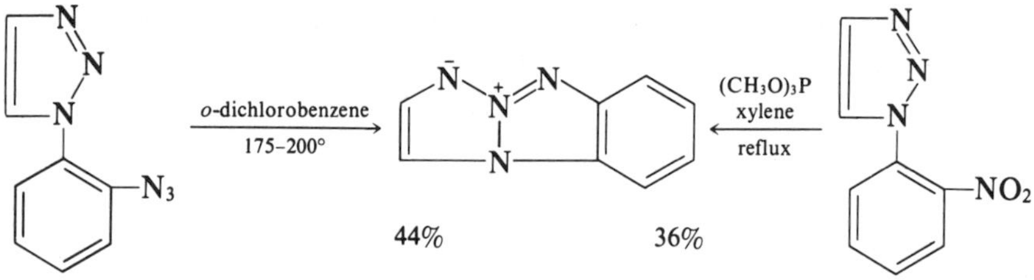

(158)

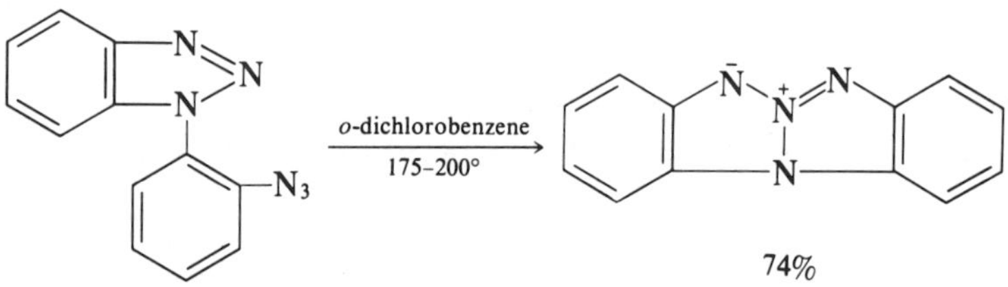

(46)

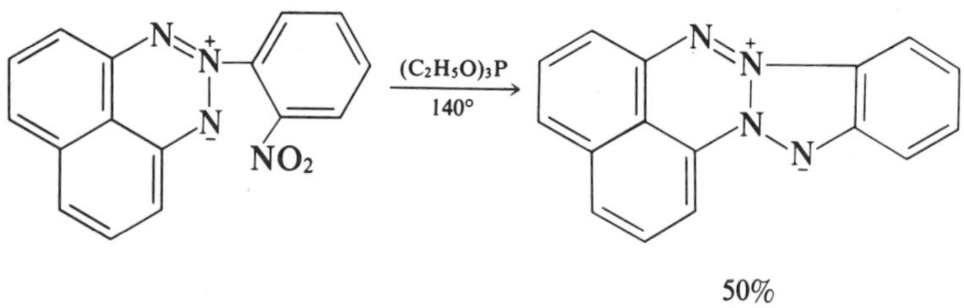

7.2. CYCLIZATIONS WITH INTACT NITROGEN CHAINS

Emphasis in this section is on ring formation with chains of three or more nitrogen atoms, rather than synthesis of nitrogen-nitrogen bonds. Cyclic high nitrogen compounds result from both inter- and intramolecular reactions with chains of three or more nitrogen atoms.

Azide supplied as ion, hydrogen azide, or other covalent azide is the source of three-nitrogen units for the synthesis of many cyclic high nitrogen compounds in both addition and displacement reations. Huisgen's concept of the concerted addition of 1,3-dipoles to dipolarophiles (135, 136) systematizes consideration of cycloadditions, including those of azides. Essential to the ring closure reactions of azide are both dipolarity and bending of the normally nearly linear azide so that bond formation to complete the ring is geometrically feasible, as discussed for both intramolecular (124) and intermolecular (185) cyclizations. LCAO calculations indicate only a small amount of energy is required to distort the straight azide (136).

Reference was made in Chapter 2 to the general synthesis of triazolines from substituted azides and olefins, in connection with their thermal stability. The reaction proceeds, usually in good yield, with unstrained and strained olefins, and with azides carrying either electron-withdrawing or electron-releasing

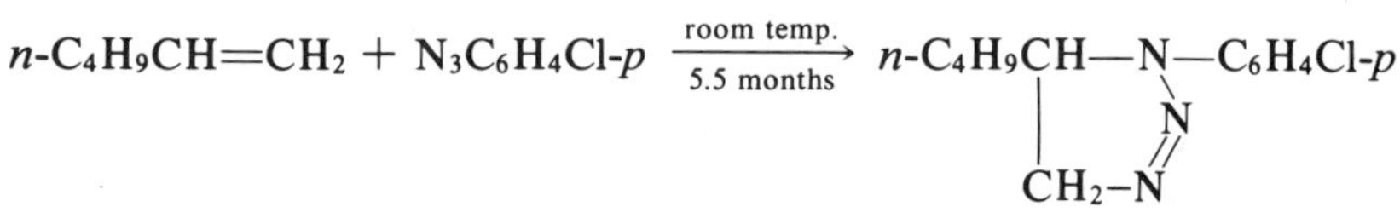

89% (259)

groups. Reactions of unstrained alkenes at room temperature are slow. Addition of azides to olefins of necessity is cis, and the olefin substituent geometry is

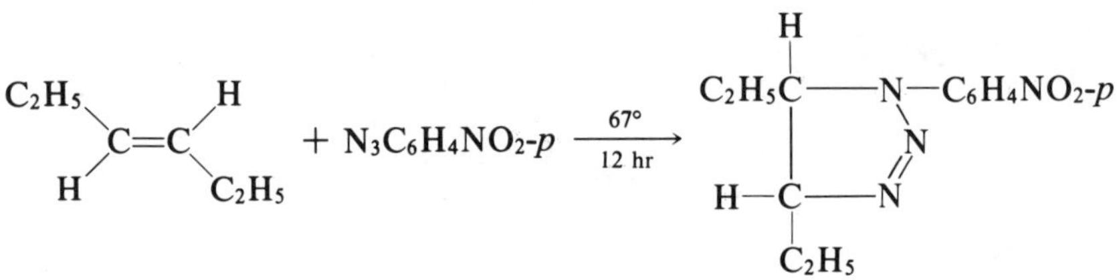

53% (259)

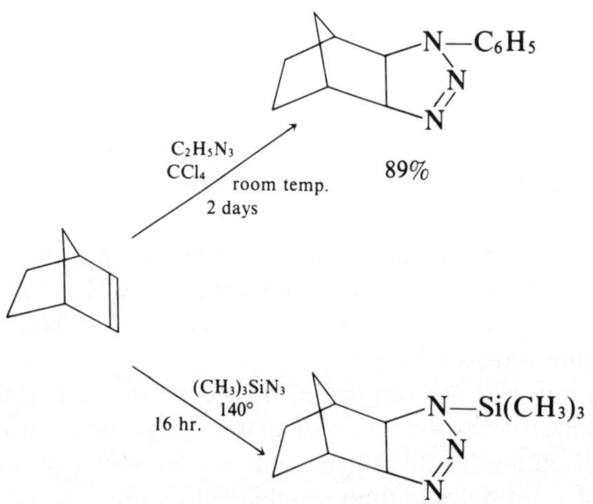

(137)

(231)

retained in the triazoline (260). Addition of azides to alkenes with electron-withdrawing substituents forms 1,4-disubstituted triazolines; alkenes with electron-releasing substituents form 1,5-disubstituted triazolines.

Phenyl azide combines with dichloroketene, generated *in situ,* to yield 4,4-dichloro-5-oxo-1-phenyltriazoline.

Alkaline condensation of phenyl azide with ethyl isopropyl ketone leads to 5-hydroxy-1-phenyltriazoline possibly through the ketone anion enolate.

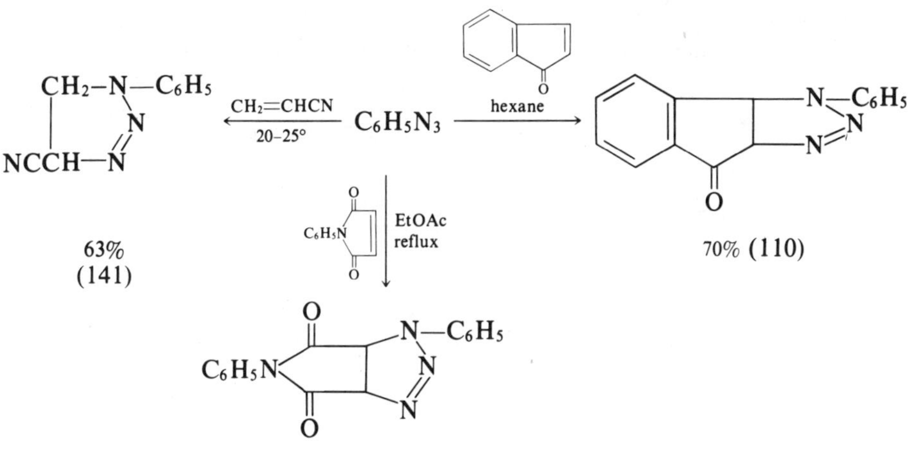

63% (141)

70% (110)

89% (211)

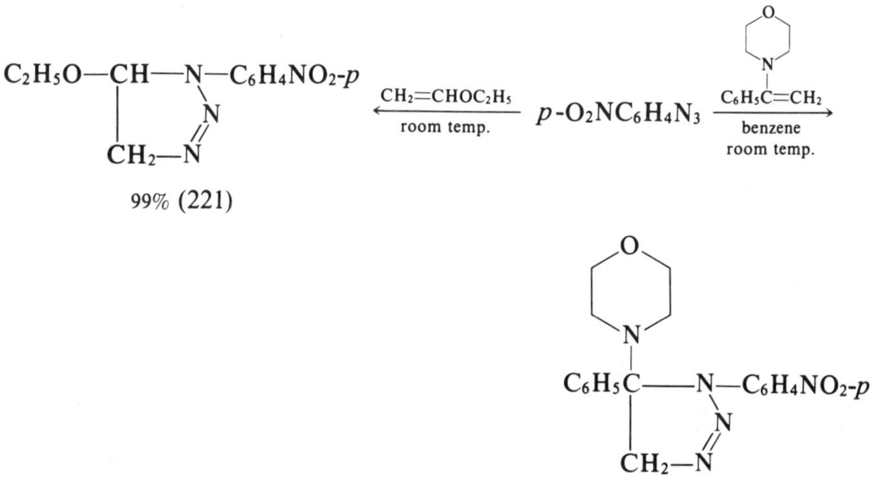

99% (221)

99% (140)

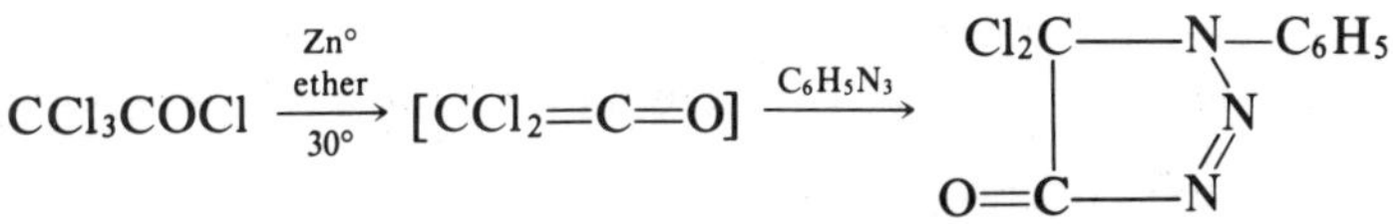

(230)

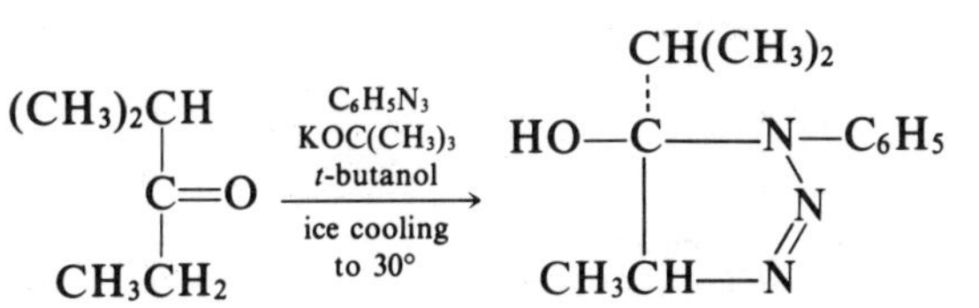

50%

(222)

Tertiary azides, substituted with alkoxycarbonyl or cyano on the tertiary carbon, are reductively cyclized to triazolines, probably through unisolated monosubstituted triazenes.

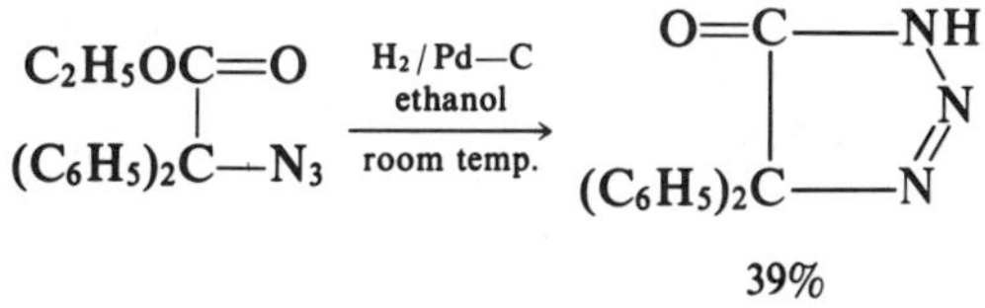
(131)

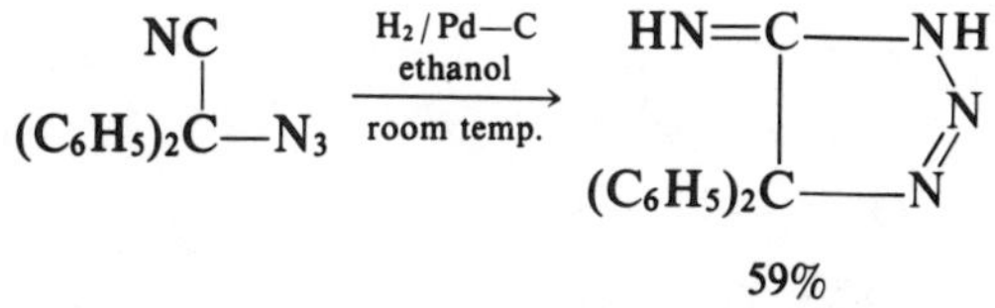
(130)

Alkyl and aryl azides, except for *p*-nitrophenyl and benzyl azides, react with 2 moles of dimethyloxosulfonium methylide to form 1-monosubstituted triazolines (95). The two methylide groups join to make the ethylene portion of the triazoline ring.

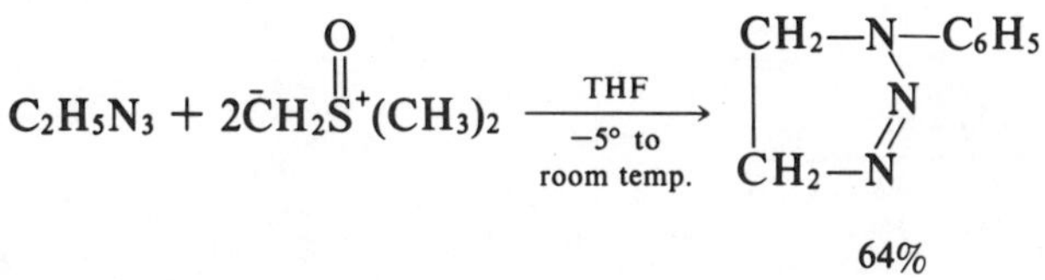
(94)

v-Triazoles are prepared by combination of azide ion, hydrogen azide, or alkyl or aryl azides with acetylenic compounds. Yields of *v*-triazoles from acetylene and its unsubstituted alkyl and aryl derivatives are lower than with acetylenes carrying more strongly electron-withdrawing or releasing substituents.

Intramolecular addition of azide to ethinyl groups leads to fused *v*-triazoles. Benzyne, generated from anthranilic acid, combines with alkyl and aryl azides forming 1-substituted benzotriazoles (252).

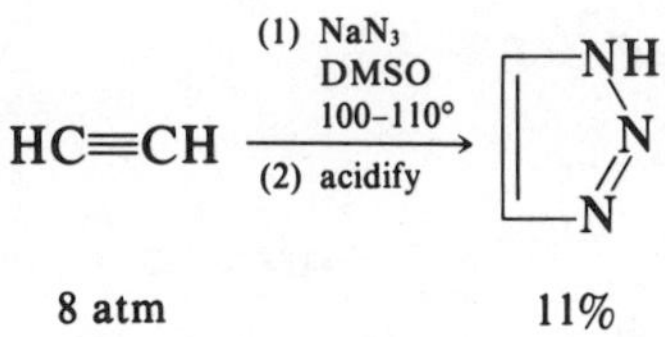
(317)

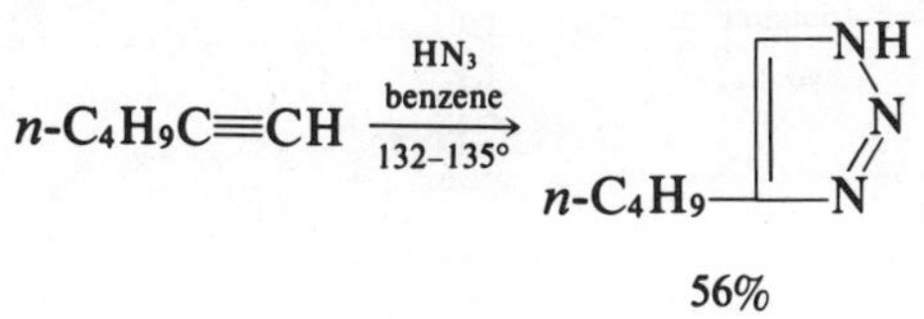
(114)

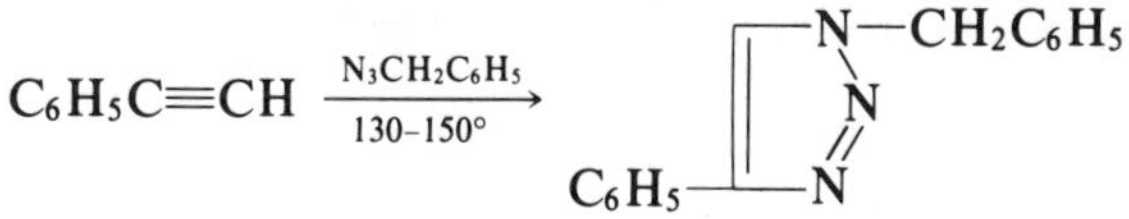 (101)

42%

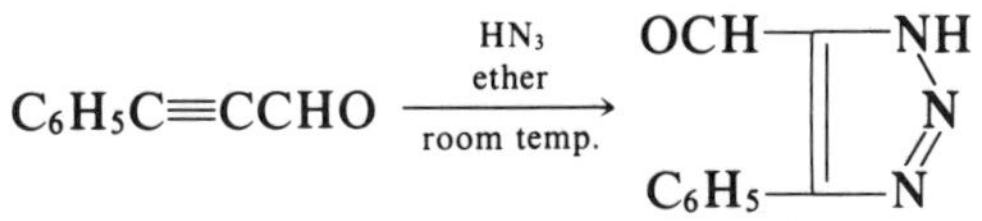 (270)

74%

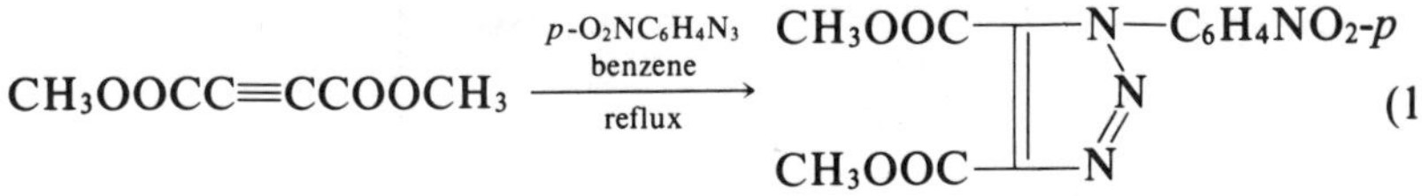 (138)

87%

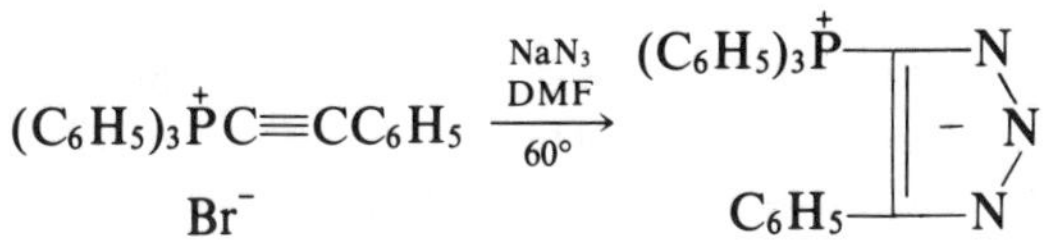 (294)

93%

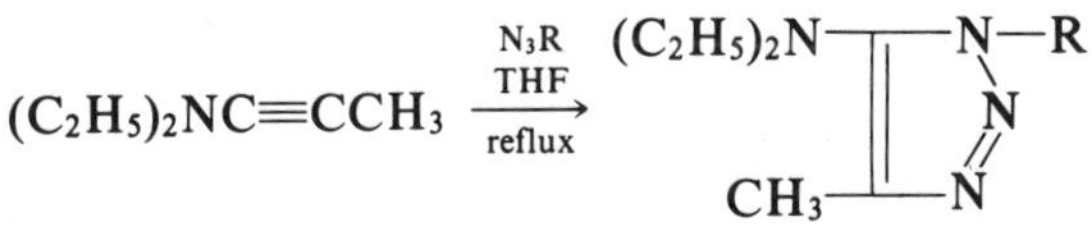 (112)

97%

R = 2,3,4,6-tetra-*O*-acetylglucopyranosyl

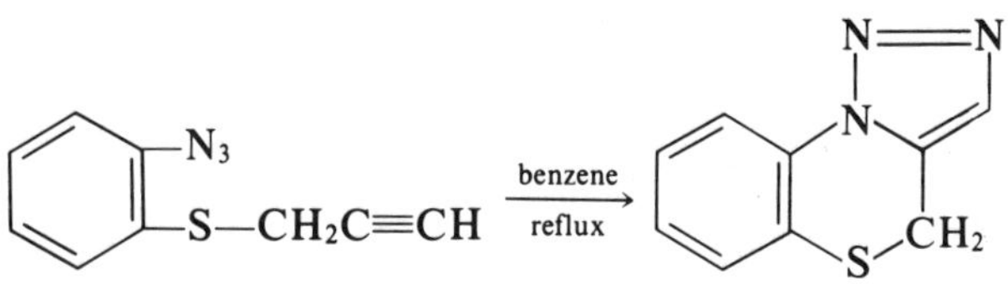 (90)

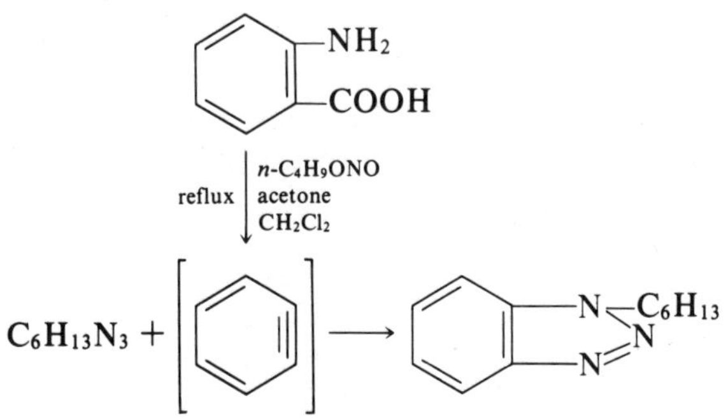 (252)

70%

Allenes also add organic azides to give ν-triazoles.

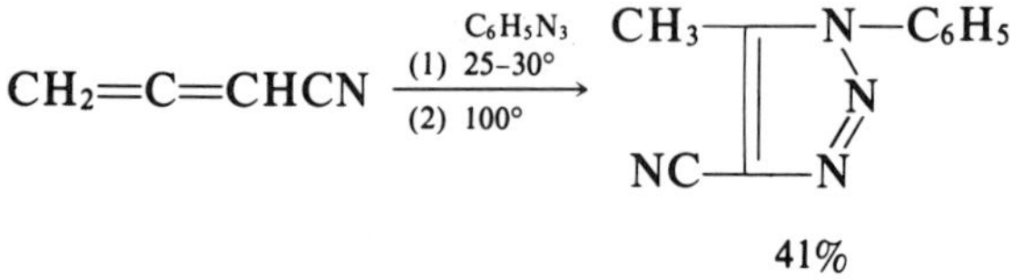

(254)

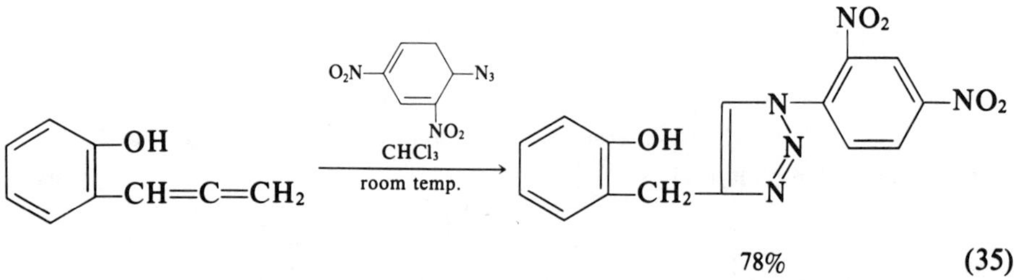

(35)

Under conditions in which a carbanion is formed on the β-carbon of a vinyl azide, cyclization to a ν-triazole occurs. Thus *trans*-2-azidovinyl *p*-tolylsulfone **(I)** is isomerized to 4-*p*-tosyl-ν-triazole **(II)** in dimethyl sulfoxide, in the presence of bases such as sodium azide, sodium *p*-toluene sulfonate, or potassium

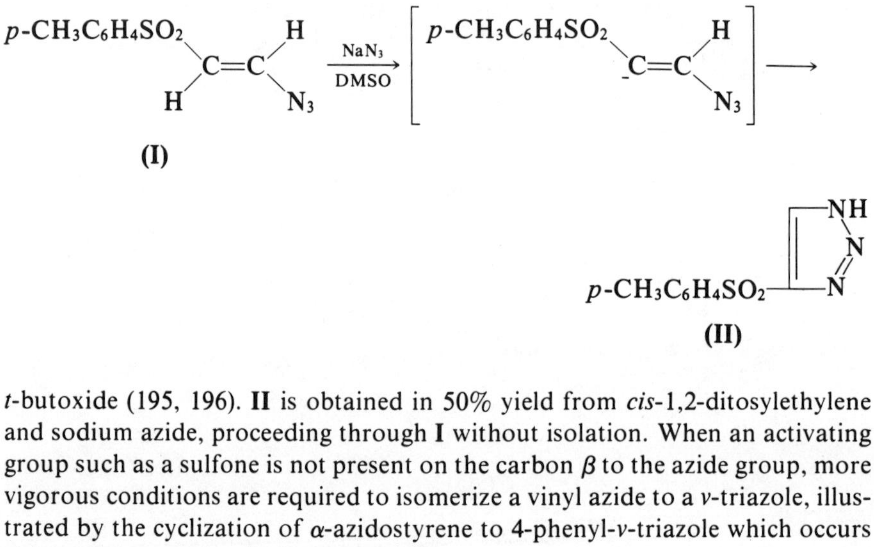

t-butoxide (195, 196). **II** is obtained in 50% yield from *cis*-1,2-ditosylethylene and sodium azide, proceeding through **I** without isolation. When an activating group such as a sulfone is not present on the carbon β to the azide group, more vigorous conditions are required to isomerize a vinyl azide to a ν-triazole, illustrated by the cyclization of α-azidostyrene to 4-phenyl-ν-triazole which occurs only under strongly alkaline conditions.

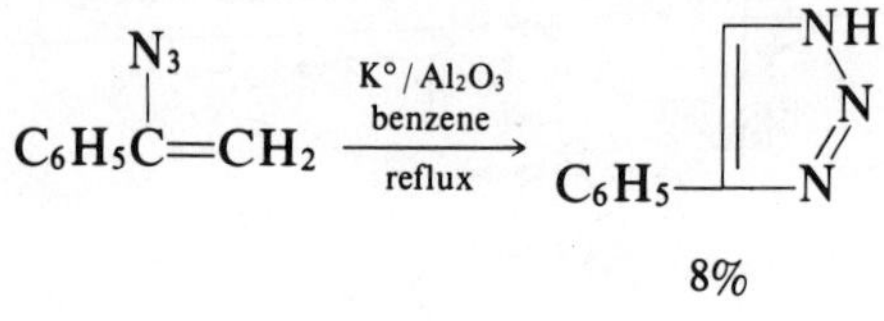

(317)

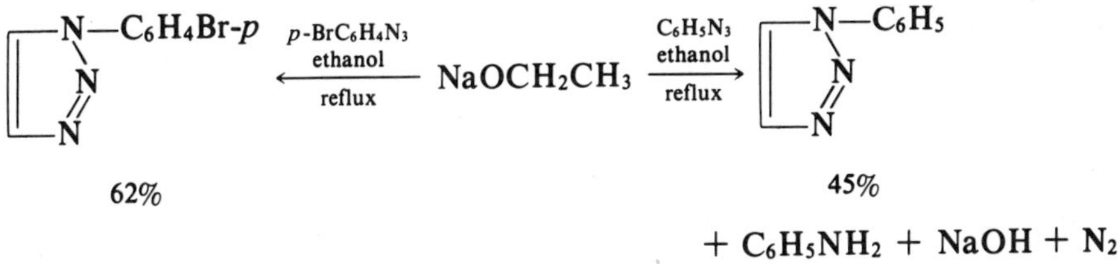

Aryl azides react with sodium alcoholates forming 1-aryl-*v*-triazoles, aryl amines, and nitrogen (30, 73).

Unsubstituted or 1-substituted *v*-triazoles are obtained from olefins carrying groups which are eliminated during or after condensation with azide ion or organic azides. Thus the olefin with a substituent X, which can be lost as HX, is a surrogate for an acetylene in the reaction with azide. Among the substituents which can be thus eliminated are alkali sulfonates, arylsulfonates (26, 196), nitro (271), cyanide (312), triphenylphosphine, and amines. These *v*-triazole syntheses pass through the corresponding triazolines, which usually are not isolated. Conversion of triazolines to *v*-triazoles under the influence of acid is discussed in Section 6.2, Reactions with Electrophiles.

Phenyl azide adds to ethylene sulfonyl chloride with simultaneous loss of sulfur dioxide and alkylation of the *v*-triazole ring with a second ethylene sulfonyl

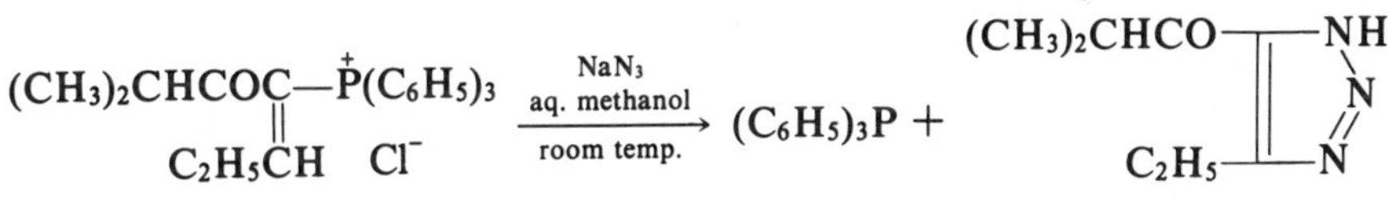

30% (324)

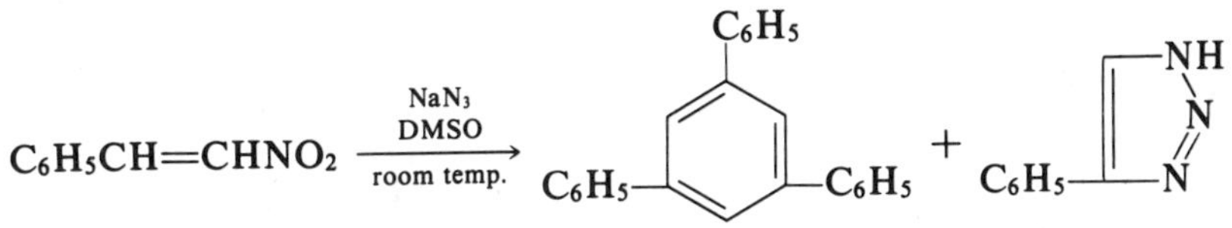

(326)

60%

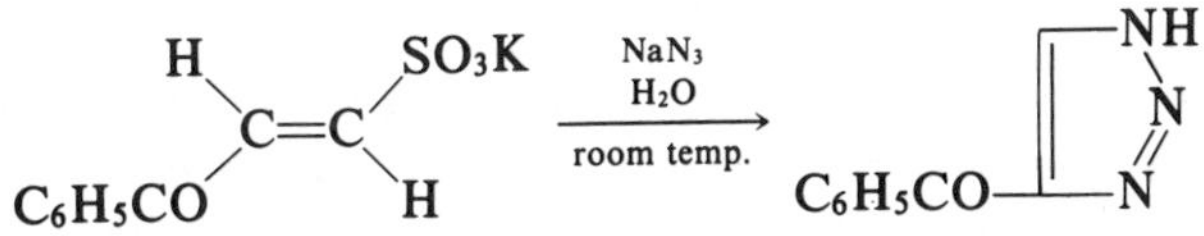

(218)

71%

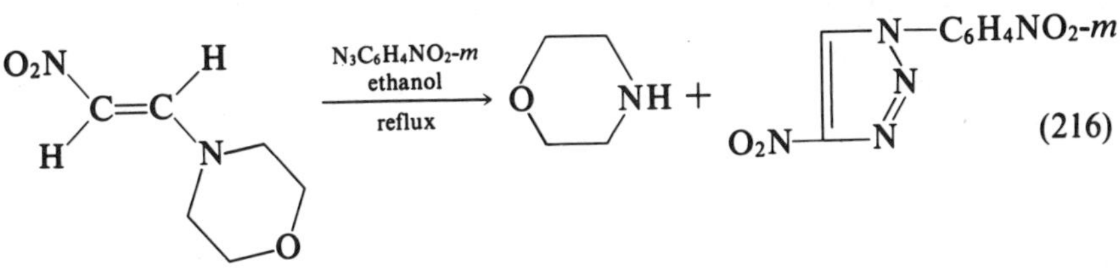

(216)

chloride molecule (256). The product hydrolyzes easily to dipolar 3-phenyl-v-triazolium-1-(2′-ethanesulfonate).

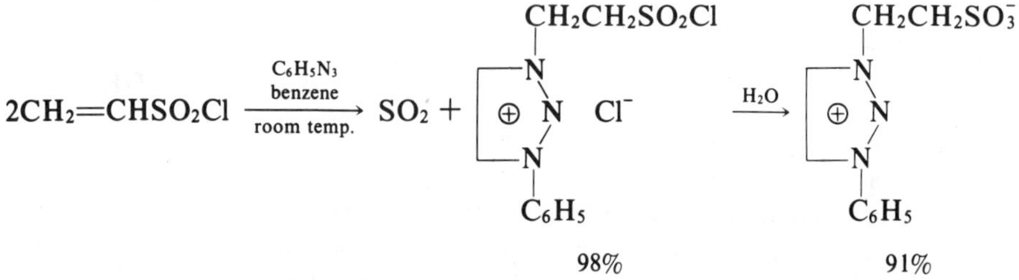

Condensation of azides with phosphorus ylides leads to v-triazoles. 4-Triphenylphosphonium v-triazole ylide is obtained by addition of azide ion to

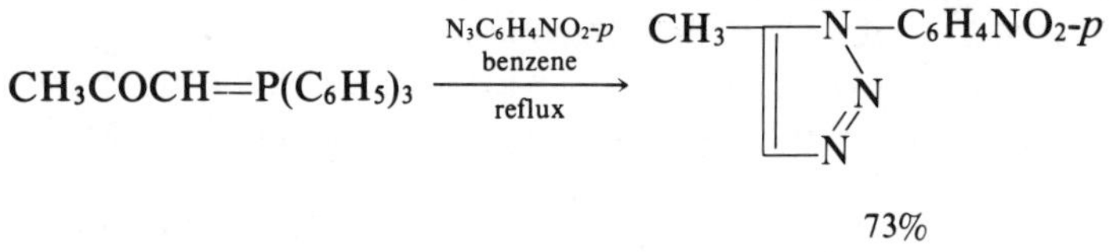 (322)

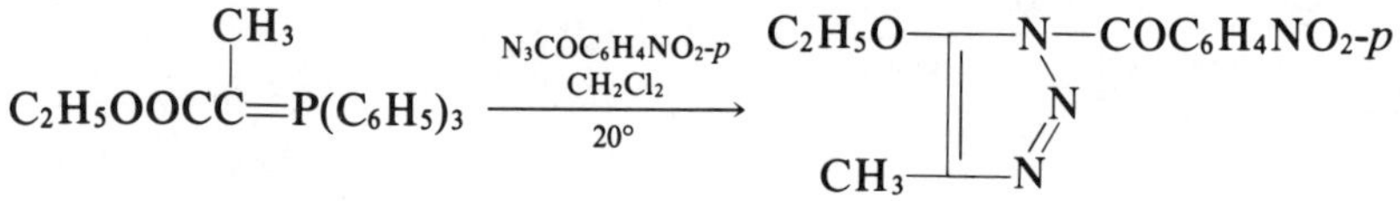 (171)

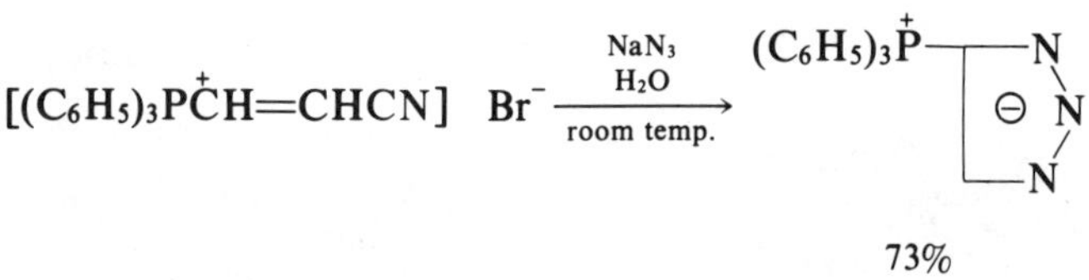 (152)

β-cyanovinyltriphenylphosphonium bromide with elimination of hydrogen cyanide.

Alkyl and aryl azides condense with the anions of compounds with methylene adjacent to electron-withdrawing groups to form v-triazoles, illustrated with ethyl acetoacetate, dimethyl malonate, and phenylacetonitrile. Ketone, alkoxycarbonyl, or cyano adjacent to the anionic carbon lead, respectively, to 5-alkyl, hydroxy, or amino-v-triazoles.

Fused ring v-triazoles are obtained from phenylacetonitrile by the same reaction when substituents present in the azide-containing molecule are capable of further reaction with the amino group generated in the condensation.

Monocyclic v-triazines result from rearrangement of cyclopropenyl and cyclopropenylium azides.

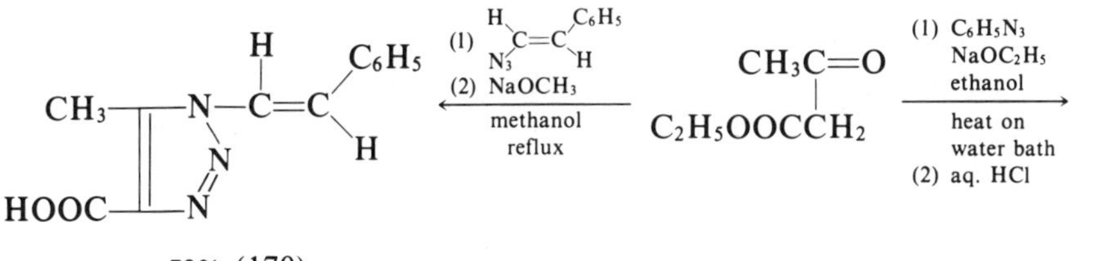

72% (170)

75–80% (62)

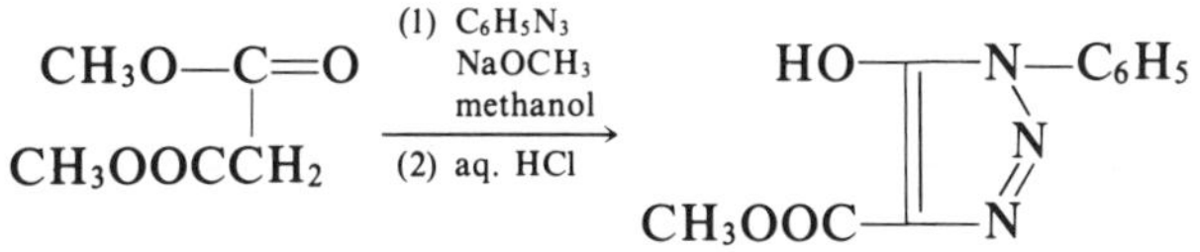

(63)

70%

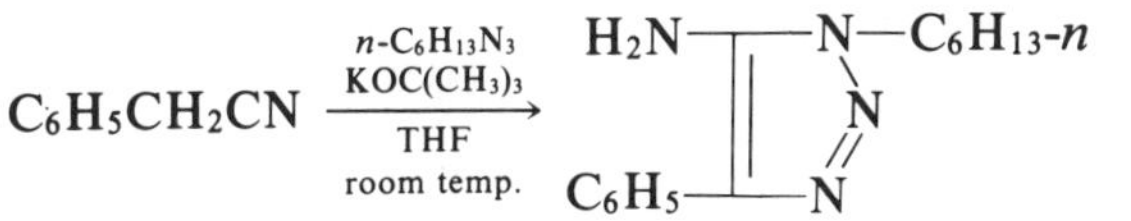

(186)

98%

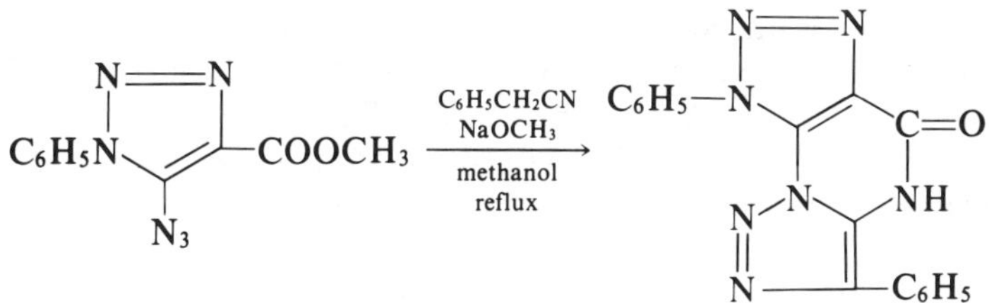

(289)

53%

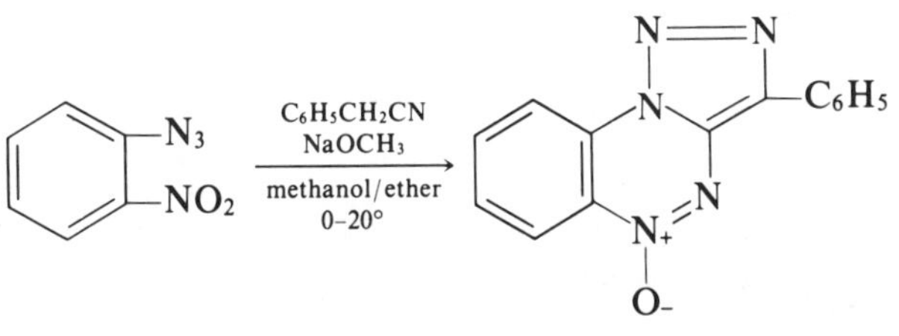

(296)

31%

Both the azide and the triazene moieties are involved in the synthesis of the pentacyclic compound **(III)** with two embedded three-nitrogen chains, formed by irradiation of 4-azidobenzotriazine (282).

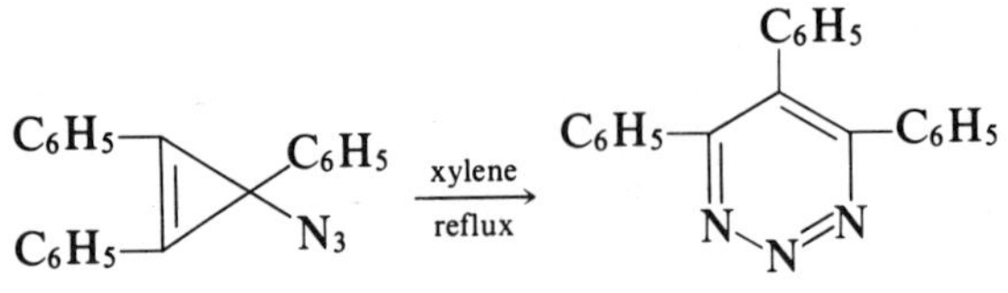

(48)

45%

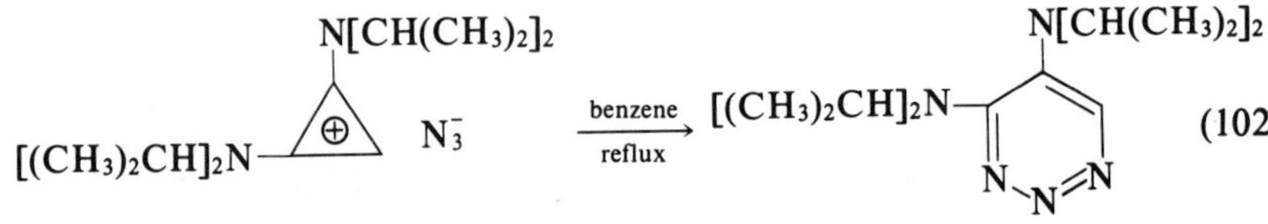

(102)

88%

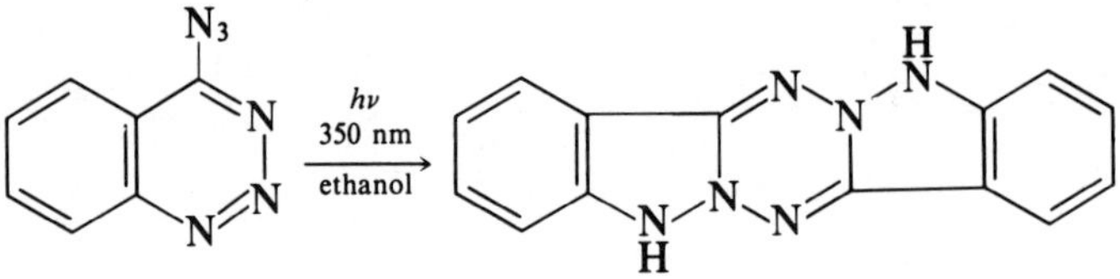

(III)

85%

5-Substituted aminothiatriazoles are obtained by addition of azides to isothiocyanates. Similarly, azide ion adds to the cumulative systems of carbon disulfide to form thiatriazolinethione and of bis(trifluoromethyl)thioketenes **(IV)** to form 5-[bis(trifluoromethyl)methyl]thiatriazole. Phenyl azide adds to **IV** to give a 2-phenylthiatriazoline, opposite to the orientation of the addition of sulfonyl isothiocyanates which leads to 4-substituted thiatriazolines.

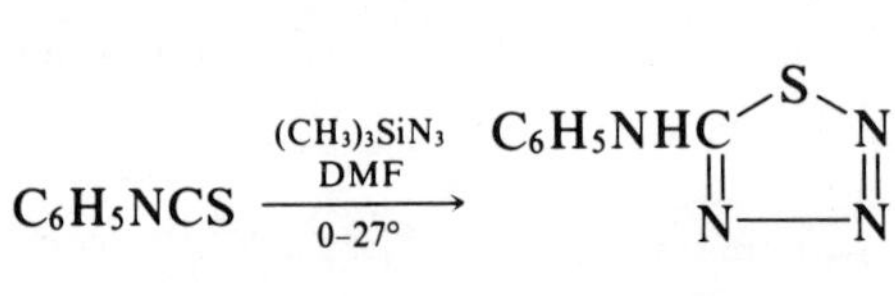

(310)

83%

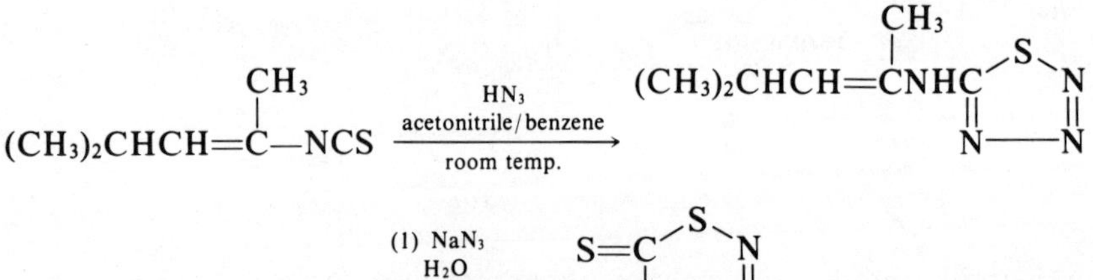

(129)

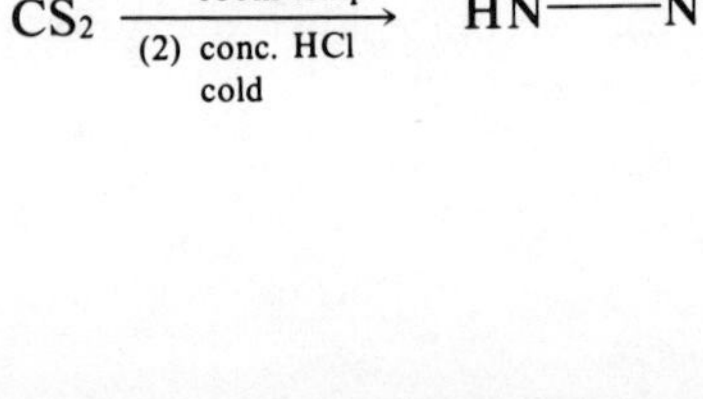

(183)

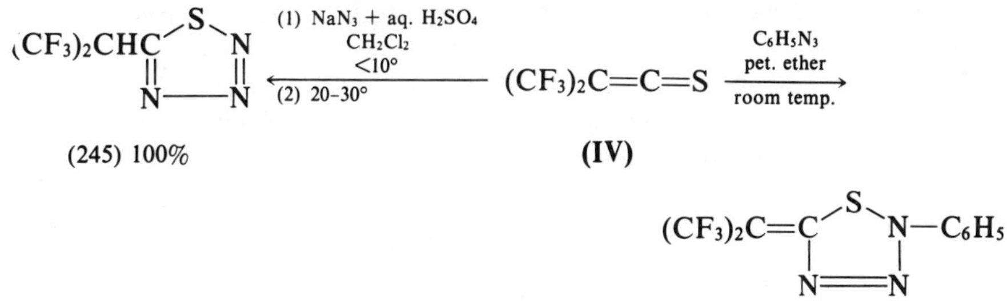

(245) 100% (IV)

26.5% (246)

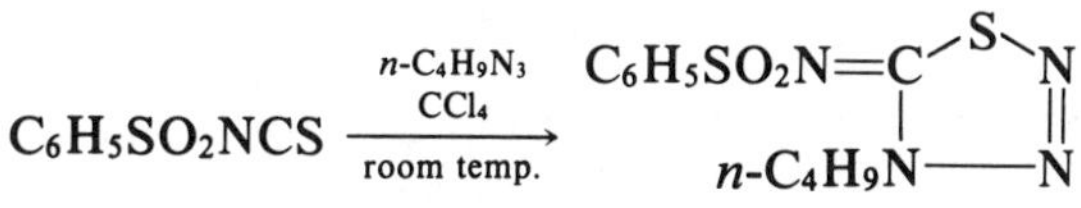

(169)

49%

Azide ion displaces chloride in chlorodithioformate esters forming 5-substituted thiothiatriazoles (243).

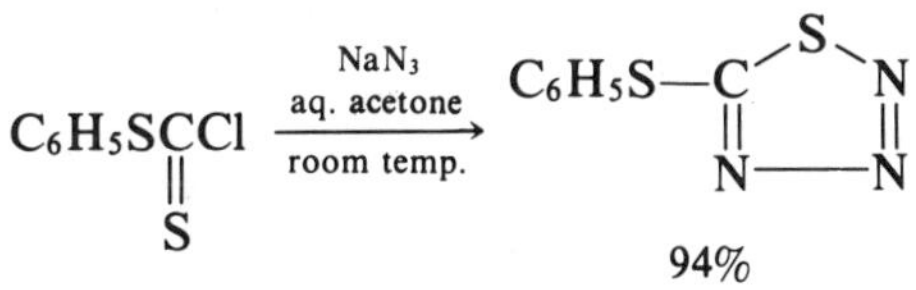

94%

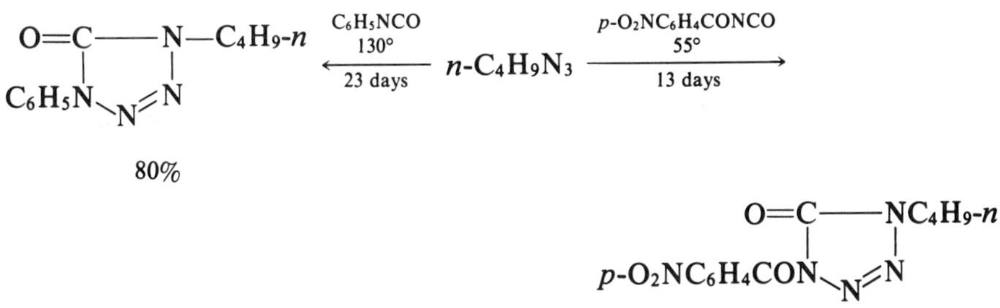

80% 65%

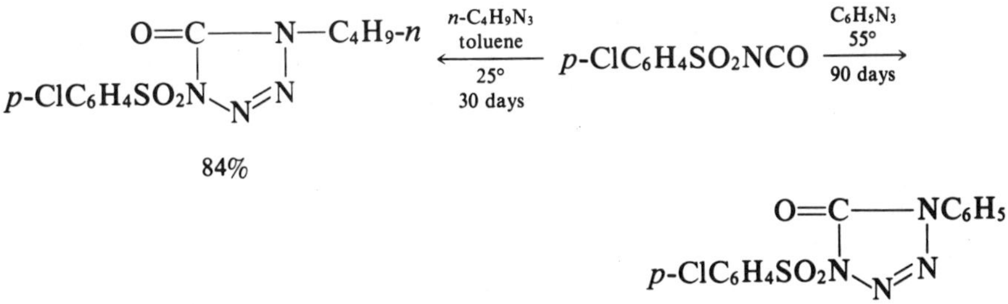

84% 60%

1,4-Tetrazolinones result from the addition of alkyl azides to aryl and acyl isocyanates and of alkyl and aryl azides to sulfonyl isocyanates (307). The reactions are slow, usually requiring from 5 to 90 days.

Chlorosulfonylisocyanate, however, reacts instantaneously with alkyl azides, forming alkyl(chlorosulfonyl)tetrazolines in high yield (288).

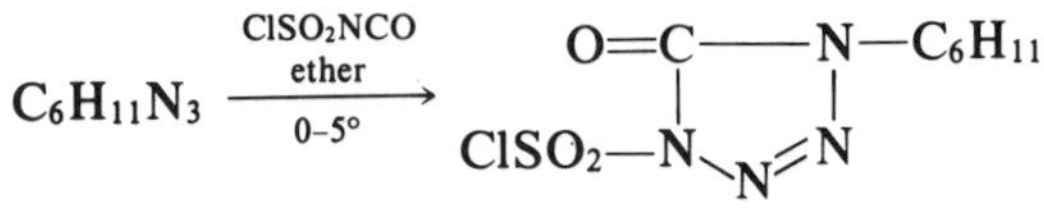

100%

Tetrazolinones with only one substituent on ring nitrogen are obtained from the action of isocyanates and aluminum azide, formed *in situ* from sodium azide and aluminum chloride. Aluminum chloride catalyzes the addition of trimethyl-

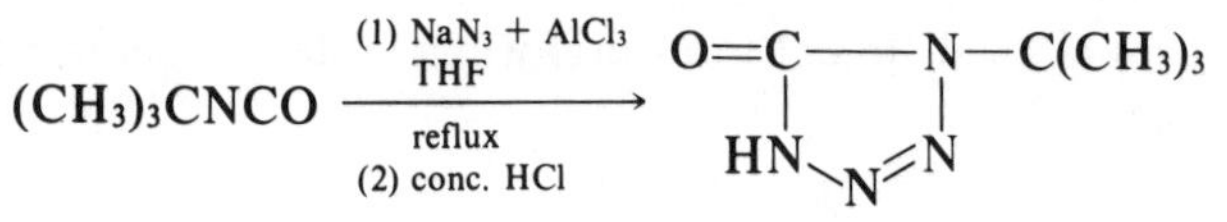

(239)

80%

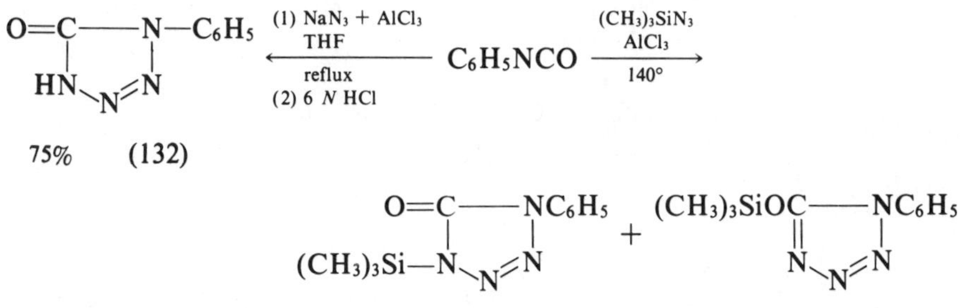

75% (132)

36% (274)

silyl azide to isocyanates. 1-Phenyltetrazolinone also results from the action of aluminum azide on benzoyl chloride, presumably through benzoyl azide and phenylisocyanate (132).

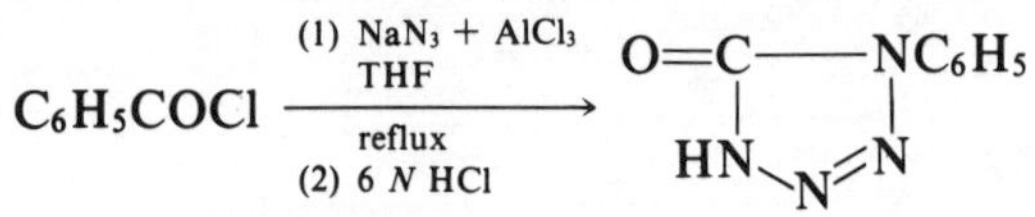

43%

Tetrazolinethiones are formed in good yields by the direct action of azide ion on isothiocyanates.

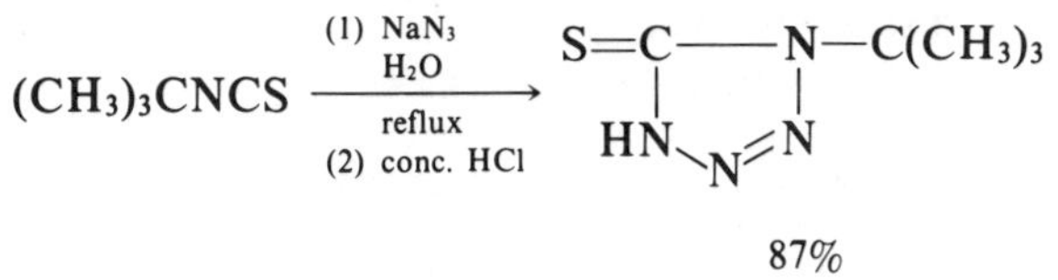 (239)

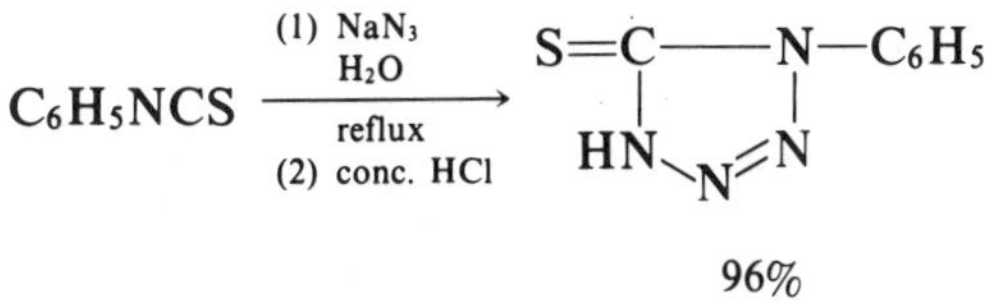 (185)

5-Monosubstituted tetrazoles are synthesized by addition of hydrogen azide or azide ion to nitriles, nitrilium salts, cyanates, thiocyanates, cyanamides, and isocyanides. Although hydrogen azide in a solvent such as benzene combines

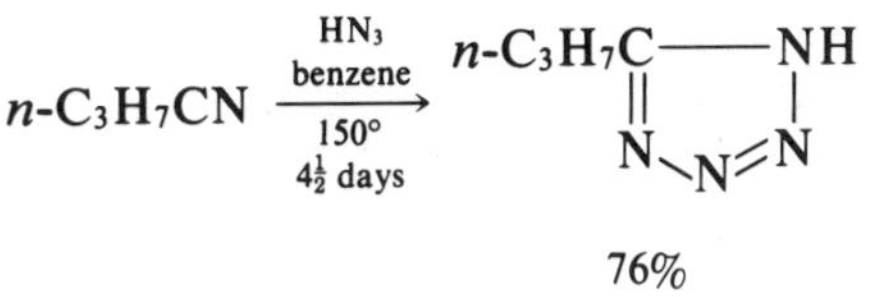 (200)

with nitriles to form tetrazoles, with the exception of nitriles carrying strongly electron-withdrawing substituents, the reaction is slow. To increase its speed, to use lower temperatures, and to avoid separate preparation of hydrogen azide solutions, alternate hydrazoic acid reagents have been sought. The data in Table 7.1 for the synthesis of 5-phenyltetrazole indicate the superiority of a combination of sodium azide and ammonium chloride containing lithium chloride catalyst (82). Slightly lower temperatures, but longer times are required

Table 7.1. Synthesis of 5-Phenyltetrazole with Alternate Hydrazoic Acid Reagents

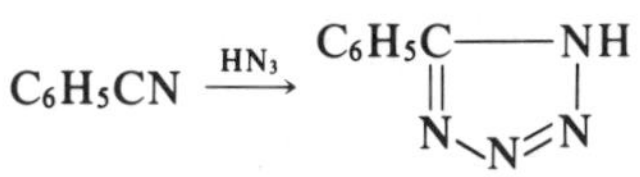

Reagent	Solvent	Temp. (°C)	Time	Yield (%)	Ref.
HN_3	Benzene	150	4 days	75	200
$NaN_3 + HOAc$	Butanol-1	118	2 days	91	126a
$NaN_3 + HOAc$	Butanol-2	99.5	2 days	84	126a
$NaN_3 + HOAc$	Isopropanol	82.5	2 days	64	126a
$NaN_3 + NH_4Cl + LiCl$	DMF	125	7 hr	100	82
$NaN_3 + NH_4Cl$	DMF	125	7 hr	59.6	82

Table 7.2. Yields of 5-Substituted Aryl Tetrazoles

5-Substituted	Yield (%)
p-$CH_3OC_6H_4$	43.2
C_6H_5	75.6
p-$O_2NC_6H_4$	96.9

with the system of sodium azide, acetic acid and n-butanol (123). The influence of substituents donating or attracting electrons in the nitrile is shown by the yields of 5-p-methoxyphenyl, phenyl, and p-nitrophenyltetrazoles (Table 7.2) on reaction with sodium azide and ammonium chloride for 3 hr at 100° in DMF (82).

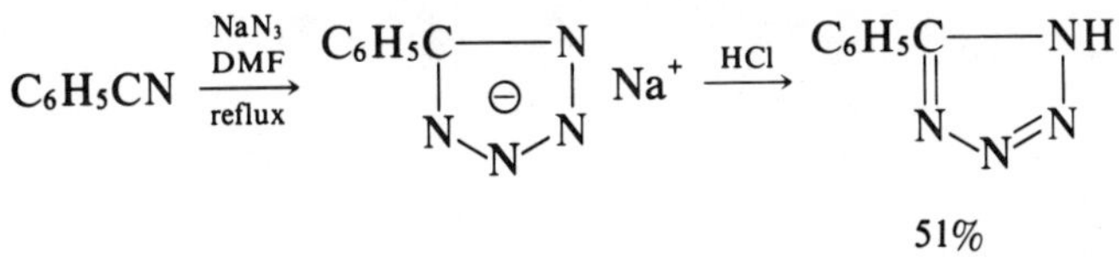

(82)

Aryl nitriles and nitriles bearing more electron-withdrawing substituents unite with sodium azide to form the sodium tetrazolates. The reaction with trifluoroacetonitrile occurs exothermally at room temperature. Electronegative

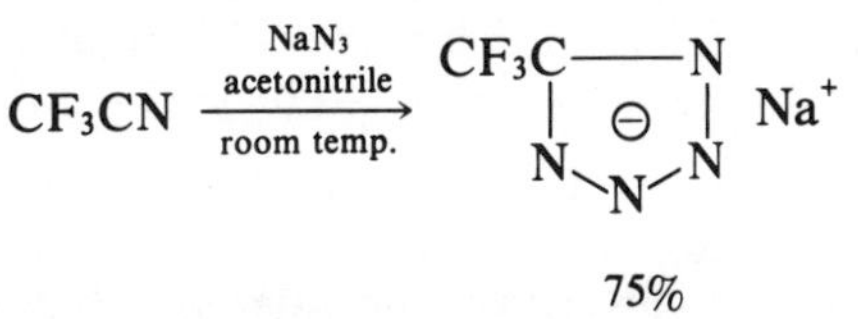

(219)

nitriles condense with alkyl, aryl, and silyl azides forming 1,5-disubstituted tetrazoles.

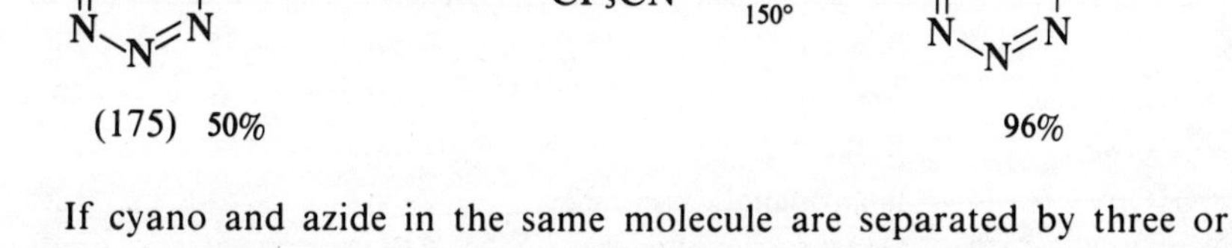

(175) (47)

If cyano and azide in the same molecule are separated by three or four atoms, intramolecular cyclization can occur, forming polycyclic tetrazoles; the reaction fails at other cyano-azide spacings (47).

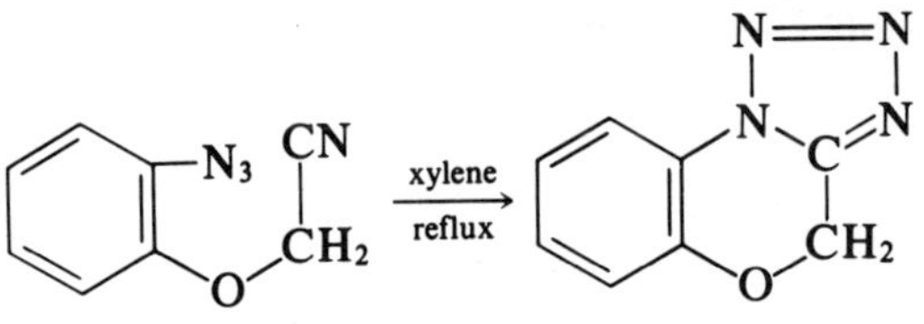

(87)

45%

1,5-Disubstituted tetrazoles result from the reaction of nitrilium salts with azide ion, usually accompanied by 1,2,4-triazolium compounds (176). Alkyl

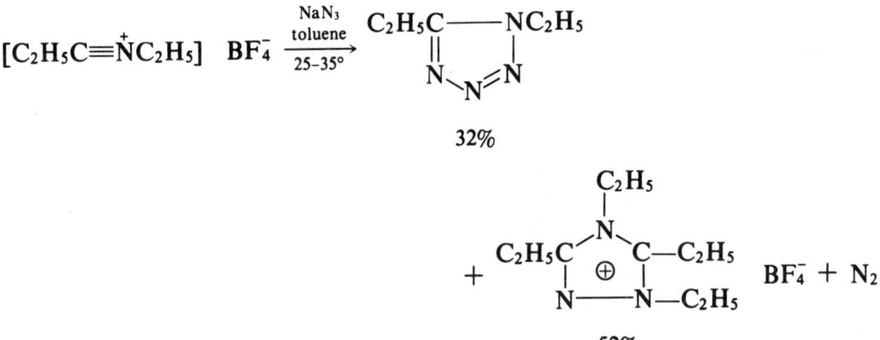

32%

52%

azides combine with nitrilum salts forming 1,4-tetrazolium salts. Nitrilium ions derived from alcohols under the influence of chlorodiphenylmethylium salts, also cyclize with azide ion to 1,5-disubstituted tetrazoles.

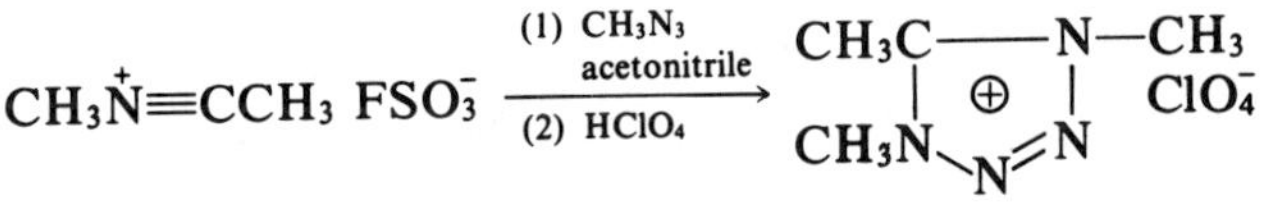

(240)

41%

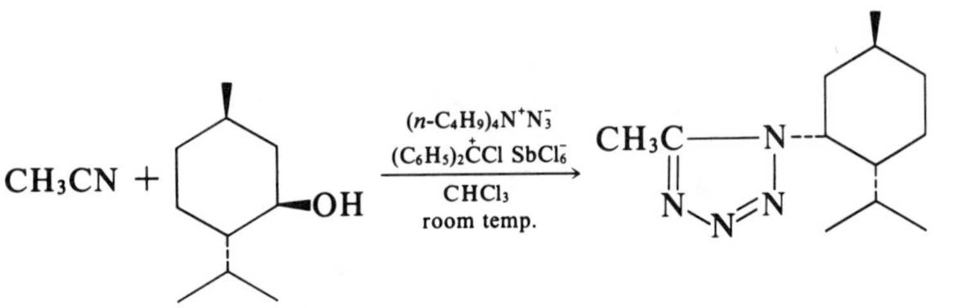

(21)

(−)-*trans*-*p*-menthane-*cis*-3-ol

50%

Tetrazoles form readily by addition of azide to cyanates and thiocyanates.

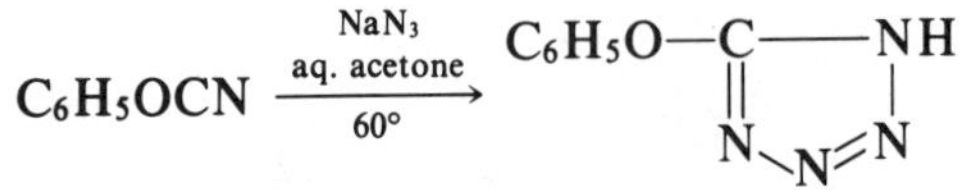

(189)

94%

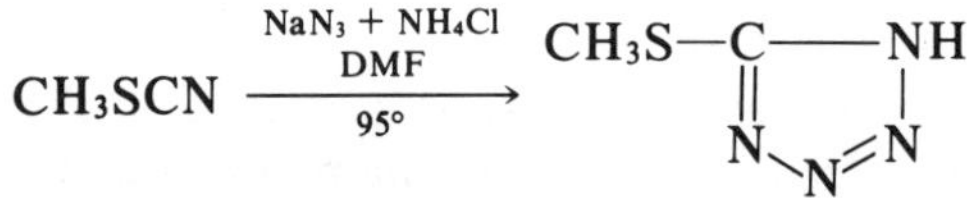

(182)

71%

N-Dialkylcyanamides add hydrogen azide or azide ion producing 5-dialkylaminotetrazoles. Alkyl- and aryl-monosubstituted cyanamides add hydrogen azide to form 5-amino 1-substituted tetrazoles.

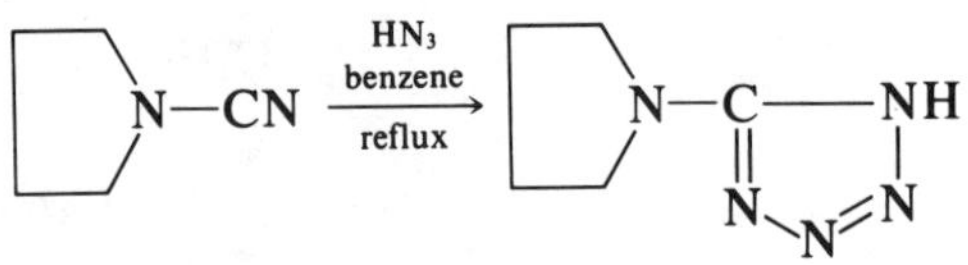

(91)

86%

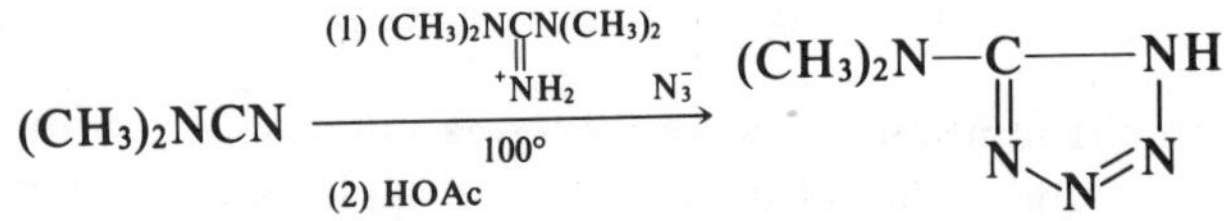

(225)

83%

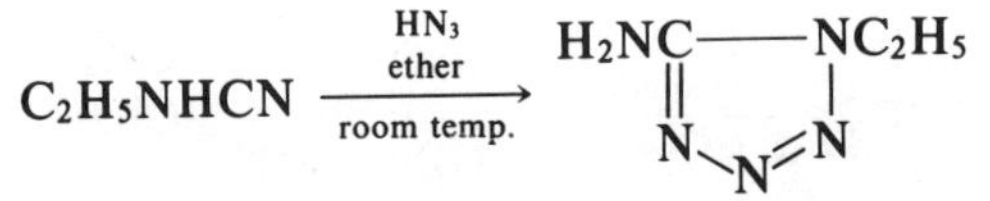

(92)

53%

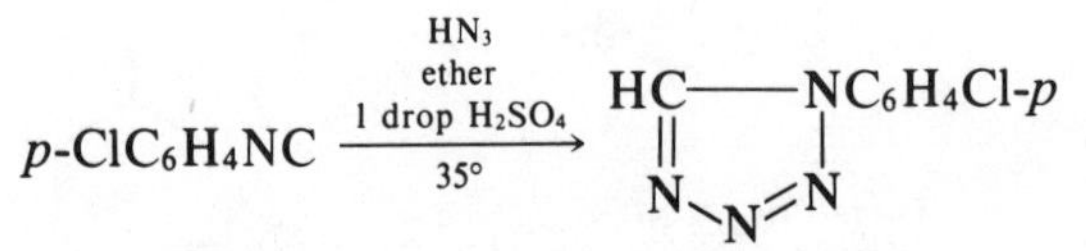

(327)

95% (35% in absence of acid)

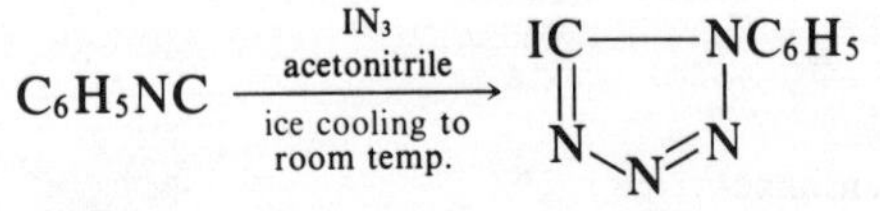

(85)

67%

Isocyanides react with hydrogen azide to synthesize 1-substituted tetrazoles, particularly rapidly in the presence of small quantities of acid. Iodine azide combines with phenylisocyanide to form 5-iodo-1-phenyltetrazole. 1-Aminotetrazole is obtained by reaction of hydrazoic acid with fulminic acid amide (206) or phenylmethylene ethoxymethylene azine, followed by hydrolysis of the phenylmethylene group (105). Hydrogen azide also adds to *N*-isocyanides of dialkylamines to form 1-disubstituted aminotetrazoles (42). 1,5-Disubstituted

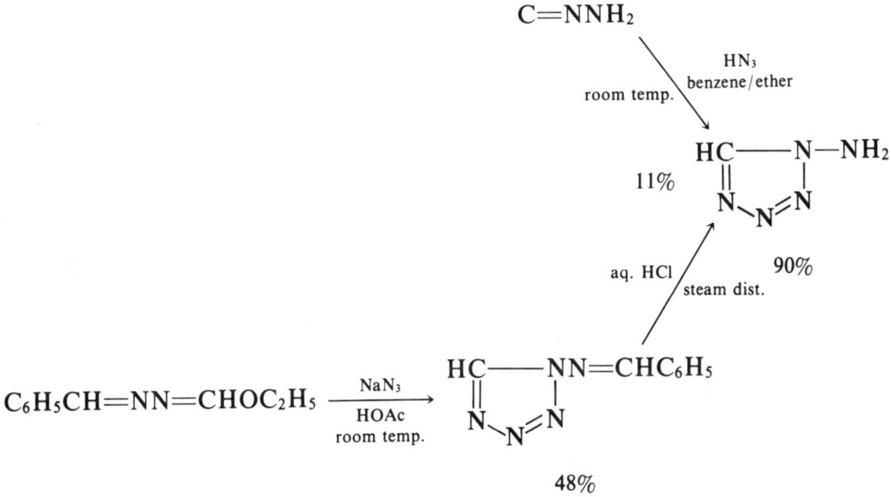

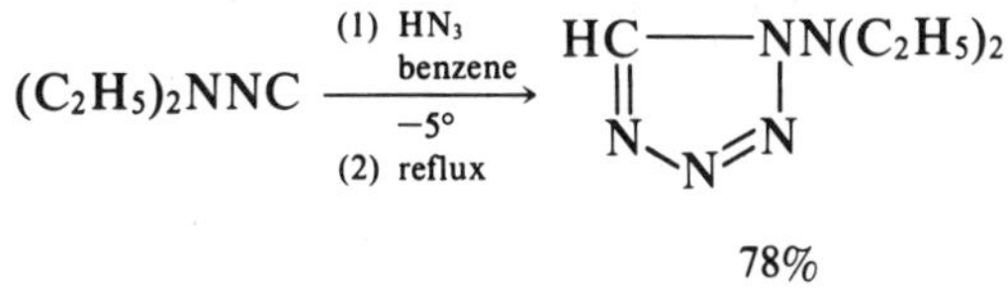

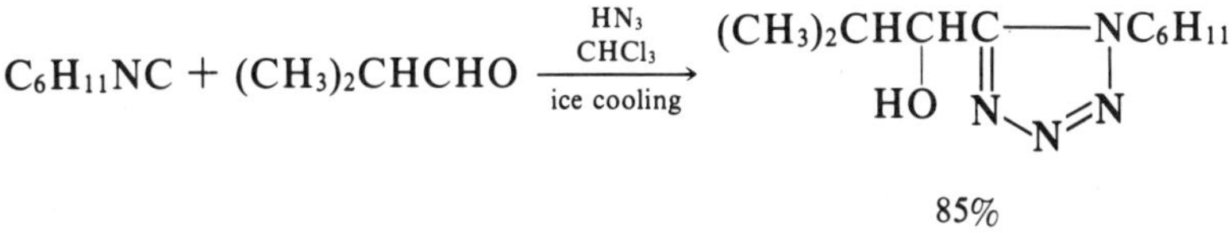

(303)

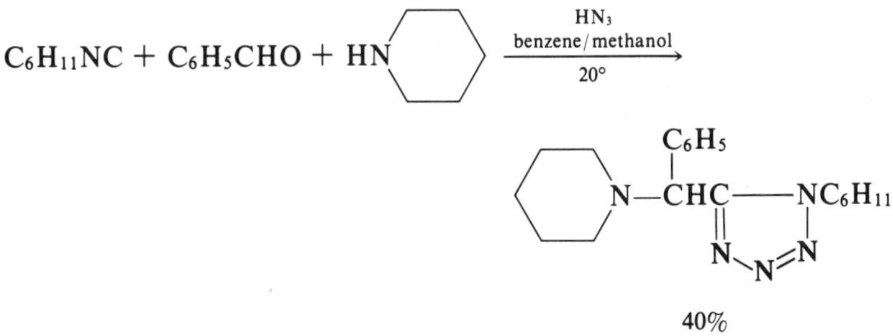

(305)

tetrazoles result from multicomponent reaction mixtures including isocyanides, hydrogen azide, and carbonyl compounds, with and without a secondary amine.

Carbodiimides and ketenimines also add hydrogen azide to form 1,5-disubstituted tetrazoles.

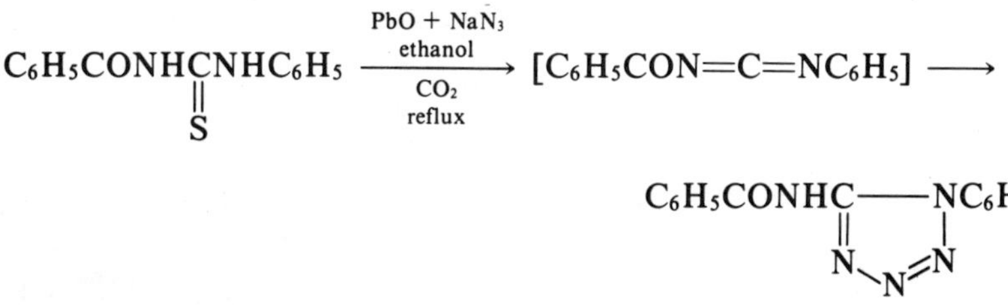

(286)

70%

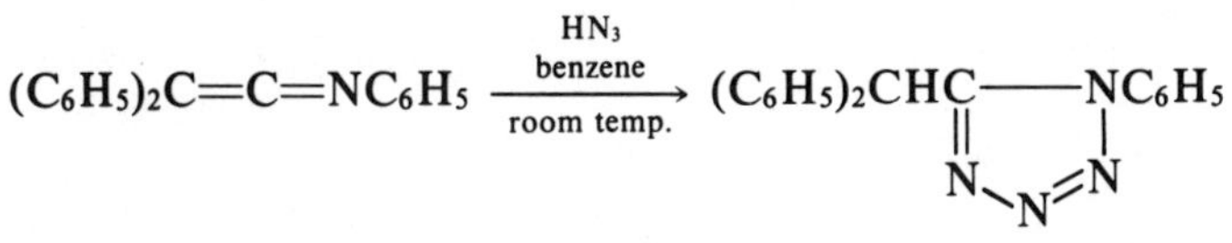

(290)

54%

1-α-Azidoarylmethyl-5-aryltetrazoles result from the combination of iodine azide with diarylazirines (223). A 1,5-disubstituted tetrazole results from combination of azide and aziridine nitrogen in ethyl 3-aryl-3-azidoaziridine-2-car-

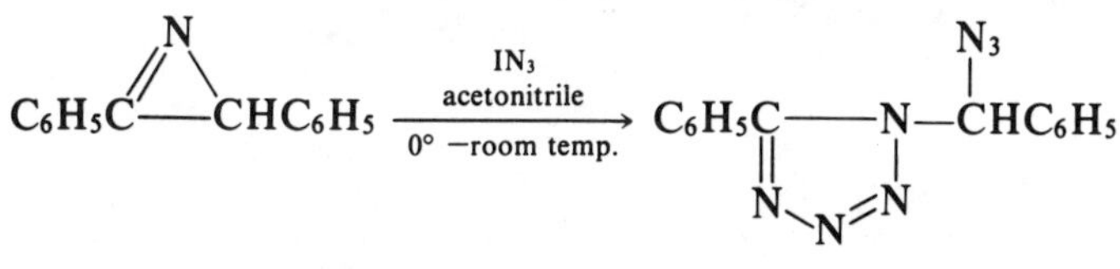

36%

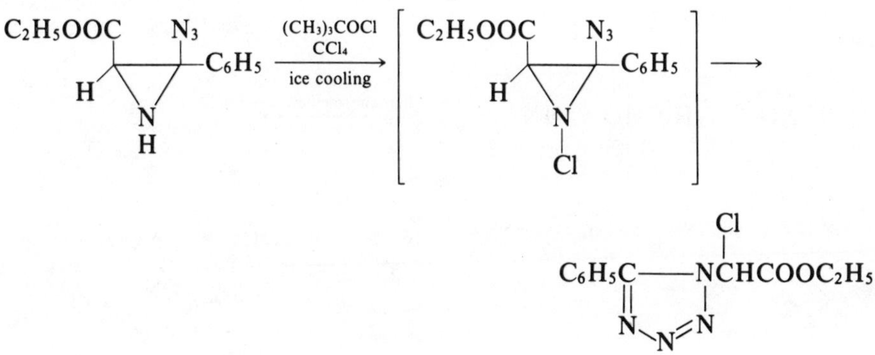

99%

boxylate on halogenation with *t*-butyl hypochlorite (291). The reaction goes through an *N*-chloroaziridine, detectable by ^{1}H NMR at $-60°$.

Iminotriphenylphosphoranes react with acetyl chloride and sodium azide to form 1-substituted 5-methyltetrazoles (325). The reaction probably involves an acetyl adduct which adds azide, subsequently losing triphenylphosphine oxide.

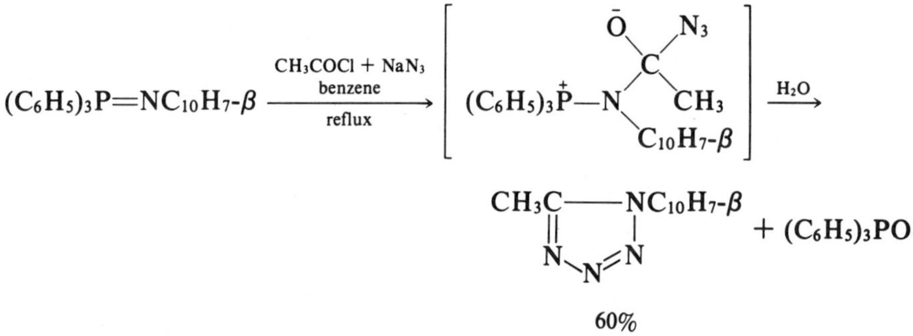

Tetrazoles are obtained by displacement of chloride, alkoxy, alkylthio, and pyridinium moieties from carbon atoms substituted with imino groups. Thus 5-chloro-1-aryltetrazoles are formed from *N*-acyldichloroazomethines. Fused tetrazolo derivatives result when the reaction is applied to heterocyclic compounds with chloro or alkoxy on carbon adjacent to ring nitrogen, illustrated with 2-chloroquinoline and 2-ethoxy-4-phenyl-1-pyrroline. Methyl dithio-*N*-monosubstituted carbamates form 1-substituted tetrazolinethiones and 2-(1′-pyridiumdihydro-1,3-oxazine)tosylate **(V)** is converted in high yield to 6,7-

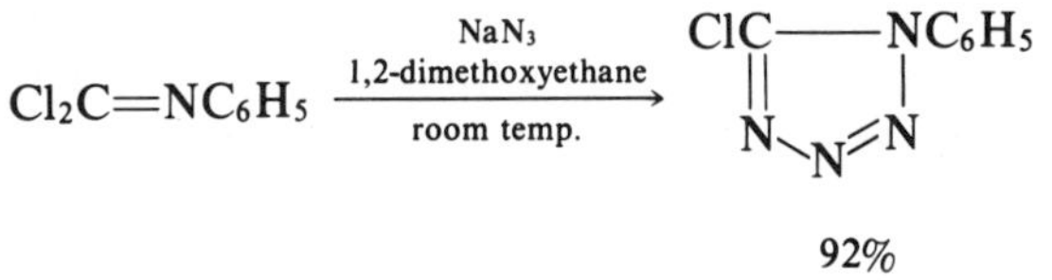

(159)

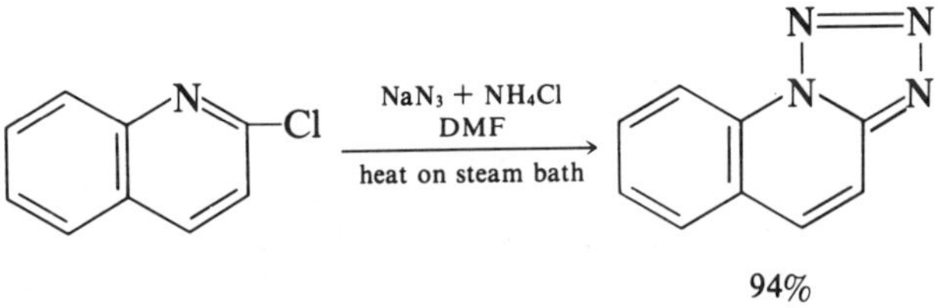

(155)

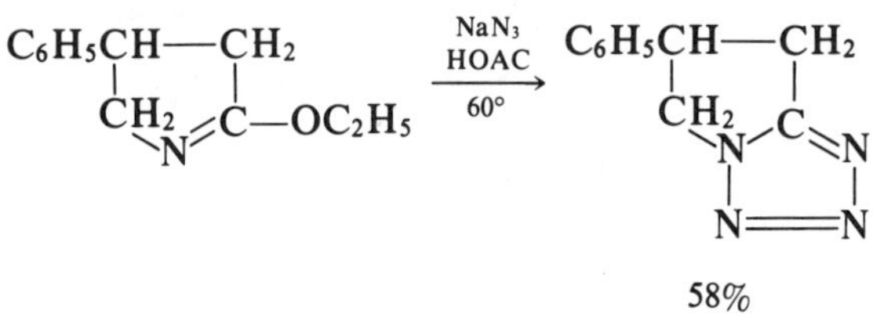

(173)

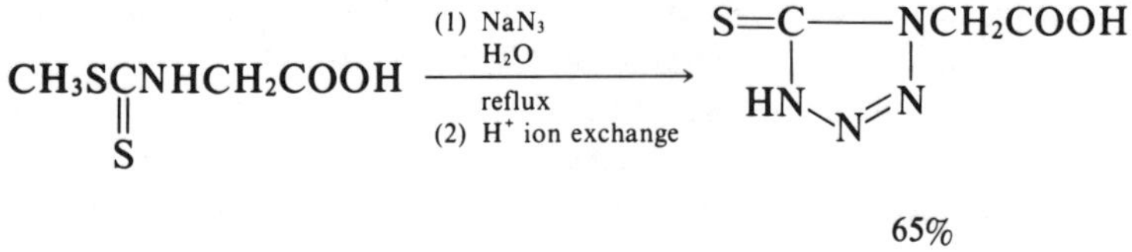
(29)

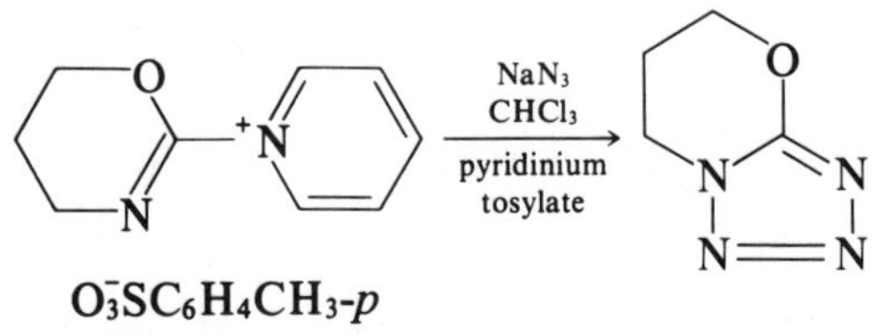
(165)

(V) 100%

dihydro-tetrazolo[5,1-*b*][1,3]oxazine. 5-[Cyano(ethoxycarbonyl)methylene]-1-methyl-4-phenyltetrazoline is obtained by displacement of the methylthio group in 1-ethoxycarbonyl-2-methylthio-2-anilinoacrylonitrile with azide ion, followed by methylation (276).

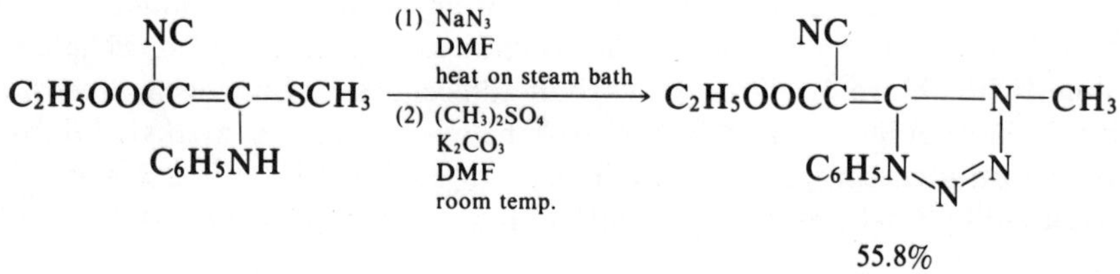

A variation on the synthesis of tetrazoles from imidoyl chlorides consists of formation of the chloride in situ from an *N*-substituted amide and its further

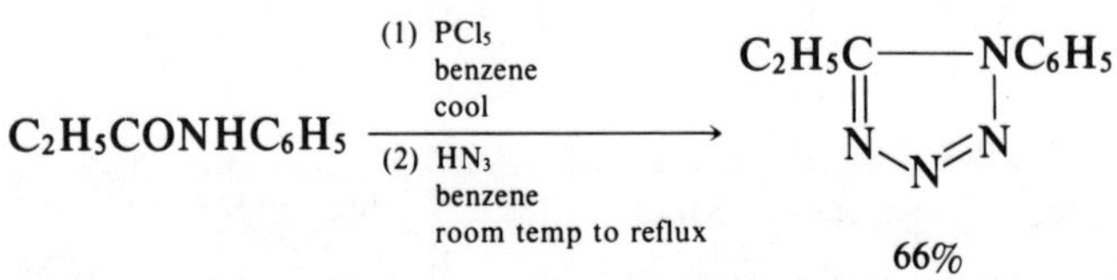
(115)

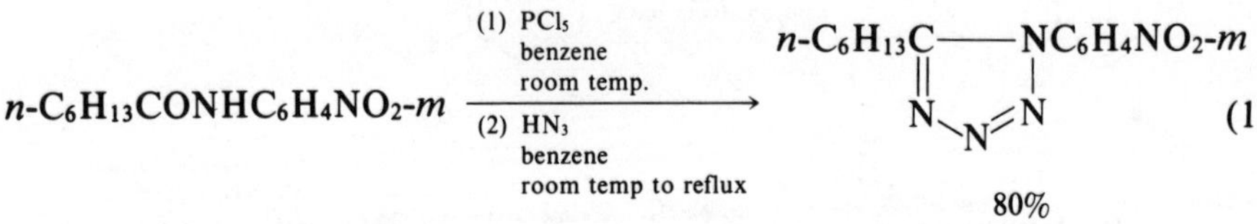
(125)

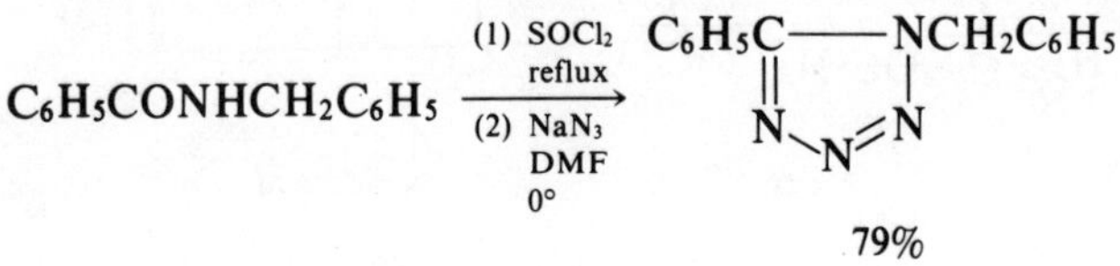
(223)

reaction, without isolation, with hydrogen azide. Reaction of hydrogen azide with the equilibrium system of cyclohexylisocyanide and acetyl chloride with their addition product yields 5-acetyl-1-cyclohexyltetrazole.

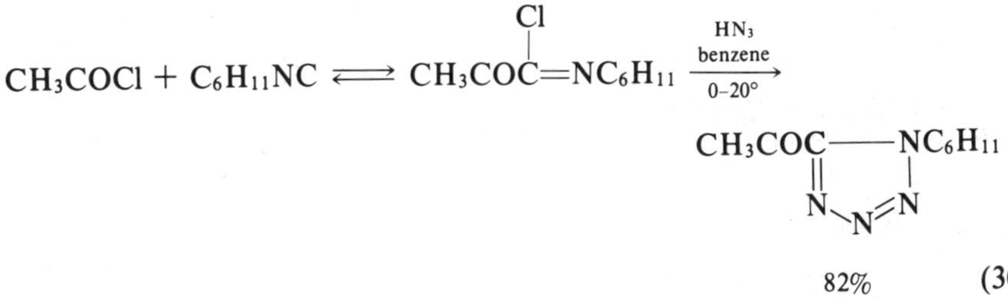

(302)

N-(*p*-Dimethylaminophenyl)nitrones react with hydrogen azide to give a mixture of tetrazoles and by-products via pathways involving reduction of the nitrone, fission to benzoyl cyanide, and addition of hydrogen azide to each product.

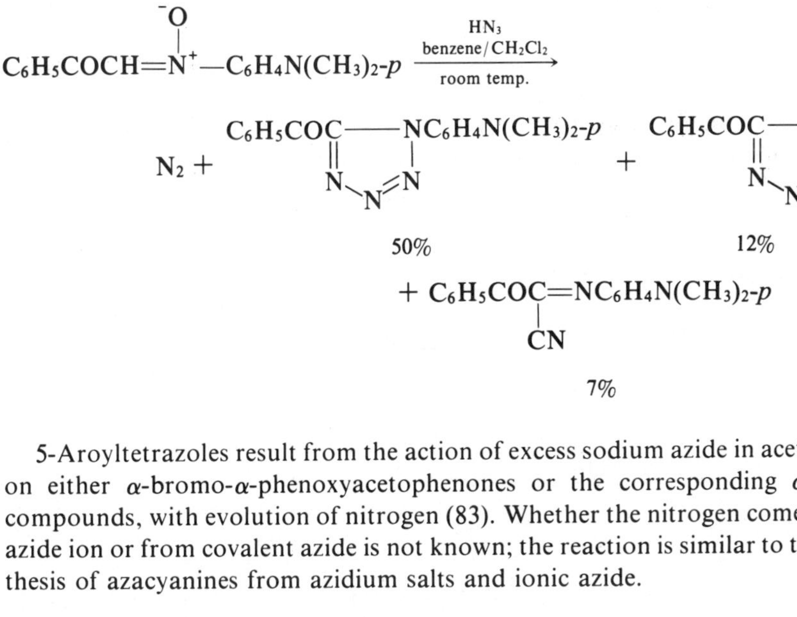

(203)

5-Aroyltetrazoles result from the action of excess sodium azide in acetic acid on either α-bromo-α-phenoxyacetophenones or the corresponding α-azido compounds, with evolution of nitrogen (83). Whether the nitrogen comes from azide ion or from covalent azide is not known; the reaction is similar to the synthesis of azacyanines from azidium salts and ionic azide.

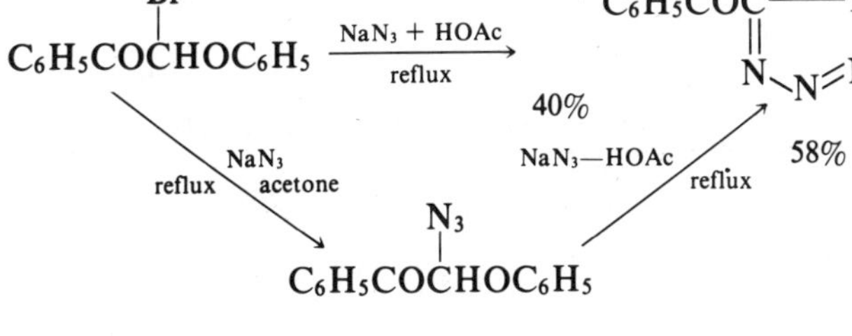

Ketones react with 2 moles of hydrazoic acid in the presence of strong acids to form tetrazoles; the action of 1 mole of hydrazoic acid produces *N*-substituted amides. These reactions and the conversion of carboxylic acids by hydrogen azide to amines are known collectively as the Schmidt reaction (319).

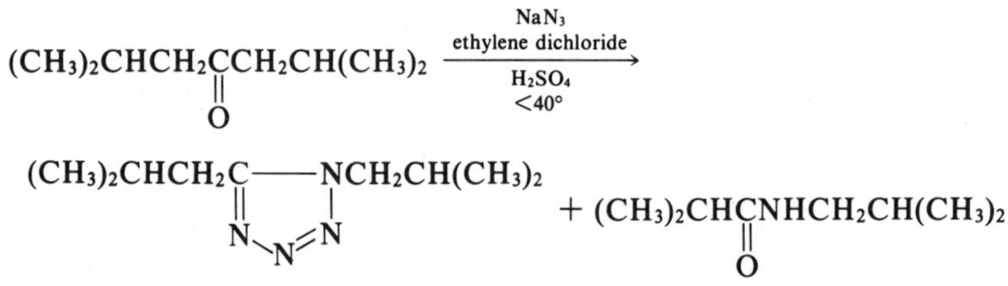

24% 46% (115)

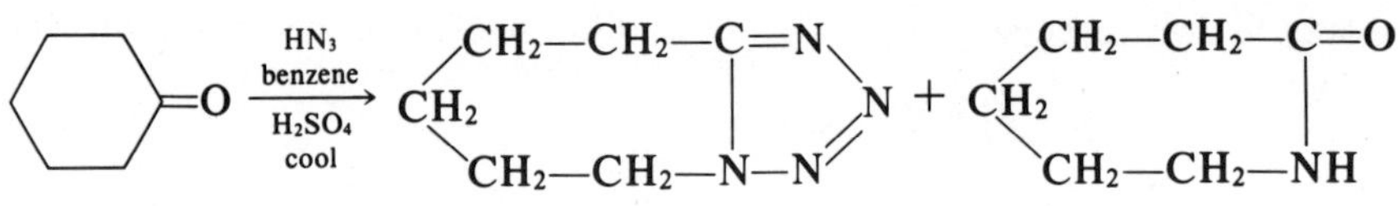

70% (262)

The reaction proceeds through combination of a carbonium ion with hydrogen azide acting as a base, rearrangement, and loss of nitrogen to form an iminocarbonium ion, and addition of a second mole of hydrogen azide (277).

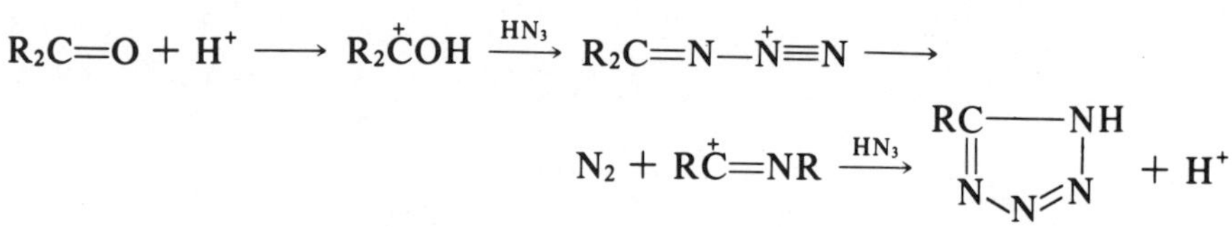

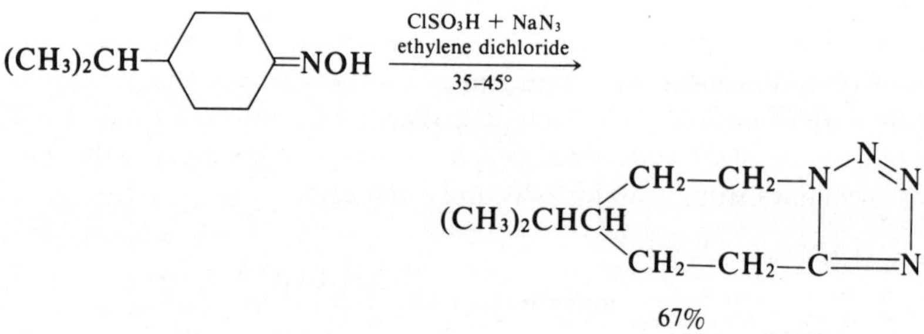

(116)

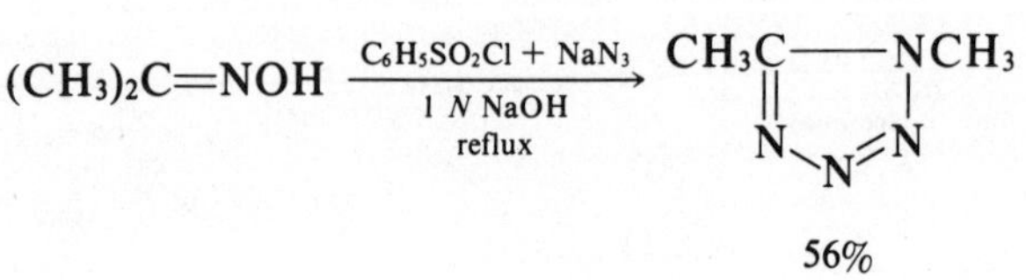

(50, 188)

If ketone oximes or oxime sulfonates are used instead of ketones, the need for 2 moles of azide is eliminated, and the reaction can be conducted to avoid acidity during combination with azide.

In a related reaction, nitriles undergo an acid-catalyzed rearrangement with hydrazoic acid to form 1-substituted 5-aminotetrazoles (41).

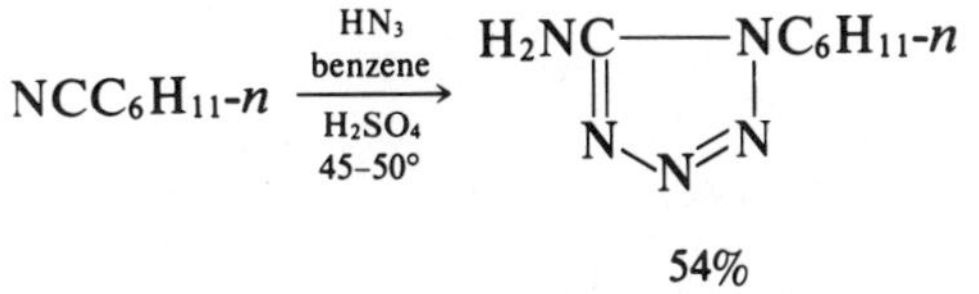

(126)

Tetraazaborolines are obtained by addition of azide to borimides with a perfluorophenyl substituent on boron (224), and by combination of diborane with organic amines and azides (103, 204).

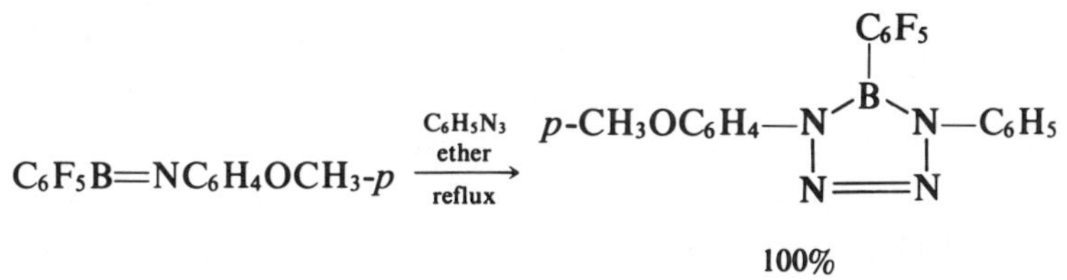

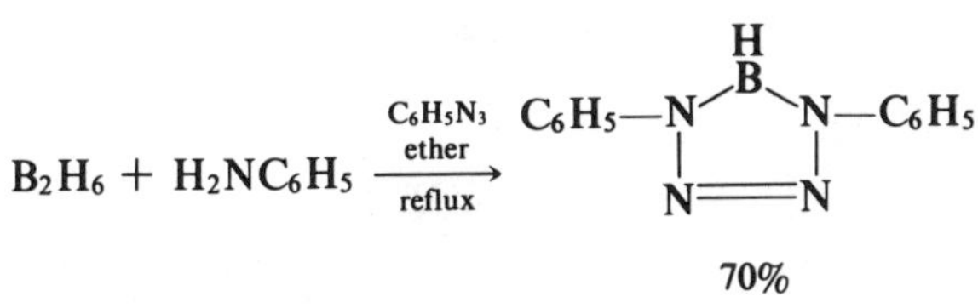

Pentazoles result by addition of azide ion to diazonium ions, as discussed in Section 2.9. The most stable pentazole, the p-dimethylaminophenyl derivative, may be prepared at $-10°$, but other members require $-35°$.

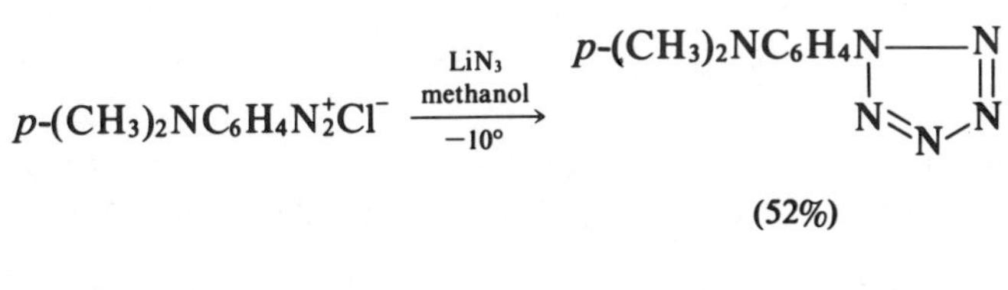

(304)

Some syntheses involve the conversion of nonazide three-nitrogen sequences to various cyclic classes. Thus heterocycles with embedded three-nitrogen chains are formed by combination of N-aminopyridazine salts or N-acylaminopyridazines with acetylene dicarboxylic methyl ester or benzyne.

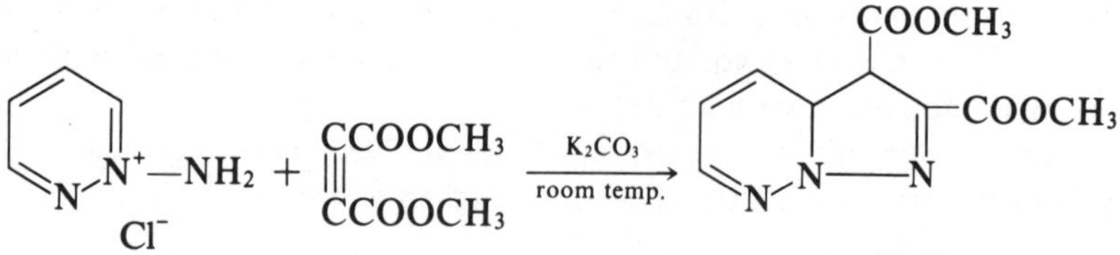

54.4% (163)

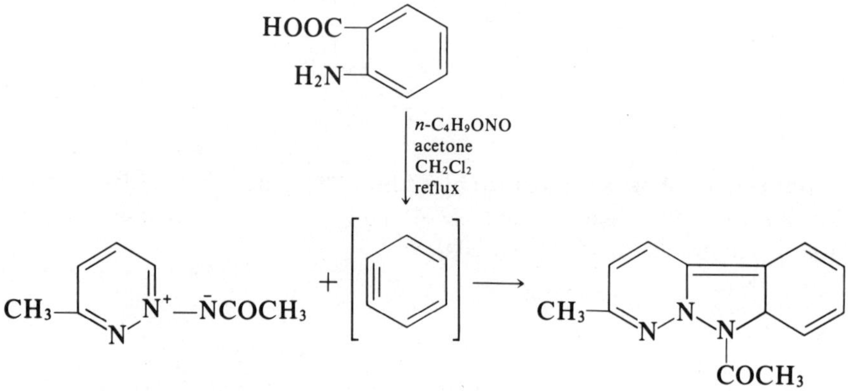

90% (117)

Heating 1-alkyl-1-nitrososemicarbazides in acid containing nitrous acid leads to 3-alkyl-oxatriazolium olates (37). Thiatriazolium olates are formed under alkaline conditions by isomerization of oxatriazolium thiolates (107).

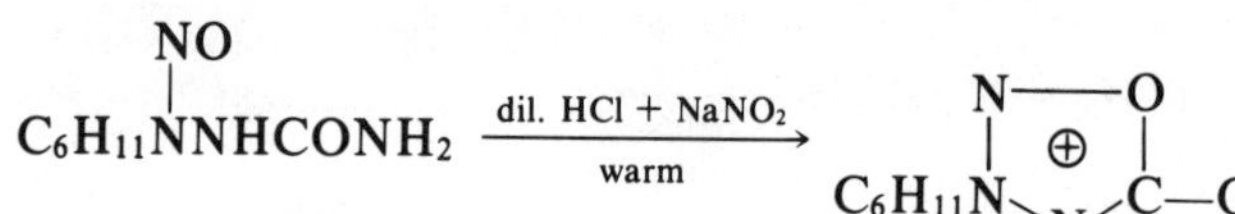

56%

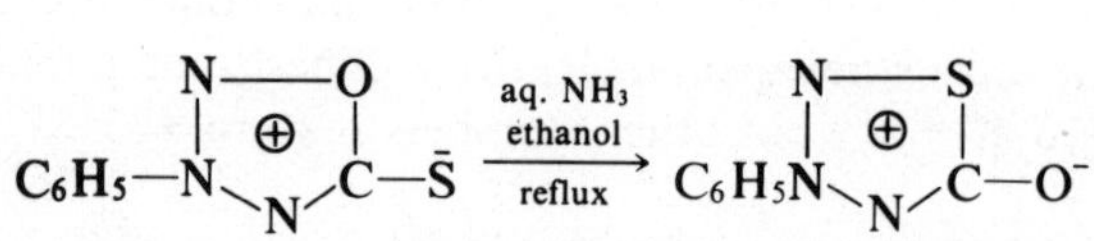

40%

The phosphorus atom in phosphorus triamides inserts between nitrogen and sulfur in the 5-aryloxythiatriazole ring to form thiatriazaphosphorines (233).

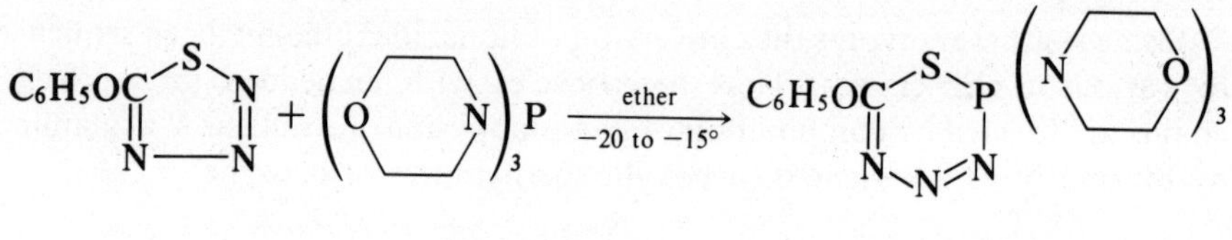

Several syntheses of tetrazole derivatives involve combination of a member of a three-nitrogen chain with another nitrogen atom within the molecule. Thus oxidation of a 2-triazenopyridine results in a tetrazolopyridium salt (198).

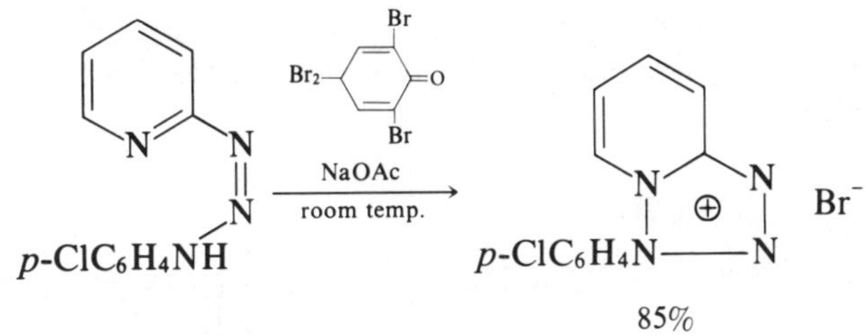

2,5-Disubstituted tetrazoles are obtained by rearrangement of 4-azo-5-hydroxy-1-substituted *v*-triazoles and 3-triazenoisoxazoles.

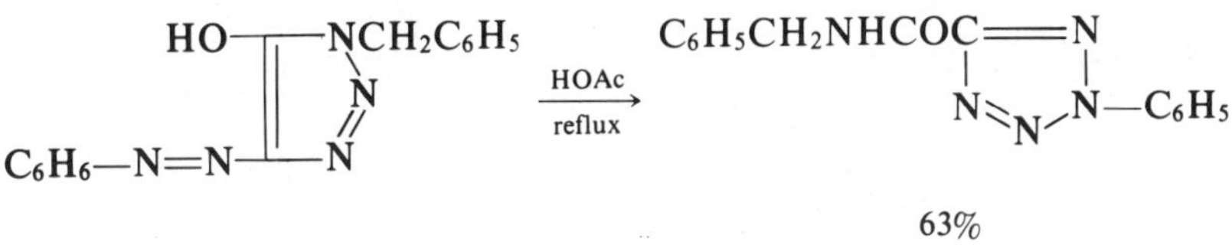

(228)

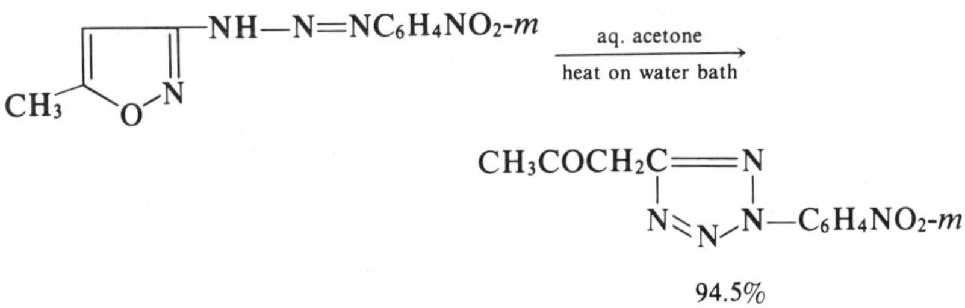

(157)

Both oxatriazolium-5-arylaminates and 5-substituted aminothiatriazoles isomerize to tetrazole derivatives.

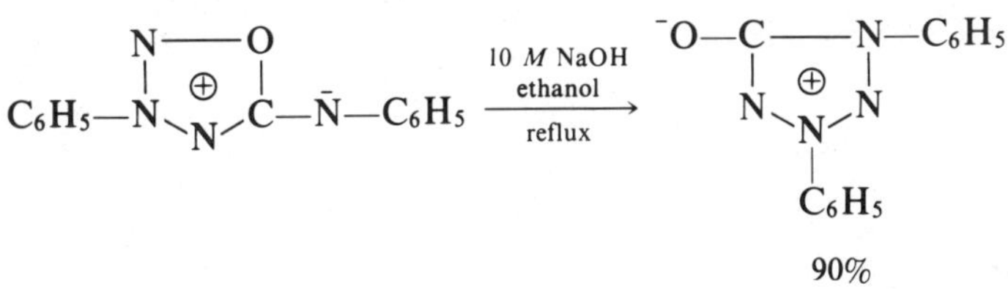

(108)

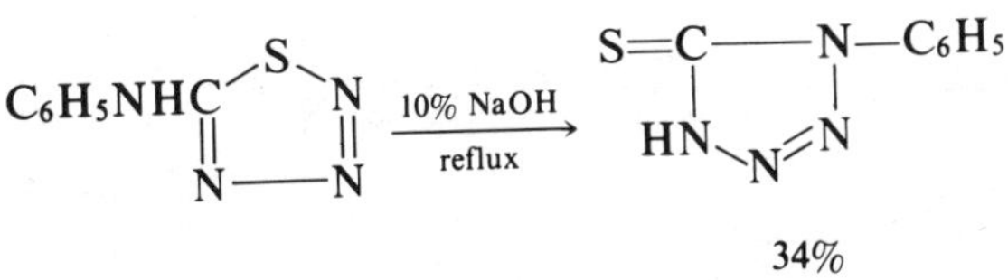

(184)

Formation of 1,4-dimethyltetrazolinone from 1,4-dimethyl-1,4-dimethoxycarbonyl-2-tetrazene by heating under nonaqueous alkaline conditions is discussed in Chapter 6, p. 512. Similarly, 1,5-disubstituted tetrazoles result by alkaline cyclization of 1-aryl-3,4-diacyl-1-tetrazenes; N-1 of the tetrazene chain combines with the carbonyl carbon at N-4. Usually the tetrazene from 1,2-diacylhydrazines and diazonium salts is converted directly to the tetrazole without isolation.

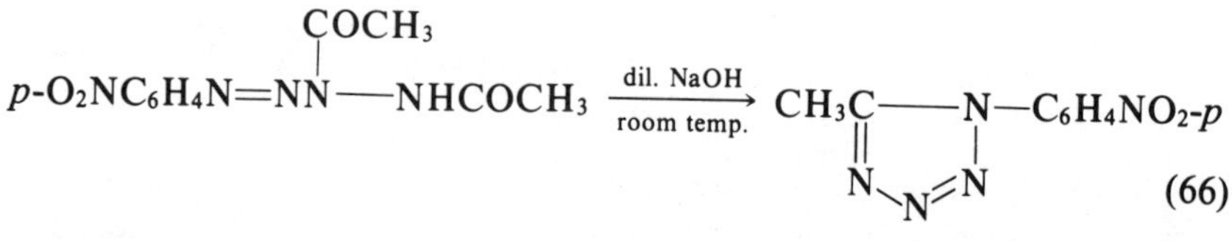

(66)

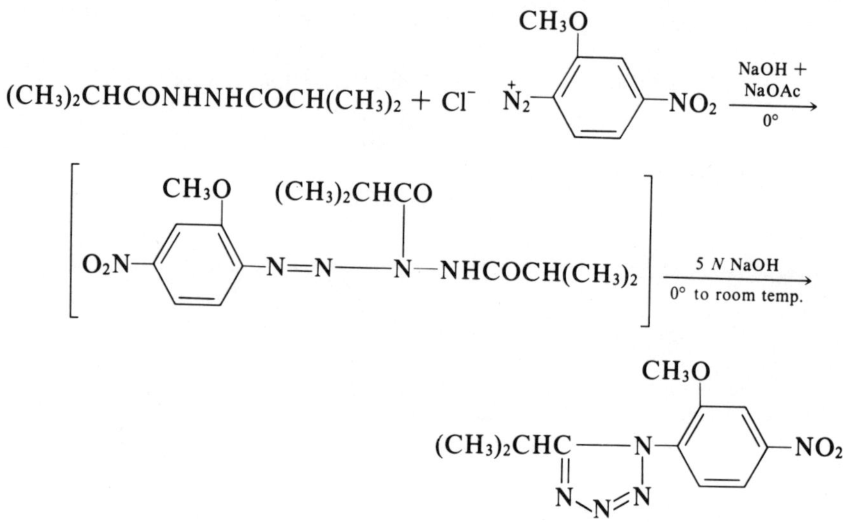

35%

(321)

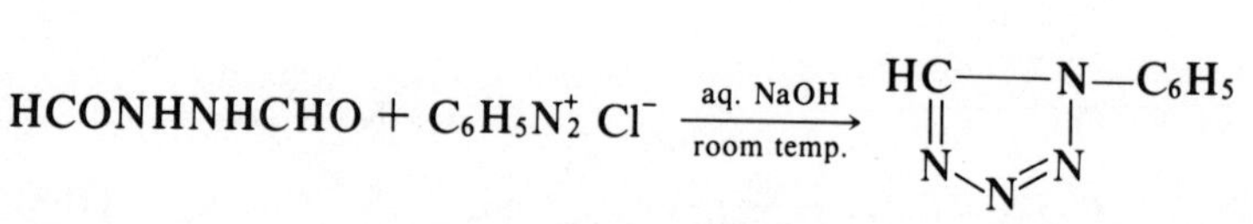

(66)

40%

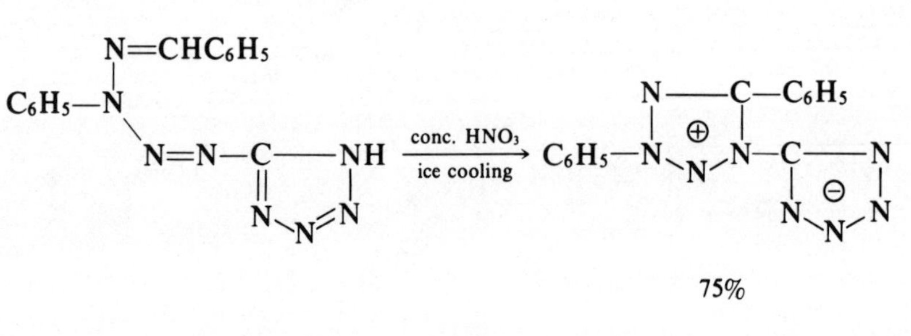

(134)

75%

1-Substituted tetrazoles can be prepared by use of 1,2-diformylhydrazine for coupling with diazonium compounds.

1-(5-Tetrazolyl)-3-aryl-4-arylmethylene-1-tetrazenes are oxidized by nitric acid to 1-tetrazolyl-3,5-diaryltetrazolium betaines.

REFERENCES

1. G. Adembri, F. Ponticelli, and P. Tedeschi, *J. Heterocycl. Chem.*, **1972,** *9,* 1219–1225.
2. B. M. Adger et al., *J. Chem. Soc., Perkin I,* **1975,** 31–40.
3. T. P. Ahern, H. Fong, and K. Vaughan, *Can. J. Chem.*, **1977,** *55,* 1701–1709.
4. R. Ahmed and J.-P. Anselme, *Can. J. Chem.*, **1972,** *50,* 1778–1780.

4a. A. Albert and K. Tratt, *Angew. Chem., Int. Ed. Engl.*, **1966,** *5,* 587.

5. A. Albini, G. F. Bettinetti, and G. Minoli, *J. Chem. Soc., Perkin I,* **1981,** 4–7.
6. N. E. Alexandrou and S. Adamopoulos, *Synthesis,* **1976,** 482–483.
7. J. N. Ashley et al., *J. Chem. Soc.*, **1953,** 3881–3888.
8. K. v. Auwers and P. Heimke, *Ann.*, **1927,** *458,* 175–220.
9. H. Balli, *Ann.*, **1961,** *647,* 11–18.
10. H. Balli and F. Kersting, *Ann.*, **1963,** *663,* 96–103.
11. O. Baltzer and H. v. Pechmann, *Ann.*, **1891,** *262,* 302–313.
12. E. Bamberger and P. DeGruyter, *Ber.*, **1893,** *26,* 2385–2397.
13. E. Bamberger and R. Hübner, *Ber.*, **1903,** *36,* 3822–3827.
14. E. Bamberger and E. Renault, *Ber.*, **1897,** *30,* 2278–2279.
15. E. Bamberger and A. Stiegelmann, *Ber.*, **1899,** *32,* 3554–3560.
16. W. R. Bamford and T. S. Stevens, *J. Chem. Soc.*, **1952,** 4735–4740.
17. N. M. Baranchik, I. V. Grachev, and D. Z. Zavel'skii, *Dokl. Akad. Nauk. SSSR,* **1954,** *94,* 701–703.
18. N. M. Baranchik, I. V. Grachev, and D. Z. Zavel'skii, *Zh. Obshch. Khim., Engl. Ed.*, **1953,** *23,* 1651–1656.
19. M. W. Barker and J. H. Gardner, *J. Heterocycl. Chem.*, **1969,** *6,* 251–252.
20. R. K. Bartlett and T. S. Stevens, *J. Chem. Soc. (C)*, **1967,** 1964–1968.
21. D. H. R. Barton et al., *J. Chem. Soc., Perkin I,* **1974,** 2101–2107.
22. R. A. Bartsch, S. Hünig, and H. Quast, *J. Am. Chem. Soc.*, **1970,** *92,* 6007–6011.
23. N. Bashir and T. L. Gilchrist, *J. Chem. Soc., Perkin I,* **1973,** 868–872.
24. H. Bauer, G. R. Bedford, and A. R. Katritzky, *J. Chem. Soc.*, **1964,** 751–755.
25. Bayer A.-G. (H. Gold and U. Claussen, inventors), Brit. Pat. 1,405,218, Sept. 10, 1975.
26. G. Beck and D. Günther, *Chem. Ber.*, **1973,** *106,* 2758–2766.
27. M. Begtrup and J. Holm, *J. Chem. Soc., Perkin I,* **1981,** 503–513.
28. H. Behringer and H. J. Fischer, *Chem. Ber.*, **1962,** *95,* 2546–2556.
29. D. A. Berges et al., *J. Heterocycl. Chem.*, **1978,** *15,* 981–985.
30. A. Bertho, *Ber.*, **1925,** *58,* 859–864.
31. J. M. Birchall, R. N. Haszeldine, and J. E. G. Kemp, *J. Chem. Soc. (C)*, **1970,** 1519–1523.
32. J. A. Bladin, *Ber.*, **1885,** *18,* 1544–1551.
33. H. Bock and K.-L. Kompa, *Angew. Chem., Int. Ed. Engl.*, **1962,** *1,* 264.
34. A. B. Boese, Jr., L. W. Jones, and R. T. Major, *J. Am. Chem. Soc.*, **1931,** *53,* 3530–3541.
35. S. Børresen and J. K. Crandall, *J. Org. Chem.*, **1976,** *41,* 678–681.

36. W. Borsche, *Ber.*, **1923,** *56,* 1488–1493.
37. J. H. Boyer and J. A. Hernandez, *J. Am. Chem. Soc.*, **1956,** *78,* 5124–5125.
38. J. H. Boyer and L. R. Morgan, Jr., *J. Am. Chem. Soc.*, **1958,** *80,* 3012–3015.
39. R. J. Boyle (American Cyanamid), U.S. Pat. 3,230,194, Jan. 18, 1966.
40. J. Bratož-Stres et al., *Tetrahedron Lett.*, **1975,** 4429–4430.
41. J. v. Braun and W. Keller, *Ber.*, **1932,** *65,* 1677–1680.
42. H. Bredereck, B. Föhlish, and K. Walz, *Ann.*, **1965,** *688,* 93–98.
43. K. Burger, S. Tremmel, and H. Schickaneder, *J. Fluorine Chem.*, **1976,** *7,* 471–480.
44. J. I. G. Cadogan et al., *J. Chem. Soc.*, **1965,** 4831–4837.
45. L. A. Carpino et al., *Org. Synth.*, **1973,** *Coll. Vol. V,* 157–159.
46. R. A. Carboni et al., *J. Am. Chem. Soc.*, **1967,** *89,* 2618–2625.
47. W. R. Carpenter, *J. Org. Chem.*, **1962,** *27,* 2085–2088.
48. E. A. Chandross and G. Smolinsky, *Tetrahedron Lett.*, **1960,** 19–22.
49. F. D. Chattaway, F. L. Garten, and G. D. Parkes, *J. Chem. Soc.*, **1924,** 1980–1991.
50. F. E. Condon et al., *Org. Prep. Proc. Int.*, **1974,** *6,* 135–139.
51. K. E. Cooper and E. H. Ingold, *J. Chem. Soc.*, **1926,** 1894–1896.
52. J. Cramer, H. Hansen, and S. Hünig, *J. Chem. Soc., Chem. Commun.*, **1974,** 264–265.
53. B. W. Cue, Jr., L. J. Czuba, and J. P. Dirlam, *J. Org. Chem.*, **1978,** *43,* 4125–4128.
54. D. Y. Curtin, R. J. Crawford, and D. K. Wedegaertner, *J. Org. Chem.*, **1962,** *27,* 4300–4305.
55. D. Y. Curtin and J. D. Druliner, *J. Org. Chem.*, **1967,** *32,* 1552–1557.
56. H. U. Daeniker and J. Druey, *Helv. Chim. Acta,* **1962,** *45,* 2441–2462.
57. R. E. Damschroder and W. D. Peterson, *Org. Synth.*, **1955,** *Coll. Vol. 3,* 106–108.
58. L. S. Davies and G. Jones, *J. Chem. Soc. (C)*, **1970,** 688–693.
59. J. De Mendoza et al., *J. Chem. Res. (M)*, **1980,** 0514–0529.
60. L. Dežman et al., *Tetrahedron,* **1977,** *33,* 2851–2855.
61. O. Diels and P. Fritzche, *Ber.*, **1911,** *44,* 3018–3027.
62. O. Dimroth, *Ber.*, **1902,** *35,* 1029–1038.
63. *Ibid.*, 4041–4060.
64. O. Dimroth and S. Merzbacher, *Ber.*, **1970,** *40,* 2402–2404.
65. *Ibid.*, **1910,** *43,* 2899–2904.
66. O. Dimroth and G. de Montmollin, *Ber.*, **1910,** *43,* 2904–2915.
67. R. S. Drago and B. R. Karstetter, *J. Am. Chem. Soc.*, **1961,** *83,* 1819–1822.
68. R. S. Drago and F. E. Paulik, *J. Am. Chem. Soc.*, **1960,** *82,* 96–98.
69. L. K. Dyall, *Aust. J. Chem.*, **1979,** *32,* 643–651.
70. G. Ege and K. Gilbert, *Tetrahedron Lett.*, **1979,** 4253–4256.
71. H. El Khadem and Z. M. El-Shafei, *J. Chem. Soc.*, **1958,** 3117–3119.
72. H. El Khadem, Z. M. El-Shafei, and M. M. Hashem, *J. Chem. Soc. (C)*, **1968,** 949–951.
73. H. El. Khadem, H. A. R. Manson, and M. H. Meshreki, *J. Chem. Soc. (C)*, **1968,** 1329–1331.
74. J. Elks and D. H. Hey, *J. Chem. Soc.*, **1943,** 441–445.
75. J. Epsztajn et al., *J. Chem. Soc., Perkin I,* **1973,** 2622–2624.
76. E. Fanghänel, *J. Prakt. Chem.*, **1975,** *317,* 631–640.
77. F. R. Farr and W. S. Wadsworth, Jr., *Proc. S. D. Acad. Sci.*, **1968,** *47,* 192–197.
78. W. V. Farrar, *J. Chem. Soc.*, **1964,** 906–907.
79. F. Fehér and K.-H. Linke, *Zt. Anorg. Allg. Chem.*, **1966,** *344,* 18–22.
80. W. G. Finnegan and R. A. Henry, *J. Org. Chem.*, **1965,** *30,* 567–575.

81. W. G. Finnegan, R. A. Henry, and E. Lieber, *J. Org. Chem.*, **1953,** *18,* 779–791.

82. W. G. Finnegan, R. A. Henry, and R. Lofquist, *J. Am. Chem. Soc.*, **1958,** *80,* 3908–3911.

83. B. F. Fisher, A. J. Tomson, and J. P. Horwitz, *J. Org. Chem.*, **1959,** *24,* 1650–1654.

84. F. Fleck, H. Balzer, and H. Aebli (Sandoz), U.S. Pat. 3,288,801, Nov. 29, 1966.

85. F. W. Fowler, A. Hassner, and L. A. Levy, *J. Am. Chem. Soc.*, **1967,** *89,* 2077–2082.

86. K. Fujinuma, Y. Hashida, and K. Matsui, *Bull. Chem. Soc. Jap.*, **1978,** *51,* 2394–2397.

87. R. Fusco, L. Garanti, and G. Zecchi, *J. Org. Chem.*, **1975,** *40,* 1906–1909.

88. S. F. Gait et al., *J. Chem. Soc., Perkin I,* **1974,** 1248–1260.

89. *Ibid.*, **1975,** 19–25.

90. L. Garanti, A. Locatelli, and G. Zecchi, *J. Heterocycl. Chem.*, **1976,** *13,* 657–659.

91. W. L. Garbrecht and R. M. Herbst, *J. Org. Chem.*, **1953,** *18,* 1003–1013.

92. *Ibid.*, 1014–1021.

93. G. Garcia-Munoz et al., *J. Heterocycl. Chem.*, **1977,** *14,* 427–430.

94. G. Gaudiano et al., *Gazz. Chim. Ital.*, **1967,** *97,* 1411–1422.

95. G. Gaudiano et al., *Tetrahedron Lett.*, **1967,** 107–111.

96. E. E. Glover and K. T. Rowbottom, *J. Chem. Soc., Perkin I,* **1974,** 1792–1794.

97. H. Gold, "Fluorescent Brightening Agents," in K. Venkataraman, Ed., *The Chemistry of Synthetic Dyes,* Vol. 5, Academic Press, New York, 1971, pp. 535–679.

98. S. Goldschmidt, *Ber.*, **1920,** *53,* 44–62.

99. S. Goldschmidt et al., *Ann.*, **1924,** *437,* 194–226.

100. S. Goldschmidt and K. Euler, *Ber.*, **1922,** *55,* 616–628.

101. R. Gompper, Chem. Ber., **1957,** *90,* 382–386.

102. R. Gompper and K. Schönafinger, *Chem. Ber.*, **1979,** *112,* 1514–1528.

103. N. N. Greenwood and J. H. Morris, *J. Chem. Soc.*, **1965,** 6205–6209.

104. C. Grundmann and H. Ulrich, *J. Org. Chem.*, **1959,** *24,* 272–274.

105. I. Hagedorn and H.-D. Winkelmann, *Chem. Ber.*, **1966,** *99,* 850–855.

106. M. Hamana and S. Kumadaki, *Chem. Pharm. Bull.*, **1974,** *22,* 1506–1518.

107. R. N. Hanley, W. D. Ollis, and C. A. Ramsden, *J. Chem. Soc., Perkin I,* **1979,** 732–735.

108. *Ibid.*, 736–740.

109. R. M. Hann and C. S. Hudson, *J. Am. Chem. Soc.*, **1944,** *66,* 735–738.

110. P. E. Hansen and K. Undheim, *J. Chem. Soc., Perkin I,* **1975,** 305–308.

111. R. J. Harder, R. A. Carboni, and J. E. Castle, *J. Am. Chem. Soc.*, **1967,** *89,* 2643–2647.

112. R. E. Harmon, R. A. Earl, and S. K. Gupta, *J. Org. Chem.*, **1971,** *36,* 2553–2556.

113. W. W. Hartman and J. B. Dickey, *Org. Synth.*, **1943,** *Coll. Vol. II,* 163–164.

114. L. W. Hartzel and F. R. Benson, *J. Am. Chem. Soc.*, **1954,** *76,* 667–670.

115. E. K. Harvill et al., *J. Org. Chem.*, **1950,** *15,* 662–670.

116. E. K. Harvill, C. W. Roberts, and R. M. Herbst, *J. Org. Chem.*, **1950,** *15,* 58–67.

117. H. Hasegawa, H. Arai, and H. Igeta, *Chem. Pharm. Bull.*, **1977,** *25,* 192–195.

118. W. T. Haskins, R. M. Hann, and C. S. Hudson, *J. Am. Chem. Soc.*, **1945,** *67,* 939–941.

119. H. Hauptmann and A. Cid M. Périssé, *Chem. Ber.*, **1956,** *89,* 1081–1094.

120. S. Hauptmann, H. Wilde, and K. Moser, *J. Prakt. Chem.*, **1971,** *313,* 882–888.

121. H. W. Heine, P. G. Willard, and T. R. Hoye, *J. Org. Chem.*, **1972,** *37,* 2980–2983.

122. G. Heller, W. Dietrich, and G. Reichardt, *J. Prakt. Chem.*, **1928,** *118,* 138–148.

123. R. A. Henry (United States of America as represented by the Secretary of the Navy), U.S. Pat. 3,096,312, July 2, 1963.

124. R. A. Henry, W. G. Finnegan, and E. Lieber, *J. Am. Chem. Soc.*, **1955,** *77,* 2264–2270.

125. R. M. Herbst et al., *J. Org. Chem.*, **1952,** *17,* 262–271.

126. R. M. Herbst, C. W. Roberts, and E. J. Harvill, *J. Org. Chem.*, **1951,** *16,* 139–149.

126a. R. M. Herbst and K. R. Wilson, *J. Org. Chem.*, **1957,** *22,* 1142–1145.

127. R. Hisada, M. Nakajma, and J.-P. Anselme, *Tetrahedron Lett.*, **1976,** 903–904.

128. L. Hoesch et al., *Helv. Chim. Acta,* **1977,** *60,* 816–830.

128a. L. Hoesch and B. Köppel, *Helv. Chim. Acta,* **1981,** *64,* 864–889.

129. S. Hoff and A. P. Blok, *Rec. Trav. Chim. Pays-Bas,* **1974,** *93,* 317–319.

130. K. Hohenlohe-Oehringen, *Monatsh.*, **1958,** *89,* 557–561.

131. *Ibid.*, 562–569.

132. J. P. Horwitz, B. E. Fisher, and A. J. Tomasewski, *J. Am. Chem. Soc.*, **1959,** *81,* 3076–3079.

133. J. P. Horwitz and V. A. Grakauskas, *J. Am. Chem. Soc.*, **1957,** *79,* 1249–1253.

134. *Ibid.*, **1958,** *80,* 926–930.

135. R. Huisgen, *J. Org. Chem.*, **1976,** *41,* 403–419.

136. R. Huisgen, *Proc. Welch Found. Conf. Chem. Res.*, **1960,** 61–95; *Proc. Chem. Soc.*, **1961,** 357–369.

137. R. Huisgen et al., *Chem. Ber.*, **1965,** *98,* 3992–4013.

138. *Ibid.*, 4014–4021.

139. R. Huisgen and H.-J. Koch, *Ann.*, **1955,** *591,* 200–231.

140. R. Huisgen, L. Möbius, and G. Szeimies, *Chem. Ber.*, **1965,** *98,* 1138–1152.

141. *Ibid.*, **1966,** *99,* 475–490.

142. S. Hünig et al., *Ann.*, **1966,** *697,* 116–139.

143. S. Hünig, H. Balli, and H. Quast, *Angew. Chem.*, **1962,** *74,* 28–29.

144. S. Hünig and H. Quast, *Ann.*, **1968,** *711,* 139–156.

145. R. Hüttel and F. Büchele, *Chem. Ber.*, **1955,** *88,* 1586–1590.

146. C. D. Hurd, "Open Chain Nitrogen Compounds," in H. Gilman, Ed., *Organic Chemistry-An Advanced Treatise,* Wiley, New York, 1938, pp. 615–616.

147. T. Ignasiak, J. Suszko, and B. Ignasiak, *J. Chem. Soc., Perkin I,* **1975,** 2122–2129.

148. S. Ito et al., *Bull. Chem. Soc. Jap.*, **1976,** *49,* 1920–1923.

149. *Ibid.*, **1978,** *51,* 953–954.

150. S. Ito and T. Fukuyama, *J. Org. Chem.*, **1971,** *36,* 2008–2009.

151. S. Ito, Y. Tanaka, and A. Kakehi, *Bull. Chem. Soc. Jap.*, **1976,** *49,* 762–766.

152. C. Ivanesies and E. Zbiral, *Monatsh.*, **1975,** *106,* 839–848.

153. M. Iwamura, P. J. Hintz, and S. F. Nelsen, *J. Am. Chem. Soc.*, **1970,** *92,* 3495–3496.

153a. J. W. A. M. Janssen et al., *J. Org. Chem.*, **1973,** *38,* 1777–1782.

154. A. Jonas and H. v. Pechmann, *Ann.*, **1891,** *262,* 277–302.

155. P. K. Kadaba, *J. Org. Chem.*, **1976,** *41,* 1073–1075.

156. P. K. Kadaba and J. O. Edwards, *J. Org. Chem.*, **1961,** *26,* 2331–2335.

157. H. Kano and E. Yamazaki, *Tetrahedron,* **1964,** *20,* 461–464.

158. J. C. Kauer and R. A. Carboni, *J. Am. Chem. Soc.*, **1967,** *89,* 2633–2637.

159. J. C. Kauer and W. A. Sheppard, *J. Org. Chem.*, **1967,** *32,* 3580–3592.

160. T. Kauffmann, S. Spaude, and D. Wolf, *Chem. Ber.*, **1964,** *97,* 3436–3441.

161. M. Keating et al., *J. Chem. Soc., Perkin I,* **1972,** 1315–1316.

162. E. Keller, R. Zweidler, and H. Hausermann (Geigy), U.S. Pat. 2,784,183, Mar. 5, 1957.

163. Y. Kobayashi, T. Kutsuma, and K. Morinaya, *Chem. Pharm. Bull.*, **1971,** *19,* 2106–2115.

164. G. Koga and J.-P., Anselme, *J. Org. Chem.*, **1970,** *35,* 960–964.
165. A. Krantz and B. Hoppe, *Tetrahedron Lett.*, **1975,** 695–698.
166. R. Kreher and H. Wissmann, *Chem. Ber.*, **1973,** *106,* 3097–3099.
167. H. Kubo, T. Nonaka, and K. Odo, *Bull. Chem. Soc. Jap.*, **1976,** *49,* 1339–1341.
168. T. Kuraishi and R. N. Castle, *J. Heterocycl. Chem.*, **1964,** *1,* 42–47.
169. G. L'abbe et al., *J. Am. Chem. Soc.*, **1974,** *96,* 3973–3978.
170. G. L'abbe and A. Hassner, *J. Heterocycl. Chem.*, **1970,** *7,* 361–366.
171. G. L'abbe, P. Ykman, and G. Smets, *Tetrahedron,* **1969,** *25,* 5421–5426.
172. M. Lacan, I. Tabakovic, and Z. Cekovic, *Tetrahedron,* **1974,** *30,* 2911–2917.
173. M. Langlois et al., *J. Heterocycl. Chem.*, **1982,** *19,* 193–200.
174. M. F. Lappert and J. S. Poland, *J. Chem. Soc.* (*C*), **1971,** 3910–3914.
175. L. A. Lazukina and V. P. Kukhar, *Zh. Org. Khim., Engl. Ed.,* **1979,** *15,* 2009–2010.
176. L. A. Lee, R. Evans, and J. W. Wheeler, *J. Org. Chem.*, **1972,** *37,* 343–347.
177. D. M. Lemal, F. Menger, and E. Coats, *J. Am. Chem. Soc.*, **1964,** *86,* 2395–2401.
178. D. M. Lemal and T. W. Rave, *J. Am. Chem. Soc.*, **1965,** *87,* 393–394.
179. C. Leuenberger, L. Hoesch, and A. S. Dreiding, *Helv. Chim. Acta,* **1981,** *64,* 1219–1233.
180. *Ibid.*, **1982,** *65,* 217–228.
181. E. Lieber et al., *Inorg. Synth.*, **1960,** *VI,* 42–46.
182. E. Lieber and T. Enkoji, *J. Org. Chem.*, **1961,** 4472–4479.
183. E. Lieber, E. Oftedahl, and C. N. R. Rao, *J. Org. Chem.*, **1963,** *28,* 194–199.
184. E. Lieber, C. N. Pillai, and R. D. Hites, *Can. J. Chem.*, **1957,** *35,* 832–842.
185. E. Lieber and J. Ramachandran, *Can. J. Chem.*, **1959,** *37,* 101–109.
186. E. Lieber, C. N. R. Rao, and T. V. Rajkumar, *J. Org. Chem.*, **1959,** *24,* 134–135.
187. R. O. Lindsay and C. F. H. Allen, *Org. Synth.*, **1955,** *Coll. Vol. III,* 710–711.
188. J. H. Markgraf, W. T. Bachman, and D. P. Hollis, *J. Org. Chem.*, **1965,** *30,* 3472–3474.
189. D. Martin and A. Weise, *Chem. Ber.*, **1966,** *99,* 317–327.
190. K. Masuda, T. Kamiya, and K. Kashiwa, *Chem. Pharm. Bull.*, **1971,** *19,* 559–563.
191. W. R. McBride and E. M. Bens, *J. Am. Chem. Soc.*, **1959,** *81,* 5546–5550.
192. W. R. McBride, W. G. Finnegan, and R. A. Henry, *J. Org. Chem.*, **1957,** *22,* 152–157.
193. A. McKillop and R. J. Kobylecki, *J. Org. Chem.*, **1974,** *39,* 2710–2714.
194. I. M. McRobbie, O. Meth-Cohn, and H. Suschitzky, *Tetrahedron Lett.*, **1976,** 925–928.
195. J. S. Meek and J. S. Fowler, *J. Am. Chem. Soc.*, **1967,** *89,* 1967.
196. J. S. Meek and J. S. Fowler, *J. Org. Chem.*, **1968,** *33,* 985–991.
197. J. Meisenheimer, O. Senn, and P. Zimmermann, *Ber.*, **1927,** *60,* 1736–1748.
198. A. Messmer and A. Gelleri, *Angew. Chem.*, **1965,** *77,* 171.
199. *Ibid.*, **1967,** *6,* 261–262.
200. J. S. Mihina and R. M. Herbst, *J. Org. Chem.*, **1950,** *15,* 1082–1092.
201. H. Minato, M. Oku, and S. H. -P. Chan, *Bull. Chem. Soc. Jap.*, **1966,** *39,* 1049–1054.
202. S. Mineo, S. Kawamura, and K. Nakagawa, *Syn. Commun.*, **1976,** *6,* 69–74.
203. D. Moderhack, *J. Heterocycl. Chem.*, **1977,** *14,* 757–763.
204. J. H. Morris and P. G. Perkins, *J. Chem. Soc.* (*A*), **1966,** 580–582.
205. W. L. Mosby and M. L. Silva, *J. Chem. Soc.*, **1965,** 1003–1012.
206. E. Müller, R. Beutler, and B. Zeeh, *Ann.*, **1968,** *719,* 72–79.
207. E. Müller and H. Haiss, *Chem. Ber.*, **1962,** *95,* 1255–1263.
208. *Ibid.*, **1963,** *96,* 570–583.

209. E. Müller and D. Ludsteck, *Chem. Ber.*, **1955,** *88,* 921–933.
210. D. B. Murphy and J. P. Picard, *J. Org. Chem.*, **1954,** *19,* 1807–1814.
211. A. Mustafa et al., *J. Am. Chem. Soc.*, **1956,** *78,* 145–149.
212. F. Musik and Z. J. Allan, *Collect. Czech. Chem. Commun.*, **1959,** *24,* 474–483.
213. M. Nakajima and J.-P. Anselme, *Tetrahedron Lett.*, **1976,** 4421–4422.
214. S. F. Nelsen et al., *J. Am. Chem. Soc.*, **1972,** *94,* 1610–1614.
215. F. A. Neugebauer and H. Fischer, *Chem. Ber.*, **1973,** *106,* 1589–1594.
216. P. N. Neuman, *J. Heterocycl. Chem.*, **1971,** *8,* 51–56.
217. H. Neunhoeffer and V. Böhnisch, *Ann.*, **1976,** 153–162.
218. Y. Nomura et al., *Bull. Chem. Soc. Jap.*, **1981,** *54,* 2779–2785.
219. W. P. Norris, *J. Org. Chem.*, **1962,** *27,* 3248–3251.
220. A. Ohsawa et al., *J. Chem. Soc. Chem. Commun.*, **1981,** 1174.
221. C. E. Olsen, *Acta Chem. Scand.*, **1974,** *B28,* 425–432.
222. C. E. Olsen and C. Pedersen, *Acta Chem. Scand.*, **1973,** *27,* 2279–2286.
223. A. Padwa et al., *J. Org. Chem.*, **1979,** *44,* 3281–3287.
224. P. I. Paetzold and G. Stohr, *Chem. Ber.*, **1968,** *101,* 2874–2880.
224a. J. P. Paolini, *J. Heterocycl. Chem.*, **1972,** *9,* 461–462.
225. A. J. Papa, *J. Org. Chem.*, **1966,** *31,* 1426–1430.
226. H. v. Pechmann and L. Frobenius, *Ber.*, **1894,** *27,* 898–900.
227. *Ibid.*, **1895,** *28,* 170–176.
228. C. Pedersen, *Acta Chem. Scand.*, **1958,** *12,* 1236–1240.
229. *Ibid.*, **1959,** *13,* 888–892.
230. R. M. Pérez Gutiérrez, G. Pieza Robles, and C. Pérez Gutiériez, *Acta Mex. Cienc. Tecnol.*, **1978,** *11–12,* 9–12.
231. W. R. Petersen, B. Arkles, and S. S. Washburne, *J. Organomet. Chem.*, **1976,** *121,* 285–291.
232. R. Pfleger, E. Garthe, and K. Rauer, *Chem. Ber.*, **1963,** *96,* 1827–1839.
233. K. Pilgram, F. Görgen, and G. Pollard, *J. Heterocycl. Chem.*, **1971,** *8,* 951–959.
234. W. H. Pirkle and P. L. Gravel, *J. Org. Chem.*, **1978,** *43,* 808–815.
235. S. Polanc et al., *J. Org. Chem.*, **1974,** *39,* 2143–2147.
236. K. T. Potts and S. Husain, *J. Org. Chem.*, **1970,** *35,* 3451–3456.
237. *Ibid.*, **1972,** *37,* 2049–2050.
238. Z. V. Pushkareva et al., *Khim. Geterotsikl. Soedin., Engl. Ed.*, **1975,** *11,* 995–998.
239. H. Quast and L. Bieber, *Chem. Ber.*, **1981,** *114,* 3253–3272.
240. H. Quast and L. Bieber, *Tetrahedron Lett.*, **1976,** 1485–1486.
241. H. Quast and E. Frankenfeld, *Angew. Chem., Int. Ed. Engl.*, **1965,** *4,* 691.
242. H. Quast and S. Hünig, *Chem. Ber.*, **1968,** *101,* 435–444.
243. A. Queen, D. M. McKinnon, and A. W. Bell, *Can. J. Chem.*, **1976,** *54,* 1906–1909.
244. A. Quilico, *Gazz. Chim. Ital.*, **1932,** *62,* 912–927.
245. M. S. Raasch, *J. Org. Chem.*, **1970,** *35,* 3470–3483.
246. *Ibid.*, **1978,** *43,* 2500–2508.
247. C. W. Rees and A. A. Sale, *J. Chem. Soc., Perkin I,* **1973,** 545–550.
248. M. Regitz, *Angew. Chem.*, **1965,** *77,* 428.
249. M. Regitz, *Chem. Ber.*, **1966,** *99,* 2918–2930.
250. H. Reimlinger, *Chem. Ber.*, **1964,** *97,* 3493–3502.
251. A. Reiser et al., *Trans. Faraday Soc.*, **1967,** *63,* 2403–2410.

252. G. A. Reynolds, *J. Org. Chem.*, **1964,** *29,* 3733–3734.

253. E. Richter and E. C. Taylor, *J. Am. Chem. Soc.*, **1956,** *78,* 5848–5852.

254. W. Ried and H. Mengler, *Ann.*, **1964,** *678,* 95–113.

255. R. Robertson and S. Thornley, *J. Chem. Soc.*, **1924,** 2169–2176.

256. C. S. Rondestvedt, Jr. and P. K. Chang, *J. Am. Chem. Soc.*, **1955,** *77,* 6532–6540.

257. K. Sakai and J.-P. Anselme, *J. Org. Chem.*, **1972,** *37,* 2351–2352.

258. T. Sakamoto, Y. Terao, and M. Sekiya, *Chem. Pharm. Bull.*, **1977,** *25,* 731–739.

259. P. Scheiner, *Tetrahedron,* **1967,** *24,* 349–356.

260. P. Scheiner, "Triazoline Decomposition," in B. S. Thyagaragan, Ed., *Selective Organic Transformations,* Wiley-Interscience, New York, 1970, pp. 327–361.

261. G. O. Schenck and H. Formanek, *Angew. Chem.*, **1958,** *70,* 505.

262. K. F. Schmidt (Knoll & Co.), U.S. Pat. 1,599,493, Sept. 14, 1926.

263. M. P. Schmidt and A. Hagenböcker, *Ber.*, **1921,** *54,* 2201–2207.

264. E. Schmitz, H. Fechner-Simon, and S. Schramm, *Ann.*, **1969,** *725,* 1–10.

265. E. Schmitz and C. Hörig, *Chem. Ber.*, **1967,** *100,* 2101–2106.

266. D. Seebach, *Chem. Ber.*, **1977,** *110,* 1879–1886.

267. D. Seebach et al., *Helv. Chim. Acta,* **1978,** *61,* 1622–1647.

268. A. S. Shawali, A. A. Fahmi, and N. F. Eweiss, *J. Heterocycl. Chem.*, **1979,** *16,* 123–128.

269. V. P. Shchipanov, *Zh. Org. Khim., Engl. Ed.*, **1979,** *15,* 1793–1798.

270. J. C. Sheehan and C. A. Robinson, *J. Am. Chem. Soc.*, **1951,** *73,* 1207–1210.

271. C. Shin, Y. Yonezawa, and J. Yoshimura, *Tetrahedron Lett.*, **1974,** 7–10.

272. H. Sieper, *Chem. Ber.*, **1967,** *100,* 1646–1654.

273. H. Sieper and P. Tavs, *Ann.*, **1967,** *704,* 161–165.

274. A. D. Sinitsa, N. A. Parkhomenko, and L. N. Markovskii, *Zh. Obsch. Khim., Engl. Ed.*, **1977,** *47,* 216.

275. H. H. Sisler and K. Utvary, *Inorg. Synth.*, **1967,** *X,* 129–131.

276. D. Smith and P. J. Taylor, *J. Chem. Soc., Perkin II,* **1979,** 1376–1386; Suppl. 22552, pp. 1–17.

277. P. A. S. Smith, *J. Am. Chem. Soc.*, **1948,** *70,* 320–323.

278. P. A. S. Smith and D. H. Kenny, *J. Org. Chem.*, **1961,** *26,* 5221–5224.

279. S. Sommer and U. Schubert, *Angew. Chem., Int. Ed. Engl.*, **1979,** *18,* 696–697.

280. B. Stanovnik et al., *J. Heterocycl. Chem.*, **1979,** *16,* 1567–1570.

281. *Ibid.*, **1982,** *19,* 577–583.

282. B. Stanovnik and M. Tišler, *J. Heterocycl. Chem.*, **1971,** *8,* 785–787.

283. H. N. E. Stevens and M. F. G. Stevens, *J. Chem. Soc. (C)*, **1970,** 765–771.

284. R. Stollé, *Ber.*, **1908,** *41,* 2811–2813.

285. R. Stollé, *J. Prakt. Chem.*, **1933,** *138,* 1–17.

286. R. Stollé and F. Henke-Stark, *J. Prakt. Chem.*, **1930,** *124,* 261–300.

287. K. B. Sukumaran, S. Satesh, and M. V. George, *Tetrahedron,* **1974,** *30,* 445–450.

288. H. Suschitzky and R. E. Walrond, *J. Chem. Soc., Perkin I,* **1977,** 47–52.

289. D. R. Sutherland and G. Tennant, *J. Chem. Soc. (C)*, **1971,** 706–713.

290. J. Svetlik and A. Martvon, *Collect. Czech. Chem. Commun.*, **1979,** *44,* 2421–2425.

291. G. Szeimes and K. Mannhardt, *Chem. Ber.*, **1977,** *110,* 2939–2946.

292. Y. Tamura et al., *J. Heterocycl. Chem.*, **1975,** *12,* 481–483.

293. Y. Tamura et al., *Tetrahedron Lett.*, **1972,** 4133–4135.

294. Y. Tanaka and S. I. Miller, *J. Org. Chem.*, **1973,** *38,* 2708–2712.

295. C. Temple, Jr., C. L. Kussner, and J. A. Montgomery, *J. Org. Chem.*, **1967,** *32,* 2241–2245.
296. G. Tennant, *J. Chem. Soc. (C)*, **1967,** 1279–1283.
297. W. Theilacker and E. C. Fintelmann, *Chem. Ber.*, **1958,** *91,* 1597–1600.
298. J. Thiele, *Ann.*, **1910,** *376,* 239–268.
299. J. Thiele, *Ber.*, **1908,** *41,* 2681–2683.
300. H.-J. Timpe, *Z. Chem.*, **1971,** *11,* 340–341.
301. O. Tsuge and H. Samura, *J. Heterocycl. Chem.*, **1971,** *8,* 707–710.
302. I. Ugi and U. Fetzer, *Chem. Ber.*, **1961,** *94,* 1116–1121.
303. I. Ugi and R. Meyr, *Chem. Ber.*, **1961,** *94,* 2229–2237.
304. I. Ugi, H. Perlinger, and L. Behringer, *Chem. Ber.*, **1958,** *91,* 2324–2329.
305. I. Ugi and C. Steinbrückner, *Chem. Ber.*, **1961,** *94,* 734–742.
306. K. Utvary, *Monatsh.*, **1968,** *99,* 1473–1482.
307. J.-M. Vandensavel, G. Smets, and G. L'Abbé, *J. Org. Chem.*, **1973,** *38,* 675–678.
308. G. Vernin et al., *Synthesis,* **1977,** 691–693.
309. J. Vilarrasa and R. Granados, *J. Heterocycl. Chem.*, **1974,** *11,* 867–872.
310. H. Vorbrüggen and K. Krolkiewicz, *Synthesis,* **1979,** 35.
311. W. S. Wadsworth, Jr., *J. Org. Chem.*, **1969,** *34,* 2994–2997.
312. E. R. Wagner, *J. Org. Chem.*, **1973,** *38,* 2976–2981.
313. R. v. Walther and W. Grieshammer, *J. Prakt. Chem.*, **1915,** *92,* 209–255.
314. N. Wiberg et al., *Chem. Ber.*, **1979,** *112,* 2718–2729.
314a. N. Wiberg and W. Uhlenbrock, *Angew. Chem. Int. Ed. Engl.*, **1970,** *7,* 70–71.
315. N. Wiberg and W. Uhlenbrock, *Chem. Ber.*, **1972,** *105,* 63–66.
316. N. Wiberg and G. Ziegleder, *Chem. Ber.*, **1978,** *111,* 2123–2129.
316a. W. Wislicenus, *Ber.*, **1892,** *25,* 2084–2087.
317. F. P. Woerner and H. Reimlinger, *Chem. Ber.*, **1970,** *103,* 1908–1917.
318. A. Wohl and H. Schiff, *Ber.*, **1900,** *33,* 2741–2758.
319. H. Wolff, *Org. React.*, **1946,** *III,* 307–336.
320. L. Wolff, *Ann.*, **1902,** *325,* 152–169.
321. D. Y. Wu and R. M. Herbst, *J. Org. Chem.*, **1952,** *17,* 1216–1227.
322. P. Ykman, G. L'Abbé, and G. Smets, *Tetrahedron,* **1971,** *27,* 845–849.
323. J. Zaloom and D. C. Roberts, *J. Org. Chem.*, **1981,** *46,* 5173–5176.
324. E. Zbiral, M. Rasberger, and H. Hengstberger, *Ann.*, **1969,** *725,* 22–28.
325. E. Zbiral and J. Stroh, *Ann.*, **1969,** *725,* 29–36.
326. N. S. Zefirov, N. K. Chapovskaya, and V. V. Kolesnikov, *Chem. Commun.*, **1971,** 1001–1002.
327. D. M. Zimmerman and R. A. Olofson, *Tetrahedron Lett.*, **1969,** 5081–5084.
328. T. Zinke and P. Schwarz, *Ann.*, **1900,** *311,* 329–340.

INDEX